T0348648

VOLUME ONE

DEVELOPMENTS IN MARINE GEOLOGY

PROXIES IN LATE CENOZOIC PALEOCEANOGRAPHY

VOLUME ONE

DEVELOPMENTS IN MARINE GEOLOGY

PROXIES IN LATE CENOZOIC PALEOCEANOGRAPHY

Edited by

CLAUDE HILLAIRE–MARCEL AND ANNE DE VERNAL
GEOTOP-Université du Québec à Montréal, Montréal, Québec Canada

ELSEVIER

Amsterdam • Boston • Heidelberg • London • New York • Oxford
Paris • San Diego • San Francisco • Singapore • Sydney • Tokyo

Elsevier
Radarweg 29, PO Box 211, 1000 AE Amsterdam, The Netherlands
The Boulevard, Langford Lane, Kidlington, Oxford, OX5 1GB, UK

First edition 2007
Reprinted 2008

Notice
No responsibility is assumed by the publisher for any injury and/or damage to persons
or property as a matter of products liability, negligence or otherwise, or from any use
or operation of any methods, products, instructions or ideas contained in the material
herein. Because of rapid advances in the medical sciences, in particular, independent
verification of diagnoses and drug dosages should be made

British Library Cataloguing in Publication Data
A catalogue record for this book is available from the British Library

Library of Congress Cataloging-in-Publication Data
A catalog record for this book is available from the Library of Congress

ISBN: 978-0444-52755-4
ISSN: 1572-5480

For information on all Elsevier publications
visit our website at www.elsevierdirect.com

Printed and bound by CPI Group (UK) Ltd, Croydon, CR0 4YY

Transferred to digital print 2013

Working together to grow
libraries in developing countries
www.elsevier.com | www.bookaid.org | www.sabre.org

ELSEVIER   BOOK AID   Sabre Foundation
            International

This volume is dedicated to the memory of Cesare Emiliani (1922–1995) and Nicholas Shackleton (1937–2006). Their research has laid the foundation of modern paleoceanography and paleoclimatology, fields of scientific endeavor that have contributed immensely to our present knowledge and understanding of the ocean-climate system.

Please note that the previous printing included a CD-ROM attached to the inside back cover of the book.

The material is now only available on the companion website: http://booksite.elsevier.com/9780444527554/

Contents

Contributors   *xiii*

Scientific Committee   *xvii*

Methods in Late Cenozoic Paleoceanography: Introduction   **1**
Claude Hillaire-Marcel and Anne de Vernal

     1. Tracers and Proxies in Deep-Sea Records   2
     2. Overview of Volume Content   3
     3. The Need for Multi-tracers and Multi-Proxy Approaches in
        Paleoceanography   7
     4. From the Geological Record to the Sedimentary Signal and the
        Properties of the Water Column   10
     5. How Far Back in Time are the Proxies Effective?   11
     6. New Perspectives and Emerging Proxies   13
     Acknowledgments   14
     References   14

Part 1:   Deep-Sea Sediment Properties

1. Deep-Sea Sediment Deposits and Properties Controlled by Currents   **19**
Ian Nicholas McCave

     1. Introduction   19
     2. Sediment Transport and Deposition by Deep-Sea Currents   30
     3. Sediment Deposition: Quaternary Records of Flow in Large-Scale Features   38
     4. Current Problems and Prospects   45
     References   54

2. Continuous Physical Properties of Cored Marine Sediments   **63**
Guillaume St-Onge, Thierry Mulder, Pierre Francus and Bernard Long

     1. Introduction   63
     2. Continuous Centimeter-Scale Measurements of Physical Properties   64
     3. Continuous Millimeter- to Micrometer-Scale Measurements of
        Physical Properties   68
     4. Recent Applications of Continuous Centimeter- to Millimeter-Scale
        Physical Properties of Marine Sediments   81
     5. Conclusion   91
     Acknowledgments   92
     References   92

3. **Magnetic Stratigraphy in Paleoceanography: Reversals, Excursions, Paleointensity, and Secular Variation**   99
Joseph S. Stoner and Guillaume St-Onge

   1. Introduction   100
   2. Background   101
   3. Soft Sediment Paleomagnetic Methods   103
   4. Magnetometers   107
   5. Measurements and Magnetizations   109
   6. Data Analysis   114
   7. Sediment Magnetism   117
   8. Development of Paleomagnetic Records   118
   9. The Paleomagnetic Record as a Stratigraphic Tool   121
   10. Some Perspectives   128
   References   130

4. **Clay Minerals, Deep Circulation and Climate**   139
Nathalie Fagel

   1. Introduction   139
   2. Methodology: The Clay Toolbox in Marine Sediments   142
   3. Applications: Clays as a Proxy for Paleocirculation   171
   4. Some Perspectives   176
   Acknowledgements   176
   References   176

5. **Radiocarbon Dating of Deep-Sea Sediments**   185
Konrad A. Hughen

   1. Introduction   185
   2. Dating Marine Sediments   187
   3. Applications of Marine ^{14}C 201
   Appendix I — Internet Resources   204
   References   206

Part 2:   Biological Tracers and Biomarkers

6. **Planktonic Foraminifera as Tracers of Past Oceanic Environments**   213
Michal Kucera

   1. Introduction   213
   2. Biology and Ecology of Planktonic Foraminifera   215
   3. Planktonic Foraminiferal Proxies   225
   4. Modifications After Death   245
   5. Perspectives   253
   WWW Resources   253
   References   254

7. **Paleoceanographical Proxies Based on Deep-Sea Benthic Foraminiferal Assemblage Characteristics**   263
    Frans J. Jorissen, Christophe Fontanier and Ellen Thomas

    1. Introduction   263
    2. Benthic Foraminiferal Proxies: A State of the Art   271
    3. Conclusions   306
    Acknowledgements   308
    4. Appendix 1.   308
    References   313

8. **Diatoms: From Micropaleontology to Isotope Geochemistry**   327
    Xavier Crosta and Nalan Koç

    1. Introduction   327
    2. Improvements in Methodologies and Interpretations   332
    3. Case Studies   350
    4. Conclusion   356
    Acknowledgments   358
    References   358

9. **Organic-Walled Dinoflagellate Cysts: Tracers of Sea-Surface Conditions**   371
    Anne de Vernal and Fabienne Marret

    1. Introduction   371
    2. Ecology of Dinoflagellates   376
    3. Dinoflagellates vs. Dinocysts and Taphonomical Processes (From the
        Biocenoses to Thanathocenoses)   377
    4. Relationships between Dinocyst Assemblages and Sea-Surface Parameters   382
    5. The Development of Quantitative Approaches for the Reconstruction
        of Hydrographic Parameters Based on Dinocysts   395
    6. The Use of Dinocysts in Paleoceanography   397
    7. Concluding Remarks   398
    References   400

10. **Coccolithophores: From Extant Populations to Fossil Assemblages**   409
    Jacques Giraudeau and Beaufort Luc

    1. Introduction   409
    2. Taxonomy   411
    3. Biogeography, Sedimentation, and Biogeochemical Significance   413
    4. Current State of Methods   414
    5. Examples of Applications   428
    Acknowledgments   432
    References   433

11. **Biomarkers as Paleoceanographic Proxies**   441
    Antoni Rosell-Melé and Erin L. McClymont

    1. Preliminary Considerations   441
    2. Methodological Approaches   443

3. Applications   466
4. Concluding Remarks   474
Acknowledgments   476
References   476

12. **Deep-Sea Corals: New Insights to Paleoceanography**   **491**
     Owen A. Sherwood and Michael J. Risk

     1. Introduction   491
     2. Methods and Interpretations   495
     3. Landmark Studies   514
     References   516

13. **Transfer Functions: Methods for Quantitative Paleoceanography Based
     on Microfossils**   **523**
     Joël Guiot and Anne de Vernal

     1. Introduction   523
     2. Methods Based on Calibration   527
     3. Methods Based on Similarity   533
     4. Comparison of Methods with a Worked Example   537
     5. Discussion and Future Developments   545
     6. The applications of Transfer Functions *Sensu Lato* in Paleoceanography   550
     7. Concluding Remarks   556
     References   557

Part 3:   Geochemical Tracers

14. **Elemental Proxies for Palaeoclimatic and Palaeoceanographic
     Variability in Marine Sediments: Interpretation and Application**   **567**
     Stephen E. Calvert and Thomas F. Pedersen

     1. Introduction   568
     2. Sedimentary Components of Marine Sediments   569
     3. Normalization of Elemental Data   569
     4. Palaeoclimatic Records from the Sea Floor   571
     5. Metalliferous Sedimentation in the Ocean   581
     6. Elemental Proxies for Palaeoproductivity   585
     7. Proxies for Redox Conditions at the Sea Floor and in Bottom Sediments   599
     8. Future Developments   621
     9. Afterword   625
     Acknowledgements   625
     References   625

15. **Isotopic Tracers of Water Masses and Deep Currents**   **645**
     Christelle Claude and Bruno Hamelin

     1. Introduction   645
     2. Present State of Methodological Approaches and Interpretations   648

3. Examples of Applications   664
4. Conclusion and Perspectives   670
References   671

16. **Paleoflux and Paleocirculation from Sediment ^{230}Th and
    ^{231}Pa/^{230}Th**   681
    Roger François

    1. Introduction   681
    2. Factors Controlling the Distribution of ^{230}Th and ^{231}Pa
       in the Ocean   684
    3. Paleoceanographic Applications   698
    4. Conclusions   712
    References   712

17. **Boron Isotopes in Marine Carbonate Sediments and the pH
    of the Ocean**   717
    Nicholas Gary Hemming and Bärbel Hönisch

    1. Introduction   717
    2. Empirical Observations and Theoretical Background   718
    3. Caveats and Complications   721
    4. Applications of the Boron Isotope Paleo-pH Proxy   726
    5. Summary and Conclusion   730
    Acknowledgments   731
    References   731

18. **The Use of Oxygen and Carbon Isotopes of Foraminifera in
    Paleoceanography**   735
    Ana Christina Ravelo and Claude Hillaire-Marcel

    1. Introduction   735
    2. Notation and Standards   736
    3. Stratigraphic and Paleoecological Use of Foraminifera   738
    4. Foraminiferal Oxygen Isotopes as Environmental Proxies   740
    5. Foraminiferal Carbon Isotopes as Environmental Proxies   751
    6. Conclusion and Summary   759
    References   760

19. **Elemental Proxies for Reconstructing Cenozoic
    Seawater Paleotemperatures from Calcareous Fossils**   765
    Yair Rosenthal

    1. Introduction   765
    2. Thermodynamic Effects on Mg Co-Precipitation
       in Calcites   766
    3. Foraminiferal Mg/Ca Paleothermometry   767
    4. Ostracode Mg/Ca Paleothermometry   777

5. Coralline Sr/Ca Paleothermometry   780
6. Contributions to Cenozoic Climate History   784
References   790

Reconstructing and Modeling Past Oceans   **799**
Katrin J. Meissner

1. A Brief Historical Overview   800
2. Classification of Climate Models   801
3. Models and Proxy Data   804
4. International Programs   807
5. Conclusion   808
References   809

Index of Taxa   **813**

Subject Index   **817**

Contributors

Luc Beaufort
CEREGE, UMR 6635 CNRS/Université P. Cézanne/IRD, Aix-en-Provence, France

Stephen E. Calvert
Department of Earth and Ocean Sciences, University of British Columbia, Vancouver, British Columbia, Canada

Christelle Claude
CEREGE, UMR 6635 CNRS/Université P. Cézanne/IRD, Aix en Provence, France

Xavier Crosta
UMR-CNRS 5805 EPOC, Université Bordeaux 1, Talence, France

Nathalie Fagel
Département de Géologie, Université de Liège, Liège, Belgium

Christophe Fontanier
Laboratoire des Bio-indicateurs Actuels et Fossiles (BIAF), Université d'Angers, Angers, France; Laboratory of Marine Bio-Indicators (LEBIM), Ile d'Yeu, France

Roger François
Department of Earth and Ocean Sciences, University of British Columbia, Vancouver, British Columbia, Canada

Pierre Francus
Institut national de la recherche scientifique, Centre Eau, Terre et Environnement (INRS-ETE), Québec, Canada; GEOTOP-UQAM & McGill, Montréal, Québec, Canada

Jacques Giraudeau
UMR-CNRS 5805 EPOC, Université Bordeaux 1, Talence, France

Joël Guiot
CEREGE, UMR 6635 CNRS/Université P. Cézanne/IRD, Aix en Provence, France

Bruno Hamelin
CEREGE, UMR 6635 CNRS/Université P. Cézanne/IRD, Aix en Provence, France

Nicholas Gary Hemming
School of Earth and Environmental Sciences, Queens College, Flushing, NY, USA

Claude Hillaire-Marcel
GEOTOP-UQAM & McGill, Université du Québec à Montréal, Québec, Canada

Bärbel Hönisch
Lamont-Doherty Earth Observatory, Columbia University, Palisades, NY, USA

Konrad A. Hughen
Department of Marine Chemistry and Geochemistry, Woods Hole Oceanographic Institution, Woods Hole, MA, USA

Frans J. Jorissen
Laboratoire des Bio-indicateurs Actuels et Fossiles, Université d'Angers, Angers, France; Laboratory of Marine Bio-Indicators (LEBIM), Ile d'Yeu, France

Nalan Koç
NPI, Polarmiljosenteret, Tromso, Norway

Michal Kucera
Institut für Geowissenschaften, Eberhard Karls Universität Tübingen, Tübingen, Germany

Bernard Long
Institut National de la Recherche Scientifique, Centre Eau, Terre et Environnement (INRS-ETE), Québec, Canada; GEOTOP-UQAM & McGill, Montréal, Québec, Canada

Fabienne Marret
Department of Geography, University of Liverpool, Liverpool, UK

Ian Nicholas McCave
Department of Earth Sciences, University of Cambridge, Cambridge, UK

Erin L. McClymont
School of Geography, Politics and Sociology, Newcastle University, Newcastle upon Tyne, UK

Katrin J. Meissner
School of Earth and Ocean Sciences, University of Victoria, Victoria, British Columbia, Canada

Thierry Mulder
Département de Géologie et Océanographie, Université Bordeaux 1, Talence, France

Thomas F. Pedersen
Department of Earth and Ocean Sciences, University of British Columbia, Vancouver, British Columbia, Canada

Ana Christina Ravelo
Ocean Sciences Department, University of California, Santa Cruz CA, USA

Michael J. Risk
School of Geography and Geology, McMaster University, Hamilton, Canada

Antoni Rosell-Melé
ICREA and Institut de Ciència i Tecnologia Ambientals, Universitat Autònoma de Barcelona, Bellaterra, Catalonia, Spain

Yair Rosenthal
Institute for Marine and Coastal Sciences, and Department of Geological Sciences, Rutgers University, NJ, USA

Guillaume St-Onge

Institut des Sciences de la Mer de Rimouski (ISMER), Université du Québec à Rimouski, Rimouski, Québec, Canada; GEOTOP-UQAM & McGill, Montréal, Québec, Canada

Owen A. Sherwood

Centre for Environmental and Marine Geology, Dalhousie University, Halifax, Canada

Joseph S. Stoner

COAS, Oregon State University, Corvallis, OR, USA

Ellen Thomas

Department of Geology and Geophysics, Yale University, New Haven, CT, USA

Anne de Vernal

GEOTOP-UQAM & McGill, Université du Québec à Montréal, Québec, Canada

Scientific Committee

Jess Adkins
Caltech, Pasadena, CA, USA

Hervé Chamley
Maison Palassimborda, Avenue de Navarre, 64250 Cambo, France

Michel Cremer
Université Bordeaux 1, Bordeaux, France

Frédérique Eynaud
Université Bordeaux 1, Bordeaux, France

Sidney Heming
Lamont-Doherty Earth Observatory, NY, USA

Gideon Henderson
Oxford University, Oxford, UK

Claude Hillaire-Marcel
GEOTOP — UQAM & McGill, Montreal, Canada

Michal Kucera
Eberhard-Karls Universität, Tübingen, Germany

Moritz Lehman
GEOTOP-UQAM & McGill, Québec, Canada

Jerry McManus
Woods Hole University, MA, USA

Reinhart Pienitz
Université Laval, Québec, Canada

David Piper
Atlantic Geoscience Centre, Dartmouth, Nova Scotia, Canada

Dan Sinclair
University of Texas in Austin, TX, USA

Ralph Schneider
Christian-Albrechts-Universität zu Kiel, Germany

Nicolas Thouveny
CEREGE & Université de Luminy, Aix-en-Provence, France

Alain Véron
CEREGE, Université Paul Cézanne, Aix-en-Provence, France

Richard Zeebe
University of Hawaii at Manoa, Honolulu, HI, USA

Patrizia Ziveri
Vrije Universiteit, Amsterdam, The Netherlands

Karin Zonneveld
Universität Bremen, Germany

METHODS IN LATE CENOZOIC PALEOCEANOGRAPHY: INTRODUCTION

Claude Hillaire-Marcel* *and* Anne de Vernal

Contents

1. Tracers and Proxies in Deep-Sea Records   2
2. Overview of Volume Content   3
3. The Need for Multi-tracers and Multi-Proxy Approaches in Paleoceanography 7
4. From the Geological Record to the Sedimentary Signal and the Properties of
   the Water Column   10
5. How Far Back in Time are the Proxies Effective?   11
6. New Perspectives and Emerging Proxies   13
Acknowledgments   14
References   14

Preamble

After more than thirty years of intensive research in paleoceanography, which has contributed considerably to our understanding of climatic changes, new avenues of research now require attention especially on internal physical and chemical processes occurring in the ocean (e.g., instabilities of thermohaline circulation, fate of anthropogenic carbon and so on). Thus, it seems timely to sum up achievements to date, take a snapshot of the state of knowledge about the past ocean, and review the methods that have allowed so much progress. In particular, the methodological approaches in paleoceanography have become extremely diversified during the last decades, as a result of wide multidisciplinary efforts from sedimentologists, geochemists, modelers, biologists, and paleontologists. In this first volume, the emphasis is placed on the most currently used methods. It aims at providing a relatively exhaustive review of the state-of-the-art approaches for reconstructing past conditions of the ocean. A second volume will address key issues about the past ocean-climate system (e.g., thermohaline circulation, the Green Ocean and carbon cycle, changes in sea-surface conditions, and sea-level) and provide climate modelers with updated reconstructions of selected past ocean synopses during the Late Cenozoic.

The present volume has been designed to provide graduate students and researchers, in any given discipline, with an overview of latest methods from the most useful tools in paleoceanography. For the specialists, the relevant chapter may seem somewhat simplified, but this is inherent in such an effort, as methods are now so complex that most of them would require a full volume each in order to be fully detailed. Students in

* Corresponding author.

Developments in Marine Geology, Volume 1
ISSN 1572-5480, DOI 10.1016/S1572-5480(07)01005-6

oceanography and marine geology should find the foundation here to a better under-standing of recording of physical, chemical, and biological processes in deep-sea se-diments.

1. TRACERS AND PROXIES IN DEEP-SEA RECORDS

The ocean dynamics is complex and the measurement of many variables is required to make proper assessments. For example, prior to assessing vertical con-vection at a given point of an ocean, temperature and salinity conditions, and thus density, have to be measured throughout the annual cycle and down the water column. Therefore, addressing the question of convection to assess thermohaline circulation patterns in the past requires the use of several tracers, which will allow the concurrent reconstruction of the sea–water density in time (annual cycle) and space (depths in the water column), with an accuracy that can be evaluated quan-titatively. Such tracers could be considered as "proxies", if we agree to define the proxy as a surrogate for quantitative estimates of given parameters (chemical, physi-cal, or biological) of past environments or climates. In this sense, proxies should provide quantitative values and include a quantitative estimate of uncertainties of the reconstructed parameter, whereas "tracers" refer to objects that are related to environmental conditions or processes, and that may provide quantitative or quali-tative insight into such conditions or processes.

PROXY (CLIMATE AND ENVIRONMENTAL DYNAMICS STUDIES)

A measurable property of an environmental/geological record which, through math-ematical or statistical treatment, can be related with a stated uncertainty to one or a combination of physical, chemical, or biological environmental factors during its formation.

A (GEOLOGICAL OR ENVIRONMENTAL) RECORD

A physical entity (sediment, ice, assemblage of fossils, chemical or biogenic mineral, growth-rings of organic remains, etc.) that has preserved properties linked to past envi-ronmental conditions with minimal or measurable diagenetic change. Its dating (in relative or absolute terms) and duration may provide a time series, allowing documen-tation of the variability of paleoenvironmental conditions based on mathematical and/or statistical tools.

Progress in paleoceanography therefore depends upon the assessment and measurement of various tracers from sediment or from biological and bio-geo-chemical recorders of environmental conditions (e.g., coral growth-rings, manganese nodules, etc.), and the development of proxies. Tracers include (i) physical properties of the sediment (grain size, mineralogy, density, and magnetic properties), (ii) biological remains preserved as calcium carbonate, silica, refractory organic matter and organic biomarkers, and (iii) geochemical and isotopic properties of detrital and biogenic material. Some of these tracers are used for defining a chronological series, some for understanding processes controlling sediment accumulation and/or diagenetic changes, and others for reconstructing the physical, biological, and chemical conditions of the past ocean environments, from the sediment–water interface to the upper water layer.

 ## 2. OVERVIEW OF VOLUME CONTENT

In this volume, most chapters focus on tracers of oceanographic conditions, many of them providing proxies of parameters such as temperature, salinity, density, current velocity, and biological productivity (Table 1), sometimes with relatively precise information on their seasonality or variability in time. Approaches developed for the use of given tracers as proxies are discussed in the respective chapters and presented in a more comprehensive manner in Chapter 13 on transfer functions.

The volume consists of three main parts, divided according to the type of tracers. The first part is concerned with deep-sea sediment physical, mineralogical, and geochemical properties that are controlled by sedimentary processes in general, such as surface and deep-sea currents (see Chapters 1, 2, and 4). Chapter 3 of this section is specifically devoted to magnetic properties of the sediment. Apart from paleoenvironmental applications, paleomagnetic properties of deep-sea sediment have been shown to provide, through paleointensity measurements, exclusive global time markers, which allow precise correlations to be made between remote sedimentary sequences, whereas most other stratigraphic tools show varying discrepancies associated with climate and ocean dynamics. This is particularly the case of radiocarbon activity in biogenic minerals, one of the most widely used chronological tools for dating deep-sea sediments, which strongly depends on sedimentary processes and deep-sea ventilation rates. The fifth chapter addresses the issue of ^{14}C-dating applications and limitations of deep-sea sediment studies.

The second part of the volume deals with biological tracers that are among the most widely used in paleoceanography. They include tracers of bottom water conditions such as benthic foraminifers (see Chapter 7) and deep corals (see Chapter 12). However, they mostly include tracers of biological productivity and of a wide range of hydrographical conditions in the upper part of the water column, such as calcareous nannofossils (coccolithophorids), planktonic foraminifers, diatoms, dinoflagellate cysts, and organic biomarkers (see Chapters 6, 8, 9, 10, and 11). There are other biological tracers, not covered in this volume due to space limitations, which too are important in paleoceanography, particularly radiolaria and ostracoda, although they are less widely used than the above-mentioned tracers.

Table 1 The Proxies and the Parameters they can Help Reconstruct.

	Temperature	Salinity	Sea ice cover	Seasonality	Current velocity	Water mass origin/age	Dissolved oxygen (Eh)	pH/alkalinity	Productivity	Stratigraphy	Remarks
Physical properties of sediment			(+)		+		(+)		(+)		
Mineralogy (clay minerals)					(+)	(+)					
Paleomagnetic properties					(+)		(+)			+[1]	1. Magnetostratigraphy (10^3 a-resolution)
^{14}C-activity of biogenic minerals						(+)[1]				+[2]	1. Ventilation through offsets with ^{230}Th-ages (e.g., deep corals); 2. Dating limit ca. 35 ka
Benthic foraminifers	(+)	(+)			(+)	(+)	(+)	(+)[2]	+	(+)[1]	1. Eco(climato)stratigraphy 2. Ratio linings vs. shells
Deep corals (& traces/isotopes)	+	+				+	+	(+)	(+)		
Planktic foraminifers	+	+	(+)	(+)	(+)	(+)		+	+	(+)[1]	1. Eco(climato)stratigraphy
Diatoms	+	+	+	(+)				+	+	(+)[1]	1. Eco(climato)stratigraphy

Method							Comments
Dinocysts	+	+	+	(+)		(+)	+[1]	1. Biostratigraphy (10^5–10^6 a pacing)
Coccoliths	+	(+)	+			(+)	+[1]	1. Biostratigraphy (10^4–10^5 a pacing)
Alkenones & biomarkers	+		(+)[1]	(+)		(+)		1. IV-insaturated are salinity/sea-ice sensitive
Elemental tracers (sediment)	(+)	(+)	(+)	(+)	+	(+)		
Radiogenic isotopes (sediment)				+				
U-series (^{230}Th–^{231}Pa)				+		(+)		
Boron isotopes in sediment					+			
^{13}C & ^{18}O in foraminifers	(+)[1]	(+)[2]	(+)[3]	(+)[3]		(+)[3]	+	1. Paleotemperature equation: T and S (or $\delta^{18}O_{water}$); 2. $\delta^{18}O$-offsets in Arctic plantics? 3. ^{13}C-data: T, ventilation rate, productivity
Traces in biogenic minerals	+							

Note: +, proxy; (+), tracer.

The third part of the volume focuses on elemental and isotopic tracers. Chapter 14 aims at providing the reader with comprehensive insights into the mechanisms and processes that modulate the concentration of various elements in sedimentary records, as well as the application of elemental sediment geochemistry to paleoceanography, particularly for the assessment of paleoproductivity and oxygen availability. The following two chapters (15 and 16) are, respectively, on radiogenic and U-series isotopes in sediment. In both cases, these isotopes have been shown to be excellent tracers of the origin of water masses (and of transported particulate matter along continental margins) and more generally provide robust indications of thermohaline circulation patterns. Chapter 17 delves into the boron isotope composition of marine carbonates, which is primarily controlled by the pH of the ocean at least during the Quaternary when the ocean volume varied by only a few percent. Determining the ocean paleo-pH allows researchers to put constraints on changes in its alkalinity and to reconstruct atmospheric pCO_2 variations due to the strong coupling between the ocean surface and the atmosphere. Finally, the last two chapters (18 and 19) examine the isotopic and geochemical properties of the tests of foraminifera. The stable isotope composition of foraminiferal calcite, its ^{18}O content in particular, is extremely important in paleoceanography, because it was used to lay down the foundations of paleoceanography by Cesare Emiliani in the 1950s and subsequently by Nicholas Shackleton in the late 1960s and early 1970s, to both of whom this volume is dedicated. However, the modern interpretation of ^{18}O data in planktonic foraminifera has evolved considerably from what was the early debate on the respective contributions of changes in ocean isotopic composition and volume, or of the ocean temperature on observed isotopic shifts between glacials and interglacials, as illustrated in Chapter 18. From this perspective, the most recent studies of minor and trace element fractionation processes in foraminiferal calcite (Chapter 19), as well as in coral carbonate skeletons (Chapter 12), have been instrumental in obtaining a better assessment of salinity vs. temperature changes in the ocean when this information is combined with oxygen isotope data. In the third section, as in the second section, several geochemical tracers have been omitted, but the merit of those chosen here lies in their wide use.

When the question of adding a conclusion to this volume arose, we thought it to be left unwritten. Methods evolve through ongoing development of technological and computational capabilities, and the need for high-resolution, high-precision studies of deep-sea sediments is still great. Thus, we instead preferred to see the volume end with a contribution, which provides a true perspective on paleoceanographic studies: the issue of the modeling of the ocean and of the potential contribution of such studies to paleoceanography. Improved predictions of future changes of the ocean-climate system are needed, not only those ensuing from global warming, but also from the constraints of the solid Earth system. This will likely require improving paleoceanographic reconstructions beyond the Late Cenozoic, but this is another challenge.

In this volume, the focus has been given on the deep-sea, from continental margins to the open ocean. The preferred sites for paleoceanographic reconstructions are found in areas of pelagic sedimentation, which are located away from the turbulence of the high-energy shelf and nearshore environments, and their predominantly terrigenous inputs. It was suggested to contributors that they address three aspects for each method: (i) its historical development and benchmark contributions, (ii) its present state of the art, and (iii) examples of modern applications or interpretations. However, because each method has its own limitations and difficulties, great flexibility was given to contributors in order to adapt each chapter to the specificity of the tracer/proxy or method described, as well as to allow them to express their personal views. Nonetheless, a scientific committee has provided an external review of each chapter. This has shown itself to be an essential step for the production of this volume, as it has provided contributors with expert feedback on the content and scope of their individual chapters.

3. The Need for Multi-tracers and Multi-Proxy Approaches in Paleoceanography

During the 1970s, the development of the first "transfer functions" to reconstruct paleotemperatures from microfossil assemblages represented a breakthrough in paleoceanography and paleoclimatology, which led to the establishment of temperature maps of the ocean of the past (Imbrie & Kipp, 1971). The CLIMAP (1976, 1981) reconstitutions of the World Ocean during the last glacial maximum still remain some of the most cited and useful contributions of the last decades. However, with the improvement of computation technology, the development of new methods, and the use of new proxies (see Chapter 13), the CLIMAP paleotemperatures have been challenged. The various reconstructions made on the basis of different proxies or different methods were often conflicting and the subject of debate. The apparent discrepancies in the paleotemperature reconstructions, for example, are now generally considered as providing complementary pictures of the ocean's state. The "sea-surface temperature" (SST) is indeed a complex parameter in time and space, with respect to seasonal changes (see Figure 1), interannual-interdecadal variability, or the vertical stratification (thus salinity) and the position of fronts (thus currents). Clearly, one proxy cannot resolve all the variables within the SST parameter as a whole. Fossil assemblages of epi- to meso-pelagic organisms generally allow the reconstruction of paleotemperatures, but each datum has temporal and bathymetric distributions that often differ from the others. Furthermore, the temporal significance of the datum is ideally seasonal, but it can be linked to decadal variability and/or exceptional blooms in low-productivity settings, for example.

Most tracers and proxies of ocean temperature or salinity are related directly or indirectly to biological productivity. They include assemblages of microfossils, geochemical or isotopic measurements made on fossil carriers, and molecular biomarkers. All of these proxies are dependent upon the physical and chemical properties of ocean waters, such as temperature and salinity, which predominantly play a role on growth rates and osmotic exchanges between cells and the ambient water. They are also subjected to ecophysiological constraints, especially the phytoplanktonic organisms that are dependent on irradiance for photosynthesis, are affected by

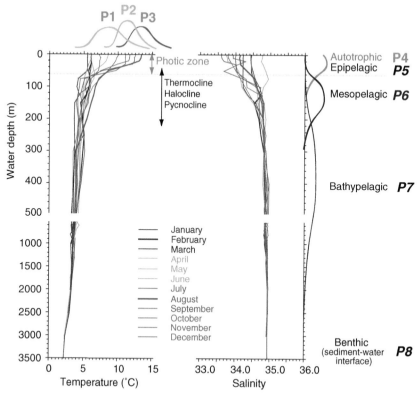

Figure 1 Monthly mean of temperature and salinity vs. depth in the water column in the northwest North Atlantic (50°N, 46°W; bathymetry = 3,450 m) drafted from data of the World Ocean Atlas (http://www.nodc.noaa.gov/OC5/pr.woa4.html). The curves P1–P9 correspond to the hypothetical ranges for nine different tracers or proxies. In the photic zone, P1, P2, and P3 might correspond to the seasonal succession of diatoms, dinoflagellates, and coccolithophorids. P4 could be any autotrophic proxy, and P5–P8 are heterotrophic proxies, either pelagic such as planktonic foraminifers (P5 and P6, for example) or radiolarians (P7), or benthic such as benthic foraminifers or many marine ostracoda (P8).

turbulence (influenced by stratification) which interferes with their mobility, and are dependent on nutrient availability. Therefore, the pelagic habitats can be defined from a turbulence-nutrient matrix along the nearshore-to-offshore continuum (see Figure 2 adapted after Smayda & Reynolds, 2001). Each biological organism has developed particular life forms and adaptive strategies, and there is a clear provincialism in the distribution of tracers and proxies. Since the primary production also determines the food availability for zooplankton and the organic carbon (particulate or dissolved) content in the water, and thus the chemistry of water masses, this provincialism impacts on the overall distribution of biological organisms, be it autotrophic, mixotrophic or heterotrophic, pelagic or benthic. From such a point of view, it is clear that none of the proxies can be universal; the simultaneous use of several proxies is necessary for a comprehensive perspective on the past ocean.

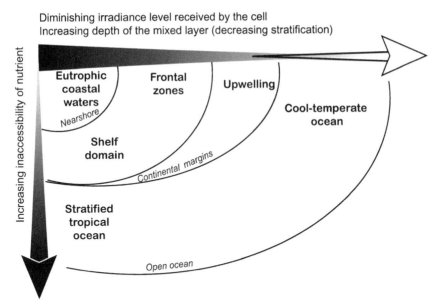

Figure 2 Simplified scheme of the provincialism of biogenic production after Smayda and Reynolds (2001). In general, the diversity in the habitats (and number of species) increases toward nearshore and eutrophic environments.

In general, the diversity in niches and habitats — and the diversity of species — is higher nearshore and along the continental margins than offshore because of more variability in biotic and abiotic parameters (see for example Chapter 7). Similarly, the pelagic realm is often more homogenous in terms of physical and chemical properties than the benthic environments, in which a high diversity of microhabitats may be found depending upon the substrate and bottom currents in addition to all other parameters. Therefore, the spatial resolution required and the precision needed from the proxies are higher for the analyses of nearshore or benthic environments, which may show mosaic-like patterns, than for offshore pelagic settings. Among the micropaleontological tracers of pelagic environments, some seem to be more appropriate for the study of nearshore and continental margin settings, whereas others seem more suitable for offshore studies. Diatoms (Chapter 8) and dinocysts (Chapter 9), which occur from freshwater to open ocean, often show high diversity of species nearshore and appear to be good tracers of hydrographical conditions along continental margins, whereas coccoliths (Chapter 10) and planktonic foraminifers (Chapter 6), which are exclusively marine and have a relatively low species diversity, seem to be more characteristic of offshore pelagic environments.

The various tracers and proxies, now available for studies in paleoceanography, yield complementary information, which should facilitate the development of models that would combine hydrographical conditions and biotic parameters to provide a comprehensive view of the past ocean.

4. FROM THE GEOLOGICAL RECORD TO THE SEDIMENTARY SIGNAL AND THE PROPERTIES OF THE WATER COLUMN

Most chapters in this volume include an in-depth examination of the difficulties associated with the reconstruction of ocean water properties from a given tracer or proxy. However, "upstream" of the use of any transfer function based on micro-paleontological assemblages, for example, there are potential sources of error linked to the transformation of the environmental signal by sedimentary processes, as well as the short- to long-term diagenetic evolution of the sedimentary record. Therefore, extracting environmental parameters from the geological archive requires a priori assessment of the discrepancies between the ocean parameters and the sedimentary record. One issue often mentioned but never fully addressed is the basic assumption that the sedimentary record represents vertical fluxes, and thus the processes and conditions that are characteristic of the overlying water column. In very open oceanic environments, such an assumption is possibly valid. However, elsewhere, lateral fluxes occur either in relation to shallow depths and the influence of currents, or through gravity displacements down continental slopes and rises, in the nepheloid layer for instance (see Chapter 1). The literature provides numerous examples of fine silt to clay particle transport over long distances (see Chapter 4). Such lateral fluxes have a strong impact on several geochemical tracers, especially the radiogenic isotopes (see Chapter 15), and Pa and Th isotopes (see Chapter 16). To some extent, micropaleontological assemblages might well be less modified by long-distance transport. It seems that a substantial part of the corresponding biological fluxes occurs with fecal pellets and mostly through marine snow (Turner, 2002). Fluxes from marine snow would indeed insure a better representativity of the micropaleontological assemblages in deep-sea sediment vs. their living counterpart in the overlying water column. Thus, the best criterion with respect to the representativity of the sedimentary record in terms of regional pelagic fluxes lies in the indications of high biogenic productivity. At sites or during intervals of low biological production, the lateral vs. vertical flux issue should be critically examined.

Aside from sedimentary processes that may be responsible for "noisy" records, early diagenetic processes may occur. They involve chemical and physical mechanisms. The preservation of biogenic minerals ($CaCO_3$ and opal) and that of organic matter (palynomorphs and biomarkers) is rarely perfect. Various dissolution indices have been proposed, mainly based on the relative state of preservation of microfossil remains. Such indices may help assess the reliability of the data or discard records with a sedimentary signature, which is too altered to be reliable. In addition, long-term diagenetic processes may occur, thus resulting in biased geochemical records.

The physical mixing of sediment is probably responsible for the most important transformation, in the mathematical sense of the term, of the sedimentary signal. It has a strong impact on the ^{14}C chronology, for example (see Chapter 5), as well as on the biological assemblages themselves and, of course, on their geochemical signature (see Chapter 18). Mixing processes have been investigated by many authors and most paleoceanographers are now familiar with benchmark papers and volume on this issue (e.g., Guinasso & Shink, 1975; Peng, Broecker, & Berger, 1979; Boudreau & Jorgensen, 2000; Bard, 2001). Except in cases in which very high sedimentation rates are ascertained, along with low biological activity on the sea floor fostered by anoxic conditions, the signal of all tracers should be considered as systematically modified or transformed by biological mixing processes. This transformation is size dependent and is highly variable in time and space. As illustrated by Smith and Schafer (1984), vertical mixing also occurs with a centimetric-scale lateral variability, which lends uncertainty to the use of deconvolution equations. There are a number of strategies for overcoming problems due to bioturbated sedimentary records, but there is no unequivocal recipe and one must acknowledge some intrinsic limitations of the paleoceanographic record. This is true for all signal carriers, and also concerns the time resolution that can be achieved, with the sole exception of deep corals, which have their own limitations (see Chapter 12).

5. HOW FAR BACK IN TIME ARE THE PROXIES EFFECTIVE?

When trying to document the past history of the ocean, uncertainty generally increases with age with respect to the accuracy of proxy reconstructions, and the use of most proxies progressively becomes impossible. In the case of biological remains, the uncertainty increases drastically with age on the Cenozoic timescale because of evolutionary processes that lead to speciation and extinction. For example, less than 10% of plankton species recovered in micropaleontological samples are common to both the early Cenozoic (Paleogene; 65.5–23.01 Myr) and the Pleistocene-Holocene epoch (1.81–0 Myr; cf. Gradstein et al., 2004). Even at the scale of the Neogene (last 23 Myr) or at the scale of the Pliocene and Pleistocene (last 5.33 Myr), the evolution of species has significantly modified the microfossil assemblages. At such time scales, the micropaleontological tracers are not accurate paleoceanographic proxies but may still constitute excellent chronostratigraphical markers. Figure 3, based on the temporal distribution of dinocyst taxa, illustrates the fact that beyond ca. 1.2 Myr ago, any transfer function with dinocysts, using their modern distribution, can be seriously questioned. One could possibly try to calibrate ecological requirements of extinct species from their co-occurrence with still existing species, but the exercise has yet to be made and its validity and usefulness demonstrated.

In addition to the uncertainty related to evolutionary processes, ecological adaptation of wide-ranging species adds another caveat. This is an issue not only for methods directly based on statistical analyses of microfossil assemblages, but also for many other proxies carried by the fossil themselves, such as the Mg/Ca ratio or ^{18}O-content of biogenic carbonates. This is a reason why, in the present volume, we have decided to include chapters on proxies that are not necessarily widely used, but do have some potential application for extending further back in time. This is the case of clay minerals or radiogenic isotopes, for example. Even in such cases, interpretations, when they go too far back in time, may differ from those based on

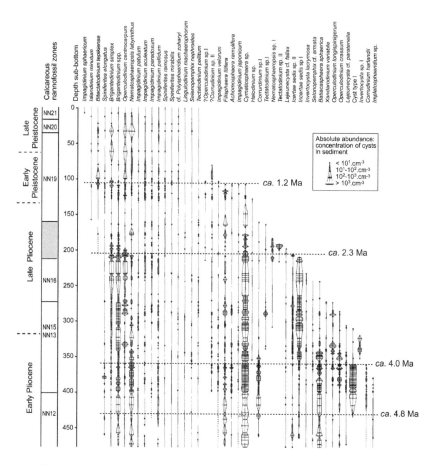

Figure 3 Biostratigraphy of major dinoflagellate taxa during the Late Cenozoic illustrating the limitation, in time, of reconstructions based on modern distribution of microorganisms. As seen here, transfer function methods using dinocyst assemblages cannot be reasonably used back in time beyond *ca.* 1.2 Myr ago (figure after de Vernal and Mudie, 1989).

the modern land–ocean geography and the general thermohaline circulation system. Major changes in the ocean's volume, tectonics, and large-scale departures from recent glacial/interglacial climate conditions have all had an impact on the concentration and residence time of geochemical tracers in the ocean.

Finally, some tracers have intrinsic temporal limits. This is the case of radioactive elements, the use of which is limited by their respective half-life. The U–series isotopes, such as ^{230}Th and ^{231}Pa (see Chapter 16), used to document sedimentary processes and water-mass properties, cannot provide robust information much beyond the last interglacial and the last glaciation, respectively. Adding to their relatively rapid decay (with half-lives of about 75 and 34 kyr, respectively), one must also take into consideration the fact that the tracer consists in both cases of their "excess" activity (over that inherited with the terrigenous carrier-fractions). Uncertainties in the geochemical properties of these fractions, linked to their source and previous cycling, particularly in soils where clay minerals show wide disequilibria between U-series isotopes, rapidly result in poorly constrained estimates of the "excess" fractions. In addition, uranium mobility in deep-sea sediments, between reduced and oxidized layers, rapidly adds some noise in records of supported/excess ratios between daughter isotopes. In a similar fashion, radiocarbon presents difficulties far back in time (see Chapter 5). Its use is not solely restricted by its decay and the analytical precision achieved (that would theoretically allow the dating of sediment back to *ca*. 50 kyr) but also by diagenetic reactions involving organic/inorganic carbon in pore-water and sediment, and sample recovery procedures; both may result in some contamination with more "active" carbon, especially of carbonates. Thus, one may reasonably consider ages beyond 35 kyr carried by foraminifer shells, for example, with much circumspection.

As seen in the above examples, using proxies far back in time raises more and more difficulties. This explains why this volume focuses on "Methods in Late Cenozoic Paleoceanography". Quantitative reconstructions of past properties of the ocean are, in practical terms, relatively attainable at the timescale of the last few hundred kyr, i.e., under the strong forcing of the eccentricity 100-kyr cycle. But they are difficult to attain in the preceding interval of the Pleistocene.

6. New Perspectives and Emerging Proxies

As a result of technological advances, new methodologies are emerging every now and then, and often provide new perspectives on the major processes that regulate the ocean. For example, direct measurement of isotopic paleotemperatures in carbonates may not be totally out of reach, as shown by Ghosh et al. (2006). In addition to the measurements of CO_2 masses 44, 45, and 46, representing various combinations of O and C isotopes, it is now possible to measure mass 47 (proportional to $^{13}C-^{18}O$ bonds in carbonate minerals, a direct function of the temperature during crystal growth), thus providing an access to the equilibrium temperature. The measurement of low abundance isotopologues should find a large array of new applications in paleoceanography, and there is hope for improvements in the near future. In a similar fashion, trace/minor element and/or isotopic analyses in single foraminifer shells by nanoSIMS, resonant laser secondary neutral mass spectrometry (Laser-SNMS) and MC-ICP-MS instruments (e.g., Rosenthal, Field, & Sherrell, 1999) are now routinely performed in many laboratories increasing the capability of geochemists to reconstruct minute differences of physical and chemical properties of the past ocean.

The wide scope of research perspectives that have opened up because of emerging technologies raises the issue of identifying important challenges and opportunities. There is no major theory to be demonstrated, as in the early days of paleoceanography when the astronomical theory of climate was still open to discussion. Now, the new challenge in this field of research would instead be to investigate why some geological records show departures from this theory, at least at the millennial timescale (e.g., Gallup, Cheng, & Edwards, 2002). There is no doubt that basic research issues are innumerable, and if the ocean is still poorly known, the paleo–ocean is even much less understood. However, investigations at sea require more international efforts, such as International Ocean Drilling Program (IODP) or IMAGES, and the use of the new technologies also entails enormous expenses. Thus, should we consider focusing future research on a few important issues and address basic problems only whenever possible? Among the many questions to examine more deeply, the following certainly require immediate attention by researchers:

- The role of the Arctic in the ocean-climate system and Thermohaline Circulation (THC) (for example, what do we know about the AMO during episodes without Arctic Ice and Greenland Ice?).
- The high-frequency instabilities and rapid changes of THC, in the recent past, not only of its AMO component, but also in upwelling areas.
- The productivity of the ocean during warmer episodes.
- The fate of carbon in the deep ocean, and the role of the nepheloid layer in geochemical fluxes to the deep ocean.
- The change in alkalinity and its impact on biomineralization processes.

Whatever the questions we choose to address on priority, efforts should be made to closely associate research in paleoceanography with modeling, in order to develop predictive tools. The second volume on "Late Cenozoic Paleoceanography" will address most of these emerging issues.

ACKNOWLEDGMENTS

The editors of this volume, as well as its contributors, would like to acknowledge the support of the scientific committee, which has been essential to the production of this volume. The help received from Ms Marie Morineaux of the UNESCO Chair for Global Change Research at UQAM, from series editor Pr. Hervé Chamley, from Ms. Patricia Wood (Montreal), and from Elsevier's staff is gratefully acknowledged.

REFERENCES

Bard, E. (2001). Paleoceanographic implications of the difference in deep-sediment mixing between large and fine particles. *Paleoceanography, 16,* 235–239.

Boudreau, B. P., & Jorgensen, B. B. (2000). *The benthic boundary layer: Transport processes and biogeochemistry.* Oxford University Press.

CLIMAP (1976). The surface of ice-age earth. *Science, 191,* 1131–1137.

Gallup, C. D., Cheng, H., & Edwards, R. L. (2002). Direct determination of the timing of sea level change during Termination II. *Science, 295,* 310–313.

Ghosh, P., Adkins, J., Affek, H., Balta, B., Guo, G., Schauble, E. A., & Schrag, D. (2006). $^{13}C-^{18}O$ bonds in carbonate minerals: A new kind of paleothermometer. *Geochimica et Cosmochimica Acta, 70,* 1439–1456.

Gradstein, F. M., Ogg, J. G., Smith, A. G., et al. (2004). *A geologic time scale 2004.* Cambridge University Press; see also the official website of the International Commission on Stratigraphy (ICS) under www.stratigraphy.org

Guinasso, N. L., & Shink, D. R. (1975). Quantitative estimates of biological mixing rates in abyssal sediments. *Journal of Geophysical Research, 80*, 3032–3043.

Imbrie, J., & Kipp, N. G. (1971). A new micropalaeontological method of quantitative palaeoclimatology: Application to a Late Pleistocene Caribbean core. In: K. K. Turekian (Ed.), *The late cenozoic glacial ages* (pp. 71–181). New Haven, CT: Yale University Press.

Peng, T. H., Broecker, W. S., & Berger, W. H. (1979). Rates of benthic mixing in deep-sea sediments as determined by radioactive tracers. *Quaternary Research, 11*, 141–149.

Rosenthal, Y., Field, P., & Sherrell, R. M. (1999). Precise determination of element/calcium ratios in calcareous samples using sector field inductively coupled plasma mass spectrometry. *Analytical Chemistry, 71*, 3248–3253.

Smayda, T. J., & Reynolds, C. S. (2001). Community assembly in marine phytoplankton: Application of recent models to harmful dinoflagellate blooms. *Journal of Plankton Research, 23*, 447–461.

Smith, J. N., & Schafer, C. T. (1984). Bioturbation processes in continental slope and rise sediments delineated by Pb-210, microfossil and textural indicators. *Journal of Marine Research, 42*, 1117–1145.

Turner, J. T. (2002). Zooplankton fecal pellets, marine snow and sinking phytoplankton blooms. *Aquatic Microbial Ecology, 27*(1), 57–102.

de Vernal, A., & Mudie, P. (1989). Pliocene to Recent Palynostratigraphy at ODP Sites 646 and 647, eastern and southern Labrador Sea. *Proceedings of the Ocean Drilling Program* 105B (pp. 401–422).

Part 1: Deep-Sea Sediment Properties

DEEP-SEA SEDIMENT DEPOSITS AND PROPERTIES CONTROLLED BY CURRENTS

Ian Nicholas McCave

Contents

1. Introduction   19
   1.1. Current indicators in deep-sea sediments   20
   1.2. Global ocean flow patterns   26
   1.3. Sediment delivery to the deep ocean   28
2. Sediment Transport and Deposition by Deep-Sea Currents   30
   2.1. Controlling factors   30
   2.2. Processes of deposition from turbulent boundary layers   33
3. Sediment Deposition: Quaternary Records of Flow in Large-Scale Features 38
   3.1. Sediment waves and drifts   38
   3.2. Applications of the sortable silt proxy: climate and flow   40
   3.3. Unmixing of size distributions into end-members   45
4. Current Problems and Prospects   45
   4.1. Grain size analysis   45
   4.2. Current or source control?   47
   4.3. Abnormally high sedimentation rates: the holy grail/poisoned chalice
        dilemma   48
   4.4. Spatial aliasing   49
   4.5. Ice-Rafted detritus   49
   4.6. Calibration   50
   4.7. Links with other flow/circulation and water mass proxies   52
References   54

1. INTRODUCTION

Sediments carry diverse types of information in their composition and grain size. An enormous number of chemical and isotopic attributes have been used to deduce environments of deposition and sediment production with particular emphasis on climatic variables. In many instances the bulk of the sediment is just

Developments in Marine Geology, Volume 1
ISSN 1572-5480, DOI 10.1016/S1572-5480(07)01006-8

seen as a carrier phase for the main information provider, often foraminifera. This attitude has led to several interpretational problems, e.g. in age modelling, which could be avoided by considering the whole sediment and the processes by which it is delivered, sorted and deposited. This topic and the interpretation of the vigour of the depositing flow are treated here.

1.1. Current Indicators in Deep-Sea Sediments

Initial recognition of the importance of deep currents in redistributing sediments came from the combination of:

1) Photographic evidence for bedforms under known currents (Heezen & Hollister, 1964),
2) Sedimentary structures in cores under strong currents (Hollister & Heezen, 1972),
3) Acoustic profiler data from echo-sounder and 3.5 kHz (Schneider et al., 1967), and
4) Seismic reflection profiles documenting large sediment bodies under current systems (Johnson & Schneider, 1969; Jones, Ewing, Ewing, & Eittreim, 1970).

1.1.1. Photographs

Systematic regional examination of photographs shows spatial variability of flow (Tucholke, Hollister, Biscaye, & Gardner, 1985) (Figure 1) and has yielded an ordinal scale of current intensity shown by photographed bedforms (McCave & Tucholke, 1986). Although these show the zonation of current speed very well by progression from subtle smoothing through longitudinal ripples to barchan ripples and crag-and-tail structure, this is virtually useless below the surface. The pervasive structure of muddy contourite sediments is bioturbation (Wetzel, 1991; Baldwin & McCave, 1999; Löwemark, Schönfeld, Werner, & Schafer, 2004 show X-radiographs). This destroys virtually all traces of original depositional structure.

1.1.2. Sedimentary structures

A few contourites show clear sedimentary structure (Figure 2), but these are relatively sandy (e.g. shown is 60–80% >63 μm). The sand (mainly foraminifera) content of many contourites is in the region of 10–20%, decreasing with increasing accumulation rate. In the case of a S.W. Indian Ocean core WIND 27K the sand percentage is 25–40% with an accumulation rate of $\sim$3 cm ka^{-1} for the period 0–60 ka. Over a 20 cm thick section shown in Figure 2, the sand percentage goes up to 80% and the structure is ripple cross-lamination. Our best estimate is that this section represents from 125 to 60 ka, a mean sedimentation rate of 0.3 cm ka^{-1}, considerably less and certainly not continuous. The current is inferred to have strengthened sufficiently to rework the section from stages 4 down to 5e, possibly in a couple of high-speed pulses lasting a few thousand years at the 5e–5d and 5a–4 transitions (McCave, Kiefer, Thornalley, & Elderfield, 2005). This demonstrates the fact that the stratigraphic resolution of sandy contourites is likely to be shot through with hiatuses. High

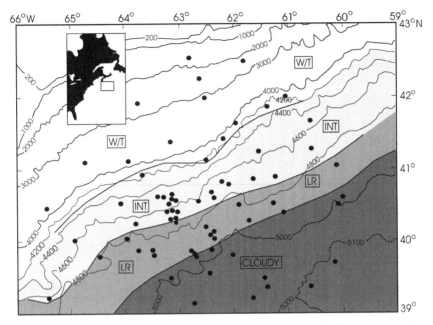

Figure 1 Current zonation of Nova Scotian Rise deduced from bedforms seen in bottom photographs. Based on data in Tucholke et al. (1985). Key to photographic indicators of increasing flow speed: W/T = weak/tranquil, INT = intermediate, LR = longitudinal ripples, Cloudy = photos showing turbid water due to resuspension. The HEBBLE instrument deployments and cores were around 40°27'N, 62°20'W.

(foram) sand contents are generally due to either dissolution of fine carbonate or winnowing removal of fine sediment.

1.1.3. Acoustic profiles

Mapping of echo character from 10 kHz sounders also showed current zonation on the margin (Hollister & Heezen, 1972), but this was superseded by the now ubiquitous 3.5 kHz whose interpretation was systematised by workers at Lamont in the 1970s (Damuth, 1980). The combination of high frequency echo sounder (10–20 kHz) and 3.5 kHz profiler can be revealing (Figure 3) (Winn, Kogler, & Werner, 1980), because the strength of the acoustic return is related to the surface sediment density (porosity) and grain size, which is a function of net accumulation rate, involving both deposition and winnowing. As can be seen from Figure 3, there is a spatially coherent response of sedimentation rate and reflectivity (mainly porosity) to current flow.

1.1.4. Seismic reflection profiles

Early seismic reflection profiling showed several drifts to be thick piles of current – deposited deep-sea sediments, e.g. Blake Outer Ridge, Eirik Drift, Feni Drift (Johnson & Schneider, 1969; Jones et al., 1970) (Figure 4). Over the years the number of drifts so recognised has increased greatly, diagnostic seismic characters have been documented (Faugères, Stow, Imbert, Viana, & Wynn, 1999) and several

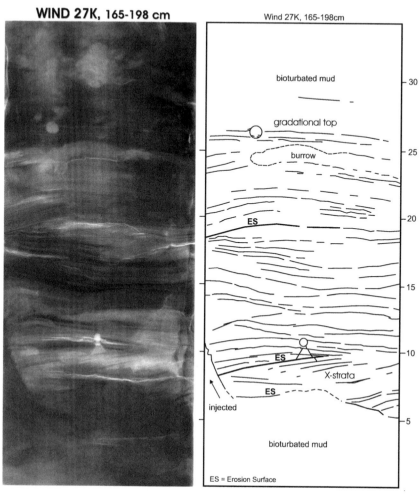

Figure 2   X-radiograph of core WIND 27K, 165-198 cm, from the Amirante Passage, equatorial western Indian Ocean, showing cross-laminated contourite muddy foraminiferal sand. Note that the base is sharp but deformed during coring and that the top is gradational in texture. Fine-scale lamination, cross lamination and erosion surfaces are evident. This 20 cm represents around 40,000 yr, demonstrating the stratigraphic penalty of too-high current speed.

edited volumes of records have been devoted to them (e.g. Rebesco & Stow, 2001; Stow, Pudsey, Howe, Faugeres, & Viana, 2002). Common large bedforms on drifts are sediment waves with wavelengths of 1–5 km, imaged by seismic reflection, to which much attention has also been paid (Mienert, Flood, & Dullo, 1994; Wynn & Stow, 2002). Some waves on continental margins are deposited under turbidity currents, some under usually geostrophic currents, thereby setting a trap for the unwary. Distinction is not easy as both usually migrate up current; but, whereas turbidite waves are generally perpendicular to flow and nearly parallel with contours while contourite waves are oblique to flow, the flow direction is often unknown and the strike of low amplitude mud waves needs high quality swath

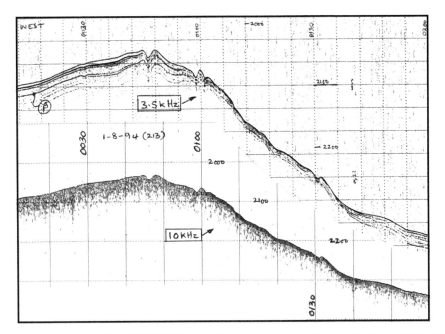

Figure 3 (Upper) 3.5 kHz WNW–ESE profile across northern Gardar Drift at 60°N, 23°W showing reduced sedimentation rate on the eastern slope of the drift by closer spaced reflectors. (Lower) 10 kHz echo sounder record along the same track showing higher amplitude reflection (redder colour) of the 'harder' sea bed more strongly affected by currents, corresponding to coarser size and slower net accumulation rate (from McCave, 1994). Coarser sediments occur on the eastern side of the drift compared with the western side where they are finer (Bianchi & McCave, 2000, their Figure 19).

bathymetry to be detected. One large fan is made of mud turbidites that have been entrained in a deep western boundary current, and thus termed a 'fan-drift' (Carter & McCave, 1994, 2002), while another is made of contourites and called a 'contourite fan' (Mezerais, Faugères, Figueiredo, & Masse, 1993). It requires a combination of seismic and 3.5 kHz profiles with cores to resolve many of these ambiguities.

1.1.5. Sediment grain size

Studies of particle size shed light on many facets of deep-sea sediments including especially depositional conditions. Because deep-sea sediments normally show few structures other than biological disturbance, grain size parameters have been used as the best indicator of relative flow speed. The mean size of medium to very coarse silt has proved to be a useful measure of the speed of the depositing flow. In the 1960s different interests led to experimental work on transport and deposition of fine-grained sediment (Einstein & Krone, 1962; Rees, 1966; White, 1970). This established critical shear stresses for erosion and deposition, but for the latter the work was done with aggregated sediment for which the control of critical deposition conditions was argued by McCave and Swift (1976) to be the (unknown) settling velocity.

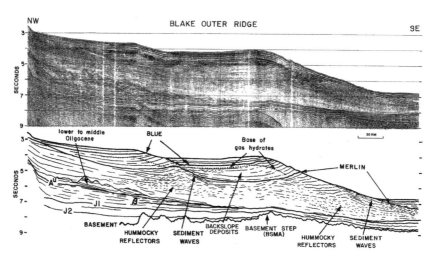

Figure 4   Multichannel seismic profile over a major drift, Blake Outer Ridge off the SE USA (Mountain & Tucholke, 1985; Figures 8–27). At several places in the edifice mud waves are observed, e.g. under the Merlin reflector, as well as episodes of erosion and deposition since erosional Horizon A^U in the late Eocene. Note the outcropping reflectors on the exposed NE face of the drift compared with deposition on the SW side.

In the deep sea the work of Ledbetter and Johnson (1976) using the Elzone counter led the way for use of fine particle size as a flow speed indicator, using the 'total silt mean size' (4–63 µm, but most workers now designate clay as <2 µm and silt as 2–63 µm grain diameter) (Figure 5). Ledbetter (1986a) demonstrated that the silt mean size was greatest under the path of the deep western boundary current (DWBC) in the Argentine Basin. It is important to remember that fine sediments need to be disaggregated for analysis, but that is not the state in which much of the material was deposited, because it involved varying degrees of particle aggregation. Dynamical inferences should be made based on the part of the size spectrum that responds directly to hydrodynamics and not from the properties of the whole disaggregated size distribution. However, as particles increase in size they become less prone to aggregation and large aggregates are more easily broken up by turbulent stresses. It was this fact that led McCave, Manighetti, and Robinson (1995b) to propose the use of the 10–63 µm silt fraction, 'Sortable Silt' (mean size denoted by $\overline{SS}$), as a flow speed indicator because the grains were more likely to have been deposited individually in response to fluid stresses.

1.1.6. Magnetic fabric and concentration

The work of Rees (1961) and Hamilton and Rees (1970) laid the basis for relating the fabric of sediments determined via the anisotropy of magnetic susceptibility (AnMS) to flow speed. Ellwood and Ledbetter (1977, and several later papers) applied this method to flow speed first in the Vema Channel (SW Atlantic), then elsewhere. The principle is that the AnMS tensor (an ellipsoid) reflects the average grain alignment and attitude (lineation, foliation and degree of anisotropy) as described by Jelinek (1981). The simple degree of anisotropy is $P = \kappa_1/\kappa_3$ (lineation

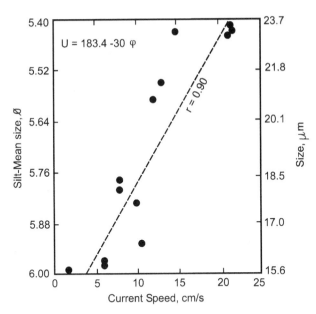

Figure 5 Ledbetter's (1986b) calibration of fine particle size to current strength based on samples and current metres in the Vema Channel area.

is κ_1/κ_2, foliation is κ_2/κ_3). His expression for the 'corrected' degree of anisotropy, related to flow strength, is

$$P' = \exp\{2((\ln \kappa_1 - K)^2 + (\ln \kappa_2 - K)^2 + (\ln \kappa_3 - K)^2)\} \tag{1}$$

where the principal susceptibilities are $\kappa_1 > \kappa_2 > \kappa_3$ on orthogonal axes, and K is the mean $(\ln\kappa_1 + \ln\kappa_2 + \ln\kappa_3)/3$. This was used by Kissel, Laj, Lehman, Labeyrie, and Bout-Roumazeilles (1997) to demonstrate clear variation in the strength of the Iceland–Scotland overflow south of Iceland with faster flow in warm stages (Figure 6).

The concentration of magnetic grains (mainly very fine magnetite) has also been interpreted as a proxy for the strength of North Atlantic Deep Water (NADW) (Kissel, 2005). This must relate to the erosive power of the lower NADW (LNADW) source flows passing over and through the basaltic Greenland–Scotland Ridge, where the magnetite is entrained. Although some have claimed a dynamically related size dependence of the magnetic grains identified by hysteresis methods (Snowball & Moros, 2003), it is difficult to envisage a hydrodynamic sorting process for such small grains (10 nm–10 μm), which must be deposited in aggregates. It is more likely that the magnetic size covaries with erosion-related parameters at source and mimics the concentration parameter of Kissel (2005).

1.1.7. Biological indicators of flow speed

Jumars (1993, p. 274) argued that 'Although species that are very sensitive and others that are very insensitive to sediment transport are known, there is as yet no agreed on a priori functional grouping of organisms into these categories'.

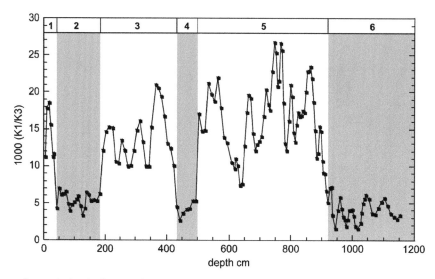

Figure 6 Variation in degree of anisotropy of magnetic susceptibility for core SU90-33 showing the marine isotope stages and low degree (corresponding to low flow speed) in cold stages 2, 4 and 6 (Kissel et al., 1997).

However, several studies have noted species assemblages, e.g. depletion of epifaunal isopods at a high energy site relative to a tranquil one (Thistle & Wilson, 1987), and individual properties, e.g. body shape allowing epifauna to enter the seabed to escape high current stress (Thistle, Yingst, & Fauchald, 1985), which are related to flow. More recently Schönfeld (2002) has found functional groups of epibenthic foraminifera that record sensitivity to flow. They occupy elevated positions above the sediment that give them better access to fast food in suspension transport. These elevated epibenthos (EEB) occur preferentially under the fast Mediterranean Outflow in the Gulf of Cadiz. A somewhat oblique means of obtaining flow speed via a speed–salinity correlation was used to calibrate the assemblages to speed, but the result does express the increase in percentage EEB of total epibenthos with location relative to the fast core of the current.

1.2. Global Ocean Flow Patterns

1.2.1. Deep ocean flow
Following Stommel's (1958) deduction of the necessity for DWBC, the validity of the theory was simultaneously demonstrated by Swallow and Worthington (1957), but was already implicit in the hydrography of Wüst (1955), all elegantly summarised by Warren (1981). Modern understanding of the thermohaline circulation was reviewed by Schmitz (1995), but this lacks detail applicable to the problem of bottom sediment transport. The geostrophic flows on the maps of Reid (1994, 1997, 2003) and bottom water properties of Mantyla and Reid (1983) reveal the bottom flow patterns that move and deposit sediment (Figure 7A). The key feature of these patterns is the localisation of currents along boundaries to the right (left) of

(a)

Figure 7A   Bottom flow in world ocean (McCave, 1986) with drift locations.

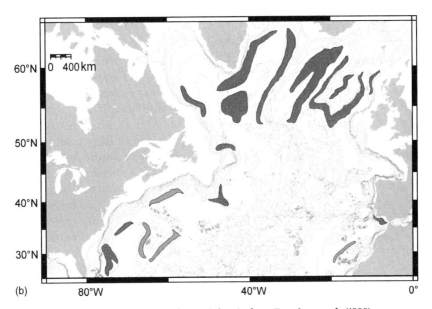

(b)

Figure 7B   Details of drift location in the N. Atlantic from Faugères et al. (1999).

the flow in the northern (southern) hemisphere. These boundaries may be in the middle of oceans, e.g. Reykjanes Ridge, SE Indian Ridge, 90-East Ridge, Pacific–Antarctic Ridge, as well as the continental margins. Drifts, however, are more common along the latter because that is where the major sediment supply points lie (Figure 7B).

1.2.2. Flow of intermediate waters

The great majority of cases of current-influenced sedimentation occur under the deep ocean current system. The influence of intermediate and surface water circulation is hard to discern because it affects sediments at fairly shallow depths on continental margins ($<\sim$1,500 m) where other slope processes intervene, particularly off-shelf fallout and down-slope transport. Convenient seamounts projecting into intermediate waters usually have locally intensified internal tides and a cap of winnowed sediment. Apart from that, intermediate waters of southern origin tend to be undersaturated in carbonate, which can result in low carbonate percentages and consequent difficulties with foraminiferal stratigraphy. For all these reasons it is hard to locate good sedimentary records of intermediate water flow, which have both satisfactory chronology and an uncontaminated record of ocean flow. Certainly many of the chemical and radioactive tracers in the current suite of proxys are prejudiced by shelf fallout, as well as grain size measures.

Nevertheless, several places have yielded sedimentary, isotopic and geochemical data at intermediate water depths; e.g. Chatham Rise, N. of Rockall Plateau, Bjorn Drift, N. Grand Banks (Nelson, Cooke, Hendy, & Cuthbertson, 1993; Pahnke et al., 2003; Lassen et al., 2002; Hall, Bianchi, & Evans, 2004; Piper & Pereira, 1992). Doubtless there are more worth searching for.

1.2.3. Surface ocean flow

It is very difficult to obtain records of the flow of the surface wind-driven circulation. Surface western boundary currents (SWBC) carry large quantities of heat polewards and so are targets for palaeoclimatology. Biogeography and sea surface temperature (SST) may reflect distributions of currents and water masses, but information on the vigour of the flow is generally lacking. This follows from the outline given above of problems with intermediate waters, only more acutely. In the major northern ocean gyres there is westward intensification that yields the Gulf Stream and Kuroshio (Stommel, 1948). These, the largest SWBCs, reach depths over 1,000 m and so influence deep-sea sediments. The Gulf Stream scours Blake Plateau at 1,000 m and sweeps sediment over into the Blake Basin. Similarly the Kuroshio sweeps over the Ryuku Ridge and deposits sediment in the southern Okinawa Trough (Wei, 2006). These areas are difficult targets for study of the surface circulation.

The biggest of all wind-driven currents is the Antarctic Circumpolar Current (ACC) (Rintoul et al., 2001) which reaches the bottom in several places, including the Scotia Sea region. Here Pudsey and Howe's (1998) grain size data suggested that the ACC was stronger during the glacials. This accords with inferences on glacial wind strengths from the record of high dust concentrations in Antarctic ice cores. However, it is of interest that the record of flow of the Florida Current (source of the Gulf Stream) (Lynch-Stieglitz, Curry, & Slowey, 1999) suggests a reduced glacial flow, but the trade winds at least were stronger then than now (Parkin, 1974).

1.3. Sediment Delivery to the Deep Ocean

1.3.1. Pelagic flux: biological, aeolian and ice-rafted detritus

The deep circulation does not deliver sediment to the ocean, but reworks it once it has arrived. The same is usually true of intermediate water flow, as it generally does not have a strong downslope component. Sediment delivery is mainly by gravity-driven processes. In the open ocean far from land, pelagic sinking flux is rapid — on a timescale of several tens of days — because of particle aggregation and biological packaging (McCave, 1975, 1984; Honjo, 1976, 1978). This affects biogenic components, ice-rafted detritus (IRD) and aeolian sediment delivered to the sea surface from the atmosphere. As a result, in areas with weak bottom currents, the pattern of even the smallest biogenic components reflects quite accurately their distribution in overlying surface waters, e.g. coccolith biogeography (McIntyre, 1967; Okada & Honjo, 1973). Moore and Heath (1978) demonstrate through careful mineralogical work that the central Pacific is marked by a belt of enhanced quartz content that reflects the position of the jet stream. Thus atmospheric patterns may also show up in sediment properties *if undisturbed by current redistribution*. Spatial distribution of IRD reflects the paths taken by icebergs (Hemming, 2004), because, as the indicator is immobile terrigenous sand, the pattern is not altered by current redistribution. The North Atlantic's 'IRD Belt' thus reflects the path of icebergs in the glacial surface circulation (plus some wind influence on trajectory).

1.3.2. Continental margins: downslope flux, climatic signals in delivered sediment

Close to continental margins, sediment is carried downslope in copious quantities by turbidity currents and debris flows that are responsible for construction of much of the continental rise. The mass failures yielding large turbidity currents and debris flows involve failure of ten to several tens of metres of sediment on the slope (Embley & Jacobi 1977), thereby mixing the products of several glacial–interglacial cycles (Weaver & Thomson 1993; Weaver, 1994). Basal erosion by turbidity-currents also produces a mixture of sediment ages and properties. Gravity flows do not, therefore, deliver sediment bearing a pure signature characteristic of conditions in the source area at the time they were triggered. Strong deep-sea currents resuspend this sediment, forming nepheloid layers, and move it to areas of spatial decrease in flow speed, causing deposition. Turbid bottom nepheloid layers (BNL) are found all over the sea bottom, with concentrations ranging from 20 to over 2,000 mg m^{-3} (McCave, 2001). Current-controlled fine sediment deposition involves transport in and removal from nepheloid layers (Figure 8). Processes acting in BNLs are responsible for sorting fine sediments.

1.3.3. Continental margins: intermediate nepheloid layers

Another class of nepheloid layers found especially at continental margins are intermediate nepheloid layers (INL) (Figure 8). These occur frequently at high levels off the upper slope and at the depth of the shelf-edge. From here they spread out across the continental margin while the sediment in them aggregates and rains out rapidly (Hill & Nowell, 1990). All nepheloid layers are principally produced by resuspension of bottom sediments. Their distribution thus indicates the dispersal of resuspended sediment. Most concentrated nepheloid layers occur on the continental shelf or deep continental margin, giving an indication of the locus of active resuspension and redeposition by strong bottom currents. On upper continental margins this is by action of internal waves and tides, and at the shelf edge by storm waves and currents (Baker & Hickey, 1986; Dickson & McCave, 1986; McPhee-Shaw, 2006). These layers may also be produced by internal wave resuspension in canyons, making canyons point sources for supply of turbid layers which mix out

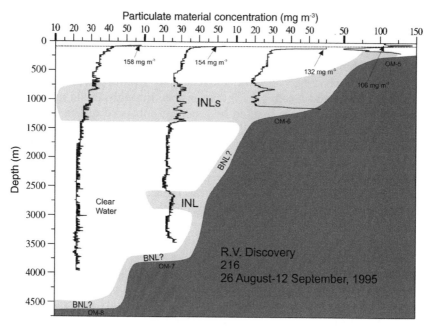

Figure 8   Turbidity distribution with depth over Goban Spur sediment trap transect during D216 (Autumn). Note the presence of intermediate nepheloid layers carrying sediment off slope. Data is cropped at 100 m to allow the structure of the deep water to be seen. Concentration estimates based on optical transmission profiles (from McCave et al., 2001).

into the ocean interior (Gardner, 1989). All these processes would have operated in the past with the added effect of lower sea level putting the whole of the shelf break well into the zone of resuspension for even moderate waves (water depths of ~50 m or less). Rivers discharged more directly to the oceans at lower sea level, so climatic signals in mineralogical composition would also have been transmitted via hyperpycnal underflows with less mixing than at sea level high-stand.

2. SEDIMENT TRANSPORT AND DEPOSITION BY DEEP-SEA CURRENTS

2.1. Controlling Factors

2.1.1. Settling velocity of grains and aggregates, w_s

Still–water settling velocity (w_s) of spheres follows Stokes Law, $w_s = \Delta\rho_p g d^2 / 18\mu$, as long as $Re_p < {\sim}0.5$ or $d < 100\,\mu m$ in water (particle Reynold's number $Re_p = w_s d/v$ where μ is the molecular viscosity and kinematic viscosity $v = \mu/\rho$, d is grain diameter, $\Delta\rho_p = (\rho_p - \rho)$ is particle minus fluid density. Reductions of up to a factor of 3 slower occur, depending on shape (Lerman, Lal, & Dacey, 1974). Viscosity variation with temperature of seawater gives nearly a factor of 2 variation in w_s (Winegard, 1970). Density exerts a strong control, particularly for *aggregates* made up of quartz-carbonate density solids of 2,500–2,900 $kg\,m^{-3}$ that, with neutrally buoyant organic matter, form particles of much lower saturated bulk density, often $<1,100\,kg\,m^{-3}$ (McCave, 1984). The observed size dependence of density has been expressed in terms of the fractal dimension, D3, the power to which diameter d_i must be raised to obtain solids mass within a floc by Logan and Wiilkinson (1990). Combined with Stokes Law one gets an equation for floc sinking speed (Hill & McCave, 2001, Chapter 4).

Particles that are not solid — aggregates, hollow particles and grains containing gas bubbles — have very variable densities. The most important classes of hollow particles are foraminifera, diatoms and radiolaria. Foraminifera on the bed are often partially sediment-filled, resulting in saturated bulk densities of 1,150–1,550 $kg\,m^{-3}$. The resulting range of $\Delta\rho_p$ from 100 to 500 (with deep seawater density of 1,050 $kg\,m^{-3}$) translates into settling velocity of 125–500 $m\,day^{-1}$ for 200 μm foraminifera — the low end for freshly sinking ones, the high end for resuspended filled ones. Aggregates typically have settling velocities of 1–2 $mm\,s^{-1}$ or 85–170 $m\,day^{-1}$. With variable size, density and viscosity, particle-sinking rates can be from 50 to 1,000 $m\,day^{-1}$. This gets material from surface to bottom in a matter of days to a few weeks.

Aggregates may be formed by physical (often referred to as flocculation, sometimes coagulation) or biological processes, frequently involving feeding and production of faeces or mucous. Particles are brought together by Brownian motion, turbulent shear or by larger, fast-sinking particles sweeping up finer ones, like rain falling through mist (McCave, 1984; Honeyman & Santschi, 1989). Organic mucous from bacteria acts as 'glue', allowing particles which get close to stick. The 'flocculation factor', the ratio of the settling velocity of an aggregate to the settling velocity of the primary particles of which it is made, can be up to 10^5 (see equation 4).

Aggregates are not stable entities. The organic membrane covering faecal pellets decays and the mucous that holds aggregates together also degrades, so particles fall apart while sinking (Honjo & Roman, 1978; Lampitt, Noji, & von Bodungen, 1990) Aggregates assembled by moderate levels of turbulence in the outer part of the boundary layer may be broken up by more energetic turbulent eddies close to the boundary. Floc diameter is inversely proportional to shear stress to some poorly known power in the range 0.25–1. Close to the bed high shear breaks large sloppy aggregates into smaller pieces (Fugate & Friedrichs, 2003).

2.1.2. Boundary layer flow structure and turbulence

In the deep sea the frictional boundary layer can extend several tens of metres above the bed. Near the bed boundary layers are intensely turbulent with the drag force exerted on the bed τ_o related to the turbulent intensity because the stress is transmitted by turbulent eddies. In the vertical plane $\tau_o = -\rho\overline{u'w'}$ where u' is the turbulent velocity component in the flow direction and w' is the vertical component, measured close to the bed. This expression is important because u' and w' are related so that $\tau_o \propto <w'>^2$ (where $<>$ denotes the rms value). This vertical turbulent velocity is responsible for keeping particles in suspension, and the turbulent stress $\rho\,\overline{u'w'}$ either causes aggregation or, at higher values, disaggregates fine particles. The term $(\tau_o/\rho)^{1/2}$ has the dimensions of a velocity called the shear or friction velocity u_*. From the above it can be seen that $u_* \propto <u'>$, (Dade, Hogg, & Boudreau, 2001). One common criterion for the ability of a flow to maintain grains in suspension is $w_s \sim <u'>$, and thus the largest grain that can be held up has $w_s \sim u_*$.

In a turbulent flow the speed decreases logarithmically towards the bed and very close to the bed becomes laminar, or at least dominated by viscosity (v), in a layer known as the *viscous sublayer* of the boundary layer. This is very thin (thickness δ_v). In water, for a flow that just moves very fine sand, ($u_* = 0.01\,\mathrm{m\,s}^{-1}$, $v = 1.5 \times 10^{-6}\,\mathrm{m^2\,s}^{-1}$), $\delta_v = 10v/u_*$ is just 1.5 mm thick. However, this is 15 times the diameter of very fine sand. The shear across this layer is large; for $u_* = 0.01\,\mathrm{m\,s}^{-1}$ the speed goes from 0 to $0.1\,\mathrm{m\,s}^{-1}$ in just 1.5 mm. Weak aggregates cannot survive this shear and break up (Hunt, 1986). Above this sublayer there is an intensely turbulent 'buffer layer', which is transitional to a region in which the flow speed varies as the logarithm of height above the bed. In deposition most particles arriving at the bed have passed through the viscous-dominated layer, which is $\sim$2–10 mm thick in the deep sea. Although viscous dominated, this layer actually has high- and low-speed streaks, and unsteady high-speed *bursts* of fluid out of the layer and *sweeps* of fluid into it from outside. These are associated with stress typically up to 10 times the average (and extremes of 30 times), so even strongly bound particles may be ripped apart on approaching the bed. Equally, weaker aggregates that have made it to the bed and are partly stuck to it may also disintegrate with their components being ejected into the main flow. These are powerful size-sorting mechanisms whereby the sublayer can retain some sizes while rejecting others.

2.1.3. Erodibility

An important feature of sediment erodability is that critical movement conditions for surface muddy sediment occur at lower stress than movement of foraminiferal sand. The surface layer of sediment is biologically processed into mobile 'fluff' that moves at $\tau_{oc} \approx 0.02\,\mathrm{Pa}$, whereas more consolidated mud moves at higher values, depending on water content, but generally $\geqslant 0.05\,\mathrm{Pa}$ (field evidence: Gross & Williams, 1991; experimental evidence: Schaaf, Grenz, Pinazo, & Lansard, 2006) (see Figure 9). This is equivalent to geostrophic flow speeds of $U_g = 9.6$–13 to 15–21 cm s^{-1} (U_g is 22–30u_*, according to the drag coefficient used (Bird, Weatherly, & Wimbush, 1982; Weatherly, 1984)). Critical conditions for partly sediment-filled fine foram sand ($d = 200\,\mu\mathrm{m}$, $\rho_s = 1{,}500\,\mathrm{kg\,m}^{-3}$) is $\tau_{oc} \geqslant 0.064\,\mathrm{Pa}$, equivalent to flow speed $U_g > 17\,\mathrm{cm\,s}^{-1}$. This means that, under conditions when most deep-sea muds can be eroded, forams are immobile, resulting in a sandy lag. To generate ripples of empty forams requires flow speeds over 18 cm s^{-1}, but sediment-filled ones would need speeds over 30 cm s^{-1}. The above is based on experimental data on erosion and bedform stability fields in non-dimensional frameworks (Miller & Komar, 1977; Southard & Boguchwal, 1990). The importance of this is that foram-moving conditions are also mud-removal conditions leading to formation of winnowed lags and reduced stratigraphic resolution.

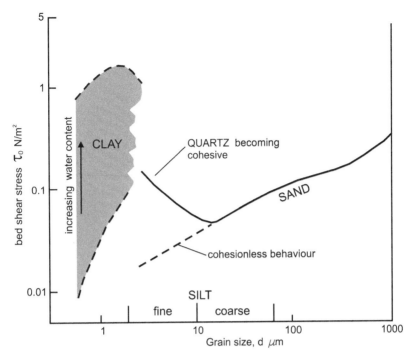

Figure 9 Critical erosion flow shear stress versus grain size showing noncohesive behaviour down to about 10 μm, and cohesive behaviour below that with a strong dependence on water content. After McCave et al. (1995a).

2.2. Processes of Deposition from Turbulent Boundary Layers

2.2.1. Rate of deposition

'Deposition' of bedload is fairly straightforward: it stops moving. This occurs in water at a shear stress only slightly lower than the critical erosion stress. Erosion stress is not measured precisely enough for single sizes (and certainly not for mixtures) to distinguish between non-cohesive erosion and deposition stresses, so they are effectively the same, though one would expect a difference analogous to the coefficients of static and sliding friction. For suspended sediment, once the stress has decreased below the suspension threshold material will sink into the near-bed region thereby increasing the concentration and causing some material to be deposited.

Experiments in flowing water by Einstein and Krone (1962) showed that for concentrations C_o less than $300 \, \mathrm{g \, m^{-3}}$, below a certain shear stress τ_d, the concentration in suspension C_t decreases exponentially with time (t):

$$C_t = C_o \exp(-w_s p t / D)$$

where D is the depth of flow (or thickness of the boundary layer) and p is the probability of deposition, the probability being given by $p = (1 - \tau_o / \tau_d)$.

2.2.2. Limiting shear stress for deposition τ_d

In the expression given above, τ_d is the limiting shear stress for deposition, the stress below which all the sediment will eventually deposit, but strictly this should be τ_{di} for particles of a given settling velocity w_{si}, as it is not to be expected that all grain sizes will deposit simultaneously. This yields:

$$R_{di} = C_{bi}w_{si}(1 - \tau_o/\tau_{di})$$

for the rate of deposition R_{di} of the ith fraction(kg m^{-2} s^{-1}) (the total rate of deposition is thus ΣR_{di} summed over all fractions). Here C_{bi} is the concentration of particles of w_{si} near the bed. If there is no flow the deposition rate reduces simply to the settling flux $C_{bi}w_{si}$.

The critical deposition stress (τ_{di}) is the stress below which particles of a given settling velocity will deposit, while those of smaller settling velocity will be ejected from the viscous sublayer, and is not well known. McCave and Swift (1976) suggested it is probably related to the settling velocity of the particles, whether aggregates or single grains. They assumed that it is given by the critical erosion stress for non-cohesive grains because below that value movement ceases and any grain reaching the bed would stop. This is shown by the non-cohesive line on the critical erosion diagram (Figure 9). These values and alternatives given by Self, Nowell, and Jumars (1989) and Dade, Nowell, and Jumars (1992) yield a range of shear stresses. Deposition of 10 µm particles occurs at stresses <0.010 Pa ($U_g<7$ to 9 cm s^{-1}) (Self), <0.015–0.03 Pa ($U_g<8$–16 cm s^{-1}) (Dade), and <0.045 Pa ($U_g<14$–20 cms^{-1}) (McCave & Swift). This range of shear stresses (0.01–0.045 Pa) is a little lower than that indicated by Hunt's (1986) experiments for the breakup of montmorillonite and illite flocs due to shear, namely 0.04–0.16 Pa. The implication is that most aggregates less than about 10 µm in diameter will survive during deposition from currents. However, above that size aggregates are increasingly likely to be broken up in the buffer layer. Hydrodynamic processes of sorting in the viscous sublayer thus tend to act on primary particles for sizes greater than 10 µm. Consequently under stronger flow this material will be size-sorted according to its disaggregated grain size whereas finer silt is not because it occurs in aggregates.

2.2.3. Sorting by selective deposition: cohesive versus sortable silt

The basis for the sortable silt flow speed proxy was set out in some detail by McCave et al. (1995b) and is reviewed at length by McCave and Hall (2006). The method proceeded from theory and observations in the High Energy Benthic Boundary Layer Experiment (HEBBLE) area (Figure 1) showing that terrigenous grain-size distributions had a pronounced mode >10 µm in places and at times when the flow speed was faster (Figure 10). The terrigenous silt fraction was accordingly divided into two fractions, 2–10 µm cohesive silt and 10–63 µm sortable silt. The mean size of the latter fraction (denoted by $\overline{SS}$) was proposed as a more sensitive indicator of the flow speed of the depositing current than the total 4–63 µm 'silt mean size' of Ledbetter (1986b), which included cohesive material. The percentage of sortable silt (SS%) of the terrigenous mud fraction was also suggested as a current index. The transition to cohesive particle behaviour occurs partly not only because clay minerals, with their charge imbalances, enter the

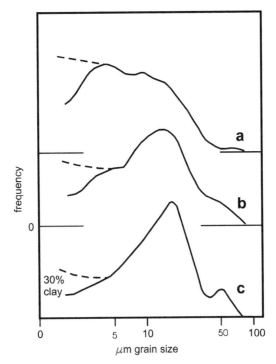

Figure 10 Size distributions of sediment from the Nova Scotian Rise measured by Coulter Counter (in McCave, 1985) showing deposits under (a) slow currents at 4,000 m water depth, (b) after moderate currents of 5–10 cm s^{-1} at 4,800 m and (c) after strong currents (10–15 cm s^{-1}) at 4,800 m.

compositional spectrum but also because Van der Waals forces become important for sizes less than ~10 µm (Figure 9). Below that size deposition is dominated by aggregates from which dynamical inferences can only be made on the basis of modelling with several assumptions (see Curran, Hill, Schell, Milligan, & Piper, 2004).

Sorting takes place through acceptance into/rejection from the viscous sublayer. There is a substantial region of overlap between the trapped and rejected populations. The controlling variables are the critical erosion stress (τ_e), the critical suspension stress (τ_s) and the critical deposition stress (τ_d). In general $\tau_d < \tau_e < \tau_s$ giving a bedload region for fine sediment between τ_e and τ_s (Dade et al., 1992). Sorting of muds thus arises mainly from *selective deposition*. Mehta and Lott (1987) also argued for sorting by selective deposition on the basis of the probable relationship between the critical deposition stress and settling velocity, and the likelihood of aggregate breakup close to the bed. Although fine-sediment distributions may have a moderately well sorted silt mode, overall most are poorly sorted. This is due to deposition of some aggregates with larger individual grains where strong flocs and grains have similar settling velocities.

Floc breakup near the bed appears to be borne out by some recent measurements in shallow water where particle size in suspension *decreases* towards the bed in

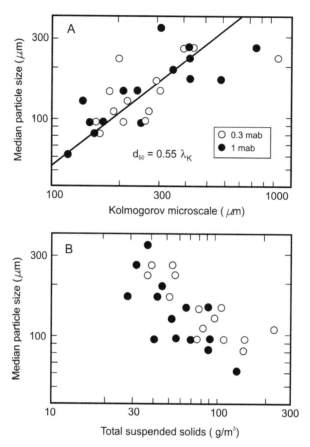

Figure 11 (A) Data from a laser *in situ* sizer (LISST) within 1 m of the bed (1 mab) showing a dependence of particle aggregate size on Kolmogorov microscale length which decreases as shear increases closer to the bed (microscale $\lambda_k = (v/\Gamma)^{\frac{1}{2}}$ where v is the kinematic viscosity and Γ is the shear rate $(\varepsilon/v)^{\frac{1}{2}}$ in which ε is the turbulent kinetic energy dissipation rate. (B) Particle size decreasing with increasing concentration low in the boundary layer because of shear breakup of aggregates (whereas increasing size would normally be expected as collisions are more frequent at higher concentration). (From Fugate & Friedrichs, 2003.)

the bottom 1.0 m of the flow (Figure 11, Fugate & Friedrichs, 2003). This is still far above the buffer layer, but is much closer to the bed than the mid–depth regions of flow in which large estuarine flocs are usually measured (van Leussen, 1999). Particles, conditioned by high near-bed turbulence are *deposited* whereas larger, weaker flocs formed higher up in the flow are *transported* but then broken up as they get close to the bed. Because flow is not steady much sediment is deposited during low flow periods.

Selective erosion (*winnowing*) is a less important sorting process for cohesive muds because cohesion means that there is little size-selective removal of particles that are prone to stick together (Winterwerp & van Kesteren, 2004). It can make the deposit somewhat coarser overall by producing intermittent erosion horizons marked by coarse silty and sandy lags, but at the cost of decreasing accumulation rate (McCave and Hall, 2006) (Figure 12). As shown in Figure 12 currents increase deposition rate up to a point beyond which winnowing sets in and accumulation decreases. Of course at a point the accumulation requires a negative spatial gradient of sediment flux (dQ_s/dx is negative). This is often accomplished in flow expansions or other spatial decreases of shear stress controlled by topography. In regions where

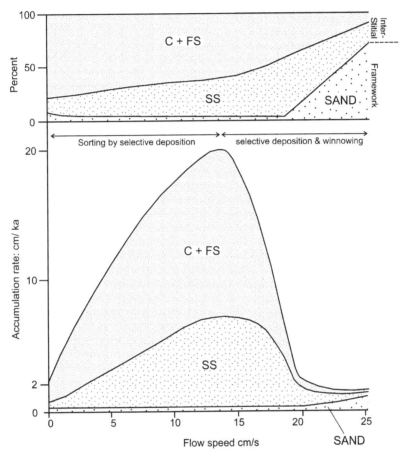

Figure 12 Hypothetical variation of accumulation rate with increasing current speed. C+FS = clay and fine silt ($< 10\,\mu m$), SS = Sortable Silt ($10–63\,\mu m$). At zero speed a pelagic rate of 2 cm ka^{-1} assumed. From zero to the peak accumulation rate the sorting process is dominantly selective deposition whereas above that point, here arbitrarily put at 13 cm s^{-1}, it is selective deposition and removal of fines (winnowing) combined. Erosional winnowing and some sand movement is assumed to occur above 20 cm s^{-1}. Above 20 cm s^{-1} mud will be mainly interstitial in sand, while above $\sim$30 cm s^{-1} (not shown) the sand will be sufficiently mobile to contain little mud at all and will be forming ripples and sand waves. The curve is shown as peaking between 10 and 15 cm s^{-1} but this is not at all well known and may well be dependent on the magnitude-frequency structure of deposition and erosion events. A peak around 10 cm s^{-1} would be entirely feasible as some records suggest the onset of surface erosion above 10–12 cm s^{-1} (Gross & Williams, 1991). (From McCave & Hall (2006).)

the currents frequently exceed about $20 \, \text{cm s}^{-1}$ the fine components are removed, leaving a lag of sand. This is most frequently foraminiferal sand with a critical erosion shear stress of 0.065–$0.078 \, \text{Pa}$ for 200–$300 \, \mu\text{m}$ sediment-filled grains (Miller and Komar, 1977) (U_g of 17–$26 \, \text{cm s}^{-1}$). Although medium foram sand has the erodibility of very coarse silt, it has the settling velocity of very fine sand and is not transported far, being largely confined to bedload. To move quartz sand of this size would take currents of $U_g = 35$–$37 \, \text{cm s}^{-1}$. The speeds required to move foraminiferal sand are only rarely exceeded frequently enough to yield a well-sorted sand, though sand dunes have been detected in places.

The significance of this is that the deposits of lateral flux driven by deep-sea currents are in the silt-clay size range and are sorted during deposition to a greater or lesser extent depending on currents. Unless the currents are very fast the proportion of foraminiferal sand is much less than in deposits resulting from vertical flux. Only the sand- and gravel-sized material may be unambiguously related to the properties of the overlying water column, as the rest is mobile.

2.2.4. Repeated sorting events under deep-sea storms

The deep-sea setting provides a situation analogous to other short-term sorting of sand by waves on beaches, because the unsteadiness of deep-sea currents, so-called 'deep-sea storms', causes periodic resuspension-transport-deposition events that sort silt (Hollister & McCave, 1984; Gross & Nowell, 1990; Gross & Williams, 1991) The response of the suspended sediment concentration field to currents is complex with long periods of flow speed below $10 \, \text{cm s}^{-1}$ during which deposition occurs (Gross & Williams, 1991; Gross & Nowell, 1990). Regions of high surface eddy kinetic energy were identified by Hollister and McCave (1984) as potential areas of benthic storms because of surface-to-bottom energy transfer, occurring particularly in western ocean basins and under the ACC. It is a key feature of the transport-deposition system because of the profound asymmetry of erosion and deposition (a few minutes' erosion can remove several weeks' deposition). Putting the material into suspension takes a short time in storms and the intervening period is then occupied by sorting through selective deposition.

3. SEDIMENT DEPOSITION: QUATERNARY RECORDS OF FLOW IN LARGE-SCALE FEATURES

3.1. Sediment Waves and Drifts

Ever since drifts began to be identified and mapped in the 1950s by Heezen (1959), and seismic reflection profiles showed them to be thick sediment piles in the 1960s (Ewing & Ewing, 1964), they have been prime coring targets. Drifts have typical dimensions of $<100 \, \text{km}$ width, several hundred kilometres length and several hundred metres thickness. Many appear to contain a record extending back through the Oligocene. Sedimentation rates are generally over $5 \, \text{cm ka}^{-1}$ and exceptionally may be over $1 \, \text{m ka}^{-1}$.

Most drifts have, somewhere on their surface, fields of sediment waves or 'mudwaves'. These features are mainly 10–60 m high with wavelengths of 1–5 km, and tend to migrate as anti-dunes, though dunes and irregular forms growing vertically are not uncommon. The most plausible theory accounts for them as the product of deposition under lee waves in the stratified lower ocean (Flood, 1988; Blumsack & Weatherly, 1989; Blumsack, 1993). The lee wave origin and anti-dune migration mode predicts minimum stress/maximum deposition on the upstream face and vice versa on the downstream face, with corresponding finer and coarser sediments. The correctness of this is borne out by the observations of Ledbetter (1993) in the great Argentine Basin mud-wave field (Figure 13), and Flood (1988) on Bahama Outer Ridge.

Many drifts and some mud-wave fields also have superimposed fields of furrows. These current-parallel depressions, maintained during surface aggradation (thus are not necessarily erosional, save at inception), also have associated grain size variations. Flood (1983) shows them to have coarser deposits in the bottom of the furrow.

Lateral migration of both waves and furrows can produce a down-core variability in grain size. Mud-wave migration is observable using 3.5 kHz profiles, but variation due to furrow movement is less obvious. The tell-tale small hyperbolic echo character of furrow fields (Damuth, 1980) are best avoided as coring targets.

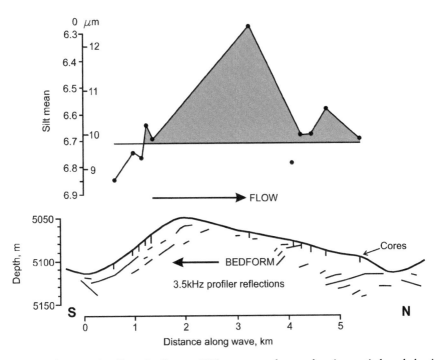

Figure 13 Silt mean size (from Ledbetter, 1993) over a mud wave showing anti-dune behaviour with the coarsest sizes on the downstream face and finer sizes indicating slower flow on the upstream slope, corresponding to deposition as shown by the acoustic profiler.

However, some well-known cored areas display these features, e.g. north-eastern Bermuda Rise (McCave, Hollister, Laine, Lonsdale, & Richardson, 1982), Bahama Outer Ridge (Hollister, Southard, Flood, & Lonsdale, 1976; Flood & Giosan, 2002).

On a much larger scale, drifts that become detached from the boundary (continental margin) to become ridges (e.g. Feni, Gardar, Eirik, Blake) have the stronger flow on the side against which Coriolis force 'presses' the current and a weaker one, often a counter-flow, on the other side (Dickson & Kidd, 1986; Bianchi & McCave, 2000). This gives rise to slower sedimentation rate, higher acoustic reflectivity and coarser grain size on the more current-stressed side (Bianchi & McCave, 2000) (Figure 3). This is unlikely to lead to any down-core ambiguity, as the lateral migration rate of drifts is very slow. However, it does mean that there is spatial variability in properties controlled by the *local* rather than global flow field. Thus positions for monitoring past flow need to be selected for representativeness as well as maximum deposition rate for high resolution.

3.2. Applications of the Sortable Silt Proxy: Climate and Flow

By now there are numerous studies employing the sortable silt mean size ($\overline{SS}$) or percentage (SS%) to show changes in the deep flow of the oceans in relation to climatic changes. These are reviewed at some length by McCave and Hall (2006), so a small selection is given here to illustrate use and problems. As will be noted below, all too few studies have employed more than one flow vigour proxy and where used they do not always agree.

3.2.1. N.E. Atlantic: last 10 ka of ISOW

Bianchi and McCave (1999) inferred changes in the current strength of Iceland Scotland Overflow Water (ISOW) from core NEAP 15K recovered from Gardar Drift (2,848 m water depth) in the South Iceland Basin. They found evidence for a quasi-periodic ~1,500-yr flow speed variability in the Holocene which they correlated with climate events in Northern Europe (Figure 14), notably faster ISOW flow during the Medieval Warm Period and slower during the Little Ice Age. These data and those of Hall et al. (2004) from core NEAP 4K farther north at 1,627 m agree with evidence of millennial periodicities in other Holocene climate proxies (Chapman & Shackleton, 2000 found ~1,660-; 1,000- and 500-yr periods in lightness — $CaCO_3$ content — and $\delta^{13}C$ in NEAP 15K). Additionally, it is notable that the IRD events of Bond et al. (1997) are neither found in NEAP 15K or −4K nor coincide with times of slower ISOW flow. The central Gardar Drift is not on any major iceberg route however.

3.2.2. Iberian margin: last 100 ka of NEADW

The sortable silt current speed proxy $\overline{SS}$ shows a consistent temporal pattern of slow flow in cold and faster in warm periods in core MD95-2040 at 2,465 m under N.E. Atlantic Deep Water on the northern Portuguese margin (Figure 15, Hall & McCave, 2000). This is currently the most unambiguous record of LNADW flow variability spanning the last glacial–interglacial cycle. The record clearly indicates

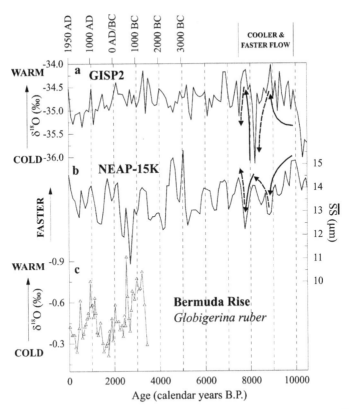

Figure 14   North Atlantic Holocene records from Bianchi and McCave (1999). (a) δ^{18}O from the central Greenland GISP2 ice core. Solid and dashed arrows between 0 and 7.5 ka BP represent periods of general warming or cooling which match relative decreases or increases in the intensity of ISOW flow, respectively, shown in (b), $\overline{SS}$ record for NEAP 15K smoothed on a 300-yr window. (c) Planktonic foraminiferal δ^{18}O data from the Sargasso Sea (Keigwin, 1996), mainly reflecting changes in sea surface temperature.

faster flow during the warm periods and slower in cold, with the strange exception of isotope in stage 2. However, another core from this margin (OMII-9K) indicates slow flow in stage 2, as do others further north, suggesting that the MD95-2040 record is anomalous at this point. Stage 2 mass accumulation rates in the latter core are high ($>20\,\mathrm{g\,cm^{-2}\,ka^{-1}}$) suggesting that the material may in part have been supplied down-slope at such a rate that the weak current could not sort it and impose an $\overline{SS}$ signal appropriate to the flow speed. This view is supported by the fact that Heinrich Layers with similar accumulation rates also show high values of $\overline{SS}$ here. The record is quite consistent with warm climate $=$ fast flow and cold $=$ slow, and thus either stage 2 or stages 4, 5b, 5d and 6 are anomalous in this core. If there is an unsorted source effect on $\overline{SS}$ in stage 2 then there is not in the other stages. It coincides with the lowest stand of sealevel. Elsewhere on this margin, this time is marked by down-slope transport with distinctive sedimentary structures (Baas, Mienert, Abrantes, & Prins, 1997). This illustrates both the utility of the method and the pitfalls associated with drifts on continental margins.

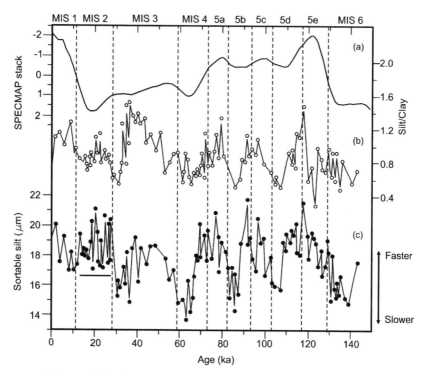

Figure 15 Sedimentological parameters for core MD95-2040 on the Iberian margin plotted against age (Hall & McCave, 2000). (a) SPECMAP stack of benthic $\delta^{18}O$ values (Martinson et al., 1987) (b) Terrigenous silt/clay ratio (wt% 2–63 µm/wt% <2 µm), (c) $\overline{SS}$ (µm); the line under the data in stage 2 indicates the part of the record believed to be unreliable as a current indicator because of possible downslope contamination. The vertical dashed lines indicate the boundaries between Marine Isotope Stages (MIS) achieved by correlating the benthic isotopic record for the core to SPECMAP. Note the clear relationship of faster flow in cold/cool periods from stages 6 to 3 with a lag of a few thousand years.

3.2.3. S.W. Pacific: Last 1 Ma of the deep western boundary current supply of CDW

ODP Site 1123 penetrates the N. Chatham Drift (Carter & McCave, 2002) on the northeast flank of Chatham Rise, east of New Zealand (3,290 m water depth) beneath the southwest Pacific DWBC. This is the largest DWBC in the world, supplying most of the Pacific's deep water. The Site 1123 $\overline{SS}$ data in Figure 16b clearly show faster flow in glacial periods and slower during interglacials over the past 1.2 Ma. Significant spectral peaks were identified at each of the orbital frequencies. These were coherent with benthic records of both oxygen and carbon isotopes at 98% (for 100 and 41 ka) and 90% confidence (for 23 ka). Hall, McCave, Shackleton, Weedon, and Harris (2001) suggest that the increased glacial DWBC flow speeds were related to greater glacial production of AABW and Circumpolar Deep Water (CDW), a feature supported by the diatom tracer data of Stickley, Carter, McCave, and Weaver (2001), and also by grain size evidence for more vigorous glacial flow of CDW seen in cores from a sediment drift in the Drake

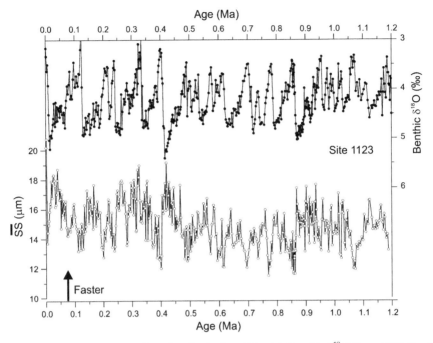

Figure 16 SW Pacific DWBC deep flow in the last 1 Ma. (a) Benthic δ^{18}O from ODP Site 1123, Chatham Rise, (b) $\overline{SS}$ at Site 1123 from Hall et al. (2001).

Passage outflow region (Pudsey & Howe, 1998; Howe & Pudsey, 1999). However, it does raise the question as to whether the enhanced flow is of the DWBC or more related to the ACC. Hall et al. (2001) therefore also measured the δ^{13}C gradient in benthic foraminifera across the Pacific, arguing that the small glacial gradient observed indicated vigorous flow across the ocean in contrast to the larger interglacial gradient.

3.2.4. Flow pulses into the Indian ocean: geostrophic transients?

McCave et al. (2005) examined bottom flow of the SW Indian Ocean DWBC through the Amirante Passage between the Madagascar and Somali Basins over the last glacial–interglacial cycle. These sediments lie at the foot of Farquhar Ridge north of the Madagascar. Sedimentation rates are rather low but core WIND 28K in the entrance to Amirante Passage (4,157 m water depth, with a sedimentation rate of 4 cm ka^{-1}) shows glacial-to-interglacial behaviour in the major deep inflow path to the Indian Ocean (Figure 17). In general the results, perhaps surprisingly, suggest bottom flows varied little between glacial and interglacials, with slightly faster flow in the last interglacial and minima in cold periods. This constancy is also picked up by the ^{231}Pa/^{230}Th flow tracer downcore (Thomas, Henderson, & McCave, 2006). However, the dominant features of the WIND 28K $\overline{SS}$ record, four pulses of substantially faster flow corresponding to positive benthic δ^{18}O shifts of 0.5–1‰, are not recorded by ^{231}Pa/^{230}Th. McCave et al. (2005) note that these pulses correspond to global cooling phases and major falls in sea level, accompanied

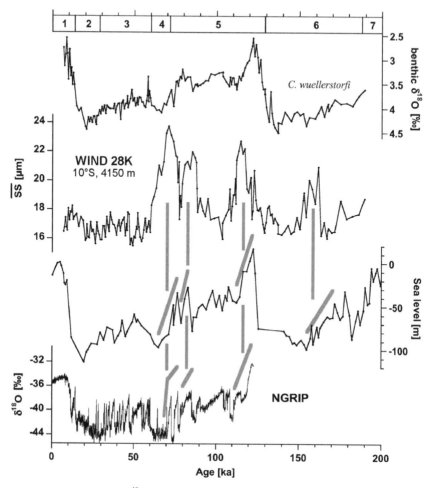

Figure 17 Benthic isotopic (δ^{18}O) and $\overline{SS}$ data for core WIND 28K plotted against age. The sea level curve of Siddall et al. (2003) based on core KL-11 from the Red Sea, and the NGRIP Greenland ice sheet δ^{18}O showing large oscillations at MIS 5a (North Greenland Ice Core Project Members, 2004) are also given. Heavy grey lines mark the sea-level and ice-core isotopic shifts corresponding to flow speed transients. (After McCave et al., 2005.)

by sharp increases of bottom-water density. This would lead to a transient local geostrophic effect in a two-layer system (with bottom inflow overlain by a south-ward outflow, Johnson, Musgrave, Warren, Ffield, & Olson, 1998). In contrast to the deep inflow to the SW Pacific (shown in Figure 16), peak flow speeds in the Amirante Passage appear limited to the periods of inferred density variation, not to the subsequent periods of uniform higher density (i.e. glacial maxima). It is also likely that they are a local, but informative, effect of the flow along a steep scarp into a narrow passage. McCave et al. (2005) conclude that during the inferred periods of density change and high flow speeds, the deep near-bed waters of the Indian Ocean inflow were strongly stratified. This process culminated in the basins filling up with higher density water, thereby removing the density contrast between the lower inflow and southward-flowing water above, resulting in slower flow speeds in the glacial maxima. Ledbetter's (1986b) Vema Channel data also indicated that flow speed maxima occurred at interglacial to glacial transitions, specifically MIS 7–6 and 5–4. Like the Amirante Passage data, this too is from flow through a choke point where perhaps transient phenomena are amplified.

3.3. Unmixing of Size Distributions into End-Members

The emphasis in this chapter is on measures based on the physics of sediment transport, but this does not repudiate the notion that sediment-size distributions can be 'unmixed' into end members (EMs), a procedure with a long history, most recently and effectively implemented statistically by Weltje (1997) and colleagues (Stuut et al., 2002; Prins et al., 2002; Weltje & Prins, 2003). This procedure is based on the whole-size distribution, whereas the SS method is based on just the dynamically sensitive fraction. For example a tropical river source would be expected to give a clay-rich end-member, whereas a more sandy or loessic source would not; even though when transported and redeposited in the deep sea, the two should have similar $\overline{SS}$ values under similar current regimes. The end-member method has most successfully been applied to marine sediments of aeolian origin in the Atlantic and Arabian sea (Holz, Stuut, & Henrich, 2004; Stuut et al., 2002) where the parent distributions have a distinctive size spectrum that can be compared with the deposit (Holz et al., 2004). Weltje and Prins (2003) show downwind fining of EMs offshore of W. Africa such that the spatial distribution of the ratio between the amounts of key EMs reveals dispersal patterns.

The method has been less successful in determining palaeoflow speeds. Prins et al. (2002) resolved size spectra from Reykjanes Ridge into four components as shown in Figure 18. end-member 2 is assigned to an IRD origin, perhaps on the basis of some sand content, but the signature with a pronounced mode at 23 µm is more likely to reflect current sorting. The component assigned a current sorted origin (end-member 3) with a mode at 8 µm (measured by laser), which is more likely to be floc-deposited fine silt and clay (Curran et al., 2004). There may be some current dependence in the ratio between end member 3 and end-member 4 used by Prins et al., but it is not clear, from a sediment mechanics perspective, how it would work.

4. CURRENT PROBLEMS AND PROSPECTS

4.1. Grain Size Analysis

Several types of instrument, based on different principles, are in use for grain size measurement. Each device has its disadvantages (McCave & Syvitski, 1991). None of them actually measures size. Presently, the Sedigraph is the instrument of choice for the study of deep-sea sediments as proxies for current intensity, since it is based on the settling velocity principle and therefore measures a 'dynamical' grain-size

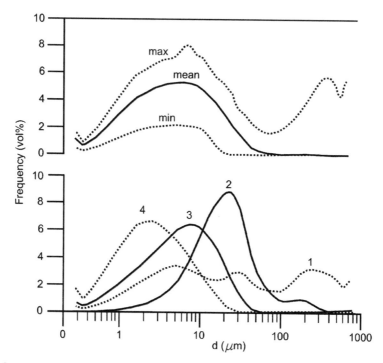

Figure 18   End-member dissection of size distributions from Reykjanes Ridge by Prins et al. (2002). A: Mean size distribution and maximum range of volume frequency for each size class; (B) modelled end members according to four-EM model, accounting for 84% of the variance. EMs 1 and 2 have been assigned to IRD and the ratio end-m/end-member 4 to flow speed (see text for alternative opinion).

distribution closely related to transport and depositional processes (Coakley & Syvitiski, 1991). Bianchi, Hall, McCave, and Joseph (1999) show that the Coulter Counter — measuring a volume-equivalent spherical diameter inferred from electrical conductivity — provides a viable alternative to the Sedigraph. Broadly, methods of size measurement may be divided into those which yield information on the whole-size distribution, and those for which the information comes from a size window with upper and lower limits. A Coulter Counter performs an analysis with a window of ~2–50% of the diameter of the orifice in which the particles are sensed. This normally means that it does not record the <2 μm clay that may comprise most of the size spectrum (Milligan & Kranck, 1991). The Coulter Counter can thus record only $\overline{SS}$, not SS%.

Laser particle sizers should be avoided for palaeocurrent reconstructions because the measured size of platey minerals is dominated by their large projected area (Konert & Vandenberghe, 1997; McCave, Hall, & Bianchi, 2006). This causes clays to be recorded as the same size as larger equant grains although they have much smaller settling velocity (Figure 19). This yields results with a weaker relationship to the dynamics of deposition. The central problem is that some coarse clay/fine silt is

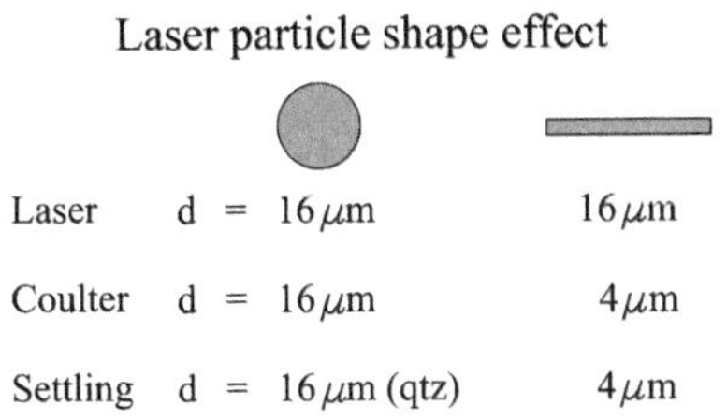

Laser particle shape effect

Laser	d	= 16 μm	16 μm
Coulter	d	= 16 μm	4 μm
Settling	d	= 16 μm (qtz)	4 μm

Figure 19 Plate versus sphere as seen by different particle size analysers.

recorded as medium to coarse silt, the key size in the 'sortable silt' means size method of inferring changes in flow speed. The performance of a laser sizer was compared with results from a Coulter Counter by McCave et al. (2006) who show that the differences in fine fraction measurement limits the application of laser sizers in determination of palaeocurrent-sensitive parameters, in particular the $\overline{SS}$ proxy. While this has not been checked on instruments other than the Malvern Multi-sizerX and 2600E, it is probably true of all those that invert the angular distribution of forward-scattered light intensity to yield the size distributions of assumed spheres, when they encounter platey particles. Laser sizers are excellent for sizing of equant grains, coarse silt and sand, and have given satisfactory results for aeolian loess (coarse to very coarse silt) where SS% is mainly >40% (e.g. Ding et al., 2002). Comparisons of instruments are given by Stein (1985); McCave, Bryant, Cook, and Coughanowr (1986); Singer et al. (1988); Syvitski, LeBlanc, and Asprey (1991); Weber, Gonthier, and Faugères (1991); Bianchi et al. (1999); McCave et al. (2006).

4.2. Current or Source Control?

One of the central problems of inferring currents from grain size is the commonly, but erroneously, presumed dominant influence of source on size. For fine sediments this is far less of a problem than for sands, where for example there can be a strong influence of source rock on grain size and size on mineralogy. Some sources obviously do not produce gravel or coarse sand, but few that supply something fail to provide mud (clay plus silt). This material comprises over 60% of all sediment on Earth (Pettijohn, 1975). Work on particle sizes of suspended sediment record a wide range. It is poorly sorted both at fluvial source (Gibbs, 1967; Potter, Heling, Shimp, & VanWie, 1975; Johnson & Kelley, 1984) and in the sea (Brun-Cottan, 1971; McCave, 1983; Kranck & Milligan, 1983). The size distribution $dV/d(\log d)$ versus log d is typically fairly flat (slope of zero ± 0.2).

Mixing of several metres to tens of metres thickness of failed sea bed in systems of sediment delivery to the deep sea (turbidity currents, debris flows) eliminates effects due to systematic temporal variability of sources, at least on short time scales (order of at least 100 ka, maybe to 1 Ma) (Embley & Jacobi, 1977; Weaver & Thomson, 1993; Weaver, 1994). Only in cases where sediment with a distinct sorted size spectrum is delivered to relatively quiescent deep sea will there be a strong source signature in the deposit.

4.3. Abnormally High Sedimentation Rates: The Holy Grail/Poisoned Chalice Dilemma

I characterised marine geologists as King Arthur's knights of the round table searching for the Holy Grail,[1] i.e. for cores with a very high sedimentation rate, to give high-resolution palaeoclimate data (McCave, 2002). The trouble is that this Grail may be a poisoned chalice. High sedimentation rates are necessary for high resolution because most marine sediments are biologically mixed on a length scale of a few centimetres (Guinasso & Schink, 1975; DeMaster, McKee, Nittrouer, Brewster, & Biscaye, 1985; Wheatcroft, 1990; Wheatcroft, Jumars, Smith, & Nowell, 1990). Cores with sedimentation rates of 50–100 cm per 1,000 yr (ka) are desired. In some cases such apparent rates are artefacts of the coring process, achieved by 'stretching' due to piston rebound (Skinner & McCave, 2003). In the North Atlantic, away from continental margins, average rates are ~2–4 cm ka^{-1} (Balsam & McCoy, 1987), and on continental margins average rates are of order 10 cm ka^{-1}. In the North Pacific they are much less than 1 cm ka^{-1}. So desirably high rates of sedimentation are anomalous. The critical uncertainty is whether cores with anomalous rates of deposition are also anomalous in other ways, e.g. in the climatic signals they contain.

Of the three scenarios for sediment deposition: (i) *pelagic settling*, (ii) *pelagic plus inert (signal-free) filler*, (iii) *lateral signal plus minor pelagic supply*, only the latter is realistic in current-controlled settings. If the sedimentation rate is 20 cm ka^{-1} and pelagic supply is 2 cm ka^{-1} then 18 cm ka^{-1} has come from somewhere upcurrent. The pelagic input provides material on which to base stratigraphy, isotopic work and inference of age, but the laterally supplied silt and clay carries an important signal indicating where it has come from (e.g. clay mineralogy, isotopic tracers indicative of provenance), or planktonic productivity or how vigorous is the bottom flow.[2] The trouble is that in most drifts scenario 3 is correct, but scenario 1 or 2 is often assumed for interpretation. If scenario 1 or 2 is correct the fine and coarse components should be of the same age, if they are not, the most likely explanation is that the filler is not inert but carries a signal similar to, but older than, that sinking from the overlying waters (Ohkouchi, Eglinton, Keigwin, & Hayes, 2002). If that is the case, some aspects of the fine fraction record at a point cannot be interpreted as a time history of events *at that point*. It is contaminated by displaced records of events at a different place with different characteristics or has integrated a signal over some distance upstream. That this was likely to be the case for alkenones was suggested by results from the Argentine Basin, which showed that modern bottom

---

[1] The Holy Grail was the chalice used by Jesus at the Last Supper and, in mediaeval legend, recovered by Joseph of Arimathea who brought it to Britain where it was lost. It was sought by the knights of whom only three (*of exceptional purity of thought*) even glimpsed, but never obtained it. (Malory, T., 1485. *Le Morte d'Arthur*. W. Caxton, London.) (Many modern editions.)

[2] Astonishingly, many paleoceanographers simply throw the fines away, thereby depriving themselves and others of the possibility of obtaining important ancilliary data. Sophisticated arguments such as "my funding agency does not pay me to keep the fines", or admission of dubious analytical practice "no-one would want the fine fractions out of my lab" have been deployed to justify this. The Lunar Receiving Lab would be insensed by a practice of crushing up Moon rock, picking out the olivines and throwing the rest away. Core repositories should similarly be incensed at profligate wastage of their material.

sediments recorded alkenone-based temperatures that were 2–6°C colder than the local sea surface (Benthien & Muller, 2000). Alkenones provide excellent SSTs, but for a *local* record the material should not be transported far. Isotopic tracers (such as Pa/Th and Nd), which rely on scavenging a signal from the water column, must integrate over some upstream extent. This conflicts with the need for high resolution records produced by copious deposition of sediment from currents. Continental margins, with high resuspension and lateral sediment supply from diverse sources are particularly prone to problems.

However, rapid mass accumulation from currents is not a problem for grain size proxies of flow speed as long as the type of flow is correctly identified, i.e. not turbidity currents.

4.4. Spatial Aliasing

With changing climate comes changing density of water masses participating in the deep circulation. Several records have demonstrated the change from dominance of 'NADW' to 'AABW'[3] between interglacials and glacials. If, for example, at the LGM there were convection of relatively fresh water south of the Arctic Polar Front as the only northern source to the Western Boundary Undercurrent (WBUC), it would be less dense than the present cocktail of NADW sources and probably shallower. At a given deep location the time trend in flow speed from glacial to interglacial would be a decrease, while at a shallow point the opposite might occur, possibly with little overall change in MOC strength (mass flux). Study of only the deep section could lead to serious error. This emphasises the need to study depth transects with flow speed proxies in conjunction with water mass tracers and density (T/S) proxies as well (currently benthic δ^{13}C and ε_{Nd} with δ^{18}O and $^{231}Pa^0_{xs}/^{230}Th^0_{xs}$ for flow vigour). A few studies have looked at this on the central N. American margin, Ledbetter and Balsam (1985), Haskell, Johnson, and Showers (1991) and Yokokawa and Franz (2002) using only one or two proxies, with all concluding that there have been shifts in the focus of current activity. In particular Haskell et al. (1991) show that the major flow on Blake Outer Ridge was shallower than 2,900 m from the LGM up to ~14 ka.

4.5. Ice-Rafted Detritus

Whether or not the input of IRD influences the $\overline{SS}$ signal depends on the rate of reworking versus the rate of supply. 10 cm ka^{-1} is ~0.6 g cm^{-2} ka^{-1}, so resuspension of this amount, 1 mm per 10 yr, is just 60 mg cm^{-2} or, for annual events removal of just 6 mg cm^{-2} per event. Benthic resuspension events typically achieve concentrations of 1–10 g m^{-3}, equivalent to 1–10 mg in a bottom mixed layer of 100 m thickness. In other words a deposition rate of 10 cm ka^{-1} can be reprocessed by just one resuspension event per year of 6 mg cm^{-2}. This is much less than the highly active HEBBLE area south of Nova Scotia. It is likely to be achieved by

---

[3] The non-genetic NSW and SSW, northern and southern source water, terms are preferable and can indicate bottom, deep or intermediate, as with modern water mass names. Thus 'glacial NSDW' rather than 'glacial NADW'.

flows reaching critical erosion conditions ($11\text{--}15\,\text{cm}\,\text{s}^{-1}$) at least once a year. For such flows the mean speed would be well under $10\,\text{cm}\,\text{s}^{-1}$.

Hass (2002) extracted a current signal from IRD-affected sediments by correlating percent sand and $\overline{\text{SS}}$, then subtracting the sand-associated part of $\overline{\text{SS}}$ from the original time series (Figure 20). The sand is assumed to be immobile and so is an index of the amount of IRD, and that the shape of the size distribution is assumed constant. Thus the sand is proxy for the amount of a constant size contaminant which, when subtracted, leaves the current-affected part of the distribution. There is significant scatter around the line. It must also be assumed that $\overline{\text{SS}}$ increases in size with percent IRD. This procedure is not rigorously justified, even though it appears to produce a credible record of flow out of the Arctic (Hass, 2002; Birgel and Hass, 2004), with flow reductions during the Younger Dryas and 8.2 ka events (Figure 20). However, if the IRD followed a crushed material Rosin–Rammler distribution, more sand would be accompanied by more silt. With a flat weight/volume distribution between 10 and 62.5 μm, its (logarithmic) mean size would be 25 μm. Values of $\overline{\text{SS}}$ tend to fall in the region 15–30 μm, so the IRD silt would not be hugely distorting but would put in more coarse silt than is usually found. So Hass's empirical correlation has some basis. Nevertheless, the major features of the record are also present in the original record that the subtraction procedure has enhanced.

Large fluxes of IRD pose the problem that the current may not be able to rework the material sufficiently rapidly to impose its characteristics on the size distribution. The rate in Hass' core PS 2837-5 is around $20\,\text{cm}\,\text{ka}^{-1}$ or 2 mm per decade and the rates of Manighetti and McCave (1995) cores are less than $10\,\text{cm}\,\text{ka}^{-1}$. Currents of moderate strength, $\sim15\,\text{cm}\,\text{s}^{-1}$ for a few days a year, should be able to process this material. Only if severely episodic dumps of sediment occur — several centimetres in a year — would these currents be unable to cope.

Data on SS and magnetic susceptibility shown by Austin and Evans (2000) for two periods in the Pliocene and early Pleistocene from south of Iceland where percentage of IRD>125 μm is up to 4% suggest little effect of IRD on the SS speed proxy (Figure 21).

4.6. Calibration

The SS proxy has yet to be calibrated in terms of flow speed and presently gives only relative changes, though Ledbetter's (1986b) total silt mean calibration (Figure 5) suggests a good relationship. At present core-top samples have been collected from locations close to long term (>1 yr) sites where current metres have been set within 100 m of the seabed (giving flow speeds from the geostrophic flow just above the boundary layer). Calibration needs to take into account the mean and variability characteristics of the currents affecting sediments. Relatively steady slow speeds may remain below the deposition threshold while more variable speeds may go up into the winnowing regime ($>12\,\text{cm}\,\text{s}^{-1}$) and occasionally into erosion ($>20\,\text{cm}\,\text{s}^{-1}$). The scalar speed (rather than vector velocity) is the key parameter to relate to size, but this is not always be strongly related to net transport. It is clearly important to know whether a change in sediment size is due to mean flow strength

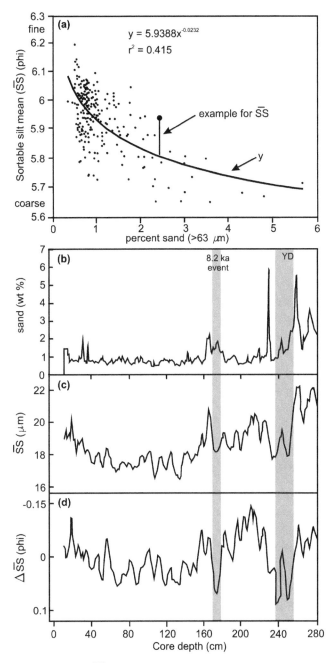

Figure 20 (A) Correlation of $\overline{SS}$ with sand percentage for the Fram Strait area (Hass, 2002). Deviations of $\overline{SS}$ from the line are presumed to have the effect of IRD on the mean removed. (B) Time trends in core PS-2837 from Yermak Plateau, E. side of Fram Strait: (a) % sand, (b) raw $\overline{SS}$, (d) $\Delta\overline{SS}$ enhancing the $\overline{SS}$ signal by the subtraction procedure shown above in (A). YD and 8.2 ka events become more apparent, but are present in the original data. (From Birgel & Hass, 2004.)

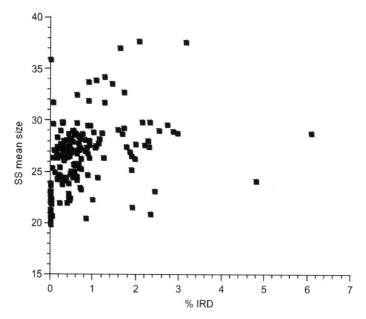

Figure 21   Mean sortable silt size (10–63 μm) (by laser) versus IRD (non-carbonate % > 125 μm). Data from ODP Site 984 south of Iceland, from Austin and Evans (2000). Due to the laser shape effect the SS% is high in these records, nevertheless, there is little correlation between IRD and $\overline{SS}$ ($r^2 = 0.0008$ or, without two very high IRD values, $r^2 = 0.134$). (From McCave & Hall, in review.)

(mean kinetic energy, K_M) or variability (eddy kinetic energy, K_E), though in many cases they are positively correlated (Dickson, 1983).

4.7. Links with other Flow/Circulation and Water Mass Proxies

There has been insufficient use of more than one flow rate proxy or more than two water mass markers from the same cores. Work needs to be done with epibenthic carbon isotopes (Mackensen & Bickert, 1999; Curry & Oppo, 2005) coupled with the phosphate proxy Cd/Ca (e.g. Boyle & Keigwin, 1982), but both are biologically mediated and can define the extent of water masses but do not act as conservative tracers capable of giving water mass mixing information. The neodymium isotope water mass tracer may provide that (Piotrowski, Goldstein, Hemming, & Fairbanks, 2004). Kissel (2005) shows that the concentration of magnetic grains (mainly very fine magnetite) may be interpreted as a proxy for the strength of NADW, perhaps related to the erosive power of the LNADW source flows passing over and through the basaltic Greenland–Scotland Ridge.

The principal flow rate or ventilation rate proxies are grain size (reviewed here) AnMS (Rees, 1961; Hamilton & Rees, 1970), carbon-14 age difference between surface and bottom waters (Broecker, Peng, Trumbore, Bonani, & Wolfli, 1990; Adkins & Boyle, 1997), $^{231}Pa_{xs}^0/^{230}Th_{xs}^0$ (Bacon & Rosholt, 1982; McManus, François, Gherardi, Keigwin, & Brown-Leger, 2004) and the density field obtained

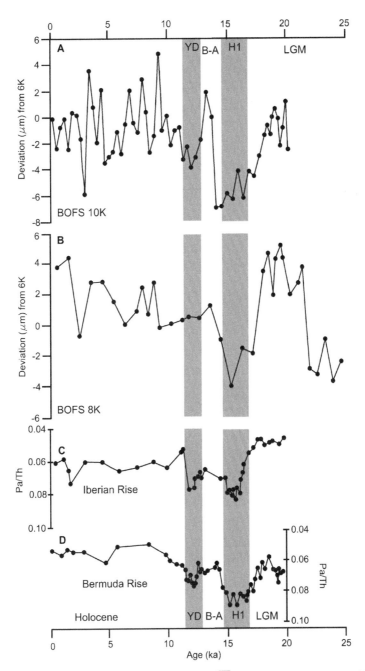

Figure 22   East Thulean Rise, differential grain size ($\Delta\overline{\text{SS}}$) for core sites BOFS 8K and 10K minus the size for BOFS 6K (presumed to be unaffected by sediment focussing or winnowing) (after Manighetti & McCave, 1995, reworked by McCave & Hall, 2006), With Pa/Th data for N. Atlantic flow rates from Gherardi et al. (2005) and McManus et al. (2004).

via benthic oxygen isotopes used to make geostrophic calculations of flow (Lynch-Stieglitz et al., 1999). Several of these have been used together with benthic δ^{13}C, but only AnMS and grain size have been used on the same cores, in one of the earliest studies in the field (Ellwood & Ledbetter, 1977). Another case produced conflicting results (Revel, Cremer, Grousset, & Labeyrie, 1996; Kissel et al., 1997) because of the use of a laser size analyser, which gives misleading results in the sortable silt range (McCave et al., 2006). The recent uses of the Pa/Th flow tracer over the deglaciation from 19 ka to the Holocene by McManus et al. (2004) and Gherardi et al. (2005) agrees with the SS grain size tracer in the N.E. Atlantic (McCave, Manighetti, & Beveridge, 1995a) (Figure 22) but on different cores.

The next step must be to use most of these tracers together on the same set of cores. It is particularly important to use both flow rate and water mass tracers together because there are numerous cases where flow rate changes are inferred to be accompanied by change in water mass. In the N. Atlantic, cores from 3,000 to 4,000 m depth sample NADW in the Holocene, but southern source water at the glacial maximum and alternations between the two EMs during deglaciation. It is essential to know what happens to the flow rate of each end-member, rather than that one was slow at a certain time while the other was fast at another time.

REFERENCES

Adkins, J. F., & Boyle, E. A. (1997). Changing atmospheric δ^{14}C and the record of deep water paleoventilation ages. *Paleoceanography, 12,* 337–344.

Austin, W. E. N., & Evans, J. R. (2000). Benthic foraminifera and sediment grain size variability at intermediate water depths in the Northeast Atlantic during the late Pliocene-early Pleistocene. *Marine Geology, 170,* 423–441.

Baas, J. H., Mienert, J., Abrantes, F., & Prins, A. (1997). Late Quaternary sedimentation on the Portuguese continental margin: Climate-related processes and products. *Palaeogeography, Palaeoclimatology, Palaeoecology, 130,* 1–23.

Bacon, M. P., & Rosholt, J. N. (1982). Accumulation rates of Th-230, Pa-231, and some transition metals on the Bermuda Rise. *Geochimica et Cosmochimica Acta, 46,* 651–666.

Baker, E. T., & Hickey, B. M. (1986). Contemporary sedimentation processes in and around an active west-coast submarine-canyon. *Marine Geology, 71,* 15–34.

Baldwin, C. T., & McCave, I. N. (1999). Bioturbation in an active deep-sea area: Implications for models of trace fossil tiering. *PALAIOS, 14,* 375–388.

Balsam, W. L., & McCoy, F. W. (1987). Atlantic sediments: Glacial/interglacial comparisons. *Paleoceanography, 2,* 531–542.

Benthien, A., & Muller, P. J. (2000). Anomalously low alkenone temperatures caused by lateral particle and sediment transport in the Malvinas current region, western Argentine Basin. *Deep-Sea Research I, 47,* 2369–2393.

Bianchi, G. G., Hall, I. R., McCave, I. N., & Joseph, L. (1999). Measurement of the sortable silt current speed proxy using the Sedigraph 5100 and Coulter Multisizer IIe: Precision and accuracy. *Sedimentology, 46,* 1001–1014.

Bianchi, G. G., & McCave, I. N. (1999). Holocene perodicity in North Atlantic climate and deep-ocean flow south of Iceland. *Nature, 397,* 515–517.

Bianchi, G. G., & McCave, I. N. (2000). Hydrography and sedimentation under the deep western boundary current on Björn and Gardar Drifts, Iceland Basin. *Marine Geology, 165,* 137–169.

Bird, A. A., Weatherly, G. L., & Wimbush, M. (1982). A study of the bottom boundary-layer over the Eastward scarp of the Bermuda Rise. *Journal of Geophysical Research, 87,* 7941–7954.

Birgel, D., & Hass, H. C. (2004). Oceanic and atmospheric variations during the last deglaciation in the Fram Strait (Arctic Ocean): A coupled high-resolution organic-geochemical and sedimentological study. *Quaternary Science Reviews*, 23, 29–47.

Blumsack, S. L. (1993). A model for the growth of mudwaves in the presence of time-varying currents. *Deep-Sea Research II*, 40, 963–974.

Blumsack, S. L., & Weatherly, G. L. (1989). Observations of the nearby flow and a model for the growth of mudwaves. *Deep-Sea Research*, 36, 1327–1339.

Bond, G., Showers, W., Cheseby, M., Lotti, R., Almasi, P., deMenocal, P., Priore, P., Cullen, H., Hajdas, I., & Bonani, G. (1997). A pervasive millennial-scale cycle in North Atlantic Holocene and glacial climates. *Science*, 278, 1257–1266.

Boyle, E. A., & Keigwin, L. D. (1982). Deep circulation of the North Atlantic over the last 200,000 years: Geochemical evidence. *Science*, 218, 784–787.

Broecker, W. S., Peng, T.-H., Trumbore, S., Bonani, G., & Wolfli, W. (1990). The distribution of radiocarbon in the glacial ocean. *Global Biogeochemical Cycles*, 4, 103–117.

Brun-Cottan, J. C. (1971). Etude de la granulometrie des particules marines, mesures éfectuées avec un compteur Coulter. *Cahiers Oceanographiques*, 23, 193–205.

Carter, L., & McCave, I. N. (1994). Structure of sediment drifts approaching an active plate margin under the S.W. Pacific deep western boundary current. *Paleoceanography*, 9, 1061–1086.

Carter, L., & McCave, I. N. (2002). Eastern New Zealand drifts: Miocene — Recent. In: D. A. V. Stow, J.-C. Faugères., J. A. Howe., C. J. Pudsey & A. R. Viana (Eds), *Deep-water contourites: modern drifts and ancient series, seismic and sedimentary characteristics*. Geological Society, London, Memoir, 22, 385–407.

Chapman, M. R., & Shackleton, N. J. (2000). Evidence of 550-year and 1000-year cyclicity in North Atlantic circulation patterns during the Holocene. *The Holocene*, 10, 287–291.

Coakley, J. P., & Syvitiski, J. P. M. (1991). Sedigraph technique. In: J. P. M. Syvitski (Ed.), *Principles, methods and application of particle size analysis* (pp. 129–142). New York: Cambridge University Press.

Curran, K. J., Hill, P. S., Schell, T. M., Milligan, T. G., & Piper, D. J. W. (2004). Inferring the mass fraction of floc-deposited mud: application to fine-grained turbidites. *Sedimentology*, 51, 927–944.

Curry, W. B., & Oppo, D. W. (2005). Glacial water mass geometry and the distribution of $\delta^{13}C$ of ΣCO_2 in the western Atlantic Ocean, *Paleoceanography*, 20, Art. No. PA1017, 1–12.

Dade, W. B., Hogg, A. J., & Boudreau, B. P. (2001). Physics of flow above the sediment-water interface. In: B. P. Boudreau & B. B. Jorgensen (Eds), *The benthic boundary layer: Transport processes and biogeochemistry* (pp. 78–103). Oxford: Oxford University Press.

Dade, W. B., Nowell, A. R. M., & Jumars, P. A. (1992). Predicting erosion resistance of muds. *Marine Geology*, 105, 285–297.

Damuth, J. E. (1980). Use of high-frequency (3.5–12 kHz) echograms in the study of near-bottom sedimentation processes in the deep-sea: A review. *Marine Geology*, 38, 51–75.

DeMaster, D. J., McKee, B. A., Nittrouer, C. A., Brewster, D. C., & Biscaye, P. E. (1985). Rates of sediment reworking at the HEBBLE site based on measurements of Th-234, Cs-137 and Pb-210. *Marine Geology*, 66, 133–148.

Dickson, R. R. (1983). Global summaries and intercomparisons: Flow statistics from long-term current meter moorings. In: A. R. Robinson (Ed.), *Eddies in marine science* (pp. 278–353). New York: Springer-Verlag.

Dickson, R. R., & Kidd, R. B. (1986). Deep circulation in the southern Rockall Trough — the oceanographic setting of Site 610. *Initial Reports of the Deep Sea Drilling Project*, 94, 1061–1074.

Dickson, R. R., & McCave, I. N. (1986). Nepheloid layers on the continental slope west of Porcupine Bank. *Deep-Sea Research*, 33, 791–818.

Ding, Z. L., Derbyshire, E., Yang, S. L., Yu, Z. W., Xiong, S. F., & Liu, T.S. (2002). Stacked 2.6-Ma grain size record from the Chinese loess based on five sections and correlation with the deep-sea $\delta^{18}O$ record. *Palaeoceanography*, 17(3), Art. No. 10335.1–5.21.

Einstein, H. A., & Krone, R. B. (1962). Experiments to determine modes of cohesive sediment transport in salt water. *Journal of Geophysical Research*, 67, 1451–1461.

Ellwood, B. B., & Ledbetter, M. T. (1977). Antarctic bottom water fluctuations in the Vema Channel: Effects of velocity changes on particle alignment and size. *Earth and Planetary Science Letters*, 35, 189–198.

Embley, R. W., & Jacobi, R. D. (1977). Distribution and morphology of large submarine slides and slumps on Atlantic continental margins. *Marine Geotechnology, 2*, 205–228.

Ewing, M., & Ewing, J. (1964). Distribution of oceanic sediments. In: K. Yoshida (Ed.), *Studies on Oceanography* (pp. 525–537). Tokyo: Tokyo University Press.

Faugères, J. C., Stow, D. A. V., Imbert, P., Viana, A. R., & Wynn, R. B. (1999). Seismic features diagnostic of contourite drifts. *Marine Geology, 162*, 1–38.

Flood, R. D. (1983). Classification of sedimentary furrows and a model for furrow initiation and evolution. *Geological Society of America Bulletin, 94*, 630–639.

Flood, R. D. (1988). A lee-wave model for deep-sea mudwave activity. *Deep-Sea Research, 35*, 973–983.

Flood, R. D., & Giosan, L. (2002). Migration history of a fine-grained abyssal sediment wave on the Bahama Outer Ridge. *Marine Geology, 192*, 259–273.

Fugate, D. C., & Friedrichs, C. T. (2003). Controls on suspended aggregate size in partially mixed estuaries. *Estuarine Coastal and Shelf Science, 58*, 389–404.

Gardner, W. D. (1989). Periodic resuspension in Baltimore Canyon by focusing of internal waves. *Journal of Geophysical Research, 94*, 18185–18194.

Gherardi, J.-M., Labeyrie, L., McManus, J. F., François, R., Skinner, L. C., & Cortijo, E. (2005). Evidence from the Northeastern Atlantic basin for variability in the rate of the meridional overturning circulation through the last deglaciation. *Earth and Planetary Science Letters, 240*, 710–723.

Gibbs, R. J. (1967). The geochemistry of the Amazon River system: Part I. The factors that control the salinity and the composition and concentration of the suspended solids. *Geological Society of America Bulletin, 78*, 1203–1232.

Gross, T. F., & Nowell, A. R. M. (1990). Turbulent suspension of sediments in the deep-sea. *Philosophical Transactions of the Royal Society of London A, 331*, 167–181.

Gross, T. F., & Williams, A. J. (1991). Characterization of deep-sea storms. *Marine Geology, 99*, 281–301.

Guinasso, N. L., & Schink, D. R. (1975). Quantitative estimates of biological mixing rates in abyssal sediments. *Journal of Geophysical Research, 80*, 3022–3043.

Hall, I. R., Bianchi, G. G., & Evans, J. R. (2004). Centennial to millennial scale Holocene climate-deep water linkage in the North Atlantic. *Quaternary Science Reviews, 23*, 1529–1536.

Hall, I. R., & McCave, I. N. (2000). Palaeocurrent reconstruction, sediment and thorium focussing on the Iberian margin over the last 140 ka. *Earth and Planetary Science Letters, 178*, 151–164.

Hall, I. R., McCave, I. N., Shackleton, N. J., Weedon, G. P., & Harris, S. E. (2001). Intensified deep Pacific inflow and ventilation during Pleistocene glacial times. *Nature, 412*, 809–812.

Hamilton, N., & Rees, A. F. (1970). The use of magnetic fabric in paleocurrent estimation. In: S. K. Runcorn (Ed.), *Paleogeophysics* (pp. 445–464). New York: Academic Press.

Haskell, B. J., Johnson, T. C., & Showers, W. J. (1991). Fluctuations in deep western North Atlantic circulation on the Blake Outer Ridge during the last deglaciation. *Paleoceanography, 6*, 21–31.

Hass, H. C. (2002). A method to reduce the influence of ice-rafted debris on a grain size record from northern Fram Strait, Arctic Ocean. *Polar Research, 21*, 299–306.

Heezen, B. C. (1959). Dynamic processes of abyssal sedimentation; erosion, transportation and re-deposition on the deep-sea floor. *Geophysical Journal of Royal Astronomical Society, 2*, 142–163.

Heezen, B. C., & Hollister, C. D. (1964). Deep sea current evidence from abyssal sediments. *Marine Geology, 1*, 141–174.

Hemming, S. R. (2004). Heinrich events: Massive late Pleistocene detritus layers of the North Atlantic and their global climate imprint. *Reviews of Geophysics, 42*, Art. No. RG1005.

Hill, P. S., & McCave, I. N. (2001). Suspended particle transport in benthic boundary layers. In: B. P. Boudreau & B. B. Jorgensen (Eds), *The benthic boundary layer: Transport processes and biogeochemistry* (pp. 78–103). New York: Oxford University Press.

Hill, P. S., & Nowell, A. R. M. (1990). The potential role of large, fast-sinking particles in clearing nepheloid layers. *Philosophical Transactions of the Royal Society of London., A331*, 103–117.

Hollister, C. D., & Heezen, B. C. (1972). Geologic effects of ocean bottom currents: Western North Atlantic. In: A. L. Gordon (Ed.), *Studies in physical oceanography* (Vol. 2, pp. 37–66). New York: Gordon and Breach.

Hollister, C. D., & McCave, I. N. (1984). Sedimentation under deep-sea storms. *Nature*, *309*, 220–225.

Hollister, C. D., Southard, J. B., Flood, R. D., & Lonsdale, P. F. (1976). Flow phenomena in the benthic boundary layer and bed forms beneath deep-current systems. In: I. N. McCave (Ed.), *The benthic boundary layer* (pp. 183–204). NewYork: Plenum Press.

Holz, C., Stuut, J.-B. W., & Henrich, R. (2004). Terrigenous sedimentation processes along the continental margin off N. W. Africa: Implications from grain-size analysis of seabed sediments. *Sedimentology*, *51*, 1145–1154.

Honeyman, B. D., & Santschi, P. H. (1989). A Brownian-pumping model for oceanic trace metal scavenging: Evidence from Th isotopes. *Journal of Marine Research*, *47*, 951–992.

Honjo, S. (1976). Coccoliths — production, transportation and sedimentation. *Marine Micropaleontology*, *1*, 65–79.

Honjo, S. (1978). Sedimentation of materials in Sargasso Sea at a 5,367 m deep station. *Journal of Marine Research*, *36*, 469–492.

Honjo, S., & Roman, M. R. (1978). Marine copepod fecal pellets — production, preservation and sedimentation. *Journal of Marine Research*, *36*, 45–57.

Howe, J. A., & Pudsey, C. J. (1999). Antarctic circumpolar deep water: A Quaternary paleoflow record from the northern Scotia Sea, South Atlantic Ocean. *Journal of Sedimentary Research*, *69*, 847–861.

Hunt, J. R. (1986). Particle aggregate breakup by fluid shear. In: A. J. Mehta (Ed.), *Estuarine cohesive sediment dynamics* (pp. 85–109). New York: Springer-Verlag.

Jelinek, V. (1981). Characterisation of the magnetic fabric of rocks. *Tectonophysics*, *79*, T63–T67.

Johnson, A. G., & Kelley, J. T. (1984). Temporal, spatial, and textural variation in the mineralogy of Mississippi River suspended sediment. *Journal of Sedimentary Petrology*, *54*, 67–72.

Johnson, G. C., Musgrave, D. L., Warren, B. A., Ffield, A., & Olson, D. B. (1998). Flow of bottom and deep water in the Amirante Passage and Mascarene Basin. *Journal of Geophysical Research*, *103*, 30973–30984.

Johnson, G. L., & Schneider, E. D. (1969). Depositional ridges in the North Atlantic. *Earth and Planetary Science Letters*, *6*, 416–422.

Jones, E. J. W., Ewing, M., Ewing, J., & Eittreim, S. L. (1970). Influence of Norwegian Sea overflow water on sedimentation in the northern North Atlantic and Labrador Sea. *Journal of Geophysical Research*, *75*, 1655–1680.

Jumars, P. A. (1993). *Concepts in Biological Oceanography* (p. 348). New York: Oxford University Press.

Keigwin, L. D. (1996). The Little Ice Age and Medieval warm period in the Sargasso Sea. *Science*, *274*, 1504–1508.

Kissel, C. (2005). Magnetic signature of rapid climatic variations in glacial North Atlantic, a review. *C. R. Geoscience*, *337*, 908–918.

Kissel, C., Laj, C., Lehman, B., Labeyrie, L., & Bout-Roumazeilles, V. (1997). Changes in the strength of the Iceland–Scotland overflow water in the last 200,000 years: Evidence from magnetic anisotropy analysis of core SU90-33. *Earth and Planetary Science Letters*, *152*, 25–36.

Konert, M., & Vandenberghe, J. (1997). Comparison of laser grain size analysis with pipette and sieve analysis a solution for the underestimation of the clay fraction. *Sedimentology*, *44*, 523–535.

Kranck, K., & Milligan, T. G. (1983). Grain size distributions of inorganic suspended river sediment. *Mitteilungen des Geologisch-Paläontologischen Institutes der Universität Hamburg*, *55*, 525–534.

Lampitt, R. S., Noji, T., & von Bodungen, B. (1990). What happens to zooplankton fecal pellets — implications for material flux. *Marine Biology*, *104*, 15–23.

Lassen, S., Kuijpers, A., Kunzendorf, H., Lindgren, H., Lindgren, H., Heinemeier, J., Jansen, E., & Knudsen, K. L. (2002). Intermediate water response during northeast Atlantic deglaciation. *Global and Planetary Change*, *32*, 111–125.

Ledbetter, M. T. (1986a). Bottom-current pathways in the Argentine Basin revealed by mean silt particle size. *Nature*, *321*, 423–425.

Ledbetter, M. T. (1986b). A late Pleistocene time-series of bottom-current speed in the Vema Channel. *Palaeogeography, Palaeoclimatology, Palaeoecology*, *53*, 97–105.

Ledbetter, M. T. (1993). Late Pleistocene to Holocene fluctuations in bottom-current speed in the Argentine Basin mudwave field. *Deep-Sea Research, 40,* 911–920.

Ledbetter, M. T., & Balsam, W. L. (1985). Paleoceanography of the deep western boundary undercurrent on the North-American continental-margin for the past 25,000 yr. *Geology, 13,* 181–184.

Ledbetter, M. T., & Johnson, D. A. (1976). Increased transport of Antarctic bottom water in the Vema Channel during last ice age. *Science, 194,* 837–839.

Lerman, A., Lal, D., & Dacey, M. F. (1974). Stokes' settling and chemical reactivity of suspended particles in natural waters. In: R. J. Gibbs (Ed.), *Suspended solids in water* (pp. 17–47). New York: Plenum Press.

van Leussen, W. (1999). The variability of settling velocities of suspended fine-grained sediment in the Ems estuary. *Journal of Sea Research, 41,* 109–118.

Logan, B. E., & Wiilkinson, D. B. (1990). Fractal geometry of marine snow and other biological aggregates. *Limnology and Oceanography, 35,* 130–136.

Löwemark, L., Schönfeld, J., Werner, F., & Schafer, P. (2004). Trace fossils as a paleoceanographic tool: Evidence from late Quaternary sediments of the southwestern Iberian margin. *Marine Geology, 204,* 27–41.

Lynch-Stieglitz, J., Curry, W. B., & Slowey, N. (1999). Weaker Gulf Stream in the Florida Straits during the last glacial maximum. *Nature, 402,* 644–648.

Mackensen, A., & Bickert, T. (1999). Stable carbon isotopes in benthic foraminifera: Proxies for deep and bottom water circulation and new production. In: G. Fischer & G. Wefer (Eds), *Use of proxies in paleoceanography* (pp. 229–254). Berlin: Springer-Verlag.

Manighetti, B., & McCave, I. N. (1995). Late glacial and Holocene palaeocurrents through South Rockall Gap, N. E. Atlantic Ocean. *Paleoceanography, 10,* 611–626.

Mantyla, A. W., & Reid, J. L. (1983). Abyssal characteristics of the World Ocean waters. *Deep-Sea Research, 30,* 805–833.

McCave, I. N. (1975). Vertical particle flux in the oceans. *Deep-Sea Research, 22,* 491–502.

McCave, I. N. (1983). Particulate size spectra, behaviour and origin of nepheloid layers over the Nova Scotian Continental Rise. *Journal of Geophysical Research, 88,* 7647–7666.

McCave, I. N. (1984). Size-spectra and aggregation of suspended particles in the deep ocean. *Deep-Sea Research, 31,* 329–352.

McCave, I. N. (1985). Stratigraphy and sedimentology of box cores from the HEBBLE site on the Nova Scotian Continental Rise. *Marine Geology, 66,* 59–89.

McCave, I. N. (1986). Local and global aspects of the bottom nepheloid layer in the world ocean. *Netherlands Journal of Sea Research, 20,* 167–181.

McCave, I. N. (1994). *Cruise Report, RRS Charles Darwin Cruise 88, NEAPACC,* Department of Earth Sciences, Cambridge University, Cambridge, U.K., 45p.

McCave, I. N. (2001). Nepheloid layers. In: J. H. Steele, S. A., Thorpe, & K. K. Turekian (Eds.), *Encyclopedia of ocean sciences* (Vol. 4, pp. 1861–1870). London: Academic Press.

McCave, I. N. (2002). A poisoned chalice? *Science, 298,* 1186–1187.

McCave, I. N., Bryant, R. G., Cook, H. F., & Coughanowr, C. A. (1986). Evaluation of laser-diffraction-size analyser for use with natural sediments. *Journal of Sedimentary Petrology, 56,* 561–564.

McCave, I. N., & Hall, I. R. (2006). Size sorting in marine muds: Processes, pitfalls and prospects for palaeoflow-speed proxies, *Geochemistry, Geophysics, Geosystems, 7,* Q10N05, doi: 10.1029/GC001284, 37pp.

McCave, I. N., Hall, I. R., & Bianchi, G. G. (2006). Laser versus settling velocity differences in silt grain size measurements: Estimation of palaeocurrent vigour. *Sedimentology, 53,* 919–928.

McCave, I. N., Hollister, C. D., Laine, E. P., Lonsdale, P. F., & Richardson, M. J. (1982). Erosion and deposition on the eastern margin of Bermuda Rise in the Late Quaternary. *Deep-Sea Research, 29,* 535–561.

McCave, I. N., Kiefer, T., Thornalley, D. J. R., & Elderfield, H. (2005). Deep flow in the Madagascar–Mascarene Basin over the last 150,000 years. *Philosophical Transactions of the Royal Society of London, 363*(1826), 81–99.

McCave, I. N., Manighetti, B., & Beveridge, N. A. S. (1995a). Changes in circulation of the North Atlantic during the last 25,000 years inferred from grain size measurements. *Nature, 374*, 149–152.

McCave, I. N., Manighetti, B., & Robinson, S. G. (1995b). Sortable silt and fine sediment size/composition slicing: parameters for palaeocurrent speed and palaeoceanography. *Paleoceanography, 10*, 593–610.

McCave, I. N., & Swift, S. A. (1976). A physical model for the rate of deposition of fine-grained sediment in the deep sea. *Geological Society of America Bulletin, 87*, 541–546.

McCave, I. N., & Syvitski, J. P. M. (1991). Principles and methods of geological particle size analysis. In: J. P. M. Syvitski (Ed.), *Principles, methods and application of particle size analysis* (pp. 3–21). New York: Cambridge University Press.

McCave, I. N., & Tucholke, B. E. (1986). Deep current-controlled sedimentation of the western North Atlantic. In: B. E. Tucholke & P. E. Vogt (Eds), *Western North Atlantic Region, The Geology of North America* (Vol. M, pp. 451–468). Boulder, CO: Geological Society of America.

McIntyre, A. (1967). Modern coccolithophoridae of Atlantic Ocean. I. Placoliths and cyrtoliths. *Deep-Sea Research, 14*, 561–1967.

McManus, J. F., François, R., Gherardi, J.-M., Keigwin, L. D., & Brown-Leger, S. (2004). Collapse and rapid resumption of Atlantic meridional circulation linked to deglacial climate changes. *Nature, 428*, 834–837.

McPhee-Shaw, E. (2006). Boundary-interior exchange: Reviewing the idea that internal-wave mixing enhances lateral dispersal near continental margins. *Deep-Sea Research II, 53*(2006), 42–59.

Mehta, A. J., & Lott, J. W. (1987). Sorting of fine sediment during deposition. In: N. C. Kraus (Ed.), *Coastal Sediments '87* (Vol. I, pp. 348–362). New York: American Society of Civil Engineers.

Mezerais, M. L., Faugères, J. C., Figueiredo, A. G., & Masse, L. (1993). Contour current accumulation off the Vema Channel mouth, southern Brazil Basin — pattern of a contourite fan. *Sedimentary Geology, 82*, 173–187.

Mienert, J., Flood, R. D., & Dullo, W. C. (1994). Research perspectives of sediment waves and drifts; monitors of global change in deepwater circulation, *Paleoceanography, 9*, 893–895 (collection of papers, 897–1094).

Miller, M. C., & Komar, P. D. (1977). Development of sediment threshold curves for unusual environments (Mars) and for inadequately studied materials (foram sands). *Sedimentology, 24*, 709–721.

Milligan, T. G., & Kranck, K. (1991). Electroresistance particle size analysers. In: J. P. M. Syvitski (Ed.), *Principles, Methods and Application of Particle Size Analysis* (pp. 109–118). New York: Cambridge University Press.

Moore, T. C., & Heath, G. R. (1978). Sea-floor sampling techniques. In: J. P. Riley, & R. Chester (Eds), *Chemical oceanography* (Vol. 7, pp. 75–126). New York: Academic Press.

Mountain, G. S., & Tucholke, B. E. (1985). Mesozoic and Cenozoic geology of the U.S. continental slope and rise. In: C. W. Poag (Ed.), *Geologic evolution of the United States Atlantic Margin* (pp. 293–341). Stroudsburg, PA: Van Nostrand-Reinhold.

Nelson, C. S., Cooke, P. J., Hendy, C. H., & Cuthbertson, A. M. (1993). Oceanographic and climatic changes over the past 160,000 years at deep-sea drilling project site-594 off southeastern New Zealand, southwest Pacific-Ocean. *Paleoceanography, 8*, 435–458.

North Greenland Ice Core Project Members (2004). High-resolution record of Northern Hemisphere climate extending into the last interglacial period. *Nature, 431*, 147–151.

Ohkouchi, N., Eglinton, T. I., Keigwin, L. D., & Hayes, J. M. (2002). Spatial and temporal offsets between proxy records in a sediment drift. *Science, 298*(5596), 1224–1227.

Okada, H., & Honjo, S. (1973). Distribution of oceanic coccolithophorids in the Pacific. *Deep-Sea Research, 20*, 355–374.

Pahnke, K., Zahn, R., Elderfield, H., & Schulz, M. (2003). 340,000-Year centennial-scale marine record of Southern Hemisphere climatic oscillation. *Science, 301*, 948–952.

Parkin, D. W. (1974). Trade-winds during the glacial cycles. *Proceedings of the Royal Society of London, 337*, 73–100.

Pettijohn, F. J. (1975). Sedimentary Rocks (3rd ed.) (p. 628). New York: Harper & Row.

Piotrowski, A. M., Goldstein, S. L., Hemming, S. R., & Fairbanks, R. G. (2004). Intensification and variability of ocean thermohaline circulation through the last deglaciation. *Earth and Planetary Science Letters, 225*, 205–220.

Piper, D. J. W., & Pereira, C. P. G. (1992). Late Quaternary sedimentation in central Flemish Pass. *Canadian Journal of Earth Sciences, 29*, 535–550.

Potter, P. E., Heling, D., Shimp, N. F., & VanWie, W. (1975). Clay mineralogy of modern alluvial muds of the Mississippi River Basin. *Bulletin du Centre de Recherches Pau-SNPA, 9*, 353–389.

Prins, M. A., Bouwer, L. M., Beets, C. J., Troelstra, S. R., Weltje, G. J., Kruk, R. W., Kuijpers, A., & Vroon, P. Z. (2002). Ocean circulation and iceberg discharge in the glacial North Atlantic; inferences from unmixing of sediment size distributions. *Geology, 30*, 555–558.

Pudsey, C. J., & Howe, J. A. (1998). Quaternary history of the Antarctic Circumpolar Current: evidence from the Scotia Sea. *Marine Geology, 148*, 83–112.

Rebesco, M., & Stow, D.A.V. (Eds). (2001). Seismic expression of contourites and related deposits. *Marine Geophysical Researches, 22*, 303–521.

Rees, A. I. (1961). The effect of water currents on the magnetic remanence and anisotropy of susceptibility of some sediments. *Geophysical Journal of Royal Astronomical Society, 5*, 235–251.

Rees, A. I. (1966). Some flume experiments with a fine silt. *Sedimentology, 6*, 209–240.

Reid, J. L. (1994). On the total geostrophic circulation of the North Atlantic ocean — flow patterns, tracers, and transports. *Progress in Oceanography, 33*, 1–92.

Reid, J. L. (1997). On the total geostrophic circulation of the Pacific Ocean: Flow patterns, tracers, and transports. *Progress in Oceanography, 39*, 263–352.

Reid, J. L. (2003). On the total geostrophic circulation of the Indian Ocean: Flow patterns, tracers, and transports. *Progress in Oceanography, 56*, 137–186.

Revel, M., Cremer, M., Grousset, F. E., & Labeyrie, L. (1996). Grain-size and Sr-Nd isotopes as tracers of paleo-bottom current strength, Northeast Atlantic Ocean. *Marine Geology, 131*, 233–249.

Rintoul, S. R., Hughes, C. W., & Olbers, D. (2001). The Antarctic circumpolar current system. In: G. Siedler, J. Church & J. Gould (Eds), *Ocean circulation and climate* (pp. 271–302). New York: Academic Press.

Schaaf, E., Grenz, C., Pinazo, C., & Lansard, B. (2006). Field and laboratory measurements of sediment erodibility: A comparison. *Journal of Sea Research, 55*, 30–42.

Schmitz, W. J. (1995). On the interbasin scale thermohaline circulation. *Reviews of Geophysics, 33*, 151–173.

Schneider, E. D., Fox, P. J., Hollister, C. D., Needham, D., & Heezen, B. C. (1967). Further evidence of contour currents in the western north Atlantic. *Earth and Planetary Science Letters, 2*, 351–359.

Schönfeld, J. (2002). A new benthic foraminiferal proxy for near-bottom current velocities in the Gulf of Cadiz, northeastern Atlantic Ocean. *Deep-Sea Research I, 49*, 1853–1875.

Self, R. F. L., Nowell, A. R. M., & Jumars, P. A. (1989). Factors controlling critical shears for deposition and erosion of individual grains. *Marine Geology, 86*, 181–199.

Siddall, M., J Rohling, E., Almogi-Labin, A., Hemleben, C., Meischner, D., Schmelzer, I., & Smeed, D. A. (2003). Sea-level fluctuations during the last glacial cycle. *Nature, 423*, 853–858.

Singer, J., Anderson, J. B., Ledbetter, M. T., Jones, K. P. N., McCave, I. N., & Wright, R. (1988). The assessment of analytical techniques for the analysis of fine-grained sediments. *Journal of Sedimentary Petrology, 58*, 534–543.

Skinner, L., & McCave, I. N. (2003). Analysis and modelling of the behaviour of gravity and piston corers based on soil mechanical principles. *Marine Geology, 199*, 181–204.

Snowball, I., & Moros, M. (2003). Saw-tooth pattern of North Atlantic current speed during Dansgaard-Oeschger cycles revealed by the magnetic grain size of Reykjanes Ridge sediments at 59 degrees N. *Paleoceanography, 18(2)*, 12. Art. No. 1026.

Southard, J. B., & Boguchwal, L. A. (1990). Bed configurations in steady unidirectional water flows. 2. Synthesis of flume data. *Journal of Sedimentary Petrology, 60*, 658–679.

Stein, R. (1985). Rapid grain-size analyses of clay and silt fraction by Sedigraph 5000D: Comparison with Coulter Counter and Atterberg methods. *Journal of Sedimentary Petrology, 55*, 590–593.

Stickley, C. E., Carter, L., McCave, I. N., & Weaver, P. P. E. (2001). Variations in the CDW flow through the S. W. Pacific gateway for the last 190 ky: Evidence from Antarctic diatoms. In: D. Seidov., B. J. Haupt & M. A. Maslin (Eds), *The oceans and rapid climate change: Past, present, and future, Geophysical Monograph* (Vol. 126, pp. 101–116). Washington, D.C.: AGU.

Stommel, H. (1948). The westward intensification of wind-driven currents. *Transactions of American Geophysical Union, 29,* 202–206.

Stommel, H. (1958). The abyssal circulation. *Deep-Sea Research, 5,* 80–82.

Stow, D. A. V., Pudsey, C. J., Howe, J. A., Faugeres, J.-C., & Viana, A. R. (Eds) (2002). Deep water contourite systems: Modern drifts and ancient series, seismic and sedimentary characteristics, *Memoirs. Geological Society London* (Vol. 22, p. 464).

Stuut, J.-B., Prins, M. A., Schneider, R., Weltje, G. J., Jansen, J. H. F., & Postma, G. (2002). A 300 kyr record of aridity and wind strength in southwestern Africa: Inferences from grain-size distributions of sediments on Walvis Ridge, S. E. Atlantic. *Marine Geology, 180,* 221–233.

Swallow, J. C., & Worthington, L. V. (1957). Measurements of deep currents in the western North Atlantic. *Nature, 179,* 1183–1184.

Syvitski, J. P. M., LeBlanc, K. W. G., & Asprey, K. W. (1991). Interlaboratory, interinstrument calibration experiment. In: J. P. M. Syvitski (Ed.), *Principles, methods and application of particle size analysis* (pp. 174–193). New York: Cambridge University Press.

Thistle, D., Yingst, J. Y., & Fauchald, K. (1985). Deep-sea benthic community exposed to strong near-bottom currents on the Scotian Rise (Western Atlantic). *Marine Geology, 66,* 91–112.

Thistle, D., & Wilson, G. D. F. (1987). A hydrodynamically modified, abyssal isopod fauna. *Deep-Sea Research, 34,* 73–87.

Thomas, A. L., Henderson, G. M., & McCave, I. N. (2006). Constant flow of AABW into the Indian Ocean over the past 140 ka? Conflict between $^{231}Pa/^{230}Th$ and sortable silt records. Abstract, Goldschmidt Conference, Melbourne.

Tucholke, B. E., Hollister, C. D., Biscaye, P. E., & Gardner, W. D. (1985). Abyssal current character determined from sediment bedforms on the Nova Scotian continental rise. *Marine Geology, 66,* 43–57.

Warren, B. A. (1981). Deep circulation of the World Ocean. In: B. A. Warren & C. Wunsch (Eds), *Evolution of Physical Oceanography* (pp. 6–41). Cambridge, MA: MIT Press.

Weatherly, G. L. (1984). An estimate of bottom frictional dissipation by Gulf-Stream fluctuations. *Journal of Marine Research, 42,* 289–301.

Weaver, P. P. E. (1994). Determination of turbidity-current erosional characteristics from reworked coccolith assemblages, Canary Basin, Northeast Atlantic. *Sedimentology, 41,* 1025–1038.

Weaver, P. P. E., & Thomson, J. (1993). Calculating erosion by deep-sea turbidity currents during initiation and flow. *Nature, 364,* 136–138.

Weber, O., Gonthier, E., & Faugères, J.-C. (1991). Analyse granulometrique de sediments fins marins: comparaison des resultats obtenus au Sedigraph et au Malvern. *Bulletin de L'Institute de Géologie du Bassin D'Aquitaine, 50,* 107–114.

Wei, K.-Y. (2006). 3. Leg 195 Synthesis: Site 1202 — Late Quaternary sedimentation and paleoceanography in the southern Okinawa Trough. *Proceedings Ocean Drilling Program, Scientific Results,* 195 (31p).

Weltje, G. J. (1997). End-member modeling of compositional data: Numerical-statistical algorithms for solving the explicit mixing problem. *Mathematical Geology, 29,* 503–549.

Weltje, G. J., & Prins, M. A. (2003). Muddled or mixed? Inferring palaeoclimate from size distributions of deep-sea clastics. *Sedimentary Geology, 162,* 39–62.

Wetzel, A. (1991). Ecologic interpretation of deep-sea trace fossil communities. *Palaeogeography, Palaeoclimatology, Palaeoecology, 85,* 47–69.

Wheatcroft, R. A. (1990). Preservation potential of sedimentary event layers. *Geology, 18*(9), 843–845.

Wheatcroft, R. A., Jumars, P. A., Smith, C. R., & Nowell, A. R. M. (1990). A mechanistic view of the particulate biodiffusion coefficient — step lengths, rest periods and transport directions. *Journal of Marine Research, 48*(1), 177–207.

White, S. J. (1970). Plane bed thresholds of fine grained sediments. *Nature, 228,* 152–154.

Winegard, C. I. (1970). Settling velocity of grains of quartz and other minerals in sea water versus pure water. *U.S. Geological Survey Professional Paper, 700-B*, pp. 161–166.

Winn, K., Kogler, F. C., & Werner, F. (1980). Simultaneous application of reflection strength recorder, sidescan sonar and sub-bottom profiler in seafloor sediment mapping. In: W. A. Kuperman & F. B. Jensen (Eds), *Bottom-interacting ocean acoustics* (pp. 85–98). New York: Plenum Press.

Winterwerp, J. C., & van Kesteren, W. G. M. (2004). *Introduction to the physics of cohesive sediment in the marine environment* (p. 466). Amsterdam: Elsevier.

Wüst, G. (1955). Stromgeschwindigkeiten im Tiefen — und Bodenwasser des Atlantischen Ozeans auf Grund dynamischer Berechnung der Meteor-Profile der Deustschen Atlantischen Expedition 1925/27, in *Bigelow Volume, supplement to Deep Sea Research*, 3, 373–397.

Wynn, R. B., & Stow, D. A. V. (Eds). (2002). Recognition and interpretation of deep-water sediment waves. *Marine Geology, 192*, 1–333.

Yokokawa, M., & Franz, S. O. (2002). Changes in grain size and magnetic fabric at Blake-Bahama Outer Ridge during the late Pleistocene (marine isotope stages 8–10). *Marine Geology, 189*, 123–144.

> **CHAPTER TWO**

Continuous Physical Properties of Cored Marine Sediments

Guillaume St-Onge*, Thierry Mulder, Pierre Francus *and* Bernard Long

Contents

1. Introduction	63
2. Continuous Centimeter-Scale Measurements of Physical Properties	64
2.1. Magnetic susceptibility	65
2.2. Gamma density	66
2.3. *P*-Wave velocity	66
2.4. Color reflectance	67
2.5. Natural gamma radiation	68
3. Continuous Millimeter- to Micrometer-Scale Measurements of Physical Properties	68
3.1. Digital core imaging	68
3.2. Digital X-ray systems	69
3.3. Computerized coaxial tomography (CAT-scan)	73
3.4. Micro X-ray fluorescence spectrometry	74
3.5. Magnetic resonance imaging (MRI) and nuclear magnetic resonance (NMR)	79
3.6. Confocal macroscopy and microscopy	79
4. Recent Applications of Continuous Centimeter- to Millimeter-Scale Physical Properties of Marine Sediments	81
4.1. Use of digital X-ray to rapidly identify turbidites in marine sediments	82
4.2. RDL in the saguenay fjord	84
4.3. CAT-Scan analysis as a millimeter-scale paleoceanographic tool	88
5. Conclusion	91
Acknowledgments	92
References	92

> ## 1. Introduction

Continuous centimeter–scale measurements of various physical properties of marine cores now form the basis of most paleoceanographic studies. These

* Corresponding author.

Developments in Marine Geology, Volume 1
ISSN 1572-5480, DOI 10.1016/S1572-5480(07)01007-X

measurements generally provide a rapid and nondestructive method for characterizing the nature and composition of long sedimentary sequences. For major paleoceanographic campaigns such as those carried out as part of the Integrated Ocean Drilling Program (IODP) or the International Marine Past Global Change Study (IMAGES), where sediment samples are used by numerous researchers for various destructive analyses, continuous and nondestructive measurements of physical properties are made quickly onboard and reflect the pristine state of the sediments. Continuous physical properties records provide the basis for stratigraphy and core correlation, the first insight into core lithology, continuous data for time series analyses and a decision tool for determining the best subsampling strategy. Physical properties that are now routinely measured continuously, both onboard and onshore, include: natural gamma radiation, gamma density, p-wave velocity, magnetic susceptibility, electrical resistivity and color reflectance. Most of these properties can now be measured continuously and automatically, on whole or split cores placed horizontally or, when the sediment-water interface must remain undisturbed, vertically. Emerging line scan systems and medical techniques may now also be used to continuously image the sediment surface on split cores or the internal structure of whole cores by high-resolution imaging, digital X-ray imaging and computerized coaxial tomography (CAT-scan). The resulting images can then be processed to obtain qualitative and quantitative information about the lithology and/or sedimentary structures of the cores. In this chapter, we first review continuous, cm-scale, nondestructive methods generally used to determine the physical properties of sediments. This is by no means an exhaustive review of all available techniques to measure physical properties of sediment cores, but rather a brief overview of the techniques most commonly used in paleoceanography. The reader is referred to the technical note by Blum (1997) for very detailed information on the measurement protocols for most of the techniques discussed in the first section of this chapter. The second part of this chapter describes emerging methods used both to image and/or determine in 2D or 3D the physical properties of long sedimentary sequences at the millimeter to micrometer-scale, allowing the reconstruction of paleoceanographic or paleoclimatological processes at temporal resolutions on the millennial- to seasonal-scale. The last part of this chapter illustrates some of these new methods using new or recently published data.

2. CONTINUOUS CENTIMETER-SCALE MEASUREMENTS OF PHYSICAL PROPERTIES

The most widely used instrument to measure continuously the physical properties of long marine cores is the Multisensor Track (MST) or Multisensor Core Logger (MSCL; Figure 1). These core loggers were all designed to measure continuously and simultaneously, at the cm-scale, several physical properties such as magnetic susceptibility, gamma density, natural gamma radiation, p-wave velocity and electrical resistivity. Although it is generally used as a hand-held technique, we will also briefly discuss the color reflectance measurement method because, as with MSCL sensors, it is fast, nondestructive, high resolution and widely used.

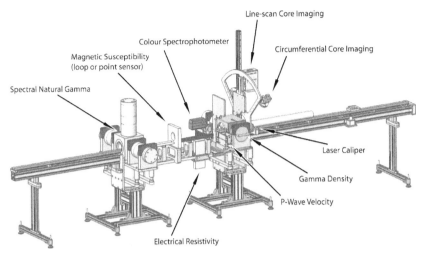

Figure 1 Schematic diagram of a GEOTEK Multisensor Core Logger, MSCL (suitable for both whole or split lined sediment cores and unlined rock cores), illustrating the range of sensor systems available (see text for details). The specific suite of sensors on any given MSCL depends on user requirements and may include a digital X-ray imaging system (not shown).

2.1. Magnetic Susceptibility

Magnetic susceptibility is one of several magnetic properties that can be measured on marine sediment cores. Other magnetic properties are the subject of various books and articles and will not be discussed in this chapter. Stoner and St-Onge (this volume) review the use of magnetic stratigraphy in paleoceanography. Most MSTs or MSCLs can measure the magnetic susceptibility of sediments (Figure 1). Data are generally expressed as uncorrected low field volumetric magnetic susceptibility (k), but may also be expressed either as corrected low field volumetric magnetic susceptibility, if the loop and core diameters are taken into account, or as mass specific magnetic susceptibility, if sediment density is simultaneously measured and taken into account. k provides a first order estimate of ferrimagnetic mineral (e.g., magnetite) abundance in sediments, but is also sensitive to grain size variations, increasing slightly with increasing grain size (e.g., Stoner, Channell, & Hillaire-Marcel, 1996). Although k is generally used to correlate cores on the basis of lithology or to identify rapidly deposited layers (RDL) such as Heinrich events, its potential applications are broader (Thompson & Oldfield, 1986; Maher & Thompson, 1999; Evans & Heller, 2003). One disadvantage of using a magnetic susceptibility loop on whole cores is the rather large response function, resulting in several centimeters of smoothing depending on the loop and core diameters. For instance, Blum (1997) estimated that axial lengths along the core corresponding to more than 99% and 50% (half-height) of the response are approximately 15 and 4.4 cm, respectively, for an 80-mm internal diameter loop. This response function would be even greater for loops of larger diameter. One way to increase the resolution is to use a point-source sensor on split cores. Such an instrument can now be used on an MSCL or a separate track (e.g., the system developed by I. Snowball at Lund University, Figure 2). According to the manufacturer, the sensitive

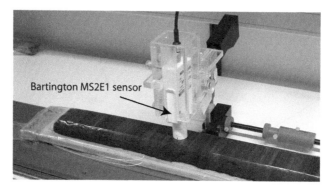

Figure 2 TAMISCAN-TS1 point-source magnetic susceptibility track. The system uses a Bartington MS2E1 sensor. A split core covered by a plastic sheet is being measured.

area of the point sensor probe (50% maximum response) is a 3.8 mm × 10.5 mm rectangle. Another alternative for increasing the resolution is to take continuous measurements on u-channel samples (2 × 2 cm, 1.5 m long plastic liner inserted in the middle of a split section for paleomagnetic measurements) using a smaller diameter loop coupled with a cryogenic magnetometer (e.g., the Gif-sur-Yvette system) or a separate track using a kappa bridge (Thomas, Guyodo, & Channell, 2003). In both cases, the response function (half-height) will be close to 3 cm.

2.2. Gamma Density

In most MSTs or MSCLs, a ^{137}Cs radioactive source and a NaI (TI) detector are used to measure sediment density based on emitted gamma ray attenuation (Figure 1). In general, a beam a few millimeters in diameter is emitted through the source collimators (2.5 or 5 mm), allowing a downcore spatial resolution at the cm-scale. To obtain precise data, the instrument must be properly calibrated using a standard of known density in distilled water. Different standards must be used for split and whole core loggers. In addition, when logging on either split or whole core systems, biases may be introduced by variations in sediment thickness or by the liner thickness itself. The GEOTEK MSCL can now measure sediment thickness variations on split cores. The gamma ray source and detector can also be placed horizontally instead of vertically to reduce possible sediment voids inside the liner due to gravity in whole cores. Clearly, the same liner used for calibration should be used during coring. Because density yields information on sediment properties such as grain size and mineralogy, the measurement of gamma density is a very widely used method.

2.3. P-Wave Velocity

P-wave velocity in marine sediments is influenced, among other things, by changes in lithology, bulk density, porosity, lithostatic pressure, degree of fracturing, degree of consolidation and/or the presence of solid gas hydrate or free gas. Combined with density measurements, the p-wave velocity is often used to calculate acoustic impedance (the product of density and acoustic velocity) in order to construct synthetic seismic profiles and to estimate the depth of seismic reflectors. P-wave velocity can be measured continuously on most available MSTs or MSCLs using a pair of transducers acting as transmitter and receiver. Earlier transducers were static and required the addition of fluid, generally water, between the transducers and the core liner. Acoustic rolling transducers were subsequently developed to ensure full contact with the liner and thus eliminate the requirement for a fluid. In this type of transducer, the center of frequency is around 230 kHz, with an accuracy of ~50 ns depending on core thickness and condition. As for gamma density, the p-wave transducer should be placed horizontally rather than vertically to minimize possible sediment voids inside the liner due to gravity in whole cores.

2.4. Color Reflectance

Color reflectance is also one of the parameters generally measured on long cores immediately after splitting. A hand-held spectrophotometer with a 3-, 4- or 8-mm aperture is generally used to measure the percent diffuse reflectance of a known light source from the sediment surface with wavelengths ranging in most cases from 400 and 700 nm (visible spectrum) at 10 nm increments. Other systems such as the one developed at Oregon State University (see below) allow measurements over a wider set of wavelengths (250–950 nm). Using the first derivative of the reflectance spectrum, it is possible to derive information on sediment mineralogy such as the presence and concentration of iron oxides such as hematite and goethite (e.g., Deaton & Baslam, 1991; Balsam & Deaton, 1991; Balsam, Damuth, & Schneider, 1997). Empirical relationships between measured geochemical parameters and color parameters, such as L^* (lightness, see below), have been derived which can be used as high-resolution geochemical proxies (e.g., Mix, Harris, & Janecek, 1995; Weber, 1998; Ortiz, Mix, Harris, & O'Connell, 1999). Reflectance data can also be converted into the widely used *Commission Internationale de l'Éclairage* (CIE) L^*, a^* and b^* color space (Nederbragt & Thurow, 2004; see also Berns, 2000, for a comprehensive discussion of color systems). L^* ranges from 0 (black) to 100 (white), whereas a^* and b^* range from +60 (red) to −60 (green) and from +60 (yellow) to −60 (blue), respectively. L^* is widely used to highlight changes in carbonate or organic carbon contents (e.g., Mix et al., 1995; Ortiz et al., 1999; Peterson, Haug, Hughen, & Röhl, 2000; St-Onge, Mulder, Piper, Hillaire-Marcel, & Stoner, 2004), whereas variations in the a^* value are often associated with changes in the concentration of red minerals such as hematite (e.g., Helmke, Schulz, & Bauch, 2002). Variations in the b^* value of anoxic sediments have been reported to closely follow variations in diatom and organic matter contents (Debret et al., 2006). Downcore variations in reflectance-derived data have also been used for stratigraphic or correlation purposes (e.g., Mix, Rugh, Pisias, Veirs, & Leg, 1992; Peterson et al., 2000). Automated color reflectance measurements are now also possible with, for instance, the GEOTEK MSCL (Figure 1) or other systems such as the Oregon State University split-core analysis track (SCAT; Mix et al., 1992, 1995; Harris, Mix, & King, 1997; Ortiz et al., 1999). Several instruments have been used by the paleoceanographic community for diffuse reflectance measurements. For instance, a few papers have discussed the use of the hand-held X-Rite Colortron (Andrews & Freeman, 1996; Boyle, 1997; Keigwin & Pickart, 1999), whereas most papers discuss the use of the more expensive Minolta hand-held spectrophotometers (e.g., Chapman & Shackleton, 2000; Ortiz & Rack, 1999 and references therein). Finally, color reflectance measurements can be correlated with results obtained using conventional tools for color description such as the Munsell chart.

2.5. Natural Gamma Radiation

Natural gamma-ray spectrometry allows estimation of elemental concentrations of K, U and Th using gamma emissions of their radioactive isotopes ^{40}K, as well as the ^{238}U and ^{232}Th series. These can be estimated by total counts or gamma spectra detected with two or more NaI scintillators and photomultiplier tubes. These detectors can be fully integrated, for instance, on the GEOTEK MSCL to perform continuous measurements (Figure 1). However, the time needed to obtain sufficient and acceptable counts is rather large compared to other measurements (Blum, Rabaute, Gaudon, & Allan, 1997), which limits the usefulness of the technique for continuous logging of sediment cores for paleoceanographic purposes. In addition, the spatial resolution of the measurements is rather low, depending in part on the NaI crystal diameter, which is generally $\sim$5–8 cm.

3. Continuous Millimeter- to Micrometer-Scale Measurements of Physical Properties

3.1. Digital Core Imaging

High quality photographs of sediment cores can be used to image, quantify, and archive sedimentological changes in marine sediment cores. One of the goals of this approach is to capture variations in sediment color and texture before oxidization of the sediment surface takes place. Consequently, photographs are taken immediately after core splitting. Until recently, most core photographs were taken on conventional film, making this approach expensive, time consuming and difficult to combine with other continuous measurements performed onboard or onshore. High quality digital cameras are now much less expensive and hence extensively used to photograph sediment cores. Unfortunately, the process is still difficult and time consuming, in part because lighting is often inadequate and further image processing is necessary. In addition, images thus obtained cannot be used effectively for direct comparison with other continuous measurements. Two commercially available systems have recently solved some of these problems: the GEOTEK Geoscan III line-scan camera and the Smartcube smartCIS camera image scanner.

The Smartcube smartCIS camera image scanner consists of an 8.2 megapixels Canon EOS 20D digital camera mounted on a moving frame equipped with lights (Figure 3). Small electric motors control both vertical and horizontal movement of the camera. Picture taking and camera movement are controlled and recorded using

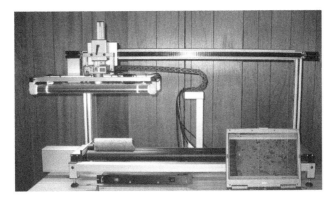

Figure 3 Smartcube smartCIS camera image scanner. In this figure, the digital image of a whole rock core was acquired, unrolled and displayed on the companion laptop. Image modified with permission from http://www.smartcube.de/indexeng.htm

a PC, and the companion software allows continuous reconstruction of an image of a core section not more than 1.5 m in length at a maximum pixel resolution of 33 μm (50 mm lens for 30–75 mm core diameters). For larger core diameters, the maximum pixel resolution is 63 μm (35 mm lens for 30–144 mm core diameters). One novel feature of this instrument is its ability to photograph whole rock cores by rotating the core, then "unrolling" the images using the software.

The Geoscan III line-scan camera is one of the latest improvements in sensors developed by GEOTEK and integrated into their automated MSCL (Figure 1). The setup consists of three individual interference filters placed in front of three 2048 Charge Coupled Device (CCD) line arrays inside the camera, one for each RGB color (red, green and blue). When properly calibrated, the resulting images can also be analyzed in terms of the three color arrays, providing downcore quantitative values of color variability (e.g., Moy, Seltzer, Rodbell, & Anderson, 2002; Nederbragt & Thurow, 2004). The standard instrument design allows the acquisition of images with a 50-μm pixel size on cores up to 10 cm wide. Pixel sizes down to 20 μm can be achieved, but for most routine applications, a 100 μm pixel size seems most practical (P. Schultheiss, personal communication, 2006). As with the Smartcube smartCIS camera image scanner, the GEOTEK MSCL can also be equipped with a new circumferential core imaging system, allowing imaging of the surface of a bare, round, whole rock core and unrolling of the resulting image to produce a complete 360° image. This makes it possible to look at geological structures and determine dip angles. A summary of the issues associated with digital core imaging is provided in Francus (2004).

3.2. Digital X-Ray Systems

X-ray radiography is a technique based on differential travel of X-rays through sediment (Bouma, 1969). During this travel, the incident X-ray beam is attenuated by various phenomena including absorption and scattering. The dominant control on beam attenuation is bulk sediment density (Holyer, Young, Sandidge, & Briggs,

1996; Jackson, Briggs, & Flint, 1996), which is in turn affected by parameters such as grain size and lithology, including carbonate and silica contents. Beam attenuation can also be affected by physical parameters such as changes in water content, compaction and porosity. The gray scale intensity of X-ray images is proportional to X-ray attenuation. Consequently, these images primarily reflect sediment density and, in theory, provide a first order picture of downcore grain size variations. However, gray level curve variations must be properly calibrated (refer to the Oregon State University system, below) or interpreted using complementary mineralogical (e.g., carbonate contents) and grain-size data obtained from the sediments using conventional sedimentological analysis methods.

Until recently, X-radiographs of marine sediment cores or slabs were obtained on conventional chemical film, making the process not only relatively time consuming but also expensive and inconvenient for image postprocessing, which required at least one additional step consisting in digitizing the film (e.g., Principato, 2004). In contrast, digital X-ray images of sediment slabs or cores can be taken quickly both onshore and onboard, and their use in image analysis is direct and straightforward. Digital X-radiography can mainly be used for high-resolution sedimentological analysis (millimeter- to centimeter-scale), but can also be used effectively for assessing bioturbation. In sedimentology, its main applications are for identifying sedimentary facies and sequences at the process scale (individual sequence) and constraining the evolution of sequences over time (nature, size, frequency, rhythms and cycles). This method is also very useful for identifying laminae or other sedimentary structures that may be unrecognizable to the naked eye, for detecting the base and top boundaries in a sedimentary sequence, for characterizing grain size trends within a given sequence, for downcore high-resolution analysis of textural and structural variations, for facies differentiation and for detecting facies changes and assessing sampling quality and position. For instance, extracted high-resolution gray curves (see Section 4.3) may be readily correlated with other physical parameters (color, gamma density, magnetic susceptibility, grain size changes) and other cores. What follows is a brief overview of some available onboard and onshore digital X-ray systems, which may be used for rapid measurement of marine cores.

The digital X-ray imaging system developed at the Université Bordeaux 1 (SCOPIX) consists of conventional X-ray equipment combined with a new radioscopy instrument developed by the Cegelec Company (Migeon, Weber, Faugères, & Saint-Paul, 1999; Lofi & Weber, 2001). The lead box (0.8 m wide, 0.8 m long, and 1.2 m high) is built on a base equipped with two motorized lateral sleeves to move the sediment samples (Figure 4). The upper part of the box contains the X-ray source (160 kV, 19 mA) with a water-cooling system. X-rays pass through the sediment and the resulting signal is amplified. Originally, X-rays were recorded by a high-resolution CCD camera (756×581 pixels) with a pixel size of 0.21 mm. This camera was used to collect the sample images shown in Figures 14 and 15. The system was recently improved with the addition of a 4096 gray level Hamamatsu ORCA camera ($1,280 \times 1,024$ pixels). The signal recorded by the camera is transferred to a computer, which converts it into gray level values. High-resolution gray level images are acquired and reconstructed using one software, while another

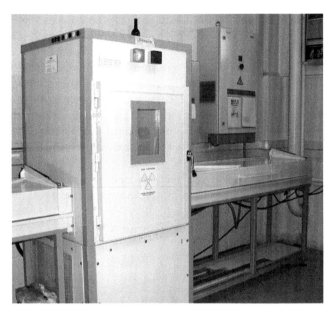

Figure 4 Photograph of the complete SCOPIX system showing the X-ray lead-protected box and the two sleeves on each side.

software is used for image processing (Migeon et al., 1999). The acquisition software records basic information about the core, whereas the processing software is used to display downcore X-ray intensity logs (gray scale values) including the mean, median, minimum and maximum values and standard deviations. Further processing of the images has been extensively described by Lofi and Weber (2001), and may include the use of several numerical filters and extractions. Filter parameters and procedure vary depending on whether the purpose is to enhance sequence boundaries or highlight internal sedimentary structures. For further examples of applications of conventional X-rays image analysis, see Francus (2004) and references therein.

Although SCOPIX measurements can be made on u-channels, split or whole cores, they are generally performed on aluminum slabs or trays. The slabs are extracted continuously from split cores using an electro-osmotic core cutter (Chmelick, 1967). Conventional slab dimensions are <7 cm wide and <1.5 m long. A thickness of ~1 cm provides the best results. Thinner slabs would yield higher precision, but would be very difficult to subsample. Uniform slab thickness prevents attenuation variations due to thickness changes.

Shipboard digital X-ray systems have several key advantages over onshore systems. Firstly, they allow rapid visualization of sediment cores prior to splitting, which may facilitate operational decision making at sea. For instance, they may help in determining whether a specific sedimentary target or stratigraphic interval was successfully cored. Secondly, they help constrain splitting, subsampling, and even logging strategies. Thirdly, archived X-ray images taken onboard reflect the pristine state of the sediments prior to any changes related to transport and splitting, such as compaction and deformation. Until recently, no commercial digital system was readily available. However, GEOTEK can now produce an X-ray system that is fully compatible with the existing range of nondestructive measurements for the MSCL. This system will provide continuous high-resolution digital X-ray images of whole or split cores, at a resolution of approximately 100–150 μm. The images will be viewed with other MSCL data, and a preview mode will be available for real-time viewing. A prototype of this X-ray system was successfully used on pressure cores by GEOTEK personnel during IODP Expedition 311 and revealed fine scale gas hydrate structures (P. Schultheiss, personal communication, 2006).

Another more portable system was recently developed by Dr. Robert Wheatcroft at Oregon State University (Figure 5). Among other features, this system uses

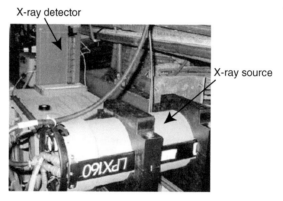

Figure 5 Oregon State University (OSU) portable shipboard digital X-ray system onboard the *R/V Garcia del Cid*. The X-ray source and detector are shown.

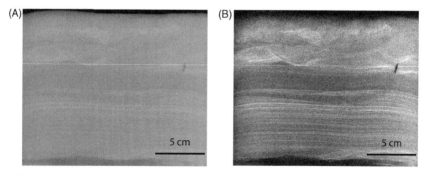

Figure 6 Example of a digital X-radiograph (negative) taken from a box core sampled in an 8-m silty sand site off the Po River using the OSU shipboard system. (A) Raw image; (B) processed image. The original bit depth was 12. The processed image is the result of applying a sharpening algorithm, followed by contrast normalization (see Wheatcroft et al., 2006 for examples). Note the very fine laminations present in the core as well as the small pelecypod in living position in the middle right.

a dpiX Flashscan 30 imager X-ray detector that can capture 29.3 × 40.6 cm images at 12-bit depth (i.e., 4,096 gray levels) and 127-μm pixel size (Figure 6). In addition, in each image, 16 sections of different glass thicknesses make it possible to calibrate image brightness to absolute density values, allowing the extraction of quantitative information. This system was recently used to identify newly deposited and highly porous sediments (Stevens, Wheatcroft, & Wiberg, 2007), as well as to map and identify flood deposits (Wheatcroft, Stevens, Hunt, & Milligan, 2006). Combined with image analysis, the system was very effective at rapidly identifying the 2000 Po River flood deposit at 33 stations, from box cores collected near the Po River Delta, Italy.

3.3. Computerized Coaxial Tomography (CAT-Scan)

Computerized axial tomography (CAT-scan) allows rapid visualization of both longitudinal and traverse sections of sediment cores. The CAT-scan method (Figure 7) uses a pixel intensity scale to quantify and map X-ray attenuation coefficients of the analyzed object on longitudinal (topogram) or transverse (tomogram) images. The resulting images are displayed in gray scale, darker and lighter zones representing lower and higher X-ray attenuation, respectively. Gray scale values are expressed as CT numbers or Hounsfield units, obtained by comparing the attenuation coefficient (μ) to that of water (μ_w):

$$CT(\text{Hounsfield units}) = (\mu/\mu_w - 1) \times 1000 \qquad (1)$$

A CT number is a complex unit related to sediment bulk density, mineralogy, as well as porosity (e.g., Boespflug, Long, & Occhietti, 1995; Crémer, Long, Desrosiers, de Montety, & Locat, 2002). This nondestructive and very high-resolution method

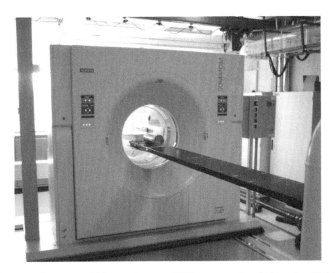

Figure 7 Siemens Somatom Volume Access CAT-scan at the *Laboratoire Multidisciplinaire de Scanographie de Québec (LMSQ).*

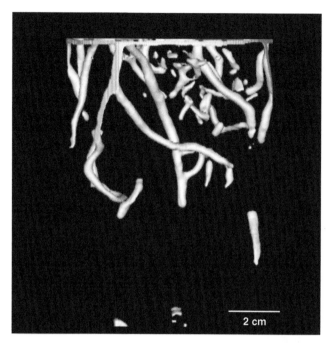

Figure 8 3D reconstruction of biogenic structures in a box core from Baie des Chaleur Bay, Eastern Canada. The structures (*Nereis* worm tubes) were highlighted by selecting a specific range of tomographic intensities using the OsiriX freeware (see Dufour et al., 2005 for method details).

($\sim$0.1–1 mm) has been used to identify sedimentary structures (Champanhet, Durand, Long, & Laberye, 1989; Holler & Kögler, 1990; Orsi, Edwards, & Anderson, 1994), determine sediment deposition mode (Crémer et al., 2002), characterize the benthic community (Mermillod-Blondin et al., 2003; Michaud et al., 2003), establish a high-resolution stratigraphy (Boespflug et al., 1995), evaluate the physical properties of sediments (Wellington & Vinegar, 1987; Kantzas, Marentette, & Jha, 1992; Orsi et al., 1994; Amos, Sutherland, Radzijewski, & Doucette, 1996) and even visualize biogenic structures in sediment cores in 3D (Figure 8, see Dufour et al., 2005 for more details). For more examples of CAT-scan applications in geoscience, refer to Mees, Swennen, Van Geet, and Jacobs (2003), as well as Duliu (1999) and Ketcham and Carlson (2001). Section 4.3 of this chapter illustrates the usefulness of CAT-scan as a paleoceanographic tool.

3.4. Micro X-Ray Fluorescence Spectrometry

A new generation of core logging instruments is currently being used in the paleoceanographic community to perform downcore mm- and even μm-scale measurements on split cores. These new instruments use micro X-ray fluorescence spectrometry to estimate continuously the elemental composition and concentration of elements from Si to U. In these instruments, an intense X-ray beam is used to irradiate the sediment surface and thus enable X-ray fluorescence analysis. Both commercially available models (Cox Analytical Systems Itrax and Avaatech XRF Core scanner) may also be combined with a line-scan camera to acquire a high-resolution RGB image of the analyzed core. The ITRAX also allows acquisition of a digital X-ray image along the center of the core for a 20 mm-wide area at a pixel size as small as 25 µm. X-ray fluorescence spectrometry can be performed at a maximum resolution of 100 µm and 1 mm, respectively, for the ITRAX and Core scanner.

Detected elements and detection limits depend upon the composition of the anode in the tube used to produce the emitted X-rays, acquisition time and the atomic number of the element being detected. For instance, using a molybdenum tube, a wide range of elements can be detected from Al to U. Detection limits for lighter elements, such as Al and Si, are much higher (in the order of a few percent) and therefore require much longer counting time. Heavier elements such as Fe or Rb can be detected in trace amounts (on the order of a few tens of ppm). For each data point, a dispersive energy spectrum is acquired and evaluated. Energy peaks for each element in the spectrum can be identified and measured. Calibration procedures are available which compare a sample of known and/or certified composition with the analyzed sample in order to produce results as concentration rather than counts per second or peak surface area. However, numerous factors such as the presence of organic matter, porosity variations, water content, grain size, crystallinity and sample topography may have an impact on the production and detection of fluorescent radiation. Caution must therefore be exercised in using such quantitative results, because the nature and physical properties of the standard are never identical to those of the sample analyzed. Elemental concentrations and ratios do however provide extremely valuable and useful information.

In paleoceanography, X-ray microfluorescence is used for two main purposes: (1) the study of sedimentation processes and (2) the study of forcing parameters affecting sedimentation (and thus the development of paleoclimatic proxies). As with X-ray analysis, this methodology can also constitute a tool for preliminary investigations and for the selection of specific intervals over which to perform conventional sedimentological analysis. In clastic sedimentology, the comparison of elements typical of the siliciclastic fraction (Al, Ti, Fe, etc.) and the biogenic fraction (Ca, Sr, etc.) can help in distinguishing periods of increased terrigeneous input into the deep oceans. For instance, in a Late Quaternary core collected off the Channel Sea, in the Armorican Deep Sea Turbidite System, both microfluorescence-X and digital X-rays measurements were used to determine changes in sedimentation type (Figure 9, Zaragosi et al., 2006). During isotopic stage 2, the Ti/Sr ratio is high and the sediment is laminated, both observations suggesting frequent and important terrigeneous inputs, probably by turbidity currents, whereas a decrease of both the Ti/Sr ratio and the number of laminae during isotopic stage 1 and, in particular, during the Holocene suggests a strongly reduced terrigeneous sediment source. The very high Ti/Sr ratio values observed between 15 and 13.8 kyr BP also clearly reflect the proximal deglacial sedimentation phase of the British Ice Sheet. This example highlights the fact that it is possible to constrain

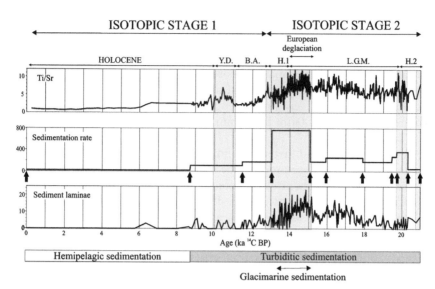

Figure 9 Microfluorescence-X (Avaatech XRF core scanner) measurements (Ti/Sr ratio) processed at the University of Bremen, and laminae count data obtained with the digital X-ray device at Université Bordeaux 1 (SCOPIX) on core MD03-2690. Modified with permission from Zaragosi et al. (2006).

sedimentation type (hemipelagic, turbiditic and glaciomarine) using only non-destructive measurements (digital X-ray and microfluorescence-X).

Microfluorescence-X has also been used for paleoceanographic purposes on cores collected off of Antarctica (Hepp, Mörz, & Grützner, 2006; Grützner, Hillenbrand, & Rebesco, 2005), Africa (Kuhlmann, Freudenthal, Helmke, & Meggers, 2004a, 2004b; Adgebie, Schneider, Röhl, & Wefer, 2003; Bozzano, Kuhlmann, & Alonso, 2002; Stuut et al., 2002a); and Chile (Lamy, Hebbeln, Röhl, & Wefer, 2001), in the Black Sea (Bahr, Lamy, Arz, Kuhlmann, & Wefer, 2005), Red Sea (Arz, Pätzold, Müller, & Moammar, 2003), Mediterranean Sea (Rothwell, Hoogakker, Thomson, & Croudace, 2006; Thomson, Croudace, & Rothwell, 2006), Gulf of California (Cheshire, Thurow, & Nederbragt, 2005), Nordic seas (Helmke, Bauch, Röhl, & Mazaud, 2005), North Pacific (Jaccard et al., 2005), SE Atlantic (Westerhold, Bickert, & Röhl, 2005; West, Jansen, & Stuut, 2004; Stuut, Prins, & Jansen, 2002b; Vidal, Bickert, Wefer, & Röhl, 2002), Equatorial Atlantic (Funk, von Dobeneck, & von Reitz, 2004a; Röhl, Brinkhuis, Fuller, Schellenberg, Stickley, & illiams, 2004b), tropical Atlantic (Arz, Gerhardt, Pätzold, & Röhl, 2001) and Southern Ocean (Röhl, Brinkhuis, & Fuller, 2004; Röhl et al., 2004; Andres, Bernasconi, McKenzie, & Röhl, 2003). This wealth of recent papers clearly highlights the increasing use of this nondestructive high-resolution method in paleoceanographic research. Furthermore, we have recently successfully employed the ITRAX on u-channel samples to gene-rate a continuous dataset including high-resolution RGB and X-ray images as well as the concentration of several important elements in paleomagnetic studies, such as Ca, Ti, and Fe (Figure 10). These data, along with images and X-ray measurements, will be useful for the interpretation of downcore magnetic properties and to assess

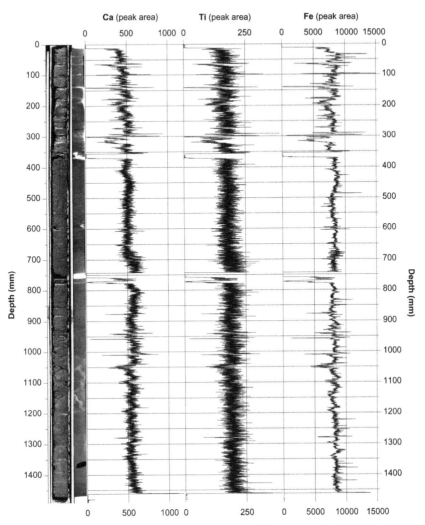

Figure 10 RGB image, digital X-ray and relative Ca, Ti and Fe counts obtained using the ITRAX core scanner on a u-channel sampled in the upper section (0–150 cm) of core 2004–804–009 from Lancaster Sound, Eastern Canadian Arctic. The downcore stability of the profiles suggests relatively low variability in the concentration of magnetic minerals, an essential criterion for paleomagnetic relative paleointensity determinations. The images also highlight sampling and coring artifacts.

the influence of coring and/or subsampling deformations on the paleomagnetic signal, because these different types of measurements can all be performed on the same u-channels. In another example (Figure 11) on fjord sediments collected in Saanich Inlet, off of Vancouver Island (British Columbia, Canada), ITRAX data clearly delineate varved sediments and were used to determine varve counts. For a detailed description of the ITRAX system, the reader is referred to Croudace, Rindby, and Rothwell (2006).

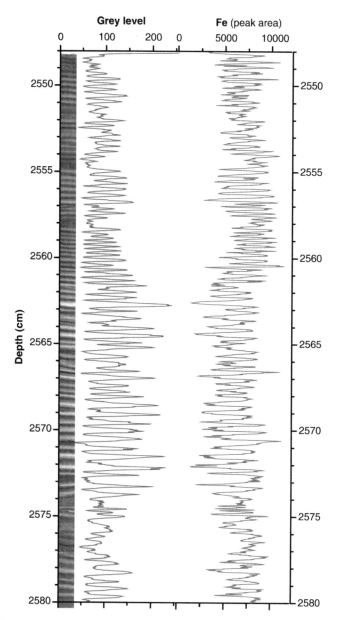

Figure 11 Digital X-ray, gray level values and relative Fe counts obtained using the ITRAX core scanner on a u-channel sampled in section XVIII (2,548–2,580 cm depth) of core MD02-2490 from Saanich Inlet, off of Vancouver Island. Each lamination is clearly resolved in both the radiograph and the Fe profile; these plots along the 32 cm long section each contain 3,200 data points.

3.5. Magnetic Resonance Imaging (MRI) and Nuclear Magnetic Resonance (NMR)

As pointed out by Rack (1998), MRI has the potential to become a very useful technique for marine sediment imaging, especially for 2D and 3D imaging of sediment cores. MRI calls upon the principles of NMR to generate images of a scanned object. Originally, this technique was called nuclear magnetic resonance imaging (NMRI), but the word "nuclear" was dropped from the acronym to remove any negative connotation. The MRI technique is used to produce images of nuclear spin density or magnetic resonance relaxation times, chemical shifts and fluid flow velocity (e.g., Ortiz and Rack, 1999). MRI has been used in petrographic studies to determine porosity, pore size distribution and/or flow and diffusion properties of various porous media such as rocks (e.g., Attard, McDonald, Roberts, & Taylor, 1994; Davies, Hardwick, Roberts, Spowage, & Packer, 1994; Mansfield & Issa, 1996; Baumann, Petsch, Fesl, & Niessner, 2002; Gingras, MacMillan, Balcom, Saunders, & Pemberton, 2002; Marica et al., 2006; Chen, Rack, & Balcom, 2006). Similarly, Kleinberg and Griffin (2005) successfully applied NMR to the quantification of the pore scale distribution of ice and the hydraulic permeability of sediments from Alaskan permafrost cores, using the Schlumberger Combinable Magnetic Resonance (CMR) tool on 15 cm of every 1-m section of core from a 438-m deep borehole. These authors also used a nearly identical NRM tool to characterize the borehole itself, allowing direct core/borehole comparison. To our knowledge, however, the study by Rack, Balcom, MacGregor, and Armstrong (1997) is the only one in which MRI was successfully used continuously to image cored sediments. In this study, the transition from Lake Agassiz proglacial to Lake Winnipeg lacustrine sediments was imaged from a core collected in Lake Winnipeg, Canada (Figure 12), and the authors showed that the resulting MRI images reflected variations in magnetic susceptibility and porosity.

3.6. Confocal Macroscopy and Microscopy

Another promising method is confocal macro- or microscopy. This method uses a confocal scanning laser (Dixon, Damaskinos, Ribes, & Beesley, 1995; Ribes, Damaskinos, & Dixon, 1995; Ribes, Damaskinos, Tiedje, Dixon, & Brodie, 1996; Dixon & Damaskinos, 1998) to image specimens ranging from $200 \times 200\,\mu m$ up to $7.5 \times 7.5\,cm$ in less than 10 s, either in reflected light or photoluminescence. In this technique, a laser beam is focused into a small aperture so as to limit the depth of field to a single plane. The resulting image is then constructed by combining several images obtained for different surface planes. In macroscopic mode, the images have a lateral resolution of $10\,\mu m$, whereas in microscopic mode, the lateral resolution is approximately $1{-}2\,\mu m$ (Ribes et al., 2000). For example, Ribes et al. (1998) and Rack et al. (1998) successfully imaged several sections of sediment cores in both macroscopic and microscopic modes. More recently, Ribes et al.

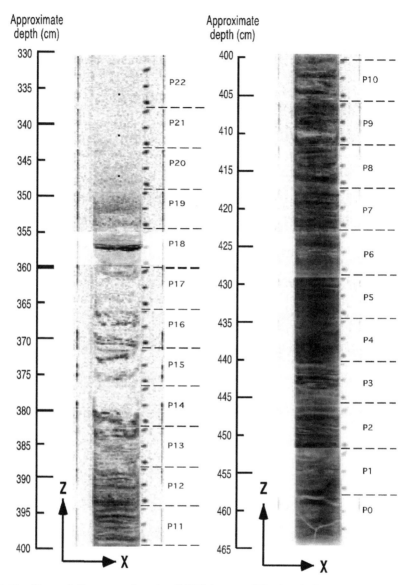

Figure 12 Magnetic Resonance Imaging (MRI) images of Section 4 of core *Namao* 94–900–122a (313–465 cm) from Lake Winnipeg, Canada. The images were obtained using the Single-Point, Ramped Imaging with T_1 Enhancement (SPRITE) technique. Individual SPRITE images are ~5 cm in length (P0 to P22). Dark bands reflect high signal intensity (associated with low porosity and/or high magnetic susceptibility), whereas lighter bands reflect lower intensity (associated with high porosity and low magnetic susceptibility). From Rack et al. (1997), with permission.

(2000) illustrated the usefulness of the method for imaging speleothems at both low and high resolution (Figure 13), allowing the clear identification of annual laminations, which may constitute a possible analogue to finely laminated marine sediments.

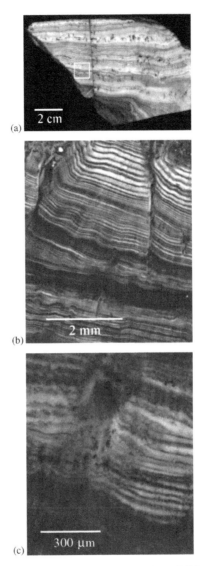

Figure 13   Photoluminescence images of speleothem specimen B104 taken with a high-resolution confocal scanning laser macroscope. (A) Overall image of the speleothem; (B and C) successive enlargements of the area highlighted in A. Modified with permission from Ribes et al. (2000).

4. RECENT APPLICATIONS OF CONTINUOUS CENTIMETER- TO MILLIMETER-SCALE PHYSICAL PROPERTIES OF MARINE SEDIMENTS

The last section of this chapter describes the use of continuous centimeter- to millimeter-scale measurements of physical properties of marine sediments and their combination with other more traditional methods to (1) identify and characterize turbidites, (2) identify and determine the trigger mechanism for RDL from the Saguenay Fjord, Québec, and (3) develop a seasonal- to millennial-scale paleoceanographic tool.

4.1. Use of Digital X-Ray to Rapidly Identify Turbidites in Marine Sediments

In 2000, a 33-cm long Barnett interface gravity core was taken in the talweg of the Capbreton Canyon (southern Bay of Biscay), at a water depth of 647 m. Digital X-ray (SCOPIX) analysis of the gravity core allowed the identification of a succession of sedimentary facies which can be interpreted as three superimposed sequences separated by hemipelagic interfaces (Mulder, Weber, Anschutz, Jorissen, & Jouanneau, 2001c; Figure 14). The two buried interfaces were paleo-seafloors and the upper interface forming the top of the core was the present-day seafloor. The top sequence showed a classical facies succession forming a turbidite sequence as defined by Bouma (1962). Micro-grain size analyses showed the classical normal grading expected for a surge-like waning flow. To validate the positional accuracy of samples used for grain size analysis, the core was once again X-radiographed after sampling.

Measurements of short-lived radiogenic isotopes provided evidence for the recent deposition of this turbidite. Values of excess ^{210}Pb (half-life $= 22.3$ yr) are very high,

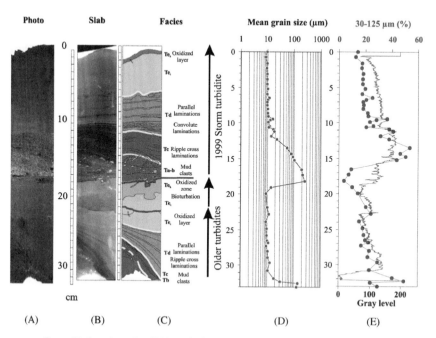

Figure 14 Core K showing the 1999 turbidite deposited in Capbreton Canyon, France. (A) Photograph, (B) X-ray image (SCOPIX), (C) facies interpretation, (D) grain size distribution curve, and (E) extracted gray levels (from the X-ray image) and abundance of the 30–125 μm fraction. Modified with permission from Mulder et al. (2001c).

and values for the interfaces both above and below the turbidite fall in the same range, which suggests very recent deposition. Excess ^{234}Th (half-life $= 24.1$ days) values confirmed this interpretation. ^{234}Th$_{exc}$ activity suggests that deposition occurred at most 6 months before sample counting (between 15/05/2000 and 02/06/2000), in the period between 05/12/1999 and 14/01/2000. During this period, the only event capable of triggering a turbulent surge was the violent "Martin" storm that hit the Bay of Biscay on December 27th, 1999. This turbulent surge could have been generated by three processes: (1) the transformation of a slide which originated at the head of Capbreton Canyon in response to excess interstitial pore pressure in the sediments resulting from 12-m high storm waves or exceptionally high swell, (2) the dissipation through the canyon of a 1–2 m high along-coast water bulge resulting from low barometric pressure, or (3) the increased coastal drift and shelf current intensities.

In the Var Canyon (Mediterranean Sea, French Riviera), digital X-ray analysis lead to the description of a new type of sedimentary sequence called hyperpycnite (Mulder, Migeon, Savoye, & Faugères, 2001a; Mulder, Migeon, Savoye, & Faugères, 2002; Figure 15). Hyperpycnites are sedimentary sequences deposited by turbulent flows generated at river mouths during floods, when the suspended

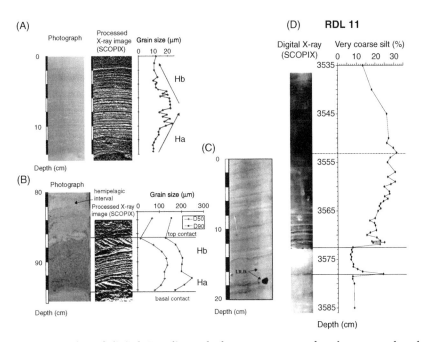

Figure 15 Examples of digital X-radiographed sequences: complete hyperpycnal turbidite sequence from the Var (A) and Zaire (B) turbidite systems. Note the superposition of coarsening upward unit Ha and fining upward unit Hb. (C) Finely laminated clays and silty clays from the northwestern Bay of Biscay associated with ice melting at the end of the last ice age, between 15–14.4 yr BP. I.R.D.: Ice-Rafted Detritus. (D) Base of the turbidite sequence of rapidly deposited layer 11 from the Saguenay Fjord, Eastern Canada. Modified with permission from Mulder, Syvitski, Migeon, Faugères, and Savoye (2003) and St-Onge et al. (2004).

sediment concentration in the fresh water is high enough for the density of river water and its suspended sediment to exceed that of seawater in the marine basin (hyperpycnal flow; Mulder & Syvitski, 1995). This sequence comprises two super-posed units. The basal, coarsening upward unit corresponds to the waning part of the hyperpycnal flow deposited during the rising portion of the flood hydrograph at the river mouth (Mulder, Syvitski, & Skene, 1998). The top, fining upward unit corresponds to the waning part of the hyperpycnal flow deposited during the falling portion of the flood hydrograph at the river mouth. The peak grain size corres-ponds to the peak of the flood at the river mouth. Numerous internal erosive or sharp surfaces are observed within this sequence. The base surface can be erosive, sharp or transitional. The top surface is usually sharp or transitional. Digital X-ray images of such sequences in the Var Giant Sedimentary Levee (Migeon, 2000; Migeon et al., 2001; Mulder, Migeon, Savoye, & Jouanneau, 2001b) and in the Zaire deep-sea turbidite system (Migeon, 2000) clearly show this type of sequence (Figure 15). They show internal sedimentary structures, the most common of which are climbing ripples, which suggest a high sediment load. In addition, numerous intrasequence erosional or sharp contact surfaces may be present. For thick hyperpycnal sequences such as the 2.2-m thick, rapidly deposited layer 11 in the Saguenay Fjord, digital X-ray imaging reveals unexpected planar horizontal lamination in the basal (waxing) unit (Figure 15).

4.2. RDL in the Saguenay Fjord

Several natural disasters have struck the Saguenay Fjord over the last few centuries. These include the 1663 ($M \approx 7.0$) and 1988 ($M = 6$) earthquakes, the 1924 Kénog-ami and 1971 Saint-Jean-Vianney landslides and the catastrophic flood of 1996, which swept more than $15 \times 10^6 \, \mathrm{m}^3$ of sediment into the Saguenay Fjord (Lapointe, Secretan, Driscoll, Bergeron, & Leclerc, 1998). Previous studies have revealed the presence of thick sediment deposits associated with these events, varying from several centimeters to several meters in thickness, in the Baie des Ha!Ha! and inner basin of the Saguenay Fjord (Smith & Walton, 1980; Syvitski & Schafer, 1996; St-Onge & Hillaire-Marcel, 2001). These deposits generally consist of light gray homogenous silty clays that contrast sharply with the dark gray, bioturbated background sediments. In this section, we illustrate how high-resolution continuous measurements were used to image and/or identify RDL associated with historic and prehistoric catastrophic events during the last 7,200 cal yr BP. For instance, the 1996 flood layer is shown in Figure 16, in which the difference between the photograph and the CAT-scan image is clearly visible. This figure also clearly highlights the laminations and internal sedi-mentary structures of the flood layer. Similarly, CAT-scan images and photographs of the four core sections of a ~4 m piston core sampled in the northern arm of the Saguenay Fjord (Figure 17) clearly show the 1996 flood layer (e.g., St-Onge & Hillaire-Marcel, 2001; Urgeles, Locat, Lee, & Martin, 2002), the 1971 Saint-Jean-Vianney (e.g., Smith & Walton, 1980) and the 1924 Kénogami (e.g., Smith & Schafer, 1987) landslide layers, as well as the top part of the 1663 turbidite. In 1999, a 38 m-long Calypso piston core (core MD99-2222) was raised from the deepest part of the Saguenay Fjord inner basin in order to identify RDLs older than the 1663

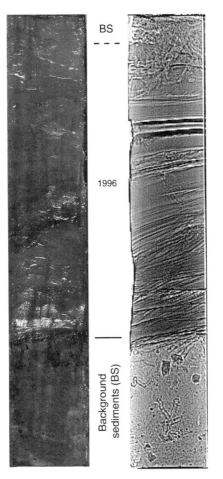

Figure 16   Digital photograph (left) and CAT-scan (right) of the 1996 Saguenay flood layer. The images were taken from a box core collected in the northern arm of the Saguenay Fjord. The push core is ~54 cm in length.

earthquake RDL. High-resolution physical (density, diffuse reflectance and digital X-radiography), magnetic (magnetic susceptibility and inclination), sedimentological (detailed description and grain size) and geochemical ($CaCO_3$) analyses revealed the presence of at least 14 RDLs, including a ~16-m thick layer associated with the 1663 AD earthquake (Figure 18). These RDLs are readily recognizable by their sharp and sandy bases, which are clearly visible on the digital X-rays images (Figure 15D) and highlighted by high density and magnetic susceptibility values. In addition, these layers are characterized by a light gray color, high $CaCO_3$ contents and low basal paleomagnetic inclinations, contrasting sharply with the dark gray bioturbated background sediments. The light gray color (higher L^* values) and the high $CaCO_3$ contents indicate the incorporation of reworked gray and slightly carbonated Laflamme Sea clays (St-Onge & Hillaire-Marcel, 2001), whereas the low paleomagnetic inclinations at the base of the RDLs indicate an energetic depositional process,

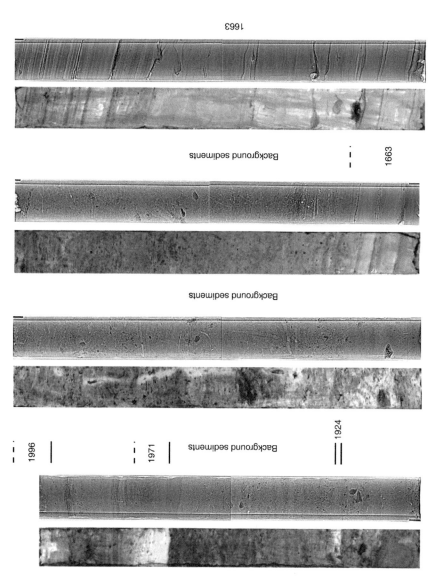

Figure 17 Digital photograph (left) and CAT-scan (right) of several historic rapidly deposited layers (RDLs) from the Saguenay Fjord. The images were taken from the four (~1 m) sections of a piston core collected in the northern arm of the Saguenay Fjord.

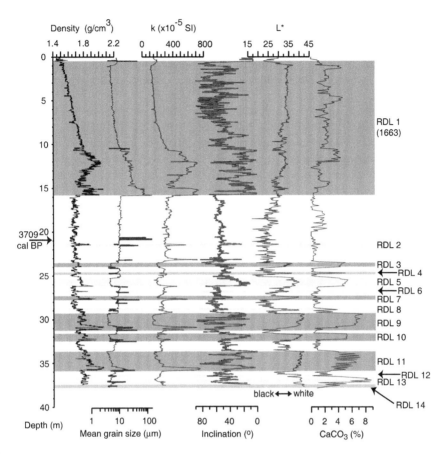

Figure 18 Physical properties of rapidly deposited layers (RDLs) in core MD99-2222. Gray zones correspond to RDLs. RDL 1 is associated with the 1663 AD earthquake. Bulk density and low field volumetric magnetic susceptibility (k) were measured using a GEOTEK MSCL onboard the *Marion Dufresne II*. L^* was measured using a CM-2002 Minolta hand-held spectrophotometer. Inclinations were calculated by principal component analysis using 4–10 alternating field (AF) demagnetization steps at peak fields of 10–80 mT, using a 2-G Enterprises model 755 cryogenic magnetometer at the University of California, Davis. The $CaCO_3$ content was analyzed using an automated Bernard calcimeter, and the grain size analyses were performed with a Malvern Supersizer "S" at the Université Bordeaux 1. Detailed data are reported in St-Onge et al. (2004).

the magnetic particles being plastered horizontally because of high flow velocity and rapid and thick sediment accumulation. These layers are interpreted either as mono-event deposits (turbidite) or bi-event deposits (turbidite+hyperpycnite). In light of the geological, sedimentological and hydrological setting of the Saguenay Fjord, St-Onge et al. (2004) concluded that these RDLs were likely all triggered, either directly or indirectly, by strong earthquakes. The mono-event RDLs resulted from earthquake-triggered slides that became turbidity currents, whereas the bi-event RDLs resulted from similar events combined with the breaching and rapid draining, during the spring freshet, of a natural dam generated by an earthquake-triggered landslide.

4.3. CAT-Scan Analysis as a Millimeter-Scale Paleoceanographic Tool

In this section, we further illustrate how CAT-scan analysis can be applied to sedimentary sequences to produce a continuous mm-scale paleoclimatic tool that can be used to identify millennial- to seasonal-scale climatic oscillations, using examples from two IMAGES Calypso piston cores from the St. Lawrence Estuary (Eastern Canada). CAT-scan analysis of the two Calypso cores was performed on 1.5 m core sections using a medical GE 7590 K Hi-speed Advantage 2.X CT/i CAT-scanner at the *Centre Hospitalier Régional de Rimouski*, Québec, Canada, in 1999. Longitudinal images were obtained using a source radiation of 120 keV and 45 mA. These images show average linear attenuation coefficients integrated over the total thickness of the scanned object (the core diameter: ~11.5 cm) and shown a plane view. The dimensions of each longitudinal image are 512×296 pixels, which corresponds to a pixel size of 1.015 mm. For each core section, five images of 300-mm long segments were thus acquired. Continuous images of both Calypso cores were then constructed using the Igor[TM] software by creating a matrix for each image and incorporating each image matrix into the final matrix of each core. Using the same software, a continuous profile of the mean pixel intensity (CT number) over a width of ~2 cm was then extracted from the newly constructed image. An artefact associated with the

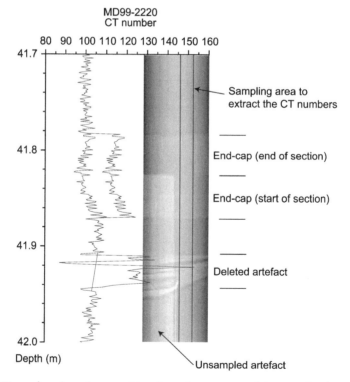

Figure 19 Examples of artefacts resulting from the presence of the examination table and the end-caps, and uncorrected (right curve) and corrected (left curve) extracted CT numbers for core 2220.

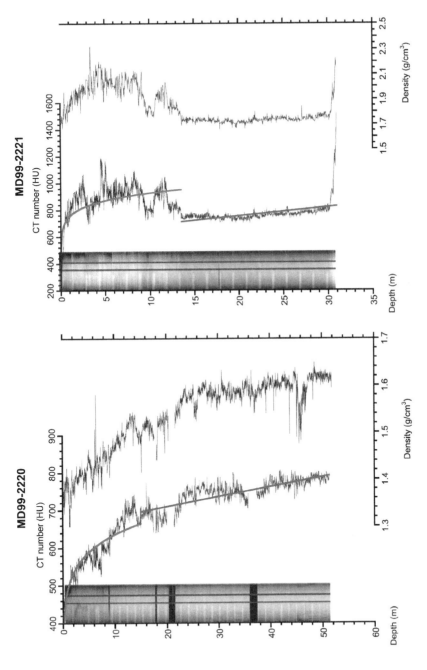

Figure 20 Longitudinal CAT-scan images and corrected extracted CT number profiles for cores 2220 (A) and 2221 (B) compared with bulk density measurements obtained onboard the *Marion Dufresne II* with a GEOTEK MSCL at 2-cm intervals. A logarithmic trend (upper part of the curve in postglacial sediments) and a linear trend (lower part of the curve in glaciomarine sediments), probably associated with sediment consolidation, are also visible. A sandy layer at the base of core 2221 was excluded from the linear fit. The areas delineated by two vertical lines on both core images are the sampling areas over which CT numbers were extracted.

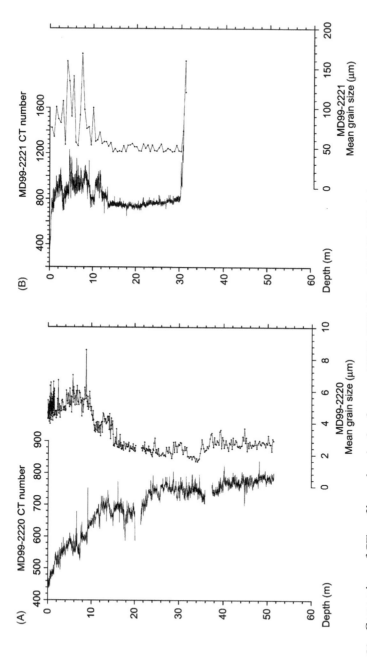

Figure 21 Comparison of CT profiles and grain size for cores MD99–2220 (A) and MD99–2221 (B). Grain size measurements were made using a Coulter Counter[TM] TAII analyzer at the Geological Survey of Canada (Atlantic), for core MD99–2220, and a Fritsch[TM] Analysette 22 laser diffraction analyzer at the INRS-ETE in Québec City, for core MD99–2221.

examination table is apparent on the first image of each core section and was only partially sampled by our procedure (Figure 19). In addition, erroneous CT numbers associated with this artefact were deleted from the continuous profile (Figure 19). Because the images were obtained in a clean hospital setting, the end-caps of the different sections could not be removed and resulted in another artefact, which slightly increased CT numbers. We corrected each core section for this small offset by visually adjusting the CT number profiles (Figure 19). Because of the CAT-scan medical software limitations, CT numbers obtained on the longitudinal images are not Hounsfield units (HU). A first-order empirical relationship between the CT numbers of the longitudinal and transverse images (in HU) at the same depth was therefore determined (Cagnat, 2004) for both Calypso cores and is as follows:

$$CT_L = 195 \ln CT_T - 1194, \quad r^2 = 0.85 \tag{2}$$

where CT_L is the CT number extracted from the longitudinal image and CT_T, the CT number, in HU, extracted from the transverse image.

In the CT number profiles of cores MD99-2220 and MD99-2221 shown in Figure 20, postglacial and glaciomarine sediments are readily distinguished (St-Onge, Stoner, & Hillaire-Marcel, 2003). Figure 20 also highlights the coarser-grained nature of postglacial sediments in core MD99-2221 (average CT number of 882 ± 109) compared to core MD99-2220 postglacial sediments (average CT number of 599 ± 70; see also Figure 21). Postglacial sediments in both cores are characterized by cyclic and high frequency oscillations superimposed on a long-term logarithmic trend, whereas glaciomarine sediments are characterized by high frequency variations superimposed on a long-term linear trend. Because these long-term trends are not seen in the grain size profiles (Figure 21), we interpret them to reflect the effect of sediment consolidation, which reduces porosity and thus increases CT number. These long-term trends are also seen in bulk density profiles (see below). In Figure 20, the CT number profiles of cores 2220 and 2221 are also compared to bulk density profiles measured onboard the RV *Marion Dufresne II* using a GEOTEK MSCL at 2 cm intervals. The overall agreement between density and CT number profiles suggests that the latter reflect variations in bulk density and could therefore be used as a mm-scale proxy (1.015-mm resolution) for sedimentological changes. Indeed, after removal of the long-term trends associated with sediment consolidation, spectral analysis of the CAT-scan data for the postglacial sediments revealed millennial- to centennial-scale oscillations, whereas spectral analysis of the glaciomarine sequences revealed decadal to annual cycles (St-Onge and Long, in review). Moreover, CAT-scan analysis of a Sangamonian sequence drilled onshore in the St. Lawrence Middle Estuary revealed similar decadal to annual cycles, as well as seasonal cycles possibly associated to tidal amplitude variations (Boespflug et al., 1995).

5. Conclusion

Continuous measurements of the physical properties of sediments now form the basis of most paleoceanographic campaign, providing quantitative information on long sediment cores at resolutions generally on the cm-scale. Most methods routinely used onboard research vessels such as the typical sensors mounted on a MSCL allow fast and nondestructive analysis and may, for instance, be used directly for core logging and stratigraphic description. The last decade has seen the development and use of methods such as high-resolution digital core imaging, micro X-ray fluorescence, CAT-scan, digital X-ray imaging, MRI and confocal macroscopy and microscopy which can achieve millimeter- to micrometer-scale downcore resolutions. These high resolutions are now used to study seasonal- to millennial-scale processes. Further technological development of methods such as CAT-scan, digital imaging and X-ray imaging will likely lead to further increase in resolution. For instance, the latest generation of medical CAT-scan instruments already produce pixel resolutions an order of magnitude higher than those of CAT-scan images presented in this paper. Several manufacturers and research groups are also working toward making some of these techniques directly available for shipboard measurements. Finally, efforts should also go toward integrating the vast amounts of data collected with this new generation of sensors into specially designed databases in order to facilitate the comparison and extraction of data, thus allowing paleoceanographers to focus on interpretation. However, these new developments are not necessarily designed by and for geoscientists, and collaboration will thus be necessary to ensure their useful transfer to the geoscience community.

ACKNOWLEDGMENTS

G. St-Onge dedicates this chapter to the memory of Gaston Desrosiers (ISMER), a friend and colleague, as well as a pioneer in the use of CAT-scan analysis in benthic ecology. We are in debt to Peter Schultheiss (GEOTEK), Frank Rack (Joint Oceanographic Institutions), Bruce Balcom (University of New Brunswick), Ian Snowball (Lund University), Jacques Labrie (INRS-ETE), Robert Wheatcroft (Oregon State University), Robert Kleinberg (Schlumberger), Kinuyo Kanamaru (University of Massachusetts), Sébastien Zaragosi (Université Bordeaux 1), and Gaston Desrosiers and Suzanne Dufour (ISMER) for generously sharing preprints, reprints or some of the figures used in this chapter. Special thanks also go to Peter Schultheiss (GEOTEK) for providing technical information about the MSCL. We also thank Jacques Locat (Université Laval) for the cores presented in Figures 16 and 17 and for the "Saguenay post-déluge" project, as well as David J.W. Piper (Geological Survey of Canada-Atlantic) for a very detailed and constructive review. We wish to sincerely thank the IPEV, Yvon Balut, the captain, officers, crew and scientific participants of several IMAGES cruises on board the R/V Marion Dufresne II. This study was supported by NSERC Discovery grants to G. St-Onge and P. Francus. This constitutes UMR CNRS 5805 EPOC contribution 1582. The authors also acknowledge the acquisition of the ITRAX and CAT-scan presented in this paper through major grants from the Canadian Foundation for Innovation (CFI) to P. Francus and B. Long (INRS-ETE), respectively. This is GEOTOP-UQAM-McGill contribution.

REFERENCES

Adgebie, A. T., Schneider, R. R., Röhl, U., & Wefer, G. (2003). Glacial millennial-scale fluctuations in central African precipitation recorded in terrigenous sediment supply and freshwater signals offshore Cameroon. Palaeogeography, Palaeoclimatology, Palaeoecology, 197, 323–333.

Amos, C. L., Sutherland, T. F., Radzijewski, B., & Doucette, M. (1996). A rapid technique to determine bulk density of fine-grained sediments by X-ray computed tomography. Journal of Sedimentary Research, 66, 1023–1039.

Andres, M. S., Bernasconi, S. M., McKenzie, J. A., & Röhl, U. (2003). Southern Ocean deglacial record supports global Younger Dryas. *Earth and Planetary Science Letters*, *216*, 515–525.

Andrews, J. T., & Freeman, W. (1996). The measurement of sediment color using the colortron spectrophotometer. *Arctic and Alpine Research*, *28*, 524–528.

Arz, H. W., Gerhardt, S., Pätzold, J., & Röhl, U. (2001). Millennial-scale changes of surface- and deep-water flow in the western tropical Atlantic linked to Northern Hemisphere high-latitude climate during the Holocene. *Geology*, *29*, 239–242.

Arz, H. W., Pätzold, J., Müller, P. J., & Moammar, M. O. (2003). Influence of Northern Hemisphere climate and global sea level rise on the restricted Red Sea marine environment during termination I. *Paleoceanography*, *18*, 1053, doi:10.1029/2002PA000864.

Attard, J. J., McDonald, P. J., Roberts, S. P., & Taylor, T. (1994). Solid state NMR imaging of irreducible water in reservoir cores for spatially resolved pore surface relaxation estimation. *Magnetic Resonance Imaging*, *12*, 355–359.

Bahr, A., Lamy, F., Arz, H., Kuhlmann, H., & Wefer, G. (2005). Late glacial to Holocene climate and sedimentation history in the NW Black Sea. *Marine Geology*, *214*, 309–322.

Balsam, W. L., Damuth, J. E., & Schneider, R. R. (1997). Comparison of shipboard vs. shore-based spectral data from Amazon fan cores: implications for interpreting sediment composition. In: R. D. Flood, D. J. W. Piper, A. Klaus & L. C. Peterson (Eds), *Proceedings of Ocean Drilling Program, Scientific Reports 155* (pp. 1–23). Texas: College Station (Ocean Drilling Program).

Balsam, W. L., & Deaton, B. C. (1991). Sediment dispersal in the Atlantic Ocean: Evaluation by visible light spectra. *Reviews in Aquatic Sciences*, *4*, 411–447.

Baumann, T., Petsch, R., Fesl, G., & Niessner, R. (2002). Flow and diffusion measurements in natural porous media using magnetic resonance imaging. *Journal of Environmental Quality*, *31*, 470–476.

Berns, R. S. (2000). *Billmeyer and Saltzman's principles of color technology* (p. 247). New York: Wiley.

Blum, P. (1997). *Physical properties handbook: A guide to the shipboard measurement of physical properties of deep-sea cores* (Ocean Drilling Program Technical Note 26). Retrieved from http://www-odp.tamu.edu/publications/tnotes/tn26/TOC.HTM.

Blum, P., Rabaute, A., Gaudon, P., & Allan, J. F. (1997). Analysis of natural gamma-ray spectra obtained from sediment cores with the shipboard scintillation detector of the Ocean Drilling Program: Example from Leg 156. *Proceedings of the Ocean Drilling Program, Scientific Results*, *156*, 183–195.

Boespflug, X., Long, B. F. N., & Occhietti, S. (1995). CAT-scan in marine stratigraphy: A quantitative approach. *Marine Geology*, *122*, 281–301.

Bouma, A. H. (1962). *Sedimentology of some flysch deposits. A graphic approach to facies interpretation* (p. 168). Amsterdam: Elsevier.

Bouma, A. H. (1969). *Methods for the Study of Sedimentary Structures* (p. 458). Wiley: New York

Boyle, E. A. (1997). Characteristics of the deep ocean carbon system during the past 150,000 years: ΣCO_2 distributions, deep water flow patterns and abrupt climate change. *Proceedings of National Academy of Sciences, U.S.A.*, *94*, 8300–8307.

Bozzano, G., Kuhlmann, H., & Alonso, B. (2002). Storminess control over African dust input to the Moroccan Atlantic margin (NW Africa) at the time of maxima boreal summer insolation: A record of the last 220 kyr. *Palaeogeography, Palaeoclimatology, Palaeoecology*, *183*, 155–168.

Cagnat, E. (2004). Étude sédimentologique de la série holocène de l'estuaire maritime du Saint-Laurent: apport de la tomodensitométrie. M.Sc. Mémoire, INRS-ETE, Québec, Canada.

Champanhet, J. M., Durand, J., Long, B., & Laberye, B. (1989). Apport du scanner à la définition géométrique des réservoirs non consolidés. *Bulletin des centres de recherche exploration-production Elf-Aquitaine*, *13*, 167–174.

Chapman, M. R., & Shackleton, N. J. (2000). Evidence of 550-year and 1000-year cyclicities in North Atlantic circulation patterns during the Holocene. *The Holocene*, *10*, 287–291.

Chen, Q., Rack, F., & Balcom, B. (2006). Quantitative magnetic resonance imaging methods for core analysis. In: R. G. Rothwell (Ed.), *New ways of looking at sediment cores and core data* (pp. 193–207). London: Geological Society (Special Publication).

Cheshire, H., Thurow, J., & Nederbragt, A. J. (2005). Late Quaternary climate change record from two long sediment cores from Guaymas Basin, Gulf of California. *Journal of Quaternary Science*, *20*, 457–469.

Chmelick, F. B. (1967). Electro-osmotic core cutting. *Marine Geology, 5,* 321–325.

Crémer, J.-F., Long, B., Desrosiers, G., de Montety, L., & Locat, J. (2002). Application de la scanographie à l'étude de la densité des sédiments et à la caractérisation des structures sédimentaires: Exemple des sédiments déposés dans la rivière Saguenay (Québec, Canada) après la crue de juillet de 1996. *Canadian Geotechnical Journal, 39,* 440–450.

Croudace, I. W., Rindby, A., & Rothwell, R. G. (2006). ITRAX: Description and evaluation of a new X-ray core scanner. In: R. G. Rothwell (Ed.), *New ways of looking at sediment cores and core data* (pp. 51–63). London: Geological Society (Special Publication).

Davies, S., Hardwick, A., Roberts, D., Spowage, K., & Packer, K. J. (1994). Quantification of oil and water in preserved reservoir rock by NMR spectroscopy and imaging. *Magnetic Resonance Imaging, 12,* 349–353.

Deaton, B. C., & Baslam, W. L. (1991). Visible spectroscopy: A rapid method for determining hematite and goethite concentrations in geological material. *Journal of Sedimentary Petrology, 61,* 628–632.

Debret, M., Desmet, M., Balsam, W., Copard, Y., Francus, P., & Laj, C. (2006). Spectrophotometer analysis of Holocene sediments from an anoxic fjord: Saanich Inlet, Bristish Columbia, Canada. *Marine Geology, 229,* 15–28.

Dixon, A. E., & Damaskinos, S. (1998). *Apparatus and method for scanning laser imaging of macroscopic samples.* US Patent 5,760,951.

Dixon, A. E., Damaskinos, S., Ribes, A., & Beesley, K. M. (1995). A new confocal scanning beam laser macroscope using a telecentric f-theta laser scan lens. *Journal of Microscopy, 178,* 261–266.

Dufour, S. C., Desrosiers, G., Long, B., Lajeunesse, P., Gagnoud, M., Labrie, J., Archambault, P., & Stora, G. (2005). A new method for three-dimensional visualization and quantification of biogenic structures in aquatic sediments using axial tomodensitometry. *Limnology and Oceanography: Methods, 3,* 372–380.

Duliu, O. (1999). Computer axial tomography in geosciences: An overview. *Earth Science Reviews, 48,* 265–281.

Evans, M.E., Heller, F. (2003). *Environmental Magnetism: Principles and Applications of Enviromagnetics* (p. 299). San Diego: Academic Press.

Francus, P. (Ed.) (2004). *Image Analysis, Sediments and Paleoenvironments* (p. 330). Dordrecht: Kluwer Academic Publishers.

Funk, J. A., von Dobeneck, T., & von Reitz, A. (2004a). Integrated rock magnetic and geochemical quantification of redoxmorphic iron mineral diagenesis in Late Quaternary sediments from the equatorial Atlantic. In: G. Wefer, S. Mulitza & V. Ratmeyer (Eds), *The South Atlantic in the Late Quaternary* (pp. 237–260). Dordrecht: Springer.

Funk, J. A., von Dobeneck, T., von Wagner, T., & Kasten, S. (2004b). Late Quaternary sedimentation and early diagenesis in the equatorial Atlantic Ocean: Pattern, trends, and processes deduced from rock magnetic and geochemical records. In: G. Wefer, S. Mulitza & V. Ratmeyer (Eds), *The South Atlantic in the Late Quaternary* (pp. 461–497). Dordrecht: Springer.

Gingras, M. K., MacMillan, B., Balcom, B., Saunders, T., & Pemberton, G. (2002). Using magnetic resonance imaging and petrographic techniques to understand the textural attributes and porosity distribution in *Macronichnus*-burrowed sandstone. *Journal of Sedimentary Research, 72,* 552–558.

Grützner, J., Hillenbrand, C.-D., & Rebesco, M. A. (2005). Terrigenous flux and biogenic silica deposition at the Antarctic continental rise during the late Miocene to early Pliocene: Implications for ice sheet stability and sea ice coverage. *Global and Planetary Change, 45,* 131–149.

Harris, S. E., Mix, A. C., & King, T. (1997). Biogenic and terrigenous sedimentation at Ceara Rise, western tropical Atlantic, supports Pliocene-Pleistocene deep-water linkage between hemispheres. In: N. J. Shackleton, W. B. Curry, C. Richter & T. J. Bralower (Eds), *Proceedings of Ocean Drilling Program, Scientific Reports 154* (pp. 331–348). Texas: College Station (Ocean Drilling Program).

Helmke, J. P., Bauch, H. A., Röhl, U., & Mazaud, A. (2005). Changes in sedimentation patterns of the Nordic seas region across the mid-Pleistocene. *Marine Geology, 215,* 107–122.

Helmke, J. P., Schulz, M., & Bauch, H. A. (2002). Sediment-color record from the Northeast Atlantic reveals patterns of millennial-scale climate variability during the Past 500,000 years. *Quaternary Research, 57,* 49–57.

Hepp, D. A., Mörz, T., & Grützner, J. (2006). Pliocene glacial cyclicity in deep-sea sediment drifts (Antarctic Peninsula Pacific Margin). *Palaeogeography, Palaeoclimatology, Palaeoecology.*, *231*, 181–198.

Holler, P., & Kögler, F. C. (1990). Computer tomography: A non destructive, high resolution technique for investigation of sedimentary structures. *Marine Geology*, *91*, 263–266.

Holyer, R. J., Young, D. K., Sandidge, J. C., & Briggs, K. B. (1996). Sediment density structure derived from textural analysis of cross-sectional X-radiographs. *Geo-Marine Letters*, *16*, 204–211.

Jaccard, S. L., Haug, G. H., Sigman, D. M., Pedersen, T. F., Thierstein, H. R., & Röhl, U. (2005). Glacial/interglacial changes in Subarctic North Pacific stratification. *Science*, *308*, 1003–1006.

Jackson, P. D., Briggs, K. B., & Flint, R. C. (1996). Evaluation of sediment heterogeneity using microresistivity imaging and X-radiography. *Geo-Marine Letters*, *16*, 219–225.

Kantzas, A., Marentette, D., & Jha, K. N. (1992). Computer-assisted tomography: From qualitative visualization to quantitative core analysis. *Journal of Petroleum Technology*, *31*, 48–56.

Keigwin, L. D., & Pickart, R. S. (1999). Slope water current over the Laurentian fan on interannual to millennial time scales. *Science*, *286*, 520–523.

Ketcham, R., & Carlson, W. A. (2001). Acquisition, optimization and interpretation of X-ray computed tomographic imagery: Applications to the geosciences. *Computers & Geosciences*, *27*, 381–400.

Kleinberg, R. L., & Griffin, D. D. (2005). NMR measurements of permafrost: Unfrozen water assay, pore scale distribution of ice, and hydraulic permeability of sediments. *Cold Regions Science and Technology*, *42*, 63–77.

Kuhlmann, H., Freudenthal, T., Helmke, P., & Meggers, H. (2004a). Reconstruction of paleoceanography off NW Africa for the last 40,000 years: Influence of local and regional factors on sediment accumulation. *Marine Geology*, *207*, 209–234.

Kuhlmann, H., Meggers, H., Freudenthal, T., & Wefer, G. (2004b). The transition of the monsoonal and the North Atlantic climate system off NW Africa during the Holocene. *Geophysical Research Letters*, 31, L22204, doi:10.1029/2004GL021267.

Lamy, F., Hebbeln, D., Röhl, U., & Wefer, G. (2001). Holocene rainfall variability in southern Chile: A marine record of latitudinal shifts of the Southern Westerlies. *Earth and Planetary Science Letters*, *185*, 369–382.

Lapointe, M. F., Secretan, Y., Driscoll, S. N., Bergeron, N., & Leclerc, M. (1998). Response of the Ha!Ha! River to the flood of July 1996 in the Saguenay Region of Quebec: Large-scale avulsion in a glaciated valley. *Water Resources Research*, *34*, 2383–2392.

Lofi, J., & Weber, O. (2001). SCOPIX–digital processing of X-ray images for the enhancement of sedimentary structures in undisturbed core slabs. *Geo-Marine Letters*, *20*, 182–186.

Maher, B. A., Thompson, R. (1999). *Quaternary climates, environments and magnetism* (p. 390). Cambridge: Cambridge University Press.

Mansfield, P., & Issa, B. (1996). Fluid transport in porous rocks. I. EPI studies and a stochastic model of flow. *Journal of Magnetic Resonance A*, *122*, 137–148.

Marica, F., Chen, Q., Hamilton, A., Hall, C., Al, T., & Balcom, B. J. (2006). Spatially resolved measurement of rock core porosity. *Journal of Magnetic Resonance*, *178*, 136–141.

Mees, F., Swennen, R., Van Geet, M., & Jacobs, P. (2003). *Applications of X-ray computed tomography in the geosciences* (p. 243). London: Geological Society (Special Publications 215).

Mermillod-Blondin, F., Marie, S., Desrosiers, G., Long, B., de Montety, L., Michaud, E., & Stora, G. (2003). Assessment of the spatial variability of intertidal benthic communities by axial tomodensitometry: Importance of fine-scale heterogeneity. *Journal of Experimental Marine Biology and Ecology*, *287*, 193–208.

Michaud, E., Desrosiers, G., Long, B., de Montety, L., Crémer, J. -F., Pelletier, E., Locat, J., Gilbert, F., & Stora, G. (2003). Use of axial tomography to follow temporal changes of benthic communities in an unstable sedimentary environment (Baie des Ha!Ha!, Saguenay Fjord). *Journal of Experimental Marine Biology and Ecology*, 285/286, 265–282.

Migeon, S. (2000). *Dunes géantes et levées sédimentaires en domaine marin profond: approche morphologique, sismique et sédimentologique*. Ph.D. Thesis, Université Bordeaux 1, p. 288.

Migeon, S., Savoye, B., Zanella, E., Mulder, T., Faugères, J.-C., & Weber, O. (2001). Detailed seismic and sedimentary study of turbidite sediment waves one the Var sedimentary ridge (SE France):

Significance for sediment transport and deposition and for the mechanism of sediment wave construction. *Marine and Petroleum Geology, 18*, 179–208.

Migeon, S., Weber, O., Faugères, J.-C., & Saint-Paul, J. (1999). SCOPIX: A new X-ray imaging system for core analysis. *Geo-Marine Letters, 18*, 251–255.

Mix, A. C., Harris, S. E., & Janecek, T. R. (1995). Estimating lithology from non intrusive reflectance spectra: Leg 138. In: N. G. Pisias, L. A. Mayer, T. R. Janecek, A. Palmer-Julson & T. H. van Andel (Eds), *Proceedings of Ocean Drilling Program, Scientific Reports 138* (pp. 413–427). Texas: College Station (Ocean Drilling Program).

Mix, A. C., Rugh, W., Pisias, N. G., Veirs, S., Leg 138 Shipboard Sedimentologists, & Leg 138 Scientific Party (1992). Color Reflectance spectroscopy: A tool for rapid characterisation of deep-sea sediments. In: N. G. Pisias, L. A. Mayer, T. R. Janecek, A. Palmer-Julson & T. H. van Andel (Eds), *Proceedings of ocean drilling program, initial reports 138* (pp. 67–77). Texas: College Station (Ocean Drilling Program).

Moy, C. M., Seltzer, G. O., Rodbell, D. T., & Anderson, D. M. (2002). Variability of El Nino/Southern Oscillation activity at millennial timescales during the Holocene epoch. *Nature, 420*, 162–165.

Mulder, T., Migeon, S., Savoye, B., Faugères, J.-C. (2001a). Inversely graded turbidite sequences in the deep Mediterranean: A record of deposits from flood-generated turbidity currents? *Geo-Marine Letters, 21*, 86–93

Mulder, T., Migeon, S., Savoye, B., & Faugères, J.-C. (2002). Inversely graded turbidite sequences in the deep Mediterranean: A record of deposits from flood-generated turbidity currents? Reply. *Geo-Marine Letters, 22*, 112–120.

Mulder, T., Migeon, S., Savoye, B., & Jouanneau, J.-M. (2001b). Twentieth century floods recorded in deep Mediterranean sediments. *Geology, 29*, 1011–1014.

Mulder, T., & Syvitski, J. P. M. (1995). Turbidity currents generated at river mouths during exceptional discharges to the world oceans. *Journal of Geology, 103*, 285–299.

Mulder, T., Syvitski, J. P. M., Migeon, S., Faugères, J.-C., & Savoye, B. (2003). Marine hyperpycnal flows: Initiation, behavior and related deposits. A review. *Marine and Petroleum Geology, 20*, 861–882.

Mulder, T., Syvitski, J. P. M., & Skene, K. I. (1998). Modelling of erosion and deposition by turbidity currents generated at river mouths. *Journal of Sedimentary Research, 68*, 124–137.

Mulder, T., Weber, O., Anschutz, P., Jorissen, F. J., & Jouanneau, J.-M. (2001c). A few months-old storm-generated turbidite deposited in the Capbreton Canyon (Bay of Biscay, S-W France). *Geomarine Letters, 21*, 149–156.

Nederbragt, A. J., & Thurow, J. W. (2004). Digital sediment colour analysis as a method to obtain high resolution climate proxy records. In: P. Francus (Ed.), *Image analysis, sediments, and paleoenvironments* (p. 330). Dordrecht: Kluwer Academic Publishers.

Orsi, T. H., Edwards, C. M., & Anderson, A. L. (1994). X-ray computed tomography: A non-destructive method for quantitative analysis of sediment cores. *Journal of Sedimentary Research A, 64*, 690–693.

Ortiz, J. D., Mix, A., Harris, S., & O'Connell, S. (1999). Diffuse spectral reflectance as a proxy for percent carbonate content in North Atlantic sediments. *Paleoceanography, 14*, 171–186.

Ortiz, J. D., & Rack, F. R. (1999). Non-invasive sediment monitoring methods: Current and future tools for high-resolution climate studies. In: F. Abrantes & A. C. Mix (Eds), *Reconstructing ocean history: A window into the future* (pp. 343–380). New York: Kluwer/Plenum.

Peterson, L. C., Haug, G. H., Hughen, K. A., & Röhl, U. (2000). Rapid changes in the hydrologic cycle of the tropical Atlantic during the last glacial. *Science, 290*, 1947–1951.

Principato, S. M. (2004). X-radiographs of sediment cores: A guide to analyzing diamicton. In: P. Francus (Ed.), *Image Analysis, Sediments and Paleoenvironments* (pp. 165–185). Dordrecht: Kluwer Academic Publishers.

Rack, F. R. (1998). *Tomorrow's technology today.* Interim report of the IMAGES standing committee on "New technologies in sediment imaging", p. 33.

Rack, F. R., Balcom, B. J., MacGregor, R. P., & Armstrong, R. L. (1997). Magnetic resonance imaging of the Lake Agassiz–Lake Winnipeg transition. *Journal of Paleolimnology, 19*, 255–264.

Rack, F. R., Ribes, A. C., Tsintzouras, G., Marshall, G., Damaskinos, S., & Dixon, A. E. (1998). Preliminary results from biomedical imaging of lake and ocean sediments. *Proceedings of the Sixth International Conference on Paleoceanography*, August 24–28, Lisbon, Portugal, p. 189.

Ribes, A. C., Damaskinos, S., & Dixon, A. E. (1995). Photoluminescence imaging of porous silicon using a confocal scanning laser macroscope/microscope. *Applied Physics Letters*, 66, 2321–2323.

Ribes, A. C., Damaskinos, S., Tiedje, H. F., Dixon, A. E., & Brodie, D. E. (1996). Reflected light, photoluminescence, and OBIC imaging of solar cells using a confocal scanning laser macroscope/ microscope. *Solar Energy Materials and Solar Cells*, 44, 439–450.

Ribes, A. C., Lundberg, J., Waldron, D. J., Vesely, M., Damaskinos, S., Guthrie, S. I., & Dixon, A. E. (2000). Photoluminescence imaging of speleothem microbanding with a high resolution confocal scanning laser macroscope. *Quaternary International*, 68–71, 253–259.

Ribes, A. C., Marshall, G., Tsintzouras, G., Damaskinos, S., Dixon, A. E., & Rack, F. R. (1998). The confocal scanning beam MACROscope/Microscope applied to imaging ocean/lake core geological specimens. *Canadian Association of Physicists (CAP) annual meeting*, June 15, 1998, Waterloo, Canada, Physics in Canada 54, 10.

Röhl, U., Brinkhuis, H., & Fuller, M. (2004). On the search for the Paleocene/Eocene boundary in the Southern Ocean: Exploring ODP Leg 189 Holes 1171D and 1172D. In: N. F. Exon, M. Malone & J. P. Kennett (Eds), *The Cenozoic Southern Ocean and climate change between Australia and Antarctica* (pp. 113–126). American Geophysical Union, Geophysical Monograph Series 181.

Röhl, U., Brinkhuis, H., Fuller, M., Schellenberg, S. A., Stickley, C. E., & Williams, G. L. (2004). Cyclostratigraphy of middle and late Eocene sediments drilled on the East Tasman Plateau (Site 1172). In: N. F. Exon, M. Malone & J. P. Kennett (Eds), *Climate evolution in the Southern Ocean and Australia's Cenozoic flight northward from Antarctica* (pp. 127–152). American Geophysical Union, Geophysical Monograph Series 181.

Rothwell, R. G., Hoogakker, B., Thomson, J., & Croudace, I. W. (2006). Turbidite emplacement on the southern Balearic Abyssal Plain (Western Mediterranean Sea) during Marine Isotope Stages 1–3: An application of XRF scanning of sediment cores in lithostratigraphic analysis. In: R. G. Rothwell (Ed.), *New ways of looking at sediment cores and core data* (pp. 79–98). London: Geological Society (Special Publication).

Smith, J. N., & Schafer, C. (1987). A 20th-century record of climatologically modulated sediment accumulation rates in a Canadian fjord. *Quaternary Research*, 27, 232–247.

Smith, J. N., & Walton, A. (1980). Sediment accumulation rates and geochronologies measured in the Saguenay Fjord using Pb-210 dating method. *Geochimica et Cosmochimica Acta*, 44, 225–240.

Stevens, A. W., Wheatcroft, R. A., & Wiberg, P. L. (2007). Seabed properties and sediment erodibility along the western Adriatic margin, Italy. *Continental Shelf Research*, 27, 400–416.

Stoner, J. S., Channell, J. E. T., & Hillaire-Marcel, C. (1996). The magnetic signature of rapidly deposited detrital layers from the deep Labrador Sea: Relationship to North Atlantic Heinrich layers. *Paleoceanography*, 11, 309–325.

St-Onge, G., & Hillaire-Marcel, C. (2001). Isotopic constraints of sedimentary inputs and organic carbon burial rates in the Saguenay Fjord, Quebec. *Marine Geology*, 176, 1–22.

St-Onge, G., & Long, B. (in review). CAT-scan analysis of sedimentary sequences: an ultrahigh-resolution paleoclimatic tool. *Engineering Geology*.

St-Onge, G., Mulder, T., Piper, D. J. W., Hillaire-Marcel, C., & Stoner, J. S. (2004). Earthquake and flood-induced turbidites in the Saguenay Fjord (Québec): A Holocene paleoseismicity record. *Quaternary Science Reviews*, 23, 283–294.

St-Onge, G., Stoner, J. S., & Hillaire-Marcel, C. (2003). Holocene paleomagnetic records from the St. Lawrence Estuary: Centennial- to millennial-scale geomagnetic modulation of cosmogenic isotopes. *Earth and Planetary Science Letters*, 209, 113–130.

Stuut, J.-B. W., Prins, M. A., & Jansen, J. H. F. (2002b). Fast reconnaissance of carbonate dissolution based on the size distribution of calcareous ooze on Walvis Ridge, SE Atlantic Ocean. *Marine Geology*, 190, 581–589.

Stuut, J.-B. W., Prins, M. A., Schneider, R. R., Weltje, G. J., Jansen, J. H. F., & Postma, G. (2002a). A 300-kyr record of aridity and wind strength in southwestern Africa: Inferences from grain-size distributions of sediments on Walvis Ridge, SE Atlantic. *Marine Geology*, 180, 221–233.

Syvitski, J. P. M., & Schafer, C. T. (1996). Evidence for earthquake-triggered basin collapse in Saguenay Fjord, Canada. *Sedimentary Geology*, *104*, 127–153.

Thomas, R. G., Guyodo, Y., & Channell, J. E. T. (2003). U channel track for susceptibility measurements. *Geochemistry, Geophysics, Geosystems*, *4*, 1050, doi:10.1029/2002GC000454.

Thomson, J., Croudace, I. W., & Rothwell, R. G. (2006). A geochemical application of the ITRAX scanner to a sediment core containing eastern Mediterranean sapropel units. In: R. G. Rothwell (Ed.), *New ways of looking at sediment cores and core data* (pp. 65–77). London: Geological Society (Special Publication).

Thompson, R., & Oldfield, F. (1986). *Environmental magnetism*. Boston: Allen and Unwin p. 227.

Urgeles, R., Locat, J., Lee, H. J., & Martin, F. (2002). The Saguenay Fjord, Quebec, Canada: integrating marine geotechnical and geophysical data for seismic slope stability and hazard assessment. *Marine Geology*, *185*, 319–340.

Vidal, L., Bickert, T., Wefer, G., & Röhl, U. (2002). Late Miocene stable isotope stratigraphy of SE Atlantic ODP Site 1085: Relation to Messinian events. *Marine Geology*, *180*, 71–85.

Weber, M. E. (1998). Estimation of biogenic carbonate an opal by continuous non-destructive measurements in deep-sea sediments: Application to the eastern Equatorial Pacific. *Deep-Sea Research I*, *45*, 1955–1975.

Wellington, S. L., & Vinegar, H. J. (1987). X-ray computed tomography. *Journal of Petroleum Technology*, *2*, 1951–1954.

West, S., Jansen, J. H. F., & Stuut, J.-B. (2004). Surface water conditions in the Northern Benguela Region (SE Atlantic) during the last 450 ky reconstructed from assemblages of planktonic foraminifera. *Marine Micropaleontology*, *51*, 321–344.

Westerhold, T., Bickert, T., & Röhl, U. (2005). Middle to late Miocene oxygen isotope stratigraphy of ODP Site 1085 (SE Atlantic): New constraints on Miocene climate variability and sea-level fluctuations. *Palaeogeography, Palaeoclimatology, Palaeoecology*, *217*, 205–222.

Wheatcroft, R. A., Stevens, A. W., Hunt, L. M., & Milligan, T. G. (2006). The large-scale distribution and internal geometry of the Fall 2000 Po River flood deposit: Evidence from digital X-radiography. *Continental Shelf Research*, *26*, 499–516.

Zaragosi, S., Bourillet, J.-F., Eynaud, F., Toucanne, S., Denhard, B., Van Toer, A., & Lanfumey, V. (2006). The impact of the last European deglaciation on the deep-sea turbidite systems of the Celtic-Armorican margin (Bay of Biscay). *Geo-Marine Letters*, *26*, 317–329.

CHAPTER THREE

Magnetic Stratigraphy in Paleoceanography: Reversals, Excursions, Paleointensity, and Secular Variation

Joseph S. Stoner* *and* Guillaume St-Onge

Contents

1. Introduction	100
2. Background	101
2.1. Geomagnetism and orientation	101
2.2. Magnetism, magnetic units, and conversions	103
3. Soft Sediment Paleomagnetic Methods	103
3.1. Sampling	103
3.2. Discrete samples	105
3.3. U-channel method	105
4. Magnetometers	107
4.1. Superconducting rock magnetometers	107
4.2. Large axis pass-through magnetometers	107
4.3. U-channel magnetometer	108
5. Measurements and Magnetizations	109
5.1. Natural remanent magnetization-AF demagnetization	109
5.2. Resolution: the response function and deconvolution	113
6. Data Analysis	114
6.1. Orthogonal projections and MAD values	115
7. Sediment Magnetism	117
7.1. The NRM recording process	117
7.2. Magnetic mineralogy	118
8. Development of Paleomagnetic Records	118
8.1. Directional records	118
8.2. Relative paleointensity determinations	119
9. The Paleomagnetic Record as a Stratigraphic Tool	121
9.1. Geomagnetic polarity time scale (GPTS)	121
9.2. Relative paleointensity stratigraphy	123
9.3. Excursions as a stratigraphic tool	126
9.4. Paleomagnetic secular variation	128
10. Some Perspectives	128
References	130

* Corresponding author.

Developments in Marine Geology, Volume 1
ISSN 1572-5480, DOI 10.1016/S1572-5480(07)01008-1

1. INTRODUCTION

The magnetic field of the Earth is thought to result from a self-exciting dynamo in the Earth's outer core (e.g., Bullard, 1949): a product of electrical currents generated through the fluid motion of an iron alloy conductor. Paleomagnetism is the study of this magnetic field as preserved in geologic material. Originally recognized in China a few thousand years ago, the study and use of Earth's magnetic field has a long history. In Europe, detailed magnetic studies were ongoing during the Middle Ages (for more details on the historical development of geomagnetism, see Jonkers, 2003). The recognition that the Earth's magnetic field was of internal (not celestial) origin and generally dipolar (like a bar magnet) goes back to at least 1600 when William Gilbert published *De Magnete*. By that time, the use of the compass for navigation was becoming increasingly established (Jonkers, 2003). The development by Carl Friedrich Gauss (1838) of the spherical harmonics expansion to characterize the geomagnetic field provided a mathematical means by which the position of the poles could be predicted and the intensity of the dipole calculated. This gave the geomagnetic field tangible and predictable elements, which paved the way for the future science of paleomagnetism.

The realization that the Earth's magnetic field was at times substantially different, even reversed, occurred at the turn of the century (Brunhes, 1906). Yet, it took more than half a century of debate before observations of magnetic reversal were universally accepted and proposals for its use as a stratigraphic synchronization tool explored (Hospers, 1955; Khramov, 1955, 1957). However, it was not until the observation of the same reversal sequence in dispersed radiometrically dated volcanic outcrops (Cox, Doell, & Dalrymple, 1963), magnetic anomalies (Vine & Wilson, 1965; Pittman & Heirtzler, 1966), and marine sediments (Opdyke, Glass, Hays, & Foster, 1966) that field reversals were fully accepted. This confirmed continental drift and began a stratigraphic revolution with the development of the geomagnetic polarity time scale (GPTS) (Doell & Dalrymple, 1966) that continues to be refined and serves as the backbone of Cenozoic stratigraphy to the present day (see Opdyke & Channell, 1996; Gradstein, Ogg, & Smith, 2004).

Since the recognition that magnetic reversals could be recorded in marine sediments (Harrison & Funnell, 1964; Opdyke et al., 1966), the GPTS has been an instrumental tool for paleoceanographic research. For example, inter-comparisons of high southern latitude radiolaria with the paleomagnetic record (Opdyke et al., 1966; Hays & Opdyke, 1967) were seminal contributions, placing late Neogene to Quaternary radiolarian events and biozones in a robust magnetostratigraphic framework. Since the 1960s, biostratigraphic researchers have been acutely aware of the importance of correlating their observations to magnetostratigraphies. The importance of this rests on the fact that magnetic reversals are globally synchronous on geological timescales, and that they are environmentally independent events that can be recorded in both deposited and thermally cooled materials alike. Much of the focus on understanding time in the geologic past has centered upon the dating of reversals and the inter-calibration of the reversal record with other chronological and stratigraphic tools (see Opdyke & Channell, 1996; Gradstein et al., 2004). For modern paleoceanography, the pioneering study of Shackleton & Opdyke (1973) presented the first example of modern marine stratigraphy with the inter-calibration of oxygen isotopes with the magnetic reversal (Matuyama/Brunhes) record.

Magnetic stratigraphy rests on the idea that the recorded magnetization of a rock reflects the behavior of the geomagnetic field. The fact that sediment deposited in water can record the geomagnetic field has been known for more than 50 yr (Johnson, Murphy, & Torreson, 1948). In the simplest case, the natural remanent magnetization (NRM) of sediment is aligned with the (geo)magnetic field and is a function of its intensity and direction at the time of deposition. In practice, many factors may work to modify the original geomagnetic input signal. Under favorable circumstances and with detailed diligence, some of these effects can be separated and others avoided so that an accurate paleomagnetic record is recovered.

Over the last decade, a significant amount of paleoceanographic research has focused on timescales much shorter than the typical interval between magnetic reversals. This has reduced the impact of magnetic polarity stratigraphy on paleoceanography, though not its fundamental importance as a stratigraphic tool (see Opdyke & Channell, 1996). Paleomagnetism has been working to keep pace and a significant new understanding of geomagnetic field behavior during times of constant polarity has begun to emerge. Essentially, it has been found that high amplitude, high frequency variations in the Earth's magnetic field occur over large spatial scales even during times of constant polarity. These changes can occur over a millennium or even less, with coherence on a regional and sometimes even global scale. New magnetostratigraphic opportunities over a range of temporal and spatial scales are emerging. Much of this contribution will outline the practical aspects of reconstructing the paleomagnetic record from marine sediments. In the latter part of this chapter, we will briefly discuss some of the recent observations on the Quaternary geomagnetic record that are being made and their uses as a stratigraphic tool for paleoceanographic research.

2. Background

2.1. Geomagnetism and Orientation

On a time-averaged basis, the geomagnetic field during times of stable polarity approximates a geocentric axial dipole (GAD). Evidence for the GAD hypothesis has been accumulated from the inclination distribution of deep-sea sediments (e.g., Opdyke & Henry, 1969; Schneider & Kent, 1990) and statistical studies of randomly distributed igneous rocks (e.g., Merrill & McElhinny, 1977; Mejia, Opdyke, Vilas, Singer, & Stoner, 2004). The dip of the field lines from horizontal in the vertical plane is known as inclination (I). Magnetic deviations from true geographic north are known as declination. Using the GAD hypothesis, inclination can be predicted as a function of latitude (λ) $\tan I = 2 \tan \lambda$ and declination would be zero. Yet, at any point in time the geomagnetic field is not a GAD. Deviations from a GAD, ~25% of the present field, are generally not considered to be stationary (this is an active topic of research) and how these change with time are known as secular variations.

The field at any point on the Earth's surface is a vector (*F*), which possesses a horizontal component (*H*), which makes an angle (*D*) between the geographical North and the magnetic meridian (Figure 1). Declination (*D*) is the angle from the geographical North measured eastward and ranging from 0° to 360°. The inclination (*I*) is the angle made by the magnetic vector with the horizontal plane. By convention, it is positive if the north-seeking vector points below the horizontal plane (present Northern Hemisphere) or negative if it points above (present Southern Hemisphere). The vertical component *Z* is positive down. The horizontal component (*H*) has two components, one to the North (*X*) and one to East (*Y*) (Figure 1).

Spherical harmonic analyses show that the geomagnetic field is almost entirely of internal origin. Approximately 90% of the present field can be explained by a dipole inclined to the Earth's axis of rotation by ~11.5°. The present magnitude of the dipole is $7.8 \times 10^{22}\,\mathrm{Am^2}$. Based on the current world magnetic model (WMM), the geomagnetic North Pole in 2005 was 79.74°N and 71.78°W and

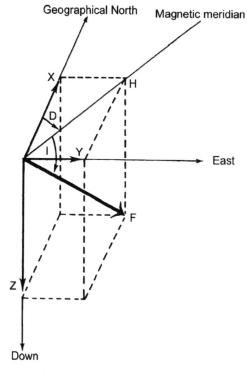

Figure 1 Earth's magnetic field. The total magnetic field is illustrated by the vector *F*. *I* = inclination; *D* = declination. The vertical component of the field is illustrated by *Z*. The horizontal component (*H*) of the field has two components, one to the North (*X*) and one to East (*Y*). Redrawn with permission from Thompson and Oldfield (1986).

the geomagnetic South Pole was 79.74°S and 108.22°E. These are based on the best fitting geocentric dipole. The great circle midway between the geomagnetic poles is called the geomagnetic equator. The actual magnetic poles (where inclination = 90°) and magnetic equator (where inclination = 0°) can deviate significantly. In 2005, based on the WMM model, the North Magnetic Pole was at 83.21°N and 118.32°W (Arctic Ocean, NW of Ellesmere Island, Canada) and South Magnetic Pole 64.53°S and 137.86°E (Southern Ocean, North of Antarctica, South of Australia).

2.2. Magnetism, Magnetic Units, and Conversions

Throughout this chapter, three types of magnetic units will be presented: one for the magnetic field, one for magnetization, and one for magnetic susceptibility. The motion of an electric charge will generate a magnetic field in the space around it. In accordance with Moskowitz (1991), as an analogy to an electron orbiting an atom, considering a loop of radius r and current i, the magnetic field is defined by $H = i/2r$ (A/m) at the center of the loop. The current loop has a magnetic moment, $m = i \times$ area (Am^2). The intensity of the magnetization (M or J), defined as a magnetic moment per unit of volume, $M = m/v$ (A/m), is the response of a material to a magnetic field passing through (note that M and H have the same units). The relationship between M in the material and the external field (H) allows the determination of the magnetic susceptibility, $k = M/H$ (dimensionless). Susceptibility is a measure of how magnetizable a material is in the presence of a magnetic field and can be used in a general way to describe the various classes of magnetic substances. The response of any material in that space is the magnetic induction (B). The unit of B is called the Tesla (T) and the total B field is the sum of the H field and the magnetization M. In the SI (Système International) system, $B = m_0(H+M)$, where the permeability of free space $m_0 = 4\pi \ 10^7 \ \text{Hm}^{-1}$ (H: Henry). Table 1 summarizes the different units in SI and cgs. SI is the standard, but because both types of units are often reported in the literature or directly given by the instruments, the relationships between the units of both systems are provided.

3. SOFT SEDIMENT PALEOMAGNETIC METHODS

3.1. Sampling

Sediment paleomagnetism is based on the requirement that the primary vector magnetization that is locked in the sediment at or shortly after the time of deposition is preserved in its original orientation, from the seafloor and through the laboratory analyses. Maintaining sediment orientation without deformation (physical or magnetic) is a non-trivial task. Assumptions are sometimes made. We will attempt to indicate, when, where, and what types of assumptions are commonly used for soft sediment paleomagnetism. For paleoceanographic uses, the first basic assumption is the original horizontality and lateral continuity of

Table 1 Magnetic Units and Relationship Between SI and CGS (centimeter–gram–second) Units.

Quantity	Symbol	SI	cgs	Relationship
Magnetic moment	m	Am^2	emu	$1\,Am^2 = 10^3\,emu$
Magnetization	M	Am^{-1}	$emu\,cm^{-3}$	$1\,Am^{-1} = 10^{-3}\,emu\,cm^{-3}$
Magnetic field	H	Am^{-1}	Oersted (oe)	$1\,Am^{-1} = 4\pi \times 10^{-3}\,oe$
Magnetic induction	B	T	Gauss (G)	$1\,T = 10^4\,G$
Permeability of free space	μ_0	Hm^{-1}	1	$4\pi\,10^7\,Hm^{-1} = 1$
Magnetic susceptibility				
Volumetric	κ	Dimensionless	$emu\,cm^{-3}\,oe$	$1\,SI = 1/4\pi\,emu\,cm^{-3}\,oe^{-1}$
Mass	χ	$m^3\,kg^{-1}$	$emu\,g^{-1}\,oe$	$1\,m^3\,kg^{-1} = 10^3/4\pi\,emu\,g^{-1}\,oe^{-1}$

Source: With permission from Tauxe (1998).
Note: Other relationships are as follows — $1\,H = kg\,m^2A^{-2}s^{-2}$, $1\,emu = 1\,G\,cm^3$, $1\,T = kg\,A^{-1}s^{-2}$.

the targeted sediment relative to the cored sediment water interface. The next assumption is that the coring process allows the acquisition of sediments that have not been stretched, compressed, or deformed in any way. Additionally, the corer should enter the sediment vertically and not rotate during penetration. Obtaining an undisturbed sample through coring or drilling is among the most difficult, yet critical components for retrieving a reliable paleomagnetic record.

Orientation control during coring is rare and/or often unreliable. Declinations are therefore most often relative. If enough time is involved, and 10,000 yr is often considered long enough (e.g., Merrill & McFadden, 2003), the assumption that the mean declination is zero is considered reasonable. One of the most critical (and easy), though sometimes neglected, measures is to split the whole core along a common plane. This can be easily done by drawing a consistent line along the core liner(s) and if there is more than one section, aligning each section together prior to insertion into the core barrel. Without this simple measure, declination must be (arbitrarily and undesirably) corrected for an unknown offset between each section, possibly degrading a critical part of the record. Nevertheless, even with these measures, the declination records are sometimes adjusted to compensate for slight rotations during coring, offsets caused by core splitting or for values close to $0°$ or $360°$. Much of the subjectivity of these corrections, except for the latter, which is just a simple scale adjustment, can be removed by measuring duplicate or overlapping material.

There are only a handful of laboratories that have magnetometers capable of running whole or half-round core sections, therefore sub-sampling is almost always required and often desired. Half or whole-round magnetometers are generally used as survey instruments with the most notable and widely used example being the pass-through magnetometer on-board the RV JOIDES Resolution, the research vessel used by the Ocean Drilling Program (ODP) and more recently by the Integrated Ocean Drilling Program (IODP). Having such a survey instrument is critical when dealing with large volumes of sediment (e.g., up to 8 km from a single ODP/IODP Leg). Initial shipboard observations provide an invaluable resource that allows later focused study of the sediments. In addition, shipboard observations are sometimes the only paleomagnetic analysis done on a sample, providing critical magnetic properties and geomagnetic polarity information that would otherwise be lost. On the other hand, sub-sampling is valued because these samples are taken from the pristine center part of the core, thus avoiding the disturbed outside part of the core section which can degrade whole or half round measurements (e.g., Acton, Okada, Clement, Lund, & Williams, 2002). In addition, measurements on sub-sampled material can be made at much higher resolution than half/whole round measurements. The width of the response function of a pass-through magnetometer is governed by the diameter of the sample measurement space. Each analysis from half/whole round magnetometer will, therefore, integrate over ~10 cm or more of stratigraphic length. More detailed studies of the NRM or laboratory magnetizations can be accomplished on sub-sampled materials at resolutions as tightly spaced as 1 cm.

3.2. Discrete Samples

Soft sediments are generally sampled using either discrete or u-channel samples (Tauxe, LaBrecque, Dodson, & Fuller, 1983). Discrete sampling of soft sediments is usually done using non-magnetic plastic cubes that are nominally 7–8 cm^3. There are several varieties of cubes with slight variations between them. If possible, truly cubic samples are more desirable. Mini-cubes of 1 cm^3 are sometimes used as they allow higher resolution sampling. However, the increased surface area to volume can result in more significant sample disturbance relative to cubes with larger volumes. Additionally, the significant reduction in material of a mini cube reduces the magnetic moment that may render weakly magnetized pelagic sediments below magnetometer sensitivity. When dealing with weakly magnetized materials, it can be advisable to demagnetize the cubes with an alternating field (AF) and measure their remanence prior to sampling so that the background can be subtracted.

3.3. U-channel Method

U-channel samples are collected by pushing rigid u-shaped plastic liners (2 × 2 cm cross-section) that are up to 1.5 m in length into the split halves of core sections (Figure 2). The u-channel is cut free using fishing line, removed from the core and capped with a matching plastic lid. End caps or waterproof polyethylene tape are often used to seal the ends. U-channels have significant advantages over discrete samples. First, u-channel sampling is much faster and introduces considerably less sediment deformation than back-to-back discrete sampling with standard 7 or 8 cm^3 plastic cubes. Second, the speed of data acquisition using u-channel samples makes it feasible to take continuous measurements on long or replicate sediment sequences that would be impractical, or even impossible, with discrete samples. Third, u-channel samples are optimal for a range of other high-resolution

Figure 2 Example of u-channel sampling. U-channels are used for continuous sampling of sediment cores by pushing the open u-channel into the split surface of the core. A lid closes the u-channel and the ends are sealed with tape.

measurements, including computerized axial tomography (CAT-scan), X-ray density scan, gamma ray attenuation density scan, and X-ray microfluorescence scan (see St-Onge, Mulder, & Farncus, this book). These not only provide useful and complementary information for downcore interpretation of the paleomagnetic data, but also allow monitoring of sediment structures and integrity. Fourth, the u-channel sample can function as a permanent archive that takes significantly less space and uses less sediment than a traditional archive half section. Because the sample is completely enclosed in plastic, dehydration is minimized and with a little additional polyethylene tape essentially eliminated. The major disadvantage of u-channel samples results from their use for continuous measurements. Though samples are measured at 1 cm intervals, each measurement is not independent and is smoothed over ~4.5 cm. This results in edge effects that compromise the measurements on the top and bottom 5 cm of each section. Gaps result in a similar effect. Sediment heterogeneities are also not easily dealt with and anomalous sediment inputs, such as tephra layers for example, will affect the measurements as much as 6 cm on either side. Careful application of deconvolution procedures can minimize these effects (Guyodo, Channell, & Thomas, 2002) and are discussed below.

4. MAGNETOMETERS

As in many fields in Earth Sciences, paleomagnetic advances have been paced by technological innovations. The rock magnetometer is the required instrument for paleomagnetic research and its development has a long history. A critical advance was made with the development of the astatic magnetometer (Blackett, 1952) which allowed weakly magnetized rocks to be accurately measured. However, the sensitivity of this instrument came with the tradeoff of requiring an extremely quite magnetic and vibration free environment. The development of spinner rock magnetometers overcame some of these limitations by using the spin to amplify the magnetic signal. The high rate of spin required was, however, disastrous for soft marine sediments. The development of the slow spin fluxgate magnetometer (Foster, 1966) was a major breakthrough for marine paleomagnetism, as it allowed soft sediment to be measured without destruction. This initiated the merger of paleomagnetism and paleoceanography that was orchestrated by Neil Opdyke at Lamont Doherty Geologic Observatory in the late 1960s and 1970s (e.g., Opdyke, 1972), essentially pioneering all marine sediment paleomagnetic work that has followed. Spinner magnetometers are still widely used, although superconducting rock magnetometers are generally the preferred instrument for marine sediments because of their great sensitivity, speed, and measurement flexibility. For more details on magnetic instrumentation the reader is referred to Collinson (1983).

4.1. Superconducting Rock Magnetometers

The development of the superconducting rock magnetometer using a Superconducting Quantum Interference Device (SQUID) designed with a radio frequencies (RF) driven weak link sensor (Goree & Fuller, 1976) revolutionized paleomagnetism since weakly magnetized samples could be rapidly measured. These systems are cooled to cryogenic temperatures using liquid helium. Precise measurements are made by reading the flux changes generated by the insertion of the sample into the pick-up coil array. This resulted in substantially increased sensitivity and dynamic range. Sediments no longer had to be rotated at high speeds or vibrated so long-cores or fragile samples could be measured. The newer Direct Current (DC) SQUIDs developed in the late 1990 s improved the sensitivity almost two orders of magnitude. The DC SQUID sensor has a total magnetic moment sensitivity of $10^{-12}\,\mathrm{Am^2}$ which, when expressed in terms of magnetization, translates to $10^{-7}\,\mathrm{A/m}$ for a $10\,\mathrm{cm^3}$ sample. Further advantages of the DC SQUIDs are the improved dynamic range and faster measurement acquisition time. The most recent improvement to superconducting rock magnetometers is the elimination of liquid helium by using a pulse tube cryocooler. This allows the system to be warmed to room temperature or cooled to cryogenic temperatures with minimal effort or consequence.

4.2. Large Axis Pass-through Magnetometers

For marine sediment studies pass-through or long-core magnetometers are most practical because of their large throughput. We will therefore focus our discussion on those types of instruments. The most widely used large axis pass-through magnetometer that can perform remanence measurements and AF demagnetization is the long-core cryogenic magnetometer (2G Enterprises model 760-R) on board the JOIDES Resolution. This instrument is equipped with a DC SQUID and has an inline AF demagnetizer capable of reaching a peak AF of 80 mT. The spatial resolution measured by the width at half-height of the pickup coils response is ~10 cm for all three axes, although they can sense a magnetization that extends up to 30 cm of core length (see Explanatory Notes Section of Channell, Sato, & Malone, 2005). The cryogenic magnetometer is sensitive to a magnetic moment of ~10^{-9} emu or 10^{-7} A/m for a 10 cm^3 rock volume. However, the practical noise level is affected by the magnetization of the core liner (~8×10^{-6} A/m) and the background magnetization of the measurement tray (~1×10^{-5} A/m).

4.3. U-channel Magnetometer

The modern workhorse of marine paleomagnetism is the automated small access pass-through u-channel superconducting rock magnetometer (Figure 3). Since its development just over a decade ago (see Weeks et al., 1993; Nagy & Valet, 1993; Roberts, Stoner, & Richter, 1996; Verosub, 1998; Sagnotti et al., 2003; Brachfeld, Kissel, Laj, & Mazaud, 2004), our knowledge of the spatial and temporal variability of the Quaternary geomagnetic field has improved dramatically. The speed of data

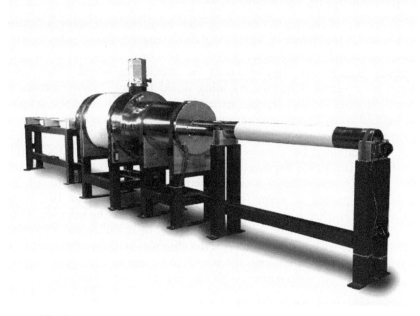

Figure 3 2G Enterprises automated small access pass-through u-channel cryogenic magnetometer. The system shown was recently installed at the University of Alberta, Canada. Picture courtesy of Bill Goree (2G Enterprises).

acquisition allows observations that could not practically be made with discrete samples. Continuous measurements at 1 cm intervals on long and/or replicate sediment sequences are becoming routine, as one measurement (e.g., one AF demagnetization step; see Section 5.1 for the use of stepwise demagnetization) of a 1.5 m long u–channel takes about 15 min.

Recent u-channel superconducting rock magnetometers fitted with DC SQUID are designed with a software controlled tracking system, in-line AF demagnetization coils capable of reaching a peak AF of 300 mT for the axial (Z) coil and 180 mT for the transverse (X, Y) coils (Nagy & Valet, 1993), in-line DC coil for acquisition of an anhysteretic remanent magnetization (ARM; see Table 2), and with a separate (or in-line) pulse magnetizer for acquisition of an isothermal remanent magnetization (IRM; see Table 2). The pick-up coils in the 2G Enterprises high-resolution cryogenic magnetometer are designed to have a narrow response function translating to a higher resolution measurement (see Weeks et al., 1993).

An important development of the long-core magnetometer is the in-line demagnetization process. A series of three AF coils are positioned in a mutually perpendicular fashion so that the sample can be demagnetized along its X, Y, and Z axes. The u-channel sample is demagnetized as it passes through the AF at a constant velocity. The peak AF is reached when the sample passes through or perpendicular to the coil and cycles down as it moves away, thus mimicking the cycling down of a standard discrete sample AF unit (Collinson, 1983). The entire process takes place in a magnetically shielded environment allowing coercivities lower than the peak AF to be randomized. Because of space restrictions, these coils cannot be operated simultaneously without interactions. Therefore, three passes (one for each perpendicular axis) are required at each demagnetization step. An in-line DC coil allows the ARM to be acquired using the axial (z) coil. It has been observed that translation speed can inversely affect the efficiency of ARM acquisition. Slow track speeds of 1 cm/s or less are most effective when acquiring an ARM on u–channel samples. Therefore, care should be taken when comparing ARM data generated with u-channel and discrete sample systems (Sagnotti et al., 2003), highlighting the need for standardization and calibration of paleomagnetic procedures.

5. MEASUREMENTS AND MAGNETIZATIONS

Several types of magnetization are now routinely imparted and measured on both discrete and u-channel samples. These include ARM, IRM, and volumetric magnetic susceptibility. Short explanations and rule of thumb downcore interpretations of these magnetizations, as well as other magnetic parameters are given in Table 2. Below, we discuss some of the requirements for acquiring high quality paleomagnetic records.

5.1. Natural Remanent Magnetization-AF Demagnetization

Even though some sediments preserve a stable magnetization after little or no demagnetization, modern standards in paleomagnetic research dictate that

Table 2 Generalized Table of Downcore Magnetic Parameters and Their Interpretations.

Parameter	Interpretation
Bulk magnetic measurements	
Natural remanent magnetization (NRM): The fossil (remanent) magnetization preserved within the sediment. NRM recorded as declination, inclination and intensity.	Dependent on mineralogy, concentration, and grain size of the magnetic material as well as mode of acquisition of remanence, and intensity and direction of the geomagnetic field.
Volumetric magnetic susceptibility (k): A measure of the concentration of magnetizable material. Defined as the ratio of induced magnetization intensity (M) per volume to the strength of the applied weak field (H): $k = M/H$.	k is a first order measure of the amount of ferrimagnetic material (e.g., magnetite). k is particularly enhanced by superparamagnetic (SP) magnetite ($<0.03\,\mu m$) and by large magnetite grains ($>10\,\mu m$). When the concentration of ferrimagnetic material is low, k responds to antiferromagnetic (e.g., hematite), paramagnetic (e.g., Fe, Mg silicates), and diamagnetic material (e.g. calcium carbonate, silica) that may complicate the interpretation.
Isothermal remanent magnetization (IRM): Magnetic remanence acquired under the influence of a strong DC field. Commonly expressed as a saturation IRM or SIRM when a field greater than 1 T is used. A back-field IRM (BIRM) is that acquired in a reversed DC field after SIRM acquisition.	SIRM primarily depends upon the concentration of magnetic, principally ferrimagnetic, material. It is grain-size dependent being particularly sensitive to magnetite grains smaller than a few tens of microns.
Anhysteretic remanent magnetization (ARM): Magnetization acquired in a biasing DC field within a decreasing alternating field. Commonly expressed as anhysteretic susceptibility (k_{ARM}) when normalized by the biasing field used.	k_{ARM} is primarily a measure of the concentration of ferrimagnetic material, however it is also strongly grain-size dependent. k_{ARM} preferentially responds to smaller magnetite grain sizes ($<10\,\mu m$), and is useful in the development of grain-size dependent ratios.
Constructed magnetic parameters	
The "hard" IRM (HIRM): This is derived by imparting a back-field of typically 0.1 or 0.3 T on a sample previously given an SIRM. The	HIRM is a measure of the concentration of magnetic material with higher coercivity than the back-field. This commonly gives information on the

Table 2. (*Continued*)

Parameter	Interpretation
resulting BIRM, which has a negative sign, is used to derive the HIRM by the formula: HIRM = (SIRM+BIRM)/2.	concentration of the antiferromagnetic (e.g. hematite) or very fine-grained ferrimagnetic (e.g. magnetite) grains depending on the back-field used.
S-ratios: These are derived by imparting a back-field of typically 0.1 or 0.3 T on a sample previously given an SIRM. The resulting BIRM, which has a negative sign, is normalized by the SIRM; S = BIRM/SIRM. This provides a measure of the proportion of saturation at the back-field applied.	The S-ratios can be used to estimate the magnetic mineralogy (e.g., magnetite or hematite). Downcore variations may be associated with changing mineralogy. Values close to −1 indicate lower coercivity and a ferrimagnetic mineralogy (e.g., magnetite); values closer to zero indicate a higher coercivity, possibly an antiferromagnetic (e.g., hematite) mineralogy. S-ratio with a back-field of 0.1 T may be sensitive to mineralogic and grain-size changes whereas the S-ratio with the 0.3 T back-field is more sensitive to mineralogical changes (e.g. proportion of magnetite to hematite).
Pseudo S-ratios: These are derived by imparting an IRM typically of 0.3 T, followed by an SIRM (e.g., 1 T). The pseudo S-ratio is determined by dividing the IRM by the SIRM.	Similarly to the classical S-ratios, the pseudo S-ratios will be used to estimate the magnetic mineralogy, with values close to 1 indicating lower coercivity and a ferrimagnetic mineralogy (e.g., magnetite) and lower values indicating a higher coercivity, possibly an antiferromagnetic (e.g., hematite) mineralogy.
k_{ARM}/k: Indicates changes in magnetic grain size, if the magnetic mineralogy is dominantly magnetite.	If the magnetic mineralogy is dominantly magnetite k_{ARM}/k varies inversely with magnetic grain size, particularly in the 1–10 μm grain-size range. However, the interpretation of this ratio may be complicated by significant amounts of SP or paramagnetic material.
SIRM/k: Indicates changes in magnetic grain size, if the magnetic mineralogy is dominantly magnetite.	If the magnetic mineralogy is dominantly magnetite, SIRM/k varies inversely with magnetic particle size. SIRM/k

Table 2. (*Continued*)

Parameter	Interpretation
	is more sensitive than k_{ARM}/k to changes in the proportion of large ($>10\,\mu m$) grains. SIRM/k may also be compromised by SP or paramagnetic material.
SIRM/k_{ARM}: Indicates changes in magnetic grain size, if the magnetic mineralogy is dominantly magnetite.	SIRM/k_{ARM} increases with increasing magnetic grain size, but is less sensitive and can be more difficult to interpret than the two ratios above. A major advantage of SIRM/k_{ARM} is that it only responds to remanence carrying magnetic material and is therefore not affected by SP or paramagnetic material.
Frequency dependent magnetic susceptibility (k_f): The ratio of low-frequency k (0.47kHz) to high-frequency k_{hf} (4.7 kHz) calculated by $k_f = 100^*(k-k_{hf})/k$	k_f is used to indicate the presence of SP material. SP material in high concentrations can compromise the grain-size interpretation made using k_{ARM}/k and SIRM/k.
Coercivity ratios: Ratios such as ARM_{20mT}/ARM_{0mT} or NRM_{30mT}/NRM_{0mT}.	These ratios provide information on the mean coercivity state of the sample, which is a reflection of its grain size and mineralogy. For example, for a magnetite-dominated mineralogy, ARM_{20mT}/ARM_{0mT} mostly reflects changes in magnetic grain size.
Median destructive field (MDF): Determined from the AF demagnetization procedure. It is the AF value needed to reduce the initial remanence by one half.	Similarly, the MDF provides information on the mean coercivity state of the sample, which is a reflection of its grain size and mineralogy. Higher MDFs indicate higher coercivity mineralogy. For a uniform mineralogy, finer grains require higher MDFs and coarser grains lower MDFs.
Hysteresis measurements[1]	
Saturation magnetization (M_s): M_s is the magnetization within a saturating field. Saturation remanence (M_{rs}): M_{rs} is the remanence remaining after removal of the saturating field.	M_{rs}/M_s decreases with increasing magnetite grain size in the submicron to few tens of microns grain-size range.

Table 2. (*Continued*)

Parameter	Interpretation
Coercive force (H_c): The back-field required to rotate saturation magnetization to zero within an applied field. Coercivity of remanence (H_{cr}): The back-field required to rotate saturation magnetization to zero remanence.	For magnetite, the ratio H_{cr}/H_c increases with increasing grain size (in the submicron to several hundred microns grain-size range) due to the strong grain-size dependence of both parameters, particularly H_c. H_{cr} is a useful guide to magnetic mineralogy.

Source: Modified with permission from Stoner, Channell, and Hillaire-Marcel (1996).
[1]Hysteresis parameters provide a means of monitoring grain-size variations in magnetite. Sediments should be homogeneous because of the small sample size (<0.05 g) typically used for hysteresis measurements. Mixed magnetic mineral assemblages greatly complicate the interpretations. The generalized interpretations listed below are based on a ferrimagnetic (e.g. magnetite) mineral assemblage.

the natural remanent magnetization (NRM) be stepwise demagnetized. This allows the component magnetization and its quality to be assessed. For u-channel samples, this is studied by progressive AF demagnetization. Because the u-channel samples are contained in plastic, thermal demagnetization techniques cannot be used. Though this limits some avenues of study, AF demagnetization does not physically alter the sediments. The sediments thus remain undisturbed for further investigation after the magnetic measurements have been completed. Typically, AF demagnetization is done using 5, 10, or even 20 mT steps between 0 and 80 mT and up to 140 mT. Discrete samples in some systems can be demagnetized at even higher levels. The magnetic mineralogy dictates the number of AF demagnetization steps. It is therefore wise to perform a pilot study prior to picking the best routine for a particular site. Generally, the routine used represents a balance between speed of data acquisition (each demagnetization and measurement step takes ~15 min) and the desire to precisely define the component magnetization of the remanence carrying grains.

5.2. Resolution: The Response Function and Deconvolution

As mentioned above, magnetic measurements on u-channel samples can be made at 1 cm intervals. However, the ~4.5 cm width (at half height) (Weeks et al., 1993) of the response function of the magnetometer pick-up coils is such that adjacent measurements at 1 cm spacing are not independent, and therefore the data are smoothed. This smoothing results in edge and other effects for a u-channel sample. Deconvolution protocols can be employed to reduce the smoothing and edge effects introduced by the response function of the magnetometer. Guyodo et al. (2002) successfully adapted the Oda and Shibuya (1996) deconvolution scheme to u-channel data from ODP sediments. Their study suggested that deconvolved resolution is comparable to that derived from 1 cm discrete samples (Figure 4). Further work supports this interpretation (Channell, 2006) and is potentially a major breakthrough for paleomagnetic research. Moreover, recent work on understanding magnetometer sensitivity as a function of position within the sensing region (Parker, 2000; Parker & Gee, 2002) may enable magnetometer calibration and future implementation of even more sophisticated corrections and deconvolution techniques.

6. DATA ANALYSIS

The amount of data generated with a u-channel cryogenic magnetometer can be substantial. Sediment cores tens or even hundreds of meters in length are

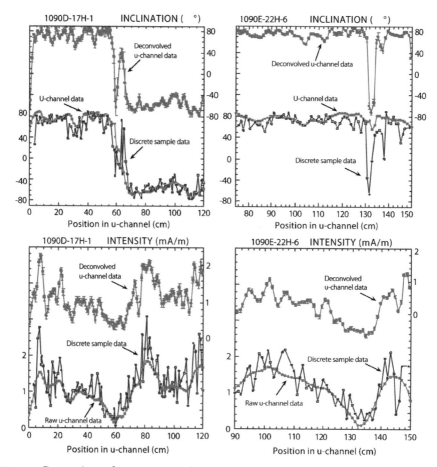

Figure 4 Comparison of component inclination (upper two graphs) and intensity (lower two graphs) from deconvolved u-channel data, raw u-channel data, and 1 cm discrete samples across a transition for low resolution (2 cm/kyr) South Atlantic ODP Site 1090. With permission from Guyodo et al. (2002).

now routinely measured with u-channel systems. Data are often collected at 1 cm intervals with each cm passing through the magnetometer 15 to more than 50 times. Careful evaluation and data reduction procedures are therefore required. In-house or commercially available programs transform the raw x, y, z magnetic moments into inclination and declination in degrees and intensity in A/m. Additional routines are often required to organize the data into the desired format for further analysis. Recently developed simple Excel spreadsheet (Mazaud, 2005) or freeware (e.g., Tauxe 1998; Jones, 2002; Cogné, 2003) can now be used to calculate and/or visualize the magnetic data in order to assess its downcore behavior. Such visualizations include the widely used Zijderveld plot (Zijderveld, 1967) for orthogonal projections of demagnetization data, normalized intensity diagrams, and stereographic projections (Figure 5). Many homegrown programs are also available and the Tauxe (1998) free software package is of particular interest for statistical analyses. The Excel spreadsheet developed by Mazaud (2005) was specifically developed for pass-through magnetometer data and is therefore particularly pertinent for sediment work. This macro allows easy calculation of component magnetizations and maximum angular deviation (MAD) values using the standard principal component analysis (PCA) of Kirschvink (1980). Additionally, the median destructive field at each interval is also provided. In this spreadsheet, the maximum number of demagnetization steps is 20, whereas the maximum number of lines is the Excel limit of 65,536 lines.

6.1. Orthogonal Projections and MAD Values

Orthogonal projections, often referred to as Zijderveld diagrams (Zijderveld, 1967), are the most commonly used approach to analyze changes in intensity and direction during demagnetization (Figure 5). The vector magnetization end points at successive demagnetization steps are plotted on both the horizontal and vertical planes. Straight-line segments indicate that the magnetic vector removed has a constant direction and the characteristic remanence is likely directed towards the origin. The component magnetization is calculated by PCA (Kirschvink, 1980), which provides a best-fit line to the demagnetization data. The MAD value provides a quantitative measurement of the precision with which the best-fit line is determined. MAD values $\geqslant 15°$ are often considered ill defined and of questionable significance (Butler, 1992; Opdyke & Channell, 1996). In the context of relative paleointensity (RPI) and paleomagnetic secular variation (PSV) of Quaternary marine sediments, much better defined components should be expected with MAD values $\geqslant 5$ to be considered suspect during times of stable polarity. High MAD values generally reflect a complex magnetization where different coercivities record a magnetization lock-in (see below) at different times and, therefore, with different characteristics. However, high MAD values are also often associated with reversals or excursions, resulting from a rapidly changing geomagnetic field relative to the time interval over which the magnetization is completely locked-in.

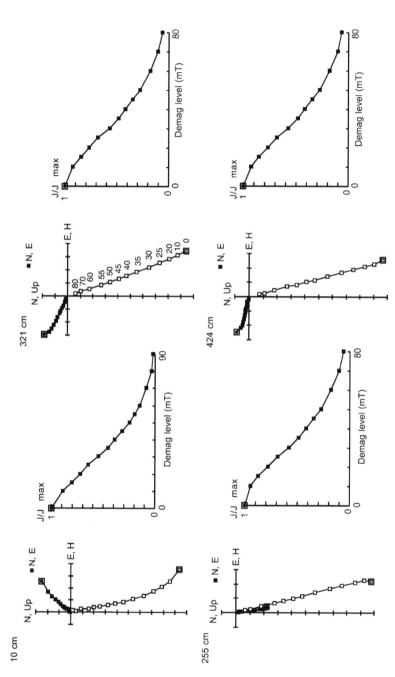

Figure 5 Typical vector endpoint diagrams (or Zijderveld plots; Zijderveld, 1967) and decay of the normalized intensity during AF demagnetization for pilot samples of core 2004-804-803, sampled in the Beaufort Sea (70°37.976′N/135°52.815′W). Closed squares represent north and east components, whereas open squares represent vertical and east components. As an example, the different steps, in mT, are illustrated in the sample at 321 cm.

7. SEDIMENT MAGNETISM

The fact that sediments can record geomagnetic field variations has been known for more than 50 yr (Johnson et al., 1948). In the simplest case, the natural remanent magnetization (NRM) of sediment is aligned with the (geo)magnetic field and is a function of its intensity and direction at the time of deposition. In practice, many factors may work to convolve the geomagnetic input signal. These include the concentration, composition, and size of the remanence-carrying magnetic grains. Non-magnetic factors also have a substantial effect including sediment composition and grain sizes, rates of deposition, and bioturbation. Preservation is affected by both biogenic and diagenetic processes, which can add, subtract, or sometimes even completely erase the NRM. In addition, magnetization can be altered during sediment retrieval, as the sample must remain both physically and magnetically undisturbed. Nevertheless, under favorable circumstances, some of these effects can be separated and others avoided so that an accurate paleomagnetic record can be recovered.

7.1. The NRM Recording Process

Our understanding of the processes involved in the magnetization of sediments remains incompletely understood even after many years of study. Apart from the redeposition experiment of Katari, Tauxe, and King (2000), which argues that the mechanical reorientation of magnetic grains after deposition is highly unlikely, it is generally assumed that sediments acquire their magnetization through a mechanism termed post-depositional remanent magnetization (PDRM). In this mechanism, the magnetization is locked-in at a specific depth below the sediment/water interface (the lock-in depth), where dewatering and compaction constrain the remobilization of the magnetic grains below that depth (e.g., Verosub, Ensley, & Ulrick, 1979). The magnetic grains thus record the intensity and direction of the ambient field at the moment they pass-through the lock-in depth or zone. This lock-in depth is dependent on the thickness of the mixed layer, where no lock-in occurs because of the mixing produced by bioturbation. Boudreau (1994, 1998) estimated a worldwide and environmentally invariant (i.e., independent of water depth or sedimentation rate) mean thickness of the mixed layer to be 9.8 ± 4.5 cm, but with variable minimum and maximum values, respectively 2 and 30 cm. On the other hand, the work of Trauth, Sarnthein, and Arnold (1997) or Smith and Rabouille (2002) suggest that particulate organic carbon fluxes to the seafloor is the primary factor controlling the depth of the mixed layer, with higher fluxes resulting in thicker mixed layers. Below the mixed layer, the magnetic grains are believed to follow a lock-in function of uncertain nature. The paleomagnetic signal measured from sediment cores thus results from the integration of geomagnetic field fluctuations, the depth of the mixed layer, and a lock-in process function. This function was previously modeled as linear (Bleil & von Dobeneck, 1999; Roberts & Winklhofer, 2004), exponential (Lovlie, 1976; Hamano, 1980; Otofuji & Sasajima, 1981; Denham & Chave, 1982; Kent & Schneider, 1995; Meynadier & Valet, 1996; Guyodo & Channell, 2002; Roberts & Winklhofer, 2004), cubic

(Roberts & Winklhofer, 2004), or sigmoidal (Channell & Guyodo, 2004) and introduces smoothing of the paleomagnetic record in the form of a low-pass filter (Teanby & Gubbins, 2000; Guyodo & Channell, 2002; Channell & Guyodo, 2004; Roberts & Winklhofer, 2004). In the recent study of Channell and Guyodo (2004), the authors showed that the magnetization was acquired in only a few centimeters when the magnetic grains passed below the mixed layer, highlighting the influence of the mixed layer thickness rather than the lock-in function itself to explain the apparent age offset between reversals in the Matuyama Chron. Moreover, according to modeling results (e.g., Vlag, Thouveny, & Rochette, 1997; Roberts & Winklhofer, 2004), the smoothing introduced by the lock-in process could remove excursions or other high frequency variations if the sedimentation rate is in the order of only a few cm/kyr. Finally, the recent study of Sagnotti, Budillon, Dinarès-Turell, Iorio, and Macrì (2005) suggests that the lock-in depth may vary through time, likely as the result of lithological factors (Bleil & von Dobeneck, 1999). Clearly, additional work is needed to fully understand the NRM recording process and its effect on the paleomagnetic record.

7.2. Magnetic Mineralogy

Among the more important factors controlling the quality of the paleomagnetic record is the magnetic mineralogy, its concentration, and preservation (e.g., Clement, Kent, & Opdyke, 1996). Magnetic minerals on the sea floor are generally derived from terrigenous or *in-situ* bacterial production, and biogenic magnetite is generally thought to be a poor paleomagnetic recorder (e.g., Schwartz, Lund, Hammond, Schwartz, & Wong, 1997). Sediments with high terrigenous content may therefore provide high quality paleomagnetic records and glaciated or previously glaciated terrains generally provide an excellent source of fine-grained magnetic minerals for paleomagnetic records. In contrast, highly organic sites may be poor targets for paleomagnetic records, though there are many complicating factors controlling the stability of magnetic minerals (e.g., Channell & Stoner, 2002; Sagnotti et al., 2003). In general, magnetite and titano-magnetite are the primary remanence carriers of high quality paleomagnetic records, especially if they have a fine and homogeneous grain size. Preservation is generally controlled by the degree of reductive diagenesis, with dissolution of magnetic oxides potentially the most important factor controlling the fidelity of a paleomagnetic record (Clement et al., 1996).

8. Development of Paleomagnetic Records

8.1. Directional Records

Inclinations and declinations can be read directly off the magnetometer or after some minor transformation of the x, y, and z magnetic moments. Yet, establishing reliability is often another story. For magnetostratigraphic data primarily applied to polarity stratigraphy, Opdyke and Channell (1996) established a set of reliability criteria following those advocated for paleomagnetic studies by Van der Voo (1990). Here, we discuss those criteria that most directly apply to marine sediment records, while emphasizing and expanding on those we believe to be the most important for Quaternary paleoceanographic studies.

For directional data, each discrete sample or individual measurement of a u-channel sample should undergo stepwise AF (thermal if appropriate) demagnetization. Orthogonal projections should be used for PCA analysis (e.g., Kirschvink, 1980) to calculate the component magnetization and the coercivity (unblocking) spectra should be determined. Blanket AF demagnetization, still relatively common for Holocene secular variation and shipboard magnetostratigraphic studies, should be used primarily for pilot studies and discouraged as definitive studies, unless no other reasonable options are available. MAD values should be used to assess the quality of the magnetization. Initially, these should be determined in a non-subjective manner, using a consistent demagnetization range and including all demagnetization steps (unless affected by a measurement error) for a particular sediment sequence. Selective use of demagnetization steps to assess the "best" component magnetization should only be done as a second step. Where possible, component inclination, declination, intensity at all demagnetization levels, and MAD values should be presented versus depth. For polarity stratigraphy, virtual geomagnetic pole (VGP) latitude should also be shown. Appropriate statistics should be shown when applicable. Evidence of magnetic mineralogy should be presented. Reversals should be antipodal. Multiple sections or replicate cores, where possible, should be studied. For high-resolution Holocene studies, attempts should be made to capture the upper sediments so that comparison and calibration with historical data (e.g., Jackson, Jonkers, & Walker, 2000) can be attempted.

8.2. Relative Paleointensity Determinations

Deriving proxy records of relative geomagnetic paleointensity from sediments involves normalizing the NRM by a magnetic parameter that compensates for change in the concentration of NRM-carrying grains. King, Banerjee, and Marvin (1983) put the use of ARM as a normalizer on a firm theoretical and empirical basis. According to these authors, the NRM/ARM ratio can be used as a paleo-intensity proxy if the NRM is a detrital remanent magnetization (DRM) and is carried by magnetite in the 1–15 μm single domain/pseudo-single domain (SD/PSD) grain-size range. The restricted grain-size range is due to the grain-size dependence of both ARM acquisition and NRM retention. Sub-micron grains and coarse multidomain (MD) magnetite grains can be efficient carriers of ARM but relatively inefficient carriers of stable DRM. According to the same authors, concentrations of magnetite should not vary downcore by more than a factor of 20–30 because of the effect of particle interactions on the efficiency of ARM acquisition. Demagnetization of both NRM and ARM in the NRM/ARM ratio serves to restrict the grain-size range contributing to both remanences.

Besides ARM, IRM and low field susceptibility (k) have been used as normalizers to derive paleointensity proxies from sediments, although the theoretical basis for the use of IRM or susceptibility (k) has not been adequately documented.

Empirically, IRM has often proved to be a better match to the NRM's coercivity than ARM and therefore a good choice as a normalizer (e.g., Stoner, Channell, Hillaire-Marcel, & Kissel, 2000). The RPI proxies can be calculated as means of NRM/IRM or NRM/ARM over a specific demagnetization range, as slopes on the plots of NRM versus ARM or IRM, or as subtracted vectors (e.g., $NRM_{25mT-50mT}/IRM_{25mT-50mT}$ or $NRM_{25mT-50mT}/ARM_{25mT-50mT}$). More complex methods have been advocated, like the pseudo-Thellier method of Tauxe, Pick, and Kok (1995). However, little difference has been found among the various methods when the sediments are high quality magnetic recorders (Valet & Meynadier, 1998), which can be assessed through studying the demagnetization behavior and is generally indicated by low MAD values. In practice, the choice of the normalizer depends on the response of the sediments to the acquisition and demagnetization of NRM, ARM, and IRM. The success of these techniques is illustrated by the proliferation of RPI studies over the last decade. This progress has been greatly accelerated by the development of the u-channel magnetometer (Weeks et al., 1993), which allows high-resolution studies as a practical undertaking and the extensive data acquisition required for RPI studies.

Determining whether normalized intensity actually reflects RPI changes is not always straightforward. Separation of geomagnetic and environmental factors may be difficult. Therefore, magnetic homogeneity has been considered a requirement for the development of trustworthy RPI records. Careful evaluation of the sediments, both in terms of their magnetic properties and the integrity of the recorded magnetization, are required. Criteria have been established that outline the types of sediments likely to provide reliable RPI records (e.g., Levi & Banerjee, 1976; King et al., 1983; Tauxe, 1993; Valet, 2003). However, it should be remembered, as with the NRM recording process, that our understanding is incomplete and that following these criteria will not guarantee a quality record, nor does it mean that only sediments that strictly follow these criteria can provide RPI estimates.

Following on the directional criteria listed above and those previously established for RPI studies (King et al., 1983; Tauxe, 1993; Valet, 2003), we will expand and add some recommendations for the construction of reliable RPI estimates. The basic idea is that the sediments should be good geomagnetic field recorders and should be geologically homogenous. Specific criteria that should be applied are: (1) the NRM of all samples should be studied by progressive AF demagnetization, permitting the removal of any viscous (low coercivity and unstable) magnetization and the determination of the component magnetization as well as the coercivity spectrum of the NRM. The stability of the magnetization and its characteristic remanence should be determined by orthogonal projections (Zijderveld plots; Figure 5) and PCA analysis. MAD values of less than 5 are preferable. Component magnetization with higher MAD values should be treated as suspect. (2) The magnetization should be carried by stable magnetite of PSD range. The mineralogy and grain size of the magnetic faction should be established (see Tauxe, 1993; Dunlop & Özdemir, 1997). (3) The sediments must be free of inclination error. This can notably be assessed by comparing the inclination record with the expected inclination for the latitude of the sampling site according to a GAD ($\tan I = 2 \tan \lambda$).

(4) Changes in magnetic concentration should vary by less than one order of magnitude, with rapid or abrupt changes treated with extra caution. These can be evaluated by looking at concentration dependent parameters such as NRM, ARM, IRM, and k. (5) The normalizer should activate the same magnetic assemblage that is responsible for the NRM acquisition. This can be assessed by demagnetization of the normalizer using the same steps as the NRM. Similar values at successive ratios provide a test of the coercivity match. The use of magnetic susceptibility as a normalizer should be carefully evaluated as it activates large MD magnetite grains and small (superparamagnetic) grains, both of which are not stable carriers of the NRM. (6) The RPI proxy should not be coherent with its normalizer and the RPI proxy should not be coherent with bulk rock magnetic parameters. This is often determined by cross-spectral analysis (Tauxe & Wu, 1990), or more recently by cross-wavelet analysis (Guyodo, Gaillot, & Channell, 2000). (7) Comparison between individual RPI records with other regional records provides replication of observations. Additional comparisons with regional stacks such as the North Atlantic paleointensity stack (NAPIS) (Laj, Kissel, Mazaud, Channell, & Beer, 2000), the South Atlantic paleointensity stack (SAPIS) (Stoner, Laj, Channell, & Kissel, 2002), or global stacks such as the high resolution global paleointensity stack (GLOPIS-75) (Laj, Kissel, & Beer, 2004), and the low resolution Sint-200 (Guyodo & Valet, 1996), Sint-800 (Guyodo & Valet, 1999), or Sint-2000 (Valet, Meynadier, & Guyodo, 2005) paleointensity stacks. Additionally, comparison with inverted cosmogenic isotope records can help assess the validity of the reconstructed paleo-intensity proxy. Finally, non-magnetic changes can also substantially affect the ability to derive a reliable RPI record. Recent improvements in non-destructive physical methods (e.g., St-Onge et al., this book) allow a rapid and precise visualisation of the cores or u-channels in order to evaluate the potential influence of cracks, coring deformation, turbidities, sand layers, or other disturbances of the paleomagnetic signal.

9. The Paleomagnetic Record as a Stratigraphic Tool

9.1. Geomagnetic Polarity Time Scale (GPTS)

The GPTS (Cande & Kent, 1995; Ogg & Smith, 2004) and magnetic polarity stratigraphy represent the stratigraphic backbone on which the geologic time-scale for the last 150 Myr is based. For a thorough review, see Opdyke and Channell (1996). Magnetic reversals are applicable to many types of geologic materials and their global synchronicity and environmental independence make them a supreme stratigraphic tool. Magnetic reversals are a non-periodic, possibly even stochastic process, resulting from an incompletely understood mechanism within the Earth's core. Reversals separate times of constant polarity with durations from 20 kyr to 50 Myr. Considering that reversals take about 5,000 yr on average (see Clement, 2004), they provide among the most precise methods of global correlation. Patterns of normal and reversed strata can be classified as polarity zones that can provide, in some circumstances, a distinct fingerprint match to the GPTS. These magnetostratigraphic polarity zones can consist of strata with a single polarity, alternating normal and reversed units or dominantly normal or reversed units with minor amounts of the opposite polarity. Correlation to the GPTS is only provided at the recorded reversal boundaries and, therefore, temporal estimates must be interpolated between reversal boundaries. In addition, reversals are non-unique, coming in either one of two forms, normal to reversed or reversed to normal. Cross-calibration is thus extremely important to uniquely identify a reversal, most often occurring with isotopic (e.g., Shackleton & Opdyke, 1973; Channell & Kleiven, 2000; Channell, Mazaud, Sullivan, Turner, & Raymo, 2002; Channell, Curtis, & Flower, 2004) or biostratigraphic data (e.g., Hays & Opdyke, 1967; Berggren, Kent, Swisher, & Aubry, 1995).

The naming of the first four magnetic polarity Chrons after prominent geomagnetists, Brunhes (normal), Matuyama (reversed), Gauss (normal), and Gilbert (reversed), spanning the past ~6 Myr, was put forward by Cox et al. (1963). These Chrons were identified based on dispersed radiometrically dated volcanic rocks, so no type localities were available. Shorter Subchrons were named after type localities, such as Jaramillo Creek in New Mexico or Olduvai Gorge in Tanzania. Beyond the Plio-Pliestocene, polarity Chrons are designated by numbers correlated to marine magnetic anomalies. The anomaly sequence of the South Atlantic was taken as a marine standard for the Late Cretaceous through the Cenozoic. The anomalies of the Cenozoic or "C-sequence" were numbered from 1 to 34 (oldest). The polarity Chron nomenclature has evolved progressively to accommodate revisions (LaBrecque, Kent, & Cande, 1977; Harland et al., 1982; Cande & Kent, 1992). The corresponding polarity Chrons (time) and polarity zones (stratigraphy) are prefaced by the letter C, with a suffix "n" denoting the younger normal polarity interval, or "r" denoting the older reversed polarity interval (e.g., Cande & Kent, 1992). When a major numbered polarity Chron is further subdivided, Subchrons are denoted by a suffix of a corresponding number polarity Chron. For the Plio-Pleistocene, the traditional names are used (Brunhes = C1n, Matuyama = C1r, Jaramillo Subchron = C1r.1n).

Starting with the pioneer work of Heirtzler, Dickson, Herron, Pitman, and Pichon (1968), the timescale for the GPTS has been derived based on assumptions that seafloor spreading at specific locations was constant or smoothly varying over long time intervals. Age calibration was achieved by fitting a smooth curve to nine-calibration levels for South Atlantic spreading history (Cande & Kent, 1992). The rest of the timescale is dated by interpolation between these tie-levels. The absolute ages of the tie-levels are much less exactly known than the relative length of the polarity intervals. The timescale has evolved with improvements in the resolution of magnetic anomalies, definition of oceanic block models, magnetostratigraphic correlation, and dating of calibration points. Recent updates, like that of Berggren et al. (1995) which was used by Cande and Kent (1995) and Ogg and Smith (2004), have incorporated improved ages for the calibration levels with many of these based on astronomical tuning (e.g., Shackleton, Berger, & Peltier, 1990; Hilgen, 1991; Lourens, Hilgen, Shackleton, Laskar, & Wilson, 2004).

9.2. Relative Paleointensity Stratigraphy

One of the most important developments in paleomagnetism over the last decade has been the demonstration that sediment normalized intensity records show globally coherent variations during times of constant polarity (e.g., Meynadier, Valet, Weeks, Shackleton, & Hegee, 1992; Tric et al., 1992; Stoner, Channell, & Hillaire-Marcel, 1995; Stoner et al., 2000; Guyodo & Valet, 1996; Guyodo & Valet, 1999; Laj et al., 2004; Valet et al., 2005; Yamazaki & Oda, 2005). The importance of these observations is that they suggest that the geomagnetic field can provide a stratigraphic tool between magnetic reversals, overcoming a major limitation of the GPTS, where correlation is only possible at reversal boundaries. In fact, high-resolution studies suggest that RPI records maybe globally coherent on timescales as short as a few thousand years (e.g., Stoner et al., 2000; Laj et al., 2004).

The principal challenge in the development of RPI stratigraphy is to define the "true" character of the record. Paleointensity cannot be predicted by theory or from numerical simulation, since the mechanisms involved in the geodynamo are not sufficiently constrained. Thus, our understanding of the record is based on continual observations and cross correlation with other dating and stratigraphic techniques. Comparison between RPI records from sediments and absolute paleointensity from thermally cooled materials (e.g., volcanic rocks and ceramic artifacts) can be used to calibrate the sediment record (e.g., Guyodo & Valet, 1999; Valet, 2003; Laj et al., 2004). In practice, however, the volcanic/sedimentary correlation is often hampered by the discontinuous nature of the volcanic record and the imprecision of available radiometric dating techniques. Comparison of sedimentary paleointensity records from different depositional environments and a detailed investigation of magnetic properties allow separation of geomagnetic and environmental signals. Distributed records from different parts of the globe are necessary to determine the characteristics of the global (as opposed to the local or regional) geomagnetic field. Stacking many individual records provides one method for determining the "true" character of the signal. Spurious features in individual records are averaged out by the stacking process. However, stacking also acts as a low pass filter, degrading the highest temporal resolution records in favor of those of lower resolution. Correlation imprecision tends to reinforce this. The use of low sedimentation rate records such as the global Sint-200 (Guyodo & Valet, 1996), Sint-800 (Guyodo & Valet, 1999), Sint-2000 (Valet et al., 2005), and regional EPAPIS-3000 (Yamazaki & Oda, 2005) paleointensity stacks have averaged out much of the high frequency variability. These records, however, do an excellent job at capturing the 10^4 yr variability of the geomagnetic field (Figure 6a), while providing excellent targets for stratigraphic correlation. Globally dispersed records from sediments that have accumulated at one order of magnitude higher (>10 cm/kyr) give a picture of the geomagnetic field with high amplitude features varying globally and generally coherently on a millennial scale (Stoner et al., 2000; Figure 7). The availability of an increased number of high resolution records and a more sophisticated stacking approach have recently been used to reduce the low pass filter effects (Laj et al., 2004) and provide GLOPIS-75 at a significantly higher resolution (Figure 6b).

One of the unique aspects of paleointensity is its relationship to cosmogenic isotopes. Assuming a constant flux of galactic cosmic rays, to a first order, the production rate of cosmogenic isotopes (e.g., ^{10}Be, ^{14}C, ^{36}Cl) reflects variations in the strength of the Earth's and Sun's magnetic fields, with stronger (weaker)

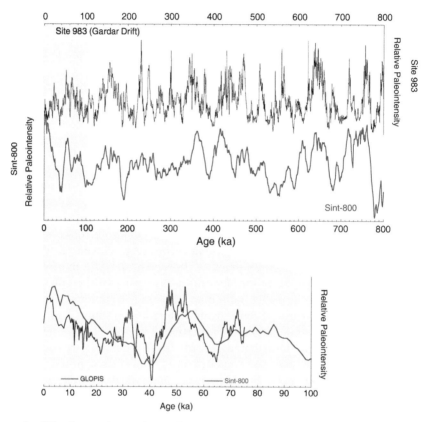

Figure 6 (A) Comparison of North Atlantic ODP site 983 high-resolution relative paleointensity record (Channell et al., 1997; Channell, Hodell, McManus, & Lehman, 1998; Channell & Kleiven, 2000) with the low-resolution SINT-800 stack (Guyodo & Valet, 1999). (B) Comparison between the high-resolution global relative paleointensity stack (GLOPIS; Laj et al., 2004) and the low-resolution SINT-800 stack. The Laschamp excursion is associated with the intensity low at ~40 ka.

Figure 7 Geomagnetic relative paleointensity records and cosmogenic isotope records. From the top to the bottom: the ^{10}Be flux from the GISP2 ice core (72°36′N/38°30′W; Finkel & Nishiizumi, 1997) and ^{36}Cl flux from the GRIP ice core (72°34′N/37°37′W; Baumgartner et al., 1998) placed on the GISP official chronology. Core MD95-2024 (50°12.26′N/45°41.14′W; Stoner et al., 2000) paleointensity record. North Atlantic ODP Site 984 (60.4°N/23.6°W; Channell 1999) paleointensity record, the North Atlantic Paleointensity Stack (NAPIS; 33–67°N/45°W–4°E; Laj et al., 2000), the Lake Baikal paleointensity record (Siberia, Russia; Peck, King, Colman, & Kravchinsky, 1996), and the South Atlantic Paleointensity Stack (SAPIS; 41–47°S/6–10°E; Stoner, Laj, Channell, & Kissel, 2002). Chronology for all records derived by tuning to the GISP2 official chronology. The offset in the NAPIS record results from an additional step of tuning to SPECMAP at ~60 kyr.

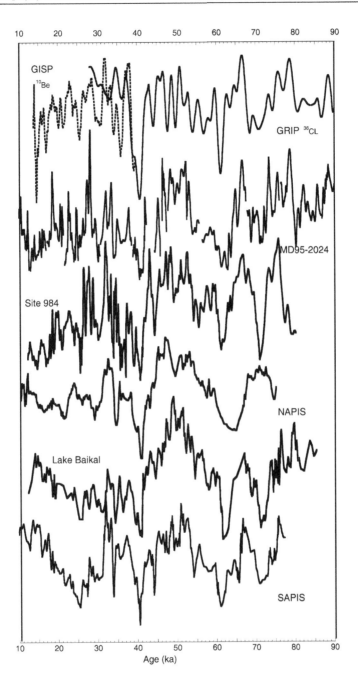

magnetic fields resulting in reduced (increased) production. In general, geo-magnetic and solar shielding of galactic cosmic rays is well known and reason-ably well understood (e.g., Lal, 1988; Masarik & Beer, 1999; Beer, 2000). What is not so well understood are the timescales of the two components. Recent comparisons between proxies of cosmogenic isotope production rates and RPI (Mazaud, Laj, & Bender, 1994; Robinson, Raisbeck, Yiou, Lehman, & Laj, 1995; Baumgartner et al., 1998; Stoner et al., 2000; Wagner et al., 2000; Beer et al., 2002; Carcaillet, Thouveny, & Bourles, 2003; Christl, Strobl, & Mangini, 2003; Hughen et al., 2004; Thouveny, Carcaillet, Moreno, Leduc, & Nérini, 2004; Muscheler, Beer, Kubik, & Synal, 2005) show substantial agreement. Beer et al. (2002) suggests that variations longer than 2,000 yr in proxy records of cosmo-genic isotope production rates can be attributed to variations in the intensity of the geomagnetic field, with higher frequency variations attributed to solar variabil-ity. Therefore, RPI provides a viable technique for correlating sediment to cosmogenic isotope records from ice cores (Mazaud et al., 1994; Stoner et al., 2000) and sediment archives (Frank et al., 1997; Hughen et al., 2004) (Figure 7). New observations from RPI records from the St. Lawrence Estuary (St-Onge, Stoner, & Hillaire-Marcel, 2003) and Scandinavia (Snowball & Sandgren, 2002, 2004) suggest that millennial and even centennial scale correlations between cosmogenic isotopes and paleointensity could be used at these timescales for stratigraphic purposes.

9.3. Excursions as a Stratigraphic Tool

Barbetti and McElhinny (1972) defined an excursion as a VGP displacement of more than 40° from the geographic pole, whereas a reversal excursion (Merrill & McFadden, 1994) reflects a VGP in the opposite hemisphere, the latter often being a more useful paleomagnetic definition. A decade ago, the existence of geomagnetic excursions within the Brunhes Chron was treated as suspect. Only two "crypto-chrons" are recognized by Cande and Kent (1992, 1995) from the Brunhes and Matuyama Chrons marine magnetic anomaly data, with estimated ages of 500 ka and 1.2 Ma (Cobb Mt. Subchron). Recent observations from sediment records suggest that they may be much more common (Langereis, Dekkers, de lange, Paterne, & van Santvoort, 1997; Lund et al., 1998; Lund, Acton, Clement, Okada, & Williams, 2001a, 2001b), with 12–17 now advocated in the Brunhes and at least another six in the Matuyama (e.g., Channell et al., 2002). The volcanic record also appears to bear this out (Singer et al., 2002), yet a strong stratigraphic understanding of most of these is presently missing. As a correlation tool, excursions have unique potential to provide short duration (<2,000 yr) "golden spikes" in the geologic record. However, only two excursions (Laschamp and Iceland Basin Events) are widely considered as globally synchronous phenomena (Channell, Stoner, Hodell, & Charles, 2000; Mazaud et al., 2002; Stoner, Channell, Hodell, & Charles, 2003), with many of the details still unknown. Many pitfalls therefore remain for their common and uncontrolled use as a stratigraphic tool without other significant stratigraphic calibration. Similarly to a reversal, the lack of any distinct fingerprint to uniquely identify an excursion is a significant stratigraphic drawback. This is important as excursions may occur in bundles closely spaced in time (Lund, Stoner, Acton, & Channell, 2006; Blanchet, Thouveny, & de Garidel-Thoron, 2006). And, if they result, as is generally thought, from non-dipole components of the geomagnetic field, then their directional signature would be distinctly different for different parts of the globe (e.g., Merrill & McFadden, 2005). However, a recent study suggests that they may be much simpler (Laj, Kissel, & Roberts, 2006), with a potentially global signature.

A more fruitful stratigraphic approach should result by combining excursions with RPI. Valet and Meynadier (1993) pointed out that lows in RPI during the Brunhes Chron from equatorial Pacific (ODP Leg 138) sediments appear synchronized with directional excursions detected elsewhere. The sedimentation rates in Leg 138 sediments and those used to construct the global stack Sint-800 (Guyodo & Valet, 1999) are, in general, too low to record geomagnetic excursions. Direct correlation of excursions and paleointensity is thus incomplete and is only convincingly achieved for a few excursions at a few locations (e.g., Lehman et al., 1996; Channell, Hodell, & Lehman, 1997; Roberts, Lehman, Weeks, Verosub, & Laj, 1997; Channell, 1999; Laj et al., 2000; Channell et al., 2000; Stoner et al., 2003; Thouveny et al., 2004; Lund, Schwartz, Keigwin, & Johnson, 2005). Therefore, our expanding knowledge of the paleointensity record provides an improving template with which to search and a guide that should be used when employing excursions as a stratigraphic tool.

An added difficulty with excursions results from the fact that they occur during times of low field intensity. The weak magnetizing field increases the likelihood that the excursional interval will be overprinted from the stronger post-excursional field (e.g., Coe & Liddicoat, 1994). In addition, it has been suggested that these short-lived phenomena are easily smoothed out of the record by the NRM recording process in low accumulation rate sediments, as discussed above and reviewed by Roberts and Winklhofer (2004). The presence of detailed oxygen isotope records has also been instrumental in improving the age estimates for excursion records in deep-sea sediments, but it will likely take a considerable international and multidisciplinary effort before a complete record of excursions within the Brunhes is successfully mapped out.

As mentioned, one excursion that can be used as a stratigraphic golden spike is the Laschamp event. Bonhommet and Babkine (1967) originally recognized excursional directions in Quaternary lava flows from the Chaine des Puys, France. The Laschamp excursion has now been recognized in many locations and represents one of the most prominent events to have affected Earth over the last 100 kyr. The Laschamp excursion has been found in sediments of the North Atlantic, South Atlantic, Indian, and Pacific oceans. It is generally associated with a broad paleointensity low. The best age estimates come from correlation of sediments to the Greenland Summit ice cores that place the event's occurrence during interstadial (IS) 10 with a GISP2 age of ~40.5 kyr. The type section lavas are independently dated at 40.4 ± 4 kyr (Guillou et al., 2004) and support the ice core derived dates. Further supporting this correlation are the observations of a major increase in the flux of cosmogenic isotopes associated with IS10 in Greenland Ice cores (e.g., Yiou et al., 1997; Wagner et al., 2000) and marine sediments (e.g., Robinson et al., 1995; Carcaillet, Bourles, & Thouveny, 2004). This event has been recently used to support correlation of paleoclimatic records to the GISP2 chronology (Mazaud et al., 2002; Lamy et al., 2004; Hill, Flower, Quinn, Hollander, & Guilderson, 2006). The fact that this event is within the radiocarbon limit has improved its stratigraphy, though its incompletely understood effect on the production of ^{14}C significantly complicates radiocarbon dating during this time interval.

9.4. Paleomagnetic Secular Variation

Observations of the historical (~the last 400 yr) geomagnetic field document continuous directional secular variations with periods on a centennial scale (Jackson et al., 2000). Paleomagnetic observations primarily from Holocene lake sediments document additional ~1,000–3,000 yr long geomagnetic directional shifts termed paleomagnetic secular variation (PSV) (Mackereth, 1971; Creer, Thompson, Moyneaux, & Mackereth, 1972; Thompson, 1973) that are regionally consistent, but different between regions (Thompson, 1984). Recent studies show that marine sediments can record these variations (Figure 8), providing a potentially continuous regional centennial to millennial stratigraphic tool (Lund & Keigwin, 1994; Lund & Schwartz, 1996; Kotilainen, Saarinen, & Winterhalter, 2000; Verosub, Harris, & Karlin, 2001; St-Onge et al., 2003; St-Onge, Piper, Mulder, Hillaire-Marcel, & Stoner, 2004; Stoner et al., 2007). Because most of the initial work was done in lake sediments, marine/terrestrial correlation (e.g., Ólafsdóttir, Stoner, Geirsdóttir, Miller, & Channell, 2005) has significant potential that, in some cases, maybe better than what can be achieved through radiocarbon dating alone (Lund, 1996; Hagstrum & Champion, 2002; Stoner et al., 2007).

Most PSV studies have been concentrated in the Holocene, though its potential impact as a correlation tool is possible for older records. For example, recent studies from Blake/Bahamas Outer Ridge (ODP Sites 1061) and the Bermuda Rise Site 1063 sediments document reproducible directional PSV records, at more than 1,000 km distance, for the interval 15,000–50,000 yr BP (Lund et al., 2005).

10. SOME PERSPECTIVES

As a community, we are just beginning to document geomagnetic change during times of constant polarity. Though there is still much to learn, progress is being made at a rapid rate. Much of this growing new understanding is fueled by the availability of new instruments (i.e., the u–channel magnetometer) that allow rapid measurements of sediment sequences, providing new observational constraints from previously untapped archives. Additionally, both numerical geodynamo and data based spherical harmonic models are also making substantial progress. Each new record is potentially an important new observation, providing new insights toward defining the "true" record, while the modeling work allows us to better interpret these observations. Continual observations and cross correlation with other dating and stratigraphic techniques allow the "true" record to be uncovered

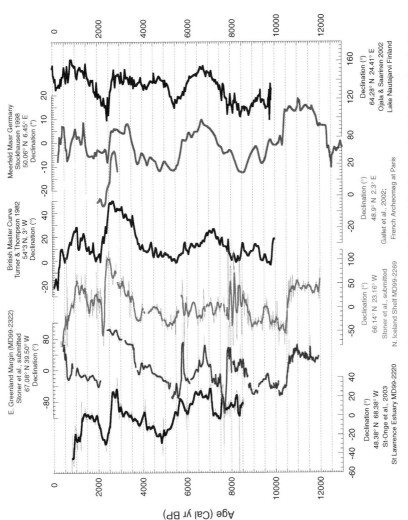

Figure 8 Comparison of Holocene paleomagnetic secular variation records from North America–North Atlantic–Europe. Only declination is shown. From left to right in each panel: St. Lawrence Estuary (MD99-2220; St-Onge et al., 2003); East Greenland Margin (MD99-2322; Stoner et al., 2007); North Iceland shelf (MD99-2269; Stoner et al., 2007); the western European archeomagnetic compilation (Bucur, 1994; Gallet, Genevey, & Le Goff, 2002); the British master curve (Thompson & Turner, 1979; Turner & Thompson, 1982); the German maar lake records (Stockhausen, 1998); and a Finish varved lake record (Ojala & Saarinen, 2002). All records are on their own chronologies and calibrated to calendar years using Stuiver et al. (1998).

piece by piece. The spatial coherence of the geomagnetic field allows prior information to be brought to each new observation and provides many stratigraphic opportunities. Yet, in many ways we are just scratching the surface. Global coverage over all timescales is still only fair, and as for high-resolution data detailed records are only found in a few restricted regions. Holocene observations suggest that if we are to understand the full dynamic range of the geomagnetic record, ultra high-resolution (>1 m/kyr) sediments, both within the Holocene and beyond, need to be targeted. As we seek to uncover the "true" geomagnetic record, our knowledge of geomagnetic stratigraphy, its applications, and its resolution will continue to improve, providing a rich future for paleomagnetic research.

REFERENCES

Acton, G. D., Okada, M., Clement, B. M., Lund, S. P., & Williams, T. (2002). Magnetic overprints in ocean sediment cores and their relationship to shear deformation caused by piston coring. *Journal of Geophysical Research, 107*(B4), EPM3-1–EPM3-15.

Barbetti, M., & McElhinny, M. (1972). Evidence of a geomagnetic excursion 30,000 yr BP. *Nature, 239*, 327–330.

Baumgartner, S., Beer, J., Masarik, J., Wagner, G., Meynadier, L., & Synal, H.-A. (1998). Geomagnetic modulation of the ^{36}Cl Flux in the GRIP Ice Core, Greenland. *Science, 279*, 1330–1332.

Beer, J. (2000). Long-term indirect indices of solar variability. *Space Science Reviews, 94*, 53–66.

Beer, J., Muscheler, R., Wagner, G., Laj, C., Kissel, C., Kubic, P. W., & Synal, H-. A. (2002). Cosmogenic nuclides during isotopic stages 2 and 3. *Quaternary Science Review, 21*, 1129–1139.

Berggren, W. A., Kent, D. V., Swisher, C. C., & Aubry, M. P. (1995). A revised Cenozoic geochronology and chronostratigraphy in time scales and global stratigraphic correlations: A unified temporal framework for an historical geology. In: W. A. Berggren, D. V. Kent, M.-P. Aubry & J. Hardenbol (Eds), *Geochronology, timescales, and stratigraphic correlation* (Vol. 54, pp. 129–212). SEPM Special Publication.

Blackett, P. M. S. (1952). A negative experiment relating to magnetism and the Earth's rotation. *Philosophical Transactions of the Royal Society of London, Series A, 245*, 309–370.

Blanchet, C. L., Thouveny, N., & de Garidel-Thoron, T. (2006). Evidence for multiple paleomagnetic intensity lows between 30 and 50 ka BP from a western Equatorial Pacific sedimentary sequence. *Quaternary Science Reviews, 25*(9–10), 1039–1052.

Bleil, U., & von Dobeneck, T. (1999). Geomagnetic events and relative paleointensity records: Clues to high-resolution paleomagnetic chronostratigraphies of the Late Quaternary marine sediments? In: G. Fisher & G. Wefer (Eds), *Use of proxies in paleoceanography: Examples from the South Atlantic* (pp. 635–654). Berlin: Springer-Verlag.

Bonhommet, N., & Babkine, J. (1967). Sur la présence d'aimantations inversées dans la Chaîne des Puys. *Comptes Rendus de l'Académie des Sciences, 264*, 92–94.

Boudreau, B. P. (1994). Is burial velocity a master parameter for bioturbation. *Geochimica et Cosmochimica Acta, 58*, 1243–1249.

Boudreau, B. P. (1998). Mean mixed depth of sediments: The wherefore and the why. *Limnology and Oceanography, 43*, 524–526.

Brachfeld, S., Kissel, C., Laj, C., & Mazaud, A. (2004). Behavior of u-channels during acquisition and demagnetization of remanence: Implications for paleomagnetic and rock magnetic measurements. *Physics of the Earth and Planetary Interiors, 145*, 1–8.

Brunhes, B. (1906). Recherches sur la direction d'aimantation des roches volcaniques. *Journal de Physique, 5*, 705–724.

Bucur, I. (1994). The direction of the terrestrial magnetic field in France during the last 21 centuries. *Physics of the Earth and Planetary Interiors, 87*, 95–109.

Bullard, E. C. (1949). The magnetic field within the Earth. *Proceedings of the Royal Society of London,* *197*(A), 433–453.

Butler, R. F. (1992). *Paleomagnetism: Magnetic domains to geologic terranes.* Oxford: Blackwell.

Cande, S. C., & Kent, D. V. (1992). A new geomagnetic polarity timescale for the late Cretaceous and Cenozoic. *Journal of Geophysical Research, 97,* 13917–13951.

Cande, S. C., & Kent, D. V. (1995). *Revised calibration of the geomagnetic polarity timescale for the Late Cretaceous and Cenozoic, 100,* 6093–6095.

Carcaillet, J. T., Bourles, D. L., & Thouveny, N. (2004). Geomagnetic dipole moment and ^{10}Be production rate intercalibration from authigenic ^{10}Be/^{9}Be for the last 1.3 Ma. *Geochemistry, Geophysics and Geosystems, 5,* Q05006, doi:10.1029/2003GC000641.

Carcaillet, J. T., Thouveny, N., & Bourles, D. L. (2003). Geomagnetic moment instability between 0.6 and 1.3 Ma from cosmonuclide evidence. *Geophysical Research Letters, 30,* 1792, doi:10.1029/2003GL017550.

Channell, J. E. T. (1999). Geomagnetic paleointensity and directional secular variation at Ocean Drilling Program (ODP) Site 984 (Bjorn Drift) since 500 ka: Comparisons with ODP Site 983 (Gardar Drift). *Journal of Geophysical Research, 104,* 22937–22951.

Channell, J. E. T. (2006). Late Brunhes polarity excursions (Mono Lake, Laschamp, Iceland Basin and Pringle Falls) recorded at ODP Site 919 (Irminger Basin). *Earth and Planetary Science Letters, 244,* 378–393.

Channell, J. E. T., Curtis, J. H., & Flower, B. P. (2004). The Matuyama-Brunhes boundary interval (500–900 ka) in North Atlantic drift sediments. *Geophysical Journal International, 158,* 489–505.

Channell, J. E. T., & Guyodo, Y. (2004). The Matuyama Chronozone at ODP Site 982 (Rockall Bank): Evidence for decimeter-scale magnetization lock-in depths. In: J. E. T Chanell, D. V. Kent, W. Lowrie & J. Meert (Eds). *Timescales of the Internal Geomagnetic Field* (pp. 205–219). AGU Geophysical Monograph Series 145.

Channell, J. E. T., Hodell, D. A., & Lehman, B. (1997). Relative geomagnetic paleointensity and ^{18}O at ODP Site 983 (Gardar Drift, North Atlantic) since 350 ka. *Earth and Planetary Science Letters, 153,* 103–118.

Channell, J. E. T., Hodell, D. A., McManus, J., & Lehman, B. (1998). Orbital modulation of the Earth's magnetic field intensity. *Nature, 394,* 464–468.

Channell, J. E. T., & Kleiven, H. F. (2000). Geomagnetic palaeointensities and astrochronological ages for the Matuyama-Brunhes boundary and the boundaries of the Jaramillo Subchron: Palaeomagnetic and oxygen isotope records from ODP Site 983. *Philosophical Transactions of the Royal Society of London, Series A, 358,* 1027–1047.

Channell, J. E. T., Mazaud, A., Sullivan, P., Turner, S., & Raymo, M. E. (2002). Geomagnetic excursions and paleointensities in the 0.9–2.15 Ma interval of the Matuyama Chron at ODP Site 983 and 984 (Iceland Basin). *Journal of Geophysical Research, 107*(B6), 10.1029/2001JB000491.

Channell, J. E. T., Sato, T., & Malone, M., et al. (2005). IODP Expedition 303, Initial Reports, Integrated Ocean Drilling Program, (in press).

Channell, J. E. T., & Stoner, J. S. (2002). Plio-Pleistocene magnetic polarity stratigraphies and diagenetic magnetite dissolution at ODP Leg 177 Sites (1089, 1091, 1093 and 1094). *Marine Micropaleontology, 45,* 269–290.

Channell, J. E. T., Stoner, J. S., Hodell, D. A., & Charles, C. D. (2000). Geomagnetic paleointensity for the last 100 kyr from the sub-Antarctic South Atlantic: A tool for interhemispheric correlation. *Earth and Planetary Science Letters, 175,* 145–160.

Christl, M., Strobl, C., & Mangini, A. (2003). Beryllium-10 in deep-sea sediments: A tracer for the Earth's magnetic field intensity during the last 200,000 years. *Quaternary Science Reviews, 22,* 725–739.

Clement, B. M. (2004). Dependence of the duration of geomagnetic polarity reversals on site latitude. *Nature, 428,* 637–640.

Clement, B. M., Kent, D. V., & Opdyke, N. D. (1996). A synthesis of magnetostratigraphic results from Pliocene-Pleistocene sediments cored using the hydraulic piston corer. *Paleoceanography, 11,* 299–308.

Coe, R. S., & Liddicoat, J. C. (1994). Overprinting of natural magnetic remanence in lake sediments by a subsequent high-intensity field. *Nature*, *367*, 57–59.

Cogné, J. P. (2003). PaleoMac: A Macintosh^TM application for treating paleomagnetic data and making plate reconstructions. *Geochemistry, Geophysics and Geosystems*, *4*, doi:10.1029/2001GC000227.

Collinson, D. W. (1983). *Methods in rock magnetism and palaeomagnetism: Techniques and instrumentation.* New York: Chapman and Hall.

Cox, A., Doell, R. R., & Dalrymple, G. B. (1963). Geomagnetic polarity epochs and Pliestocene geochronometry. *Nature*, *198*, 1049–1051.

Creer, K. M., Thompson, R., Moyneaux, L., & Mackereth, F. J. H. (1972). Geomagnetic secular variation recorded in the stable magnetic remanence of recent sediments. *Earth Planetary Science Letters*, *14*, 115–127.

Denham, C. R., & Chave, A. D. (1982). Detrital remanent magnetization: Viscosity theory of the lock-in zone. *Journal of Geophysical Research*, *87*, 7126–7130.

Doell, R. R., & Dalrymple, G. B. (1966). Geomagnetic polarity epochs: A new polarity event and the age of the Brunhes-Matuyama boundary. *Science*, *152*, 1060–1061.

Dunlop, D. J., & Özdemir, Ö. (1997). *Rock magnetism, fundamentals and frontiers. Cambridge studies in magnetism.* Cambridge: Cambridge University Press.

Finkel, R. C., & Nishiizumi, K. (1997). Beryllium 10 concentrations in the Greenland Ice Sheet Project 2 ice core from 3–40 ka. *Journal of Geophysical Research*, *102*, 26699–26706.

Foster, J. H. (1966). A paleomagnetic spinner magnetometer using a flux gate gradiometer. *Earth and Planetary Science Letters*, *1*, 463–466.

Frank, M., Schwarz, B., Baumann, S., Kubik, P., Suter, M., & Mangini, A. (1997). A 200 kyr record of cosmogenic radionuclide production rate and geomagnetic field intensity from ^{10}Be in globally stacked deep-sea sediments. *Earth and Planetary Science Letters*, *149*, 121–129.

Gallet, Y., Genevey, A., & Le Goff, M. (2002). Three millennia of directional variations of the Earth's magnetic field in western Europe as revealed by archeological artifacts. *Physics of the Earth and Planetary Interiors*, *131*, 81–89.

Goree, W. S., & Fuller, M. D. (1976). Magnetometers using r-f driven SQUIDs and their application in rock magnetism and paleomagnetism. *Reviews of Geophysics and Space Physics*, *14*, 591–608.

Gradstein, F. M., Ogg, J. G., & Smith, A. G. (2004). *A geologic time scale 2004.* Cambridge, UK: Cambridge University Press.

Guillou, H., Singer, B. S., Laj, C., Kissel, C., Scaillet, S., & Jicha, B. R. (2004). On the age of the Laschamp geomagnetic excursion. *Earth and Planetary Science Letters*, *227*, 331–341.

Guyodo, Y., & Channell, J. E. T. (2002). Effects of variable sedimentation rates and age errors on the resolution of sedimentary paleointensity records. *Geochemistry, Geophysics and Geosystems*, *3*, doi: 10.1029/2001GC000211.

Guyodo, Y., Channell, J. E. T., & Thomas, R. (2002). Deconvolution of u-channel paleomagnetic data near geomagnetic reversals and short events. *Geophysical Research Letters*, *29*, 1845, doi:10.1029/2002GL014963.

Guyodo, Y., Gaillot, P., & Channell, J. E. T. (2000). Wavelet analysis of relative geomagnetic paleointensity at ODP Site 983. *Earth and Planetary Science Letters*, *184*, 109–123.

Guyodo, Y., & Valet, J.-P. (1996). Relative variations in geomagnetic intensity from sedimentary records: The past 200,000 years. *Earth and Planetary Science Letters*, *143*, 23–36.

Guyodo, Y., & Valet, J. P. (1999). Global changes in intensity of the Earth's magnetic field during the past 800 kyr. *Nature*, *399*, 249–252.

Hagstrum, J. T., & Champion, D. E. (2002). A Holocene paleosecular variation record from ^{14}C-dated volcanic rocks in western North America. *Journal of Geophysical Research*, *107*, doi: 10.1029/2001JB000524.

Hamano, Y. (1980). An experiment on the post-depositional remanent magnetization in artificial and natural sediments. *Earth and Planetary Sciences Letters*, *51*, 221–232.

Harland, W. B., Cox, A. V., Llewellyn, P. G., Picton, G. A. G., Smith, A. G., & Walters, R. (1982). *A geologic time scale.* New York, USA: Cambridge University Press.

Harrison, C. G. A., & Funnell, B. M. (1964). Relationship of paleomagnetic reversal and micro-paleontology in two late Cenozoic cores from the Pacific Ocean. *Nature, 204*, 566.

Hays, J. D., & Opdyke, N. D. (1967). Antarctic radiolaria, magnetic reversals, and climate change. *Science, 158*, 1001–1011.

Heirtzler, J. R., Dickson, G. O., Herron, E. M., Pitman, W. C., III, & Le Pichon, X. (1968). Marine magnetic anomalies, geomagnetic field reversals, and motions of the ocean floor and continents. *Journal of Geophysical Research, 73*, 2119–2136.

Hilgen, F. J. (1991). Astronomical calibration of Gauss to Matuyama sapropels in the Mediteranean and implication for the geomagnetic polarity time scale. *Earth and Planetary Science Letters, 104*, 226–244.

Hill, H. W., Flower, B. P., Quinn, T. M., Hollander, D. J., & Guilderson, T. P. (2006). Laurentide Ice Sheet meltwater and abrupt climate change during the last glaciation. *Paleoceanography, 21*, PA1006, doi:10.1029/2005PA001186.

Hospers, J. (1955). Rock magnetism and polar wandering. *Journal of Geology, 63*, 59–74.

Hughen, K., Lehman, S., Southon, J., Overpeck, J., Marchal, O., Herring, C., & Turnbull, J. (2004). ^{14}C activity and global carbon cycle changes over the past 50,000 years. *Science, 303*, 202–207.

Jackson, A., Jonkers, A. R. T., & Walker, M. R. (2000). Four centuries of geomagnetic secular variation from historical records. *Philosophical Transactions of the Royal Society of London, Series A, 358*, 957–990.

Johnson, E. A., Murphy, T., & Torreson, O. W. (1948). Pre-history of the Earth's magnetic field. *Terrestrial Magnetism and Atmospheric Electricity, 53*, 349–372.

Jones, C. H. (2002). User-driven integrated software lives: "PaleoMag" paleomagnetics analysis on the MacIntosh. *Computers and Geosciences, 28*, 1145–1151.

Jonkers, A. R. T. (2003). *Earth's magnetism in the age of sail.* Baltimore, London: The Johns Hopkins University Press.

Katari, K., Tauxe, L., & King, J. (2000). A reassessment of post-depositional remanent magnetism: Preliminary experiments with natural sediments. *Earth and Planetary Science Letters, 183*, 147–160.

Kent, D. V., & Schneider, D. A. (1995). Correlation of paleointensity variation records in the Brunhes/Matuyama polarity transition interval. *Earth and Planetary Sciences Letters, 129*, 135–144.

Khramov, A. N. (1955). Study of remanent magnetization and the proble of stratigraphic correlation and subdivision of non-fossiliferous strata. *Akademiya Nauk SSR, 100*, 551–551 (in Russian).

Khramov, A. N. (1957). Paleomagnetism as a basis for a new technique of sedimentary rock correlation and subdivision. *Akademiya Nauk SSR, 112*, 849–852 (in Russian).

King, J. W., Banerjee, S. K., & Marvin, J. A. (1983). A new rock magnetic approach to selecting sediments for geomagnetic paleointensity studies: Application to paleointensity for the last 4000 years. *Journal of Geophysical Research B, 88*, 5911–5921.

Kirschvink, J. L. (1980). The least-squares line and plane and the analysis of paleomagnetic data. *Geophysical Journal of the Royal Astronomical Society, 62*, 699–718.

Kotilainen, A. T., Saarinen, T., & Winterhalter, B. (2000). High-resolution paleomagnetic dating of sediments deposited in the central Baltic Sea during the last 3000 years. *Marine Geology, 166*, 51–64.

LaBrecque, J. L., Kent, D. V., & Cande, S. C. (1977). Revised magnetic polarity time-scale for the Late Cretaceous and Cenozoic time. *Geology, 5*, 330–335.

Laj, C., Kissel, C., & Beer, J. (2004). High resolution global paleointensity stack since 75 kyr (GLOPIS-75) calibrated to absolute values. In: J. E. T. Channell, D. V. Kent, W. Lowrie & J. Meert, (Eds). *Timescales of the internal geomagnetic field* (pp. 255–266). AGU Geophysical Monograph Series 145.

Laj, C., Kissel, C., Mazaud, A., Channell, J. E. T., & Beer, J. (2000). North Atlantic paleointensity stack since 75 ka (NAPIS-75) and the duration of the Laschamp event. *Philosophical Transactions of the Royal Society of London, Series A, 358*, 1009–1025.

Laj, C., Kissel, C., & Roberts, A. P. (2006). Geomagnetic field behavior during the Iceland Basin and Laschamp geomagnetic excursions: A simple transitional field geometry? *Geochemistry, Geophysics and Geosystems, 7*, Q03004, doi:10.1029/2005GC001122.

Lal, D. (1988). Theoretically expected variations in the terrestrial cosmic-ray production rates of isotopes. In: G. C. Castagnoli (Ed.), *Theoretically expected variations in the terrestrial cosmic-ray production rates of isotopes* (Vol. XCV, pp. 215–233). Amsterdam: North-Holland.

Lamy, F. J., Kaiser, U., Ninnemann, D., Hebbeln, H. W., Arz, H. W., & Stoner, J. S. (2004). Antarctic timing of surface water changes off Chile and Patagonian ice-sheet response. *Science, 304,* 1959–1962.

Langereis, C. G., Dekkers, M. J., de lange, G. J., Paterne, M., & van Santvoort, P. J. M. (1997). Magnetostratigraphy and astronomical calibration of the last 1.1 Myr from an eastern Mediterranean piston core and dating of short events in the Brunhes. *Geophysical Journal International, 129,* 75–94.

Lehman, B., Laj, C., Kissel, C., Mazaud, A., Paterne, M., & Labeyrie, L. (1996). Relative changes of the geomagnetic field intensity during the last 280 kyear from piston cores in the Azores area. *Physics of the Earth and Planetary Interiors, 93,* 269–284.

Levi, S., & Banerjee, S. K. (1976). On the possibility of obtaining relative paleointensities from lake sediments. *Earth and Planetary Science Letters, 29,* 219–226.

Lourens, L., Hilgen, F., Shachleton, N. J., Laskar, J., & Wilson, D. (2004). The Neogene Period. In: F. M. Gradstein, J. Ogg & A. G. Smith (Eds), *A geologic time scale 2004* (pp. 409–440). Cambridge, UK: Cambridge University Press.

Lovlie, R. (1976). The intensity pattern of post-depositional remanence acquired in some marine sediments deposited during a reversal of the external magnetic field. *Earth and Planetary Sciences Letters, 30,* 209–214.

Lund, S. P. (1996). A comparison of paleomagnetic secular variation records from North America. *Journal of Geophysical Research, 101,* 8007–8024.

Lund, S. P., Acton, G., Clement, B., Hastedt, M., Okada, M., Williams, T. & ODP Leg 172 Scientific Party (1998). Geomagnetic field excursions occurred often during the last million years. *EOS Transactions, AGU, 79*(14), 178–179.

Lund, S. P., Acton, G. D., Clement, B., Okada, M., & Williams, T. (2001a). Brunhes chron magnetic field excursions recovered from Leg 172 sediments. In: L. D. Keigwin, D. Rio, G. D. Acton & E. Arnold (Eds). *Proceedings of the Ocean Drilling Project,* Scientific Results (Vol. 172) (http://www-odp.tamu.edu/publications/172_SR/chap_10/chap_10.htm).

Lund, S. P., Acton, G., Clement, B., Okada, M., & Williams, T. (2001b). Paleomagnetic records of Stage 3 excursions from ODP Leg 172. In: L. D. Keigwin, D. Rio, G. D. Acton & E. Arnold (Eds). *Proceedings of the Ocean Drilling Project,* Scientific Results (Vol. 172) (http://www-odp.tamu.edu/publications/172_SR/chap_11/chap_11.htm).

Lund, S. P., & Keigwin, L. (1994). Measurement of the degree of smoothing in sediment paleomagnetic secular variation records: An example from late Quaternary deep-sea sediments of the Bermuda Rise, western North Atlantic Ocean. *Earth and Planetary Science Letters, 122,* 317–330.

Lund, S. P., & Schwartz, M. (1996). Paleomagnetic secular variation as a high-resolution chronostratigraphic tool in Quaternary paleoclimatic studies of lacustrine and deep-sea sediments. *GSA Annual Meeting, Abstracts, 28,* A-232.

Lund, S. P., Schwartz, M., Keigwin, L., & Johnson, T. (2005). Deep-sea sediment records of the Laschamp geomagnetic field excursion (41,000 calendar years before present). *Journal of Geophysical Research, 110,* B04101, doi:10.1029/2003JB002943.

Lund, S., Stoner, J. S., Acton, G., & Channell, J. E. T. (2006). Brunhes paleomagnetic field variability recorded in Ocean Drilling Program Cores. *Physics of the Earth and Planetary Interiors, 156,* 194–204.

Mackereth, F. (1971). On the variation in the direction of the horizontal component of magnetization in lake sediments. *Earth and Planetary Science Letters, 12,* 332–338.

Masarik, J., & Beer, J. (1999). Simulation of particle fluxes and cosmogenic nuclide production in the Earth's atmosphere. *Journal of Geophysical Research D, 104*(10), 12099–12111.

Mazaud, A. (2005). User-friendly software for vector analysis of the magnetization of long sediment cores. *Geochemistry, Geophysics and Geosystems, 6,* doi:10.1029/2005GC001036.

Mazaud, A., Laj, C., & Bender, M. (1994). A geomagnetic chronology for Antarctic ice accumulation. *Geophysical Research Letters, 21,* 337–340.

Mazaud, A., Sicre, M. A., Ezat, U., Pichon, J. J., Duprat, J., Laj, C., Kissel, C., Beaufort, L., Michel, E., & Turon, J.-L. (2002). Geomagnetic-assisted stratigraphy and sea surface temperature changes in core MD94-103 (Southern Indian Ocean): Possible implications for North-South climatic relationships around H4. *Earth and Planetary Science Letters, 201,* 159–170.

Mejia, V., Opdyke, N. D., Vilas, J. F., Singer, B. S., & Stoner, J. S. (2004). Paleomagnetic secular variation from Pliocene-Pleistocene lavas from Southern Patagonia. *Geochemistry, Geophysics and Geosystems, 5,* Q03H08, doi:10.1029/2003GC000633.

Merrill, R. T., & McElhinny, M. (1977). Anomalies in the time-averaged paleomagnetic field and their implications for the loser mantle. *Reviews of Geophysics and Space Physics, 15,* 309–323.

Merrill, R. T., & McFadden, P. L. (1994). Geomagnetic field stability: Reversal events and excursions. *Earth and Planetary Science Letters, 121,* 57–69.

Merrill, R. T., & McFadden, P. L. (2003). The geomagnetic axial dipole field assumption. *Physics of the Earth and Planetary Interiors, 139,* 171–185.

Merrill, R. T., & McFadden, P. L. (2005). The use of magnetic field excursions in stratigraphy. *Quaternary Research, 63,* 232–237.

Meynadier, L., & Valet, J.-P. (1996). Post-depositional realignement of magnetic grains and asymmetrical saw-tooth patterns of magnetization intensity. *Earth and Planetary Sciences Letters, 140,* 123–132.

Meynadier, L., Valet, J.-P., Weeks, R., Shackleton, N. J., & Hagee, V. L. (1992). Relative geomagnetic intensity of the field during the last 140 ka. *Earth and Planetary Science Letters, 114,* 39–57.

Moskowitz, B. M. (1991). *Hitchhikers Guide to Magnetism.* http://www.irm.umn.edu/hg2m/hg2m_index.html

Muscheler, R., Beer, J., Kubik, P. W., & Synal, H-A. (2005). Geomagnetic field intensity during the last 60,000 years based on ^{10}Be, and ^{36}Cl from the Summit ice cores and ^{14}C. *Quaternary Science Reviews, 24,* 1849–1860.

Nagy, E. A., & Valet, J.-P. (1993). New advances for paleomagnetic studies of sediment cores using U-channels. *Geophysical Research Letters, 20,* 671–674.

Oda, H., & Shibuya, H. (1996). Deconvolution of long-core paleomagnetic data of Ocean Drilling Program by Akaike's Bayesian criterion minimization. *Journal of Geophysical Research, 101,* 2815–2834.

Ogg, J. C., & Smith, A. G. (2004). The geomagnetic polarity time scale. In: F. M. Gradstein, J. Ogg & A. G. Smith (Eds), *A geologic time scale 2004* (pp. 63–86). Cambridge, UK: Cambridge University Press.

Ojala, A., & Saarinen, T. (2002). Palaeosecular variation of the Earth's magnetic field during the last 10,000 years based on the annually laminated sediment of Lake Nautajärvi, central Finland. *The Holocene, 12,* 391–400.

Ólafsdóttir, S., Stoner, J. S., Geirsdóttir, Á., Miller, G. H., & Channell, J. E. T. (2005). High-resolution Holocene paleomagnetic secular variation records from Iceland: Towards marine-terrestrial synchronization. *EOS Transactions, AGU, 86*(52), U43A-0820.

Opdyke, N. D. (1972). Paleomagnetism of deep sea cores. *Reviews of Geophysics and Space Physics, 10,* 213–249.

Opdyke, N. D., & Channell, J. E. T. (1996). *Magnetic stratigraphy.* San Diego, CA: Academic Press.

Opdyke, N. D., Glass, B., Hays, J. D., & Foster, J. (1966). Paleomagnetic study of Antarctic Deep-Sea cores. *Science, 154,* 349–357.

Opdyke, N. D., & Henry, K. W. (1969). A test of the dipole hypothesis. *Earth and Planetary Science Letters, 6,* 139–151.

Otofuji, Y., & Sasajima, S. (1981). A magnetization process of sediments: Laboratory experiments on post-depositional remanent magnetization. *Geophysical Journal of the Royal Astronomical Society, 66,* 241–259.

Parker, R. L. (2000). Calibration of the pass-through magnetometer-I: Theory. *Geophysical Journal International, 142,* 371–383.

Parker, R. L., & Gee, J. (2002). Calibration of the pass-through magnetometer-II: Application. *Geophysical Journal International, 150,* 140–152.

Peck, J. A., King, J. W., Colman, S. M., & Kravchinsky, V. A. (1996). An 84 kyr paleomagnetic record from the sediments of Lake Baikal. *Journal of Geophysical Research, 101,* 11365–11385.

Pittman, W. C., III, & Heirtzler, J. R. (1966). Magnetic anomalies over the Pacific-Antarctic ridge. *Science*, *154*, 1164–1171.

Roberts, A. P., Lehman, B., Weeks, R. J., Verosub, K. L., & Laj, C. (1997). Relative paleointensity of the geomagnetic field over the last 200,000 years from ODP Sites 883 and 884, North Pacific Ocean. *Earth and Planetary Science Letters*, *152*, 11–23.

Roberts, A. P., Stoner J. S., & Richter, C. (1996). Coring-induced magnetic overprints and limitations of the long-core paleomagnetic measurement technique: Some observations from ODP Leg 160, Eastern Mediterranean Sea, In: K.-C. Emeis, A. H. F. Robertson, C. Richter et al. (Eds). *Proceedings of the Ocean Drilling Program*, Initial Reports, *160*, 497–504.

Roberts, A. P., & Winklhofer, M. (2004). Why are geomagnetic excursions not always recorded in sediments? Constraints from post-depositional remanent magnetization lock-in modelling. *Earth and Planetary Science Letters*, *227*, 345–359.

Robinson, C., Raisbeck, G. M., Yiou, F., Lehman, B., & Laj, C. (1995). The relationship between ^{10}Be and geomagnetic field strength records in central North Atlantic sediments during the last 80 ka. *Earth and Planetary Science Letters*, *136*, 551–557.

Sagnotti, L., Budillon, F., Dinarès-Turell, J., Iorio, M., & Macrì, P. (2005). Evidence for a variable paleomagnetic lock-in depth in the Holocene sequence from the Salerno Gulf (Italy): Implications for "high-resolution" paleomagnetic dating. *Geochemistry, Geophysics and Geosystems*, *6*, doi: 10.1029/2005GC001043.

Sagnotti, L., Rochette, P., Jackson, M., Vadeboin, F., Dinarès- Turell, J., Winkler, A. & "Mag-Net" Science Team (2003). Inter-laboratory calibration of low field magnetic and anhysteretic susceptibility measurements. *Physics of the Earth and Planetary Interiors*, *138*, 25–38.

Schneider, D. A., & Kent, D. V. (1990). The time-averaged paleomagnetic field. *Reviews of Geophysics*, *28*, 71–96.

Schwartz, M., Lund, S. P., Hammond, D. E., Schwartz, R., & Wong, K. (1997). Early sediment diagenesis on the Blake/Bahama Outer Ridge, North Atlantic Ocean, and its effects on sediment magnetism. *Journal of Geophysical Research*, *102*, 7903–7914.

Shackleton, N. J., Berger, A., & Peltier, W. R. (1990). An alternative astronomical calibration of the lower Pleistocene timescale based on ODP Site 677. *Transactions of the Royal Society Edinburgh*, *81*, 251–261.

Shackleton, N. J., & Opdyke, N. D. (1973). Oxygen isotope and paleomagnetic stratigraphy of Equatorial Pacific core V28-238: Oxygen isotope temperatures and ice volumes on a 10^5 year and 10^6 year scale. *Quaternary Research*, *3*, 39–55.

Singer, B. S., Relle, M. K., Hoffman, K. A., Battle, A., Laj, C., Guillou, H., & Carracedo, J. C. (2002). Ar/Ar ages from transitionally magnetized lavas on La Palma, Canary Islands, and the geomagnetic instability timescale. *Journal of Geophysical Research*, *107*(B11), 2307, doi:10.1029/2001JB001613.

Smith, C. R., & Rabouille, C. (2002). What controls the mixed-layer depth in deep-sea sediments? The importance of POC flux. *Limnology and Oceanography*, *47*, 418–426.

Snowball, I., & Sandgren, P. (2002). Geomagnetic field variations in northern Sweden during the Holocene from varved lake sediments and their implications for cosmogenic nuclide production rates. *The Holocene*, *12*, 517–530.

Snowball, I., & Sandgren, P. (2004). Geomagnetic field intensity changes in Sweden between 9000 and 450 cal BP: Extending the record of "archaeomagnetic jerks" by means of lake sediments and the pseudo-Thellier technique. *Earth and Planetary Science Letters*, *227*, 361–376.

Stockhausen, H. (1998). Geomagnetic paleosecular variation (0–13,000 yr BP) as recorded in sediments from the three maar lakes from West Eifel (Germany). *Geophysical Journal International*, *135*, 898–910.

Stoner, J. S., Channell, J. E. T., & Hillaire-Marcel, C. (1995). Late Pleistocene relative geomagnetic paleointensity from the deep Labrador Sea: Regional and global correlations. *Earth and Planetary Science Letters*, *134*, 237–252.

Stoner, J. S., Channell, J. E. T., & Hillaire-Marcel, C. (1996). The magnetic signature of rapidly deposited detrital layers from the deep Labrador sea: Relationship to North Atlantic Heinrich layers. *Paleoceanography*, *11*, 309–325.

Stoner, J. S., Channell, J. E. T., Hillaire-Marcel, C., & Kissel, C. (2000). Geomagnetic paleointensity and environmental record from Labrador Sea core MD99-2024: Global marine sediment and ice core chronostratigraphy for the last 110 kyr. *Earth and Planetary Science Letters, 183*, 161–177.

Stoner, J. S., Channell, J. E. T., Hodell, D. A., & Charles, C. D. (2003). A ∼580 kyr paleomagnetic record from the sub-Antarctic South Atlantic (ODP Site 1089). *Journal of Geophysical Research, 108*(B5), 2244, doi:10.1029/2001JB001390.

Stoner, J. S., Jennings, A., Kristjansdottir, G. B., Dunhill, G., Andrews, J. T., & Hardadottir, J. (2007). paleomagnetic approach toward refining Holocene radiocarbon-based chronologies: Paleoceanographic records from North Iceland (MD99-2269) and east Greenland (MD99-2322) margins. *Paleoceanography, 22*, PA1209, doi:10.1029/2006PA001285.

Stoner, J. S., Laj, C., Channell, J. E. T., & Kissel, C. (2002). South Atlantic (SAPIS) and North Atlantic (NAPIS) geomagnetic paleointensity stacks (0–80 ka): Implications for inter-hemispheric correlation. *Quaternary Science Reviews, 21*, 1141–1151.

St-Onge, G., Mulder, T., & Francus, P. (this book). Continuous physical properties of cored marine sediments. In: C. Hillaire-Marcel & A. de Vernal (Eds), *Proxies in Late Cenozoic paleoceanography.* Amsterdam: Elsevier.

St-Onge, G., Piper, D. J. W., Mulder, T., Hillaire-Marcel, C., & Stoner, J. S. (2004). Earthquake and flood-induced turbidites in the Saguenay Fjord (Québec): A Holocene paleoseismicity record. *Quaternary Science Reviews, 23*, 283–294.

St-Onge, G., Stoner, J. S., & Hillaire-Marcel, C. (2003). Holocene paleomagnetic records from the St. Lawrence Estuary: Centennial- to millennial-scale geomagnetic modulation of cosmogenic isotopes. *Earth and Planetary Science Letters, 209*, 113–130.

Stuiver, M., Reimer, P. J., Bard, E., Beck, J. W., Burr, G. S., Hughen, K. A., Kromer, B., McCormac, G., Van der Plicht, J., & Spurk, M. (1998). INTCAL98 radiocarbon age calibration, 24,000–0 cal BP. *Radiocarbon, 40*, 1041–1083.

Tauxe, L. (1993). Sedimentary records of relative paleointensity of the geomagnetic field: Theory and practice. *Reviews of Geophysics, 31*, 319–354.

Tauxe, L. (1998). *Paleomagnetic principles and practice.* Dordrecht, The Netherlands: Kluwer Academic Publishers.

Tauxe, L., LaBrecque, J. L., Dodson, R., & Fuller, M. (1983). U-channels: A new technique for paleomagnetic analysis of hydraulic piston cores. *EOS Transactions, AGU, 64*, 219.

Tauxe, L., Pick, T., & Kok, Y. S. (1995). Relative paleointensity in sediments: A pseudo-Thellier approach. *Geophysical Research Letters, 22*, 2885–2888.

Tauxe, L., & Wu, G. (1990). Normalized remanence in sediments of the western equatorial Pacific: Relative paleointensity of the geomagnetic field? *Journal of Geophysical Research, 95*, 12337–12350.

Teanby, N., & Gubbins, D. (2000). The effects of aliasing and lock-in processes on paleosecular variation records from sediments. *Geophysical Journal International, 142*, 563–570.

Thompson, R. (1973). Palaeolimnology and Paleomagnetism. *Nature, 242*, 182–184.

Thompson, R. (1984). A global review of paleomagnetic results from wet lake sediments. In: E. Y. Haworth & J. W. G. Lund (Eds), *Lake sediments and environmental history* (pp. 145–165). Minneapolis, MN: University of Minnesota Press.

Thompson, R., & Oldfield, F. (1986). *Environmental magnetism.* London, UK: Allen & Unwin.

Thompson, R., & Turner, G. M. (1979). British geomagnetic master curve 10,000–0 yr BP for dating European sediments. *Geophysical Research Letters, 6*, 249–252.

Thouveny, N., Carcaillet, J., Moreno, E., Leduc, G., & Nérini, D. (2004). Geomagnetic moment variation and paleomagnetic excursions since 400 kyr BP: A stacked record from sedimentary sequences of the Portuguese margin. *Earth and Planetary Science Letters, 219*, 377–396.

Trauth, M. H., Sarnthein, M., & Arnold, M. (1997). Bioturbational mixing depth and carbon flux at the seafloor. *Paleoceanography, 12*, 517–526.

Tric, E., Valet, J.-P., Tucholka, P., Paterne, M., Labeyrie, L., François, G., Tauxe, L., & Fontugne, M. (1992). Paleointensity of the geomagnetic field during the last 80,000 years. *Journal of Geophysical Research, 97*, 9337–9351.

Turner, G. M., & Thompson, R. (1982). Detransformation of the British geomagnetic secular variation record for Holocene times. *Geophysical Journal of the Royal Astronomical Society, 70*, 789–792.

Valet, J.-P. (2003). Time variations in geomagnetic intensity. *Reviews of Geophysics, 41,* doi: 10.1029/2001RG000104.

Valet, J.-P., & Meynadier, L. (1993). Geomagnetic field intensity and reversals during the past four million years. *Nature, 366,* 234–238.

Valet, J.-P., & Meynadier, L. (1998). A comparison of different techniques for relative paleointensity. *Geophysical Research Letters, 25,* 89–92.

Valet, J.-P., Meynadier, L., & Guyodo, Y. (2005). Geomagnetic dipole strength and reversal rate over the past two million years. *Nature, 435,* 802–805.

Van der Voo, R. (1990). Phanerozoic paleomagnetic poles from Europe and North America and comparisons with continental reconstructions. *Reviews of Geophysics, 28,* 167–206.

Verosub, K. L. (1998). Faster is better. *Science, 281,* 1297–1298.

Verosub, K. L., Ensley, R. A., & Ulrick, J. S. (1979). The role of water content in the magnetization of sediments. *Geophysical Research Letters, 6,* 226–228.

Verosub, K. L., Harris, A. H., & Karlin, R. (2001). Ultra high-resolution paleomagnetic record from ODP Leg 169S, Saanich Inlet, British Columbia: Initial results. *Marine Geology, 174,* 79–93.

Vine, F. J., & Wilson, J. T. (1965). Magnetic anomalies over a young ocean ridges off Vancouver Island. *Science, 150,* 485–489.

Vlag, P., Thouveny, N., & Rochette, P. (1997). Synthetic and sedimentary records of geomagnetic excursions. *Geophysical Research Letters, 24,* 723–726.

Wagner, G., Beer, J., Laj, C., Kissel, C., Mazarik, J., Muscheler, R., & Synal, H.-A. (2000). Chlorine-36 evidence for the Mono Lake event in the Summit GRIP ice core. *Earth and Planetary Science Letters, 181,* 1–6.

Weeks, R. J., Laj, C., Endignoux, L., Fuller, M. D., Roberts, A. P., Manganne, R., Blanchard, E., & Goree, W. (1993). Improvements in long core measurement techniques: Applications in palaeomagnetism and palaeoceanography. *Geophysical Journal International, 114,* 651–662.

Yamazaki, T., & Oda, H. (2005). A geomagnetic paleointensity stack between 0.8 and 3.0 Ma from equatorial Pacific sediment cores. *Geochemistry, Geophysics and Geosystems, 6,* Q11H20, doi:10.1029/2005GC001001.

Yiou, F., Raisbeck, G. M., Baumgartner, S., Beer, J., Hammer, C., Johnsen, S., Jouzel, J., Kubik, P. W., Lestringuez, J., Stievenard, M., Suter, M., & Yiou, P. (1997). Beryllium 10 in Greenland Ice Core Project ice core at Summit, Greenland. *Journal of Geophysical Research, 102*(C12), 26783–26794.

Zijderveld, J. D. A. (1967). A.C. demagnetization of rocks: Analysis of results. In: D. W. Collinson, K. M. Creer & S. K. Runcorn (Eds), *Methods in paleomagnetism* (pp. 245–286). Amsterdam: Elsevier.

CHAPTER FOUR

Clay Minerals, Deep Circulation and Climate

Nathalie Fagel

Contents

1. Introduction	139
2. Methodology: The Clay Toolbox in Marine Sediments	142
2.1. Clay mineral groups in deep-sea sediments	142
2.2. Formation of clay minerals	143
2.3. The origin of clays in deep-sea sediments	145
2.4. Clay particle transport mechanisms	147
2.5. Clay mineral distribution in the world ocean basins	150
2.6. The significance of clays in cenozoic marine sediments	158
2.7. Provenance of detrital inputs	160
2.8. Relationship between clay mineralogy and ocean circulation	163
2.9. Relationship between clay minerals and climate	168
3. Applications: Clays as a Proxy for Paleocirculation	171
3.1. Clay distribution in Arctic Ocean surface sediments	171
3.2. Clay distribution in holocene and last glacial sediments in the North Atlantic	172
3.3. Clay distribution since the last glacial in the southeast Indian Ocean	175
4. Some Perspectives	176
Acknowledgements	176
References	176

1. Introduction

The detrital fraction of deep-sea sediments may carry important information about conditions on adjacent continents and about the mechanisms by which material is transported from land to sea (Figure 1). Clay minerals are the main constituents of recent deep-sea or abyssal sediments. Since the late sixties, their role as paleoclimatic and paleoceanographic indicators has been investigated worldwide by X-ray diffraction techniques (Yeroshchev–Shak, 1964; Biscaye, 1965; Berry & Johns, 1966; Chamley, 1967; Griffin, Windom, & Goldberg, 1968; Rateev, Gorbunova,

Developments in Marine Geology, Volume 1
ISSN 1572-5480, DOI 10.1016/S1572-5480(07)01009-3

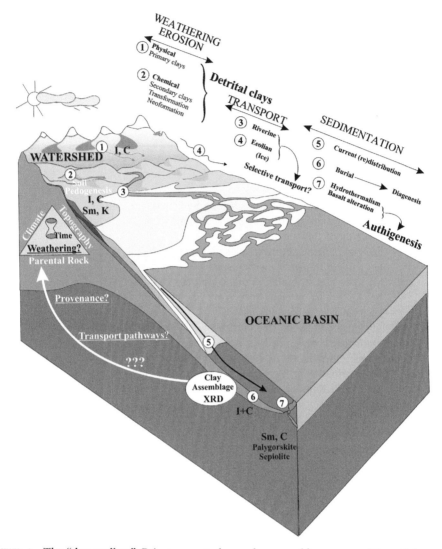

Figure 1 The "clay toolbox". Primary controls on clay assemblage composition of deep-sea sediments. I: illite, C: chlorite, Sm: smectite, K: kaolinite, XRD: X-ray diffraction technique. In marine sediments, only detrital clays can carry *indirect* information regarding weathering conditions within the watershed (clays = climate proxy) or changes in provenance (clays = transport pathway proxy).

Lisitzyn, & Nosov, 1969; Chamley, 1974; Kolla, Kostecki, Robinson, Biscaye, & Ray, 1981; Petschick, Kuhn, & Gingele, 1996; Gingele, Schmieder, Petschick, von Dobeneck, & Rühlemann , 1999; Robert, Diester-Haass, & Paturel, 2005). In marine sediments, clays are mainly detrital (~90%; Velde, 1995) and their abundances provide abiotic *proxy data*, which may be used to decipher either (1) climate changes in the source area on adjacent landmasses, (2) changes in the intensity of the transport agent, or (3) changes in the ocean currents that disperse the terrigenous input (Gingele, De Deckker, Girault, & Guichard, 2004).

In essence, clay mineralogy is a useful tool to constrain the provenance of fine-grained terrigenous sediments (Biscaye, 1965). After much discussion in the literature about the usefulness of clay minerals in delineating suspended sediment dispersal routes and constraining the provenance of fine-grained marine deposits (Biscaye, 1965; Hein, Bouma, Hampton, & Ross, 1979), clay mineralogy has been widely used as a *tracer of provenance* in studies of the world's oceans. In these studies, the mineralogy of the fine-grained detrital fraction generally reflects the intensity of continental weathering in the source areas (Biscaye, 1965; Griffin et al., 1968; Rateev et al., 1969). Combining clay mineralogy and radiogenic isotope measurements (Sr, Nd and/or Pb) provides further constraints on the identification of the source area (e.g., Grousset, Biscaye, Zindler, Prospero, & Chester, 1988; Fagel, Innocent, Stevenson, & Hillaire-Marcel, 1999; Walter, Hegner, Diekmann, Kuhn, & Rutgers Van Der Loeff, 2000; Rutberg, Goldstein, Hemming, & Anderson, 2005).

Clays are sensitive indicators of their environment of formation, and their composition may be used to constrain climatic variations which do not affect other size fractions (Moriarty, 1977). The composition of terrestrial clay mineral reflects the prevailing weathering regimes which control the nature and intensity of pedogenetic processes in continental source areas. Weathering depends primarily on climate zonation, which determines the intensity of physical and/or chemical weathering (e.g., Chamley, 1989; Weaver, 1989).

Clay minerals are eroded from soils by rivers, wind or ice and carried into shallow and deep water masses of the surrounding seas. Their modern distribution pattern on the sea floor provides insight into their propagation by ocean currents (Gingele, De Deckker, & Hillenbrand, 2001b): the dispersal of detrital clays constrains the transport pathway of suspended fine-grained particles. As such, clays are *indirect tracers of water masses*. The interpretation of down-core changes in mineral assemblages depends on available information on the modern distribution and sources, as well as independent climate proxy data about the source area. Once the relationship between a specific clay mineral assemblage and a given source, water mass or current system is established, variations of this assemblage in down-core profiles may be used to detect fluctuations in the propagation of the water mass or current system (Diekmann et al., 1996; Gingele et al., 1999).

As the distribution of clay minerals in modern oceans appears to be controlled by contemporaneous climates, marine clays in Cenozoic (and Mesozoic) sediments have been widely used to reconstruct *paleoclimates* (Chamley, 1981). Singer (1984), and later Thiry (2000), questioned the relevance of using marine detrital clays as a continental paleoclimate proxy: indeed, changes in marine clay mineral assemblages do not systematically reflect changes in weathering conditions in the continental source area, but rather changes in source areas or transport media. For older sedimentary sequences, diagenesis can completely erase the paleoenvironmental memory of clay minerals, initially diversified clay assemblages evolving with burial towards an anonymous illite- and chlorite-bearing association with intermediate irregular and regular mixed-layers (e.g., Hower, Eslinger, Hower, & Perry, 1976; Nadeau, Wilson, McHardy, & Tait, 1985; Inoue, 1987). The dominant diagenetic change is the progressive reaction of smectite to illite, which depends primarily on temperature (Hower et al., 1976; Hoffman & Hower, 1979; Srodon & Eberl, 1984). Other than temperature, the two most important factors are time and fluid chemistry (Eberl & Hower, 1977; Roberson & Lahann, 1981; Ramsayer & Boles, 1986). In sedimentary deposits characterized by a normal geothermal gradient (~30°C/km), burial diagenesis operates at depths greater than 2,500–3,000 m (i.e., at temperatures >80°C, Chamley, 1989).

In this chapter, we review the literature in order to describe the relationships between (1) clay mineral abundance in deep-sea sediments and deep circulation, and (2) clay variability and short- or long-term climate changes. As the present volume is dedicated to *Proxies in Paleoceanography*, we focus on the interpretation of marine clay mineral data. The identification and quantification of clay minerals using X-ray diffraction methods are described in other methodological textbooks (e.g., Brindley & Brown, 1980; Moore & Reynolds, 1989).

2. METHODOLOGY: THE CLAY TOOLBOX IN MARINE SEDIMENTS

2.1. Clay Mineral Groups in Deep-Sea Sediments

2.1.1. Definition and identification of clay mineral groups

The mineral component of the "clay-sized" fraction (<2 μm, or even <1 μm) of deep-sea sediments comprises the common clay minerals montmorillonite (or smectite), illite, kaolinite and chlorite, with lesser amounts of quartz and feldspars (Windom, 1976). In addition, the mixed-layer minerals palygorskite and sepiolite have been also identified in this size fraction (Windom, 1976). Clay minerals are identified by their characteristic basal X-ray diffraction maxima (i.e., "reflection" or "peak"; Brindley & Brown, 1980; Moore & Reynolds, 1989). The term "clay minerals" refers to mineral groups rather than specific mineralogical species. "Illite" is used "as a general term for the clay mineral constituent of argillaceous sediments belonging to the mica group" (definition from Grim, Bray, & Bradley, 1937). Illites are characterized by a 10 Å reflection which is affected neither by glycolation nor heating. The term is used in the same general sense as "montmorillonite" or "kaolinite", that is, to refer to a group of minerals. Any material which expands to 17 Å upon glycolation is assigned to the montmorillonite group (Biscaye, 1965). The term "smectite" is often used in more recent works, and encompasses pure smectite but also smectite–illite mixed layers with variable abundances of smectite layers. Chlorites are defined by a 14 Å reflection which is affected neither by solvatation nor, usually, by heating (except if vermiculite is present). Kaolinite is identified by a 7 Å reflection that disappears above 500°C. Minor quantities of non-clay minerals are present in the terrigenous fine-grained component of deep-sea sediments.

2.1.2. Clay mineral classification

Clay minerals belong to the phyllosilicate family (Brindley & Brown, 1980; Caillière, Hénin, & Rautureau, 1982; Moore & Reynolds, 1989). Phyllosilicates are hydrous silicate minerals essentially composed of two-dimensional Si-bearing tetrahedral, and Al-bearing octahedral sheets stacked in a regular array (White, 1999). The fundamental classification criteria retained by the AIPEA (Association Internationale Pour l'Etude des Argiles) are the tetrahedral–octahedral sheet combination type (layer type, i.e., 1/1, 2/1 or 2/1/1), the cation content of the octahedral sheets (dioctahedral or trioctahedral), the layer charge and the type of interlayer material (Bailey, 1980).

Kaolinite consists of a combination of one tetrahedral layer and one octahedral layer (layer type 1/1). It is a dioctahedral mineral with two-thirds of the octahedral sites filled by a trivalent cation (Al or possibly Fe^{3+}). There is no interlayer material other than H_2O.

Illite comprises a combination of two tetrahedral sheets and one intercalated octahedral sheet (layer type 2/1). It is a dioctahedral mineral characterized by the presence of K cations in the interlayer which compensate the charge deficit resulting from Al^{3+} substitution by Si^{4+} in the tetrahedral sheet.

Smectite is also a 2/1 layer type mineral, but is characterized by variable amounts of the exchangeable cations Na and/or Ca in the interlayer, as well as one or two layers of water. Smectites are divided into two subgroups. Montmorillonites *sensu lato* comprise all dioctahedral smectites, i.e., montmorilllonite *sensu stricto*, characterized by an octahedral Mg^{2+} charge, beidellite, characterized by a tetrahedral Al^{3+} charge, and its Fe-rich end-member nontronite (tetrahedral Fe^{3+} charge). Saponites *sensu lato* consist of a trioctahedral mineral which includes either an octahedral charge (Mg^{2+} in stevensite, Li^+ and F^+ in hectorite) or an octahedral Mg^{2+} charge and a tetrahedral Al^{3+} charge in saponite *sensu stricto*.

Chlorite is also composed of three sheets, two tetrahedral ones and an octahedral one, but its interlayer space is taken up by a hydroxide layer (layer type 2/1/1). In sedimentary environments, although chlorites are mainly trioctahedral with 3/3 octahedral sites filled by bivalent cations such as Fe, Mg, Mn or, more rarely, monovalent cations (Li, Ni), some dioctahedral species are possible (e.g., sudoite). Their chemical composition is complex and varied due to numerous possible substitutions in both the tetrahedral and/or octahedral layers and in the interlayer.

Mixed layers are formed by the regular or irregular stacking of two or more layer types. Not all possible combinations are observed in sedimentary environments (see reported occurrences in Thorez, 1986).

Palygorskite and sepiolite are fibrous clay minerals comprising two discontinuous tetrahedral sheets and one continuous octahedral sheet (modulated 2/1 layer type). They are characterized by a ribbon structure with a large space available for water, hydroxyl and exchangeable cations, more specifically Mg in sepiolite and Al in palygorskite (Caillière et al., 1982).

2.2. Formation of Clay Minerals

Sedimentary clays are divided into two main categories according to their origin (Figure 1). Clay minerals are either formed by *in situ* precipitation from a concentrated solution in closed continental or marine sedimentary basins (authigenic clays) or by weathering processes on the continent (detrital clays). Authigenesis is a minor process which accounts for less than 10% of clay minerals (Velde, 1995), although it may be important locally. In deep-sea environments, authigenic clays result from the weathering of basaltic oceanic crust or hydrothermal alteration (Chamley, 1989; Weaver, 1989). The nature and chemical composition of clays are strongly controlled by reaction temperatures (Velde, 1992). Authigenic deep-sea clays are mainly Fe- and/or Mg-rich minerals, including various species of smectites (nontronite, saponite), micas (celadonite), chlorites and fibrous clays (palygorskite, sepiolite; Velde, 1995).

Whereas authigenic clays provide insight into the geochemical environment, detrital clays hold a record of weathering conditions in the adjacent landmass (Velde, 1995). Detrital clays result from the physical breakdown of parental rocks (primary minerals) or from the selective dissolution of pre-existing clay and non-clay minerals and growth of secondary minerals in soils (Allen, 1997). Under normal pH conditions, water attack or *hydrolysis*, is the main chemical weathering process involved (Chamley, 1989). During pedogenesis, secondary clays form by transformation processes or neoformation. Transformed clays are the result of cation loss due to the opening of the sheet structure (degradation) or, conversely, cation addition to repair slightly altered clay (rejuvenation) or to form a new mineral (aggradation; Thorez, 1989). Neoformed clays are formed by recombination and precipitation of ions from a leachate. The intensity of chemical weathering, and hence the nature of the secondary minerals, are primarily controlled by climate conditions. The temperature, water availability and precipitation regimes affect the extent of cation, silica and alumina leaching (Pédro, 1968; Chamley, 1989; Allen, 1997). In cold or arid climate, water availability is too low for chemical weathering to operate. Clays, mainly illite and chlorite, are inherited from parental rocks (*heritage* process). For moderate leaching under temperate conditions, cations released by the breakdown parental minerals promote the formation of cation-bearing clay minerals comprised of two tetrahedral Si-layers and one octahedral Al-layer, such as illite and smectite (*bisiallitisation* process). With increasing hydrolysis under tropical conditions, more silica is released in solution and only one tetrahedral Si-layer and one octahedral Al-layer combine to form kaolinite (*monosiallitisation* process). More intense leaching under warm and humid equatorial climate precludes the formation of clays; the secondary minerals are oxides (gibbsite, *allitisation* process). Although the global weathering pattern shows a zonal or latitudinal control (Allen, 1997), the relationship between climate parameters and clay mineral formation is not always straightforward. For instance, clay mineral products formed under different climate conditions may show mimetism (Singer, 1984). Important factors other than climate include parental material, topography and time (Singer, 1984; Velde, 1992). The acidic or basic nature of parental rocks controls the composition of secondary minerals. Mineral grain size determines specific area, and hence reaction kinetics. Topography controls runoff conditions, making the recombination of ions possible only in depressed areas. Tectonic stability determines the ratio between physical and chemical weathering. Pedogenesis is a relatively slow process the intensity of which increases with time.

2.3. The Origin of Clays in Deep-Sea Sediments

2.3.1. Detrital clays: illite and kaolinite

Studies over large oceanic areas (Biscaye, 1965; Griffin et al., 1968; Rateev et al., 1969; Lisitzin, 1996) show that the interaction of continental sources and ocean distribution mechanisms can account for the major clay mineral abundance patterns (Moriarty, 1977). In particular, for Atlantic Ocean sediments, Biscaye (1965) demonstrated striking geographical distribution patterns for the main clay minerals; more specifically, recent clay distribution in the Atlantic Ocean is controlled by climatic and weathering zonations on adjacent land masses, implying that most clay minerals are terrigenous (Biscaye, 1965; Chamley, 1989). Moriarty (1977) points out that some authors (Berry & Johns, 1966) have proposed that alteration and authigenesis are important mechanisms, but what limited chemical data are available point to no alteration and minimal authigenic inputs, especially in sediments characterized by high terrigenous accumulation rates (McDougall & Harris, 1969; Darby, 1975). In later studies, the occurrence of illite, chlorite and kaolinite in soil-sized eolian dust from the Eastern margins of the Atlantic Ocean is used as evidence for a land-derived origin for these clays (Chester, Elderfield, Griffin, Johnson, & Padgham, 1972). Griffin et al. (1968) claim that clay minerals in the $<2\,\mu m$ fraction of deep-sea sediments are useful indicators of marine sedimentary processes, particularly those involving the transport of land-derived solids to and within the oceans. However, according to Behairy, Chester, Griffiths, Johnson, and Stoner (1975), little is known about the large-scale distribution of major clays in particulate material in the world's oceans. This material represents an intermediate stage between the initial transport of solid components to the ocean and their deposition in bottom sediments. It is during this stage that mixing of various components is expected to begin. Clay mineral concentrations in particulate material from surface $(<5\,m)$ waters of the eastern margins of the Atlantic Ocean have been determined (Behairy et al., 1975). Clay minerals in deep-sea sediments and surface particulate matter may be either detrital or authigenic (Behairy et al., 1975). However, clay minerals in eolian dusts have a purely detrital origin. The average concentrations (40% illite, 25% kaolinite) and the latitudinal distribution of illite and kaolinite in surface seawater particulate material between 60°N and 40°S are similar to those of eolian dusts: the relative abundance of illite decreases towards low latitudes, and is compensated by an increase in kaolinite abundance. This observation constitutes strong evidence for the detrital origin of these clays in particulate matter and the underlying deep-sea sediments, which record similar distributional trends (Biscaye, 1965).

As underlined by Biscaye (1965), the strongest evidence yet for the detrital origin of illite is provided by K/Ar measurements from recent North Atlantic deep-sea sediments (Hurley, Heezen, Pinson, & Fairbairn, 1963). Ages on the order of hundreds of millions of years indicate that a major component of the sediments must be derived from old continental rocks and their weathering products. Illite is a widespread mineral in many types of rocks and soils and is quite resistant to chemical weathering (Gradusov, 1974).

Given its elemental composition and the pH conditions required for its formation (Arrhenius, 1963), it is unlikely that kaolinite forms in the oceans (Windom, 1976).

One exception to this is the reported occurrence of kaolinite authigenesis in deep hydrothermal-volcanogenic sediments (Karpoff, Peterschmitt, & Hoffert, 1980).

2.3.2. Mixed detrital and authigenic clays: smectite, chlorite and fibrous clays

Distinguishing authigenic clays from detrital clays can be difficult (Singer, 1984). Because it can be either detrital or authigenic, the origin of smectite in marine sediments was, at one point, a matter of debate (e.g., Biscaye, 1965; Griffin et al., 1968). Authigenic smectites derive mainly from volcanism, hydrothermal activity or diagenetic processes (Chamley, 1989). Submarine alteration of basaltic volcanic glasses and volcanic rock fragments is one of the most important authigenic smectite-forming processes (Windom, 1976; Chamley, 1989; Weaver, 1989). Fe-rich beidellites and nontronites commonly derive from hydrothermal alteration of basalts, and they are particularly abundant in active areas of mid-oceanic ridges (Haggerty & Baker, 1967; McMurthy, Wang, & Yeh, 1983; Parra, Delmont, Ferragne, Latouche, & Puechmaille, 1985, 1986; Parra, Puechmaille, Dumon, Delmont, & Ferragne, 1986). In contrast, detrital smectites are sourced on adjacent continents, and they primarily form by hydrolysis under temperate to semi-arid climate conditions (Pédro, 1968; Chamley, 1989).

In the North Atlantic, similar smectite abundances in particulate material and deep-sea sediments argue for a primarily detrital origin (Behairy et al., 1975). In the South Altantic, however, the abundance of smectite in particulate matter is too low to account for smectite abundances in deep-sea sediments. This smectite enrichment (by up to a factor of 2) in South Atlantic sediments suggests the presence of authigenically-derived smectites (Behairy et al., 1975). On the whole, however, the lack of correlation between smectite, zeolites and volcanic shards abundances, as well as the low abundance of smectites in Atlantic Ocean surface sediments point to a primarily detrital origin for these clays (Biscaye, 1965).

The question regarding whether smectite is essentially authigenic or detrital is particularly important when it comes to the Pacific Ocean, where the large extent of oceanic area and the occurrence of sediment-trapping deep-sea trenches seem, a priori, to argue against an important terrigenous input (Chamley, 1986). In such an oceanic environment (low sedimentation rate, strong volcanic-hydrothermal activity, deposition of metalliferous sediments and other exchange processes at the water/sediment interface), synsedimentary and diagenetic processes would be expected, especially in intraplate basins. Griffin and Goldberg (1963) concluded that smectite in recent South Pacific sediments is of volcanic origin: it represents the most abundant clay type within the $<2\,\mu m$ fraction and occurs in close association with zeolites and volcanic shards. However, Deep-Sea Drilling Project (DSDP) data from the Shatsky Rise ($>6,000\,m$ water depth) rather suggest that the pelagic red clays, which are diversified and show no evidence of lithological control, are mainly detrital (Chamley, 1986). Thus, a terrigenous influence is particularly important close to continental margins in the North Pacific Ocean, but may also be a defining factor of clay assemblages in pelagic red clays in the West Pacific Ocean (Shatsky Rise), which is characterized by relatively high sedimentation rates.

Authigenic formation of chlorite in marine sediments requires specific conditions; hence, most chlorites in deep-sea sediments are detrital (Windom, 1976). Chlorite is a widespread primary constituent of low-grade metamorphic, magmatic and terrigenous sedimentary rocks and a secondary weathering product of other clay minerals (illite). Chlorite is relatively uniformly distributed on the continents (Gradusov, 1974). For instance, its detrital origin in the Atlantic Ocean is supported by a higher abundance in the particulate matter (20%) than in the underlying sediments (Behairy et al., 1975). However, authigenic chlorite may form during alteration of oceanic crust or by contact metamorphism of sediments intruded by basaltic lavas (Chamley, 1989).

The fibrous clays palygorskite and sepiolite have been identified in Cenozoic sediments of the Atlantic, Pacific and Indian Oceans (Chamley, 1989 and references therein). Couture (1977) suggested that these clays mainly form authigenically in hydrothermal environments. Although experimental studies suggest the possible *in situ* precipitation of sepiolite from seawater, sepiolite is rare in hydrothermal environments. On the other hand, palygorskite has never been synthesized under such conditions, but it is often associated with metalliferous deposits (Kastner, 1981). Fibrous clays are also formed in calcareous soils or precipitated from concentrated freshwater (lake) or saline water (permarine lagoon) in arid climates (Weaver, 1989; Velde, 1995). The occurrence of abundant palygorskite in oceanic areas adjacent to deserts (Gulf of Aden, Red Sea and Arabian Sea; Kolla et al., 1981; Fagel, Debrabant, De Menocal, & Demoulin, 1992a; Windom, 1976) and the occurrence of sepiolite in dust samples from New Zealand are strong evidence for eolian transport from adjacent continents (Windom, 1976).

In marine sediments, trace element chemistry can be a useful indicator of sedimentary material provenance (e.g., Murray, Buchholtz Ten Brink, Jones, Gerlach, & Russ, 1990; Nath, Roelandts, Sudhakar, & Pluger, 1992; Fagel, André, Chamley, Debrabant, & Jolivet, 1992b, 1997a; Li & Schoonmaker, 2003). In particular, rare earth element (REE) patterns may help in distinguishing between detrital and authigenic clays (Desprairies & Bonnot-Courtois, 1980; Chamley & Bonnot-Courtois, 1981). Indeed, authigenic clays are characterized by a seawater-type REE profile (i.e., negative Ce anomaly and depletion in light REE relative to heavy REE; Piper, 1974). A seawater-like Sr isotopic signature may also constitute an indicator of authigenesis (Clauer, O'Neil, Bonnot-Courtois, & Holtzapffel, 1990; Clauer & Chaudhuri, 1995).

2.4. Clay Particle Transport Mechanisms

Due to their small particle size, clay minerals might be expected to have a long residence time in seawater and, consequently, to get well-mixed (Behairy et al., 1975). This assumption is, however, not valid because clay mineral distributions in deep-sea sediments are, in general, similar to those on surrounding land masses. This implies that the clay minerals present in the upper layers of the seawater column are transported to the bottom relatively rapidly. Sedimentary material initially deposited on continental margins is subsequently transported to deep-sea areas by turbidity currents, and further redistributed by deep currents. As such, clay distribution in deep-sea sediments must be constrained by seafloor topography. For instance, bathymetry and the abyssal plain outline exert a strong control on the clay mineral abundance contours of the South Indian Basin (Moriarty, 1977). In contrast, the westward decrease in kaolinite abundance observed in the Atlantic Ocean off Africa and across the Mid–Atlantic Ridge is independent of seafloor topography.

The transport of clay particles to the deep sea is controlled by the complex interplay of atmospheric, hydrographic, glaciogenic and topographic conditions (Biscaye, 1965; Petschick et al., 1996). In the arid regions of northern and southwestern Africa, wind transport is the main process which supplies terrigenous matter to the ocean (Prospero, 1981). Along the African coast, the distributions of suspended clayey particles (Behairy et al., 1975) and aerosols (Chester et al., 1972) show a clear relationship with clay mineral provinces in source regions on land. If humid conditions prevail, the nearshore clay input is controlled by river systems (e.g., Amazon River; Gibbs, 1977). At high latitudes, clay minerals are also supplied by glaciomarine processes at the Antarctic continental margin (Windom, 1976; Anderson, Kurtz, Domack, & Balshaw, 1980; Ehrmann, Melles, Kuhn, & Grobe, 1992) and at the Arctic margin (Wahsner et al., 1999).

Different processes exert a control on the transport of sediment from shelves into deep ocean basins (Wahsner et al., 1999): (1) suspension along different current systems which leads to dispersion of clay minerals over large distances within the water column; (2) gravitational flows (turbidity currents and debris flows); (3) at high latitudes, sinking over the continental margin of suspended clay-rich cold and saline water masses formed on the shelf during ice formation (sea–ice or iceberg transport; Figure 2).

Near continental slopes, abundant clayey material can be derived from turbidity currents and local dense thermohaline underflows (Kuhn & Weber, 1993). With increasing distance from the source areas, advection of fine-grained particles by deep-water currents becomes the most important mode of detrital sediment transport to the deep sea (Biscaye, 1965; Petschick et al., 1996). For instance, in the North

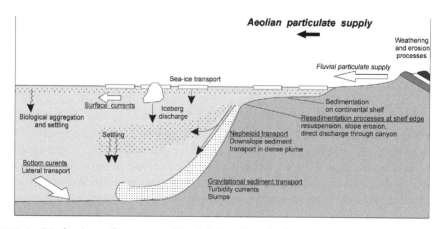

Figure 2 Mechanisms of transport of detrital clays into the deep sea. Modified with permission from Stow (1994) and Wahsner et al. (1999).

Atlantic, erosion products from Iceland are transported into basins by gravity currents, then cut off by Norwegian Bottom Current (NWBC) and transported southwards as a fine-grained suspension (Grousset, Latouche, & Maillet, 1983). Southwest of the Grand Banks of Newfoundland, Alaf (1987) showed evidence for large amounts of detrital clay being injected into the system by turbidity currents through (Laurentian) channels, canyons and surface sediment plumes, the latter being, in turn, strongly affected by surface circulation (i.e., locally, the Labrador Current). Fine-grained sediments injected into the system by turbidity currents through the Laurentian Channel are probably trapped by deep circulation. In the Weddell Sea, it is likely that high proportions of silt and clay are mainly carried within permanent or temporary bottom currents prior to deposition, whereas sand is transported by ice rafting (Diekmann & Kuhn, 1999).

Bottom current activity causes resuspension of these minerals to form nepheloid layers as well as horizontal drift on the scale of several thousands of miles (Biscaye, 1965). The resuspended particulate loads in the nepheloid layer of basins west of the Mid-Atlantic Ridge, which result from the interaction of abyssal currents with the bottom, range from $\sim 2 \times 10^6$ tons in the equatorial Guyana basin, to $\sim 50 \times 10^6$ tons in the North American basin (Biscaye & Eittreim, 1977). Offshore of Africa, the weight of clays accounts on average for almost 81% of suspended particulate matter in the near-bottom layer (McMaster, Betzer, Carder, Miller, & Eggimann, 1977). The total resuspended particulate load in the western basins (111×10^6 tons) is almost an order of magnitude greater than those in the basins located east of the Mid-Atlantic Ridge (13×10^6 tons). The net northward flux of resuspended particles carried in Antarctic bottom waters (AABW) drops from $\sim 8 \times 10^6$ tons/yr between the southern and northern ends of the Brazil basin and remains at $\sim 1 \times 10^6$ tons/yr across the Guyana basin. The extension of the nepheloid layer in the North Atlantic was constrained by comparing mineralogical analyses of subsurface sediments from different cores (Bout-Roumazeilles, Cortijo, Labeyrie, & Debrabant, 1999). The nepheloid is enriched in peculiar illite–vermiculite mixed layers. Those clays are not commonly observed in sediments, and hence constitute a good indicator of particle sources. They are primarily derived from moderate pedogenic processes in the Appalachian Mountains of Canada and transported by run-off to the Labrador shelf area. The nepheloid layer forms in shallow areas where it is enriched in illite–vermiculite mixed layers by erosion of the shelf. It flows eastwards along the slope and into the basin at intermediate water depths, to the mid-oceanic ridge, which prevents its propagation into the northeastern Atlantic basin. The current flows primarily southwards, following the main pattern of deep-water circulation (Bout-Roumazeilles et al., 1999) and carrying illite–vermiculite mixed layers at least as far south as New Jersey (Deconinck & Vanderaveroet, 1996). A dense nepheloid layer flowing at intermediate water depth and following the general pattern of intermediate and deep-water circulation is seasonally documented in the modern Labrador Sea (Biscaye & Eittreim, 1977).

During transport, detrital clay assemblages may undergo differentiation processes (e.g., selective erosion of the soil source) by size sorting or differential flocculation (Singer, 1984). The dispersal pathways of clay minerals during transport from source to deposition site are dependent on the transport agent (wind, ice or river; Singer, 1984).

There has been much debate in the literature about the lateral evolution of clay mineral assemblages from river mouths to shelf and slope sediments (Chamley, 1989). The alteration of clay minerals due to chemical disequilibrium in seawater was the first hypothesis put forth (Grim & Johns, 1954). In a study of clay minerals on the Brazilian continental shelf, Gibbs (1977) demonstrated that the dominant mechanism accounting for lateral changes in clay mineral composition is the physical sorting of sediments by grain size. Mineral composition changes from the mouth of the Amazon River over a distance of 1,400 km along the shelf, the proportion of montmorillonite increasing from 27% to 40%, as those of kaolinite and micas decrease from 36% to 32% and 28% to 18%, respectively. Gibbs (1977) pointed out that all these clays are characterized by similar flocculating properties due to organic and metallic coatings.

2.5. Clay Mineral Distribution in the World Ocean Basins

2.5.1. Literature review of clay distribution patterns in the world oceans

With the expansion of oceanographic programmes in the late sixties, the distribution of clay minerals in recent deep-sea sediments from the world oceans was studied by X-ray diffraction (Table 1, Figure 3). Biscaye (1965) determined the relative abundance of the four main clay mineral groups (i.e., montmorillonite, illite, kaolinite and chlorite) within the carbonate-free, $<2\,\mu m$ fraction of 500 sediments of the Atlantic Ocean and adjacent seas. For this author, the "clay fraction" refers to the silicate fraction of deep-sea sediments. The term is to be distinguished from "clay minerals", which refers to specific mineral constituents of "clay", and from "clay-size", which is a particle size range (Windom, 1976; Weaver, 1989). Using contour maps, Biscaye (1965) reported the relative mineral composition of the $<2\,\mu m$ size fraction assuming that montmorillonite, illite, kaolinite and chlorite account for 100% of the mineralogical composition in that size fraction. Using Biscaye's quantification method to obtain comparative values, Griffin et al. (1968) presented some 350 new analyses, primarily from the Pacific and Indian Oceans. Incorporating Biscaye's (1965) database for the Atlantic Ocean, Griffin et al. (1968) and Rateev et al. (1969) reported quite similar distributions of clay minerals in the world oceans, in spite of the fact that they did not analyse the same size fraction. Rateev et al. (1969) analysed the $<1\,\mu m$ fraction of some 380 additional surface samples from the Pacific, Antarctic and Indian Oceans. Compiling new and available (Biscaye, 1965) data, they published the distribution patterns of clay minerals in surface sediments in all of the world oceans, with the exception of the Arctic Ocean. Later, Windom (1976) integrated all those data to produce average mineral assemblages (Table 2) and maps of the distribution of the four main clay mineral groups in the world oceans. Lisitzin (1972) presented a compilation of the numerous Russian investigations of deep-sea clay minerals, but this work, initially published in Russian, was only translated in 1996. The contour maps drawn by Russian scientific teams (Yeroshchev-Shak, 1964; Rateev et al., 1969; Lisitzin, 1996) have the advantage of presenting clay mineral distributions in both deep-sea sediments and potential source areas on land (see simplified Figure 4a–d, which show the distributions of illite, chlorite, smectite and kaolinite,

Table 1 Literature Review of Clay Minerals from Surface Sediments in Basins of the World Oceans.

Ocean	Oceanic Area	References
	World ocean	Griffin et al. (1968); Rateev et al. (1969); Windom (1976); Lisitzin (1996)
Pacific Ocean	All basins	Griffin et al. (1968)
	Central Arctic	Darby (1975); Stein, Grobe, and Washner (1994); Wahsner et al. (1999) and references therein
	Alaska, Chukchi Sea and East Siberian seas	Hein et al. (1979); Naidu et al. (1982); Naidu and Mowatt (1983); Moser and Hein (1984)
	Barents and Kara seas	Elverhøi, Pfirman, Solheim, and Larssen (1989); Nürnberg, Levitan, Pavlidis, and Shelkhova (1995); Gorbunova (1997)
	Laptev Sea	Serova and Gorbunova (1997)
	North East Pacific	Carson and Arcaro (1983)
	West Pacific basins	Chamley (1981)
Atlantic Ocean	All basins	Biscaye (1965)
	Labrador, Irminger and Iceland basins	Zimmerman (1982); Grousset et al.,(1983); Bout-Roumazeilles (1995); Fagel et al. (1996)
	Western North Atlantic margin	Zimmerman (1972); Piper and Slatt (1977)
	Eastern North and South Atlantic margin	Behairy et al. (1975)
	Equatorial Atlantic	Bremner and Willis (1993)
	South Atlantic	Petschick et al. (1996); Diekmann et al. (2000, 2003)
	Antarctic shelf	Ehrmann et al. (1992); Ehrmann et al. (2005); Hillenbrand and Ehrmann (2005)
	Mediterranean Sea	Venkatarathnam and Ryan (1971)
Indian Ocean	North Indian	Bouquillon et al. (1989)
	Arabian Sea	Kolla et al. (1981)
	Central Indian Basin	Rao and Nath (1988)
	South East Indian Ocean	Kolla and Biscaye (1973); Moriarty (1977); Kolla, Henderson, Sullivan et al. (1978); Gingele et al. (2001b)
	West Indian Ocean	Kolla et al. (1976)

respectively). As underlined by Chamley (1989), these maps are for "large-scale reference" purposes only.

2.5.2. Clay distribution in Atlantic Ocean surface sediments

Like gibbsite, kaolinite abundance decreases both northwards and southwards from the equatorial Atlantic region (Biscaye, 1965). The strong correlation between

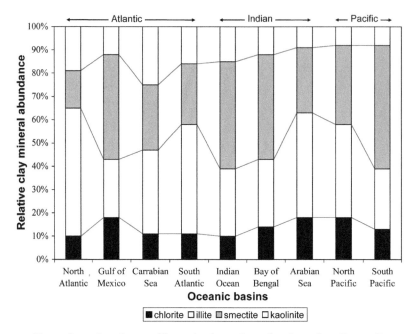

Figure 3 Mean clay mineral assemblages in the <2 μm fraction of surface sediments from basins of the world oceans compiled by Windom (1976) (Table 1). Illite and smectite are the two most variable clays. Inherited from continental landmasses, illite tends to decrease southwards, consistent with the asymmetrical distribution of landmasses. The climate-derived (or zonal) distribution of smectite is locally masked by the presence of additional authigenetic clays produced by alteration of volcanic material (including basaltic oceanic crust) and/or hydothermalism. In addition, major fluviatile inputs locally dilute deep-sea clay assemblages (e.g., smectite-rich inputs from the Mississippi River into the Gulf of Mexico; illite- and chlorite-rich inputs from the Indus into the Arabian Sea).

Table 2 Mean Clay Mineral Assemblages in the <2 μm Fraction of Surface Sediments from Basins of the World Oceans.

Ocean area	Illite	Chlorite	Smectite	Kaolinite	Nb. samples
North Atlantic	55	10	16	19	181–193
Gulf of Mexico	25	18	45	12	38
Carrabian Sea	36	11	28	25	49–56
South Atlantic	47	11	26	16	196–214
North Pacific	40	18	34	8	170
South Pacific	26	13	53	8	140–151
Indian Ocean	29	10	46	15	129–245
Bay of Bengal	29	14	45	12	25–51
Arabian Sea	45	18	28	9	29

Note: Data expressed as relative abundance in the clay-size fraction (i.e., values are summed to 100%). Data from Windom (1976).

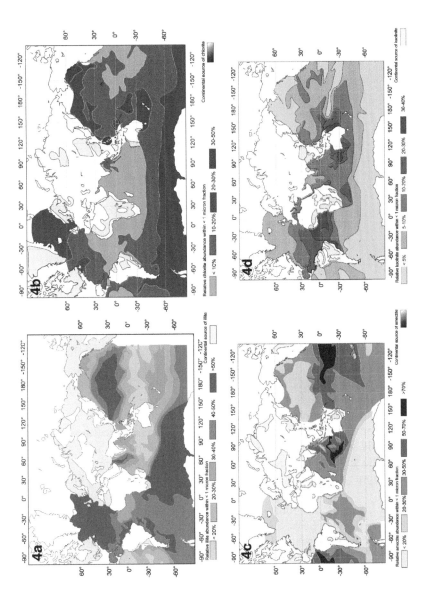

Figure 4 (A) Distribution of illite in the "clay" (<1 μm) fraction of surface marine sediments. Continental areas characterized by high abundance of illite (i.e., >50%) in soils are also reported. (B) Distribution of chlorite in the "clay" (<1 μm) fraction of surface marine sediments. Major continental sources of chlorite are reported. (C) Distribution of smectite in the "clay" (<1 μm) fraction of surface marine sediments. Major continental areas characterized by high abundances of smectite (i.e., >50%) in soils are also reported. (D) Distribution of kaolinite in the "clay" (<1 μm) fraction of surface marine sediments. Continental areas characterized by occurrence of kaolinite or kaolinite and gibbsite in soils are also reported. Simplified from Lisitzin (1996), with permission.

kaolinite and gibbsite abundances and proximity to continental areas undergoing intense tropical weathering seems to constitute conclusive evidence for their continental origin (Biscaye, 1965). Kaolinite constitutes 20–30% of the clay fraction in the eastern sector of the equatorial Atlantic but, due to input from the Niger River draining kaolinite-rich African equatorial soils and to wind transport from North African deserts, it reaches up to 50% in the western sector (Windom, 1976).

In Atlantic sediments, chlorite abundances increase towards the polar regions (Windom, 1976). This mineral is more prevalent in soils at higher latitudes where chemical weathering is less intense (Gradusov, 1974). The strong correlation between the continental and deep-sea abundances of chlorite supports a detrital origin for this mineral (Biscaye, 1965).

Illite is the dominant clay mineral in the $<2\,\mu m$ fraction of Atlantic Ocean sediments, showing minor variations compared to other clay minerals (Biscaye, 1965). Rivers play a major role in the asymmetrical distribution of illites in deep-sea Atlantic sediments (Rateev et al., 1969). There is an obvious illite zonation in the North Atlantic, which is supplied by large drainage areas within temperate and cold zones (Windom, 1976). In the South Atlantic, drainage areas are small or non-existent (Windom, 1976). The concentration of continental masses in the northern hemisphere supports a detrital origin for illites, a conclusion which is confirmed by the old K/Ar ages of these clays minerals (see above, Biscaye, 1965). Additional illite-rich inputs are transported by wind from desert soils of southwest Africa to the South Atlantic (Windom, 1976). For instance, the high concentration of illite measured in the suspended load of rivers draining South America contributes to the high illite content in South Atlantic sediments at the mouths of the Amazon, the Parana, the Orinoco, etc. In contrast, smectite-rich particulate inputs from the Mississippi dilute the illite content in the Gulf of Mexico. The lack of correlation between smectite, zeolites and volcanic shards, as well as the low abundance of smectites in surface Atlantic Ocean sediments point to a primarily detrital origin for smectite (Biscaye, 1965). Smectite abundance is low in the North Atlantic, where it is diluted by high inputs of illite and chlorite (Windom, 1976). The Mississippi River supplies smectite-rich particulate matter to the Gulf of Mexico and the Caribbean Sea. The Amazon River also supplies smectites to the southern North Atlantic which are mainly detrital, except in the vicinity of the Lesser Antilles Arc. The high smectite contents in the South Atlantic have been attributed to eolian inputs or alteration of Mid-Atlantic Ridge (MAR) material (Windom, 1976).

The resulting clay mineral distribution patterns highlight the fact that most recent deep-sea clays are derived from surrounding continents (Biscaye, 1965). The mineralogical analysis of the fine-grained fraction of Atlantic deep-sea sediments is therefore a useful indicator of sediment provenance. In the North Atlantic, the topographic control exerted by the MAR contributes to the importance of bottom current transport. In the equatorial Atlantic Ocean, detrital clays are transported by rivers draining South America and Africa, and by winds from Africa. Biscaye (1965) emphasizes the importance of AABW in the northward transport of clays in the South Atlantic.

2.5.3. Clay distribution in Pacific Ocean surface sediments

Griffin et al. (1968) related the relative amounts of chlorite, montmorillonite or smectite, kaolinite and illite in pelagic sediments to the sources and transport paths of solid particles from the continents to the oceans and to injections of volcanic materials into marine environments, especially in the South Pacific Ocean. According to Rateev et al. (1969), wind is the primary transport agent in the North and South Pacific.

Windom (1976) provides a detailed discussion of the maps presented by Griffin et al. (1968). Kaolinite is abundant in low latitude marine sediments. Pacific Ocean sediments have low kaolinite contents due to minimal terrigenous supply from river discharge, especially in the South Central and East Pacific. Along the eastern Pacific coast, relatively arid source areas with low run-off supply very little kaolinite. Regions of intense chemical weathering and high run-off along the western coast act as important sources of kaolinite. In the South Pacific, kaolinite-rich particles are supplied by wind from the Australian desert.

Although more uniform, the distribution of chlorite in deep-sea Pacific sediments is inversely related to that of kaolinite. The highest chlorite abundances occur in sediments adjacent to the Alasakan and Canadian coasts, reflecting suspended particle compositions in major rivers draining into this area. The Asian continent also supplies chlorite to the northwestern Pacific, which is subsequently redistributed by surface current. The high abundance of chlorite in the south-western Pacific has been attributed to eolian transport from Australian soils.

An illite-rich belt observed in North Pacific sediments has been attributed to eolian inputs. It is likely that illite enrichment in the South Pacific results from dust supplied from Australia deserts. It is interesting to note that sepiolite has been observed in a few samples from the South Pacific, including dust from New Zealand (Windom, 1976).

The highest smectite abundances observed are from the Pacific Ocean. Smectite is uniformly distributed and undergoes local dilution by continental run-off or wind-transported, continentally-derived material. In the South Pacific, high abundances of smectite are observed in the vicinity of volcanically active areas (Windom, 1976). The high abundance of smectite and its close association with zeolites and volcanic glasses are evidences for its volcanogenic origin (Griffin & Goldberg, 1963).

2.5.4. Clay distribution in Indian Ocean surface sediments

The kaolinite contents of recent Indian Ocean sediments are intermediate between those of the Atlantic and Pacific Oceans (Windom, 1976). A clear trend of increasing kaolinite abundance towards Western Australia, which is attributable to wind transport, is observed.

Indian Ocean sediments are depleted in chlorite, except in the Arabian Sea, where chlorite is supplied by rivers (e.g., the Indus) or wind (Windom, 1976). Illite contents are also low, decreasing towards the Indian coast as a result of dilution by smectite-rich fluvial detrital inputs from the Deccan traps. The exceptionally high illite abundance in the Bay of Bengal is related to Ganges run-off (Bouquillon, Chamley, & Frolich, 1989; Fagel, Debrabant, & André, 1994), whereas the high smectite content of the Indian Ocean is linked to river discharge (Windom, 1976). In addition, abundant authigenic smectites are observed locally in the eastern and southwestern Indian Ocean (Biscaye, 1965; Windom, 1976). Palygorskite has been observed in the Gulf of Aden, the Red Sea and the Arabian Sea (Windom, 1976; Kolla et al., 1981; Fagel et al., 1992a).

2.5.5. General trends in recent clay distributions in the world oceans

As previously highlighted for the Atlantic Ocean (Biscaye, 1965), the spatial distribution of clay minerals in the world oceans shows some significant regular patterns (Rateev et al., 1969). The relative abundances of the different clay minerals in bottom sediments show a pronounced latitudinal control. In bottom sediments of the world oceans, two groups of clay minerals can be distinguished based on their distribution patterns:

(1) kaolinite, gibbsite and smectite are concentrated in the tropical-humid zone, and their abundances decrease to both the north and south; this is called the equatorial-type distribution;
(2) chlorite and illite are concentrated at moderate and high latitudes, mainly within the cold, moderately humid and glacial zones; this is called the bipolar-type distribution.

To explain the latitudinal distribution of clay minerals, Rateev et al. (1969) called upon several factors related to the nature of the continental catchment basins (climatic zonation, type of continental weathering, and intensity of denudation), ocean hydrodynamics (current directions and the overall movement of water masses), the influence of volcanism, etc. It is possible to assess the influence of catchment basins on the distribution of clay minerals by comparing the distribution of clay minerals with the distribution of the different types of weathering and the intensity of mechanical denudation on the continents (Strakhov, 1960). The location of the kaolinite- and smectite-rich belt coincides with an area of intense chemical weathering in the tropics (35°N–35°S). The northern concentration maxima of illite (60–80%) and chlorite (>30%) are associated with the belt of normal weathering which encompasses the cold, moderately humid zones of Siberia and Canada. The distribution of clay minerals in world ocean sediments is closely related to latitude or climate. Kaolinite, smectite and gibbsite are formed in areas of tropical lateritic soil development. Illite and chlorite are mainly formed in soils of moderate and high latitudes. A gradual decrease in the abundance of equatorial minerals towards the poles and in the abundance of bipolar minerals towards the equator is the results of hydrodynamic scattering.

Deviations from the normal picture are associated with the asymmetry of catchment basins (e.g., the southern hemisphere), the effect of vertical climatic zonation on the delivery of clayey particles (India), and/or irregular patterns of ocean currents. A relationship between the distribution of minerals and hydrodynamics may be established by comparing compiled maps of clay mineral distributions with maps of currents in the world oceans. Rateev et al. (1969) noted the relationship between an area of high kaolinite abundances (40–60%) in the Indian and Pacific Oceans and the south trade wind currents, and the fact that the area characterized by 20–40%

kaolinite in the Atlantic Ocean is limited by the north trade wind and North Atlantic currents. In the southern hemisphere, westerly winds affect clay mineral distribution. Ocean currents which usually display a latitudinal direction play an important role in promoting the latitudinal zonation of clay minerals in ocean sediments by retaining them within their latitudinal belts. The influence of volcanism on the distribution of clay minerals in ocean sediments is evidenced by the occurrence of aclimatic and azonal areas of high smectite contents which are superposed on latitudinal climatic belts, thus creating a more complex zonal distribution.

Clay mineral patterns in deep-sea sediments match the overall latitudinal distribution of clay minerals in soils worldwide (600 samples; Gradusov, 1974). Regions with soils characterized by high concentrations of illites ($>70\%$) in the C horizon primarily occur in the northern hemisphere, with minor regions occurring in the southern hemisphere. These areas include morainic deposits in mountainous regions, shields in cold and temperate humid climates, as well as low altitude areas covered by alluvial and glacial deposits. All these areas are characterized by a cold climate. In the southern hemisphere, high illite contents are mainly observed in soils of mountainous regions. In parental material of the equatorial belt and bordering regions, illite abundance is minimal ($<30\%$). The illite content is usually lower in upper eluvial horizon A. Smectite and mixed-layers abundance maxima are observed in soils formed from alteration products of basic parental rocks in subtropical and tropical, arid and semi-arid climates. Smectite abundances tend to increase from the upper to the lower horizons of soil profiles. Soils with low smectite contents predominate in mountainous areas and shields characterized by free drainage and cold or temperate climates.

Later regional studies have mainly served to confirm the overall distribution patterns of clays in the world oceans (see Table 1). For instance, Naidu and Mowatt (1983) compiled clay mineral assemblage data from 700 sediment samples from the Gulf of Alaska and the South Bering Sea in the North Pacific Ocean. Although the data compiled were obtained by various investigators using different sample preparation, analytical and quantification techniques, which increases uncertainties (Naidu, Creager, & Mowatt, 1982), the results are consistent with the general latitudinal trends, controlled by climate and associated pedogenic processes, previously identified for clay mineral distribution in the world oceans (Biscaye, 1965; Griffin et al., 1968; Rateev et al., 1969). The distribution patterns of clay minerals in all of the marginal seas of Alaska (except the Beaufort Sea) are related to the various contemporaneous terrigenous sources of clay minerals and subsequent clay mineral dispersal by prevailing currents. Conversely, clay mineral distribution may be used to infer net current patterns. Naidu and Mowatt (1983) do, however, point to the higher kaolinite abundances and kaolinite/chlorite ratios in the north Bering, Chukchi and Beaufort Seas. This local kaolinite enrichment is attributed to reworking of kaolinite-bearing sedimentary rocks of northern Alaska, and subsequent transport to the adjacent Arctic seas. In the outer continental shelf region of the Beaufort Sea, the lack of recognizable dispersal patterns in clay distribution may be due to peculiarities of the sedimentary regime resulting from ice sediment transport (Naidu & Mowatt, 1983).

Exceptions to the latitudinal zonation of clay minerals in the oceans suggest that climate is not the only factor responsible for the distribution of terrigenous clay minerals (Thiry, 2000). Around continental masses, marine clay patterns are clearly controlled by terrigenous inputs, but the presence of abundant smectite in the centre of ocean basins must result from other processes, among which authigenesis and/or differential settling along advective marine currents are the most commonly mentioned (Thiry, 2000).

2.6. The Significance of Clays in Cenozoic Marine Sediments

Only the detrital clay component of marine sediments is indicative of weathering conditions in the continental formation environment. Detrital clays also provide insight into provenance and transport processes (Velde, 1995). Many papers dealing with the mineralogy of present-day marine environments have shown that clay minerals are, in general, continentally-derived (Biscaye, 1965; Griffin et al., 1968; Rateev et al., 1969). This also appears to be the case for older deposits. For instance, Chamley (1981) studied the long term Cenozoic clay deposition trends in DSDP material from different marine environments of the Mediterranean Sea, the North and South Atlantic and the northwestern Pacific, and concluded that the clay assemblages of recent deep-sea deposits are largely derived from the terrigenous supply of rocks and soil materials (Biscaye, 1965; Griffin et al., 1968). This is evidenced by comparing the average latitudinal zonation of clay mineral abundance in continental soils and in marine sediments, and by demonstrating a relationship between the principal geological provinces and clay sedimentary provinces. Mineralogical and geochemical studies point to the existence of similar conditions during past geological times in the same areas. The dominant nature of the terrigenous component in most Cenozoic deep-sea clays is suggested by: (1) the absence of continuous changes in clay content related to diagenetic modifications with the depth of burial; (2) the occurrence of a variety of clay assemblages, as in present-day deposits, including species derived from surficial and thermodynamically immature environments (e.g., poorly crystallized chlorite, irregular mixed-layers) which are not genetically associated with diagenetic minerals like clinoptilolite or opal CT; (3) the overall lack of a relationship between clay mineralogy and lithology; (4) the lack of a relationship between smectite abundance and volcanic activity, except in oceanic basalts and volcanogenic sediments; and (5) the occurrence of fragile fibrous clay minerals in reworked sediments. Long-term marine clay sequences are therefore potential tools in deciphering continental paleoenvironments expressed by changing pedogenesis, marginal sedimentation, erosion, transport and sedimentation.

Because of their continental provenance, clays have been used to reconstruct oceanic paleoenvironments over the Cenozoic period. The interpretation of down-core changes in mineral concentrations depends on the best possible information on the modern distribution and sources, as well as independent climate *proxy data* about the source area (Gingele et al., 2004). The significance of clay minerals for paleoenvironmental reconstructions varies with site location as well as sediment age. For instance, Ehrmann et al. (1992) studied different geological settings in the Antarctic Ocean to assess the significance of clay mineral assemblages in reconstructing the glacial history of the Antarctica Peninsula, the paleoceanographic history of the Antarctic Ocean, and the sedimentary processes operating at the Antarctic continental margin. They concluded that clay mineral assemblages in late Mesozoic to Paleogene sediments are sensitive tools for reconstructing climatic conditions. For example, the shift from smectite-rich to illite- and chlorite-dominated assemblages in the earliest Oligocene reflects the transition from chemical weathering conditions under a warm and humid climate to physical weathering under cooler conditions. Submarine plateaus (e.g., Maud Rise, Kerguelen Plateau) provide the best record for direct paleoclimatic studies. On proximal sites of the continental slope and shelf, as well as in the deep sea, paleoclimatic information is often masked by a variety of processes resulting in sediment redistribution. At those sites, however, clay mineral assemblages carry a wealth of information on different sedimentary processes. Following the establishment of the continental East Antarctic ice sheet, physical weathering prevailed and variations in the clay mineral record predominantly reflect the influence of different sediment sources controlled by different glacial, hydrographic or gravitational transport processes. Because these sedimentary processes are generally linked to climatic variations, clay mineral assemblages in most of Neogene and Quaternary sediments in Antarctic deposits provide indirect paleoclimatic information (also see Robert, 1982; Robert, Caulet, & Maillot, 1988). These processes are best documented by clay mineral composition in those areas where changes in source regions with distinct petrographic properties are expected and where distances from the source region are short.

Clay mineral assemblages in the approximately 1,600 m thick Cenozoic sedimentary succession off of Cape Roberts on the McMurdo Sound shelf in Antarctica were analysed in order to reconstruct the paleoclimate and glacial history of this part of Antarctica (Ehrmann, Setti, & Marinoni, 2005). The Cenozoic sediments exhibit a large number of unconformity-bound, fining-upwards cycles attributed to sea level variations or changes in the proximity of the ice masses. The lack of correlation between sedimentary facies and clay mineral assemblages, and the absence of systematic differences in clay signature between proximal and distal glaciomarine sediments suggest that the clay mineral distribution pattern is not influenced by transport, sedimentation and reworking processes, but that it is rather primarily controlled by climate conditions and changes in source area.

In the North Atlantic, this approach has enabled paleohydrological reconstructions based on clays from different periods, including the Quaternary (Latouche & Parra, 1979), the Cenozoic (Latouche, 1978) and the Mesozoic (Chamley, 1981). In some environments, the use of clay data for sedimentological reconstruction purposes may be difficult. For instance, in volcanic environments, volcanic products and their by-products (i.e., authigenic minerals from aerial and sub-marine environments) mix with terrigenous inputs, making it difficult to differentiate the two types of materials (Grousset et al., 1983). Moreover, in the North Atlantic, source rocks yield relatively constant clay mineral assemblages through time with minimum variability between glacial and interglacial stages (Fagel & Hillaire-Marcel, 2006). Elsewhere, processes such as rapid tectonic uplift, as in the Himalayas, may result in progressive changes through time (e.g., Bouquillon et al., 1989; and other references in Chamley, 1989).

2.7. Provenance of Detrital Inputs

Detrital clay mineral assemblages in deep-sea sediments have been widely used to determine the provenance of detrital inputs. Clay mineralogy has been used as a tracer of provenance and transport mechanisms in studies of the world oceans in which the mineralogy of the fine detrital fraction ($<20\,\mu m$) generally reflects the intensity of continental weathering in the source areas (Biscaye, 1965; Griffin et al., 1968). The effectiveness of clay proxies is best documented in areas characterized by distinct mineralogical provinces (e.g., the southwest Pacific, Gingele, De Deckker, & Hillenbrand, 2001a; the southeast Indian Ocean, Moriarty, 1977; the Arabian Sea, Fagel et al., 1992a; the South China Sea, Liu et al., 2003; the eastern Mediterranean Sea, Venkatarathnam & Ryan, 1971). For instance, clay mineral assemblages in 166 surface sediments from the southwest Pacific, between Indonesia and NW Australia, show a strong relationship with the geology and weathering regime of the adjacent hinterland and allow the distinction of four clay mineral provinces (Gingele et al., 2001a). Likewise, the study of 245 surface sediments from the southeast Indian Ocean and southwest Pacific allows the delineation of four geographical provinces (Moriarty, 1977).

The alternation of two distinct clay mineral assemblages as expressed by their clay mineral ratios has been used to trace seasonal transport direction changes related to monsoon circulation in the South China Sea (Liu et al., 2003) and the Arabian Sea (Fagel et al., 1992a), for instance. In the northern South China Sea, present-day illite and chlorite mainly come from northern source areas (Taiwan and the Yangtze River), and are carried by the surface currents which prevail during the winter monsoon. In contrast, smectites are mainly derived from southern volcanic source areas (Luzon and the Indonesian islands) and are transported by summer monsoon circulation (Liu et al., 2003). In the Arabian Sea, southwesterly summer monsoon winds carry palygorskite-rich clay assemblages from the Somalian and Arabian desert regions whereas the winter northeasterly monsoon winds bring illite-rich material from the Indus River (Fagel et al., 1992a). It should be pointed out that comparing two clay mineral components using their ratio reduces the effect of dilution by other components, and hence facilitates interpretation (Gingele et al., 2001b).

In general, the presence of multiple sources and transport processes makes it difficult to assign one main source area to a given clay mineral assemblage. For instance, Carson and Arcaro (1983) emphasized the fact that constraining the areal distribution patterns of clay minerals is not sufficient to define their provenance. Clay mineral studies require analysis of mineralogy-size variables to assess the relative contribution of provenance and selective transport. Carson and Arcaro (1983) studied Late Pleistocene and Holocene sediments from the Cascadia Basin and Juan de Fuca abyssal plain in the northeast Pacific Ocean. They noted that clay mineralogy is size-dependent, with a grain-size mineralogy relationship similar to that reported by Gibbs (1977) for sediments along the Amazon shelf, namely that the fine fraction is enriched in montmorillonite. The increase in montmorillonite in Holocene sediments cannot be explained by a change of source. For the American continental shelf, Nittrouer (1978) established that selective transport resulted in the preferential deposition of sands in nearshore waters and silts to silty clays on the middle to outer continental shelf. If the shoreline and associated river mouths are immediate sources of clays, a change in the position of the shoreline would result in a change in the distance to any deposition site. Moreover, selective transport is a function of current velocity: a decrease in bottom velocity could lead to progressively smaller particles being deposited. Given the size dependency of the clay mineralogical content, less vigorous currents than those that were active during the Pleistocene could produce smectite-rich Holocene deposits without any change in provenance.

Further constraints on sediment sources may be provided by complementary analyses. In some cases, detailed studies of X-ray patterns and cation saturations can yield additional information. For instance, the link between smectite composition in sediments from the northern North Atlantic and the Labrador Sea and deep circulation allows a clear distinction between the various dominant terrigenous sources associated with the main components of the modern Western Boundary Undercurrent (Fagel, Robert, Preda, & Thorez, 2001).

The radiogenic isotopic signature of the detrital sedimentary fraction has been used to trace sediment provenance. Huon, Jantschik, Kubler, and Fontignie (1991) measured the isotopic K/Ar signature of the clay-size fraction of NE Atlantic sediments to constrain the sources of detrital material. In ice-rafted layers, K/Ar ages reflect an enrichment in old detrital material which is consistent with erosion of Precambrian rocks. Because the contribution of radiogenic ^{40}Ar in recent rocks is negligible, it was difficult to pin down the influence of recent volcanically derived material. Sm/Nd systematics provide better insight into the mixing of crustal and volcanically derived material. Goldstein and O'Nions (1981) have shown that detrital clays in marine sediments retain their Nd isotopic signature throughout the sedimentary cycle, that is, from continental weathering, through sediment transport to diagenesis. Nd isotopic analyses of these terrigenous fractions can therefore provide direct and quantitative information on sediment provenance (McCulloch & Wasserburg, 1978; Jones, Halliday, Rea, & Owen, 1994). In order to constrain complex sedimentary mixing, Nd isotopic analysis may be coupled with other radiogenic isotope measurements. For instance, Nd and Pb isotopes have proven to be powerful tracers of the origin and provenance of deep-sea sediments in subduction areas (e.g., White, Dupré, & Vidal, 1985; White & Dupré, 1986; Vroon, Van Bergen, Klaver, & White, 1995), in worldwide turbidites (McLennan, McCulloch, Taylor, & Maynard, 1989) or, more regionally, in the Indian (Dia, Dupré, & Allègre, 1992), the Pacific (Jones et al., 1994) and the South Atlantic Oceans (Bayon, German, Nesbitt, Bertrand, & Schneider, 2003). In particular, these isotopes constitute suitable tracers of the origin of fine-grained deep-sea sediments of the northwestern North Atlantic (Labrador Sea), where, at present, mid-Atlantic mantle sources interact with old crustal North American inputs (Innocent, Fagel, Stevenson, & Hillaire-Marcel, 1997; Fagel, Innocent, Gariépy, & Hillaire-Marcel, 2002; Fagel et al., 2004). Nd and Pb isotopes have been measured in the clay-size fraction of Late Glacial and Holocene deep-sea sediments recovered from two Labrador Sea piston cores. These data provide better constraints on the different source areas that supplied clay-size material into the Labrador Sea. Changes in their relative contribution through time further constrain the deep circulation pattern from the last deglaciation until 8.6 ka (Fagel et al., 2002).

Although Sr isotopes are less conservative than Nd isotopes during weathering processes (Goldstein, 1988; Allègre, Dupré, & Négrel, 1996), sedimentary Sr isotope signatures have been successfully related to continental sources of the detrital fraction in the North Atlantic (Dasch, 1969; Biscaye, Chesselet, & Prospero, 1974), the equatorial Atlantic (Hemming et al., 1998) and the northern (Asahara, Tanaka, Kamioka, & Nishimura, 1999) and western Pacific (Mahoney, 2005). Nd and Sr have been widely used to trace detrital provenance in the North Atlantic (Grousset et al., 1988; Revel, Sinko, Grousset, & Biscaye, 1996a), in the South Atlantic (Walter et al., 2000; Bayon et al., 2003) and in the Arctic Oceans (Tütken, Eisenhauer, Wiegand, & Hansen, 2002).

Although such multiproxy methods remain scarce in the literature, coupling radiogenic isotope analyses and clay mineralogy is a powerful approach. The interpretation of isotopic data is made easier with prior knowledge of the main potential sources of detrital particles obtained from recent clay distribution, especially in complex environments involving multiple transport agents. For instance, Bayon et al. (2003) have analysed the Nd isotopic composition of the entire detrital fractions (mainly $<20\,\mu m$) of four sediment cores from the SE Atlantic Ocean to investigate variations in the accumulation of particles transported by Circumpolar Deep Water (CDW) and North Atlantic Deep Water (NADW) to this region during the Late Quaternary. Their data reproduce the complex modern-day hydrography of the SE Atlantic Ocean and confirm the relative variations in NADW and CDW at the Last Glacial Maximum (LGM). On the basis of Nd concentrations and isotopic compositions of terrigenous sediment in the Cape Basin, Bayon et al. (2003) demonstrated that the supply of terrigenous material derived from the southwest Atlantic region was greater during glacial periods than during interglacials. These authors suggested that the increase in the flux of clays from the southwest Atlantic region was the result of increased transport into the Cape Basin by CDW. The glacial-related change in deep circulation was questioned by Rutberg et al. (2005) who called instead upon higher glacial inputs from Patagonia (Diekmann et al., 2000; Walter et al., 2000). Indeed, clay mineralogy records from Cape Basin sediments at ODP 1089 show systematic climate-related changes (Rutberg et al., 2005). Kuhn and Diekmann (2002) showed that kaolinite/chlorite ratios varied by more than a factor of two, low values corresponding to glacial periods. They determined that Patagonia was the principal source of chlorite input into Cape Basin sediments. In fact, chlorite enrichment (Kuhn & Diekmann, 2002) and the shift in Nd isotopic signature (Bayon et al., 2003) suggest a greater contribution from South America or other western sources during the LGM rather than any change in the deep current regime (Rutberg et al., 2005).

Walter et al. (2000) have also combined clay mineralogy and Sr–Nd isotopic data for Late Quaternary surface sediment and sediment cores from the South Atlantic and southeast Pacific to reconstruct past circulation changes in the Austral Ocean. The glacial/interglacial shift in sources may be due to either the decreasing influence of NADW during glacial times or a larger contribution of glaciogenic detritus from southern South America. Hemming et al. (1998) combined clay mineralogy and Sr isotopic data of the detrital bulk sedimentary fraction of western equatorial Atlantic sediments to constrain particle provenance. The geographic gradients in clay mineralogy and Sr isotopic composition of the detrital bulk fraction suggest mixing between Andean and Brazilian shield-derived sediments transported by the nepheloid layer from more kaolinite-rich sources, probably located in the Brazil Basin.

Isotopic measurements can be performed on the bulk detrital fraction (e.g., Walter et al., 2000; Bayon et al., 2003) or on specific size fractions ($<63\,\mu m$, Hemming et al., 1998; $10–63\,\mu m$ and $<10\,\mu m$, Revel, Cremer, Grousset, & Labeyrie, 1996b; $<2\,\mu m$, Innocent et al., 1997; Fagel et al., 1999). Walter et al. (2000) analysed the Sr isotopic compositions of bulk, silt and clay fractions and pointed out that this approach allows additional constraints to be placed on transport processes. Innocent, Fagel, and Hillaire-Marcel (2000) showed that the clay-size fractions ($<2\,\mu m$) yielded different Sm-Nd signatures than the cohesive silt fraction ($2–10\,\mu m$). Of these different size fractions, the clay-size fraction appears to be of particular interest for tracing sedimentary inputs transported by deep currents (Fagel et al., 2002). Silt fractions mainly record the influence of proximal supplies, whereas clay-size fractions are more sensitive to sedimentary input advected by deep currents (Innocent et al., 2000). Clay-size particles may be transported by deep currents then deposited when current strength decreases. However, because of their cohesive behaviour, they are less sensitive to further current winnowing following deposition (McCave, Manighetti, & Beveridge, 1995). More investigations of grain-size dependence are required, because Tütken et al. (2002) provided evidence for the grain-size-dependent fractionation of Sr isotopes, a behaviour which Nd isotopes do not appear to display.

2.8. Relationship between Clay Mineralogy and Ocean Circulation

2.8.1. General statement

Clay mineral assemblages have been used to trace ocean currents at the ocean scale (e.g., the Atlantic Ocean) since the sixties (Biscaye, 1965) and, more recently, at the regional scale as, for instance, in the North Atlantic (Grousset & Latouche, 1983), the South Atlantic (Gingele, 1996; Petschick et al., 1996; Figure 5), the East Indian Ocean (Gingele et al., 2001a), the southeast Indian Ocean and the southwest Pacific Basin (Moriarty, 1977). In given water mass, fine-grained clay minerals can be advected over considerable distances to finally settle far away from their original source. There has been much discussion in the literature about the usefulness of clay minerals in delineating suspended sediment dispersal routes and in indicating the provenance of marine fine-grained deposits (e.g., Biscaye, 1965; Knebel, Conomos, & Commeau, 1977). Hein et al. (1979) showed that clay minerals in estuarine and shelf deposits in south central Alaska provide a record of their source areas and patterns of dispersal. Surface clay mineral distributions have allowed the reconstruction of a mean regional oceanic circulation pattern which is consistent with independent current, salinity and temperature data.

A prerequisite for the reconstruction of transport pathways is the identification of specific source areas on the adjacent landmasses (Hillenbrand & Ehrmann, 2005). Clay mineral assemblages in surface ocean sediments, which differ significantly from those in local sources, have already been used successfully to outline the extent and propagation of water masses (e.g., Petschick et al., 1996, for the South

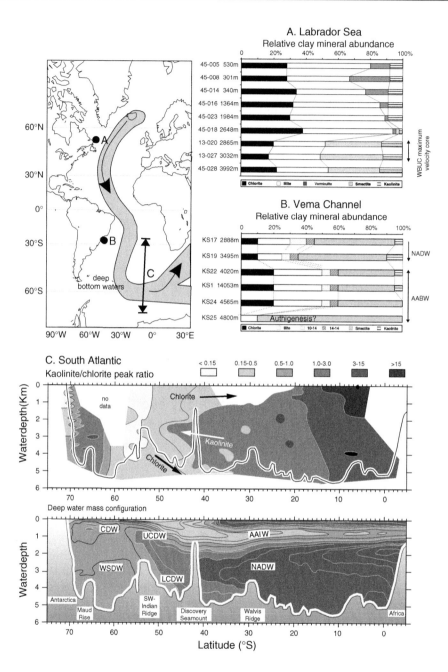

A. Labrador Sea
Relative clay mineral abundance

B. Vema Channel
Relative clay mineral abundance

C. South Atlantic
Kaolinite/chlorite peak ratio

Deep water mass configuration

Atlantic; Gingele et al., 1999, for the eastern subtropical Atlantic; Fagel, Robert, & Hillaire-Marcel, 1996, for the northern North Atlantic; Gingele et al., 2001a, for the southeast Indian Ocean). Over glacial/interglacial cycles, changes in marine clay mineral assemblages usually record changes in source areas or transport media (Gingele et al., 2001a). Once the relationship between a specific clay mineral suite and a certain source, water mass or current system is established, variations of this suite in down-core profiles may be used to constrain fluctuations in the propagation of the water mass or the current system (Diekmann et al., 1996; Gingele, Schmieder et al., 1999; Gingele et al., 2004).

2.8.2. Surface sediments
Many studies have demonstrated a relationship between clay mineral assemblages in surface sediments and water depth. Venkatarathnam and Ryan (1971) mapped the distribution of clay minerals in surface sediments in the deep-water basins of the eastern Mediterranean Sea to trace the sources and understand the mechanisms of sediment dispersal in this body of water by correlating the distribution of clay minerals with water circulation patterns. These authors concluded that bottom currents are important in transporting illite and chlorite-rich sediments into the deep parts of the Ionian Basin.

In the South Atlantic, advection of chlorite with southern AABW was already suggested by Biscaye (1965). For the southwest Pacific, Moriarty (1977) interpreted the high chlorite content in the South Tasmanian Basin to have resulted from the

Figure 5 The relationship between clay mineral assemblages and deep oceanic circulation. Evidence for clay mineral evolution with water depth reported for three areas (labelled A, B, C) located along the present-day general circulation path of cold deep bottom water generated in the northern North Atlantic. (A) SW-NE transect across the Canadian margin, Labrador Sea, northern North Atlantic. The clay mineral assemblage changes with increasing water depth (data from Fagel et al., 1996, with permission). Note the sharp change in clay mineral composition and abundances between 2,648 and 2,865 m. Abundant vermiculite-like clays are only reported in sediments from the continental shelf. The disappearance of vermiculite coincides with massive amounts of smectite in the deepest sediment cores. The highest smectite abundance occurs at depths consistent with the maximum velocity core of the WBUC. (B) NW-SE transect along the western flank of the Rio Grande Rise, Vema channel, Equatorial Atlantic (data from Melguen et al., 1978, with permission). A significant change in clay mineral assemblage is observed between 3,495 and 4,020 m, i.e., at the lower limit of NADW and the upper limit of AABW. Chlorite abundance in sediments collected between 4,000 and 4,565 m is twice as high as that observed in shallower sediments (<3,495 m). Chlorite was widely used as a proxy for AABW-transported inputs. The significant change in chlorite abundance in the deepest sediments is striking; such smectite-dominated clay assemblages may reflect the presence of authigenetic clays, although this is not confirmed by any geochemical anomaly in the bulk sediments (Melguen et al., 1978). (C) Latitudinal transect through the South Atlantic, from Antarctica to the equator including data collected from 20°W to 20°E and projected along the Greenwich meridian. Data from Petschick et al. (1996), with permission. Upper panel: relationship between the kaolinite/chlorite ratio (3.58 Å/3.54 Å) and water depth. Lower panel: water mass distribution. This figure, simplified from Petschick et al. (1996), clearly demonstrates the strong link between clay mineral assemblage and the nature of the intermediate or deep water mass: at present, kaolinite-rich inputs are transported by NADW, whereas chlorite-enriched inputs are carried by Antarctic Intermediate Water and Lower Circum Antarctic Deep Water (LCDW, i.e., previously AABW).

reworking of sediment from the Tasman Fracture zone by AABW bottom current. In the southeast Indian Ocean, dispersal of both the clay and quartz components in deep basins is also largely controlled by AABW circulation (Kolla, Henderson, Sullivan, & Biscaye, 1978). Quartz is primarily continent-derived, whereas clay abundances may reflect both continental derivation and the alteration of submarine basalts. Not only are clay abundances high in zones of high quartz abundances, but clays also extend farther into deeper, more seaward areas than quartz. Clays are also more readily dispersed than the coarse-grained quartz and are characterized by more extensive, uniform distribution of high abundances. Clay abundances cannot be used as a reliable indicator of the extent of influx of continental detritus. To do this, data on clay mineralogy and clay accumulation rates are required (Kolla et al., 1978).

The use of deep-sea clay mineral assemblages as indicators of water mass fluctuations (i.e., NADW versus AABW) was well documented for the Vema Channel in the southeast Atlantic (Petschick et al., 1996). Based on their strong latitudinal and opposite distribution patterns in surface sediments of the South Atlantic, kaolinite and chlorite were inferred to be useful tracers of the major deep water masses in the vicinity of the Rio Grande Rise. Advection of chlorite with southern AABW deep water was already suggested by Biscaye (1965). Melguen et al. (1978) demonstrated that the clay mineralogy of six cores from the Vema Channel largely depends on the provenance of the water masses (AABW or NADW) that transported the inherited products of continental weathering. They concluded that AABW transports chlorite and crystalline illite northwards. In contrast, the distribution of kaolinite and poorly crystallized illite is related to southwards flowing NADW (Figure 5).

More recently, the advection of kaolinite with NADW was documented from numerous cores (> 900 samples) from the South Atlantic (Diekmann et al., 1996; Petschick et al., 1996). Alternatively, assuming that the suspended load of NADW in the western South Atlantic is too small to account for significant kaolinite enrichment, Jones (1984) proposed that kaolinite is advected along isopycnals from resuspended kaolinite-rich deposits of the Sao Paulo Plateau. Kaolinite/chlorite ratios of surface sediment samples provide evidence for both the advection and isopycnal transport models (Gingele et al., 1999). On the one hand, the symmetrical nature of the kaolinite enrichment at mid-depths on either side of the Rio Grande Rise is consistent with advective transport from low latitude. Consequently, similar kaolinite/chlorite ratios should be observed at comparable depths along the Middle Atlantic Ridge. However, only a slight increase in kaolinite/chlorite values is recorded near the surface, which may result from minor advection of kaolinite from low latitude. On the other hand, the transport of kaolinite to the western slope of the Rio Grande Rise along isopycnals, from the Sao Paulo Plateau may explain the kaolinite enrichment on the western side of the Rio Grande Rise, but not on its eastern side. To explain the symmetrical nature of the kaolinite enrichment on the Rio Grande Rise, Gingele et al. (1999) suggested the injection of suspended kaolinite at intermediate depths of the NADW off the mouth of the Rio Doce, and short-distance transport southwards instead of long-distance advection from lower latitude.

In the Labrador Sea, the clay mineralogy of deep-sea sediments has been used to trace the outflow of the Western Boundary Undercurrent (WBUC), the dominant lateral transport agent of suspended load in the North Atlantic (Eittreim & Ewing, 1974). Surface sediments from both margins of the Labrador Sea provide evidence for a link between smectite abundance in the clay fraction and the depth of the high velocity axis of the WBUC (Fagel et al., 1996; Figure 5). Based on a significant shift in the Nd isotope signature (Innocent et al., 1997), the WBUC is thought to be responsible for erosion and transport of clay particles from the smectite-rich western North Atlantic Iceland and Irminger basins, followed by redeposition in the Labrador Sea.

2.8.3. Late Cenozoic sediments

Grousset and Latouche (1983) have suggested that smectite-rich advective inputs in northwest Atlantic basins derived from erosion of Icelandic products were probably transported by Norwegian bottom currents. In the northeast Atlantic, changes in the smectite/illite ratio and variations in smectite and illite fluxes have been used to monitor variations of the WBUC outflow and, consequently, to constrain the NADW evolution during the last glacial/interglacial transition (Fagel et al., 1997b). This interpretation derived from clay mineralogy was later confirmed by estimating the relative contribution of proximal and distal current-driven clayey inputs, which are characterized by distinct Nd isotopic signatures (Fagel et al., 1999).

The clay mineral record in South Atlantic sediments (ODP Site 1089) has been used to reconstruct regional current systems in response to climate variability over the last 590 ka (Kuhn & Diekmann, 2002). Clay minerals indicate the source of terrigeous mud and trace the dispersal of fine-grained suspensions by water-mass advection within oceanic currents (Diekmann et al., 2003). Terrigenous sediments mainly originate from African sources with minor contributions from distal southern sources, and are supplied by circumpolar water masses, NADW, and surface currents of the Agulhas Current (Kuhn & Diekmann, 2002). Surface distribution maps and depth transects clearly indicate that kaolinite is especially abundant in the deep waters of the South Atlantic basins, and closely associated with NADW circulation (Petschick et al., 1996). Changes in clay mineralogy are best displayed by the quartz/feldspar and kaolinite/chlorite ratios of the clay-size fraction. Mineralogical variations reflect both the northward displacement of NADW injection into the Antarctic Circumpolar Current and a weakening of the Agulhas Current during glacial stages, sub-stages and stadials. During glacial stages, the influence of NADW decreases and allows the spreading of southern-source water masses in the area. The LGM map shows that low kaolinite/chlorite ratios in the northeast migrate towards lower latitudes and shallower water depths on topographic highs because of the larger extent of chlorite-bearing southern-source water masses and the reduced influx of kaolinite-bearing NADW (Diekmann et al., 2003).

In the southeast Indian Ocean, three cores taken below the path of the present day Leeuwin Current in the Timor Passage, offshore of the Australian Northwest Shelf and of the North West Cape of Western Australia, provide a Late Quaternary record of environmental changes (Gingele et al., 2001b). Today, kaolinite and chlorite are transported into the Timor Passage by the Indonesian Throughflow, while illite is locally derived from Timor. Where the Leeuwin Current leaves the Timor Passage, it bears a characteristic clay mineral signature acquired in the Indonesian Archipelago (kaolinite, chlorite and illite). The uptake of clay minerals along its path through the Timor Sea (e.g., illite from the Kimberley area) changes this signature. South of North West Cape, chlorite injected by rivers of the Pilbara region into the path of the Leeuwin Current is a prominent constituent of surface sediments at water depths less than 1,000 m and delineates the present-day flow of the current. During the last glacial period, the volume of the Indonesian Through-flow decreased, and less kaolinite and chlorite reached the Timor Passage. Offshore from North West Cape, a reduction in the amount of chlorite during the last glacial may indicate a decrease in the intensity of the Leeuwin Current or its absence, and/ or a lower chlorite input due to drier conditions on land. An illite maximum in Holocene and recent sediments offshore of North West Cape is the result of ma-terial input from rivers which periodically drain the adjacent hinterland. Lower illite abundances point to a drier climate in the area during the last glacial stage.

Recently, Barker and Thomas (2004) highlighted the fact that their "knowledge of the development and palaeoclimatic significance of the ACC will be best served by (…) mineralogical studies of clays as a way of examining provenance and therefore surface and bottom current directions and the existence of interocean connections".

2.9. Relationship between Clay Minerals and Climate

2.9.1. General statement

The paleoclimatic interpretation of clay mineral data requires knowledge of the potential source areas (Moriarty, 1977), as well as of the mode and strength of transport processes (Diekmann et al., 1996; Gingele et al., 1999). Variations in down-core clay mineral distribution in deep-sea sediments have been interpreted in terms of changes in the climatic conditions prevailing in the continental source area (Chamley, 1967, 1989; Robert & Maillot, 1983; Clayton, Pearce, & Pzeterson, 1999; Foucault & Mélières, 2000) and, potentially, subsequent changes in sea level (Robert et al., 2005). The paleoclimatic interpretation of clays is based on the assumptions that they are detrital and have not been significantly altered by diagenesis, that their source areas can be identified, and that they represent secondary products of continental weathering (Singer, 1984). Based on the latitudinal climate-driven clay-mineral distribution trends in recent deep-sea sediments (Biscaye, 1965; Griffin et al., 1968; Rateev et al., 1969), variations in vertical marine clay mineral distribution patterns have been interpreted in terms of shifts in the climatic conditions prevailing in the continental source areas of the detrital clay minerals (Singer, 1984) and have been widely used to reconstruct paleoclimates.

However, the interpretation of past climate change based on clay mineralogy is complicated by several factors described below (Moriarty, 1977; Singer, 1984; Chamley, 1989; Thiry, 2000):

(1)   Clay minerals go through several stages between their development in soils and their final deposition in an ocean basin. For instance, clay minerals may persist

in soils through climate changes (Caroll, 1970), especially in regions where humid and arid weathering conditions alternate.

(2) The signal may be obliterated by erosion and transport processes as erosion products from bedrock and different soil horizons undergo mixing. Soils formed during interglacial periods in the Arctic can be eroded during the following glacial period, and become mixed with glacial "rock flour" and loess (Darby, 1975; Caroll, 1970).

(3) Marine sediment may also contain ancient recycled clay minerals which hold no paleoclimatic information regarding the period of interest (Singer, 1984). Unexpectedly high amounts of kaolinite at high latitudes in the Arctic Ocean are the result of reworking of kaolinite-bearing Mesozoic sedimentary rocks from surrounding continental masses (Singer, 1984). Although the deposition of kaolinite in recent sediments of the world oceans is consistent with proximity to wet climatic zones, this is partly due to the fact that kaolinitic paleosurfaces coincide with tropical areas where kaolinite is still developing at present (Thiry, 2000).

(4) There is a lag time between the formation of the soil-derived clay assemblages on the continents and their arrival in ocean basins. Clay sedimentation may occur several thousands of years and more after the formation of soils on the continent, and climate may have considerably changed during that time. The resolution of the paleoclimatic record in marine clays is therefore limited by soil formation rates, which are sometimes on the order of 1 or 2 Ma (Moriarty, 1977; Thiry, 2000).

Although clay assemblages in Cenozoic DSDP sedimentary records are primarily controlled by climate, other factors, such as selective distribution of clay minerals during erosion and transport (Singer, 1984), may readily modify them. Rivers are the principal carriers of fine-grained sediments from continental areas into the ocean. With increasing distance from the source areas, the sediment load decreases, but the extent of this decrease is a function of clay mineral type. Later deposition of fine-grained marine sediments is controlled by current flow strength and direction, and seasonal meteorological and oceanic factors (winter storm, periodic river discharge, seasonal upwelling, etc.). Although these factors are related to climate, their relationship to it is complex (Singer, 1984). The relationships between climate and paleoceanography will be illustrated in the last section of this chapter (see Section 3, "Applications").

According to Thiry (2000), clay mineral assemblages provide an integrated record of overall climatic impacts. Paleoclimatic interpretations of marine clay assemblages are "yielding nothing more than rather broad paleoclimatic information". However, short-term changes in clay mineral assemblages have been observed in many oceanic basins. These were not directly related to climate-induced changes in the adjacent landmasses, but were interpreted as resulting from either changing pathways and/or strength of transport processes (Latouche & Parra, 1979; Kissel, Laj, Lehman, Labeyrie, & Bout-Roumazeilles, 1997; Liu et al., 2003), local changes in glacial erosion related to ice sheet advance and retreat and in the supply of source rocks (Vanderaveroet, Averbuch, Deconinck, & Chamley, 1999; Hillenbrand & Ehrmann, 2005), fluctuations in river discharge between humid and arid phases

(Gingele et al., 1999) or sea level change and the related variation in shoreline position (Carson & Arcaro, 1983). These processes are direct or indirect consequences of climate change. According to Gingele et al. (1999), multiple sources and transport processes as well as the dilution of different clay minerals make it difficult to relate changes in the down-core record of any one component to changes in specific paleoclimatic parameters. In general, comparing two clay mineral components using the ratio of one to the other reduces the effect of dilution by other clays, which facilitates interpretation (Gingele et al., 1999).

2.9.2. Glacial/interglacial fluctuations of clay mineral assemblages

Short-term clay mineralogical (and geochemical) changes (i.e., over glacial/interglacial periods) were observed in four cores of Quaternary sediments from the Northwest Atlantic Mid-Ocean Canyon (Latouche & Parra, 1979). The mineralogical evolution over the last climate cycle was explained "both by climatic changes and the nature of the sedimentary sources which successively contributed to the deposits". During the glacial period, illite- and chlorite-rich sediments are autochthonous, carried from the nearby continental platform by turbidity currents. The lack of other inputs is explained by the shutdown of the main circulation of North Atlantic water masses during the glacial period. During the interglacials, sedimentary material inputs were more complex, the variability of sedimentary materials reflecting the establishment of circulation patterns similar to present-day patterns. Detrital deposits are mixed with smectite-rich material which may have originated from volcanic areas such as Iceland, the Reykjanes Ridge or the Gibbs fracture and have been carried by Norwegian deep-sea waters. Furthermore, the Gulf Stream seems to account for the transport of montmorillonitic clay mineral inputs from southern areas towards the north. Thus, clay mineral assemblages provide indirect information about hydrological features of the study area over the last 120,000 yr. Parra (1982) attributed mineralogical differences between glacial and interglacial periods primarily to changes in erosion and alteration conditions on the continent, while recognizing that hydrological changes may also play a role. The distribution of clay minerals in Late Quaternary sediments between the Gibbs fracture and the Greenland basin is attributed to glacial/interglacial cycles (Grousset et al., 1982), although this relationship does not result from climate-induced clay formation in the continental source areas, but from the relative influence of different transport and deposition agents, such as bottom currents in warm periods and ice rafting during cold periods (Singer, 1984). Also in the northeast Atlantic basins, Kissel et al. (1997) combined the study of clay minerals, and magnetic anisotropy and susceptibility for one core (SU90-33) from 2,400 m water depth along the Iceland–Scotland overflow water (ISOW), a branch of NADW. The magnetic susceptibility record indicates down-core variations in the amount of magnetite which co-vary with changes in the proportion of smectite in the clay fraction (Bout-Roumazeilles, 1995). Both parameters are "climatically" controlled, glacial periods being associated with lower values than interglacial periods. Down-core changes in the degree of anisotropy have been interpreted to be due to changes in the strength of bottom water circulation. Indeed, the strength of the contour current associated with transport in the ISOW appears to have been significantly greater during climatic stages 5, 3 and 1 than during stages 6, 4 and 2.

Gingele et al. (1999) analysed sedimentary clay mineral assemblages from the eastern terrace of the Vema Channel, the western flank of the Rio Grande Rise and the Brazilian continental slope in the South Atlantic. Variations in kaolinite/chlorite ratios are believed to record variations in the Rio Doce discharge and to reflect humidity conditions on the adjacent South American hinterland. The long-term decrease in smectite content and kaolinite/chlorite ratios from 1,500 ka to the present is believed to document a trend towards more arid and cooler climate conditions for subtropical southern latitudes of South America.

Regarding the southwest Indian Ocean, Gingele et al. (2004) studied a sediment core from a submarine rise on the continental slope offshore of southern Sumatra in order to reconstruct the paleoceanographic evolution of the South Java Current over the past 80 ka. This core shows variations in clay mineral assemblages over glacial/interglacial cycles related to alternating changes in atmospheric and oceanographic circulation dominated by the Northern Hemisphere East Asian Monsoon system and the Southern Hemisphere Australian Monsoon system. During low sea level glacial periods, the Sunda Strait was closed and kaolinite-rich terrigenous supply from that source ceased. The SE Winter Monsoon reached its maximum between 20 and 12 ka and intensified the westward-flowing South Java Current which carries smectite-rich particles originating south of Java. A similar, albeit much weaker, regime prevailed from 74 to 70 ka. During most of the glacial period from 70 to 20 ka, strong northeasterly winds associated with the East Asian Winter Monsoon intensified the Indian Monsoon Current and the eastward-flowing South Java Current, and may also have carried dust. The monsoonal system as we know it today, with its distinct dry and wet seasons, may not have been active before ~12 ka.

As for the Pacific Ocean, Liu et al. (2003) also provided evidence for strong glacial/interglacial cyclicity, northern South China Sea sediments showing high illite, chlorite and kaolinite abundances associated with glacial periods and high smectite and mixed-layers mineral abundances associated with interglacials. Liu et al. (2003) interpreted fluctuations in glacial/interglacial clay mineral assemblages at Site 1146 to have resulted from variations in the intensity of transport processes. In the South China Sea, the intensity of ocean surface currents is controlled by seasonal monsoon variability. Based on these present-day regional mineralogical provinces, (smectite +mixed-layers)/(illite+chlorite) ratios were used to constrain alternating summer and winter surface circulation patterns. Higher smectite abundances at Site 1146 reflect stronger southwesterly currents. Therefore, higher (smectite+mixed-layers)/(illite+chlorite) ratios during interglacials indicate enhanced summer monsoon and weakened winter monsoon circulation. In contrast, lower clay ratios indicate strongly intensified winter monsoon and weakened summer monsoon during glacial periods.

3. Applications: Clays as a Proxy for Paleocirculation

3.1. Clay Distribution in Arctic Ocean Surface Sediments

The identification of source areas and transport pathways of terrigenous material was illustrated by an extensive study of clay minerals from the Arctic Ocean and adjacent Eurasian shelf areas (Wahsner et al., 1999). Detailed distribution maps of clay minerals were derived from the study of the fine fraction (<2 m) of a large number (~470) of surface sediment samples, including new analyses and data from the literature. All semi-quantitative estimates were obtained using the same quantification method or were recalculated. Results for other size fraction (e.g., Russian data on clays from the <1 µm fraction) were excluded from this compilation.

In the Arctic Ocean, clay minerals in surface sediments are purely detrital, reflecting the mineralogical composition of surrounding landmasses. Illite (>50%) and chlorite are the main clay minerals. Except in some parts of the Kara Sea, illite is the dominant clay mineral. The highest concentrations of illite (>70%) are observed in the East Siberian Sea and around Svalbard. Chlorite abundances show the least variations in the Eastern Arctic, ranging from 10% to 25%. Smectite and kaolinite occur in minor amounts (<20%) in the Central Arctic, but show strong variations in shelf areas. Wahsner et al. (1999) emphasized the fact that "these differences can be used to reconstruct pathways of terrigenous matter and circulation patterns in Arctic Ocean". Indeed, the Kara Sea and the western part of the Laptev Sea are enriched in smectite, the highest values (upto 70%) being observed in the Ob and Yenisey deltas. Using a triangular diagram, it is therefore possible to distinguish Kara Sea (highest smectite values) from East Siberian Sea clay mineral assemblages (lowest smectite and kaolinite values and highest illite content).

Clay mineral data also yield information on the transport mechanisms of fine particulate material. According to Wahsner et al. (1999), the high smectite content of the Kara and Laptev Seas surface sediments is related to sea-ice formation and surface current transport. In contrast, smectite-poor sediments in the Eurasian Arctic Ocean are related to turbidity and deep oceanic currents. The high kaolinite content of Nansen Basin sediments is related to sediment transport by turbidity currents. Chlorite abundance is relatively uniform in the Eurasian Arctic Ocean and therefore does not constitute a useful source indicator. Wahsner et al. (1999) concluded that clay mineral abundances do not seem to be a reliable indicator of the extent of sediment supply by sea-ice in Arctic Ocean surface sediments.

3.2. Clay Distribution in Holocene and Last Glacial Sediments in the North Atlantic

Following Biscaye's (1965) seminal paper, the distribution of clay minerals in the North Atlantic and their evolution through glacial/interglacial cycles have been the subject of numerous studies (e.g., Zimmerman, 1972; Parra, 1982; Parra et al., 1985; Grousset et al., 1983; Bout-Roumazeilles, 1995; Fagel, Robert et al., 1996; Fagel et al., 1997b; Kissel et al., 1997; Bout-Roumazeilles et al., 1999). Among these studies, Zimmerman (1982) analysed the clay mineral composition of sediments from 88 sites in the North Atlantic and compared the distribution of the different clays for both the Holocene and the Last Glacial (Figure 6). This study illustrated the spatial variability of clay mineral distribution at the oceanic basin scale and provided significant insight into the relationship between regional deep circulation patterns and NADW.

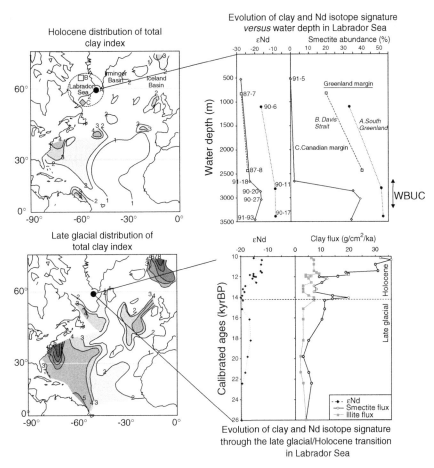

Figure 6 Comparison of Holocene (upper left panel) and Last Glacial (lower left panel) Total Clay Index distributions in the North Atlantic. Data from Zimmerman (1982), with permission. Upper right panel --- The influence of WBUC on clay transport in recent sediments from the Labrador Sea is highlighted. A significant increase in relative smectite abundance is systematically observed in WBUC-bathed surface sediments. The most pronounced shift is recorded in sediments from the Canadian margin. Mineralogical data for transects A and C are from Fagel et al. (1996), transect B data are from Fagel et al. (2001). The influence of eastern inputs at WBUC depth is also marked by contrasting Nd isotope signatures as a function of water depth (data from Innocent et al., 1997, with permission). Lower right panel --- Clay mineralogical evolution through the Last Glacial/Holocene was used as a proxy for WBUC intensity (Fagel et al., 1997b). The smectite flux evolved significantly through the Last Glacial/Holocene transition, whereas the illite flux remained more or less constant. The evolution of the Nd isotope signature of the clay fraction (data from Fagel et al., 1999) is consistent with higher inputs from the eastern basins into the Labrador Sea, which is consistent with the WBUC pathway.

According to Zimmerman's (1982) distribution maps, both the Iceland and Irminger basins show low present-day illite abundances. However, the evolution of clay mineral assemblages in these basins during glacial periods differs, illite abundance decreasing in the Irminger basin, whereas it increases in the Iceland basin.

Zimmerman (1982) speculated that large amounts of illite were brought into the glacial Iceland basin by bottom current transport from the Norwegian Sea. The influence of WBUC on clay fraction transport from the Eastern basins into the western North Atlantic is further evidenced by specific clay mineral and Nd isotopic signatures recorded from a depth corresponding to the WBUC maximum velocity axis in the Labrador Sea (Fagel et al., 1996; Innocent et al., 1997; see Figure 6).

In addition to clay distribution maps, Zimmerman (1982) also estimated the abundance of clay minerals within bulk sediments. He defined a "Total Clay Index" (TCI) based on the sum of the areas above background under the curve for the 17, 10 and 7 Å peaks on solvated XRD patterns, after multiplying by Biscaye's (1965) factors. According to Zimmerman (1982), the TCI defines mineral provinces that are dependent on source, current transport, basin topography and clay deposition rate. In the North Atlantic, three areas are characterized by TCI values greater than three (Figure 6, upper panel). A high Holocene TCI characterizes the continental rise off of the east coast of North America. It is explained by an important depocenter of clay minerals transported southwards by WBUC (Zimmerman, 1972). The locally high TCI at the base of the Great Banks in the Newfoundland basin suggests a high clay deposition rate behind the Newfoundland barrier, which obstructs the south-flowing WBUC. A third area of locally high TCI off of northeastern South America reflects Orinoco River related deposition.

During the Last Glacial, the areas of high TCI were much wider, including the entire North American basin, the TCI reaching values of up to eight off of the east American margin (Figure 6, lower panel). This pattern suggests a more active dispersal system resulting from important continental erosion and active currents. For instance, a latitudinal pattern with high quartz content and dilution of total clay contents occurred along the 45°N parallel. Coinciding with the position of the polar front, this pattern reflects the ice-rafted origin of marine sediments in this area, an interpretation consistent with more recent studies of clay minerals (Bout-Roumazeilles et al., 1999). In contrast, for the finer fraction, Zimmerman (1982) suggested that sediments are injected into WBUC by active turbidity currents, enhanced by glacial low sea-level stands. This transport mechanism is consistent with the proposed sediment dispersal mode (Heezen, Hollister, & Ruddiman, 1966). The relatively high abundance of smectite in Last Glacial sediments along the northwest Atlantic margin (e.g., off the Grand Banks, Piper & Slatt, 1977; or in the Labrador Sea, Fagel et al., 1997b) have been interpreted to result from the persistent, albeit sluggish, influence of WBUC on clay transport during the Last Glacial. At the southern tip of Greenland, variations with time of clay mineral assemblages suggest gradually increasing sedimentary fluxes and WBUC intensity since the Late Glacial Maximum. The smectite flux increases throughout the 2/1 transition whereas the illite flux remains approximately constant (Figure 6). The increasing WBUC transport through the Last Glacial/Holocene transition is further confirmed by a closely related significant change in the Nd isotope signature of the clay fraction (Fagel et al., 1999).

3.3. Clay Distribution since the Last Glacial in the Southeast Indian Ocean

Gingele et al. (2001a) analysed the clay mineral distribution in 166 core-top samples from between Indonesia and NW Australia to trace modern current pathways. In addition, the recent distribution of clay minerals was used to reconstruct the evolution of the Leeuwin Current through the Late Quaternary (Gingele et al., 2001b). The Indonesian Island Arc and Australia, which are characterized by contrasting geology and climates, constitute an ideal provenance case study. Gingele et al. (2001a) identified four clay mineral provinces closely related to the geology of the adjacent hinterland. Although they are all subjected to tropical conditions, the samples from the Indonesian Island Arc display significant differences which allowing three clay mineral provinces to be distinguished. Illite is the major clay mineral of the Western and Eastern Provinces, whereas the Central Province abounds with smectite. The fourth province comprises the northwest and west Australian shelf, slope and offshore plateaus and is dominated by kaolinite. Unlike the latitudinal zonation of clay mineral assemblages observed in other parts of the world oceans, samples from offshore of northwestern Australia show little variation in clay mineral composition (Windom, 1976; see Figure 4). Gingele et al. (2001a) emphasized the fact that the observed clay mineral assemblages do not reflect changes in weathering conditions of the terrigenous source, because weathering profiles develop over long periods of time (Thiry, 2000), particularly in Australia. Rather, the clay assemblages record changes in source areas or transport media. Transport of clays by surface and subsurface ocean currents is observed within the provinces (e.g., with the Leeuwin and West Australian Currents in the NW Australian Province, and with the outflow of low-salinity water through the Sunda and Lombok straits in the Central Province). Kaolinite and chlorite are transported into the Timor Passage by the Indonesian Throughflow, while illite is brought in locally from Timor. The Leeuwin Current leaves the Timor Passage with a characteristic clay mineral signature acquired in the Indonesian Archipelago (kaolinite, chlorite and illite). The uptake of clay minerals along its path through the Timor Sea (e.g., illite from the Kimberley area) changes this signature. South of North West Cape, chlorite injected by rivers of the Pilbara region into the path of the Leeuwin Current is a major constituent of surface sediments in water depths of less than 1,000 m and delineates the present-day flow of the current. The transport of clays across province boundaries is inhibited by strong salinity fronts, with the exception of the boundary between the Central and Eastern Province. It should be pointed out that the relationship between clay assemblage and current is not always obvious. For instance, Gingele et al. (2001a) emphasized the fact that clay mineral patterns do not provide decisive information regarding the transport of suspended matter from the Timor Passage along the path of the northern branch of the Leeuwin Current. Although high illite contents around Timor provide a distinct signal, the main Indonesian Throughflow through the Timor Passage cannot be traced by clay mineral distribution alone. Northeasterly current directions during the major runoff periods prevent the development of an expected illite-rich tongue of sediment to the southwest.

Clay mineral evolution from three cores taken from beneath the path of the present day Leeuwin Current in the Timor Passage was studied in order to constrain changes in the strength and/or path of the current system since the Last Glacial (Gingele et al., 2001b). Based on the lower amounts of kaolinite and chlorite that reached the Timor Passage, Gingele et al. (2001b) inferred a decrease in the volume of the Indonesian Throughflow during the Last Glacial. Offshore, a reduction in the amount of chlorite during the Last Glacial is interpreted to have resulted either from a decrease in the intensity of the Leeuwin Current or its absence, and/or a lower chlorite input due to drier conditions on land. The latter hypothesis is consistent with the synchronous decrease in illite abundance. An illite maximum in Holocene and recent sediments offshore of North West Cape is the result of material input from rivers which periodically drain the adjacent hinterland. Lower illite abundances point to a drier climate in the area during the last glacial stage.

4. Some Perspectives

Most marine sediments comprise a complex mix of clay minerals derived from different source areas and/or formed by different processes. The interpretation of marine clay mineral data could be facilitated by a multiproxy approach. For instance, combining radiogenic isotopic analysis and clay mineralogy would help to distinguish between variations in particulate inputs and variations in paleoclimate and/or paleoenvironmental conditions in the hinterland.

In addition, the clay mineralogy approach is hindered by non-uniform sample preparation methods (pre treatments, clay-size fraction extraction methods) and interpretation of X-ray diffraction patterns, such that comparison of clay data sets produced by different labs or countries remains tricky. Intercalibrations must be developed. Ongoing developments in computer-assisted clay quantification methods could contribute to improving this delicate step in the interpretation of the significance of clay minerals.

ACKNOWLEDGEMENTS

The author wishes to thank R. Petschick for kindly providing data used in the drafting of Figure 4c, François De Vleeschouwer and Gaetan Renson for figure design and Hervé Chamley for editing this chapter.

REFERENCES

Alaf, M. (1987). Late Quaternary plume, nepheloid and turbidite sedimentation and effect of the Gulf Stream near the tail of the Grand Banks, Newfoundland. *Marine Geology, 74*, 277–290.
Allègre, C. J., Dupré, B., & Négrel, R. (1996). Sr-Nd-Pb isotopes systematics in Amazon and Congo River systems: Constraints about erosion processes. *Chemical Geology, 131*, 93–112.
Allen, P. A. (1997). *Earth surface processes* (p. 404). Oxford: Blackwell.
Anderson, J. B., Kurtz, D. D., Domack, E. W., & Balshaw, K. M. (1980). Glacial and glacial marine sediments of the Antarctic continental shelf. *Journal of Geology, 88*, 399–414.

Arrhenius, G. (1963). Pelagic sediments. In: M. N. Hill (Ed.), *The Sea* (Vol. III, pp. 655–727). New York: Interscience Publication.

Asahara, Y., Tanaka, T., Kamioka, H., & Nishimura, A. (1999). Asian continental nature of ^{87}Sr/^{86}Sr ratios in north central Pacific sediments. *Earth Planetary Sciences Letters, 133*, 105–116.

Bailey, S. W. (1980). Summary of recommendations of AIPEA nomenclature committee. *Clay and Clay Minerals, 28*, 73–78.

Barker, A. P. F., & Thomas, E. (2004). Origin, signature and palaeoclimatic influence of the Antarctic Circumpolar Current. *Earth-Science Reviews, 66*, 143–162.

Bayon, G., German, C. R., Nesbitt, R. W., Bertrand, P., & Schneider, R. R. (2003). Increased input of circumpolar deep water-borne detritus to the glacial SE Atlantic Ocean. *Geochemical Geophysical Geosystems, 4*,1025, doi:10.1029/2002GC000371.

Behairy, A. K., Chester, R., Griffiths, A. J., Johnson, L. R., & Stoner, J. H. (1975). The clay mineralogy of particulate material from some surface seawaters of the Eastern Atlantic Ocean. *Marine Geology, 18*, M45–M56.

Berry, R. W., & Johns, W. D. (1966). Mineralogy of the clay-sized fractions of some North Atlantic–Arctic ocean bottom sediments. *Geological Society American Bulletin, 77*, 183–196.

Biscaye, P. E. (1965). Mineralogy and sedimentation of recent deep-sea clay in the Atlantic Ocean and adjacent seas and oceans. *Geological Society American Bulletin, 76*, 803–832.

Biscaye, P. E., Chesselet, R., & Prospero, J. M. (1974). Rb-Sr, ^{87}Sr/^{86}Sr isotope system as an index of provenance of continental dusts in the open Atlantic Ocean. *Journal Atmospheric Research, 8*, 819–829.

Biscaye, P. E., & Eittreim, S. L. (1977). Suspended particulate loads and transports in the nepheloid layer of the abyssal Atlantic Ocean. *Marine Geology, 23*, 155–172.

Bouquillon, A., Chamley, H., & Frolich, F. (1989). Sédimentation argileuse au Cénozoïque de l'Océan Indien Nord Oriental. *Oceanologica Acta, 12*, 133–147.

Bout-Roumazeilles, V. (1995). Relations entre variabilités minéralogiques et climatiques enregistrées dans les sédiments de l'Atlantique Nord depuis les 8 derniers stades glaciaires-interglaciaires. Université de Lille I, 1995 (180 pp).

Bout-Roumazeilles, V., Cortijo, E., Labeyrie, L., & Debrabant, P. (1999). Clay mineral evidence of nepheloid layer contributions to the Heinrich layers in the northwest Atlantic. *Palaeogeography, Palaeoclimatology, Palaeoecology, 146*, 211–228.

Bremner, J. M., & Willis, J. P. (1993). Mineralogy and geochemistry of the clay fraction of sediments from the Namibian continental margin and the adjacent hinterland. *Marine Geology, 115*, 85–116.

Brindley, G. W., & Brown, G. (1980). *Crystal structures and their X-ray identification* (p. 495). London: Mineralogical Society.

Caillière, S., Hénin, S., & Rautureau, M. (1982). *Minéralogie des argyles. 2. Classification et nomenclature* (p. 189). Paris: Masson.

Caroll, D. (1970). Clay minerals in Arctic Ocean sea floor sediments. *Journal Sedimentary Petrology, 45*, 272–279.

Carson, B., & Arcaro, N. P. (1983). Control of clay-mineral stratigraphy by selective transport in late Pleistocene-Holocene sediments of Northern Cascadia Basin-Juan de Fuca Abyssal Plain: Implications for studies of clay-mineral provenance. *Journal Sedimentary Petrology, 53*, 395–406.

Chamley, H. (1967). Possibilités d'utilisation de la cristallinité d'un minéral argileux (illite) comme témoin climatique dans les sédiments récents. *Compte Rendu Académie des Sciences Paris Série D, 265*, 184–187.

Chamley, H. (1974). Place des argiles marines parmi divers indicateurs paléoclimatiques. In: Coll. Int. CNRS 219, Les méthodes quantitatives d'étude des variations du climat au cours du Pléistocènes, Gif-sur-Yvette, 1973 (pp. 35–37).

Chamley, H. (1981). Long-term trends in clay deposition in the ocean. *Oceanologica Acta Proceedings 26th International Geological Congress, Geology of oceans symposium*, Paris, July 7–17, 1980 (pp. 105–110).

Chamley, H. (1986). Continental and marine paleoenvironments reflected by west Pacific clay sedimentation. *Geologische Rundschau, 75*, 271–285.

Chamley, H. (1989). *Clay sedimentology* (p. 623). Berlin: Springer.

Chamley, H., & Bonnot-Courtois, C. (1981). Argiles authigènes et terrigènes de l'Atlantique et du Pacifique NW (Leg 11 et 58 DSDP): Apport des terres rares. *Oceanologica Acta, 4*, 229–238.

Chester, R., Elderfield, H., Griffin, J. J., Johnson, L. R., & Padgham, R. C. (1972). Eolian dust along the eastern margins of the Atlantic Ocean. *Marine Geology*, *13*, 91–105.

Clauer, N., & Chaudhuri, S. (1995). *Clays in crustal environments. Isotope dating and tracing* (p. 359). Berlin: Springer-Verlag.

Clauer, N., O'Neil, J. R., Bonnot-Courtois, C., & Holtzapffel, T. (1990). Morphological, chemical, and isotopic evidence for an early diagenetic evolution of detrital smectite in marine sediments. *Clay and Clays Minerals*, *38*, 33–46.

Clayton, T., Pearce, R. B., & Pzeterson, L. C. (1999). Indirect climatic control of the clay mineral composition of Quaternary sediments from the Cariaco Basin, northern Venezuela (ODP Site 1002). *Marine Geology*, *161*, 191–206.

Couture, R. A. (1977). Composition and origin of palygorskite-rich and montmorillonite-rich zeolite-containing sediments from the Pacific Ocean. *Chemical Geology*, *19*, 113–130.

Darby, D. A. (1975). Kaolinite and other clay minerals in Arctic Ocean sediments. *Journal Sedimentary Petrology*, *45*, 272–279.

Dasch, E. J. (1969). Sr isotopes in weathering profiles, deep-sea sediments and sedimentary rocks. *Geochimica et Cosmochimica Acta*, *33*, 1521–1552.

Deconinck, J. F., & Vanderaveroet, P. (1996). Eocene to Pleistocene clay mineral sedimentation off New-Jersey, western north Atlantic (ODP Leg 150, sites 903 and 905). *Proceedings of ODP Science Results*, *150*, 147–170.

Desprairies, A., & Bonnot-Courtois, C. (1980). Relation entre la composition des smectites d'altération sous-marine et leur cortège de terres rares. *Earth Planetary Science Letters*, *48*, 124–130.

Dia, A., Dupré, B., & Allègre, C. J. (1992). Nd isotopes in Indian Ocean used as a tracer of supply to the ocean and circulation paths. *Marine Geology*, *103*, 349–359.

Diekmann, B., Fütterer, D. K., Grobe, H., Hillenbrand, C. D., Kuhn, G., Michels, K., Petschick, R., & Pirrung, M. (2003). Terrigenous sediment supply in the polar to temperate South Atlantic: Land-ocean links of environmental changes during the Late Quaternary. In: G. Wefer, S. Mulitza & V. Ratmeyer (Eds), *The South Atlantic in the Late Quaternary: Reconstruction of material budgets and current systems* (pp. 375–399). Berlin: Springer-Verlag.

Diekmann, B., & Kuhn, G. (1999). Provenance and dispersal of glacial-marine surface sediments in the Weddell Sea and adjoining areas, Antarctica: Ice-rafting versus current transport. *Marine Geology*, *158*, 209–231.

Diekmann, B., Kuhn, G., Rachold, V., Abelmann, A., Brathauer, U., Futterer, D. K., Gersonde, R., & Grobe, H. (2000). Terrigenous sediment supply in the Scotia Sea (Southern Ocean): Response to Late Quaternary ice dynamics in Patagonia and on the Antarctic Peninsula. *Palaeogeography, Palaeoclimatology, Palaeoecology*, *162*, 357–387.

Diekmann, B., Petschick, R., Gingele, F. X., Fütterer, D. K., Abelmann, A., Brathauer, U., Gersonde, R., & Mackensen, A. (1996). Clay mineral fluctuations in Late Quaternary sediments of the southeastern South Atlantic: Implications for past changes of deep water advection. In: G. Wefer, et al. (Eds), *The South Atlantic: Present and past circulation* (pp. 621–644). Springer-Verlag.

Eberl, D., & Hower, J. (1977). The hydrothermal transformation of sodium and potassium smectite into mixed-layer clay. *Clays and Clay Minerals*, *26*, 327–340.

Ehrmann, W., Setti, M., & Marinoni, L. (2005). Clay minerals in Cenozoic sediments off Cape Roberts (McMurdo Sound, Antarctica) reveal palaeoclimatic history. *Palaeogeography, Palaeoclimatology, Palaeoecology*, *229*, 187–211.

Ehrmann, W. U., Melles, M., Kuhn, G., & Grobe, H. (1992). Significance of clay mineral assemblages in the Antarctic Ocean. *Marine Geology*, *107*, 249–273.

Eittreim, S. L., & Ewing, M. (1974). Turbidity distribution in the deep waters of the western Atlantic trough. In: R. J. Gibbs (Ed.), *Suspended solids in water* (pp. 213–225). New York, NY: Plenum Press.

Elverhøi, A., Pfirman, S. L., Solheim, A., & Larssen, B. B. (1989). Glaciomarine sedimentation in epicontinental seas exemplified by the Northern Barents sea. *Marine Geology*, *85*, 225–250.

Fagel, N., André, L., Chamley, H., Debrabant, P., & Jolivet, L. (1992b). Clay sedimentation in the Japan Sea since the Early Miocene: influence of source-rock and hydrothermal activity. *Sedimentary Geology*, *80*, 27–40.

Fagel, N., André, L., & Debrabant, P. (1997a). The geochemistry of pelagic clays: Detrital versus non-detrital signals?. *Geochimica et Cosmochimica Acta, 61,* 989–1008.

Fagel, N., Debrabant, P., & André, L. (1994). Clay supplies in the Central Indian Basin since the Late Miocene: Climatic or tectonic control?. *Marine Geology, 122,* 151–172.

Fagel, N., Debrabant, P., De Menocal, P., & Demoulin, B. (1992a). Utilisation des minéraux sédimentaires argileux pour la reconstitution des variations paléoclimatiques à court terme en Mer d'Arabie. *Oceanologica Acta, 15,* 125–137.

Fagel, N., & Hillaire-Marcel, C. (2006). Glacial/interglacial instabilities of the Western Boundary Undercurrent during the last 360 kyr from Sm/Nd signatures of sedimentary clay fractions at ODP-Site 646 (Labrador Sea). *Marine Geology, 232,* 87–99.

Fagel, N., Hillaire-Marcel, C., Humblet, M., Brasseur, R., Weis, D., & Stevenson, R. (2004). Nd and Pb isotope signatures of the clay-size fraction of Labrador Sea sediments during the Holocene: Implications for the inception of the modern deep circulation pattern. *Paleoceanography, 19,* doi:10.1029/2003PA000993, PA1029 1–16.

Fagel, N., Hillaire-Marcel, C., & Robert, C. (1997b). Changes in the Western Boundary Undercurrent outflow since the Last Glacial Maximum, from smectite/illite ratios in deep Labrador Sea sediments. *Paleocanography, 12,* 79–96.

Fagel, N., Innocent, C., Gariépy, C., & Hillaire-Marcel, C. (2002). Sources of Labrador Sea sediments since the last glacial maximum inferred from Nd-Pb isotopes. *Geochimica et Cosmochimica Acta, 66,* 2569–2581.

Fagel, N., Innocent, C., Stevenson, R. K., & Hillaire-Marcel, C. (1999). Nd isotopes as tracers of paleocurrents: a high resolution study of Late Quaternary sediments from the Labrador Sea. *Paleoceanography, 14,* 777–788.

Fagel, N., Robert, C., & Hillaire-Marcel, C. (1996). Clay mineral signature of the North Atlantic Boundary undercurrent. *Marine Geology, 130,* 19–28.

Fagel, N., Robert, C., Preda, M., & Thorez, J. (2001). Smectite composition as a tracer of deep circulation: The case of the Northern North Atlantic. *Marine Geology, 172,* 309–330.

Foucault, A., & Mélières, F. (2000). Palaeoclimatic cyclicity in central Mediterranean Pliocene sediments: The mineralogical signal. *Palaeogeography, Palaeoclimatology, Palaeoecology, 158,* 311–323.

Gibbs, R. J. (1977). Clay mineral segregation in the marine environment. *Journal Sedimentary Petrology, 47,* 237–243.

Gingele, F. X. (1996). Holocene climate optimum in Southwest Africa — evidence from the marine clay record. *Paleogeography, Paleoecology, Paleoceanography, 122,* 77–87.

Gingele, F. X., De Deckker, P., Girault, A., & Guichard, F. (2004). History of the South Java Current over the past 80 ka. *Palaeogeography, Palaeoclimatology, Palaeoecology, 183,* 247–260.

Gingele, F. X., De Deckker, P., & Hillenbrand, C. D. (2001a). Late Quaternary fluctuations of the Leeuwin Current and palaeoclimates on the adjacent land masses: Clay mineral evidence. *Australian Journal of Earth Sciences, 48,* 867–874.

Gingele, F. X., De Deckker, P., & Hillenbrand, C. D. (2001b). Clay mineral distribution in surface sediments between Indonesia and NW Australia — source and transport by ocean currents. *Marine Geology, 179,* 135–146.

Gingele, F. X., Schmieder, F., Petschick, R., von Dobeneck, T., & Rühlemann, C. (1999). Terrigenous flux in the Rio Grande Rise area during the past 1500 ka: Evidence of deepwater advection or rapid response to continental rainfall patterns? *Paleoceanography, 14,* 84–95.

Goldstein, S. L. (1988). Decoupled evolution of Nd and Sr isotopes in the continental crust and mantle. *Nature, 336,* 733–738.

Goldstein, S. L., & O'Nions, R. K. (1981). Nd and Sr isotopic relationships in pelagic clays and ferromanganese deposits. *Nature, 292,* 324–327.

Gorbunova, Z. N. (1997). Clay-sized minerals in the Kara Sea sediments. *Oceanology, 37,* 709–712.

Gradusov, B. P. (1974). A tentative study of clay mineral distribution in soils of the world. *Geoderma, 12,* 49–55.

Griffin, J. J., & Goldberg, E. D. (1963). Clay mineral distribution in the Pacific Ocean. In: M. N. Hill (Ed.), *The sea* (Vol. 3, pp. 728–741). New York: Interscience.

Griffin, J. J., Windom, H. L., & Goldberg, E. D. (1968). The distribution of clay minerals in the World Ocean. *Deep-Sea Research, 15*, 433–459.

Grim, R. E., Bray, R. H., & Bradley, W. F. (1937). The mica in argillaceous sediments. *American Mineralogist, 22*, 813–829.

Grim, R. E., & Johns, W. D. (1954). Clay mineral investigations of sediments in the northern Gulf of Mexico. Clays Clay Mineral, 2nd National Conference, Pergamon, New York (pp. 81–103).

Grousset, F., & Latouche, C. (1983). Rôles respectifs de l'advection et de la décantation sur la Ride Médio-Atlantique (40° à 50°N). *Chemical Geology, 40*, 225–249.

Grousset, F., Latouche, C., & Maillet, N. (1983). Clay minerals as indicators of wind and current contribution to post-glacial sedimentation on the Azores/Iceland Ridge. *Clay Minerals, 18*, 65–75.

Grousset, F., Latouche, C., & Parra, M. (1982). Late Quaternary sedimentation between the Gibbs Fracture and the Greenland Basin: Mineralogical and geochemical data. *Marine Geology, 47*, 303–330.

Grousset, F. E., Biscaye, P. E., Zindler, A., Prospero, J., & Chester, R. (1988). Neodymium isotopes as tracers in marine sediments and aerosols: North Atlantic. *Earth Planetary Science Letters, 87*, 367–378.

Haggerty, S. E., & Baker, I. (1967). The alteration of olivine in basaltic and associated lavas. Part II: Intermediate and low temperature alteration. Contribution. *Mineralogy Petrology, 16*, 258–273.

Heezen, B. C., Hollister, C. D., & Ruddiman, W. F. (1966). Shaping of the continental rise by deep geostrophic contour currents. *Science, 152*, 502–508.

Hein, J. R., Bouma, A. H., Hampton, M. A., & Ross, C. R. (1979). Clay mineralogy, fine-grained sediment dispersal, and inferred current patterns, lower Cook Inlet and Kodiak shelf, Alaska. *Sedimentary Geology, 24*, 291–306.

Hemming, S. R., Biscaye, P. E., Broecker, W. S., Hemming, N. G., Klas, M., & Hajdas, I. (1998). Provenance change coupled with increased clay flux during deglacial times in the western equatorial Atlantic. *Paleogeography, Paleoecology, Paleoceanography, 142*, 217–230.

Hillenbrand, C. D., & Ehrmann, W. (2005). Late Neogene to Quaternary environmental changes in the Antarctic Peninsula region: Evidence from drift sediments. *Global Planetary Change, 45*, 165–191.

Hoffman, J., & Hower, J. (1979). Clay mineral assemblages as low-grade metamorphic geothermometers: Application to the thrust-faulted disturbed belt of Montana, USA. In: P. A. Scholle & P. R. Schluger (Eds), *Aspects of diagenesis* (Vol. 26, pp. 55–79). Society of Economic Paleontologists and Mineralogists Publication.

Hower, J., Eslinger, E. V., Hower, M. E., & Perry, E. A. (1976). Mechanisms of burial metamorphism of argillaceous sedimentary: 1. Mineralogical and chemical evidence. *Geological Society American Bulletin, 87*, 727–737.

Huon, S., Jantschik, R., Kubler, B., & Fontignie, D. (1991). Analyses K-AR, Rb-Sr et minéralogiques des fractions argileuses de sédiments quaternaires, Atlantique N-E: Résultats préliminaires. *Schweizerische Mineralogische and Petrographische Mitteilungen, 71*, 275–280.

Hurley, P. M., Heezen, B., Pinson, C., & Fairbairn, H. W. (1963). K-Ar age values in pelagic sediments of the North Atlantic. *Geochimica et Cosmochimica Acta, 27*, 393–399.

Innocent, C., Fagel, N., & Hillaire-Marcel, C. (2000). Sm–Nd isotope systematics in deep-sea sediments: Clay-size versus coarser fractions. *Marine Geology, 168*, 79–87.

Innocent, C., Fagel, N., Stevenson, R. K., & Hillaire-Marcel, C. (1997). Sm-Nd signature of modern and late Quaternary sediments from the northwest North Atlantic: Implications for deep current changes since the Last Glacial Maximum. *Earth Planetary Science Letters, 146*, 607–625.

Inoue, A. (1987). Conversion of smectite to chlorite by hydrothermal and diagenetic alterations, Hokuroko Kuroko mineralization area, Northeast Japan. *Proceedings International Clay Conference,* Denver, 1985 (pp. 158–164).

Jones, C. E., Halliday, A. N., Rea, D. K., & Owen, R. M. (1994). Neodymium isotopic variations in the North Pacific modern silicate sediment and their significance of detrital REE contributions to seawater. *Earth Planetary Science Letters, 127*, 55–66.

Jones, G. A. (1984). Advective transport of clay minerals in the region of the Rio Grande Rise. *Marine Geology, 58*, 187–221.

Karpoff, A. M., Peterschmitt, I., & Hoffert, M. (1980). Mineralogy and geochemistry of sedimentary deposits on Emperor seamounts, sites 430, 431, and 432: Authigenesis of silicates, phosphates and ferromanganese oxides. In: E. D. Jackson, I. Koisumi, et al. (Eds), *Initial Reports of the Deep Sea Drilling Project*, 55, Washington (U.S. Gov. Print. Off.): 463–489.

Kastner, M. (1981). Authigenic silicates in deep-sea sediments: Formation and diagenesis. In: C. Emiliani (Ed.), *The sea* (Vol. 7, pp. 915–980). New York: Wiley.

Kissel, C., Laj, C., Lehman, B., Labeyrie, L., & Bout-Roumazeilles, V. (1997). Changes in the strength of the Iceland–Scotland Overflow Water in the last 200,000 years: Evidence from magnetic anisotropy analysis of core SU90-33. *Earth and Planetary Science Letters*, *152*, 25–36.

Knebel, H. J., Conomos, T. J., & Commeau, J. A. (1977). Clay-mineral variability in the suspended sediments of the San Fransisco bay system, California. *Journal Sedimentary Petrology*, *47*, 229–236.

Kolla, V., & Biscaye, P. E. (1973). Clay mineralogy and sedimentation in the eastern Indian Ocean. *Deep-Sea Research*, *20*, 727–738.

Kolla, V., Henderson, L., & Biscaye, P. E. (1976). Clay mineralogy and sedimentation in the western Indian Ocean. *Deep-Sea Research and Oceanographic Abstracts*, *23*, 949–961.

Kolla, V., Henderson, L., Sullivan, L., & Biscaye, P. E. (1978). Recent sedimentation in the southeast Indian Ocean with special reference to the effects of Antarctic Bottom Water circulation. *Marine Geology*, *27*, 1–17.

Kolla, V., Kostecki, J. A., Robinson, F., Biscaye, P. E., & Ray, P. K. (1981). Distributions and origins of clay minerals and quartz in surface sediments of the Arabian Sea. *Journal of Sedimentary Petrology*, *51*, 563–569.

Kuhn, G., & Diekmann, B. (2002). Late Quaternary variability of ocean circulation in the southeastern South Atlantic inferred from the terrigenous sediment record of a drift deposit in the southern Cape Basin (ODP Site 1089). *Paleogeography, Paleoecology, Paleoceanography*, *182*, 287–303.

Kuhn, G., & Weber, M. (1993). Acoustical characterization of sediments by Parasound and 3.5 KHz systems: Related sedimentary processes on the southeastern Weddell Sea continental slope, Antarctica. *Marine Geology*, *113*, 201–217.

Latouche, C. (1978). Clay minerals as indicators of the Cenozoic evolution of the North Atlantic Ocean. In: M. M. Mortland & V. C. Farmer (Eds), *International Clay Conference, 1978, Developments in Sedimentology* (Vol. 27, pp. 271–279). Amsterdam: Elsevier.

Latouche, C., & Parra, M. (1979). La sédimentation au Quaternaire récent dans le "Northwest Atlantic Mid-Ocean Canyon" — Apport des données minéralogiques et géochimiques. *Marine Geology*, *29*, 137–164.

Li, Y. H., & Schoonmaker, J. E. (2003). Chemical composition and mineralogy of marine sediments. In: F. T. Mackenzie (Ed.), *Sediments, diagenesis, and sedimentary rocks. Treatise on geochemistry* (Vol. 7). Amsterdam: Elsevier.

Lisitzin, A. P. (1972). Sedimentation in the world ocean. *Society of Economic Paleontologists and Mineralogists Special Publication*, *17*, 218.

Lisitzin, A. P. (1996). *Oceanic sedimentation. Lithology and geochemistry* (p. 400). Washington: American Geophysical Union.

Liu, Z., Trentesaux, A., Clemens, S. C., Colin, C., Wang, P., Huang, B., & Boulay, S. (2003). Clay mineral assemblages in the northern South China Sea: Implications for East Asian monsoon evolution over the past 2 million years. *Marine Geology*, *201*, 133–146.

Mahoney, J. B. (2005). Nd and Sr isotopic signatures of fine-grained clastic sediments: A case study of western Pacific marginal basins. *Sedimentary Geology*, *182*, 183–199.

McCave, I. N., Manighetti, B., & Beveridge, N. A. S. (1995). Circulation in the glacial North Atlantic inferred from grain-size measurements. *Nature*, *374*, 149–151.

McCulloch, M. T., & Wasserburg, G. J. (1978). Sm-Nd and Rb-Sr chronology of continental crust formation. *Science*, *200*, 1003–1011.

McDougall, J. D., & Harris, R. C. (1969). The geochemistry of an Arctic watershed. *Canadian Journal of Earth Sciences*, *6*, 305–315.

McLennan, S. M., McCulloch, M. T., Taylor, S. R., & Maynard, J. B. (1989). Effects of sedimentary sorting on neodymium isotopes in deep-sea turbidites. *Nature*, *337*, 547–549.

McMaster, R. L., Betzer, P. R., Carder, K. L., Miller, L., & Eggimann, D. W. (1977). Suspended particle mineralogy and transport in water masses of the West Africa Shelf adjacent to Sierra Leone and Liberia. *Deep-Sea Research, 24*, 651–665.

McMurthy, G. M., Wang, C. H., & Yeh, H. W. (1983). Chemical and isotopic investigations into the origin of clay minerals from the Galapagos hydrothermal mounds field. *Geochimica et Cosmochimica Acta, 47*, 475–489.

Melguen, M., Debrabant, P., Chamley, H., Maillot, H., Hoffert, M., & Courtois, C. (1978). Influence des courants profonds sur les faciès sédimentaires du Vema Channel (Atlantique Sud) à la fin du Cénozoïque. *Bulletin Société Géologique de France, XX*, 121–136.

Moore, D. M., & Reynolds, R. C. (1989). *X-ray diffraction and the identification and analysis of clay minerals* (p. 332). Oxford: Oxford University Press.

Moriarty, K. C. (1977). Clay minerals in southeast Indian Ocean sediments, transport mechanisms and depositional environments. *Marine Geology, 25*, 149–174.

Moser, F. C., & Hein, R. H. (1984). Distribution of clay minerals in the suspended and bottom sediments from the Northern Bering Sea shelf area, Alaska. *U.S. Geological Survey Bulletin, 1624*, 1–19.

Murray, R. W., Buchholtz Ten Brink, M. R., Jones, D. L., Gerlach, D. C., & Russ, G. P. I. (1990). Rare earth elements as indicators of different marine depositional environments in chert and shale. *Geology, 18*, 268–271.

Nadeau, P. H., Wilson, M. J., McHardy, W. J., & Tait, J. M. (1985). The conversion of smectite to illite during diagenesis: Evidence from some illitic clays from bentonites and sandstones. *Mineral Magazine, 49*, 393–400.

Naidu, A. S., Creager, J. S., & Mowatt, T. C. (1982). Clay mineral dispersal patterns in the north Bering and Chukchi Seas. *Marine Geology, 47*, 1–5.

Naidu, A. S., & Mowatt, T. C. (1983). Sources and dispersal patterns of clay minerals in surface sediments from the continental-shelf areas off Alaska. *Geological Society American Bulletin, 94*, 841–854.

Nath, B. N., Roelandts, I., Sudhakar, M., & Pluger, W. L. (1992). Rare earth element patterns of the Central Indian Basin sediments related to their lithology. *Geophysical Research Letters, 19*, 1197–1200.

Nittrouer, C. A. (1978). *The process of detrital sediment accumulation in a continental shelf environment: Examination of the Washington Shelf*. Unpublished Ph.D. thesis, Seattle, University of Washington, 243 pp.

Nürnberg, D., Levitan, M. A., Pavlidis, J. A., & Shelkhova, E. S. (1995). Distribution of clay minerals in surface sediments from the eastern Barents and southwestern Kara seas. *Geologische Rundschau, 84*, 665–682.

Parra, M. (1982). North Atlantic sedimentation and paleohydrology during the late quaternary — mineralogical and geochemical data. *Oceanologica Acta, 5*, 241–247.

Parra, M., Delmont, P., Ferragne, A., Latouche, C., & Puechmaille, C. (1985). Origin and evolution of smectites in recent marine sediments of the NE Atlantic. *Clay Minerals, 20*, 335–346.

Parra, M., Puechmaille, C., Dumon, J. C., Delmont, P., & Ferragne, A. (1986). Geochemistry of tertiary alterite clay phases on the Iceland-Faeroe ridge (Northeast Atlantic), Leg 38, Site 336. *Chemical Geology, 54*, 165–176.

Pédro, U. (1968). Distribution des principaux types d'altération chimique à la surface du globe. Présentation d'une esquisse géographique. *Géographie Physique Géologie Dynamique, 2*, 457–470.

Petschick, R., Kuhn, G., & Gingele, F. X. (1996). Clay mineral distribution in surface sediments of the South Atlantic — sources, transport and relation to oceanography. *Marine Geology, 130*, 203–229.

Piper, D. J. W., & Slatt, R. M. (1977). Late Quaternary clay-mineral distribution on the eastern continental margin of Canada. *Geological Society American Bulletin, 88*, 267–272.

Piper, D. Z. (1974). Rare earth elements in the sedimentary cycle: A summary. *Chemical Geology, 14*, 285–304.

Prospero, J. (1981). Eolian transport to the world ocean. In: C. Emiliani (Ed.), *The sea, the oceanic lithosphere* (Vol. 7, pp. 801–874). New York: Wiley.

Ramsayer, K., & Boles, J. R. (1986). Mixed-layer illite/smectite minerals in Tertiary sandstones and shales San Joaquin Basin, California. *Clays and Clay Minerals*, *29*, 129–135.

Rao, V. P., & Nath, B. N. (1988). Nature, distribution and origin of clay minerals in grain size fractions of sediments from manganese nodule field, Central Indian Basin. *Indian Journal of Marine Sciences*, *17*, 202–207.

Rateev, M. A., Gorbunova, Z. N., Lisitzyn, A. P., & Nosov, G. L. (1969). The distribution of clay minerals in the oceans. *Sedimentology*, *13*, 21–43.

Revel, M., Cremer, M., Grousset, F. E., & Labeyrie, L. (1996b). Grain-size and Sr-Nd isotopes as tracer of paleo-bottom current strength, Northeast Atlantic Ocean. *Marine Geology*, *131*, 233–249.

Revel, M., Sinko, J. A., Grousset, F. E., & Biscaye, P. E. (1996a). Sr and Nd isotopes as tracers of North Atlantic lithic particles: Paleoclimatic implications. *Paleoceanography*, *11*, 95–113.

Roberson, H. E., & Lahann, R. W. (1981). Smectite-to-illite conversion rates, effects of solution chemistry. *Clays and Clay Minerals*, *29*, 129–135.

Robert, C. (1982). *Modalité de la sédimentation argileuse en relation avec l'histoire géologique de l'Atlantique Sud*. Thèse de doctorat, Univ. Aix-Marseille II, 141 pp.

Robert, C., Caulet, J. P., & Maillot, H. (1988). Evolution climatique et hydrologique en Mer de Ross (Site DSDP 274) au Néogène, d'après les associations de radiolaires, la minéralogie des argiles et al géochimie minérale. *Comptes rendus de l'Academie des sciences. Serie Paris, II*, *306*, 437–442.

Robert, C., Diester-Haass, L., & Paturel, J. (2005). Clay mineral assemblages, siliciclastic input and paleoproductivity at ODP Site 1085 off Southwest Africa: A late Miocene–early Pliocene history of Orange river discharges and Benguela current activity, and their relation to global sea level change. *Marine Geology*, *216*, 221–238.

Robert, C., & Maillot, H. (1983). Paleoenvironmental significance of clay mineralogical and geochemical data, southwest Atlantic, Deep Sea Drilling Project Legs 36 and 71. In: W. J. Ludwig & V. A. Krasheninnikov (Eds), *Initial Reports of the Deep Sea Drilling Project 71* (pp. 379–397). Washington: U.S. Government Printing Office.

Rutberg, R. L., Goldstein, S. L., Hemming, S. R., & Anderson, R. F. (2005). Sr isotope evidence for sources of terrigenous sediment in the southeast Atlantic Ocean: Is there increased available Fe for enhanced glacial productivity? *Paleoceanography*, *20*, PA1018, doi:10.1029/2003PA000999, PA1018 1–10.

Serova, V. V., & Gorbunova, Z. N. (1997). Mineral composition of soils, aerosols, suspended matter, and bottom sediments of the Lena river and the Laptev Sea. *Oceanology*, *37*, 121–125.

Singer, A. (1984). The paleoclimatic interpretation of clay minerals in sediments — a review. *Earth Science Reviews*, *21*, 251–293.

Srodon, J., & Eberl, D. D. (1984). Illite. In: S. W. Bailey (Ed.), *Micas, Review in Mineralogy* (Vol. 13, pp. 495–544). Mineralogical Society of America.

Stein, R., Grobe, H., & Washner, M. (1994). Organic carbon, carbonate and clay mineral distributions in eastern central Arctic Ocean surface sediments. *Marine Geology*, *119*, 269–285.

Stow, D. A. V. (1994). Deep sea processes of sediment transport and deposition. In: K. Pye (Ed.), *Sediment transport and depositional processes* (pp. 257–291). Oxford: Blakwell.

Strakhov, N. M., (1960). Principles and theories of lithogenesis, 1. Akad. Nauk. S.S.S.R., Moscow (pp. 6–24). [In: Rateev, et al., 1969.]

Thiry, M. (2000). Paleoclimatic interpretation of clay minerals in marine deposits: an outlook from the continental origin. *Earth Science Review*, *49*, 201–221.

Thorez, J. (1986). Argillogenesis and the hydrolysis index. *Mineralogica et Petrographica Acta*, *29*, 313–338.

Thorez, J. (1989). Argilloscopy of weathering and sedimentation. *Bulletin Société Géologique de Belgique*, *98*, 245–267.

Tütken, T., Eisenhauer, A., Wiegand, B., & Hansen, B. T. (2002). Glacial-interglacial cycles in Sr and Nd isotopic composition of Arctic marine sediments triggered by the Salbard/Barents Sea ice sheet. *Marine Geology*, *182*, 351–372.

Vanderaveroet, P., Averbuch, O., Deconinck, J. F., & Chamley, H. (1999). A record of glacial/ interglacial alternations in Pleistocene sediments off New Jersey expressed by clay mineral, grain-size and magnetic susceptibility. *Marine Geology*, *159*, 79–92.

Velde, B. (1992). *Introduction to clay minerals* (p. 198). London: Chapman and Hall.

Velde, B. (1995). *Origin and mineralogy of clays. Clays and the environment* (p. 356). Berlin: Springer-Verlag.

Venkatarathnam, K., & Ryan, W. B. F. (1971). Dispersal patterns of clay minerals in the sediments of the eastern Mediterranean Sea. *Marine Geology, 11,* 261–282.

Vroon, P. Z., Van Bergen, M. J., Klaver, G. J., & White, W. M. (1995). Strontium, neodymium, and lead isotopic and trace element signatures of the East Indonesian sediments: Provenance and implications for Banda Arc magma genesis. *Geochimica et Cosmochimica Acta, 59,* 2573–2598.

Wahsner, M., Müller, C., Stein, R., Ivanov, G., Levitan, M., Shelekhova, E., & Tarasov, G. (1999). Clay-mineral distribution in surface sediments of the Eurasian Arctic Ocean and continental margin as indicator for source areas and transport pathways — a synthesis. *Boreas, 28,* 215–233.

Walter, H. J., Hegner, E., Diekmann, B., Kuhn, G., & Rutgers Van Der Loeff, M. M. (2000). Provenance and transport of terrigenous sediment in the South Atlantic Ocean and their relations to glacial and interglacial cycles: Nd and Sr isotopic evidence. *Geochimica et Cosmochimica Acta, 64,* 3813–3827.

Weaver, C. E. (1989). *Clays, muds, and shales. Developments in sedimentology, SEPM 44* (p. 819). Amsterdam: Elsevier.

White, R. E. (1999). *Principles and practice of soil science. The soil as a natural resource* (p. 348). Oxford: Blackwell.

White, W. M., & Dupré, B. (1986). Sediment subduction and magma genesis in the Lesser Antilles: Isotopic and trace element constraints. *Journal of Geophysical Research, 91,* 5927–5941.

White, W. M., Dupré, B., & Vidal, P. (1985). Isotope and trace element geochemistry of sediments from the Barbados Ridge–Demerara Plain region, Atlantic Ocean. *Geochimica et Cosmochimica Acta, 49,* 1875–1886.

Windom, H. L. (1976). Lithogenous material in marine sediments. In: J. P. Riley & R. Chester (Eds), *Chemical oceanography* (Vol. 5, pp. 103–135). New York: Academic Press.

Yeroshchev-Shak, V. A. (1964). Clay minerals of the Atlantic Ocean, Soviet. *Oceanography, 30,* 90–105.

Zimmerman, H. B. (1972). Sediments of the New England Continental Rise. *Geololgical Society American Bulletin, 83,* 3709–3724.

Zimmerman, H. B. (1982). Fine-grained sediment distribution in the late Pleistocene/Holocene North Atlantic. *Bulletin Institut Géologique du Bassin d'Aquitaine, 31,* 337–357.

CHAPTER FIVE

RADIOCARBON DATING OF DEEP-SEA SEDIMENTS

Konrad A. Hughen

Contents

1. Introduction	185
2. Dating Marine Sediments	187
2.1. Calculation of ^{14}C ages	187
2.2. Measurement techniques	189
2.3. Contamination/Sample materials	190
2.4. Calibration of the radiocarbon timescale	194
2.5. Marine ^{14}C reservoir age	196
2.6. Bioturbation-Abundance effects	198
3. Applications of Marine ^{14}C	201
Appendix I — Internet Resources	204
Radiocarbon journal	204
Calibration programs	204
Data sets	205
References	206

1. INTRODUCTION

Marine sediments are valuable archives for paleoclimate reconstructions. Deep-sea sediments typically provide continuous deposition with few hiatuses, despite large shifts in climate and depositional regime. In addition, marine sediment archives offer opportunities for application of multiple proxies representing different climatic and oceanographic systems. Over the past 50,000 yr, radiocarbon dating has served as a powerful tool for accurate dating of marine records, although caution must be exercised in sample selection and interpretation of ages in the earliest part of the timescale. Nevertheless, radiocarbon measurements have evolved into a universal and precise dating tool, which at the same time can be used as a geochemical tracer of climate and carbon cycle changes.

Several factors contribute to the unique role of ^{14}C in paleoceanographic studies. Since any material containing carbon has the potential to be dated, the method is applicable to materials and situations throughout the world oceans,

Developments in Marine Geology, Volume 1
ISSN 1572-5480, DOI 10.1016/S1572-5480(07)01010-X

including all organisms and widely-occurring inorganic carbonate minerals, as well as inorganic carbon dissolved in water. Although the production of ^{14}C, like any cosmogenic isotope, is controlled by the local magnetic dipole moment of the Earth and hence varies with latitude, the relatively fast mixing of the troposphere results in a spatially homogeneous input of ^{14}C into the ocean (e.g., Broecker, Sutherland, Smethie, Peng, & Ostlund, 1995). The amount of carbon in those 'active' reservoirs, primarily the oceans, exchanging with the atmosphere over the lifetime of ^{14}C is large (Figure 1), attenuating strongly any high-frequency variability from the production rate of ^{14}C. Finally, although the natural abundance ratio of ^{14}C/^{12}C is low (10^{-12} in a modern pre-bomb sample), high-precision techniques exist to measure ^{14}C routinely even in very small samples (less than a milligram). Nevertheless, there are several caveats of which researchers must be aware when applying ^{14}C dating techniques to marine paleoclimate records. These include the need for calibration of the ^{14}C timescale, potential variability in marine ^{14}C reservoir ages, and age biases resulting from variable fluxes of ^{14}C carriers to the sediment together with differential mixing due to sediment bioturbation. In addition, in the case of fossil carbonates, diagenetic evolution may result in unreliable chronologies beyond ca. 35,000 yr.

Deriving ages from ^{14}C measurements of carbon-bearing fossils in deep sea sediments requires a detailed understanding of the carbon cycle. From this view point, the injection of large quantities of nuclear-bomb ^{14}C into the atmosphere in the late 1950s and early 1960s, and its subsequent transfer into active carbon reservoirs, has yielded valuable information on carbon cycling and residence times

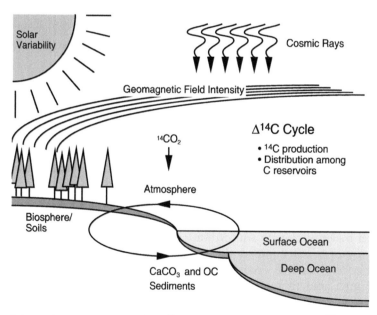

Figure 1 Schematic representation of the ^{14}C cycle, depicting cosmogenic ^{14}C production and distribution of ^{14}C within the global carbon cycle. The global carbon cycle includes only active ^{14}C reservoirs that exchange carbon on timescales of relevance to radiocarbon (i.e., 10^1–10^4 yr).

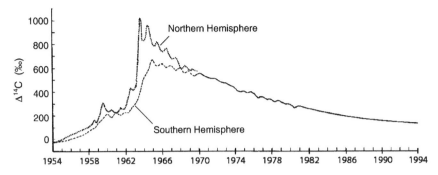

Figure 2   Atmospheric radiocarbon concentrations for the northern and southern hemispheres, resulting from atmospheric nuclear weapons testing. This large signal provides a valuable tracer for global carbon cycle studies. Figure from Broecker and Peng (1994).

(e.g., Broecker & Peng, 1994; Figure 2). Although results from such studies are needed to take into account ^{14}C chronologies in deep-sea sediments, they will not be examined in detail in this chapter, as the focus here is on the geochronological use of natural ^{14}C activities. In the second volume of the series on Late Cenozoic Ocean, a chapter will address the issue of the carbon cycle.

2. DATING MARINE SEDIMENTS

The radiocarbon isotope ^{14}C was the first cosmogenic isotope to be detected (Rubin & Kamen, 1941; Libby, 1946). ^{14}C is produced by cosmic radiation interacting in an (n, p) reaction with atmospheric nitrogen (Masarik & Beer, 1999) and enters the global carbon cycle as ^{14}CO$_2$, which is well mixed within the atmosphere. From here the two main fields of radiocarbon dating originate: (1) photosynthesis fixes CO_2, and hence ^{14}CO$_2$, into plant organic matter, allowing dating of organic carbon samples, and (2) CO_2 dissolves in sea water to form dissolved inorganic carbon (DIC), which can be dated directly or following incorporation into marine carbonates.

2.1. Calculation of ^{14}C Ages

^{14}C dating is based on the radioactive decay law:

$$t = \frac{1}{\lambda} \times \ln\left(\frac{N_0}{N_t}\right) = \frac{1}{\lambda} \times \ln\left(\frac{A_0}{A_t}\right) \tag{1}$$

where λ = decay constant of ^{14}C (1/mean life); N_t and A_t = remaining ^{14}C concentration (N) or activity (A) of a sample as measured today (standardized to AD 1950 and corrected for carbon isotope fractionation; see Broecker & Olson, 1959; Olson & Broecker, 1961; and N_0 and A_0 = initial ^{14}C concentration (N) or activity (A) at the time when the sample ceased exchange of ^{14}C with the source reservoir. The ^{14}C decay constant originally measured by Engelkemeir et al. (1949) yielded a half-life of $5,568 \pm 30$ yr (mean life 8,033 yr). Later on this quantity was measured more accurately, and a consensus was reached fixing the half-life at $5,730 \pm 40$ yr (Godwin, 1962). Concentrations are usually measured with reference to the most abundant stable isotope of carbon (^{12}C), whereas activities are expressed with reference to total carbon content of a given sample or material (i.e., ^{12}C+^{13}C).

By convention, standards for reporting ^{14}C ages have been adopted since the 1972 Radiocarbon Conference (see also Stuiver & Pollach, 1977) to ensure that more recently reported values remain consistent with previous studies. *Conventional radiocarbon ages* are therefore defined to meet the following conditions:

1) Use of the 5,568 yr (Libby) half-life.
2) Use of the National Bureau of Standards (NBS) Oxalic acid standard directly or indirectly as the "modern" radiocarbon standard.
3) Correction for sample isotopic fractionation normalized to $\delta^{13}C = -25$ per mil relative to a standard (cf. the Pee Dee Belemnite, or VPDB; see Coplen, 1996).
4) Standard and sample activity are both reported relative to AD 1950, established as the year 0 before present (BP). Hence the date of the measurement does not enter calculations, except for carrying back any measured activity to its value in 1950.
5) Past atmospheric ^{14}C concentration is assumed to have remained unchanged through time.

The ^{14}C activity (depletion) of a sample relative to the standard is given by:

$$\delta^{14}C = \left(\frac{A_{\text{sample}}}{A_{\text{standard}}} - 1 \right) \times 1000 \tag{2}$$

where $\delta^{14}C$ = standard-corrected ^{14}C activity, expressed in per mil (‰); A_{sample} = original measured activity; and A_{standard} = 95% of the NBS oxalic acid standard (equal to 1890 wood used originally by Libby to estimate atmospheric ^{14}C activity, prior to its dilution by "dead" CO_2 from the industrial combustion of fossil fuel and coal). In practical terms, $A_{\text{standard}} = 13.56$ dpm/gC (disintegration per minute and per gram of carbon) corresponding to an isotopic ratio ^{14}C/C of 1.176×10^{-12}, values often referred to as those of "modern" carbon.

In addition, to obtain a ^{14}C date based on Equation (1), the measured ^{14}C activity has to be corrected for isotopic fractionation. Normally, one would assume that during the lifetime of the specimen its ^{14}C activity was identical to that of the source reservoir, and that any decrease measured in the remaining ^{14}C activity today is solely due to radioactive decay. However, the transport, uptake, and respiration of carbon, as well as laboratory processing, may discriminate (isotopically fractionate) against the heavier carbon isotopes ^{13}C and ^{14}C compared to ^{12}C. This may result in a systematic difference from the source reservoir, which mimics radioactive decay for ^{14}C. Hence, a ^{14}C sample, which is depleted in ^{14}C owing to isotope fractionation would appear older than its true age.

Fortunately, the effect of isotope fractionation on the ^{14}C activity can be corrected by determining the abundance of the stable carbon isotope ^{13}C, compared to the source material. This is because isotopic fractionation depends on the mass difference with respect to ^{12}C; hence, for ^{14}C it is twice that of ^{13}C (Craig, 1954; for details, see Wigley & Muller, 1981). ^{13}C isotope abundances are calculated relative to the VPDB standard, according to:

$$\delta^{13}C = \left(\frac{(^{13}C/^{12}C)_{sample}}{(^{13}C/^{12}C)_{VPDB}} - 1\right) \times 1000 \qquad (3)$$

Terrestrial plants following the photosynthetic cycle of Calvin (often referred to as the C3 cycle since the first sugar to be formed contains 3 atoms of carbon), and notably the 1890 wood used by Libby to determine the "standard" ^{14}C activity value, have $\delta^{13}C$ close to –25‰ (e.g., Smith & Epstein, 1971). Thus, all ^{14}C isotopic fractionation corrections are normalized to this value, according to:

$$D^{14}C = \delta^{14}C - 2(\delta^{13}C + 25)\left(1 + \frac{\delta^{14}C}{1000}\right) \qquad (4)$$

where $D^{14}C$ is standard and fractionation-corrected ^{14}C concentration in per mil of "modern" carbon; $\delta^{13}C$ for the sample is measured relative to the PDB standard; and the value 25 arises from the $\delta^{13}C$ value of the 1890 wood (-25‰ vs. VPDV) used in the early steps of the development of the method by Libby.

2.2. Measurement Techniques

The ^{14}C activity or ratio of ^{14}C to total C in a sample compared to that in a standard is either measured by detection of its radioactive decay (radiometric techniques), or by counting carbon ions using accelerator mass spectrometry (AMS). From a user's perspective the main difference between the two techniques is in the amount of carbon required for analysis. Decay counting over one week utilizes only a minute fraction of the ^{14}C atoms present in a sample, roughly given by the ratio of one week over the true half-life (8,267 yr). On the other hand, direct counting of ^{14}C ions, even accounting for losses in various stages of an AMS system, results in five orders of magnitude higher sensitivity and reductions in sample size (1 milligram) and measurement time (1 h).

Radiometric techniques were developed to a high level of sophistication during the period between 1960 and 1980 (for reviews, see Kromer & Münnich, 1992; Cook & Harkness, 1994). Today the most common techniques used in this field are CO_2 gas counting and the liquid scintillation counting of benzene. Sample sizes range from 0.5 to 10 g of carbon. Using counting times of up to one week, high-precision laboratories have demonstrated routine precision of 0.2%, which is useful in minimizing the contribution of ^{14}C error to uncertainties in ^{14}C age calibration data sets (Reimer et al., 2004), and for studies of ocean ventilation (Stuiver et al., 1996; Schlosser et al., 1997).

The AMS technique was developed initially by adaptation of accelerators to the detection of rare isotopes (for review, see Gove, 1992), but today many machines are dedicated to ^{14}C (Figure 3). Briefly, a cesium sputter ion source ionizes sample carbon in the form of graphite, producing negative ion beams of the three isotopes

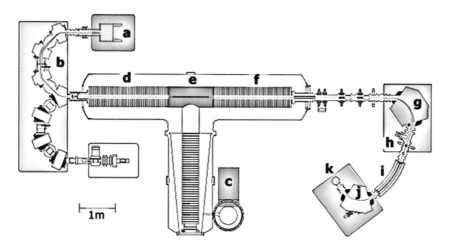

Figure 3 Schematic of a 3 MV AMS in operation at the National Ocean Sciences AMS Facility (NOSAMS), at Woods Hole Oceanographic Institution. Indicated are the (a) ion source, (b) recombinator region, (c) power supply, (d & f) accelerator tubes, (e) stripper canal, (g & j) analyzer magnets, (h) Faraday cups, (i) electrostatic deflector, and (k) ^{14}C detector.

of carbon (mass 12, 13, and 14). The negative ionization permits the elimination of inferences from isobaric ^{14}N. The beams are mass filtered to remove ions with masses smaller than 12 and larger than 14, and the largest ^{12}C beam is reduced to the same intensity as the ^{13}C beam. The three ion beams are accelerated and passed through a tube containing argon gas, which acts as an electron stripper, converting the incoming negative carbon ions to positive ions. The resulting carbon ions are then accelerated to their highest energy and passed through an electrostatic filter to select a given ion charge (usually C^{3+}), then through a spectrometer magnet to disperse the beams of different masses. The abundant isotopes ^{12}C and ^{13}C are measured in Faraday cups, while the vastly smaller ^{14}C ion beam is analyzed using an ion counter. Additional filtering devices serve to screen out any remaining particles matching the kinematic properties of the ^{14}C ions, such as molecular ^{12}CH$_2$ ions. The relative intensity ratios of the three carbon isotope beams provide a direct measure of the sample ^{14}C concentration and ^{13}C/^{12}C isotopic composition due to natural and analytical fractionation.

The typical sample size for AMS measurement is in the order of 0.3 mg carbon, and samples smaller than 50 µg carbon are routinely measured using specialized techniques and standards. The routine precision of the AMS technique has been demonstrated to be in the 0.3–0.4% range. Although AMS has been developed into a mature and precise technique, it is continually being improved to provide even greater sensitivity (lower sample size).

2.3. Contamination/Sample Materials

A central issue with the ^{14}C dating method is the question of context, i.e., the time interval between the event dated by the method and the event of interest for the user. Ideally, both events should be well defined and identical or close in time, compared to the age resolution of the dating method. The radiocarbon dating method is capable of an age resolution in the order of a few decades; hence, it is crucial to keep in mind that the event to be dated is the cessation of exchange with the respective ^{14}C reservoir. In marine sciences, however, the focus is usually on dating the sample burial, which may come later, particularly in the event of sediment remobilization and focusing. Hence, the interpretation of any ^{14}C date must take into account any potential gap between the date reported by the laboratory and the date of the deposition of the sample.

A second general issue relevant for ^{14}C dating is the potential extent of contamination, either after burial and before recovery of the sample, or in the dating laboratory. In addition to the specific information given below, a few general remarks are possible for carbon samples. Over the past 50 yr extensive research has been devoted to obtaining reliable results from various fractions in carbonaceous samples, and chemical procedures have been developed for decontamination and pretreatment of most materials (methodologies for many different materials are published as individual papers in the proceedings of International Radiocarbon Conferences). The relative contribution to age determination from any contamination source depends on the amount of contaminating carbon and its activity (age) difference from the sample. Here, samples older than a few half-lives, and very small samples, which leave less room for rigorous pretreatment, are of concern.

A few general handling precautions can be given: prior to submittal, organic samples must be stored in a suitable way to prevent attack of fungi (i.e., they should be dried or frozen). Most organic samples are pretreated with acid and alkali solutions in the ^{14}C laboratory (e.g., van Klinken & Hedges, 1998); hence contamination from washing with tap water and touching a sample usually is not a concern. Milligram-sized samples to be submitted for AMS dating require the most care, as microgram levels of contamination (especially modern CO_2 with high radiocarbon concentrations) must be avoided here.

2.3.1. Seawater

Atmospheric ^{14}C enters the ocean as dissolved CO_2, $CO_2(aq)$, which reacts with water to form carbonic acid, and then dissociates into bicarbonate and carbonate ion forms according to:

$$CO_2(aq) + H_2O \rightleftharpoons H_2CO_3 \tag{5}$$

$$H_2CO_3 \rightleftharpoons HCO_3^- + H^+ \tag{6}$$

$$HCO_3^- \rightleftharpoons CO_3^{2-} + H^+ \tag{7}$$

The total of all marine aquatic carbonate species ($CO_2(aq)+H_2CO_3+HCO_3^-+CO_3^{2-}$), leaving aside low abundance more complex ions, is referred to as dissolved inorganic carbon (DIC), also called ΣCO_2. In the ocean, DIC spreads from the surface to depths carried by the flux of sinking and laterally spreading advective ocean currents. During the relatively long transport time, the ^{14}C activity of DIC is changed by radioactive decay, as well as mixing with water of different

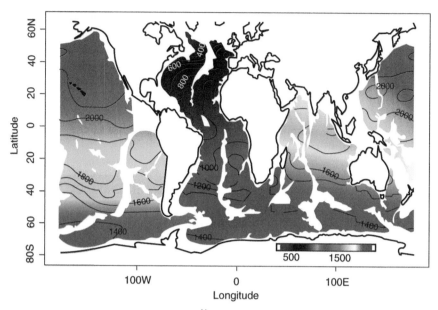

Figure 4 Objectively mapped conventional ^{14}C age of natural radiocarbon on the 3,500 m level. For this figure, Δ^{14}C was first converted to ^{14}C age, which was then objectively mapped (i.e., converted then mapped, not mapped then converted). Contours are 100 yr apart. Figure from Matsumoto and Key (2004).

origin. Hence, from the ^{14}C-activity distribution of carbonate dissolved in seawater, the patterns and rates of deep-water fluxes can be inferred (Figure 4). This application of ^{14}C analyses has led to large-scale programs in oceanography, both on global and regional scales.

2.3.2. Marine carbonates

The base ^{14}C activity of the biogenic carbonate pool of marine fauna is set by the mixed-layer carbonate species. For most applications, the carbonate material to be dated is assumed to be in equilibrium with the seawater in which it grew (accounting for fractionation and constant offset "vital" effects). The main systematic uncertainty then comes from the variation of the marine reservoir correction. For a wide range of sites, the marine reservoir correction has been determined from close to modern (pre-bomb and generally pre-industrial time) marine samples of known age, archived in an online database at http://www.calib.org. In general terms, this database allows users to estimate apparent ages of mixed-layer DIC and shallow marine carbonates under oceanic conditions close to modern, notably with reference to thermohaline circulation patterns including intensity and location of deep water formation and upwelling. Extrapolating information from this database or other source to estimate DIC age and reservoir corrections back in time thus requires caution.

Foraminifera. Due to their near-ubiquitous occurrence, foraminifera hand-picked and ^{14}C dated by AMS have become a powerful tool in dating of marine sediments.

For planktonic species originating from the ocean surface layer, the respective marine reservoir correction can be applied with some reservation as explained above, and quoted with the result. The reservoir correction for benthic species is set by the apparent ^{14}C age of the bottom waters at the site. From the ^{14}C age difference of pairs of benthic vs. planktonic foraminifera, found at the same depth in a sediment core, attempts at determining the transit times of the bottom water (from sea surface to the site), also called ventilation age, have been made (Broecker et al., 1988; Keigwin & Schlegel, 2002). As the ocean ventilation is directly tied to oceanic heat transfer, and hence climate change, this field is of considerable importance in paleoclimate research. However, direct rapid export of organic carbon to the sea floor (e.g., Bauer, Reimers, Druffel, & Williams, 1995), as well as oxidation of older organic matter and the subsequent release of ^{14}C-depleted CO_2 in sediment pore water, may result in ^{14}C-offsets independent of ventilation rates, and possible differences between endo- and epibenthic fauna (e.g., Wu & Hillaire-Marcel, 1994).

Corals.   Coral skeletons accurately record the ^{14}C activity of DIC in seawater, allowing corals to be applied to a wide range of radiocarbon research. The high concentration of U in aragonitic corals reduces the potential impact of any diagenetic U mobility (e.g., Pons-Branchu, Hillaire-Marcel, Deschamps, Ghaleb, & Sinclair, 2005), also provides the potential for U-series dating methods to be applied to these materials with great precision (Edwards & Chen, 1986). Paired ^{14}C-U/Th dates from reef-building corals have been used to calibrate the radiocarbon time scale beyond tree rings (Bard et al., 1998; Fairbanks et al., 2005) as well as to record changes in surface ocean (and atmospheric) ^{14}C activity as a function of changing climate (Edwards et al., 1993).

In addition to shallow reef-building corals, solitary corals from the deep ocean have been measured for radiocarbon to provide information about ^{14}C activity and changes in rates of deep ocean ventilation (Adkins et al., 1998; Robinson et al., 2005).

Coral aragonite is stable in seawater, but unstable when tectonic uplift exposes reef terraces to meteoric rainfall and groundwater. Modern atmospheric carbon can contaminate fossil coral skeletal material during recrystallization (usually in secondary calcite), and selective dissolution (up to 40%) is needed to successfully eliminate the contaminated fraction (Burr et al., 1998). Worthy of mention here is the fact that diagenetic effects and recrystallization generally have a large impact on U-series ages as well. Although correction models have been proposed (e.g., Thompson, Spiegelman, Goldstein, & Speed, 2003), most studies discard samples containing more than a few tenths of 1% calcite for U-series determination (Hughen et al., 2004a; Fairbanks et al., 2005), reducing the number of samples suitable for ^{14}C vs. U-series cross-dating and documenting changes in DIC ^{14}C activity back in time (e.g., Pons-Branchu et al., 2005).

Molluscs, bivalves.   Carbonate samples, including the older parts of mollusc or bivalve shells, may exchange isotopically with seawater carbonate and/or CO_2. Often the contamination (towards younger ages in seawater) is confined to the outer layers of a sample. Hence, it is customary to remove 10–30% of a sample by etching the surface in weak acid. The thickest part of the shell, or umbo in the case of bivalves, is often targeted as the most robust for extensive leaching treatment.

2.3.3. Organic/molecular compounds

In some circumstances, the ^{14}C dating of recognizable plant or carbonate macro-fossils is not possible owing to low productivity or lack of preservation, and researchers turn to the pool of organic carbon as material for ^{14}C dating. Total organic carbon (TOC) in sediments is extremely heterogeneous, representing sources ranging in age from modern plant remains to fossil kerogen (Eglinton et al., 1997), and thus dates on bulk TOC should be avoided if possible. Certain classes of organic compounds can be traced to specific sources, such as aquatic algae and terrestrial plants. These 'biomarker' compounds can be isolated and concentrated for AMS ^{14}C analysis, allowing ^{14}C dating of previously undatable sediments, as well as providing a window into details of carbon cycling at the molecular level (Huang et al., 1996; Currie et al., 1999; McNichol et al., 2000; Pearson & Eglinton, 2000).

As with all materials used for ^{14}C dating, the source and history of organic molecules must be understood to interpret biomarker ^{14}C ages in the context of a sediment horizon. For example, compounds derived from terrestrial vegetation can be ablated directly from leaf surfaces via eolian transportation, or slowly leached from storage in forest litter and soils via fluvial transport. These scenarios have different implications for dating applications. To date, most studies have focused on ^{14}C as a constraint on determining biomarker sources and transport pathways (Ohkouchi, Eglinton, Keigwin, & Hayes, 2002; Mollenhauer et al., 2003), although their geochronological value is high.

The identification and isolation of organic biomarkers requires careful organic chemistry techniques, including solvent extraction of sediments, removal or substitution of functional groups, column chromatography to isolate compounds, and gas chromatographic (GC) analysis of compound recovery and purity, followed by preparative capillary GC (PCGC) to separate and concentrate individual compounds (for reviews, see Eglinton et al., 1996; Hedges & Oades, 1997). Many of these procedures have become automated and the routine performance of compound-specific ^{14}C analyses is becoming increasingly common.

2.4. Calibration of the Radiocarbon Timescale

As discussed above, conventional ^{14}C ages assume that initial ^{14}C concentration has remained constant. However, studies show that atmospheric and surface ocean ^{14}C concentration has changed notably through time. As a result of changing ^{14}C, radiocarbon ages have deviated significantly from calendar ages in the past, primarily owing to changes in either the rate of ^{14}C production in the atmosphere (a function of geomagnetic field intensity and solar variability; Stuiver & Braziunas, 1993; Masarik & Beer, 1999; Laj et al., 2002), or the distribution of ^{14}C between different reservoirs in the global carbon cycle (primarily deep ocean ventilation; Siegenthaler & Sarmiento, 1993) (Figure 1).

The Earth's geomagnetic field serves to shield the atmosphere from incoming cosmic rays, and when the magnetic field strength increases, ^{14}C production decreases (and vice versa). Similarly, the solar wind distorts the Earth's geomagnetic field in a way that decreases ^{14}C production, and a *rise* in solar activity will cause a *decline* in ^{14}C production. In addition, the Earth's global carbon cycle contains several reservoirs which exchange carbon with the atmosphere on timescales relevant to the lifetime of ^{14}C (10^1–10^4 yr). Within the deep ocean in particular, ^{14}C is sequestered from atmospheric exchange for 1,000 yr on average, long enough for decay to reduce the deep ocean ^{14}C activity significantly (Broecker & Peng, 1982). Changes in the rate of exchange between the deep ocean and the atmosphere through fluctuations in the meridional overturning circulation (i.e., the North Atlantic component of global thermohaline circulation belt), can strongly influence atmospheric ^{14}C activity (Hughen et al., 1998).

Tree-ring dendrochronologies provide the most accurate and highest resolution data for ^{14}C calibration (Reimer et al., 2004), but currently are limited to the past 12.4 ka (Friedrich et al., 2001). In addition, for applications to calibration of marine dates, tree-ring ^{14}C ages must be modeled to derive equivalent ocean mixed-layer ages (Hughen et al., 2004a). High-resolution measurements of planktonic foraminifer in layer-counted varved sediments of the Cariaco Basin have extended the marine ^{14}C calibration curve beyond tree rings back to 14.7 ka (Hughen, Southon, Lehman, & Overpeck, 2000). Additional marine calibration data parallel to and beyond the varved sediment record are available through extensive measurements from U/Th-dated corals around the world (Edwards et al., 1993; Bard et al., 1998; Burr et al., 1998, 2004; Cutler et al., 2004), but at lower (approximately centennial) resolution (Hughen et al., 2004a). Beyond the limit of the IntCal04 data set at 26 ka, a comparison of data sets from previous studies shows considerable scatter and no single ^{14}C calibration curve is available for this interval (Van der Plicht et al., 2004).

Recently, marine-based calibration data back to 50 ka have been provided by ^{14}C and ^{230}Th-dated coral results with irregular sample spacing (Fairbanks et al., 2005), and at higher resolution from sediments of the Cariaco Basin (Hughen et al., 2004b). Cariaco Basin sediments are only intermittently laminated beyond ~14.7 ka, but they show distinct millennial-scale variability in sedimentological and geochemical records, which can be reliably correlated with Dansgaard-Oeschger (D-O) events in Greenland ice cores (Hughen et al., 2004b) and ^{230}Th-dated Hulu Cave speleothems (Hughen, Southon, Lehman, Bertrand, & Turnbull, 2006). These correlations have been used to transfer the calendar chronologies onto the Cariaco ^{14}C series to provide calibration data sets. Where Cariaco Basin data, linked to the ^{230}Th Hulu Cave chronology of Wang et al. (2001), overlap with coral data sets they agree well (Figure 5), indicating that the higher resolution of the ^{14}C dating in the Cariaco Basin can be used to accurately interpolate the calibration between lower resolution coral data (Hughen et al., 2006). Other records retroactively linked to the NGRIP Greenland ice core layer chronology (Rasmussen et al., 2006; Johnsen et al., 2001), including marine sediments from the Iberian Margin (Bard, Rostek, & Menot-Combes, 2004) and the Iceland Sea (Voelker et al., 2000), as well as revised ^{230}Th ages for Bahama speleothems (Beck et al., 2001; Richards et al., 2006), appear to agree with Cariaco and coral data over this interval. The long-term trend in ^{14}C vs. calendar age over the past ~40,000 yr leads

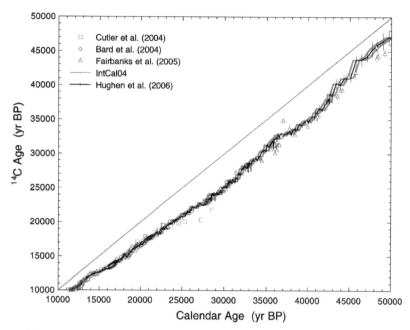

Figure 5 ^{14}C calibration from 0 to 50 calendar kyr BP. High-resolution ^{14}C calibration data from IntCal04 are shown as a black line. Cariaco Basin ^{14}C ages are plotted vs. Hulu Cave ^{230}Th calendar age model as small blue circles. Thin gray lines show 1-sigma error envelope for Cariaco record due to calendar age uncertainty. Paired ^{230}Th/^{14}C dates from corals are shown as solid circles. All error bars are 1-sigma. Figure from Hughen et al. (2006).

to conventional ^{14}C ages being lower than true calendar ages. A monotonously varying trend would not affect adversely the transformation of ^{14}C age intervals into calendar age ranges, but the fluctuations observed in high-resolution ^{14}C data sets lead to ambiguities in calibration. For example, comparing ^{14}C vs. calendar age shows plateaus or rapid reversals with ^{14}C ages remaining the same, or actually becoming older, with decreasing calendar age. Such details of ^{14}C variability, including brief periods in which samples of different true age may show identical measured ^{14}C activity, exemplify the need for high-resolution ^{14}C calibration data sets. The observed convergence of data sets from dispersed archives and geographic locales will likely provide, in the near future, the basis for an extended ^{14}C calibration back to 50 ka.

2.5. Marine ^{14}C Reservoir Age

The ^{14}C activity of the ocean can vary independently from the atmosphere, and in the surface ocean, upwelling of deep water and gas exchange with the atmosphere maintain an intermediate ^{14}C level, producing an apparent "reservoir age" where surface waters appear older than the contemporaneous atmosphere. Interpretation or calibration of ^{14}C dates from these systems should ideally involve carbon cycle models including the effects of changing local reservoir ages. The world oceans hold ca. 93% of all readily exchangeable carbon (i.e., those reservoirs with residence times relevant to ^{14}C timescales) in the form of carbon ion species (CO_2, HCO_3^- and CO_3^{2-}). The residence time of ^{14}C in the deep ocean, with respect to gas exchange with the atmosphere, is on the order of several hundred to 1,600 yr (Broecker & Peng, 1982). Hence, sufficient decay can occur in the ^{14}C activity of the deep ocean (depending on depth and location) to fall between 80% and 90% of the atmosphere. Currently, stratified surface waters at low latitudes (Tropical Warm Pools) are well equilibrated with the atmosphere and have ^{14}C activity of ca. 96% compared to the atmosphere, equivalent to a 400 yr reservoir age. These areas typically have younger reservoir ages than at high latitudes, where older water outcrops at the surface.

This pattern is influenced by the global deep 'conveyor-belt' circulation in which young surface water is advected into the high-latitude North Atlantic Ocean (reservoir age ~400 yr) and the oldest deep waters return to the surface in the Northeast Pacific region (reservoir age ~700 yr). Reservoir ages are also affected by localized conditions, such as local or regional (equatorial, coastal) upwelling, wind speed (CO_2 gas exchange), and extent of sea ice. Efforts to calibrate marine ^{14}C dates back in time must therefore take into account the potential for changes in regional ocean circulation and surface conditions, as well as atmospheric ^{14}C concentration. From modelling of the carbon cycle (Siegenthaler et al., 1980), the variation at low-latitude open ocean sites is expected to be moderate, because even drastic changes in these parameters affect the surface activity to only a few percent, equivalent to less than ca. ± 150 yr in the marine reservoir correction.

For high latitudes and shallow water sites, the reservoir variability may be substantially higher. For example, there there is evidence for large (factor of two or greater) reservoir age shifts in the past, during late Glacial and deglacial times in the high-latitude North Atlantic (+400 to 800 yr) (Bard et al., 1994; Austin, Bard, Hunt, Kroon, & Peacock, 1995; Bondevik, Birks, Gulliksen, & Mangerud, 1999; Eiriksson, Knudsen, Haflidason, & Heinemeier, 2000; Waelbroeck et al., 2001), Mediterranean Sea (+500 yr) (Siani et al., 2001), and New Zealand region (+1,200 to 2,000 yr) (Sikes, Samson, Guilderson, & Howard, 2000). Fortunately, the warm pools of the tropics may have remained relatively stable throughout the past, and calibration data based on tropical surface reservoir ages (corals, planktonic foraminifera) may provide accurate constraints for atmospheric values. For example, Cariaco basin ^{14}C ages agree closely with anchored tree-ring ages from 10.5 to 12.4 cal kyr BP across the large climatic shifts of the Younger Dryas (Hughen, Southon, Bertrand, Frantz, & Zermeno, 2004c), exhibiting no evidence of significant reservoir variability associated with local upwelling. Nevertheless, a Cariaco comparison with floating tree-ring sections indicates the possibility that reservoir age increased by up to 250 yr during the Allerød (Kromer et al., 2004). Therefore, both within single locations and between regions, some changes in reservoir correction through time may be apparent—either as slight trends or increased/decreased variability (Figure 6). Many of these changes reflect real shifts in regional or local oceanography, such as surface circulation and advection, meridional overturning, or local upwelling, rather than reflecting analytical uncertainties because of sample diagenesis or laboratory error.

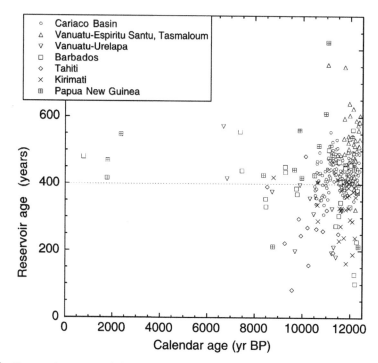

Figure 6 Reservoir age variability for the marine locations used in IntCal04 and Marine04 calibration data sets, including Cariaco Basin (Hughen et al., 2004c); Vanuatu (Burr et al., 1998; Cutler et al., 2004); Barbados (Bard et al., 1998; Fairbanks et al., 2005); Tahiti (Bard et al., 1998); Kirimati (Fairbanks et al., 2005); and Papua New Guinea (Edwards et al., 1993; Cutler et al., 2004). Slight but coherent changes in reservoir age through time are likely due to real changes in oceanographic conditions, rather than laboratory uncertainty. Figure from Hughen et al. (2004a).

Quantifiable records of changes in regional oceanographic conditions adequate for predicting and correcting past reservoir variability are presently lacking. Thus, a certain degree of scatter in site-specific reservoir values through time cannot be avoided, and must instead be characterized as reservoir uncertainty. This uncertainty incorporates all sources of error in reservoir measurement and calculation and is likely an overestimation of true oceanic variability. Increased data density in the future may allow us to identify spatial and temporal patterns of reservoir variability, increasing precision for calibration as well as our understanding of ocean circulation change.

2.6. Bioturbation-Abundance Effects

Biases in apparent ^{14}C ages measured on foraminiferal shells can occur through the combined effects of bioturbation and abrupt changes in ^{14}C-carrier abundances and size (Broecker, Matsumoto, Clark, Hajdas, & Bonani, 1999; Keigwin & Schlegel, 2002; Bard, 2001). In circumstances where sedimentation rates are low, abrupt shifts in overall foraminiferal fluxes, or changes in abundance among different species, can

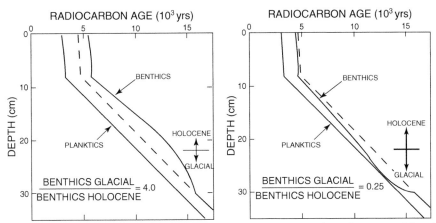

Figure 7 Model calculations showing the perturbation in the radiocarbon age of benthic foraminifera due to the bioturbation-abundance couple. For these calculations the concentration of planktonics is assumed to have remained constant with time. In each case, the age trend for benthics assuming no change in abundance is shown for reference by the dashed line; (left) the concentration of benthics is assumed to have abruptly decreased by a factor of four at the close of glacial time; (right) the concentration of benthics is assumed to have abruptly increased by a factor of four at the close of glacial time. The bioturbation depth (8 cm), the sedimentation rate (2 cm/kyr) and the deep water-surface water age difference (1,600 yr) are assumed to have remained constant with time. Figure from Broecker et al. (1984).

result in significant biases in the final vertical redistribution of shells due to bioturbation (Figure 7). Aside from resulting in variable smoothing of proxy records in deep-sea sediments (e.g., Guinasso & Shink, 1975), complex, size-dependent bioturbation and variable fluxes of ^{14}C-carriers to the sea floor can lead to distortions in ^{14}C chronologies independent of the cycling of this isotope in the atmosphere and ocean.

There are a number of strategies for overcoming problems due to this bioturbation-abundance couple. Selecting cores with high sedimentation rates, and particularly, high and steady foraminiferal fluxes, will substantially minimize any biases. In addition, a useful approach is to date only those samples occurring within peaks in foraminiferal abundance (Broecker et al., 1999; Keigwin & Schlegel, 2002), although this may not solve problems linked to distinct redistributions of ^{14}C due to complex mixing processes (e.g., Guinasso & Shink, 1975). In situations where multiple species may need to be combined for dating, and relative abundances between the two species vary sharply with depth, both species should be dated independently in a number of test samples to test for coherency (Duplessy et al., 1986). Abrupt climate shifts often result in complete replacement of dominant foraminiferal species in sediments. Rather than switching between abundant species across the climate transition, ^{14}C dating should be restricted to species that exist during both climatic regimes, despite being lower in abundance (e.g., Robinson et al., 2005).

Other potential biases may occur through differential dissolution of foraminiferal shells after burial. Biases may occur in part due to the fact that shallower dwelling species often have thinner shells, as well as higher ^{14}C activity. Dissolution would preferentially remove these 'younger' species and bias the sample toward older ages. A solution would be, whenever possible, to restrict dating to species with relatively narrow depth ranges (e.g., *Globigerina bulloides, Globigerinoides ruber*) rather than those known to migrate to greater depths during the life cycle (e.g., *Neogloboquadrina dutertrei*). Unfortunately, in high-latitude marine environments of the Northern Hemisphere, glacial climate intervals yielded practically monospecific assemblages of *N. pachyderma* (left coiling), and thus the issue of variable ^{14}C-age distortions across the abrupt deglaciation remains a potential problem.

Finally, one should remember that shallow marine settings with highly variable environmental and sedimentological conditions may present specific problems with respect to the establishment of reliable ^{14}C chronologies. Examples from the Ross Ice Shelf area in the Antarctic (Andrews et al., 1999), or from the post-glacial Champlain Sea in eastern Canada (Richard & Occhietti, 2005), illustrate such difficulties.

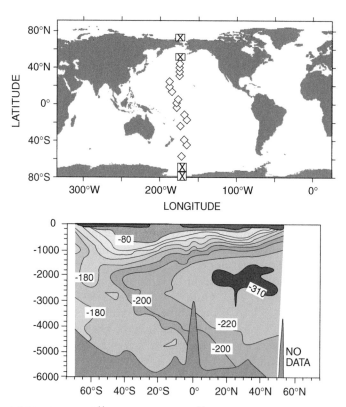

Figure 8 Variability in natural ^{14}C concentration, $\Delta^{14}C$, (in ‰) of DIC from a vertical section (a) along the Western Pacific GEOSECS cruise track as estimated (b) from observations by Broecker et al. (1995). Figure from Orr (1998).

3. APPLICATIONS OF MARINE ¹⁴C

Aside from the establishment of chronostratigraphies in deep-sea sediments, ¹⁴C studies provide essential information about changes in carbon cycling in relation to climate. Moreover, they also yield direct information on variability of past ocean circulation.

Due to the complex patterns and long timescale of surface and deep ocean circulation, the ¹⁴C activity of DIC varies systematically throughout the world ocean. The formation of North Atlantic Deep Water through the Atlantic Meridional Overturning and the global "conveyor belt" circulation results in young DIC ¹⁴C ages in the abyssal North Atlantic (500 yr) and older ages in the North Pacific (>2,000 yr) (Figure 4). Surface reservoir ages are typically lowest in the well-stratified tropical Warm Pools with minimum upwelling of older deep water, whereas upwelling and outcropping of deep water at high-latitudes results in maximum reservoir ages (Figure 8). In the high-latitude North Atlantic region, advection of high ¹⁴C-activity surface water from the tropical Atlantic, together with

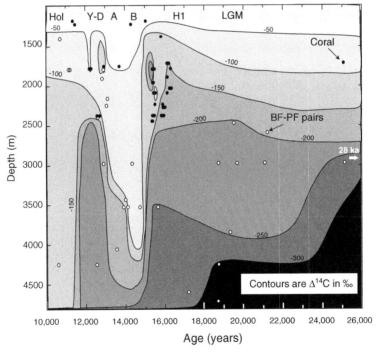

Figure 9 Deep ventilation changes in the deglacial western North Atlantic Ocean. Contour plot shown for coral (closed circles) and benthic-planktonic foraminifera (open circles) data from 26 to 10 kyr BP and deeper than 1,000 m. Data are plotted as marine Δ¹⁴C (‰), relative to atmospheric Δ¹⁴C inferred from IntCal04 (Reimer et al., 2004), Cariaco Basin ODP core 1002D (Hughen et al., 2004b) and Bahama speleothem (Beck et al., 2001). Dark colors are low in radiocarbon and light colors are rich in radiocarbon. Figure from Robinson et al. (2005).

limited upwelling, maintains a low reservoir age (300–400 yr). The oldest reservoir ages are found in the Southern Ocean and NW Pacific, due to upwelling of the most ancient abyssal waters.

Changes in surface reservoir age can be used to reconstruct past changes in ocean circulation including deep ventilation. For example, during the Younger Dryas interval, high-latitude North Atlantic reservoir ages increased sharply (+400 to 800 yr) due to reduced northward advection of warm pool surface water and increased deep mixing associated with a slowdown of NADW formation (Bard et al., 1994; Waelbroeck et al., 2003). Similarly, records of deep and intermediate water ^{14}C age relative to the surface ocean provide a direct measure of ventilation rates and in theory can be used to constrain past circulation changes (Broecker et al., 1984; Keigwin & Schlegel, 2002). As described above, however, it can be difficult to obtain undisturbed benthic and planktic specimens in proper abundances from the same sediment samples, due to the influences of bioturbation and sharp changes in abundance (e.g., Broecker et al., 1999).

In addition to its use for dating, ^{14}C concentration can also be expressed as initial activity, Δ^{14}C, following the convention of Stuiver and Pollach (1977)

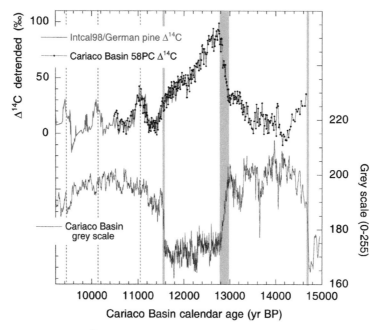

Figure 10 Atmospheric Δ^{14}C during the last deglaciation. (Upper) Purple line is detrended atmospheric Δ^{14}C from IntCal98 and German Pine data, blue line with solid circles is detrended Δ^{14}C measured in Cariaco Basin core PL07-58PC, using a constant reservoir correction of 420 yr. (Lower) Red line is Cariaco Basin gray scale, showing climate transitions of the last deglacial transition (vertical gray bars). Vertical dashed lines indicate century-scale anomalies common to both cosmogenic ^{10}Be and ^{14}C, seen throughout the Holocene and attributed to solar variability. Large Δ^{14}C peak at onset of Younger Dryas, ~12.8 kyr BP, is consistent with reduced NADW formation and deep ocean ventilaton at that time. Figure from Hughen et al. (2000).

according to:

$$\Delta^{14}C = [(F_{\mathrm{m}} \times e - \lambda_t) - 1] \times 1000 \tag{8}$$

where $\Delta^{14}C$ is age, standard and fractionation-corrected ^{14}C concentration in per mil; F_{m} the fraction of modern ^{14}C, λ the true ^{14}C decay constant ($8{,}267^{-1}$ yr) and t the calendar age of the sample.

$\Delta^{14}C$ is a versatile parameter for reconstructions, because if the calendar age is known, it can provide a measure of the apparent ^{14}C concentration of marine DIC back through time. Paired ^{14}C-^{230}Th ages from solitary deep corals in the North Atlantic region have been used to document rapid changes in deep ocean ventilation during the last deglaciation (Adkins et al., 1998). Deep coral $\Delta^{14}C$ data, together with paired benthic-planktic ^{14}C measurements, allow the mapping of deep ocean ventilation history throughout the past (Robinson et al., 2005) (Figure 9). One should keep in mind the consideration that diagenetic effects may alter U-series systematics more than the ^{14}C content of biogenic carbonates, and result in erroneous age offsets in paired ^{14}C-^{230}Th ages.

Nevertheless, $\Delta^{14}C$ from well-stratified low-latitude surface ocean sites can be used to reconstruct past changes in mixed-layer and atmospheric $\Delta^{14}C$ (Druffel & Suess, 1983). Over the last century, high-resolution ($\sim$monthly) time series of

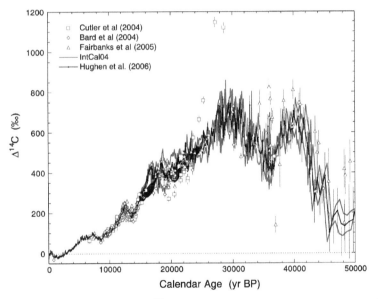

Figure 11 Marine-based atmospheric $\Delta^{14}C$ over the past 50,000 yr. High-resolution $\Delta^{14}C$ data from IntCal04 are shown as a black line. Cariaco Basin $\Delta^{14}C$ (vs. Hulu Cave ^{230}Th calendar age) is plotted as small blue circles. The error bars show $\Delta^{14}C$ uncertainty in the Cariaco-Hulu record due to independent 1-sigma errors in ^{14}C age measurement alone, whereas the thin gray lines show the 1-sigma error envelopes due to calendar age uncertainties alone. Paired ^{230}Th/^{14}C dates from corals are shown as solid circles with 1-sigma error bars. Comparison of Cariaco-Hulu record with $\Delta^{14}C$ from fossil corals shows excellent agreement back to 33 cal ka ($r = 0.79$), and a continued strong correlation from 33 to 50 cal ka ($r = 0.58$), despite increased scatter. Figure from Hughen et al. (2006).

Δ^{14}C from the tropical Pacific show strong interannual variability related to the seasonal upwelling of Δ^{14}C-depleted water and provide a means of correlating El Niño-Southern Oscillation (ENSO) variability and thermocline structure (Guilderson & Schrag, 1998; Guilderson et al., 1998). During the last deglaciation, Δ^{14}C from the Cariaco Basin shows an abrupt increase coincident with the onset of the Younger Dryas interval (Figure 10) (Hughen et al., 2000). Similar magnitude changes are not seen in other cosmogenic isotopes such as ^{10}Be (Finkel & Nishiizumi, 1997), and the Δ^{14}C results are most consistent with reduced global deep ocean ventilation as a result of a slowdown in North Atlantic Deep Water (NADW) formation (Hughen et al., 1998). Over the past 50,000 yr, longer records of surface ocean Δ^{14}C shows elevated values and large variability during the Glacial period, implying significantly greater ^{14}C production and possibly diminished uptake of ^{14}C by the ocean compared to the Holocene (Figure 11).

APPENDIX I — INTERNET RESOURCES

Radiocarbon journal

The Radiocarbon journal (http://www.radiocarbon.org) is the central, peer-reviewed publication of the radiocarbon community. It publishes the proceedings of International Radiocarbon Conferences, held every three years, as well as papers on all aspects of the technique. The web site provides links to a wide range of aspects relevant to radiocarbon.

Calibration programs

The task of calibration of a conventional ^{14}C date is done by transformation of the ^{14}C date, and its error probability distribution, to the calendar date, using the calibration data set as transformation function. A simple manual approach is based on finding the calendar age intersect(s) of a ^{14}C date and its associated error ranges using the most suitable calibration data set (Aitchison & Scott, 1987). With the intersect approach the shape of probability distribution is ignored, and any calendar age within the interval(s) of intersection is considered equally probable. As potentially valuable detail is lost by this approach, computer programs, which more accurately transform the probability distribution, are available. Programs such as OxCal (Ramsey, 1995) in particular are especially useful when a series of ^{14}C dates are to be calibrated, including independent information such as the relative sequence in time (e.g., from stratigraphy), which may substantially reduce the calendar age uncertainty (Ramsey, 1998).

Special care should be taken to reduce confusion in reporting *calibrated* ^{14}C ages. In addition to the uncalibrated ^{14}C date (the conventional ^{14}C or Libby's age), the calibrated result is reported separately and denominated '*cal* BP' or '*cal* BC/AD'. In view of the many continuing refinements in calibration work, the ^{14}C calibration data sets and the calibration technique (e.g., name of the computer program) should ideally also be quoted. The format of the calibrated result varies according to the program, and its interpretation depends on the nature of the sample.

A number of calibration programs can be obtained through the Internet, and some sites offer online calibration. The calibration data sets are available and can be accessed at these sites. Updated links may be found at www.radiocarbon.org.

CALIB (http://calib.qub.ac.uk/calib/) is available for a wide range of platforms, and also in an online version. This program allows selection from a number of calibration data sets, primarily the IntCal04 and Marine04 data sets back to 26 cal kyr BP for terrestrial and marine samples, ratified by the International Radiocarbon Conference community. High precision single-year calibration is available back to 350 ^{14}C BP, as is Southern Hemisphere sample calibration (SHCal04). In addition, global surface ocean reservoir ages can be accessed and local reservoir ages for individual sites can be determined using the Marine Reservoir Correction database.

OxCal (http://www.rlaha.ox.ac.uk/orau/index.htm) has been developed for Windows and an online version at the Oxford radiocarbon laboratory. It allows selection of the IntCal04 data sets described above. This program is particularly powerful and provides extensive support for calibrating a series of ^{14}C dates within a known floating context, such as stratigraphy, to constrain more precisely the calibration result.

WinCal25 (http://www.cio.phys.rug.nl/HTML-docs/cio-us/frb10.htm) is available as a Windows version from the University of Groningen, and has been updated with the IntCal04 data sets.

CalPal (http://www.calpal.de/) has been developed for Windows with an online version at the University of Cologne. This program provides calibration beyond IntCal04 back to 50,000 cal BP, and allows the user to select from a large range of data sets. However, many of these data sets are compilations of published data, which have been subjectively modified by the CalPal authors, so caution is strongly urged in the program Scientific Disclaimer.

Marine Radiocarbon Calibration (http://radiocarbon.ldeo.columbia.edu/research/radcarbcal.htm) is available in an online version from Columbia University. This program is based on paired ^{14}C-^{230}Th ages from fossil corals, and provides marine calibration beyond IntCal04 back to 50,000 cal BP.

Data sets

The community recommended calibration data sets INTCAL04 (for terrestrial data) and MARINE04 (for marine data) are available at http://www.calib.org as part of the CALIB program package. The primary data upon which IntCal04 and Marine04 are based are available at http://www.radiocarbon.org/IntCal04.htm as supplemental data to the IntCal04 issue of Radiocarbon. In addition to the ^{14}C age data, Δ^{14}C is included in the text files. Currently, these data sets do not extend beyond 26 cal ka BP.

The Cariaco Basin marine calibration data sets, including high-resolution data for the last deglaciation and data back to 50,000 cal BP, are available online at the NOAA World Data Center for Paleoclimatology (http://www.ncdc.noaa.gov/paleo/data.html).

The Fairbanks0805 marine calibration data set based on paired ^{14}C-^{230}Th ages from fossil corals back to 50,000 cal BP is available at http://radiocarbon.ldeo. columbia.edu/research/radcarbcal.htm

REFERENCES

Adkins, J. F., Cheng, H., Boyle, E. A., Druffel, E. R. M., & Edwards, R. L. (1998). Deep-sea coral evidence for rapid change in ventilation of the North Atlantik 15,400 years ago. *Science, 280*, 725–728.

Aitchison, T. C., & Scott, E. M. (1987). A review of the methodology for calibrating radiocarbon dates into historical ages. *BAR International Series, 333*, 187–201.

Andrews, J. T., Domack, E. W., Cunningham, W. L., Leventer, A., Licht, K. J., Jull, A. J. T., DeMaster, D. J., & Jennings, A. E. (1999). Problems and possible solutions concerning radiocarbon dating of surface marine sediments, Ross Sea, Antarctica. *Quaternary Research, 52*, 206–216.

Austin, W., Bard, E., Hunt, J., Kroon, D., & Peacock, J. (1995). The ^{14}C age of the Icelandic Vedde Ash: Implications for Younger Dryas marine reservoir age corrections. *Radiocarbon, 37*, 53–62.

Bard, E. (2001). Paleoceanographic implications of the difference in deep-sea sediment mixing between large and fine particles. *Paleoceanography, 16*, 235–239.

Bard, E., Arnold, M., Mangerud, J., Paterne, M., Labeyrie, L., Duprat, J., Melieres, M.-A., Sonstegaard, E., & Duplessy, J.-C. (1994). The North Atlantic atmosphere-sea surface ^{14}C gradient during the Younger Dryas climatic event. *EPSL, 126*, 275–287.

Bard, E., Arnold, M., Hamelin, B., Tisnerat, L. N., & Cabioch, G. (1998). Radiocarbon calibration by means of mass spectrometric ^{230}Th/^{234}U and ^{14}C ages of corals; an updated database including samples from Barbados, Mururoa and Tahiti. *Radiocarbon, 403*, 1085–1092.

Bard, E., Rostek, F., & Menot-Combes, G. (2004). Radiocarbon calibration beyond 20,000 ^{14}C yr B.P. by means of planktonic foraminifera of the Iberian Margin. *Quaternary Research, 61*, 204–214.

Bauer, J. E., Reimers, C. E., Druffel, E. R. M., & Williams, P. M. (1995). Isotopic constraints on carbon exchange between deep ocean sediments and sea water. *Nature, 373*, 686–689.

Beck, J. W., Richards, D. A., Edwards, R. L., Silverman, B. W., Smart, P. L., Donahue, D., Hererra-Osterheld, S., Burr, G., Calsoyas, L., Jull, A. J. T., & Biddulph, D. (2001). Extremely large variations of atmospheric ^{14}C concentration during the last glacial period. *Science, 292*, 2453–2458.

Bondevik, S., Birks, H., Gulliksen, S., & Mangerud, J. (1999). Late Weichselian marine ^{14}C reservoir ages at the western coast of Norway. *Quaternary Research, 52*, 104–114.

Broecker, W., Andree, M., Bonani, G., Wolfli, W., Oeschger, H., Klas, M., Mix, A., & Curry, W. (1988). Preliminary estimates for the radiocarbon age of deep water in the glacial ocean. *Paleoceanography, 3*, 659–669.

Broecker, W., Matsumoto, K., Clark, E., Hajdas, I., & Bonani, G. (1999). Radiocarbon age difference between coexisting foraminiferal species. *Paleoceanography, 14*, 431–436.

Broecker, W., Mix, A., Andree, M., & Oeschger, H. (1984). Radiocarbon measurements on coexisting benthic and planktic foraminifera shells: Potential for reconstructing ocean ventilation times over the past 20,000 years. *Nuclear Instruments and Methods, B5*, 331–339.

Broecker, W. S., & Olson, E. A. (1959). Lamont Radiocarbon Measurements VI. *American Journal of Science Radiocarbon Supplement, 1*, 111–132.

Broecker, W. S., & Peng, T.-H. (1982). *Tracers in the sea*. New York: Lamont-Doherty Earth Observatory.

Broecker, W. S., & Peng, T.-H. (1994). Stratospheric contribution to the global bomb radiocarbon inventory: Model versus observation. *Global Biogeochemical Cycles, 8*, 377–384.

Broecker, W. S., Sutherland, S., Smethie, W., Peng, T.-H., & Ostlund, G. (1995). Oceanic radiocarbon: Separation of the natural and bomb components. *Global Biogeochemical Cycles, 9*, 263–288.

Burr, G. S., Beck, J. W., Taylor, F. W., Recy, J., Edwards, R. L., Cabioch, G., Correge, T., Donahue, D. J., & O'Malley, J. M. (1998). A high-resolution radiocarbon calibration between 11,700 and 12,400 calendar years BP derived from ^{230}Th ages of corals from Espiritu Santo Island, Vanuatu. *Radiocarbon*(403), 1093–1105.

Burr, G., Galang, C., Taylor, F., Gallup, C., Edwards, R. L., Cutler, K., & Quirk, B. (2004). Radiocarbon results from a 13-kyr BP coral from the Huon Peninsula, Papua New Guinea. *Radiocarbon*, 46, 1211–1224.

Cook, G., Harkness, D., MacKenzie, A., Miller, B., & Scott, M. (1994). *Liquid scintillation spectrometry.* Tucson, AZ: Radiocarbon.

Coplen, T. B. (1996). New guidelines for the reporting of stable hydrogen, carbon, and oxygen isotope ratio data. *Geochimica et Cosmochimica Acta*, 60, 3359.

Craig, H. (1954). Carbon 13 in plants and the relationship between carbon 13 and carbon 14 variations in nature. *Journal of Geology*, 62, 115–149.

Currie, L. A., Klouda, G. A., Benner, B. A., Garrity, K., & Eglinton, T. I. (1999). Isotopic and molecular fractionation in combustion; three routes to molecular marker validation, including direct molecular "dating" GC/AMS. *Atmospheric Environment*, 33, 2789–2806.

Cutler, K., Gray, S. C., Burr, G., Edwards, R. L., Taylor, F. W., Cabioch, G., Beck, J. W., Cheng, H., & Moore, J. (2004). Radiocarbon calibration and comparison to 50 kyr BP with paired ^{14}C and ^{230}Th dating of corals from Vanuatu and Papua New Guinea. *Radiocarbon*, 46, 1127–1160.

Druffel, E. R. M., & Suess, H. E. (1983). On the radiocarbon record in banded corals: Exchange parameters and net transport of $^{14}CO_2$ between atmosphere and surface ocean. *Journal of Geophysical Research*, 88(C2), 1271–1280.

Duplessy, J.-C., Arnold, M., Maurice, P., Bard, E., Duprat, J., & Moyes, J. (1986). Direct dating of the oxygen-isotope record of the last deglaciation by ^{14}C accelerator mass spectrometry. *Nature*, 320, 350–352.

Edwards, R. L., Beck, J. W., Burr, G. S., Donahue, D. J., Chappell, J. M. A., Bloom, A. L., Druffel, E. R. M., & Taylor, F. W. (1993). A large drop in atmospheric $^{14}C/^{12}C$ and reduced melting in the Younger Dryas, documented with ^{230}Th ages of corals. *Science*, 260, 962–968.

Edwards, R. L., Chen, J. H., & Wasserburg, G. J. (1986). ^{238}U-^{234}U-^{230}Th-^{232}Th systematics and the precise measurement of time over the past 500,000 years. *Earth and Planetary Science Letters*, 812–3, 175–192.

Eglinton, T. I., Benitez-Nelson, B. C., Pearson, A., McNichol, A. P., Bauer, J. E., & Druffel, E. R. M. (1997). Variability in radiocarbon ages of individual organic compounds from marine sediments. *Science*, 277, 796–799.

Eglinton, T. I., McNichol, A., Benitez-Nelson, C., Pearson, A., von Reden, K., Schneider, R., Bauer, J., & Druffel, E. (1996). Isolation of individual organic compounds for AMS radiocarbon analysis: A novel approach. *Radiocarbon*, 38, 26.

Eiriksson, J., Knudsen, K., Haflidason, H., & Heinemeier, J. (2000). Chronology of late Holocene climatic events in the northern North Atlantic based on AMS ^{14}C dates and tephra markers from the volcano Hekla, Iceland. *Journal of Quaternary Science*, 15, 573–580.

Engelkemeir, A. G., Hamill, W. H., Inghram, M. G., & Libby, W. F. (1949). The half-life of radiocarbon C14. *Physical Review*, 75, 1825–1833.

Fairbanks, R., Mortlock, R., Chiu, T., Cao, L., Kaplan, A., Guilderson, T., Fairbanks, T., Arthur, L., Grootes, P., & Nadeau, M. (2005). Radiocarbon calibration curve spanning 0 to 50,000 years BP based on paired ^{230}Th/^{234}U/^{238}U and ^{14}C dates on pristine corals. *Quaternary Science Reviews*, 24, 1781–1796.

Finkel, R. C., & Nishiizumi, K. (1997). Beryllium 10 concentrations in the Greenland Ice Sheet Project 2 ice core from 3–40 ka. *JGR*, 102(C12), 26699–26706.

Friedrich, M., Kromer, B., Kaiser, K. F., Spurk, M., Hughen, K. A., & Johnsen, S. J. (2001). High-resolution climate signals in the Bolling-Allerod Interstadial (Greenland Interstadial 1) as reflected in European tree-ring chronologies compared to marine varves and ice-core records. *Quaternary Science Reviews*, 20, 1223–1232.

Godwin, H. (1962). Radiocarbon dating. *Nature*, 195, 943–945.

Gove, H. E. (1992). The history of AMS, its advantages over decay counting: Applications and prospects. In: R. E. Taylor, A. Long & R. S. Kra (Eds), *Radiocarbon after Four Decades*. New York: Springer.

Guilderson, T. P., & Schrag, D. P. (1998). Abrupt shift in subsurface temperatures in the eastern tropical Pacific associated with recent changes in El Niño. *Science, 281*, 240–243.

Guilderson, T. P., Schrag, D. P., Kashgarian, M., & Southon, J. (1998). Radiocarbon variability in the western equatorial Pacific inferred from a high-resolution coral record from Nauru Island. *Journal of Geophysical Research, 10311*, 24641–24650.

Guinasso, N. L., & Shink, D. R. (1975). Quantitative estimates of biological mixing rates in abyssal sediments. *Journal of Geophysical Research, 80*, 3032–3043.

Hedges, J. I., & Oades, J. M. (1997). Comparative organic geochemistries of soils and marine sediments. *Organic Geochemistry, 27*, 319–361.

Huang, Y., Bol, R., Harkness, D. D., Ineson, P., & Eglinton, G. (1996). Post-glacial variations in distributions, ^{13}C and ^{14}C contents of aliphatic hydrocarbons and bulk organic matter in three types of British acid upland soils. *Organic Geochemistry, 243*, 273–287.

Hughen, K. A., Lehman, S. J., Southon, J., Overpeck, J. T., Marchal, O., Herring, C., & Turnbull, J. (2004b). ^{14}C activity and global carbon cycle changes over the past 50,000 years. *Science, 303*, 202–207.

Hughen, K. A., Overpeck, J. T., Lehman, S. J., Kashgarian, M., Southon, J., Peterson, L. C., Alley, R., & Sigman, D. M. (1998). Deglacial changes in ocean circulation from an extended radiocarbon calibration. *Nature, 391*, 65–68.

Hughen, K. A., Baillie, M., Bard, E., Beck, J. W., Bertrand, C., Blackwell, P., Buck, C., Burr, G. S., Cutler, K., Damon, P., Edwards, R. L., Fairbanks, R., Friedrich, M., Guilderson, T., Kromer, B., McCormac, G., Manning, S., Ramsey, C., Reimer, P., Reimer, R., Remmele, S., Southon, J., Stuiver, M., Talamo, S., Taylor, F., van der Plicht, J., & Weyhenmeyer, C. (2004a). MARINE04 marine radiocarbon age calibration, 26–0 ka BP. *Radiocarbon, 46*, 1059–1086.

Hughen, K. A., Southon, J. R., Bertrand, C., Frantz, B., & Zermeno, P. (2004c). Cariaco basin calibration update: Revisions to calendar and ^{14}C chronologies for Core PL07-58PC. *Radiocarbon, 46*, 1161–1188.

Hughen, K. A., Southon, J. A., Lehman, S. J., Bertrand, C. J. H., & Turnbull, J. (2006). Revised 14C activity record from the Cariaco basin for the past 60,000 years. *Quaternary Science Reviews, 25*, 3216–3227.

Hughen, K. A., Southon, J. R., Lehman, S. J., & Overpeck, J. T. (2000). Synchronous radiocarbon and climate shifts during the last deglaciation. *Science, 290*, 1951–1954.

Johnsen, S., Dahl-Jensen, D., Gundestrup, N., Steffensen, J. P., Clausen, H., Miller, H., Masson-Delmotte, V., Sveinbjornsdottir, A., & White, J. (2001). Oxygen isotope and palaeotemperature records from six Greenland ice-core stations; Camp Century, Dye-3, GRIP, GISP2, Renland and NorthGRIP. *Journal of Quaternary Science, 16*, 299–307.

Keigwin, L. D., & Schlegel, M. A. (2002). Ocean ventilation and sedimentation since the glacial maximum at 3 km in the western North Atlantic. *Geochemistry, Geophysics, Geosystems, 3*, 10.

van Klinken, G. J., & Hedges, R. E. M. (1998). Chemistry strategies for organic ^{14}C samples. *Radiocarbon, 401–2*, 51–56.

Kromer, B., & Münnich, K.-O. (1992). CO_2 gas proportional counting in radiocarbon dating — Review and perspective. In: R. E. Taylor, A. Long & R. S. Kra (Eds), *Radiocarbon after four decades* (pp. 184–197). New York: Springer.

Kromer, B., Friedrich, M., Hughen, K. A., Kaiser, K. F., Remmele, S., Schaub, M., & Talamo, S. (2004). Late glacial ^{14}C ages from a floating 1270-ring pine chronology. *Radiocarbon, 46*, 1203–1210.

Laj, C., Kissel, C., Mazaud, A., Michel, E., Muscheler, R., & Beer, J. (2002). Geomagnetic field intensity, North Atlantic deep water circulation and atmospheric D^{14}C during the last 50 kyr. *Earth and Planetary Science Letters, 200*, 177–190.

Libby, W. F. (1946). Atmospheric helium-3 and radiocarbon from cosmic radiation. *Physical Review, 69*, 671–672.

Masarik, J., & Beer, J. (1999). Simulation of particle fluxes and cosmogenic nuclide production in the Earth's atmosphere. *Journal of Geophysical Research, 104*(D10), 12099–12111.

Matsumoto, K. & Key, R. M. (2004). Natural radiocarbon distribution in the deep ocean. In: M. Shiyomi, et al. (Eds), *Global environmental change in the ocean and on land*, pp. 45–58.

McNichol, A. P., Ertel, J. R., et al. (2000). The radiocarbon content of individual lignin-derived phenols: Technique and initial results. *Radiocarbon*(422), 219–227.

Mollenhauer, G., Eglinton, T. I., Ohkouchi, N., Schneider, R. R., Müller, P. J., Grootes, P. M., & Rullkötter, J. (2003). Asynchronous alkenone and foraminifera records from the Benguela Upwelling System. *Geochimica et Cosmochimica Acta*, *6712*, 2157–2171.

Ohkouchi, N., Eglinton, T. I., Keigwin, L. D., & Hayes, J. M. (2002). Spatial and temporal offsets between proxy records in a sediment drift. *Science*, *298*, 1224–1227.

Olson, E. A., & Broecker, W. S. (1961). Lamont natural radiocarbon measurements VII. *Radiocarbon*, *3*, 141–175.

Orr, J. C. (1998). Ocean carbon-cycle model intercomparison project (OCMIP): Phase I (1995–1997). *IGBP/GAIM REPORT SERIES*, 7, 16.

Pearson, A., & Eglinton, T. I. (2000). The origin of *n*-alkanes in Santa Monica Basin surface sediment: A model based on compound-specific $D^{14}C$ and $D^{13}C$ data. *Organic Geochemistry*(3111), 1103–1116.

Pons-Branchu, E., Hillaire-Marcel, C., Deschamps, P., Ghaleb, B., & Sinclair, D. J. (2005). Early diagenesis impact on precise U-series dating of deep-sea corals: Example of a 100–200-year old *Lophelia pertusa* sample from the northeast Atlantic. *Geochimica et Cosmochimica Acta*, *69*, 4865–4879.

Ramsey, C. B. (1995). Radiocarbon and analysis of stratigraphy: The OxCal program. *Radiocarbon*(372), 425–430.

Ramsey, C.B., 1998. Probability and dating. Radiocarbon, (401–2): 461–474

Rasmussen, S. O., Andersen, K. K., Svensson, A. M., Steffensen, J. P., Vinther, B. M., Clausen, H. B., Siggaard-Andersen, M.-L., Johnsen, S. J., Larsen, L. B., Dahl-Jensen, D., Bigler, R., Rothlisberger, R., Fischer, H., Goto-Azuma, K., Hansson, M., & Ruth, U. (2006). A new Greenland ice core chronology for the last glacial termination. *Journal of Geophysical Research*, *111*, D06102, doi:10.1029/2005JD006079.

Reimer, P., Baillie, M., Bard, E., Bayliss, A., Beck, J. W., Bertrand, C., Blackwell, P., Buck, C., Burr, G. S., Cutler, K., Damon, P., Edwards, R. L., Fairbanks, R., Friedrich, M., Guilderson, T., Hogg, A., Hughen, K., Kromer, B., McCormac, G., Manning, S., Ramsey, C., Reimer, R., Remmele, S., Southon, J., Stuiver, M., Talamo, S., Taylor, F., van der Plicht, J., & Weyhenmeyer, C. (2004). INTCAL04 terrestrial radiocarbon age calibration, 26–0 ka BP. *Radiocarbon*, *46*, 1029–1058.

Richard, P. J. H., & Occhietti, S. (2005). ^{14}C chronology for ice retreat and inception of Champlain Sea in the St. Lawrence lowlands, Canada. *Quaternary Research*, *63*, 353–358.

Richards, D., Hoffmann, D. L., Beck, J. W., Smart, P., Paterson, B., & Mattey, D. (2006). Exploring the potential causes of atmospheric $\delta^{14}C$ variation using multi-proxy evidence from Bahamian speleothems (34 to 45 KA), 19th International Radiocarbon Conference, April 3–7, Keble College, Oxford, UK.

Robinson, L., Adkins, J., Keigwin, L., Southon, J., Fernandez, D. P., Wang, S., & Scheirer, D. (2005). Radiocarbon variability in the western north Atlantic during the last deglaciation. *Science*, *310*, 1469–1473.

Rubin, S., & Kamen, M. D. (1941). Long-lived radioactive carbon: ^{14}C. *Physical Review*, *59*, 349–354.

Schlosser, P., Kromer, B., et al. (1997). The first trans-Arctic ^{14}C section: Comparison of the mean ages of the deep waters in the Eurasian and Canadian basins of the Arctic Ocean. *Nuclear Instruments and Methods in Physics Research*, B123, 431–437.

Siani, G., Paterne, M., Michel, E., Sulpizio, R., Sbrana, A., Arnold, M., & Haddad, G. (2001). Mediterranean sea surface radiocarbon reservoir age changes since the last glacial maximum. *Science*, *294*, 1917–1920.

Siegenthaler, U., & Sarmiento, J. L. (1993). Atmospheric carbon dioxide and the ocean. *Nature*, *365*, 119–125.

Siegenthaler, U., Heimann, M., et al. (1980). ^{14}C variations caused by changes in the global carbon cycle. *Radiocarbon*, *222*, 177–191.

Sikes, E., Samson, C., Guilderson, T., & Howard, W. (2000). Old radiocarbon ages in the southwest Pacific Ocean during the last glacial period and deglaciation. *Nature*, *405*, 555–559.

Smith, B. N., & Epstein, S. (1971). Two categories of $^{13}C/^{12}C$ ratios for higher plants. *Plant Physiology, 47*, 380–384.

Stuiver, M., & Braziunas, T. F. (1993). Sun, ocean, climate and atmospheric $^{14}CO_2$: An evaluation of causal and spectral relationships. *The Holocene, 34*, 289–305.

Stuiver, M., Östlund, H. G., et al. (1996). Large-volume WOCE radiocarbon sampling in the Pacific Ocean. *Radiocarbon, 383*, 519–560.

Stuiver, M., & Pollach, H. (1977). Discussion: Reporting of ^{14}C data. *Radiocarbon, 19*, 355–363.

Thompson, W. G., Spiegelman, M. W., Goldstein, S. L., & Speed, R. C. (2003). An open-system model for U-series age determinations of fossil corals. *Earth and Planetary Science Letters, 210*, 365–381.

Van der Plicht, J., Beck, J. W., Bard, E., Baillie, M. G. L., Blackwell, P. G., Buck, C. E., Friedrich, M., Guilderson, T. P., Hughen, K. A., Kromer, B., McCormac, F. G., Ramsey, C. B., Reimer, P. J., Reimer, R. W., Remmele, S., Richards, D. A., Southon, J. R., Stuiver, M., & Weyhenmeyer, C. E. (2004). NotCal04 — comparison/calibration ^{14}C records 26–50 cal kyr BP. *Radiocarbon, 46*, 1225–1238.

Voelker, A. H. L., Grootes, P. M., et al. (2000). Radiocarbon levels in the Iceland Sea from 25–53 Kyr and their link to the earth's magnetic field intensity. *Radiocarbon, 42*, 25–36.

Waelbroeck, C., Duplessy, J., Michel, E., Labeyrie, L., Paillard, D., & Duprat, J. (2001). The timing of the last deglaciation in north Atlantic climate records. *Nature, 412*, 724–727.

Waelbroeck, C., Parrenin, F., Ferrer, M., Chappellaz, J., Levi, C., Grimalt, J., Labeyrie, L., Jouzel, J. (2003). How closely are millennial scale benthic ^{18}O and Antarctic ice records related? *Congress of the International Union for Quaternary Research, 16*, 90

Wang, Y. J., Cheng, H., Edwards, R. L., An, Z. S., Wu, J. Y., Shen, C. C., & Dorale, J. A. (2001). A high-resolution absolute-dated late pleistocene monsoon record from Hulu Cave, China. *Science, 294*, 2345–2348.

Wigley, T. M. L., & Muller, A. B. (1981). Fractionation corrections in radiocarbon dating. *Radiocarbon, 232*, 173–190.

Wu, G. P., & Hillaire-Marcel, C. (1994). Accelerator mass spectrometry radiocarbon stratigraphies in deep Labrador Sea cores: Paleoceanographic implications. *Canadian Journal of Earth Sciences, 31*, 38–47.

PART 2: BIOLOGICAL TRACERS AND BIOMARKERS

PLANKTONIC FORAMINIFERA AS TRACERS OF PAST OCEANIC ENVIRONMENTS

Michal Kucera

Contents

1. Introduction	213
2. Biology and Ecology of Planktonic Foraminifera	215
2.1. Cellular structure, reproduction, and shell formation	215
2.2. Classification and species concept	219
2.3. Ecology and distribution	221
3. Planktonic Foraminiferal Proxies	225
3.1. Census data	228
3.2. Shell morphology	235
3.3. Planktonic foraminifera as substrate for geochemical studies	244
4. Modifications After Death	245
4.1. Settling through the water column	245
4.2. Calcite dissolution	247
5. Perspectives	253
WWW Resources	253
References	254

1. INTRODUCTION

Paleoceanography has always been closely connected with the study of planktonic foraminifera. The prolific production and excellent preservation of foraminiferal fossils in oceanic sediments (Figure 1) has produced probably the best fossil record on Earth, providing unparalleled archives of morphological change, faunal variations, and habitat characteristics. Planktonic foraminifera are the most common source of paleoceanographic proxies, be it through the properties of their fossil assemblages or as a substrate for extraction of geochemical signals. The steady rain of foraminiferal shells is responsible for the deposition of a large portion of deep-sea biogenic carbonate. Vincent and Berger (1981) estimated that over a period of 500 years planktonic foraminifera deposit a mass of carbon equal to that of the entire biosphere. Fossilized planktonic foraminifera form the backbone of Cenozoic bio-stratigraphy (Berggren, Kent, Swisher, & Aubry, 1995) and have been instrumental in the study of rates and patterns of evolution (Norris, 2000).

Developments in Marine Geology, Volume 1
ISSN 1572-5480, DOI 10.1016/S1572-5480(07)01011-1

Figure 1 Light-microscope image of the sand-fraction residue from a tropical deep-sea sediment sample. The residue is dominated by planktonic foraminiferal shells representing ∼20 species. The foraminifera are well preserved and illustrate the large variation in shell sizes typical for tropical assemblages. (Photo: Wilfried Rönnfeld.)

The potential for planktonic foraminifera to be used as tracers of surface-water properties was first noted by Murray (1897), who recognized that extant species in the plankton and in sea floor sediments are distributed in global belts related to surface-water temperatures. Schott (1935) pioneered the use of quantitative census counts and discovered that fossil assemblages in short deep-sea cores changed between glacial and interglacial times. The prominent role of planktonic foraminifera in reconstructions of Pleistocene climate variation has been established since the birth of paleoceanography. Pfleger (1948) and Arrhenius (1952) used planktonic foraminifera to describe Quaternary climate cycles in the first long piston cores recovered from the deep sea by the Swedish Deep Sea Expedition with the four-mast schooner *Albatross* in 1947–1948. In less than 20 years, enormous progress has been made in the understanding of the biology and ecology of planktonic foraminifera, culminating in the development of the first sophisticated transfer function by Imbrie and Kipp (1971), that laid the foundation for the grandest virtual time-travelling exercise of its time: the reconstruction of the surface of the Earth at the time of the last glacial maximum (CLIMAP, 1976).

The value of foraminiferal calcite as a recorder of chemical and isotopic signals was recognized by Emiliani (1954a, 1954b). Stable isotopic signals extracted from planktonic foraminifera soon became a standard tool for the recognition of glacial cycles and eventually facilitated the recognition of orbital pacing of the ice-ages (Shackleton & Opdyke, 1973; Hays, Imbrie, & Shackleton, 1976). The chemical composition of foraminiferal calcite proved to be a fertile ground for the development of proxies: almost every trace element and stable or radiogenic isotope imaginable has been, or is being, measured and calibrated in an effort to reconstruct past seawater chemistry and biogeochemical cycles (Henderson, 2002).

Early work on the biology and ecology of planktonic foraminifera has been treated comprehensively in the reviews by Hedley and Adams (1974, 1976), Bé (1977), Vincent and Berger (1981), and Hemleben, Spindler, and Anderson (1989). This chapter will thus focus on the work of the previous 20 years with the objective of highlighting the most common and most promising foraminiferal proxies, and put them in the context of modern biological knowledge. The reader should be aware that stable–isotopic and geochemical proxies, as well as transfer functions, are treated comprehensively in separate chapters of this volume (Chapters 7 and 13, respectively). The use of planktonic foraminifera as tracers of ocean properties is a mature field of science. As a result, we know a great deal about the limitations of foraminiferal proxies and the circumstances in which they can or cannot be applied, and these are well covered in this review. This sign of maturity of the field should not be interpreted by the reader as an argument against the use of planktonic foraminiferal proxies. Planktonic foraminifera continue to play a central role in paleoceanography, providing the science with robust and reliable proxies, and will continue to do so for some time. These inconspicuous organisms and their tiny shells are the true heroes of our quest to reveal the past of our planet.

2. BIOLOGY AND ECOLOGY OF PLANKTONIC FORAMINIFERA

2.1. Cellular Structure, Reproduction, and Shell Formation

Planktonic foraminifera are marine heterotrophic protists that surround their unicellular body with elaborate calcite shells[1]. Cytoplasm inside the shells contains typical eukaryotic cellular organelles, supplemented by the so-called fibrillar bodies, which are unique to planktonic foraminifera and may act to control buoyancy (Hemleben et al., 1989). Outside the shell, the cytoplasm is stretched into thin, anastomosing strands (rhizopodia), which may extend several shell-diameter lengths away from the shell. The external rhizopodial network serves to collect food particles and transport them toward the primary opening of the shell (aperture). Inside the shell, food particles are digested and stored as lipids and starches in specialized vacuoles.

Planktonic foraminifera exhibit a range of trophic behaviors from indiscriminate omnivory to selective carnivory (Hemleben et al., 1989). Herbivorous and omnivorous species consume phytoplankton, mainly diatoms and dinoflagellates, while carnivorous species prey on copepods, ciliates, and other similarly sized zooplankton (Hemleben et al., 1989). Species that inhabit the photic zone often harbor intracellular algal symbionts (dinoflagellates or chrysophycophytes). A symbiotic relationship with photosynthesizing algae is particularly advantageous in warm oligotrophic waters, where nutrients and food are scarce but light is abundant. Typical population

---

[1] The correct technical term for foraminiferal skeleton is *test*, from Latin *testa* = shell, however, this term has an English homonym with a very different meaning. To avoid confusion, the term *shell* will be used throughout this chapter.

densities of planktonic foraminifera range from $>1,000$ individuals/m^3 in polar ocean blooms to <100 individuals/m^3 in oligotrophic gyres (Schiebel & Hemleben, 2005). Given their low population densities and low nutrient/weight ratio (due to the shells), it is not surprising that no selective predators of planktonic foraminifera have been discovered. Instead, planktonic foraminifera appear to be indiscriminately ingested by filter-feeding planktotrophs (Lipps & Valentine, 1970; Hemleben et al., 1989).

Except for the Antarctic species *Neogloboquadrina pachyderma*, which overwinters in brine channels in sea ice (Spindler & Diekmann, 1986), all extant species of planktonic foraminifera are holoplanktonic, spending their entire life freely floating in surface waters. The mixed layer and the upper thermocline are the most densely populated, while virtually no living individuals are found at depths below 1,000 m (Vincent & Berger, 1981). Laboratory observations indicate that some individuals survive when placed on the sediment surface (Hilbrecht & Thierstein, 1996), but there have been no reports of living (or resting) planktonic foraminifera on the sea floor.

Although benthic foraminifera exhibit a complex life cycle including an array of reproductive strategies, solely sexual reproduction has been observed among planktonic foraminifera (Hemleben et al., 1989). Given the lack of morphological dimorphism, which is often indicative of multiple reproductive strategies in foraminifera, it is most likely that all fossil species reproduced exclusively sexually as well. During reproduction, the cytoplasm is divided into hundreds of thousands of biflagellated isogametes that are released into the environment. In order to maximize the chances of gametes from different individuals finding each other, the reproduction has to be synchronized in space and time. Indeed, most shallow-water species appear to reproduce in pace with the synodic lunar cycle (*Hastigerina pelagica*, *Globigerinoides sacculifer*, *Globigerina bulloides*) or half-synodic lunar cycle (*Globigerinoides ruber*) (Spindler, Hemleben, Bayer, Bé, & Anderson, 1979; Bijma, Erez, & Hemleben, 1990a; Schiebel, Bijma, & Hemleben, 1997), and lunar pacing appears important for carbonate production in the tropical oceans (Kawahata, Nishimura, & Gagan, 2002). Recently, the prevalence of the lunar reproductive cycle became a matter of debate (Lončarić, Brummer, & Kroon, 2005). Deep-dwelling species like *Globorotalia truncatulinoides* may follow longer, perhaps yearly, reproductive cycles (Hemleben et al., 1989) and individuals of *N. pachyderma* isolated from Antarctic sea-ice were kept in culture for 230 days (Spindler, 1996).

During their life, individual species are known to migrate vertically within the water column and release gametes at well-defined, species-specific depths, often close to the pycnocline (Schiebel & Hemleben, 2005). The need for deep oceanic waters to complete their life cycles is perhaps the reason why planktonic foraminifera avoid neritic waters over continental shelves (Schmuker, 2000) and resist every effort made to reproduce them under laboratory conditions (Hemleben et al., 1989).

Following gamete fusion, shell growth is facilitated by the sequential addition of chambers, gradually increasing the dimensions of the shell. The process of shell formation and calcification is described in detail by Hemleben et al. (1989). The external rhizopodial network forms the outline of the new chamber and secretes the primary organic membrane (POM) that acts as the nucleation centre for

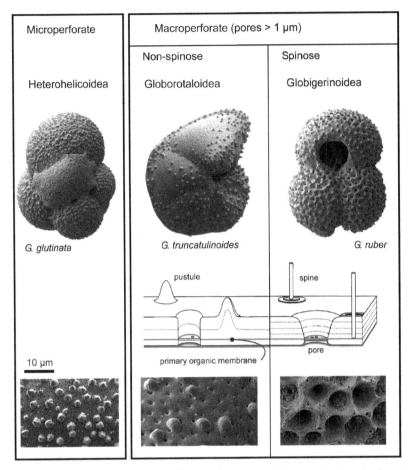

Figure 2 Classification scheme of the three main groups of extant planktonic foraminifera. Representative specimens of the three groups (not to scale) document typical morphology and wall ornament (enlarged sections, all to the same scale). Shell walls are layered, perforated by pores and the outer surface features either pustules or spines (diagram modified from Schiebel & Hemleben, 2005).

calcification (Figure 2). With the exception of the monolamellar Hastigerinidae, calcite layers are added on both sides of the POM, with the external layer extending across the entire outer surface of the shell. Pores are formed within the early stages of wall calcification, while surface ornament including pustules and ridges are formed simultaneously. Spines are plugged into pre-formed cavities in the outer shell layer. They are solid and can be repeatedly shed, or resorbed and regrown. The exact mechanism of foraminiferal calcification is not fully understood. However, laboratory observations on benthic foraminifera indicate that the calcification is extra-cellular and mediated through cation enrichment and transport of seawater in specialized vacuoles (Erez, 2003) and that two separate processes producing different mineral phases may be involved (Bentov & Erez, 2005).

During growth, the shape of a planktonic foraminiferal shell may change dramatically (Brummer, Hemleben, & Spindler, 1986; Hemleben et al., 1989). Adult characteristics, important for the identification of species, develop late in the ontogeny, making classification of juvenile stages next to impossible. Transitions between ontogenetic stages may be linked to changes in trophic behavior and the onset of symbiont infestation. Algal symbionts play an important role in the calcification process, providing extra energy to the host and modulating the chemical microenvironment by lowering dissolved CO_2 concentration. Laboratory experiments show that specimens that were grown in darkness or without symbionts produce substantially smaller shells (Bé, Spero, & Anderson, 1982). The metabolic activity of algal symbionts alters the stable isotopic composition of foraminiferal calcite (Spero & Deniro, 1987) and this distinctive signature (Figure 3) can be used to detect the presence of photosymbiosis in fossil species (Norris, 1996). Significant changes to the shell are associated with reproduction. Some species deposit an

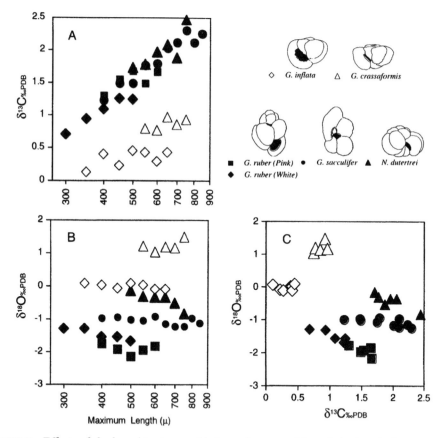

Figure 3 Effects of algal symbionts on stable isotopic composition of planktonic foraminiferal shells. Symbiont-bearing species (filled symbols) show a distinct increase with size toward heavier carbon signature, reflecting an increasing rate of removal of light carbon by the photosymbionts. The distinct isotopic signatures can be used to trace the presence of symbionts in the past (adapted from Norris (1996). Copyright 1996, The Paleontological Society).

additional thick layer of gametogenic calcite prior to reproduction, while others shed spines and *H. pelagica* even resorbs its inner septa (Hemleben et al., 1989). The final chamber of the shell may be disfigured and dislocated. The significance of these kummerform chambers is not fully understood (Berger, 1970b; Hecht & Savin, 1972), but some are likely to represent the products of residual cytoplasm still active after gamete release (Hemleben et al., 1989).

2.2. Classification and Species Concept

Classification of planktonic foraminifera is based entirely on the properties of their shells. At the highest taxonomic level, late Cenozoic planktonic foraminifera are subdivided into three superfamilies: Globigerinoidea, Globorotaloidea, and Heterohelicoidea (Figure 2). The taxonomic position of the family Hastigerinidae, which produces monolamellar shells, remains unclear (Schiebel & Hemleben, 2005). There are ~50 living species of planktonic foraminifera in the modern oceans, but only ~20 of these are sufficiently abundant in the larger sediment fractions to be used for paleoenvironmental reconstructions (Table 1). Planktonic foraminiferal faunas have remained relatively uniform throughout the entire late Cenozoic. As a result, paleoceanographers working with Quaternary climate change can normally get away with the knowledge of only a few dozen taxa. Species-level classification of Quaternary planktonic foraminifera follows the concept developed by Parker (1962). For a detailed overview of foraminifera taxonomy, the reader is referred to the compilations of Bé (1967), Saito, Thompson, and Breger (1981), Vincent and Berger (1981), and Hemleben et al. (1989).

Following the pioneering work by Darling, Kroon, Wade, and Leigh Brown (1996), the taxonomy of extant planktonic foraminifera could be tested using molecular genetic data. Thus, Darling, Wade, Kroon, and Leigh Brown (1997) and de Vargas, Zaninetti, Hilbrecht, and Pawlovski (1997) were able to confirm that the three major clades of planktonic foraminifera, defined by shell ultrastructure, are monophyletic, although it appears that they may have originated from different benthic lineages. It was further shown that every consistently recognized species and morphotype proved to be genetically distinct. This holds true even for forms whose taxonomic status was long unclear, such as the biologically distinct but morphologically identical types of *Globigerinella siphonifera* (Huber, Bijma, & Darling, 1997), the pink and white varieties of *G. ruber* (Darling, Wade, Kroon, Leigh Brown, & Bijma, 1999), and the two coiling forms of *N. pachyderma* (Darling et al., 2000; Darling, Kucera, Kroon, & Wade, 2006; Bauch et al., 2003). However, the genetic data also showed that morphology has not always been effective in describing the diversity of planktonic foraminifera. Apart from confirming the status of existing taxonomic groups, molecular data also revealed the presence of distinct genetic types within planktonic foraminiferal morphospecies, where no intraspecific clusters were suspected (Kucera & Darling, 2002; de Vargas, Sáez, Medlin, & Thierstein, 2004).

Many of the cryptic genetic types recognized within species of planktonic foraminifera show a considerable degree of genetic separation, comparable to that seen among morphologically defined species. In addition, molecular clock estimates suggest that these cryptic species diverged hundreds of thousands to millions of

Table 1 Ecologically Important Species of Extant Planktonic Foraminifera.

Species	Symbionts	Reproduction	Habitat depth	Genetic types	Dissolution resistance
Orbulina universa	Obligatory	Monthly	Surface	3	Moderate
Globigerinoides ruber (pink)	Obligatory	?Bi-weekly	Surface		Susceptible
Globigerinoides ruber (white)	Obligatory	Bi-weekly	Surface	3	Susceptible
Globigerinoides sacculifer	Obligatory	Monthly	Surface		Susceptible
Globigerinella siphonifera	Obligatory	Monthly	Surface to subsurface	3	Susceptible
Globigerina bulloides	None	Monthly	Surface	6	Moderate
Turborotalita quinqueloba	Facultative	Monthly	Surface	5	Moderate
Neogloboquadrina pachyderma	None	?Monthly	Surface to subsurface	5	Resistant
Neogloboquadrina incompta	None	Monthly	Surface to subsurface	2	Moderate
Neogloboquadrina dutertrei	Facultative	Monthly	Surface to subsurface	3	Resistant
Pulleniatina obliquiloculata	Facultative	Monthly	Subsurface		Resistant
Globorotalia inflata	Facultative	Monthly	Subsurface		Resistant
Globorotalia truncatulinoides	None	?Annual	Deep	4	Moderate
Globorotalia hirsuta	None	?Annual	Deep		Moderate
Globorotalia scitula	None	Monthly	Subsurface		Moderate
Globorotalia menardii	Facultative	Monthly	Subsurface		Resistant
Globorotalia tumida	None	?Annual	Deep	n.n.	Resistant
Globigerinita glutinata	Facultative	Monthly	Surface to subsurface		Moderate
Globigerinita uvula	None	Monthly	Surface to subsurface		Moderate

Source: Modified from Hemleben et al. (1989), Schiebel and Hemleben (2005) and Kucera and Darling (2002).

years ago (Darling et al., 1999; Darling, Kucera, Wade, von Langen, & Pak, 2003; Darling, Kucera, Pudsey, & Wade, 2004; de Vargas, Bonzon, Rees, Pawlowski, & Zaninetti, 2002; de Vargas, Norris, Zaninetti, Gibb, & Pawlowski, 1999; de Vargas, Renaud, Hilbrecht, & Pawlowski, 2001). Although biological species concepts are difficult to apply to planktonic foraminifera, due to their reluctance to complete their reproductive cycle in laboratory conditions, it appears reasonable to assume that at least some of the genetically identified types represent distinct species. This conclusion is further supported by the distinct biogeographic distribution of these cryptic genetic types, which appears to follow trophic regimes (de Vargas et al., 1999) and surface-water properties (Figure 4; Darling et al., 2004; de Vargas et al., 2001, 2002). Many morphologically defined species are thus in fact lumping ecologically distinct taxa, increasing the amount of noise in foraminifera-based paleoceanographic proxies (Kucera & Darling, 2002).

2.3. Ecology and Distribution

Extant species of planktonic foraminifera can be grouped into five main assemblages that define the tropical, subtropical, temperate, subpolar, and polar provinces (Bradshaw, 1959; Bé & Tolderlund, 1971). Almost two-thirds of the world oceans are covered by the warm-water provinces (Figure 5). The boundary between the warm subtropical and colder transitional province is marked by the annual isotherm of 18°C (Figure 5), which corresponds approximately to the latitude of balanced radiative heat budget (Vincent & Berger, 1981). Most extant species are cosmopolitan within their preferred bioprovince, although three Indopacific (*Globigerinella adamsi*, *Globoquadrina conglomerata*, *Globorotaloides hexagonus*) and one Atlantic tropical species (*G. ruber* pink) are endemic. The ubiquitous distribution of foraminiferal morpho-species is not mirrored by the cryptic genetic types. Although some of these do occur globally (Darling et al., 1999, 2000), many show a considerable degree of endemism (Kucera & Darling, 2002; Kucera et al., 2005).

The distribution and abundance of planktonic foraminifera species is strongly linked to surface-water properties. Sea-surface temperature (SST) appears to be the single most important factor controlling assemblage composition (Figure 5; Morey, Mix, & Pisias, 2005), diversity (Rutherford, D'Hondt, & Prell, 1999), and shell size (Schmidt, Renaud, Bollmann, Schiebel, & Thierstein, 2004a). Both laboratory experiments (Bijma, Faber, & Hemleben, 1990b) and sediment-trap observations (Zaric, Donner, Fischer, Mulitza, & Wefer, 2005) indicate that planktonic fora-minifera species survive under a considerable range of SST, but that their optimum ranges, defined by highest relative and absolute abundances, are typically narrow and distinct (Figure 5). At present, the polar waters of both hemispheres are dom-inated by a single small species (*N. pachyderma*), while the highest diversity and largest sizes are found in the oligotrophic subtropical gyres. The increase in SST toward the equator is accompanied by a proportional increase in surface-water stratification. The strength of vertical gradients in the ocean determines the number of physical niches available for passively floating plankton, and may thus control their diversity and morphological disparity (as reflected by size range) (Rutherford et al., 1999; Schmidt et al., 2004a).

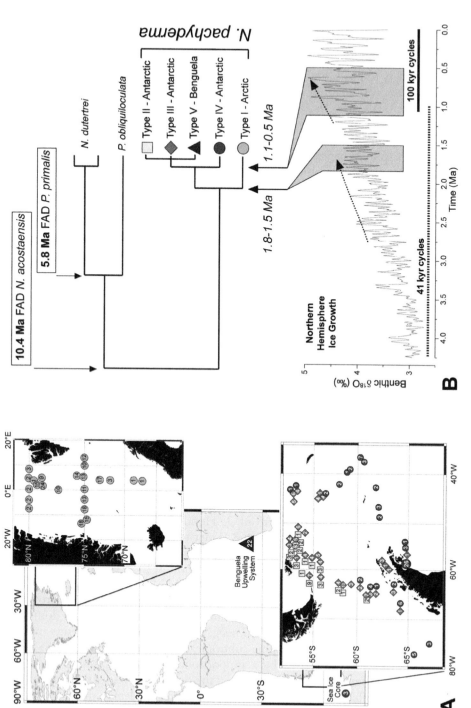

Figure 4   Distribution (A) and phylogeny (B) of molecular genetic types of *N. pachyderma* in the Atlantic Ocean. The data show that genetic diversification within the species began in early Quaternary and that the genetic types show a high degree of endemism (adapted from Darling et al. (2004). Copyright 2004, National Academy of Sciences).

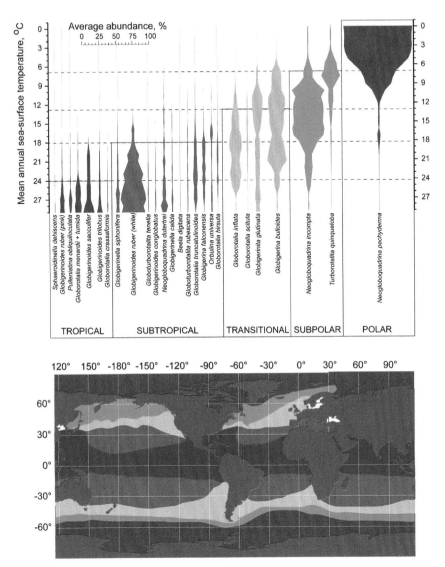

Figure 5 Planktonic foraminiferal provinces in the modern ocean. The distribution of the provinces (Bé, 1977; Vincent & Berger, 1981) follows sea-surface temperature gradients, reflecting the strong relationship between SST and species abundances. The abundance plots are based on surface-sediment data from the Atlantic Ocean (Kucera et al., 2005), averaged at one degree centigrade intervals.

The general trend toward higher diversity and larger sizes with increasing SST is reversed in equatorial and coastal upwelling zones, which are characterized by higher population densities of smaller species (Rutherford et al., 1999; Schmidt et al., 2004a). Large, symbiont-bearing carnivorous specialists are adapted to oligotrophic conditions, and in high-productivity regimes they are easily outnumbered by omnivorous and herbivorous species such as *G. bulloides* and *Globigerinita glutinata*.

These opportunists can rapidly react to organic particle redistribution and phyto-plankton blooms following nutrient entrainment (Schiebel, Hiller, & Hemleben, 1995; Schiebel, Wanilk, Bork, & Humleben, 2001; Schiebel & Hemleben, 2005).

Episodic pulses of primary productivity, coupled with the seasonal SST cycle result in predictable successions of planktonic foraminifera species, which react to the changing environmental conditions according to their ecological prefer-ences. Such successions have been documented in numerous sediment-trap studies (Figure 6) and their understanding is of great importance for geochemical proxies. Schiebel and Hemleben (2005) give an excellent summary of the main seasonal production patterns. In general, the flux rate of planktonic foraminiferal shells follows primary productivity cycles with a lag of several weeks. In polar oceans, the

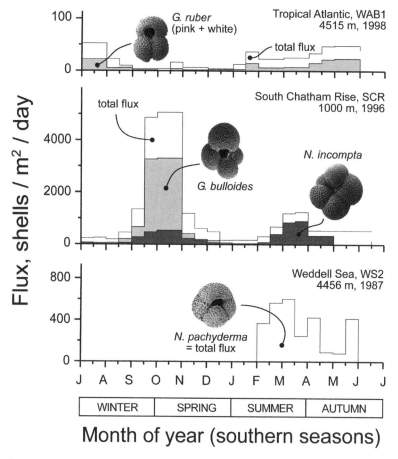

Figure 6 Typical patterns of annual cycle of planktonic foraminifera shell flux in polar, temperate, and tropical oceans. The flux in the polar ocean is limited to ice-free conditions; in temperate oceans it is typically focused into two seasonal peaks, each dominated by different species, and the oligotrophic tropical waters are characterized by extremely low and even fluxes throughout the year. Flux data are from different studies as reported in the compilation by Zaric et al. (2005). Note that all sediment-traps are from the southern hemisphere.

flux peak is observed during the summer, whereas in temperate oceans the spring flux maximum is often followed by a smaller autumn peak. Tropical and subtropical oceans are characterized by a steady rain of foraminiferal shells throughout the year (Figure 6).

Within the range of normal marine conditions (33–36‰), salinity does not appear to exert any significant influence on planktonic foraminifera (Hemleben et al., 1989). Laboratory experiments indicate that some species can tolerate a remarkable range of salinities (*G. ruber*: 22–49‰) and that salinity tolerances differ among species (Bijma et al., 1990b). In nature, no planktonic foraminifera are known to live under hyposaline conditions. *N. pachyderma* is known to avoid low salinity (<32‰) surface layers in the Arctic (Carstens, Hebbeln, & Wefer, 1997; Simstich, Sarnthein, & Erlenkeuser, 2003; Hillaire-Marcel, De Vernal, Polyak, & Darby, 2004) and the low-salinity surface water associated with the Zaire River plume is inhabited by a distinct assemblage dominated by *G. ruber* pink and *Neogloboquadrina dutertrei* (Ufkes, Jansen, & Brummer, 1998). At the other end of the spectrum, planktonic foraminifera inhabiting the Red Sea live at salinities in excess of 40‰ and the Antarctic *N. pachyderma* live in sea-ice where brine salinities exceed 80‰ (Dieckmann, Spindler, Lange, Ackley, & Eicken, 1991; Spindler, 1996). The upper salinity limit for tropical foraminifera determined by laboratory experiments seem to correspond well with observations from glacial Red Sea sediments, where aplanktonic zones correspond with paleosalinities determined from hydrological models (Fenton, Geiselhart, Rohling, & Hemleben, 2000; Siddall et al., 2003). The influence of ecological factors other than temperature, salinity, and fertility is difficult to disentangle, because most surface-water properties are strongly inter-correlated.

3. Planktonic Foraminiferal Proxies

Many properties of individual organisms and whole ecological systems are affected by the physical and chemical parameters of their habitat. If we knew how the environment modifies the basic genetic design of organisms and how it controls their spatial and temporal distribution, we could use the fossil record of such organisms to reconstruct the state and variation of past environments. The various signals locked in fossils are only rarely directly related to individual environmental parameters. Therefore, paleoenvironmental reconstructions rely on recipes and algorithms describing ways of how to relate measurements and observations made on fossils and other geological material to past environmental variables.

Planktonic foraminifera are by far the most important signal carriers in paleoceanography. The physical and chemical properties of foraminiferal shells provide a multitude of paleoproxies, based on the chemical composition and morphology of their shells as well as species abundance patterns. This chapter will deal with the physical properties of foraminiferal shells and proxies that can be derived from them (Table 2). Chemical properties of foraminiferal shells and their use as proxies are treated in detail in chapters on stable isotopes and trace elements. Unlike their

Table 2   Physical Properties of Foraminiferal Shells that can be Used for Paleoproxies.

Census	Counts of taxa, ecological or functional types
	Presence/absence
	Semi-quantitative abundance estimates
	Absolute abundances
	Relative abundances
Biometry	**Measurements made on individual specimens**
	Morphology — size, shape
	Shell ultrastructure
	Colour and weight
Modification	**Changes to fossils incurred after deposition**
	Preservation indices
	Fragmentation

chemical composition, the physical properties of foraminiferal shells can be determined relatively easily and with high precision. In addition, physical and chemical properties of foraminiferal shells follow different taphonomic pathways and proxies based on these properties can be used to derive independent estimates of paleoceanographic parameters from the same fossil assemblage, thus providing a unique opportunity to assess the robustness of such paleoenvironmental reconstructions. However, the processes that contribute to the physical form of a foraminiferal shell are very complex, and, as a result, reconstructions based on physical proxies are often considered less reliable and more difficult to interpret.

In an ideal world, one would wish to have a full mechanistic understanding of why and how each proxy works. In reality, mechanistic understanding of most paleoproxies is rare, particularly of those based on fossils and their physical form. As soon as life with all its complexities enters the equation, paleoceanographers have to resort to the process of empirical calibration. Here, the relationship between a physical parameter of a fossil and an environmental variable is derived by observing and describing the distribution of taxa and their properties in the present-day ocean. This process is methodologically relatively simple but it involves a number of assumptions that limit the applicability of each empirically calibrated proxy.

An empirical calibration requires a database of measurements or recordings of a physical parameter of an extant organism, with simultaneous observations of the desired environmental variable(s) associated with its habitat and a mathematical tool or algorithm to determine the form of the relationship between the two types of data. In paleoceanography, this relationship is then applied on data extracted from the fossil record. The main assumptions of this process are that the target environmental variable exerts a significant influence on the measured property, and that the mathematical tool is able to describe this relationship effectively.

There are two additional issues that need to be considered when applying an empirical calibration to a fossil sample: how far back in time can a calibration be used, and how do we recognize that a proxy is extrapolating into the unknown, yielding reconstructions of unknown reliability? Both issues reflect the basic assumption of empirical calibration: the stationarity principle. This principle states that the properties of fossils and the relationships among them and the environment must remain identical throughout the range of the application of the proxy.

The exact range in time of each calibration depends on the rate of evolution in the group of fossils on which the calibration is based. The reliability is highest in samples derived from the same time frame as the calibration data set and decreases until the time when the signal carrier or its components first evolved (Figure 7). Molecular-clock ages of the most recent divergences among cryptic species in planktonic foraminifera (Darling et al., 2003, 2004; de Vargas et al., 1999, 2001), as well as morphometric observations on changes in their ecological preferences (Schmidt, Renaud, & Bollmann, 2003), suggest that an ecological calibration based on modern planktonic foraminifera can be used with high confidence during the last glacial cycle, while its application in samples older than 1 Myr would very likely suffer significantly from the breakdown of the stationarity assumption. For example, Kucera and Kennett (2002) discovered a small but distinct shift in the morphology of Pacific *N. pachyderma* at ~1 Myr and showed that this inconspicuous event coincides with a major change in the habitat of this species. A similar example from the Neogene *Fohsella* lineage is given by Norris, Corfield, and Cartlidge (1996). Apparent taxonomic uniformity of Cenozoic planktonic foraminifera has led to attempts to apply calibrated proxies in deep time (e.g., Andersson, 1997). However, morphological similarity does not imply ecological similarity, and morphology alone is clearly not sufficient to describe the true diversity in planktonic foraminifera. Therefore, caution has to be exercised whenever observations and assumptions derived from extant species are transferred to the fossil record.

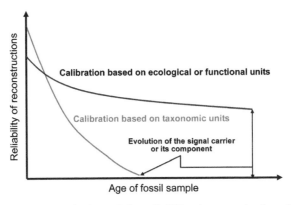

Figure 7 Gradual decrease with time of the reliability (accuracy) of a calibration based on taxonomic units and a calibration based on ecological or functional units. The reason for the decrease is the increasing possibility that ecological preferences of the constituent units of the calibration may have changed. Significant shifts in habitat and ecology within planktonic foraminiferal lineages can occur without noticeable changes to shell morphology.

Recognition of situations where a proxy yields unreliable reconstructions is a more difficult issue. In the case of transfer functions (see below), this uncertainty is embodied in the concept of the no-analog condition (Hutson, 1977). Such conditions occur when the measured value of a fossil property (size of a shell, abundance of a species) exceeds the variation in the calibration data set, or when the oceanographic situation in the past has no analog in the present. The latter is particularly difficult to detect and thus it is fair to say that all paleoceanographic reconstructions based on fossils have to be critically scrutinized. If in doubt, one should abstain from interpreting the absolute value of the proxy signal and use only its qualitative content (like the sign of a change). For obvious reasons, proxies delivering only qualitative information (warmer, colder) are much more robust and can be applied further back in time than rigorous quantitative proxies.

3.1. Census Data

Deep-sea sediments deposited above the calcite compensation depth abound with shells of planktonic foraminifera. Typically, one gram of deep-sea calcareous ooze contains thousands to tens of thousands of specimens larger that 0.150 mm. Such high abundances combined with the ease of their extraction from the sediment make planktonic foraminifera particularly suitable for quantitative analyses. In addition, as we have shown in Section 2.3, species composition of foraminiferal assemblages is extremely sensitive to surface-water properties, particularly SST (Morey et al., 2005). Data from laminated sediments from the California Margin (Field, Baumgartner, Charles, Ferreira-Bartrina, & Ohman, 2006) show that even inter-annual SST variations are recorded in the composition of foraminiferal assemblages deposited onto the sea floor. Thus it is not surprising that the analysis of census counts of species abundances and assemblage composition is the most common source of planktonic foraminiferal proxies.

The determination of a foraminiferal assemblage census typically involves the counting of 300–500 specimens in random sub-samples of the >0.150 mm fraction. This standard procedure has been developed within the CLIMAP project and was motivated on the one hand by the need for rapid acquisition of large amounts of census data, and on the other hand by statistical reproducibility (CLIMAP, 1976). Subsequent studies validated this optimization (e.g., Pflaumann, Duprat, Pujol, & Labeyrie, 1996; Fatela and Taborda, 2002) and the continuation of this method can be safely recommended for routine data acquisition in normal environments. However, if a particular rare species is of interest, the census size has to be increased in order to be able to distinguish statistically significant changes in the relative abundance of that species (Pflaumann et al., 1996). Alternatively, the absolute abundance of a rare species can be expressed in terms of its accumulation rate (Kucera, 1998), rather than its proportion in the total assemblage.

Smaller-sized planktonic foraminifera are generally difficult to identify and time-consuming to count. In order to minimize errors due to taxonomic uncertainty, CLIMAP project members (1976) recommended the use of the >0.150 mm size fraction. This practice has been followed ever since and enormous quantities of data continue to be generated using this procedure (Kucera et al., 2005). Although such a standard is essential for studies using environmental calibration based on species abundance, such as transfer functions (see below), many planktonic foraminiferal shells are smaller than 0.150 mm and the introduction of this artificial size limit inevitably biases the assemblage composition. Planktonic foraminiferal size decreases toward the poles (Schmidt et al. 2004a) and the use of the same lower size limit thus leads to an artificial decrease in diversity in high-latitude assemblages. The loss of information in census counts from such regions based on the >0.150 mm fraction led some researchers to consider the use of smaller sieve sizes (normally >0.125 mm). The value of analyzing smaller size fractions has been demonstrated for plankton samples from the Fram Strait (Carstens et al., 1997), as well as for reconstructions of the glacial–interglacial palaeoceanography of the polar Arctic Ocean (Kandiano & Bauch, 2002). Although there can be no doubt that the use of smaller fractions affords a more appropriate characterization of polar and subpolar assemblages, the use of counts based on this procedure in established environmental calibrations requires additional assumptions (Hendy & Kennett, 2000), or the development of new calibration datasets (Niebler & Gersonde, 1998).

The quality of census data relies on correct identification of the counted taxonomic units. Given the low overall number of planktonic foraminifera species and the common practice of lumping similar forms (e.g., G. tumida and G. menardii; Pflaumann et al., 1996; Kucera et al., 2005), there is no a priori reason to doubt the general applicability of environmental calibrations based on data generated by a range of researchers. Taxonomic schemes that are commonly used to generate census counts of planktonic foraminifera (Kucera et al., 2005) reflect a compromise between accuracy, speed, and reproducibility. However, no objective analysis of the errors resulting from different taxonomic opinions has ever been performed and every practitioner would agree that some species, such as G. bulloides and G. falconensis, are easily confused. Authors producing census counts are thus encouraged to consider their taxonomic concept carefully and note explicitly which species were recognized and which taxa were not being distinguished.

3.1.1. Indicator species and assemblages

The simplest types of foraminiferal proxies are based on abundances of ecologically significant (indicator) species. For example, the deep-dwelling G. truncatulinoides exhibits a yearly reproductive cycle during which it descends to considerable depth in the ocean (Schiebel & Hemleben, 2005). It can only complete its reproductive cycle if the waters at all depths it passes through are oxygenated. Following this argument, Casford et al. (2003) interpreted the presence of G. truncatulinoides in eastern Mediterranean sapropels as evidence for intermittent relaxation of intermediate-water anoxia during the sapropel-deposition events. Similarly, the affinity of Globorotalia scitula for sub-thermocline waters rich in organic debris allowed Schiebel, Schmuker, Alves, and Hemleben (2002) to trace the position of the Azores Front during the late Pleistocene.

The most commonly used index species is the shallow-dwelling opportunistic G. bulloides, which is known to thrive in high-productivity regimes, making it a good indicator of upwelling intensity (Thiede, 1975; Conan, Ivanova, & Brummer, 2002; Figure 8). Naidu and Malmgren (1995) and Gupta, Anderson, and Overpeck (2003)

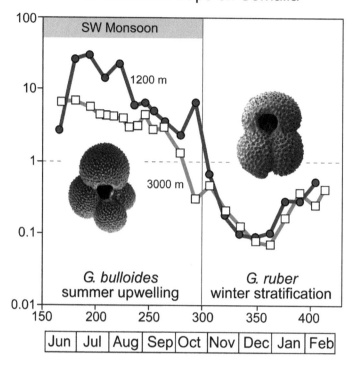

Figure 8 Increased flux of *G. bulloides* shells during upwelling off Somalia. Data from two sediment-traps moored at different depths show a consistent picture of dominance of this species during the south-western Monsoon, whereas *G. ruber* dominates the assemblage during the low-productivity winter period (modified from Conan et al., 2002).

used the abundance of this species to reconstruct sub-Milankovitch oscillations in Holocene monsoon-driven upwelling in the Arabian Sea, and Black et al. (1999) showed that its accumulation rate in laminated sediments from the Cariaco Basin correlates with historical records of wind strength and resulting upwelling in the region.

The ecology of individual species is often complex and poorly understood. Grouping ecologically similar species into "indicator assemblages" often provides a more robust approach. Indeed, the proportion between warm- and cold-water species could be used to track the thermal history of the ocean's surface on geological time scales (Ingle, 1977), as well as with centennial resolution (Rohling, Mayewski, Hayes, Abu-Zied, & Casford, 2002). A specific fauna is known to be associated with the warm Agulhas Current off South Africa (Giraudeau, 1993) and past changes in the abundance of this "Agulhas fauna" can be used to reconstruct the history of surface-water exchange between the Indian and Atlantic Oceans on glacial–interglacial time scales (Peeters et al., 2004) (Figure 9).

In the North Atlantic, planktonic foraminiferal assemblages allow an indirect reconstruction of sea-ice distribution. Kucera et al. (2005) showed that surface sediment samples deposited below the modern Arctic Domain surface water in the North Atlantic always contain a small but detectable fraction (0.3–1%) of subpolar species, including *T. quinqueloba*, *G. bulloides*, and *G. glutinata*, whereas in the perennially ice-covered regions these species are totally absent. The subpolar intruders are strictly bound to seasonally open conditions guaranteeing food and light supply, in particular the symbiont-bearing *T. quinqueloba* (Schiebel & Hemleben, 2005). Thus, presence of even small portions of subpolar species can be taken as a proxy of seasonally ice-free conditions. This approach has been used by Kucera et al. (2005) to reconstruct the extent of seasonally ice-free conditions in the Nordic Seas during the last glacial maximum (Figure 10).

Species and assemblage abundance proxies are simple and effective, but their main limitation is that they only deliver qualitative reconstructions. The understanding of many oceanographic processes requires knowledge of the absolute values of environmental parameters or the magnitudes of their changes. In order to derive such information from census data, the abundances of planktonic foraminiferal species have to be empirically calibrated to environmental variables. Such calibration is the subject of the so-called transfer functions.

3.1.2. Transfer functions

The strong relationship between environmental variables, notably SST, and assemblage composition of planktonic foraminifera has long tempted workers to make qualified guesses of absolute values of past environmental conditions (see review in Vincent & Berger, 1981). The art of qualified guessing eventually evolved into mathematical formalization of the ecological relationships. This transition can be exemplified by the weighted-average optimum temperature method of Berger (1969). This technique combines a somewhat intuitive ecological analysis of species with a rigorous formula:

$$T_{est} = \frac{\sum(p_i \cdot t_i)}{\sum p_i}$$

where p_i is the proportion of species i and t_i the "optimal" temperature for species i.

The choice of t_i is informed by the distribution of foraminiferal assemblages in modern oceans and sea-floor samples and represents a simple kind of empirical calibration. Such multi-dimensional empirical calibrations of species abundances and environmental parameters are called "transfer functions" in paleoceanography. Generally speaking, transfer functions can be defined as empirically calibrated mathematical formulas or algorithms that serve to optimally extract the general relationship between faunal composition in sediment samples and environmental conditions reflected by the fauna. This relationship is then applied to census data from fossil samples (Figure 11). Like any other empirical calibration, transfer functions rely on a number of assumptions. For more details, the

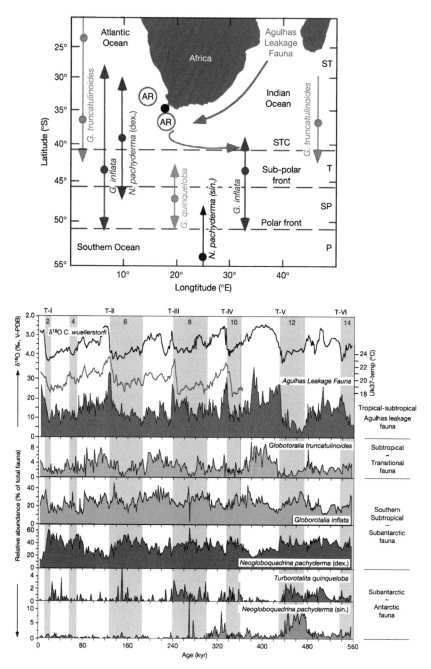

Figure 9 Changes in the abundance of assemblages of planktonic foraminifera in a core taken off Cape of Good Hope (solid circle). Present-day distribution of individual species is shown by vertical arrows in the upper panel; AR = Agulhas Rings. Subtropical species are carried into the region by the warm Agulhas Current and their abundance reflects the intensity of the current in the past (after Peeters et al. (2004). Copyright 2004, Nature).

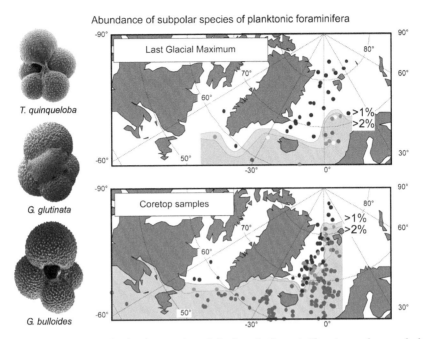

Abundance of subpolar species of planktonic foraminifera

Figure 10 Distribution of subpolar species of planktonic foraminifera in modern and glacial North Atlantic sediments. The presence of subpolar species in glacial Norwegian Sea indicates that this region must have been seasonally ice-free during the glacial period. Yellow, orange and red colours indicate samples with more than 2, 5 and 10% of subpolar species, respectively (data from Kucera et al., 2005).

reader is referred to the reviews in Birks (1995), Kucera et al. (2005), and Chapter 13 of this book.

The struggle for improvement in the precision of transfer function reconstructions has caused researchers to resort to more complex, often computer-intensive methods. A good review of early work on planktonic foraminiferal transfer function techniques is given by Hutson (1977). A true breakthrough in this field came with the so-called Imbrie-Kipp Transfer Function method (Imbrie & Kipp, 1971) which used Q-mode principal component analysis to decompose the variation in the faunal data into a smaller number of variables that were then regressed upon the known physical parameters. The Imbrie-Kipp method was the foundation for the pivotal effort of the CLIMAP group to reconstruct the sea–surface temperature field of the last glacial maximum ocean (CLIMAP, 1976). Foraminiferal transfer functions have seen a recent revival, fuelled mainly by the development and application of new computational techniques (Kucera et al., 2005, Chapter 13). Despite its caveats and limitations, the method is extremely important since its reconstructions are independent of geochemical proxies.

Although most planktonic foraminiferal transfer functions have dealt with SST, there is no reason to exclude the possibility that other environmental variables can be meaningfully extracted from the census data. Anderson and Archer (2002) used the modern analog technique (see Chapter 13) to reconstruct calcite saturation of

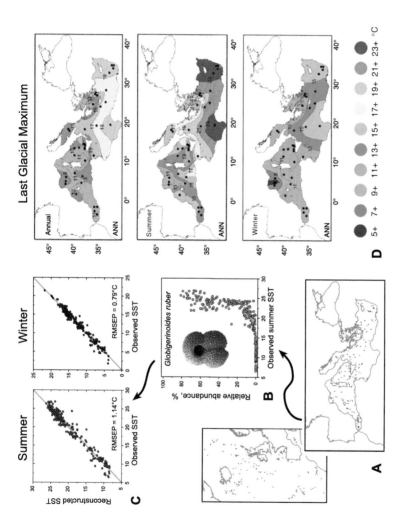

Figure 11 Foraminiferal transfer function for reconstruction of sea surface temperature in the Mediterranean. The calibration is based on 145 Mediterranean and 129 Atlantic core-top samples (A). In each sample, the abundance of 23 species of planktonic foraminifera was determined (B) and an artificial neural network algorithm was used to extract the relationship between species abundance and SST (C). The strong relationship between species abundance and SST is exemplified by the tropical species *G. ruber*. The estimated errors of SST reconstruction (RMSEP) range around 1 °C. The algorithms were then used to reconstruct SST in 37 cores representing the Last Glacial Maximum (D) (modified from Hayes, Kucera, Kallel, Sbaffi, & Rohling, 2005).

glacial bottom waters, and Ivanova et al. (2003) reconstructed paleoproductivity in the Arabian Sea using a variant of the Imbrie-Kipp method. The reason why most of the focus has been on SST is clearly shown in the analysis of Morey et al. (2005). These authors used a multivariate statistical technique known as canonical correspondence analysis in order to determine which of a series of 35 environmental parameters showed a strong and independent relationship with assemblage composition. SST came out as the single most significant factor, followed by a weak and diffuse relationship with a combination of parameters that Morey et al. (2005) interpreted as indicative of surface-water fertility.

The greatest challenge of foraminiferal transfer functions is generalization. Through a combination of increasing size of calibration data sets and increasing complexity of mathematical techniques, the apparent prediction errors of transfer functions have been brought to below 1°C. However, it is imperative to constantly remind ourselves that transfer functions are not developed to reproduce present-day faunal patterns. They are being devised to describe the general relationship between fauna and environmental forcing. Only those transfer functions that are capable of extracting the general relationship between fauna and environment will be robust to no-analog situations and can be meaningfully applied to the past (Hutson, 1977; Kucera et al., 2005).

3.2. Shell Morphology

All organisms are affected during growth by the state of their physical environment. The phenotype of each species thus reflects the combined action of the genetically stored information overprinted by ecological effects. If the nature of the action of environment on the phenotype was known, the morphology of an organism could be used to reconstruct the environmental conditions to which it was exposed during its life. This is the basic premise of proxies based on the morphology of planktonic foraminiferal shells. Because of the inherent complexity of the factors affecting shell morphology, quantitative empirical calibrations of morphological properties are rare and often imprecise. On the other hand, physical properties of foraminiferal shells can be determined unambiguously and accurately. Normally, measurements made on 20–50 specimens are enough to characterize the average state of a morphological variable in a fossil assemblage. This leads to the curious situation whereby morphological variability can be reconstructed with great accuracy but the resulting interpretations remain qualitative (Renaud & Schmidt, 2003; Schmidt et al., 2003). The difficulty in developing quantitative morphological proxies is also the reason why such proxies are rarely applied, especially when compared to the widespread use of transfer functions.

3.2.1. Shell size

The dimensions of planktonic foraminiferal shells found in sediments may vary by as much as two orders of magnitude. Some of the variation can be attributed to ontogenetic growth, but shells vary in size considerably even among adult individuals. In part, this variation is linked to taxonomy: the tiniest modern species build shells consistently smaller than 0.1 mm, while the giants can reach sizes well over 1 mm (Figure 1). Larger size is typically associated with warm-water species and Schmidt et al. (2004a) showed how this relationship is manifested in a spectacular expansion of the size range in foraminiferal assemblages from the poles toward the equator (Figure 12). The exact cause of this pattern is difficult to disentangle, as most of the involved variables are highly inter-correlated. Most likely a combination of higher carbonate saturation, faster metabolic rates, higher light intensity, and greater niche diversity (due to stronger stratification) can promote growth to larger and heavier shell sizes in the warm subtropical and tropical oceans (Schmidt et al., 2004a; de Villiers, 2004).

Temperature-related effects appear to control shell size in planktonic foraminifera even at the species level. Kennett (1976) and Hecht (1976) noticed that abundance and size maxima of many taxa tend to occur at specific temperatures. Bé, Harrison, and Lott (1973) documented a marked decrease in shell size of *O. universa* south

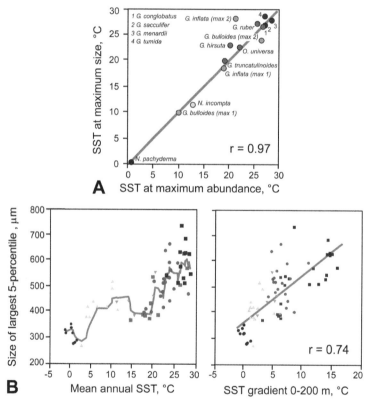

Figure 12 Relationship between shell size and temperature in planktonic foraminifera. Individual species achieve maximum size where they are most abundant (A), indicating that large size signals optimum ecological conditions. Size of the entire assemblage, expressed as the value dividing the 5% largest specimens from the rest, shows a gradual increase toward the tropics, interrupted at oceanographic fronts (B). Large assemblage size correlate with stronger vertical gradients in SST, indicating a greater number of niches. Symbols and shading distinguish samples representative of the five bioprovinces shown in Figure 5 (modified from Schmidt et al., 2004a).

of 30°S in the Indian Ocean and Hecht (1976) showed that in the North Atlantic, *G. bulloides* reach their largest sizes around 50°N, whereas in the subtropical to tropical *G. ruber*, maximum sizes occur around 10°N. The fact that temperature ranges leading to the largest size coincide with the highest relative abundance of individual species indicates that the largest size is reached under optimum environmental conditions, facilitating faster growth (Figure 12, Schmidt et al., 2004a). This model appears robust: it has been reproduced in all other oceanic basins and confirmed by laboratory experiments (Hemleben, Spindler, Breitinger, & Ott, 1987; Caron, 1987a).

Malmgren and Kennett (1978a, 1978b) pioneered the use of shell size as a proxy for SST. Their records from the southern Indian Ocean (Figure 13) revealed systematic shifts in mean shell size of *G. bulloides* that followed isotopically and faunally defined glacial stages. In these records, size was negatively correlated with temperature. However, one must realize that this is only because these cores are located near the present-day ecological optimum of this species and during glacial times surface ocean conditions shifted toward colder temperatures, away from the ecological optimum. If the core was located in a water mass which was warmer than the ecological optimum of *G. bulloides*, glacial cooling would have caused the ambient water mass to appear closer to the ecological optimum, leading to larger shell sizes (Figure 13). The range of possible responses of shell size as well as assemblage size to environmental forcing has been documented and extensively discussed by Schmidt et al. (2003).

An additional factor affecting size is food availability. Intuitively, a greater availability of food particles should lead to less energy being spent on foraging resulting in faster growth. However, Schmidt et al. (2004a) showed that this relationship only holds up to an "optimum primary productivity" of about 150 gC/m^2/yr. At high primary productivity, the shell size decreases. This pattern is mirrored in assemblage size range data, which show distinct minima at the position of major frontal systems (Figure 12), a finding consistent with the observation of Ortiz, Mix, and Collier (1995). Large, warm-water species are adapted to oligotrophic open ocean conditions. As discussed in the previous section, in high-productivity regimes, such species are outcompeted by generalists such as *G. bulloides*, which tend to be smaller. In sediments from the Arabian Sea, Naidu and Malmgren (1995) found a relationship between Holocene upwelling intensity and the shell size of four planktonic foraminiferal species, and speculated that this relationship may reflect changes in primary productivity rather than SST.

The study of Schmidt et al. (2003) showed that for the analyzed species, the relationship between shell size and SST remained stationary throughout the last 300 kyr. However, on evolutionary time scales, it appears that assemblage size of planktonic foraminifera followed the development of thermal gradients in the oceans and that late Neogene (including current) tropical oceans harbor unusually large planktonic foraminifera (Schmidt, Thierstein, Bollmann, & Schiebel, 2004b).

In summary, shell size is a potentially interesting variable, since it is objective, easy to determine and existing records show that it is highly sensitive to environmental forcing. Although the interpretation of the observed changes could be complex, the possibility of simultaneously analyzing several species as well as entire

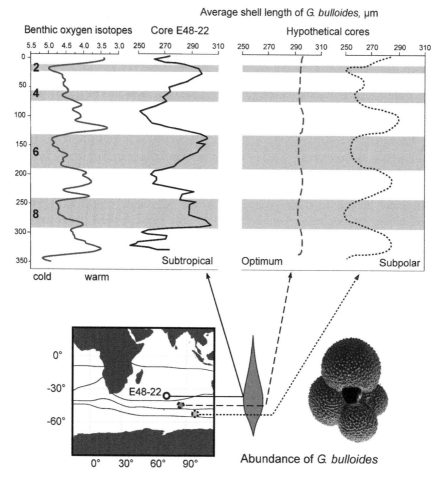

Figure 13 Variation in shell size of *G. bulloides* during the late Quaternary. The record from Core E48-22 (data from Malmgren & Kennett, 1978a, 1978b) in the subtropical province shows that larger sizes occurred during glacial periods (marine isotope stages 2, 4, 6, and 8), when the colder water masses expanded. This reflects the phenomenon that largest sizes correspond to ecological optimum: a change away from the optimum results in smaller sizes. Hypothetically, a core located near the ecological optimum would show a muted signal, whereas a core located in colder water should record largest sizes during warm intervals and smaller sizes during glacial periods. The boundaries between present-day foraminiferal bioprovinces are indicated in the map.

assemblages highlights the potential of foraminiferal shell size as a proxy for surface-water conditions throughout the late Quaternary.

3.2.2. Coiling direction

Planktonic foraminiferal shells with trochospirally arranged chambers can exhibit either dextral (right-handed) or sinistral (left-handed) coiling (Figure 14). Some species show a strong preference (bias) for either right-handed or left-handed

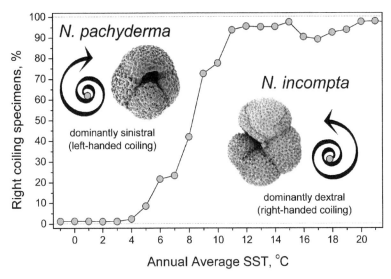

Figure 14   Changes in coiling direction of shells of the high-latitude species of *Neogloboquadrina*. The proportion of right-coiling specimens increases dramatically between 6 and 10°C, reflecting the replacement of *N. pachyderma*, which produces mainly sinistral shells by *N. incompta*, which produces mainly dextral shells (modified from Darling et al., 2006).

coiling, while other species exhibit mixed coiling proportions. Brummer and Kroon (1988) noticed that biased coiling is associated with non-spinose macro-perforate species, whereas proportionate coiling was typical for spinose species and microperforate species.

The ratio between the two coiling types of a species can vary through time and/ or space. Patterns of distinct shifts in coiling preference have been recognized and quantitatively characterized among several species of modern planktonic foraminifera (Table 3). Assuming that coiling direction in such species reflected ecophenotypic response to environmental parameters, mainly sea-surface temperature (Ericson, 1959; Ericson, Wollin, & Wollin, 1954; Boltovskoy, 1973), coiling ratios became an important early tool to reconstruct past marine environments. The determination of coiling ratios is rapid, accurate, and reproducible. A census of 50–100 specimens is normally sufficient to detect environmentally significant changes in coiling direction. Useful reviews of earlier work on coiling direction in planktonic foraminifera are given in Kennett (1976), Vincent and Berger (1981), and Hemleben et al. (1989).

Ericson (1959) and Bandy (1960) developed the most widely used coiling direction proxy. It is based on the remarkably strong and consistent relationship between coiling direction and sea-surface temperature in high-latitude species of the genus *Neogloboquadrina* (Figure 14). Earlier workers explained this behavior by temperature controlling the coiling direction in a single species, *N. pachyderma*, but Darling et al. (2000, 2004, 2006) demonstrated that the pattern reflects the presence of two distinct species with opposite coiling preferences. Polar waters of both hemispheres are inhabited by *N. pachyderma*, with dominantly sinistral shells, whilst

Table 3   Coiling Direction Proxies in Late Cenozoic Planktonic Foraminifera.

Species	Application	References
Neogloboquadrina pachyderma	Sinistral coiling associated with cold temperatures	Ericson (1959); Bandy (1960)
Globorotalia truncatulinoides	Sinistral coiling associated with cold temperatures or low salinity	Ericson et al. (1954); Thiede (1971)
	Holocene biostratigraphy in the North Atlantic	Pujol (1980); Zaragosi et al. (2000)
Globigerina bulloides	Sinistral coiling associated with colder temperatures or higher fertility	Boltovskoy (1973); Naidu and Malmgren (1996)
Globorotalia hirsuta	Holocene biostratigraphy in the North Atlantic	Duprat (1983); Zaragosi et al. (2000)
Pulleniatina spp.	Neogene biostratigraphy of tropical Atlantic and Pacific	Saito (1976)

the dominantly dextrally coiled *N. incompta* thrives in subpolar and temperate regions (Figure 14).

Similarly, Ericson et al. (1954) noted the presence of distinct regions in the North Atlantic characterized by different coiling directions of *G. truncatulinoides*. Early interpretations of this pattern again focused on phenotypic response to water temperature, but a molecular genetic study by de Vargas et al. (2001) revealed that dextral coiling in this species is associated with one of the four distinct genetic types they identified; the remaining three types showing sinistral coiling. These recent discoveries support earlier work by Brummer and Kroon (1988), who concluded that coiling direction in planktonic foraminifera is likely to be a genetically determined binary trait and that coiling direction changes were not driven by environmental factors.

The discovery of a link between coiling preference and genetic distinction implies that any qualitative or quantitative proxy based on coiling direction in planktonic foraminifera is only applicable as long as the ecological and coiling preferences of the genetic types remain unchanged. Morphological and molecular genetic data suggest that in the case of high-latitude *Neogloboquadrina*, coiling direction can only be used as a proxy for sea-surface temperature during the last 1 Myr (Kucera & Kennett, 2002; Darling et al., 2004). In contrast, the ecophenotypic hypothesis allowed a more extensive application of the proxy throughout the geological past.

As in other organisms, the genetic control on body (shell) symmetry is not 100% efficient. Darling et al. (2006) showed that a low level ($<3\%$) of aberrant coiling is associated with both *N. pachyderma* and *N. incompta*, indicating that coiling direction cannot be taken as an absolute discriminator among genetic types. While species with biased coiling seem to deliver a consistent picture, the control on coiling direction in species with proportionate coiling remains unclear. Boltovskoy (1973), Malmgren and Kennett (1976), and Naidu and Malmgren (1996) showed that *G. bulloides* exhibits a significant bias toward sinistral coiling which seems to be linked to temperature. Darling et al. (2003) confirmed the presence of this bias, which does not seem to be linked to genetic distinction, but no relationship between coiling direction and temperature was observed.

In summary, shifts of coiling ratios in planktonic foraminifera species are almost certainly a signature of distinct genetic types, which are revealed through their opposite coiling directions. If such genetic types are linked to different ecological preferences, the coiling ratio can be used as a meaningful paleoenvironmental signal. The abundance of coiling types in species exhibiting systematic shifts in coiling preference should thus be recorded separately for the purpose of environmental calibration and where foraminiferal calcite is used as substrate for geochemical proxies.

3.2.3. Shape

Compared to other groups of marine microzooplankton, planktonic foraminifera show a surprisingly low diversity. However, all of the ~50 living species exhibit a remarkable degree of morphological plasticity. The origin of this morphological plasticity has been traditionally attributed to ecophenotypic response to environmental forcing. Scott (1974) and Kennett (1976) give excellent summaries of early work on planktonic foraminiferal morphology and its relationships with environmental parameters. Kennett (1968a) documented a morphological gradation in *N. pachyderma* in surface sediments from the South Pacific. Shell morphology of this species, including variables such as the number of chambers in the last whorl, was shown to follow surface ocean hydrography in the region. Malmgren and Kennett (1972) later reanalyzed the same data using multivariate statistics and identified four distinct latitudinal clusters. Similarly, Kennett (1968b) showed compelling evidence for the relationship between shell morphology and SST in *G. truncatulinoides* (Figure 15), while Hecht (1974) found that the rate of chamber expansion in *G. ruber* from the North Atlantic appears correlated with surface salinity. Malmgren and Kennett (1976) noted a systematic relationship between SST and shell compression and aperture size in *G. bulloides* from the southern Indian Ocean.

Two factors hamper the application of foraminiferal morphological variation in reconstructions of Quaternary paleoceanography. First, virtually nothing is known about the functional morphology of planktonic foraminiferal shells, and thus any environmental calibration inevitably remains a black box. Secondly, the assumption that morphological variability is linked to ecophenotypy must be supported by independent means. Although the former constraint remains, advances in molecular genetics of foraminifera have made it possible to test the latter assumption. Recent molecular genetic studies have shown a high level of genetic diversity among morphospecies of planktonic foraminifera and suggested a significant genetic component in the morphological variation of planktonic foraminiferal species. Darling et al. (2004) showed that the southern ocean is inhabited by a series of distinct genetic types of *N. pachyderma*, whose distribution follows that of the

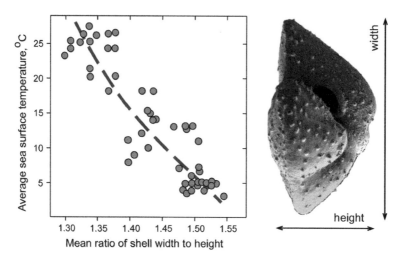

Figure 15 Variation in shell morphology of modern G. *truncatulinoides* (modified from Kennett, 1968a). The apparent ecophenotypic increase in shell height toward the tropics in fact reflects the varying proportions of four different genetic types (de Vargas et al., 2001).

morphological clusters identified by Kennett (1968a) and Malmgren and Kennett (1972). Even stronger evidence was presented in the study of G. *truncatulinoides* by de Vargas et al. (2001), who found a direct link between genetic distinction and shell morphology in this morphospecies. This discovery provides an elegant explanation for the presence of morphological and stable isotopic "subpopulations" in this species (Healy-Williams et al., 1985). Similarly, the pattern of changes in the morphology (including coiling direction) of this species in late Quaternary Atlantic sediments (Lohmann & Malmgren, 1983) can be explained in terms of habitat tracking among cryptic species with different ecological preferences (Renaud & Schmidt, 2003).

Most geochemical proxies rely on species-specific empirical calibrations. The specificity reflects the interference of metabolic processes and ecological behavior with the incorporation of chemical signals into foraminiferal calcite. If morphological variation is linked to genetic distinction, different morphotypes of planktonic foraminiferal species could be used to detect the presence of genetic diversity and assist in the interpretation of geochemical signals. This potential is not a theoretical conjunction; Bijma, Hemleben, Huber, Erlenkeuser, and Kroon (1998) found geochemical differences between the two genetically distinct types of G. *siphonifera* and recent studies link apparent intraspecific variability to different stable isotopic signatures in G. *bulloides* (Bemis, Spero, Lea, & Bijma, 2000) and isotopic and trace-element differences in G. *ruber* (Wang, 2000; Steinke et al., 2005). Both of these morphospecies are known to consist of several distinct genetic types (Kucera & Darling, 2002). Clearly, an understanding of the origin and significance of the morphological variation in individual species has a great potential for increasing the capacity of planktonic foraminifera to produce accurate and reliable information on past sea-surface conditions.

3.2.4. Wall ultrastructure

Several features of planktonic foraminiferal shell wall have been considered to be related to environmental parameters; thorough reviews of this topic are given in Kennett (1976), Vincent and Berger (1981), and Hemleben et al. (1989). Kennett (1968a, 1970) noted an increase in wall thickness of several globorotalid species in colder waters and Srinivasan and Kennett (1974) showed that surface ornament in *N. pachyderma* varied in pace with Neogene climate oscillations in temperate waters of the South Pacific. Although the final thickness of the shell wall appears to be controlled chiefly by carbonate ion concentration, both during life and after the deposition (see Section 4.2.2 "Shell weight"), the functional significance of changes in surface ornament in planktonic foraminifera remains obscure. Vincent and Berger (1981) caution against the use of surface ornament. Given the exposure of this part of the shell, and its particularly large surface-to-volume ratio, the preservation of the ornament ought to be particularly prone to dissolution. Indeed, Dittert and Henrich (2000) were able to devise an SEM-based wall-texture index for *G. bulloides* that could have been used to measure calcite dissolution intensity.

Of all properties of the shell wall, by far the maximum attention has been given to porosity. The size and frequency of pores is easy to measure and a number of studies have shown a close relationship between porosity and temperature in surface sediments both among species (Figure 16; Bé, 1968) and within species of macroperforate planktonic foraminifera (Frerichs, Heiman, Borgman, & Bé, 1972; Bé et al., 1973). Wiles (1967) and Hecht, Bé, and Lott (1976) showed that porosity could be used to trace Quaternary climatic cycles. Based on experimental observations and

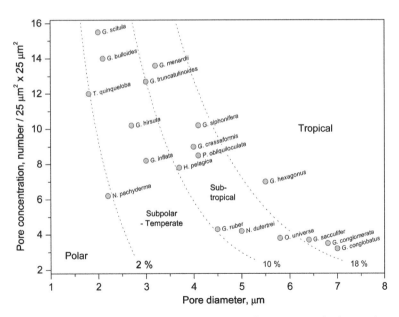

Figure 16   Variation in shell porosity among 19 macro-perforate extant planktonic foraminifera species. Tropical species show the highest porosity values whereas polar species show the lowest porosities (modified from Bé, 1968).

mitochondrial distribution (Hemleben et al., 1989), it appears that pores are related to gas exchange. This hypothesis is supported by laboratory experiments which indicate that increase in temperature correlates with larger pore diameter in *G. sacculifer* and *Orbulina universa* (Caron, Faber, & Bé, 1987a, 1987b; Bijma et al., 1990b): lower gas solubility under higher temperature promotes the growth of larger pores.

The use of shell porosity as a proxy of surface-water properties is complicated by several factors. Porosity in planktonic foraminifera changes dramatically during ontogeny (e.g., Huber et al., 1997) implying that a meaningful comparison is only possible among individuals of the same maturity. Next, calcite dissolution leads to an apparent increase in pore diameter (Bé, Morsem, & Harrison, 1975), although pore density is immune to this effect. Finally, although the environmental relationship between porosity and temperature is undisputed, it is clear that some of the variation in porosity is linked to cryptic genetic diversity. Huber et al. (1997) have shown that porosity is the only morphological character discriminating genetic Types I and II of *G. siphonifera*. Interestingly, in a survey of porosity in five species of planktonic foraminifera from sedimentary samples, Frerichs et al. (1972) found no relationship between latitude and porosity in this species. Similarly, a link between shell porosity in *O. universa* (Bé et al., 1973; Hecht et al., 1976) and genetic diversity has been proposed by de Vargas et al. (1999), although this hypothesis remains to be verified by direct observations on genotyped specimens. Despite these limitations, pore properties in planktonic foraminifera in well-preserved sediments are an underestimated source of useful environmental information, particularly because of the known functional significance of these structures and the verification of environmental observations by laboratory experiments.

3.3. Planktonic Foraminifera as Substrate for Geochemical Studies

Foraminiferal calcite has been used as a passive recorder of surface-water composition, as well as for the monitoring of kinetically and metabolically mediated fractionations (Henderson, 2002). Isotopic and trace element proxies have been treated comprehensively in the reviews by Rohling and Cook (1999), Lea (1999), and Schiebel and Hemleben (2005), and are the subjects of Chapters 18, 16, and 17 of this book. The use of planktonic foraminifera as substrate for geochemical proxies increasingly relies upon a detailed knowledge of the ecology of the signal carrier (Rohling et al., 2004). It may be useful to remind ourselves of the main factors influencing the incorporation of geochemical signatures into foraminiferal shells.

The chemical composition of foraminiferal shells derives primarily from the chemistry of the ambient seawater. Proxies that are known to be passively incorporated into the shells, require knowledge of the habitat and ecology of the analyzed species. The fossil assemblage is biased toward shells deposited during the time of maximum production. Depending on the species and the ecological circumstances, a signal measured on a group of specimens may either represent the yearly average, the spring, the summer, or the autumn (Figure 6). The vertical migration pattern and depth of calcification determine what level in the water column the chemical signal represents. Size may be used to estimate the ontogenetic stage indicating what part of the vertical migration cycle the analyzed specimens represent.

Proxies monitoring kinetic or metabolic fractionations must further consider the microhabitat of the analyzed species. Symbionts alter the chemical microenvironment of the foraminifera, while changes in growth rates related to food availability or sub-optimum ecological conditions affect the rate of metabolic processes and associated fractionations. All proxies using species-specific ecology require correct identification of taxa whose behavior does not change through their geographical range. It is increasingly obvious that the presence of multiple cryptic genetic types in common species of planktonic foraminifera has been an underestimated source of noise in paleoceanographical reconstructions.

Like physical proxies, geochemical proxies can only be applied as far back in time as the relationship between species ecology and biology and the target chemical process remain stationary. Even if the metabolic or biochemical pathway responsible for incorporation of the chemical signal remains the same, any change in calcification depth or production season will alter the meaning of the recon-structed record. As we have shown in Section 3, planktonic foraminiferal species may have kept to a similar niche for several hundreds of thousands of years. Environmental calibrations derived from extant species cannot be applied to fo-raminifera extracted from early Quaternary and Neogene sediments, without con-sidering the possibility of non-stationary behavior. Such applications are possible in principle, but require elaborate matching and cross-calibration of the ecology of fossil species.

4. Modifications After Death

4.1. Settling through the Water Column

Unlike ciliates or flagellated protozoans, planktonic foraminifera have no active means of propulsion. Their calcite shells cause negative buoyancy, which is com-pensated for by the production of low-density lipids or gasses during metabolism and which allows the foraminifera to control their vertical position in the water column. Subsequent to premature death or following sexual reproduction, the positive buoyancy is lost and planktonic foraminiferal shells, empty or filled with residual cytoplasm, begin to descend to the sea floor. Schiebel (2002) estimated the global export production of foraminiferal calcite at 100 m depth to 1.3–3.2 Gigatonnes/yr, equivalent to 25–50% of the total pelagic carbonate flux. Of this amount, 0.36–0.88 Gigatonnes/yr arrive on the sea floor, making up 30–80% of the deep-sea biogenic carbonate (Figure 17).

Clearly, a large amount of foraminiferal calcite is lost during settling. This loss is attributed to biogeochemical cycling in the water column. The intensity of these processes is a function of the residence time of individual shells in the water column, which is directly proportional to settling velocity. Berger and Piper (1972) conducted the first modern settling experiments, followed by the work of Fok-Pun and Komar (1983) and Takahashi and Bé (1984). Furbish and Arnold (1997) investigated in detail the effect of spine geometry on sinking. All studies show that the majority of foraminiferal shells have Reynolds numbers larger than 0.5 and that

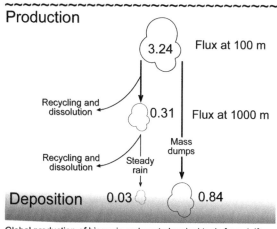

Global production of biogenic carbonate by planktonic foraminifera
in Gigatonnes per year

Figure 17   The contribution of planktonic foraminifera to global biogenic carbonate flux in the ocean. Note that only about one-quarter of the export flux from the surface layer is deposited on the sea floor. A majority of this carbonate is transported to the sea floor during mass deposition events (blooms). Modified from Schiebel (2002).

they consequently do not settle in the Stoke's region. Further, it was shown that the settling velocity of empty shells of planktonic foraminifera depends on the shell size, weight, shape, and presence of residual cytoplasm, and on the physical properties of ambient seawater. In general, large, spineless, gametogenic specimens are expected to sink at speeds between 500 and 3,000 m/day, i.e., reaching the sea floor within a week, while small specimens and specimens with spines sink more slowly (> 200 m/day), with a residence time of two weeks or more.

The varying settling velocities of foraminiferal shells imply that the time of deposition on the sea floor, or in sediment traps, lags behind the time of reproduction and growth. Prolonged exposure during settling of small specimens leads to their preferential dissolution and disproportionate amounts of adult shells thus appear to accumulate in deep-sea sediments (Peeters et al., 1999; Schiebel & Hemleben, 2005). Schiebel (2002) argues that mass dumping of fast-settling particles during export production peaks is in fact responsible for the majority of foraminiferal shells that reach the sea floor (Figure 17). In addition, fast-sinking specimens are less likely to be expatriated during settling. Expatriation is a process of lateral advection of passively floating organisms associated with surface or subsurface currents (Berger, 1970b; Weyl, 1978) or storm events (Schiebel et al., 1995). Expatriation affects living specimens as well as empty shells and leads to an apparent expansion of the geographical ranges of species away from their ecological optima. This effect is particularly obvious in sediments deposited near major fronts or margins of biogeographical provinces. Increasing steepness of surface ocean gradients, such as during the glacial expansion of polar waters, may be manifested in unusual "mixed" faunal assemblages that may prevent environmental reconstructions using transfer functions (see Section 3.1 "Census Data").

4.2. Calcite Dissolution

The deep waters of the world oceans are undersaturated with respect to calcium carbonate (Berger, 1970a), and the decomposition of organic matter during settling and the acidic environments in predator guts create corrosive microenvironments conducive to calcite dissolution throughout the water column (Schiebel, 2002). Therefore, empty shells of planktonic foraminifera settling onto the sea floor are subject almost immediately after death to dissolution (Berger, 1971). The intensity of dissolution depends upon the final settling depth and residence time in the water column and on the sea floor. Although recent estimates suggest that only about 25% of planktonic foraminiferal shells reach the sea floor (Schiebel, 2002), the rate of water-column dissolution is too low in comparison with the rate of carbonate supply, and the weight loss to dissolution is not manifested in the carbonate content of the sediment (Berger, Bonneau, & Parker, 1982). However, several hundred meters above the calcite compensation depth (CCD), the dissolution intensity rapidly increases and begins to affect the bulk composition of the sediment in favor of insoluble constituents (clay, opal). This level is known as the foraminiferal lysocline (Berger 1970a, Figure 18). In regions with high primary productivity, degradation of excess organic carbon delivered to the sea floor causes increased CO_2 concentration in pore waters that may result in carbonate dissolution above the lysocline (Peterson & Prell, 1985; Adler, Hensen, Wenzhöfer, Pfeifer, & Schulz,

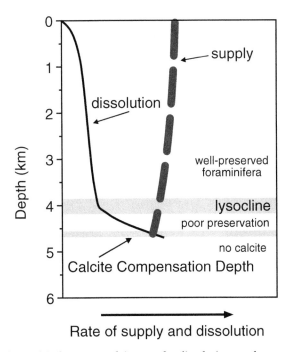

Rate of supply and dissolution

Figure 18 The relationship between calcite supply, dissolution, and preservation on the sea floor. The lysocline is defined by a sudden increase in dissolution rate, reflected by poor preservation of planktonic foraminifera in surface sediments (modified from Berger, 1970a).

2001). The position of the lysocline and the CCD reflect bottom water carbonate ion concentration, which varies in response to changes in oceanic circulation and redistribution of carbon among the main global reservoirs. Although changes in the position of the CCD cannot be effectively monitored in one sediment core, the proximity of the lysocline and the intensity of dissolution are well reflected in the preservation of planktonic foraminiferal shells. Therefore, preservation of planktonic foraminifera can be used as a proxy for bottom water chemistry and indirectly for deep circulation and carbon cycling.

4.2.1. Quantifying dissolution intensity

In order to reconstruct dissolution intensity, we need to devise an effective quantitative measure of foraminiferal preservation. Following studies by Ruddiman and Heezen (1967) and Berger (1968, 1970a) based on core-top samples from depth transects, as well as experimental investigations by Bé et al. (1975), the effects of dissolution on foraminiferal shells are well understood (Figure 19, Table 4). With increasing dissolution, the thickness of the shell decreases, corresponding to the mass of calcite lost to dissolution. Early stages of dissolution are seen as distinct etching patterns (Bé et al., 1975; Dittert & Henrich, 2000), but eventually, thinner, or more exposed parts of the shell become so weakened that the shell disintegrates into fragments. In pelagic oozes, where planktonic foraminiferal shells are the main constituent of the coarse fraction, this process leads to a gradual decrease in the sand content of the sediment. Due to the different morphology of their shells some planktonic foraminifera species are more prone to dissolution (Ruddiman & Heezen, 1967; Berger, 1968). This means that assemblages affected by dissolution are enriched in species resistant to dissolution (Table 1). This effect can be used to quantify dissolution intensity by determining the excess abundance of resistant species. Similarly, deep-sea benthic foraminifera have normally thick and smooth shells that are more resistant to dissolution than those of planktonic foraminifera and the ratio between whole shells of the two groups (the P/B ratio) decreases with

Increasing intensity of calcite dissolution

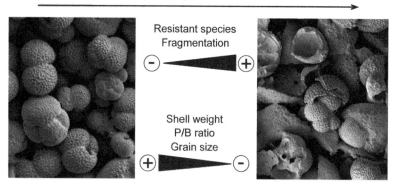

Figure 19 A well-preserved assemblage of planktonic foraminifera compared to an assemblage affected by calcite dissolution and a summary of the various effects of calcite dissolution on foraminiferal shells.

Table 4 Common Indices Used for Estimating Calcite Dissolution Intensity from Planktonic Foraminifera.

Index	Remarks
Degree of fragmentation $F(\%) = 100 \times F/(C+F)$; $F =$ number of fragments; $C =$ number of complete shells; Berger (1970a)	The most commonly used dissolution index, considered to be reliable and robust. Its determination is subjective and it is therefore not suitable for quantitative calibrated reconstructions.
$F_{LS}(\%) = 100 \times (F/8)/((F/8)+C)$; Le and Shackleton (1992)	A modification used to improve the linearity of the relationship of the index with dissolution intensity.
Shell weight $\Delta CO_3^{2-} = [G.\ sacculifer\ (355–425\ \mu m)$ shell weight $(\mu g) - 20.27]/0.70$; Broecker and Clark (2001a) $\Delta CO_3^{2-} = [G.\ ruber\ (300–355\ \mu m)$ shell weight $(\mu g)-12.6]/022$; de Villiers (2005) $\Delta CO_3^{2-} =$ Calcite saturation state at the sediment–water interface	Both calibrations are based on data from Indian and Pacific Ocean coretops and assume that all of the shell weight variation is due to calcite dissolution. Unlike fragmentation, this index is objective and thus better suited for quantitative calibration.
Percentage of resistant species $res(\%) = 100 \times r/(r+s)$; $r =$ number of shells of resistant species; $s =$ number of shells of susceptible species; Ruddiman and Heezen (1967)	Requires regional calibration and an *a priori* knowledge of species dissolution susceptibility. It is only sensitive at low dissolution intensity.
Assemblage dissolution index $FDX = \Sigma(p_i R_i)/\Sigma p_i$; $R_i =$ rank of species i; $p_i =$ percentage of species i; Berger (1968)	Requires regional calibration and an *a priori* knowledge of species dissolution susceptibility. It has a wider sensitivity range than the previous index.
Loss of susceptible species $L(\%) = 100 \times (1-r_o/r)$; $r_o =$ percentage of resistant species in an unaltered sample; $r =$ percentage of resistant species in a sediment sample; Berger, 1971	Requires one unaltered sample from the investigated region and an *a priori* knowledge of species dissolution susceptibility. It is only sensitive at low dissolution intensity.
P/B — plankton to benthos ratio $B(\%) = 100 \times B/(B+P)$; $B =$ number of benthic foraminiferal shells; $P =$ number of planktic foraminiferal shells (Arrhenius, 1952)	Shells of benthic foraminifera are more resistant to dissolution than those of planktic foraminifera, but there are many other factors influencing the abundance of both groups. This index is robust and sensitive even at high dissolution intensity, but it can be difficult to interpret unambiguously.

Source: Modified from Conan et al., 2002.

increasing dissolution. The latter can be used as a dissolution proxy only in abyssal sediments. In shallower settings, the P/B ratio is primarily controlled by food availability.

The phenomena described above have been used to devise a number of numerical or ordinal indices (Table 4). Although they may differ in their sensitivity at various intervals of the dissolution intensity range, their reproducibility, and their spatial and temporal applicability, all of these indices are often highly correlated, and the choice is thus primarily guided by the specific circumstances of each application. Below, we will discuss two measures of dissolution intensity. For a more thorough review of other indices, the reader is referred to the volume by Sliter, Bé and Berger (1975) and to the reviews in Thunell (1976), Berger et al. (1982), Hemleben et al. (1989), Dittert et al. (1999), and Conan et al. (2002).

4.2.2. Shell weight

The use of size-normalized weight of planktonic foraminiferal shells as an indicator of dissolution intensity was first proposed by Lohmann (1995); a recent review of the method is given by de Villiers (2005). Broecker and Clark (2001a, 2001b) showed that the weight loss of foraminiferal shells due to dissolution is strongly correlated with bottom water carbonate ion concentration and showed how this relationship could be used to reconstruct past ocean chemistry. The method is based on empirical calibration of bottom water carbonate ion concentration with average weight of clean, empty shells of selected species picked from narrow size ranges in core-top sample transects (Figure 20). The application of this proxy rests on the assumption that the "initial weight" of foraminiferal shells of a certain size is

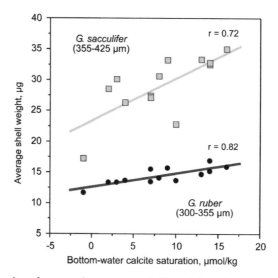

Figure 20 Progressive decrease in average shell weight of two species of planktonic foraminifera with decreasing calcite saturation of ambient bottom waters in Pacific and Indian Ocean core-top samples. The decrease in shell weight reflects gradual thinning of the shell wall. The relationship can be used to reconstruct bottom-water chemistry in the past (data from de Villiers, 2005).

constant through both time and within the geographical range of each species. However, Barker and Elderfield (2002) noticed a significant variability in the size-normalized weight of foraminiferal shells and attributed it to the influence on shell growth of carbonate saturation in the surface waters. This conclusion is supported by laboratory experiments (Spero, Bijma, Lea, & Bemis, 1997) and by glacial–interglacial shell weight variability consistent with lowered glacial atmospheric CO_2 (Barker & Elderfield, 2002).

The shell weight technique is elegant in its simplicity and potentially powerful, but one must not forget that its applicability is limited by the assumptions of the species-specific empirical relationships upon which it is based. Core-top calibrations may be valid for the modern ocean, but their application in the geological past and across evolutionary time scales is questionable. In addition, the technique is still in its infancy, and a considerable number of additional factors that potentially influence shell weight remain to be explored (de Villiers, 2004, 2005).

4.2.3. Fragmentation
The proportion between fragments and complete shells of planktonic foraminifera is the most commonly used proxy for calcite dissolution intensity. It is based on visual identification and counting of the abundance of fragments and complete shells, typically out of 300 particles. Because every researcher has to develop one's own criteria for discrimination between fragments and whole shells, the index is highly subjective and not easily reproducible. This subjectivity is particularly significant when the index is used for quantitative calibration with bottom water carbonate ion concentration (de Villiers, 2005). Nevertheless, the degree of fragmentation of planktonic foraminifera has been used in numerous studies, which indicates that in low-to-moderate dissolution regimes it represents an excellent index of calcite dissolution (Thunell, 1976; Le & Shackleton, 1992). This proxy has a number of advantages: it is rapid and simple to determine, independent of species composition of the analyzed assemblage, and as a semi-quantitative index it can be applied to pelagic sediments with planktonic foraminifera of any age.

4.2.4. Effect of calcite dissolution on foraminiferal proxies
Calcite dissolution has the potential to modify the chemical composition of foraminiferal shells and affect proxies using foraminiferal calcite as substrate. The most significant effect is seen in the Mg/Ca ratio of foraminiferal calcite. The Mg-rich carbonate phase is more soluble than pure calcite and foraminiferal shells that are subject to dissolution will thus appear depleted in Mg. This effect is discussed in detail by Brown and Elderfield (1996) and a recent review is given in Barker, Cacho, Benway, and Tachikawa (2005). A similar effect is to be expected for other trace elements, especially as many of these are not distributed evenly throughout the foraminiferal shell (Eggins, DeDeckker, & Marshall, 2003, 2004). Progressive solution of outer layers can thus alter the bulk composition of the entire shell even if the studied trace element dissolves out at the same rate as Ca. Berger and Killingley (1977) showed that calcite dissolution may influence even the carbon and oxygen stable isotopic composition of foraminiferal calcite. In this case, a shift in the isotopic composition can be linked to preferential removal of outer layers of the shell that store

geochemical and isotopic signatures of the adult habitat of each species. The effects of calcite dissolution on geochemical proxies become most apparent at depths approaching the foraminiferal lysocline. At such depths, calcite dissolution must be considered as a potential source of noise in any paleoceanographic application.

The effects of calcite dissolution are not limited to geochemical proxies. Dissolution increases fragmentation and alters the size spectrum of the assemblage. In addition, the abundance of dissolution-prone species becomes gradually reduced and the residual assemblages appear to represent colder conditions (Berger, 1968; Vincent & Berger 1981; Thunell & Honjo, 1981). This phenomenon reflects the tendency of dissolution-prone species to be more common in tropical faunas (Figure 21). Although there have been attempts to compensate for dissolution-related assemblage composition bias, the best recommendation would seem to refrain from basing environmental reconstructions on samples that bear signs of moderate to severe dissolution. Although calcite dissolution has a potentially significant impact on foraminiferal proxies, it is easy to recognize and its effects are understood. Paleoceanographers are well aware of this problem and take great care to limit their studies to appropriately preserved fossil material. As a result, dissolution is rarely an issue when considering the reliability of foraminiferal proxy results.

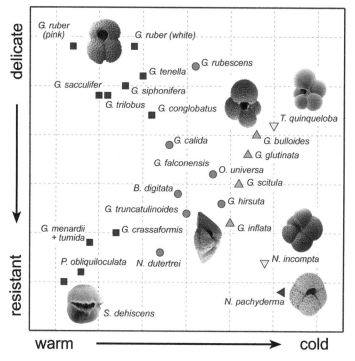

Figure 21 A comparison of temperature and dissolution susceptibility ranking of 26 extant species of planktonic foraminifera. Symbols and shading indicate species characteristic of the five foraminiferal bioprovinces as shown in Figure 5. Warm-water species tend to be delicate and easily dissolved, whereas most cold-water species are resistant (data from Figure 5 and Hemleben et al., 1989).

5. PERSPECTIVES

Planktonic foraminifera are the main provider of paleoceanographic proxies and there is every reason to believe that they will continue to represent our main source of information on the state of past oceans. The enormous research effort of the last two decades has helped to elucidate many aspects of foraminiferal life, ecology, and shell chemistry, so that the proxies we are using today are more precise and reliable then ever. At the same time, the huge progress that has been made in the research on planktonic foraminifera over the last 20 years has also helped to highlight a number of basic issues that continue to hamper the use of these organisms for paleoceanographical applications. It is especially the increasing effort to achieve a more mechanistic understanding of proxies that calls for urgent action in the three following areas:

- It is increasingly obvious that further development of geochemical proxies is only tenable if the mechanism of foraminiferal biomineralization is understood. The studies of Erez (2003) and Bentov and Erez (2005) reveal how little we know, and how many of our assumptions may need to be revised, and the necessity to clarify nannoscale mineralogy of foraminiferal shells is highlighted in the analyses made by Eggins, Sadekov, and De Deckker (2003, 2004).
- Geochemical and physical proxies alike suffer from our insufficient knowledge of the microhabitat and natural behavior of planktonic foraminifera, including the functional morphology of their shells. This ignorance is epitomized in our inability to have these organisms reproduce in laboratory cultures, a technical constraint that severely restricts work on geochemical calibrations, ecological experiments, and genetic analyses.
- As if there were not enough other issues to tackle, molecular genetics seems to have shattered the very basis of foraminiferal proxies — morphologically defined species are hiding a manifold of cryptic genetic types that appear to have distinct ecologies (de Vargas et al., 1999; Darling et al., 2004) and introduce noise into environmental calibrations (Kucera & Darling, 2002). The nature, origin, and ecological significance of this cryptic diversity have to be clarified, so that proxies requiring species-specific calibration can be adjusted and applied appropriately.

WWW RESOURCES

http://www.emidas.org/
The Electronic Microfossil Image Database System (EMIDAS) offers access to digital images of planktonic foraminifera and includes an annotated key to their modern species.
http://www.maureenraymo.com/taxonomy.php
The taxonomic plates from Bé (1977), reprinted with permission from Elsevier.
http://www.nmnh.si.edu/paleo/foram/

The foraminiferal site of the Smithsonian Institution, including access to the largest collection of type specimens.

http://portal.chronos.org/

The Chronos project provides electronic access to stratigraphic distributions of species through the Neptune database and includes excellent taxonomic tools for identification of fossil planktonic foraminifera.

http://www.pangaea.de/Projects/MARGO/

The MARGO project houses the definitive collection of quantitative counts of planktonic foraminifera from core-tops and last glacial maximum samples. It includes the CLIMAP data set and the Brown University Foraminiferal database.

http://www.ngdc.noaa.gov/mgg/geology/hh1996.html

A useful graphical compilation of the relationships between extant planktonic foraminiferal species and surface-water properties in the Atlantic and Indian Oceans, using the CLIMAP data set.

http://www.cushmanfoundation.org/

The official site of the Cushman Foundation for Foraminiferal Research, which publishes the Journal of Foraminiferal Research.

http://www.tmsoc.org/

The official site of The Micropalaeontological Society, which promotes research on all microfossils, including foraminifera, and publishes the Journal of Micropalaeontology. The site contains an excellent collection of links to online resources.

REFERENCES

Adler, M., Hensen, C., Wenzhöfer, F., Pfeifer, K., & Schulz, H. D. (2001). Modeling of calcite dissolution by oxic respiration in supralysoclinal deep-sea sediments. *Marine Geology, 177*, 167–189.

Andersson, C. (1997). Transfer function vs. modern analog technique for estimating Pliocene sea-surface temperatures based on planktic foraminiferal data, western Equatorial Pacific Ocean. *Journal of Foraminiferal Research, 27*, 123–132.

Anderson, D. M., & Archer, D. (2002). Glacial-interglacial stability of ocean pH inferred from foraminifer dissolution rates. *Nature, 416*, 70–73.

Arrhenius, G. (1952). Sediment cores from the East Pacific. *Reports of the Swedish Deep-Sea Expedition, 1947–1948, 5*, 1–228.

Bandy, O. L. (1960). The geologic significance of coiling ratios in the foraminifer *Globigerina pachyderma* (Ehrenberg). *Journal of Paleontology, 34*, 671–681.

Barker, S., Cacho, I., Benway, H., & Tachikawa, K. (2005). Planktonic foraminiferal Mg/Ca as a proxy for past oceanic temperatures: A methodological overview and data compilation for the Last Glacial Maximum. *Quaternary Science Reviews, 24*, 821–834.

Barker, S., & Elderfield, H. (2002). Foraminiferal calcification response to glacial-interglacial changes in atmospheric CO_2. *Science, 297*, 833–836.

Bauch, D., Darling, K. F., Simstich, J., Bauch, H. A., Erlenkeuser, H., & Kroon, D. (2003). Palaeoceanographic implications of genetic variation in living North Atlantic *N. pachyderma*. *Nature, 424*, 299–302.

Bé, A. W. H. (1967). Foraminifera, families: Globigerinidae and Globorotaliidae, Fiche no. 108. In: J. H. Frasier (Ed.). Fiches d'intification du zooplancton. Charlottenlund, Denmark: Conseil International pour l'Exploration de la Mer.

Bé, A. W. H. (1968). Shell porosity of Recent planktonic foraminifera as a climatic index. *Science, 161*, 881–884.

Bé, A. W. H. (1977). An ecological, zoogeographic and taxonomic review of Recent planktonic foraminifera. In: A. T. S. Ramsey (Ed.), *Oceanic micropaleontology* (Vol. 1, pp. 1–100). London: Academic Press.

Bé, A. W. H., Harrison, S. M., & Lott, L. (1973). *Orbulina universa* d'Orbigny in the Indian Ocean. *Micropaleontology, 19,* 150–192.

Bé, A. W. H., Morsem J. W., & Harrison, S. M. (1975). Progressive dissolution and ultrastructural breakdown in planktonic foraminifera. Special Publication 13, *Cushman Foundation for Foraminiferal Research,* 27–55.

Bé, A. W. H., Spero, H. J., & Anderson, O. R. (1982). Effects of symbiont elimination and reinfection on the life processes of the planktonic foraminifer *Globigerinoides sacculifer*. *Marine Biology, 70,* 73–86.

Bé, A. W. H., & Tolderlund, D. S. (1971). Distribution and ecology of planktonic foraminifera. In: B. M. Funnell & W. R. Riedel (Eds), *The micropaleontology of oceans* (pp. 105–150). London: Cambridge University Press.

Bemis, B. E., Spero, H. J., Lea, D. W., & Bijma, J. (2000). Temperature influence on the carbon isotopic composition of *Globigerina bulloides* and *Orbulina universa* (planktonic foraminifera). *Marine Micropaleontology, 38,* 213–228.

Bentov, S., & Erez, J. (2005). Novel observations on biomineralization processes in foraminifera and implications for Mg/Ca ratio in the shells. *Geology, 33,* 841–844.

Berger, W. H. (1968). Planktonic foraminifera: Selective solution and paleoclimatic interpretation. *Deep-Sea Research, 15,* 31–43.

Berger, W. H. (1969). Ecologic patterns of living planktonic foraminifera. *Deep-Sea Research, 16,* 1–24.

Berger, W. H. (1970a). Planktonic foraminifera: Selective solution and the lysocline. *Marine Geology, 8,* 111–138.

Berger, W. H. (1970b). Planktonic foraminifera: Differential production and expatriation off Baja California. *Limnology and Oceanography, 15,* 183–204.

Berger, W. H. (1971). Sedimentation of planktic foraminifera. *Marine Geology, 11,* 325–358.

Berger, W. H., Bonneau, M.-C., & Parker, F. L. (1982). Foraminifera on the deep-sea floor: Lysocline and dissolution rate. *Oceanologia Acta, 5,* 249–258.

Berger, W. H., & Killingley, J. S. (1977). Glacial-Holocene transition in deep-sea carbonates: Selective dissolution and the stable isotope signal. *Science, 197,* 563–566.

Berger, W. H., & Piper, D. J. W. (1972). Planktic foraminifera: Differential settling, dissolution, and redeposition. *Limnology and Oceanography, 17,* 275–287.

Berggren, W. A., Kent, D. V., Swisher III, C. C., & Aubry, M. P. (1995). A revised Cenozoic geochronology and chronostratigraphy. *Society of Economic Paleontologists and Mineralogists Special Publication, 54,* 129–212.

Bijma, J., Erez, J., & Hemleben, C. (1990a). Lunar and semi-lunar reproductive cycles in some spinose planktonic foraminifers. *Journal of Foraminiferal Research, 20,* 117–127.

Bijma, J., Faber, W. W., & Hemleben, C. (1990b). Temperature and salinity limits for growth and survival of some planktonic foraminifers in laboratory cultures. *Journal of Foraminiferal Research, 20,* 95–116.

Bijma, J., Hemleben, C., Huber, B. T., Erlenkeuser, D., & Kroon, D. (1998). Experimental determination of the ontogenetic stable isotope variability in two morphotypes of *Globigerinella siphonifera* (d'Orbigny). *Marine Micropaleontology, 35,* 141–160.

Birks, H. J. B. (1995). Quantitative palaeoenvironmental reconstructions. In: D. Maddy & J. S. Brew (Eds), *Statistical modelling of Quaternary science data: Technical guide 5* (pp. 161–254). Cambridge: Quaternary Research Association.

Black, D. E., Peterson, L. C., Overpeck, J. T., Kaplan, A., Evans, M. N., & Kashgarian, M. (1999). Eight centuries of North Atlantic ocean atmosphere variability. *Science, 286,* 1709–1713.

Boltovskoy, E. (1973). Note on the determination of absolute surface water temperature by means of the foraminifer *Globigerina bulloides* d'Orbigny. *Paläontologische Zeitschrift, 47,* 152–155.

Bradshaw, J. S. (1959). Ecology of living foraminifera in the North and Equatorial Pacific Ocean. *Cushman Foundation for Foraminiferal Research: Contributions, 10,* 25–64.

Broecker, W. S., & Clark, E. (2001a). An evaluation of Lohmann's foraminifera weight dissolution index. *Paleoceanography, 16,* 531–534.

Broecker, W. S., & Clark, E. (2001b). Glacial-to-Holocene redistribution of carbonate ion in the deep sea. *Science, 294,* 2152–2154.

Brown, S. J., & Elderfield, H. (1996). Variations in Mg/Ca and Sr/Ca ratios of planktonic foraminifera caused by postdepositional dissolution: Evidence of shallow Mg-dependent dissolution. *Paleoceanography, 11,* 543–551.

Brummer, G.-J. A., Hemleben, C., & Spindler, M. (1986). Planktonic foraminiferal ontogeny and new perspectives for micropalaeontology. *Nature, 319,* 50–52.

Brummer, G.-J. A., & Kroon, D. (1988). *Planktonic foraminifers as tracers of ocean-climate history.* Amsterdam: Free University Press.

Caron, D. A., Faber, W. W. J., & Bé, A. W. H. (1987a). Effects of temperature and salinity on the growth and survival of the planktonic foraminifer *Globigerinoides sacculifer. Journal of the Marine Biological Association of the United Kingdom, 67,* 323–342.

Caron, D. A., Faber, W. W. J., & Bé, A. W. H. (1987b). Growth of the spinose planktonic foraminifer *Orbulina universa* in laboratory culture and the effect of temperature on life processes. *Journal of the Marine Biological Association of the United Kingdom, 67,* 343–358.

Carstens, J., Hebbeln, G., & Wefer, G. (1997). Distribution of planktic foraminifera at the ice margin in the Arctic (Fram Strait). *Marine Micropaleontology, 29,* 257–269.

Casford, J. S. L., Rohling, E. J., Abu-Zied, R. H., Jorissen, F. J., Leng, M., & Thomson, J. (2003). A dynamic concept for eastern Mediterranean circulation and oxygenation during sapropel formation. *Palaeogeography, Palaeoclimatology, Palaeoecology, 190,* 103–119.

CLIMAP Project Members (1976). The surface of the ice-age Earth. *Science, 191,* 1131–1137.

Conan, S. M.-H., Ivanova, E. M., & Brummer, G.-J. A. (2002). Quantifying carbonate dissolution and calibration of foraminiferal dissolution indices in the Somali Basin. *Marine Geology, 182,* 325–349.

Darling, K. F., Kroon, D., Wade, C. M., & Leigh Brown, A. J. (1996). Molecular phylogeny of the planktonic foraminifera. *Journal of Foraminiferal Research, 26,* 324–330.

Darling, K. F., Kucera, M., Kroon, D., & Wade, C. M. (2006). A resolution for the coiling direction paradox in *Neogloboquadrina pachyderma. Paleoceanography,* 21, PA2011, doi:10.1029/2005PA001189.

Darling, K. F., Kucera, M., Pudsey, C. J., & Wade, C. M. (2004). Molecular evidence links cryptic diversification in polar plankton to Quaternary climate dynamics. *Proceedings of the National Academy of Sciences, U.S.A., 101,* 7657–7662.

Darling, K. F., Kucera, M., Wade, C. M., von Langen, P., & Pak, D. (2003). Seasonal occurrence of genetic types of planktonic foraminiferal morphospecies in the Santa Barbara Channel. *Paleoceanography, 18,* 1032, doi:10.1029/2001PA000723.

Darling, K. F., Wade, C. M., Kroon, D., & Leigh Brown, A. J. (1997). Planktonic foraminiferal molecular evolution and their polyphyletic origins from benthic taxa. *Marine Micropaleontology, 30,* 251–266.

Darling, K. F., Wade, C. M., Kroon, D., Leigh Brown, A. J., & Bijma, J. (1999). The diversity and distribution of modern planktonic foraminiferal small subunit ribosomal RNA genotypes and their potential as tracers of present and past ocean circulations. *Paleoceanography, 14,* 3–12.

Darling, K. F., Wade, C. M., Steward, I. A., Kroon, D., Dingle, R., & Leigh Brown, A. J. (2000). Molecular evidence for genetic mixing of Arctic and Antarctic subpolar populations of planktonic foraminifers. *Nature, 405,* 43–47.

Dieckmann, G. S., Spindler, M., Lange, M. A., Ackley, S. F., & Eicken, H. (1991). Antarctic sea ice: A habitat for the foraminifer *Neogloboquadrina pachyderma. Journal of Foraminiferal Research, 21,* 182–189.

Dittert, N., Baumann, K.-H., Bickert, T., Henrich, R., Huber, R., Kinkel, H., & Meggers, H. (1999). Carbonate dissolution in the deep-sea: Methods, quantification and paleoceanographic application. In: G. Fischer & G. Wefer (Eds), *Use of proxies in paleoceanography: Examples from the South Atlantic* (pp. 255–284). Berlin: Springer-Verlag.

Dittert, N., & Henrich, R. (2000). Carbonate dissolution in the South Atlantic Ocean: Evidence from ultrastructure breakdown in *Globigerina bulloides. Deep-Sea Research, 47,* 603–620.

Duprat, J. (1983). Les foraminiferes planctoniques du Quaternaire terminal d'un domaine pericontinental (Golfe de Gascogne, cotes ouest-iberiques, mer d'Alboran): Ecologie-biostratigraphie. *Bulletin de l'Institut de Geologie du Bassin d'Aquitaine, 33,* 71–150.

Eggins, S., DeDeckker, P., & Marshall, A. T. (2003). Mg/Ca variation in planktonic foraminfera tests: Implications for reconstructing palaeo-seawater temperature and habitat migration. *Earth and Planetary Science Letters, 212*, 291–306.

Eggins, S. M., Sadekov, A., & De Deckker, P. (2004). Modulation and daily banding of Mg/Ca in *Orbulina universa* tests by symbiont photosynthesis and respiration: A complication for seawater thermometry?. *Earth and Planetary Science Letters, 225*, 411–419.

Emiliani, C. (1954a). Depth habitats of some species of pelagic foraminifera as indicated by oxygen isotopes ratios. *American Journal of Science, 252*, 149–158.

Emiliani, C. (1954b). Temperatures of Pacific bottom waters and polar superficial waters during the Tertiary. *Science, 119*, 853–855.

Erez, J. (2003). The source of ions for biomineralization in foraminifera and their implications for paleoceanographic proxies. *Reviews in Mineralogy and Geochemistry, 54*, 115–149.

Ericson, D. B. (1959). Coiling direction of *Globigerina pachyderma* as a climatic index. *Science, 130*, 219–220.

Ericson, D. B., Wollin, G., & Wollin, J. (1954). Coiling direction of *Globorotalia truncatulinoides* in deep-sea cores. *Deep Sea Research, 2*, 152–158.

Fatela, F., & Taborda, R. (2002). Confidence limits of species proportions in microfossil assemblages. *Marine Micropaleontology, 45*, 169–174.

Fenton, M., Geiselhart, S., Rohling, E. J., & Hemleben, Ch. (2000). Aplanktonic zones in the Red Sea. *Marine Micropaleontology, 40*, 277–294.

Field, D. B., Baumgartner, T. R., Charles, C. D., Ferreira-Bartrina, V., & Ohman, M. D. (2006). Planktonic foraminifera of the California Current reflect 20th-century warming. *Science, 311*, 63–66.

Fok-Pun, L., & Komar, P. D. (1983). Settling velocities of planktonic foraminifera: Density variations and shape effects. *Journal of Foraminiferal Research, 13*, 60–68.

Frerichs, W. E., Heiman, M. E., Borgman, L. E., & Bé, A. W. H. (1972). Latitudal variations in planktonic foraminiferal test porosity. Part 1: Optical studies. *Journal of Foraminiferal Research, 2*, 6–13.

Furbish, D. J., & Arnold, A. J. (1997). Hydrodynamic strategies in the morphological evolution of spinose planktonic foraminifera. *Geological Society of America Bulletin, 109*, 1055–1072.

Giraudeau, J. (1993). Planktonic foraminiferal assemblages in surface sediments from the southwest African continental margin. *Marine Geology, 110*, 47–62.

Gupta, A. K., Anderson, D. M., & Overpeck, J. T. (2003). Abrupt changes in the Asian southwest monsoon during the Holocene and their links to the North Atlantic Ocean. *Nature, 421*, 354–357.

Hayes, A., Kucera, M., Kallel, N., Sbaffi, L., & Rohling, E. J. (2005). Glacial Mediterranean sea surface temperatures based on planktonic foraminiferal assemblages. *Quaternary Science Reviews, 24*, 999–1016.

Hays, J. D., Imbrie, J., & Shackleton, N. J. (1976). Variations in the Earth's orbit: Pace maker of the ice ages. *Science, 194*, 1121–1132.

Healy-Williams, N., Ehrlich, R., & Williams, D. F. (1985). Morphometric and stable isotopic evidence for subpopulations of Globorotalia truncatulinoides. *Journal of Foraminiferal Research, 15*, 242–253.

Hecht, A. D. (1974). Intraspecific variation in Recent populations of *Globigerinoides ruber* and *Globigerinoides trilobus* and their application to paleoenvironmental analysis. *Journal of Palaeontology, 48*, 1217–1234.

Hecht, A. D. (1976). An ecologic model for test size variation recent planktonic foraminifera: Application to the fossil record. *Journal of Foraminiferal Research, 6*, 295–311.

Hecht, A. D., Bé, A. W., & Lott, L. (1976). Ecologic and paleoclimatic implications of morphologic variation of *Orbulina universa* in the Indian Ocean. *Science, 194*, 422–424.

Hecht, A. D., & Savin, S. M. (1972). Phenotypic variation and oxygen isotope ratios in Recent planktonic foraminifera. *Journal of Foraminiferal Research, 2*, 55–67.

Hedley, R. G., & Adams, C. G. (Eds) (1974). *Foraminifera* (Vol. 1, p. 276). London: Academic Press.

Hedley, R. G., & Adams, C. G. (Eds) (1976). *Foraminifera* (Vol. 2, p. 265). London: Academic Press.

Hemleben, C., Spindler, M., & Anderson, O. R. (1989). *Modern planktonic foraminifera* (p. 363). New York: Springer.

Hemleben, Ch., Spindler, M., Breitinger, I., & Ott, R. (1987). Morphological and physiological responses of *Globigerinoides sacculifer* (Brady) under varying laboratory conditions. *Marine Micropaleontology*, 12, 305–324.

Henderson, G. M. (2002). New oceanic proxies for paleoclimate. *Earth and Planetary Science Letters*, 203, 1–13.

Hendy, I. L., & Kennett, J. P. (2000). Dansgaard-Oeschger cycles and the California Current System: Planktonic foraminiferal response to rapid climate change in Santa Barbara Basin, Ocean Drilling Program hole 893A. *Paleoceanography*, 15, 30–42.

Hilbrecht, H., & Thierstein, H. R. (1996). Benthic behavior of planktic foraminifera. *Geology*, 4, 200–202.

Hillaire-Marcel, C., De Vernal, A., Polyak, L., & Darby, D. (2004). Size-dependent isotopic composition of planktic foraminifers from Chukchi Sea vs. NW Atlantic sediments: Implications for the Holocene paleoceanography of the western Arctic. *Quaternary Science Reviews*, 23, 245–260.

Huber, B. T., Bijma, J., & Darling, K. (1997). Cryptic speciation in the living planktonic foraminifer *Globigerinella siphonifera* (d'Orbigny). *Paleobiology*, 23, 33–62.

Hutson, W. H. (1977). Transfer functions under no-analog conditions: Experiments with Indian Ocean planktonic foraminifera. *Quaternary Research*, 8, 355–367.

Imbrie, J., & Kipp, N. (1971). A new micropaleontological method for quantitative paleoclimatology: Application to a late Pleistocene Caribbean core. In: K. K. Turekian (Ed.), *The Late Cenozoic glacial ages* (pp. 71–181). New Haven, Connecticut: Yale University Press.

Ingle, J. C., Jr. (1977). Late Neogene marine events and the Pliocene-Pleistocene boundary in the marginal North Pacific. *Giornale di Geologia*, 44, 359–374.

Ivanova, E., Schiebel, R., Singh, A. D., Schmiedl, G., Niebler, H.-S., & Hemleben, C. (2003). Primary production in the Arabian Sea during the last 135,000 years. *Palaeogeography, Palaeoclimatology, Palaeoecology*, 197, 61–82.

Kandiano, E. S., & Bauch, H. A. (2002). Implications of planktic foraminiferal size fractions for the glacial-interglacial paleoceanography of the polar North Atlantic. *Journal of Foraminiferal Research*, 32, 245–251.

Kawahata, H., Nishimura, A., & Gagan, M. K. (2002). Seasonal change in foraminiferal production in the western equatorial Pacific warm pool: Evidence from sediment trap experiments. *Deep-Sea Research II*, 49, 2783–2800.

Kennett, J. P. (1968a). Latitudinal variation in *Globigerina pachyderma* (Ehrenberg) in surface sediments of the southwest Pacific Ocean. *Micropaleontology*, 14, 305–318.

Kennett, J. P. (1968b). *Globorotalia truncatulinoides* as a paleo-oceanographic index. *Science*, 159, 1461–1463.

Kennett, J. P. (1970). Pleistocene paleoclimates and foraminiferal biostratigraphy in Subantarctic deep-sea cores. *Deep-Sea Research*, 17, 125–140.

Kennett, J. P. (1976). Phenotypic variation in some Recent and late Cenozoic planktonic foraminifera. In: R. H. Hedley & C. G. Adams (Eds), *Foraminifera* (vol. 2, pp. 111–170). New York: Academic Press.

Kucera, M. (1998). Biochronology of the mid-Pliocene *Sphaeroidinella* event. *Marine Micropaleontology*, 35, 1–16.

Kucera, M., & Darling, K. F. (2002). Cryptic species of planktonic foraminifera: Their effect on palaeoceanographic reconstructions. *Philosophical Transaction of the Royal Society of London Series A*, 360, 695–718.

Kucera, M., & Kennett, J. P. (2002). Causes and consequences of a middle Pleistocene origin of the modern planktonic foraminifer *Neogloboquadrina pachyderma* sinistral. *Geology*, 30, 539–542.

Kucera, M., Weinelt, M., Kiefer, T., Pflaumann, U., Hayes, A., Weinelt, M., Chen, M.-T., Mix, A. C., Barrows, T. T., Cortijo, E., Duprat, J., Juggins, S., & Waelbroeck, C. (2005). Reconstruction of sea-surface temperatures from assemblages of planktonic foraminifera: Multi-technique approach based on geographically constrained calibration datasets and its application to glacial Atlantic and Pacific Oceans. *Quaternary Science Reviews*, 24, 951–998.

Le, J., & Shackleton, N. J. (1992). Carbonate dissolution fluctuations in the western Equatorial Pacific during the late Quaternary. *Paleoceanography*, 7, 21–42.

Lea, D. W. (1999). Trace elements in foraminiferal calcite. In: B. K. Sen Gupta (Ed.), *Modern foraminifera* (pp. 259–280). Dordrecht: Kluwer.

Lipps, J. H., & Valentine, J. W. (1970). The role of foraminifera in the trophic structure of marine communities. *Lethaia, 3,* 279–286.

Lohmann, G. P. (1995). A model for variation in the chemistry of planktonic foraminifera due to secondary calcification and selective dissolution. *Paleoceanography, 10,* 445–457.

Lohmann, G. P., & Malmgren, B. A. (1983). Equatorward migration of *Globorotalia truncatulinoides* ecophenotypes through the Late Pleistocene: Gradual evolution or ocean change? *Paleobiology, 9,* 414–421.

Lončarić, N., Brummer, G.-J. A., & Kroon, D. (2005). Lunar cycles and seasonal variations in deposition fluxes of planktic foraminiferal shell carbonate to the deep South Atlantic (central Walvis Ridge). *Deep-Sea Research I, 52,* 1178–1188.

Malmgren, B. A., & Kennett, J. P. (1972). Biometric analysis of phenotypic variation: *Globigerina pachyderma* (Ehrenberg) in the South Pacific Ocean. *Micropaleontology, 18,* 241–248.

Malmgren, B. A., & Kennett, J. P. (1976). Biometric analysis of phenotypic variation in recent *Globigerina bulloides* d'Orbigny in the Southern Indian Ocean. *Marine Micropaleontology, 1,* 3–25.

Malmgren, B., & Kennett, J. P. (1978a). Test size variation in *Globigerina bulloides* in response to Quaternary palaeoceanographic changes. *Nature, 275,* 123–124.

Malmgren, B. A., & Kennett, J. P. (1978b). Late Quaternary paleoclimatic applications of mean size variations in *Globigerina bulloides* d'Orbigny in the Southern Indian Ocean. *Journal of Paleontology, 52,* 1195–1207.

Morey, A. E., Mix, A. C., & Pisias, N. G. (2005). Planktonic foraminiferal assemblages preserved in surface sediments correspond to multiple environmental variables. *Quaternary Science Reviews, 24,* 925–950.

Murray, J. (1897). On the distribution of the pelagic foraminifera at the surface and on the floor of the ocean. *Natural Science (Ecology), 11,* 17–27.

Naidu, P. D., & Malmgren, B. A. (1995). Monsoon upwelling effects on test size of some planktonic foraminifera species from the Oman Margin, Arabian Sea. *Paleoceanography, 10,* 117–122.

Naidu, P. D., & Malmgren, B. A. (1996). Relationship between late Quaternary upwelling history and coiling properties of *Neogloboquadrina pachyderma* and *Globigerina bulloides* in the Arabian Sea. *Journal of Foraminiferal Research, 26,* 64–70.

Niebler, H. S., & Gersonde, R. (1998). A planktic foraminiferal transfer function for the southern South Atlantic Ocean. *Marine Micropaleontology, 34,* 213–214.

Norris, R. D. (1996). Symbiosis as an evolutionary innovation in the radiation of Paleocene planktic foraminifera. *Paleobiology, 22,* 461–480.

Norris, R. D. (2000). Pelagic species diversity, biogeography, and evolution. *Paleobiology, 26*(Suppl.), 236–258.

Norris, R. D., Corfield, R. M., & Cartlidge, J. (1996). What is gradualism? Cryptic speciation in globorotalid foraminifera. *Paleobiology, 22,* 386–405.

Ortiz, J. D., Mix, A. C., & Collier, R. W. (1995). Environmental control of living symbiotic and asymbiotic foraminifera of the California Current. *Paleoceanography, 10,* 987–1009.

Parker, F. L. (1962). Planktonic foraminifera species in Pacific sediments. *Micropaleontology, 8,* 219–254.

Peeters, F. J. C., Acheson, R., Brummer, G.-J. A., de Ruijter, W. P. M., Schneider, R. R., Ganssen, G. M., Ufkes, E., & Kroon, D. (2004). Vigorous exchange between the Indian and Atlantic Ocean at the end of the past five glacial periods. *Nature, 430,* 661–665.

Peeters, F., Ivanova, E., Conan, S., Brummer, G.-J., Ganssen, G., Troelstra, S., & van Hinte, J. (1999). A size analysis of planktic foraminifera from the Arabian Sea. *Marine Micropalaeontology, 36,* 31–36.

Peterson, L. C., & Prell, W. L. (1985). Carbonate dissolution in recent sediments of the Eastern Equatorial Indian Ocean: Preservation patterns and carbonate loss above the lysocline. *Marine Geology, 64,* 259–290.

Pflaumann, U., Duprat, J., Pujol, C., & Labeyrie, L. (1996). SIMMAX: A modern analog technique to deduce Atlantic sea surface temperatures from planktonic foraminifera in deep-sea sediments. *Paleoceanography, 11,* 15–35.

Pfleger, F. B. (1948). Foraminifera of a submarine core from the Caribbean Sea. *Göteborgs Kungliga Vetenskaps-och Vitterhets-samhälles Handlingar, 6B*, 5, 3–9.

Pujol, C. (1980). Les foraminifères planctoniques de l'Atlantique Nord au Quaternaire: Ecologie–stratigraphie–environnement. *Memoires de l'Institut de Geologie du Bassin d'Aquitaine, 10*, 1–254.

Renaud, S., & Schmidt, D. N. (2003). Habitat tracking as a response of the planktic foraminifer *Globorotalia truncatulinoides* to environmental fluctuations during the last 140 kyr. *Marine Micropaleontology, 49*, 97–122.

Rohling, E. J., & Cook, S. (1999). Stable oxygen and carbon isotopes in foraminiferal carbonate shells. In: B. K. Sen Gupta (Ed.), *Modern foraminifera* (pp. 239–258). Dordrecht: Kluwer.

Rohling, E. J., Mayewski, P. A., Hayes, A., Abu-Zied, R. H., & Casford, J. S. L. (2002). Holocene atmosphere–ocean interactions: Records from Greenland and the Aegean Sea. *Climate Dynamics, 18*, 587–593.

Rohling, E. J., Sprovieri, M., Cane, T. R., Casford, J. S. L., Cooke, S., Bouloubassi, I., Emeis, K. C., Schiebel, R., Rogerson, M., Hayes, A., Jorissen, F. J., & Kroon, D. (2004). Reconstructing past planktic foraminiferal habitats using stable isotope data: A case history for Mediterranean sapropel S5. *Marine Micropaleontology, 50*, 89–123.

Ruddiman, W. F., & Heezen, B. C. (1967). Differential solution of planktic foraminifera. *Deep-Sea Research, 14*, 801–808.

Rutherford, S., D'Hondt, S., & Prell, W. (1999). Environmental controls on the geographic distribution of zooplankton diversity. *Nature, 400*, 749–753.

Saito, T. (1976). Geologic significance of coiling direction in the planktonic foraminifera *Pulleniatina*. *Geology, 4*, 305–309.

Saito, T., Thompson, R. R., & Breger, D. (1981). *Systematic index of Recent and Pleistocene planktonic foraminifera* (pp. 1–190). Tokyo: University of Tokyo Press.

Schiebel, R. (2002). Planktic foraminiferal sedimentation and the marine calcite budget. *Global Biogeochemical Cycles, 16*, 1065, doi:10.1029/2001GB001459.

Schiebel, R., Bijma, J., & Hemleben, C. (1997). Population dynamics of the planktic foraminifer *Globigerina bulloides* from the eastern North Atlantic. *Deep-Sea Research Part II, 44*, 1701–1713.

Schiebel, R., & Hemleben, C. (2005). Extant planktic foraminifera: A brief review. *Paläontogische Zeitschrift, 79*, 135–148.

Schiebel, R., Hiller, B., & Hemleben, C. (1995). Impacts of storms on Recent planktic foraminiferal test production and $CaCO_3$ flux in the North Atlantic at 47 degrees N, 20 degrees W (JGOFS). *Marine Micropaleontology, 26*, 115–129.

Schiebel, R., Schmuker, B., Alves, M., & Hemleben, C. (2002). Tracking the Recent and late Pleistocene Azores front by the distribution of planktic foraminifers. *Journal of Marine Systems, 37*, 213–227.

Schiebel, R., Waniek, J., Bork, M., & Hemleben, C. (2001). Planktic foraminiferal production stimulated by chlorophyll redistribution and entrainment of nutrients. *Deep-Sea Research Part I, 48*, 721–740.

Schmidt, D. N., Renaud, S., & Bollmann, J. (2003). Response of planktic foraminiferal size to late Quaternary climate change. *Paleoceanography, 18*, 10.1029/2002PA000831.

Schmidt, D. N., Renaud, S., Bollmann, J., Schiebel, R., & Thierstein, H. R. (2004a). Size distribution of Holocene planktic foraminifer assemblages: Biogeography, ecology and adaptation. *Marine Micropaleontology, 50*, 319–338.

Schmidt, D. N., Thierstein, H. R., Bollmann, J., & Schiebel, R. (2004b). Abiotic forcing of plankton evolution in the Cenozoic. *Science, 303*, 207–210.

Schmuker, B. (2000). The influence of shelf vicinity on the distribution of planktic foraminifera south of Puerto Rico. *Marine Geology, 166*, 125–143.

Schott, W. (1935). Die Foraminiferen aus dem aequatorialen Teil des Atlantischen Ozeans. *Deutsch Atlanta Expeditie "Meteor" 1925–1927, 3*, 34–134.

Scott, G. H. (1974). Biometry of the foraminiferal shell. In: R. H. Hedley & C. G. Adams (Eds), *Foraminifera* (Vol. 1, pp. 55–151). New York: Academic Press.

Shackleton, N. J., & Opdyke, N. D. (1973). Oxygen isotope and paleomagnetic stratigraphy of equatorial Pacific core V28–238: Oxygen isotope temperatures and ice volumes on a 10^5 year and 10^6 year scale. *Quaternary Research, 3*, 39–55.

Siddall, M., Rohling, E. J., Almogi-Labin, A., Hemleben, Ch., Meischner, D., Schmelzer, I., & Smeed, A. (2003). Sea-level fluctuations during the last glacial cycle. *Nature, 423,* 853–858.

Simstich, J., Sarnthein, M., & Erlenkeuser, H. (2003). Paired $\delta^{18}O$ signals of *Neogloboquadrina pachyderma* (s) and *Turborotalita quinqueloba* show thermal stratification structure in Nordic Seas. *Marine Micropaleontology, 48,* 107–125.

Sliter, W. V., Bé, A. W. H., & Berger, W. H. (Eds). (1975). Dissolution of deep sea carbonates. Special Publication 13. *Cushman Foundation for Foraminiferal Research,* 159 pp.

Spero, H. J., Bijma, J., Lea, D. W., & Bemis, B. E. (1997). Effect of seawater carbonate concentration on foraminiferal carbon and oxygen isotopes. *Nature, 390,* 497–500.

Spero, H. J., & Deniro, M. J. (1987). The influence of symbiont photosynthesis on the $\delta^{18}O$ and $\delta^{13}C$ values of planktonic foraminiferal shell calcite. *Symbiosis, 4,* 213–228.

Spindler, M. (1996). On the salinity tolerance of the planktonic foraminifera *Neogloboquadrina pachyderma* from Antarctic sea ice. *Proceedings of the NIPR Symposium on Polar Biology, 9,* 85–91.

Spindler, M., & Diekmann, G. S. (1986). Distribution and abundance of the planktic foraminifer *Neogloboquadrina pachyderma* in Sea Ice of the Weddell Sea (Antarctica). *Polar Biology, 5,* 186–191.

Spindler, M., Hemleben, C., Bayer, U., Bé, A. W. H., & Anderson, O. R. (1979). Lunar periodicity of reproduction in the planktonic foraminifer *Hastigerina pelagica. Marine Ecology Progress Series, 1,* 61–64.

Srinivasan, M. S., & Kennett, J. P. (1974). Secondary calcification of the planktonic foraminifer *Neogloboquadrina pachyderma* as a climatic index. *Science, 186,* 630–631.

Steinke, S., Chiu, H.-Y., Yu, P.-S., Shen, C.-C., Löwemark, L., Mii, H.-S., & Chen, M.-T. (2005). Mg/Ca ratios of two *Globigerinoides ruber* (white) morphotypes: Implications for reconstructing past tropical/subtropical surface water conditions. *Geochemistry, Geophysics and Geosystems, 6,* Q11005, doi:10.1029/2005GC000926.

Takahashi, K., & Bé, A. W. H. (1984). Planktonic foraminifera: Factors controlling sinking speeds. *Deep Sea Research, 31,* 1477–1500.

Thiede, J. (1971). Variations in coiling ratios of Holocene planktonic foraminifera. *Deep-Sea Research, 18,* 823–831.

Thiede, J. (1975). Distribution of foraminifera in coastal waters of an upwelling area. *Nature, 253,* 712–714.

Thunell, R. C. (1976). Optimum indices of calcium carbonate dissolution in deep-sea sediments. *Geology, 4,* 525–528.

Thunell, R. C., & Honjo, S. (1981). Calcite dissolution and the modification of planktic foraminiferal assemblages. *Marine Micropaleontology, 6,* 169–182.

Ufkes, E., Jansen, J. H. F., & Brummer, G. J. (1998). Living planktonic foraminifera in the eastern South Atlantic during spring: Indicators of water masses, upwelling and Congo (Zaire) River plume. *Marine Micropaleontology, 33,* 27–53.

de Vargas, C., Bonzon, M., Rees, N., Pawlowski, J., & Zaninetti, L. (2002). A molecular approach to biodiversity and ecology in the planktonic foraminifera *Globigerinella siphonifera* (d'Orbigny). *Marine Micropalaeontology, 45,* 101–116.

de Vargas, C., Norris, R., Zaninetti, L., Gibb, S. W., & Pawlovski, J. (1999). Molecular evidence of cryptic speciation in planktonic foraminifers and their relation to oceanic provinces. *Proceedings of the National Academy of Sciences, USA, 96,* 2864–2868.

de Vargas, C., Renaud, S., Hilbrecht, H., & Pawlovski, J. (2001). Pleistocene adaptive radiation in *Globorotalia truncatulinoides*: Genetic, morphologic, and environmental evidence. *Paleobiology, 27,* 104–125.

de Vargas, C., Sáez, A. G., Medlin, L. K., & Thierstein, H. R. (2004). Super-species in the calcareous plankton. In: H. R. Thierstein & J. R. Young (Eds), *Coccolithophores: From molecular processes to global impact.* Berlin: Springer-Verlag.

de Vargas, C., Zaninetti, L., Hilbrecht, H., & Pawlovski, J. (1997). Phylogeny and rates of molecular evolution of planktonic foraminifera: SSU rDNA sequences compared to the fossil record. *Journal of Molecular Evolution, 45,* 285–294.

de Villiers, S. (2004). Occupation of an ecological niche as the fundamental control on the shell-weight of calcifying planktonic foraminifera. *Marine Biology, 144,* 45–50.

de Villiers, S. (2005). Foraminiferal shell-weight evidence for sedimentary calcite dissolution above the lysocline. *Deep-Sea Research I, 52*, 671–680.

Vincent, E., & Berger W. H. (1981). Planktonic foraminifera and their use in Paleoceanography. In: C. Emiliani (Ed.), *The oceanic lithosphere. The sea* (Vol. 7, pp. 1025–1119). Hoboken, N. J.: Wiley-Interscience.

Wang, L. (2000). Isotopic signals in two morphotypes of *Globigerinoides ruber* (white) from the South China Sea: Implications for monsoon climate change during the last glacial cycle. *Palaeogeography, Palaeoclimatology, Palaeoecology, 161*, 381–394.

Weyl, P. K. (1978). Micropaleontology and ocean surface climate. *Science, 202*, 475–481.

Wiles, W. W. (1967). Pleistocene changes in the pore concentration of a planktonic foraminiferal species from the Pacific Ocean. *Progress in Oceanography, 4*, 53–160.

Zaragosi, S., Auffret, G. A., Faugeres, J.-C., Garlan, T., Pujol, C., & Cortijo, E. (2000). Physiography and recent sediment distribution of the Celtic Deep-Sea Fan, Bay of Biscay. *Marine Geology, 169*, 207–237.

Zaric, S., Donner, B., Fischer, G., Mulitza, S., & Wefer, G. (2005). Sensitivity of planktic foraminifera to sea surface temperature and export production as derived from sediment trap data. *Marine Micropaleontology, 55*, 75–105.

CHAPTER SEVEN

Paleoceanographical Proxies Based on Deep-Sea Benthic Foraminiferal Assemblage Characteristics

Frans J. Jorissen*, Christophe Fontanier *and* Ellen Thomas

Contents

1. Introduction	263
1.1. General introduction	263
1.2. Historical overview of the use of benthic foraminiferal assemblages	266
1.3. Recent advances in benthic foraminiferal ecology	267
2. Benthic Foraminiferal Proxies: A State of the Art	271
2.1. Overview of proxy methods based on benthic foraminiferal assemblage data	271
2.2. Proxies of bottom water oxygenation	273
2.3. Paleoproductivity proxies	285
2.4. The water mass concept	301
2.5. Benthic foraminiferal faunas as indicators of current velocity	304
3. Conclusions	306
Acknowledgements	308
4. Appendix 1	308
References	313

1. Introduction

1.1. General Introduction

The most popular proxies based on microfossil assemblage data produce a quantitative estimate of a physico-chemical target parameter, usually by applying a transfer function, calibrated on the basis of a large dataset of recent or core-top samples. Examples are planktonic foraminiferal estimates of sea surface temperature (Imbrie & Kipp, 1971) and reconstructions of sea ice coverage based on radiolarian (Lozano & Hays, 1976) or diatom assemblages (Crosta, Pichon, & Burckle, 1998). These

---

* Corresponding author.

Developments in Marine Geology, Volume 1
ISSN 1572-5480, DOI 10.1016/S1572-5480(07)01012-3

methods are easy to use, apply empirical relationships that do not require a precise knowledge of the ecology of the organisms, and produce quantitative estimates that can be directly applied to reconstruct paleo-environments, and to test and tune global climate models.

Assemblage-based proxy methods that do not yield fully quantitative results have become less popular over the last decennia, mainly because semi-quantitative or qualitative proxy methods implicitly admit a considerable degree of uncertainty. In many allegedly quantitative proxies, however, the error may be as large, but is concealed by the numerical aspect of the estimates, commonly expressed with of 1 or 2 numbers behind the decimal point, suggesting a highly precise and trustworthy reconstruction of the target parameter.

The decreased popularity of assemblage-based proxy methods affects proxies based on benthic foraminifera, although they have been used to reconstruct a wide range of oceanographic parameters, including water depth, water mass properties, bottom water oxygen content, and the extent and/or seasonality of the organic flux to the ocean floor. Although the use of benthic foraminiferal assemblage compositions has become less common, the use of benthic foraminiferal tests as carriers of geochemical proxy methods (stable isotopes, Mg/Ca, Sr/Ca, etc.) has never been so widespread (e.g., Wefer, Berger, Bijma, & Fischer, 1999; Lea, 2004; Lynch-Stieglitz, 2004; Ravizza & Zachos, 2004; Sigman & Haug, 2004).

All geochemical proxies based on the remains of micro-organisms rely on a thorough knowledge of their ecology that determines when, where, and exactly under which conditions the proxy value is fixed in the microfossil test. In the case of benthic foraminifera, their strongly increased application in geochemical studies has led to a renewed interest in ecological studies in field situations, but also under controlled laboratory conditions (e.g., Heinz, Schmiedl, Kitazato, & Hemleben, 2001; Ernst & Van der Zwaan, 2004; Geslin, Heinz, Hemleben, & Jorissen, 2004). Due to these studies and to studies affiliated with the Joint Global Ocean Flux Studies (JGOFS), benthic foraminiferal ecology is much better known than 20 years ago, and it has become clear that the composition of benthic foraminiferal faunas is controlled by a limited number of closely interrelated environmental parameters.

In spite of the significant advances in knowledge of foraminiferal ecology, only hesitant attempts to develop or improve proxy methods based on benthic foraminiferal assemblage composition have been undertaken in recent years. The further elaboration of these proxy methods is hampered by a number of problems, most of which are not unique to benthic foraminifera, and concern most commonly used proxy methods:

1) Many of the controlling parameters that the proxy methods aim to reconstruct are strongly interdependent in the present oceans, making the reconstruction of individual parameters particularly difficult.

2) Recent ecosystems cover only a fraction of the environmental conditions encountered in past oceans. The successful reproduction of non-analog conditions is a critical point for all proxy methods, particularly because usually we cannot test the validity of the proxy estimates.

3) Laboratory experiments may remedy these two problems by observing faunal responses to changes in single environmental parameters, and to environmental conditions not encountered in recent ecosystems (e.g., Chandler, Williams, Spero, & Xiaodong, 1996; Wilson-Finelli, Chandler, & Spero, 1998; Toyofuku & Kitazato, 2005). Unfortunately, experiments with benthic foraminiferal faunas from open ocean environments are complicated (e.g., Hintz et al., 2004). It appears very difficult to artificially create environments in which deep-sea foraminiferal species feed, grow new chambers, and reproduce. Progress in this promising field of research requires important investments and is consequently rather slow.

4) Our knowledge of taphonomical processes is insufficient. Taphonomical processes are responsible for important compositional differences between living faunas, the subject of ecological studies, and fossil faunas as used by proxy methods to reconstruct past environmental parameters. As for many other proxies, a major effort is needed to study the transformations and losses taking place during the transition from a living to a sub-recent (core-top) fauna and finally to a fossil fauna.

5) A successful proxy method requires calibration with a very large dataset, encompassing as many different ecosystems and ecological settings as possible. The best way to develop such a dataset is by cooperation of a large group of scientists, adopting the same methodology and putting their data in a common database.

Wefer et al. (1999) give an overview of all commonly used paleoceanographical proxies, and their relatively limited attention to benthic foraminiferal assemblage studies reflects their decreased popularity. Following the important progress in foraminiferal ecology in recent years, several papers (e.g., Gooday, 1994; Jorissen, 1999a; Van der Zwaan, Duijnstee, Den Dulk, Ernst, & Kouwenhoven, 1999; Murray, 2001; Smart, 2002) investigated the possibility to apply these new findings to paleoceanography. In a very thorough review, Gooday (2003) presents new ecological concepts, and gives an updated overview of the methods using benthic foraminiferal assemblage studies for paleoceanographical reconstructions. The use of benthic foraminiferal tests as carriers of geochemical proxies is reviewed in several publications (e.g., Lea, 2004; Lynch-Stieglitz, 2004; Ravizza & Zachos, 2004; Sigman & Haug, 2004; papers in Elderfield, 2004).

In the present chapter on paleoceanographical proxies based on deep-sea benthic foraminiferal assemblage characteristics we do not aim to present a complete treatment of all existing proxy methods, for which we strongly recommend Gooday (2003). Instead, we will give a rather personal view of the three proxy relationships that in our opinion are most promising: those between benthic foraminiferal faunas and benthic ecosystem oxygenation, export productivity and deep-sea water mass characteristics. For each, we will concentrate on the advantages and shortcomings of the existing methods, and try to indicate possible pathways for future improvements that may lead to a better application of this potentially useful group of micro-organisms. As a consequence of this approach, we will highlight the many problems encountered when developing proxies based on benthic foraminiferal assemblage data, but want to stress that most of these problems are not unique to proxies based on faunal data. We will limit our treatment to the reconstruction of environmental parameters in open marine ecosystems, and will not consider more coastal ecosystems, such as estuaries, inner continental shelf areas, or reef ecosystems.

1.2. Historical Overview of the Use of Benthic Foraminiferal Assemblages

The practical application of benthic foraminifera started in the beginning of the 20th century, when Joseph Cushman and his co-workers developed their use as biostratigraphic markers, providing age control (Cushman, 1928), and leading to their wide-spread application in oil exploration. Natland (1933) first used benthic foraminifera in a strictly paleo-environmental way to determine the depth of deposition of sediments of the Ventura Basin. Sediments originally interpreted as outer shelf deposits later turned out to be turbidity sequences (Natland & Kuenen, 1951), confirming Natland's original paleobathymetric estimates. For the next 40 years, benthic foraminifera were used extensively to determine paleobathymetry. At first, efforts concentrated on the recognition of isobathyal species, supposed to have the same depth distribution in all oceans (e.g., Bandy, 1953a, 1953b; Bandy & Arnal, 1957; Bandy & Echols, 1964; Bandy & Chierici, 1966; Pujos-Lamy, 1973). In more coastal areas, the ratio between hyaline, porcellaneous and agglutinated taxa was used as an indicator of water depth and salinity (e.g., Bandy & Arnal, 1957; Sliter & Baker, 1972; Murray, 1973; Greiner, 1974). The ratio of planktonic and benthic foraminifera was also proposed as a paleobathymetrical proxy (Grimsdale & Van Morkhoven, 1955).

In this early period multivariate statistical methods were not yet widely available, and only relations between individual species or species groups and single environmental parameters could be analysed, leading to major oversimplification of the complex natural situation in which a host of environmental factors interact and control foraminiferal ecology. Pflum and Frerichs (1976) noted that some "delta-depressed" species in the Gulf of Mexico had a lower upper depth limit in front of the Mississippi delta, than in other areas, whereas others, "delta-elevated" taxa, showed the opposite tendency. They suggested that these differences in bathymetrical range were caused by specific redox conditions, resulting from the input of organic matter by river outflow (Pflum & Frerichs, 1976), and were probably first to recognise the influence of organic input on the bathymetrical range of benthic foraminiferal species.

The early 1970s saw the advent of the water mass concept: different deep-sea water masses are characterised by a specific combination of temperature, salinity and pH, and were supposed to be inhabited by highly characteristic faunas (e.g., Streeter, 1973; Schrader, 1974; Lohmann, 1978a). On the basis of this hypothesis it should have become possible to track past variations of the extension of the water masses bathing the ocean floor, thus deep-sea circulation patterns.

From the 1950s on, research on low oxygen basins off California (e.g., Smith, 1964; Phleger & Soutar, 1973; Douglas & Heitman, 1979) and on Mediterranean sapropels (e.g., Parker, 1958), led to the recognition that recent as well as ancient low oxygen environments were inhabited by specific faunas, generally with a low diversity and strongly dominated by a few species adapted to these apparently hostile environments (e.g., Cita & Podenzani, 1980; Mullineaux & Lohmann, 1981; Parisi & Cita, 1982; Van der Zwaan, 1982). Bernhard (1986) showed that taxa inhabiting low oxygen environments are characterised by a specific morphology, often with a maximum surface-to-volume ratio. As a consequence, bottom water oxygenation became gradually accepted as a major environmental factor in many ecosystems.

The work of Lutze's group on the northwest Africa upwelling area (Lutze, 1980; Lutze & Coulbourn, 1984; Lutze, Pflaumann, & Weinholz, 1986) probably caused the breakthrough that led to the recent opinion that the organic flux to the ocean floor is the most important parameter controlling benthic life in deep, open ocean ecosystems. Since the 1980s, many efforts have been made to develop reliable proxies of bottom water oxygenation and various aspects of the organic flux to the ocean floor.

1.3. Recent Advances in Benthic Foraminiferal Ecology

It is outside the scope of this paper to give a complete overview of our present knowledge of deep-sea foraminiferal ecology (recent reviews in Murray, 1991; Sen Gupta, 1999; Van der Zwaan et al., 1999; Gooday, 2003), but we will highlight some of the most important advances since the 1980s, which have significantly changed our view of the parameters controlling foraminiferal faunas, thus the potential use of fossil faunas as paleoceanographical proxies.

1) Since the key papers of Basov and Khusid (1983) and Corliss (1985), and the numerous ecological studies confirming their observations (e.g., Mackensen & Douglas, 1989; Corliss & Emerson, 1990; Corliss, 1991; Barmawidjaja, Jorissen, Puskaric, & Van der Zwaan, 1992; Rathburn & Corliss, 1994; Kitazato, 1994; Rathburn, Corliss, Tappa, & Lohmann, 1996; Ohga & Kitazato, 1997; De Stigter, Jorissen, & Van der Zwaan, 1998; Jorissen, Wittling, Peypouquet, Rabouille, & Relexans, 1998; Kitazato et al., 2000; Schmiedl et al., 2000; Fontanier et al., 2002; Licari, Schumacher, Wenzhöfer, Zabel, & Mackensen, 2003), we know that deep-sea benthic foraminifera do not exclusively live at or on the sediment surface, but are present alive in the upper 1–10 cm of the sediment, in microhabitats that become increasingly oxygen-depleted from the sediment surface downwards. Elaborating ideas proposed by Shirayama (1984), Corliss and Emerson (1990), and Loubere, Gary, and Lagoe (1993), Jorissen, De Stigter, and Widmark (1995) proposed the so-called TROX model, which explains that the depth of the foraminiferal microhabitat in oligotrophic ecosystems is limited by the availability of food particles within the sediment, whereas in eutrophic systems a critical oxygen level decides down to what depth in the sediment most species can live (Figure 1). Some authors have suggested that oxygen concentration is not a major limiting factor for many taxa (e.g., Rathburn & Corliss, 1994; Moodley, Heip, & Middelburg, 1997, 1998a, 1998b), and that competition and predation may interfere (Buzas, Collins, Richardson, & Severin, 1989; Mackensen & Douglas, 1989; Van der Zwaan et al., 1999; Gooday, 2003), but the general validity of the TROX concept has been confirmed in many

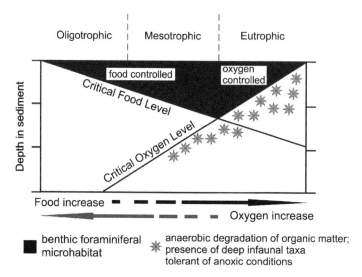

Figure 1   Trox-model (Jorissen et al., 1995, with permission from Springer), explaining the depth limits of the foraminiferal microhabitat by a combination of oxygen penetration and food availability in the sediment. In oligotrophic environments, the microhabitat depth is limited by the low amount of food in the sediment; in eutrophic ecosystems, the penetration depth of most taxa is limited by the shallow level at which no oxygen is present. In mesotrophic areas, microhabitat depth is maximal. Some deep infaunal taxa are not limited by the zero oxygen level, but may participate in anaerobic pathways of organic matter remineralisation at several cm depth in the sediment.

studies, and Carney (2005) suggested that the TROX model may also successfully explain many macrofaunal distribution patterns.

The foraminiferal niche is thus much more variable than thought previously, with different species inhabiting a wide range of biogeochemically different micro-environments, from the sediment-water interface to several cm depth in the sediment. Therefore, the isotopic and trace element composition ($\delta^{18}O$, $\delta^{13}C$, Mg/Ca, etc.) of the foraminiferal test can be interpreted adequately only if the microhabitat (and calcification depth) of each investigated species is precisely known. The study of site-specific differences in these parameters opens up new possibilities to reconstruct former redox conditions at and below the sediment-water interface, as well as their controlling parameters.

2) Since the early 1990s, Rose Bengal stained foraminifera have been reported to occur in anoxic environments below the sediment-water interface (e.g., Bernhard, 1989; Bernhard & Reimers, 1991; Loubere et al., 1993; Alve, 1994; Rathburn & Corliss, 1994; Rathburn et al., 1996; Jannink, Zachariasse, & Van der Zwaan, 1998; Jorissen et al., 1998; Fontanier et al., 2002). Rose Bengal is a protein stain (Walton, 1952) which has been widely used for the recognition of living foraminifera. Rose Bengal, however, will also stain protoplasm in a more or less advanced state of decay (e.g., Bernhard, 1988; Corliss & Emerson, 1990; Hannah & Rogerson, 1997; Jorissen, 1999a), and thus its reliability for recognition of

living foraminifera has been seriously questioned, especially in anoxic environments where protoplasm decay may be very slow (e.g., Corliss & Emerson, 1990). Studies using more specific vital stains have confirmed that some foraminiferal species can indeed live, and be active, in anoxic sediments (Bernhard & Reimers, 1991; Bernhard, 1993; Bernhard & Alve, 1996). The abundant evidence that populations may be active in low oxygen conditions suggests that oxygen concentration is a critical factor only below a certain threshold, which for most species is surprisingly low, i.e., below 1 ml/l or even less (Jorissen et al., 1995; Bernhard et al., 1997; Van der Zwaan et al., 1999; Levin et al., 2001). This suggests that the quantitative reconstruction of bottom water oxygenation may be feasible only for values below 1 ml/l, and will probably be much more complicated or impossible, at higher concentrations (Murray, 2001).

3) Since publication of the key papers of Lutze and co-workers (Lutze, 1980; Lutze & Coulbourn, 1984; Lutze et al., 1986) it has become clear that the flux of organic matter to the deep-sea floor is the main parameter structuring benthic foraminiferal faunas by controlling their density and species composition. Many foraminiferal taxa appear to have an optimum range with respect to organic input, in which their competitiveness is maximal, and under which conditions they become dominant faunal elements. If the organic input falls below, or exceeds the optimal flux levels, they are replaced by taxa that are more competitive under more oligotrophic or more eutrophic conditions. Changes in the bathymetrical range of foraminiferal species under different organic flux regimes were first documented by Pflum and Frerichs (1976) in front of the Mississippi delta, more recently by De Rijk, Jorissen, Rohling, and Troelstra (2000), who showed a progressive shoaling of the bathymetrical ranges of many species along a West–East transect through the Mediterranean, coincident with a transition from eutrophic to very oligotrophic ecosystems.

The fact that many foraminiferal species depend strongly upon the organic flux must open up new pathways for the reconstruction of paleoproductivity, but as we will explain in Section 2.3, the development of such methods is hampered by calibration problems (Altenbach et al., 1999).

4) Carney (1989) proposed that two independent organic matter remineralisation cycles exist at the sea floor, fueled by two types of organic input. Labile organic matter (marine organic matter, phytodetritus), is rapidly remineralised at the sediment-water interface using aerobic pathways, whereas more refractory organic matter (terrestrial organic matter, laterally advected material) is remineralised much more slowly in the dysaerobic ecosystems deeper in the sediment. Since this key publication, several authors have speculated on the importance of organic matter quality as a controlling ecological parameter. Most superficially living taxa probably participate in aerobic remineralisation, whereas deeper living taxa may contribute to dysaerobic mineralization of lower quality organic matter (Fontanier et al., 2002, 2005). In the eastern Atlantic Cap Ferret Canyon, for instance, where there is lateral input of large quantities of refractory organic matter, a rich deep infauna co-exists with much poorer surface faunas (Fontanier et al., 2005).

These observations suggest that it should be possible to use the benthic foraminiferal faunal composition to reconstruct not only the quantity, but also the
quality of the organic input. On the basis of the succession of fossil faunas in a core
off NW Africa, Caralp (1984, 1989) suggested that *Bulimina exilis* has the advantage
over other species when there is an input of fresh, labile organic matter with a high
phaeopigment/glucide ratio, whereas *Melonis barleeanus* was more abundant when
the organic input consisted of more refractory, laterally advected organic material.
Goldstein and Corliss (1994) observed that diatom frustules were ingested by
U. peregrina, but not by *Globobulimina* spp., suggesting adaptation to different food
types. On the basis of fatty acid analysis in foraminiferal protoplasma, Suhr, Pond,
Gooday, and Smith (2003) concluded that some taxa are more selective than others.
For instance, *Globocassidulina subglobosa* may preferentially ingest fresh diatoms,
whereas *Thurammina albicans* feeds on degraded material. The shallow infaunal taxa
Uvigerina akitaensis and *Bulimina aculeata* show a higher carbon assimilation rate than
the deeper infaunal species *Textularia kattegatensis* and *Chilostomella ovoidea* in a field
feeding experiment with ^{13}C-labeled algae (Nomaki, Heinz, Hemleben, & Kitazato,
2005). All these observations suggest that different species respond differently to the
input of various types of food particles, and that food preferences exist.

5) From long-term observations based on sediment trap deployments, we know
    that in most areas the organic matter flux to the ocean floor strongly varies
    seasonally and/or interannually (e.g., Billett, Lampitt, Rice, & Mantoura, 1983;
    Berger & Wefer, 1990; Lohrenz et al., 1992). In some areas, episodic events are
    responsible for most of the organic input to the seafloor, and specialised benthic
    foraminiferal faunas may rapidly colonise phytodetritus deposits, reacting
    by accelerated growth and/or reproduction (e.g., Gooday, 1988, 1993, 1996;
    Gooday & Lambshead, 1989; Gooday & Rathburn, 1999; Gooday & Hughes,
    2002). These phenomena have been documented most conclusively after
    episodic phytodetritus falls in otherwise oligotrophic deep oceanic basins, but
    also may control foraminiferal population dynamics in more eutrophic areas
    with a less variable organic input (e.g., Ohga & Kitazato, 1997; Kitazato et al.,
    2000; Rathburn et al., 2001; Fontanier, Jorissen, Anschutz, & Chaillou, 2003,
    2006). This strong response of some foraminiferal taxa (e.g., *Epistominella
    exigua*) has led to attempts to use these taxa as indicators of pulsed organic input
    (Loubere, 1998; Smart, King, Gooday, Murray, & Thomas, 1994; Thomas,
    Booth, Maslin, & Shackleton, 1995; Thomas & Gooday, 1996).

6) Since the early publications of Buzas and co-workers (Buzas & Gibson, 1969;
    Buzas et al., 1989), it has become increasingly clear that spatial variability at a
    micro- or mesoscale (patchiness) may be important in the deep ocean (e.g.,
    Gooday & Rathburn, 1999; Hughes & Gooday, 2004). This phenomenon may
    cause large differences between living faunas collected with multi-corers on one
    hand, and fossil samples in which patchiness has been eliminated by time
    averaging on the other hand. Taphonomical processes may be responsible for a
    partial or even total loss of many foraminiferal taxa and will further increase the
    differences between fossil and living faunas. Ecological observations of living

faunas are essential for a better understanding of proxy relationships, but we think that a successful calibration of a proxy relation is very difficult, if not impossible, to establish on the basis of living faunas only, because of the above mentioned processes which create large differences between living and fossil faunas. We argue that once the underlying mechanisms of proxy relations are fully understood, proxy calibration should be performed on the basis of sub-recent faunas in which the transformation of the living fauna to the fossil fauna due to early diagenetic processes has been concluded.

2. Benthic Foraminiferal Proxies: A State of the Art

2.1. Overview of Proxy Methods Based on Benthic Foraminiferal Assemblage Data

In this chapter, we will describe the present state of the art of the most commonly used proxy methods based on benthic foraminiferal assemblage data. Benthic foraminiferal assemblage data may be of several types: presence/absence data of various taxa, measures of faunal density, of biodiversity, or data on the morphology of dominant taxa. All of these have been proposed as paleoceanographical proxies. The reconstructed environmental parameters fall into two groups:

a) Physico-chemical parameters, such as temperature, salinity, carbonate saturation, hydrodynamics, or oxygen concentration of the bottom water. Such parameters may be expected to act as limiting factors, determining whether a foraminiferal species can live somewhere, and if so, if it can feed actively, grow, calcify and reproduce. If all limiting factors fall within the optimum range for a specific taxon, it will attain maximum competitiveness, and may be expected to reach its maximum abundance. Such a maximum abundance may be very high for opportunistic species, but very low for highly specialised K-selected taxa (e.g., Levinton, 1970; Pianka, 1970; Dodd & Stanton, 1990).

b) Resource parameters, such as the quality and quantity of the organic matter flux that directly influence the amount of food available to specific taxa. Resources are expected to act directly on the density of foraminiferal populations, but will also determine what species will dominate the faunas, because most taxa have maximal competitiveness within a specific flux range.

Traditionally, benthic foraminiferal assemblage data have been used to reconstruct water depth. In the deep-sea, water depth itself is not a controlling ecological parameter, but most other parameters change with water depth, thus causing the well-established bathymetrical successions seen in many ocean basins. Although elaborate bathymetrical distribution schemes have been proposed from the 1950s on, more recent data consistently show important differences in bathymetrical ranges between ocean basins, which appear to be at least partially caused by differences in organic flux regime, so that reliable reconstruction of paleo waterdepth by the use of benthic foraminiferal presence–absence patterns is probably not possible. Alternative methods, such as the ratio between planktonic and benthic foraminifera (e.g., Van der Zwaan, Jorissen, & De Stigter, 1990; Van Hinsbergen, Kouwenhoven, & Van der Zwaan, 2005), or modern analogue techniques (Hayward, 2004) continue to be explored.

Foraminiferal assemblage characteristics have rarely been used for the reconstruction of bottom water temperature and salinity in open ocean environments, because the variability in these parameters in most oceanic basins is probably too limited to cause a significant faunal response. In shallow and coastal water environments, on the contrary, where strong temperature and salinity gradients exist, foraminifera have been used successfully to reconstruct these parameters. The combination of stable oxygen isotope and Mg/Ca values is considered the most promising method to reconstruct temperature and salinity in deep oceanic environments (e.g., Lear, Elderfield, & Wilson, 2000; Lear, Rosenthal, Coxall, & Wilson, 2004).

The combination of bottom water temperature and salinity determines the specific density of sea water, and the stratification of the oceanic water masses (e.g., Schmitz, 1992). Until the early 1980s, the benthic foraminiferal assemblage composition was thought to mainly reflect these water mass characteristics, and it was thought possible to reconstruct the geographical and depth extent of specific water masses on the basis of the distributional patterns of benthic foraminiferal marker species (e.g., Streeter, 1973; Schnitker, 1974, 1979, 1980; Gofas, 1978; Lohmann, 1978b; Lutze, 1979; Streeter & Shackleton, 1979; Corliss, 1979b, 1983a; Peterson, 1984; Caralp, 1987). With the realisation that the spatial and bathymetrical distribution of most recent foraminiferal taxa is predominantly based on organic flux rates, many scientists today doubt the validity of the so-called "water mass concept". Although small changes in temperature and salinity in deep ocean ecosystems appear to have a minimal influence on benthic foraminiferal assemblages, other parameters related to water masses, such as carbonate saturation (alkalinity) may have a profound influence on bio-calcification, and in this way structure the live benthic foraminiferal assemblages (e.g., Mackensen, Grobe, Kuhn, & Fütterer, 1990; Mackensen, Schmiedl, Harloff, & Giese, 1995), and the preservation of their tests. Carbonate saturation becomes especially important close to the carbonate compensation depth (CCD) (e.g., Bremer & Lohmann, 1982).

The reconstruction of the extent of former water masses is still of paramount importance, and today a range of geochemical proxies, such as Cd/Ca ratios in calcite, Nd isotopes in fish teeth and manganese nodules, Pb isotopes in manganese nodules, or carbon isotopes in carbonates are used (e.g., Lynch-Stieglitz, 2004), and the foraminifera-based "water mass" concept can not be discarded without a serious discussion (Section 2.4).

The organic flux to the ocean floor, its quantity, quality and periodicity, is today considered to be the prevailing environmental parameter structuring deep-sea benthic foraminiferal faunas, so that our best hopes to develop paleoceanographic proxies on the basis of deep-sea benthic foraminiferal assemblages lay in the field of paleo-productivity. We will treat this subject in Section 2.3, where we will also focus on proxies reconstructing the periodicity of the organic input, and the presence/absence of episodic events.

Although many modern studies suggest that bottom water oxygen concentration is for most species less critical than thought previously, several foraminiferal proxies of bottom water oxygenation have been proposed, and yield promising results. In Section 2.2 we will discuss the existing methods, and indicate possible pathways for future improvement.

The observation that a specific assemblage of benthic foraminifera occurs in areas with elevated current velocities (Schönfeld, 1997, 2002a, 2002b) has led to the tentative development of a proxy for bottom current velocity. This promising new field of proxy development could inform us about past variations of the intensity of deep-water circulation and is briefly treated in Section 2.5.

2.2. Proxies of Bottom Water Oxygenation

2.2.1. Introduction

Since the 1960s, abundant live foraminiferal faunas have been reported from low oxygen environments (e.g., Smith, 1964; Phleger & Soutar, 1973; Douglas & Heitman, 1979). Recent review papers (e.g., Bernhard, 1986; Sen Gupta & Machain-Castillo, 1993; Bernhard & Sen Gupta, 1999) agree that faunas from such environments have a characteristic taxonomic composition, and are generally strongly dominated by bolivinids, buliminids, globobuliminids and some other taxa, at least when oxygen concentrations reach concentrations below ~1 ml/l. If the relative proportion of these taxa would increase with decreasing bottom water oxygenation, it should be possible to use the composition of fossil faunas to reconstruct ancient bottom water oxygen concentrations. It has been argued, however, that two groups of fundamentally different taxa, with very different life strategies, may profit from the disappearance of less resistant taxa at low bottom water oxygen concentrations: (1) deep infaunal taxa, which at the onset of bottom water hypoxia migrate from the deeper sediment layers to the sediment surface (Jorissen, 1999a), and (2) epifaunal or shallow infaunal taxa which have developed adaptations to or tolerance for low oxygen conditions. Unlike the first group, the second group of taxa are usually rare in environments with well-oxygenated bottom waters, to become abundant under severely hypoxic conditions, at least in some locations.

Before looking at the available proxy methods for paleo bottom water oxygenation based on foraminiferal assemblages, we must discuss the terminology used to describe environments where bottom and/or pore waters are under-saturated in oxygen. Unfortunately, there is considerable confusion in the literature concerning the exact meaning of the various descriptors. Figure 2 presents an inventory of the terms most often encountered in the foraminiferal literature, and their meaning according to various authors. There is considerable confusion about the exact meaning of the terms "dysoxic" and "suboxic", which have been used for environments with entirely different oxygen concentrations. The apparently precise term anoxic (a = without) is used for environments where oxygen concentrations are below the detection limit (dependent upon techniques used), whereas others use it for concentrations below 0.1 ml/l. Bernhard and Sen Gupta (1999) differentiate between anoxic environments with and without sulphate reduction, the latter being termed postoxic. In order to avoid further confusion, we will use the term hypoxic for all environments (without giving a precise range of oxygen concentration) where foraminiferal faunas may potentially be influenced by low oxygen

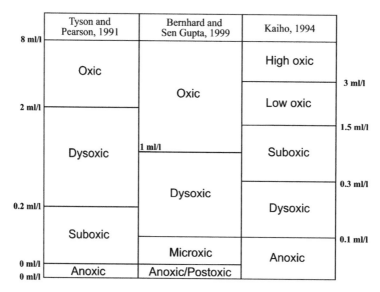

Figure 2 Classification of environments with different oxygen concentrations, according to Tyson and Pearson (1991), Kaiho (1994), and Bernhard and Sen Gupta (1999).

concentrations, and we will use anoxic for all environments without detectable oxygen.

2.2.2. Potential problems for paleo-oxygenation proxies

Several problems must be considered when benthic foraminiferal assemblages are used to reconstruct bottom water oxygen concentration, some of which (points 1, 2, 4, and 5 in the following list) are specific for bottom water oxygenation proxies, whereas others are of a more general nature, and concern most proxies based on microfossil remains:

1. In most benthic ecosystems oxygen concentration is subject to a strong gradient. Starting in the more or less well-oxygenated bottom waters, oxygen concentrations decrease first at the sediment-water interface, followed by a rapid downward decrease in the superficial sediment layer, to become zero at a few mm or cm depth. In continental slope and margin environments the depth of the oxic sediment layer rarely exceeds 5 cm. Extreme oxygen penetration, down to 1 m or more, has been observed in the Angola Basin, Weddell Sea and in the West Equatorial Pacific (Rutgers van der Loeff, 1990). In contrast, oxygen penetration is limited to the topmost mm of the sediment in many of the world's oxygen minimum zones (OMZs) (e.g., Miao & Thunell, 1993; Levin, 2003).

   In environments where oxygen concentrations change considerably over short vertical distances within the sediment and/or over short time periods, it must be made very clear what the bottom water oxygenation proxy exactly intends to reconstruct: bottom water oxygenation (at some tens of centimeter above the sediment-water interface), oxygen concentration at the sediment-water interface, or the depth of oxygen penetration into the sediment. These three elements each tell part of the story, and only the knowledge of all three (which is probably unrealistic to expect from proxies) gives a complete picture of benthic ecosystem oxygenation.

2. Benthic foraminiferal faunas commonly inhabit a several cm deep superficial sediment layer, coinciding with the complete range from well oxygenated to strongly hypoxic conditions. Therefore, most faunas collected in environments with well-oxygenated bottom waters contain a mixture of taxa that inhabit fully oxic to strongly hypoxic, or even anoxic microhabitats. Although individual species usually have a preference for a specific depth interval (e.g., Corliss, 1985; Jorissen et al., 1998; Fontanier et al., 2002), even taxa considered "epifaunal" do not exclusively live at the sediment-water interface, but may migrate through the topmost mm or cm of the sediment, and will, as all other taxa, experience a wide range of oxygen concentrations during their life time (e.g., Barmawidjaja et al., 1992; Ohga & Kitazato, 1997).

3. Geological samples, the basis for paleoceanographic reconstructions, are always time-averaged. In the ideal case (without addition or loss of tests due to lateral transport), a sample will contain a mixture of all faunas that inhabited the site during several decennia. Even in laminated sediments with annual laminae, the fossil fauna found in a single lamina will represent the average conditions during a complete season, i.e., ~6 months. If environmental conditions were stable during the period under consideration, the composition of the fauna may indeed record bottom water oxygenation precisely. However, bottom water oxygen concentration (and oxygen penetration depth into the sediment) will very often experience important short-term variability, for instance when strong water column stratification, or pulsed phytodetritus deposits, cause seasonal hypoxia/anoxia. Such events may be annual, but can also be highly episodic, occurring once every 10 years, or even less often. In such cases it will be extremely difficult (if not impossible) to extract the precise oxygenation history from the time-averaged faunas. For instance, in cases of anoxic conditions (without fauna) interrupted by short periods with oxic bottom waters, characterised by rich pioneer faunas, a foraminifer-based bottom water oxygenation proxy will probably overestimate the average long-term bottom water oxygen concentration. In other settings the duration and severity of the periodical hypoxia/anoxia may easily be overestimated, for instance when oligotrophic, well-oxygenated ecosystems are affected by short-term anoxia leading to the explosive development of a few low-oxygen tolerant taxa.

4. The critical values at which the oxygen concentration starts to have a negative impact on the organisms are probably very low, below 1 ml/l or even below 0.5 ml/l (Levin & Gage, 1998). Many elongated, commonly biserial or triserial taxa, have been described as abundant at very low oxygen concentrations. In several papers (e.g., Bernhard, 1986; Corliss & Chen, 1988; Corliss & Fois, 1990; Corliss, 1991) it has been suggested that their morphology corresponds to an infaunal microhabitat. Not only elongated uniserial, biserial and triserial taxa, however, but a wide range of planoconvex to biconvex,

planispiral and trochospiral taxa have been found alive in low oxygen environments (Table 1). For example, *Cibicidoides wuellerstorfi*, usually described as a strictly epifaunal taxon and considered typical of well oxygenated bottom waters, is present in significant numbers in several stations in the Sulu Sea where bottom water oxygen concentrations are well below 2 ml/l (Rathburn & Corliss, 1994). Many of the planoconvex to biconvex taxa listed in Table 1 (e.g., *Cassidulina carinata*, *Cibicides ungerianus*, *Cibicidoides wuellerstorfi*, *Gavelinopsis*

Table 1 Observations of Planoconvex Planispiral and Trochospiral Taxa Generally Considered as Typical of Well-Oxygenated Bottom Waters in Low Oxygen Environments; the Second Column Gives the Lowest Bottom Water Oxygen Concentration at Which Living (Rose Bengal stained) Specimens have been Described; and the Third Column Gives the Reference: 1. Alve (1990); 2. Bernhard (1992); 3. Douglas and Heitman (1979); 4. Jannink et al. (1998); 5. Jorissen et al. (1992); 6. Maas (2000); 7. Mackensen and Douglas (1989); 8. Sen Gupta et al. (1997); 9. Smith (1964).

Ammonia batavus	0.3 ml/l	1	*Epistominella smithi*	0.1 ml/l	3
Anomalina sp.	0.2 ml/l	4	*Epistominella vitrea*	0.8 ml/l	5
Cancris auriculus	0.3 ml/l	6	*Eponides antillarum*	0.9 ml/l	9
Cancris inaequalis	0.2 ml/l	7	*Eponides leviculus*	0.2 ml/l	3
Cancris oblongus	0.2 ml/l	4	*Eponides regularis*	0.1 ml/l	6
Cancris panamensis	0.6 ml/l	9	*Gavelinopsis lobatulus*	2.1 ml/l	4
Cancris sagra	0.4 ml/l	9	*Gavelinopsis translucens*	0.0 ml/l	8
Cassidulina carinata	0.3 ml/l	4	*Gyroidina io*	0.2 ml/l	3
Cassidulina crassa	0.3 ml/l	6	*Gyroidina lamarckiana*	0.1 ml/l	6
Cassidulina cushmani	0.3 ml/l	9	*Gyroidina multilocula*	0.4 ml/l	9
Cassidulina delicata	0.1 ml/l	3,7	*Gyroidina parva/pulchra*	0.9 ml/l	4
Cassidulina depressa	0.5 ml/l	4	*Gyroidina rotundimargo*	0.3 ml/l	9
Cassidulina laevigata	0.2 ml/l	7	*Gyroidina umbonata*	0.4 ml/l	4
Cassidulina limbata	1.0 ml/l	3	*Gyroidinoides neosoldanii*	0.5 ml/l	2
Cassidulina minuta	0.6 ml/l	9	*Gyroidinoides soldanii*	0.1 ml/l	6
Cassidulina oblonga	0.6 ml/l	9	*Hanzawaia boueana*	0.8 ml/l	5
Cassidulina sgarellae	0.3 ml/l	6	*Hanzawaia concentrica*	0.3 ml/l	6
Cassidulina subcarinata	0.2 ml/l	3	*Hoeglundina elegans*	0.2 ml/l	3
Cassidulina subglobosa	0.2 ml/l	4	*Hyalinea balthica*	0.3 ml/l	4
Cassidulina teretis	0.5 ml/l	3	*Islandiella subglobosa*	0.2 ml/l	3
Cassidulina tumida	0.4 ml/l	9	*Islandiella* sp.	0.3 ml/l	6
Cibicides bradyi	0.3 ml/l	4	*Lenticulina articulata*	0.3 ml/l	6
Cibicides fletcheri	0.4 ml/l	7	*Oridorsalis umbonatus*	0.4 ml/l	4
Cibicides refulgens	0.3 ml/l	2	*Osangularia culter*	0.3 ml/l	4
Cibicides ungerianus	2.1 ml/l	4	*Osangularia rugosa*	0.0 ml/l	8
Cibicidoides wuellerstorfi	1.6 ml/l	2	*Planulina ariminensis*	0.2 ml/l	7
Elphidium albiumbilicatulum	0.0 ml/l	1	*Planulina exorna*	0.6 ml/l	9
Elphidium excavatum	0.1 ml/l	1	*Planulina limbata/ornata*	0.3 ml/l	9
Elphidium incertum	0.1 ml/l	1	*Pullenia* sp.	0.4 ml/l	4
Elphidium tumidum	2.4 ml/l	9	*Valvulineria araucana*	0.2 ml/l	3
Epistominella decorata	0.1 ml/l	6	*Valvulineria javana*	0.2 ml/l	4
Epistominella exigua	0.3 ml/l	6	*Valvulineria oblonga*	0.3 ml/l	9

translucens, Hoeglundina elegans), have been found as shallow infauna or epifauna in well-oxygenated environments. In some papers (e.g., Kaiho, 1994; Jannink et al., 2001) it was therefore implicitly suggested that these species need elevated oxygen concentrations, but observations of these taxa in strongly hypoxic environments suggest that they are much less sensitive to low oxygen concentrations than generally assumed. Their preference for superficial sediment niches may well be explained by their dependence on a supply of high quality food particles. It should also be realised that no foraminiferal species occur exclusively in low-oxygen environments (Sen Gupta & Machain-Castillo, 1993), and that most taxa indicative for strongly hypoxic conditions can also be found at much higher bottom water oxygenation values. Murray (2001) thus suggested that foraminifera can only be used as proxies for oxygen levels below concentrations of 1 or perhaps 2 ml/l, and above these values, there is no relation between oxygen levels and the composition of the foraminiferal faunas (Murray, 2001).

5. In recent benthic ecosystems there is a complex interplay between export productivity (organic flux to the sea floor) and oxygenation of bottom and pore waters. The competitive ability of most species appears to be determined by one or both of these parameters. Jorissen et al. (1995) suggested that deep infaunal taxa, often considered indicative of hypoxic conditions, are only present when the organic flux is sufficiently high to have organic detritus within the sediment that can be used metabolically. In areas with lower flux rates most organic matter is consumed at the sediment-water interface, and deeper sediment layers are too poor in organic matter to be inhabited by deep infaunal taxa (Figure 1). This means that a sudden appearance, or a relative frequency increase, of these deep infaunal taxa in the fossil record may very well be the result of an increased organic flux, and not of lowered bottom water oxygen concentrations. Oxygen concentrations have a direct impact on the faunal composition only in areas where bottom water concentrations become so low that they cause a diminished competitive ability and/or reproductive potential of the more superficially living taxa. In such cases, less resistant taxa will disappear, and deep infaunal taxa that are perfectly adapted to live in low oxygen environments will take over niches at the sediment surface and dominate the foraminiferal assemblages (Jorissen, 1999b). Such conditions will occur only when the oxygen concentrations fall below a threshold level critical for the surface dwelling taxa ($\leqslant 1$ ml/l; Murray, 2001). Finally, even faunas that are strongly dominated by deep infaunal taxa may form in environments with well-oxygenated bottom waters. For instance, at a 2,800 m deep station in Cap Ferret Canyon (north-eastern Atlantic) bottom waters are well oxygenated (5 ml/l), but deep infaunal taxa are dominant, probably because of intensive degradation of large quantities of low quality organic matter by anaerobic bacterial stocks (Fontanier et al., 2005).

6. The living fauna always differs considerably from the fossil faunas accumulating at the same site (e.g., Murray, 1991; Jorissen & Wittling, 1999; Walker & Goldstein, 1999). This difference is caused by early diagenetic processes, causing a preferential loss of the more fragile taxa, but also by biological parameters, such

as large interspecific differences in test production, or selective predation on some foraminiferal species (e.g., Culver & Lipps, 2003). Additionally, transport by bottom currents may be responsible for the addition of allochthonous foraminifera and/or the removal of small autochthonous elements. Since paleoceanographic reconstructions are based on fossil material that has been seriously modified by taphonomic processes, it is probably not a good practice to use living faunas for the calibration of paleoceanographical proxies. Individuals living at any specific moment of sample collection do not necessarily reflect a fauna as averaged over 1 year or a full seasonal cycle, for instance. We think that proxy relationships between faunal characteristics and oceanographical parameters should rather be calibrated by a comparison of the recent values of the target parameter with "subfossil" faunas that represent recent environmental conditions but have fully concluded the suit of diagenetic processes and will undergo no more post-mortem compositional changes. Unfortunately, in most ocean basins, it takes several thousands of years before a fauna has passed through the biologically active top part of the sediment, where most of the diagenetic processes and losses of foraminiferal tests take place (e.g., Walker & Goldstein, 1999). It is very improbable that the Holocene fossil faunas collected below this topmost sediment layer where early diagenesis is taking place, can directly be compared with the recent (anthropogenically-influenced) environmental conditions, a procedure that cannot be avoided when a proxy relationship has to be calibrated. These problems can be at least partly solved by a careful, high-resolution study of a succession of Holocene faunas in areas with high sedimentation rates.

7. Most oceanic environments are much less stable then previously thought: even abyssal environments experience episodic phytodetritus inputs, provoking a strong response of some highly opportunistic taxa that may rapidly attain high densities (e.g., Gooday, 1988, 1993; Gooday & Lambshead, 1989). Other unpredictable events such as whale falls or pulsed food input from hydrothermal vents may also have a profound impact on benthic foraminiferal faunas, although hydrothermal vent regions and cold seep areas do not contain endemic benthic assemblages (e.g., Rathburn et al., 2003; Barberi & Panieri, 2004). At present, it cannot be excluded that in many parts of the ocean an important part of the fossil faunas is formed during ephemeral events. If true, the fossil fauna should not be considered representative of average conditions, but for the very specific conditions associated with the short periods of high productivity.

In spite of these problems, there appear to be significant differences in tolerance for low oxygen levels between species, and therefore oxygen concentration must have an impact on the assemblage composition, at least at low to very low oxygenation levels. The possibility to estimate former oxygen levels with a high precision in the 0–1 ml/l range is important, because it will open up the possibility to reconstruct the history (including extent and severity) of OMZs. These continental margin environments potentially play an important role as organic carbon sinks in the global carbon cycle, and the reconstruction of the variability of intensity of OMZs through time may give insight into the dynamics of intermediate and deeper water masses.

2.2.3. Existing proxies of bottom water oxygenation

The proposed methods for the reconstruction of ancient bottom water oxygenation concentrations on the basis of benthic foraminiferal assemblage characteristics fall into four categories:

1. Most often, a number of taxa is considered indicative of hypoxic conditions, and the relative frequency of these taxa (expressed as a percentage of the total benthic foraminiferal fauna) is used as a semi-quantitative index of bottom water oxygenation. The best example of this method is the widely applied benthic foraminiferal oxygen index (BFOI) of Kaiho (1991, 1994, 1999). In this index, a subdivision of taxa into categories of dysoxic, suboxic or oxic indicators is made rather arbitrarily, mainly based on test morphology, and differs between the 1991 paper that deals with Paleogene faunas, and the 1994 and 1999 papers that try to calibrate the method on the basis of recent faunas. It is surprising to see that for many species adult individuals ($\geq 350\,\mu$m) are considered as "oxic" indicators, whereas smaller individuals ($< 350\,\mu$m) of the same species are considered as "suboxic" indicators (Kaiho, 1994, 1999). The hypothesis underlying the method is that in well-oxygenated bottom waters, dysoxic indicators live in poorly oxygenated deep infaunal microhabitats. In the case of hypoxic conditions at the sediment-water interface, less resistant taxa disappear and the "dysoxic indicators" become dominant. However, only a few of these "dysoxic" indicators (21 taxa listed by Kaiho, 1994) have actually been observed alive in intermediate or deep infaunal microhabitats, and the suggested microhabitat separation between oxic and suboxic indicators (Kaiho, 1994, Figure 1) is not supported by data. In its latest version (Kaiho, 1999), Kaiho's oxygen index distinguishes between five classes of bottom water oxygenation (Figure 1), and at levels above 3.2 ml/l is no longer correlated with bottom water oxygenation. Other authors have slightly modified the attribution of taxa to the categories of dysoxic, suboxic or oxic indicators (e.g., Baas, Schönfeld, & Zahn, 1998), in order to better fit the method (and results!) to their ideas.

2. The method proposed by Jannink et al. (2001) follows an inverse approach: taxa that are found consistently living in the topmost sediment are considered oxiphylic, and their cumulative percentage is considered a proxy for bottom water oxygenation. The rationale behind this procedure is that oxygen penetration into the sediment increases with increasing bottom water oxygenation, leading to an increased volume of the niche potentially occupied by oxiphylic taxa. The problem with this method is the determination of which species are oxiphylic. A species found living close to the sediment-water interface may prefer this microhabitat because it does not tolerate the lower oxygen concentrations deeper in the sediment, but also because it prefers high quality food particles that are most concentrated at the sediment-water interface, as indicated by an increasing amount of data (e.g., Kitazato, Nomaki, Heinz, & Nakatsuka, 2003; Fontanier et al., 2003, 2006; Ernst, Bours, Duijnstee, & Van der Zwaan, 2005; Nomaki et al., 2005).

3. Several authors (e.g., Loubere, 1994, 1996; Morigi, Jorissen, Gervais, Guichard, & Borsetti, 2001) have attempted to link sub-recent faunal assemblages to

bottom water oxygenation values by multivariate statistical methods, with some encouraging results for recent faunas. It is very difficult to know, however, whether application of these relationships to fossil material would yield reliable results, because the investigated datasets contain a rather limited array of combinations of oxygen concentration and organic flux (and other environmental parameters). Non-analog conditions thus may not be correctly reconstructed. An additional problem is the fact that in order to be robust, the multivariate statistical methods need many more data (samples) than variables (taxa) (e.g., Tabachnick & Fidell, 1983), and at present the size of the analysed datasets is small in comparison with the number of species.

4. Schmiedl et al. (2003) proposed a method based on a combination of the relative proportion of low-oxygen tolerant marker species and a diversity index, counting each for 50% of the final score of their oxygen index, because all hypoxic environments show a lower biodiversity than well-oxygenated settings. By adding a factor independent of taxonomic composition the proxy method may become more robust, and applicable in an array of areas with different faunal compositions.

Unfortunately, all four procedures suffer from one or more of the pitfalls outlined before. The main problem is the fact that all species simultaneously respond to bottom and pore water oxygenation and to the organic flux level. Many taxa that have been proposed to be low oxygen indicators should probably be considered high productivity markers instead. Although these taxa are indeed abundantly present in many low oxygen environments, they also dominate faunas in high productivity areas with fairly high bottom water oxygen concentrations, such as the upwelling area off NW Africa (e.g., Jorissen et al., 1995; Morigi et al., 2001). The reverse also may occur: benthic environments with low oxygen concentrations are not necessarily dominated by "low oxygen indicators" in areas where export production is low (e.g., Rathburn & Corliss, 1994).

2.2.4. Future developments of bottom water oxygenation proxies

Some of the best examples of past changes in bottom water oxygenation have been preserved in Mediterranean sapropels (e.g., Jorissen, 1999b; Schmiedl et al., 2003). A detailed study of the faunal succession at the transition from homogeneous sediments characteristic of oxygenated bottom waters to the laminated sediments associated with anoxic conditions can yield essential information about the tolerance levels of the various species. Rapid faunal changes took place at the onset of the anoxic periods that lasted for several thousands of years (Figure 3). The faunas that lived during the centuries immediately preceding the onset of anoxia (resulting in azoic sediments) have a low diversity and are strongly dominated by taxa (e.g., *Globobulimina* spp., *Chilostomella* spp.), which in recent ecosystems have been described in deep infaunal microhabitats. Surface dwellers adapted to strongly hypoxic conditions, such as those in the Santa Barbara Basin (e.g., *Epistominella smithi, Nonionella stella* and *Cassidulina delicata*; Mackensen & Douglas, 1989; Bernhard et al., 1997) were absent in the late Quaternary Mediterranean.

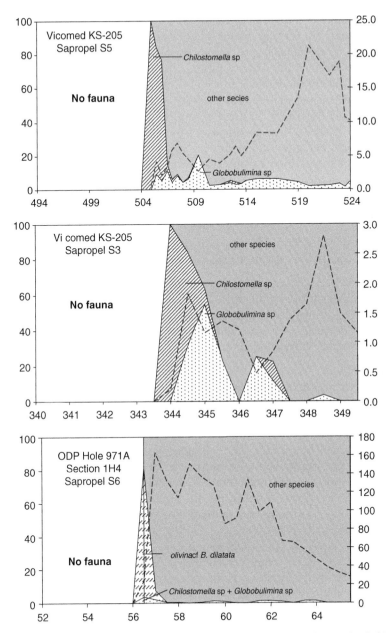

Figure 3 Relative frequency of deep infaunal taxa (percentage, scale on the left), in the centuries before deposition of sapropels S5, S3 (both from core Vicomed KS205, 38°11,86′N, 18°08,04′E, 2,345 m) and S6 (ODP Hole 971A, section 1H4, 33°42,19′N, 24°42.81′E, 2,143 m) and total faunal density (interrupted line, in number of foraminifera per gram dry weight, scale on the right), in function of depth in the core (in centimeter, scale below). Note the abrupt frequency increase to a total dominance in the last centimeter below azoic sediments, which coincides with the disappearance of almost all other taxa. In all three cases this frequency increase is accompanied by a strong drop in faunal density. Sapropel S5 but not S6 shows a preliminary frequency increase (at 510 cm) ~5 cm below this ultimate frequency increase.

Figure 3 shows the relative frequencies of deep infaunal taxa immediately below sapropels S5 and S6. In the rather oligotrophic environments of the central Mediterranean, *Globobulimina* spp., *Chilostomella* spp. and other deep infaunal taxa are rare or even absent at this water depth (De Rijk et al., 2000). These taxa increase dramatically in relative abundance in the 1–2 cm below the azoic sediments (here reflecting a time period of 1–2 centuries), to attain dominance in the last fauna-containing sample (Figure 3). In these examples, the increase in deep infaunal taxa to percentages of 80–100% is accompanied by a strong decrease in the number of foraminiferal tests per gram dry sediment, suggesting that the increase in relative abundance is mainly caused by the disappearance of other taxa, which are less resistant to low oxygen concentrations.

In some but not all sapropels the takeover of the benthic faunas by deep infaunal taxa is preceded by a preliminary increase in relative abundance of these taxa (as seen at 510 cm for sapropel S5, 347 cm for sapropel S3; but not for sapropel S6). This first increase of deep infaunal taxa is not accompanied by a disappearance of less resistant taxa. Two explanations are possible:

1) an increased organic flux to the ocean floor may have opened up niches deeper in the sediment for deep infaunal taxa.
2) a short-term strong decrease in bottom water oxygenation may have caused mixing of deep infaunal taxa with the more diverse faunas deposited before and/or after this short hypoxic event.

The first frequency increase of deep infaunal taxa at both S3 and S5 is accompanied by a temporary decrease of overall faunal density, so that the second possibility seems most probable.

The curves in Figure 4 present different scenarios during a strong decrease of bottom water oxygenation. In Figure 4a the decrease of bottom water oxygenation is not accompanied by an increase in organic flux, and could occur because of stratification of the water column, or a long-term lack of renewal of deep waters. The percentage of deep infaunal taxa will initially remain stable, until a critical point (indicated by arrow; here arbitrarily placed at 1 ml/l), at which taxa less resistant to hypoxic conditions start to disappear. With a further decrease of bottom water oxygenation, deep infaunal taxa rapidly take over hypoxic niches at the sediment-water interface, and their abundance increases exponentially, reaching 100% at ~0.01 ml/l. The initial percentage of deep infaunal taxa (the horizontal lines to the right of the threshold point in Figure 4a) is determined by the background organic flux level. Curve (1) depicts an oligotrophic setting without deep infaunal taxa where only the sediment-water interface is inhabited. Curve (2) characterises more eutrophic settings where the organic flux can sustain a fauna living at several cm depth in the sediment. Curve (1) corresponds to the situation for sapropel S6, ODP hole 971A (Figure 3); curve (2) corresponds to the succession below sapropel S5, core Vicomed KS-205 (Figure 3).

In Figure 4b the decrease in bottom water oxygenation is at least partially accompanied by an increase of the organic flux, as in places where hypoxia is caused by eutrophication. Deep infaunal taxa are then expected to show a first increase in abundance due to an increased food availability within the sediment, but with a

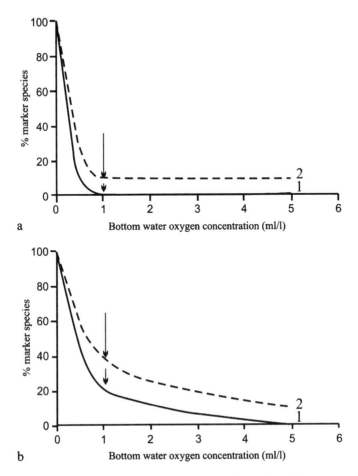

Figure 4 Hypothetical curves showing the percentage of deep infaunal species as a function of bottom water oxygen concentration and organic flux to the ocean floor. (a) The decrease of bottom water oxygenation is not accompanied by an increase in organic flux. (b) The decrease of bottom water oxygenation is accompanied by an increase of organic flux to the ocean floor. (a/b), Curve 1: oligotrophic setting without deep infaunal taxa in the background fauna. Curve 2: More eutrophic setting with ~10% deep infaunal taxa in the background fauna.

further decrease of bottom water oxygenation, less resistant taxa will start to disappear (vertical arrows in Figure 4), and the increase of deep infaunal taxa will strongly accelerate. From this threshold point on the percentage of deep infaunal taxa is no longer determined by the organic flux, but mainly by bottom water oxygenation.

This theoretical example illustrates that the interplay between organic flux and bottom water oxygenation, which controls the percentage of deep infaunal taxa, may be complex. As soon as oxygenation levels fall below a certain threshold value, further increase of the percentage of deep infaunal taxa seems to be largely due to the disappearance of taxa less resistant to low oxygen conditions, and therefore bottom water oxygenation, and it thus should be possible to use the percentage of deep infaunal taxa to estimate bottom water oxygenation. For this approach to work, it must be determined precisely at what oxygenation level the various less resistant taxa start to disappear from the ecosystem, and the percentage of deep infaunal taxa starts its final increase. In Figure 4, this level has arbitrarily been placed at 1 ml/l, but a different threshold value would not significantly change the shape of the curves. At bottom water oxygenation levels above this critical boundary, on the contrary, the percentage of deep infaunal taxa is determined by a combination of bottom water concentration and organic flux level, and the reconstruction of bottom water oxygenation is more difficult. However, when longer records are available (such as the sapropel records, Figure 3), it may be possible to constrain the organic flux level by other methods (Section 2.3).

Summarising, there is potential to develop a quantitative proxy method for bottom water oxygenation in spite of the complexity of the interplay between bottom water oxygenation and organic matter flux, on the basis of the relative proportions of a group of indicator species that replace taxa less resistant to low oxygen concentrations below a critical threshold value. The calibration of such a proxy has to be based on the rigorous application of the following conditions:

1) Indicator species for low oxygen conditions should be selected on the basis of observations in recent low oxygen ecosystems. All deep infaunal taxa (e.g., *Chilostomella* spp., *Globobulimina* spp.) qualify as marker species. Other taxa observed to be abundant in low oxygen settings have a planoconvex morphology (e.g., *Epistominella smithi*, *Nonionella stella*, and *Cassidulina delicata*) that suggests an adaptation to a microhabitat close to the sediment-water interface (e.g., Mackensen & Douglas, 1989; Bernhard et al., 1997).
2) In order to take taphonomical processes into account, the proxy method should be calibrated by a comparison between upper Holocene fossil faunas (if existing) and the recent value of the proxy parameter. The calibration should be performed in many different settings (specifically different productivity regimes) with oxygenation values between 0 and 2 ml/l. Above 2 ml/l, benthic foraminiferal assemblage composition is probably not influenced by bottom water oxygenation (Murray, 2001).
3) It should be tested whether the addition of a diversity index (Schmiedl et al., 2003) can make the proxy method more robust, and more globally applicable. However, diversity indices may decrease in response to an increase in organic flux as well as to a decrease in oxygen concentration.

If the relation between the percentage of deep infaunal taxa, bottom water oxygen concentration and downward organic flux can be successfully calibrated for a wide range of recent and/or sub-recent environments, a precise proxy of bottom water oxygenation may be obtained for the range of 0–1 ml/l. Potential success of this approach is suggested by Martinez et al. (1999), who documented that changes in the percentage of deep infaunal taxa mirrored changes in values of Mo/Al (Molybdenum/Aluminium), a geochemical oxygenation index. If we succeed in the development of a reliable proxy for bottom water oxygenation, the next challenge will be the reconstruction of the redox conditions within the superficial sediments.

2.3. Paleoproductivity Proxies

2.3.1. Introduction

Knowledge of past changes in primary production and the subsequent transport of organic matter to the ocean floor is essential for the understanding of the response of the biological carbon pump to climate change. Paleoproductivity proxies, which address these complex phenomena, may have several objectives: in some cases they attempt to reconstruct surface water primary production, whereas in other cases a less ambitious, but equally valuable approach is followed, aiming to reconstruct the flux of organic carbon to the ocean floor. A major complication is the fact that most paleoproductivity proxies based on responses to the organic carbon flux to the ocean floor (e.g., sedimentary organic carbon flux, benthic ecosystem response, benthic $\delta^{13}C$ values) will not only be influenced by the quantity of organic matter arriving at the sea floor, but also by other aspects of the organic flux, such as the quality of the organic matter, and the timing (constant versus pulsed) of the organic input. Another problem is that the mechanisms of organic material transport from the surface waters to the ocean floor are still badly known. If we are able to gain insight into these secondary aspects of paleoproductivity, we will also improve our understanding of the functioning of surface water, pelagic phytoplankton communities and the biological carbon pump in past oceans.

The use of benthic foraminifera as markers of export productivity to the sea floor has become conceivable since the 1970s, when it became progressively clear that not water depth, temperature and salinity ("water masses"), but organic input and bottom water oxygenation are the main environmental parameters controlling faunal patterns in open ocean benthic ecosystems. In the majority of open ocean benthic ecosystems, the flux of organic matter from the productive surface waters to the ocean floor constitutes the main food source for benthic organisms. Although hydrocarbon seeps and hydrothermal vents have received much attention in the past decennia because of their rich and often spectacular faunas, in the present oceans their cumulative surface area is probably small in comparison with sea floor areas, which depend on the input of organic detritus produced in the surface waters. Unfortunately, there is still considerable uncertainty about the processes of transport of particulate organic carbon to the ocean floor. Since the late 1970s, sediment trap data have been used to develop equations describing the downward organic flux through the water column (e.g., Eppley & Peterson, 1979; Suess, 1980; Betzer et al., 1984; Martin, Knauer, Karl, & Broenkow, 1987; Pace, Knauer, Karl, & Martin, 1987; Berger, Smetacek, & Wefer, 1989; Berger & Wefer, 1990). In all these equations, the downward organic flux is described simply as a function of primary production in the surface waters and water depth. More recent data, however, suggest that many other parameters may interfere:

- Several authors (e.g., Eppley & Peterson, 1979; Berger et al., 1989; Berger & Wefer, 1990) have suggested that the organic flux to the ocean floor constitutes a higher percentage of total productivity in unstable settings, such as ecosystems with a high seasonal variability in primary production. Recently, however, François, Honjo, Krishfield, and Manganini (2002) described a decreasing particle flux with higher seasonality.

- The amount of particulate organic carbon transported from the surface to deeper layers seems to depend on the structure and functioning of the pelagic food webs (e.g., Wassmann, 1993). For instance, Boyd and Newton (1995) describe a twofold increase of the primary organic carbon flux as a function of community structure. In general, vertical transport of large particles, such as marine snow (Turley, 2002), diatom aggregates (e.g., Kemp, Baldauf, & Pearce, 1995, 2000) or zooplankton fecal pellets, is much faster and more efficient than transport of small particles.
- Ballasting of organic particles with siliceous or carbonate tests (François et al., 2002), terrigenous dust (Ittekot, 1993) or tunicate feeding structures (Robinson, Reisenbichler, & Sherlock, 2005) may significantly increase the efficiency of organic matter transport to deeper parts of the water column. Armstrong, Lee, Hedges, Honjo, and Wakeham (2002) suggest that ballast minerals are largely responsible for the deep-water (>1,800 m) POC fluxes.
- Particle organic carbon fluxes through the water column appear to be strongly modulated by biological, physical and chemical transformation processes. Unfortunately, the various mechanisms of POC degradation in the water column are still poorly known (Jackson & Burd, 2002), and the potentially important role of dissolved organic carbon (DOC) is not taken into account in many organic flux studies.
- In many parts of the ocean, and especially in ocean margin settings, lateral advection by intermediate or deep water currents or by slope failure and turbidity currents is a major factor responsible for the transport of particulate organic carbon to the ocean floor (e.g., Antia, Von Bodungen, & Peinert, 1999). Laterally advected organic matter is often aged, with the more labile, easily consumable components stripped off. As a consequence, laterally advected organic matter will be only partially remineralised in the oxic ecosystems at the sediment-water interface. The more refractory parts, however, may trigger intensive early diagenetic processes in the anaerobic ecosystems deeper in the sediment (Carney, 1989; Fontanier et al., 2005).

In spite of these many complicating factors, an important part of the particle rain appears to fall vertically (e.g., Berelson et al., 1997; Nelson et al., 2002), and the POC flux varies as a function of primary production in the surface waters at least down to ~1,000 depth (Fischer, Ratmeyer, & Wefer, 2000). The above-mentioned flux equations could therefore be useful as a first approximation of the quantity of particulate organic matter arriving at the ocean floor.

One of the major problems inhibiting a more precise knowledge of the quantity (and quality) of organic matter arriving at the ocean floor is the difficulty to obtain reliable measurements. Empirical flux formulae are useful, but can only give a rough, long-term estimate of the organic flux to the ocean floor. More precise estimates can be obtained by measuring the amount of organic matter remineralisation in the benthic ecosystem, for instance by *in situ* measurements of oxygen consumption by benthic landers (e.g., Reimers, 1987; Gundersen & Jørgensen, 1990; Epping & Helder, 1998) or in benthic chambers (e.g., Tengberg et al., 1995). Onboard ship measurements in multi-cores are easier to perform, but generally considered less reliable. In order to have a more complete picture of the extent of organic matter degradation, including mineralisation under anaerobic conditions, vertical profiles of important redox species (e.g., nitrate, manganese, reactived iron-oxides and sulphate) are measured in the superficial sediment and pore waters (Froelich et al., 1979). Models integrating the downcore concentration profiles of all major redox species provide accurate estimates of the total amount of organic matter remineralised in the benthic ecosystem (e.g., Jahnke, Emerson, & Murray, 1982; Rabouille & Gaillard, 1991; Soetaert, Herman, & Middelburg, 1996; Kelly-Gerreyn, Hydes, & Waniek, 2005).

Although these approaches may provide reliable data about the remineralisation of labile organic matter, the data reflect rather short periods of time (days to weeks), and do not necessarily reflect the long-term, average conditions represented by a fossil benthic foraminiferal fauna, or by any other proxy carrier. Furthermore, these measurements mostly represent the labile, highly reactive part of the organic matter, but not the more refractory components that are degraded in the anaerobic ecosystems deeper in the sediment, over time spans of months to years, or are fossilised, to become sedimentary organic matter.

Problems concerning the quantification of the downward organic flux and the relative importance of laterally advected material do not only complicate reconstructions based on benthic foraminiferal faunas, but concern most paleoproductivity proxies. If we want to apply benthic foraminiferal assemblages successfully as quantitative paleoproductivity proxies we need to understand in detail how recent assemblages are structured by the input of various amounts of organic matter of different quality. In addition, we have to understand how low oxygen concentrations may modify the foraminiferal response to different organic flux regimes.

Several foraminiferal proxies for paleo-export production have been proposed during the last decennia, based on three different approaches:

1) Most proxies are based on supposed relations between faunal composition and organic carbon flux rates. In several areas, preferred flux regimes have been tentatively determined for dominant benthic foraminiferal taxa (e.g., Altenbach et al., 1999, 2003; De Rijk et al., 2000; Schönfeld & Altenbach, 2005). For example, *Uvigerina mediterranea* and *U. peregrina* are typical for flux rates above 2.5 g/m^2/yr, whereas *Cibicides wuellerstorfi*, *Gyroidina altiformis*, and *G. orbicularis* are almost exclusively found below this flux range (Altenbach et al., 1999; de Rijk et al., 2000). However, important inconsistencies appear to exist between studied areas (Schönfeld & Altenbach, 2005), and some important taxa, such as *Epistominella exigua* (Altenbach et al., 1999) or *Melonis* spp. (Altenbach et al., 1999; de Rijk et al., 2000) occur in areas with very different flux regimes, and appear to be not very dependent on a specific organic flux level. These inconsistencies may be due to inadequate estimates of export production, but also to the impact of other ecological factors that may interfere with the foraminiferal dependence on export production. Until today, observed flux regimes for individual taxa have not been used systematically to reconstruct paleo-productivity.

2) Multivariate statistical methods have been applied in order to investigate the response of the foraminiferal assemblage composition to different flux regimes

(e.g., Loubere, 1991, 1994, 1996, 1998; Fariduddin & Loubere, 1997; Kuhnt et al., 1999; Loubere & Fariduddin, 1999; Wollenberg & Kuhnt, 2000; Morigi et al., 2001). Loubere (1998) and Loubere and Fariduddin (1999) presented multiple regression functions that relate the foraminiferal assemblage composition quantitatively to primary production in the surface waters. Others (e.g. Kuhnt et al., 1999; Wollenburg, Kuhnt, & Mackensen, 2001) used the statistical relations observed in recent ecosystems to reconstruct paleoproductivity on the basis of fossil faunas.

3) Herguera and Berger (1991) and Herguera (1992) proposed a paleoproductivity proxy based on the accumulation rate of the benthic foraminiferal fossil fauna larger than 150 μm. Their data from the western equatorial Pacific show that the benthic foraminiferal accumulation rate (BFAR) varied linearly with export production for stations with primary production ranging from 40 to 135 $g/m^2/yr$, so that for every mg of organic carbon reaching the ocean floor, one foraminiferal shell larger than 150 μm is deposited. The striking simplicity of this idea is appealing but also discouraging, because somehow we feel that nature cannot be that simple. It is difficult to imagine that faunas subject to the much higher flux regimes characteristic of more eutrophic areas will have test production rates per unit of organic carbon identical to those described by Herguera (1992), and a non-linear response to increasing export production has indeed been suggested by Schmiedl and Mackensen (1997). Naidu and Malmgren (1995) showed that in low oxygen environments on the Oman margin OMZ, BFAR does not reflect surface-water productivity. In spite of these problems, BFARs have tentatively been applied to reconstruct paleo-productivity semi-quantitatively (e.g., Den Dulk, Reichart, van Heyst, Zachariasse, & Van der Zwaan, 2000; Herguera, 2000; Wollenburg & Kuhnt, 2000). No objective tests are available to check the reliability of the results, a problem with all paleo-productivity proxies, so that we are still in doubt about the reliability of this method.

2.3.2. Paleoproductivity proxies based on flux-dependence of individual species

The realisation that foraminiferal species depend on specific organic flux rates dates from the 1970s, when Lutze and co-workers (Lutze, 1980; Lutze & Coulbourn, 1984) conclusively showed that some foraminiferal taxa (e.g., *Uvigerina* spp.) have high relative densities in areas with high organic input. Until today, only very few studies have tried to accurately quantify the flux-dependence of the various benthic foraminiferal taxa.

In a study of the morphology of benthic foraminifera in core-top samples from the Norwegian Sea, Corliss and Chen (1988) observed a rapid shift from faunas dominated by morphotypes considered typical of epifaunal lifestyles (trochospiral, milioline) to faunas dominated by morphotypes considered typical of infaunal microhabitats (rounded planispiral, flattened ovoid, tapered cylindrical, spherical, tapered flattened) at a water depth of ~1,500 m. They estimated that this change in morphotype dominance takes place at an organic flux level of 3–6 $gC/m^2/yr$. Buzas, Culver, and Jorissen (1993) calculated that microhabitat assignments on the basis of morphology have only a 75% accuracy, but the data of Corliss and Chen (1988) clearly document the existence of a major faunal change related to a specific organic flux level.

In a new analysis of the large datasets generated by the Lutze-team, Altenbach et al. (1999) investigated the dependence on specific organic flux rates for various foraminiferal taxa. Rather disappointingly, the range of most foraminiferal taxa appears to be very large: correlation coefficients between the percentage in the total fauna of almost all taxa and the organic flux rates at the stations where they are found are very weak. Altenbach et al. (1999) concluded that down to 2,000 m the succession of faunal species seems to be related mainly to bathymetry, whereas only below 2,000 m a relation with export production becomes visible, as confirmed for faunas from the Gulf of Guinea (Altenbach, Lutze, Schiebel, & Schönfeld, 2003). Altenbach et al. (1999) concluded that the presence or absence of a taxon does not seem to be a valuable measure for the reconstruction of flux rates. In spite of these rather disappointing results, they suggested that an organic flux rate of $\sim$2–3 gram organic carbon per meter square per year ($gC/m^2/yr$) is an important threshold limit for many species, as confirmed by Jian et al. (1999), De Rijk et al. (2000) and Weinelt et al. (2001). De Rijk et al. (2000) observed that such a faunal threshold occurs at an estimated input of labile organic matter of $\sim$2.5 g/m^2/yr in the Mediterranean, which occurs at $\sim$1,500 m water depth in the eutrophic Western Mediterranean, but shoals to $\sim$400 m depth in the much more oligotrophic eastern Mediterranean. Morigi et al. (2001) used multivariate statistical analysis of recent foraminiferal thanatocoenoses in the upwelling area off Cap Blanc, a part (19–27°N) of the much larger area (Arctic Ocean, Norwegian-Greenland Sea, NW Africa, and Guinea Basin) investigated by Altenbach et al. (1999), to subdivide frequent taxa into six groups, each with a preference for a rather narrow range of organic flux rates.

In view of these studies, we think that Altenbach et al. (1999), who studied a huge area, may have been too pessimistic when they concluded that the presence or absence of foraminiferal species cannot be used to reconstruct organic flux levels. Taxa with a preference for a rather narrow range of flux levels will necessarily display a weak correlation coefficient with organic flux when their percentage distribution is compared with the total range of observed flux rates over a huge area.

The direct application of these observed flux-dependencies is hampered by three problems:

1) In all previously mentioned studies, the estimated organic flux values to the ocean floor were based on a combination of (usually satellite-image derived) primary productivity (PP) values and empirical flux equations. As a consequence, the resulting J_z (organic flux arriving at a water depth of z meters) values are very approximate, and it is difficult to compare flux-dependencies between different regions. The obvious solution would be to calibrate these calculated flux values by a comparison with actual flux data observed in sediment traps in the same areas. Unfortunately, sediment trap data are still scarce, in many areas non-existent, and we do not have measured flux data for larger sets of stations. However, sediment trap data for only one or two stations in each investigated area

would already allow us to verify the correctness of the order of magnitude of values provided by the flux equations, and, when necessary, to apply corrections for nearby stations. Successful proxy calibration depends on the acquisition of reliable flux data for many open ocean and continental margin areas.

2) Presently most scientists agree that the organic flux is the main parameter structuring open ocean benthic faunas, but evidently it is not the only one. Other factors may modify the foraminiferal response to varying flux levels, especially the response as expressed in presence/absence patterns. Therefore, proxy calibration efforts should consider not only the relations between faunal characteristics and J_z, but should take other environmental factors (such as bottom water oxygenation, current velocity or sediment grain-size) into account.

3) Foraminifera have a wide range of ecological strategies. Species vary from K-selected, highly specialised taxa, occupying very specific and narrow ecological niches, to much more opportunistic, r-selected taxa, occupying a wide range of ecological niches. The group of r-selected taxa shows a strong response to highly episodic phytodetritus deposits, but will probably be weakly dependent upon annual flux rates. Good marker species for annual organic flux rates are expected to be K-selected species.

These problems are serious, but we do not think that they are insoluble. Tentative quantifications of preferred flux regimes for individual species are still scarce, but in our opinion, the first results are encouraging. A large research effort is needed to improve the quantification of the flux of organic matter to the benthic ecosystem, and to determine upper and lower tolerance limits as well as the optimum flux range for many foraminiferal taxa. We are convinced that such an inventory could be the basis of a reliable paleoproductivity proxy.

2.3.3. Paleoproductivity proxies based on flux-dependency revealed by multivariate statistical methods

Since the early 1990s Loubere and co-workers (Loubere, 1991, 1994, 1996, 1998; Fariduddin & Loubere, 1997; Loubere & Fariduddin, 1999) built up a large set of core-top data on benthic foraminiferal assemblage distribution in the Atlantic, Indian and Pacific Ocean, and developed a multiple regression function linking the composition of the foraminiferal faunas to the primary production in the over-lying surface waters. For their samples from a fairly narrow bathymetrical range (2,800–3,500 m), the correlation between bottom water oxygenation and primary production was very weak, so that apparent reactions to different organic flux rates cannot be caused by co-varying oxygen concentrations. In the 1994 and 1996 papers, Loubere presents a Pacific Ocean dataset of 74 samples. The multiple regression analysis shows a very strong correlation coefficient to surface water primary production values. Fariduddin and Loubere (1997) present a similar analysis for an 84 station Atlantic Ocean dataset, where productivity also seems to be the most important ecological factor. In 1998, Loubere presented a paper in which the Pacific multiple regression function was tested with Indian Ocean core-top material (water depth 2,400–3,500 m). Although in general the Pacific regression function reproduced the Indian Ocean primary production values rather well, the Indian Ocean benthic foraminiferal faunas differed by much higher percentages of *Epistominella*, interpreted as a response to the more seasonal aspect of the organic matter flux in the Indian Ocean. In 1999, Loubere and Fariduddin presented a multiple regression analysis of a 207 station sample set from the Pacific, Indian and Atlantic Oceans, from water depths between 2,300 and 3,700 m. Again, the correlation coefficients between primary production, seasonality and other parameters are weak.

Kuhnt et al. (1999) applied correspondence analysis to data on benthic foraminiferal faunas in 43 box cores from the South China Sea. The first factor of their correspondence analysis has a very strong correlation to the estimated organic carbon flux to the ocean floor. This relation was then used to reconstruct the organic carbon flux to the ocean floor for the last glacial–interglacial cycle. In a study of 37 Rose-bengal stained samples from the Laptev Sea (Arctic Ocean), Wollenburg and Kuhnt (2000) found a strong positive correlation between their first factor and the organic flux to the ocean floor, which was used by Wollenburg et al. (2001) to reconstruct a paleoproductivity record on the basis of faunal successions found in two piston cores from the Arctic Ocean.

These multivariate statistical approaches are promising, but they suffer from the general weaknesses of such methods. First, multiple regression analyses ideally need about 20 times more samples than variables in order to be reliable (Tabachnick & Fidell, 1983). Even the 207 station dataset of Loubere and Fariduddin (1999) is far too small in view of the large number of variables (species abundances). Next, all described regression functions are strictly empirical equations, thus not based on understanding of the ecological mechanisms controlling the faunal distribution. With other words, the statistical relationships are black boxes, with no underlying logic to explain the results. Recurrent combinations between environmental parameters observed in today's oceans (e.g., low temperature and high primary production, or high organic flux and low oxygen concentrations) may not have existed or may have been different in the past, causing different responses of faunal composition to the target parameter. The application of statistical relations observed in recent ecosystems becomes especially hazardous when extrapolation is used to reconstruct conditions not present in the dataset used for calibration. Such non-analog conditions include extreme primary production regimes (higher or lower than today), or combinations of environmental parameters (organic flux, oxygenation, quality of organic matter, etc.) not encountered in the recent oceans. For instance, in the geological past deep ocean temperatures were considerably higher than today, and it has been speculated that the resulting increased metabolic activity would lead to increased organic matter mineralisation in the water column (and a lower flux to the ocean floor). Benthic faunas would probably have required higher food fluxes to sustain a similar biomass (Thomas, 2007).

Finally, as for most proxies, we can only check the validity of the results by comparing them to estimates obtained by independent (e.g., geochemical) proxy methods. In spite of our instinctive mistrust when confronted with methods not based on a mechanistical understanding of observed ecological patterns, we agree that multivariate statistical methods are a promising pathway to a useful paleoproductivity proxy. It is therefore important to continue the efforts to increase the size of available datasets, and to extend this type of study to a wider range of continental slope and open ocean environments.

2.3.4. Paleoproductivity proxies based on the benthic foraminiferal accumulation rate (BFAR)

In 1991, Herguera and Berger proposed the BFAR (number of benthic foraminifera per unit of area per unit of time) as a paleoproductivity proxy. The underlying hypothesis of the method is that for every mg of organic carbon reaching the ocean floor, a fixed number of fossil foraminiferal tests is deposited. In their 1991 benchmark paper, Herguera and Berger indicated that for such a concept to work, four conditions must be met: (1) a linear relation must exist between the organic matter flux and the number of fossilised foraminifera; (2) the flux of organic matter arriving at the ocean floor must in a linear way depend on surface water primary production; (3) the sedimentation rate must be invariable or known in sufficient detail; and (4) there must be no significant carbonate dissolution. The authors suggest that probably none of these assumptions is fully met.

Herguera and Berger (1991) investigated two series of cores in the western equatorial Pacific, from oligotrophic to mesotrophic areas with primary production rates of 65–80, and 105–135 g/m^2/yr, respectively. After calculation of the flux of organic matter to the ocean floor by applying a flux equation based on sediment trap data, and a comparison with the accumulation rates of foraminifera in coretops, they arrived at the conclusion that for every mg of organic carbon reaching the ocean floor, ~1 foraminiferal test larger than 150 µm is fossilised. Herguera (1992) exploits various flux formulae, and concludes that the BFAR is not reliable below a water depth of ~4,000 m due to carbonate dissolution. Naidu and Malmgren (1995), who tested the BFAR in the Oman Margin OMZ, concluded that the proxy does not work in a low oxygen setting.

Although BFAR has been applied widely as a semi-quantitative proxy of paleoproductivity (e.g., Thomas et al., 1995; Nees, Altenbach, Kassens, & Thiede, 1997; Schmiedl & Mackensen, 1997; Den Dulk et al., 2000; Herguera, 2000; Wollenburg & Kuhnt, 2000), most authors have not used the BFAR concept to obtain quantitative estimates of primary production, or even of the organic flux to the ocean floor, because of the problems outlined by Herguera and Berger (1991), and the absence of a reliable calibration, either in the western equatorial Pacific or in regions with different patterns of productivity.

Such a calibration is needed because it is very improbable that in different flux regimes a similar number of foraminifera will be deposited per gram organic carbon arriving at the ocean floor (Schmiedl & Mackensen, 1997). One would expect that in eutrophic areas, with a high export production, faunas would be dominated by rather opportunistic taxa which may produce much more offspring per mg organic carbon than the more K-selected taxa encountered in more oligotrophic areas. The scheme presented in Figure 5 presents the successive steps for the calibration of a paleoproductivity proxy based on the BFAR.

The first problem (A in Figure 5) is the quantitative relation between primary production in the surface waters (PP), the quantity of organic matter that leaves the

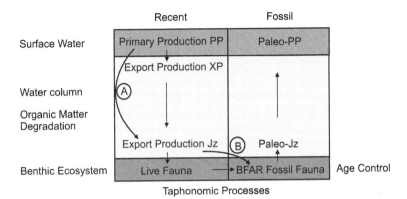

Figure 5   Scheme indicating the various steps and potential problems involved in the calibration of the BFAR as a paleoproductivity proxy. See text for further explanation.

productive surface layer (Export Production, XP), and the flux of organic matter to the sea floor at a water depth of z meters (J_z). As outlined in Section 2.3.1, equations quantitatively describing these relations have been developed on the basis of a comparison of primary production values measured in surface waters with sediment trap data. In paleoceanography, once the paleo-J_z (flux to the sea floor) has been reconstructed, these flux equations can theoretically be used to calculate paleo-PP (past primary production). However, all attempts to translate a reconstructed paleo-J_z into a paleo-PP value should accept a wide error margin, due to the uncertainties of the flux equations and the possibility of a significant contribution of laterally advected material. Only a very small portion (0.01–1%) of the total primary production is transported to the ocean floor (e.g., Murray et al., 1996), so that even small changes in the transport mechanism may have a large impact on the amount of organic matter arriving at the sea floor. It seems therefore judicious to limit the use of the BFAR concept to the reconstruction of the organic flux to the ocean floor (paleo-J_z) for the time being.

For calibration purposes, the characteristics of the benthic fauna have to be compared quantitatively with recent J_z values, usually calculated by introducing satellite-derived PP-values into the afore-mentioned flux equations, adding another source of uncertainty. Difficulties to obtain precise J_z values, needed to quantify its impact on the foraminiferal faunas, have seriously hampered the development of the BFAR concept.

A next problem is the calibration of the BFAR by comparison with recent J_z values (B in Figure 5). When more or less accurate estimates of J_z are available, they can easily be compared with the density of a living foraminiferal fauna (e.g., Fontanier et al., 2002), but it is much more complicated to compare J_z with the accumulation rate of fossil foraminifera. In order to do so one must have a precise knowledge of the sediment accumulation rate, requiring a very detailed age control for the topmost sediment layer that in ideal cases and for the last ~100 years may be obtained by downcore profiles of excess ^{210}Pb and/or [237]Cs profiles (e.g., Nittrouer, DeMaster, McKee, Cutshall, & Larsen, 1984). However, bioturbation activities and/or sediment resuspension after deposition may significantly alter the ^{210}Pb-profiles, which can lead to incorrect calculation of the sediment accumulation rates. In aquatic environments, assumptions must also be made about the exact transfer process of the radionuclides from atmosphere to sea floor (e.g., Appleby & Oldfield, 1992). Next, the loss of foraminiferal tests due to taphonomical processes should be extremely well constrained. Unfortunately, our present knowledge of taphonomical processes is far too fragmentary to have any realistic hope to calibrate the BFAR method by comparing J_z with recent or core-top foraminiferal faunas, for which taphonomical processes have not been finished. An additional problem is the fact that recent faunas may show important seasonal variability, and are therefore not always representative for longer periods.

We thus are left with the option to calibrate the BFAR on the basis of (sub-) recent material, for which two essential conditions have been met: (1) taphonomical processes have been terminated, and (2) the fauna is representative of recent environmental conditions. Unfortunately, it is very difficult, if not impossible, to be certain that both conditions are fulfilled. In recent or very modern faunas, faunal transformations due to taphonomical processes have not been finished (as discussed above). For fossil faunas we do not know whether the environmental parameters by which they were structured were similar to the recent conditions observed at the site. The best solution seems to calibrate the proxy relationship by using faunas deposited during the early Holocene or other recent interglacial periods under environmental conditions (as far as we know them) relatively comparable to modern ones. A considerable uncertainty will be the result, thus paleo-J_z estimates resulting from BFAR will always have a large error margin.

A final problem is the quality of the organic flux to the ocean floor. Fresh marine organic matter (phytodetritus) is directly consumed by benthic foraminifera, and its availability in large quantities may lead to the production of a relatively large number of produced tests per mg of organic carbon (e.g., , 1988, 1993). Aged, or continental organic matter, on the contrary, will contain more refractory components, and only a small part can be directly consumed by the benthic fauna. As a consequence, fewer foraminifera will be produced per mg organic carbon when refractory organic carbon accounts for an important part of the input. There are as yet no methods available that are routinely used to describe the nutritious value of organic matter, and it may be particularly complicated to take this parameter into account.

Below, we use data from three piston cores (Table 2) from the continental margin off Cap Blanc (NW Africa) to illustrate some of the calibration problems. Core Sed17aK was sampled in an oligotrophic area, at 2,975 m depth, Core KS04 at 1,000 m depth in an area of strong coastal upwelling, whereas core Sed20bK comes from a site at 1,445 m depth, presently outside the direct influence of upwelling. For all three cores age control is based on stable isotope analyses of planktonic foraminifera (G. bulloides for Sed17aK and Sed20bK; G. inflata for KS04) and correlation to an orbitally tuned age model. We will use core Sed20bK, which has the highest time resolution, as an example (Figure 6).

In all three cores the topmost sediment layer has been lost and only the lower part of the Holocene is present, as is often the case for piston cores. In core Sed20bK (as well as in the other cores) the BFAR (of the >150 μm size fraction of

Table 2 Cores KS04, Sed20bK, and Sed 17aK: Geographical Position, Water Depth, Age at the Bottom of the Core (Indicated as MIS: Marine Isotopic Stage), Number of Age Control Points, Primary Production (According to Schemainda et al., 1975), Total Estimated Organic Carbon Flux to the Sea Floor, and its Labile Component (Calculated According to Equation (1), After Herguera (1992)).

Core	Latitude	Longitude	Water depth	Age at bottom core	Number of age control points	Primary production (g/m^2/yr)	J_z total C$_{org}$ flux (g/m^2/yr)	J_z labile component (g/m^2/yr)
KS04	20°34.70'N	18°08.80'W	1,000 m	MIS7	8	200	7.9	5.65
Sed20bK	25°01.70'N	16°39.02'W	1,445 m	MIS5d	12	100	2.7	1.3
Sed17aK	25°16.80'N	17°06.45'W	2,975 m	MIS6	15	65	1.1	0.35

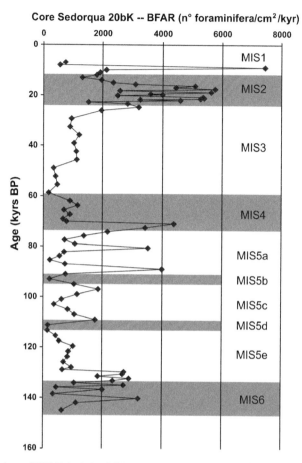

Figure 6 Variation of BFAR in NW Africa core Sed20bK (25°01.70′N, 16°39.02′W, 1,445 m), over the last two glacial–interglacial cycles. After Guichard et al. (1999).

benthic foraminifers, following Herguera & Berger, 1991) shows a strong variability, with peak values ~10 times higher than the baseline (Figure 6). Such peaks are present in glacial and interglacial periods, and may be due to rapid changes in sediment accumulation rate which are not recognised due to the imperfect time resolution, or may be real, due to the occurrence of long-lived benthic foraminiferal bloom periods in response to upwelling events that have resulted in periods with large-scale phytodetritus deposition. Because the topmost sediment layer has not been recovered, we do not know whether the present organic flux to the ocean floor corresponds to (1) a peak of BFAR, (2) Holocene background BFAR values, or (3) some intermediate BFAR value. The only acceptable solution is to consider the two extreme situations, and use those to define an error envelope. Guichard, Jorissen, and Peypouquet (1999) decided to compare recent J_z values with BFAR values of Marine Isotope Stage (MIS) 5a, an interglacial period with climatic conditions more or less comparable to the Holocene. With the exception of the peak value observed at ~9,000 BP, BFAR values observed in MIS 5a cover the range of values observed in the lower Holocene part of the core.

Today, primary production at site SED20bK is ~100 gC/m^2/yr (Schemainda, Nehring, & Schultz, 1975). For the calculation of the flux to the sea floor, we use the flux equation proposed by Herguera (1992):

$$ J_z = \left(2\sqrt{PP} \times \frac{PP}{z}\right) + \left(\frac{5}{\sqrt{PP}} \times \frac{PP}{\sqrt{z}}\right) \tag{1} $$

in which J_z is the total organic flux to the sea floor at water depth of z meter, and PP primary production in the surface waters. We prefer this flux equation over other ones (review in Herguera, 1992), because it distinguishes between a labile component (first term) that rapidly diminishes with water depth, and a more refractory component (second term) that is much more constant with water depth.

For site KS20bK, the estimated total J_z is 2.7 gC/m^2/yr; the labile component is estimated at 1.4 gC/m^2/yr, the more refractory component 1.3 gC/m^2/yr. Three different hypotheses can now be tested. Estimated recent flux rates:

1. correspond to the background (minimal) BFAR values observed in MIS 5a,
2. correspond to the average BFAR values during MIS 5a
3. correspond to the peak values during MIS 5a, which seems improbable in view of the moderate recent primary production value.

Next, for these three different hypotheses (recent PP corresponds to minimum, average or peak BFAR) the BFAR values can be compared with the estimated values of the total organic flux to the sea floor, and with its labile component. The result is the number of foraminifera deposited per mg of organic carbon (total or labile) arriving at the sea floor (Table 3); for core Sed20bK this number varies from 2.9 to 28 foraminifera per mg for total organic carbon, and from 5.6 to 54 foraminifer per mg if only the labile component is considered.

This first estimate produces values that differ by an order of magnitude, but it seems very improbable that the recent primary production value (very moderate for the Cap Blanc region) corresponds to peak BFAR values (Figure 6), which probably represent periods of intensified upwelling. It seems more probable that recent PP values correspond to a BFAR value between the minimum and average value, which would mean that for every mg of organic carbon arriving at the ocean floor, 2.9–5.6 fossil foraminifera >150 μm are preserved (or 5.6–10.7 per mg labile organic carbon).

For cores KS04 (water depth 1,000 m) and Sed17aK (water depth 2,975 m) (Guichard et al., 1997), a similar procedure was followed (Table 3). The flux rates were estimated by introducing primary production measurements of Schemainda et al. (1975) into flux equation (1) (Herguera, 1992). Next, the number of fossil foraminifera per mg organic carbon was calculated for each of the three hypotheses, once using the total organic flux, and once using only the labile component (Table 3).

From these BFAR calibrations we observe:

1) The tentative value of 1 foraminifer larger than >150 μm deposited per 1 mg organic matter reaching the sea floor (Herguera & Berger, 1991) has the right order of magnitude, but appears too low for our continental margin cores.

Table 3 Cores Sed2obK, KS04, and Sed17aK: Comparison of Observed Benthic Foraminiferal Accumulation Rates (BFAR, According to Three Hypotheses) with the Estimated Flux of Organic Carbon to the Sea Floor (Total Flux and Labile Component).

Core Sed20bK–1,445 m	**BFAR (forams/cm^2/kyr)**	**Total C$_{org}$ flux (2.7 gC/m^2/year)**	**C$_{org}$, labile component (1.4 gC/m^2/yr)**
Background BFAR	780	2.9 forams/mg C$_{org}$	5.6 forams/mg labile C$_{org}$
Average BFAR	1,500	5.6 forams/mg C$_{org}$	10.7 forams/mg labile C$_{org}$
Peak values BFAR	4,000–7,500	15–28 forams/mg C$_{org}$	28–54 forams/mg labile C$_{org}$
Core KS04–1,000 m	**BFAR (forams/m^2/yr)**	**Total C$_{org}$ flux (7.9 gC/m^2/yr)**	**C$_{org}$, labile component (5.65 gC/m^2/yr)**
Background BFAR	17,000	2.2 forams/mg C$_{org}$	3.0 forams/mg labile C$_{org}$
Average BFAR	24,800	3.1 forams/mg C$_{org}$	4.4 forams/mg labile C$_{org}$
Peak values BFAR	50,000	6.3 forams/mg C$_{org}$	8.8 forams/mg labile C$_{org}$
Core Sed17aK–2,975 m	**BFAR (forams/m^2/yr)**	**Total C$_{org}$ flux (1.1 gC/m^2/yr)**	**C$_{org}$, labile component (0.35 gC/m^2/yr)**
Background BFAR MIS5a	450	0.4 forams/mg C$_{org}$	1.3 forams/mg labile C$_{org}$
Peak value BFAR MIS1	4,000	3.6 forams/mg C$_{org}$	11 forams/mg labile C$_{org}$

Note: Three scenarios are tested, and the numbers of foraminifera>150 μm preserved in the sediment are given for each case.

If we do not consider the rather improbable numbers based on Hypothesis 3 (recent situation corresponds to peak BFAR), we find that 0.4–5.6 fossilised foraminifera >150 μm are deposited per mg total organic carbon arriving at the sea floor off NW Africa.

2) The values for our deepest core (Sed17aK; Table 3) correspond rather well to the values proposed by Herguera and Berger (1991), whereas our two continental margin cores, representing much more eutrophic ecosystems, show much higher values (2.2–5.6 foraminifera deposited per mg total organic carbon arriving at the ocean floor). This suggests that the relationship between J_z and BFAR is not linear. More eutrophic areas appear indeed to be inhabited by more opportunistic taxa that produce more tests per unit of organic carbon.

3) By only considering the labile component of the downward organic flux, the differences between our deep oligotrophic site, and our shallower, more eutrophic sites diminish slightly (Table 3). This suggests that it may be judicious to consider only the labile part of the organic flux to the ocean floor.

We think that this example documents that BFAR is a promising proxy for paleo-J_z. We need to develop a large data-basis of BFAR records in well-dated cores from regions with different patterns of PP (oligotrophic to eutrophic, highly seasonal to less seasonal), in order to determine how BFAR responds to various organic flux regimes.

2.3.5. Reconstructing the quality of the downward organic flux

Benthic foraminifera, as well as most other benthic organisms, will mainly respond to the flux to the ocean floor of labile organic particles that are easily metabolised. Terrigenous or old marine organic particles (refractory organic matter) have only a limited nutritious value, and will commonly bypass the oxic niches at the sediment-water interface. Only in the anaerobic environments deeper in the sediment part of this material is recycled (e.g., Fenchel & Finlay, 1995), whereas the remainder will be preserved as sedimentary organic matter. The absolute quantities and relative proportions of these two types of organic matter (labile versus refractory) differ as a function of the amount of terrigenous input (usually determined by the distance from land) and the local biological and sedimentological conditions. At sites with an influx of predominantly labile organic matter, most will be consumed by the benthic fauna at the sediment-water interface, whereas at sites with more refractory components, a larger part of the organic matter will become preserved in the sediments. It thus seems unlikely that sedimentary organic flux rates (the amount of fossilised C_{org} per surface area per year) can provide a reliable paleoproductivity proxy (e.g., Wefer et al., 1999), although the sedimentary organic carbon content has often been used to reproduce past biological production (e.g., Müller & Suess, 1979; Sarnthein, Winn, Duplessy, & Fontugne, 1987, 1988).

In our opinion, the differential response of benthic foraminiferal faunas to influx of these two types of organic matter could be used to gain insight into the character of past organic particle supply, but more information is required about the dependence of various species on food quality. Surface-dwelling taxa have been said to rely on labile organic matter, whereas intermediate and deep infaunal taxa tolerate more refractory food particles made available by partial recycling under anaerobic conditions (e.g., Jorissen, 1999a; Fontanier et al., 2005).

In order to gain insight into the organic matter quality, Guichard et al. (1999) combined data on BFAR and sedimentary organic matter flux (Figure 7). The underlying idea is that the BFAR varies as a function of the labile organic carbon input, but is rather insensitive to the input of refractory organic carbon. The quality of the organic matter flux probably changed considerably over time: during periods with intensified upwelling, large amounts of labile organic matter reached the sea floor where they undoubtedly triggered biological production, and high BFARs. During prolonged periods without sustained upwelling (as today), on the contrary, the quality of the organic input was probably much lower, but a larger proportion was preserved in the sediment, leading to a relatively low BFAR with respect to the sedimentary C_{org} flux.

Sedimentary C_{org} flux is positively correlated to BFAR, but there is a wide scatter around regression line (1), which represents the average ratio between the BFAR and sedimentary C_{org} flux (of 15 foraminiferal tests per 1 mg sedimentary organic carbon), and thus an average quality of organic matter. Samples with minimal BFAR values probably represent periods when the quality of the organic influx was poor (i.e., minimal labile fraction). Points high above the regression line represent samples with a very high BFAR relative to the sedimentary organic carbon flux, suggesting an organic matter flux dominated by labile components, in this region possibly representing periods with strongly intensified upwelling. The variability of BFAR is not only a response to a varying organic input, but may be partially caused by short-term variability of the sediment accumulation rate, not recognised due to the limits on the resolution of the age model. In spite of these complications, we think that this approach deserves to be further explored.

2.3.6. Reconstructing the periodicity of the downward organic flux

Studies on the Porcupine abyssal plane (Lampitt, 1985; Gooday, 1988; Lambshead & Gooday, 1990) have conclusively shown the importance of the massive phytoplankton

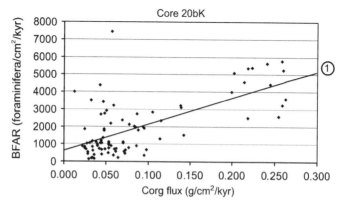

Figure 7 Core Sed20bK; comparison between BFAR and sediment Corg flux (after Guichard et al., 1999); 88 samples; regression line (1): BFAR = 15,160 × C_{org}+616.5 (r^2 = 0.42).

deposits in the deep ocean, to which the benthic fauna responds by a period of intense biological activity, for many taxa leading to accelerated growth and/or reproduction (Pfannkuche & Lochte, 1993). Benthic foraminifera are among the organisms with the strongest response to such events (e.g., Gooday, 1988, 1993, 1996; Gooday, Levin, Linke, & Heeger, 1992; Gooday & Hughes, 2002). Several small taxa (e.g., *Alabaminella weddellensis*, *Epistominella exigua*, *Epistominella pusilla*, *Fursenkoina* spp., *Globocassidulina subglobosa*) as well as soft-shelled forms rapidly colonise the phyto-detritus, producing abundant offspring, leading to the prediction that these taxa will probably become dominant elements in the fossil assemblages at sites with episodic phytodetritus deposits. Smart et al. (1994) applied this concept to Miocene and Quaternary deep-sea cores; Thomas et al. (1995) to two mid-latitude North Atlantic sediment cores (last 45 kyr). They interpreted the abundance of phytodetritus species and increased BFAR during the last deglaciation, as the result of increased surface primary production, and subsequent transport of phytodetritus to the ocean floor in response to a northward migration of the polar front (Thomas et al., 1995). Thomas and Gooday (1996) argued that phytodetritus deposition increased strongly at the establishment of the Antarctic ice sheet (earliest Oligocene), leading to the occur-rence of lower diversity assemblages with dominant phytodetritus species at high latitudes. Several other papers follow a similar approach to reconstruct past phyto-detritus flux events (e.g., Nees, Armand, De Deckker, Labracherie, & Passlow, 1997, 1999; Nees & Struck, 1999; Ohkushi, Thomas, & Kawahata, 2000).

A potential problem is the fact that most of the indicator species of these pulsed phytodetritus deposits are small (63–150 μm), have therefore not been studied very often, and may be sensitive to dissolution. The *E. exigua–A. weddellensis* assemblage was positively correlated with the seasonality of primary production, whereas individual species of this assemblage did not show such a relationship (Sun, Corliss, Brown, & Showers, 2006). In fact, all opportunistic taxa will be advantaged by unstable conditions such as episodic transport of phytodetritus to the ocean floor, but some opportunistic species can respond strongly to phytodetritus input in an oligotrophic deep ocean setting, whereas other taxa may be typical for similar conditions in a more eutrophic continental slope context. For instance, at 550 m water depth in the Bay of Biscay large (adult specimens >150 μm) *Uvigerina peregrina* and *Uvigerina mediterranea* show (together with *E. exigua*) a marked increase in abundance in response to phytodetritus deposits following phytoplankton blooms (Fontanier et al. (2003). These observations are in contrast with the situation in the open-ocean environments of the north-western Pacific (Ohkushi et al., 2000), where *Uvigerina* spp. reacted very differently from *E. exigua*. These results show that we can not validly extrapolate from observations at one location to large areas, and that more observations of faunal variability over time are needed to further understand the periodicity of the past organic matter flux.

2.4. The Water Mass Concept

As described in Section 2.1, the 'water mass concept' implied that the differences in physico-chemical parameters between various water masses were responsible for the fact that they could be characterised by a specific benthic foraminiferal fauna.

Below we will discuss arguments that led to the rejection of the fixed bathymetry concept, then consider the pros and cons of the water mass concept, and discuss present ideas on the relations between foraminiferal assemblage structure and the physico-chemical characteristics of bottom waters.

2.4.1. Deep-sea foraminifera as indicators of bathymetry

Attempts to compare bathymetrical species distributions in various oceans led to the conclusion that for many species important differences in bathymetrical distribution existed. For instance, Pujos-Lamy (1973) compared the bathymetrical species succession for shelf to abyssal environments on the French continental margin (Bay of Biscay, NE Atlantic; Caralp, Lamy, & Pujos, 1970) with successions in other ocean basins such as the western Pacific (Polski, 1959), the north-western Pacific (Stschedrina, 1957), the north Pacific (Saidova, 1961), the Californian margin (Bandy, 1961), and the Gulf of Mexico (Phleger, 1960). At water depths of less than 2,000 m at least some cosmopolitan taxa showed comparable depth distributions in the Pacific and Atlantic Oceans, but below that depth the successive faunal associations occurred at significantly different depths. Pujos-Lamy (1973) (in agreement with Boltovskoy, 1965) concluded that these differences in bathymetrical distribution could be explained by the existence of different oceanic provinces, thereby preparing the ground for the water mass theory.

Many subsequent papers (e.g., Streeter, 1973; Schnitker, 1974, 1980; Lohmann, 1978a; Douglas & Woodruff, 1981; Bremer & Lohmann, 1982; Corliss, 1983b; Mackensen et al., 1990, 1995; Schmiedl, Mackensen, & Müller, 1997) confirmed that the upper and lower depth limits of many foraminiferal species show significant differences between various ocean basins, although bathymetrical species successions may be very similar. Many papers on the distribution of living and fossil deep-sea foraminiferal faunas (e.g., Corliss, 1985, 1991; Woodruff, 1985; Thomas, 1986; Mackensen & Douglas, 1989; Woodruff & Savin, 1989; Corliss & Emerson, 1990; Barmawidjaja et al., 1992; Rathburn & Corliss, 1994; Kitazato, 1994; Rathburn et al., 1996; Ohga & Kitazato, 1997; de Stigter et al., 1998; Jorissen et al., 1998; Jannink et al., 1998; Kitazato et al., 2000; Schmiedl et al., 2000; Fontanier et al., 2002; Licari et al., 2003; Hess, Jorissen, Venet, & Abu-Zied, 2005) show that cosmopolitan species, such as *Nuttallides umboniferus*, *Epistominella exigua* or *Cibicidoides wuellerstorfi*, thrive in all deep-sea basins, without showing a clear preference for a specific bathymetrical range or bio-province.

2.4.2. Deep-sea foraminifera as water mass indicators?

In the 1970s and early 1980s deep-sea foraminiferal assemblages were related to specific water masses, and various authors (e.g., Streeter, 1973; Schnitker, 1974, 1980; Lagoe, 1977; Lohmann, 1978a, 1978b; Gofas, 1978; Corliss, 1978, 1979; Streeter & Shackleton, 1979; Osterman & Kellogg, 1979; Belanger & Streeter, 1980; Miller & Lohmann, 1982; Streeter, Belanger, Kellogg, & Duplessy, 1982; Bremer & Lohmann, 1982; Weston, 1982; Corliss, 1983a, 1983b; Peterson, 1984; Woodruff, 1985; Murray, Weston, Haddon, & Powell, 1986) investigated whether their past spatial distribution could be used to reconstruct deep-sea circulation patterns (i.e., the geographical distribution of water masses). In most of these papers, the presence of foraminiferal species in some deep-sea environments and their absence in others was explained by their dependence on water mass characteristics rather than by their bathymetrical preferences. Several "index species", or a "specific combination of benthic foraminiferal species" were related to the specific physical and chemical properties of water masses, in some cases rather subjectively, in others by using statistical methods (Appendix 1). Most of these data pertained to foraminiferal thanatocoenoses sampled in core-tops (of piston, gravity, trigger-weight and grab cores), considered representative of the living faunas. However, the use of thanatocoenoses to determine a possible correlation between foraminiferal distribution and the geographical extension of water masses is highly problematic (e.g., Mackensen et al., 1990; Douglas & Woodruff, 1981). In most cases the uppermost material does not contain recent assemblages, because the sediment-water interface is only rarely sampled by these methods. Taphonomical processes, such as carbonate dissolution below the lysocline or in organic-rich sediments, differential disintegration of fragile tests, deposition of reworked material, or winnowing of autochthonous species may all cause large differences between living and fossil faunas (e.g., Corliss & Honjo, 1981; Mackensen et al., 1990, 1995). In some of the studies documenting the water mass concept (Lohmann, 1978a; Schnitker, 1979; Streeter & Shackleton, 1979), agglutinated and porcellaneous foraminifera were excluded from the faunal counts, leading to less reliable correlations between water mass distribution and foraminiferal assemblages. Moreover, as underlined by several authors (e.g., Thomas et al., 1995; Gooday, 2003), the use of different size fractions in most of these studies (>125, >150, or >250 μm) created important methodology-related faunal discrepancies in datasets from the various study areas.

One of the major problems of the water mass concept is the fact that in most of the classical papers water mass characteristics are presented very simplistically. For instance, the physico-chemical definition of North Atlantic deep Water (NADW) is very complex (e.g., Frew, Dennis, Heywood, Meredith, & Boswell, 2000; Van Aken, 2000). NADW can be subdivided in a number of regional water masses (e.g., LSW, DSOW, ISOW, NEADW, NWADW) with a large variability in physico-chemical parameters (temperature, salinity, nutrient concentrations, oxygen concentration, etc.). In the "Southern Ocean", some physico-chemical water mass properties such as oxygenation, alkalinity, nutrient concentration and corrosiveness progressively change in watermasses as they move along their pathways along the ocean floor (e.g., Peterson, 1984; Mackensen et al., 1995). Appendix 1 thus does not show clear-cut physico-chemical differences between the various water masses, and one cannot pinpoint the exact water mass properties that limit the occurrence of a specific group of taxa. Furthermore, many environmental parameters defining a specific water mass are cross-correlated (temperature, salinity, pressure, oxygen concentration, nutrient concentration, alkalinity/acidity, current velocity, sedimentological patterns, etc.), posing another obstacle to efforts to determine which physico-chemical parameter(s) control the distribution of the foraminiferal assemblages. We are therefore of the opinion that the water mass concept has lost much of its credibility over the last 20 years.

The alleged relation between faunal composition and water masses has been used to reconstruct the Quaternary history of bottom water circulation (e.g., Streeter, 1973; Schnitker, 1974, 1976, 1979, 1980; Gofas, 1978; Lohmann, 1978b; Streeter & Shackleton, 1979; Corliss, 1979, 1983a; Caralp, Grousset, Moyes, Peypouquet, & Pujol, 1982; Peterson & Lohmann, 1982; Peterson, 1984; Caralp, 1984; Caralp, 1987; Murray et al., 1986; Murray, 1988). In our opinion, the interpretations of the data presented in these studies are no longer valid.

2.4.3. Deep-sea foraminifera as indicators of complex environmental conditions

A number of ecological studies, mainly based on living (Rose Bengal stained) foraminiferal faunas (e.g., Mackensen et al., 1990, 1995; Schmiedl et al., 1997; Murray, 2001), has shown that the composition of deep-sea benthic foraminiferal faunas is related to a complex of environmental parameters. The faunal composition appears to be mainly determined by the food supply, the characteristics of the sediment and of the overlying surface waters, whereas water mass properties appear to play at best only a minor role. For example, Mackensen et al. (1990) show that the distribution of live foraminiferal faunas in the eastern Weddell Sea is related to parameters such as sediment granulometry, current velocity, organic matter deposition and the corrosiveness of bottom and interstitial waters. The interplay of four main environmental parameters controls the distribution of benthic foraminiferal taxa in a large South Atlantic Ocean database (Mackensen et al., 1995):

1. organic carbon content of the sediment and the exported organic matter flux to the sea-floor,
2. hydrodynamical properties of the benthic environment, and the related sediment grain size,
3. bottom water oxygenation, and
4. bottom water carbonate saturation.

The abundance of *Fontbotia* (= *Cibicidoides*) *wuellerstorfi* is related to the occurrence of young and well-ventilated (NADW-like) bottom waters, where the organic carbon flux does not exceed $1 \, \mathrm{gC/m^2/yr}$. *Cassidulina laevigata*, *Uvigerina peregrina*, and some buliminids and bolivinids are correlated to higher organic carbon flux rates ($>3 \, \mathrm{gC/m/yr}$), whereas assemblages dominated by *Epistominella exigua* occur in the low salinity core of NADW, where primary production shows a large seasonal contrast. *Angulogerina angulosa* is well correlated with coarse-grained sediment in environments characterised by strong bottom currents. *Nuttallides umboniferus* is found above the CCD, but below the lysocline (Mackensen et al., 1995). Similarly, Schmiedl et al. (1997) related the spatial distribution of seven living and dead assemblages to the dissolved oxygen content of the bottom water, the organic carbon flux, the organic carbon content of the surface sediment, the nature of the substrate and the related porosity.

2.5. Benthic foraminiferal faunas as indicators of current velocity

Intensified bottom water currents (contour, tidal or slope currents) have been shown to influence the microhabitats and composition of the benthic foraminiferal fauna (Lutze & Altenbach, 1988; Lutze & Thiel, 1989; Linke & Lutze, 1993;

Schönfeld. 1997, 2002a, 2002b). Under high current velocities (20–50 cm/s), some species may live preferentially on elevated substrates, or within sedimentary objects such as pteropod shells. Elevated substrates include large biogenic objects such as coarse shell debris, sponges, crinoids, hydroids, tube worms (Sen Gupta, Smith, & Lobegeier, 2006), or living arborescent benthic foraminifera (e.g., *Rhabdammina* spp.), as well as large terrigenous particles (pebbles). At some mm or cm above the sediment-water interface, specialised epibenthic foraminifera may catch suspended organic matter particles (suspension-feeders). Mackensen et al. (1995) recognised a faunal association dominated by *Angulogerina angulosa*, correlated with very coarse-grained sediments (sand-gravel) and putatively strong bottom currents. On the southern Portuguese continental margin, Schönfeld (1997) described common *Cibicides lobatulus*, *Planulina ariminensis*, *Discanomalina* spp., and *Epistominella exigua* living on elevated substrates, and suggested that these taxa thus maximise the acquisition of suspended organic compounds. A similar assemblage of sessile epifaunal and/or epibenthic suspension feeders was found on the Spanish continental margin, influenced by high velocity bottom currents related to the Mediterranean Outflow Water (Schönfeld, 2002a). Using data from Gulf of Cadiz and the southern Portuguese margin, Schönfeld (2002b) showed a linear correlation between the percentage of taxa thought typical for elevated microhabitats and bottom water current velocity, and used this relation to reconstruct the Holocene history of bottom current velocity.

The use of the composition of benthic foraminiferal assemblages as a proxy for past current regimes is complicated by the fact that active currents may constitute one of the main taphonomical factors, causing resuspension and/or transport at the sediment-water interface, leading to winnowing of or addition of components. Schönfeld (1997) recommended the use of the >250 µm fraction benthic foraminifera in a proxy for bottom current regimes because tests in this large size fraction may be least affected by transport.

The development of such a current velocity proxy requires more ecological studies based on living and subfossil faunas, in order to understand taphonomical modifications, and the record of the faunal response to current regimes in the sediments (Schönfeld, 1997, 2002b). The calibration of such proxies is hampered by the lack of precise and highly detailed measurements of bottom current velocities in regions where information on benthic foraminiferal assemblages is available. For example, no precise current measurements are available for the Bay of Biscay, where foraminiferal ecology has been well studied (e.g., Fontanier et al., 2002, 2003, 2005, 2006; Hess et al., 2005). In the absence of current measurements, various non-faunal proxies for current activity could be used, such as grain size analysis, X-ray photographs and ^{210}Pb and/or ^{234}Th profiles of the upper sediment layers (Fontanier et al., 2005). Direct observations of sedimentary structures in core material, photographic/video surveys of the seafloor and parasound records, may also give information on sedimentary processes and the associated current flow regimes (Schönfeld, 1997). Finally, a potential problem of this proxy method is the fact that we do not have observations that the benthic foraminifera living on elevated substrates are indeed suspension feeders. At least some species considered to be indicators of high current velocities (Schönfeld, 1997, 2002a, 2002b) also occur in shallow infaunal microhabitats (e.g., *Cibicidoides* spp. in Corliss, 1985; Rathburn & Corliss, 1994; Fontanier et al., 2002; *Planulina ariminensis* in De Stigter et al., 1998). Species such as *Cibicides lobatulus*, *Planulina ariminensis*, and especially *Epistominella exigua* are common in areas where we have no indications of significant bottom currents. Most of these species appear to be typical of rather oligotrophic environments, where a microhabitat close to the sediment-water interface and perhaps a suspension-feeding life strategy, may be advantageous.

We thus suggest that the method proposed by Schönfeld should be used with caution, and may not be useful in all open ocean environments. In a well-constrained local context, on the contrary, this method could be used to provide valuable information about past current regimes, but more research is needed.

3. CONCLUSIONS

The main environmental parameters structuring benthic foraminiferal faunas appear to be the organic flux to the ocean floor (its quantity, quality, and periodicity), and bottom water oxygenation (especially at very low ranges). To a lesser degree, sediment grain size and current velocity may act as limiting factors (e.g., Miller & Lohmann, 1982; Lutze & Coulbourn, 1984; Mackensen et al., 1990, 1995; Schmiedl et al., 1997). Conservative water mass properties such as salinity and temperature play only a minor role. Only in the deepest part of ocean basins, where strongly oligotrophic conditions prevail, the corrosiveness of the bottom waters (highest in waters such as AABW), may control the distribution of a few cosmopolitan taxa, such as *Nuttallides umboniferus* or *Oridorsalis tener* (e.g., Bremer & Lohmann, 1982; Mackensen et al., 1990, 1995; Schmiedl et al., 1997).

One of the first conclusions of this overview of the most important paleoceanographic proxy methods based on benthic foraminiferal assemblage characteristics could be that the scientific community working on benthic foraminiferal assemblages has been too modest. Many recent ecological studies show conclusively that under most circumstances the composition of deep-sea benthic foraminiferal assemblages is controlled by a rather limited number of environmental factors. Our understanding of the often complex interactions between these factors, and of the way in which they structure the faunal composition, has made much progress over the last 20 years. Once we understand how environmental parameters influence the faunal composition, it should be possible to use these relations in the reverse way, i.e., to use fossil faunas for the reconstruction of the controlling environmental parameters in the past, although more research is needed in order to establish for which part of earth history conditions resembled those in the present oceans sufficiently for this approach to work.

Methods to reconstruct past values of the essential oceanographic parameters of organic flux to the sea floor and bottom water oxygenation have largely remained qualitative, or at best semi-quantitative. In spite of the indisputable presence of a number of problems (many of which are shared with other, e.g., geochemical, proxy methods), some of the proposed proxies are based on firm ecological observations. Other physico-chemical parameters, such as current velocity, water corrosiveness to $CaCO_3$, temperature and/or salinity appear less important, except perhaps in some specific environmental contexts where benthic foraminiferal proxies may provide adequate reconstructions of their past variability.

The available proxies based on benthic foraminiferal assemblage composition show that they have major potential, but further research is needed to add or improve the quantitative aspects (Table 4). In many cases (e.g., bottom water oxygenation, and C_{org} flux to the ocean floor) this can be done by significantly increasing the size of existing databases. In others (e.g., periodicity of the organic flux), time series observations are necessary. A major obstacle is our insufficient

Table 4 Overview of Proxy Methods Based on Benthic Foraminiferal Assemblage Characteristics, Their Problems and Possible Remedies.

Parameter	Method	Problems and remedies
Bottom water oxygenation	Marker species approach	Needs a more objective choice of marker species and a better calibration, probably only feasible for low O_2 concentrations
Bottom water oxygenation	Marker species+faunal diversity index	Needs further testing
Organic matter flux to the sea floor (J_z)	Flux dependency of species or species groups	Needs better estimates of J_z, needs calibration in many more areas
Organic matter flux to the sea floor (J_z)	BFAR	Needs calibration, using a large number of cores from different PP regimes
Primary production (PP)	BFAR, flux dependency	Needs better knowledge of relations between PP and J_z
Quality of organic input	Marker species approach	Needs much more data on the geochemical composition of organic matter
Quality of organic input	BFAR/Corg flux	Needs more research
Periodicity of the organic flux		Needs more time series studies in various productivity contexts
Physico-chemical characteristics of water masses		May work in oligotrophic abyssal environments; more research needed
Current velocity		More current velocity measurements needed; probably only feasible in areas with high current velocities

knowledge of the differences between recent and fossil faunas due to taphonomical alterations. This phenomenon, of importance for all paleoceanographic proxies, can to some extent be solved relatively easily in the case of foraminiferal assemblages by detailed studies of their vertical succession in sediments deposited in the last 5,000 years, when environmental conditions were probably rather invariable in many areas. Unfortunately, such taphonomical studies are extremely time-consuming. Finally, we want to stress that scientists working with benthic foraminiferal assemblages should try to quantify the very complex relations observed in nature. If a single proxy reconstruction has a large degree of uncertainty, a multi-proxy approach, with as many independent proxies as possible, may successfully reduce the uncertainty related to each single method. Proxies based on foraminiferal assemblage composition are fundamentally different from all geochemical proxies, and thus may provide independent reconstructions of essential oceanographic parameters. We conclude that benthic foraminiferal proxies deserve to be much more widely applied than they are today.

ACKNOWLEDGEMENTS

This paper could not exist without the many prolific scientific discussions we had with numerous colleagues; we would especially like to mention the contributions of Joan Bernhard, Gérald Duchemin, Emmanuelle Geslin, Andy Gooday, Sophie Guichard, Johann Hohenegger, Hélène Howa, Andreas Mackensen, Gerhard Schmiedl, Stefanie Schumacher, and Bert van der Zwaan. We (FJ, CF) acknowledge funding of the Foramprox program by the French national programs PNEDC and PROOF; ET acknowledges funding by NSF and USSSP.

4. APPENDIX 1

Summary of early studies dealing with putative relationship between foraminiferal assemblages and water mass properties and/or bathymetry. Note the differences between papers concerning the water mass characteristics and the faunas inhabiting specific water masses. References are numbered in the first column and listed below. Water mass nomenclature is also detailed below. "nbd" means "no bathymetric data" available in the related note. Taxonomic names are not homogenised between studies. Reference list: (1) Caralp et al. (1970) and Pujos-Lamy (1973); (2) Streeter (1973); (3) Schnitker (1974); (4) Lagoe (1977); (5) Gofas (1978); (6) Gofas (1978); (7) Corliss (1978a); (8) Corliss (1978b, 1979); (9) Osterman and Kellogg (1979); (10) Schnitker (1980); (11) Belanger and Streeter (1980) and Streeter et al. (1982); (12) Burke (1981); (13) Corliss (1981, 1983b); (14) Miller and Lohmann (1982); (15) Peterson (1984); (16) Weston (1982), and Murray et al. (1986). Water mass nomenclature: CPDW/CDW/Circumpolar Deep Water; NADW/North Atlantic Deep Water; AABW/Antarctic Bottom Water; ABW/ Arctic Bottom Water; NSOW/Norwegian Sea Overflow Water; AAIW/Antarctic Intermediate Water; IDW/Indian Deep Water; IBW/Indian Bottom Water; PBW/ Pacific Bottom Water; PDW/Pacific Deep Water; MW/Mediterreanean Water; NEADW/North East Atlantic Deep Water.

Reference	Study area	Water mass/depth	Temperature (°C)	Salinity (psu)	Oxygenation(ml/l)	Dominent species
1	Northeast Atlantic Ocean, Bay of Biscay	100–250 m	–	–	–	*Cibicides lobatulus, Cibicides refulgens, Gaudryana* spp., *Textularia sagittula*
	Northeast Atlantic Ocean, Bay of Biscay	200–1,700 m	–	–	–	*Trifarina angulosa, Trifarina bradyi, Bulimina buchiana, Uvigerina peregrina (s.l.), Epistominella umbonifera*
	Northeast Atlantic Ocean, Bay of Biscay	1,700–3,000 m	–	–	–	*Bulimina inflata, Bulimina alazanensis, Uvigerine peregrina dirupta, Hyperammina* div. sp., *Planulina wuellerstorfi, Nonion pompilioides*
	Northeast Atlantic Ocean, Bay of Biscay	2,500–4,500 m	–	–	–	*Epistominella exigua, Bulimina alazanensis, Planulina wuellerstorfi, Eponides* spp., *Oridorsalis umbonatus*
2	North Atlantic	NADW, <2,500 m	>3,0	>35.00	–	*Nummoloculina irregularis, Cibicides kullenbergi*
	North Atlantic	NADW, 2,500–4,000 m	2.0 to 3.0	From 34.90 to 35.00	–	*Planulina wuellerstorfi, Epistominella exigua*
3	North Atlantic	AABW, >4,000 m	<2.0	<34.90	–	*Epistominella umbonifera*
	Western North Atlantic Ocean	Lower NADW (nbd)	From 0 to 4.0		–	*Hoeglundina* sp., *Uvigerina* sp., *Gyroidina* sp.
	Western North Atlantic Ocean	ABW (NSOW) (nbd)	1.9		–	*Epistominella exigua*
	Western North Atlantic Ocean	AABW (nbd)	1.5		–	*Epistominella umbonifera*
4	Central Arctic Ocean	<2,500 m (permanently under Arctic pack ice)	−0.5	~35.00	6.5	*Epistominella arctica, Buliminella elegans* var. *hensoni, Cassidulina teretis, Valvulineria arctica*
	Central Arctic Ocean	2,000–2,500 m (permanently under Arctic pack ice)	−0.5	~35.00	6.5	*Eponides tener, Eponides tumidulus* var. *horvathi, Quinqueloculina akneriana, Planulina wuellerstorfi, Ceratobulimina arctica*
	Central Arctic Ocean	2,500–3,700 m (permanently under Arctic pack ice)	−0.5	~35.00	6.5	*Stetsonia horvathi, Eponides tener, Triloculina frigida*

(*Continued*)

Appendix 1   *(Continued)*

Reference	Study area	Water mass/depth	Temperature (°C)	Salinity (psu)	Oxygenation(ml/l)	Dominent species
5	Western South Atlantic Ocean	NADW, 3,000–3,500 m	From 2.0 to 2.7	~34.90	~5.7	*Miliolidea, Cibicides wuellerstorfi, Gyroidina altiformis, Pyrgo murrhina*
	Western South Atlantic Ocean	Transition to AABW, 3,500–4,000 m	From 1.0 to 2.0	From 34.75 to 34.90	From 5.0 to 5.7	*Epistominella umbonifera*
	Central Pacific	PBW (upper layers), ~4,500	~1.0	~34.71	–	*Cibicides wuellerstorfi, Favocassidulina favus, Pulleria bulloides, Melonis sphaeroides, Melonis bradyi, Pyrgo murrhina*
	Central Pacific	PBW (lower layers), 4,500–5,000 m	From 1.0 to 1.2	~34.71	–	*Epistominella umbonifera, Globocassidulina subglobosa*
6	Western South Atlantic Ocean	Upper NADW/AAIW (= upper branch of CPDW), <2,200 m	~3.0	<34.90	<5.6	*Uvigerina peregrina, Planulina wuellerstorfi, Globocassidulina subglobosa*
	Western South Atlantic Ocean	NADW, 2,200–4,000 m (+ lower branch of CPDW, ~4,200 m)	From 1.0 to 3.0	From 34.75 to 34.90	From 5.0 to 5.6	*Planulina wuellerstorfi, Hoeglundina elegans, Pyrgo spp., Quinqueloculina spp., Nummoloculina irregularis*
	Western South Atlantic Ocean	AABW, >4,000 m	<1.0	<34.75	~5.0	*Nuttallides umbonifera, Oridorsalis tener*
7	Southeast Indian Ocean	IBW (variable depth)	From 0.8 to 1.2	From 34.72 to 34.74	From 4.4 to 4.8	*Uvigerina* spp., *Epistominella exigua*
	Southeast Indian Ocean	Mixing between IBW and AABW (variable depths)	From 0.4 to 0.8	From 34.70 to 34.72	From 4.8 to 5.0	*Epistominella umbonifera, Planulina wuellerstorfi, Globocassidulina subglobosa*
	Southeast Indian Ocean	AABW (variable depths)	From −0.2 to 0.8	From 34.68 to 34.70	From 5.0 to 5.6	*Globocassidulina subglobosa, Planulina wuellerstorfi, Astrononion echolsi,*
8	Southeast Indian Ocean	AABW1 (variable depths)	From 0.6 to 0.8	–	–	*Oridorsalis tener, Gyroidinoides soldanii, Pulleria bulloides*
	Southeast Indian Ocean	AABW2 (variable depths)	From −0.2 to 0.4	–	–	*Epistominella umbonifera, Oridorsalis tener, Pullenia wuellerstorfi*
9	Ross Sea	Eastern shelf (almost permanently under pack ice), 400–2,000 m	–	–	–	*Cyclammina* spp., *Hormosina ovicula gracilis, Milliammina arenacea, Reophax nodulosus*

	Ross Sea	Western shelf (seasonally under pack ice), 500–800 m	—	—	—	Globocassidulina subglobosa, Ehrenbergina glabra
	Ross Sea	Continental slope	—	—	—	Trifarina earlandi, Cibicides lobatulus
10	Western North Atlantic Ocean	NADW (nbd)	From 2.0 to 4	—	—	Globocassidulina subglobosa, Elphidium incertum, Urigerina peregrina
	Western North Atlantic Ocean	(NSOW) (nbd)	1.9	—	—	Epistominella exigua
	Western North Atlantic Ocean	AABW (nbd)	1.4	—	—	Osangularia umbonifera
11	Norwegian-Greenland Sea	Source of Atlantic Bottom Water, 600–1,200 m	>−1.0	34.90	>6.7	Melonis barleeanus, Pullenia bulloides, Islandiella norcrossi
	Norwegian-Greenland Sea	Source of Atlantic Bottom Water, 950–1,500 m	>−1.0	34.90	>6.7	Cassidulina teretis, Trifarina angulosae
	Norwegian-Greenland Sea	Source of Atlantic Bottom Water, 1,250–3,200 m	−1.0	34.90	From 6.7 to 7.1	Cibicides wuellerstorfi, Oridorsalis tener
	Norwegian-Greenland Sea	Source of Atlantic Bottom Water, >2,900 m	−1.0	34.90	From 6.7 to 7.1	Oridorsalis tener, Epistominella exigua, Cibicides wuellerstorfi
12	Western Equatorial Pacific Ontong java Plateau	Deep oxygen minimum	From 1.9 to 3.9	From 34.58 to 34.65	From 2.8 to 3.9	Siphougerina interrupta, Cibicides pseudoungerianus, Astrononion stelligerum, Hoeglundina elegans, Gyroidina lamarckiana
	Western Equatorial Pacific Ontong java Plateau	PDW, 2,500–3,000 m (above lysocline)	From 1.4 to 1.9	34.74	From 4.0 to 4.4	Epistominella exigua, Pullenia bulloides, Melonis affinis, Melonis pompilioides, Nuttallides umbonifera
	Western Equatorial Pacific Ontong java Plateau	PBW, 3,000–4,300 m (below lysocline)	From 1.5 to 1.75	34.70	From 3.4 to 4.6	Nuttallides umbonifera, Epistominella exigua
13	Southwest Indian Ocean	CDW (nbd)	—	—	—	Uvigerin sp.
	Southwest Indian Ocean	NADW, 1,600–3,800 m	From 0.8 to 2.6	From 34.66 to 34.72	—	Planulina wuellerstorfi, Globocassidulina subglobosa, Astrononion echolsi, Pullenia bulloides

(Continued)

Appendix 1 *(Continued)*

Reference	Study area	Water mass/depth	Temperature (°C)	Salinity (psu)	Oxygenation (ml/l)	Dominent species
	Southwest Indian Ocean	AABW, 3,600–4,800 m	From −0.3 to 0.8	From 34.66 to 34.72	–	*Epistominella umbonifera*
14	Northeast Atlantic Ocean	<700 m	>4.5	>35.10	<6.0	*Globobulimina* spp., *Bulimina* spp.
	Northeast Atlantic Ocean	700–2,500 m	From 3.0 to 4.5	~35.10	From 5.5 to 6.0	*Uvigerine peregrina*
	Northeast Atlantic Ocean	2500–4,000 m	From 1.5 to 3	From 35.1 to 35.4	~6.0	*Hoeglundina elegans*
15	Eastern Equatorial Indian Ocean	IDW, 2,000–3,800 m	From 1.2 to 2.6	From 34.72 to 34.76	From 3.1 to 4.2	*Globocassidulina subglobosa, Pyrgo* spp., *Uvigerine peregrina, Eggerella brady, Cibicidoides wuellerstorfi*
	Eastern Equatorial Indian Ocean	IBW, 3,800–4,600 m (below lysocline)	~1.2	~34.72	From 4.1 to 4.4	*Epistominella exigua, Cibicidoides wuellerstorfi, Pullenia bulloidoides, Oridorsalis umbonatus*
	Eastern Equatorial Indian Ocean	IBW, 4,200–5,000 m (below lysocline)	From 1.2 to 1.4	~34.72	From 3.9 to 4.4	*Nuttallides umbonifera, Cibicidoides wuellerstorfi, Pullenia bulloides, Oridorsalis umbonatus*
16	Northeast Atlantic Ocean	MW	>4.0	>35.00	~4.0	*Cassiduline obtusa, Globocassidulina subglobosa*
	Northeast Atlantic Ocean	NEADW (or upper NADW)	From 3.0 to 4.0	From 34.92 to 34.97	6.0	*Epistominella exigua*
	Northeast Atlantic Ocean	NADW	From 2.5 to 4.0	From 34.95 to 35.00	From 5.2 to 5.6	*Planulina wuellerstorfi, Globocassidulina subglobosa, Cibicidoides kullenbergi, Oridorsalis umbonatus*
	Northeast Atlantic Ocean	AABW	<2.0	34.90	>6.0	*Osangularia umbonifera*

REFERENCES

Altenbach, A. V., Lutze, G. F., Schiebel, R., & Schönfeld, J. (2003). Impact of interrelated and interdependent ecological controls on benthic foraminifera: An example from the Gulf of Guinea. *Palaeogeography, Palaeoclimatology, Palaeoecology, 197*, 213–238.

Altenbach, A. V., Pflaumann, U., Schiebel, R., Thies, A., Timm, S., & Trauth, M. (1999). Scaling percentages and distributional patterns of benthic foraminifera with flux rates of organic carbon. *Journal of Foraminiferal Research, 29*, 173–185.

Alve, E. (1990). Variations in estuarine foraminiferal biofacies with diminishing oxygen conditions in Drammensfjord, SE Norway. In: C. Hemleben, M. A. Kaminski, W. Kuhnt & D. B. Scott (Eds), *Paleoecology, biostratigraphy, paleoceanography and taxonomy of agglutinated foraminifera* (pp. 661–694). The Netherlands: Kluwer Academic Publishers.

Alve, E. (1994). Opportunistic features of the foraminifer *Stainforthia fusiformis* (Williamson): Evidence from Frierfjord, Norway. *Journal of Micropaleontology, 13*, 24.

Antia, A. N., Von Bodungen, B., & Peinert, R. (1999). Particle flux accross the mid-European continental margin. *Deep-Sea Research, I, 46*, 1999–2024.

Appleby, P. G., & Oldfield, F. (1992). Application of lead-210 to sedimentation studies. In: M. Ivanovich & R. S. Harmon (Eds), *U-series disequilibrium: Applications to earth, marine and environmental studies* (pp. 731–778). Oxford: Clarendon Press.

Armstrong, R. A., Lee, C., Hedges, J. I., Honjo, S., & Wakeham, S. G. (2002). A new, mechanistic model for organic carbon fluxes in the ocean based on the quantitative association of POC with ballast minerals. *Deep-Sea Research II, 49*, 219–236.

Baas, J. H., Schönfeld, J., & Zahn, R. (1998). Mid-depth oxygen drawdown during Heinrich events: evidence from benthic foraminiferal community structure, trace-fossil tiering, and benthic $\delta^{13}C$ at the Portuguese Margin. *Marine Geology, 152*, 25–55.

Bandy, O. L. (1953a). Ecology and paleoecology of some California Foraminifera, Part I. The frequency distribution of Recent Foraminifera off California. *Journal of Paleontology, 27*, 161–192.

Bandy, O. L. (1953b). Ecology and paleoecology of some California Foraminifera, Part II. Foraminiferal evidence of subsidence rates in the Ventura Basin. *Journal of Paleontology, 27*, 200–203.

Bandy, O. L. (1961). Distribution of foraminifera, radiolaria and diatoms in sediments of the Gulf of California. *Micropaleontology, 7*, 1–26.

Bandy, O. L., & Arnal, R. (1957). Distribution of Recent foraminifera off the west coast of Central America. *American Association of Petroleum Geologists Bulletin, 41*, 2037–2053.

Bandy, O. L., & Chierici, M. (1966). Depth-temperature evaluation of selected California and Mediterranean bathyal foraminifera. *Marine Geology, 4*, 254–271.

Bandy, O. L., & Echols, J. (1964). Antarctic foraminiferal zonation. *American Geophysical Union, Antarctic Research Series, 1*, 73–91.

Barberi, R., & Panieri, G. (2004). How are benthic foraminiferal faunas influenced by cold seeps? Evidence from the Miocene of Italy. *Palaeogeography, Palaeoclimatology, Palaeoecology, 204*, 257–275.

Barmawidjaja, D. M., Jorissen, F. J., Puskaric, S., & Van der Zwaan, G. J. (1992). Microhabitat selection by benthic foraminifera in the northern Adriatic Sea. *Journal of Foraminiferal Research, 22*, 297–317.

Basov, I. A., & Khusid, T. A. (1983). Biomass of benthic foraminifera in the sediments of the Sea of Okhotsk. *Oceanology, 33*, 489–495.

Belanger, P. E., & Streeter, S. S. (1980). Distribution and ecology of benthic foraminifera in the Norwegian Greenland Sea. *Marine Micropaleontology, 5*, 401–428.

Berelson, W., Anderson, R., Dymond, J., DeMaster, D., Hammond, D., Collier, R., Honjo, S., Leinen, M., McManus, J., Pope, R., Smith, C., & Stephens, M. (1997). Biogenic budgets of particle rain, benthic remineralization and sediment accumulation in the equatorial pacific. *Deep-Sea Research, II, 44*, 2251–2282.

Berger, W., Smetacek, V., & Wefer, G. (1989). Ocean productivity and paleoproductivity—An overview. In: W. Berger, V. Smetacek & G. Wefer (Eds), *Productivity of the Ocean: Present and Past, Dahlem Workshop Reports* (Vol. 44, pp. 1–34). Chichester: Wiley.

Berger, W., & Wefer, G. (1990). Export production: Seasonality and intermittency, and paleocea-nographic implications. *Palaeogeography, Palaeoclimatology, Paleoecology, 89*, 245–254.

Bernhard, J. M. (1986). Characteristic assemblages and morphologies of benthic foraminifera from anoxic, organic-rich deposits: Jurassic through Holocene. *Journal of Foraminiferal Research, 16*, 207–215.

Bernhard, J. M. (1988). Postmortem vital staining in benthic foraminifera: Duration and importance in population and distributional studies. *Journal of Foraminiferal Research, 18*, 143–146.

Bernhard, J. M. (1989). The distribution of benthic foraminifera with respect to oxygen concentration and organic carbon levels in shallow-water Antarctic sediments. *Limnology and Oceanography, 34*, 1131–1141.

Bernhard, J. M. (1992). Benthic foraminiferal distribution and biomass related to pore water oxygen: Central California Continental slope and rise. *Deep Sea Research, 39*, 586–605.

Bernhard, J. M. (1993). Experimental and field evidence of Antarctic foraminiferal tolerance to anoxia and hydrogen sulphide. *Marine Micropaleontology, 20*, 203–213.

Bernhard, J. M., & Alve, E. (1996). Survival, ATP pool, and ultrastructural characterization of benthic foraminifera from Drammensfjord (Norway): Response to anoxia. *Marine Micropaleontology, 28*, 5–17.

Bernhard, J. M., & Reimers, C. E. (1991). Benthic foraminiferal population fluctuations related to anoxia: Santa Barbara Basin. *Biogeochemistry, 15*, 127–149.

Bernhard, J. M., & Sen Gupta, B. K. (1999). Foraminifera of oxygen-depleted environments. In: B. K. Sen Gupta (Ed.), *Modern Foraminifera* (pp. 201–216). New York, NY: Kluwer Academic Press.

Bernhard, J. M., Sen Gupta, B. K., & Borne, P. F. (1997). Benthic foraminiferal proxy to estimate dysoxic bottom-water oxygen concentrations; Santa Barbara basin, U.S. Pacific continental margin. *Journal of Foraminiferal Research, 27*, 301–310.

Betzer, P. R., Showers, W. J., Laws, E. A., Winn, C. D., Di Tullio, G. R., & Kroopnick, P. P. (1984). Primary productivity and particle fluxes on a transect of the equator at 153°W in the Pacific Ocean. *Deep-Sea Research, 31*, 1–11.

Billett, D. S. M., Lampitt, R. S., Rice, A. L., & Mantoura, R. F. C. (1983). Seasonal sedimentation of phytoplankton to the deep-sea benthos. *Nature, 302*, 520–522.

Boltovskoy, E. (1965). *Los Foraminiferos recientes, Editorial* (pp. 1–510). Argentina: Universitaria de Buenos Aires..

Boyd, P. W., & Newton, P. (1995). Evidence of the potential influence of planktonic community structure on the interannual variability of particulate organic carbon flux. *Deep-Sea Research, I, 42*, 619–637.

Bremer, M. L., & Lohmann, G. P. (1982). Evidence of primary control of the distribution of certain Atlantic Ocean benthonic foraminifera by degree of carbonate saturation. *Deep-Sea Research, 29*, 987–998.

Burke, S. C. (1981). Recent benthic foraminifera of the Ontong Java Plateau. *Journal of Foraminiferal Research, 11*, 1–19.

Buzas, M. A., Collins, L. S., Richardson, S. L., & Severin, K. P. (1989). Experiments on predation, substrate preference, and colonization of benthic foraminifera at the shelfbreak off the Ft. Pierce Inlet, Florida. *Journal of Foraminiferal Research, 19*, 146–152.

Buzas, M. A., Culver, S. J., & Jorissen, F. J. (1993). A statistical evaluation of the microhabitats of living (stained) infaunal benthic foraminifera. *Marine Micropaleontology, 20*, 311–320.

Buzas, M. A., & Gibson, T. G. (1969). Species diversity: Benthonic foraminifera in the western North Atlantic. *Science, 163*, 72–75.

Caralp, M. H. (1984). Quaternary calcareous benthic foraminifers. *Initial Reports of the Deep Sea Drilling Project, 80*, 725–755.

Caralp, M. H. (1987). Deep-sea circulation in the northeastern Atlantic over the past 30,000 years: the benthic foraminiferal record. *Oceanologica Acta, 10*, 27–40.

Caralp, M. H. (1989). Abundance of *Bulimina exilis* and *Melonis barleeanum*: Relationship to the quality of marine organic matter. *Geo-Marine Letters, 9*, 37–43.

Caralp, M., Grousset, F., Moyes, J., Peypouquet, J.-P., & Pujol, C. (1982). L'environnement confiné du Golfe de Gascogne avant le maximum glaciaire. Actes Colloque International CNRS,

Bordeaux, Sept. 1981, Bulletin de l'Institut Géologique du Bassin d'Aquitaine, Bordeaux (Vol. 31, pp. 411–422).

Caralp, M., Lamy, A., & Pujos, M. (1970). Contribution à la connaissance de la distribution bathymétrique des foraminifères dans le Golfe de Gascogne. *Revista Española De Micropaleontologia, 2*, 55–84.

Carney, R. S. (1989). Examining relationship between organic carbon flux and deep-sea deposit feeding. In: G. Lopez, G. Taghon & J. Levinton (Eds), *Ecology of marine deposit feeders* (Vol. 31, pp. 24–58). *Lecture Notes on Coastal and Estuarine Studies*, New York: Springer.

Carney, R. S. (2005). Zonation of deep biota on continental margins. Oceanography and Marine Biology. *Annual Review, 43*, 211–278.

Chandler, G. T., Williams, D. F., Spero, H. J., & Xiaodong, G. (1996). Sediment microhabitat effects on carbon stable isotopic signatures of microcosm-cultured benthic foraminifera. *Limnology and Oceanography, 41*, 680–688.

Cita, M. B., & Podenzani, M. (1980). Destructive effects of oxygen starvation and ash falls on benthic life: A pilot study. *Quaternary Research, 13*, 230–241.

Corliss, B. H. (1978). Recent deep-sea benthonic foraminiferal distributions in the southeast Indian ocean: Inferred bottom water routes and ecological implications. *Marine Geology, 31*, 115–138.

Corliss, B. H. (1979). Quaternary Antarctic bottom-water history: Deep-sea benthonic foraminiferal evidence from the Southeast Indian Ocean — Inferred bottom-water routes and ecological implications. *Marine Geology, 31*, 115–138.

Corliss, B. H. (1983a). Quaternary circulation of the Antarctic Circumpolar Current. *Deep-Sea Research, 30*, 47–61.

Corliss, B. H. (1983b). Distribution of Holocene deep-sea benthonic foraminifera in the southwest Indian Ocean. *Deep-Sea Research, 30*, 95–117.

Corliss, B. H. (1985). Microhabitats of benthic foraminifera within deep-sea sediments. *Nature, 314*, 435–438.

Corliss, B. H. (1991). Morphology and microhabitat preferences of benthic foraminifera from the northwest Atlantic Ocean. *Marine Micropaleontology, 17*, 195–236.

Corliss, B. H., & Chen, C. (1988). Morphotype patterns of Norwegian Sea deep-sea benthic foraminifera and ecological implications. *Geology, 16*, 716–719.

Corliss, B. H., & Emerson, S. (1990). Distribution of Rose Bengal stained deep-sea benthic foraminifera from the Nova Scotia continental margin and Gulf of Maine. *Deep-Sea Research, 37*, 381–400.

Corliss, B. H., & Fois, E. (1990). Morphotype analysis of deep-sea benthic foraminifera from the northwest Gulf of Mexico. *Palaios, 5*, 589–605.

Corliss, B. H., & Honjo, S. (1981). Dissolution of deep-sea benthonic foraminifera. *Micropaleontology, 27*, 356–378.

Crosta, X., Pichon, J. J., & Burckle, L. H. (1998). Application of the modern analog technique to marine Antarctic diatoms reconstruction of maximum sea-ice extent at the Last Glacial Maximum. *Paleoceanography, 13*, 284–297.

Culver, S. J., & Lipps, J. (2003). Predation on and by Foraminifera. In: P. H. Kelley, M. Kowalewski & T. A. Hansen (Eds), *Predator–prey interactions in the fossil record* (pp. 7–32). New York, NY: Kluwer Academic Press/Plenum Publishers.

Cushman, J. A. (1928). Foraminifera: Their classification and economic use. *Cushman Laboratory for Foraminiferal Research, Special Publications* (Vol. 1, p. 401).

De Rijk, S., Jorissen, F. J., Rohling, E. J., & Troelstra, S. R. (2000). Organic flux control on bathymetric zonation of Mediterranean benthic foraminifera. *Marine Micropaleontology, 40*, 151–166.

De Stigter, H. C., Jorissen, F. J., & Van der Zwaan, G. J. (1998). Bathymetric distribution and microhabitat partitioning of live (Rose Bengal stained) benthic foraminifera along a shelf to deep-sea transect in the southern Adriatic Sea. *Journal of Foraminiferal Research, 28*, 40–65.

Den Dulk, M., Reichart, G. J., van Heyst, S., Zachariasse, W. J., & Van der Zwaan, G. J. (2000). Benthic foraminifera as proxies of organic matter flux and bottom water oxygenation? A case history from the northern Arabian Sea. *Palaeogeography, Palaeoclimatology, Palaeoecology, 161*, 337–359.

Dodd, J. R., & Stanton, R. J. (1990). *Paleoecology: Concepts and applications* (p. 502). New York, NY: Wiley.

Douglas, R. G., & Heitman, H. L. (1979). Slope and basin benthic foraminifera of the California Borderland. *Society of Economic Paleontologists and Mineralogists, 27*, 231–246.

Douglas, R. G., & Woodruff, F. (1981). Deep-sea benthic foraminifera. The oceanic Lithosphere. In: C. Emiliani (Ed.), *The sea* (7, pp. 1233–1327). New York, NY: Wiley-Interscience.

Elderfield, H. (Ed.) (2004). *Treatise on geochemistry: The oceans and marine geochemistry* (Vol. 6). Amsterdam: Elsevier.

Epping, E., & Helder, W. (1998). Oxygen budgets calculated from in situ oxygen microprofiles for northern Adriatic sediments. *Continental Shelf Research, 17*, 1737–1764.

Eppley, R., & Peterson, B. (1979). Particulate organic matter flux and planktonic new production in the deep ocean. *Nature, 282*, 677–680.

Ernst, S. R., & Van der Zwaan, G. J. (2004). Effects of experimentally induced raised levels of organic flux and oxygen depletion on a continental slope benthic foraminiferal community. *Deep-Sea Research I, 51*, 1709–1739.

Ernst, S. R., Bours, I., Duijnstee, Y., & Van der Zwaan, G. J. (2005). Experimental effects of an organic matter pulse and oxygen depletion on a benthic foraminiferal shelf community. *Journal of Foraminiferal Research, 35*, 177–197.

Fariduddin, M., & Loubere, P. (1997). The surface ocean productivity response of deeper water benthic foraminifera in the Atlantic Ocean. *Marine Micropaleontology, 32*, 289–310.

Fenchel, T., & Finlay, B. J. (1995). *Ecology and evolution in anoxic worlds* (p. 276). Oxford: Oxford studies in Ecology and Evolution, New York: Oxford University Press.

Fischer, G., Ratmeyer, V., & Wefer, G. (2000). Organic carbon fluxes in the Atlantic and the Outhern Ocean: Relationship to primary production compiled from satellite radiometer data. *Deep-Sea Research II, 47*, 1961–1997.

Fontanier, C., Jorissen, F. J., Anschutz, P., & Chaillou, G. (2006). Seasonal variability of foraminiferal faunas at 1000 m depth in the Bay of Biscay. *Journal of Foraminiferal Research, 36*, 61–76.

Fontanier, C., Jorissen, F. J., Chaillou, G., Anschutz, P., Grémare, A., & Griveaud, C. (2005). Live foraminiferal faunas from a 2800 m deep lower canyon station from the Bay of Biscay: Faunal response to focusing of refractory organic matter. *Deep-Sea Research I, 52*, 1189–1227.

Fontanier, C., Jorissen, F. J., Chaillou, G., David, C., Anschutz, P., & Lafon, V. (2003). Seasonal and interannual variability of benthic foraminiferal faunas at 550 m depth in the Bay of Biscay. *Deep-Sea Research I, 50*, 457–494.

Fontanier, C., Jorissen, F. J., Licari, L., Alexandre, A., Anschutz, P., & Carbonel, P. (2002). Live benthic foraminiferal faunas from the Bay of Biscay: Faunal density, composition, and microhabitats. *Deep-Sea Research I, 49*, 751–785.

François, R., Honjo, S., Krishfield, R., & Manganini, S. (2002). Factors controlling the flux of organic carbon to the bathypelagic zone of the ocean. *Global Biogeochemical Cycles, 16*, 1087–1097.

Froelich, P. N., Klinkhammer, G. P., Bender, M. L., Luedke, N. A., Heath, G. R., Cullen, D., dauphin, P., Hammond, D., Hartman, B., & Maynard, V. (1979). early oxidation of organic matter in pelagic sediments of the eastern Equatorial Atlantic: Suboxic diagenesis. *Geochimica et Cosmochimica Acta, 43*, 1075–1090.

Frew, R. D., Dennis, P. F., Heywood, K. J., Meredith, P. H., & Boswell, S. M. (2000). The oxygen isotope composition of water masses in the northern North Atlantic. *Deep-Sea Research I, 47*, 2265–2286.

Geslin, E., Heinz, P., Hemleben, Ch., & Jorissen, F. J. (2004). Migratory response of deep-sea benthic foraminifera to variable oxygen conditions: Laboratory investigations. *Marine Micropaleontology, 53*, 227–243.

Gofas, S. (1978). *Une approche du paléoenvironnement océanique: les foraminifères benthiques calcaires traceurs de la circulation abyssale.* Ph.D. thesis, University of Western Brittany, France.

Goldstein, S. T., & Corliss, B. H. (1994). Deposit feeding in selected deep-sea an shallow-water benthic foraminifera. *Deep-Sea Research I, 41*, 229–241.

Gooday, A. J. (1988). A response by benthic Foraminifera to the deposition of phytodetritus in the deep-sea. *Nature, 332*, 70–73.

Gooday, A. J. (1993). Deep-sea benthic foraminifera species which exploit phytodetritus: Characteristic features and controls on distribution. *Marine Micropaleontology, 22*, 187–205.

Gooday, A. J. (1994). The biology of deep-sea foraminifera: A review of some advances and their applications in paleoceanography. *Palaios, 9*, 14–31.

Gooday, A. J. (1996). Epifaunal and shallow infaunal foraminiferal communities at three abyssal NE Atlantic sites subject to differing phytodetritus input regimes. *Deep-Sea Research I, 43*, 1395–1421.

Gooday, A. J. (2003). Benthic foraminifera (Protista) as tools in deep-water palaeoceanography: Environmental influences on faunal characteristics. *Advances in Marine Biology, 46*, 1–90.

Gooday, A. J., & Hughes, J. A. (2002). Foraminifera associated with phytodetritus deposits at a bathyal site in the northern Rockall Trough (NE Atlantic): Seasonal contrasts and a comparison of stained and dead assemblages. *Marine Micropaleontology, 46*, 83–110.

Gooday, A. J., & Lambshead, P. J. D. (1989). Influence of seasonally deposited phytodetritus on benthic foraminiferal populations in the bathyal northeast Atlantic: The species response. *Marine Ecology Progress Series, 5*, 53–67.

Gooday, A. J., Levin, L. A., Linke, P., & Heeger, T. (1992). The role of benthic foraminifera in deep-sea food webs and carbon cycling. In: G. T. Rowe & V. Pariente (Eds), *Deep-sea food chains and the global carbon cycle* (pp. 63–91). New York, NY: Kluwer Academic Publishers.

Gooday, A. J., & Rathburn, A. E. (1999). Temporal variability in living deep-sea foraminifera: A review. *Earth Sciences Reviews, 46*, 187–212.

Greiner, G. O. G. (1974). Environmental factors controlling the distribution of benthonic foraminifera. *Breviora, 420*, 1–35.

Grimsdale, T. F., & Van Morkhoven, F. P. C. M. (1955). The ratio between pelagic and benthonic foraminifera as a means of estimating depth of deposition of sedimentary rocks. *Proceedings of 4th World Petroleum Congress, Section I/D4* (pp. 473–491).

Guichard, S., Jorissen, F. J., Bertrand, P., Gervais, A., Martinez, P., Peypouquet, J.-P., Pujol, C., & Vergnaud-Grazzini, C. (1997). Foraminifères benthiques et paléoproductivité : réflexions sur une carotte de l'upwelling (NW Africain). Comptes rendus de l'Académie des Sciences, Paris. *Sciences de la Terre et des Planètes, 325*, 65–70.

Guichard, S., Jorissen, F. J., & Peypouquet, J. P. (1999). Late Quaternary benthic foraminiferal records testifying lateral variability of the Cape Blanc upwelling signal. Comptes Rendus de l'Académie des Sciences, Paris. *Sciences de la Terre et des Planètes, 329*, 295–301.

Gundersen, J. K., & Jørgensen, B. B. (1990). Microstructure of diffusive boundary layers and oxygen uptake of the sea floor. *Nature, 345*, 604–607.

Hannah, F., & Rogerson, A. (1997). The temporal and spatial distribution of foraminiferans in marine benthic sediments of the Clyde Sea area, Scotland. *Estuarine, Coastal and Shelf Science, 44*, 377–383.

Hayward, B. W. (2004). Foraminifera-based estimates of paleobathymetry using Modern Analogue Technique, and the subsidence history of the early Miocene Waitemata Basin, New Zealand. *Journal of Geology and Geophysics, 47*, 749–767.

Heinz, P., Schmiedl, G., Kitazato, H., & Hemleben, C. (2001). Response of deep-sea benthic foraminifera from the Mediterranean Sea to simulated phytoplankton pulses under laboratory conditions. *Journal of Foraminiferal Research, 31*, 210–227.

Herguera, J. C. (1992). *Deep-sea benthic foraminifera and biogenic opal: Glacial to postglacial productivity changes in the western equatorial Pacific, Marine Micropaleontology, 18*, 79–98.

Herguera, J. C. (2000). Last glacial paleoproductivity patterns in the eastern equatorial Pacific: Benthic foraminifera records. *Marine Micropaleontology, 40*, 259–275.

Herguera, J. C., & Berger, W. (1991). Paleoproductivity from benthonic foraminifera abundance: Glacial to postglacial change in the west-equatorial Pacific. *Geology, 19*, 1173–1176.

Hess, S., Jorissen, F. J., Venet, V., & Abu-Zied, R. (2005). Benthic foraminiferal recovery after recent turbidite deposition in Cap Breton Canyon, Bay of Biscay. *Journal of Foraminiferal Research, 35*, 114–129.

Hintz, C. J., Chandler, G. T., Bernhard, J. M., McCorkle, D. C., Havach, S. M., Blanks, J. K., & Shaw, T. J. (2004). A physicochemically constrained seawater culturing system for production of benthic foraminifera. *Limnology and Oceanography: Methods, 2*, 160–170.

Hughes, J. A., & Gooday, A. J. (2004). Associations between living benthic foraminifera and dead tests of *Syringammina gracilissima* (Xenophyophorea) in the Darwin Mounds region (NE Atlantic). *Deep-Sea Research, I, 51*, 1741–1758.

Imbrie, J., & Kipp, N. (1971). A new micropaleontological method for quantitative paleoclimatology: Application to a late Pleistocene Caribbean core. In: K. K. Turekian (Ed.), *The late Cenozoic glacial ages* (pp. 71–181). New Haven, CT: Yale University Press.

Ittekot, V. (1993). The abiotically driven biological pump in the ocean and short-term fluctuations in atmospheric CO_2 contents. *Global and Planetary Change, 8*, 17–25.

Jackson, G. A., & Burd, A. B. (2002). A model for the distribution of particle flux in the mid-water column controlled by subsurface biotic interactions. *Deep-Sea Research II, 49*, 193–217.

Jahnke, R. A., Emerson, S. R., & Murray, J. W. (1982). A model of oxygen reduction, denitrification, and organic matter mineralization in marine sediments. *Limnology and Oceanography, 27*, 610–623.

Jannink, N. T., Van der Zwaan, G. J., Almogi-Labin, A., Duijnstee, I., & Jorissen, F. J. (2001). A transfer function for the quantitative reconstruction of oxygen contents in marine paleo-environments. *Geologica Ultraiectina, 203*, 161–169.

Jannink, N. T., Zachariasse, W. J., & Van der Zwaan, G. J. (1998). Living (Rose Bengal stained) benthic foraminifera from the Pakistan continental margin (northern Arabian Sea). *Deep-Sea Research I, 45*, 1483–1513.

Jian, Z., Wang, L., Kienast, M., Sarnthein, M., Kuhnt, W., Lin, H., & Wang, P. (1999). Benthic foraminiferal paleoceanography of the South China Sea over the last 40,000 years. *Marine Geology, 156*, 159–186.

Jorissen, F. J. (1999a). Benthic foraminiferal microhabitats below the sediment-water interface. In: B. K. Sen Gupta (Ed.), *Modern Foraminifera* (pp. 161–179). Dordrecht: Kluwer Academic Publishers.

Jorissen, F. J. (1999b). Benthic foraminiferal successions across late Quaternary Mediterranean sapropels. In E. J. Rohling (Ed.), *Fifth decade of Mediterranean paleoclimate and sapropel studies, Marine Geology, 153*, 91–101.

Jorissen, F. J., Barmawidjaja, D. M., Puskaric, S., & Van der Zwaan, G. J. (1992). Vertical distribution of benthic Foraminifera in the Northern Adriatic Sea. The relation with high organic flux. *Marine Micropaleontology, 19*, 131–146.

Jorissen, F. J., De Stigter, H. C., & Widmark, J. G. V. (1995). A conceptual model explaining benthic foraminiferal microhabitats. *Marine Micropaleontology, 22*, 3–15.

Jorissen, F. J., & Wittling, I. (1999). Ecological evidence from taphonomical studies; living-dead comparisons of benthic foraminiferal faunas off Cape Blanc (NW Africa). *Palaeogeography, Palaeoclimatology, Palaeoecology, 149*, 151–170.

Jorissen, F. J., Wittling, I., Peypouquet, J. P., Rabouille, C., & Relexans, J. C. (1998). Live benthic foraminiferal faunas off Cap Blanc, NW Africa: Community structure and microhabitats. *Deep-Sea Research I, 45*, 2157–2188.

Kaiho, K. (1991). Global changes of Paleogene aerobic/anaerobic benthic foraminifera and deep-sea circulation. *Palaeogeography, Palaeoclimatology, Palaeoecology, 83*, 65–85.

Kaiho, K. (1994). Benthic foraminiferal dissolved-oxygen index and dissolved-oxygen levels in the modern ocean. *Geology, 22*, 719–722.

Kaiho, K. (1999). Evolution in the test size of deep-sea benthic foraminifera during the past 120 m.y.. *Marine Micropaleontology, 37*, 53–65.

Kelly-Gerreyn, B. A., Hydes, D. J., & Waniek, J. J. (2005). Control of the diffusive boundary layer on benthic fluxes: A model study. *Marine Ecology Progress Series, 292*, 61–74.

Kemp, A. E. S., Baldauf, J. G., & Pearce, R. B. (1995). Origins and paleoceanographic significance of laminated diatom ooze from the eastern equatorial Pacific. *Proceedings Ocean Drilling Program, 138*, 641–663.

Kitazato, H. (1994). Diversity and characteristics of benthic foraminiferal microhabitats in four marine environments around Japan. *Marine Micropaleontology, 24*, 29–41.

Kitazato, H., Nomaki, H., Heinz, P., & Nakatsuka, T. (2003). The role of benthic foraminifera in deep-sea food webs at the sediment-water interface: Results from in situ feeding experiments in Sagami Bay. *Frontier Research on Earth Evolution, 1*, 227–232.

Kitazato, H., Shirayama, Y., Nakatsuka, T., Fujiwara, S., Shimanaga, M., Kato, Y., Okada, Y., Kanda, J., Yamaoka, A., Masukawa, T., & Suzuki, K. (2000). Seasonal phytodetritus deposition and responses of bathyal benthic foraminiferal populations in Sagami Bay, Japan: Preliminary results from "Project Sagami 1996–1999". *Marine Micropaleontology, 40*, 135–149.

Kuhnt, W., Hess, S., & Jian, Z. (1999). Quantitative composition of benthic foraminiferal assemblages as a proxy indicator for organic carbon flux rates in the South China Sea. *Marine Geology, 156*, 123–157.

Lagoe, M. L. (1977). Recent benthic foraminifera from the Central Artic Ocean. *Journal of Foraminiferal Research, 7*, 106–129.

Lambshead, P. J. D., & Gooday, A. J. (1990). The impact of seasonally deposited phytodetritus on epifaunal and shallow infaunal benthic foraminiferal populations in the bathyal northeast Atlantic: The assemblage response. *Deep-Sea Research I, 37*, 1263–1283.

Lampitt, R. S. (1985). Evidence for seasonal deposition of detritus to deep sea floor and its subsequent resuspension. *Deep-Sea Research, 32*, 885–897.

Lea, D. L. (2004). Elemental and Isotopic Proxies of Past Ocean Temperatures. In: H. Elderfield (Ed.), *Treatise on geochemistry* (Vol. 6, pp. 365–390). The Oceans and Marine Geochemistry, Chapter 14. Amsterdam: Elsevier.

Lear, H., Elderfield, H., & Wilson, P. A. (2000). Cenozoic deep-sea temperatures and global ice volumes from Mg/Ca in benthic foraminiferal calcite. *Science, 287*, 269–272.

Lear, C. H., Rosenthal, Y., Coxall, H. K., & Wilson, P. A. (2004). Late Eocene to early Miocene ice sheet dynamics and the global carbon cycle. *Paleoceanography, 19*, PA4015, doi:10.1029/2004PA001039, 2004.

Levin, L. A. (2003). Oxygen minimum zone benthos: Adaptation and community response to hypoxia. *Oceanography and Marine Biology: An Annual Review, 41*, 1–45.

Levin, L. A., Etter, R. J., Rex, M. A., Gooday, A. J., Smith, C. R., Pineda, J., Stuart, C. T., Hessler, R. R., & Pawson, D. (2001). Environmental influences on regional deep-sea species diversity. *Annual Review of Ecology and Systematics, 132*, 51–93.

Levin, L. A., & Gage, J. D. (1998). Relationships between oxygen, organic matter and the diversity of bathyal macrofauna. *Deep-Sea Research, 45*, 129–163.

Levinton, J. S. (1970). The paleoecological significance of opportunistic species. *Lethaia, 3*, 69–78.

Licari, L. N., Schumacher, S., Wenzhöfer, F., Zabel, M., & Mackensen, A. (2003). Communities and microhabitats of living benthic foraminifera from the tropical East Atlantic: Impact of different productivity regimes. *Journal of Foraminiferal Research, 33*, 10–31.

Linke, P., & Lutze, G. F. (1993). Microhabitats preferences of benthic foraminifera – A static concept or a dynamic adaptation to optimise food acquisition? In: M. R. Langer (Ed.), *Foraminiferal Microhabitats. Marine Micropaleontology, 20*, 215–234.

Lohmann, G. P. (1978a). Abyssal benthonic foraminifera as hydrographic indicators in the western South Atlantic ocean. *Journal of Foraminiferal Research, 8*, 6–34.

Lohmann, G. P. (1978b). Response of the deep sea to ice ages. *Oceanus, 21*, 58–64.

Lohrenz, S. E., Knauer, G. A., Asper, V. L., Tuel, M., Michaels, A. F., & Knap, A. H. (1992). Seasonal variability in primary production and particle flux in the northwestern Sargasso Sea: U.S. JGOFS Bermuda Atlantic Time-series Study. *Deep-Sea Research, 39*, 1373–1391.

Loubere, P. (1991). Deep sea benthic foraminiferal assemblage response to a surface ocean productivity gradient: A test. *Paleoceanography, 6*, 193–204.

Loubere, P. (1994). Quantitative estimation of surface ocean productivity and bottom water oxygen concentration using benthic foraminifera. *Paleoceanography, 9*, 723–737.

Loubere, P. (1996). The surface ocean productivity and bottom water oxygen signals in deep water benthic foraminiferal assemblages. *Marine Micropaleontology, 28*, 247–261.

Loubere, P (1998). The impact of seasonality on the benthos as reflected in the assemblages of deep-sea foraminifera. *Deep-Sea Research I, 45*, 409–432.

Loubere, P., & Fariduddin, M. (1999). Benthic foraminifera and the flux of organic carbon to the seabed. In: B. K. Sen Gupta (Ed.), *Modern foraminifera* (pp. 181–199). Dordrecht: Kluwer.

Loubere, P., Gary, A., & Lagoe, M. (1993). Benthic foraminiferal microhabitats and the generation of a fossil assemblage: Theory and preliminary data. *Marine Micropaleontology, 20*, 165–181.

Lozano, J. A., & Hays, J. D. (1976). Relationship of radiolarian assemblages to sediment types and physical oceanography in the Atlantic and western Indian oceanic sectors of the Antarctic Ocean. In R. M. Cline, J. D. Hays (Eds), *Investigations of Late Quaternary Paleoceanography and Paleoclimatology. Geological Society of America, Memoirs, 145*, 303–336.

Lutze G. F. (1979). Benthic foraminifers at Site 397: fluctuations and ranges in the Quaternary. In: U. von Rad, W. B. F. Ryan et al. (Eds), *Init. Repts. DSDP* (Vol. 47 (Pt. 1), pp. 419–431), Washington: U.S. Government Printing Office.

Lutze, G. F. (1980). Depth distribution of benthic foraminifera on the continental margin off NW Africa. *Meteor Forschungs-Ergebnisse, C32*, 31–80.

Lutze, G. F., & Altenbach, A. (1988). *Rupertina stabilis* (WALLICH), a highly adapted, suspension feeding foraminifer. *Meyniana, 40*, 55–69.

Lutze, G. F., & Coulbourn, W. T. (1984). Recent benthic foraminifera from the continental margin of northwest Africa: Community structure and distribution. *Marine Micropaleontology, 8*, 361–401.

Lutze, G. F., Pflaumann, U., & Weinholz, P. (1986). Jungquartäre Fluktuationen der benthischen Foraminiferenfaunen in Tiefsee-Sedimenten vor NW-Afrika — Eine Reaktion auf Productivitätsänderungen im Oberfläschenwasser. *Meteor Forschungs-Ergebnisse, C40*, 163–180.

Lutze, G. F., & Thiel, H. (1989). Epibenthic foraminifera from elevated microhabitats: *Cibicidoides wuellerstorfi* and *Planulina ariminensis. Journal of Foraminiferal Research, 19*, 153–158.

Lynch-Stieglitz, J. (2004). Tracers of past ocean circulation. In: H. Elderfield (Ed.), *Treatise on geochemistry* (Vol. 6, pp. 433–451). The Oceans and Marine Geochemistry, Chapter 16. Amsterdam: Elsevier.

Maas, M. (2000). Verteilung lebendgefärbter bentischer Foraminiferen in einer intensivierten Sauerstoffminimumzone, Indo-Pacifischer Kontinentalrand, nördliches Arabisches Meer. *Meyniana, 52*, 101–129.

Mackensen, A., & Douglas, R. G. (1989). Down-core distribution of live and dead deep-water benthic foraminifera in box cores from the Weddell Sea and the California continental borderland. *Deep-Sea Research, 36*, 879–900.

Mackensen, A., Grobe, H., Kuhn, G., & Fütterer, D. K. (1990). Benthic foraminiferal assemblages from the eastern Weddell Sea between 68 and 73°S: Distribution, ecology and fossilization potential. *Marine Micropaleontology, 16*, 241–283.

Mackensen, A., Schmiedl, G., Harloff, J., & Giese, M. (1995). Deep-sea foraminifera in the South Atlantic Ocean: Ecology and assemblage generation. *Micropaleontology, 41*, 342–358.

Martin, J. H., Knauer, G. A., Karl, D. M., & Broenkow, W. W. (1987). VERTEX: Carbon cycling in the north-east pacific. *Deep-Sea Research, 34*, 267–285.

Martinez, P., Bertrand, P., Shimmield, G. B., Cochrane, K., Jorissen, F. J., Foster, J., & Dignan, M. (1999). Upwelling intensity and ocean productivity changes off Cape Blanc (northwest Africa) during the last 70,000 years: Geochemical and micropalaeontological evidence. *Marine Geology, 158*, 57–74.

Miao, Q., & Thunell, R. C. (1993). Recent deep-sea benthic foraminiferal distributions in the South China and Sulu Seas. *Marine Micropaleontology, 22*, 1–32.

Miller, K. G., & Lohmann, G. P. (1982). Environmental distribution of Recent benthic foraminifera on the northeast United States continental slope. *Geological Society of America, Bulletin, 93*, 200–206.

Moodley, L., Heip, C. H. R., & Middelburg, J. J. (1998b). Benthic activity in sediments of the northwestern Adriatic Sea: Sediment oxygen consumption, macro- and meiofauna dynamics. *Journal of Sea Research, 40*, 263–280.

Morigi, C., Jorissen, F. J., Gervais, A., Guichard, S., & Borsetti, A. M. (2001). Benthic foraminiferal faunas in surface sediments off NW Africa: Relationship with the organic flux to the ocean floor. *Journal of Foraminiferal Research, 31*, 350–368.

Müller, P. J., & Suess, E. (1979). Productivity, sedimentation rate, and sedimantary organic matter in the oceans. I. Organic carbon preservation. *Deep-Sea Research, 26*, 1347–1362.

Mullineaux, L. S., & Lohmann, G. P. (1981). Late Quaternary Stagnations and recirculation of the eastern Mediterranean: Changes in the deep water recorded by fossil benthic foraminifera. *Journal of Foraminiferal Research, 11*, 20–39.

Murray, J. W. (1973). *Distribution and ecology of living benthic foraminiferids*. London: Heinemann, 288 pp.

Murray, J. W. (1988). Neogene bottom water-masses and benthic foraminifera in the NE Atlantic Ocean. *Journal of the Geological Society, London, 145*, 125–132.

Murray, J. W. (1991). *Ecology and paleoecology of benthic foraminifera* (p. 397). Harlow: Longman.

Murray, J. W. (2001). The niche of benthic foraminifera, critical thresholds and proxies. *Marine Micropaleontology, 41*, 1–7.

Murray, J. W., Weston, J. F., Haddon, C. A., & Powell, A. D. J. (1986). Miocene to recent bottom water masses of the north-east Atlantic: An analysis of benthic foraminifera. In: C. P. Summerhayes & N. J. Shackleton (Eds), *North Atlantic Palaeoceanography* (Vol. 21, pp. 219–230). London: Special Publication of the Geological Society.

Murray, J., Young, J., Newton, J., Dunne, J., Chapin, T., Paul, B., & McCarthy, J. J. (1996). Export flux of particulate organic carbon from the central equatorial Pacific determined using a combined drifting trap-[234]Th approach. *Deep Sea Research II, 43*, 1095–1132.

Naidu, P. D., & Malmgren, B. A. (1995). Do benthic foraminifer records represent a productivity index in oxygen minimum zone areas? An evaluation from the Oman Margin, Arabian Sea. *Marine Micropaleontology, 26*, 49–55.

Natland, M. L. (1933). The temperature- and depth-distribution of some Recent and fossil Foraminifera in the Southern California Region. *Scripps Institute of Oceanography Bulletin, Technical Series, 3*, 225–230.

Natland, M. L., & Kuenen, P. H. (1951). Sedimentary history of the Ventura Basin, California, and the action of turbidity currents. In: J. L. Hough (Ed.), *Turbidity currents and the transportation of coarse sediments to deep water. A symposium* (Vol. 2, pp. 76–107). Tulsa, OK: Society of Economic Paleontologists and Mineralogists, Special Publication.

Nees, S., Altenbach, A. V., Kassens, H., & Thiede, J. (1997). High-resolution record of foraminiferal response to Late Quaternary sea-ice retreat in the Norwegian-Greenland Sea. *Geology, 25*, 659–662.

Nees, S., Armand, L., De Deckker, P., Labracherie, M., & Passlow, V. (1999). A diatom and benthic foraminiferal record from the South Tasman Rise (southeastern Indian Ocean): Implications for palaeoceanographic changes for the last 200,000 years. *Marine Micropaleontology, 38*, 69–89.

Nees, S., & Struck, U. (1999). Benthic foraminiferal response to major oceanographic changes. In: F. Abrantes & A. Mix (Eds), *Reconstructing ocean history: A window into the future* (pp. 195–216). Dordrecht: Kluwer Academic/Plenum Publishers.

Nelson, D. M., Anderson, R. F., Barber, R. T., Brzezinski, M. A., Buesseler, K. O., Chase, Z., Collier, R. W., Dickson, M. L., François, R., Hiscock, M. R., Honjo, S., Marra, J., Martin, W. R., Sambrotto, R. N., Sayles, F. L., & Sigmon, D. E. (2002). Vertical budgets for organic carbon and biogenic silica in the Pacific sector of the Southern Ocean, 1996–1998. *Deep Sea Research Part II: Topical Studies in Oceanography, 49*(9-10), 1645–1674.

Nittrouer, C. A., DeMaster, D. J., McKee, B. A., Cutshall, N. H., & Larsen, I. L. (1984). The effect of sediment mixing on Pb-210 accumulation rates for the Washington continental shelf. *Marine Geology, 54*, 201–221.

Nomaki, H., Heinz, P., Hemleben, C., & Kitazato, H. (2005). Behavior and response of deep-sea benthic foraminifera to freshly supplied organic matter: A laboratory feeding experiment in microcosm environments. *Journal of Foraminiferal Research, 35*, 103–113.

Ohga, T., & Kitazato, H. (1997). Seasonal changes in bathyal foraminiferal populations in response to the flux of organic matter (Sagami Bay, Japan). *Terra Nova, 9*, 33–37.

Ohkushi, K., Thomas, E., & Kawahata, H. (2000). Abyssal benthic foraminifera from the Northwestern Pacific (Shatsky Rise) during the last 298 kyr. *Marine Micropaleontology, 38*, 119–147.

Osterman, L. E., & Kellogg, T. B. (1979). Recent benthic foraminiferal distribution from the Ross Sea, Antarctica: Relation to ecological and oceanographic conditions. *Journal of Foraminiferal Research, 9*, 250–269.

Pace, M. L., Knauer, G. A., Karl, D. M., & Martin, J. H. (1987). Primary production, new production, and vertical flux in the eastern Pacific Ocean. *Nature, 325*, 803–804.

Parisi, E., & Cita, M. B. (1982). Late Quaternary paleoceanograpic changes recorded by deep-sea benthos in the western Mediterranean Ridge. *Geografia Fisica e Dinamica Quaternaria, 5*, 102–114.

Parker, F. L. (1958). Eastern Mediterranean Foraminifera. Reports of the Swedish deep-sea expedition 1947–1948, VIII: *Sediment cores from the Mediterranean and the Red Sea, 4,* 283.

Peterson, L. C. (1984). Recent abyssal benthic foraminiferal biofacies of the eastern Equatorial Indian Ocean. *Marine Micropaleontology, 8,* 479–519.

Peterson, L. C., & Lohmann, G. P. (1982). Major change in Atlantic Deep and Bottom Waters 700,000 yr Ago: Benthonic Foraminiferal Evidence from the South Atlantic. *Quaternary Research, 17,* 26–38.

Pfannkuche, O., & Lochte, K. (1993). Open ocean pelago-benthic coupling: Cyanobacteria as tracers of sedimenting salp faeces. *Deep-Sea Research, 40,* 727–737.

Pflum, C. E., & Frerichs, W. E. (1976). Gulf of Mexico Deep-Water Foraminifers. *Cushman Foundation for Foraminiferal Research, Special Publication, 14,* 125.

Phleger, F. L. (1960). *Ecology and distribution of recent Foraminifera* (p. 297). Baltimore, MD: Johns Hopkins Press.

Phleger, F. L., & Soutar, A. (1973). Production of benthic foraminifera in three east Pacific oxygen minima. *Micropaleontology, 19,* 110–115.

Pianka, E. R. (1970). On r and K selection. *American Naturalist, 104,* 592–597.

Polski, W. (1959). Foraminiferal biofacies off North Asiatic Coast. *Journal of Paleontology, 33,* 569–587.

Pujos-Lamy, A. (1973). Répartition bathymétrique des foraminifères benthiques du Golfe de Gascogne. *Revista Española De Micropaleontologia, 5,* 213–234.

Rabouille, C., & Gaillard, J. F. (1991). A coupled model representing the deep-sea organic carbon mineralization and oxygen consumption in surficial sediments. *Journal of Geophysical Research, 96,* 2761–2776.

Rathburn, A. E., & Corliss, B. H. (1994). The ecology of living (stained) deep-sea benthic foraminifera from the Sulu Sea. *Paleoceanography, 9,* 87–150.

Rathburn, A. E., Corliss, B. H., Tappa, K. D., & Lohmann, K. C. (1996). Comparison of the ecology and stable isotopic compositions of living (stained) benthic foraminifera from the Sulu and South China Seas. *Deep-Sea Research I, 43,* 1617–1646.

Rathburn, A. E., Perez, M. E., & Lange, C. (2001). Benthic-pelagic coupling in The Southern California bight: Relationships between sinking organic material, diatoms and benthic foraminifera. *Marine Micropaleontology, 43,* 261–271.

Rathburn, A. E., Perez, M. E., Martin, J. B., Day, S. E., Mahn, C., Gieskes, J., Ziebis, W., Williams, D., & Bahls, A., (2003). Relationships betweeb the distribution and stable isotopic composition of living benthic foraminifera and cold methane seeps biogeochemistry in Monterey Bay, California. *Geochemistry, Geophysics, Geosystems, 4*(12), doi:10.1029/2003GC000595.

Ravizza, G. E., & Zachos, J. C. (2004). Records of Cenozoic Ocean Chemistry. In: H. Elderfield (Ed.), *Treatise on geochemistry* (Vol. 6, pp. 551–581). The Oceans and Marine Geochemistry, Chapter 20. Amsterdam: Elsevier).

Reimers, C. E. (1987). An in situ microprofiling instrument for measuring interfacial pore water gradients: Methods and oxygen profiles from the North Pacific Ocean. *Deep-Sea Research, 34,* 2019–2035.

Robinson, B. H., Reisenbichler, K. R., & Sherlock, R. E. (2005). Giant larvacean houses: Rapid carbon transport to the deep sea floor. *Science, 308,* 1609–1611.

Rutgers van der Loeff, M. M. (1990). Oxygen in pore waters of deep-sea sediments. *Philosophical Transactions Royal Society London, A, 331,* 69–84.

Saidova, K. M. (1961). Ecology of the foraminifera and paleoceanography of the Far Eastern seas of the USSR and the northwestern part of the Pacific Ocean. *Akademie Nauk SSSR, Instituta Okeanologii, 45,* 65–71 (in Russian).

Sarnthein, M., Winn, K., Duplessy, J. C., & Fontugne, M. R. (1988). Global variations of surface ocean productivity in low and mid latitudes: Influence on CO_2 reservoirs of the deep ocean and atmosphere during the last 21,000 years. *Paleoceanography, 3,* 361–399.

Schemainda, R., Nehring, D., & Schultz, S. (1975). Ozeanologische untersuchungen zum produktionspotential der Nordwestafrikanischen wasserauftriebsregion 1970–1973. *Nationalkomitee für Geodäsie und Geophysik bei der Akademie der Wissenschaften der Deutschen Demokratischen Republik, IV/16,* 85.

Schmiedl, G., De Bovée, F., Buscail, R., Charrière, B., Hemleben, C., Medernach, L., & Picon, P. (2000). Trophic control of benthic foraminiferal abundance and microhabitat in the bathyal Gulf of Lions, western Mediterranean Sea. *Marine Micropaleontology*, 40, 167–188.

Schmiedl, G., & Mackensen, A. (1997). Late Quaternary paleoproductivity and deep water circulation in the eastern South Atlantic Ocean: Evidence from benthic foraminifera. *Palaeogeography, Palaeoclimatology, Palaeoecology*, 130, 43–80.

Schmiedl, G., Mackensen, A., & Müller, P. J. (1997). Recent benthic foraminifera from the eastern South Atlantic Ocean: Dependence on food supply and water masses. *Marine Micropaleontology*, 32, 249–287.

Schmiedl, G., Mitschele, A., Beck, S., Emeis, K. C., Hemleben, C., Schulz, H., Sperling, M., & Weldeab, S. (2003). Benthic foraminiferal record of ecosystem variability in the eastern Mediterranean Sea during times of sapropel S$_5$ and S$_6$ deposition. *Palaeogeography, Palaeoclimatology, Palaeoecology*, 190, 139–164.

Schmitz, W. J. (1992). On the intrabasinal thermohaline circulation. *Reviews of Geophysics*, 33, 151–173.

Schnitker, D. (1974). West abyssal circulation during the past 120,000 years. *Nature*, 248, 385–387.

Schnitker, D. (1976). Structure and cycles of the western North Atlantic bottom water, 24,000 years BP to present. *EOS*, 57, 257–258.

Schnitker, D. (1979). The deep waters of the western North Atlantic during the past 24,000 years and the re-initiation of the Western Boundary Undercurrent. *Marine Micropaleontology*, 4, 264–280.

Schnitker, D. (1980). Quaternary deep-sea benthic foraminifers and bottom water masses. *Annual Review of Earth and Planetary Science*, 8, 343–370.

Schönfeld, J. (1997). The impact of the Mediterranean Outflow Water (MOW) on benthic foraminiferal assemblages and surface sediments at the southern Portuguese continental margin. *Marine Micropaleontology*, 29, 211–236.

Schönfeld, J. (2002a). Recent benthic foraminiferal assemblages in deep high-energy environments from the Gulf of Cadiz (Spain). *Micropaleontology*, 44, 141–162.

Schönfeld, J. (2002b). A new benthic foraminiferal proxy for near-bottom current velocities in the Gulf of Cadiz, northeastern Atlantic Ocean. *Deep-Sea Research I*, 49, 1853–1875.

Schönfeld, J., & Altenbach, A. V. (2005). Late Glacial to Recent distribution pattern of deep-water *Uvigerina* species in the north-eastern Atlantic. *Marine Micropaleontology*, 57, 1–24.

Schrader, H. J. (1974). Cenozoic marine planktonic diatom stratigraphy of the tropical Indian Ocean. In: R. L. Fisherm & E. T. Bunce (Eds), *Initial Reports of the Deep Sea Drilling Project* (Vol. 24, pp. 887–968). Washington: U. S. Government Printing Office.

Sen Gupta, B. (Ed.) (1999). *Modern foraminifera* (p. 371). Dordrecht: Kluwer Academic Publishers.

Sen Gupta, B. K., & Machain-Castillo, M. L. (1993). Benthic foraminifera in oxygen-poor habitats. *Marine Micropaleontology*, 20, 3–4.

Sen Gupta, B. K., Platon, E., Bernhard, J. M., & Aharon, P. (1997). Foraminiferal colonization of hydrocarbon-seep bacterial mats and underlying sediment, Gulf of Mexico Slope. *Journal of Foraminiferal Research*, 27, 292–300.

Sen Gupta, B. K., Smith, L. E., & Lobegeier, M. K. (2006). Attachment of Foraminifera to vestimentiferan tubeworms at cold seeps Refuge from sea-floor hypoxia and sulfide toxicity. *Marine Micropaleontology*, 62(1), 1–6.

Shirayama, Y. (1984). The abundance of deep sea meiobenthos in the Western Pacific in relation to environmental factors. *Oceanologica Acta*, 7, 113–121.

Sigman, D. M., & Haug, G. H. (2004). The biological pump in the past. In: H. Elderfield (Ed.), *Treatise on geochemistry* (Vol. 6, pp. 491–528) The Oceans and Marine Geochemistry, Chapter 18. Amsterdam: Elsevier.

Sliter, W. V., & Baker, R. A. (1972). Cretaceous bathymetric distribution of benthic foraminifera. *Journal of Foraminiferal Research*, 2, 167–183.

Smart, C. W. (2002). Environmental applications of deep-sea benthic foraminifera. In: S. K. Haslett (Ed.), *Quaternary Environmental Micropalaeontology* (pp. 14–58). London: Arnold Publishers.

Smart, C. W., King, S. C., Gooday, A. J., Murray, J. W., & Thomas, E. (1994). A benthic foraminiferal proxy of pulsed organic matter paleofluxes. *Marine Micropaleontology*, 23, 89–99.

Smith, P. B. (1964). Recent foraminifera off Central America. Ecology of benthic species. *USGS Professional Paper, 429B,* 1–55.

Soetaert, K., Herman, P. M. J., & Middelburg, J. J. (1996). A model of early diagenetic processes from the shelf to abyssal depths. *Geochimica et Cosmochimica Acta, 60,* 1019–1040.

Streeter, S. S. (1973). Bottom water and benthonic foraminifera in the North Atlantic-glacial-interglacial cycles. *Quaternary Research, 3,* 131–141.

Streeter, S. S., Belanger, P. E., Kellogg, T. B., & Duplessy, J. C. (1982). Late Pleistocene paleo-oceanography of the Norwegian-Greenland sea: Benthic foraminiferal evidence. *Quaternary Research, 18,* 72–90.

Streeter, S. S., & Shackleton, N. J. (1979). Paleocirculation of the deep North Atlantic: 150,000 year record of benthic foraminifera and oxygen-18. *Science, 203,* 168–171.

Stschedrina, Z. G. (1957). Some regularities in the distribution of recent foraminifera. Trudy Leningradskogo Obshch. *Estest, 73,* 99–106.

Suess, E. (1980). Particulate organic carbon flux in the ocean-surface productivity and oxygen utilization. *Nature, 280,* 260–263.

Suhr, S. B., Pond, D. W., Gooday, A. J., & Smith, C. R. (2003). Selective feeding by benthic foraminifera on phytodetritus on the western Antarctic Peninsula shelf: Evidence from fatty acid biomarker analysis. *Marine Ecology Progress Series, 262,* 153–162.

Sun, X., Corliss, B. H., Brown, C. W., & Showers, W. J. (2006). The effect of primary productivity and seasonality on the distribution of deep-sea benthic foraminifera in the North Atlantic. *Deep-Sea Research I, 53,* 28–47.

Tabachnick, B. G., & Fidell, L.S. (1983). *Using multivariate statistics* (p. 509). New York, NY: Harper & Row Publishers.

Tengberg, A., De Bovee, F., Hall, P., Berelson, W., Chadwick, D., Cicceri, G., Crassous, P., Devol, A., Emerson, S., Gage, J., Glud, R., Graziottin, F., Gundersen, J., Hammond, D., Helder, W., Hinga, K., Holby, O., Jahnke, R., Khripounoff, A., Lieberman, S., Nuppenau, V., Pfannkuche, O., Reimers, C., Rowe, G., Sahami, A., Sayles, F., Schuster, M., Smallman, D., Wehrli, B., & De Wilde, P. (1995). Benthic chamber and profiling landers in oceanography—A review of design, technical solutions and functioning. *Progress in Oceanography, 35,* 253–294.

Thomas, E. (1986). Changes in composition of Neogene benthic foraminiferal faunas in equatorial Pacific and north Atlantic. *Palaeogeography, Palaeoclimatology, Palaeoecology, 53,* 47–61.

Thomas, E. (2007). Cenozoic mass extinctions in the deep sea: What perturbs the largest habitat on earth? In: S. Monechi, M. Rampino & R. Coccioni (Eds), *Large ecosystem perturbations: Causes and consequences.* Geological Society of America Special paper, in press.

Thomas, E., Booth, L., Maslin, M., & Shackleton, N. J. (1995). Northeastern Atlantic benthic foraminifera during the last 45,000 years: Changes in productivity seen from the bottom up. *Paleoceanography, 10,* 545–562.

Thomas, E., Gooday, A. J. (1996). Cenozoic deep-sea benthic foraminifers: Tracers for changes in oceanic productivity? *Geology, 24,* 355–358

Toyofuku, T., & Kitazato, H. (2005). Micromapping of Mg/Ca values in cultured specimens of the high-magnesium benthic foraminifera. *Geochemistry, Geophysics, and Geosystems, 6,* Q11P05.

Turley, C. (2002). *The importance of 'marine snow': Microbiology Today, 29,* 177–179.

Tyson, R. V., & Pearson, T. H. (1991). Modern and ancient continental shelf anoxia: An overview. In: Tyson, R. V., Pearson, T. H., (Eds), *Modern and Ancient Continental Shelf Anoxia* (Vol. 58, pp. 1–24). London: Geological Society of London Special Publication.

Van Aken, H. M. (2000). The hydrography of the mid-latitude northeast Atlantic Ocean I: The deep water masses. *Deep-Sea Research I, 47,* 757–788.

Van der Zwaan, G. J. (1982). Paleoecology of Late Miocene Mediterranean Foraminifera. *Utrecht Micropaleontological Bulletins, 25,* 202.

Van der Zwaan, G. J., Duijnstee, I. A. P., Den Dulk, M., Ernst, S. R., & Kouwenhoven, N. T. (1999). Benthic foraminifers: proxies or problems? A review of paleoecological concepts. *Earth Sciences Reviews, 46,* 213–236.

Van der Zwaan, G. J., Jorissen, F. J., & De Stigter, H. (1990). The depth dependency of planktonic/benthic foraminiferal ratios: Constraints and applications. *Marine Geology, 95,* 1–16.

Van Hinsbergen, D. J. J., Kouwenhoven, T. J., & Van der Zwaan, G. J. (2005). Paleobathymetry in the backstripping procedure: Correction for oxygenation effects on depth estimates. *Palaeogeography, Palaeoclimatology, Palaeoecology, 221*, 245–265.

Walker, S. E., & Goldstein, S. T. (1999). Taphonomic tiering: Experimental field taphonomy of molluscs and foraminifera above and below the sediment-water interface. *Palaeogeography, Palaeoclimatology, Palaeoecology, 149*, 227–244.

Walton, W. R. (1952). Techniques for recognition of living Foraminifera. *Contributions Cushman Foundation for Foraminiferal Research, 3*, 56–60.

Wassmann, P. (1993). Regulation of vertical export of particulate organic matter from the euphotic zone by planktonic heterotrophs in eutrophicated aquatic environments. *Marine Pollution Bulletin, 26*, 636–643.

Wefer, G., Berger, W. H., Bijma, J., & Fischer, G. (1999). Clues to ocean history: A brief overview of proxies. In: G. Fischer & G. Wefer (Eds), *Uses of proxies in Paleoceanography: Examples from the South Atlantic* (pp. 1–68). Berlin, Heidelberg: Springer-Verlag.

Weinelt, M., Kuhnt, W., Sarnthein, M., Altenbach, A., Costello, O., Erlenkeuser, H., Pflaumann, U., Simstich, J., Thies, A., Trauth, M. H., & Vogelsand, E. (2001). Paleoceanographic proxies in the northern North Atlantic. In: P. Schäfer, W. Ritzrau, M. Schlüter & T. Thiede (Eds), *The northern North Atlantic: A changing environment* (pp. 319–352). Berlin, Heidelberg: Springer-Verlag.

Weston, J. F. (1982). *Distribution and ecology of Recent Deep Sea benthic foraminifera in the Northeast Atlantic Ocean*. Ph.D. thesis, University of Exeter, England.

Wilson-Finelli, A., Chandler, G. T., & Spero, H. J. (1998). Stable isotope behavior in paleoceanographically important benthic foraminifera: Results from microcosm culture experiments. *Journal of Foraminiferal Research, 28*, 312–320.

Wollenburg, J. E., & Kuhnt, W. (2000). The response of benthic foraminifers to carbon flux and primary production in the Arctic Ocean. *Marine Micropaleontology, 40*, 189–231.

Wollenburg, J. E., Kuhnt, W., & Mackensen, A. (2001). Changes in Arctic Ocean paleoproductivity and hydrography during the last 145 kyr: The benthic foraminiferal record. *Paleoceanography, 16*, 65–77.

Woodruff, F. (1985). Changes in Miocene deep-sea benthic foraminiferal distribution in the Pacific Ocean: Relationship to paleoceanography. In: Kennett, J. P. (Ed.), *The Miocene Ocean: Paleoceanography and Biogeography* (Vol. 163, pp. 131–175). Boulder, Memoir: Geological Society of America.

Woodruff, F., & Savin, S. M. (1989). Miocene deepwater oceanography. *Paleoceanography, 4*, 87–140.

CHAPTER EIGHT

Diatoms: From Micropaleontology to Isotope Geochemistry

Xavier Crosta* *and* Nalan Koç

Contents

1. Introduction	327
1.1. Classification of diatoms	327
1.2. Biology of diatoms	329
1.3. Ecology of diatoms	329
1.4. Diatoms in surface sediments	330
1.5. Conceptual progress in diatom methods	331
2. Improvements in Methodologies and Interpretations	332
2.1. Micropaleontology	332
2.2. Isotope geochemistry	344
3. Case Studies	350
3.1. SST in the north atlantic	350
3.2. Sea-ice in the southern ocean	352
3.3. C, n, and si isotopes in the southern ocean	353
4. Conclusion	356
Acknowledgments	358
References	358

1. Introduction

1.1. Classification of Diatoms

Kingdom: Protista

Phylum: Chrysophyta

Class: Bacillariophyceae

Orders: Centrales and Pennales

* Corresponding author.

Developments in Marine Geology, Volume 1
ISSN 1572-5480, DOI 10.1016/S1572-5480(07)01013-5

Diatoms are unicellular organisms in which the cell is encapsulated in an amorphous silica box, called the frustule, composed of two intricate valves. Diatom size
varies from 2 μm to 1–2 mm, and diatom shape exhibits any variation from round
(Centrales) to needle-like (Pennales). The frustule is highly ornamented with pores
(areolae), processes (labiate, strutted, internal or external, with or without exten-
sions), spines, costae, horns, hyaline areas, and other distinguishing features. Diatom
taxonomy is historically based on the shape and ornamentation of the frustule
(Pfitzer, 1871; Schütt, 1896; Simonsen, 1979; Round, Crawford, & Mann, 1990;
Hasle & Syversten, 1997).

Centric diatoms are separated into three suborders, based on the presence or
absence of the marginal ring of processes and the polarity of the symmetry, while
Pennate diatoms are separated into two suborders based on the presence or absence
of the raphe, observed as an elongated fissure or pair of fissures through the valve
wall (Figure 1) (see Anonymous (1975) for more details on the distinguishing
features of diatoms). Recently, molecular investigations provided a new understanding of species determination and showed that several species may share the

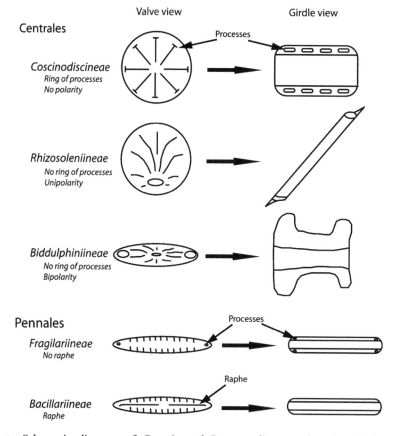

Figure 1 Schematic diagram of Centric and Pennate diatom sub-orders (redrawn with
permission from Hasle & Syvertsen, 1997).

same morphology (Graham & Wilcox, 2000). The question of whether this genetic variability is related to environmental conditions and may be useful for paleoclimatic investigations is still under debate. To date, the classification system developed by Simonsen (1979) is still the most widely accepted.

1.2. Biology of Diatoms

Diatoms are photosynthetic organisms possessing yellow–brown chloroplasts with pigments including chlorophyll a and c, β-carothene, fucoxanthin, diatoxanthin, and diadinoxanthin (Jeffrey, Mantoura, & Wight, 1997). This large set of pigments enables diatoms to capture a wide range of wavelengths and to live at low light levels, for example under sea-ice that filters most of the solar energy.

Diatoms generally reproduce through vegetative fission at a rate of 0.1–8 times per day. This vegetative reproduction allows diatoms to build a very high biomass, which is at the origin of diatomite, when the preservation process allows it. Vegetative reproduction involves the formation of two new hypovalves in the parent diatom's frustule, which progressively reduce the average size of diatom frustules in the population. At a given threshold, diatoms undergo sexual reproduction through gamete fusion and the formation of an auxospore that renews a full-sized vegetative cell (Round, 1972). Some species have another peculiar reproductive stage, the resting spore. The spore is formed under unfavorable conditions (depleted nutrient levels, low light levels, etc.) and allows the diatom to survive until better conditions return.

1.3. Ecology of Diatoms

About 285 genera and 12,000 species of diatoms have been identified (Round et al., 1990). Diatoms are found in almost every aquatic environment, including fresh and marine waters. They are nonmotile and restricted to the photic zone. Diatoms may be solitary as well as colonial. In the marine environment diatoms are generally planktonic, although some benthic or pseudo-benthic species attached to macro-algae or sea-ice are also encountered.

The relationships between abiotic and biotic factors and diatom distribution in surface water are poorly understood. Many factors interact to determine the distribution of planktonic diatoms in any given oceanic region, but the most important factors are sea-surface temperatures (SSTs) (Neori & Holm-Hansen, 1982), sea-ice conditions (SIC) (Horner, 1985), macro- and micronutrient levels (Fitzwater, Coale, Gordon, Johnson, & Ondrusek, 1996), stability of the surface water layer (Leventer, 1991), light levels (El Sayed, 1990), and grazing (El Sayed, 1990). Salinity may also exert a major role on diatom distribution, especially in coastal regions and regions of the Artic Ocean influenced by sea-ice, where strong gradients in salinity exist (Campeau, Pienitz, & Héquette, 1998; Licursi, Sierra, & Gomez, 2006). Similarly, many factors interact to determine the distribution of benthic diatoms, the most important being the biotope category (Aleem, 1950), the substrate (Round et al., 1990), and the water depth (Campeau, Pienitz, & Héquette, 1999), perhaps associated with irradiance penetration. Benthic diatoms are generally restricted to environments shallower than 100 m.

In the world ocean, diatoms are restricted to cold, nutrient-rich regions where silicic acid is not limiting, such as in the polar regions, the coastal and equatorial upwelling systems, and in the coastal areas. In other regions, diatoms are outcompeted by carbonate organisms that have lower nutrient requirements.

Some diatom species thrive in very narrow ranges of conditions and are encountered in specific regions. For example, some *Fragilariopsis* species occur in both polar regions, while others occur only in upwelling systems. This specificity can be extreme and some diatoms are endemic to a single region. Several species are, for example, restricted to the Antarctic Ocean, such as *Fragilariopsis kerguelensis* and *F. curta*. Species thriving in a limited range of conditions are obviously much more useful than widely distributed species for paleoceanographic reconstructions. Although it is difficult to talk about diatom zonation for the world ocean, clear zonations are evident in specific areas. Different ecological preferences lead to gradients of different diatom species abundances in surface waters (Heiden & Kolbe, 1928; Hendey, 1937; Hustedt, 1958; Hasle, 1969) and generally in surface sediments (Sancetta, 1992; DeFelice & Wise, 1981; Abrantes, 1988a; Koç-Karpuz & Schrader, 1990; Armand, Crosta, Romero, & Pichon, 2005; Crosta, Romero, Armand, & Pichon, 2005a; Romero, Armand, Crosta, & Pichon, 2005). Understanding diatom ecology in the study area is therefore essential for paleoceanographic investigations.

1.4. Diatoms in Surface Sediments

The distribution of diatoms in surface sediments is related to secondary processes that modify the surface water assemblages, except for autochthonous benthic diatom assemblages. Sedimentation type (Schrader, 1971; von Bodungen, Smetacek, Tilzer, & Zeitzschel, 1985; Smetacek, 1985), lateral transport (Leventer, 1991), and dissolution in the water column and at the water–sediment interface (Kamatani, Ejiri, & Tréguer, 1988; Shemesh, Burckle, & Froelich, 1989) are major processes determining diatom flux to the seafloor. Generally, 1–10% of the diatoms produced in surface waters reach the sediment (Kozlova, 1971; Ragueneau et al., 2000). Although the surface water assemblages, which bear the ecological and climatic signal, are altered during settling to the seafloor and burying, it has been shown that the residual sedimentary assemblages are still indicative of surface conditions in different oceanic regions such as the North Pacific (Sancetta, 1992), the Southern Ocean (Armand et al., 2005; Crosta et al., 2005a; Romero et al., 2005), the Benguela upwelling system (Pokras & Molfino, 1986), the Equatorial Atlantic (Treppke et al., 1996), and the high North Atlantic (Koç-Karpuz & Schrader, 1990; Andersen, Koç, Jennings, & Andrews, 2004a). Diatoms can therefore be used to infer past oceanographic and climatic changes in these regions (Sancetta, 1979; DeFelice & Wise, 1981; Burckle, 1984a, 1984b; Pokras & Molfino, 1987; Pichon et al., 1992; Koç, Jansen, & Haflidason, 1993; Zielinski & Gersonde, 1997).

Autochthonous benthic diatom assemblages result from the ecological preferences of benthic diatoms as described above, and can therefore be used as quantitative paleodepth indicators in coastal areas (Campeau et al., 1999; Jiang, Seidenkrantz, Knudsen, & Eiriksson, 2001).

1.5. Conceptual Progress in Diatom Methods

Diatoms have been known and identified since the beginning of the 18th century, but they have only recently been used to investigate past oceanographic and climatic changes. Three main applications can be described: biostratigraphy for age dating, micropaleontology and geochemistry for paleoceanography.

Fossil diatoms were initially used for biostratigraphic purposes. Biostratigraphy is the science of dating rocks or sediments by using the fossils they contain. Usually, the objective of biostratigraphy is basin-wide correlations when other stratigraphic methods are lacking. The fossil species used must be geographically widespread and have short life spans. Diatom species that achieve these two requirements are key stratigraphic markers. In the 1970s, it was shown that diatom sequences in large regions were similar through time although the sediment composition and texture could be completely different. The different diatom units were tied to paleomagnetic stratigraphy or other biostratigraphy to define Epoch boundaries that one could extrapolate to other records representing the same units, thus ascribing age control at a basin-wide scale. This science is in constant evolution and diatom units are continuously refined due to changing diatom taxonomy, investigations of high-resolution records and better dating techniques. Some key studies are mentioned below, as no further mention of biostratigraphy is made in this chapter.

In the North Pacific, where sediments are mainly barren of $CaCO_3$, diatoms are the prime biostratigraphic tool. Neogene diatom biostratigraphy was developed there in the 1970s (Koizumi, 1977) as a complement to paleomagnetic and tephra chronology. Nineteen diatom zones are currently determined for the Neogene and Pleistocene epochs and are valid for the entire North Pacific (Akiba, 1985; Sancetta & Silvestri, 1984; Akiba & Yanagisawa, 1985). DSDP/ODP cruises (Legs 38, 94, 104, 151, 152, and 162) have shown that the main biogenic component of the Tertiary sediments of the North Atlantic Ocean and the Norwegian-Greenland Sea are the siliceous microfossil group diatoms, and that the area was primarily a silica ocean until the onset of Northern Hemisphere glaciations during the late Miocene. Diatom species show very rapid evolution through the Cenozoic, and this has made it possible to establish a high-resolution biostratigraphy for the area. There is a well-established diatom biostratigraphy for the North Atlantic (Baldauf, 1984, 1987), which has recently been refined (Koç & Flower, 1998; Koç, Hodell, Kleiven, & Labeyrie, 1999; Koç, Labeyrie, Manthé, Flower, & Hodell, 2001). Most of the fossil diatom species of the Norwegian-Greenland Sea are endemic to the area. Therefore, a separate diatom biostratigraphy had to be developed for the Norwegian-Greenland Sea. Based on the DSDP Leg 38 material, Dzinoridze et al. (1978) and Schrader and Fenner (1976) proposed a diatom biostratigraphy for the area. Meanwhile, development of drilling techniques and availability of reliable paleomagnetic stratigraphies enabled the development of a new Neogene (Koç & Scherer, 1996) and late Paleogene (Scherer & Koç, 1996) diatom biostratigraphy for the Norwegian-Greenland Sea based on the ODP Leg 151 material. In the Equatorial Pacific, seven diatom datum levels for the Neogene and Pleistocene epochs were identified and related to the paleomagnetic reversal record (Burckle,1972). Diatom zones are characterized by unique floral assemblages that have proved useful for basin-wide correlations. In the Southern Ocean, McCollum (1975) defined zonal schemes for most of the Tertiary. They were subsequently extended to the Cenozoic (Gersonde, 1990; Fenner, 1991; Gersonde & Barcena, 1998) and recently improved (Zielinski & Gersonde, 2002). Stratigraphic markers from the late MIS 6 at 135 kyr BP, late MIS 7 at 190 kyr BP, and early MIS 8 at 290 kyr BP are again essential to confirm oxygen isotope stratigraphy (Burckle, Clarke, & SHackleton, 1978; Zielinski & Gersonde, 2002).

In the 1980s, paleoceanographers realized that it would be possible to extrapolate the relationships between diatom assemblages in surface sediments and modern parameters to down-core fossil assemblages in order to document past changes in oceanography, in siliceous productivity and ultimately in climate. The starting hypothesis is that a given diatom assemblage is produced and preserved under specific modern conditions. If the same assemblage is found down-core, then oceanographic and climatic conditions may have been the same in the past as they are now. A great number of surface parameters, ecologically important for diatom development, were thus reconstructed: SSTs, SIC, hydrology, productivity events, etc. Initially, investigations were based on down-core variations of given species, or groups of given species, of a known ecology, but it became rapidly apparent that only statistical methods (factor analysis) could provide full understanding of the down-core assemblages and therefore produce better paleoceanographic reconstructions (Burckle, 1984a). The ultimate step was taken in the 1990s with the appearance of transfer functions that provided quantitative estimates of surface properties (Koç-Karpuz & Schrader, 1990; Pichon et al., 1992) and water depth (Campeau et al., 1999), which are essential to constrain or verify paleoclimatic models.

In the 1990s, with the rapid development of isotope geochemistry, it became possible to analyze stable isotopic ratios of light elements in diatoms to track changes in surface water properties. Isotope geochemistry was first applied to foraminifera (Emiliani, 1955, 1966) and then similarly applied to diatoms where carbonated organisms were lacking. Several different and complementary isotopes can be measured in diatoms. Two groups of isotopes detected in diatoms can be differentiated: (1) oxygen (O) and silicon (Si) isotopes that are carried by the diatom frustule, and (2) carbon (C) and nitrogen (N) isotopes that are carried by the organic matrix. This organic matrix is called diatom-intrinsic organic matter (DIOM) and is intimately embedded into the silica lattice where it directs biomineralization (Kröger & Sumper, 1998). Analyzing DIOM rather than bulk organic matter provides a more direct picture of surface water nutrient cycling because the DIOM is protected from remineralization and diagenesis by the silica matrix (Sigman, Altabet, François, McCorkle, & Gaillard, 1999a).

2. IMPROVEMENTS IN METHODOLOGIES AND INTERPRETATIONS

2.1. Micropaleontology

Analysis of microfossil assemblage census counts is one of the principal tools of paleoceanographic studies because distribution of individual organisms and whole ecological systems are affected by the physico-chemical parameters of their habitat. Diatoms are microscopic organisms and should be observed under the microscope at a strong magnification, usually of 1,000. Diatoms must therefore be glued to a permanent medium embedded in between a slide and a cover-slide. Sample preparation is the first laboratory step that will guarantee (or not) a high-quality and reliable study. Additionally, diatomists must follow the same taxonomic references and the same counting rules in an effort of harmonization.

2.1.1. Slide preparation

There are several ways to clean, concentrate, and mount diatom slides, even though all techniques emanate from the original protocol described below (Schrader, 1973). All of them must achieve random settling and random diatom distribution to ensure a good representativity of the sedimentary assemblages.

Generally, the protocol starts by leaching the dry raw sediment with H_2O_2 to remove the organic matter coating the valves, and with HCl to remove carbonates, at a temperature of 65°C. Complete removal has occurred when the bubbling stops. Diatom valves are then concentrated by eliminating the clays through a fractionated settling technique, in which the residue is allowed to sediment for 90 min in a given volume of distilled water. The water, containing not only clays but also some small diatoms, is subsequently removed using a vacuum pump. The settling step is repeated eight times (Schrader & Gersonde, 1978). The final residue is transferred to a 50 ml Nalgene bottle for storage. From this bottle, a given volume is taken after thorough shaking and transferred to another 50 ml Nalgene bottle that serves as a dilution step. A subsample of 0.2 ml is taken from the second bottle after homogenization, and spread on a wet cover-slide hosted in a Petri dish. The water is then evaporated in an oven at 45°C. Permanent mount is achieved by adding a few drops of resin dissolved in xylene or toluene and evaporated on a hot plate.

Variations from this technique include:

- The raw material is freeze-dried instead of dried in the oven. The benefit is that sediment porosity is preserved (Zielinski, 1993).
- Additional boiling of the raw sediment in benzene or tetrasodium diphosphate to stimulate the dispersion of diatom valves (Koç-Karpuz & Schrader, 1990; Pichon, Labracherie, Labeyrie, & Duprat, 1987).
- Transfer of the whole aliquot from the dilution bottle into the Petri dish to ensure better diatom distribution, and use of a paper towel (Koç-Karpuz & Schrader, 1990) or "vaccum" (Scherer, 1995) to suck the water out of the Petri dish after diatom settling.
- Centrifugation of the solution containing the diatom valves instead of the fractionated settling technique to ensure better recovery of small diatoms, transfer of 1–2 drops of the final diluted residue into a prefilled Petri dish, and use of a wool wire to evacuate the water (Pichon et al., 1987; Rathburn, Pichon, Ayress, & DeDeckker, 1997).
- Use of three cover-slides with glue-covered surfaces per sample in one Petri dish (Koç-Karpuz & Schrader, 1990; Zielinski, 1993) or in three different Petri dishes

to avoid artifacts on subsamples during processing (Pichon et al., 1987; Rathburn et al., 1997).

Each protocol presents its own advantages and disadvantages in that the fractionated settling technique may underestimate small diatoms, while the centrifugation technique may destroy very fragile diatoms. Similarly, evaporation of the water contained in the Petri dish yields a better diatom distribution, but takes more time than sucking the water out with a wire, which may displace small diatoms if cover-slides do not have glue-covered surfaces. Finally, using three mounting slides per sample in three different Petri dishes may be statistically more relevant because processing artifacts cannot possibly occur on each subsample.

Recently, a new method based on the different hydrodynamical behavior of diatoms and mineral grains was recently developed (Rings, Lücke, & Schleser, 2004). This method employs split-flow thin fractionation (SPLITT) as a tool for separating diatom frustules from other sedimentary particles. The principle of SPLITT fractionation is the gradual separation of particles in a laminar flow within a tunnel/cell with a field of gravity force applied perpendicular to the flow, the carrier liquid being deionized water. As a result of different sinking velocities, which depend on the density, shape, and size of the particles, particles are separated into two fractions with diatoms escaping the SPLITT by the upper outlet and sedimentary particles flowing through the lower outlet. Length, breadth, and height of the SPLITT channel can be adjusted to obtain the best separation whatever the sediment composition. Advantages of SPLITT fractionation over other techniques are good reproducibility, minimum loss of diatoms, and minimum contamination of diatoms by terrigenous particles and sponge spicules (Rings et al., 2004).

2.1.2. Diatom counts

It is absolutely necessary to follow a few counting rules in order for diatom abundances to be directly compared from site to site and from laboratory to laboratory. A reference convention was developed by Schrader and Gersonde (1978).

Generally, more than half of the valve must be seen to count one specimen (Figure 2). However, some diatom types have particularities, and the reference convention needed to be amended. For example, *Rhizosolenia* type specimens are centric diatoms, and they can reach extreme lengths by increasing their number of girdle bands, but the valve itself is short and circular, and has a spine-like proboscis that it is absolutely necessary to identify in order to count one specimen (Armand & Zielinski, 2001). If only the girdle bands or a part of the proboscis are observed, no specimen is identified (Figure 2). *Thalassiothrix* type specimens are very long and narrow pennate diatoms, and can be broken into hundreds of pieces in the sediment. *Thalassiothrix* relative abundances were estimated from the number of fragments (Pichon et al., 1992), but it was rapidly understood that there is a weak relationship between the number of fragments and the number of valves, since valves can randomly break into few or numerous pieces. Only apices can give an idea of the number of valves, as two apices represent one valve. The number of apices counted in one sample is therefore divided by two to calculate *Thalassiothrix* relative abundance while intermediate fragments are rejected (Armand, 1997) (Figure 2).

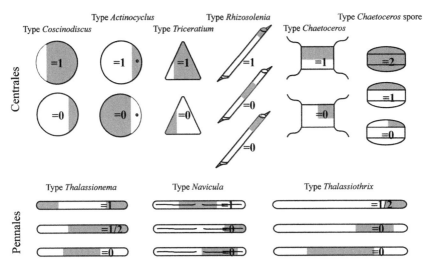

Figure 2 Counting convention for the main diatom groups. The shaded area represents diatom fragments that can be encountered in slides. Redrawn and modified with permission from Schrader and Gersonde (1978) and Armand (1997).

Chaetoceros is another complex genus in which vegetative valves are readily identified but barely preserved in the sediment, particularly in the case of *Hyalochaete* specimens, and resting spores are difficult to identify but are sometimes very abundant in coastal sediments (Leventer, 1991; Crosta, Pichon, & Labracherie, 1997; Hay, Pienitz, & Thompson, 2003). The picture is also often complicated by the presence of numerous pieces of setae. The same rule applies to *Chaetoceros* vegetative cells and resting spores just as for other diatoms, i.e., that more than half of the valve should be present to be counted as one, except that different species are generally lumped together in a *Phaeoceros* group and a *Hyalochaete* group. However, some particularities arise since full resting spore cells have two valves and setae are not counted.

Generally, more than 300–400 specimens should be counted to ensure a good statistical reproducibility. When *Chaetoceros* resting spores are overwhelming (>40–50%), 300–400 specimens other than *Chaetoceros* should be counted to provide an accurate picture of the diatom diversity, and therefore provide better confidence in the paleoceanographic reconstructions (Allen, Pike, Pudsey, & Leventer, 2005).

2.1.3. Diatom assemblages: from presence/absence to statistics

Fossil diatom assemblages can be used to track past environmental changes if (1) modern assemblages are representative of the environmental conditions in which they grow and (2) that diatom ecology has not changed through time.

Relationships with surface parameters. Many papers have shown that diatom assemblages in surface water generally respond to local-to-regional parameters such as nutrient content, water dynamics (currents, hydrological fronts, stratification, etc.), SST, and SIC. In upwelling systems, the main environmental parameter is the intensity of the upwelling that dictates the nutrient input from deep waters and the subsequent nutrient gradient in surface waters. As deep waters are colder than surface waters it also creates a temperature gradient. Diatoms are then distributed in relation to the nutrient and SST gradients. Diatoms having high nutrient requirements thrive closer to the upwelling cell than diatoms having low nutrient requirements. For example, *Chaetoceros* thrives in tropical to polar waters of very high productivity (Hendey, 1937; Pokras & Molfino, 1986; Abrantes, 1988a; Leventer, 1991), while *Fragilariopsis doliolus* thrives in tropical to temperate waters of low to moderate productivity (Simonsen, 1974; Romero, Fischer, Lange, & Wefer, 2000), and *Roperia tessalata* thrives in warm waters of low to moderate productivity (Hasle & Syvertsen, 1997; Semina, 2003).

Most of the time, fossil diatoms preserved in surface sediments have geographical distributions in relation to their ecological preferences. High relative abundances of a given species are found in sediments underlying their maximum production zone in surface waters, where an optimal set of environmental conditions allows the species to develop. Fossil diatoms therefore experience distribution in gradients from high abundances indicating favorable overlying conditions, to low abundances indicating unfavorable conditions. In upwelling systems, favorable conditions are adequate nutrient concentrations and temperatures (Pokras & Molfino, 1986), while in the polar oceans favorable conditions are temperatures and sea-ice cover (DeFelice & Wise, 1981; Koç-Karpuz & Schrader, 1990; Zielinski & Gersonde, 1997). In the Southern Ocean diatoms generally show north–south gradients of increasing or decreasing abundances depending upon their ecological preferences for warmer or colder temperatures, whereas in the Nordic Seas they mainly display east–west gradients ranging from the warm Atlantic current in the east to the sea-ice in the west.

In the Southern Ocean, *F. curta*, the main sea-ice diatom (Armand et al., 2005), reaches its highest relative abundances of ~70% at very cold SSTs between −1°C and 1°C, and heavy sea-ice cover between 8 and 11 months per year (Figure 3). Relative abundances of this species sharply drop to zero at warmer SSTs and lower sea-ice cover. *F. kerguelensis*, the main open ocean diatom (Crosta et al., 2005a), reaches maximum relative abundances of ~80% at SSTs between 1°C and 7°C and low sea-ice cover between 0 and 3 months per year (Figure 3). Relative abundances of *F. kerguelensis* sharply drop to zero at lower SSTs, but drop more gently towards warmer SSTs where it is replaced by species thriving in warmer waters. *F. kerguelensis* also shows an inverse relationship with sea-ice cover that inhibits its production but promotes sea-ice diatom (*F. curta*) production. The *Azpeitia tabularis* group, one of the main warm water diatoms in the Southern Ocean (Romero et al., 2005), reaches highest relative abundances at SSTs between 11°C and 14°C and no sea-ice cover (Figure 3). Relative abundances of this group decrease towards both colder and warmer SSTs.

In most of the cases, maximum abundances of fossil diatoms reflect narrow ranges of environmental conditions (Figure 3). Additionally, overlaps of diatom gradients are common with maximum abundances of species 1 occurring during a decreasing

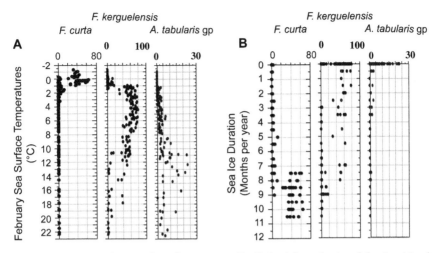

Figure 3 Relative abundances of *Fragilariopsis curta*, *Fragilariopsis kerguelensis* and the *Azpeitia tabularis* group in 228 surface sediment samples from the Southern Ocean versus sea-surface temperatures (A) and sea-ice presence (B). Modified with permission from Armand et al. (2005), Crosta et al. (2005a), and Romero et al. (2005).

trend of species 2 and an increasing trend of species 3. For example, the sharp decrease in *F. kerguelensis* maximum abundances centered at 1°C occurs concomitantly to the appearance of *F. curta*, and the decreasing abundances of *F. kerguelensis* towards warmer waters occur concomitantly to the increasing abundances trend of the *A. tabularis* group (Figure 3). These are some of the specificities that paleoceanographers use to reconstruct past changes. At a given site, down-core changes in the relative abundances of one or several diatom species indicate changes in the environmental conditions. The main challenge is to quantify the type and the magnitude of the changes.

One must keep in mind that preserved fossil assemblages and diatom biogeography is not a direct picture of surface conditions since most of the information is lost during settling to the seafloor. Dissolution, grazing, winnowing, transport, reworking, and bioturbation may deeply alter the surface water assemblages. Sedimentary assemblages therefore represent average surface conditions. The average time covered by the sedimentary assemblage depends on the sedimentation rate, about few centimeters per thousands of years in the open ocean, to a few meters per thousands of years in upwelling systems, coastal areas, and fjords. Some laminated records from exceptional sites allow reconstruction of seasonal signals (Kemp, 1995; Kemp, Baldauf, & Pearce, 1995; Stickley et al., 2005). Still, it is possible to use what we know about regional diatom ecology to reconstruct environmental changes in the past. It can be done by looking at down-core records of a single species, of species ratios or of the total assemblage through statistical methods.

Single species-based reconstructions. Investigation of down-core records of a single species provides information on very specific parameters in restricted areas.

This approach is obviously limited to the range of the species distribution relative to the parameters and requires a very good knowledge of the ecology and distribution of the species. Indeed, diatoms may have different behavior in different environments. Extrapolation of the regional behavior of a given species to another area may lead to spurious interpretation of past changes. Additionally, a resistant species may be concentrated by dissolution, transport, and reworking during settling and burying. One should therefore be careful when using down-core records of a single species to infer paleoceanography and paleoclimate. An example of this type of dichotomic ecology in different environments may be found in the Southern Ocean. Based on extensive investigations of time-series sediment-traps and diatom distribution in surface sediment of the Weddell Sea, Gersonde and Zielinski (2000) showed that relative abundances of the *F. curta* group (*F. curta* and *F. cylindrus*), which were greater than 3% of the total diatom assemblage, indicated the presence of winter sea-ice. They also showed that relative amounts of *F. obliquecostata*, which were greater than 3%, indicated the presence of summer sea-ice. Comparisons of winter and summer sea-ice extents at the last glacial maximum (LGM), estimated by the single species proxies with winter and summer sea-ice extents estimated through a transfer function approach, provide very similar results in the South Atlantic sector, while some discrepancies arise in the Indian sector of the Southern Ocean (Gersonde, Crosta, Abelmann, & Armand, 2005). Reasons behind the inter-basin discrepancy between the two micropaleontological methods are still not fully understood, but it seems that different species ecology in the two sectors and specific transport and dissolution in the Indian sector are the two most likely explanations. Variations in the *F. curta* group were further used down-core to track past changes in sea-ice extent. Application of this proxy to several cores from the Atlantic sector of the Southern Ocean showed rapid sea-ice retreats during deglaciations in phase with SST warming (Bianchi & Gersonde, 2002, 2004).

Species ratios-based reconstructions. Investigation of down-core records of species ratios also provides very specific information in restricted areas, as it does for single species. A very good knowledge of the ecology of the species used in the ratio is absolutely necessary. Ratios can involve different species (Shemesh et al., 1989), different varieties of a single species (Fryxell & Prasad, 1990), different stages of a single species or species group (Leventer et al., 1996), or number of fragments to full cells of a single species (Abrantes, 1988a, 1988b).

Based on the observation of modern diatom distribution and dissolution in laboratory experiments, Shemesh et al. (1989) showed a depletion of *F. kerguelensis* (K) relative to *Thalassiosira lentiginosa* (L) when dissolution increases. The preservation index calculated as K/(K+L) gives information on the relative extent of dissolution. Application of the preservation index to Holocene and LGM samples from the Southern Ocean indicated that Holocene and LGM diatoms from the Atlantic sector are equally preserved while Holocene diatoms from the Indian sector are better preserved than LGM diatoms.

Another preservation index, called the fragmentation index, was identified in the upwelling system off Portugal on the basis of diatom fragmentation (Abrantes, 1988a). Application of this fragmentation index, calculated as the number of diatom fragments to the number of full diatom valves, to cores from the upwelling off Portugal indicated variable temporal and spatial diatom dissolution with greater dissolution during Marine Isotope Stage 3 (MIS 3) than during MIS 2, and greater dissolution at the outer upwelling fringe (Abrantes, 1991).

Investigations of the modern distribution of *Eucampia antarctica* in the phyto-plankton have shown this species to form morphologically different summer and winter stages, morphologically different terminal and intercalary valves, and mor-phologically different warm and cold varieties (Fryxell & Prasad, 1990; Fryxell, 1991). The ratio of summer to winter valves in down-core records potentially gives qualitative information on SSTs. Greater ratio values indicate prominence of the summer stage versus the winter stage and therefore warmer annual temperatures (Fryxell, 1991). Similarly, the ratio of terminal to intercalary valves can be used to track sea-ice extent. A lower ratio indicates greater winter diatom production and therefore less sea-ice. Application of this ratio to a sediment core off the Kerguelen Islands in the Indian sector of the Southern Ocean indicates repetitive sea-ice waning and waxing over the last 800,000 yr, in phase with Milankovitch oscillations of Earth obliquity (Kaczmarska, Barbrick, Ehrman, & Cant, 1993). Another productivity index was built on the concomitant presence of *Chaetoceros* resting spores and *Chaetoceros* vegetative cells in the sediment. Resting spores are formed in the vegetative valves when a strong bloom depletes surface water nutrients (Hargraves & French, 1975; Harrison, Conway, Holmes, & Davis, 1977). Higher values of the spores to vegetative valves ratio indicate higher productivity and subsequent nutrient depletion (Leventer, 1991). The down-core record of this ratio shows repetitive changes during the Holocene with a 200–300 yr cyclicity, suggesting that the siliceous productivity in the Antarctic Peninsula region is primarily controlled by solar activity (Leventer et al., 1996).

More regional paleo-reconstructions are generally based on multispecies investi-gations that provide a greater spatial coverage and a better characterization of surface water parameters. A set of diatom species covers a wider range of conditions than a single species, with each species covering a small range of conditions (Figure 3). A set of diatom species is also more representative of the phytoplank-tonic production and is less prone to dissolution and reworking artefacts. This approach, due to the complexity of dealing with many variables, calls for a statistical analysis of the assemblages.

Statistics-based reconstructions. Statistical treatments are used to reduce the number of variables (species) by grouping species exhibiting similar ecological re-sponses together, and used to detect structure in the relationships between variables. The most common method is the Q-mode factor analysis (QFA). The QFA starts with a principal component analysis (PCA) and is followed by a varimax rotation of the selected principal components (PC). The PCA method involves a mathematical procedure that transforms a number of possibly correlated variables into a smaller number of uncorrelated variables called "principal components." The first PC accounts for as much of the variability in the data as possible, and each succeeding component accounts for as much of the remaining variability as possible (Pielou, 1984). In this way, one can find directions in which the data set has the most significant amounts of variation. Species grouping is obtained from the projection of the species squared weights on the space defined by the PC. The QFA is used to study the patterns of relationship among many dependent variables, with the goal of discovering something about the nature of the independent variables that affect them. The inferred independent variables are called "factors." The extraction of factors amounts to a variance maximizing (varimax) rotation of the original variable space defined by the PC. This type of rotation is called variance maximizing because the goal of the rotation is to maximize the variance of the "new" variable (factor), while minimizing the variance around the new variable (Imbrie & Kipp, 1971). The QFA provides two matrices; first, the varimax score matrix that presents the variance accounted for by the factors in each sample and second, the varimax score factor matrix that represents the species weight in each factor. In the varimax factor matrix, the sum of the squares of the factor loadings, defined as the communality of the sample, provides a way of testing the significance of the statistical treatment applied, while the cumulated factor loadings of each factor, called the variance, indicates the significance of each factor in the total data set. Samples belong to the factor in which they reach the highest factor loading. Mapping factor loadings gives information on the geographical extent of the factors. The varimax score factor matrix is most useful to draw preliminary relationships between the factors and environmental conditions based on the ecology of the species included in each factor.

The QFA input data are generally relative abundances of diatom species, but it is sometimes necessary to transform the percent data to reduce the overrepresentation of some species. Class ranking (Pichon et al., 1992) or logarithmic transformation of the relative abundances (Zielinski, Gersonde, Sieger, & Fütterer, 1998) may be used to this effect.

In order to develop a calibration set for paleo-reconstructions, the first step is to apply the QFA to modern samples to extract and map factors and to draw a relationship between the factors and modern environmental conditions. Generally, not all of the species present in the surface sediments are used. Rare diatom species (less than 2% of the total assemblage), reworked species, widely distributed species and benthic species are eliminated because they do not highlight specific surface conditions. Input or not of a diatom species obviously depends on the parameters to be reconstructed. The second step is to apply the same statistical treatment to the same species counted down-core. The same factors are extracted for each fossil assemblage. From the down-core evolution of factor dominance it is possible to infer past oceanographic changes at the core location.

Such a statistical approach has been widely used in the 1980s. Sancetta (1979) applied a QFA treatment to diatom assemblages in 62 core-top samples from the North Pacific that resolved Subtropical, Transitional, Subarctic, Production, and Okhotsk factors with clear relationships to regional water types and currents. The five factors accounted for 96% of the total variance. The QFA treatment of diatom assemblages in a series of cores indicated a strong cooling of surface and deep waters and higher productivity in the northwestern Pacific during the last glacial. A QFA analysis of diatom assemblages in 59 core-tops from the Eastern Equatorial Atlantic derived Tropical — Moderate Productivity, High Productive, Runoff, Subtropical

— Low Productivity and Antarctic Displaced factors (Pokras & Molfino, 1986). Each factor presents a different dominant diatom species or species association. The five factors accounted for 95% of the original variance. When applied to a set of cores, the QFA approach indicated strong variations of the factors in phase with climate changes. Higher diatom productivity in the Equatorial Atlantic during glacial MIS 2, 4, and 6, and low diatom productivity during the warm substage 5.5 were observed (Pokras & Molfino, 1987). Based on very low scores of the Antarctic Displaced factor, influx of Antarctic Bottom Water was supposed insignificant throughout the last 160,000 yr. Factor analysis of diatoms in 55 core-tops from the Southern Ocean produced Sea-Ice, Polar Front Zone, and Antarctic Zone factors, the latter one being encompassed by the two first factors (Burckle, 1984a). The three factors accounted for 97% of the total variance. A QFA of the same 27 diatom species in 51 fossil samples showed the distribution of these factors during the LGM. For each factor, high factor loadings were generally located more to the north than their modern distribution, indicating a northward migration of the Polar Front Zone, of the winter sea-ice and a strengthening of the Weddell Gyre in relation with colder temperatures and stronger winds.

2.1.4. Transfer functions

Presence/absence of a diatom species, relative abundance variations of one or several species, and QFA on many species provide qualitative interpretation of past environmental changes. Transfer functions go a step further and produce quantitative estimates of surface physico-chemical parameters, such as SSTs in degree Celsius, thanks to the development of advanced computational methods. Such quantitative estimates are essential because they are independent of geochemical proxies and are most useful to constrain or validate paleoclimatic models. They provide a range of values in which model results may fall if the physics are correctly computed (Kucera, Rosell-Melé, Schneider, Walbroeck, & Weilnet, 2005).

A transfer function must be understood as any kind of mathematical approach that analyses census counts of fossil assemblages to produce absolute values of surface properties by comparing fossil samples to a subset of modern samples having definite modern conditions. Transfer functions can work on reduced species data sets but generally between 20 and 40 diatom species are used. Reduced species data sets can perform better than raw data sets because the high variability of the diatom assemblages is smoothed (Racca, Gregory-Eaves, Pientiz, & Prairie, 2004). Similarly, although it is possible to work on limited surface sample data sets, it may be best to work on extended data sets that cover broader modern conditions, hence reducing the possibility of nonanalog conditions. There are several types of transfer functions, each one based on different mathematics. The most common ones are the Imbrie and Kipp Method (IKM; Imbrie & Kipp, 1971), the Modern Analog Technique (MAT; Hutson, 1980), the Weighted Averaging Partial Least Square (WA-PLS; ter Braak & Juggins, 1993), Maximum Likelihood (ML; Birks & Koç, 2002), and the Artificial Neural Network method (ANN; Malmgren & Nordlund, 1997; Malmgren, Kucera, Nyberg, & Waelbroeck, 2001). The General Additive Model (GAM; Armand, 1997) and the Revised Analog Method (RAM;

Waelbroeck et al., 1998) are variations of the IKM and MAT approaches, respectively.

All transfer functions operate within the same framework. Whatever the algorithms and the techniques used, they all start with three databases. First, the modern species database that displays the chosen diatom species present in core-top sediments (Figure 4). This data set is the same as the one used in the QFA mentioned above. Second, the modern parameter database that gives quantitative values of surface properties extracted from *in situ* measurements, generally compiled in numerical atlases. Values are extracted at the vertical of the core-top samples, as it is impossible to cope with lateral advection of sinking particles in extended databases. Third, the fossil species database that includes diatom census counts of the same species in down-core samples. Whatever the algorithms and the techniques used, most transfer functions work in three steps. First, the calibration step compares the modern species database to the modern parameter database to determine species–environment relationships between the two sets (Figure 4). Second, the comparison step correlates the fossil database to the modern database to detect similarities between the two sets. Third, the estimation step produces the quantitative estimate based on the two first steps.

Each technique is dependent upon, but differently affected by, the quality of the three databases and therefore upon diatom taxonomy, core-top coverage, and the extraction of the modern parameters. Databases are thus validated through an auto-run of the modern data sets to check whether modern surface properties are accurately estimated. The modern database serves therefore as both the reference database and the fossil database. Good databases and appropriate transfer functions will provide paleoenviromental estimates close to the modern environmental values

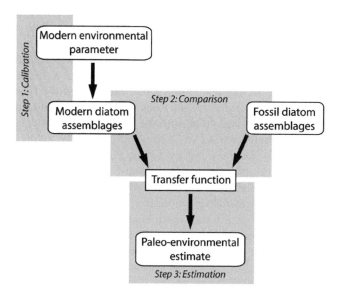

Figure 4   Schematic protocol of a transfer function highlighting the databases and the three-step mathematical technique.

associated with each sample. Linear regression between observed and estimated values must yield a correlation coefficient and a slope close to 1, low residuals and low standard errors on the estimates. When it is validated, the whole package, including the databases and the mathematical approach, can be applied to fossil samples.

The IKM is based on a QFA of the core-top diatom assemblages (program CABFAC) and a regression between the calculated factors and the modern parameters (program REGRESS) that builds the paleoecological equation (Figure 4, step 1). The fossil diatom assemblages are also reduced through a QFA (Figure 4, step 2). Factor loadings of each fossil assemblage are then introduced into the paleoecological equation (program THREAD) to produce the quantitative paleoenvironmental estimate (Figure 4, step 3). The IKM is possibly the best technique to apply on restricted modern databases because it calculates a mathematical function between the core-tops samples and the parameters, thus coping with the lack of samples. The program CABFAC provides information on the biogeography of the modern factors and on the representativity of the chosen species in the assemblages. The communality is a good tool to discard core-tops that are possibly affected by dissolution, reworking, or are just not represented by the chosen factors. The IKM allows extrapolation, i.e., paleoenvironmental estimates outside the range of values covered by the modern parameter database. Conversely, the paleo-equation also has important flaws in that (1) it only provides a mean standard error on the equation, (2) it smoothes the estimates, and (3) it is affected by addition of any modern sample that will subsequently change the estimates. Moreover, this technique is strongly influenced by species with high relative percentages, at least in Southern Ocean sediments. This dependency upon overrepresented species should be alleviated through normalization of the diatom relative abundances using a system of class ranking (Pichon et al., 1987) or logarithmic transformation (Zielinski et al., 1998). Different systems of normalization induce different estimates.

The MAT is a simple comparison between fossil assemblages and modern ones. There is no calibration step besides plots of species relative abundances in core-top samples versus associated parameters. For each fossil sample, a dissimilarity coefficient, which measures the difference between the fossil assemblage and the modern assemblages, is calculated using the square chord distance (Hutson, 1980). The MAT then chooses the x less dissimilar analogs to calculate the paleoenvironmental estimate. This calculation can be a simple average of the x quantitative values associated with the chosen analogs (Prell, 1985), or an average of the x values weighted by the geographical distance of the analogs to the fossil sample (Pflaumann, Duprat, Pujol, & Labeyrie, 1996), or weighted by the dissimilarity coefficients (Guiot, 1990). This approach generally works with relative percentages and does not require normalization of the relative abundances, because rare species with low abundances are as equally important as dominant species. As the estimate is a simple average of the core-top parameter values, the MAT provides a root mean square error of prediction (RMSEP) for each fossil sample, and therefore a point-by-point control of the paleoenvironmental estimate. Any new core-top sample can be added to the modern database and can contribute to the result of any fossil samples without changing the whole set of estimates. The MAT provides the location of the chosen analogs that may give further environmental information than the quantitative estimate. However, the MAT has several flaws. It requires an extended core-top database to provide reliable analogs to any fossil sample. It is very sensitive to the number of chosen analogs and to the maximum value of dissimilarity above which the analog is rejected and not used in the calculation of the paleoenvironmental estimate. Estimates are restricted to the range of values covered by the modern databases.

WA-PLS can be regarded as the unimodal-based equivalent of multiple linear regressions (ter Braak & Juggins, 1993). This means that a species has an optimal abundance along the environmental gradient being investigated. As with the IKM method, WA-PLS uses several components in the final transfer function. These components are however selected to maximize the covariance between the environmental variables to be reconstructed and hence the better predictive power of the method, whereas in the IKM method the components are chosen irrespective of their predictive value to capture the maximum variance within the biological data.

The ANN works using a back propagation (BP) neural network, which relies on the hypothesis that there is a relationship between the distribution of modern assemblages and the physical and chemical properties of the environment. The ANN is based on an algorithm that has the ability of autonomous "learning" of a relationship between two groups of numbers (Malmgren & Nordlund, 1997), by exchanging information between the interconnected processing units composing the network. The learning persists as long as the prediction error for each sample in the calibration data set decreases and provides a calibration equation calculated on the modern databases. The ANN is best when relationships between core-top assemblages and surface properties are nonlinear. It is not dependent upon the size of the modern database and it allows extrapolation similarly to the IKM. Nevertheless, this technique has several flaws. The ANN calibration is more or less a black box and it is extremely time-consuming because of the learning period. Different architectures of the network yield different estimates.

2.2. Isotope Geochemistry

Isotope analyses were first developed for bulk sediment (N isotopes) or for organisms other than diatoms (C and N isotopes). They were eventually applied to diatoms to cope with important diagenetic problems or wherever carbonate organisms were not present. Up until now, four isotope ratios are routinely measured in the diatom organic-intrinsic matter (C and N) and in the diatom frustule (O and Si). Specific protocols that are developed in the following paragraphs were built to extract and purify diatoms from the bulk sediment.

2.2.1. Rationale behind the isotopes

Diatoms preferentially assimilate light isotopes (^{12}C, ^{14}N, ^{16}O and ^{28}Si) to build the organic matter and biomineralize the frustules, thus leaving the nutrient pool in surface waters enriched in heavy isotopes (^{13}C, ^{15}N, ^{18}O and ^{30}Si). As the initial nutrient pools are consumed during biomass production, their nutrient light to heavy isotope ratios progressively increase. This progressive increase is transferred to the biogenic material subsequently produced using the enriched pool, thus leading to a parallel isotope enrichment of the organic material (Figure 5). Stable isotope ratios of the particulate organic matter and of the buried organic matter reflect the proportion of nutrients assimilated during phytoplankton development as a measure of the balance between supply to the surface waters and biological uptake. Therefore, they do not represent an absolute value of the assimilation but rather a relative uptake of the nutrient.

The isotopic signal, noted δ, provides a way to visualize the isotopic enrichment of the source and of the product. Additionally, δ is calibrated versus reference values used worldwide that allow for intercomparisons. Standard notation for δ is depicted in Equation (1) where E is the element, H is the heavy isotope and L is the light isotope (Figure 5). δ may therefore be understood as a deviation from the reference isotopic ratio values.

The isotopic enrichment between the organic product and the dissolved nutrients is calculated as fractionation factor α that measures the reactivity of an organism to the various isotopes of an element. The fractionation factor is determined at equilibrium and is dependent upon physico-chemical and environmental factors. Because it is expressed as the ratio of heavy to light isotope ratios in the source and the product, α is very close to 1. Isotope geochemists therefore prefer to use the fractionation, ε_p, which represents the deviation from 1. The higher ε_p is, the less heavy isotopes are assimilated, which results in more depleted δ values in the biogenic material (Figure 5 and equation (2)). In Rayleigh's model, a constant ε_p yields at any moment an instant product δ (dotted gray line) depleted by ε_p regarding the source δ (black line) (Figure 5 and equation (3)). The integrated

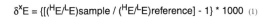

$$\delta^X E = \{[(^HE/^LE)sample \, / \, (^HE/^LE)reference] - 1\} * 1000 \quad (1)$$

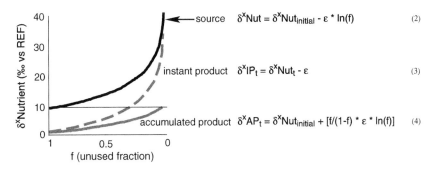

source $\delta^X Nut = \delta^X Nut_{initial} - \varepsilon * \ln(f)$ (2)

instant product $\delta^X IP_t = \delta^X Nut_t - \varepsilon$ (3)

accumulated product $\delta^X AP_t = \delta^X Nut_{initial} + [f/(1-f) * \varepsilon * \ln(f)]$ (4)

Figure 5 Simulation of nutrient fractionation during biogenic material formation by diatoms. Curves depict changes in the delta of the source (black line), of the instant biogenic product (dotted gray line) and of the accumulating biogenic product (gray line). Nut = nutrient; ε = fractionation; f = unused nutrient fraction; E = element; H = heavy isotope; and L = light isotope.

product δ has the same value as the source initial δ value when all nutrients are used (Figure 5 and equation (4)).

2.2.2. Carbon isotope ratios in diatoms

On wide oceanic scales, the $\delta^{13}C_{org}$ is anticorrelated with the concentration of molecular dissolved CO_2 ($CO_{2(aq)}$) in surface waters (Rau, Froelich, Takahashi, & Marais, 1989, 1991a). The $CO_{2(aq)}$ is dependent upon physical processes (SST and salinity, diffusivity, wind intensity) and biological processes (carbon uptake). It was believed that passive diffusion into phytoplankton cells was the primary carbon acquisition pathway (Laws, Popp, Bidigare, Kennicutt, & Macko, 1995), and therefore $\delta^{13}C_{org}$ down-core records were tentatively used to reconstruct past CO_2 concentrations in surface waters (Jasper & Hayes, 1990; Bentaleb & Fontugne, 1998). However, the anticorrelation between $\delta^{13}C_{org}$ and $CO_{2(aq)}$ is not consistently observed regionally within a given ocean system when other factors may become dominant, such as growth rate, community structure (Popp et al., 1999), cell size/shape fraction (Pancost, Freeman, Wakeham, & Robertson, 1997; Popp et al., 1998; Burkhardt, Riebesell, & Zondervan, 1999; Trull & Armand, 2001), and nondiffusive carbon uptake through carbon concentration mechanisms (Rau, 2001; Tortell, Rau, & Morel, 2000; Tortell & Morel, 2002; Cassar, Laws, Bidigare, & Popp, 2004; Woodworth et al., 2004). These processes, by strongly affecting the carbon isotopic fractionation (ε_p), weaken the relationship between $\delta^{13}C_{org}$ and $CO_{2(aq)}$, and may account for the discrepancy between marine $\delta^{13}C_{org}$-based pCO_2 reconstructed from low-latitude records and Vostok CO_2 (Kienast, Calvert, Pelejero, & Grimalt, 2001).

The cleaning procedure to isolate DIOM follows the method described by Singer and Shemesh (1995), which involves a decarbonation, a stepwise physical washing and sieving at $32 \, \mu m$ in order to separate the diatom fraction from the bulk sediment, a heavy liquid step to remove the heavy minerals, and an oxidation of the labile organic matter of the diatom fraction $< 32 \, \mu m$ to remove the labile organic matter coating the diatom valves. The advantages of using the fraction $< 32 \, \mu m$ are that (1) it generally accounts for the largest amount of the whole diatom assemblage, (2) the same species generally dominate down-core records, and (3) no radiolarians or sponge spicules are present.

Analyses of DIOM-$\delta^{13}C_{org}$ are therefore performed on a restricted diatom size fraction that may limit the influence of community structure and cell size/shape changes, thus providing a more direct link to $CO_{2(aq)}$ and phytoplankton carbon uptake as a mirror of paleoproductivity changes (Shemesh, Macko, Charles, & Rau, 1993; Singer & Shemesh, 1995; Rosenthal, Dahan, & Shemesh, 2000; Crosta & Shemesh, 2002). Analyses of DIOM-$\delta^{13}C_{org}$ are also conducted on a specific organic matter, mainly composed of proteins (Kröger & Sumper, 1998; Kröger, Bergsdorf, & Sumper, 2002), which directs biomineralization of the frustule (Kröger, Deutzmann, & Sumper, 1999). This organic matter is protected from alteration and diagenesis by the silica matrix (Sigman et al., 1999a; Crosta, Shemesh, Salvignac, Gildor, & Yam, 2002), again providing a more faithful picture of processes occurring in surface waters. It is, however, important to keep in mind that analysis of the DIOM only limits the issues mentioned above, and that many unknowns still exist.

Up until now, investigations of DIOM-$\delta^{13}C_{org}$ were exclusively conducted in the Southern Ocean to document past changes in productivity and nutrient cycling in relation to oceanographic and climate changes. Comparison of several down-core records of DIOM-$\delta^{13}C_{org}$ with other productivity proxies (Mortlock et al., 1991; Kumar, Gwiazda, Anderson, & Froelich, 1993, 1995; Anderson et al., 1998; Bareille et al., 1998; Frank et al., 2000; Dézileau, Reyss, & Lemoine, 2003) have shown a glacial drop in productivity south of the Antarctic Polar Front, a glacial increase in productivity in the Subantarctic Zone, and no glacial changes in productivity in the Subtropical Zone (Shemesh et al., 1993; Singer & Shemesh, 1995; Rosenthal et al., 2000; Crosta & Shemesh, 2002; Crosta et al., 2005b).

2.2.3. Nitrogen isotope ratios in diatoms

In many oceanic regions, the $\delta^{15}N_{org}$ of sinking bulk organic matter is correlated to the relative uptake of nitrate in surface waters (Rau, Sullivan, & Gordon, 1991b; Altabet & François, 1994; Sigman, Altabet, McCorkle, François, & Fisher, 1999b). The higher the consumption, the heavier the $\delta^{15}N_{org}$ becomes. More enriched nitrogen isotopes in glacial sediments of the Antarctic Indian Ocean were therefore taken to indicate greater nutrient use during cold periods (François, Altabet, & Burckle, 1992; François et al., 1997), although it was long known that several other factors may influence the sedimentary $\delta^{15}N_{org}$ records. Indeed, bacterial remineralization during sinking and burial preferentially removes ^{14}N, leaving the fossilized organic matter enriched in ^{15}N relative to the organic matter produced in surface waters (Altabet & François, 1994). Early diagenesis similarly leads to the preservation of ^{15}N-enriched organic matter. Such alteration of the surface water signal may be different from place to place and, more importantly, may not be constant through time in a given place. Enrichment can be up to 2–5‰ (Altabet & François, 1994) and is mainly dependent upon the flux and speed of sinking organic matter (Lourey, Trull, & Sigman, 2003), and on the redox conditions at the water–sediment interface (Ganeshram, Pedersen, Calvert, McNeill, & Fontugne, 2000). Analysis of DIOM-$\delta^{15}N_{org}$ allows us to deal with the impact of remineralization and diagenesis because the DIOM is protected from alteration by the frustule (Sigman et al., 1999a; Crosta & Shemesh, 2002). It also reduces the potential impact of community changes, diatom size fraction (Karsh, Trull, Lourey, & Sigman, 2003), and contamination by continental organic matter (Huon, Grousset, Burdloff, Bardoux, & Mariotti, 2002). We are still far from fully understanding bulk $\delta^{15}N_{org}$ and DIOM-$\delta^{15}N_{org}$ signals in the modern ocean because of species-dependent isotopic fractionation factors (Sigman & Casciotti, 2001) and of different nutrient sources (Lourey et al., 2003). Uncertainties are even higher for the past oceans due to the preservation state of the organic matter and the diatoms.

Although bulk $\delta^{15}N_{org}$ measurements have been conducted in many places (François et al., 1997; Kienast, Calvert, & Pedersen, 2002; Higginson, Maxwell, & Altabet, 2003; Galbraith, Kienast, Pedersen, & Calvert, 2004, and references cited therein; Higginson & Altabet, 2004), DIOM-$\delta^{15}N_{org}$ investigations are restricted to the Southern Ocean (Shemesh et al., 1993, 2002; Sigman et al., 1999a; Sigman

& Boyle, 2000; Hodell et al., 2001; Crosta & Shemesh, 2002; Robinson, Brunelle, & Sigman, 2004, 2005; Crosta et al., 2005b). The cleaning procedure follows the one described above for the DIOM-$\delta^{13}C_{org}$ analysis. Combustion-based measurement of DIOM-$\delta^{15}N_{org}$ is generally performed simultaneously to the DIOM-$\delta^{13}C_{org}$ analysis (Crosta & Shemesh, 2002), although DIOM-$\delta^{15}N_{org}$ can be measured alone on the IRMS to gain sensitivity and reproducibility. Another technique involving conversion of DIOM nitrogen to nitrate and denitrification of the resulting nitrate into N_2, which is subsequently introduced into the IRMS, was recently developed (Sigman et al., 2001). This method reduces the amount of N_{org} necessary to attain the detection level and alleviates the potential air contamination introduced during the combustion-based protocol (Robinson et al., 2004). The persulfate-denitrifier method leads to different results in the Antarctic Zone, but to similar results in the Subantarctic Zone relative to the combustion-based method (Robinson et al., 2004, 2005; Crosta et al., 2005b). Why these discrepancies exist between the two methods is still under debate.

DIOM-$\delta^{15}N_{org}$ investigations, coupled with other paleoproductivity proxies, indicate increased relative uptake of nitrate in the Antarctic and Subantarctic Zones and no changes in uptake in the Subtropical Zone during the last glacial period. The reason for increased relative uptake of nitrate is regionally different. South of the Antarctic Polar Front, heavier DIOM-$\delta^{15}N_{org}$ values result from reduced nutrient supply in the surface waters, certainly in relation to stratification of surface waters by greater glacial sea-ice melting (François et al., 1997). In the Subantarctic Zone, heavier DIOM-$\delta^{15}N_{org}$ values result from an increase in glacial productivity and iron fertilization promoting the N/Si uptake ratio by diatoms (Crosta et al., 2005b; Robinson et al., 2005).

2.2.4. Silicon isotope ratios in diatoms

Laboratory-culture experiments and *in situ* investigations have shown that the $\delta^{30}Si$ of diatoms is correlated to the relative uptake of silicic acid ($Si(OH)_4$) by diatoms in surface water (De la Rocha, Brzezinski, DeNiro, & Shemesh, 1998; Varela, Pride, & Brzezinski, 2004). From the few studies made, it seems that silicon fractionation is independent of temperature and diatom species, although silicon ε_p measured in low-temperature waters of the Southern Ocean was twice as high (Varela et al., 2004) compared to temperate culture batches (De la Rocha, Brzezinski, & DeNiro, 1997). Additional investigations are required to fill in several gaps in our knowledge. Also, it seems that $Si(OH)_4$ is the only silicon source and that frustule dissolution does not modify the sedimentary isotopic silicon composition of diatoms (De la Rocha et al., 1998), thus facilitating paleoceanographic interpretations.

The analytical protocol to measure $\delta^{30}Si$ in diatoms involves the recovery and purification of the silicon as SiO_2 and the fluorination of the purified silica to form SiF_4 gas, which is subsequently injected into the IRMS (De la Rocha et al., 1997). However, strong leaching with HF and laser heating render this technique tedious and dangerous. New techniques to measure silicon isotopes by MC-ICP-MS using dry plasma conditions are under development (De la Rocha, 2002; Cardinal et al., 2003). This new method provides better accuracy than the IRMS technique (less than 0.1‰), which is appreciable when silicon ε_p is 1‰.

Most of δ^{30}Si studies are from the Southern Ocean and more particularly from the Antarctic Zone (De la Rocha et al., 1998; Brzezinski et al., 2002; Beucher, Brzezinski, Crosta, & Tréguer, 2006). In the Antarctic and Subantarctic Zones, δ^{30}Si signals are anticorrelated to DIOM-δ^{15}N$_{org}$ signals, indicating less silicon uptake and more nitrate uptake during the last glacial period relative to modern times. This shows that surface water stratification is not the only process affecting nutrient cycling and biological uptake. Iron fertilization by dust input or vertical supply is necessary to decouple Si(OH)$_4$ and NO$_3^-$ consumption by diatoms (Hutchins & Bruland, 1998; Takeda, 1998; Crosta et al., 2002). In the Subtropical Zone, δ^{30}Si signals are correlated to DIOM-δ^{15}N$_{org}$ signals, both indicating almost no change in Si(OH)$_4$ and NO$_3^-$ consumption by diatoms over the last 50,000 yr.

2.2.5. Oxygen isotope ratios in diatoms

The δ^{18}O of diatoms is dependent upon the SST and the isotopic composition of the water in which diatoms formed their frustule (Juillet-Leclerc & Labeyrie, 1987). It seems that the isotopic signal is free of species effect, although more laboratory-culture experiments are necessary to confirm preliminary results. The isotopic composition of the water is tied to salinity (Craig & Gordon, 1965). Equations linking diatom δ^{18}O and SST have been developed. These paleotemperature equations show slopes different than the ones developed for carbonate δ^{18}O, thus allowing the reconstruction of both SST and isotopic composition of the water when foraminifera and diatoms grow in the same water mass (Moschen, Lücke, & Schleser, 2005).

Measurement of δ^{18}O is difficult because of the exchangeable nature of a fraction of oxygen atoms included in the silica matrix. Approximately 10–20% of oxygen is labile, which explains the lack of reproducibility during early investigations (Labeyrie & Juillet, 1982). The fraction of nonexchangeable oxygen is stable over thousands of years and retains the surface water isotopic composition after burial (Shemesh, Charles, & Fairbanks, 1992). The goal of the protocol is to exchange the labile oxygen fraction with oxygen of known isotopic composition under controlled conditions of temperature and water isotopic ratio (Labeyrie & Juillet, 1982). For example, Shemesh, Burckle, and Hays (1995) let pure diatoms react with water vapor at 40‰ during 6 h at 200°C. Pure diatom samples are obtained using a method similar to that used for DIOM-δ^{13}C$_{org}$, except for a stronger leaching and additional settling and heavy liquid steps in order to completely remove the organic matter, clays, and heavy minerals that alter the isotopic composition of diatom silica (Juillet-Leclerc, 1984; Shemesh et al., 1995). After the exchange, diatoms are recrystallized. The extraction of the oxygen and its conversion to CO$_2$ is carried out by fluorination. The CO$_2$ is then analyzed for its oxygen isotopic composition in the IRMS with reproducibility better than 0.2‰. A new technique was recently developed for the determination of oxygen isotope composition in biogenic silica. The inductive high-temperature carbon reduction method (iHTR) is based on the reduction of silica by carbon, using temperatures as high as 1,830°C, to produce carbon monoxide for isotope analysis. Details of this method are presented in Lücke, Moschen, and Schleser (2005). The amount of material necessary is 1.5 mg of biogenic silica, and the reproducibility for natural samples is better than 0.15‰.

Most studies of diatom $\delta^{18}O$ have been conducted in the Atlantic sector of the Southern Ocean where they indicate melt-water input during the LGM (Shemesh, Burckle, & Hays, 1994) and the last deglaciation (Shemesh et al., 2002). Diatom $\delta^{18}O$, although difficult to analyze, is a particularly suitable tool to document melt-water events in the Southern Ocean, for example whether MWP 1A originates from Antarctica (Weaver, Saenko, Clarck, & Mitrovica, 2003), since foraminifera are often not present in the sediments during these events.

 ## 3. CASE STUDIES

3.1. SST in the North Atlantic

In the North Atlantic and the North Pacific Oceans, diatom-based SST estimates have been generally provided via IKM transfer functions (Sancetta, 1979; Sancetta, Heusser, Labeyrie, Naidu, & Robinson, 1985; Koç et al., 1993; Koç, Jansen, Hald, & Labeyrie, 1996; Andersen, Koç, & Moros, 2004a, 2004b; Jiang, Eiricksson, Schulz, Knudsen, & Seidenkrantz, 2005). Even though the QFA is certainly a good method to cope with the huge range of environmental conditions encountered at high northern latitudes, both WA-PLS and ML are being used more and more often (Birks & Koç, 2002). Transfer functions in the high-latitude North Atlantic have been applied primarily to the deglaciation (Koç-Karpuz & Jansen, 1992) and the Holocene (Koç et al., 1993; Andersen et al., 2004a, 2004b; Jiang et al., 2005). Due to sea-ice cover during the glacial periods it has been almost impossible to obtain long and continuous diatom records from the high-latitude North Atlantic.

As a result of societal pressure in the context of global warming, the focus today is on understanding the frequency and origin of Holocene climate variability in the region of the North Atlantic. This ocean is a key region in modulating the global climate through the thermohaline circulation. More locally, complicated atmospheric and oceanic circulation patterns regulate the amount of heat transported to northern North America and northern Europe. Intensive quantitative reconstructions of climate parameters in this region will help us to understand how atmospheric and oceanic circulation patterns have evolved and interacted during the Holocene, and can then be used to forecast their behavior in the future.

In the example below, an IKM transfer function was applied to diatom fossil assemblages of core MD95-2011 from the Voring Plateau in the Norwegian Sea (66°58.18′N — 07°38.36′E — 1,048 m) (Birks & Koç, 2002; Andersen et al., 2004a). The modern database was composed of 139 core-top samples. The modern species database was composed of 52 species. The QFA calculated eight factors defined by specific diatom assemblages. The eight factors, accounting for 95% of the total variance, are described in detail in Andersen et al. (2004a). The down-core diatom relative abundances were transformed into the same eight factors, which were subsequently introduced into the equation to calculate paleotemperatures. The IKM technique produced in this example a coefficient of determination (R^2)

of 0.9 and a RMSEP of 1.25 °C. The chronology of the core was based on 10 AMS dates calibrated to calendar ages using CALIB 4.3 software (Stuiver et al., 1998) after removing the reservoir age, and one tephra layer.

The diatom-based SSTs indicate a division of the Holocene into three periods: first, the Holocene Climate Optimum (HCO) between 9,500 yr B.P. and 6,500 yr B.P. with SSTs around 15 °C, i.e., 4 °C warmer than the modern SSTs at the core location (Figure 6); second, the Holocene Transition Period (HTP) between 6,500 yr B.P. and 3,000 yr B.P. displaying a SST decrease towards modern values; and third, the Cool Late Holocene Period (CLPH) between 3,000 yr B.P. and 0 yr B.P. with temperature around the modern SST value of 10–11 °C. The timing of the CLPH onset at 3,000 yr B.P. seems in phase with the initiation of the global Neoglacial cool period, similarly detected at high Southern latitudes (Leventer, Dunbar, & DeMaster, 1993; Brachfeld, Banerjee, Guyodo, & Acton, 2002; Shevenell & Kennett, 2002).

The reconstructed cooling trend is in agreement with other reconstructions from the same region and from other regions of the North Atlantic (Bauch et al., 2001; Jennings, Knudsen, Hald, Hansen, & Andrews, 2002; Andrews & Giraudeau, 2003). It is also in step with the decreasing Northern Hemisphere summer insolation since the last 10,000 yr, indicating a strong orbital-driven impact on Holocene climate evolution. There are, however, regional discrepancies in the timing and duration of the HCO in particular, indicating complex atmospheric and oceanic responses to the insolation forcing. Specifically, it is believed that an intense cold East Greenland Current associated with greater sea-ice presence led to delayed HCO warming in the western North Atlantic relative to the eastern North Atlantic (Andersen et al., 2004b). A large-scale North Atlantic Oscillation signature influencing wind strength and direction could also explain the reconstructed zonal SST differences.

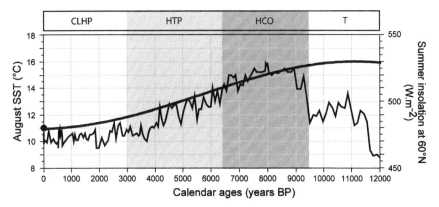

Figure 6 Sea-surface temperatures in core MD95-2011 as estimated by a diatom-based IKM transfer function (modified from Andersen et al., 2004a, 2004b). Modern August SST value is ~11 °C (black point), which matches well with the core-top estimate of 10.1 °C. Stratigraphy of the core is based on 10 AMS-14C dates and one tephra layer. T: Transition; HCO: Holocene Climatic Optimum; HTP: Holocene Transition Period; CLPH: Cool Late Holocene Period.

3.2. Sea-ice in the Southern Ocean

In the Southern Ocean, diatom-based SST estimates are generally given through IKM and MAT approaches, depending on the modern data set available (Pichon et al., 1992; Zielinski et al., 1998; Crosta, Sturm, Armand, & Pichon, 2004). Estimation of sea-ice winter and summer extents are provided by the *F. curta* and *Fragilariopsis obliquescostata* indexes, respectively (Gersonde & Zielinski, 2000) and by MAT transfer functions (Crosta, Pichon, & Burckle, 1998a; Crosta et al., 2004). Sea-ice reconstructions were initiated by CLIMAP (1981) and subsequently fell into disuse until the concomitant development of the above-mentioned qualitative and quantitative approaches and performing paleoclimatic models. It was additionally shown that CLIMAP sea-ice estimates in the Nordic seas were spurious (Weilnet et al., 1996), thus creating a resurgence in interest in the LGM period and in Antarctic SIC as important boundary conditions. Records of past SIC focus not only on the LGM (Crosta, Pichon, & Burckle, 1998a, 1998b; Gersonde et al., 2003, 2005), but also on long records (Crosta et al., 2004) and on high-resolution records of rapid climate changes (Hodell et al., 2001; Shemesh et al., 2002; Bianchi & Gersonde, 2002, 2004; Nielsen, Koç, & Crosta, 2004).

The example below represents the only study ever to reconstruct winter and summer sea-ice extents around Antarctica based on the combination of the *Fragilariopsis* proxies and the MAT approach (Gersonde et al., 2005). This comprehensive study conducted on sediments dated from the LGM provides more insight into the seasonal SIC than previous studies (Crosta et al., 1998a; Gersonde & Zielinski, 2000) because of the new modern data set used to estimate sea-ice, the greater area covered, and the comparison of the outputs of both methods whenever possible. SSTs at the LGM were concurrently estimated (Gersonde et al., 2005). It was also the first time that a set of quality controls on the modern data sets, the fossil data sets, and the estimates were provided.

The *F. curta* and *F. obliquecostata* proxies, calibrated by sediment-trap investigations in the South Atlantic (Gersonde and Zielinski, 2000), were applied to 45 LGM samples from the Atlantic and eastern Indian sector of the Southern Ocean (Gersonde et al., 2003). The MAT approach was applied to 73 LGM samples from the Atlantic, Indian, and eastern Pacific sectors of the Southern Ocean (Crosta et al., 1998a). The modern data set was composed of 204 surface sediment samples of recent to subrecent age and involved 31 diatom species. The transfer function is referenced hereafter as $MAT_5204/31$. Modern sea-ice input data were from Schweitzer's (1995) numerical atlas. The $MAT_5204/31$ accurately reconstructed the modern sea-ice distribution of yearly presence and winter concentration/extent with correlation coefficients of 0.97 and 0.96, slopes of the linear regression of 0.96 and 0.93, and mean RMSEP of 0.6 months per year and 6%, respectively. The $MAT_5204/31$ was less efficient in reconstructing the modern sea-ice distribution of summer concentration/extent with a correlation coefficient of 0.8, a slope of 0.6 indicating an overestimation of the estimates, and a mean RMSEP of 4%. Mean root mean square errors of prediction were almost twice as high when samples showing no sea-ice were discarded from the regressions.

The winter sea-ice extent at the LGM, as estimated from diatom assemblages, was 5–10° of latitude of its modern location (Figure 7), thus doubling the winter sea-ice area. This limit, calculated on diatom floral assemblages, is in good agreement with CLIMAP (1981) winter sea-ice limit, which served many years as a reference for paleoclimatic models. Models that compute LGM sea-ice as a consolidated cap calculate a direct effect of Antarctic sea-ice on atmospheric CO_2 concentration of $\sim$70 ppm (Stephens & Keeling, 2000). Models that compute LGM sea-ice as concentration gradients estimated by diatom transfer functions (Crosta et al., 1998b) calculate a pCO_2 drop of 10–30 ppm, thus attributing a lesser direct role to Antarctic sea-ice on atmospheric CO_2 (Morales-Maqueda & Rahmstorf, 2002).

The summer sea-ice extent at the LGM, as estimated from diatom assemblages, was much more extended in the South Atlantic sector of the Southern Ocean, but had a similar extent to today's one in the Indian and eastern Pacific sectors (Figure 7). Although the LGM database does not cover the western Pacific sector, it is believed that LGM summer sea-ice was more extended there relative to the modern cover, as the Ross Sea acts as an ice factory identically to the Weddell Sea. All in all, LGM summer sea-ice cover was obviously greater than the modern one, but diatom-based estimates argue against an area multiplied by 5 as reconstructed by CLIMAP (1981) and Cooke and Hays (1982), who positioned the glacial summer limit around the modern winter sea-ice limit. The new reconstruction implies much less perennial sea-ice cover, which has great implication for the albedo, the CO_2 efflux at the ocean–atmosphere interface, the productivity (Moore, Abbott, Richman, & Nelson, 2000), the hydrological cycle in redistributing salt (Shin, Liu, Otto-Bliesner, Kutzbach, & Vavrus, 2003), and surface water stratification (François et al., 1997). It is absolutely essential to arrive at a more accurate and comprehensive LGM summer sea-ice limit in the near future in order to better constrain paleoclimatic models.

3.3. C, N, and Si Isotopes in the Southern Ocean

The Southern Ocean is one of the largest high-nutrient, low-chlorophyll (HNLC) regions of the world in which low concentrations of trace metals, such as iron, limit productivity (Boyd et al., 2000; Boyd, 2002). Martin (1990) hypothesized that greater iron-bearing dust input during glacial times promoted phytoplanktonic productivity, therefore leading to the recorded reduced glacial atmospheric iron concentrations (Sarmiento & Toggweiler, 1984). However, investigations of deep-sea sedimentary records failed to clearly show the glacial period's increase in productivity, with contradictory results according to the study area and the proxy used (Mortlock et al., 1991; Kumar et al., 1995; Bareille et al., 1998). A decoupling between accumulation rates of organic carbon and biogenic silica has been shown (Anderson et al., 1998). In this context, the question of productivity changes in the Southern Ocean might have been abandoned if the "silicic leakage" hypothesis had not recently been proposed. Laboratory experiments showed that the uptake ratios of N/Si and C/Si by diatoms were greater when iron was not limiting (Hutchins & Bruland, 1998; Takeda, 1998; Brzezinski et al., 2002). *In situ*, it is believed that a

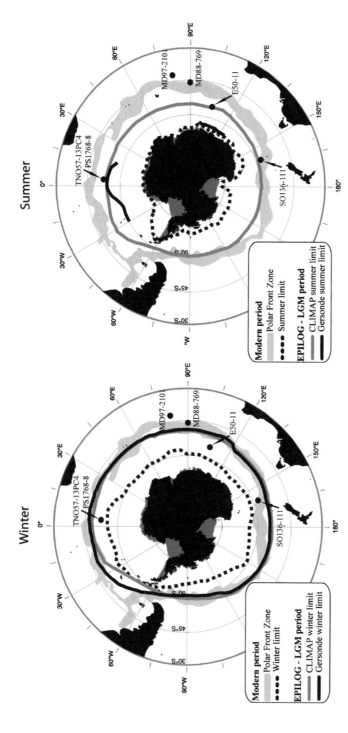

Figure 7 Sea-ice extent at the Last Glacial Maximum during the winter and summer seasons in the context of the MARGO program (modified from Gersonde et al., 2005). Gersonde's winter sea-ice limit was calculated from the *F. curta* group proxy (Gersonde & Zielinski, 2000) and the MAT approach (Crosta et al., 1998a), while Gersonde's summer limit was estimated from the *F. obliquecostata* proxy (Gersonde et al., 2005) and the MAT approach (Crosta et al., 1998a). CLIMAP's winter and summer limits were located at the faunally identified winter and summer 0°C isotherms, respectively (CLIMAP, 1981), and lithogenic tracers such as changes in the sedimentation rate, presence of ice-rafted debris, and geographical contact between diatom oozes and clayey sediments (Cooke & Hays, 1982). Position of the modern sea-ice limits is from Schweitzer (1995), and location of the modern hydrographic fronts is from Orsi, Whitworth, and Nowlin (1995).

higher N/Si uptake ratio during glacial times, stimulated by greater eolian iron input (Petit et al., 1999), led to surface water enrichment in silicic acid. Subantarctic Mode Waters (SAMW) thus supplied more silicic acid to low-latitude phytoplankton, stimulating the growth of siliceous organisms at the expense of carbonate organisms, thus reducing atmospheric CO_2 by decreasing the $CaCO_3/C_{org}$ ratio (Archer, Winguth, Lea, & Mahowaki, 2000). However, little geological evidence supports this hypothesis. In this context, diatom isotopic ratios may provide the best test of the silicic acid hypothesis as they are not affected by remineralization and diagenesis.

DIOM-$\delta^{13}C_{org}$, DIOM-$\delta^{15}N_{org}$, and $\delta^{30}Si$ were analyzed in the same three cores from the Indian sector of the Southern Ocean. Core SO136-111/E50-11 is from the Antarctic Zone, core MD88-769 is from the Subantarctic Zone, and core MD97-2101 is from the southern Subtropical Zone (Figure 7). Stratigraphy of the cores is based on AMS-^{14}C dates subsequently calibrated to calendar ages using CALIB 4.3 software (Stuiver et al., 1998) after removing the reservoir age. In the two southernmost cores, DIOM-$\delta^{15}N_{org}$ and $\delta^{30}Si$ are anticorrelated while DIOM-$\delta^{13}C_{org}$ and $\delta^{30}Si$ are positively correlated (Figure 8, APF and SAF boxes). Discarding potential modification of the nitrate to silicic acid supply ratio by the upwelling of Circumpolar Deep Water, anticorrelated DIOM-$\delta^{15}N_{org}$ and $\delta^{30}Si$ records argues for different relative uptake of nitrate and silicic acid through time, with greater uptake of nitrate during the last glacial and greater uptake of silicic acid during the Holocene. These changes are in agreement with the dust concentration record of Vostok (Petit et al., 1999), indicating a potential role of iron deficiency alleviation on diatom physiology during glacial times (Crosta et al., 2005b; Robinson et al., 2005; Beucher et al., 2006). The remnant nutrient pool in Antarctic surface waters was therefore enriched in nitrate during the Holocene and enriched in silicic acid during the last glacial period. Antarctic surface waters circulate around Antarctica within the Antarctic Circumpolar Current but have a northward component due to the Coriolis force and cross-frontal eddy diffusion. One should expect similarly anticorrelated Si and N isotopic signals in the northernmost core located on the Southern Subtropical Front if Antarctic Surface Waters were advected as far north. Nitrogen and silicon isotopic records in core MD97-2101 are, however, correlated (Figure 8, SSTF box). This indicates that N/Si uptake ratios by diatoms remained almost constant over the last 50,000 yr, which has two implications. First, no iron supply modified diatom physiology in the northern part of the Indian sector of the Southern Ocean, in agreement with the reconstruction of dust deposition at the LGM (Andersen, Armengaud, & Genthon, 1998). Second, no Antarctic surface waters reached the latitude of the core location, but they were seemingly transformed in SAMW that fueled low-latitude upwelling systems with waters enriched in silicic acids (Crosta et al., 2005b; Beucher et al., 2006). Investigations of diatom isotopic ratios thus provide support to the silicic acid hypothesis, although it is still impossible to quantify the amplitude of N/Si changes in the SAMW, and therefore the real impact on atmospheric CO_2 concentrations (Matsumoto, Sarmiento, & Brzezinski, 2002). Investigations of diatom isotopic ratios in low-latitude upwelling systems may help in resolving this issue.

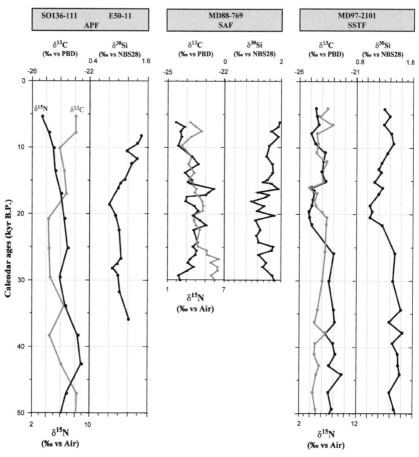

Figure 8   Carbon, nitrogen, and silicon isotopes in diatoms versus age from the Antarctic Polar Front Zone (APF), the Subantarctic Front (SAF), and the Southern Subtropical Front (SSTF) (De La Rocha et al., 1998; Crosta et al., 2005b; Beucher et al., 2006). Positions of the cores are shown in Figure 7.

As illustrated above, several methods using diatoms have been developed during the last couple of decades, providing scientists with exceptional tools to reconstruct past ocean conditions. There is no doubt that further application of these methods to high-latitude, upwelling and coastal regions of the world's oceans will enhance our understanding of climate change at all time scales.

 ## 4. Conclusion

Due to their distribution from oceanic to fresh waters and their ability to synthesize a siliceous test that is generally well preserved in the sediments, diatoms have received a great amount of attention. Thanks to the constant advancement of the analytical techniques, the improving understanding of the ecology and physiology of diatoms, and the drilling of better/longer sedimentary deep-sea cores, diatoms are now used to investigate past changes of a wide range of oceanographic parameters. Reconstruction of SSTs, sea-ice cover, water mass fluxes, fresh water input, productivity, and nutrient cycling represent only a small part of what it is potentially possible to do using diatoms, in both the fields of micropaleontogy and sediment geochemistry. These same parameters are also the most important in paleoclimate modeling, and can be used as boundary conditions or as an independent external check of model output. Paleoclimatic models are generally coupled with thermodynamic sea-ice models that, although fully interactive, may not provide accurate seasonal sea-ice extent and concentration. In this context, extensive reconstructions of sea-ice cover at key periods and at Milankovitch scale are greatly needed. Diatoms are the main tool to achieve quantitative estimates of Antarctic sea-ice extent and concentration, although chemical content of ice cores may provide complementary information (Wolff et al., 2006).

Future developments of the diatom methods will mainly concern geochemistry. Currently, a great effort is being made to obtain reliable radiocarbon dates from the organic compounds intrinsic to diatom frustules (Ingalls, Anderson, & Pearson, 2004), in order to help in dating sediments devoid of foraminifera, to measure silicon isotopes (Beucher et al., 2006) and oxygen isotopes (Swann, Maslin, Leng, Sloane, & Haug, 2006), to quantify production and dissolution of diatoms (Nelson, Tréguer, Brzezinski, Leynaert, & Quéguiner, 1995; Tréguer et al., 1995), and to understand biomineralization of the frustule (Martin-Jézéquel, Hildebrand, & Brzezinski, 2000) in order to help reconstruct modern and past silica cycles.

However, developments regarding micropaleontology should not be ignored. Slide preparation and diatom census counts are very long processes that would greatly benefit from automatic microfossil recognition. Although we know that diatom intra-specific variability is high, and that specimens may be present in different states of preservation, or even be tilted on the permanent slide, it might be possible in the future to have access to the most prominent species or group of species with acceptable taxonomic precision. Laminated sediments were recently discovered in different oceanic regions and can be used to reconstruct climate changes at a seasonal scale over the whole Holocene period. Two aspects, nevertheless, have to be addressed. First, modern diatom ecology needs to be better understood at the seasonal scale in the regions where such exceptional records exist: the coastal Antarctic (Leventer, Domack, Barkoukis, McAndrews, & Murray, 2002; Maddison et al., 2005; Stickley et al., 2005; Denis et al., 2006), Baja California (Pike & Kemp, 1999; Kemp et al., 2000), fjords along the Canadian West Coast (Hay et al., 2003), and the Red Sea (Seeberg-Elverfeldt, Lange, Arz, Pätzold, & Pike, 2004). Sediment-trap investigations along the Canadian West Coast have already begun to provide more insights into diatom seasonal succession and ecology (Hay et al., 2003; Dallimore et al., 2005). Second, techniques to accurately determine diatom species and sediment composition on thin sections need to be further developed (Pike & Kemp, 1996; Pike, Bernhard, Moreton, & Butler, 2001; Denis et al., 2006).

ACKNOWLEDGMENTS

International colleagues are thanked for discussions and for providing in press or unpublished data contained in this chapter. This is an UMR EPOC publication no. 1602.

REFERENCES

Abrantes, F. (1988a). Diatom assemblages as upwelling indicators in surface sediments off Portugal. *Marine Geology, 85,* 15–39.

Abrantes, F. (1988b). Diatom productivity peak and increased circulation during the latest quaternary: Alboran Basin (Western Mediterranean). *Marine Micropaleontology, 13,* 79–96.

Abrantes, F. (1991). Increased upwelling off Portugal during the last glaciation: Diatom evidence. *Marine Micropalentology, 17,* 285–310.

Akiba, F. (1985). Middle Miocene to Quaternary diatom biostratigraphy in the Nankai Trough and Japan Trench, and modified Lower Miocene through Quaternary diatom zones for middle-to-high latitudes of the North Pacific. In: H. Kagami, D. E. Karig & W. T. Coulbourn, et al. (Eds), *Initial reports of the deep sea drilling project* (Vol. 87, pp. 393–481). Washington, D.C.: U.S. Government Printing Office.

Akiba, F., & Yanagisawa, Y. (1985). Taxonomy, morphology and phylogeny of the Neogene diatom zonal marker species in the middle-to-high latitudes of the North Pacific. In: H. Kagami, D. E. Karig & W. T. Coulbourn, et al. (Eds), *Initial reports of the deep sea drilling project* (Vol. 87, pp. 483–553). Washington, D.C.: U.S. Government Printing Office.

Aleem, A. A. (1950). Distribution and ecology of British Littoral diatoms. *Journal of Ecology, 38.*

Allen, C. S., Pike, J., Pudsey, C. J., & Leventer, A. (2005). Submillennial variations in ocean conditions during deglaciation based on diatom assemblages from the southwest Atlantic. *Paleoceanography, 20,* PA2012, doi:10.1029/2004PA001055.

Altabet, M. A., & François, R. (1994). Sedimentary nitrogen isotopic ratio as a recorder for surface ocean nitrate utilization. *Global Biogeochemical Cycles, 8*(1), 103–116.

Andersen, C., Koç, N., Jennings, A., & Andrews, J. T. (2004a). Nonuniform response of the major surface currents in the Nordic Seas to insolation forcing: Implications for the Holocene climate variability. *Paleoceanography, 19,* PA2003, doi:10.1029/2002PA000873.

Andersen, C., Koç, N., & Moros, M. (2004b). A highly unstable Holocene climate in the subpolar North Atlantic: Evidence from diatoms. *Quaternary Science Reviews, 23,* 2155–2166.

Andersen, K. K., Armengaud, A., & Genthon, C. (1998). Atmospheric dust under glacial and interglacial conditions. *Geophysical Research Letters, 25*(13), 2281–2284.

Anderson, R. F., Kumar, N., Mortlock, R. A., Froelich, P. N., Kubik, P., Dittrich-Hannen, B., & Suter, M. (1998). Late-Quaternary changes in productivity of the Southern Ocean. *Journal of Marine Systems, 17,* 497–514.

Andrews, J. T., & Giraudeau, J. (2003). Multi-proxy records showing significant Holocene environmental variability: The inner N Iceland Shelf (Hunafloi). *Quaternary Science Reviews, 22,* 175–193.

Anonymous (1975). Proposals for a standardization of diatom terminology and diagnoses. *Nova Hedwigia, 53,* 323–354.

Archer, D. E., Winguth, A., Lea, D., & Mahowaki, N. (2000). What caused the glacial/interglacial atmospheric pCO_2 cycles? *Reviews of Geophysics, 38*(2), 159–189.

Armand, L., Crosta, X., Romero, O., & Pichon, J. J. (2005). The biogeography of major diatom taxa in Southern Ocean sediments. 1. Sea-ice related species. *Palaeogeography, Palaeoclimatology, Palaeoecology, 223,* 93–126.

Armand, L. (1997). *The use of diatom transfer functions in estimating sea-surface temperature and sea-ice in cores from the southeast Indian Ocean.* Ph.D thesis, Australian National University, Canberra, Australia.

Armand, L. K., & Zielinski, U. (2001). Diatom species of the genus *Rhizosolenia* from Southern Ocean sediments: Distribution and taxonomic notes. *Diatom Research, 16*(2), 259–294.

Baldauf, J. G. (1984). Cenozoic diatom biostratigraphy and paleoceanography of the Rockall Plateau region, North Atlantic, Deep Sea Drilling Project, Leg 81. In: (Eds), *Initial reports of the Deep Sea Drilling Project* (Vol. 81, pp. 439–478). Washington, D.C.: U.S. Government Printing Office.

Baldauf, J. G. (1987). Diatom biostratigraphy of the middle and high latitude North Atlantic Ocean, Deep Sea Drilling Project, Leg 94. In: (Eds), *Initial reports of the Deep Sea Drilling Project* (Vol. 94, pp. 729–762). Washington, D.C.: U.S. Government Printing Office.

Bareille, G., Labracherie, M., Bertrand, P., Labeyrie, L., Lavaux, G., & Dignan, M. (1998). Glacial-interglacial changes in the accumulation rates of major biogenic components in Southern Indian Ocean sediments. *Journal of Marine Systems, 17*, 527–539.

Bauch, H. A., Erlenkeuser, H., Spielhagen, R. F., Struck, U., Matthiessen, J., Thiede, J., & Heinemeier, J. (2001). A multiproxy reconstruction of the evolution of deep and surface waters in the sub-Arctic Nordic seas over the last 30.000 years. *Quaternary Science Reviews, 20*, 659–678.

Bentaleb, I., & Fontugne, M. (1998). The role of the Southern Indian Ocean in the glacial to interglacial atmospheric CO_2 change: Organic carbon isotope evidences. *Global and Planetary Change, 16–17*, 25–36.

Beucher, C., Brzezinski, M. A., Crosta, X., & Tréguer, P. (2006). Nutrient dynamics in the subantarctic and subtropical zones of the Southern Ocean from the Last Ice Age to the Holocene revealed by silicon isotope composition. ASLO Ocean Science Meeting, Hawaiï, USA, February.

Bianchi, C., & Gersonde, R. (2002). The Southern Ocean surface between marine isotope stages 6 and 5d: Shape and timing of climate changes. *Palaeogeography, Palaeoclimatology, Palaecology, 187*, 151–177.

Bianchi, C., & Gersonde, R. (2004). Climate evolution at the last deglaciation: The role of the Southern Ocean. *Earth and Planetary Science Letters, 228*, 407–428.

Birks, C. J. A., & Koç, N. (2002). A high-resolution diatom record of late-Quaternary sea-surface temperatures and oceanographic conditions from the eastern Norwegian Sea. *Boreas, 31*(4), 323–344.

von Bodungen, B., Smetacek, V. S., Tilzer, M. M., & Zeitzschel, B. (1985). Primary production and sedimentation during spring in the Antarctic Peninsula region. *Deep Sea Research, 33*, 177–194.

Boyd, P. W. (2002). The role of iron in the biogeochemistry of the Southern Ocean and equatorial Pacific: A comparison of *in situ* iron enrichments. *Deep Sea Research, Part II, 49*, 1803–1821.

Boyd, P. W., Watson, A. J., Law, C. S., Abraham, E. R., Trull, T., Murdoch, R., Bakker, D. C. E., Bowie, A. R., Buesseler, K. O., Chang, H., Charette, M., Croot, P., Downing, K., Frew, R., Gall, M., Hadfield, M., Hall, J., Harvey, M., Jameson, G., LaRoche, J., Liddicoat, M., Ling, R., Maldonado, M. T., McKay, R. M., Nodder, S., Pickmere, S., Pridmore, R., Rintoul, S., Safi, K., Sutton, P., Strzepzk, R., Tanneberger, K., Turner, S., Waite, A., & Zeldis, J. (2000). A mesoscale phytoplankton bloom in the polar Southern Ocean stimulated by iron fertilization. *Nature, 407*, 695–702.

ter Braak, C. J. F., & Juggins, S. (1993). Weighted averaging partial least squares regression (WA-PLS): An improved method for reconstructing environmental variables from species assemblages. *Hydrobiologia, 269–270*, 485–502.

Brachfeld, S. A., Banerjee, S. K., Guyodo, Y., & Acton, G. D. (2002). A 13200 year history of century to millenial-scale paleoenvironmental change magnetically recorded in the Palmer Deep, western Antarctic Peninsula. *Earth and Planetary Science Letters, 194*, 311–326.

Brzezinski, M. A., Pride, C. J., Franck, V., Sigman, D. M., Sarmiento, J. L., Matsumoto, K., Gruber, N., Rau, G. H., & Coale, K. H. (2002). A switch from $Si(OH)_4$ to NO_3^- depletion in the glacial Southern Ocean. *Geophysical Research Letters, 29*(12), doi:10.1029/2001GL014349.

Burckle, L. H. (1972). Late Cenozoic planktonic diatom zones from the Eastern Equatorial Pacific. *Nova Hedwigia, 39*, 217–246.

Burckle, L. H. (1984a). Diatom distribution and paleoceanographic reconstruction in the Southern Ocean — present and Last Glacial Maximum. *Marine Micropaleontology, 9*, 241–246.

Burckle, L. H. (1984b). Ecology and paleoecology of the marine diatom *Eucampia antarctica* (Cast.) Mangin. *Marine Micropaleontology, 9*, 77–86.

Burckle, L. H., Clarke, D. B., & SHackleton, N. J. (1978). Isochronous last-abundant-appearance datum (LAAD) of the diatom *Hemidiscus karstenii* in the sub-Antarctic. *Geology, 6*, 243–246.

Burkhardt, S., Riebesell, U., & Zondervan, I. (1999). Effects of growth rate, CO_2 concentration, and cell size on the stable carbon isotope fractionation in marine phytoplankton. *Geochimica et Cosmochimica Acta, 63*(22), 3729–3741.

Campeau, S., Pienitz, R., & Héquette, A. (1998). Diatom from the Beaufort Sea coast, southern Arctic Ocean (Canada): Modern analogues for reconstructing Late Quaterny environments and relative sea levels. *Bibliotheca Diatomologica, 42*, 244.

Campeau, S., Pienitz, R., & Héquette, A. (1999). Diatom as quantitative paleodepth indicators in coastal areas of the southeastern Beaufort Sea, Arctic Ocean. *Palaeogeography, Palaeoclimatology, Palaeoecology, 146*, 67–97.

Cardinal, D., Alleman, L. Y., Jong, J. d., Ziegler, K., & André, L. (2003). Isotopic composition of silicon measured by multicollector plasma source mass spectrometry in dry plasma mode. *Journal of Analytical Atomic Spectrometry, 18*, 213–218.

Cassar, N., Laws, E. A., Bidigare, R. R., & Popp, B. N. (2004). Bicarbonate uptake by Southern Ocean phytoplankton. *Global Biogeochemical Cycles, 18*, doi:10.1029/2003GB002116.

CLIMAP, 1981. Seasonal reconstructions of the Earth's surface at the last glacial maximum. *Geological Society of America*. Map Chart Series MC-36.

Cooke, D. W., & Hays, J. D. (1982). Estimates of Antarctic Ocean seasonal sea-ice cover during glacial intervals. In: C. Craddock (Ed.), *Antarctic Geoscience, International union of Geological Sciences* (pp. 1017–1025). Madison, WI: The Union of Wisconsin Press.

Craig, H., & Gordon, L. I. (1965). Deuterium and oxygen 18 variations in the ocean and the marine atmosphere. In: E. Tongiorgi (Ed.), *Stable isotopes in oceanogaphic studies*. Spoleto, Italy: CNR.

Crosta, X., Pichon, J. J., & Burckle, L. H. (1998a). Application of modern analog technique to marine Antarctic diatoms: Reconstruction of the maximum sea-ice extent at the last glacial maximum. *Paleoceanography, 13*(3), 284–297.

Crosta, X., Pichon, J. J., & Burckle, L. H. (1998b). Reappraisal of seasonal Antarctic sea-ice extent at the last glacial maximum. *Geophysical Research Letters, 25*(14), 2703–2706.

Crosta, X., Pichon, J. J., & Labracherie, M. (1997). Distribution of *Chaetoceros* resting spores in modern peri-antarctic sediments. *Marine Micropaleontology, 29*, 283–299.

Crosta, X., Romero, O., Armand, L., & Pichon, J. J. (2005a). The biogeography of major diatom taxa in Southern Ocean sediments. 2. Open ocean related species. *Palaeogeography, Palaeoclimatology, Palaeoecology, 223*, 66–92.

Crosta, X., & Shemesh, A. (2002). Reconciling down-core anti-correlation of diatom carbon and nitrogen isotopic ratios from the Southern Ocean. *Paleoceanography, 17*(1), 1010, doi:10.1029/2000PA000565.

Crosta, X., Shemesh, A., Etourneau, J., Yam, R., Billy, I., & Pichon, J. J. (2005b). Nutrient cycling in the Indian sector of the Southern Ocean over the last 50,000 years. *Global Biogeochemical Cycles, 19*, GB3007, doi:10.1029/2004GB002344.

Crosta, X., Shemesh, A., Salvignac, M. E., Gildor, H., & Yam, R. (2002). Late Quaternary variations of elemental ratios (C/Si and N/Si) in diatom-bound organic matter from the Southern Ocean. *Deep Sea Research, Part II, 49*, 1939–1952.

Crosta, X., Sturm, A., Armand, L., & Pichon, J. J. (2004). Late Quaternary sea-ice history in the Indian sector of the Southern Ocean as recorded by diatom assemblages. *Marine Micropaleontology, 50*, 209–223.

De la Rocha, C. L., (2002). Measurement of silicon stable isotope natural abundances via multicollector inductively coupled plasma mass spectrometry (MC-ICP-MS). *Geochemistry, Geophysics, Geosystems, 3*(8), 10.1029/2002GC000310.

De la Rocha, C. L., Brzezinski, M. A., & DeNiro, M. J. (1997). Fractionation of silicon isotopes by marine diatoms during biogenic silica formation. *Geochimica et Cosmochimica Acta, 61*(23), 5051–5056.

De la Rocha, C. L., Brzezinski, M. A., DeNiro, M. J., & Shemesh, A. (1998). Silicon-isotope composition of diatom as an indicator of past oceanic change. *Nature, 395*, 680–683.

Defelice, D. R., & Wise, S. W. (1981). Surface lithofacies, biofacies, and diatom diversity patterns as models for delineation of climatic change in the Southeast Atlantic Ocean. *Marine Micropaleontology, 6*, 29–70.

Denis, D., Crosta, X., Zaragosi, S., Romero, O., Martin, B., & Mas, V. (2006). Seasonal and sub-seasonal climate changes recorded in laminated diatom ooze sediments, Adélie Land, East Antarctica. *The Holocene, 16*(8), 1137–1147.

Dézileau, L., Reyss, J. L., & Lemoine, F. (2003). Late Quaternary changes in biogenic opal fluxes in the Southern Indian Ocean. *Marine Geology, 202,* 143–158.

Dzinoridze, R. N., Jousé, A. P., Koroleva-Golikova, G. S., Kozlova, G. E., Nagaeva, G. S., Petrushevskaya, M. G., & Strelnikova, N. I. (1978). Diatom and radiolarian Cenozoic stratigraphy, Norwegian Basin, DSDP Leg 38. In: M. Talwani & G. Udintsev (Eds), *Initial reports of the Deep Sea Drilling Project* (Vol. 38, pp. 289–428). Washington, D.C.: U.S. Government Printing Office.

El Sayed, S. Z. (1990). Plankton. In: G. P. Glasby (Ed.), *Antarctic sector of the Pacific* (pp. 95–125). Amsterdam: Elsevier.

Emiliani, C. (1955). Pleistocene temperatures. *Journal of Geology, 63,* 538–578.

Emiliani, C. (1966). Paleotemperature analysis of Caribbean cores, P6304-8 and P6304-9 and a generalized temperature curve for the past 425,000 years. *Journal of Geology, 74,* 109–126.

Fenner, J. M. (1991). Late Pliocene-Quaternary quantitative diatom stratigraphy in the Atlantic sector of the Southern Ocean. In: P. F. Ciesielski, Y. Kristoffersen, et al. (Eds), *Proceedings of the Ocean Drilling Program, scientific results,* College Station, TX (Vol. 114, pp. 97–121).

Fitzwater, S. E., Coale, K. H., Gordon, R. M., Johnson, K. H., & Ondrusek, M. E. (1996). Iron deficiency and phytoplankton growth in the equatorial Pacific. *Deep Sea Research II, 43*(4–6), 995–1015.

François, R., Altabet, M. A., & Burckle, L. H. (1992). Glacial to interglacial changes in surface nitrate utilization in the Indian sector of the Southern Ocean as recorded by sediment $\delta^{15}N$. *Paleoceanography, 7*(5), 589–606.

François, R., Altabet, M. A., Yu, E. F., Sigman, D. M., Bacon, M. P., Frank, M., Bohrmann, G., Bareille, G., & Labeyrie, L. D. (1997). Contribution of Southern Ocean surface-water stratification to low atmospheric CO_2 concentrations during the last glacial period. *Nature, 389,* 929–935.

Frank, M., Gersonde, R., van der Loeff, M. R., Bohrmann, G., Nürnberg, C. C., Kubik, P. W., Suter, M., & Mangini, A. (2000). Similar glacial and interglacial export bioproductivity in the Atlantic sector of the Southern Ocean: Multiproxy evidence and implications for glacial atmospheric CO_2. *Paleoceanography, 15*(6), 642–658.

Fryxell, G. A. (1991). Comparison of winter and summer growth stages of the diatom *Eucampia Antarctica* from the Kerguelen Plateau and south of the Antarctic Convergence Zone, In: J. Barron & B. Larsen (Eds), *Proceedings of the Ocean Drilling Program, scientific results,* College Station, TX (Vol. 119, pp. 675–685).

Fryxell, G. A., & Prasad, A. K. S. K. (1990). *Eucampia antarctica* var. *recta* (Mangin) stat. nov. (Biddulphiaceae, Bacillariophyceae): Life stages at the Weddell sea-ice edge. *Phycologia, 29*(1), 27–38.

Galbraith, E. D., Kienast, M., Pedersen, T. F., & Calvert, S. E. (2004). Glacial-interglacial modulation of the marine nitrogen cycle by high-latitude O_2 supply to the global thermocline. *Paleoceanography, 19*(4), PA4007, doi:10.1029/2003PA001000.

Ganeshram, R., Pedersen, T. F., Calvert, S. E., McNeill, G. W., & Fontugne, M. R. (2000). Glacial-interglacial variability in denitrification in the world's oceans: Causes and consequences. *Paleoceanography, 15,* 361–376.

Gersonde, R. (1990). Taxonomy and morphostructure of Neogene diatoms from the Southern Ocean, ODP Leg 113, In: P. F. Barker, J. P. Kennett, et al. (Eds), *Proceedings of the Ocean Drilling Program, scientific results,* College Station, TX (Vol. 113, pp. 791–802).

Gersonde, R., Abelmann, A., Brathauer, U., Becquey, S., Bianchi, C., Cortese, G., Grobe, H., Khun, G., Niebler, H. S., Segl, M., Sieger, R., Zielinski, U., & Fütterer, D. K. (2003). Last glacial sea surface temperatures and sea-ice extent in the Southern Ocean (Atlantic-Indian sector): A multiproxy approach. *Paleoceanography, 18*(3), 1061, doi:10.1029/2002PA000809.

Gersonde, R., & Barcena, M. A. (1998). Revision of the upper Pliocene — Pleistocene diatom biostratigraphy for the northern belt of the Southern Ocean. *Micropaleontology, 44*(1), 84–98.

Gersonde, R., Crosta, X., Abelmann, A., & Armand, L. K. (2005). Sea surface temperature and sea-ice distribution of the Southern Ocean at the EPILOG last glacial maximum — A circum-Antarctic view based on siliceous microfossil records. *Quaternary Science Reviews, 24,* 869–896.

Gersonde, R., & Zielinski, U. (2000). The reconstruction of late quaternary Antarctic sea-ice distribution — the use of diatoms as a proxy for sea-ice. *Palaeogeography, Palaeoclimatology, Palaeoecology, 162*, 263–286.

Graham, L. E., Wilcox, L. W. (2000). *Algae* (p. 650). Indianapolis, IN: Prentice Hall.

Guiot, J. (1990). Methodology of the last climatic cycle reconstructions in France from pollen data. *Palaeogeography, Palaeoclimatology, Palaeoecology, 80*, 49–69.

Hargraves, P. E., & French, F. W. (1975). Observation of the survival of diatom resting spores. *Nova Hedwigia, 53*, 229–238.

Harrison, P. J., Conway, H. L., Holmes, R. W., & Davis, C. O. (1977). Marine diatoms grown in chemostats under silicate or ammonium limitation. III. Cellular chemical composition and morphology of *Chaetoceros debilis*, *Skeletonema costatum* and *Thalassiosira gravida*. *Marine Biology, 43*, 19–31.

Hasle, G. R. (1969). *An analysis of the phytoplancton of the Pacific Southern Ocean: Abundance, composition, and distribution during the Brategg Expedition, 1947–1948* (p. 168). Oslo, Norway: Universitetsforlaget.

Hasle, G. R., & Syversten, E. E. (1997). Marine diatoms. In: C. R. Tomas (Ed.), *Identifying marine diatoms and dinoflagellates* (pp. 5–385). San Diego, CA: Academic Press.

Hay, M. B., Pienitz, R., & Thompson, R. E. (2003). Distribution of diatom surface sediment assemblages within Effingham inlet, a temperate fjord on the west coast of Vancouver Island (Canada). *Marine Micropaleontology, 48*, 291–320.

Heiden, H., & Kolbe, R. W. (1928). Die marinen diatomeen des Deutschen Südpolar Expedition 1901–1903. *Deutschen Südpolar Expedition, 8*, 450–714.

Hendey, N. I. (1937). The plankton diatoms of the Southern Seas. *Discovery Reports, XVI*, 151–364.

Higginson, M. J., & Altabet, M. A. (2004). Inital test of the silicic acid leakage hypothesis using sedimentary biomarkers. *Geophysical Research Letters, 31*, L18303, doi:10.1029/2004GL020511.

Higginson, M. J., Maxwell, J. R., & Altabet, M. A. (2003). Nitrogen isotopes and chlorin paleoproductivity records from the northern South China Sea: Remote vs local forcing of millenial- and orbital-scale variability. *Marine Geology, 201*, 223–250.

Hodell, D. A., Kanfoush, S. L., Shemesh, A., Crosta, X., Charles, C. D., & Guilderson, T. P. (2001). Abrupt cooling of Antarctic surface waters and sea-ice expansion in the South Atlantic sector of the Southern Ocean at 5000 cal yr BP. *Quaternary Research, 56*, 191–198.

Horner, R. A. (1985). Ecology of sea ice microalgae. In: R. H. Horner (Ed.), *Sea Ice Biota* (pp. 83–104). Boca Raton, FL: CRC Press.

Huon, S., Grousset, F., Burdloff, D., Bardoux, G., & Mariotti, A. (2002). Sources of fine-sized organic matter in North Atlantic Heinrich layers: Delta C-13 and delta N-15 tracers. *Geochimica et Cosmochimica Acta, 66*, 223–239.

Hustedt, F. (1958). Diatomeen aus der Antarktis und dem Südatlaktik. Reprinted from "Deutsche Antarktishe Expedition 19838/1939" Band II, Geographische-Kartographische Anstalt "Mundus", Hamburg, p. 191.

Hutchins, D. A., & Bruland, K. W. (1998). Iron-limited diatoms growth and Si:N uptake ratios in a coastal upwelling regime. *Nature, 393*, 561–564.

Hutson, W. H. (1980). The Agulhas current during the Late Pleistocene: Analysis of aodern faunal analogs. *Science, 207*, 64–66.

Imbrie, J., & Kipp, N. G. (1971). A new micropaleontological method for quantitative paleoclimatology: Application to a Late Pleistocene Carribean core. In: K. K. Turekian, (Ed.), *Late Cenozoïc glacial ages* (pp. 71–181). New Haven, CT: Yale University Press.

Ingalls, A. E., Anderson, R. F., & Pearson, A. (2004). Radiocarbon dating of diatom-bound organic compounds. *Marine Chemistry, 92*, 91–105.

Jasper, J. P., & Hayes, J. M. (1990). A carbon-isotopic record of CO_2 levels during the last quaternary. *Nature, 347*, 462–464.

Jeffrey, S. W., Mantoura, R. F. C., & Wight, S. W. (1997). *Phytoplankton pigments in oceanography: Guidelines to modern methods* (p. 661). Mouflon, Belgium: UNESCO Publishing.

Jennings, A., Knudsen, K. L., Hald, M., Hansen, C. V., & Andrews, J. T. (2002). A mid-Holocene shift in Arctic sea-ice variability on the East Greenland Shelf. *Holocene, 12*, 49–58.

Jiang, H., Eiricksson, J., Schulz, M., Knudsen, K. L., & Seidenkrantz, M. S. (2005). Evidence for solar forcing of sea-surface temperature on the North Icelandic Shelf during the Late Holocene. *Geology, 33*(1), 73–76.

Jiang, H., Seidenkrantz, M. S., Knudsen, K. L., & Eiricksson, J. (2001). Diatom surface sediment assemblages around Iceland and their relationships to oceanic environmental variables. *Marine Micropaleontology, 41*, 73–96.

Juillet-Leclerc, A. (1984). Cleaning process for diatomaceous samples, In: Ricard, M. (Ed.), *8th Symposium on Living and Fossil Diatoms* (pp. 733–763). Koetlz Scientific Books, Koenigstein, Germany.

Juillet-Leclerc, A., & Labeyrie, L. (1987). Temperature dependence of the oxygen isotopic fractionation between diatom silica and water. *Earth and Planetary Science Letters, 84*, 69–74.

Kaczmarska, I., Barbrick, N. E., Ehrman, J. M., & Cant, G. P. (1993). *Eucampia* Index as an indicator of the Late Pleistocene oscillations of the winter sea-ice extent at the ODP Leg 119 Site 745B at the Kerguelen Plateau. *Hydrobiologia, 269/270*, 103–112.

Kamatani, A., Ejiri, N., & Tréguer, P. (1988). The dissolution kinetics of diatom ooze from the Antarctic area. *Deep Sea Research, 35*(7), 1195–1203.

Karsh, K. L., Trull, T. W., Lourey, M. J., & Sigman, D. M. (2003). Relationship of nitrogen isotope fractionation to phytoplankton size and iron availability during the Southern Ocean Iron RElease Experiment (SOIREE). *Limnology and Oceanography, 48*(3), 1058–1068.

Kemp, A. E. S. (1995). Laminated sediments from coastal and open ocean upwelling zones: What variability do they record. In: C. P. Summerhayes & K. C. Emeis, et al. (Eds), *Upwelling in the cean: Modern processes and ancient records* (pp. 239–257). Southampton, UK: Wiley.

Kemp, A. E. S., Pike, J., Pearce, R. B., & Lange, C. B. (2000). The "Fall dump" — A new perspective on the role of a "shade flora" in the annual cycle of diatom production and export flux. *Deep Sea Research II, 47*, 2129–2154.

Kemp, A. E. S., Baldauf, J. G., & Pearce, R. B. (1995). Origins and paleoceanographic significance of laminated diatom ooze from the eastern equatorial Pacific Ocean, In: N. G. Pisias, L. A. Mayer, et al. (Eds), *Proceedings of the Ocean Drilling Program, scientific results*, College Station, TX (Vol. 138, pp. 641–645).

Kienast, M., Calvert, S. E., Pelejero, C., & Grimalt, J. (2001). A critical review of marine sedimentary $\delta^{13}C_{org}$-pCO_2 estimates: New paleorecords from South China Sea and a revisit of other low-latitude $\delta^{13}C_{org}$-pCO_2 records. *Global Biogeochemical Cycles, 15*(1), 113–127.

Kienast, S. S., Calvert, S. E., & Pedersen, T. F. (2002). Nitrogen isotope and productivity variations along the northeast Pacific margin over the last 120 kyr: Surface and subsurface paleoceanography. *Paleoceanography, 17*(4), 1055, doi:10.1029/2001PA000650.

Koç, N., & Flower, B. (1998). High-resolution Pleistocene diatom biostratigraphy and paleoceanography of Site 919 from the Irminger Basin. In: A. D. Saunders, H. C. Larsen, S. W. J. Wise (Eds), *Proceedings of the Ocean Drilling Program, scientific results*, College Station, TX (Vol. 152, pp. 209–219).

Koç, N., Hodell, D. A., Kleiven, H., & Labeyrie, L. (1999). High-resolution Pleistocene diatom biostratigraphy of Site 983 and correlations to isotope stratigraphy, In: E. Jansen, M. Raymo, & P. Blum (Eds), *Proceedings of the Ocean Drilling Program, scientific results*, College Station, TX (Vol. 162, pp. 51–62).

Koç, N., Jansen, E., & Haflidason, H. (1993). Paleoceanographic reconstructions of surface ocean conditions in the Greenland, Iceland and Norwegian Seas through the last 14 ka based on diatoms. *Quaternary Science Reviews, 12*, 115–140.

Koç, N., Jansen, E., Hald, M., & Labeyrie, L. (1996). Late glacial-Holocene sea surface temperatures and gradients between the North Atlantic and the Norwegian Seas: Implications for the Nordic heat pump. In: J. T. Andrews & W. E. N., Austin, et al. (Eds), *Late Quaternary palaeoceanography of the North Atlantic margins* (Vol. 111, pp. 177–185). London: Geological Society Special Publication.

Koç, N., Labeyrie, L., Manthé, S., Flower, B. P., & Hodell, D. A. (2001). The last occurrence of *Proboscia curvirostris* in the North Atlantic marine isotope stages 9–8. *Marine Micropaleontology, 41*, 9–23.

Koç, N., & Scherer, R. (1996). Neogene diatom biostratigraphy of the Iceland Sea Site 907, In: (Eds), *Proceedings of the Ocean Drilling Program, scientific results*, College Station, TX (Vol. 151, pp. 61, 74).

Koç-Karpuz, N., & Jansen, E. (1992). A high resolution diatom record of the last deglaciation from the SE Norwegian Sea: Documentation of rapid climatic changes. *Paleoceanography*, 7, 499–520.

Koç-Karpuz, N., & Schrader, H. (1990). Surface sediment diatom distribution and Holocene paleotemperature variations in the Greenland, Iceland and Norwegian Sea. *Paleoceanography*, 5(4), 557–580.

Koizumi, I. (1977). Diatom biostratigraphy in the North Pacific region. *Proceedings of the First International Congress on Pacific Neogene Stratigraphy*, Tokyo, Japan (pp. 235–253).

Kozlova, O. G. (1971). The main feature of diatoms and silicoflagellate distribution in the Indian Ocean. In: B. M. Funnel & W. R. Riedel (Eds), *Micropaleontology of the oceans* (p. 271). Cambridge: Cambridge University Press.

Kröger, N., Bergsdorf, C., & Sumper, M. (2002). Frustulins: Domain conservation in a protein family associated with diatom cell walls. *European Journal of Biochemistry*, 239, 259–264.

Kröger, N., Deutzmann, R., & Sumper, M. (1999). Polycationic peptides from diatom biosilica that direct silica nannospheres formation. *Science*, 286, 1129–1132.

Kröger, N., & Sumper, M. (1998). Diatom cell wall proteins and the cell biology of silica biomineralization. *Protist*, 149, 213–219.

Kucera, M., Rosell-Melé, A., Schneider, R., Walbroeck, C., & Weilnet, M. (2005). Mutliproxy approach for the reconstruction of the glacial ocean surface (MARGO). *Quaternary Science Reviews*, 24, 813–819.

Kumar, N., Anderson, R. F., Mortlock, R. A., Froelich, P. N., Kubik, P., Dittrich-Hannen, B., & Suter, M. (1995). Increased biological productivity and export production in the glacial Southern Ocean. *Nature*, 378, 675–680.

Kumar, N., Gwiazda, R., Anderson, R. F., & Froelich, P. N. (1993). ^{231}Pa/^{230}Th ratios in sediments as a proxy for past changes in Southern Ocean productivity. *Nature*, 362, 45–48.

Labeyrie, L. D., & Juillet, A. (1982). Oxygen isotopic exchangeability of diatom valve silica: Interpretation and consequences for paleoclimate studies. *Geochimica et Cosmochimica Acta*, 46, 967–975.

Laws, E. A., Popp, B. N., Bidigare, R. P., Kennicutt, M. C., & Macko, S. A. (1995). Dependence of phytoplankton carbon isotopic composition on growth rate and $(CO_2)_{aq}$: Theoritical considerations and experimental results. *Geochimica et Cosochimica Acta*, 59(6), 1131–1138.

Leventer, A. (1991). Sediment trap diatom assemblages from the northern Antarctic Peninsula region. *Deep Sea Research*, 38, 1127–1143.

Leventer, A., Domack, E., Barkoukis, A., McAndrews, B., & Murray, J. (2002). Laminations from the Palmer Deep: A diatom-based interpretation. *Paleoceanography*, 17(2), doi:10.1029/2001PA000624.

Leventer, A., Domack, E. W., Ishman, S. E., Brachfeld, S., McClennen, C. E., & Manley, P. (1996). Productivity cycles of 200–300 years in the Antarctic Peninsula region: Understanding linkages among the sun, atmosphere, oceans, sea-ice, and biota. *Geological Society of America - Bulletin*, 108(12), 1626–1644.

Leventer, A., Dunbar, R. B., & DeMaster, D. J. (1993). Diatom evidence for the Late Holocene climatic events in Granite Harbor, Antarctica. *Paleoceanography*, 8, 373–386.

Licursi, M., Sierra, M. V., & Gomez, N. (2006). Diatom assemblages from a turbid coastal plain estuary: Rio de la Plata (South America). *Journal of Marine Systems*, 61(1-2), 35–45.

Lourey, M. J., Trull, T. W., & Sigman, D. M. (2003). Sensitivity of δ^{15}N of nitrate, surface suspended and deep sinking particulate nitrogen to seasonal nitrate depletion in the Southern Ocean. *Global Biogeochemical Cycles*, 17(3), 1081, doi:10.1029/2002GB001973.

Lücke, A., Moschen, R., & Schleser, G. H. (2005). High-temperature carbon reduction of silica: A novel approach for oxygen isotope analysis of biogenic opal. *Geochimica et Cosmochimica Acta*, 69(6), 1423–1433.

Maddison, E. J., Pike, J., Leventer, A., & Domack, E. (2005). Deglacial seasonal and sub-seasonal diatom record from Palmer Deep, Antarctica. *Journal of Quaternary Science*, 20(5), 435–446.

Malmgren, B. A., Kucera, M., Nyberg, J., & Waelbroeck, C. (2001). Comparison of statistical and artificial neural network techniques for estimating past sea-surface temperatures from planktonic foraminifera census data. *Paleoceanography, 16*(5), 520–530.

Malmgren, B. A., & Nordlund, U. (1997). Application of artificial neural networks to paleoceanographic data. *Palaeogeography, Palaeoclimatology, Palaeoecology, 136*, 359–373.

Martin, J. (1990). Glacial-interglacial CO_2 change: The iron hypothesis. *Paleoceanography, 5*, 1–13.

Martin-Jézéquel, V., Hildebrand, M., & Brzezinski, M. A. (2000). Silicon metabolism in diatoms: Implications for growth. *Journal of Phycology, 36*, 821–840.

Matsumoto, K., Sarmiento, J. L., & Brzezinski, M. A. (2002). Silicic acid leakage from the Southern Ocean: A possible explanation for glacial atmospheric pCO_2. *Global Biogeochemical Cycles, 16*(3), doi:10.1029/2001GB001442.

McCollum, D. W. (1975). Diatom stratigraphy of the Southern Ocean. In: D. Hays & L. Frakes, et al. (Eds), *Initial reports of the Deep Sea Drilling Project* (Vol. 28, pp. 515–571). Washington D.C.: U.S. Government Printing Office.

Moore, J. K., Abbott, M. R., Richman, J. G., & Nelson, D. M. (2000). The Southern Ocean at the Late Glacial maximum: A strong sink for atmospheric CO_2. *Global Biogeochemical Cycles, 14*(1), 455–475.

Morales-Maqueda, M. A., & Rahmstorf, S. (2002). Did Antarctic sea-ice expansion cause glacial CO_2 decline? *Geophysical Research Letters, 29*(1), 1011, doi:10.1029/2001GL013240.

Mortlock, R. A., Charles, C. D., Froelich, P. N., Zibello, M. A., Saltzman, J., Hays, J. D., & Burckle, L. H. (1991). Evidence for lower productivity in the Antarctic Ocean during the last glaciation. *Nature, 351*, 220–222.

Moschen, R., Lücke, A., & Schleser, G. H. (2005). Sensitivity of biogenic silica oxygen isotopes to changes in surface water temperature and palaeoclimatology. *Geophysical Research Letters, 32*, LO7708, doi:10.1029/2004GL22167.

Nelson, D. M., Tréguer, P., Brzezinski, M. A., Leynaert, A., & Quéguiner, B. (1995). Production and dissolution of biogenic silica in the ocean: Revised global estimates, comparison with regional data and relationship to biogenic sedimentation. *Global Biogeochemical Cycle, 9*(3), 359–372.

Neori, A., & Holm-Hansen, O. (1982). Effect of temperature on rate of photosynthesis in Antarctic phytoplankton. *Polar Biology, 1*, 33–38.

Nielsen, S. H. H., Koç, N., & Crosta, X. (2004). Holocene climate in the Atlantic sector of the Southern Ocean: Controlled by insolation or oceanic circulation. *Geology, 32*(4), 317–320.

Orsi, A. H., Whitworth, T., & Nowlin, W. D. (1995). On the meridonal extent and fronts of the Antarctic Circumpolar Current. *Deep Sea Research, 42*(5), 641–673.

Pancost, R. D., Freeman, K. H., Wakeham, S. G., & Robertson, C. Y. (1997). Controls on carbon isotope fractionation by diatoms in the Peru upwelling region. *Geochimica et Cosmochimica Acta, 61*(23), 1983–4991.

Petit, J. R., Raynaud, D., Barkov, N. I., Barnola, J. M., Basile, I., Benders, M., Chappellaz, J., Davis, M., delaygue, G., Delmotte, M., kotlyakov, V. M., Legrand, M., Lipenkov, V. Y., Lorius, C., Pepin, L., Ritz, C., Saltzman, E., & Stievenard, M. (1999). Climate and atmospheric history of the past 420.000 years from the Vostok ice core, Antarctica. *Nature, 399*, 429–436.

Pfitzer, E. (1871). Untersuchungen über Bau und Entwicklung der Bacillariaceen (Diatomaceen). In: H. Hanstein (Ed.), *Botanische Abhandlungen aus dem Gebiet der Morphologie, 1*(2), 1–189.

Pflaumann, U., Duprat, J., Pujol, C., & Labeyrie, L. D. (1996). SIMMAX: A modern analog technique to deduce Atlantic sea-surface temperatures from planktonic foraminifera in deep-sea sediments. *Paleoceanography, 11*, 15–35.

Pichon, J. J., Labeyrie, L. D., Bareille, G., Labracherie, M., Duprat, J., & Jouzel, J. (1992). Surface water temperature changes in the high latitudes of the southern hemisphere over the last glacial–interglacial cycle. *Paleoceanography, 7*, 289–318.

Pichon, J. J., Labracherie, M., Labeyrie, L. D., & Duprat, J. (1987). Transfer function between diatom assemblages and surface hydrology in the Southern Ocean. *Palaeogeography, Palaeoclimatology, Palaeoecology, 61*, 79–95.

Pielou, E. C. (1984). *The interpretation of ecological data: A primer on classification and ordination* (p. 263). New York: Wiley.

Pike, J., Bernhard, J. M., Moreton, S. G., & Butler, I. B. (2001). Microbioirrigation of marine sediments in dysoxic environments: Implications for early sediment fabric formation and diagenetic processes. *Geology, 29*(10), 923–926.

Pike, J., & Kemp, A. E. S. (1996). Preparation and analysis techniques for studies of laminated sediments. In: A. E. S. Kemp (Ed.), *Palaeoclimatology and Palaeoceanography from Laminated Sediments*, (Vol. 116, pp. 37–48). London: Geological Society Special Publication.

Pike, J., & Kemp, A. E. S. (1999). Diatom mats in Gulf of California sediments: Implications for the paleoenvironmental interpretation of laminated sediments and silica burial. *Geology, 27*(4), 311–314.

Pokras, E. M., & Molfino, B. (1986). Oceanographic control of diatom abundances and species distribution in surface sediment of the Tropical and Southeast Atlantic. *Marine Micropaleontology, 10*, 165–188.

Pokras, E. M., & Molfino, B. (1987). Diatom record of Late Quaternary climatic change in the eastern equatorial Atlantic and tropical Africa. *Paleoceanography, 2*(3), 273–286.

Popp, B. N., Laws, E. A., Bidigare, R. R., Dore, J. E., Hanson, K. L., & Wakeham, S. G. (1998). Effect of phytoplankton cell geometry on carbon isotopic fractionation. *Geochimica et Cosmochimica Acta, 62*(1), 69–77.

Popp, B. N., Trull, T., kenig, F., Wakeham, S. G., Rust, T. M., Tilbrook, B., Griffiths, F. B., Wright, S. W., Marchant, H. J., Bidigare, R. R., & Laws, E. A. (1999). Controls on the carbon isotopic composition of Southern Ocean phytoplankton. *Global Biogeochemical Cycles, 13*(4), 827–843.

Prell, W. L. (1985). *The stability of low-latitude sea-surface temperatures: An evaluation of the CLIMAP reconstruction with emphasis on the positive SST anomalies*. Report TR 025, US Department of Energy Washington D.C., pp. 1–60.

Racca, J. M. J., Gregory-Eaves, I., Pientiz, R., & Prairie, Y. (2004). Tailoring palaeolimnological, diatom-based transfer functions. *Canadian Journal of Fisheries and Aquatic Sciences, 61*, 1–15.

Ragueneau, O., Tréguer, P., Leynaert, A., Anderson, R. F., Brezinski, M. A., DeMaster, D. J., Dugdale, R. C., Dymont, J., Fisher, G., François, R., Heinze, C., Maier-Reimer, E., Martin-Jézéquel, V., Nelson, D. M., & Quéguiner, B. (2000). A review of the Si cycle in the modern ocean: Recent progress and missing gaps in the application of biogenic opal as a paleoproductivity proxy. *Global and Planetary Change, 26*, 317–365.

Rathburn, A. E., Pichon, J. J., Ayress, M. A., & DeDeckker, P. (1997). Microfossil and stable-isotope evidence for changes in Late Holocene paleoproductivity and paleoceanographic conditions in the Prydz Bay region of Antarctica. *Palaeogeography, Palaeoclimatology, Palaeoecology, 131*, 485–510.

Rau, G. H. (2001). Plankton $^{13}C/^{12}C$ variations in Monterey Bay, California: Evidence of non-diffusive inorganic carbon uptake by phytoplankton in an upwelling environment. *Deep Sea Research I, 48*, 79–94.

Rau, G. H., Froelich, P. N., Takahashi, T., Marais, D. J. D. (1991a). Does sedimentary organic $\delta^{13}C$ record variations in quaternary ocean [$CO_{2(aq)}$]? *Paleoceanography, 6*(3), 335–347.

Rau, G. H., Sullivan, C. W., & Gordon, L. I. (1991b). $\delta^{13}C$ and $\delta^{15}N$ variations in Weddell Sea particulate organic matter. *Marine Chemistry, 35*, 355–369.

Rings, A., Lücke, A., & Schleser, G. H. (2004). A new method for the quantitative separation of diatom frustules from lake sediments. *Limnology and Oceanography: Methods, 2*, 25–34.

Robinson, R. S., Brunelle, B. G., & Sigman, D. M. (2004). Revisiting nutrient utilisation in the glacial Antarctic: Evidence from a new method for diatom-bound N isotopic signal. *Paleoceanography, 19*, PA3001, doi:10.1029/2003PA000996.

Robinson, R. S., Sigman, D., DiFiore, P. J., Rohde, M. M., Mashiotta, T. A., & Lea, D. W. (2005). Diatom-bound 15N/14N: New support for enhanced nutrient consumption in the ice age subantarctic. *Paleoceanography, 20*(3), PA3003, doi:10.1029/2004PA001114.

Romero, O., Armand, L. K., Crosta, X., & Pichon, J. J. (2005). The biogeography of major diatom taxa in Southern Ocean sediments: 3. Tropical/subtropical species. *Palaeogeography, Palaeoclimatology, Palaeoecology, 223*, 49–65.

Romero, O., Fischer, G., Lange, C. B., & Wefer, G. (2000). Siliceous phytoplankton of the western equatorial Atlantic: Sediment traps and surface sediments. *Deep Sea Research II, 47*, 1939–1959.

Rosenthal, Y., Dahan, M., & Shemesh, A. (2000). The glacial Southern Ocean: A source of atmospheric CO_2 as inferred from carbon isotopes in diatoms. *Paleoceanography, 15*(1), 65–75.

Round, F. E. (1972). The problem of reduction of cell size during diatom cell division. *Nova Hedwigia, 23*, 291–303.

Round, F. E., Crawford, R. M., & Mann, D. G. (1990). *The diatoms. Biology and morphology of the genera* (p. 747). Cambridge: Cambridge University Press.

Sancetta, C. (1979). Oceanography of the North Pacific during the last 18.000 years: Evidence from fossil diatoms. *Marine Micropaleontology, 4*, 103–123.

Sancetta, C. (1992). Comparison of phytoplankton in sediment trap series and surface sediments along a productivity gradient. *Paleoceanography, 7*(2), 183–194.

Sancetta, C., Heusser, L., Labeyrie, L., Naidu, A. S., & Robinson, S. W. (1985). Wisconsin-Holocene paleoenvironment of the Bering Sea: Evidence from diatoms, pollen, oxygen isotopes and clay minerals. *Marine Geology, 62*, 55–68.

Sancetta, C., & Silvestri, S. (1984). Diatom stratigraphy of the Late Pleistocene (Brunhes) Subarctic Pacific. *Marine Micropaleontology, 9*, 263–274.

Sarmiento, J. L., & Toggweiler, J. R. (1984). A new model for the role of the oceans in determining atmospheric pCO_2. *Nature, 308*, 621–624.

Scherer, R. (1995). A new method for the determination of absolute abundances of diatoms and other silt-sized sedimentary particles. *Journal of Paleolimnology, 12*, 171–179.

Scherer, R., & Koç, N. (1996). Late Paleogene diatom biostratigraphy of the northern Norwegian-Greenland Sea, ODP Leg 151. In: J. Thiede & A. Myhre (Eds), *Proceedings of the Ocean Drilling Program, scientific results*, College Station, TX (Vol. 151, pp. 75–99).

Schrader, H. J. (1971). Fecal pellets: Role in sedimentation of pelagic diatoms. *Science, 174*, 55–77.

Schrader, H. (1973). Cenozoic diatoms from the northeast Pacific, Leg 18. *Proceedings of the Ocean Drilling Program, Initial Reports* (Vol. 18, pp. 673–797). Washington, D.C.: U.S. Government Printing Office.

Schrader, H. J., & Fenner, J. (1976). Norwegian Sea Cenozoic diatom biostratigraphy and taxonomy, In: M. Talwani, G. Udintsev, et al. (Eds), *Proceedings of the Ocean Drilling Program, initial reports*, College Station, TX (Vol. 38, pp. 921–1099).

Schrader, H. J., & Gersonde, R. (1978). Diatoms and silicoflagellates. In: A. Zachariasse, et al. (Eds), Micropaleontological counting methods and techniques — An exercise on an eight meters section of the Lower Pliocene of Capo Rossello, Sicily. *Utrecht Micropaleontology Bulletin, 17*, 129–176.

Schütt, F. (1896). Bacillariales. In: P. Engler (Ed.), *Die natürlichen Pflanzenfamilien* (pp. 31–153). Leipzig, Germany: Engelmann, W.

Schweitzer, P. N. (1995). *Monthly averaged polar sea-ice concentration.* U.S. Geological Survey Digital Data Series, Virginia.

Seeberg-Elverfeldt, I. A., Lange, C. B., Arz, H. W., Pätzold, J., & Pike, J. (2004). The significance of diatoms in the formation of laminated sediments of the Shaban Deep, Northern Red Sea. *Marine Geology, 209*, 279–301.

Semina, H. J. (2003). SEM-studied diatoms of different regions of the World Ocean. *Iconographia Diatomologica, 10*, 1–362.

Shemesh, A., Burckle, L. H., & Froelich, P. N. (1989). Dissolution and preservation of Antarctic diatoms and the effect on sediment thanatocenoses. *Quaternary Research, 31*, 288–308.

Shemesh, A., Burckle, L. H., & Hays, J. D. (1994). Meltwater input to the Southern Ocean during the last glacial maximum. *Science, 266*, 1542–1544.

Shemesh, A., Burckle, L. H., & Hays, J. D. (1995). Late Pleistocene oxygen isotope records of biogenic silica from the Atlantic sector of the Southern Ocean. *Paleoceanography, 10*(2), 179–196.

Shemesh, A., Charles, C. D., & Fairbanks, R. G. (1992). Oxygen isotopes in biogenis silica: Global changes in ocean temperature and isotopic composition. *Science, 256*, 1434–1436.

Shemesh, A., Hodell, D., Crosta, X., Kanfoush, S., Charles, C., & Guilderson, T. (2002). Sequence of events during the last deglaciation in Southern Ocean sediments and Antarctic ice cores. *Paleoceanography, 17*(4), 1056, doi:10.1029/2000PA000599.

Shemesh, A., Macko, S. A., Charles, C. D., & Rau, G. H. (1993). Isotopic evidence for reduced productivity in the Glacial Southern Ocean. *Science, 262*, 407–410.

Shevenell, A. E., & Kennett, J. P. (2002). Antarctic Holocene climate change: A benthic foraminiferal stable isotope record from Palmer Deep. *Paleoceanography, 17*(3), doi:10.1029/2000PA000596.

Shin, S. I., Liu, Z., Otto-Bliesner, B. L., Kutzbach, J. E., & Vavrus, S. (2003). Southern Ocean sea-ice control of the glacial North Atlantic thermohaline circulation. *Geophysical Research Letters, 30*(2), 1096, doi:10.1029/2002GL015513.

Sigman, D. M., Altabet, M. A., François, R., McCorkle, D. C., & Gaillard, J. F. (1999a). The isotopic composition of diatom-bound nitrogen in the Southern Ocean sediments. *Paleoceanography, 14*(2), 118–134.

Sigman, D. M., Altabet, M. A., McCorkle, D. C., François, R., & Fisher, G. (1999b). The $\delta^{15}N$ of nitrate in the Southern Ocean: Consumption of nitrate in surface waters. *Global Biogeochemical Cycles, 13*(4), 1149–1166.

Sigman, D. M., & Boyle, E. A. (2000). Glacial/interglacial variations in atmospheric carbon dioxide. *Nature, 407*, 859–869.

Sigman, D., & Casciotti, K. L. (2001). Nitrogen isotopes in the ocean. In: J. H. Steele, K. K. Turekian & S. A. Thorpe (Eds), *Encyclopedia of ocean sciences* (p. 2449). London, UK: Academic Press.

Sigman, D., Casciotti, K. M., Andreani, M., Barford, C., Galanter, M., & Bohlke, J. K. (2001). A bacterial method for the nitrogen analysis of nitrate in seawater and freshwater. *Analytical Chemistry, 73*, 4145–4153.

Simonsen, R. (1974). The diatom plankton of the Indian Ocean Expedition of R/V Meteor 1964/ 1965. *Meteor Forschergeb Reihe D, 19*, 1–107.

Simonsen, R. (1979). The diatom system: Ideas on phylogeny. *Bacillaria, 2*, 9–71.

Singer, A. J., & Shemesh, A. (1995). Climatically linked carbon isotope variation during the past 430,000 years in Southern Ocean sediments. *Paeloceanography, 10*(2), 171–177.

Smetacek, V. S. (1985). Role of sinking in diatom life-history cycles: Ecological, evolutionary and geological significance. *Marine Biology, 84*, 239–251.

Stephens, B. B., & Keeling, R. F. (2000). The influence of Antarctic sea-ice on glacial-interglacial CO_2 variations. *Nature, 404*, 171–174.

Stickley, C. E., Pike, J., Leventer, A., Dunbar, R., Domack, E. W., Brachfeld, S., Manley, P., & McClennan, C. (2005). Deglacial ocean and climate seasonality in laminated diatom sediments, Mac.Robertson Shelf, Antarctica. *Palaeogeography, Palaeoclimatology, Palaeoecology, 227*, 290–310.

Stuiver, M., Reimer, P. J., Bard, E., Beck, J. W., Burr, G. S., Hughen, K. A., Kromer, B., McCormac, G., van den Plicht, J., & Spurk, M. (1998). INTCAL98 radiocarbon age callibration, 24.000-0 cal BP. *Radiocarbon, 40*(3), 1041–1083.

Swann, G. E., Maslin, M. A., Leng, M. J., Sloane, H. J., & Haug, G. H. (2006). Diatom $\delta^{18}O$ evidence for the development of the modern halocline system in the subarctic northwest Pacific at the onset of major Northern Hemisphere glaciation. *Paleoceanography, 21*, PA1009, doi:10.1029/ 2005PA001147.

Takeda, S. (1998). Influence of iron availability on nutrients consumption ratio of diatoms in oceanic waters. *Nature, 393*, 774–777.

Tortell, P. D., & Morel, F. M. M. (2002). Sources of inorganic carbon for phytoplankton in the eastern Subtropical and Equatorial Pacific Ocean. *Limnology and Oceanography, 47*(4), 1012–1022.

Tortell, P. D., Rau, G. H., & Morel, F. M. M. (2000). Inorganic carbon acquisition in coastal Pacific phytoplankton communities. *Limnology and Oceanography, 45*(7), 1485–1500.

Tréguer, P., Nelson, D. M., Van Bennekom, A. J., DeMaster, D. J., Leynaert, A., & Quéguiner, B. (1995). The silica balance in the world ocean: A reestimate. *Science, 268*, 375–379.

Treppke, U. F., Lange, C. B., Donner, B., Fischer, G., Ruhland, G., & Wefer, G. (1996). Diatom and silicoflagellate fluxes at the Walvis Ridge: An environment influenced by coastal upwelling in the Benguela system. *Journal of Marine Research, 54*, 991–1016.

Trull, T. W., & Armand, L. (2001). Insights into Southern Ocean carbon export from the $\delta^{13}C$ of particles and dissolved inorganic carbon during the SOIREE iron release experiment. *Deep Sea Research, Part II, 48*, 2655–2680.

Varela, D. E., Pride, C. J., & Brzezinski, M. A. (2004). Biological fractionation of silicon isotopes in Southern Ocean surface waters. *Global Biogeochemical Cycles, 18*, GB1047, doi:10.1029/ 2003GB002140.

Waelbroeck, C., Labeyrie, L., Duplessy, J. C., Guiot, J., Labracherie, M., Lelaire, H., & Duprat, J. (1998). Improving past sea surface temperature estimates based on planktonic fossil faunas. *Paleoceanography, 13*(3), 272–283.

Weaver, A. J., Saenko, O. A., Clarck, P. U., & Mitrovica, J. X. (2003). Meltwater pulse 1A from Antarctica as a trigger of the Bolling-Allerod warm interval. *Science, 299*, 1709–1713.

Weilnet, M., Sarnthein, M., Pflaumann, U., Schulz, H., Jung, S., & Erlenkeuser, H. (1996). Ice-free Nordic seas during the Last Glacial Maximum? Potential sites of deepwater formation. *Paleoclimates, 1*, 283–309.

Wolff, E. W., et al. (2006). Southern Ocean sea-ice extent, productivity and iron flux over the past eight glacial cycles. *Nature, 440*, 491–496.

Woodworth, M., Goni, M., Tappa, E., Tedesco, K., Thunell, R., Astor, Y., Varela, R., Diaz-Ramos, J. R., & Müller-Karger, F. (2004). Oceanographic controls on the carbon isotopic compositions of sinking particles from the Cariaco Basin. *Deep Sea Research I, 51*, 1955–1974.

Zielinski, U. (1993). *Quantitative estimation of palaeoenvironmental parameters of the Antarctic Surface Water in the Late Quaternary using transfer functions with diatoms*. Ph.D. thesis, Alfred Wegener Institute for Polar and Marine Research, Bremerhaven, Germany, p. 171.

Zielinski, U., & Gersonde, R. (1997). Diatom distribution in Southern Ocean surface sediments (Atlantic Sector): Implications for paleoenvironmental reconstructions. *Paaleogeography, Palaeoclimatology, Palaeoceanography, 129*, 213–250.

Zielinski, U., & Gersonde, R. (2002). Plio-Pleistocene diatom biostratigraphy from ODP Leg 177, Atlantic sector of the Southern Ocean. *Marine Micropaleontology, 45*, 225–268.

Zielinski, U., Gersonde, R., Sieger, R., & Fütterer, D. (1998). Quaternary surface water temperature estimations: Calibration of a diatom transfer function for the Southern Ocean. *Paleoceanography, 13*(4), 365–384.

CHAPTER NINE

ORGANIC-WALLED DINOFLAGELLATE CYSTS: TRACERS OF SEA-SURFACE CONDITIONS

Anne de Vernal* *and* Fabienne Marret

Contents

1. Introduction	371
2. Ecology of Dinoflagellates	376
3. Dinoflagellates vs. Dinocysts and Taphonomical Processes (From the Biocenoses to Thanathocenoses)	377
3.1. Living vs. fossil dinoflagellates	377
3.2. Biogeography of motile cells vs. distribution of the cysts in sediments	379
3.3. Sedimentation	381
3.4. Preservation	382
4. Relationships between Dinocyst Assemblages and Sea-Surface Parameters	382
4.1. Nearshore vs. offshore distribution, sea-level and continentality indices	383
4.2. Salinity	386
4.3. Sea-Surface temperature and seasonality	386
4.4. Sea-Ice cover	387
4.5. Productivity, upwelling and polynyas	390
4.6. Environmental quality and eutrophication	392
4.7. Red tides and harmful algal blooms (HAB)	394
5. The Development of Quantitative Approaches for the Reconstruction of Hydrographic Parameters Based on Dinocysts	395
5.1. A brief history	395
5.2. A few caveats	396
6. The Use of Dinocysts in Paleoceanography	397
7. Concluding Remarks	398
References	400

1. INTRODUCTION

Dinoflagellates are microscopic unicellular organisms belonging to the division of Dinoflagellata (Fensome et al., 1993). They inhabit most types of aquatic

* Corresponding author.

Developments in Marine Geology, Volume 1
ISSN 1572-5480, DOI 10.1016/S1572-5480(07)01014-7

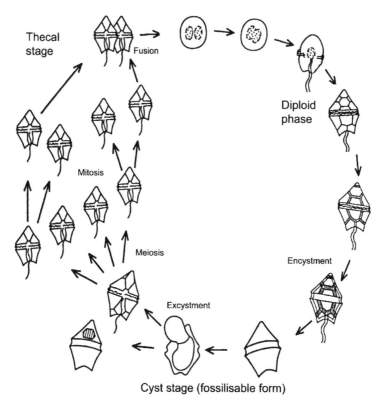

Thecal
stage

Fusion

Diploid
phase

Mitosis

Meiosis

Encystment

Excystment

Cyst stage (fossilisable form)

Figure 1   Diagram of the life cycle of a dinoflagellate showing the alternation of the motile stage
(cannot be fossilized) and the cyst stage (yielding fossil remains).

environments, from lakes to open ocean, and occur at all latitudes from the Equator
to Polar seas. During their motile or vegetative phase, dinoflagellates bear two
flagella used for swimming with a spiral-like motion that is at the origin of their
name (from the Greek word *dinos* meaning whirling). Together with diatoms
and coccolithophorids, dinoflagellates constitute important primary producers.
Approximately half of the extant dinoflagellate taxa are autotrophic; others are
heterotrophic, mixotrophic, parasitic or symbiotic. Some dinoflagellates, such as
those belonging to the genus *Symbiodinium*, and commonly referred to as zoo-
xanthellae, are symbionts of various organisms including corals, radiolarians and
foraminifera (Pawlowski, Holzmann, Fahrni, Pochon, & Lee, 2001; Pochon,
LaJeunesse, & Pawlowski, 2004).

The motile cell of dinoflagellates does not yield remains that can be fossilized.
However, many dinoflagellates have a complex life cycle involving several stages,
asexual and sexual, motile and non-motile (Figure 1). During the course of sexual
reproduction, some species form a diploid cell protected within a cyst, which
permits survival of the organism during a dormancy period of variable length
(*cf.* Wall & Dale, 1968; Fensome et al., 1993). A few dinoflagellate taxa produce
calcareous cysts (Zonneveld et al., 1999), and approximately 10–20% of the species

produce cysts composed of highly resistant organic matter (e.g., Dale, 1976; Head, 1996). The organic cyst material is called dinosporin: it has been compared to the sporopollenin of pollen grains, but shows distinct chemical macromolecular composition, which varies depending upon the taxon (Fensome et al., 1993; Kokinos et al., 1998; Versteegh & Blokker, 2004). The organic-walled cysts, also known as "dinocysts," are typically 15 to 100 μm in diameter. They are routinely observed at magnifications ranging from 250 × to 1,000 × in palynological slides, prepared following laboratory procedures used for pollen analyses, which involve treatments with hydrofluoric and hydrochloric acids. The study of dinocysts is often considered to be a sub-discipline of palynology.

Fossil dinocysts are mainly known from marine sediments, and appear to be particularly abundant along continental margins (estuaries, continental shelves and slopes, epicontinental seas). Dinocysts are widely used in biostratigraphy and paleoecology of the Mesozoic and Tertiary (e.g., Powell, 1992; Fensome & Williams, 2004). In the field of Late Cenozoic paleoceanography and paleoecology, the study of dinocysts is of growing interest. Because they are composed of resistant organic matter, dinocysts are generally well preserved in most sediment, unlike calcareous or siliceous biological remains that can be affected by dissolution. Moreover, unlike many biological tracers that are stenohaline and thus restricted to open ocean, dinoflagellates dwell in a wide range of salinities and permit investigations in nearshore environments, epicontinental seas or estuaries, in addition to full oceanic environments. However, the concentration of dinocysts in sediment decreases significantly offshore, in low productivity ocean gyre areas (e.g., de Vernal, Turon, & Guiot, 1994; Marret, 1994a; Vink et al., 2000). Thus, dinocyst assemblages seem to be more useful along continental margins and are often viewed as complementary to other tracers more adapted to open ocean conditions, such as calcareous dinoflagellates (e.g., Vink, Brune, Holl, Zonneveld, & Willems, 2002; Vink, 2004), coccoliths (e.g., Winter, Jordan, & Roth, 1994) and planktonic foraminifera (e.g., Bé & Tolderlund, 1971).

Since the early work of Wall, Dale, Lohmann, and Smith (1977), many studies have documented the geographical distribution of modern dinocysts on the sea floor. There are now regional data sets for the North Atlantic and the Arctic Oceans, the circum-Antarctic Ocean, the low latitudes of the Atlantic Ocean, the eastern and western Pacific Ocean margins (see for example, synthesis by Rochon, de Vernal, Turon, Matthiessen, & Head, 1999; de Vernal et al., 2001, 2005a; Marret & Zonneveld, 2003; see also Figure 2; Table 1). These data sets were used to define relationships between the distribution of dinocyst assemblages and hydrographic parameters, notably the temperature, salinity, sea-ice cover and productivity or eutrophication (cf. Section 4). In parallel to studies dealing with modern distribution of assemblages, the number of paleoceanographical studies using dinocysts has increased significantly during the last two decades. Many of these studies present empirical interpretations of the assemblages in terms of productivity, salinity or temperature changes. Quantitative reconstructions of oceanographical parameters from dinocyst assemblages were tentatively made using various approaches, including the modern analogue technique (de Vernal et al., 1994). It has been used in particular to estimate late Quaternary temperature and salinity in the North

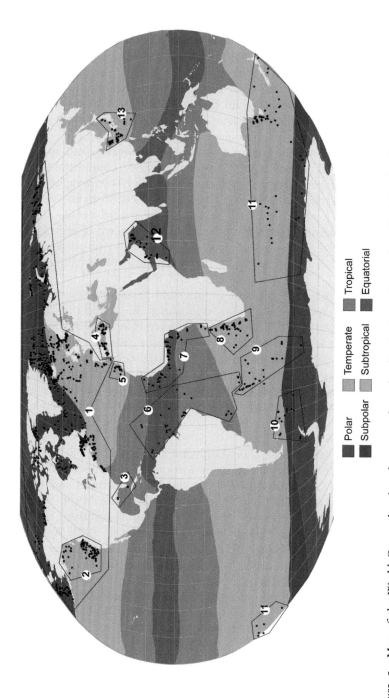

Figure 2 Map of the World Ocean showing the areas where the distributions of organic-walled dinoflagellate cysts in sediment have been documented from the study of assemblages in surface sediments. Climatic provinces after Gross and Gross (1994). For references, see Table 1.

Table 1 List of Publications Documenting the Distribution of Organic-Walled Dinocysts in Surface Sediments.

Data set no.	Location	References
1	Northern Hemisphere	de Vernal et al. (1994); de Vernal, Rochon, Turon, and Matthiessen, (1997); de Vernal et al. (2001, 2005a); Rochon and de Vernal (1994); Rochon et al. (1999); Radi et al. (2001); Voronina, Polyak, de Vernal, and Peyron (2001); Kunz-Pirrung (2001); Grosfjeld and Harland (2001); Mudie and Rochon (2001); Hamel, de Vernal, Gosselin, and Hillaire-Marcel (2002)
2	NE Pacific Ocean	Radi and de Vernal (2004)
3	Gulf of Mexico, Mississippi Sound, Florida Bay	Edwards in Marret and Zonneveld (2003)
4	Mediterranean Sea	Mangin (2002)
5	Canary Islands	Targarona, Warnaar, Boessenkool, Brinkhuis, and Canals (1999)
6	Western tropical and equatorial Atlantic Ocean	Vink, Zonneveld, and Willems (2000)
7	Eastern equatorial Atlantic Ocean	Marret (1994a), Marret and Dupont in Marret and Zonneveld (2003)
8	Namibian margin and SW tropical Atlantic Ocean	Zonneveld, Hoek, Brinkhuis, and Willems (2001a)
9	Southern Atlantic Ocean	Esper, Zonneveld, and Willems (2002)
10	Antarctic and Scotia Sea	Harland, Pudsey, Howe, and Fitzpatrick (1998)
11	Southern Ocean	Marret and de Vernal (1997); Marret, de Vernal, Benderra, and Harland (2001a)
12	Arabian Sea	Zonneveld (1997)
13	W Pacific (Japan)	Matsuoka (1981, 1985); Matsuoka et al. (1999); Kobayashi, Matsuoka, and Iizuka (1986); Cho and Matsuoka (1999)

Source: Updated from Marret and Zonneveld (2003).

Atlantic (e.g., Hillaire-Marcel, de Vernal, Bilodeau, & Weaver, 2001; Solignac, de Vernal, & Hillaire-Marcel, 2004) and in the Southern Ocean (Marret et al., 2001a), as well as to estimate sea-ice cover extent in the North Atlantic during the Last Glacial Maximum (LGM) (de Vernal et al., 2005a; de Vernal, Hillaire-Marcel, Turon, & Matthiessen, 2000). The use of dinocysts in paleoceanography therefore seems useful and unique with respect to some hydrographic parameters (e.g., sea-ice cover). In the present paper, we discuss the state of the art knowledge that allows for the use of dinocysts in paleoceanography, with an emphasis on the strengths and weaknesses of the methods, and we present a few examples of applications selected from the literature.

2. ECOLOGY OF DINOFLAGELLATES

Dinoflagellates live in various types of aquatic environments, including lakes, estuaries, epicontinental seas and oceans, from equatorial to polar settings (e.g., Taylor & Pollingher, 1987; Matthiessen et al., 2005). However, most dinoflagellates are marine: about 2,000 species are known to live in the modern marine waters, whereas only a few hundred species are known to live in freshwater. In marine environments, and within the planktonic communities, dinoflagellates show particularly high species diversity, together with high variability in morphology, and adaptation to a wide range of environments (e.g., Sournia, 1995; Smayda & Reynolds, 2003).

Dinoflagellates are usually abundant in neritic environments, including the estuaries, epicontinental seas and continental shelves. This is due to the high tolerance of many species toward low salinity, in addition to nutrient availability and stratification of water masses (e.g., Pemberton, Rees, Miller, Raine, & Joint, 2004; Gowen & Stewart, 2005). Dinoflagellates can also be abundant offshore in the open ocean (e.g., Pitcher, Walker, Mitchell-Innes, & Moloney, 1991; Veldhuis, de Baar, Kraay, Van Bleijswijk, & Baars, 1997; Smayda & Reynolds, 2003). Particularly high species diversity can be observed near the shelf edge where both low-salinity tolerant taxa and oceanic species can occur together or successively throughout the year.

In addition to salinity gradients, the nearshore to offshore distribution of dino-flagellates can be related to many parameters that include nutrients and water mass stratification. Smayda and Reynolds (2001, 2003) examined the ecology of dino-flagellates and the relationships between their living populations and abiotic parameters. They distinguished nine pelagic habitats characterized by specific life forms and assemblages, which correspond to a nearshore–offshore gradient of decreasing nutrients, reduced mixing and increasingly deeper euphotic zone. Turbulence, in particular, is an important parameter in dinoflagellate distribution because it has an inhibitory effect on dinoflagellate cyst growth and reproduction, which may vary depending on the species (e.g., Gibson & Thomas, 1995; Gibson, 2000).

Dinoflagellates are adapted to a wide range of temperatures. Although particu-larly high species diversity is observed in intertropical areas, assemblages show a relatively high number of species in polar environments, where some taxa appear tolerant toward extensive sea-ice cover (e.g., Matthiessen et al., 2005). The ability of dinoflagellates to encyst in relation to sexual reproduction, or to make temporary cysts in case of adverse conditions possibly explains their occurrence even in extremely harsh environments (e.g., Graham & Wilcox, 2000).

The feeding strategies of dinoflagellates are diverse. Many are phototrophic, with peridinin as the dominant pigment, and account for an important part of the primary production together with coccolithophorids and diatoms (e.g., Parsons, Takahashi, & Hargrave, 1984; Taylor & Pollingher, 1987). In general, dinoflagellates produce their blooms after diatoms, which have the ability to reproduce much more rapidly than dinoflagellates. Many dinoflagellate species are heterotrophic or mixotrophic, feeding on other organisms or on dissolved organic substances (e.g., Taylor & Pollingher, 1987; Gaines & Elbrachter 1987). Most families actually include both autotrophic and heterotrophic taxa (Schnepf & Elbrächter, 1999; Smayda & Reynolds, 2003). Dinoflagellates as a group belong to either to phytoplankton or to microzooplankton. They are usually recovered together in plankton samples collected in the upper 50 m (e.g., Dodge & Harland, 1991) or 100 m (e.g., Raine, White, & Dodge, 2002) of the water column. Their living depth is relatively shallow (down to the bottom of the euphotic zone) because the autotrophic taxa are dependant upon light penetration, and because the habitat of the heterotrophic species appears to be closely coupled to diatoms, on which they particularly feed, and/or to the maximum chlorophyll zone (e.g., Gaines & Elbrächter, 1987).

Dinoflagellates are mobile in the water column. They have two flagella, one around the cingulum and the other longitudinal (Figure 3), which allow swimming with a "whirling" motion at a speed ranging from a few centimeters to a few meters per hour. They use their flagella together with physiological adjustments of buoyancy to migrate vertically on a diurnal basis in the upper waters, in order to optimize their metabolic and feeding activities. Despite their ability to move vertically, dinoflagellates generally inhabit a relatively thin and shallow surface layer, especially in stratified marine environments, because most of the taxa cannot migrate across the pycnocline, which represents an important physical barrier (cf. Levandowsky & Kaneta, 1987).

The reproduction of dinoflagellates is most commonly asexual by mitosis. In bloom periods, vegetative cell division occurs at a rate of about one per day. Sexual reproduction has been observed for many species. During blooms, dinoflagellates can be responsible for "red tides," so-called because the very large number of cells in the surface water induces a color change. Some dinoflagellates are bioluminescent and cause sparkling of the sea at night. A few dinoflagellate species produce neurotoxins that may be bioconcentrated by filtering organisms, notably shellfish, which then become poisonous and dangerous for human consumption, as well as to the animals feeding on them.

3. DINOFLAGELLATES VS. DINOCYSTS AND TAPHONOMICAL PROCESSES (FROM THE BIOCENOSES TO THANATHOCENOSES)

3.1. Living vs. Fossil Dinoflagellates

The relationships between living populations and the cyst assemblages in sediment are difficult to establish for several reasons. (1) Dinocysts represent only a fragmentary

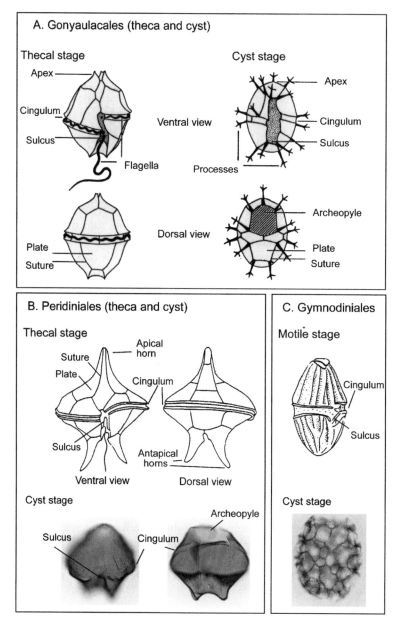

Figure 3   Schematic illustration of the main types of dinoflagellate cysts routinely recovered in marine sediments. A: Illustration of the morphology of Gonyaulacales (theca of *Gonyaulax* sp. and corresponding cyst belonging to the genus *Spiniferites*). B: Illustration of the morphology of Peridiniales (theca of *Protoperidinium* sp. and photograph of the cyst of *Quinquecuspis concreta*). C: Illustration of Gymnodiniales, which are athecate by definition (motile stage and cyst of *Polykrikos schwartzii*).

picture of the original dinoflagellate populations, as only 10–20% of the species produce cysts that can be fossilized (Dale, 1976; Head, 1996; Matthiessen et al., 2005; cf. Table 2). Amongst dinoflagellates, there are three main groups producing organic-walled microfossils: the Gymnodiniales and the Peridiniales that mostly include heterotrophic taxa, and the Gonyaulacales that are autotrophic (Figure 3, Table 2). Many of the dinoflagellate taxa producing organic-walled cysts belong to life-form types that are characteristic of nearshore and continental margin habitats (cf. Smayda & Reynolds, 2001). (2) The morphology of dinoflagellate cells and that of their cysts often differ significantly, which led to the development of different nomenclatures by biologists and paleontologists. (3) Finally, because the biological affinities of many Quaternary and most pre-Quaternary dinocysts are still uncertain (cf. Head, 1996; cf. Figure 3, Table 2 and Plates), the ecological affinities of most taxa cannot be documented easily from that of their living counterparts.

Laboratory cultures have permitted the life cycle of many dinoflagellate species to be reproduced to the motile cells starting from cysts recovered in sediment, and from living populations to the cyst stage. These studies have shown that cysts are viable many years after burial and sampling of the sediments (Lewis, Harris, Jones, & Edmonds, 1999; McQuoid, Godhe, & Nordberg, 2002), and have helped to confirm the taxonomical affinities of many taxa (e.g., Wall & Dale, 1968; Head, 1996; Ellegaard, Daugbjerg, Rochon, Lewis, & Harding, 2003). They also showed that cysts of the same motile species may differ morphologically depending on salinity and temperature (e.g., Ellegaard, Lewis, & Harding, 2002; Zonneveld & Susek, 2006).

3.2. Biogeography of Motile Cells vs. Distribution of the Cysts in Sediments

There are limited studies coupling information on planktonic dinoflagellates and fluxes of dinocysts to the sea floor. Nevertheless, it seems that the biogeographical distributions of cyst-forming dinoflagellates in surface waters and that of dinocysts in sediments are generally consistent with each other (e.g., Dodge, 1994). There is a general consistency with respect to nearshore–offshore patterns and latitudinal gradients, which closely relate to salinity and temperature controlled by current patterns (e.g., Dodge & Harland, 1991; Dodge, 1994; Raine et al., 2002). However, the detailed comparison of observations based on motile populations in the water column, sediment-trap contents, and cyst assemblages on the sea floor does not yield a perfect correspondence. This can be due in part to the fact that motile dinoflagellates sampled from the plankton correspond to an instantaneous time interval, whereas the cysts in the upper first centimeter of the sediments may represent several years, decades or even centuries of sedimentary fluxes. The analyses of traps collected throughout annual cycles illustrate seasonal patterns of cyst production, with the maximum recorded during summer, and species-specific duration of cyst production periods; they also show variations from one year to another in the overall fluxes (Montresor, Zingone, & Sarno, 1998).

Table 2   List of Main Dinocyst Taxa Recovered in Recent Marine Sediments and General Affinities.

Dinocyst species name	Motile affinity	Order
Ataxiodinium choane	*Gonyaulax* sp.	Gonyaulacales
Bitectatodinium spongium	Unknown	Gonyaulacales
Bitectatodinium tepikiense	*Gonyaulax* sp.	Gonyaulacales
Brigantedinium cariacoense	*Protoperidinium avellanum*	Peridiniales
Brigantedinium simplex	*Protoperidinium conicoides*	Peridiniales
Dalella chathamensis	*Gonyaulax* sp.	Gonyaulacales
Dubridinium caperatum	*Preperidinium meunieri*	Perdiniales
Echinidinium aculeatum	Unknown	Peridiniales
Echinidinium bispiniformum	Unknown	Peridiniales
Echinidinium delicatum	Unknown	Peridiniales
Echinidinium granulatum	Unknown	Peridiniales
Echinidinium karaense	Unknown	Peridiniales
Echinidinium transparantum	Unknown	Peridiniales
Cysts of *Gymnodinium catenatum*	*Gymnodinium catenatum*	Gymnodiniales
Impagidinium aculeatum	*Gonyaulax* sp.	Gonyaulacales
Impagidinium pacificum	*Gonyaulax* sp.	Gonyaulacales
Impagidinium pallidum	*Gonyaulax* sp.	Gonyaulacales
Impagidinium paradoxum	*Gonyaulax* sp.	Gonyaulacales
Impagidinium patulum	*Gonyaulax* sp.	Gonyaulacales
Impagidinium plicatum	*Gonyaulax* sp.	Gonyaulacales
Impagidinium sphaericum	*Gonyaulax* sp.	Gonyaulacales
Impagidinium strialatum	*Gonyaulax* sp.	Gonyaulacales
Impagidinium variaseptum	*Gonyaulax* sp.	Gonyaulacales
Impagidinium velorum	*Gonyaulax* sp.	Gonyaulacales
Islandinium brevispinosum	*Protoperidinium* sp.	Peridiniales
Islandinium? cezare	*Protoperidinium* sp.	Peridiniales
Islandinium minutum	*Protoperidinium* sp.	Peridiniales
Leipokatium invisitatum	Unknown	Peridiniales
Lejeunecysta oliva	Unknown	Peridiniales
Lejeunecysta Sabrina	*Protoperidinium leone*	Peridiniales
Lingulodinium machaerophorum	*Lingulodinium polyedrum*	Gonyaulacales
Nematosphaeropsis labyrinthus	*Gonyaulax spinifera*	Gonyaulacales
Nematosphaeropsis rigida	*Gonyaulax* sp.	Gonyaulacales
Operculodinium centrocarpum-short processes form	*Protoceratium reticulatum*	Gonyaulacales
Operculodinium short processes	*Protoceratium reticulatum*	Gonyaulacales
Operculodinium israelianum	*Protoceratium* sp.	Gonyaulacales
Operculodinium janduchenei	Unknown	Gonyaulacales
Operculodinium longispinigerum	Unknown	Gonyaulacales
Cysts of *Pentapharsodinium dalei*	*Pentapharsodinium dalei*	Peridiniales
Cysts of *Pheopolykrikos hartmannii*	*Pheopolykrikos hartmannii*	Gymnodiniales
Cyst of *Polykrikos* sp.-Arctic morphotype	*Polykrikos* sp.	Gymnodiniales
Cysts of *Polykrikos kofoidii*	*Polykrikos kofoidii*	Gymnodiniales
Cysts of *Polykrikos schwartzii*	*Polykrikos schwartzii*	Gymnodiniales
Polysphaeridium zoharyi	*Pyrodinium bahamense*	Gonyaulacales

Table 2 (*Continued*)

Dinocyst species name	Motile affinity	Order
Cysts of *Protoperidinium americanum*	*Protoperidinium americanum*	Peridiniales
Cysts of *Protoperidinium nudum*	*Protoperidinium nudum*	Peridiniales
Cysts of *Protoperidinium stellatum*	*Protoperidinium stellatum*	Peridiniales
Pyxidinopsis psilata	Unknown	Gonyaulacales
Pyxidinopsis reticulata	Unknown	Gonyaulacales
Quinquecuspis concreta	*Protoperidinium leone*	Peridiniales
Cysts of cf. *Scrippsiella trifida*	*Scippsiella trifida*	Gonyaulacales
Selenopemphix antarctica	Unknown	Peridiniales
Selenopemphix nephroides	*Protoperidinium subinerme*	Peridiniales
Selenopemphix quanta	*Protoperidinium conicum*	Peridiniales
Spiniferites alaskensis	*Gonyaulax* sp.	Gonyaulacales
Spiniferites belerius	*Gonyaulax scrippsae*	Gonyaulacales
Spiniferites bentorii	*Gonyaulax digitale*	Gonyaulacales
Spiniferites bulloideus	*Gonyaulax scrippsae*	Gonyaulacales
Spiniferites cruciformis	*Gonyaulax* sp.	Gonyaulacales
Spiniferites delicatus	*Gonyaulax* sp.	Gonyaulacales
Spiniferites elongatus	*Gonyaulax* sp.	Gonyaulacales
Spiniferites frigidus	*Gonyaulax* sp.	Gonyaulacales
Spiniferites hyperacanthus	*Gonyaulax* sp.	Gonyaulacales
Spiniferites lazus	*Gonyaulax* sp.	Gonyaulacales
Spiniferites membranaceus	*Gonyaulax* sp.	Gonyaulacales
Spiniferites mirabilis	*Gonyaulax spinifera*	Gonyaulacales
Spiniferites pachydermus	*Gonyaulax* sp.	Gonyaulacales
Spiniferites ramosus	*Gonyaulax* sp.	Gonyaulacales
Stelladinium reidii	Unknown	Peridiniales
Stelladinium robustum	Unknown	Peridiniales
Tectatodinium pellitum	*Gonyaulax spinifera*	Gonyaulacales
Trinovantedinium applanatum	*Protoperidinium pentagonum*	Peridiniales
Trinovantedinium variabile	Unknown	Peridiniales
Tuberculodinium vancampoae	*Pyrophacus steinii*	Gonyaulacales
Votadinium calvum	*Protoperidinium oblongum*	Peridiniales
Votadinium spinosum	*Protoperidinium claudicans*	Peridiniales
Xandarodinium xanthum	*Protoperidinium divaricatum*	Peridiniales

Notes: Grey highlighting: heterotrophic or mixotrophic taxa; no highlighting: autotrohic taxa.

3.3. Sedimentation

Part of the differences observed between plankton populations and microfossil contents of deep-sea sediments could be attributed to some lateral transport. The sinking rate of microscopic remains such as dinocysts, or any small size microfossil, is difficult to estimate, but the settling of individual cells or cysts is slow (in the order of meters per day). Incorporation within fecal pellets and marine "snow" particles, which consist of cohesive aggregates of microscopic organic material having a density high enough to sink in the water column, are mechanisms likely to explain the vertical fluxes of pelagic particles over hundreds or thousand of meters (see review by Turner, 2002). The dinocyst assemblages on the sea floor are most probably related to production in the upper part of the water column above the site location, and to subsequent sedimentation within marine snow and fecal pellets. A few trap studies in high productivity areas, comparing assemblages at different depths, indeed suggest that vertical sedimentation of dinocysts is very rapid and that lateral transport is limited (e.g., Zonneveld & Brummer, 2000). However, some lateral transport with surface, intermediate or bottom currents cannot be ruled out, especially in areas of low productivity where any reworking or lateral input can weigh significantly on the "local" assemblages.

3.4. Preservation

Dinocysts are made of highly resistant organic polymers (e.g., Kokinos et al., 1998; Versteegh & Blokker, 2004) and are usually preserved very well in marine sediments, unlike diatoms or foraminifera that are prone to dissolution of opal silica or calcium carbonate, respectively. From this point of view, dinocysts represent an extremely useful proxy of ocean changes in regions of the world's oceans where calcium carbonate dissolution occurs (because of a shallow lysocline and oxidation of organic matter-rich sediments that create a high pCO_2). However, it has been noticed that some taxa may be affected by strong oxidation: laboratory treatments show that acetolysis or strong oxidants may selectively destroy some taxa, in particular *Protoperidinium* species (e.g., Dale, 1976; Marret, 1993; Brenner & Biebow, 2001). In a field experiment, Zonneveld, Versteegh, and deLange (1997, 2001b) classified a number of taxa according to their sensitivity to oxygen availability in bottom waters (Table 3). Selective degradation of the cyst wall of some dinocyst taxa could be a limitation, especially in areas characterized by oxygen-rich bottom waters and low sedimentation rates, preventing rapid burial of the organic matter.

4. RELATIONSHIPS BETWEEN DINOCYST ASSEMBLAGES AND SEA-SURFACE PARAMETERS

Despite uncertainties about the biological affinities of organic-walled dinoflagellate cysts and caveats concerning lateral dispersion and preservation, the distribution of dinocyst assemblages on the sea floor is illustrative of relationships with hydrographical parameters and productivity in surface waters.

On a qualitative or quantitative basis, the species composition of assemblages appears informative at different levels: sea-level changes, biogeography, sea-surface conditions (sea-ice cover duration, temperature, salinity and productivity), hydrological fronts, etc. The pioneering work of Wall et al. (1977) enabled us to view dinocysts as potential paleoceanographical proxies, prompting a number of paleoenvironmental studies. Consequently, numerous studies based on the present-day distribution of dinocysts have allowed for the definition of environmental indices.

Table 3 Organic-Walled Cysts Grouped in Relation to Their Sensitivity to Oxygen Availability in Bottom Waters.

Sensitivity to oxygen	Organic–walled cysts
Extremely sensitive	Cysts from *Protoperidinium* species *Echinidinium* species
Moderately sensitive	*Lingulodinium machaerophorum* *Operculodinium centrocarpum* *Pyxidinopsis reticulata* *Spiniferites* species (*S. bentorii, S. mirabilis, S. pachydermus, S. ramosus*) *Nematosphaeropsis labyrinthus*
Resistant	*Impagidinium* species (*I. aculeatum, I. paradoxum, I. patulum, I. plicatum, I. Sphaericum*) *Operculodinium israelianum* Cysts of *Pentapharsodinium dalei* *Polysphaeridium zoharyi*

Source: Data collected from Zonneveld et al. (2001b).

In the early 1980s, Turon (1980) and Harland (1983) were the first to produce distribution maps of dinocysts on the sea-floor. Harland (1983) published the relative abundance of 42 taxa for the North Atlantic Ocean and adjacent seas, initiating an international interest. Since the 1980s, numerous studies from around the world have permitted the definition of distribution patterns at regional levels, and have led to the determination of relationships between cyst assemblages and sea-surface conditions (e.g., Matsuoka, 1985; Baldwin, 1987; Bint, 1988; Dodge & Harland, 1991; Edwards & Andrle, 1992; Mudie, 1992; de Vernal et al., 1994; Sun & Mcminn, 1994; Ellegaard, Christensen, & Moestrup, 1994; Marret, 1994a; Matthiessen, 1994; Mudie & Harland, 1996; Marret & de Vernal, 1997; Sonneman & Hill, 1997; Zonneveld, 1997; Harland et al., 1998; Rochon et al., 1999; Godhe, Karunasagar, Karunasagar, & Karlson, 2000; Persson, Godhe, & Karlson, 2000; Vink et al., 2000; Grosfjeld & Harland, 2001; Mudie & Rochon, 2001; Zonneveld et al., 2001a; Cho, Kim, Moon, & Matsuoka, 2003; Debenay et al., 2003; Morquecho & Lechuga-Devèze, 2003; Azanza et al., 2004; Kawamura, 2004; Orlova, Morozova, Gribble, Kulis, & Anderson, 2004; Radi & de Vernal, 2004; Sangiorgi, Fabbri, Comandini, Gabbianelli, & Tagliavini, 2005; *cf.* Table 1).

4.1. Nearshore vs. Offshore Distribution, Sea-Level and Continentality Indices

The study of Wall et al. (1977) illustrated a biogeographic zonation reflecting the distance to the shore based on the identification of inner neritic, outer neritic and oceanic taxa. Thus, the species composition of assemblages and the relative abundance of taxa have been used as indicators of sea level, assuming that lateral and down-slope transports are minimal and that preservation bias is negligible. The

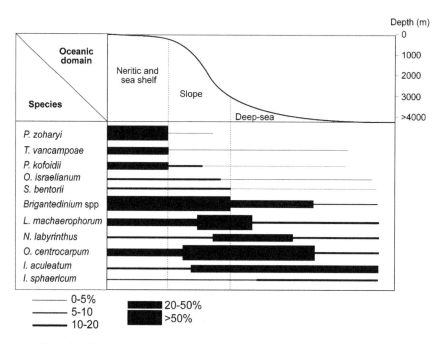

Figure 4   Relative abundance vs. water depth of the most common dinocyst taxa in temperate to equatorial regions (40°N to 40°S). Data compiled from Dale et al. (2002), Marret and Zonneveld (2003), and de Vernal et al. (2005).

schematic profile in Figure 4 clearly shows that some taxa have a specific distribution in the oceanic domain, hence they are commonly used to assess sea-level changes and to make biogeographical reconstructions (e.g., Morzadec-Kerfourn, 1992; Harland & Long, 1996; Marret, Scourse, & Austin, 2004b). However, the relationship between sea level and dinocyst assemblages is indirect, and reflects a combination of many parameters that often characterize nearshore–offshore gradients, notably the turbulence, water mass stratification and nutrient availability. Because these gradients are changing along frontal zones and upwellings, dinocysts should be used cautiously as a sea-level indicator.

Dinocysts are relatively abundant in estuarine environments, which are generally characterized by stratified waters and strong salinity gradients because of dilution with freshwater discharge. Estuarine regions play an important role in the global carbon cycle as they procure nutrient enrichment that triggers high primary productivity. In low latitude environments, deep-sea fans are ideal archives for past environments and land–ocean interaction records. River discharge events in past deep-sea sediments can be appraised by: (1) increase of freshwater algae flux (i.e., *Pediastrum, Botryococcus*); (2) increase of neritic or lagunal species (such as *Tuberculodinium vancampoae* and *Polysphaeridium zoharyi*) in the case of West African sequences; (3) higher ratio of the terrestrial vs. marine palynomorph content; (4) occurrence peaks of old palynomorphs reworked from sedimentary outcrops. Pollen grains are commonly found in marine sediments. Along continental margins, pollen concentration is high

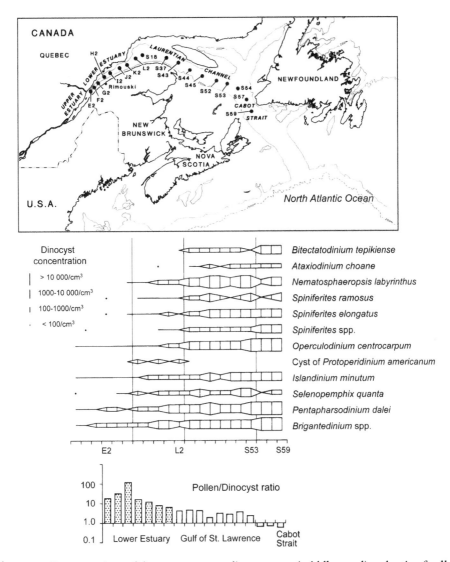

Figure 5 Concentrations of the most common dinocyst taxa (middle panel) and ratio of pollen vs. dinocyst concentration (lower panel) from the lower St. Lawrence Estuary to the Cabot Strait at the edge of the Atlantic Ocean, where sea-surface salinity in August ranges from 24 (E2) to 31 (S59). The figure is redrafted with permission from de Vernal and Giroux (1991).

and decreases rapidly toward open oceanic environments. Transport mechanisms involve eolian and fluvial processes, and pollen flux is often considered to be an in-dicator of the dynamics of the transport mechanism (e.g., Melia, 1984; Hooghiemstra & Agwu, 1986; Dupont & Agwu, 1991; Dupont, 1999; Dupont & Wyputta, 2003). The (pollen+spores)/dinocyst (or P/D) ratio enables the qualification of the origin of the organic flux, i.e. terrestrial vs. marine, with values tending to zero toward the open ocean (de Vernal & Giroux, 1991; see Figure 5). This "continentality" index reflects the proportion of terrestrial microfossils transported by rivers or in the atmosphere, although production at the source, storage and reworking, differential preservation of both terrestrial and marine palynomorphs may also influence this ratio.

4.2. Salinity

Dinocysts are found from freshwater to hypersaline environments, with only a few species occurring within a restricted salinity range (e.g., Wall et al., 1977; Dodge & Harland, 1991; de Vernal et al., 1994; Dale, 1996; Marret & Zonneveld, 2003). Oceanic taxa, such as *Impagidinium*, are only found where fully saline conditions occur, whereas most of the taxa seem to be euryhaline. However, brackish environments, such as the Baltic Sea, the Black sea and other Central Asian seas, contain assemblages characterized by a high morphological variability and very low diversity (less than eight taxa) (Dale, 1996; Mudie, Aksu, & Yasar, 2001a; Mudie, Harland, Matthiessen, & de Vernal, 2001b; Marret, Leroy, Chalie, & Gasse, 2004a). The cruciform shape of the cyst body typical in these regions, or the process length, had been initially attributed to low salinity or environmental stress by Wall and Dale (1973). Mudie, Rochon, Aksu, and Gillespie (2002) attempted to associate morphotypes of *Spiniferites cruciformis* with a specific salinity range; although the relationship was unclear, they observed that extreme morphotypes (i.e., the most and the least developed processes) seemed to be dominant at the extremities of the salinity range. Lewis and Hallett (1997) documented that *L. machaerophorum*, a euryhaline species, may develop shorter processes with low saline conditions, although all shapes have been found with a constant salinity. Other species, such as *Operculodinium centrocarpum* (Nehring, 1994, 1997) or cysts of *Gonyaulax baltica* (Ellegaard et al., 2002), also present considerable variations in process length that have been attributed to low saline conditions. However, the relationship between variations of processes and salinity is not clearly demonstrated on a quantitative basis for many species, and other environmental factors such as turbulence, temperature or density may affect the development of the cysts and their morphological attributes (Kokinos & Anderson, 1995; Zonneveld & Susek, 2006; Lewis, personal communication).

In areas characterized by a large amplitude gradient of salinity such as estuaries and continental margins, a relationship can be established between the assemblages of dinocysts and sea–surface salinity. This can be shown on a qualitative basis, in the Estuary and Gulf of St. Lawrence for example (*cf.* de Vernal & Giroux, 1991; Figure 5). This has been illustrated using multivariate analyses, in the northwest North Atlantic, for example (e.g., de Vernal et al., 1994; Rochon et al. 1999). However, although quantitative relationships between dinocyst assemblages and salinity are clear in some marine environments, they cannot be extrapolated unequivocally on a hemispheric or global scale.

4.3. Sea-Surface Temperature and Seasonality

Dinocysts are found in marine sediment from polar to tropical environments. In general, the diversity of species decreases from the tropics to the poles, although the dinocyst concentration can be very high in Arctic seas and subpolar seas (e.g., de Vernal et al., 2001, 2005a). More than 60 taxa have been recorded in the Northern Hemisphere, but only 10 to 12 taxa are common in the Arctic seas. They include opportunistic or ubiquitous taxa, such as *Operculodinium centrocarpum* and *Brigantedinium* spp. They also include a few taxa that seem adapted to particularly cold conditions (notably *Impagidinium pallidum*, cyst of *Polykrikos* sp. Arctic morphotype, *Islandinium* spp.). In the southern hemisphere also, there is a gradient of decreasing diversity toward the pole where a few taxa seem to characterize cold conditions (*Selenopemphix antarctica* in particular; Marret & de Vernal, 1997). From high to low latitudes, the increase in the number of species is related to increased temperature (Figure 6). The relative abundance of many taxa also varies in relation with sea-surface temperatures (SST), either annually or seasonally (Edwards & Andrle, 1992; Marret, 1994a; Marret & Zonneveld, 2003). Multivariate analyses further support the determining influence of SST in taxa and assemblage distribution, both at regional or hemispheric scales (e.g., Marret, 1994a; Marret & Zonneveld, 2003; Marret et al., 2001a; Rochon et al., 1999; de Vernal et al., 1994, 1997, 2001, 2005a; Radi & de Vernal, 2004).

The existence of a relationship between SST and dinocyst assemblages is unquestionable. However, this relationship is most probably season-dependent. In high latitudes, the bloom of dinoflagellates, which is often followed by encystment, most frequently occurs during summer after the diatom bloom (e.g., Matthiessen et al., 2005). Therefore, it can be assumed that dinocyst assemblages are mainly related to summer SSTs. However, dinoflagellates and their cyst population also appear dependent upon the temperature changes over the annual cycle. The overall life cycle of dinoflagellates and cyst production can take place over a few weeks to over a few months, depending upon the taxon. In polar seas where the ice-free season is very short, only the species having the ability to form cysts in a short time can develop. The seasonal constraint is low in intertropical environments, but possibly plays an important role on determining the abundance of taxa in temperate regions. Actually, the distribution of dinocyst assemblages in the North Atlantic and adjacent subpolar and polar seas suggests a relationship between dinocyst assemblages and the seasonality, as expressed by the difference between the coldest and warmest months (e.g., de Vernal et al., 1994, 2001; Rochon et al., 1999). For example, some taxa, such as *Bitectatodinium tepikiense*, apparently require a high summer temperature but tolerate freezing winter conditions, whereas many other thermophilic taxa do not tolerate a wide amplitude temperature gradient from winter to summer (e.g., *Impagidinium aculeatum* or *Spiniferites mirabilis*). Seasonality is also probably a parameter that plays a major role in dinoflagellate distribution, ecology and cyst production (Figure 7).

4.4. Sea-Ice Cover

Biogeographical distribution of dinocysts shows that some taxa have affinities for polar environments, and some may be used as sea-ice indicators. In their study, Matthiessen et al. (2005) presented a general review of Arctic taxa and their ecological significance. Only two cyst-forming species are known to dwell in the

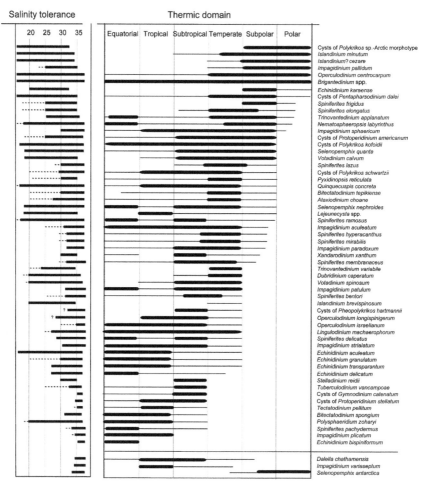

Figure 6 Diagram showing the known distribution of the main dinocyst taxa in surface sediments according to bioclimatic domains and their salinity tolerance. The compilation was made from Marret and Zonneveld (2003), Radi and de Vernal (2004) and de Vernal et al. (2005a). Note that *Dalella chathamensis*, *Selenopemphix antarctica* and *Impagidinium variaseptum* are taxa exclusively reported to occur in the Southern Ocean. A few taxa also seem to occur exclusively in middle- to high latitudes of the northern Hemisphere: *Polykrikos* sp. — Arctic morphotype, *Spiniferites elongatus*, *Spiniferites frigidus*, and *Trinovantedinium variabile*.

ice-pack environment, *Polarella glacialis* and *Peridiniella catenata*. However, their cysts were not recovered in sediment. The sea-floor sediments of areas characterized by multiyear perennial pack-ice are usually barren of dinocysts (e.g., Rochon et al., 1999; de Vernal et al., 2005a). Nevertheless, there are a few dinocyst taxa that are known to occur in sediments of areas marked by seasonal sea-ice. In both hemispheres, and especially in Arctic and subarctic seas, *Islandinium* species are often abundant (e.g., Pienkowski-Furze, 2004; de Vernal et al., 2005a). *Selenopemphix antarctica* is also characteristic of seasonal sea-ice cover, but its occurrence is

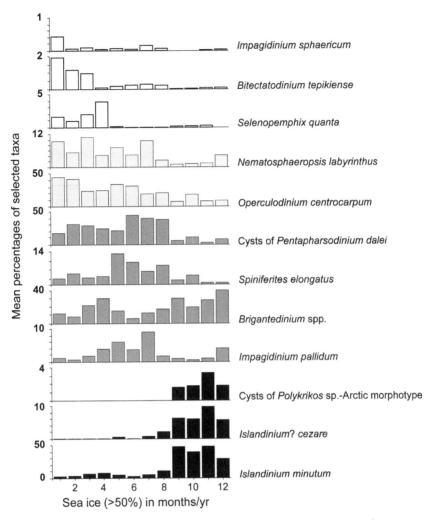

Figure 7   Mean percentage of selected dinocyst taxa vs. sea-ice extent in months per year with more than 50% of ice concentration. The compilation was made from the North Atlantic and Arctic Ocean database (*cf.* de Vernal et al., 2001).

restricted to the circum Antarctic Ocean (Marret & de Vernal, 1997; de Vernal et al., 2001). Similarly, the cyst of *Polykrikos* sp. — Arctic morphotype exclusively occurs in ice-covered marine environments of the Northern Hemisphere. Many other taxa appear tolerant to sea-ice cover and may occur in high proportions in sea-ice environments. This is particularly the case of the ubiquitous taxa *Operculodinium centrocarpum* and *Brigantedinium* spp. Other taxa such as *Pentaphar-sodinium dalei*, *Spiniferites elongatus-frigidus* and *Impagidinium pallidum* have affinities for subarctic environments and often characterize areas of winter sea-ice (e.g., de Vernal et al., 2001; Figure 7).

4.5. Productivity, Upwelling and Polynyas

Dinocyst assemblages that include both phototrophic and heterotrophic taxa seem to reflect the trophic characteristics of the upper water mass (e.g., Devillers & de Vernal, 2000). Paradoxically, high productivity areas such as upwellings or polynyas are usually characterized by the dominance of heterotrophic dinocysts (e.g., Lewis, Dodge, & Powell, 1990; Hamel et al., 2002; Radi & de Vernal, 2004). This can be attributed to competition among primary producers, diatoms being dominant in high nutrient contexts because they reproduce at a much faster rate (up to eight cell divisions per day) than dinoflagellates (about one division per day). Therefore, in upwellings and polynyas, diatoms are advantaged in as much as nutrients are not limiting. In these environments, heterotrophic dinoflagellates feeding on diatoms can be very abundant. The main dinocyst taxa commonly used for reconstructing the strength of upwelling regimes belong to the heterotrophic *Protoperidinium* group. The ratio Protoperidiniaceae/Gonyaulacaceae (P/G) was first documented by Lewis et al. (1990) to illustrate changes in the upwelling intensity along the Peruvian margins. This ratio has been broadly used and adapted, with the inclusion of heterotrophic species such as *Polykrikos* (Marret, 1994a) or *Echinidinium* (Esper, Versteegh, Zonneveld, & Willems, 2004) species, to look into past variations of upwelling systems along the west African margin (Susek, Zonneveld, Fischer, Versteegh, & Willems, 2005).

Over the last decade, a number of studies have documented qualitatively and quantitatively the cyst distribution in recent sediments from many high productivity regions such as NW European shelves (Dale, 1996; Nehring, 1997; Grosfjeld & Harland, 2001; Marret & Scourse, 2002; Marret et al., 2004b), Arctic and subarctic areas (de Vernal et al., 2001; Hamel et al., 2002), and coastal upwelling areas from the west African margin, NW America, the Peruvian margin and the Arabian Sea (e.g., Marret, 1994b; Blanco, 1995; Biebow, 1996; Zonneveld, 1997; Zonneveld et al., 2001a; Dale, Dale, & Jansen, 2002; Marret & Zonneveld, 2003; Radi & de Vernal, 2004; Sprangers, Dammers, Brinkhuis, Van Weering, & Lotter, 2004; Joyce, Pitcher, du Randt, & Monteiro, 2005; Susek et al., 2005). In general, dinocyst assemblages from high productivity environments are dominated by heterotrophic taxa (Figure 8). However, the species composition of assemblages differs from one site to another in relation to latitude (Table 4), and shelf areas from the northeast North Atlantic and North Pacific that are not under upwelling influence contain high proportions of autotrophic gonyaulacoid taxa (*Operculodinium centrocarpum*, *Spiniferites* species), accompanied by cysts of *Pentapharsodinium dalei* (e.g., Voronina et al., 2001; Marret & Zonneveld, 2003; Radi, Pospelova, de Vernal, & Vaughn, 2007).

The relationship between productivity and dinocyst assemblages has been analyzed statistically using regional data sets (e.g., Devillers & de Vernal, 2000; Radi & de Vernal, 2004, 2006). The results suggest that quantitative reconstruction would be feasible with an accuracy comparable to the uncertainty of estimates from satellite observations. To date, however, only qualitative estimates of past productivity are available based on dinocyst fluxes, P/G ratios or percentages of indicator species. For example, studies of marine sequences collected from high productivity regions of the South Atlantic Ocean have allowed the identification of periods of

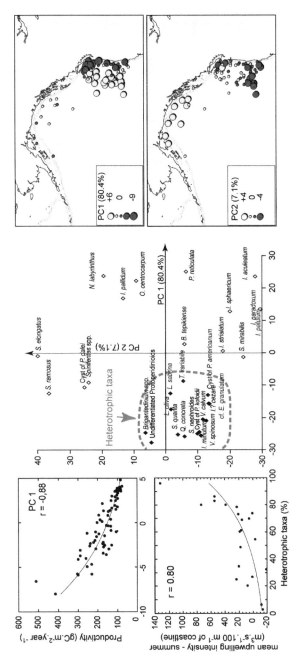

Figure 8 Illustration of the relationships between dinocyst assemblages, heterotrophic taxa, productivity and upwelling in the northeast North Pacific (with permission from Radi & de Vernal, 2004). Right: Maps showing the geographical distribution of principal component 1 and 2 (PC1 and PC2) scores (right panels). Middle: Ordination of taxa on the axes of PC1 and PC2; white diamonds represent cysts of autotrophic taxa and black diamonds represent cysts of heterotrophic taxa. Left: Relationship between the percentages of heterotrophic taxa and productivity and mean upwelling intensity (UI) in summer (average over June, July, August and September; UI can be generated only at coastal sites; the coefficient of correlation is calculated based on a logarithmic relationship). Relationship between PC1 and the annual primary productivity (the coefficient of correlation is calculated based on a logarithmic relationship).

Table 4 Classification of Dinocyst Taxa in Upwelling and River Induced-Upwelling Systems of the Eastern Sides of the Atlantic, Pacific Oceans and Arabian Sea Based on Their Relative Abundance in Recent Sediments.

Hydrological context	Region	Taxa
Seasonal coastal upwelling	Northeastern Pacific	*Brigantedinium* spp., *O. centrocarpum*, Cysts of *P. dalei*, *P. reticulata*, *S.* quanta, *S. nephroides*
	NW Africa and Iberia	*L. machaerophorum*, *Protoperidinium* spp., *P. schwartzii*
	Somalia/Arabian Sea	*Brigantedinium* spp. with *Echinidinium* spp.
	West equatorial Africa	*Brigantedinium* spp., *S. nephroides*, *P. kofoidii*, *L. machaerophorum*, *O. israelianum*, *N. labyrinthus*, *S. delicatus*, *S. ramosus*, *S. membranaceus*
Permanent coastal upwelling	SW Africa (Benguela)	*Brigantedinium* spp. with *Echinidinium* spp., *O. centrocarpum*, *P. americanum*, *S. quanta*, *N. labyrinthus*, *S. pachydermus*
River-induced upwelling with cells of coastal upwelling	Off Congo River	*O. centrocarpum*, *S. delicatus*, *Brigantedinium* spp.
	Off SW Africa	*Brigantedinium* spp., with *Echinidinium* spp., *O. israelianum*, *P. kofoidii*, *P. schwartzii*, *P. americanum*, *S. quanta*, *T. applanatum*

Source: Data collected from Zonneveld et al. (2001a), Dale et al. (2002), Marret and Zonneveld (2003), Radi and de Vernal (2004), Sprangers et al. (2004).

enhanced upwelling regime concurrent with increased atmospheric circulation strength during glacial stages (e.g., Marret, 1994b; Dupont, Marret, & Winn, 1998; Holl, Zonneveld, & Willems, 2000). An example from the east equatorial Atlantic Ocean is illustrated in Figure 9. It shows the similarity between the cyst concentration and flux variability, and the export productivity based on TOC and foraminiferal assemblages over the last climatic cycle (Struck, Sarnthein, Westerhausen, Barnola, & Raynaud, 1993; Marret, 1994b).

4.6. Environmental Quality and Eutrophication

Increasing numbers of studies are using dinocysts to characterize eutrophication and/or human-induced impacts on fjords, harbors, or bays (Lewis & Hallett, 1997;

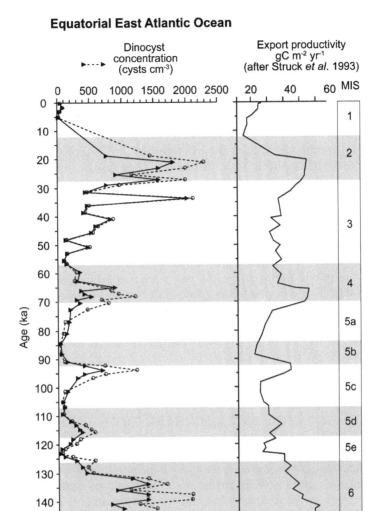

Equatorial East Atlantic Ocean

Figure 9 Comparison of dinocyst concentration and flux curve with quantified export productivity from site GIK16772, Gulf of Guinea (Marret, 1994b).

Thorsen & Dale, 1998; Dale, Thorsen, & Fjellså, 1999; Matsuoka, 1999; Dale, 2001; Dale & Dale, 2002; Pospelova, Chmura, Boothman, & Latimer, 2002; Harland, Nordberg, & Filipsson, 2004; Sangiorgi & Donders, 2004; Brenner, 2005). In particular, increased abundance of L. *machaerophorum* in sediments has been used as an indicator of eutrophication in European fjords and lochs, in association with increased levels of nutrients due to human activities. As shown above, it seems clear that dinocyst assemblages reflect the trophic characteristics of surface waters. However, whether they allow us to distinguish human-induced eutrophication from natural variations in productivity is another matter, and one which would certainly require more investigation.

Beyond the relationship between productivity and dinocyst assemblages, some studies have examined the links that could be established with nutrients, including phosphate, silica and nitrate (Devillers & de Vernal, 2000; Radi et al., 2007). The analyses suggested some relationships between distribution of dinocysts and nutrients, especially nitrate. However, the relationships are regional and cannot be extrapolated on a hemispheric scale. Moreover, they remain unclear due to the interdependency of nutrient content in water and hydrographical parameters.

4.7. Red Tides and Harmful Algal Blooms (HAB)

Although HAB and red tides are natural phenomena that have occurred throughout geological times, their increased frequency over the last two decades has become a concern for the public health and economic resources (Hallegraeff, Anderson, & Cembella, 2003). The bloom-species group is dominated by dinoflagellates, with approximately 200 taxa (or 10% of the total number) creating toxic blooms or red tides (Smayda & Reynolds, 2003), and of these only 3% are reported to be harmful (60 taxa). The majority seems to be photosynthetic, with the exception of *Noctiluca* and *Dinophysis*. Red tides usually occur when one non-toxic bloom-forming species proliferates in huge numbers (up to 20 million cells per liter), generating anoxic conditions and a discoloration of water (red or brown). For instance, a luminescent red tide is caused by *Lingulodinium polyedrum* (cyst producer). During HABs, toxins are released and affect at different levels the marine ecosystem and the food web (see Table 5).

An explanation for a global increase in HABs has been attributed to four main causes: (1) increased scientific awareness of toxic species; (2) increased use of coastal waters for fisheries and shellfish farms; (3) occurrence of algal blooms due to cultural eutrophication and/or unusual climatic conditions; (4) transport of cysts from toxic species either in ship ballast or from translocation of shellfish stocks from one region to another one (Hallegraeff et al., 2003).

The viability of cysts in sediments raises the question of "seed" banks that could generate toxic blooms. Laboratory experiments and field observations document that some cyst taxa may still germinate after a number of years. For instance, McQuoid et al. (2002) managed to generate a viable population of *Lingulodinium polyedra* from 20- to 55-year old cysts collected in the Koljö Fjord (Sweden). Mizuchima and Matsuoka (2004) successfully germinated cysts of *Alexandrium* species that were at least eight years old. Monitoring eutrophication and/or bloom events in coastal regions and estuaries would require a survey of dinocyst records from high sedimentation rate cores covering the last few decades.

Among HAB taxa, only a few species produce organic-walled cysts that are recovered in sediment, and very few studies have thus provided evidence of past "red tides." Thorsen and Dale (1998) suggested that the increase in cysts of *Gymnodinium catenatum* is related to climate warming in Skagerrak-Kattegat.

Table 5 List of Some Dinoflagellate Species that Produce Toxins.

Paralytic shellfish poisoning (PSP)	Diarrhetic shellfish poisoning (DSP)	Neurotoxic shellfish poisoning (NSP)	Ciguatera fish poisoning
Alexandrium catenella[1]	*Dinophysis acuminata,*	*Karenia brevis* (Florida)	*Gambierdiscus toxicus*
Alexandrium minutum[1]	*Dinophysis acuta*	*K. papilionaceae*	*Ostreopsis siamensis*
Alexandrium tamarense[1]	*Dinophysis fortii*	*K. selliformis*	*Coolia monotis*
Gymnodinium catenatum[1]	*Dinophysis norvegica*		
Pyrodinium bahamense[1]	*Prorocentrum lima*	*K. bicuneiformis* (New Zealand)	

Note: Data collected from Hallegraeff et al. (2003). It is of note that the only species that yield known dinocysts in the sedimentary record with unquestionable taxonomical affinities are *Gymnodinium catenatum* and *Pyrodinium bahamense* (cyst name *Polysphaeridium zoharyi*).
[1] Cyst-forming species.

5. THE DEVELOPMENT OF QUANTITATIVE APPROACHES FOR THE RECONSTRUCTION OF HYDROGRAPHIC PARAMETERS BASED ON DINOCYSTS

5.1. A Brief History

Dinocysts can be used as proxies for various environmental parameters because the species occurrence, taxa abundance and assemblage distribution show relationships with sea–surface temperature, seasonality, salinity, sea–ice cover, as well as with productivity, and possibly nutrients (*cf.* Section 4). On this basis, quantitative approaches were developed or adapted for the reconstruction of past oceanographical conditions using dinocyst data. The earliest attempts for quantitative reconstructions of paleoceanographical conditions based on dinocyst data were made by Edwards, Mudie, and de Vernal (1991), Mudie (1992) and Edwards and Andrle (1992). They were based either on indices (for example, the relative abundance of thermophilous *Impagidinium* species vs. more ubiquitous taxa) or on the Imbrie and Kipp (1971) transfer function. These early studies were done using compilations of the few dinocyst data available in the literature. They showed the potential of dinocysts for estimating past oceanographical parameters and proposed regressions for the calculation of winter temperature, summer temperature or salinity (Edwards et al., 1991; Mudie, 1992). The development of standardized databases with respect to laboratory procedures and taxonomy (e.g., de Vernal et al., 1994, 1997) made it possible to apply the best analogue method of Guiot (1990), which allows the simultaneous reconstruction of several oceanographical parameters. The best analogue approach was then used to quantitatively reconstruct sea–surface temperature in summer and winter, salinity and sea–ice cover for several late Quaternary sequences of high latitudes, in the North Atlantic (e.g., de Vernal, Guiot, & Turon, 1993; de Vernal et al. 1994, 2001; de Vernal, Hillaire-Marcel, & Bilodeau, 1996; Rochon, de Vernal, Sejrup, & Haflidason, 1998; Eynaud et al., 2002; Levac, 2001),

Arctic and subarctic seas (Levac & deVernal, 1997; Levac, de Vernal, & Blake, 2001; Voronina et al., 2001; de Vernal et al., 2005a; de Vernal, Hillaire-Marcel, & Darby, 2005b), the northeastern North Pacific (de Vernal & Pedersen, 1997; Marret, de Vernal, Pedersen, & McDonald, 2001b) and of the Southern Ocean (Marret et al., 2001a). Alternative methods based on regression techniques (e.g., de Vernal et al., 1994; Marret et al., 2001a) or neural network approaches (e.g., Peyron & de Vernal, 2001) were also tested and have yielded comparable results. The advantages and limitations of each transfer function approach used for quantitative paleoceano-graphic reconstruction are discussed in depth by Guiot and de Vernal (Chapter XIII, this volume). Reference dinocyst databases are under development and various approaches are currently being tested, not only to reconstruct SSTs, salinity and sea-ice, but also primary productivity.

5.2. A Few Caveats

One of the advantages of dinocyst compared to other paleoceanographic proxies is the possibility of simultaneously reconstructing various parameters, such as salinity or sea-ice in addition to temperature. This is possible because of the wide range of dinoflagellate occurrence with respect to hydrographic conditions, and because the reference databases include data from epicontinental and estuarine settings in addition to open ocean, which results in various combinations of salinity vs. temperature, or summer vs. winter temperature. However, in some marine environments, co-variance does occur among the above-mentioned parameters. In the open North Atlantic Ocean, there is generally co-variance of seasonal SSTs and of SST vs. salinity. Similarly, there is a relationship between sea-ice cover and SSTs. Also, large differences between summer and winter SSTs often characterize near-shore areas, epicontinental seas and estuaries as the result of low thermal inertia in surface waters due to the stratification of a buoyant low salinity surface layer. Therefore, depending upon the marine setting, there is interdependency between hydrographic parameters (salinity, stratification, seasonal SSTs) and it is not always obvious to assess which is the one that is the most determinant and best recon-structed. Nevertheless, the overall reconstructions from dinocysts provide an in-tegrated picture of the diverse hydrographical conditions, including situations that are not represented by proxies occurring exclusively in open ocean environments.

Another caveat concerns the influence of nutrient availability and the structure of plankton populations on the dinoflagellate and cyst assemblages, which could possibly interfere in the relationships with hydrographical conditions. As shown by assemblages in polynyas and upwelling areas (cf. Section 4.5), there is a relationship between dinocyst assemblages and productivity, which can be the most determinant parameter at regional scales. Moreover, it should be stressed that the differences in the assemblages of the Pacific and Atlantic Oceans could be due to differences with respect to nutrient limitations.

Because of the above mentioned caveats, regression-based techniques such as the Imbrie and Kipp (1971) method or artificial neural networks have to be used with much caution. The relationship between assemblages and a given parameter, and the equation describing this relationship, may differ significantly depending upon the initial calibration data set. The problem is not as acute with the modern analogue technique, which does not assume any relationship but simply supposes that identical assemblages should be the result of identical abiotic and biotic conditions. However, by using the modern analogue technique, spatial autocorrelation may possibly result in an underestimation of the prediction error (Telford, 2006; see discussion by Guiot and de Vernal, Chapter XIII, this volume).

The analogue techniques and their variants appear more appropriate than calibration approaches. However, they can also yield ambiguous results. One potential problem concerns false analogues, notably when assemblages are characterized by the dominance of ubiquitous taxa. This is why the accompanying taxa, which often have more narrow environmental tolerances than ubiquitous taxa, are more heavily weighted through logarithmic transformation prior to statistical comparison of assemblages (cf. de Vernal et al., 2001). Another issue concerns non-analogue situations, which are relatively frequent when dealing with some interval of the past, such as the LGM (cf. de Vernal et al., 2005a). In such cases, the distance between the best modern analogue and the fossil assemblages can be used to define a confidence level to estimate the reliability or quality of reconstructions.

6. THE USE OF DINOCYSTS IN PALEOCEANOGRAPHY

The use of dinocysts in paleoceanography is relatively recent. Until the mid-1960s, the biological affinities of dinocysts were unknown. Dinocysts were associated with hystrichospheres. They were used as biostratigraphical markers but virtually ignored in the fields of paleoecology and paleoceanography. Moreover, the study of dinocysts is usually more time consuming than that of other proxies, such as planktonic foraminifera because of treatments involving repeated chemical (HCl, HF) and mechanical (wet sieving) manipulations prior to any observation. It should also be mentioned that the taxonomy of modern dinocysts is still under development, with new species and new morphotypes being regularly described during the course of surface sediment analyses for the development of reference databases.

During the 1980s and 1990s, the development of modern databases led to an improvement in the understanding of dinocyst distribution in relation to environmental parameters, which permitted their use as proxies in paleoceanography. As mentioned above, and further illustrated by a few examples below, dinocysts appear useful in documenting various aspects of changes in the ocean, notably productivity variations at low latitudes and sea-surface temperature, salinity and sea-ice at high latitudes.

In the tropical and equatorial domains of the Atlantic Ocean, dinocyst studies in late Quaternary cores have provided insight into the variability of a number of past sea-surface parameters in relation to climate changes. Periods of enhanced oceanic upwelling, in particular during glacial stages, were characterized by increased productivity and a higher abundance of dinocysts (see Figure 9). Along the SW African coasts, dinocyst assemblages in marine cores have yielded evidence for a coupling between sea-level changes and nutrient-enrichment induced by coastal upwellings and/or river discharges (Marret, 1994b; Holl et al., 2000; Vink et al., 2001; Dupont et al., 1998; Marret, Scourse, Versteegh, Jansen, & Schneider, 2001c; Shi, Schneider, Beug, & Dupont, 2001).

In the Southern Ocean, investigations of dinocysts allowed for the reconstruction of past SST in the Chatham Region, notably at DSDP Site 594, which is a well-studied late Quaternary record off the coast of New Zealand (e.g., Marret et al., 2001a; Wells & Okada, 1997). This record has provided evidence for large amplitude SSTs changes following the glacial–interglacial cycles based on several proxies, including foraminiferal assemblages. At this site, the dinocyst-based SST reconstructions yielded results coherent with those obtained using planktonic foraminifera (Figure 10), with, however, larger amplitude of temperature variations. Dinocyst-based estimates tend to indicate colder conditions during the isotope stage 2 and higher SSTs during isotope stage 5 than foraminifera. These discrepancies might result from different sensitivity of dinoflagellates and planktonic foraminifera toward seasonal temperatures.

In the mid to high latitudes of the North Atlantic and adjacent polar and subpolar seas, many studies were done based on dinocysts to document late Quaternary changes of SSTs, sea-ice cover and salinity. As an example, the studies of de Vernal et al. (1996), Rochon et al. (1998) and Grosfjeld et al. (1999) have shown very extensive sea-ice cover along the southeastern Canadian margin and along the Scandinavian coast during the Younger Dryas. As another example, an in-depth study of the LGM suggests that the northern North Atlantic was characterized by sea-surface conditions much different from the modern ones: sea-ice cover was more extensive, salinity was lower, and SSTs apparently recorded much larger winter to summer amplitudes (cf. de Vernal et al., 2000, 2005a). The reconstruction of the North Atlantic SSTs at LGM based on dinocysts shows significant discrepancies with estimates based on planktonic foraminifera (cf. CLIMAP, 1976; Kucera et al., 2005; de Vernal et al., 2006). These discrepancies are evidence of the complexity of the biotic responses to climate changes, and most probably result from a combination of factors, including salinity and water mass stratification, seasonality, nutrient concentrations and productivity, in addition to mean summer temperature.

7. CONCLUDING REMARKS

Dinocysts are paleoceanographical proxies that are complementary to other microfossils in many respects:

1. They are composed of refractory organic matter. Thus, their preservation is not affected by dissolution processes, which can be a problem in the case of siliceous microfossils because of under-saturation of the water in opal silica, and in the case of calcareous remains below the lysocline. However, dinocysts can be sensitive to oxidation.
2. Dinoflagellates are ubiquitous and their cysts are particularly abundant in epicontinental seas and along continental margins, whereas the open ocean

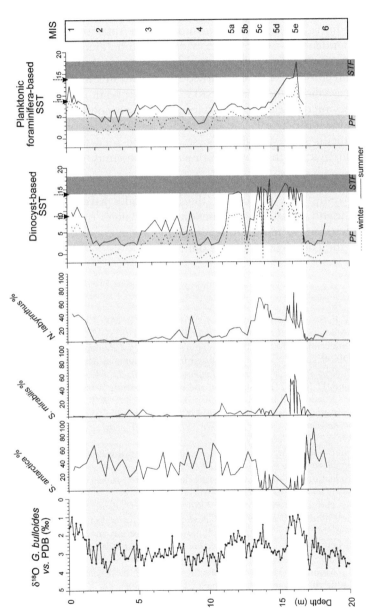

Figure 10 Comparison of winter and summer dinocyst-based (with permission from Marret et al., 2001a) and foraminifera-based SSTs (Wells & Okada, 1997) from site DSDP 594, southwest Pacific The percentages of a subantarctic taxon (*Selenopemphix antarctica*) and of cool temperate and warm temperate taxa (*Nematosphaeropsis labyrinthus* and *Spiniferites mirabilis*, respectively) are illustrated. Vertical shaded bands correspond to the thermic domains of the polar and subtropical fronts (PF = Polar front and STF = Subtropical Front). Arrows indicate present-day winter and summer SST at the site.

areas are often marked by low cyst concentrations. It is thus believed that dinocysts are very useful proxies in neritic areas, on sea floor slopes and rises, but that they are of more limited use in central gyre areas of the open ocean domain.

3. Dinocyst assemblages show a relatively high species diversity in polar seas and constitute one of the rare tracers for the paleoceanographical investigations and sea-ice cover reconstruction in the Arctic and circum-Antarctic seas.

4. Dinocysts are abundant nearshore where wide amplitude gradients of salinity and large seasonal contrasts of temperature, due to stratification of the upper water mass, are recorded. They can provide information on hydrographical parameters, such as salinity and seasonal SSTs. From this point of view, they are complementary to micropaleontological tracers such as coccoliths or planktonic foraminifera that are more stenohaline and restricted to the open ocean domain.

5. Recent developments suggest that dinocysts could be used to quantitatively reconstruct productivity in the ocean and upwelling intensity.

Our understanding of dinocysts as paleoceanographical tracers is still developing. Nevertheless, the investigations made over the last decades have clearly shown that dinocysts are a useful proxy for documenting the changes of biotic (trophic character) and abiotic (seasonal temperature, salinity, sea-ice) conditions in surface waters, especially in the continental margins domain.

REFERENCES

Azanza, R. V., Siringan, F. P., San Diego-Mcglone, M. L., Yñiguez, A. T., Macalalad, N. H., Zamora, P. B., Agustin, M. B., & Matsuoka, K. (2004). Horizontal dinoflagellate cyst distribution, sediment characteristics and benthic flux in Manila Bay, Philippines. *Phycological Research, 52*(4), 376–386.

Baldwin, R. P. (1987). Dinoflagellate resting cysts isolated from sediments in Marlborough Sounds, New Zealand. *New Zealand Journal of Marine and Freshwater Research, 21,* 543–553.

Bé, A. W. H., & Tolderlund, D. S. (1971). Distribution and ecology of living planktonic foraminifera in surface waters of the Atlantic and Indian oceans. In: B. M. Funnell & W. R. Riedel (Eds), *The micropaleontology of oceans* (pp. 105–149). Cambridge, UK: Cambridge University Press.

Biebow, N. (1996). *Dinoflagellatenzysten als indikatoren der spät- und postglazialen entwicklung des auftriebgeshehens vor Peru* (p. 130). Geomar Report 57.

Bint, A. N. (1988). Recent dinoflagellate cysts from Mermaid Sound, northwestern Australia. *Memoir of the Association of Australasian Palaeontologists, 5,* 329–341.

Blanco, J. (1995). Cyst production in four species of neritic dinoflagellates. *Journal of Plankton Research, 17*(1), 165–182.

Brenner, W. W. (2005). Holocene environmental history of the Gotland Basin (Baltic Sea): A micropalaeontological model. *Palaeogeography, Palaeoclimatology, Palaeoecology, 220,* 227–241.

Brenner, W. W., & Biebow, N. (2001). Missing autofluorescence of recent and fossil dinoflagellate cysts—an indicator of heterotrophy? *Neues Jahrbuch Fur Geologie Und Palaontologie-Abhandlungen, 219*(1–2), 229–240.

Cho, H.-J., Kim, C.-H., Moon, C.-H., & Matsuoka, K. (2003). Dinoflagellate cysts in recent sediments from the Southern Coastal waters of Korea. *Botanica Marina, 46,* 332–337.

Cho, H.-J., & Matsuoka, K. (1999). *Dinoflagellate cyst composition in the surface sediments from the Yellow Sea and the East China Sea. Proceedings of 2nd International Workshop on Oceanography and Fisheries.* East China Sea (pp. 73–81).

CLIMAP (1976). The surface of ice-age earth. *Science, 191,* 1131–1137.

Dale, B. (1976). Cyst formation, sedimentation, and preservation: Factors affecting dinoflagellate assemblages in recent sediments from Trondheimsfjord, Norway. *Review of Palaeobotany and Palynology, 22*, 39–60.

Dale, B. (1996). Dinoflagellate cyst ecology: Modeling and geological applications. In: J. Jansonius, & D. C. McGregor (Eds), *Palynology: Principles and applications* (Vol. 3, pp. 1249–1275). Salt Lake City, UT: American Association of Stratigraphic Palynologist.

Dale, B. (2001). The sedimentary record of dinoflagellate cysts: Looking back in the future of phytoplankton blooms. *Scienta Marina, 65*(2), 257–272.

Dale, B., & Dale, A. L. (2002). Environmental applications of dinoflagellate cysts and acritarchs. In: S. K. Haslett (Ed.), *Quarternary environmental micropalaeontology* (pp. 208–240). London: Arnold.

Dale, B., Dale, A. L., & Jansen, J. H. F. (2002). Dinoflagellate cysts as environmental indicators in surface sediments from the Congo deep-sea fan and adjacent regions. *Palaeogeography, Palaeoclimatology, Palaeoecology, 185*(3–4), 309–338.

Dale, B., Thorsen, T. A., & Fjellså, A. (1999). Dinoflagellate cysts as indicators of cultural eutrophication in the Oslofjord, Norway. *Estuarine, Coastal and Shelf Science, 48*, 371–382.

Debenay, J. P., Carbonel, P., Morzadec-Kerfourn, M. T., Cazaubon, A., Denefle, M., & Lezine, A. M. (2003). Multi-bioindicator study of a small estuary in Vendee (France). *Estuarine Coastal and Shelf Science, 58*(4), 843–860.

Devillers, R., & de Vernal, A. (2000). Distribution of dinoflagellate cysts in surface sediments of the northern North Atlantic in relation to nutrient content and productivity in surface waters. *Marine Geology, 166*, 103–124.

Dodge, J. D. (1994). Biogeography of marine armoured dinoflagellates and dinocysts in the NE Atlantic and North Sea. *Review of Palaeobotany and Palynology, 84*, 169–180.

Dodge, J. D., & Harland, R. (1991). The distribution of planktonic dinoflagellates and their cysts in the eastern and northeastern Atlantic Ocean. *New Phytologist, 118*(4), 593–603.

Dupont, L. (1999). Pollen and spores in marine sediments from the East Atlantic—a view form the ocean into the African continent. In: G. Fisher & G. Wefer (Eds), *Use of proxies in paleoceanography: Examples from the South Atlantic* (pp. 523–546). Berlin Heidelberg: Springer-Verlag.

Dupont, L., & Agwu, C. (1991). Environmental-control of pollen grain distribution patterns in the Gulf of guinea and offshore NW-Africa. *Geologische Rundschau, 80*(3), 567–589.

Dupont, L., Marret, F., & Winn, K. (1998). Land-sea correlation by means of terrestrial and marine palynomorphs from the equatorial East Atlantic: Phasing of SE trade winds and the oceanic productivity. *Palaeogeography, Palaeoclimatology, Palaeoecology, 142*(1–2), 51–84.

Dupont, L., & Wyputta, U. (2003). Reconstructing pathways of aeolian pollen transport to the marine sediments along the coastline of SW Africa. *Quaternary Science Reviews, 22*(2–4), 157–174.

Edwards, L. E., & Andrle, V. A. S. (1992). Distribution of selected dinoflagellate cysts in modern marine sediments. In: M. J. Head & J. H. Wrenn (Eds), *Neogene and quaternary dinoflagellate cysts and acritarchs* (pp. 259–288). Texas, TX: American Association of Stratigraphic Palynologists Foundation.

Edwards, L. E., Mudie, P. J., & de Vernal, A. (1991). Pliocene paleoclimatic reconstruction using dinoflagellate cysts: Comparison of methods. *Quaternary Science Reviews, 10*(2–3), 259.

Ellegaard, M., Christensen, N. F., & Moestrup, Ø. (1994). Dinoflagellate cysts from Recent Danish marine sediments. *European Journal of Phycology, 29*, 183–194.

Ellegaard, M., Daugbjerg, N., Rochon, A., Lewis, J., & Harding, I. (2003). Morphological and LSU rDNA sequence variation within the *Gonyaulax spinifera-Spiniferites* group (Dinophyceae) and proposal of *G. elongata* comb. nov. and *G. membranacea* comb. nov. *Phycologia, 42*(2), 151–164.

Ellegaard, M., Lewis, J., & Harding, I. (2002). Cyst-theca relationship, life cycle and effetcs of temperature and salinity on the cyst morphology of *Gonyaulax baltica* sp. nov. (Dinophyceae) from the Baltic Sea area. *Journal of Phycology, 38*, 775–789.

Esper, O., Versteegh, G. J. M., Zonneveld, K. A. F., & Willems, H. (2004). A palynological reconstruction of the Agulhas retroflection (South Atlantic Ocean) during the Late Quaternary. *Global and Planetary Change, 41*, 31–62.

Esper, O., Zonneveld, K. A. F., & Willems, H. (2002). Distribution of organic-walled dinoflagellate cysts in surface sediments of the southern Ocean (Atlantic sector) between the Subtropical Front and Weddell Gyre. *Marine Micropaleontology, 46*, 177–208.

Eynaud, F., Turon, J. L., Matthiessen, J., Kissel, C., Peypouquet, J. P., de Vernal, A., & Henry, M. (2002). Norwegian sea-surface palaeoenvironments of marine oxygen-isotope stage 3: The paradoxical response of dinoflagellate cysts. *Journal of Quaternary Science*, 17(4), 349–359.

Fensome, R. A., Taylor, F. J. R., Norris, G., Sarjeant, W. A. S., Wharton, D. I., & Williams, G. L. (1993). A classification of living and fossil dinoflagellates. American Museum of Natural History. *Micropaleontology*, Special publication number 7, pp. 1–351.

Fensome, R. A., & Williams, G. L. (2004). The Lentin and Williams Index of fossil dinoflagellates 2004 edition. American Association of Stratigraphic Palynologists Foundation Contribution Series 42.

Gaines, G., & Elbrachter, M. (1987). Heterotrophic nutrition. In: F. J. R. Taylor (Ed.), *The biology of dinoflagellates* (pp. 224–281). Oxford: Blackwell Scientific.

Gibson, C. H. (2000). Laboratory and ocean studies of phytoplankton response to fossil turbulence. *Dynamics of Atmospheres and Oceans*, 31, 295–306.

Gibson, C. H., & Thomas, W. H. (1995). Effects of turbulence intermittency on growth inhibition of a red tide dinoflagellate, *Gonyaulax polyedra* Stein. *Journal of Geophysical Research*, 100, 24841–24846.

Godhe, A., Karunasagar, I., Karunasagar, I., & Karlson, B. (2000). Dinoflagellate cysts in recent marine sediments from SW India. *Botanica Marina*, 43(1), 39–48.

Gowen, R. J., & Stewart, B. M. (2005). The Irish Sea: Nutrient status and phytoplankton. *Journal of Sea Research*, 54, 36–50.

Graham, L. E., & Wilcox, L. W. (2000). *Algae*. Upper Saddle River: Prentice Hall.

Grosfjeld, K., & Harland, R. (2001). Distribution of modern dinoflagellate cysts from inshore areas along the coast of southern Norway. *Journal of Quaternary Science*, 16(7), 651–659.

Grosfjeld, K., Larsen, E., Sejrup, H. P., de Vernal, A., Flatebo, T., Vestbo, M., Haflidason, H., & Aarseth, I. (1999). Dinoflagellate cysts reflecting surface-water conditions in Voldafjorden, western Norway during the last 11,300 years. *Boreas*, 28(3), 403–415.

Gross, M. G., & Gross, E. (1994). *Oceanography, a view of the Earth* (7th ed.). Englewood Cliffs, NJ: Prentice-Hall.

Guiot, J. (1990). Methodology of the Last Climatic Cycle reconstruction in France from pollen data. *Palaeogeography, Palaeoclimatology, Palaeoecology*, 80(1), 49–69.

Hallegraeff, C. M., Anderson, D. M., & Cembella, A. D. (2003). *Manual on harmful marine microalgae. Monographs on Oceanographic Methodology*. Paris: UNESCO.

Hamel, D., de Vernal, A., Gosselin, M., & Hillaire-Marcel, C. (2002). Organic-walled microfossils and geochemical tracers: Sedimentary indicators of productivity changes in the North Water and northern Baffin Bay during the last centuries. *Deep-Sea Research Part II-Topical Studies in Oceanography*, 49(22–23), 5277–5295.

Harland, R. (1983). Distribution maps of recent dinoflagellate cysts in bottom sediments from the North-Atlantic Ocean and adjacent seas. *Palaeontology*, 26, 321–387.

Harland, R., & Long, D. (1996). A Holocene dinoflagellate cyst record from offshore north-east England. *Proceedings of the Yorkshire Geological Society*, 51(1), 65–74.

Harland, R., Nordberg, K., & Filipsson, H. L. (2004). The seasonal occurrence of dinoflagellate cysts in surface sediments from Koljo Fjord, west coast of Sweden—a note. *Review of Palaeobotany and Palynology*, 128(1–2), 107–117.

Harland, R., Pudsey, C., Howe, J., & Fitzpatrick, M. (1998). Recent dinoflagellate cysts in a transect from the Falkland trough to the Weddell Sea, Antarctica. *Palaeontology*, 41, 1093–1131.

Head, M. J. (1996). Modern dinoflagellate cysts and their biological affinities. In: J. Jansonius, & D. C. McGregor (Eds), Palynology: Principles and applications (Vol. 3, pp. 1197–1248). Salt Lake City, UT: American Association of Stratigraphic Palynologists Foundation.

Hillaire-Marcel, C., de Vernal, A., Bilodeau, G., & Weaver, A. J. (2001). Absence of deep water formation in the Labrador Sea during the last interglacial period. *Nature*, 410, 1073–1077.

Holl, C., Zonneveld, K. A. F., & Willems, H. (2000). Organic-walled dinoflagellate cyst assemblages in the tropical Atlantic Ocean and oceanographical changes over the last 140 ka. *Palaeogeography, Palaeoclimatology, Palaeoecology*, 160(1–2), 69–90.

Hooghiemstra, H., & Agwu, C. O. C. (1986). Distribution of palynomorphs in marine sediments—a record for seasonal wind patterns over NW Africa and adjacent Atlantic. *Geologische Rundschau*, 75(1), 81–95.

Imbrie, J., & Kipp, N. G. (1971). A new micropalaeontological method of quantitative palaeoclimatology: Application to a Late Pleistocene Caribbean core. In: K. K. Turekian (Ed.), *The Late Cenozoic Glacial Ages* (pp. 71–181). New Haven, CT: Yale University Press.

Joyce, L. B., Pitcher, G. C., du Randt, A., & Monteiro, P. M. S. (2005). Dinoflagellate cysts from surface sediments of Saldanha Bay, South Africa: An indication of the potential risk of harmful algal blooms. *Harmful Algae, 4*, 309–318.

Kawamura, H. (2004). Dinoflagellate cyst distribution along a shelf to slope transect of an oligotrophic tropical sea (Sunda Shelf, South China Sea). *Phycological Research, 52*(4), 355.

Kobayashi, S., Matsuoka, K., & Lizuka, S. (1986). Distribution of dinoflagellate cysts in surface sediments of Japanese coastal waters. I. Omura Bay, Kyushu (in Japanese). *Bulletin of Plankton Society of Japan, 33*, 81–93.

Kokinos, J. P., & Anderson, D. M. (1995). Morphological development of resting cysts in cultures of the marine dinoflagellate *Lingulodinium polyedrum* (= *L. machaerophorum*). *Palynology, 19*, 143–166.

Kokinos, J. P., Eglinton, T. I., Goni, M. A., Boon, J. J., Martoglio, P. A., & Anderson, D. M. (1998). Characterization of a highly resistant biomacromolecular material in the cell wall of a marine dinoflagellate resting cyst. *Organic Geochemistry, 28*(5), 265–288.

Kucera, M., Weinelt, M., Kiefer, T., Pflaumann, U., Hayes, A., Weinelt, M., Chen, M. T., Mix, A. C., Barrows, T. T., Cortijo, E., Duprat, J., Juggins, S., & Waelbroeck, C. (2005). Reconstruction of sea-surface temperatures from assemblages of planktonic foraminifera: Multi-technique approach based on geographically constrained calibration data sets and its application to glacial Atlantic and Pacific Oceans. *Quaternary Science Reviews, 24*(7–9), 951–998.

Kunz-Pirrung, M. (2001). Dinoflagellate cyst assemblages in recent sediments of the Laptev Sea (Arctic Ocean) and their relation to hydrographic conditions. *Journal of Quaternary Science, 16*, 637–650.

Levac, E. (2001). High resolution Holocene palynological record from the Scotian Shelf. *Marine Micropaleontology, 43*(3–4), 179–197.

Levac, E., & de Vernal, A. (1997). Postglacial changes of terrestrial and marine environments along the Labrador coast: Palynological evidence from cores 91-045-005 and 91-045-006, Cartwright Saddle. *Canadian Journal of Earth Sciences, 34*(10), 1358–1365.

Levac, E., de Vernal, A., & Blake, W. (2001). Sea-surface conditions in northernmost Baffin Bay during the Holocene: Palynological evidence. *Journal of Quaternary Science, 16*(4), 353–363.

Levandowsky, M., & Kaneta, P. (1987). Behaviour in dinoflagellates. In: F. J. R. Taylor (Ed.), *The biology of dinoflagellates* (pp. 330–397). Oxford: Blackwell Scientific.

Lewis, J., Dodge, J. D., & Powell, A. J. (1990). Quaternary dinoflagellate cysts from the upwelling system offshore Peru, Hole 686B, ODP LEG 112. In: E. Suess, R. von Huene, et al. (Eds), *Proceedings of the Ocean Drilling Program, Scientific Results* (Vol. 112, pp. 323–327).

Lewis, J., & Hallett, R. (1997). *Lingulodinium polyedrum* (*Gonyaulax polyedra*) a blooming dinoflagellate. *Oceanography and Marine Biology; an Annual Review, 35*, 97–161.

Lewis, J., Harris, A. S. D., Jones, K. J., & Edmonds, R. L. (1999). Long-term survival of marine planktonic diatoms and dinoflagellates in stored sediment samples. *Journal of Plankton Research, 21*(2), 343–354.

Mangin, S. (2002). *Distribution actuelle des kystes de dinoflagellés en Méditerranée occidentale et application aux fonctions de transfert* (p. 34). Memoir of DEA, University of Bordeaux 1.

Marret, F. (1993). Les effets de l'acétolyse sur les assemblages des kystes de dinoflagellés. *Palynosciences, 2*, 267–272.

Marret, F. (1994a). Distribution of dinoflagellate cysts in recent marine sediments from the east equatorial Atlantic (Gulf of Guinea). *Review of Palaeobotany and Palynology, 84*(1–2), 1–22.

Marret, F. (1994b). *Evolution paléoclimatique et paléohydrologique de l'Atlantique est-équatorial et du proche continent au Quaternaire terminal. Contribution palynologique (kystes de dinoflagellés, pollen et spores)*. Doctoral thesis, Bordeaux I. 271pp.

Marret, F., & de Vernal, A. (1997). Dinoflagellate cyst distribution in surface sediments of the southern Indian Ocean. *Marine Micropaleontology, 29*(3–4), 367–392.

Marret, F., de Vernal, A., Benderra, F., & Harland, R. (2001a). Late Quaternary sea-surface conditions at DSDP Hole 594 in the southwest Pacific Ocean based on dinoflagellate cyst assemblages. *Journal of Quaternary Science, 16*(7), 739–751.

Marret, F., de Vernal, A., Pedersen, T. F., & McDonald, D. (2001b). Middle Pleistocene to Holocene palynostratigraphy of Ocean Drilling Program Site 887 in the Gulf of Alaska, northeastern North Pacific. *Canadian Journal of Earth Science, 38*(3), 373–386.

Marret, F., Leroy, S., Chalié, F., & Gasse, F. (2004a). New organic-walled dinoflagellate cysts from recent sediments of Central Asian seas. *Review of Palaeobotany and Palynology, 129*(1–2), 1–20.

Marret, F., & Scourse, J. D. (2002). Control of modern dinoflagellate cyst distribution in the Irish and Celtic seas by seasonal stratification dynamics. *Marine Micropaleontology, 47*(1–2), 101–116.

Marret, F., Scourse, J. D., & Austin, W. E. N. (2004b). Holocene shelf-sea seasonal stratification dynamics: A dinoflagellate cyst record from the Celtic Sea, NW European shelf. *Holocene, 14*(5), 689–696.

Marret, F., Scourse, J., Versteegh, G., Jansen, J., & Schneider, R. (2001c). Integrated marine and terrestrial evidence for abrupt Congo River palaeodischarge fluctuations during the last deglaciation. *Journal of Quaternary Science, 16*(8), 761–766.

Marret, F., & Zonneveld, K. A. F. (2003). Atlas of modern organic-walled dinoflagellate cyst distribution. *Review of Palaeobotany and Palynology, 125*(1–2), 1–200.

Matsuoka, K. (1981). Dinoflagellate cysts and pollen in pelagic sediments of the northern part of the Philippine Sea. *Bulletin of Faculty of Liberal Arts, Nagasaki University (Natural Science), 21*, 59–70.

Matsuoka, K. (1985). Organic-walled dinoflagellate cysts from surface sediments of Nagasaki Bay and Senzaki Bay, West Japan. *Bulletin of Faculty of Liberal Arts, Nagasaka University, (Natural Science), 25*(2), 21–115.

Matsuoka, K. (1999). Eutrophication process recorded in dinoflagellate cyst assemblages—a case of Yokohama Port, Tokyo Bay, Japan. *The Science of the Total Environment, 231*, 17–35.

Matsuoka, K., Saito, Y., Katayama, H., Kanai, Y., Chen, J., & Zho, H. (1999). Marine palynomorphs found in surface sediments and a core sample collected from of Chanjang River, western part of the East China Sea. *Proceedings of the 2nd International Workshop on Oceanography and Fisheries. East China Sea* (pp. 195–207).

Matthiessen, J. (1994). Distribution Patterns of dinoflagellate cysts and other organic-walled microfossils in recent Norwegian-Greenland sea sediments. *Marine Micropaleontology, 24*(3–4), 307–334.

Matthiessen, J., de Vernal, A., Head, M. J., Okolodkov, Y. B., Zonneveld, K. A. F., & Harland, R. (2005). Modern organic-walled dinoflagellate cysts in Artic marine environments and their (paleo) environmental significance. *Paläontologische Zeitschrift, 79*(1), 3–51.

McQuoid, M. R., Godhe, A., & Nordberg, K. (2002). Viability of phytoplankton resting stages in the sediments of a coastal Swedish fjord. *European Journal of Phycology, 37*(2), 191–201.

Melia, M. (1984). The distribution and relationship between palynomorphs in aerosols and deep-sea sediments off the coast of northwest Africa. *Marine Geology, 58*(3–4), 345–371.

Mizuchima, K., & Matsuoka, K. (2004). Vertical distribution and germination ability of *Alexandrium* spp. cysts (Dinophyceae) in the sediments collected from Kure Bay of the Seto Inland Sea, Japan. *Phycological Research, 52*(4), 408–413.

Montresor, M., Zingone, A., & Sarno, D. (1998). Dinoflagellate cyst production at a coastal Mediterranean site. *Journal of Plankton Research, 20*(12), 2291–2312.

Morquecho, L., & Lechuga-Devèze, C. H. (2003). Dinoflagellate cysts in recent sediments from Bahía Concepción, Gulf of California. *Botanica Marina, 46*, 132–141.

Morzadec-Kerfourn, M. T. (1992). Estuarine dinoflagellate cysts among oceanic assemblages of Pleistocene deep-sea sediments from the West African margin and their palaeoenvironmental significance. In: M. J. Head & J. H. Wrenn (Eds), *Neogene and Quaternary dinoflagellate cysts and acritarchs* (pp. 133–146). Dallas, TX: American Association of Stratigraphic Palynologists Foundation.

Mudie, P. J. (1992). Circum-arctic Quaternary and Neogene marine palynofloras: Paleoecology and statistical analysis. In: M. J. Head & J. H. Wrenn (Eds), *Neogene and Quaternary dinoflagellate cysts and acritarchs* (pp. 347–390). Dallas, TX: American Association of Stratigraphic Palynologists Foundation.

Mudie, P. J., Aksu, A. E., & Yasar, D. (2001a). Late Quaternary dinoflagellate cysts from the black, Marmara and Aegean seas: variations in assemblages, morphology and paleosalinity. *Marine Micropaleontology, 43*, 155–178.

Mudie, P. J., & Harland, R. (1996). Aquatic Quaternary. In: J. Jansonius & D. C. McGregor (Eds), *Palynology: Principles and applications* (pp. 843–878). Salt Lake City, UT: American Association of Stratigraphic Palynologist Foundation.

Mudie, P. J., Harland, R., Matthiessen, J., & de Vernal, A. (2001b). Dinoflagellate cysts and high latitude Quaternary paleoenvironmental reconstructions: An introduction. *Journal of Quaternary Science, 16,* 595–602.

Mudie, P. J., & Rochon, A. (2001). Distribution of dinoflagellate cysts in the Canadian Arctic marine region. *Journal of Quaternary Science, 16*(7), 603.

Mudie, P. J., Rochon, A., Aksu, A. E., & Gillespie, H. (2002). Dinoflagellate cysts, freshwater algae and fungal spores as salinity indicators in Late Quaternary cores from Marmara and Black seas. *Marine Geology, 190*(1–2), 203.

Nehring, S. (1994). Spatial distribution of dinoflagellate resting cysts in recent sediments of Kiel Bight, Germany (Baltic Sea). *Ophelia, 39*(2), 137–158.

Nehring, S. (1997). Dinoflagellate resting cysts from recent German coastal sediments. *Botanica Marina, 40,* 307–327.

Orlova, T. Y., Morozova, T. V., Gribble, K. E., Kulis, D. M., & Anderson, D. M. (2004). Dinoflagellate cysts in recent marine sediments from the east coast of Russia. *Botanica Marina, 47*(3), 184.

Parsons, T. R., Takahashi, M., & Hargrave, B. (1984). *Biological oceanographic processes.* Oxford: Pergamon Press.

Pawlowski, J., Holzmann, M., Fahrni, J. F., Pochon, X., & Lee, J. J. (2001). Molecular identification of algal endosymbionts in large miliolid foraminifera. 2. Dinoflagellates. *Journal of Eukaryotic Microbiology, 48,* 368–373.

Pemberton, K., Rees, A. P., Miller, P. I., Raine, R., & Joint, I. (2004). The influence of water body characteristics on phytoplankton diversity and production in the Celtic Sea. *Continental Shelf Research, 24*(17), 2011.

Persson, A., Godhe, A., & Karlson, B. (2000). Dinoflagellate cysts in recent sediments from the west coast of Sweden. *Botanica Marina, 43*(1), 69–79.

Peyron, O., & de Vernal, A. (2001). Application of artificial neural networks (ANN) to high-latitude dinocyst assemblages for the reconstruction of past sea-surface conditions in Arctic and sub-Arctic seas. *Journal of Quaternary Science, 16*(7), 699–709.

Pienkowski-Furze, A. (2004). *Dinoflagellates and their cysts from the Weddell Sea.* Ph. D. thesis, University of Wales-Bangor. 185pp.

Pitcher, G. C., Walker, D. R., Mitchell-Innes, B. A., & Moloney, C. L. (1991). Short-term variability during an anchor station study in the southern Benguela upwelling system: Phytoplankton dynamics. *Progress in Oceanography, 28,* 39–64.

Pochon, X., LaJeunesse, T. C., & Pawlowski, J. (2004). Biogeographic partitioning and host specialization among foraminiferan dinoflagellate symbionts (*Symbiodinium*; Dinophyta). *Marine Biology, 146,* 17–27.

Pospelova, V., Chmura, G. L., Boothman, W. S., & Latimer, J. S. (2002). Dinoflagellate cyst records and human disturbance in two neighboring estuaries, New Bedford Harbor and Apponagansett Bay, Massachusetts (USA). *Science of the Total Environment, 298*(1–3), 81.

Powell, A. J. (1992). *A stratigraphic Index of Dinoflagellate cysts.* British Micropaleontological Society Publication Series, London: Chapman & Hall.

Radi, T., Pospelova, V., de Vernal, A., & Vaughn, B. J. (2007). Dinoflagellate cysts as indicators of water quality and productivity in estuarine environments of British Columbia, *Marine Micropaleontology, 62,* 269–297.

Radi, T., & de Vernal, A. (2004). Dinocyst distribution in surface sediments from the northeastern Pacific margin (40–60°N) in relation to hydrographic conditions, productivity and upwelling. *Review of Palaeobotany and Palynology, 128*(1–2), 169.

Radi, T., & de Vernal, A. (2006). *Transfer functions for the reconstruction of primary productivity in marine environments and paleoenvironments based on organic-walled dinoflagellate cysts.* GAC-MAC Annual meeting, Montreal, May 2006, Abstract SS-14.

Radi, T., de Vernal, A., & Peyron, O. (2001). Relationships between dinocyst assemblages in surface sediment and hydrographic conditions in the Bering and Chukchi seas. *Journal of Quaternary Science, 16*(7), 667–681.

Raine, R., White, M., & Dodge, J. D. (2002). The summer distribution of net plankton dinoflagellates and their relation to water movements in the NE Atlantic Ocean, west of Ireland. *Journal of Plankton Research, 24*, 1131–1147.

Rochon, A., & de Vernal, A. (1994). Palynomorph distribution in recent sediments from the Labrador Sea. *Canadian Journal of Earth Sciences, 31*, 115–127.

Rochon, A., de Vernal, A., Sejrup, H. P., & Haflidason, H. (1998). Palynological evidence of climatic and oceanographic changes in the North Sea during the last deglaciation. *Quaternary Research, 49*(2), 197–207.

Rochon, A., de Vernal, A., Turon, J.-L., Matthiessen, J., & Head, M. J. (1999). *Distribution of dinoflagellate cyst assemblages in surface sediments from the North Atlantic Ocean and adjacent basins and quantitative reconstructions of sea-surface parameters.* Series Dallas, TX: American Association of Stratigraphic Palynologists Foundation. Special contribution no. 35.

Sangiorgi, F., & Donders, T. H. (2004). Reconstructing 150 years of eutrophication in the north-western Adriatic Sea (Italy) using dinoflagellate cysts, pollen and spores. *Estuarine, Coastal and Shelf Science, 60*(1), 69.

Sangiorgi, F., Fabbri, D., Comandini, M., Gabbianelli, G., & Tagliavini, E. (2005). The distribution of sterols and organic-walled dinoflagellate cysts in surface sediments of the North-western Adriatic Sea (Italy). *Estuarine, Coastal and Shelf Science, 64*(2–3), 395.

Schnepf, E., & Elbrächter, M. (1999). Dinophyte chloroplasts and phylogeny—A review. *Grana, 38*, 81–97.

Shi, N., Schneider, R., Beug, H., & Dupont, L. (2001). Southeast trade wind variations during the last 135 kyr: Evidence from pollen spectra in eastern South Atlantic sediments. *Earth and Planetary Science Letters, 187*(3–4), 311–321.

Smayda, T. J., & Reynolds, C. S. (2001). Community assembly in marine phytoplankton: application of recent models to harmful dinoflagellate blooms. *Journal of Plankton Research, 23*, 447–461.

Smayda, T. J., & Reynolds, C. S. (2003). Strategies of marine dinoflagellate survival and some rules of assembly. *Journal of Sea Research, 49*(2), 95.

Solignac, S., de Vernal, A., & Hillaire-Marcel, C. (2004). Holocene sea-surface conditions in the North Atlantic—Contrasted trends and regimes in the western and eastern sectors (Labrador Sea vs. Iceland Basin). *Quaternary Science Reviews, 23*(3–4), 319.

Sonneman, J. A., & Hill, D. R. A. (1997). A taxonomic survey of cyst-producing dinoflagellates from recent sediments of Victorian Coastal Waters, Australia. *Botanica Marina, 40*, 149–177.

Sournia, A. (1995). Red tide and toxic marine phytoplankton of the world ocean: An inquiry into biodiversity. In: P. Lassus, G. Arzul, E. Erard, P. Gentien & B. Marcaillou (Eds), *Harmful marine algal blooms* (pp. 103–112). Paris: Lavoisier.

Sprangers, M., Dammers, N., Brinkhuis, H., Van Weering, T. C. E., & Lotter, A. F. (2004). Modern organic-walled dinoflagellate cyst distribution offshore NW Iberia; tracing the upwelling system. *Review of Palaeobotany and Palynology, 128*(1–2), 97.

Struck, U., Sarnthein, M., Westerhausen, L., Barnola, J. M., & Raynaud, D. (1993). Ocean-atmosphere carbon exchange—impact of the biological pump in the Atlantic equatorial upwelling belt over the last 330,000 years. *Palaeogeography, Palaeoclimatology, Palaeoecology, 103*(1–2), 41–56.

Sun, X. K., & Mcminn, A. (1994). Recent dinoflagellate cyst distribution associated with the subtropical convergence on the Chatham Rise, east of New-Zealand. *Marine Micropaleontology, 23*(4), 345–356.

Susek, E., Zonneveld, K. A. F., Fischer, G., Versteegh, G. J. M., & Willems, H. (2005). Organic-walled dinoflagellate cyst production in relation to upwelling intensity and lithogenic influx in the Cape Blanc region (off north-west Africa). *Phycological Research, 53*(2), 97.

Targarona, J., Warnaar, J., Boessenkool, K. P., Brinkhuis, H., & Canals, M. (1999). Recent dinoflagellate cyst distribution in the North Canary Basin, NW Africa. *Grana, 38*, 170–178.

Taylor, F. J. R., & Pollingher, U. (1987). The ecology of Dinoflagellates. In: F. J. R. Taylor (Ed.), *The biology of Dinoflagellates* (pp. 398–529). Oxford: Blackwell Scientific Publications.

Telford, R. J. (2006). Limitations of dinoflagellate cyst transfer functions. *Quaternary Science Reviews, 25*(13–14), 1375–1382.

Thorsen, T. A., & Dale, B. (1998). Climatically influenced distribution of *Gymnodinium catenatum* during the past 2000 years in coastal sediments of southern Norway. *Palaeogeography, Palaeoclimatology, Palaeoecology, 143*.

Turon, J. L. (1980). Recent pollen assemblages from deep-sea cores and provenance of Reykjanes Ridge and Gibbs Fracture Sediments. *Comptes Rendus Hebdomadaires des Séances De L'Académie ses Sciences-Série D, 291*(5), 453–456.

Turner, J. T. (2002). Zooplankton fecal pellets, marine snow and sinking phytoplankton blooms. *Aquatic Microbial Ecology, 27*(1), 57–102.

Veldhuis, M. J. W., de Baar, H. J. W., Kraay, G. W., Van Bleijswijk, D. L., & Baars, M. A. (1997). Seasonal and spatial variability in phytoplankton biomass, productivity and growth in the northwestern Indian Ocean: The southwest and northeast monsoon, 1992–1993. *Deep-Sea Research I, 44*, 425–449.

de Vernal, A., Eynaud, F., Henry, M., Hillaire-Marcel, C., Londeix, L., Mangin, S., Matthiessen, J., Marret, F., Radi, T., Rochon, A., Solignac, S., & Turon, J. (2005a). Reconstruction of sea-surface conditions at middle to high latitudes of the Northern Hemisphere during the Last Glacial Maximum (LGM) based on dinoflagellate cyst assemblages. *Quaternary Science Reviews, 24*(7–9), 897–924.

de Vernal, A., & Giroux, L. (1991). Distribution of organic-walled microfossils in recent sediments from the estuary and Gulf of St Lawrence. *Canadian Special Publication of Fisheries and Aquatic Sciences, 113*, 189–199.

de Vernal, A., Guiot, J., & Turon, J.-L. (1993). Postglacial evolution of environments in the Gulf of St. Lawrence: Palynological evidence. *Géographie physique et Quaternaire, 47*, 167–180.

de Vernal, A., Henry, M., Matthiesen, J., Mudie, P. J., Rochon, A., Boessenkool, K. P., Eynaud, F., Grøsfjeld, K., Guiot, J., Hamel, D., Harland, R., Head, M. J., Kunz-Pirrung, M., Levac, E., Loucheur, V., Peyron, O., Pospelova, V., Radi, T., Turon, J.-L., & Voronina, E. (2001). Dinoflagellate cyst assemblages as tracers of sea-surface conditions in the Northern North Atlantic, Arctic and sub-Arctic seas: The new '*n* = 677' data base and its application for quantitative palaeoceanographic reconstruction. *Journal of Quaternary Science, 16*(7), 681.

de Vernal, A., Hillaire-Marcel, C., & Bilodeau, G. (1996). Reduced meltwater outflow from the Laurentide ice margin during the Younger Dryas. *Nature, 381*, 774–777.

de Vernal, A., Hillaire-Marcel, C., & Darby, D. (2005b). Variability of sea ice cover in the Chukchi Sea (western Arctic Ocean) during the Holocene. *Paleoceanography, 20*, PA4018, doi:10.1029/2005PA001157.

de Vernal, A., Hillaire-Marcel, C., Turon, J.-L., & Matthiessen, J. (2000). Reconstruction of sea-surface temperature, salinity, and sea-ice cover in the northern North Atlantic during the last glacial maximum based on dinocyst assemblages. *Canadian Journal of Earth Sciences, 37*(5), 725–750.

de Vernal, A., & Pedersen, T. F. (1997). Micropaleontology and palynology of core PAR87A-10: A 23,000 year record of paleoenvironmental changes in the Gulf of Alaska, northeast North Pacific. *Paleoceanography, 12*(6), 821–830.

de Vernal, A., Rochon, A., Turon, J.-L., & Matthiessen, J. (1997). Organic-walled dinoflagellate cysts: Palynological tracers of sea-surface conditions in middle to high latitude marine environments. *GEOBIOS, 30*, 905–920.

de Vernal, A., Rosell-Melé, A., Kucera, M., Hillaire-Marcel, C., Eynaud, F., Weinelt, M., Dokken, T., & Kageyama, M. (2006). Testing the validity of proxies for the reconstruction of LGM sea-surface conditions in the northern North Atlantic. *Quaternary Science Reviews, 25*(21–22), 2820–2834.

de Vernal, A., Turon, J. L., & Guiot, J. (1994). Dinoflagellate cyst distribution in high-latitude marine environments and quantitative reconstruction of sea-surface salinity, temperature, and seasonality. *Canadian Journal of Earth Sciences, 31*(1), 48–62.

Versteegh, G. J. M., & Blokker, P. (2004). Resistant macromolecules of extant and fossil microalgae. *Phycological Research*, *52*, 325–339.

Vink, A. (2004). Calcareous dinoflagellate cysts in South and equatorial Atlantic surface sediments: Diversity, distribution, ecology and potential for palaeoenvironmental reconstruction. *Marine Micropaleontology*, *50*(1–2), 43.

Vink, A., Brune, A., Holl, C., Zonneveld, K. A. F., & Willems, H. (2002). On the response of calcareous dinoflagellates to oligotrophy and stratification of the upper water column in the equatorial Atlantic Ocean. *Palaeogeography, Palaeoclimatology, Palaeoecology*, *178*(1–2), 53–66.

Vink, A., Ruhlemann, C., Zonneveld, K. A. F., Mulitza, S., Huls, M., & Willems, H. (2001). Shifts in the position of the North Equatorial Current and rapid productivity changes in the western Tropical Atlantic during the last glacial. *Paleoceanography*, *16*(5), 479–490.

Vink, A., Zonneveld, K. A. F., & Willems, H. (2000). Organic-walled dinoflagellate cysts in western equatorial Atlantic surface sediments: Distributions and their relation to environment. *Review of Palaeobotany and Palynology*, *112*(4), 247–286.

Voronina, E., Polyak, L., de Vernal, A., & Peyron, O. (2001). Holocene variations of sea-surface conditions in the southeastern Barents Sea, reconstructed from dinoflagellate cyst assemblages. *Journal of Quaternary Science*, *16*(7), 717–726.

Wall, D., & Dale, B. (1968). Modern dinoflagellate cysts and the evolution of the Peridiniales. *Micropaleontology*, *14*, 265–304.

Wall, D., & Dale, B. (1973). Palaeosalinity relationships of dinoflagellates in the late Quaternary of the Black Sea—a summary. *Geosciences Man*, *VII*, 95–102.

Wall, D., Dale, B., Lohmann, G. P., & Smith, W. K. (1977). The environmental and climatic distribution of dinoflagellate cysts in the North and South Atlantic Oceans and adjacent seas. *Marine Micropaleontology*, *2*, 121–200.

Wells, P., & Okada, H. (1997). Response of nannoplankton to major changes in sea-surface temperature and movements of hydrological fronts over Site DSDP 594 (south Chatham Rise, southeastern New Zealand), during the last 130 kyr. *Marine Micropaleontology*, *32*(3–4), 341–363.

Winter, A., Jordan, R. W., & Roth, P. H. (1994). Biogeography of living coccolithophores in ocean waters. In: A. Winter & W. G. Siesser (Eds), *Coccolithophores* (pp. 161–177). Cambridge, UK: Cambridge University Press.

Zonneveld, K. A. F. (1997). Dinoflagellate cyst distribution in surface sediments from the Arabian Sea (northwestern Indian Ocean) in relation to temperature and salinity gradients in the upper water column. *Deep-Sea Research Part II-Topical Studies in Oceanography*, *44*(6–7), 1411–1443.

Zonneveld, K. A. F., & Brummer, G. J. A. (2000). (Palaeo-)ecological significance, transport and preservation of organic-walled dinoflagellate cysts in the Somali Basin, NW Arabian Sea. *Deep-Sea Research Part II-Topical Studies in Oceanography*, *47*(9–11), 2229–2256.

Zonneveld, K. A. F., Hoek, R. P., Brinkhuis, H., & Willems, H. (2001a). Geographical distributions of organic-walled dinoflagellate cysts in surficial sediments of the Benguela upwelling region and their relationship to upper ocean conditions. *Progress in Oceanography*, *48*(1), 25–72.

Zonneveld, K. A. F., Holl, C., Janofske, D., Karwath, B., Kerntopf, B., Rühlemann, C., & Willems, H. (1999). Calcareous dinoflagellate cysts as palaeo-environmental tools. In: G. Fischer & G. Wefer (Eds), *Use of proxies in paleoceanography: Examples from the South Atlantic* (pp. 145–167). Berlin: Springer.

Zonneveld, K. A. F., & Susek, E. (2006). Effect of temperature, light and salinity on cyst production and morphology of *Tuberculodinium vancampoae* (Rossignol 1962) Wall 1967 (*Pyrophacus steinii* (Schiller 1935) Wall et Dale 1971). *Review of Palaeobotany and Palynology* (in press).

Zonneveld, K. A. F., Versteegh, G. J. M., & deLange, G. J. (1997). Preservation of organic-walled dinoflagellate cysts in different oxygen regimes: A 10,000 year natural experiment. *Marine Micropaleontology*, *29*(3–4), 393–405.

Zonneveld, K. A. F., Versteegh, G. J. M., & de Lange, G. J. (2001b). Palaeoproductivity and post-depositional aerobic organic matter decay reflected by dinoflagellate cyst assemblages of the Eastern Mediterranean S1 sapropel. *Marine Geology*, *172*(3–4), 181–195.

CHAPTER TEN

Coccolithophores: From Extant Populations to Fossil Assemblages

Jacques Giraudeau* *and* Luc Beaufort

Contents

1. Introduction	409
2. Taxonomy	411
3. Biogeography, Sedimentation, and Biogeochemical Significance	413
4. Current State of Methods	414
4.1. Pleistocene biostratigraphy	415
4.2. Advances in sample preparation, observation, and census counts	417
4.3. Coccolithophore contribution to carbonate production and sedimentation	422
4.4. Coccolith-Based transfer functions	424
5. Examples of Applications	428
5.1. Spatio-Temporal variability of extant coccolithophore populations in the tropical pacific	428
5.2. Reconstruction of equatorial indo-pacific ocean primary production variability	431
Acknowledgments	432
References	433

1. Introduction

Coccolithophores form a major group of marine algae occurring in modern oceans. They are, consequently, of great interdisciplinary interest. As one of the main ocean primary producers, as well as the main source of calcareous deep-sea oozes, they have fostered an enormous amount of research initiatives from marine biologists and geologists since the pioneering study by Ehrenberg (1836) of chalk formations in the Baltic Sea. This was followed by the first record of extant populations during the first expedition around the globe of the HMS *Challenger*.

More importantly, in the context of climate research, coccolithophores are unique among the marine biota as they contribute to the three major forcing functions by which pelagic organisms influence the global climate system over

* Corresponding author.

Developments in Marine Geology, Volume 1
ISSN 1572-5480, DOI 10.1016/S1572-5480(07)01015-9

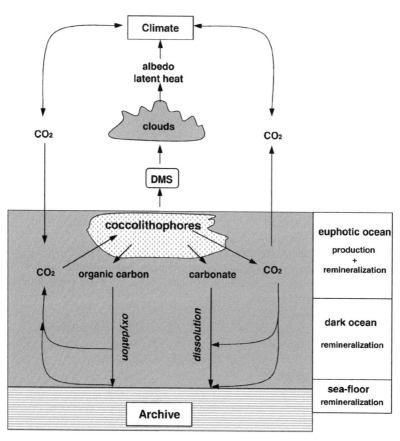

Figure 1   A conceptual representation of the global biogeochemical impacts of coccolithophore bloom events (after Westbroek et al., 1993).

sub-millennial timescales (Figure 1; Holligan, 1992; Westbroek et al., 1993): (1) the organic carbon pump (creation, export, and remineralization of organic carbon), (2) the carbonate pump (creation, sinking, and partial dissolution of particulate inorganic carbon — $CaCO_3$), (3) large-scale albedo effects (creation of highly reflecting clouds following gaseous emissions of dimethyl sulphide — DMS). The global biogeochemical significance of coccolithophores was put to the forefront of the scientific community when CZCS and AVHHR satellite images reported recurrent large-scale blooms (mainly induced by the cosmopolitan species *Emiliania huxleyi*) covering annually $1.5 \times 10^6 \, km^2$, the subpolar latitudes accounting for 71% of this area (Brown & Yoder, 1994). This fostered a series of interdisciplinary and international initiatives, such as the Global Emiliania Modelling initiative (Westbroek et al., 1993), the European project EHUX (Green & Harris, 1996) and the European network CODENET (Thierstein & Young, 2004), from which major advances in the ecology, biology, taxonomy, and paleoceanographic significance of these organisms have been achieved.

> ## 2. TAXONOMY

Kingdom	Protista
Division	Haptophyta
Class	Prymnesiophyceae
Orders	Isochrysidales and Coccolithales

The Class Prymnesiophyceae includes both calcifying (e.g., *Emiliania, Gephyrocapsa, Coccolithus*) and non-calcifying taxa (e.g., *Phaeocystis, Prymnesium*). Although cytological and biochemical evidence suggest that both groups were derived from coccolithophores (Green, Perch-Nielsen, & Westbroek, 1989; Green & Leadbeater, 1994), in this review we chose to limit the term "coccolithophores" to organisms belonging to the Orders Isochrysidales and Coccolithales, which form minute calcareous plates (average length = 7–8 μm), the coccoliths, because of their value for paleoceanographic studies. Coccoliths interlock to form a spherical external skeleton, the coccosphere (average diameter 20 μm), which surrounds the living cell. In the broad group of coccolithophores, we included some genera of unknown taxonomic affiliation (e.g., *Florisphaera*) which produce calcareous platelets — nannoliths — of the size of coccoliths, and which have been traditionally observed and studied with fossil and extant calcareous nannoplankton.

The taxonomy of coccolithophore species relies historically on the morphological characteristics of coccoliths preserved in the sedimentary record (e.g., Tappan, 1980; Perch-Nielsen, 1985). Culture techniques as well as recent advances in transmitted (TEM) or scanning electron microscope (SEM) observations have provided vital information on the biology and systematics of living and fossil forms. This entails the determination of various life cycles, motile and nonmotile (e.g., Billard, 1994), which for certain species, such as *Coccolithus pelagicus* (Parke & Adams, 1960), involves the production of morphologically different coccoliths/ holococcoliths vs. heterococcoliths, depending on the phase of their life cycle. Heterococcoliths, which are formed from calcite crystals of varying sizes and shapes, are only produced during the passive floating phase of coccolithophore life cycles, whereas holococcoliths, made of calcite crystals with identical size and shape, may only be synthesized during their motile stage. The fragile structure of holococcoliths renders them particularly vulnerable to disintegration during the process of sedimentation, and therefore, although common among living populations, they are barely represented in the fossil record.

SEM observations of extant coccolithophores revealed an additional complication in that some species, such as most of the *Syracosphaera* species (e.g., Kleijne, 1993), are dimorphic, i.e., they possess two or more morphologically different coccoliths.

The taxonomy of coccolithophores, carried out under the International Code of Botanical Nomenclature, is therefore in constant evolution. The phylogenetic schemes and classifications initially proposed by Jordan and Kleijne (1994), Green and Jordan (1994), and Jordan and Green (1994), and recently amended by Edvardsen et al. (2000) and Saez et al. (2004) on the grounds of genetic analyses, are considered to be the best up-to-date references in the field.

Most of the late Quaternary species are also currently thriving in the world's oceans. Up to 200 extant coccolithophore species have been described so far (e.g., Winter & Siesser, 1994), of which 30–40 are common in the sedimentary record. Of the 13–15 classified extant coccolithophore families, 6, whose representative species are both ecologically significant and abundant in the fossil record, are commonly used in late Quaternary paleoceanographic studies (Figure 2).

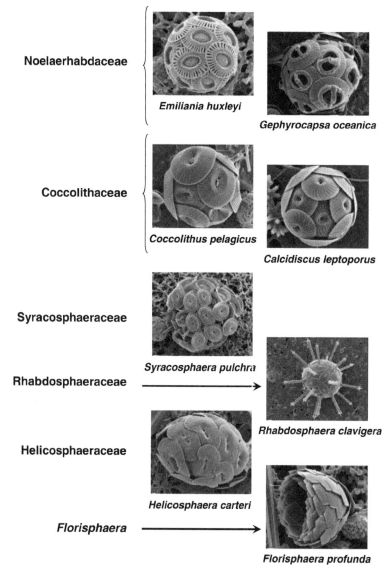

Figure 2   SEM views of the most common extant coccolithophore families (and representative species). All images courtesy Young et al. (2003).

3. BIOGEOGRAPHY, SEDIMENTATION, AND BIOGEOCHEMICAL SIGNIFICANCE

The primary interest in coccolithophores for paleoceanographic reconstruction lies in the biogeographical distribution and habitat of the most common late Quaternary species, as inferred from the now-standard surface water studies by McIntyre and Bé (1967) and by Okada and Honjo (1973) (see the review by Winter, Jordan, & Roth, 1994), as well as the synoptic surface sediment studies by Geitzenauer, Roche, and McIntyre (1977) and by McIntyre, Bé, and Roche. (1970), as summarized by Roth (1994). Both data sets (plankton and surface sediment) highlight a broad latitudinal distribution of coccolithophore/coccolith assemblages according to four floral zones (subpolar, temperate, subtropical, and tropical) in relation to major water masses. The coherence between surface water and surface sediment within biogeographical zones, as shown by e.g. McIntyre and Bé (1967), and in a more recent study by Baumann, Andruleit, & Samtleben (2000), is an indication of the rapid transport of coccoliths from the photic layer of the ocean to the seafloor (Honjo, 1976). This initiated the development of a suite of quantitative ocean-wide paleoecological transfer functions for Late Pleistocene paleotemperature reconstructions (Geitzenauer, Roche, & McIntyre, 1976; Molfino, Kipp, & Morley, 1982; Giraudeau & Pujos, 1990), with sea-surface temperature being viewed as the most discriminating parameter of water masses on a large scale. However, such a synoptic zonal distribution is of limited interest when one is interested in specific environments characterized by large gradients in biological, chemical, and physical parameters over restricted areas, where the response of the marine realm to paleo-environmental changes is particularly amplified. Consequently, most recent investigations on the distribution of extant and sediment surface assemblages have concentrated on small-scale, regional characterizations such as coastal upwelling systems (Baumann, Cepek, & Kinkel, 1999; Giraudeau, 1992), equatorial upwelling (Kinkel, Baumann, & Cepek, 2000), the subantarctic (Eynaud, Giraudeau, Pichon, & Pudsey, 1999; Findlay & Giraudeau, 2000) and subarctic domains (Baumann et al., 2000), the Indonesian (Kleijne, 1990) and Japan seas (Tanaka, 1991). A revised biogeography of Holocene coccoliths, which takes into account both large- and small-scale investigations of surface sediment assemblages, has recently been proposed by Ziveri, Baumann, Böckel, Bollmann, and Young (2004) for the Atlantic Ocean, and suggests that primary productivity and related trophic conditions might explain, in addition to sea-surface temperatures, the taxonomic composition of coccolith assemblages. This ocean-wide inference confirms previous results obtained from observations in restricted areas characterized by large gradients in nutrient conditions (Giraudeau & Bailey, 1995; Andruleit & Rogalla, 2002).

Investigations on seasonal and annual variations in coccolithophore production and species diversity (see the review by Baumann, Böckel, Geisen, & Kinkel, 2005) have been the focus of recent water column studies considering the importance of such information for (1) refining the ecological niches of key taxa used in paleoceanographic studies, (2) a better understanding of sedimentation processes in the water column and at the water-sediment interface, and (3) assessing the coccolith contribution to the global budget of biogenic carbonate. Studies on long-term sediment trapping, such as those done by Honjo, Manganini, and Cole (1982) and by Steinmetz (1991), and time-series trapping studies (see the comprehensive studies by Knappertsbusch & Brummer, 1995; Samtleben et al., 1995; Andruleit, 1997; Ziveri, Broerse, Van Hinte, Westbroek, & Honjo, 2000; Broerse, Ziveri, Van Hinte, & Honjo, 2000a; Beaufort & Heussner, 2001; among others) have quantified the coccolith contribution to total carbonate flux as ranging from 20 to 80%, depending on specific marine settings, with an average of 60% (Honjo, 1996). Sediment trapping was also crucial in investigating the mechanisms of vertical transport of coccoliths to the seafloor. For example, the rapid preferential transfer of coccoliths from surface waters to the seafloor by zooplankton fecal pellets, and various macroscopic organic and inorganic macroaggregates, as initially proposed by Honjo (1976), was further quantified by sediment trap studies, with resulting settling velocities in the order of 200 m/day, which explains the excellent preservation of these delicate skeletons even at depths below the lysocline (Steinmetz, 1994; Knappertsbusch & Brummer, 1995). This mechanism of transport has been shown to considerably enhance the transfer efficiency of organic matter produced by the photosynthetic coccolithophores; coccolith-$CaCO_3$ is acting as a ballast mineral to particulate organic carbon (Van der Wal, Kempers, & Veldhuis, 1995). Such a phenomenon, which is particularly active in bloom conditions, is thought to reduce the calcification to photosynthesis ratio, leading to a net sink of CO_2 to the deep ocean (Buitenhuis, van der Wal, & de Baar, 2001). Another application of coccolith investigations in sediment trap material is their use as tracers of particle transfer processes across passive continental margins, supporting the dominant role of lateral transport in the downslope sedimentation of silt and clay fractions (Beaufort & Heussner, 1999; Giraudeau, Bailey, & Pujol, 2000).

4. CURRENT STATE OF METHODS

Major advances in our knowledge of coccolithophore ecology and in our ability to retrieve paleoceanographic data from fossil records are highly dependent upon progress in methods for accurate extraction, observation, and census counts of extant and fossil assemblages. The increasing interest of the scientific community in the investigation of the mechanisms and implications of rapid and often subtle (paleo-) climate changes calls for a rapid, highly reproducible and accurate data collection. The same prerequisites concern other aspects of coccolith-based investigations, such as ongoing studies on the phylogeny and the biogeochemical impact of coccolithophores on present and past climate changes (i.e., contribution to carbonate sedimentation, DMS production), as well as recent initiatives in the construction and/or calibration of geochemical paleoproxies from both their organic and inorganic remains in sediments.

Coccolith researchers have been on the forefront of the progress in methods of microbiology and micropaleontology during the last 10 yr. This section will review these advances and their implications for understanding present and past physical, chemical, and biological aspects of the ocean's surface. Ongoing developments on some of the most exciting new geochemical proxies linked with coccolithophore production, such as biomarkers, stable isotopes, and trace elements are summarized in accompanying chapters by Rosell-Mele and Rosenthal or have been the subject of recent reviews (e.g., Stoll & Ziveri, 2004), and will therefore not be discussed in the present chapter.

4.1. Pleistocene Biostratigraphy

The unquestionable importance of coccoliths in biostratigraphy is due to their abundance and diversity in marine sediments, their wide geographic distribution, and their rapid evolution since their first appearance in the Late Triassic. Therefore, they have been extensively used during the past 40 yr for precise stratigraphic delineations within the Mesozoic and the Cenozoic. Their success lies also with the ease of sample preparation for routine work, as smears of bulk sediment are often sufficient to investigate the presence of biostratigraphically significant index species by using a standard light microscope. This makes the establishment of a preliminary age frame of sedimentary archives possible as soon as they are recovered, as it is routinely done during drilling and coring expeditions, such as the Ocean Drilling Program (ODP).

The last 10 yr have seen substantial progress in the refining of the standard zonation of Martini (1971) and of Okada and Bukry (1980). Improvements in quantitative techniques (see Section 4.2), as well as high resolution sampling, and the calibration of bio-events through astronomical tuning (Lourens et al., 1996; Berggren et al., 1995) have all led to a high-quality definition of the timing of the first and last occurrences of index species (= datums), which define the standard biostratigraphic zonations. The most spectacular advances concern the Pleistocene period, where micropaleontologists benefited from the increased availability of high-quality, high-sedimentation rate deep-sea cores, associated with high-resolution stable isotope (oxygen) and magnetostratigraphic records (Wei, 1993; Raffi, Backman, Rio, & Shackleton, 1993). A special effort was made to investigate the degree of isochrony and diachrony of bio-horizons from globally distributed low- to high-latitude deep-sea successions (Flores, Gersonde, Sierro, & Niebler, 2000; Raffi, 2002; Maiorano & Marino, 2004), developing the relationships between stratigraphic patterns and environmental changes. The use of dominance patterns of taxa from the Noelaerhabdaceae family was equally important and crucial in refining the Pleistocene biostratigraphic zonal scheme. This family includes the genera *Emiliania* and *Gephyrocapsa*, whose species are known to have alternatively dominated the calcareous phytoplankton from the early Pleistocene to present times. The most recent dominance or "acme" event is attributed to the species *E. huxleyi*, the single most abundant coccolithophore species in today's oceans, which has also dominated the sediment assemblages in the world's oceans for approximately the last 90 kyr (Thierstein, Geitzenauer, Molfino, & Shackleton, 1977). Prior to this interval, all acme intervals are related to morphologically differentiated representatives of *Gephyrocapsa* (e.g., Pujos, 1988). Most of these dominance changes in gephyrocapsid coccoliths have a global stratigraphic significance (Pujos & Giraudeau, 1993; Weaver, 1993), and are thought to be related to evolutionary adaptation rather than to global environmental changes (Bollmann, Baumann, & Thierstein, 1998).

The resulting coccolith-based biostratigraphic framework, which has considerably improved the stratigraphic resolution of the Pleistocene interval, has been used successfully in preliminary investigations during deep-sea coring expeditions (Figure 3).

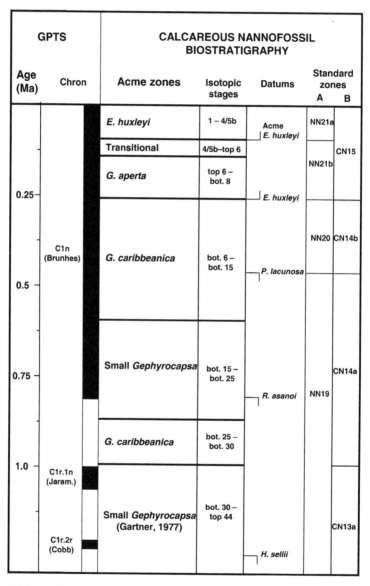

Figure 3 High-resolution coccolithophore zonal scheme adopted for the mid- to late Quaternary biostratigraphy of sediment cores recovered by ODP Leg 175 (Shipboard Scientific Party, 1998). Geomagnetic-polarity timescale (GPTS) after Berggren et al. (1995). Standard zones A and B from Martini (1971) and Okada and Bukry (1980), respectively.

4.2. Advances in Sample Preparation, Observation, and Census Counts

4.2.1. Sample preparation: from relative to absolute abundance

The ease of sample preparation using smears of raw sediment and subsequent observation with a light microscope was part of the success of coccoliths for bio-stratigraphical purposes, as well as for paleoceanographical studies based on their relative abundance. However, this simple method was rapidly found to be inadequate for the purposes of assessing various processes, such as characterizing transport processes through the water column to the sediment, providing an estimate of bulk or species-specific (paleo-) productivity and (paleo-) fluxes, or quantifying the contribution of coccoliths to carbonate sedimentation and, therefore, to the global carbon budget in the present and in the past. In addition, the interpretation of down-core records of coccolith species using their relative abundances was often biased by the so-called "closed sum" problem, which is particularly important in calcareous nannofossils assemblages in which a large number of ecologically significant species are often subordinate to a few dominant taxa. Specialists of both extant and fossil coccolithophores have therefore developed, during the last 10 yr, a series of methods for absolute census counts, which can be ranked into three categories: the "spiking with microbeads" method, the random settling method, and the filtration method, as summarized in Figure 4.

Okada (1992) was the first to apply the "spiking with microbeads" method to coccolith studies, by adding to raw sediment material a known weight of soda-ash spheres (initially manufactured to ensure the nighttime reflectivity of paints used in road markings). Accordingly, the estimates of the ratio of coccoliths to microbeads (2–10 µm size range), as obtained from light-microscopy census counts, is translated into absolute abundances of particle elements. This method, initially introduced by palynologists (Benninghoff, 1962), was reassessed by Bollmann, Brabec, Cortés, and Geisen (1999) who benefited from recently available microbeads with better constrained density and diameter variation, allowing them to infer absolute abundance of coccoliths with an acceptable standard deviation. Bollmann et al. (1999), coupled this method to a spraying device ("spray gun") in order to ensure a homogeneous distribution of particles on the target (slide or cover slip), following various steps of suspension in alcohol and ultrasonication to ensure a homogeneous mixing of microbeads and microfossils. The accuracy and reproducibility of the so-called SMS ("spiking with microbeads and spraying") method was subsequently illustrated by Herrle and Bollmann (2004) using comparisons with estimates obtained from the random settling and filtration techniques.

However, the use of microbeads for estimating coccolith absolute abundance suffers from the lack of standardization, among scientists, in the kinds of microbeads used, as well as in the equipment used for homogenizing the sediment/tracer mixture and slide preparation. Nonetheless, this method is popular for regularly checking the accuracy and reproducibility of absolute abundance calculations based on alternate techniques (e.g., Herrle & Bollmann, 2004).

The random settling technique, first introduced for coccolith studies by Beaufort (1991), as a cheaper alternative to the SMS method described above, is much more popular for absolute quantification. Its prime feature is that it ensures an

Devices	Equation for Cabs (number of coccoliths/weight unit)	Selected references	Pitfalls
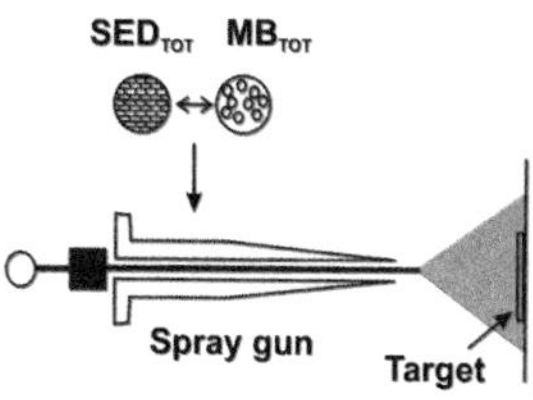 **SED_TOT MB_TOT** **Spray gun Target**	$$Cabs = \frac{Ccount}{MBcount} \times \frac{SEDtot}{MBtot}$$ Ccount: coccoliths counted MBcount: microbeads counted SED tot: sample weight MBtot: microbeads weight	Okada (1992) Bollmann et al. (1999)	1) lack of standardisation in microbead type and preparation devices. 2) expensive setup.
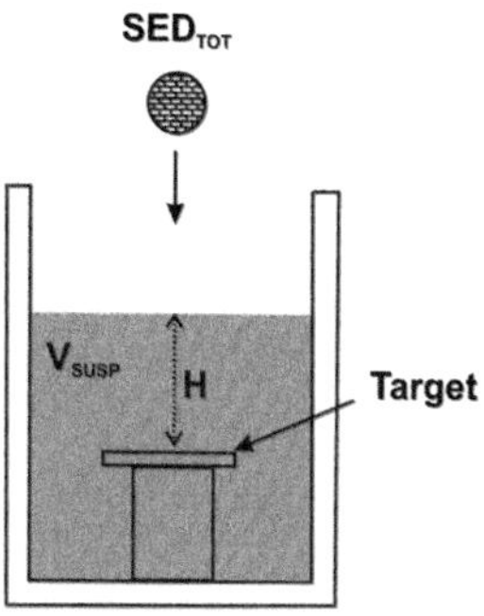 **SED_TOT** **V_SUSP H Target**	$$Cabs = \frac{Ccount \times Vsusp}{SEDtot \times F \times A \times H}$$ Ccount: coccoliths counted Vsusp: volume of suspension SED tot: sample weight F: number of field of views observed A: surface of one field of view H: height of water column above slide	Beaufort (1991) Flores and Sierro (1998) Geisen et al.(1999)	1) anomalies due to convective currents within the settling device. 2) anomalies due to elevated target support (Geisen et al., 1999). 3) time consuming if based on evaporation of the suspension.
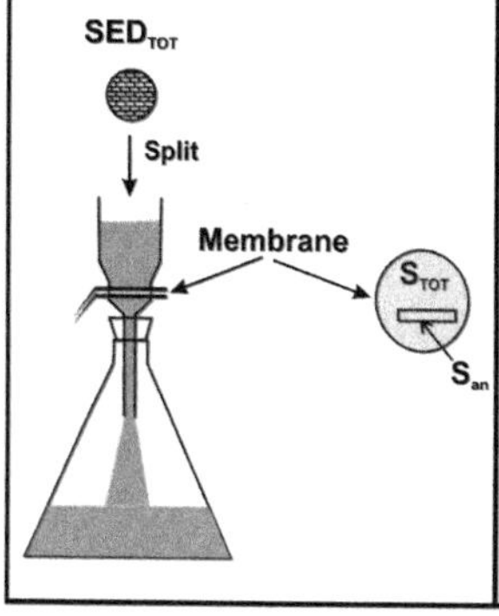 **SED_TOT** **Split Membrane S_TOT S_an**	$$Cabs = \frac{Stot \times Ccount \times Split}{San \times SEDtot}$$ Stot: surface of effective filtration Ccount: coccoliths counted Split: split factor Sana: analysed surface SED tot: sample weight	Andruleit (1996) Andruleit and Baumann (1998) Herrle and Bollmann (2004)	1) need to adjust sediment loading to avoid clogging of the membrane. 2) errors induced by successive splitting. 3) risks of uneven distribution of material on the membrane attributed to flow velocities, air locking, structure of the support sieve (Bollmann et al., 2002).

Figure 4   Most common methods and associated mathematical equations for the calculation of coccolithophore absolute abundance (*Cabs*) in marine sediments. From top to bottom: "spiking and spraying", "random settling", and "filtration" methods.

optimal and even distribution of particles on slides, which are later used for census counts. The principle of this method is to allow suspended particles to gently settle on a mounting support (slides or stubs), which is fixed in a cylinder in which the suspension containing the sedimentary material is poured. Preliminary devices (Williams & Bralower, 1995; Flores & Sierro, 1998) were based on the time-consuming process of evaporation, but a new generation of devices including a drain valve greatly reduced preparation time (Geisen, Bollmann, Herrle, Mutter-lose, and Young, 1999). These improved devices were successfully used in analyzing

Mesozoic and Cenozoic material (Mattioli & Pittet, 2002; Westphal Munnecke, Pross, & Herrle, 2004). As discussed by Geisen et al. (1999), the settling technique might, however, induce an artificial reduction or enrichment of the resulting particle loadings on the mounting support. Such a bias is supposed to be mainly caused by a convection current within the settling device, a function of the water temperature and of the volume of the chamber. The reproducibility of the absolute abundance estimates between various studies and associated experimental settling devices is therefore questionable.

While the three above-described methods for absolute quantification were initially developed for the investigation of marine sediments, the filtration technique applied to coccolithophores was first introduced for the quantitative analyses of living calcareous phytoplankton (McIntyre & Bé, 1967; Okada & Honjo, 1973), then adapted to sediment traps (e.g., Ziveri, Thunell, & Rio, 1995; Knappertsbusch & Brummer, 1995) and sediment samples (Andruleit, 1996; Andruleit & Baumann, 1998). Standard techniques for sediment samples involve a series of splits and dilutions of the suspended material, followed by funnel filtration through 47 mm-diameter membrane filters (0.8 or 0.45 μm pore-size), air- or oven-drying of the membrane, and mounting of approximately 20 mm^2 pieces on SEM stubs or between slides and cover slips. A cleaning step involving chemical oxidation and ultrasonification, as initially described by Bairbakhish, Bollmann, Sprengel, & Thierstein (1999) for sediment trap material, might be implemented prior to funnel filtration when dealing with sediment samples rich in organic matter. The accuracy and reproducibility of the filtration technique is highly dependent upon the dilution/splitting method used to adjust the amount of material to be filtered, to ensure an even distribution of the particles on the filter and avoid clogging. Whereas splitting errors induced by basic series of dilutions and divisions with a pipette of suspended material (Okada, 2000) are unknown, it is assumed that the use of the now standard rotary wet splitter induces a splitting error smaller than 1% (Andruleit, von Rad, Bruns, & Ittekot, 2000). Herrle and Bollmann (2004) additionally argued that the reproducibility and accuracy of coccolith census counts were better achieved when an in-line-filtration, instead of a funnel-filtration device, is used.

This section on sample preparation would not be complete without mentioning recent developments in methods allowing the separation of morphologically and/or taxonomically homogeneous coccolith fractions. Such methods are essential for geochemical analyses of calcareous nannofossils (stable isotopes and trace elements), in which the partitioning of minor elements as well as non-equilibrium effects in isotope fractionations are dependent upon changes in the relative carbonate contribution of the different species which compose the fossil assemblage (Stoll & Schrag, 2000; Ziveri et al., 2003). Standard mechanical separations using small-opening sieves or filters are not satisfactory due to clogging problems and limited precision in size separation. Accordingly, Paull and Thierstein (1987) successfully separated five compositional fractions of fine marine sediments (<38 μm) with a custom-made automated device for repeated decanting. Each fraction was shown to be dominated by 50–60% of individuals of a single species or a few species. This method was afterwards refined by Stoll and Ziveri (2002) using coupled repeated decanting and density-stratified settling columns. The density-stratified method was found to be particularly effective for separating the smaller coccoliths (2–5 μm) with a resulting 80–90% dominance of single species in separate sediment fractions. An alternate method developed by Minoletti, Gardin, Nicot, Renard, and Spezzaferri (2001) involves a series of ultrasonic disintegrations, low-speed centrifugations, and cascade microfiltrations on polycarbonate membranes. According to the authors, this time-consuming analytical procedure resulted in size-fractions composed of more than 80% calcareous nannofossils, and was found to be particularly effective for separating discoasters from other coccoliths in Neogene marine sediments.

As summarized by Stoll and Ziveri (2004), it must be recognized that none of the above-described methods were able to effectively separate coccoliths from similarly sized non-carbonate or carbonate particles present in sediments, which is a limitation for the use of size-fractions in trace element or stable isotope geochemistry of marine sediments rich in lithogenic material, or highly affected by carbonate dissolution.

4.2.2. Observation and census counts: from manual to automated analyses

Improved microscope technology, computing power, and software developments have, in recent years, opened the way to partially or fully automated coccolithophore analyses. While still developed and routinely used by a limited number of specialists, such initiatives are likely to become standard techniques considering their interest in sedimentological and paleoenvironmental studies (Francus, 2004). Potential applications in micropaleontology, and more specifically in coccolithophore studies, are numerous and linked with the two main technical developments: image acquisition and coccolith recognition.

One of the main objectives of automated image acquisition from transmitted light or transmitted and scanning electron microscopes is to allow reproducible species identifications and counts by different specialists. The reproducibility of species identification and census counts is indeed often hampered by the variety of preparation and observation techniques and taxonomical concepts used by the various researchers (see above section). Given the size of coccoliths, double checking of species determinations and relative abundances of species assemblages can only be achieved by the reexamination of stored digitized overview or particle-specific images, and subsequent manual processing on computer screens. While the new generation of scanning electron microscopes (SEMs) are routinely equipped with digital imaging units and storage capacity, the capture of images with transmitted light microscopes is more difficult, as it necessitates a sophisticated camera system able to cope with the low-light conditions of crossed nicols (crossed polarized light), as well as specific objectives for a homogeneous illumination of the observed field of view. Given these constraints and the developments in optical and camera settings, the processing of manually acquired images proved to be very effective in the morphometric analyses of coccolithophores for taxonomical (e.g., Young & Westbroek, 1991) and paleoecological/paleoenvironmental purposes (e.g., Knappertsbusch, Cortés, & Thierstein, 1997; Bollmann, 1997; Colmenero-Hidalgo, Flores, & Sierro, 2002). A further step was recently achieved by the automatic capture of images using computer-controlled robots for the motion and focus of the microscope (*X*, *Y*, *Z* directions). This provides the advantages of speeding up the process of collecting microfossil images and of producing consistently homogeneous high-quality digital outputs (Bollmann et al., 2004). It also opens the way to automatic coccolith recognition.

Coccolithophore specialists are at the forefront of research initiatives for fully automatic microfossil recognition and subsequent census counts. Compared to other microfossil groups, coccolith quantification is indeed particularly tedious due to their size, and the fact that most assemblages are overwhelmingly dominated by one or two species, making the confident quantification of the less abundant species more time-consuming. To date, two versions of automated coccolith recognition have been developed and tested: SYRACO (Système de Reconnaissance Automatique de Coccolithes) introduced by Dollfus and Beaufort (1999) and recently modified by Beaufort and Dollfus (2004), and COGNIS (Computer Guided Nannofossil Identification System) developed by Bollmann et al. (2004). Contrary to other existing techniques of sediment particle recognition, based on a complex set of algorithms, which resolves specific shape or textural features, both systems are based on the application of artificial neural networks (ANNs). These self-learning systems are particularly suited and flexible for the investigation of coccoliths characterized by high intraspecific variability in shape, size, and preservation, as well as for the analysis of microfossil slides which contain a variable amount of non-coccolith particles. The procedures used in these ANNs imply a series of learning steps on previously acquired images of individual species, and subsequent applications to the analyses of digitally acquired view fields of sedimentary material. As an example of performance, the modified version of SYRACO (Figure 5; Beaufort

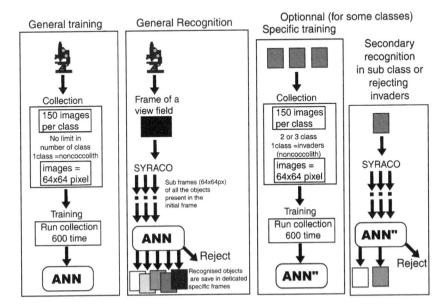

Figure 5 Flow chart of the SYRACO artificial neural network system for automatic image acquisition and coccolith recognition (after Dollfus & Beaufort, 1999; Beaufort & Dollfus, 2004).

& Dollfus, 2004), which includes a dynamic view of the object analyzed, presently extracts 11 classes of Pleistocene coccoliths (at the genus or species level) with a 96% level of reliability. It is able to complete the recognition and quantification of coccolith-rich sediment samples (100 view fields per sample) at a rate of 30–50 samples per day. Finally, as with COGNIS, SYRACO produces an output of digitized frames of classified objects (coccoliths and non-coccoliths), allowing the investigator to refine, if necessary, the automated classification, as well as to conduct species-specific morphometric analyses.

4.3. Coccolithophore Contribution to Carbonate Production and Sedimentation

Coccolithophores stand as major, if not dominant, contributors to carbonate fluxes in the open ocean. Given the impact of biocalcification on the global carbon cycle, collecting quantitative data on coccolith-carbonate export to the seafloor, and subsequent mass burial, has become a major research concern over the last 10 yr. Increasingly popular studies on the biogeochemical impact of bloom events, such as those conducted within JGOFS (Joint Global Ocean Flux Study) (Holligan et al., 1993; Van der Wal et al., 1995), as well as paleoceanographic investigations on the role of calcareous plankton production and dissolution in past CO_2 variations (Archer, Winguth, Lea, & Mahowald, 2000; Ridgwell, Watson, & Archer, 2002), among other research topics, are highly dependent upon accurate measurements of the amount of calcite produced by coccolithophores in the photic layers and the proportion of this calcite that becomes buried.

Estimates of coccolith contribution to bulk particle and sediment mass fluxes in sediment traps and sediment cores are commonly assessed by weighing the carbonate content of the fine fraction (Haidar, Thierstein, & Deuser, 2000; Ziveri et al., 2000). While easily determined using standard coulometric or inductively coupled plasma spectrometric methods, the resulting measurements are very approximate, as they commonly provide an overestimation of coccolith contribution to bulk carbonate contents. Fine fractions in particles settling throughout the water column and in buried sediments include a non-negligible amount of non-coccolith fragments, such as detrital microcalcite, calcareous dinophytes or foraminiferal fragments (Paull, Hills, & Thierstein, 1988; Beaufort & Heussner, 1999). Comparisons with more sophisticated methods such as those described below showed, for instance, increased discrepancies with water depth between the fine fraction and the coccolithophore-$CaCO_3$ in high- and middle-latitude settings due to dissolution-alteration processes (Broerse, Ziveri, & Honjo, 2000b; Ziveri et al., 2000).

A more accurate and elaborate determination of coccolith-carbonate content is based on estimates of the average weight of coccolith units. The total coccolith contribution to bulk carbonate content of sinking and buried material can thus be assessed using data on absolute coccolithophore abundances as estimated using the methods described in the above section. Recent progress in microscope technology and image capture systems (see above section), as well as development of software for morphometric analyses, has opened the way to this method based on coccolith geometry. The first step, as introduced by Honjo (1976) and later on applied by

Samtleben and Bickert (1990), Ziveri et al. (1995), and Knappertsbusch and Brummer (1995), was to use a unique or a couple of estimated coccolith weights as standard values for all extant and late Pleistocene species. Although attractive, this method is far from accurate; in particular, high interspecific variability in coccolith volume (and related carbonate-weight) considerably affects the accuracy of the estimates when analyzing material from low to middle latitudes with usually a high species diversity. The following step was therefore needed to provide a species-specific data set of mean weight values, which took into account the large range of coccolith sizes and shapes among the various extant and late Pleistocene species. This was accurately done by Beaufort and Heussner (1999) and by Young and Ziveri (2000) through the collection of precise morphometric measurements (length, width, and thickness) on a large set of SEM or light microscope-based digitized images, as well as the application of a correction factor for volume calculation, which takes into account the species-specific shape of the coccoliths. According to these authors, coccolith weight (CW), expressed in picograms, can be estimated from the following equations:

$$CW = V \times K \times \text{calcite density} \qquad (1)$$

where V is the mean coccolith volume calculated from mean length, width, and thickness (all expressed in µm) of coccolith species, and K the species-specific correction factor as given in Beaufort and Heussner (1999).

$$CW = l^3 \times k_s \times \text{calcite density} \qquad (2)$$

where l is the mean length of coccolith species, and k_s the species-specific correction factor as given in Young and Ziveri (2000).

The data sets of Beaufort and Heussner (1999) and of Young and Ziveri (2000) compare relatively well, and have since been extensively used in subsequent sediment trap and sediment core studies for local, regional, or basin-wide investigations of the coccolithophore component of the biogeochemical system, or in calculations of paleo-flux rates (e.g., Broerse et al., 2000a; Ziveri et al., 2000; Andrews & Giraudeau, 2003; Giraudeau, Jennings, & Andrews, 2004; Baumann, Böckel, & Frenz, 2004). While providing a revised assessment of the contribution of coccolith carbonate to bulk carbonate fluxes, these studies also illustrated the importance of the effect of assemblage composition on this calculated contribution; the range of species-specific mean weights, with two degrees of magnitudes between the smaller (*E. huxleyi*: 2–3.5 pg) and the larger species (*C. pelagicus*: ca. 150 pg), can indeed imply large differences in coccolith-carbonate content of sinking or buried materials, otherwise characterized by the same total concentration of coccolith scales.

Although confident of the mathematical method used in their study, Young and Ziveri (2000) stressed that the resulting data set of species-specific weights, which is based on a selected set of samples from a North Atlantic sediment trap, has to be used with extreme caution in subsequent work due to intraspecific size variability. This variation, common to most coccolithophore species (e.g., *E. huxleyi*: Young, 1994; *Calcidiscus leptoporus*: Knappertsbusch et al., 1997), results from genotypic variability or varying growth response to different ecological conditions. This pitfall implies that an accurate use of Equations (1) and (2) for bulk coccolith-mass weight estimates necessitates re-investigations of the size range of the most abundant species in any subsequently studied sedimentary material.

A totally novel approach for weight estimates of coccoliths was recently proposed by Beaufort (2005), as a way to both accelerate data collecting and bypass the pitfalls of the above-described methods which are related to inter- and intra-specific size variations. This method uses the optical properties of calcite by translating the brightness of individual coccoliths when viewed in cross-polarized light, into calcite weight. Making full use of automated image capture systems and computer-controlled robots for focus and motion of the microscope, and post-calibration of the calcite birefringence (gray level) with known carbonate values, the author has managed to provide weight estimates of coccolith species on the same order of magnitude of the mean species weight estimates using their geometry (see above-described methods). Besides being rapid, the main interest of this method is that it automatically measures the carbonate weight of any individual particles (among them coccoliths) distributed on a microscope slide. Combined with an automatic microfossil recognition system such as SYRACO (see above section), the calculated individual coccolith weight estimates can be translated into bulk or species-specific coccolith-carbonate contribution to the studied sediment material. According to Beaufort (2005), this method is of broad appeal for current studies in paleoceanography for which calcite production and dissolution are important.

4.4. Coccolith-Based Transfer Functions

Coccolith-based transfer functions for quantitative estimates of key physical, chemical, or biological parameters of surface waters was introduced in the 1970s and 1980s mostly within the auspices of the CLIMAP project (CLIMAP, 1976, 1981). Most initiatives were based on multivariate statistical analyses (factor analyses, and regression or canonical analyses) of fossil assemblages in a basin-wide set of surface sediments for the derivation of paleoecological equations for estimating surface water temperature and salinities (see the review by Roth, 1994). While theoretically justified by the broad biogeographic patterns of numerous coccolith species according to dominant surface water masses (McIntyre et al., 1970), this method, initially used with more or less success by, for example, Geitzenauer et al. (1976), Molfino et al. (1982) and Giraudeau and Pujos (1990) for paleoceanographic reconstructions of late Pleistocene surface circulation, has barely been developed during the last 15 yr.

Beside limited progress in recent years to refine the ocean-wide coccolithophore biogeography, some major pitfalls limit the use of standard multivariate analyses of surface sediment coccolith-assemblages for the derivation of paleoecological equations: the alternative dominance of single coccolith species throughout the Pleistocene, although useful for biostratigraphic purposes (see section on "Biostratigraphy" above), creates non-analogous situations between recent and past assemblages, and necessitates exclusion of these species from the analysis and/or log-transformation of raw census counts to increase the importance of rare but ecologically significant species (Geitzenauer et al., 1976; Giraudeau & Pujos, 1990). Also, although relatively sustained by the large-scale biogeography of coccolith species, considering a low taxonomic resolution, the dominant influence of sea-surface temperature upon species distribution is far from obvious when considering specific marine settings, such as continental margin or frontal areas where additional biotic or abiotic parameters (salinity, macronutrient content, primary productivity, among others) are as important in explaining the composition of species assemblages. This pitfall is exacerbated when considering higher taxonomic resolution, morphologically and/or genetically differentiated sub-species of single broad taxa being known to occupy distinct habitats (Ziveri et al., 2004). Accordingly, the most recent developments in coccolith-based quantitative paleoecological reconstructions have made full use of these various limits to introduce new methodological concepts.

A first step was introduced by Giraudeau and Rogers (1994) by restricting the application of a standard multivariate analysis (Imbrie & Kipp, 1971) to a geographically limited set of sediment samples representative of a specific oceanographic process: the southwest African margin and the associated Benguela upwelling process. Species assemblages as derived from a factor analysis of the floral census counts were essentially related to the range of chlorophyll *a* concentrations in surface waters (Figure 6). The paleoecological equation, as given by a stepwise

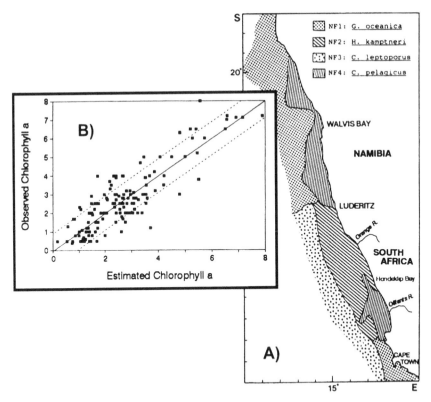

Figure 6 Transfer function for phytoplankton biomass estimates in the Benguela coastal upwelling system (after Giraudeau & Rogers, 1994). (A) Biogeography of the factor assemblages (with dominant species) as produced by Q-mode factor analysis. (B) Scatter plot of observed mean annual chlorophyll *a* concentrations (mg/m^3) versus estimates produced by the regression equation. Dashed lines indicate the interval of standard error of estimates (± 0.93 mg/m^3).

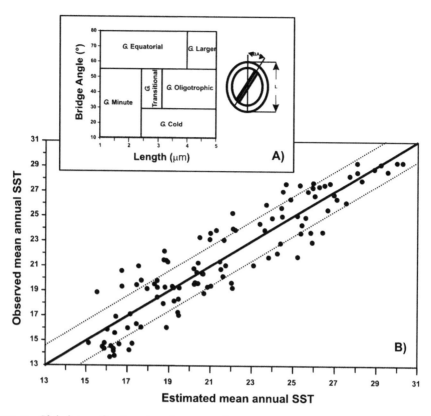

Figure 7 Global transfer function for sea-surface temperature (SST) estimates based on abundance of *Gephyrocapsa* morphotypes (after Bollmann et al., 2002). (A) Morphological associations of *Gephyrocapsa* as determined in Holocene sediment assemblages. (B) Scatter plot of observed mean annual SST (°C) vs. estimates produced by the regression equation. Dashed lines indicate the interval of standard error of estimates ($\pm 1.78°$C).

multiple regression, reproduced mean annual phytoplankton biomass with a standard error of 0.9 mg (chl *a*)/m³ over a chlorophyll spectrum of 0.2–8 mg/m³ (Giraudeau & Rogers, 1994).

More recently Bollmann, Henderiks, and Brabec (2002) provided a calibration of *Gephyrocapsa* coccoliths for paleotemperature assessment. This was made possible after a thorough analysis of the world-wide biogeographic distribution of six morphotypes belonging to this genus (Bollmann, 1997), and differentiated on the basis of coccolith length and bridge angle (Figure 7A). Standard multiple regression, applied to a set of 110 surface sediment samples from the Pacific, Indian and Atlantic Oceans covering a surface temperature gradient ranging from 13.6° to 29.3°C, yielded a paleoecological model for paleotemperature reconstruction with a standard error of estimates ($< 1.8°$C) comparable to values obtained from other micropaleontological proxies such as planktonic foraminifera (Figure 7B). Given the observed distinct biogeographic distribution of morphotypes from other species/taxa such as *C. leptoporus* (Knappertsbusch et al., 1997), the method proposed by

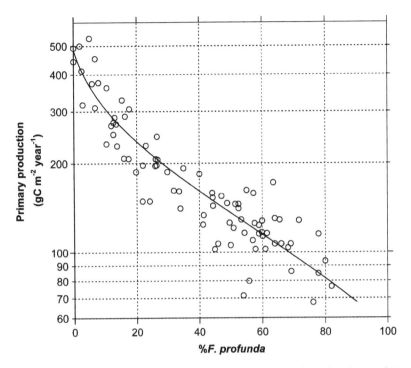

Figure 8   Transfer function for primary production estimates based on abundance of *F. profunda* (Beaufort et al., 1997, 2001): scatter plot of wt% *F. profunda* in surface sediment samples of the Indian Ocean vs. measured total yearly primary production at the sample locations.

Bollmann et al. (2002) might well soon become a standard procedure in subsequent studies for quantitative estimates of biotic or abiotic surface water parameters.

A very straightforward method of coccolithophore-based quantitative paleoecology was proposed by Beaufort et al. (1997), based on the relative abundance of a single coccolith species (Figure 8). Following the assumption of Molfino and McIntyre (1990) that nutricline depth variations could be monitored by the relative abundance of the deep living species *Florisphaera profunda*, Beaufort et al. (1997) used an Indian Ocean core-top data set and modern measurements of phytoplankton biomass to derive the following equation for primary productivity estimates:

$$PP = 316 \times \log\left(\%Fp + 3\right)$$

where PP is primary productivity (expressed in gC/m^2/yr), and %Fp the relative abundance of *F. profunda*.

The correlation and standard deviation of the residuals between the estimated and observed productivity in the calibration data set were $r = 0.94$ and ± 26 gC/m^2/yr, respectively. This transfer function has been shown to be reliable in the equatorial Atlantic (Henriksson, 2000) and the Pacific Ocean (Beaufort, de Garidel-Thoron, Mix, & Pisias, 2001). In samples external to the calibration set, results based on %Fp compared favorably with several other primary productivity proxies, such as radiolarian and foraminiferal-based productivity indexes, organic carbon or alkenone concentrations (Beaufort et al., 2001). An application of this method for studying the response of the tropical Indian and Pacific Oceans to insolation forcing is given in the following section.

5. Examples of Applications

The following results are based on two ongoing projects conducted by the authors in the tropical ocean realm. They both make full use of the recent advances in sample preparation techniques, as well as in coccolith-based paleoecological equations as illustrated by the second example. The rationale behind the first example is to better understand the creation and transformation of sedimentary proxies (in our case "coccoliths"), which are used to reconstruct past changes in key oceanic processes by integrating them into modern process studies. The second example will illustrate how information gained on calibrated coccolithophore relationships with biotic or abiotic parameters can procure an original view of the mechanisms involved in the response of the tropical ocean to orbitally driven insolation cycles.

5.1. Spatio-Temporal Variability of Extant Coccolithophore Populations in the Tropical Pacific

The Geochemistry, Phytoplankton and Color of Ocean project (GeP & CO; Dandonneau et al., 2004) was undertaken in 1999 to describe the spatio-temporal variability of phytoplankton populations at the ocean's surface in relation to the observed variability of the ocean's physical and chemical conditions (http://www.lodyc.jussieu.fr/gepco). GeP & CO stands, together with the UK biannual Atlantic Meridional Transects (AMT; Aiken et al., 2000), as one of a few series of cruises that have sampled the ocean on long tracks on a seasonal basis. All water samples were collected en route by a commercial ship between Le Havre (France) and Nouméa (New Caledonia), between November 1999 and June 2002, from the outlet of a thermosalinograph installed in the engine room at the intake of the cooling system. Results presented herein concern the Pacific part of four successive cruises and illustrate the seasonal and regional changes of coccolithophore production in the top 10 m of the surface waters.

The Pacific transect crosses four biogeochemical provinces as described by Longhurst (1998) on the basis of their differences in oceanic physics and surface chlorophyll fields (Figure 9A). Both the Pacific North Equatorial Countercurrent (PNEC) and the Pacific Equatorial Divergence (PEQD) provinces show weak seasonality in mixed layer depth, primary productivity, and phytoplankton biomass and are defined as HNLC (high nutrient low chlorophyll) regions. The oligotrophic South Pacific Subtropical Gyre (SPSG) shows a slight winter increase in productivity with slight deepening of the mixed layer, while nutrient limitation occurs 8–10 months a year. Winter mixing is much enhanced in the Southern Subtropical Convergence (SSTC) province, which shows higher seasonality in primary production rates than found in SPSG.

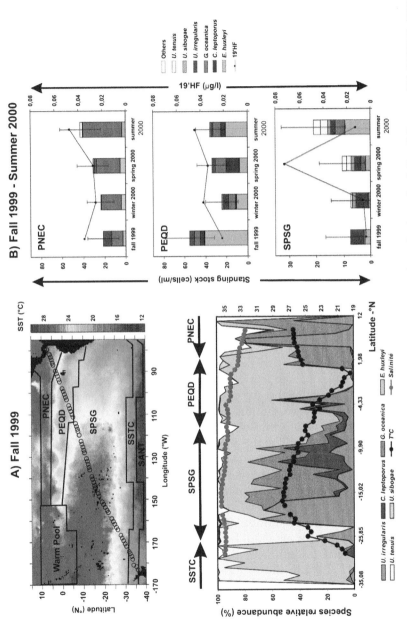

Figure 9 Biogeography of coccolithophore species in surface waters of the tropical Pacific along the GeP & CO route. (A) AVHRR (NOAA) map of sea-surface temperature in November 1999, sample locations (open dots), and related distribution of species weight % according to the biogeochemical provinces defined by Longhurst (1998): Pacific North Equatorial Countercurrent (PNEC), Pacific Equatorial Divergence (PEQD), South Pacific Subtropical Gyre (SPSG), South Subtropical Convergence (SSTC). The right Y axis on the bottom plot refers to sea-surface temperatures (°C) and salinities (‰) at sample locations. (B) Average coccolithophore standing stocks (boxes) and concentrations of 19'hexanoyloxyfucoxanthin (19'HF, black line) in the PNEC, PEQD and SPSG provinces from fall 1999 to summer 2000 (boreal seasons). The vertical error bars are the standard deviation of measured total standing stocks within each biogeochemical province.

This partitioning in biogeochemical provinces is very well expressed by spatial changes in coccolithophore species diversity as shown in Figure 9A. Though mean annual standing stocks in PEQD and PNEC are relatively comparable (Figure 9B), they strongly differ by their dominant species, with populations in the Equatorial Divergence dominated by the opportunistic species *E. huxleyi*, while those in the North Equatorial Countercurrent are dominated by *G. oceanica*. The SPSG province carries a low abundance but highly diverse population, as is the characteristic of other subtropical gyre domains of the Atlantic and Indian Oceans (Winter et al., 1994).

This data set of coccolithophore distribution pattern across the equatorial and SPSG domains is essential, as it supplements the unique record published by Okada and Honjo (1973) for the North and Equatorial Pacific: the succession of species groups identified along the GeP & Co route is symmetrical to the one observed in the Northern Hemisphere. The pattern of coccolithophore distribution given in Figure 9A seems therefore valid for the whole tropical and subtropical Pacific.

The seasonal evolution of coccolithophore standing stocks within the biogeochemical provinces follows the changes affecting the mixed layer depth and associated nutrient content of the surface layer (Figure 9B). Cell concentrations peak during the boreal summer in the PNEC and SPSG provinces as a response, respectively, to a strengthening of the North Equatorial Countercurrent, which enhances the Ekman divergence, and to the austral winter mixing. The maximum standing stock observed in the PEQD province during the boreal fall is induced by the seasonal reinforcement of trade winds, which intensify upwelling along the equator from July to December (Wyrtki & Kilonsky, 1984).

Marker pigments were routinely analyzed during the GeP & CO experiment, offering an opportunity to test the correlation between the concentration of diagnostic pigments and the abundance of phytoplankton groups. The photosynthetic pigment 19'hexanoyloxyfucoxanthin (19'HF) is considered as a marker pigment of haptophytes, of which coccolithophores are the main open-ocean contributors (Letelier et al., 1993). The seasonal evolution of 19'HF concentration within each biogeochemical province shows a poor relationship with cell concentrations (Figure 9B). This general disagreement between both the variables is possibly caused by changes in pigment ratio in different coccolithophore species (Jeffrey & Wright, 1994). Recent results indicate that variations in coccolithophore pigment composition have an evolutionary origin, with similarities increasing towards the lower taxonomic level (Van Lenning, Probert, Latasa, Estrada, & Young, 2004). The seasonal evolution of the nearly monospecific (*G. oceanica*) populations in the PNEC province shows a good match with the 19'HF concentration pattern, suggesting that species diversity has to be considered in this respect. Another cause of disagreement might be induced by organic-scale producing haptophytes, such as *Phaeocystis*, known to contain 19'HF (Belviso, Claustre, & Marty, 2001), but whose abundance is barely measurable using standard light microscope instruments for calcifying coccolithophores.

While the absence of major hydrological changes during the four consecutive GeP & CO years supposedly minimized the observed inter-annual variability of coccolithophore standing stocks and species diversity across the tropical Pacific (Dandonneau, Montel, Blanchot, Giraudeau, & Neveux, 2006), the above-described biogeography of species assemblages, as well as their quick response to changes in the physics and chemistry of the photic layer, suggest that this species group might be particularly suited to assess the present and past dynamics of El Niño climate anomalies.

5.2. Reconstruction of Equatorial Indo-Pacific Ocean Primary Production Variability

Little is known about large regional patterns of past primary production variability. In particular, it is crucial to describe the Glacial/Interglacial change in area, which could contribute significantly to the carbon cycle. The tropical ocean is a huge oceanic area, which produces a large part of the total phytoplanktonic production, but which also is particularly able to produce even more because its average primary production (PP) per unit area is relatively moderate due to the surface water stratification.

Beaufort et al. (2001) published the estimates of PP from nine cores from Indian and Pacific Equatorial Oceans spanning the last 250,000 yrs. When compared to each other, these PP records show two main modes: the first one, common to all cores, is related to global climate variability; the second, that opposes the western Pacific Warm Pool to the rest of the tropics on the precession band, is specific to the tropics and resembles present El Nino Southern Oscillation (ENSO) cycles but pulsing on a much longer timescale (Beaufort et al., 2001). This array of cores also provides the opportunity to study the pattern of variability of PP in this area (Beaufort, 2006). Each PP record is considered to be representative of the area in which the core has been retrieved. These areas are represented by the squares in Figure 10. For each of these areas, the total yearly primary production (TYPP) has been estimated using the map of primary production estimates from satellite imagery (Antoine, Andre, & Morel, 1995). For each record of primary production, its average Holocene PP (PPh) value has been subtracted from all the PP values (PPs). These new values have been divided by the variance of the original record (PPv), and then multiplied by the total PP of the zone in which the records have been taken.

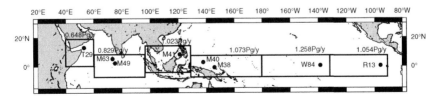

Figure 10   Location of the cores used to estimate the total primary production (PP) variability in the Equatorial Indian Ocean. The squares represent the area used in preparing the stack of PP. The numbers in the square are the total integrated yearly primary production in the area (after Antoine, André, & Morel, 1995).

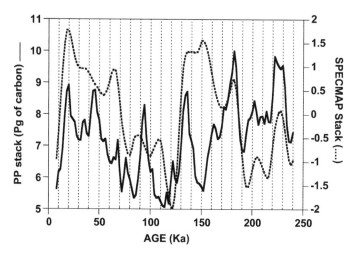

Figure 11 Record of low latitude Indo-Pacific Primary Production in Pg of carbon per year produced by adding the weighted variability of the different records. The weighing factor corresponds to the annual production of the area where the core has been retrieved (solid line). The dotted line is the δ^{18}O SPECMAP stack.

The new record of variations of total yearly production of each area (TYPPa) is given by the following equation: .

$$\text{TYPPs} = \text{TYPP} \times \text{PPv} \times (\text{PPs} - \text{PPh})$$

In areas where more than one core was studied, the average of PP between those cores was used for the calculations presented above. The sum of the six different TYPPs provides the first estimate of the total yearly production of the low-latitude Indo-Pacific Ocean for the last 250 kyr (Figure 11). The importance of precession is clearly visible, even if some cycles are missing. The obliquity also induces a strong response in the Tropical PP. This is confirmed by spectral analysis of the PP stack (Beaufort, 2006). Primary production is highest during times of high precession and high obliquity. The fact that obliquity is important in this record is a surprise, because this orbital parameter has little influence on low-latitude insolation. However, obliquity strongly influences high- to low-latitude contrasts, and can therefore play a significant role on the depth of the thermocline at low latitudes (Philander & Fedorov, 2003). The stack of all PP records shows that primary production was 50% (3.5 Pg of carbon) higher during glacials than during interglacials in the Equatorial Indo-Pacific Ocean. This increase in carbon produced at low latitudes could explain a significant (although small) part of the low pCO_2 recorded in ice cores in glacial times.

ACKNOWLEDGMENTS

The authors thank Patrizia Ziveri for reviewing this chapter and providing useful comments, and Jeremy Young for allowing reproduction of SEM images of extant coccolithophores. This is UMR 5805 EPOC contribution no. 1603.

REFERENCES

Aiken, J., Rees, N., Hooker, S., Holligan, P., Bale, A., Robins, D., Moore, G., Harris, R., & Pilgrim, D. (2000). The Atlantic Meridional Transect: Overview and synthesis of data. *Progress in Oceanography, 45*, 257–312.

Andrews, J. T., & Giraudeau, J. (2003). Multi-proxy records showing significant Holocene environmental variability: the inner N, Iceland shelf (Hunafloi). *Quaternary Science Reviews, 22*, 175–193.

Andruleit, H. (1996). A filtration technique for quantitative studies of coccoliths. *Micropaleontology, 42*, 403–406.

Andruleit, H. (1997). Coccolithophore fluxes in the Norwegian-Greenland Sea: Seasonality and assemblage alterations. *Marine Micropaleontology, 31*, 45–64.

Andruleit, H., & Baumann, K. H. (1998). History of the last deglaciation and holocene in the Nordic Seas as revealed by coccolithophore assemblages. *Marine Micropaleontology, 35*, 179–201.

Andruleit, H., von Rad, U., Bruns, A., & Ittekot, V. (2000). Coccolithophore fluxes from sediment traps in the northeastern Arabian Sea off Pakistan. *Marine Micropaleontology, 38*, 285–308.

Andruleit, H., & Rogalla, U. (2002). Coccolithophores in surface sediments of the Arabian Sea in relation to environmental gradients in surface waters. *Marine Geology, 31*, 1–22.

Antoine, D., André, J.-M., & Morel, A. (1995). Oceanic primary production 2. Estimation at global scale from satellite (coastal zone color scanner) chlorophyll. *Global Biogeochemical Cycles, 10*, 57–69.

Archer, D., Winguth, A., Lea, D., & Mahowald, N. (2000). What caused the Glacial/Interglacial atmospheric pCO2 cycles. *Review of Geophysics, 38*, 159–189.

Bairbakhish, A. N., Bollmann, J., Sprengel, C., & Thierstein, H. R. (1999). Disintegration of aggregates and coccospheres in sediment trap samples. *Marine Micropaleontology, 37*, 219–223.

Baumann, K. H., Andruleit, H. A., & Samtleben, C. (2000). Coccolithophores in the Nordic Seas: comparison of living communities with surface sediment assemblages. *Deep-Sea Research II, 47*, 1743–1772.

Baumann, K. H., Böckel, B., & Frenz, M. (2004). Coccolith contribution to South Atlantic carbonate sedimentation. In: H. R. Thierstein & J. Young (Eds), *Coccolithophores: From molecular processes to global impact* (pp. 367–402). Heidelberg: Springer-Verlag.

Baumann, K.-H., Böckel, B., Geisen, M., & Kinkel, H. (2005). The significance of extant coccolithophores as indicators of ocean water masses, surface water temperature, and paleoproductivity: A review. *Palaeontologische Zeitschrift, 79*(1), 93–112.

Baumann, K. H., Cepek, M., & Kinkel, H. (1999). Coccolithophores as indicators of ocean water masses, surface water temperature, and paleoproductivity: Examples from the South Atlantic. In: G. Fisher & G. Wefer (Eds), *Use of proxies in paleoceanography: Examples from the South Atlantic* (pp. 117–144). Heidelberg: Springer-Verlag.

Beaufort, L. (1991). Adaptation of the random settling method for quantitative studies of calcareous nannofossils. *Micropaleontology, 37*, 415–418.

Beaufort, L. (2005). Weight estimates of coccoliths using the optical properties (birefringence) of calcite. *Micropaleontology, 51*, 289–297.

Beaufort, L. (2006). Precession and ENSO-like variability in the Equatorial Indo-Pacific Ocean. In: H. Kawahata & Y. Awaya (Eds), *Global climate change and response of carbon cycle in the equatorial Pacific and Indian Oceans and adjacent landmasses*. Utrecht: Elsevier.

Beaufort, L., & Dollfus, D. (2004). Automatic recognition of coccoliths by dynamical neural networks. *Marine Micropaleontology, 51*, 57–73.

Beaufort, L., de Garidel-Thoron, T., Mix, A. C., & Pisias, N. G. (2001). ENSO-like forcing on oceanic primary production during the late Pleistocene. *Science, 293*, 2440–2444.

Beaufort, L., & Heussner, S. (1999). Coccolithophorids on the continental slope of the Bay of Biscay:Production, transport and contribution to mass fluxes. *Deep-Sea Research II, 46*, 2147–2174.

Beaufort, L., & Heussner, S. (2001). Seasonal dynamics of calcareous nannoplankton on a West European continental margin: The Bay of Biscay. *Marine Micropaleontology, 43*, 27–55.

Beaufort, L., Lancelot, Y., Camberlin, P., Cayre, O., Vincent, E., Bassinot, F., & Labeyrie, L. (1997). Insolation cycles as a major control of Indian Ocean primary productivity. *Science, 278*, 1451–1454.

Belviso, S., Claustre, H., & Marty, J.-C. (2001). Evaluation of the utility of chemotaxonomic pigments as a surrogate for particulate DMSP. *Limnology and Oceanography*, *46*, 989–995.

Benninghoff, W. S. (1962). Calculation of pollen and spores density in sediments by addition of exotic pollen in known quantities. *Pollen Spores*, *4*, 332–333.

Berggren, W. A., Hilgen, F. J., Langereis, C. G., Kent, D. V., Obradovich, J. D., Raffi, I., Raymo, M. E., & Shackleton, N. J. (1995). Late Neogene chronology: New perspectives in high-resolution stratigraphy. *Geological Society of America Bulletin*, *107*, 1272–1287.

Billard, C. (1994). Life cycles. In: J. C. Green & B. S. C. Leadbeater (Eds), *The haptophyte algae. The Systematics Association. Special Volume 51* (pp. 167–186). Oxford: Clarendon Press.

Bollmann, J. (1997). Morphology and biogeography of *Gephyrocapsa* coccoliths in Holocene sediments. *Marine Micropaleontology*, *29*, 319–350.

Bollmann, J., Baumann, K.-H., & Thierstein, H. R. (1998). Global dominance of *Gephyrocapsa* coccoliths in the late Pleistocene: Selective dissolution, evolution, or global environmental change? *Paleoceanography*, *13*, 517–529.

Bollmann, J., Brabec, B., Cortés, M. Y., & Geisen, M. (1999). Determination of absolute abundances in deep-sea sediments by spiking with microbeads and spraying (SMS-method). *Marine Micropaleontology*, *38*, 29–38.

Bollmann, J., Henderiks, J., & Brabec, B. (2002). Global calibration of *Gephyrocapsa* coccolith abundance in Holocene sediments for paleotemperature assessment. *Paleoceanography*, *17*(3) 10.1029/2001PA000742.

Bollmann, J., Quinn, P. S., Vela, M., Brabec, B., Brechner, S., Cortés, M. Y., Hilbrecht, H., Schmidt, D. N., Schiebel, R., & Thierstein, H. R. (2004), Automated particle analysis: Calcareous microfossils. In: P. Francus (Ed.), *Image Analysis, Sediments and Paleoenvironments*. Series: Developments in Paleoenvironmental Research (Vol. 7, p. 330). New York: Springer.

Broerse, A. T. C., Ziveri, P., & Honjo, S. (2000b). Coccolithophore (-CaCO₃) flux in the Sea of Okhotsk: Seasonality, settling and alteration processes. *Marine Micropaleontology*, *39*, 179–200.

Broerse, A. T. C., Ziveri, P., Van Hinte, J. E., & Honjo, S. (2000a). Coccolithophore export production, species composition, and coccolith-CaCO₃ fluxes in the NE Atlantic (34°N 21°W and 48°N 21°W). *Deep-Sea Research II*, *47*, 1877–1906.

Brown, C. W., & Yoder, J. A. (1994). Coccolithophorid blooms in the global ocean. *Journal of Geophysical Research*, *99*, 7467–7482.

Buitenhuis, E. T., van der Wal, P., & de Baar, H. J. W. (2001). Blooms of *Emiliania huxleyi* are sinks of atmospheric carbon dioxide: A field and mesocosm study derived simulation. *Global Biogeochemical Cycles*, *15*, 577–587.

CLIMAP project members (1976). The surface of the ice age earth. *Science*, *191*, 1131–1137.

CLIMAP project members (1981). Seasonal reconstructions of the earth surface at the last glacial maximum. *Geological Society of America. Map and Chart Series*, *MC-36*.

Colmenero-Hidalgo, E., Flores, J.-A., & Sierro, F. J. (2002). Biometry of *Emiliania huxleyi* and its biostratigraphic significance in the Eastern North Atlantic Ocean and Western Mediterranean Sea in the last 20,000 years. *Marine Micropaleontology*, *46*, 247–263.

Dandonneau, Y., Deschamps, P.-Y., Nicolas, J.-M., Loisel, H., Blanchot, J., Montel, Y., Thieuleux, F., & Bécu, G. (2004). Seasonal and interannual variability of ocean colour and composition of phytoplankton communities in the North Atlantic, Equatorial Pacific and South Pacific. *Deep-Sea Research II*, *51*, 303–318.

Dandonneau, Y., Montel, Y., Blanchot, J., Giraudeau, J., & Neveux, J. (2006). Temporal variability in phytoplankton pigments, picoplakton and coccolithophores along a transect through the North Atlantic and tropical Southwestern Pacific. *Deep-Sea Research I*, *53*, 689–712.

Dollfus, D., & Beaufort, L. (1999). Fat neural network for recognition of position-normalised objects. *Neural Networks*, *12*, 553–560.

Edvardsen, B., Eikrem, W., Green, J. C., Andersen, R. A., Moon-Van der Staay, S. Y., & Medlin, L. K. (2000). Phylogenetic reconstructions of the Haptophyta inferred from rRNA sequences and available morphological data. *Phycologia*, *39*, 19–35.

Ehrenberg, C. G. (1836). Bemrkungen uber feste mikroscopische anogarnische Formen in den erdigen und derben Mineralien. Berict. Verh. K. Preuss. Akad. Wiss. Berlin, pp. 84–85.

Eynaud, F., Giraudeau, J., Pichon, J.-J., & Pudsey, C. J. (1999). Sea-surface distribution of cocco-lithophores, diatoms, silicoflagellates and dinoflagellates in the South Atlantic Ocean during the late austral summer 1995. *Deep-Sea Research I, 46*, 451–482.

Findlay, C. S., & Giraudeau, J. (2000). Extant calcareous nannoplankton in the Australian sector of the Southern Ocean (austral summers 1994 and 1995). *Marine Micropaleontology, 40*, 417–439.

Flores, J. A., Gersonde, R., Sierro, F. J., & Niebler, H. S. (2000). Southern Ocean Pleistocene calcareous nannofossil events: Calibration with isotope and geomagnetic stratigraphies. *Marine Micropaleontology, 40*, 377–402.

Flores, J. A., & Sierro, F. J. (1998). Revised technique for calculation of calcareous nannofossil accumulation rates. *Micropaleontology, 43*, 321–324.

Francus, P. (Ed.) (2004). *Image analysis, sediments and paleoenvironments*. Series: Developments in Paleoenvironmental Research (Vol. 7, p. 330). New York: Springer.

Geisen, M., Bollmann, J., Herrle, J. O., Mutterlose, J., & Young, J. (1999). Calibration of the random settling technique for calculation of absolute abundances of calcareous nannoplankton. *Micropaleontology, 45*, 437–442.

Geitzenauer, K. R., Roche, M. B., & McIntyre, A. (1976). Modern Pacific coccolith assemblages: Derivation and application to late Pleistocene paleotemperature analysis. In: R. M. Cline & J. D. Hays (Eds), *Investigation of Late Quaternary Paleoceanography and Plaeoclimatology* (Vol. 145, pp. 423–448). Geological Society of America. Mem.

Geitzenauer, K. R., Roche, M. B., & McIntyre, A. (1977). Coccolith biogeography from North Atlantic and Pacific surface sediments. In: A. T. S. Ramsey (Ed.), *Oceanic Micropaleontology* (pp. 973–1008). New York: Academic Press.

Giraudeau, J. (1992). Distribution of Recent nannofossils beneath the Benguela system: southwest African continental margin. *Marine Geology, 108*, 219–237.

Giraudeau, J., & Bailey, G. W. (1995). Spatial dynamics of coccolithophore communities during an upwelling event in the Southern Benguela system. *Continental Shelf Research, 15*, 1825–1852.

Giraudeau, J., Bailey, G. W., & Pujol, C. (2000). A high-resolution time series analyses of particle fluxes in the Northern Benguela coastal upwelling system: Carbonate record of changes in biogenic production and particle transfer processes. *Deep-Sea Research II, 47*, 1999–2028.

Giraudeau, J., Jennings, A. E., & Andrews, J. T. (2004). Timing and mechanisms of surface and intermediate water circulation changes in the Nordic Seas over the last 10,000 cal years: A view from the North Iceland shelf. *Quaternary Science Reviews, 23*, 2127–2139.

Giraudeau, J., & Pujos, A. (1990). Fonction de Transfert basée sur les nannofossiles calcaires du Pléistocène des Caraïbes. *Oceanologica Acta, 13*, 453–469.

Giraudeau, J., & Rogers, J. (1994). Phytoplankton biomass and sea-surface temperature estimates from sea-bed distribution of nannofossils and planktonic foraminifera in the Benguela upwelling system. *Micropalaeontology, 40*, 275–285.

Green, J. C., & Harris, R. (1996). EHUX (*Emiliania huxleyi*). *Journal of Marine Systems, 9*(1-2), 136.

Green, J. C., & Jordan, R. W. (1994). Systematic history and taxonomy, In: J. C. Green & B. S. C. Leadbeater (Eds), *The haptophyte algae*. Systematics Association (Spec. 51, pp. 1–21). Oxford: Clarendon Press.

Green, J. C., & Leadbeater, B. S. C. (1994). *The haptophyte algae*. Systematics Association (Spec. 51, pp. 1–46). Oxford: Clarendon Press.

Green, J. C., Perch-Nielsen, K., & Westbroek, P. (1989). Prymnesiophita. In: L. Margulis (Ed.), *Handbook of Protoctista* (pp. 293–317). Boston, MA: Jones and Bartlett.

Haidar, A. T., Thierstein, H. R., & Deuser, W. G. (2000). Calcareous phytoplankton standing stocks, fluxes and accumulation in Holocene sediments off Bermuda (N. Atlantic). *Deep-Sea Research II, 47*, 1907–1938.

Henriksson, A. S. (2000). Coccolithophore response to oceanographic changes in the equatorial Atlantic during the last 200,000 years. *Palaeogeography, Palaeoclimatology, Palaeoecology, 156*, 161–173.

Herrle, J. O., & Bollmann, J. (2004). Accuracy and reproducibility of absolute nannoplankton abundances using the filtration technique in combination with a rotary sample splitter. *Marine Micropaleontology, 53*, 389–404.

Holligan, P. M. (1992). Do marine phytoplankton influence global climate? In: P. G. Falkowski & A. D. Woodhead (Eds), *Primary productivity and biogeochemical cycles in the sea* (pp. 487–501). New York: Plenum Press.

Holligan, P. M., Fernandez, E., Aiken, J., Balch, W. M., Boyd, P., Burkill, P. H., Finch, M., Groom, S. B., Malin, G., Muleer, K., Purdie, D. A., Robinson, C., Trees, C. C., Turner, S. M., & Van der Wal, P. (1993). A biogeochemical study of the coccolithophore, *Emiliania huxleyi*, in the North Atlantic. *Global Biogeochemical Cycles, 7*, 879–900.

Honjo, S. (1976). Coccoliths: Production, transportation and sedimentation. *Marine Micropaleontology, 1*, 65–79.

Honjo, S. (1996). Fluxes of particles to the interior of the open oceans. In: V. Ittekot, et al. (Eds), *Particle flux in the ocean* (pp. 91–154). Chichester: Wiley.

Honjo, S., Manganini, S. J., & Cole, J. J. (1982). Sedimentation of biogenic matter in the deep ocean. *Deep-Sea Research, 29*, 609–625.

Imbrie, J., & Kipp, N. G. (1971). A new micropaleontological method for quantitative paleoclimatology: Application to a Late Pleistocene Caribbean core. In: K. K. Turekian (Ed.), *Late cenozoic gliacia ages* (pp. 71–181). New Haven, CT: Yale University Press.

Jeffrey, S. W., & Wright, S. W. (1994). Photosynthetic pigments in the haptophyta. In: J. Green & B. Leadbeater (Eds), *The haptophyte algae* (pp. 111–132). Oxford: Oxford University Press.

Jordan, R. W., & Green, J. C. (1994). A check-list of the extant haptophyta of the world. *Journal of the Marine Biological Association UK, 74*, 149–174.

Jordan, R. W., & Kleijne, A. (1994). A classification system for living coccolithophores. In: A. Winter & W. G. Siesser (Eds), *Coccolithophores* (pp. 83–106). Cambridge: Cambridge University Press.

Kinkel, H., Baumann, K. H., & Cepek, M. (2000). Coccolithophores in the equatorial Atalntic Ocean: Response to seasonal and late quaternary surface water variability. *Marine Micropaleontology, 39*, 87–112.

Kleijne, A. (1990). Distribution and malformation of extant calcareous nannoplankton in the Indonesian seas. *Marine Micropaleontology, 16*, 293–316.

Kleijne, A. (1993). *Morphology, taxonomy and distribution of extant coccolithophorids (calcareous nannoplankton)*. Ph.D. thesis, Free University of Amsterdam, FEBO, Enschede (p. 321).

Knappertsbusch, M., & Brummer, G.-J. A. (1995). A sediment trap investigation of sinking coccolithophorids in the North Atlantic. *Deep-Sea Research I, 42*, 1083–1109.

Knappertsbusch, M., Cortés, M. Y., & Thierstein, H. R. (1997). Morphologic variability of the coccolithophorid *Calcidiscus leptoporus* in the plankton, surface sediments and from the early pleistocene. *Marine Micropaleontology, 30*, 293–317.

Letelier, R. M., Bidigare, R. R., Hebel, D. V., Ondrusek, M., Winn, C. D., & Karl, D. M. (1993). Temporal variability of phytoplankton community structure based on pigment analysis. *Limnology and Oceanography, 38*, 1420–1437.

Longhurst, A. (1998). *Ecological geography of the sea* (p. 398). London: Academic Press.

Lourens, L. J., Antonarakou, A., Hilgen, F. J., Van Hoof, A. A. M., Vergnaud-Grazzini, C., & Zachariasse, W. J. (1996). Evaluation of the plio-pleistocene astronomical timescale. *Paleoceanography, 11*, 391–413.

Maiorano, P., & Marino, M. (2004). Calcareous nannofossil bioevents end environmental control on temporal and spatial patterns at the early-middle pleistocene. *Marine Micropaleontology, 53*, 405–422.

Martini, E. (1971). Standard Tertiary and Quaternary calcareous nannoplankton zonation. In: Farinacci, A. (Ed.), *Proceedings of 2nd International Conference on Planktonic Microfossils Roma*, Rome (Ed. Tednosci, Vol. 2, 739–785).

Mattioli, E., & Pittet, B. (2002). Contribution of calcareous nannoplankton to carbonate deposition: A new approach applied to the Lower Jurassic of central Italy. *Marine Micropaleontology, 45*, 175–190.

McIntyre, A., & Bé, A. W. H. (1967). Modern coccolithophoridae of the Atlantic Ocean — I: Placoliths and cyrtoliths. *Deep-Sea Research, 14*, 561–597.

McIntyre, A., Bé, A. W. H., & Roche, M. B. (1970). Modern pacific coccolithophorida: A paleontological thermometer. *Transactions of NewYork Academy of Sciences, 32*, 720–731.

Minoletti, F., Gardin, S., Nicot, E., Renard, M., & Spezzaferri, S. (2001). Mise au point d'un protocole expérimental de séparation granulométrique d'assemblages de nannofossiles calcaires : Applications paléoécologiques et géochimiques. *Bulletin de la Société Géologique de France, 172,* 437–446.

Molfino, B., Kipp, N. G., & Morley, J. J. (1982). Comparison of foraminiferal, coccolithophorid, and radiolarian paleotemperature equations: Assemblage coherency and estimate concordancy. *Quaternary Research, 17,* 279–313.

Molfino, B., & McIntyre, A. (1990). Precessional forcing of nutricline dynamics in the equatorial Atlantic. *Science, 249,* 766–769.

Okada, H. (1992). Use of microbeads to estimate the absolute abundance of nannofossils. *International Nannoplankton Association, 14,* 96–97.

Okada, H. (2000). An improved filtering technique for calculation of calcareous nannofossil accumulation rates. *Journal of Nannoplankton Research, 22,* 203–204.

Okada, H., & Bukry, D. (1980). Supplementary modification and introduction of code numbers to the low-latitude coccolith biostratigraphic zonation (Bukry, 1973; 1975). *Marine Micropaleontology, 5,* 321–325.

Okada, H., & Honjo, S. (1973). The distribution of oceanic coccolithophorids in the Pacific. *Deep-Sea Research, 20,* 355–374.

Parke, M., & Adams, I. (1960). The motile (*Crystallolithus hyalinus* Gaarder and Markali) and non-motile phases in the life history of *Coccolithus pelagicus* (Wallich) Schiller. *Journal of the Marine Biological Association, 39,* 263–274.

Paull, C. K., Hills, S. J., & Thierstein, H. R. (1988). Progressive dissolution of fine carbonate particles in pelagic sediments. *Marine Geology, 81,* 27–40.

Paull, C. K., & Thierstein, H. R. (1987). Stable isotopic fractionation among particles in quaternary coccolith-sized deep-sea sediments. *Paleoceanography, 2,* 423–429.

Perch-Nielsen, K. (1985). Cenozoic calcareous nannofossils. In: H. M. Bolli, J. B. Saunders & K. Perch-Nielsen (Eds), *Plankton Stratigraphy* (Volume 1, pp. 427–554). Cambridge: Cambridge University Press.

Philander, G., & Fedorov, A. V. (2003). Role of tropics in changing the response to Milankovich forcing some three million years ago. *Paleoceanography, 18,* 1045.

Pujos, A. (1988). Spatio-temporal distribution of some quaternary coccoliths. *Oceanologica Acta, 11,* 65–78.

Pujos, A., & Giraudeau, J. (1993). Distribution of Noelaerhabdaceae (calcareous nannofossils) in the upper and middle quaternary of the Atlantic and Pacific oceans. *Oceanologica Acta, 16,* 349–362.

Raffi, I. (2002). Revision of the early-middle pleistocene calcareous nannofossil biochronology (1.75–0.85 Ma). *Marine Micropaleontology, 45,* 25–55.

Raffi, I., Backman, J., Rio, D., & Shackleton, N. J. (1993). Plio-Pleistocene nannofossil biostratigraphy and calibration to oxygen isotopes stratigraphies from Deep Sea Drilling Project Site 607 and Ocean Drilling Program Site 677. *Paleoceanography, 8,* 387–408.

Ridgwell, A. J., Watson, A. J., & Archer, D. E. (2002). Modelling the response of the oceanic Si inventory to perturbation, and consequences for atmospheric CO_2. *Global Biogeochemical Cycles, 16*(4), 1071, doi: 10.1029/2002GB001877

Roth, P. H. (1994). Distribution of coccoliths in oceanic sediments. In: A. Winter & W. G. Siesser (Eds), *Coccolithophores* (pp. 199–218). Cambridge: Cambridge University Press.

Saez, A. G., Probert, I., Young, J. R., Edvardsen, B., Eikrem, W., & Medlin, L. K. (2004). A review of the phylogeny of the haptophyta. In: H. R. Thierstein & J. Young (Eds), *Coccolithophores: From molecular processes to global impact* (pp. 251–269). Heidelberg: Springer-Verlag.

Samtleben, C., & Bickert, T. (1990). Coccoliths in sediment traps from the Norwegian Sea. *Marine Micropaleontology, 16,* 39–64.

Samtleben, C., Schäfer, P., Andruliet, H., Baumann, A., Baumann, K. H., Kohly, A., Matthiessen, J., Schröder-Ritzrau, A. & Synpal Working Group (1995). Plankton in the Norwegian-Greenlnad Sea: From living communities to sediment assemblages — an actualistic approach. *Geologische Rundschau, 84,* 108–136.

Shipboard Scientific Party. (1998). Explanatory notes. In: G. Wefer, W. H. Berger, C. Richter, et al. *Proceedings of Ocean Drilling Program, Initial Reports, 175.* College Station, TX (Ocean Drilling Program) (pp. 27–46).

Steinmetz, J. C. (1991). Calcareous nannoplankton biocoenosis: Sediment trap studies in the equatorial Atlantic, central Pacific, and Panama Basin. In: S. Honjo (Ed.), *Ocean Biocoenosis Series 1* (p. 85). Woods Hole Oceanographic Inst. Press.

Steinmetz, J. C. (1994). Sedimentation of coccolithophores. In: A. Winter & W. G. Siesser (Eds), *Coccolithophores* (pp. 179–198). Cambridge: Cambridge University Press.

Stoll, H. M., & Schrag, D. P. (2000). Coccolith Sr/Ca as a new indicator of coccolithophorid calcification and growth rate. *Geochemistry, Geophysics, And Geosystems, 1,* 1–24.

Stoll, H. M., & Ziveri, P. (2002). Separation of monospecific and restricted coccolith assemblages from sediments using differential settling velocity. *Marine Micropaleontology, 46,* 209–221.

Stoll, H. M., & Ziveri, P. (2004). Coccolithophorid-based geochemical paleoproxies. In: H. R. Thierstein & J. Young (Eds), *Coccolithophores — from molecular processes to global impact* (pp. 529–562). Heidelberg: Springer-Verlag.

Tanaka, Y. (1991). Calcareous nannoplankton thanatocoenoses in surface sediments from seas around Japan. *Science Reports of the Tôhoku University, 61,* 127–198.

Tappan, H. (1980). *The paleobiology of plant protists.* San Francisco, CA: W.H. Freeman.

Thierstein, H. R., Geitzenauer, K., Molfino, B., & Shackleton, N. J. (1977). Global synchroneity of late Pleistocene coccolith datum levels, validation by oxygen isotopes. *Geology, 5,* 400–404.

Thierstein, H. R., & Young, J. (2004). *Coccolithophores — from molecular processes to global impact* (p. 565). Heidelberg: Springer-Verlag.

Van der Wal, P., Kempers, R. S., & Veldhuis, M. J. W. (1995). Production and downward flux of organic matter and calcite in a North Sea bloom of the coccolithophore *Emiliania huxleyi. Marine Ecology Progress Series, 126,* 247–265.

Van Lenning, K., Probert, I., Latasa, M., Estrada, M., & Young, J. (2004). Pigment diversity of coccolithophores in relation to taxonomy, phylogeny and ecological preferences. In: H. R. Thierstein & J. Young (Eds), *Coccolithophores — from molecular processes to global impact* (pp. 51–73). Heidelberg: Springer-Verlag.

Weaver, P. P. E. (1993). High resolution stratigraphy of marine Quaternary sequences. In: I. A. Hailwood & R. B. Kidd (Eds), *High resolution stratigraphy* (Vol. 70, pp. 137–153). London: Geological Society Special Publication.

Wei, W. (1993). Calibration of upper Pliocene-lower Pleistocene nannofossil events with oxygen isotope stratigraphy. *Paleoceanography, 8,* 85–89.

Westbroek, P., Brown, C. W., van Bleijswijk, J., Brownlee, C., Brummer, G. J., Conte, M., Egge, J., Fernandez, E., Jordan, R., Knappertsbusch, M., Stefels, J., Veldhuis, M., van der Wal, P., & Young, J. (1993). A model system approach to biological climate forcing. The example of *Emiliania huxleyi. Global and Planetary Change, 8,* 27–46.

Westphal, H., Munnecke, A., Pross, J., & Herrle, J. O. (2004). Multiproxy approach to understanding the origin of Cretaceous pelagic limestone–marl alternations (DSDP site 391, Blake-Bahama Basin). *Sedimentology, 51,* 109–126.

Williams, J. R., & Bralower, T. J. (1995). Nannofossil assemblages, fine fraction stable isotopes, and the paleoceanography of the Valanginian-Barremian (Early Cretaceous) North Sea Basin. *Paleoceanography, 10,* 815–839.

Winter, A., Jordan, R. W., & Roth, P. H. (1994). Biogeography of living coccolithophores in ocean waters. In: A. Winter & W. G. Siesser (Eds), *Coccolithophores* (pp. 161–178). Cambridge: Cambridge University Press.

Winter, A., & Siesser, W. G. (1994). Atlas of living coccolithophores. In: A. Winter & W. G. Siesser (Eds), *Coccolithophores* (pp. 107–160). Cambridge: Cambridge University Press.

Wyrtki, K., & Kilonsky, B. (1984). Mean water and current structure during the Hawaii-to-Tahiti Shuttle Experiment. *Journal of Physical Oceanography, 14,* 242–254.

Young, J. R. (1994). Variation in *Emiliania huxleyi* coccolith morphology in samples from the Norwegian EHUX experiment, 1992. *Sarsia, 79/4,* 417–425.

Young, J. R., Geisen, L., Cros, L., Kleijne, A., Sprengel, C., Probert, I., & Ostergaard, J. (2003). A guide to extant coccolithophore taxonomy. *Journal of Nannoplankton Research*, *1*(Special Issue), 125.

Young, J. R., & Westbroek, P. (1991). Genotypic variation in the coccolithophorid species *Emiliania huxleyi*. *Marine Micropaleontology*, *18*, 5–23.

Young, J. R., & Ziveri, P. (2000). Calculation of coccolith volume and its use in calibration of carbonate flux estimates. *Deep-Sea Research II*, *47*, 1679–1700.

Ziveri, P., Baumann, K. H., Böckel, B., Bollmann, J., & Young, J. (2004). Biogeography of selected Holocene coccoliths in the Atlantic Ocean. In: H. R. Thierstein & J. Young (Eds), *Coccolithophores — from molecular processes to global impact* (pp. 403–428). Heidelberg: Springer-Verlag.

Ziveri, P., Broerse, A. T. C., Van Hinte, J. E., Westbroek, P., & Honjo, S. (2000). The fate of coccoliths at 48°N 21°W, northeastern Atlantic. *Deep-Sea Research II*, *47*, 1853–1876.

Ziveri, P., Stoll, H., Probert, I., Klaas, C., Geisen, M., Ganssen, G., & Young, J. (2003). Stable isotope "vital effects" in coccolith calcite. *Earth and Planetary Science Letters*, *210*, 137–149.

Ziveri, P., Thunell, R. C., & Rio, D. (1995). Export production of coccolithophores in an upwelling region: Results from San Pedro Basin, Southern California Borderlands. *Marine Micropaleontology*, *24*, 335–358.

CHAPTER ELEVEN

Biomarkers as Paleoceanographic Proxies

Antoni Rosell-Melé* *and* Erin L. McClymont

Contents

1. Preliminary Considerations	441
2. Methodological Approaches	443
2.1. Sea surface temperature	445
2.2. Salinity	456
2.3. Export productivity of algal class types	459
2.4. Dissolved carbon dioxide	460
2.5. Land-Ocean correlations: changes in continental vegetation	463
3. Applications	466
3.1. Sea-Surface temperature	467
3.2. Marine carbon cycle: CO_2 and phytoplankton production	470
3.3. Land vegetation changes	473
4. Concluding Remarks	474
Acknowledgments	476
References	476

1. Preliminary Considerations

In palaeoceanography the term biomarker is used to name organic molecules found in sediments, initially produced by a variety of organisms either on land or in the aquatic environment. A key characteristic of biomarkers is that after their biosynthesis, and the death of the source organisms, they survive deposition to sediments in a recognizable form in terms of their original structure and sterical configuration (i.e., spatial distribution of the atoms). They can thus be considered chemical fossils (Eglinton & Calvin, 1967). The usefulness of organic components as palaeoproxies largely depends on their resilience to early degradation processes during sedimentation and after incorporation into the sediment. Some molecules appear very resistant to degradation (e.g., aliphatic hydrocarbons), whereas others are barely preserved in the sedimentary record (e.g., carotenoids and amino acids). The overall effect of diagenesis will be a reduction of the absolute amounts within all compound classes with increasing water column and sediment depth (e.g., Wakeham, Hedges, Lee, Peterson, & Hernes, 1997; Wakeham & Lee, 1993).

---

* Corresponding author.

Developments in Marine Geology, Volume 1
ISSN 1572-5480, DOI 10.1016/S1572-5480(07)01016-0

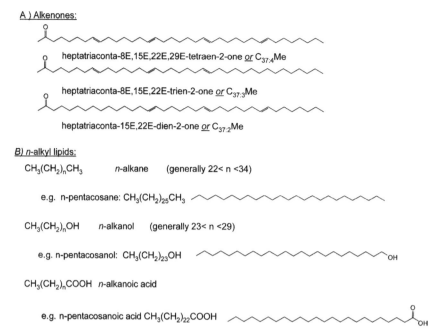

A) Alkenones:

heptatriaconta-8E,15E,22E,29E-tetraen-2-one or $C_{37:4}$Me

heptatriaconta-8E,15E,22E-trien-2-one or $C_{37:3}$Me

heptatriaconta-15E,22E-dien-2-one or $C_{37:2}$Me

B) n-alkyl lipids:

$CH_3(CH_2)_nCH_3$ n-alkane   (generally 22< n <34)

e.g. n-pentacosane: $CH_3(CH_2)_{25}CH_3$

$CH_3(CH_2)_nOH$ n-alkanol   (generally 23< n <29)

e.g. n-pentacosanol: $CH_3(CH_2)_{23}OH$

$CH_3(CH_2)_nCOOH$ n-alkanoic acid

e.g. n-pentacosanoic acid $CH_3(CH_2)_{22}COOH$

Figure 1 Structures of commonly applied biomarkers, their IUPAC nomenclature, and shorthand notations. (A) the C_{37} alkenones, produced by the Haptophyte algae. Structures determined by de Leeuw et al. (1980) and Rechka and Maxwell (1988); (B) n-alkyl lipids, found in a wide variety of organisms but associated with terrestrial higher plant inputs in marine sediments.

The most common biomarkers used as climate proxies belong to a few compound types, namely C_{37} alkenones and n-alkyl lipids (Figure 1). The key factors that make them suitable as proxies are that they are extremely common in sediments, their sources are identified, and their study provides information on key environmental variables. They are also more refractory to degradation than other components, so that sediments become relatively enriched as other compounds are being degraded (Madureira, Conte, & Eglinton, 1995; Prahl, de Lange, Lyle, & Sparrow, 1989a; Sinninghe Damsté, Rijpstra, & Reichart, 2002a; Wakeham et al., 1997), and their molecular structures remain unchanged in immature or shallow sediments.

Biomarkers may be transported to sediments as part of the remains of the original organism (e.g., leaf debris, marine snow), its digested remains (e.g., faecal pellets), or adsorbed to mineral particles (i.e., ballast minerals: silicate and carbonate biominerals, and dust), which will eventually settle on the ocean bottom through gravity. Allochthonous materials are carried to the site of deposition by wind and water (i.e., rivers and currents) (e.g., Fahl & Stein, 1999; Poynter, Farrimond, Robinson, & Eglinton, 1989; Prahl, 1985; Simoneit, Chester, & Eglinton, 1977). Materials can be further redistributed laterally and vertically by sediment reworking, through erosion and redeposition by sediment-gravity flow processes, bottom water currents, tidal movement, or contour currents, which are common processes in the continental margins and abyssal depths (Thomsen & Gust, 2000; Weaver, Wynn, Kenyon, & Evan, 2000). In areas and time periods of intense glaciomarine sedimentological activity, sea–ice and icebergs can also be a vehicle for the transport of continental organic matter to the sea-floor via ice rafted debris or dropstones, resuspension of marine shelf deposits by iceberg scouring, and gravimetric flows caused by advancing ice-sheets (Fahl & Stein, 1999; Rosell-Melé & Koç, 1997; Rosell-Melé, Maslin, Maxwell, & Schaeffer, 1997; Thomsen, Schulz-Bull, Petrick, & Duinker, 1998). Different organic components may also be associated with particles of different size (Thompson & Eglinton, 1978; Wade & Quinn, 1979). During particle dispersion, spatial fractionation of the various chemicals may take place according to the hydrodynamic properties of the carrier particles (Prahl, 1985). Clearly, organic components are subjected to a range of modes of transport, which can operate on different timescales.

It cannot be taken for granted that in the same stratigraphic horizon all biomarkers will have the same age, especially those with different origins, as it has been demonstrated through ^{14}C dating of phytoplanktonic, archaeal, and higher-plant biomarkers in marine sediments (Eglinton et al., 1997; Mollenhauer et al., 2003; Smittenberg et al., 2004). Moreover, consideration must be given to the age of the biomarkers in a sample in relation to the age of other sediment constituents that are used to investigate other proxy parameters, particularly foraminifera that are typically used to establish the age models of a sediment record. In high sedimentation sites, the ages of alkenones have been reported to be older in comparison to co-occurring foraminifera, by up to 7,000 yr in Bermuda Rise drift sediments (Ohkouchi, Eglinton, Keigwin, & Hayes, 2002), 4,500 yr in the Namibian slope, or 1,000 yr in Chilean margin sediments (Mollenhauer et al., 2005; Figure 2). In these settings lateral advection of (recycled) biomarker-bearing material may explain age decoupling between biomarkers and other proxy parameters as a function of particle size (Benthien & Müller, 2000). In contrast, alkenones and foraminifera have been found to be of similar age in NW African and South China Sea sediments (Mollenhauer et al., 2005). In a site off Northeastern Japan, the alkenones were found to be younger than the planktonic foraminifera (Uchida et al., 2005). In this latter case, the area studied is characterized by low accumulation rates of sediments, and the offsets between sediment constituents might be caused by bioturbation. The association of biomarkers and microfossils to particles of different sizes may lead to a decoupling of different palaeo records obtained from the same core, because the extent of bioturbation depends on particle size (Bard, 2001).

2. METHODOLOGICAL APPROACHES

The interpretation of biomarker data relies on determining the presence, absolute and relative contents, and isotopic composition of specific components. In this chapter only detailed attention will be given to those biomarker proxies that the authors consider to be well established, or that show considerable potential and represent significant recent contributions to the paleoceanographer tool kit. The list

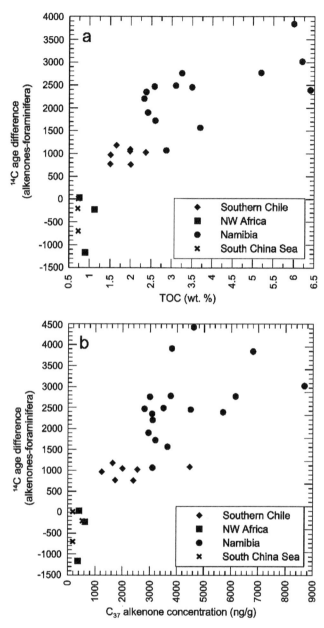

Figure 2 Differences in radiocarbon age (as radiocarbon years) measured between alkenones and foraminifera and their relationship to (a) total organic carbon (TOC) content and (b) C_{37} alkenone concentration in sediment samples from four different continental margin settings. Note that age offsets between alkenones and foraminifera are largest where TOC content and alkenone concentration are highest, and seem to be negligible where sediments are relatively poor in TOC and alkenones. *Source*: Mollenhauer et al. (2005). With permission from American Geophysical Union.

of paleoceanographic proxies derived from biomarkers is thus longer than those considered in this paper. The reader is directed to seek elsewhere examples of approaches proposed to reconstruct photic zone anoxia (Menzel, Hopmans, van Bergen, de Leeuw, & Sinninghe Damsté, 2002; Passier et al., 1999; Sinninghe Damsté, Wakeham, Kohnen, Hayes, & de Leeuw, 1993), sea-level rise (Kim, Dupont, Behling, & Versteegh, 2005; Scourse et al., 2005; Versteegh et al., 2004), phytoplankton communities (Dahl, Repeta, & Goericke, 2004; Schubert et al., 1998; Werne, Hollander, Lyons, & Peterson, 2000), water masses (Rosell-Melé, Weinelt, Koç, Jansen, & Sarnthein, 1998; Versteegh, Bosch, & de Leeuw, 1997; Versteegh, Jansen, de Leeuw, & Schneider, 2000), and biomass burning (Elias, Simoneit, Cordeiro, & Turcq, 2001; Simoneit & Elias, 2000; Simoneit et al., 1999) amongst other developments.

2.1. Sea Surface Temperature

2.1.1. The $U^{K'}_{37}$ index

The definition of an alkenone proxy to reconstruct sea surface temperature (SST) was first investigated by the group at the Organic Geochemistry Unit in the University of Bristol in the early 1980s. It is based on the measurement of the relative content in sediments of alkenones with 37 carbon atoms and 2, 3, or 4 double bonds (abbreviated as e.g., 37:4 or $C_{37:4}$, which stands for C_{37} alkenone with 4 double bonds). As described in Brassell (1993), the observation that the number of unsaturations in C_{37} alkenones varied in marine sediments from different locations led to the investigation of a possible climate control on the degree of unsaturations of the molecules, given that it was already known that the biological production of some lipids was influenced by the growth temperatures of the source organism. An unsaturation index was first defined in Brassell, Eglinton, Marlowe, Pflaumann, and Sarnthein (1986) based on extensive culture and sediment work (Marlowe, 1984):

$$U^K_{37} = \frac{C_{37:2} - C_{37:4}}{C_{37:2} + C_{37:3} + C_{37:4}} \tag{1}$$

U^K_{37} stands for unsaturated ketones with 37 carbon atoms. The work in the largely unpublished thesis by Marlowe (1984) suggests that the formula of U^K_{37} was derived by an empirical trial and error process until an index was obtained that afforded the best linear fit with growth temperature from cultured algae. The tetraunsaturated alkenone, $C_{37:4}$, is not common outside locations in the high latitudes and some coastal sites. This soon prompted the definition of the simplified alkenone index, $U^{K'}_{37}$, without $C_{37:4}$ (Prahl & Wakeham, 1987):

$$U^{K'}_{37} = \frac{C_{37:2}}{C_{37:2} + C_{37:3}} \tag{2}$$

Towards colder temperatures the number of unsaturations in the alkenones increases, whereas in some tropical settings the index tends towards 1 and the $C_{37:2}$ is often the only component that can be detected (Figure 3).

Physiological role of alkenones. Alkenones are biosynthesized by some algae of the class Haptophyceae (also known as Prymnesiophyceae) (Marlowe, Brassell,

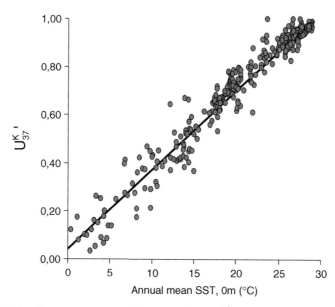

Figure 3 Global sediment core-top calibrations of the $U^{K'}_{37}$ index to mean annual sea-surface temperatures at 0 m depth. The solid line corresponds to the regression equation indicated in the graph. *Source*: Adapted from Müller et al. (1998).

Eglinton, & Green, 1984a; Marlowe et al., 1984b). This includes the cocco-lithophorid *Emiliania huxleyi*, which is the most abundant and widespread cocco-lithophorid in the oceans (Okada & Honjo, 1973; Okada & McIntyre, 1977; Winter, Jordan, & Roth, 1994) and consequently, it is probably the main producer of alkenones found in recent sediments in addition to contributions from the other notably abundant coccolithophorid *Gephyrocapsae oceanica* (Conte, Thompson, Eglinton, & Green, 1995; Conte, Volkman, & Eglinton, 1994; Marlowe et al., 1984a, 1984b; Volkman, Barrett, Blackburn, & Sikes, 1995; Volkman, Eglinton, Corner, & Forsberg, 1980a; Volkman, Eglinton, Corner, & Sargent, 1980b). Deter-mination of the biosynthetic role and cellular location of the alkenones remains uncertain. Initial proposals that alkenones were membrane lipids and played a role in maintaining membrane fluidity (Brassell et al., 1986; Prahl, Muehlhausen, & Zahnle, 1988) have not been endorsed by subsequent studies. Eltgroth, Watwood, and Wolfe (2005) have suggested that alkenones could be synthesized in chloroplasts and present in lipid bodies, and in agreement with others (Bell & Pond, 1996; Epstein, D'Hondt, & Hargraves, 2001) have proposed that alkenones are metabolic storage lipids. Sawada and Shiraiwa (2004) have suggested, in contrast, that the alkenones together with their related alkenoates might be associated with the function of the coccolith-producing compartment and play a role as buoyancy controllers for the heavy cells bearing coccoliths.

Alkenone production. Given their phytoplanktonic source, the alkenone "export production" must originate from somewhere in the euphotic zone, and it will eventually be incorporated into sediments. Direct measurements of alkenones in the upper water column indicate that the zone of maximum alkenone production originates in the isothermal surface mixed layer (SML) rather than within a deep chlorophyll maximum layer (DCML) as suggested on occasions (Ohkouchi, Kawamura, Kawahata, & Okada, 1999; Prahl, Collier, Dymond, Lyle, & Sparrow, 1993; Prahl, Pilskaln, & Sparrow, 2001). For instance, Conte, Weber, King, and Wakeham (2001) showed that over a 7 yr period in the western Sargasso Sea alkenone concentrations in the surface mixed layer (0–20 m) were generally 2–4 times higher than in the deep fluorescent maximum (75–110 m), consistent with *E. huxleyi* concentration profiles (Haidar & Thierstein, 2001). Using *in situ* incubation of water samples labeled with ^{13}C, the maximum in the alkenone production rate in the euphotic zone of Sagami Bay, Japan, was observed in a narrow layer in the upper euphotic zone, particularly around 5 m depth, above the DCML (Hamanaka, Sawada, & Tanoue, 2000). In the North Pacific subtropical gyre, based on a bi-annual sediment trap collection, bi-seasonal measurement of alkenones depth profiles in the water column, and ^{13}C labeling incubation experiments, Prahl, Popp, Karl, and Sparrow (2005) concluded that the depth of maximum alkenone export in winter and summer is within the SML rather than the DCLM.

The production of alkenones throughout the year, or seasonality, develops with maxima in export fluxes occurring once or twice a year. The timing of maximum production differs between oceanic regions. Spring alkenone maximum fluxes have been reported in the NE Atlantic and the NE Pacific (Prahl et al., 1993; Rosell-Melé, Comes, Müller, & Ziveri, 2000). Summer alkenone "blooms" have been reported off north-eastern United States (Prahl et al., 2001) and in the Southern Ocean (Indian Ocean sector) (Ternois et al., 1998). An autumn bloom of maximum alkenone production occurs in the Mediterranean (Sicre, Ternois, Miquel, & Marty, 1999; Ternois, Sicre, Boireau, Marty, & Miquel, 1996), in the central equatorial Pacific (Harada, Handa, Harada, & Matsuoka, 2001), and in the Norwegian Sea (Thomsen et al., 1998). Biannual blooms have been reported in the North Pacific subtropical gyre (Prahl et al., 2005), the western Mediterranean Sea (Ternois et al., 1996), and the NE Norwegian Sea (Thomsen et al., 1998). Conte, Weber, and Ralph (1998b) have shown that in the Sargasso Sea erratic high fluxes of alkenones occur which are not always associated to the annual spring bloom. In the coastal upwelling region off Cape Blanc, alkenone fluxes have shown considerable inter-annual variations and no consistent seasonality over a 4 yr period (Müller & Fischer, 2001). In the central Arabian Sea particulate fluxes of alkenones displayed distinct maxima at the start and end of the NE and SW Monsoons (Prahl, Dymond, & Sparrow, 2000). In the Gulf of California high fluxes of alkenones tend to occur during the post-upwelling (warmer SST) periods, past the expected spring bloom (Goñi, Hartz, Thunell, & Tappa, 2001). Studies with multiannual trap collections show that interannual and interseasonal variability in the intensity of the fluxes can be several fold (Müller & Fischer, 2001; Prahl et al., 2005; Sicre et al., 1999). Hence, the studies carried out to date clearly indicate that the seasonality of annual-maximum alkenone fluxes vary markedly across the oceans, depending on the characteristics of the local oceanographic settings.

Calibration of $U_{37}^{K'}$. Initial calibrations were based on measuring $U_{37}^{K'}$ in a laboratory culture of a Pacific strain of *E. huxleyi* (8–25 °C) (Prahl et al., 1988; Prahl & Wakeham, 1987). Subsequent work on marine surface sediments (Doose, Prahl, & Lyle,

1997; Herbert et al., 1998; Pelejero & Grimalt, 1997; Pichon, Sikes, Hiramatsu, & Robertson, 1998; Rosell-Melé, Eglinton, Pflaumann, & Sarnthein, 1995; Sikes, Farrington, & Keigwin, 1991; Sonzogni et al., 1997) led to the conclusion that the initial culture equation was equivalent to that derived from a global core-top calibration ($60°N$–$60°S$; 0–$29°C$) using annual mean temperatures at $0\,m$ depth from a climatological database (Müller, Kirst, Ruhland, von Storch, & Rosell-Melé, 1998; Figure 3), and further studies on surface marine sediments around the world have confirmed this observation (Bentaleb, Fontugne, & Beaufort, 2002; Ohkouchi et al., 1999; Conte et al., 2006; Prahl, Mix, & Sparrow, 2006). The current calibration of $U_{37}^{K'}$ is thus derived empirically and does not directly reflect the temporal and spatial variations of the production of alkenones in the oceans, both in terms of seasonality and depth of production. It provides an integrated "global" signal of patterns of alkenone accumulation in bottom sediments. This is quite remarkable because studies following that of Prahl et al. (1988) have revealed considerable variability in the calibration equations of $U_{37}^{K'}$ using algal cultures (Conte et al., 1995; Conte, Thompson, Lesley, & Harris, 1998a; Grimalt et al., 2000; Sawada, Handa, Shiraiwa, Danbara, & Montani, 1996; Volkman et al., 1995) or regionally, employing water column particulate matter (Conte & Eglinton, 1993; Harada, Shin, Murata, Uchida, & Nakatani, 2003; Sicre, Bard, Ezat, & Rostek, 2002; Sikes & Sicre, 2002; Sikes & Volkman, 1993; Ternois et al., 1998; Ternois, Sicre, Boireau, Conte, & Eglinton, 1997). There is also a systematic difference in the slope of the $U_{37}^{K'}$– SST relationship derived from global water column particulate organic matter (POM) samples and the data from the culture by Prahl & Wakeham (1987) and from core-tops by Müller et al. (1998). Water column POM $U_{37}^{K'}$ values tend to be lower than those found in their underlying surface sediments, particularly in the range ~5–$15°C$ (Bendle & Rosell-Melé, 2004; Conte et al., 2001; Rosell-Melé et al., 1995; Conte et al., 2006). Furthermore, a number of studies highlight a degree of nonlinearity in the relationship of alkenones to SST at high ($>25°C$) and low ($<8°C$) temperature extremes (Conte et al., 1995, 2001, 2006; Pelejero & Calvo, 2003; Rosell-Melé, 1998; Rosell-Melé et al., 1995; Sikes & Volkman, 1993; Sonzogni et al., 1997). Therefore it is apparent that in certain contexts or regions absolute temperatures derived from the "recommended" Prahl & Wakeham (1987) or Müller et al. (1998) equations are unrealistic. It has been suggested that in the Nordic Seas a calibration based on the original U_{37}^{K} index, which incorporates the $C_{37:4}$ compound, gives more accurate results than $U_{37}^{K'}$ (Bendle & Rosell-Melé, 2004; Rosell-Melé, 1998). However, precise environmental control of the abundance of the tetraunsaturated alkenone still remains to be clarified and thus this may constrain $U_{37}^{K'}$ as a proxy to estimate SST in low temperature locations. In fact, in the Nordic Seas, below $6°C$ neither the $U_{37}^{K'}$ or U_{37}^{K} index measured in surface sediments is correlated to SST. The breakdown of the sedimentary $U_{37}^{K'}$ – SST relationship in the Nordic Seas is suggested to be due to both biological limitations and inclusion of an autochthonous alkenone component in surface marine sediments (Bendle & Rosell-Melé, 2004). This contrasts with the cold waters of the Southern Ocean, where $U_{37}^{K'}$ is well correlated to SST down to $\sim3°C$, or in the South Atlantic where the correlation reaches $0°C$ (Sikes, Volkamn, Robertson, & Pichon, 1997; Müller et al., 1998).

The observed discrepancies between predicted and measured alkenone ratios at a given temperature, based on the calibration equations of Prahl/Müller is explained in some studies by ecological or physiological factors, or some combination of the two. As already mentioned, the quantitative response of $U_{37}^{K'}$ to growth temperature can differ significantly between species and even between strains of the same species of alkenone-producing haptophytes (Conte et al., 1998a). It is also recognized that non-thermal factors such as nutrient and light availability can alter the $U_{37}^{K'}$ value set by alkenone producers growing under isothermal conditions in laboratory conditions (Epstein et al., 2001; Epstein, D'Hondt, Quinn, Zhang, & Hargraves, 1998; Popp, Kenig, Wakeham, Laws, & Bidigare, 1998a; Prahl et al., 2006; Prahl, Wolfe, & Sparrow, 2003; Versteegh, Riegman, de Leeuw, & Jansen, 2001). However, it is not clear how the laboratory observations of non-thermal effects on $U_{37}^{K'}$ translate to the interpretation of field observations and, at present, quantification of ecological biases in the accuracy of $U_{37}^{K'}$ in sediment samples cannot be determined confidently (Prahl et al., 2006). For paleoceanographic applications, where sediment samples rarely have annual or interannual resolution, paleotemperature assessments based on $U_{37}^{K'}$ are unlikely to be impaired by the higher frequency temporal variability in the surface water ecology of alkenone producers (Prahl et al., 2005).

Occurrence of alkenones in the sedimentary record. Di- and tri-unsaturated alkenones are widespread components of modern sediments from all the oceans and marginal seas (e.g., for surveys, see Brassell, 1993; Müller et al., 1998), and occur continuously in sediments spanning the Cenozoic, at least since the Eocene (Lutz, Gieren, Luckge, Wilkes, & Littke, 2000; Marlow, Lange, Wefer, & Rosell-Melé, 2000; Marlowe, Brassell, Eglinton, & Green, 1990; Pagani, Freeman, & Arthur, 1999b). A single C_{40} alkadien-3-one has been found in sediments deposited during the Paleocene (Yamamoto, Ficken, Baas, Bosch, & deLeeuw, 1996). In mid-Cretaceous sediments only di-unsaturated C_{37}–C_{39} alkenones (Albian ~105 Ma, Farrimond, Eglinton, & Brassell, 1986; Aptian ~120.5 Ma, Brassell & Dumitrescu, 2004) have been reported. Brassell and Dumitrescu (2004) have suggested that, given that the earliest occurrence of alkatrienones is in the Eocene at high latitudes (Marlowe et al., 1990), temperature controls on alkenone unsaturation may represent a biological response triggered by global cooling in the Paleogene. This may indicate that the application of $U_{37}^{K'}$ to sediments may not be justified in sediments pre-dating the Cenozoic, assuming that the calibration principles employed in assessment of contemporary characteristics are equally valid as a paleotemperature proxy throughout the Cenozoic. This is based on the biochemical response of *E. huxleyi* to growth temperature, given that it is the most abundant species of coccolithophorid at present and the calibration of $U_{37}^{K'}$ is based on modern surface sediments and a culture of this organism (Müller et al., 1998; Prahl et al., 1988). In fact, *E. huxleyi* only emerged in the sedimentary record *ca.* 260 ka, becoming dominant after *ca.* 85–77 ka (Hay, 1977; Thierstein, Geitzenauer, & Molfino, 1977). Therefore, successful application of the alkenone palaeothermometer relies on past alkenone-synthesisers responding to growth-temperature changes in a similar manner to the modern alkenone producers, chiefly *E. huxleyi*. Combined analyses of coccolithophore assemblages and alkenones from sediments in the south-east and north-east

Atlantic suggest that this assumption is correct at least for the last 1.5 Myr (e.g., Figure 4; McClymont, Rosell-Melé, Giraudeau, Pierre, & Lloyd, 2005; Müller, Cepek, Ruhland, & Schneider, 1997; Villanueva, Flores, & Grimalt, 2002; Weaver et al., 1999). Significant changes to the coccolithophore assemblages have occurred with no detectable impact on $U_{37}^{K'}$ values, and vice versa. The fossil assemblages also tended to be dominated by species of *Gephyrocapsa*, which are abundant in the modern ocean. These are also contributors to the modern data set of alkenones in surface sediments and hence to the core-top $U_{37}^{K'}$–SST calibration; similar $U_{37}^{K'}$/SST

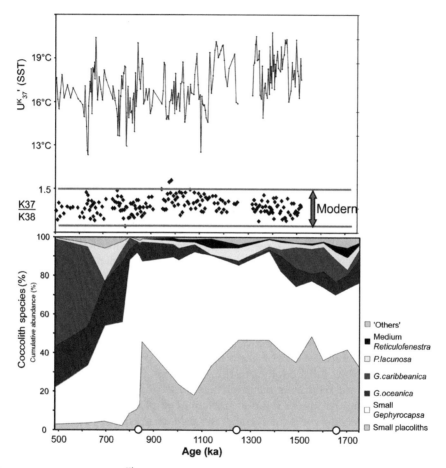

Figure 4   Comparison of $U_{37}^{K'}$ derived SSTs and coccolith extinction events during the early- and mid-Pleistocene. (A) $U_{37}^{K'}$ derived SST record; (B) ratio between C_{37} and C_{38} alkenones, showing the modern range in sediments and cultures with strong $U_{37}^{K'}$-SST correlation; (C) relative abundances of the dominant coccolith species groups. Open circles along the *x*-axis denote last appearance datums for three coccolith species identified in ODP Shipboard biostratigraphy (*Calcidiscus macintyrei* at ca. 1,670 ka, *Hellicosphaera selli* at ca. 1,250 ka, *Reticulofenestra asanoi* at ca. 830 ka). Note absence of change in K37/K38 ratio despite changes in coccolith assemblages, which supports the application of the modern $U_{37}^{K'}$-SST calibration. Adapted from McClymont et al. (2005).

sensitivities to that of *E. huxleyi* in the ocean are thus expected in the sedimentary records to enable the application of $U_{37}^{K'}$ as a paleo-thermometer.

Alkenone diagenesis and $U_{37}^{K'}$. The transport of organic carbon to marine sediments must also compete with alteration and remineralization within the water column, and both during and after deposition. The export flux of alkenones to sediments in pelagic regions is generally less than 1% of the total production in the euphotic zone (Conte, Eglinton, & Madureira, 1992; Müller & Fischer, 2001; Prahl et al., 2000; Wakeham et al., 1997). Alkenones in settling POM and at the sediment–water interface have been shown to be at least or more labile relative to total organic carbon (TOC) and undergo considerable alteration and remineralization. Only in a few instances have alkenones been reported as being less labile than TOC (in the Gulf of Maine, Prahl et al., 2001) possibly owing to the suboxic water column. In highly productive and near-coastal sites the fraction preserved of alkenones in sediments has been reported to be between 25 and 100% of the euphotic zone production (Goñi et al., 2001; Prahl et al., 1993, 2001). However alkenones have been found to be refractory relative to nearly all other functionalized lipids (Conte et al., 1992; Freeman & Wakeham, 1992; Gong & Hollander, 1999; Madureira et al., 1995; Prahl et al., 1993, 2000; Prahl & Muehlhausen, 1989). This may be caused by the uncommon *E* or *trans* configuration of the double bonds in the chain (Rechka & Maxwell, 1988), which may contribute to make them more resilient to bacterial attack (Brassell, 1993).

Diagenetic alteration of the $U_{37}^{K'}$ signal has been a concern since the definition of the proxy, owing to the increased functionality in $C_{37:3}$ and hence increased reactivity from the additional double bond compared to $C_{37:2}$. In fact, some researchers have argued that the tri-unsaturated component degrades preferentially during deposition (Gong & Hollander, 1999; Hoefs, Versteegh, Rijpstra, de Leeuw, & Sinninghe-Damsté, 1998). However, to date a demonstration of such a mechanism widely affecting the alkenone record, and hence biasing $U_{37}^{K'}$, is lacking. No sediment trap study has shown a shift in $U_{37}^{K'}$ with depth related to degradation (Conte et al., 1992; Müller & Fischer, 2001; Sawada, Handa, & Nakatsuka, 1998). Experiments show that the $U_{37}^{K'}$ index is unaffected by enzyme-mediated digestion and faecal pellet production (Grice et al., 1998; Volkman et al., 1980b), and the negative bias resulting from photodegradation and aerobic microbial degradation in the upper water column is small (<0.05; Rontani, Cuny, Grossi, & Beker, 1997). The $U_{37}^{K'}$ signal was also unaffected by diagenesis of organic carbon (including alkenones) in mm- and cm-scale sediment profiles from the Peru margin (McCaffrey, Farrington, & Repeta, 1990) and the Biscay Abyssal Plain (Madureira et al., 1995). Laboratory simulations of microbial activity provide the only direct measurements of the effects of biological activity in surface sediments. Teece, Getliff, Leftley, Parkes, and Maxwell (1998) identified no significant bias to $U_{37}^{K'}$ between slurries of sediment and denatured *E. huxleyi* incubated with bacteria under oxic, sulphate reducing, and methanogenic conditions. Different bottom water redox conditions from three sites in the Indian Ocean were not associated with any diagenetic bias in $U_{37}^{K'}$ (Sinninghe Damsté et al., 2002a), and although severe post-depositional degradation of alkenones in the oxic layer across acute redox boundaries in the turbidites of the Madeira Abyssal Plain was observed by Prahl et al. (1989a) and Hoefs et al. (1998)

only the latter study found evidence for a significant positive bias to $U_{37}^{K'}$ in the oxic layer. Grimalt et al. (2000) suggested that this may be the result of low alkenone concentrations in the older (Miocene and Pliocene) turbidites of Hoefs et al. (1998) introducing analytical bias to the measured $U_{37}^{K'}$ value (Rosell-Melé et al., 2001). It is not clear whether Hoefs et al. (1998) circumvented this problem by extracting larger amounts of sample and/or having smaller analyte dilution volumes than Prahl et al. (1989a) and thus whether the discrepancy represents analytical or diagenetic bias in the $U_{37}^{K'}$ index.

As with many other sedimentary components, alkenones will undergo sedimentary diagenesis with increasing depth of burial and temperature. Increasing burial leads to secondary sedimentary processes, which tend to be controlled by increased anoxia and reducing conditions and at greater depths by thermal alteration, all of which could lead to a $U_{37}^{K'}$ bias. Marlowe et al. (1990) noted that in these environments the alkenones are likely to be the limiting reactant and the diagenetic products that are produced should be preserved in the sedimentary matrix. For example, high sedimentary accumulation rates will lead to enhanced burial and sub-oxic conditions causing exposure to reactive species such as inorganic sulphides and hydrogenating conditions. Possible mechanisms for the diagenetic removal of alkenones from the "free" solvent extractable fraction to the "bound" fraction have been indicated by the presence of alkenones and C_{37} n-alkanes and related products released after chemical and physicochemical degradation of sedimentary organic matter (Hoefs et al., 1998; Koopmans et al., 1997; Sinninghe Damsté, Rijpstra, Kock-van Dalen, de Leeuw, & Schenck, 1989). However, there have not been many published investigations on the fate of the $U_{37}^{K'}$ signal during these types of secondary sedimentary processes. Marlow (2001) addressed this issue by undertaking a detailed study of the potential role of geomacromolecular formation in the post-depositional fate of alkenones, by chemically and physicochemically (by hydropyrolysis) degrading the bound macromolecular organic matter of sediments from the Benguela upwelling system. Minor amounts of alkenones and potentially related diagenetic products were released by hydropyrolysis and chemolysis compared to the abundance of free alkenones. In a worst-case scenario the post-depositional bias in $U_{37}^{K'}$ was estimated to be 0.4°C in sediments spanning the Pleistocene and Pliocene.

2.1.2. The TEX$_{86}$ index

In 2002, Schouten, Hopmans, Schefuss, and Sinninghe Damsté (2002a) proposed a new tool to reconstruct past seawater temperatures based on measuring the average number of cyclopentane rings found in glycerol dialkyl glycerol tetraethers (GDGTs; Figure 5). After analyzing GDGTs distribution in 40 marine surface sediments from 15 locations worldwide and comparing them with annual mean SSTs from a climatological database they found that the best fit was obtained with the TEX$_{86}$ index (TetraEther indeX of tetraethers consisting of 86 carbon atoms; Figure 6):

$$TEX_{86} = \frac{[IV] + [V] + [VI]}{[III] + [IV] + [V] + [VI]} \tag{3}$$

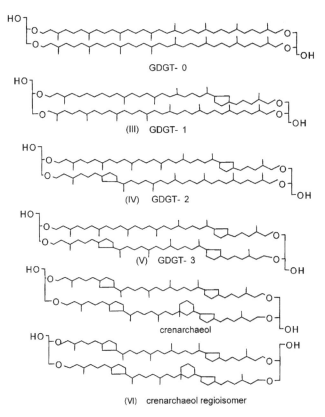

Figure 5 Structures of the glycerol diphytanyl glycerol tetraethers (GDGTs) produced by Archaea and used in the generation of the TEX_{86} index.

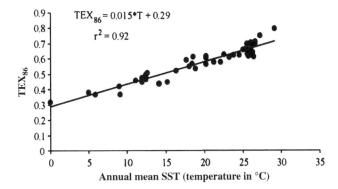

Figure 6 Correlation between the TEX_{86} index in core top samples to the mean annual SST at a variety of sites (number of samples in each site between brackets): from Skan Bay (1), Saanich Inlet (1), Washington margin (7), Santa Monica Basin (2), Peru Margin (2), Cariaco Basin (1), Halley Bay Station (1), Angola Basin (24), Iberian Margin (2), Wadden Sea (3), Skagerrak and Drammensfjord (2), Aegean Sea (1), Black Sea (1), Arabian Sea (13), and Kau Bay (1). *Source*: Wuchter et al. (2004). With permission from American Geophysical Union.

The roman numerals refer to the GDGT isomers shown in Figure 5, which contain 1 (III), 2 (IV), 3 (V), or 4 (VI) cyclopentane rings (Figure 5). GDGTs are archaeabacterial membrane lipids. They have been detected in sediments more than 112 Myr old (Kuypers et al., 2001). Culture experiments of hyperthermophilic archaea have shown that the relative distribution of cyclopentane rings in the GDGTs strongly depends on culture temperatures (Derosa, Gambacorta, Nicolaus, Sodano, & Bulock, 1980; Uda, Sugai, Itoh, & Itoh, 2001). The rigid structure of the cyclopentane ring makes the membrane less fluid and the archaeal tetraether lipids seem to be the simplest and most stable membranes in living organisms (Itoh, Sugai, Uda, & Itoh, 2001). GDGTs III–VI, the components in TEX_{86}, are expected to derive chiefly from organisms of the kingdom of Crenarchaeota. This was formerly thought to consist only of hyperthermophilic organisms living at temperatures $>60°C$. However, phylogenetically related Crenarchaeota are also present in oceans and lakes with environmental temperatures ranging from 0 to $30°C$ (Delong, Wu, Prezelin, & Jovine, 1994; Fuhrman, McCallum, & Davis, 1992; Hershberger, Barns, Reysenbach, Dawson, & Pace, 1996; Karner, DeLong, & Karl, 2001; MacGregor, Moser, Alm, Nealson, & Stahl, 1997). Phylogenetically diverse crenarchaeal 16S rRNA phylotypes have also been identified in sediments estuaries (Abreu, Jurgens, De Marco, Saano, & Bordalo, 2001), in continental shelf anoxic sediments (Vetriani, Reysenbach, & Dore, 1998), and the deep sea (Vetriani, Jannasch, MacGregor, Stahl, & Reysenbach, 1999), but the biogeochemical role of Crenarchaeota in these sedimentary environments is unknown. The so-called "cold" Crenarchaeota from the marine water column biosynthesize similar GDGTs as those encountered in cultured hyperthermophilic archaea with the exception of crenarchaeol, present only in Crenarchaeota, which is uniquely characterized by the presence of a cyclohexane ring (Sinninghe Damsté, Schouten, Hopmans, van Duin, & Geenevasen, 2002b; Figure 5). Pelagic Crenarchaeota represent one of the ocean's single most abundant cell types (Karner et al., 2001), with plankton archaeal assemblages dominated by a few cosmopolitan phylotypes (Massana, DeLong, & Pedros-Alio, 2000).

The occurrence and diversity of marine pelagic Crenarcheota were noted in the 1990s. Non-extremophilic archaea (Crenarchaeota and Euryarchaeota) are now recognized to be widespread in the ocean (Fuhrman et al., 1992), but our current understanding of their physiology and biogeochemical function remains largely speculative and not much is known on their ecology and distribution in the oceans because of the lack of comprehensive surveys and difficulty to isolate single species (Francis, Roberts, Beman, Santoro, & Oakley, 2005). Consequently, the metabolic requirements of planktonic archaea remain a major unsolved issue (Herndl et al., 2005). Marine crenarchaeotes have been found in different geographic areas and at different depths in the water column, and are considered ubiquitous, abundant, and often the dominant archaeal phylotypes in most marine waters (Massana et al., 2000). In the same study it was determined that in temperate regions (Pacific Ocean, Atlantic Ocean, and Mediterranean Sea) although marine Crenarchaeota occurred throughout the water column they were more predominant at depth, in the aphotic zone. In a monitoring program, consisting of monthly sampling conducted throughout the water column (surface to 4,750 m) from September 1997 to

December 1998 at the Hawaii's Ocean Timeseries station ALOHA in the North Pacific subtropical gyre, fluorescence *in situ* hybridization (FISH) data indicate that in open ocean sites the upper 500 m of the ocean are characterized by constantly high numbers of pelagic Crenarchaeota; that they only drop below this level and comprise a large fraction of total marine picoplankton below the euphotic zone, at depths greater than 1,000 m (Karner et al., 2001). The fraction of Crenarchaeota increased with depth reaching 39% of total DNA-containing picoplankton detected. While bacteria decreased in relative abundance with increasing depth, the relative abundance of the pelagic Crenarchaeota increased sharply at the 250 m depth layer and below 1,000 m they were as common as bacteria. In surface layers, pelagic Crenarchaeota were only present sporadically, and never abundant numerically. Pelagic Crenarchaeota are thus considered a dominant component of the deep ocean and that an adaptive strategy has allowed them to radiate throughout nearly the entire oceanic water column (Karner et al., 2001). Herndl et al. (2005) have determined the abundance of Archaea in the mesopelagic and bathypelagic North Atlantic along a south–north transect of more than 4,000 km. It was determined that planktonic archaea are actively growing in the dark ocean although at lower growth rates than bacteria and might play a significant role in the oceanic carbon cycle. The contribution of Crenarchaeota increased with depth from 18.5% in the 100 m layer to 26.4% of the picoplankton abundance in the NADW, although it decreased in absolute terms with depth with maximum cell counts towards the surface (Herndl et al., 2005).

Despite the widespread abundance of Crenarchaeota in the water column the sedimentary signal of GDGTs seems to originate primarily from the upper part, *ca.* 100 m, of the water column and represent annual mean fluxes (Schouten et al., 2002a; Wuchter, Schouten, Wakeham, & Sinninghe Damsté, 2005). This is in apparent contradiction with the observation that the GDGT depth distribution has maxima below 100 m, which then decrease with depth (Wuchter et al., 2005). In addition, analyses of water samples in the Arabian Sea at depths varying between 35 and 1,500 m show that concentrations of GDGTs I–VI are on average similar for samples from 35 to 500 m depths but lower at 1,000–1,500 m depths (Sinninghe Damsté et al., 2002). It has been argued (Schouten, Hopmans, Schefuss, Wuchter, & Sinninghe Damsté, 2002b; Wakeham, Lewis, Hopmans, Schouten, & Sinninghe Damsté, 2003; Wuchter et al., 2005), based on the comparison of estimated water temperatures from TEX$_{86}$ in surface sediments and mean water temperatures from climatological databases, and the fact that TEX$_{86}$ values in suspended organic matter in the water column are relatively constant over depth, that the exported GDGT signal is better represented by crenarchaeotal cell material derived from the photic zone where an active food web exists, and transported with fast sinking aggregates that reach the seafloor. These authors have proposed that living suspended deep sea Crenarchaeota do not substantially influence the sedimentary GDGT signal probably because of the lack of aggregation processes and transport mechanisms to the seafloor.

Archaea abundances also appear to vary seasonally although information on this issue is even more scarce. Molecular biological studies and the analysis of membrane lipids have shown that marine Crenarchaeota abundance in the water column off

Antarctica, California, and The Netherlands increases after blooms in surface phytoplankton and is negatively correlated with chlorophyll concentrations (Murray et al., 1999; Murray et al., 1998; Wuchter et al., 2005). These studies show that Archaea are not abundant in surface waters during periods of high primary production, although data from the Northwest Atlantic indicates that these observations cannot be generalized too widely (Wuchter et al., 2005).

Calibration of TEX_{86} against SST has been undertaken by comparing values of TEX_{86} in surface sediments from a variety of locations with water temperatures in the upper water column (0–100 m depth). When TEX_{86} values from POM in the water column are compared with *in situ* temperatures only a weak correlation is observed (Wuchter et al., 2005). The TEX_{86} in surface sediment and POM correlates most strongly with annual mean temperature at the ocean surface, 0–10 m depth (Schouten et al., 2002a; Wuchter et al., 2005).

$TEX_{86} = 0.015 \times T+0.29$; $n = 61$, $r^2 = 0.92$ (core-tops) (Wuchter, Schouten, Coolen, & Sinninghe Damsté, 2004)

$TEX_{86} = 0.015 \times T+0.030$; $n = 118$, $r^2 = 0.70$ (POM) (Wuchter et al., 2005)

Incubation in a mesocosm of seawater containing marine crenarcheota has also confirmed that water temperature is the main control in the number of cyclopentane moieties in the crenarchaeotal membrane lipids (Wuchter et al., 2004).

TEX_{86} also seems to be resistant to diagenetic alteration (Schouten, Hopmans, & Sinninghe Damsté, 2004) although GDGTs themselves are not more refractory to degradation than other lipids and under oxic conditions they degrade at similar rates as other phytoplanktonic lipids (Sinninghe Damsté et al., 2002a). A restriction of the application of TEX_{86} is that it cannot be used in sites with high rates of anaerobic oxidation of methane because high amounts of GDGTs II–IV have been found in these sediments that are derived from anaerobic methanotrophic archaea (e.g., Pancost, Hopmans, Sinninghe Damsté, & Party, 2001; Wakeham et al., 2003). Moreover, GDGTs can also be produced by archaeabacteria thriving in soils and peats, and thus the terrestrial contribution to aquatic sediments, particularly in lakes, may bias the autochthonous TEX_{86} signal (Weijers, Schouten, van den Linden, van Geel, & Sinninghe Damsté, 2004).

2.2. Salinity

At present there are no well-established sea surface salinity (SSS) proxies available to paleoceanographers. In the northern latitudes in the Atlantic and Pacific, some studies have highlighted an empirical relationship between percent $C_{37:4}$ (% $C_{37:4}$) and SSS (Harada et al., 2003; Rosell-Melé, 1998; Rosell-Melé, Jansen, & Weinelt, 2002; Sicre et al., 2002). In the Baltic Sea and Skagerrak, increasing values of % $C_{37:4}$ and generally decreasing abundances of total C_{37} alkenones correlate with a decreasing salinity gradient (~33–8 psu) (Blanz, Emeis, & Siegel, 2005; Schulz, Schoner, & Emeis, 2000). Moreover, some have applied the % $C_{37:4}$ measurement down core as a tentative proxy to infer paleosalinity variations (Bard, Rostek, Turon, & Gendreau, 2000; Rosell-Melé, 1998; Rosell-Melé et al., 2002; Seki et al., 2005). On a global scale it has been demonstrated, however, that there is no discernable relationship of % $C_{37:4}$ to SSS (Bendle, Rosell-Melé, & Ziveri, 2005;

Sikes & Sicre, 2002).Thus, it is likely then that the relationship of % C$_{37:4}$ to SSS is restricted to individual basins. Bendle et al. (2005) have assessed the distribution of % C$_{37:4}$ in the Nordic Seas and North Atlantic POM and showed that although it is closely associated with SSS, the relationship displays considerable scatter and takes the form of a cluster of high % C$_{37:4}$ values associated with polar waters and a low value cluster with Atlantic waters, with mixing between these two groups in the intermediate water masses (Figure 7). They concluded, however, that whether an environmental factor or combination of factors promotes the production of relatively high % C$_{37:4}$ values in polar or polar influenced ocean waters remains to be established and results to date do not support the use of % C$_{37:4}$ to estimate, quantitatively, past variations in SSS. However, the data obtained do support earlier suggestions to use % C$_{37:4}$ as a proxy to estimate the relative influence of water masses in the North Atlantic and to infer local variations in the salinity budget of the surface ocean. Hence, a similar interpretation of % C$_{37:4}$ as a qualitative salinity proxy could be applied elsewhere where contrasting changes in surface salinity may have taken place in the past.

Isotopic fractionation between deuterium (^{2}H) and hydrogen (^{1}H) in water (expressed as δD) is affected by condensation and evaporation processes. A strong correlation exists between δD and precipitation (Dansgaard, 1964). This information was initially exploited in a variety of sedimentary settings, including ice cores, speleothems, and tree-ring cellulose (see Sauer, Eglinton, Hayes, Schimmelmann, & Sessions, 2001 for further references). The interpretation of δD values in bulk materials is confounded by the possible mixture of inputs from sources with

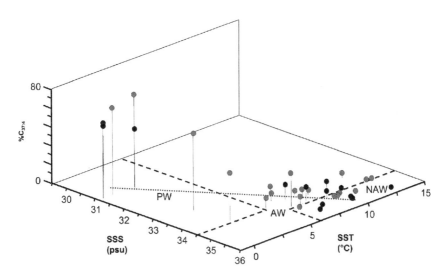

Figure 7 Comparison of the C$_{37:4}$ alkenone abundance, expressed as a percentage of total C$_{37}$ alkenones, to temperature and salinity in the Nordic Seas. The major water masses are delimited by dashed lines, based on published values (Hansen and Østerhus, 2000; Johanessen, 1986; Swift, 1986). Water masses: Polar Water (PW), Arctic Water (AW), and North Atlantic Water (NAW). The dotted line shows a linear regression between SSS and SST: $y = 1.5x + 44.2$, $r^2 = 0.54$ ($n = 50$).
Source: Bendle et al. (2005). With permission from American Geophysical Union.

potentially different δD signatures, different biosynthetic influences over δD values, and potential diagenetic alteration of bulk δD values. This can also be addressed through the application of a molecular approach given that recent advances in isotopic ratio mass spectrometric techniques have enabled the determination of very precise and accurate δD measurements from individual compounds (Sessions, Sylva, Summons, & Hayes, 2004). Consequently, biomarkers that can record past seawater δD values show great potential for reconstructing past changes in surface ocean salinity (Schouten et al., 2006). Bulk δD analysis of Mediterranean sapropels demonstrated the potential for marine sediments to record changes in surface salinities (Krishnamurthy, Meyers, & Lovan, 2000), but it is unclear what effect diagenetic exchange of hydrogen could have had on the record (e.g., Schimmelmann, Lewan, & Wintsch, 1999). Only a few studies testing compound-specific δD composition in marine lipids have been performed. Initial cultures of the alkenone-producing Haptophyte *E. huxleyi* revealed a consistent isotopic depletion in the alkenones compared to the culture media between 232 and 225‰ (Englebrecht & Sachs, 2005; Paul, 2002). Subsequent culturing of *E. huxleyi* and *G. oceanica* has identified a wider range of fractionation between the alkenones and the water in which they were grown, in the range -175 to -261‰, and a depletion in δD for *G. oceanica* relative to *E. huxleyi* of $\sim$30‰ (Schouten et al., 2006). Salinity was observed to be a strong control over both δD in the meteoric water and in the alkenones from both algal species (Figure 8). However, the relationship is complex, and in addition to the difference in isotopic depletion between the two species of Haptophyte, Schouten et al. (2006) also observed an influence of growth rate over δD-alkenone values, and noted that the δD-alkenone signal amplified the changing δD values in the water associated with changing salinity. Thus, the application of δD-alkenone analyses to reconstruct past changes in salinity require growth rates to be

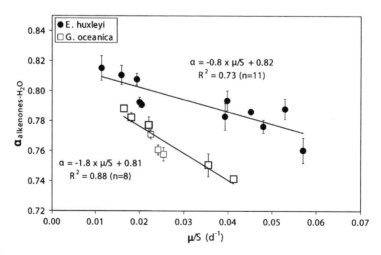

Figure 8 The isotopic fractionation of deuterium and hydrogen in C_{37} alkenones compared to the ratio between growth rate (m) and salinity (S) from culture experiments. Two different regressions are shown, for *E. huxleyi* and *G. oceanica*. *Source*: With permission from Schouten et al. (2006).

constrained, and for further research to determine the species-effect over δD fractionation.

2.3. Export Productivity of Algal Class Types

The sedimentary concentration of a biomarker can be related to the flux of the component from its source to the burial site to reconstruct changes in the productivity of the source organism (Prahl & Muehlhausen, 1989). Changes in productivity of an algal group or the total algal assemblage can be related to factors such as nutrient abundance and ecological structure, which in turn may relate to climatic conditions. For example, chlorins are biomarkers for chlorophyll and have been proposed as proxies of primary productivity (Harris et al., 1996; Rosell-Melé, 1994; Summerhayes et al., 1995). There exists a correlation between surface sediment chlorin concentrations and productivity in overlying waters which further suggest the use of chlorins as proxies for past production (Higginson, 1999; Shankle, Goericke, Franks, & Levin, 2002). However, quantitative reconstructions of past total production and the relative importance of different phytoplankton groups using biomarker accumulation rates are prone to error unless they take into account other processes which influence the content of a biomarker in the sediment, namely degradation during transport and burial of the biomarker and dilution by mineral matter. At present there is no reliable method to measure or estimate for past environments the fraction of a compound degraded or transformed during transport, deposition, and burial in order to quantify past production in the surface ocean. Variations in the processes that control biomarker degradation rates (e.g., oxygen contents, enzymatic reactions, particle surface areas, sedimentation rate) can lead to important variations in accumulation rates, even at a local scale, resulting in an unknown degree of uncertainty in the interpretation of the data (Mayer, 1994; Sinninghe Damsté et al., 2002a). A "Chlorin Index" has recently been proposed as a method for determining the extent of post-depositional diagenesis of organic matter in marine sediments (Schubert, 2005). The index uses the fluorescence intensity of sediment extracts to record the relative importance of "fresh" chlorophylls versus their degradation products. The study demonstrated promising results in a number of different sedimentary settings, and also demonstrated that the technique could be applied to palaeoceanographic archives to determine not only the accumulation rates of organic material but also its "freshness" (Schubert, 2005). The results also compared well with existing degradation indices using amino acid compositions, including the Dauwe Index (Dauwe, Middelburg, Herman, & Heip, 1999) and the β-alanine/total hydrolysable amino acid ratio (Cowie & Hedges, 1992). Changes to palaeo-redox conditions in the water column may also be determined using relative concentrations of various biomarkers (Poynter & Eglinton, 1991; Schulte, Rostek, Bard, Rullkötter, & Marchal, 1999; Sinninghe Damsté, Kuypers, Schouten, Schulte, & Rullkotter, 2003). However, it is known that degradation rates are different for different biomarkers, as indicated for instance by the relatively refractory nature of the alkenones relative to other organic components (see discussion above) and variability in biomarker oxidation rates in surface sediments (Marchand, Marty, Miquel, & Rontani, 2005). During the interpretation of changes in the relative concentrations of different algal biomarkers the possibility that the results may reflect preferential preservation rather than production must be acknowledged. This issue may in part be addressed by comparing the organic and inorganic remains of the producers in question, e.g., using coccolith and alkenone accumulation rates to determine past Haptophyte production (Weaver et al., 1999). As discussed in Section 1, issues of biomarker advection causing accumulation rate variability can also be assessed using compound-specific radiocarbon dating and a comparison to the ages of other sedimentary components in each horizon (e.g., Mollenhauer et al., 2005; Ohkouchi et al., 2002). A recent study has also demonstrated that δD signatures of biomarkers can be used to trace water masses in the North Atlantic and to detect allochthonous inputs of biomarkers to sedimentary archives (Englebrecht & Sachs, 2005). In conclusion, biomarker mass accumulation rates in sediments cannot be applied to quantify absolute past levels of phytoplanktonic production in the surface ocean and should be used with care to determine the relative contributions of different producers. However, valuable information may be gained to understand carbon cycling processes within the marine realm, and in particular the biologically driven sequestration of organic carbon into marine sediments that has been proposed to account at least in part for the glacial reductions in atmospheric CO_2 (Kohfeld, Le Quere, Harrison, & Anderson, 2005).

2.4. Dissolved Carbon Dioxide

Photosynthetic fixation of carbon by marine phytoplankton utilizes dissolved inorganic carbon (as CO_2 or bicarbonate) from the surrounding waters, and as with terrestrial plants it will discriminate strongly against ^{13}C relative to ^{12}C (Degens, Guillard, Sackett, & Hellebus, 1968; Smith & Epstein, 1971). Correlation between $\delta^{13}C$ in marine sediments, POM, and phytoplankton to surface ocean CO_2 concentrations (Bentaleb et al., 1996; Freeman & Wakeham, 1992; Rau, Takahashi, & Marais, 1989) suggested that this technique could be used to determine past CO_2 concentrations in the ocean and the atmosphere. The degree of isotopic fractionation within an organism (εp) generally shows a positive correlation to the reciprocal of dissolved CO_2 concentrations, $[CO_2(aq)]$ (Bidigare et al., 1997; Laws, Popp, Bidigare, Kennicutt, & Macko, 1995; Pagani, Arthur, & Freeman, 1999a; Popp et al., 1998b). Using Henry's Law, $[CO_2(aq)]$ can be converted to pCO_2 values with knowledge of the SST and salinity (defined as α) contemporaneous to the $[CO_2(aq)]$ measurement (Benthien et al., 2002; Pagani et al., 1999a; Weiss, 1974):

$$pCO_2 = \frac{CO_2(aq)}{\alpha} \tag{4}$$

However, the interpretation of the sedimentary record of $\delta^{13}C$ and its relationship to $[CO_2(aq)]$ is complex, due to the physiological factors that can affect εp, including cell geometry (Burkhardt, Riebesell, & Zondervan, 1999; Pancost, Freeman, Wakeham, & Robertson, 1997; Popp et al., 1998b), cellular growth rates (Bidigare et al., 1997; Burkhardt et al., 1999; Laws et al., 1995; Pancost et al., 1997), and the response of the organism to changing nutrient supplies (Eek, Whiticar, Bishop, & Wong, 1999). These factors affect the intracellular concentration of CO_2 and thus influence the degree of fractionation that takes place, and may differ between phytoplankton species and algal groups, e.g., between diatoms and haptophytes (Pancost et al., 1997; Popp et al., 1998b) and even between different compounds isolated from the same species (Riebesell, Revill, Holdsworth, & Volkman, 2000; Schouten et al., 1998). Furthermore, it has emerged that fractionation may also differ depending upon whether the phytoplankton utilize CO_2 or bicarbonate (HCO_3^-) for photosynthesis (Burkhardt et al., 1999; Gonzalez, Riebesell, Hayes, & Laws, 2001). Preferential loss of isotopically depleted (e.g., lipids) or enriched (e.g., carbohydrates, proteins) organic matter may also account for diagenetic alteration of isotope values (Frishman & Behrens, 1971; Macko, 1981). These issues make interpretation of bulk sedimentary $\delta^{13}C$ signals a difficult task on top of the uncertainties introduced by the mixed composition of bulk organic matter. Compound-specific stable isotope analyses (CSIA) can address a number of these issues, because where the source of the biomarkers is known, attention can be targeted towards understanding the physiological controls over the carbon isotope fractionation, and [CO_2(aq)] can potentially be reconstructed.

To use $\delta^{13}C$ to reconstruct past pCO_2 there is therefore a need to find biomarkers of a restricted and well-defined origin. The highly source-specific nature of the alkenones as biomarkers for the Haptophyte algae, their global distribution in surface sediments, and post-depositional stability, discussed above, makes them a promising proxy for pCO_2. Furthermore, the $\delta^{13}C$-alkenone values can be coupled to $U_{37}^{K'}$-SST values, and coccolith carbonate $\delta^{13}C$, to give pCO_2 estimates via Henry's Law with the knowledge that all information has been gained from the same source organisms. The $\delta^{13}C$ values of the alkenones have been investigated in algae cultures (Popp et al., 1998a, 1998b; Riebesell et al., 2000; Schouten et al., 1998), in the water column POM (Eek et al., 1999; Pancost et al., 1997), and in surface sediments (Benthien et al., 2002; Benthien & Müller, 2000; Pagani, Freeman, Ohkouchi, & Caldeira, 2002) to try and elucidate the major controls over carbon isotope fractionation in the Haptophytes. The relatively limited range in cell geometry and size of *E. huxleyi* and *G. oceanica* mean that the dominant controls over $\delta^{13}C_{\text{alkenone}}$ in cultures and particulate matter are [CO_2(aq)] and nutrient-driven changes in growth rates (Bidigare et al., 1997; Eek et al., 1999; Pancost et al., 1997; Popp et al., 1998a, 1998b; Riebesell et al., 2000). However, the conditions under which the algae are cultured simplify and perhaps even distort the real environmental controls over isotopic fractionation and preservation of the $\delta^{13}C$ signal, which will ultimately determine the application of $\delta^{13}C$ as a proxy. It is therefore necessary to examine and calibrate the sedimentary relationship between $\delta^{13}C$, [CO_2(aq)], and nutrient availability. Surface sediment alkenone $\delta^{13}C$ values from the equatorial and South Atlantic are positively correlated to 1/[CO_2(aq)], but also show a strong and negative correlation to phosphate and nitrate contents in the overlying waters, suggesting that nutrient-limited growth rates are the dominant control over $\delta^{13}C_{\text{alkenones}}$ (Benthien et al., 2002; Schulte, Benthien, Müller, & Ruhlemann, 2004; Figure 9). However, where the nutrient concentrations can be estimated, for example using phosphate concentrations in surface waters, [CO_2(aq)] can be calculated (Pagani et al., 2002). This study also highlighted the need to know the depth of alkenone production, given that this may vary considerably within the surface mixed layer, but showed that light limited growth and/or active carbon uptake had minimal effect over reconstructed [CO_2(aq)]. One potential proxy for Haptophyte growth rates that has recently emerged is the application of Sr/Ca ratios in coccolith calcites, as the uptake of Sr in coccolith calcite appears to be determined by growth rate (Stoll, Klaas, Probert, Encinar, & Alonso, 2002a; Stoll & Schrag, 2000; Stoll, Ziveri, Geisen, Probert, & Young, 2002b; Figure 14). Recent advances in cleaning and separation techniques for coccolith calcite also enable Sr/Ca ratios from individual species of Haptophytes to be analyzed (Stoll & Ziveri,

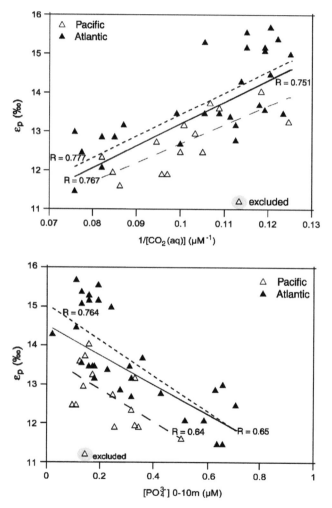

Figure 9 Relationships between the isotopic composition of $C_{37:2}$ alkenone from surface sediments (εp) to concentrations of (a) CO_2 and (b) phosphate in seawater. South Atlantic data is shown in solid triangles, and the central Pacific data is shown in open triangles. Solid line shows regression of the complete data set excluding the marked outlier. *Source*: Schulte et al. (2004). With permission from American Geophysical Union.

2002). Where information on nutrient contents and/or growth rates are not available in conjunction with $\delta^{13}C_{alkenones}$ to reconstruct pCO_2, the degree of isotopic fractionation may still give information on levels of productivity in the Haptophytes, which can also be used to infer variations in palaeo-nutrient supplies (Benthien et al., 2002; Schulte et al., 2004).

The effect of post-depositional degradation on molecular isotopic composition is not clear. Cretaceous-aged porphyrins have displayed minimal diagenetic imprint (Hayes, Freeman, Popp, & Hoham, 1990), and it is generally assumed that the refractory nature of the alkenones makes them resistant to diagenetic alteration of both $U_{37}^{K'}$ and $\delta^{13}C$ (Pagani et al., 1999a). The alkenone $\delta^{13}C$ signal also survives herbivory by zooplankton (Grice et al., 1998). However, laboratory incubations under oxic and anoxic conditions have also demonstrated that the $\delta^{13}C_{alkenone}$ may be depleted by up to 6‰ (Sun et al., 2004).

2.5. Land-Ocean Correlations: Changes in Continental Vegetation

Photosynthesis in higher plants occurs through either the Calvin-Benson (C_3) or the Hatch-Slack (C_4) pathway. The carbon isotopic fractionations associated with these pathways differ significantly. In the bulk tissues of plants, $\delta^{13}C$ commonly range from -25 to $-30‰$ in C_3 plants and -10 to 16% in C_4 plants (Brassell, 1993; Collister, Rieley, Stern, Eglinton, & Fry, 1994; Hadley & Smith, 1989; O'Leary, 1981; Smith & Epstein, 1971). C_3 plants include trees, shrubs, and cool-climate grasses (Cerling et al., 1997). C_4 plants include grasses and sedges found predominantly in tropical savannahs, temperate grasslands, and semideserts (Cerling et al., 1997). Past vegetation changes can be determined from the stratigraphic record of pollen and spores but it is not possible, however, to distinguish between C_3 from C_4 plants using pollen records.

Marine sediments contain organic molecular components that have an unambiguous terrestrial origin. Some of these are land plant contributions indicated by the presence of C_{27}, C_{29}, and C_{31} n-alkanes, or C_{24}, C_{26}, and C_{28} n-alkanoic acids (Cranwell, 1973; Eglinton & Hamilton, 1963; Rieley, Collier, Jones, & Eglinton, 1991). Leaf lipids record isotopic differences produced by different photosynthetic pathways (Collister et al., 1994; O'Leary, 1981). Determination of the $\delta^{13}C$ signatures of these terrigenous n-alkyl lipids can be used to determine the relative contributions of C_3 and C_4 plants on land to aquatic sediments and establish changes in vegetation patterns through time.

Homologous series of long-chain n-alkyl compounds (n-alkanes, n-alkanols, n-alkanoic acids, and wax esters) are major components of epicuticular waxes from vascular plant leaves that serve as protection against desiccation and bacterial attack (Eglinton & Calvin, 1967). Plant-wax n-alkanes typically contain between 25 and 35 carbons, with a strong predominance of odd- over even-carbon number chain lengths and a strong predominance of odd-carbon number homologues over even-numbered ones (the most common being C_{27}, C_{29}, C_{31}, and C_{33}; Eglinton & Calvin, 1967; Eglinton, Gonzalez, Hamilton, & Raphael, 1962; Kolattukudy, 1976). The n-alkanols and n-alkanoic acids typically occur in higher plants as the C_{16}–C_{36} homologues with a strong even-over-odd predominance, reflecting their biosynthesis from acetyl moieties; the most common homologues are the C_{22}, C_{24}, C_{26}, C_{28}, and C_{30} components (Eglinton & Hamilton, 1963; Kolattukudy, 1976).

N-alkyl lipids have been identified in recent and ancient terrestrial and marine sediments. Oceanic sediments, however, commonly include both marine and terrigenous organic material, and both categories can include reworked fossil carbon. Eglinton et al. (1997) and Pearson and Eglinton (2000) determined the ^{14}C composition of n-alkanes in surface sediments and showed that n-alkanes in marine sediments can be derived from either soil or weathered shales, although the input from ancient sources is minor. To overcome this issue some authors have focussed on n-alkanoic acids because of their abundance in sediments and because they are not subject to contamination by fossil hydrocarbons (Dahl et al., 2005).

The carbon preference index (CPI) can help to establish the likelihood that n-alkyl lipids in a given sample derive from higher plants. For the n-alkanes, this determines the relative chain length magnitude with odd or even carbon numbers. N-alkane distributions from modern plants typically have CPI numbers higher than 4 (4.3–40.3; Collister et al., 1994). Algal inputs are characterized by C_{17} n-alkane or C_{12}, C_{14}, and C_{16} n-alkanoic acids (e.g., Cranwell, 1978), although these components are produced by all plants albeit in lower amounts than in algae. Thus, low CPI values are usually attributed to algal inputs (Clark & Blumer, 1967). Low CPI numbers can also be related to degradation, microbial inputs, and contamination with petroleum. The likelihood of the sources of n-alkanols and n-alkanoic acids, be it from land plants or bacterial/algal inputs, can be also established with the CPI (Freeman & Collarusso, 2001), or the relative abundance of higher molecular-weight components (Bourbonniere & Meyers, 1996).

Higher plant material in sediments from most marine settings can be assumed to be predominately terrestrially derived (Figure 10). The modes of transport to marine sediments of n-alkyl higher plant lipids are basically two: aeolian and/or riverine transport (Ikehara et al., 2000; Kawamura, 1995; Ohkouchi, Kawamura, Kawahata, & Taira, 1997a; Schefuss, Ratmeyer, Stuut, Jansen, & Sinninghe Damsté, 2003a; Schefuss, Schouten, Jansen, & Sinninghe Damsté, 2003b; Simoneit et al., 1977). Aquatic vascular higher plants only occur widely in a limited number of marine settings (e.g., shallow shelves and lagoons; De Leeuw, Frewin, van Bergen, Sinninghe Damsté, & Collinson, 1995). In sediments near coastal sites and river mouths leaf lipids are associated to fluvial inputs. Only a small fraction (ca. 20% or less) of riverine suspended matter is deposited in deep-sea sediments (Berner & Berner, 1996; McKee, Aller, Allison, Bianchi, & Kineke, 2004). As a result, much of the particulate terrigenous organic matter transported by rivers to the oceans is deposited in continental margin sediments (Burdige, 2005). But, the controls on biomarker transport and preservation during fluvial delivery to marine sediments are poorly understood (McKee, 2003). As noted in Pancost and Boot (2004), the mechanisms of organic matter delivery can be highly variable and biases may be associated with proximity to watersheds, differential degradation of biomarkers, and particle trapping or sorting in rivers and continental margins, which they could all exert primary controls on both the quantity and nature of organic matter delivered from fluvial processes.

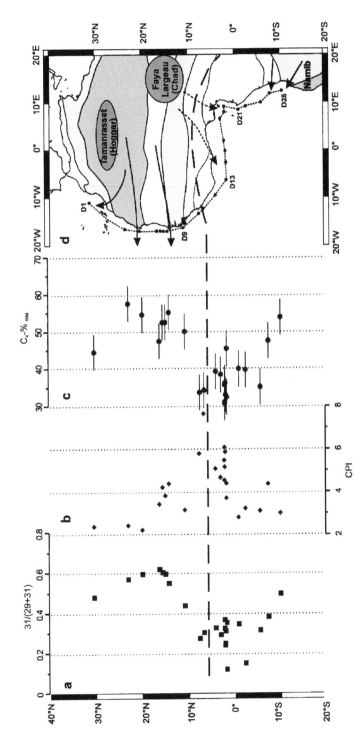

Figure 10 Comparison between *n*-alkane distributions and their isotopic composition in aerosols, and the dominant vegetation zones onshore. The ship track during aerosol collection is shown, and the horizontal line indicates the ITCZ position. (a) ratio between two dominant *n*-alkane homologues; (b) carbon preference index; (c) isotopically determined ($\delta^{13}C$) percentage of C_4 plant derived *n*-alkanes; and (d) vegetation zones. Mixed C_3/C_4 vegetation in light shading. Sparse vegetation but dominated by C_4 plants in dark shading. *Source:* Schefuss et al. (2003a). With permission from Elsevier.

In open ocean settings, aeolian transport mechanisms are predominant. An exception can occur when the organic matter is transported as ice rafted debris, and for instance the glaciomarine sedimentation of terrestrial organic matter in the Northern North Atlantic is well documented during glacial periods and polar regions (Fahl & Stein, 1999; Wagner & Henrich, 1994). Higher plant biomarkers from leaf wax lipids can be relatively abundant in abyssal sediments and atmospheric dust. Typical terrestrial biomarkers in atmospheric dust are long-chain n-alkanes, n-alkanols, and n-alkanoic acids (Simoneit, 1977). Waxes on plant leaf surfaces are thought to be removed by wind and dust by ablation and become airborne, or plant organic matter in soils can be lifted during dust storms (Conte & Weber, 2002; Hadley & Smith, 1989; Simoneit, 1977). Ablated waxes accumulate in air as particles with a molecular composition broadly similar to that of their source vegetation (Rogge, Hildemann, Mazurek, Cass, & Simoneit, 1993; Simoneit, 1989) and air masses moving across the land surface can potentially accumulate and integrate waxes from vegetation over large continental areas. Long-range transport of windborne terrestrial organic matter has been well documented, and in areas under the influence of the trade winds is particularly important (Conte & Weber, 2002). Off the Eastern African continent and in the Arabian Sea higher plant biomarker distributions and their $\delta^{13}C$ values in surface sediments and dust samples reflect the modern vegetation type along their transport pathway (Dahl et al., 2005; Huang, Dupont, Sarnthein, Hayes, & Eglinton, 2000; Rommerskirchen et al., 2003; Schefuss et al., 2003a), indicating that leaf lipids found in sediments are derived from contemporary vegetation sources on the neighbouring continents (Figure 10). Similarly, surface sediments from a Central Pacific latitudinal transect (Ohkouchi et al., 1997a) and aerosols from the North Western Pacific also contained terrestrially aeolian-derived n-alkyl compounds (Bendle, Kawamura, & Yamazaki, 2006). Their presence and variability were interpreted as reflecting different wind regimes at high and low latitudes transporting wax compounds from Asia and Oceania to the North Pacific by westerlies or delivered by trade winds from Central and South America to the tropical Pacific. These findings validate the use of n-alkyl lipids $\delta^{13}C$ signatures in ancient offshore marine sediments for the reconstruction of past African, Asian, and American climates through the reconstruction of past vegetation changes. In combination with pollen data it represents potentially a powerful approach to be applied in the reconstruction of past continental phytogeographic developments.

3. APPLICATIONS

Arguably, some of the most notable contributions to paleoceanographic reconstruction during the Cenozoic using biomarker proxies have been to determine the thermal history of the surface oceans during the Pleistocene, the evolution of land plants in Africa, and changes in the marine carbon cycle through the reconstruction of export productivity and pCO_2 in pre-Quaternary climates. Most of these applications, detailed below, relate in fact to the use of alkenone proxies and n-alkyl lipids. So far, published TEX_{86} applications have been restricted to continental sites and pre-Cenozoic reconstructions (Jenkyns, Forster, Schouten, & Sinninghe Damsté, 2004; Powers et al., 2005; Schouten et al., 2003).

3.1. Sea-Surface Temperature

Alkenone paleothermometry has played an important role in quantifying changes in SST over a range of timescales and time-frequencies in all the ocean basins. One of the earliest contributions, which came with the definition of the proxy, was to show that changes in foraminiferal oxygen isotopes during the Quaternary were paralleled by changes in SST in the NE Atlantic (Brassell et al., 1986). Consequently, a number of studies demonstrated that tropical and sub-tropical SSTs varied coherently with Milankovitch insolation cycles and in-phase with the precessional component of insolation through changes to the intensity and zonality of wind systems (Herbert et al., 2001; Lyle, Prahl, & Sparrow, 1992; Schneider, Müller, & Ruhland, 1995; Schneider et al., 1996). In relation to this, the aim of a range of studies has been to determine if changes in upwelling strength are in phase with the glacial/interglacial cycles, and the role of these systems in drawing down atmospheric CO_2 from the atmosphere. Clearly, the drop in SSTs during glacials in comparison to interglacials is a recurrent feature in most oceanic locations, although changes in productivity associated to these changes is more difficult to determine (Hinrichs, Schneider, Müller, & Rullkötter, 1999; Lyle et al., 1992; Sicre et al., 2000; Summerhayes et al., 1995). Quantification of the glacial/interglacial cyclicity in SST has been undertaken now in all major ocean basins and marginal seas, although it is usually restricted to sites relatively close to coastal basins as shown in a global compilation of the alkenone–SST glacial–Holocene anomaly (Rosell-Melé et al., 2004). In this context, quantification of tropical SSTs have been a topic of debate for some time, and several studies have contributed to resolve that the magnitude of the change in tropical SSTs worldwide at the last glacial maximum was of just a few degrees centigrade (Ohkouchi, Kawamura, Nakamura, & Taira, 1994; Pelejero, Grimalt, Heilig, Kienast, & Wang, 1999; Sikes & Keigwin, 1994; Sonzogni, Bard, & Rostek, 1998).

The $U_{37}^{K'}$ index has also proved useful to obtain evidence of rapid climate change affecting the ocean at centennial or lower timescales (Eglinton et al., 1992). One of the early applications of $U_{37}^{K'}$ showed that in an annually layered record off Southern California major historical El Niño related warmings along the California Coast stood out in most cases as warm $U_{37}^{K'}$ anomalies (Kennedy & Brassell, 1992). The impacts of the Dansgaard-Oeschger cycles (Cacho et al., 1999; Martrat et al., 2004; Sachs & Lehman, 1999) (Figure 11) or the Heinrich events over the hydrology of the North Atlantic have both been examined using $U_{37}^{K'}$ and the % $C_{37:4}$ as a proxy for relative SSS variability (Bard et al., 2000). The impact of Heinrich events in the tropical Atlantic was also established using $U_{37}^{K'}$ estimates and a significant warming was documented for Heinrich event HI and the Younger Dryas, showing that changes in the tropical and high-latitude North Atlantic are out of phase and suggesting that the thermohaline circulation was the important trigger for these rapid climate changes (Rühlemann, Mulitza, Müller, Wefer, & Zahn, 1999). Alkenone paleothermometry has been used to establish the relative timing of the

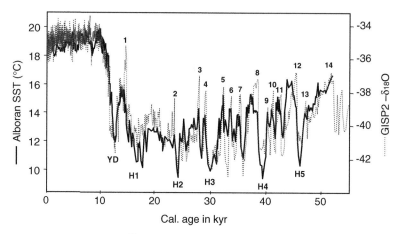

Figure 11 Rapid changes in $U^{K'}_{37}$-SST in the Alboran Sea during the last 50 kyr, and their correlation to events recorded in the Greenland ice core. Note that during the Heinrich events of iceberg rafting into the North Atlantic (H1, H2 etc.) and the cool but short-lived Younger Dras stadial (YD), SSTs in the Alboran Sea show rapid declines. *Source*: Cacho et al. (1999). With permission from American Geophysical Union.

last deglacial warming in the Southern and Northern hemispheres through the quantification of tropical temperatures during the last glacial cycle in high-resolution sediment cores recovered from the tropical Indian Ocean (Bard, Rostek, & Sonzogni, 1997). This study established that inferred initial SST warming in the tropical Indian Ocean were in phase with Northern Hemisphere sea and air temperature changes, but lagged Antarctic warming. In the tropical Atlantic, Schneider et al. (1996) and Schneider, Müller, and Acheson (1999) identified early SST warming that preceded northern hemisphere temperature change and global ice volume decline at glacial terminations, and highlighted the potential role of the tropics in amplifying the climate response to insolation forcing.

$U^{K'}_{37}$ has been used to reconstruct the longer-term thermal evolution of the oceans during the Pleistocene and Pliocene. In the north-west Pacific, an explanation for the development of the northern hemisphere glaciations *ca.* 2.7 Ma as relating to surface ocean stratification and atmospheric moisture transports (Haug et al., 2005). A continuous time series of SST for the past 4.5 Myr has been provided by $U^{K'}_{37}$ that showed how the climate has shifted from the relatively stable global warmth during the mid-Pliocene to the high-amplitude glacial/interglacial cycles of the Late Quaternary (Marlow et al., 2000; Figure 12). The role of the tropical SSTs in driving or modulating global climate change and the interplay between the high and low latitudes during the Pliocene–Pleistocene has been shown in several studies using biomarkers (Haywood, Dekens, Ravelo, & Williams, 2005; McClymont & Rosell-Melé, 2005). These reconstructions show that SST variability analogous to the modern El Niño/Southern Oscillation, as represented by SST gradients across the equatorial Pacific, can have global climate repercussions. Long range teleconnections are a prominent feature in climate dynamics, and in addition to some of the cases described so far, alkenone-derived sea-surface

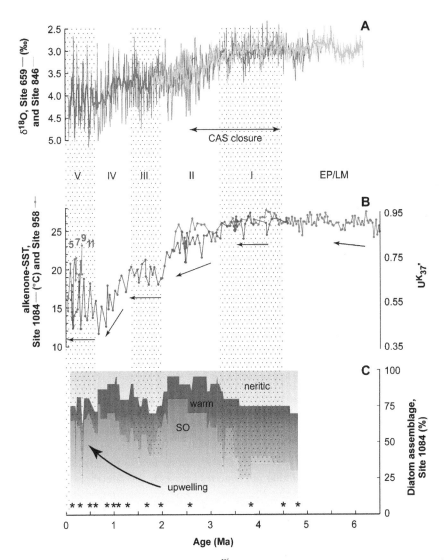

Figure 12 A Pleistocene–Pliocene record of $U_{37}^{K'}$-SST changes and upwelling intensity from the Benguela upwelling region. (A) benthic oxygen isotope record of global ice volume (Tiedemann et al., 1994); (B) Benguela SSTs (red) and a comparison to an additional eastern boundary current system, the Canary Current (blue; Herbert and Schuffert, 1998); (C) diatom indices for upwelling, Southern Ocean, warm and neritic surface waters in the Benguela. Note the shift towards upwelling species coincides with the drop in SSTs after 3.2 Ma. *Source*: Adapted with permission from Marlow et al. (2000).

temperature (SST) records have been used to establish that these occur in millennial timescales during the Holocene as reflected in the temperature trends in SST during this period worldwide (Kim et al., 2004; Rimbu, Lohmann, Lorenz, Kim, & Schneider, 2004).

3.2. Marine Carbon Cycle: CO_2 and Phytoplankton Production

The application of CSIA for the reconstruction of past pCO_2 using $\delta^{13}C_{alkenones}$ shows great potential for assessing the interaction between the production and export of organic carbon from the surface ocean (the "biological pump"), and the 80–100 ppmv variability of pCO_2 recorded in the ice cores between glacial and interglacial cycles (e.g., Kohfeld et al., 2005). The first reconstruction was produced by Jasper and Hayes (1990) who demonstrated that the technique could be used to broadly identify the Vostok ice core variations in pCO_2 over the last glacial/interglacial cycle, variations that could not be estimated from bulk $\delta^{13}C$ values at this site due to the additional inputs of terrestrial carbon into the marine sediments. Subsequently, Jasper, Hayes, Mix, and Prahl (1994) generated a 250 ka record of pCO_2 from the equatorial Pacific using $\delta^{13}C_{alkenone}$, which broadly showed the same trends to the Vostok record but misplaced a pCO_2 maximum at 80 ka instead of 130 ka. Correction of the pCO_2 data for changes in coccolithophore growth rate variations using Sr/Ca gave a much closer reconstruction to Vostok (Stoll & Schrag, 2000; Figure 13). The most extensive reconstruction of pCO_2 has come from the work of Pagani and co-workers, who have used the $\delta^{13}C_{alkenone}$ method to reconstruct a 40 Myr record of pCO_2 that spans the Eocene, Oligocene, and Miocene (Pagani et al., 1999a, 1999b; Pagani, Zachos, Freeman, Tipple, & Bohaty, 2005; Figure 14). The strength of this work lies in the selection of samples from long-term nutrient-limited oligotrophic gyres, where changes in growth rates are assumed to have been minimal through geological time, and thus the relationship to $[CO_2(aq)]$ and pCO_2 should dominate the $\delta^{13}C_{alkenone}$ signature. An alternative method to reconstructing past pCO_2 over the last 160 Myr was presented by Freeman and Hayes (1992), who used $\delta^{13}C$ values in sedimentary porphyrins as a proxy. It should be noted that despite the recent emergence of very long records of atmospheric gas composition from Antarctic ice cores (e.g., EPICA community members, Augustin et al., 2004), there is a large gap in the reconstruction of pCO_2 using molecular approaches between the late Pleistocene studies of Jasper and co-workers and the Miocene–Eocene reconstructions of Pagani and co-workers. This gap spans a number of significant events in the palaeoclimate records where pCO_2 has been highlighted as a potentially significant driver or feedback mechanism, including the onset of northern hemisphere glaciation, the mid–Pleistocene climate transition, and the Stage 11 problem (e.g., Raymo, 1994).

An alternative application of the $\delta^{13}C_{alkenone}$ method has been to use the reconstructed $[CO_2(aq)]$ values to assess particular oceanographic regions as sinks or sources through the late Pleistocene, where the reconstructed $[CO_2(aq)]$ values differ from those provided by the ice core records. These results have demonstrated that the central equatorial Pacific has remained a source of CO_2 to the atmosphere over the last 230 kyr, but that the sea-to-air flux was greatest during glacial periods (Jasper et al., 1994). Similar results were gained from the upwelling region of the Angola Current in the eastern South Atlantic, which has also been a source of CO_2 to the atmosphere over the last 200 kyr, particularly during glacial times (Andersen, Müller, Kirst, & Schneider, 1999). Low glacial εp values in the oligotrophic regions of the South Atlantic (Brazil Current, South Atlantic tropical gyre, and the western

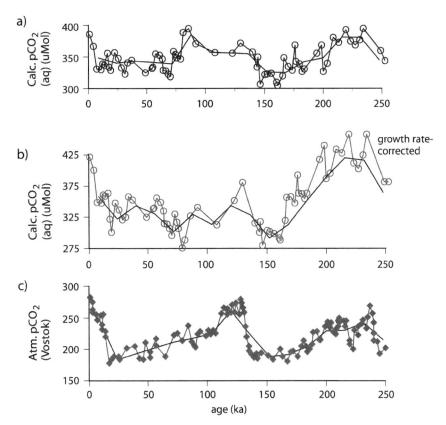

Figure 13 A comparison between calculated atmospheric CO_2 concentrations using $\delta^{13}C$ alkenones to the Vostok ice core record of Petit et al. (1999). Smoothing at 16 kyr intervals is shown by the black lines. (a) Constant algal growth rates assumed, from Jasper et al. (1994); (b) corrections for algal growth rate variations made using Sr/Ca values in calcite; (c) Vostok pCO_2 (Petit et al. 1999). *Source*: Stoll and Schrag (2000). With permission from American Geophysical Union.

tropical Atlantic) also reveal these areas as either small sources of CO_2 to the atmosphere or that glacial Haptophyte growth rates have biased the εp signal (Benthien et al., 2002). The implication of these records is the need to find the sinks of CO_2 in the climate system and in particular during glacial times, in order to assess the effectiveness of the biological pump (or other carbon sequestration mechanisms) in drawing down $pCO2$.

Where growth rates are found to dominate the $\delta^{13}C$ signal, rather than [$CO_2(aq)$], or where palaeo-nutrient information is not available, isotopic analysis has been used to give valuable information about growth rates and thus palaeo-nutrient compositions and phytoplankton productivity (Benthien et al., 2005; Schulte et al., 2004). Changes in phytoplankton production may also be assessed using compound-specific $\delta^{13}C$ in a variety of biomarkers. Only small changes to $\delta^{13}C$ in alkenones (Haptophytes) and loliolide/isololiolide (diatoms) were found

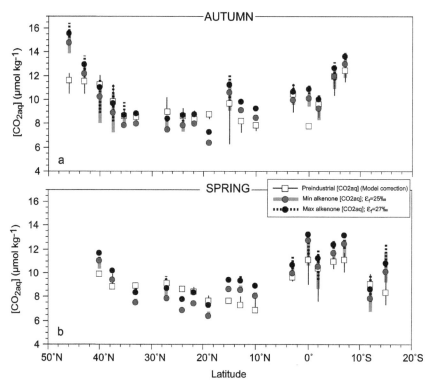

Figure 14   Comparison between measured CO_2 concentrations in surface ocean waters at the depth of haptophyte production, and those calculated using $\delta^{13}C$ values of the $C_{37:2}$ alkenone, in two seasons. Measured CO_2 concentrations had to be corrected for anthropogenic CO_2 that elevated values above those calculated from the alkenones. *Source*: Pagani et al. (2002). With permission from American Geophysical Union.

associated with a Pliocene sapropel in the eastern Mediterranean, despite high increases in biomarker accumulation rates during sapropel deposition that suggest increased biological productivity at these times (Menzel, Schouten, van Bergen, & Sinninghe Damsté, 2004). It was proposed that the $\delta^{13}C$ values were instead a reflection of recycled and isotopically depleted CO_2 being advected to the photic zone during storms when there existed a generally shallow chemocline, and that this isotopically depleted CO_2 source probably counter-balanced high growth-rate induced isotopic enrichment (Menzel et al., 2004).

Acknowledging the caveats described above using biomarker accumulation rates to determine production and changes to plankton assemblages, a number of studies have demonstrated the potential for biomarkers to give a more detailed under-standing of organic matter production and preservation in the marine realm. Good correlations have been determined at a number of sites between chlorin accumu-lation rates and other palaeo-productivity proxies down-core, proposed to indicate a production-driven signal to the organic matter accumulation rates (Harris et al., 1996; Lyle et al., 1992; Summerhayes et al., 1995). For instance, Schulte et al. (1999) analyzed two cores in the central Arabian Sea for the last 330 kyr, and identified precession-related cycles in alkenones, dinosterol, and brassicasterol (for haptophytes, dinoflagellates, and diatoms) superimposed upon longer, glacial/interglacial changes to bottom water oxygen contents. The latter process was also responsible for preferential preservation of dinosterol under anoxic conditions in one of the cores, and here it was recognized that the signal could not be used to address production changes alone (Schulte et al., 1999).

3.3. Land Vegetation Changes

Determination of the timing and intensity of inputs to n-alkyl lipids from leaf waxes to marine sediments has been used, for example, to identify wind borne or riverine contributions of land plants in marine sediments (Hinrichs et al., 1999; Ohkouchi, Kawamura, & Taira, 1997b; Pelejero, Grimalt, Sarnthein, Wang, & Flores, 1998; Poynter et al., 1989; Prahl, MuehlhauSen, & Lyle, 1989b; Sicre et al., 2000; Ternois, Kawamura, Keigwin, Ohkouchi, & Nakatsuka, 2001). Stratigraphic determination of the $\delta^{13}C$ in n-alkyl lipids is a relatively recent development and only more recently has been applied to marine sequences. One of the first studies was carried out by Yamada and Ishiwatari (1999) who identified an increase in n-alkane $\delta^{13}C$ values in Japan Sea sediments about 10 ka before present. They attributed this shift to either enhanced terrestrial productivity or a change in dominant plant species in the source areas of the terrigenous inputs. Most of the studies have as an aim to determine the timing of the expansion of C_4 plants on land and any possible links with climate. CSIA data in paleosols and Bengal Fan sediment provided evidence of the expansion of C_4 grasslands on the Indian subcontinent prior to 6 Ma, consistent with a relatively rapid transition from dominantly C_3 vegetation to an ecosystem dominated by C_4 plants typical of semi-arid grasslands (Freeman & Collarusso, 2001). The African continent has been the focus of several studies on this question. In the eastern South Atlantic, Schefuss et al. (2003b) composed a record of African C_4 plant abundance between 1.2 and 0.45 Myr ago using the carbon isotope of wind-transported terrigenous plant waxes (Figure 15). They concluded that large-scale changes in African vegetation were linked closely to SSTs in the tropical Atlantic Ocean because the latter influenced the strength of the African monsoon and in turn aridity on the African continent. Similarly, the effect on varying monsoon strength, rather than pCO_2, was also proposed to be the main forcing that explained observed changes in southern African vegetation 2.5 Ma, in Marine Isotope Stages 101/100, through the analysis of n-alkane abundances and $\delta^{13}C$ values (Denison, Maslin, Boot, Pancost, & Ettwein, 2005). Significant changes in vegetation during the late Neogene have also been established for Northeast Africa, where large and repeated oscillations between more open and more closed landscapes have been proposed as being an important aspect of northeast African vegetation change during the past 4 Myr (Feakins, deMenocal, & Eglinton, 2005). The link between vegetation changes and climate seems obvious, but plants may not respond immediately to climate oscillations. In a record in the Cariaco basin (Hughen, Eglinton, Xu, & Makou, 2004), high-resolution vegetation records have been reconstructed using the molecular isotopic approach. These workers have determined that while climate shifts in both high-latitude and tropical North

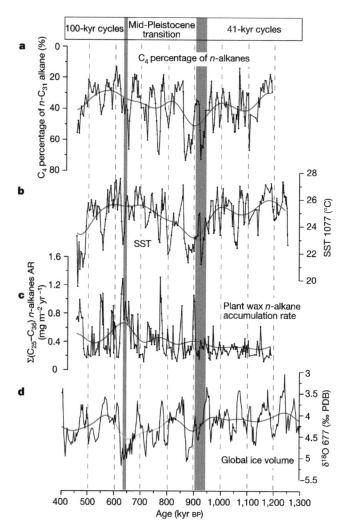

Figure 15 Combining isotopic composition of higher plant lipids and $U_{37}^{K'}$ derived SSTs to investigate land–ocean correlation during the mid-Pleistocene. (a) C_4 vegetation percentage onshore based on $\delta^{13}C$ of the C_{31} *n*-alkane; (b) $U_{37}^{K'}$-derived SSTs; (c) plant wax lipid accumulation rates; (d) benthic $\delta^{18}O$ record of global ice volume, from Shackleton et al. (1990). *Source*: Schefuss et al. (2003b). With permission from Nature Publishing group.

Atlantic regions were synchronous, tropical vegetation change lagged abrupt climatic shifts by several decades.

4. Concluding Remarks

The various approaches described in the preceding sections are at distinct stages in their development and application. Arguably, $U_{37}^{K'}$ is the most established proxy and is held in the same standing as other SST estimators (Kucera, Rosell-Melé, Schneider, Waelbroeck, & Weinelt, 2005). Investigation to constraint its application will continue in the years to come, but not to disprove its value, rather to enhance the reliability of the reconstructions. Hopefully, the TEX_{86} will also prove to be an equally useful proxy in marine settings. At the time this review has been produced there is plenty of evidence to suggest that the comparative analysis of $U^{K'}_{37}$ and TEX_{86}, in parallel to using other SST proxies, will facilitate the reconstruction of the thermal structure of the upper part of the water column with increased accuracy. It is important for modeling to have absolute values of climatic variables, and $U^{K'}_{37}$ and TEX_{86} make an important contribution to obtain quantitative reconstruction of past water temperatures. Other applications as outlined in Section 3, have had a significant impact in the field but have been less widely applied probably because the number of groups worldwide capable of carrying out in-house measurements is still low. Other approaches, noted in the introduction to Section 2, are also yet to be fully exploited.

Most applications have been undertaken using gas chromatography, sometimes hyphenated to mass spectrometry. An exception is the measurement of TEX_{86}, which is undertaken by liquid chromatography-mass spectrometry, and chlorins that can be measured by light-absorption or fluorescence techniques. As new analytical techniques are developed, and multidisciplinary investigations are conducted, it is likely that new applications will keep being proposed and applied. Clearly CSIA still has much to offer to provide in the future some of the best insights into the climatic evolution of the Earth, particularly as sensitivity is increased in isotope ratio mass spectrometers and new techniques are implemented e.g., the study of larger and thermally labile biomolecules via the hyphenation of liquid chromatography and isotopic ratio mass spectrometry, enabling routine CSIA of isotopes other than those of carbon (e.g., $\delta^{15}N$ in chlorophyll).

Biomarker proxies are usually placed in one bag, i.e., the one containing the proxies derived from organic molecules. This is equivalent to considering foraminiferal and coccolith proxies, or diatom and radiolarian approaches to be somehow related just because the fossil tests are made of carbonate or silica. Hopefully the current review will contribute to making the case further that there are more biomarker proxies than just $U^{K'}_{37}$, and that the information derived from various approaches is diverse and in occasions unique. Thus, changes in key climatic parameters and the carbon cycle can be inferred from studying biomarkers in sediments. Biomarkers can be employed to gain quantitative reconstructions on past SST and pCO_2, and insights into ocean and atmospheric circulation, ice–ocean interactions, salinity, sea level, marine phytoplankton ecology and production, and onshore vegetation. The latter is particularly important because biomarker analyses in marine sediment cores allow undertaking an "integrated past reconstruction approach" and investigate land–ocean interactions (see Figure 15). Furthermore, investigation of marine and terrestrial carbon fluxes to marine sediments, associated to specific plant functional types and modes of transport, permits reconstruction of the marine carbon cycle and its terrestrial components and helps us to decipher the role of biogeochemical cycles in climate change.

ACKNOWLEDGMENTS

The comments and feedback provided by colleagues to put together this review are very much appreciated, with special thanks to Marina Escala and Alfredo Martínez, and to Ralph Schneider for reviewing the manuscript. Thanks to colleagues that provided data or granted permission to show their published figures in this manuscript.

REFERENCES

Abreu, C., Jurgens, G., De Marco, P., Saano, A., & Bordalo, A. A. (2001). Crenarchaeota and Euryarchaeota in temperate estuarine sediments. *Journal of Applied Microbiology, 90*, 713–718.

Andersen, N., Müller, P. J., Kirst, G., & Schneider, R. R. (1999). Alkenone $\delta^{13}C$ as a proxy for past PCO_2 in surface waters: Results from the late Quaternary Angola Current. In: G. Fischer & G. Wefer (Eds), *Use of proxies in paleoceanography: Examples from the South Atlantic* (pp. 469–488). Berlin: Springer-Verlag.

Augustin, L., Barbante, C., Barnes, P. R. F., Barnola, J. M., Bigler, M., Castellano, E., Cattani, O., Chappellaz, J., DahlJensen, D., Delmonte, B., Dreyfus, G., Durand, G., Falourd, S., Fischer, H., Fluckiger, J., Hansson, M. E., Huybrechts, P., Jugie, R., Johnsen, S. J., Jouzel, J., Kaufmann, P., Kipfstuhl, J., Lambert, F., Lipenkov, V. Y., Littot, G. V. C., Longinelli, A., Lorrain, R., Maggi, V., Masson-Delmotte, V., Miller, H., Mulvaney, R., Oerlemans, J., Oerter, H., Orombelli, G., Parrenin, F., Peel, D. A., Petit, J. R., Raynaud, D., Ritz, C., Ruth, U., Schwander, J., Siegenthaler, U., Souchez, R., Stauffer, B., Steffensen, J. P., Stenni, B., Stocker, T. F., Tabacco, I. E., Udisti, R., van de Wal, R. S. W., van den Broeke, M., Weiss, J., Wilhelms, F., Winther, J. G., Wolff, E. W., & Zucchelli, M. (2004). Eight glacial cycles from an Antarctic ice core. *Nature, 429*, 623–628.

Bard, E. (2001). Paleoceanographic implications of the difference in deep-sea sediment mixing between large and fine particles. *Paleoceanography, 16*, 235–239.

Bard, E., Rostek, F., & Sonzogni, C. (1997). Interhemispheric synchrony of the last deglaciation inferred from alkenone paleothermometry. *Nature, 385*, 707–710.

Bard, E., Rostek, F., Turon, J. L., & Gendreau, S. (2000). Hydrological impact of Heinrich events in the subtropical northeast Atlantic. *Nature, 289*, 1321–1324.

Bell, M. V., & Pond, D. (1996). Lipid composition during growth of motile and coccolith forms of *Emiliania huxleyi. Phytochemistry, 41*, 465–471.

Bendle, J., & Rosell-Melé, A. (2004). Distributions of U^K_{37} and $U^{K'}_{37}$ in the surface waters and sediments of the Nordic Seas: Implications for paleoceanography. *Geochemistry, Geophysics and Geosystems, 5*, Q11013.

Bendle, J., Rosell-Melé, A., & Ziveri, P. (2005). Variability of unusual distributions of alkenones in the surface waters of the Nordic Seas. *Paleoceanography, 20*, PA2001.

Bendle, J. A., Kawamura, K., & Yamazaki, K. (2006). Seasonal changes in stable carbon isotopic composition of *n*-alkanes in the marine aerosols from the western North Pacific: Implications for the source and atmospheric transport. *Geochimica et Cosmochimica Acta, 70*, 13–26.

Bentaleb, I., Fontugne, M., & Beaufort, L. (2002). Long-chain alkenones and $U^{K'}_{37}$ variability along a south–north transect in the Western Pacific Ocean. *Global and Planetary Change, 34*, 173–183.

Bentaleb, I., Fontugne, M., Descolas-Gros, C., Girardin, C., Mariotti, A., Pierre, C., Brunet, C., & Poisson, A. (1996). Organic carbon isotopic composition of phytoplankton and sea-surface pCO_2 reconstructions in the Southern Indian Ocean during the last 50,000 yr. *Organic Geochemistry, 24*, 399–410.

Benthien, A., Andersen, N., Schulte, S., Müller, P. J., Schneider, R. R., & Wefer, G. (2002). Carbon isotopic composition of the $C_{37:2}$ alkenone in core top sediments of the South Atlantic Ocean: Effects of CO_2 and nutrient concentrations. *Global Biogeochemical Cycles, 16*, 1012.

Benthien, A., Andersen, N., Schulte, S., Müller, P. J., Schneider, R. R., & Wefer, G. (2005). The carbon isotopic record of the $C_{37:2}$ alkenone in the South Atlantic: Last Glacial Maximum (LGM) vs. Holocene. *Palaeogeography, Palaeoclimatology, Palaeoecology, 221*, 123–140.

Benthien, A., & Müller, P. J. (2000). Anomalously low alkenone temperatures caused by lateral particle and sediment transport in the Malvinas Current region, western Argentine Basin. *Deep-Sea Research Part I — Oceanographic Research Papers, 47*, 2369–2393.

Berner, E. K., & Berner, R. A. (1996). *Global environment: Water, air, and geochemical cycles.* Upper Saddle River, NJ: Prentice-Hall.

Bidigare, R. R., Fluegge, A., Freeman, K. H., Hanson, K. L., Hayes, J. M., Hollander, D., Jasper, J. P., King, L. L., Laws, E. A., Milder, J., Miller, F. J., Pancost, R., Popp, B. N., Steinberg, P. A., & Wakeham, S. G. (1997). Consistent fractionation of ^{13}C in nature and in the laboratory: Growth-rate effects in some haptophyte algae. *Global Biogeochemical Cycles, 11*, 279–292.

Blanz, T., Emeis, K. C., & Siegel, H. (2005). Controls on alkenone unsaturation ratios along the salinity gradient between the open ocean and the Baltic Sea. *Geochimica et Cosmochimica Acta, 69*, 3589–3600.

Bourbonniere, R. A., & Meyers, P. A. (1996). Sedimentary geolipid records of historical changes in the watersheds and productivities of Lakes Ontario and Erie. *Limnology and Oceanography, 41*, 352–359.

Brassell, S. C. (1993). Applications of biomarkers for delineating marine palaeoclimatic fluctuations during the Pleistocene. In: M. H. Engel & S. A. Macko (Eds), *Organic geochemistry principles and applications* (pp. 699–738). New York: Plenum Press.

Brassell, S. C., & Dumitrescu, M. (2004). Recognition of alkenones in a lower Aptian porcellanite from the west-central Pacific. *Organic Geochemistry, 35*, 181–188.

Brassell, S. C., Eglinton, G., Marlowe, I. T., Pflaumann, U., & Sarnthein, M. (1986). Molecular stratigraphy: A new tool for climatic assessment. *Nature, 320*, 129–133.

Burdige, D. J. (2005). Burial of terrestrial organic matter in marine sediments: A re-assessment. *Global Biogeochemical Cycles, 19*, GB4011, doi:10.1029/2004GB002368.

Burkhardt, S., Riebesell, U., & Zondervan, I. (1999). Stable carbon isotope fractionation by marine phytoplankton in response to daylength, growth rate, and CO_2 availability. *Marine Ecology Progress Series, 184*, 31–41.

Cacho, I., Grimalt, J. O., Pelejero, C., Canals, M., Sierro, F. J., Flores, J. A., & Shackleton, N. (1999). Dansgaard-Oeschger and Heinrich event imprints in Alboran Sea paleotemperatures. *Paleoceanography, 14*, 698–705.

Cerling, T. E., Harris, J. M., MacFadden, B. J., Leakey, M. G., Quade, J., Eisenmann, V., & Ehleringer, J. R. (1997). Global vegetation change through the Miocene/Pliocene boundary. *Nature, 389*, 153–158.

Clark, R. C., & Blumer, M. (1967). Distribution of *n*-paraffins in marine organisms and sediment. *Limnology and Oceanography, 12*, 79–87.

Collister, J. W., Rieley, G., Stern, B., Eglinton, G., & Fry, B. (1994). Compound-specific $\delta^{13}C$ analyses of leaf lipids from plants with differing carbon-dioxide metabolisms. *Organic Geochemistry, 21*, 619–627.

Conte, M. H., & Eglinton, G. (1993). Alkenone and alkenoate distributions within the euphotic zone of the eastern north Atlantic: Correlation with production temperature. *Deep-Sea Research I, 40*, 1935–1961.

Conte, M. H., Eglinton, G., & Madureira, L. A. S. (1992). Long-chain alkenones and alkyl alkenoates as palaeotemperature indicators: Their production, flux and early sedimentary diagenesis in the Eastern North Atlantic. *Organic Geochemistry, 19*, 287–298.

Conte, M. H., Sicre, M.-A., Rühlemann, C., Weber, J. C., Schulte, S., Schulz-Bull, D., & Blanz, T. (2006). Global temperature calibration of the alkenone unsaturation index ($U_{37}^{K'}$) in surface waters and comparison with surface sediments. *Geochemistry, Geophysics and Geosystems, 7*, Q02005, doi:10.1029/2005GC001054.

Conte, M. H., Thompson, A., Eglinton, G., & Green, J. C. (1995). Lipid biomarker diversity in the coccolithophorid *Emiliania huxleyi* (Prymnesiophyceae) and the related species *Gephyrocapsa oceanica*. *Journal of Phycology, 31*, 272–282.

Conte, M. H., Thompson, A., Lesley, D., & Harris, R. P. (1998a). Genetic and physiological influences on the alkenone/alkenoate versus growth temperature relationship in *Emiliania huxleyi* and *Gephyrocapsa oceanica*. *Geochimica et Cosmochimica Acta, 62*, 51–68.

Conte, M. H., Volkman, J. K., & Eglinton, G. (1994). Lipid biomarkers of the Haptophyta. In: B. Leadbeater & J. C. Green (Eds), *The haptophyte algae* (pp. 351–377). Oxford: Clarendon.

Conte, M. H., & Weber, J. C. (2002). Plant biomarkers in aerosols record isotopic discrimination of terrestrial photosynthesis. *Nature, 417*, 639–641.

Conte, M. H., Weber, J. C., King, L. L., & Wakeham, S. G. (2001). The alkenone temperature signal in western North Atlantic surface waters. *Geochimica et Cosmochimica Acta, 65*, 4275–4287.

Conte, M. H., Weber, J. C., & Ralph, N. (1998b). Episodic particle flux in the deep Sargasso Sea: An organic geochemical assessment. *Deep-Sea Research I, 45*, 1819–1841.

Cowie, G. L., & Hedges, J. I. (1992). Sources and reactivities of aminoacids in a coastal marine-environment. *Limnology and Oceanography, 37*, 703–724.

Cranwell, P. A. (1973). Chain-length distribution of *n*-alkanes from lake sediments in relation to post-glacial environmental change. *Freshwater Biology, 3*, 259–265.

Cranwell, P. A. (1978). Extractable and bound lipid components in a freshwater sediment. *Geochimica et Cosmochimica Acta, 42*, 1523–1532.

Dahl, K. A., Oppo, D. W., Eglinton, T. I., Hughen, K. A., Curry, W. B., & Sirocko, F. (2005). Terrigenous plant wax inputs to the Arabian Sea: Implications for the reconstruction of winds associated with the Indian Monsoon. *Geochimica et Cosmochimica Acta, 69*, 2547–2558.

Dahl, K. A., Repeta, D. J., & Goericke, R. (2004). Reconstructing the phytoplankton community of the Cariaco Basin during the Younger Dryas cold event using chlorin steryl esters. *Paleoceanography, 19*, PA1006, doi:10.1029/2003PA000907.

Dansgaard, W. (1964). Stable isotopes in precipitation. *Tellus, 16*, 436–468.

Dauwe, B., Middelburg, J. J., Herman, P. M. J., & Heip, C. H. R. (1999). Linking diagenetic alteration of amino acids and bulk organic matter reactivity. *Limnology and Oceanography, 44*, 1809–1814.

Degens, E. T., Guillard, R. R., Sackett, W. M., & Hellebus, Ja. (1968). Metabolic fractionation of carbon isotopes in marine plankton. I. Temperature and respiration experiments. *Deep-Sea Research, 15*, 1.

De Leeuw, J. W., Frewin, N. L., van Bergen, P. F., Sinninghe Damsté, J. S., & Collinson, M. E. (1995). Organic carbon as a palaeoenvironmental indicator in the marine realm. In: D. W. J. Bosence & P. A. Allison (Eds), *Marine palaeoenvironmental analysis from fossils* (pp. 43–71). London: Geological Society.

Delong, E. F., Wu, K. Y., Prezelin, B. B., & Jovine, R. V. M. (1994). High abundance of archaea in Antarctic marine picoplankton. *Nature, 371*, 695–697.

Denison, S. M., Maslin, M. A., Boot, C., Pancost, R. D., & Ettwein, V. J. (2005). Precession-forced changes in south west African vegetation during marine isotope stages 101–100 (similar to 2.56–2.51 Ma). *Palaeogeography, Palaeoclimatology, Palaeoecology, 220*, 375–386.

Derosa, M., Gambacorta, A., Nicolaus, B., Sodano, S., & Bulock, J. D. (1980). Structural regularities in tetraether lipids of Caldariella and their biosynthetic and phyletic implications. *Phytochemistry, 19*, 833–836.

Doose, H., Prahl, F. G., & Lyle, M. W. (1997). Biomarker temperature estimates for modern and last glacial surface waters of the California Current system between 33 degrees and 42 degrees N. *Paleoceanography, 12*, 615–622.

Eek, M. K., Whiticar, M. J., Bishop, J. K. B., & Wong, C. S. (1999). Influence of nutrients on carbon isotope fractionation by natural populations of Prymnesiophyte algae in NE Pacific. *Deep-Sea Research II, 46*, 2863–2876.

Eglinton, G., Bradshaw, S. A., Rosell, A., Sarnthein, M., Pflaumann, U., & Tiedemann, R. (1992). Molecular record of secular sea surface temperature changes on 100-year timescales for glacial terminations I, II and IV. *Nature, 356*, 423–426.

Eglinton, G., & Calvin, M. (1967). Chemical fossils. *Scientific American, 261*, 32–43.

Eglinton, G., Gonzalez, A. G., Hamilton, R. J., & Raphael, R. A. (1962). Hydrocarbon constituents of the wax coatings of plant leaves: A taxonomic survey. *Phytochemistry, 1*, 89–102.

Eglinton, G., & Hamilton, R. J. (1963). The distribution of alkanes. In: T. Swain (Ed.), *Chemical plant taxonomy* (pp. 187–208). London: Academic Press.

Eglinton, T. I., Benitez-Nelson, B. C., Pearson, A., Nichol, A. P., Bauer, J. E., & Druffel, E. R. M. (1997). Variability in radiocarbon ages of individual organic compounds from marine sediments. *Science, 277*, 796–799.

Elias, V. O., Simoneit, B. R. T., Cordeiro, R. C., & Turcq, B. (2001). Evaluating levoglucosan as an indicator of biomass burning in Carajas, Amazonia: A comparison to the charcoal record. *Geochimica et Cosmochimica Acta, 65,* 267–272.

Eltgroth, M. L., Watwood, R. L., & Wolfe, G. V. (2005). Production and cellular localization of neutral long-chain lipids in the haptophyte algae *Isochrysis galbana* and *Emiliania huxleyi. Journal of Phycology, 41,* 1000–1009.

Englebrecht, A. C., & Sachs, J. P. (2005). Determination of sediment provenance at drift sites using hydrogen isotopes and unsaturation ratios in alkenones. *Geochimica et Cosmochimica Acta, 69,* 4253–4265.

Epstein, B. L., D'Hondt, S., & Hargraves, P. E. (2001). The possible metabolic role of C$_{37}$ alkenones in *Emiliania huxleyi. Organic Geochemistry, 32,* 867–875.

Epstein, B. L., D'Hondt, S., Quinn, J. G., Zhang, J., & Hargraves, P. E. (1998). An effect of dissolved nutrient concentrations on alkenone-based temperature estimates. *Paleoceanography, 13,* 122–126.

Fahl, K., & Stein, R. (1999). Biomarkers as organic-carbon and environmental indicators in the Late Quaternary Arctic Ocean: Problems and perspectives. *Marine Chemistry, 63,* 293–309.

Farrimond, P., Eglinton, G., & Brassell, S. C. (1986). Alkenones in Cretaceous black shales, Blake-Bahama Basin, western North Atlantic. *Organic Geochemistry, 10,* 897–903.

Feakins, S. J., deMenocal, P. B., & Eglinton, T. I. (2005). Biomarker records of late Neogene changes in northeast African vegetation. *Geology, 33,* 977–980.

Francis, C. A., Roberts, K. J., Beman, J. M., Santoro, A. E., & Oakley, B. B. (2005). Ubiquity and diversity of ammonia-oxidizing archaea in water columns and sediments of the ocean. *Proceedings of the National Academy of Sciences of the United States of America, 102,* 14683–14688.

Freeman, K. H., & Collarusso, L. A. (2001). Molecular and isotopic records of C-4 grassland expansion in the late Miocene. *Geochimica et Cosmochimica Acta, 65,* 1439–1454.

Freeman, K. H., & Hayes, J. M. (1992). Fractionation of carbon isotopes by phytoplankton and estimates of ancient CO$_2$ levels. *Global Biogeochemical Cycles, 6,* 185–198.

Freeman, K. H., & Wakeham, S. G. (1992). Variations in the distributions and isotopic compositions of alkenones in Black Sea particles and sediments. *Organic Geochemistry, 19,* 277–285.

Frishman, S. A., & Behrens, E. W. (1971). Stable carbon isotopes in blue-green algal mats. *Journal of Geology, 79,* 95–100.

Fuhrman, J. A., McCallum, K., & Davis, A. A. (1992). Novel major archaebacterial group from marine plankton. *Nature, 356,* 148–149.

Gong, C., & Hollander, D. J. (1999). Evidence for differential degradation of alkenones under contrasting bottom water oxygen conditions: Implication for paleotemperature reconstruction. *Geochimica et Cosmochimica Acta, 63,* 405–411.

Goñi, M. A., Hartz, D. M., Thunell, R. C., & Tappa, E. (2001). Oceanographic considerations for the application of the alkenone-based paleotemperature U$^{K'}_{37}$ index in the Gulf of California. *Geochimica et Cosmochimica Acta, 65,* 545–557.

Gonzalez, E. L., Riebesell, U., Hayes, J. M., & Laws, E. A. (2001). Effects of biosynthesis and physiology on relative abundances and isotopic compositions of alkenones. *Geochemistry, Geophysics and Geosystems, 2,* U1–U9.

Grice, K., Klein-Breteler, W. C. M. K., Schouten, S., Grossi, V., de Leeuw, J. W., & Sinninghe-Damsté, J. S. (1998). Effects of zooplankton herbivory on biomarker proxy records. *Paleoceanography, 13,* 686–693.

Grimalt, J. O., Rullkötter, J., Sicre, M.-A., Summons, R., Farrington, J., Harvey, H. R., Goñi, M., & Sawada, K. (2000). Modifications of the C$_{37}$ alkenone and alkenoate composition in the water column and sediment: Possible implications for sea surface temperature estimates in paleoceanography. *Geochemistry, Geophysics and Geosystems, 1,* 2000GC000053.

Hadley, J. L., & Smith, W. K. (1989). Wind erosion of leaf surface wax in alpine timberline conifers. *Arctic and Alpine Research, 21,* 392–398.

Haidar, A. T., & Thierstein, H. R. (2001). Coccolithophore dynamics off Bermuda (N. Atlantic). *Deep-Sea Research Part II-Topical Studies in Oceanography, 48,* 1925–1956.

Hamanaka, J., Sawada, K., & Tanoue, E. (2000). Production rates of C$_{37}$ alkenones determined by C$_{13}$-labeling technique in the euphotic zone of Sagami Bay, Japan. *Organic Geochemistry, 31,* 1095–1102.

Harada, N., Handa, N., Harada, K., & Matsuoka, H. (2001). Alkenones and particulate fluxes in sediment traps from the central equatorial Pacific. *Deep-Sea Research Part I— Oceanographic Research Papers*, *48*, 891–907.

Harada, N., Shin, K. H., Murata, A., Uchida, M., & Nakatani, T. (2003). Characteristics of alkenones synthesized by a bloom of *Emiliania huxleyi* in the Bering Sea. *Geochimica et Cosmochimica Acta*, *67*, 1507–1519.

Harris, P. G., Zhao, M., Rosell-Melé, A., Tiedemann, R., Sarnthein, M., & Maxwell, J. R. (1996). Chlorin accumulation rate as a proxy for Quaternary marine primary productivity proxy. *Nature*, *383*, 63–65.

Haug, G. H., Ganopolski, A., Sigman, D. M., Rosell-Melé, A., Swann, G. E. A., Tiedemann, R., Jaccard, S. L., Bollmann, J., Maslin, M. A., Leng, M. J., & Eglinton, G. (2005). North Pacific seasonality and the glaciation of North America 2.7 million years ago. *Nature*, *433*, 821–825.

Hay, W. W. (1977). Calcareous nannofossils. In: A. T. S. Ramsay (Ed.), *Oceanic micropaleontology* (pp. 1055–1200). London: Academic Press.

Hayes, J. M., Freeman, K. H., Popp, B. N., & Hoham, C. H. (1990). Compound-specific isotopic analyses: A novel tool for reconstruction of ancient biogeochemical processes. *Organic Geochemistry*, *16*, 1115–1128.

Haywood, A. M., Dekens, P., Ravelo, A. C., & Williams, M. (2005). Warmer tropics during the mid-Pliocene? Evidence from alkenone paleothermometry and a fully coupled ocean-atmosphere GCM. *Geochemistry, Geophysics and Geosystems*, *6*, Q03010.

Herbert, T. D., Schuffert, J. D., Andreasen, D., Heusser, L., Lyle, M., Mix, A., Ravelo, A. C., Stott, L. D., & Herguera, J. C. (2001). Collapse of the California Current during glacial maxima linked to climate change on land. *Science*, *293*, 71–76.

Herbert, T. D., Schuffert, J. D., Thomas, D., Lange, C., Weinheimer, A., Peleo-Alampay, A., & Herguera, J.-C. (1998). Depth and seasonality of alkenone production along the California margin inferred from a core top transect. *Paleoceanography*, *13*, 263–271.

Herndl, G. J., Reinthaler, T., Teira, E., van Aken, H., Veth, C., Pernthaler, A., & Pernthaler, J. (2005). Contribution of archaea to total prokaryotic production in the deep Atlantic Ocean. *Applied and Environmental Microbiology*, *71*, 2303–2309.

Hershberger, K. L., Barns, S. M., Reysenbach, A. L., Dawson, S. C., & Pace, N. R. (1996). Wide diversity of Crenarchaeota. *Nature*, *384*, 420.

Higginson, M. J. (1999). "Chlorin pigment stratigraphy as a new and rapid palaeoceanographic proxy in the Quaternary." Doctoral thesis, University of Bristol.

Hinrichs, K. U., Schneider, R. R., Müller, P. J., & Rullkötter, J. (1999). A biomarker perspective on paleoproductivity variations in two Late Quaternary sediment sections from the Southeastern Atlantic Ocean. *Organic Geochemistry*, *30*, 341–366.

Hoefs, M. J. L., Versteegh, G. J. M., Rijpstra, W. I. C., de Leeuw, J. W., & Sinninghe-Damsté, J. S. (1998). Postdepositional oxic degradation of alkenones: Implications for the measurement of palaeo sea surface temperatures. *Paleoceanography*, *13*, 42–49.

Huang, Y. S., Dupont, L., Sarnthein, M., Hayes, J. M., & Eglinton, G. (2000). Mapping of C-4 plant input from North West Africa into North East Atlantic sediments. *Geochimica et Cosmochimica Acta*, *64*, 3505–3513.

Hughen, K. A., Eglinton, T. I., Xu, L., & Makou, M. (2004). Abrupt tropical vegetation response to rapid climate changes. *Science*, *304*, 1955–1959.

Ikehara, M., Kawamura, K., Ohkouchi, N., Murayama, M., Nakamura, T., & Taira, A. (2000). Variations of terrestrial input and marine productivity in the Southern Ocean (48 degrees S) during the last two deglaciations. *Paleoceanography*, *15*, 170–180.

Itoh, Y. H., Sugai, A., Uda, I., & Itoh, T. (2001). The evolution of lipids. Space life sciences: Living organisms, biological processes and the limits of life. *Advances in Space Research*, *28*, 719–724.

Jasper, J. P., & Hayes, J. M. (1990). A carbon isotope record of CO_2 levels during the late Quaternary. *Nature*, *347*, 462–464.

Jasper, J. P., Hayes, J. M., Mix, A. C., & Prahl, F. G. (1994). Photosynthetic fractionation of C^{13} and concentrations of dissolved CO_2 in the Central Equatorial Pacific during the last 255,000 years. *Paleoceanography*, *9*, 781–798.

Jenkyns, H. C., Forster, A., Schouten, S., & Sinninghe Damsté, J. S. (2004). High temperatures in the Late Cretaceous Arctic Ocean. *Nature*, *432*, 888–892.

Karner, M. B., DeLong, E. F., & Karl, D. M. (2001). Archaeal dominance in the mesopelagic zone of the Pacific Ocean. *Nature*, *409*, 507–510.

Kawamura, K. (1995). Land-derived lipid class compounds in the deep-sea sediments and marine aerosols from North Pacific. In: H. Sakai & Y. Nozaki (Eds), *Biogeochemical processes and ocean flux in the Western Pacific* (pp. 31–51). Tokyo: Terra Scientific Publishing Company.

Kennedy, J. A., & Brassell, S. C. (1992). Molecular stratigraphy of the Santa Barbara basin: Comparison with historical records of annual climate change. *Organic Geochemistry*, *19*, 235–244.

Kim, J. H., Dupont, L., Behling, H., & Versteegh, G. J. M. (2005). Impacts of rapid sea-level rise on mangrove deposit erosion: Application of taraxerol and Rhizophora records. *Journal of Quaternary Science*, *20*, 221–225.

Kim, J. H., Rimbu, N., Lorenz, S. J., Lohmann, G., Nam, S. I., Schouten, S., Ruhlemann, C., & Schneider, R. R. (2004). North Pacific and North Atlantic sea-surface temperature variability during the holocene. *Quaternary Science Reviews*, *23*, 2141–2154.

Kohfeld, K. E., Le Quere, C., Harrison, S. P., & Anderson, R. F. (2005). Role of marine biology in glacial-interglacial CO_2 cycles. *Science*, *308*, 74–78.

Kolattukudy, P. E. (1976). *The chemistry and biochemistry of natural waxes*. Amsterdam: Elsevier.

Koopmans, M. P., SchaefferReiss, C., deLeeuw, J. W., Lewan, M. D., Maxwell, J. R., Schaeffer, P., & Sinninghe Damsté, J. S. (1997). Sulphur and oxygen sequestration of n-C_{37} and n-C_{38} unsaturated ketones in an immature kerogen and the release of their carbon skeletons during early stages of thermal maturation. *Geochimica et Cosmochimica Acta*, *61*, 2397–2408.

Krishnamurthy, R. V., Meyers, P. A., & Lovan, N. A. (2000). Isotopic evidence of sea-surface freshening, enhanced productivity, and improved organic matter preservation during sapropel deposition in the Tyrrhenian Sea. *Geology*, *28*, 263–266.

Kucera, M., Rosell-Melé, A., Schneider, R., Waelbroeck, C., & Weinelt, M. (2005). Multiproxy approach for the reconstruction of the glacial ocean surface (MARGO). *Quaternary Science Reviews*, *24*, 813–819.

Kuypers, M. M. M., Blokker, P., Erbacher, J., Kinkel, H., Pancost, R. D., Schouten, S., & Sinninghe Damsté, J. S. (2001). Massive expansion of marine archaea during a mid-Cretaceous oceanic anoxic event. *Science*, *293*, 92–94.

Laws, E. A., Popp, B. N., Bidigare, R. R., Kennicutt, M. C., & Macko, S. A. (1995). Dependence of phytoplankton carbon isotopic composition on growth rate and [CO_2]aq: Theoretical considerations and experimental results. *Geochimica et Cosmochimica Acta*, *59*, 1131–1138.

Lutz, R., Gieren, B., Luckge, A., Wilkes, H., & Littke, R. (2000). Composition of organic matter in subducted and unsubducted sediments off the Nicoya peninsula, Costa Rica (ODP Leg 170, Sites 1039 and 1040). *Organic Geochemistry*, *31*, 1597–1610.

Lyle, M. W., Prahl, F. G., & Sparrow, M. A. (1992). Upwelling and productivity changes inferred from a temperature record in the central equatorial Pacific. *Nature*, *355*, 812–815.

MacGregor, B. J., Moser, D. P., Alm, E. W., Nealson, K. H., & Stahl, D. A. (1997). Crenarchaeota in Lake Michigan sediment. *Applied and Environmental Microbiology*, *63*, 1178–1181.

Macko, S. A. (1981). "Stable nitrogen isotope ratios as tracers of organic geochemical processes." Doctoral thesis, University of Texas.

Madureira, L. A. S., Conte, M. H., & Eglinton, G. (1995). Early diagenesis of lipid biomarker compounds in North Atlantic sediments. *Paleoceanography*, *10*, 627–642.

Marchand, D., Marty, J. C., Miquel, J. C., & Rontani, J. F. (2005). Lipids and their oxidation products as biomarkers for carbon cycling in the northwestern Mediterranean Sea: Results from a sediment trap study. *Marine Chemistry*, *95*, 129–147.

Marlow, J. R. (2001). "Application of $U^{K'}_{37}$ for long-term (Pliocene-Pleistocene) palaeoclimate reconstruction." Doctoral thesis, Newcastle upon Tyne.

Marlow, J. R., Lange, C., Wefer, G., & Rosell-Melé, A. (2000). Upwelling intensification as part of the Pliocene-Pleistocene climate transition. *Science*, *290*, 2288–2291.

Marlowe, I. (1984). "Lipids as palaeoclimatic indicators." Doctoral thesis, University of Bristol.

Marlowe, I. T., Brassell, S. C., Eglinton, G., & Green, J. C. (1984a). Long chain unsaturated ketones and esters in living algae and marine sediments. *Organic Geochemistry*, *6*, 135–141.

Marlowe, I. T., Brassell, S. C., Eglinton, G., & Green, J. C. (1990). Long-chain alkenones and alkyl alkenoates and the fossil coccolith record of marine sediments. *Chemical Geology*, *88*, 349–375.

Marlowe, I. T., Green, J. C., Neal, A. C., Brassell, S. C., Eglinton, G., & Course, P. A. (1984b). Long chain (n-C$_{37}$-C$_{39}$) alkenones in the Prymnesiophyceae: Distribution of alkenones and other lipids and their taxonomic significance. *British Phycological Journal, 19,* 203–216.

Martrat, B., Grimalt, J. O., Lopez-Martinez, C., Cacho, I., Sierro, F. J., Flores, J. A., Zahn, R., Canals, M., Curtis, J. H., & Hodell, D. A. (2004). Abrupt temperature changes in the Western Mediterranean over the past 250,000 years. *Science, 306,* 1762–1765.

Massana, R., DeLong, E. F., & Pedros-Alio, C. (2000). A few cosmopolitan phylotypes dominate planktonic archaeal assemblages in widely different oceanic provinces. *Applied and Environmental Microbiology, 66,* 1777–1787.

Mayer, L. M. (1994). Surface-area control of organic-carbon accumulation in continental-shelf sediments. *Geochimica et Cosmochimica Acta, 58,* 1271–1284.

McCaffrey, M. A., Farrington, J. W., & Repeta, D. J. (1990). The organic geochemistry of Peru margin surface sediments: I. A comparison of the C$_{37}$ alkenone and historical El Niño records. *Geochimica et Cosmochimica Acta, 54,* 1671–1682.

McClymont, E. L., & Rosell-Melé, A. (2005). Links between the onset of modern Walker circulation and the mid-Pleistocene climate transition. *Geology, 33,* 389–392.

McClymont, E. L., Rosell-Melé, A., Giraudeau, J., Pierre, C., & Lloyd, J. M. (2005). Alkenone and coccolith records of the mid-Pleistocene in the south-east Atlantic: Implications for the U$^{K'}_{37}$ index and South African climate. *Quaternary Science Reviews, 24,* 1559–1572.

McKee, B. (2003). RiOMar: The transport, transformation and fate of Carbon in river-dominated ocean margins. Report of the RiOMar workshop. New Orleans: Tulane University.

McKee, B. A., Aller, R. C., Allison, M. A., Bianchi, T. S., & Kineke, G. C. (2004). Transport and transformation of dissolved and particulate materials on continental margins influenced by major rivers: Benthic boundary layer and seabed processes. *Continental Shelf Research, 24,* 899–926.

Menzel, D., Hopmans, E. C., van Bergen, P. F., de Leeuw, J. W., & Sinninghe Damsté, J. S. (2002). Development of photic zone euxinia in the eastern Mediterranean Basin during deposition of Pliocene sapropels. *Marine Geology, 189,* 215–226.

Menzel, D., Schouten, S., van Bergen, P. F., & Sinninghe Damsté, J. S. (2004). Higher plant vegetation changes during Pliocene sapropel formation. *Organic Geochemistry, 35,* 1343–1353.

Mollenhauer, G., Eglinton, T. I., Ohkuchi, N., Schneider, R. R., Müller, P. J., Grootes, P. M., & Rullkotter, J. (2003). Asynchronous alkenone and foraminifera records from the Benguela Upwelling System. *Geochimica et Cosmochimica Acta, 67,* 2157–2171.

Mollenhauer, G., Kienast, M., Lamy, F., Meggers, H., Schneider, R. R., Hayes, J. M., & Eglinton, T. I. (2005). An evaluation of ^{14}C age relationships between co-occurring foraminifera, alkenones, and total organic carbon in continental margin sediments. *Paleoceanography, 20.*

Müller, P. J., Cepek, M., Ruhland, G., & Schneider, R. R. (1997). Alkenone and coccolithophorid species changes in Late Quaternary sediments from the Walvis Ridge: Implications for the alkenone paleotemperature method. *Palaeogeography, Palaeoclimatology, Palaeoecology, 135,* 71–96.

Müller, P. J., & Fischer, G. (2001). A 4-year sediment trap record of alkenones from the filamentous upwelling region off Cape Blanc, NW Africa and a comparison with distributions in underlying sediments. *Deep-Sea Research Part I — Oceanographic Research Papers, 48,* 1877–1903.

Müller, P. J., Kirst, G., Ruhland, G., von Storch, I., & Rosell-Melé, A. (1998). Calibration of the alkenone paleotemperature index U$^{K'}_{37}$ based on core-tops from the eastern South Atlantic and the global ocean (60°N–60°S). *Geochimica et Cosmochimica Acta, 62,* 1757–1772.

Murray, A. E., Blakis, A., Massana, R., Strawzewski, S., Passow, U., Alldredge, A., & DeLong, E. F. (1999). A time series assessment of planktonic archaeal variability in the Santa Barbara Channel. *Aquatic Microbial Ecology, 20,* 129–145.

Murray, A. E., Preston, C. M., Massana, R., Taylor, L. T., Blakis, A., Wu, K., & DeLong, E. F. (1998). Seasonal and spatial variability of bacterial and archaeal assemblages in the coastal waters near Anvers Island, Antarctica. *Applied and Environmental Microbiology, 64,* 2585–2595.

Ohkouchi, N., Eglinton, T. I., Keigwin, L. D., & Hayes, J. M. (2002). Spatial and temporal offsets between proxy records in a sediment drift. *Science, 298,* 1224–1227.

Ohkouchi, N., Kawamura, K., Kawahata, H., & Okada, H. (1999). Depth ranges of alkenone production in the Central Pacific Ocean. *Global Biogeochemical Cycles, 13,* 695–704.

Ohkouchi, N., Kawamura, K., Kawahata, H., & Taira, A. (1997a). Latitudinal distributions of terrestrial biomarkers in the sediments from the Central Pacific. *Geochimica et Cosmochimica Acta*, *61*, 1911–1918.

Ohkouchi, N., Kawamura, K., Nakamura, T., & Taira, A. (1994). Small changes in the sea-surface temperature during the last 20,000 years: Molecular evidence from the western tropical Pacific. *Geophysical Resarch Letters*, *21*, 2207–2210.

Ohkouchi, N., Kawamura, K., & Taira, A. (1997b). Fluctuations of terrestrial and marine biomarkers in the western tropical Pacific during the last 23,300 years. *Paleoceanography*, *12*, 623–630.

Okada, H., & Honjo, S. (1973). The distribution of oceanic coccolithophorids in the Pacific. *Deep-Sea Research*, *20*, 355–374.

Okada, H., & McIntyre, A. (1977). Modern coccolithophores of the Pacific and North Atlantic oceans. *Micropaleontology*, *23*, 1–55.

O'Leary, M. H. (1981). Carbon isotope fractionation in plants. *Phytochemistry*, *20*, 553–567.

Pagani, M., Arthur, M. A., & Freeman, K. H. (1999a). Miocene evolution of atmospheric carbon dioxide. *Paleoceanography*, *14*, 273–292.

Pagani, M., Freeman, K. H., & Arthur, M. A. (1999b). Late Miocene atmospheric CO_2 concentrations and the expansion of C-4 grasses. *Science*, *285*, 876–879.

Pagani, M., Freeman, K. H., Ohkouchi, N., & Caldeira, K. (2002). Comparison of water column [CO_2aq] with sedimentary alkenone-based estimates: A test of the alkenone-CO_2 proxy. *Paleoceanography*, *17*, 1069, doi:10.1029/2002PA000756.

Pagani, M., Zachos, J. C., Freeman, K. H., Tipple, B., & Bohaty, S. (2005). Marked decline in atmospheric carbon dioxide concentrations during the Paleogene. *Science*, *309*, 600–603.

Pancost, R. D., & Boot, C. S. (2004). The palaeoclimatic utility of terrestrial biomarkers in marine sediments. *Marine Chemistry*, *92*, 239–261.

Pancost, R. D., Freeman, K. H., Wakeham, S. G., & Robertson, C. Y. (1997). Controls on carbon isotope fractionation by diatoms in the Peru upwelling region. *Geochimica et Cosmochimica Acta*, *61*, 4983–4991.

Pancost, R. D., Hopmans, E. C., Sinninghe Damsté, J. S., & Party, T. M. S. S. (2001). Archaeal lipids in the Mediterranean cold seeps: Molecular proxies for anaerobic methane oxidation. *Geochimica et Cosmochimica Acta*, *65*, 1611–1627.

Passier, H. F., Bosch, H. J., Nijenhuis, I. A., Lourens, L. J., Bottcher, M. E., Leenders, A., Sinninghe Damsté, J. S., de Lange, G. J., & de Leeuw, J. W. (1999). Sulphidic Mediterranean surface waters during Pliocene sapropel formation. *Nature*, *397*, 146–149.

Paul, H. (2002). "Application of novel stable isotope methods to reconstruct paleoenvironments: Compound specific hydrogen isotopes and pore-water oxygen isotopes". Doctoral theisis, Swiss Federal Institute of Technology.

Pearson, A., & Eglinton, T. I. (2000). The origin of *n*-alkanes in Santa Monica Basin surface sediment: A model based on compound-specific Delta C-14 and Delta C-13 data. *Organic Geochemistry*, *31*, 1103–1116.

Pelejero, C., & Calvo, E. (2003). The upper end of the $U^{K'}_{37}$ temperature calibration revisited. *Geochemistry, Geophysics and Geosystems*, *4*, Art. No. 1014 Feb 15.

Pelejero, C., & Grimalt, J. O. (1997). The correlation between the U^{K}_{37} index and sea surface temperatures in the warm boundary: The South China Sea. *Geochimica et Cosmochimica Acta*, *61*, 4789–4797.

Pelejero, C., Grimalt, J. O., Heilig, S., Kienast, M., & Wang, L. (1999). High-resolution $U^{K'}_{37}$ temperature reconstructions in the South China Sea over the past 220 kyr. *Paleoceanography*, *14*, 224–231.

Pelejero, C., Grimalt, J. O., Sarnthein, M., Wang, L., & Flores, J.-A. (1998). Molecular biomarker record of sea surface temperature and climatic change in the South China Sea during the last 140,000 yrs. *Marine Geology*, *156*, 109–121.

Pichon, J. J., Sikes, E. L., Hiramatsu, C., & Robertson, L. (1998). Comparison of $U^{K'}_{37}$ and diatom assemblage sea surface temperature estimates with atlas derived data in holocene sediments from the southern west Indian Ocean. *Journal of Marine Systems*, *17*, 541–554.

Popp, B. N., Kenig, F., Wakeham, S. G., Laws, E. A., & Bidigare, R. R. (1998a). Does growth rate affect ketone unsaturation and intracellular carbon isotopic variability in *Emiliani huxleyi*? *Paleoceanography*, *13*, 35–41.

Popp, B. N., Laws, E. A., Bidigare, R. R., Dore, J. E., Hanson, K. L., & Wakeham, S. G. (1998b). Effect of phytoplankton cell geometry on carbon isotopic fractionation. *Geochimica et Cosmochimica Acta, 62,* 69–77.

Powers, L. A., Johnson, T. C., Werne, J. P., Castanada, I. S., Hopmans, E. C., Sinninghe Damsté, J. S., & Schouten, S. (2005). Large temperature variability in the southern African tropics since the Last Glacial Maximum. *Geophysical Research Letters, 32,* L08706.

Poynter, J. G., & Eglinton, G. (1991). The biomarker concept: Strength and weaknesses. *Fresenius Journal of Analytical Chemistry, 339,* 725–731.

Poynter, J.-G., Farrimond, P., Robinson, N., & Eglinton, G. (1989). Aeolian derived higher plant lipids in the marine sedimentary record: Links with paleoclimate. In: M. Leinen & M. Sarnthein (Eds), *Paleoclimatology and paleometeorology: Modern and past patterns of global atmospheric transport* (pp. 435–462). Dordrecht, The Netherlands: Kluwer, Academic Press.

Prahl, F. G. (1985). Chemical evidence of differential particle dispersal in the southern Washington coastal environment. *Geochimica et Cosmochimica Acta, 49,* 2533–2539.

Prahl, F. G., Collier, R. B., Dymond, J., Lyle, M., & Sparrow, M. A. (1993). A biomarker perspective on prymnesiophyte in the northeast Pacific Ocean. *Deep-Sea Research I, 40,* 2061–2076.

Prahl, F. G., de Lange, G. J., Lyle, M., & Sparrow, M. A. (1989a). Post-depositional stability of long-chain alkenones under contrasting redox conditions. *Nature, 341,* 434–437.

Prahl, F. G., Dymond, J., & Sparrow, M. (2000). Annual biomarker record for export production in the central Arabian Sea. *Deep-Sea Research II, 47,* 1581–1604.

Prahl, F. G., Mix, A. C., & Sparrow, M. A. (2006). Alkenone paleothermometry: Biological lessons from marine sediment records off western South America. *Geochimica et Cosmochimica Acta, 70,* 101–117.

Prahl, F. G., & Muehlhausen, L. A. (1989). Lipid biomarkers as geochemical tools for paleoceano-graphic study. In: W. H. Berger, V. S. Smetaceck & G. Wefer (Eds), *Productivity of the ocean: Present and past* (pp. 271–289). New York: John Wiley & Sons.

Prahl, F. G., MuehlhauSen, L. A., & Lyle, M. (1989b). An organic geochemical assessment of oceanographic conditions at MANOP site over the past 26,000 years. *Paleoceanography, 4,* 495–510.

Prahl, F. G., Muehlhausen, L. A., & Zahnle, D. I. (1988). Further evaluation of long-chain alkenones as indicators of paleoceanographic conditions. *Geochimica et Cosmochimica Acta, 52,* 2303–2310.

Prahl, F. G., Pilskaln, C. H., & Sparrow, M. A. (2001). Seasonal record for alkenones in sedimentary particles from the Gulf of Maine. *Deep-Sea Research Part I — Oceanographic Research Papers, 48,* 515–528.

Prahl, F. G., Popp, B. N., Karl, D. M., & Sparrow, M. A. (2005). Ecology and biogeochemistry of alkenone production at Station ALOHA. *Deep-Sea Research Part I — Oceanographic Research Papers, 52,* 699–719.

Prahl, F. G., & Wakeham, S. G. (1987). Calibration of unsaturation patterns in long-chain ketone compositions for palaeotemperature assessment. *Nature, 320,* 367–369.

Prahl, F. G., Wolfe, G. V., & Sparrow, M. A. (2003). Physiological impacts on alkenone paleothermo-metry. *Paleoceanography, 18,* Art. No. 1025 Apr 9.

Rau, G. H., Takahashi, T., & Marais, D. J. D. (1989). Latitudinal variations in plankton Delta-C-13: Implications for CO_2 and productivity in past oceans. *Nature, 341,* 516–518.

Raymo, M. E. (1994). The initiation of Northern-Hemisphere Glaciation. *Annual Review of Earth and Planetary Sciences, 22,* 353–383.

Rechka, J. A., & Maxwell, J. R. (1988). Characterisation of alkenone temperature indicators in sediments and organisms. *Organic Geochemistry, 13,* 727–734.

Riebesell, U., Revill, A. T., Holdsworth, D. G., & Volkman, J. K. (2000). The effects of varying CO_2 concentration on lipid composition and carbon isotope fractionation in *Emiliania huxleyi*. *Geochimica et Cosmochimica Acta, 64,* 4179–4192.

Rieley, G., Collier, R. J., Jones, D. M., & Eglinton, G. (1991). The biogeochemistry of Ellesmere Lake, UK-I: Source correlation of leaf wax inputs to the sedimentary lipid record. *Organic Geochemistry, 17,* 901–912.

Rimbu, N., Lohmann, G., Lorenz, S. J., Kim, J. H., & Schneider, R. R. (2004). Holocene climate variability as derived from alkenone sea surface temperature and coupled ocean-atmosphere model experiments. *Climate Dynamics, 23,* 215–227.

Rogge, W. F., Hildemann, L. M., Mazurek, M. A., Cass, G. R., & Simoneit, B. R. T. (1993). Sources of fine organic aerosol. 4: Particulate abrasion products from leaf surfaces of urban plants. *Environmental Science and Technology, 27,* 2700–2711.

Rommerskirchen, F., Eglinton, G., Dupont, L., Guntner, U., Wenzel, C., & Rullkotter, J. (2003). A north to south transect of Holocene southeast Atlantic continental margin sediments: Relationship between aerosol transport and compound-specific Delta C-13 land plant biomarker and pollen records. *Geochemistry, Geophysics and Geosystems, 4,* Art. No. 1101 Dec 10.

Rontani, J. F., Cuny, P., Grossi, V., & Beker, B. (1997). Stability of long-chain alkenones in senescing cells of *Emiliania huxleyi*: Effect of photochemical and aerobic microbial degradation on the alkenone unsaturation ratio ($U_{37}^{K'}$). *Organic Geochemistry, 26,* 503–509.

Rosell-Melé, A. (1994). "Long-chain alkenones, alkyl alkenoates and total pigment abundances as climatic proxy-indicators in the Northeastern Atlantic." Doctoral thesis, University of Bristol.

Rosell-Melé, A. (1998). Interhemispheric appraisal of the value of alkenone indices as temperature and salinity proxies in high-latitude locations. *Paleoceanography, 13,* 694–703.

Rosell-Melé, A., Bard, E., Emeis, K. C., Grieger, B., Hewitt, C., Müller, P. J., & Schneider, R. R. (2004). Sea surface temperature anomalies in the oceans at the LGM estimated from the alkenone-$U_{37}^{K'}$ index: Comparison with GCMs. *Geophysical Research Letters, 31,* L03208.

Rosell-Melé, A., Bard, E., Emeis, K. C., Grimalt, J. O., Müller, P., Schneider, R., Bouloubassi, I., Epstein, B., Fahl, K., Fluegge, A., Freeman, K., Goni, M., Guntner, U., Hartz, D., Hellebust, S., Herbert, T., Ikehara, M., Ishiwatari, R., Kawamura, K., Kenig, F., de Leeuw, J., Lehman, S., Mejanelle, L., Ohkouchi, N., Pancost, R. D., Pelejero, C., Prahl, F., Quinn, J., Rontani, J. F., Rostek, F., Rullkotter, J., Sachs, J., Blanz, T., Sawada, K., Schutz-Bull, D., Sikes, E., Sonzogni, C., Ternois, Y., Versteegh, G., Volkman, J. K., & Wakeham, S. (2001).Precision of the current methods to measure the alkenone proxy $U_{37}^{K'}$ and absolute alkenone abundance in sediments: Results of an interlaboratory comparison study. *Geochemistry, Geophysics and Geosystems, 2,* Art. No. 2000GC000141 Jul 6.

Rosell-Melé, A., Comes, P., Müller, P. J., & Ziveri, P. (2000). Alkenone fluxes and anomalous $U_{37}^{K'}$ values during 1989–1990 in the Northeast Atlantic (48 degrees N 21 degrees W). *Marine Chemistry, 71,* 251–264.

Rosell-Melé, A., Eglinton, G., Pflaumann, U., & Sarnthein, M. (1995). Atlantic core-top calibration of the U_{37}^{K} index as a sea-surface palaeotemperature indicator. *Geochimica et Cosmochimica Acta, 59,* 3099–3107.

Rosell-Melé, A., Jansen, E., & Weinelt, M. (2002). Appraisal of a molecular approach to infer variations in surface ocean freshwater inputs to the North Atlantic during the last glacia. *Global and Planetary Change, 34,* 143–152.

Rosell-Melé, A., & Koç, N. (1997). Paleoclimatic significance of the stratigraphic occurrence of photosynthetic biomarker pigments in the Nordic Seas. *Geology, 25,* 49–52.

Rosell-Melé, A., Maslin, M., Maxwell, J. R., & Schaeffer, P. (1997). Biomarker evidence for "Heinrich" events. *Geochimica et Cosmochimica Acta, 61,* 1671–1678.

Rosell-Melé, A., Weinelt, M., Koç, N., Jansen, E., & Sarnthein, M. (1998). Variability of the Arctic front during the last climatic cycle: Application of a novel molecular proxy. *Terra Nova, 10,* 86–89.

Rühlemann, C., Mulitza, S., Müller, P. J., Wefer, G., & Zahn, R. (1999). Warming of the tropical Atlantic Ocean and lowdown of thermohaline circulation during the last deglaciation. *Nature, 402,* 511–514.

Sachs, J. P., & Lehman, S. J. (1999). Subtropical North Atlantic temperatures 60,000 to 30,000 years ago. *Science, 286,* 756–759.

Sauer, P. E., Eglinton, T. I., Hayes, J. M., Schimmelmann, A., & Sessions, A. L. (2001). Compound-specific D/H ratios of lipid biomarkers from sediments as a proxy for environmental and climatic conditions. *Geochimica et Cosmochimica Acta, 65,* 213–222.

Sawada, K., Handa, N., & Nakatsuka, T. (1998). Production and transport of long-chain alkenones and alkyl alkenoates in a sea water column in the northwestern Pacific off central Japan. *Marine Chemistry, 59,* 219–234.

Sawada, K., Handa, N., Shiraiwa, Y., Danbara, A., & Montani, S. (1996). Long-chain alkenones and alkyl alkenoates in the coastal and pelagic sediments of the northwest North Pacific, with special reference to the reconstruction of *Emiliania huxleyi* and *Gephyrocapsa oceanica* ratios. *Organic Geochemistry, 24*, 751–764.

Sawada, K., & Shiraiwa, Y. (2004). Alkenone and alkenoic acid compositions of the membrane fractions of *Emiliania huxleyi*. *Phytochemistry, 65*, 1299–1307.

Schefuss, E., Ratmeyer, V., Stuut, J. B. W., Jansen, J. H. F., & Sinninghe Damsté, J. S. (2003a). Carbon isotope analyses of *n*-alkanes in dust from the lower atmosphere over the central eastern Atlantic. *Geochimica et Cosmochimica Acta, 67*, 1757–1767.

Schefuss, E., Schouten, S., Jansen, J. H. F., & Sinninghe Damsté, J. S. (2003b). African vegetation controlled by tropical sea surface temperatures in the mid-Pleistocene period. *Nature, 422*, 418–421.

Schimmelmann, A., Lewan, M. D., & Wintsch, R. P. (1999). D/H isotope ratios of kerogen, bitumen, oil, and water in hydrous pyrolysis of source rocks containing kerogen types I, II, IIS, and III. *Geochimica et Cosmochimica Acta, 63*, 3751–3766.

Schneider, R. R., Müller, P. J., & Acheson, R. (1999) Atlantic alkenone sea surface temperature records: Low versus mid latitudes and differences between hemispheres. In F. Abrantes & A. C. Mix (Eds), *Reconstructing ocean history: A window to the future* (pp. 33–56). New York, Dordrecht: Kluwer Academic/Plenum Publishers.

Schneider, R. R., Müller, P. J., & Ruhland, G. (1995). Late Quaternary surface circulation in the east equatorial South Atlantic: Evidence from alkenone sea surface temperatures. *Paleoceanography, 10*(2), 197–220, doi:10.1029/94PA03308.

Schneider, R. R., Müller, P. J., Ruhland, G., Meinecke, G., Schmidt, H., & Wefer, G. (1996). Late Quaternary surface temperatures and productivity in the east-equatorial South Atlantic: Response to changes in trade/monsoon wind forcing and surface water advection. In: G. Wefer, W. H. Berger, G. Siedler & D. J. Webb (Eds), *The South Atlantic: Present and past circulation* (pp. 527–551). Berlin, Heilderberg: Springer.

Schouten, S., Hopmans, E. C., & Sinninghe Damsté, J. S. (2004). The effect of maturity and depositional redox conditions on archaeal tetraether lipid palaeothermometry. *Organic Geochemistry, 35*, 567–571.

Schouten, S., Hopmans, E. C., Forster, A., van Breugel, Y., Kuypers, M. M. M., & Sinninghe Damsté, J. S. (2003). Extremely high sea-surface temperatures at low latitudes during the middle Cretaceous as revealed by archaeal membrane lipids. *Geology, 31*, 1069–1072.

Schouten, S., Hopmans, E. C., Schefuss, E., & Sinninghe Damsté, J. S. (2002a). Distributional variations in marine Crenarchaeotal membrane lipids: A new tool for reconstructing ancient sea water temperatures? *Earth and Planetary Science Letters, 204*, 265–274.

Schouten, S., Hopmans, E. C., Schefuss, E., Wuchter, C., & Sinninghe Damsté, J. S. (2002b). Geochemical and environmental importance of marine archaea. *Geochimica et Cosmochimica Acta, 66* A687.

Schouten, S., Klein-Breteler, W. C. M. K., Blokker, P., Schogt, N., Rijpstra, W. I. C., Grice, K., Baas, M., & Sinninghe Damsté, J. S. (1998). Biosynthetic effects on the stable carbon isotope compositions of algal lipids: Implications for deciphering the carbon isotopic biomarker record. *Geochimica et Cosmochimica Acta, 62*, 1397–1406.

Schouten, S., Ossebaar, J., Schrieber, K., Kienhuis, M. V. M., Langer, G., Benthien, A., & Bijma, J. (2006). The effect of temperature, salinity and growth rate on the stable hydrogen isotopic composition of long-chain alkenones produced by *Emiliania huxleyi* and *Geophyrocapsa oceanica*. *Biogeosciences, 3*, 113–119.

Schubert, C. J. (2005). Chlorin Index: A new parameter for organic matter freshness in sediments. *Geochemistry, Geophysics and Geosystems, 6*, Art. No. Q03005 Mar 8.

Schubert, C. J., Villanueva, J., Calvert, S. E., Cowie, G. L., von Rad, U., Schulz, H., Berner, U., & Erlenkeuser, H. (1998). Stable phytoplankton community structure in the Arabian Sea over the past 20,000 years. *Nature, 394*, 563–566.

Schulte, S., Benthien, A., Müller, P. J., & Ruhlemann, C. (2004). Carbon isotopic fractionation (εp) of C_{37} alkenones in deep-sea sediments: Its potential as a paleonutrient proxy. *Paleoceanography, 19*.

Schulte, S., Rostek, F., Bard, E., Rullkötter, J., & Marchal, O. (1999). Variations of oxygen-minimum and primary productivity recorded in sediments of the Arabian Sea. *Earth and Planetary Science Letters, 173*, 205–221.

Schulz, H. M., Schoner, A., & Emeis, K. C. (2000). Long-chain alkenone patterns in the Baltic Sea: An ocean-freshwater transition. *Geochimica et Cosmochimica Acta, 64*, 469–477.

Scourse, J., Marret, F., Versteegh, G. J. M., Jansen, J. H. F., Schefuss, E., & van der Plicht, J. (2005). High-resolution last deglaciation record from the Congo fan reveals significance of mangrove pollen and biomarkers as indicators of shelf transgression. *Quaternary Research, 64*, 57–69.

Seki, O., Kawamura, K., Sakamoto, T., Ikehara, M., Nakatsuka, T., & Wakatsuchi, M. (2005). Decreased surface salinity in the Sea of Okhotsk during the last glacial period estimated from alkenones. *Geophysical Research Letters, 32*, L08710.

Sessions, A. L., Sylva, S. P., Summons, R. E., & Hayes, J. M. (2004). Isotopic exchange of carbon-bound hydrogen over geologic timescales. *Geochimica et Cosmochimica Acta, 68*, 1545–1559.

Shankle, A., Goericke, R., Franks, P., & Levin, L. (2002). Chlorin distribution and degradation in sediments within and below the Arabian Sea oxygen minimum zone. *Deep-Sea Research I, 49*, 953–969.

Sicre, M. A., Bard, E., Ezat, U., & Rostek, F. (2002). Alkenone distributions in the North Atlantic and Nordic Sea surface waters. *Geochemistry, Geophysics and Geosystems, 3*, 10.1029/2001GC000159.

Sicre, M. A., Ternois, Y., Miquel, J. C., & Marty, J. C. (1999). Alkenones in the Northwestern Mediterranean Sea: Interannual variability and vertical transfer. *Geophysical Research Letters, 26*, 1735–1738.

Sicre, M. A., Ternois, Y., Paterne, M., Boireau, A., Beaufort, L., Martinez, P., & Bertrand, P. (2000). Biomarker stratigraphic records over the last 150 kyears off the NW African coast at 25 degrees N. *Organic Geochemistry, 31*, 577–588.

Sikes, E. L., Farrington, J. W., & Keigwin, L. D. (1991). Use of the alkenone unsaturation ratio $U^{K'}_{37}$ to determine past sea surface temperatures: Core-top SST calibrations and methodology considerations. *Earth and Planetary Science Letters, 104*, 36–47.

Sikes, E. L., & Keigwin, L. D. (1994). Equatorial Atlantic sea surface temperatures for the last 30 kyr: A comparison of $U^{K'}_{37}$, $\delta^{18}O$ and foraminiferal assemblage temperature estimates. *Paleoceanography, 9*, 31–45.

Sikes, E. L., Sicre, M. A. (2002). Relationship of the tetra-unsaturated C_{37} alkenone to salinity and temperature: Implications for paleoproxy applications. *Geochemistry, Geophysics and Geosystems, 3*, Art. No. 1063 Nov 6.

Sikes, E. L., & Volkman, J. K. (1993). Calibration of alkenone unsaturation ratios ($U^{K'}_{37}$) for paleotemperature estimation in cold polar waters. *Geochimica et Cosmochimica Acta, 57*, 1883–1889.

Sikes, E. L., Volkamn, J. K., Robertson, L., & Pichon, J.-J. (1997). Alkenones and alkenes in surface waters and sediments of the Southern Ocean: Implications for paleotemperature estimation in polar regions. *Geochimica et Cosmochimica Acta, 61*, 1495–1505.

Simoneit, B. R. T. (1977). Organic matter in eolian dusts over the Atlantic Ocean. *Marine Chemistry, 5*, 443–464.

Simoneit, B. R. T. (1989). Organic-matter of the troposphere. 5: Application of molecular marker analysis to biogenic emissions into the troposphere for source reconciliations. *Journal of Atmospheric Chemistry, 8*, 251–275.

Simoneit, B. R. T., Chester, R., & Eglinton, G. (1977). Biogenic lipids in particulates from the lower atmosphere over the eastern Atlantic. *Nature, 267*, 682–685.

Simoneit, B. R. T., & Elias, V. O. (2000). Organic tracers from biomass burning in atmospheric particulate matter over the ocean. *Marine Chemistry, 69*, 301–312.

Simoneit, B. R. T., Schauer, J. J., Nolte, C. G., Oros, D. R., Elias, V. O., Fraser, M. P., Rogge, W. F., & Cass, G. R. (1999). Levoglucosan, a tracer for cellulose in biomass burning and atmospheric particles. *Atmospheric Environment, 33*, 173–182.

Sinninghe Damsté, J. S., Kuypers, M. M. M., Schouten, S., Schulte, S., & Rullkotter, R. (2003). The lycopane/C_{31} *n*-alkane ratio as a proxy to assess palaeoxicity during sediment deposition. *Earth and Planetary Science Letters, 209*, 215–226.

Sinninghe Damsté, J. S., Rijpstra, W. I. C., Hopmans, E. C., Prahl, F. G., Wakeham, S. G., & Schouten, S. (2002). Distribution of membrane lipids of planktonic Crenarchaeota in the Arabian Sea. *Applied and Environmental Microbiology, 68*, 2997–3002.

Sinninghe Damsté, J. S., Rijpstra, W. I. C., Kock-van Dalen, A. C., de Leeuw, J. W., & Schenck, P. A. (1989). Quenching of labile functionalised lipids by inorganic sulphur species: Evidence for the formation of sedimentary organic sulphur compounds at the early stages of diagenesis. *Geochimica et Cosmochimica Acta, 53,* 1343–1355.

Sinninghe Damsté, J. S., Rijpstra, W. I. C., & Reichart, G. J. (2002a). The influence of oxic degradation on the sedimentary biomarker record II. Evidence from Arabian Sea sediments. *Geochimica et Cosmochimica Acta, 66,* 2737–2754.

Sinninghe Damsté, J. S., Schouten, S., Hopmans, E. C., van Duin, A. C. T., & Geenevasen, J. A. J. (2002b). Crenarchaeol: The characteristic core glycerol dibiphytanyl glycerol tetraether membrane lipid of cosmopolitan pelagic Crenarchaeota. *Journal of Lipid Research, 43,* 1641–1651.

Sinninghe Damsté, J. S., Wakeham, S., Kohnen, M. E. L., Hayes, J. M., & de Leeuw, J. W. (1993). A 6000-year sedimentary molecular record of chemocline excursions in the Black Sea. *Nature, 362,* 827–829.

Smith, B. N., & Epstein, S. (1971). Two categories of C-13/C-12 ratios for higher plants. *Plant Physiology, 47,* 380–384.

Smittenberg, R. H., Hopmans, E. C., Schouten, S., Hayes, J. M., Eglinton, T. I., Sinninghe Damsté, J. S. (2004). Compound-specific radiocarbon dating of the varved Holocene sedimentary record of Saanich Inlet, Canada. *Paleoceanography, 19,* PA2012, doi:10.1029/2003PA000927.

Sonzogni, C., Bard, E., & Rostek, F. (1998). Tropical sea surface temperatures during the last glacial period: A view based on alkenones in Indian Ocean sediments. *Quaternary Science Reviews, 17,* 1185–1201.

Sonzogni, C., Bard, E., Rostek, F., Lafont, A., Rosell-Melé, A., & Eglinton, G. (1997). Core-top calibration of the alkenone index versus sea surface temperature in the Indian Ocean. *Deep-Sea Research, 44,* 1445–1460.

Stoll, H. M., Klaas, C. M., Probert, I., Encinar, J. R., & Alonso, J. I. G. (2002a). Calcification rate and temperature effects on Sr partitioning in coccoliths of multiple species of coccolithophorids in culture. *Global and Planetary Change, 34,* 153–171.

Stoll, H. M., & Schrag, D. P. (2000). Coccolith Sr/Ca as a new indicator of coccolithophorid calcification and growth rate. *Geochemistry, Geophysics and Geosystems, 1,* Paper number 10.1029/1999GC000015.

Stoll, H. M., & Ziveri, P. (2002). Separation of monospecific and restricted coccolith assemblages from sediments using differential settling velocity. *Marine Micropaleontology, 46,* 209–221.

Stoll, H. M., Ziveri, P., Geisen, M., Probert, I., & Young, J. R. (2002b). Potential and limitations of Sr/Ca ratios in coccolith carbonate: New perspectives from cultures and monospecific samples from sediments. *Philosophical Transactions of the Royal Society of London Series a — Mathematical Physical and Engineering Sciences, 360,* 719–747.

Summerhayes, C. P., Kroon, D., Rosell-Melé, A., Jordan, R. W., Schrader, H. J., Hearn, R., Grimalt, J. O., & Eglinton, G. (1995). Variability in the Benguela Current upwelling system over the past 70,000 years. *Progress in Oceanography, 35,* 207–251.

Sun, M. Y., Zou, L., Dai, J. H., Ding, H. B., Culp, R. A., & Scranton, M. I. (2004). Molecular carbon isotopic fractionation of algal lipids during decomposition in natural oxic and anoxic seawaters. *Organic Geochemistry, 35,* 895–908.

Teece, M. A., Getliff, J. M., Leftley, J. W., Parkes, R. J., & Maxwell, J. R. (1998). Microbial degradation of the marine prymnesiophyte *Emiliania huxleyi* under oxic and anoxic conditions as a model for early diagenesis: Long chain alkadienes, alkenones and alkyl alkenoates. *Organic Geochemistry, 29,* 863–880.

Ternois, Y., Kawamura, K., Keigwin, L., Ohkouchi, N., & Nakatsuka, T. (2001). A biomarker approach for assessing marine and terrigenous inputs to the sediments of Sea of Okhotsk for the last 27,000 years. *Geochimica et Cosmochimica Acta, 65,* 791–802.

Ternois, Y., Sicre, M. A., Boireau, A., Beaufort, L., Miquel, J. C., & Jeandel, C. (1998). Hydrocarbons, sterols and alkenones in sinking particles in the Indian Ocean sector of the Southern Ocean. *Organic Geochemistry, 28,* 489–501.

Ternois, Y., Sicre, M. A., Boireau, A., Conte, M. H., & Eglinton, G. (1997). Evaluation of long-chain aleknones as paleo-temperature indicators in the Mediterranean Sea. *Deep-Sea Research I, 44,* 271–286.

Ternois, Y., Sicre, M.-A., Boireau, A., Marty, J.-C., & Miquel, J.-C. (1996). Production pattern of alkenones in the Mediterranean Sea. *Geophysical Research Letters, 23,* 3171–3174.

Thierstein, H. R., Geitzenauer, K. R., & Molfino, B. (1977). Global synchroneity of Late Quaternary coccolith datum levels: Validation by oxygen isotopes. *Geology, 5,* 400–404.

Thompson, S., & Eglinton, G. (1978). The fractionation of a recent sediment for organic geochemical analysis. *Geochimica et Cosmochimica Acta, 42,* 199–207.

Thomsen, C., Schulz-Bull, D. E., Petrick, G., & Duinker, J. C. (1998). Seasonal variability of the long-chain alkenone flux and the effect on the $U_{37}^{K'}$ index in the Norwegian Sea. *Organic Geochemistry, 28,* 311–323.

Thomsen, L., & Gust, G. (2000). Sediment erosion thresholds and characteristics of resuspended aggregates on the western European continental margin. *Deep-Sea Research Part I — Oceanographic Research Papers, 47,* 1881–1897.

Uchida, M., Shibata, Y., Ohkushi, K., Yoneda, M., Kawamura, K., & Morita, M. (2005). Age discrepancy between molecular biomarkers and calcareous foraminifera isolated from the same horizons of Northwest Pacific sediments. *Chemical Geology, 218,* 73–89.

Uda, I., Sugai, A., Itoh, Y. H., & Itoh, T. (2001). Variation in molecular species of polar lipids from *Thermoplasma acidophilum* depends on growth temperature. *Lipids, 36,* 103–105.

Versteegh, G. J. M., Bosch, H. J., & de Leeuw, J. W. (1997). Potential palaeoenvironmental information of C-24 to C-36 mid-chain diols, keto-ols and mid-chain hydroxy fatty acids: A critical review. *Organic Geochemistry, 27,* 1–13.

Versteegh, G. J. M., Jansen, J. H. F., de Leeuw, J. W., & Schneider, R. R. (2000). Mid-chain diols and keto-ols in SE Atlantic sediments: A new tool for tracing past sea surface water masses? *Geochimica et Cosmochimica Acta, 64,* 1879–1892.

Versteegh, G. J. M., Riegman, R., de Leeuw, J. W., & Jansen, J. H. F. (2001). $U_{37}^{K'}$ values for *Isochrysis galbana* as a function of culture temperature, light intensity and nutrient concentrations. *Organic Geochemistry, 32,* 785–794.

Versteegh, G. J. M., Schefuss, E., Dupont, L., Marret, F., Sinninghe Damsté, J. S., & Jansen, J. H. F. (2004). Taraxerol and *Rhizophora* pollen as proxies for tracking past mangrove ecosystems. *Geochimica et Cosmochimica Acta, 68,* 411–422.

Vetriani, C., Jannasch, H. W., MacGregor, B. J., Stahl, D. A., & Reysenbach, A. L. (1999). Population structure and phylogenetic characterization of marine benthic archaea in deep-sea sediments. *Applied and Environmental Microbiology, 65,* 4375–4384.

Vetriani, C., Reysenbach, A. L., & Dore, J. (1998). Recovery and phylogenetic analysis of archaeal rRNA sequences from continental shelf sediments. *FEMS Microbiology Letters, 161,* 83–88.

Villanueva, J., Flores, J. A., & Grimalt, J. O. (2002). A detailed comparison of the $U_{37}^{K'}$ and coccolith records over the past 290 kyr: Implications to the alkenone paleotemperature method. *Organic Geochemistry, 33,* 897–905.

Volkman, J. K., Barrett, S. M., Blackburn, S. I., & Sikes, E. L. (1995). Alkenones in *Gephyrocapsa oceanica*: Implications for studies of paleoclimate. *Geochimica et Cosmochimica Acta, 59,* 513–520.

Volkman, J. K., Eglinton, G., Corner, E. D. S., & Forsberg, T. E. V. (1980a). Long-chain alkenes and alkenones in the marine coccolithophorid *Emiliania huxleyi*. *Phytochemistry, 19,* 2619–2622.

Volkman, J. K., Eglinton, G., Corner, E. D. S., & Sargent, J. R. (1980b). Novel unsaturated straight-chain C_{37}-C_{39} methyl and ethyl ketones in marine sediments and coccolithophore *Emiliania huxleyi*. In: A. G. Douglas & J. R. Maxwell (Eds), *Advances in Organic Geochemistry 1979* (pp. 219–227). Oxford: Pergamon.

Wade, T., & Quinn, J. G. (1979). Geochemical distributions of hydrocarbons in sediments from mid-Narrangasett Bay, Rhode Island. *Organic Geochemistry, 1,* 157–167.

Wagner, T., & Henrich, R. (1994). Organofacies and lithofacies of glacial-interglacial deposits in the Norwegian-Greenland Sea-Responses to paleoceanographic and paleoclimatic changes. *Marine Geology, 120,* 335–364.

Wakeham, S. G., Hedges, J. I., Lee, C., Peterson, M. L., & Hernes, P. J. (1997). Compositions and transport of lipid biomarkers through the water column and surficial sediments of the equatorial Pacific Ocean. *Deep-Sea Research II, 44,* 2131–2162.

Wakeham, S. G., & Lee, C. (1993). Production, transport, and alteration of particulate organic matter in the marine water column. In: M. H. Engel & S. A. Macko (Eds), *Organic geochemistry: Principles and applications* (pp. 145–169). New York: Plenum Press.

Wakeham, S. G., Lewis, C. M., Hopmans, E. C., Schouten, S., & Sinninghe Damsté, J. S. (2003). Archaea mediate anaerobic oxidation of methane in deep euxinic waters of the Black Sea. *Geochimica et Cosmochimica Acta*, *67*, 1359–1374.

Weaver, P. P. E., Chapman, M. R., Eglinton, G., Zhao, M., Rutledge, D., & Read, G. (1999). Combined coccolith, foraminiferal, and biomarker reconstruction of paleoceanographic conditions over the last 120 kyr in the northern North Atlantic (59°, 23°W). *Paleoceanography*, *14*, 336–349.

Weaver, P. P. E., Wynn, R. B., Kenyon, N. H., & Evan, J. (2000). Continental margin sedimentation, with special reference to the north-east Atlantic margin. *Sedimentology*, *47*, 239–256.

Weijers, J., Schouten, W. H. S., van den Linden, M., van Geel, B., & Sinninghe Damsté, J. S. (2004). Water table related variations in the abundance of intact archaeal membrane lipids in a Swedish peat bog. *FEMS Microbiology Letters*, *239*, 51–56.

Weiss, R. F. (1974). Carbon dioxide solubility in water and seawater: The solubility of a non-ideal gas. *Marine Chemistry*, *2*, 203.

Werne, J. P., Hollander, D. J., Lyons, T. W., & Peterson, L. C. (2000). Climate-induced variations in productivity and planktonic ecosystem structure from the Younger Dryas to Holocene in the Cariaco Basin, Venezuela. *Paleoceanography*, *15*, 19–29.

Winter, A., Jordan, R., & Roth, P. (1994). Biogeography of living coccolithophores in ocean waters. In: A. Winter & W. G. Siesser (Eds), *Coccolithophores* (pp. 161–177). Cambridge: Cambridge University Press.

Wuchter, C., Schouten, S., Coolen, M. J. L., & Sinninghe Damsté, J. S. (2004). Temperature-dependent variation in the distribution of tetraether membrane lipids of marine Crenarchaeota: Implications for TEX_{86} paleothermometry. *Paleoceanography*, *19*.

Wuchter, C., Schouten, S., Wakeham, S. G., Sinninghe Damsté, J. S. (2005). Temporal and spatial variation in tetraether membrane lipids of marine Crenarchaeota in particulate organic matter: Implications for TEX_{86} paleothermometry. *Paleoceanography*, *20*, PA3013, doi:10.1029/2004PA001110.

Yamamoto, M., Ficken, K., Baas, M., Bosch, H. J., & deLeeuw, J. W. (1996). Molecular palaeontology of the earliest Danian at Geulhemmerberg (The Netherlands). *Geologie En Mijnbouw*, *75*, 255–267.

CHAPTER TWELVE

DEEP-SEA CORALS: NEW INSIGHTS TO PALEOCEANOGRAPHY

Owen A. Sherwood* *and* Michael J. Risk

Contents

1. Introduction	491
1.1. Overview of past work on geochemistry of deep-sea corals	492
1.2. Advantages of the deep-sea coral archive	495
2. Methods and Interpretations	495
2.1. Growth and sclerochronology in deep-sea scleractinians	495
2.2. Growth and sclerochronology in horny corals	498
2.3. Fossil preservation	500
2.4. Sources of carbon to deep-sea corals	501
2.5. Biocalcification models	502
2.6. Stable isotopic disequilibria in deep-sea corals	503
2.7. Overcoming isotopic disequilibria: the "lines technique"	505
2.8. Trace element vital effects	506
2.9. U-series dating of deep-sea corals	508
2.10. Radiocarbon dating and paleo-$\Delta^{14}C$	511
2.11. Surface signals from the organic skeletons of horny corals	511
3. Landmark Studies	514
References	516

1. INTRODUCTION

Deep-sea corals were first described from the coast of Norway, and the first species identified by Linnaeus (1758). Since then, at least until recently, interest in the group has waxed and waned, seemingly in concert with the frequency of their relationship with fishing efforts on the continental shelves.

In recent years, two critical concerns have heightened interest in deep-sea corals. Worldwide decline of coastal fisheries, especially off the shelves on eastern North America, has prompted a re-examination of the role of deep-sea corals as habitat for fishes. In addition, concern about the rate and scale of climate change has combined with the growing realization that this process can be measured by

* Corresponding author.

Developments in Marine Geology, Volume 1
ISSN 1572-5480, DOI 10.1016/S1572-5480(07)01017-2

using the records encoded in coral skeletons. Instrumental records are not sufficiently widespread or long-standing to allow predictive models to be constructed,
and proxy records have to be consulted. Other than deep-sea corals, few archives
offer the combination of global distribution (Figure 1), temporal resolution, and
duration of record. In this chapter, when we speak of deep-sea "corals," we are
referring not only to the true corals, the Scleractinia (which date from the Triassic)
but any coelenterate that makes a hard skeleton capable of recording climate data
(Figure 2). This would include several orders that secrete a horny skeleton, such as
Octocorals, (Ordovician-recent), Antipatharians (Miocene-recent) and Zoanthids
(recent).

1.1. Overview of Past Work on Geochemistry of Deep-Sea Corals

Previous attempts several decades ago to decipher the climate record in deep-sea
corals from stable isotope analyses were overshadowed by the upsurge in interest in
records from shallow-water reef corals. Weber (1973) found that deep-sea corals
were generally less depleted in $\delta^{18}O$ and $\delta^{13}C$ than were reef corals; Emiliani,
Hudson, Shinn, George, and Lidz (1978) described a series of analyses of subsamples of corals from the Blake Plateau, off Florida, and found a progressive up-polyp
closer approach to equilibrium values.

Since the early 1970s, paleoceanographic studies have relied heavily on cores
from reef corals. Two examples, from hundreds of papers, involve our understanding of El Niño-Southern Oscillation (ENSO) events. Carriquiry and Risk (1988)
were the first to establish the record of the timing and intensity of such an event,
working up the stable isotopic record in reef corals from the west coast of Costa
Rica. Tudhope et al. (2001) analyzed cores from the superbly-exposed and preserved reef terraces of Papua New Guinea, and found that ENSO events have
existed for at least the past 130,000 yr, and that they continued even through glacial
times. They also found that ENSO events of the 20th century were stronger than in
past glacial or interglacial times. Cores from reef corals can frequently span centuries; Hendy et al. (2002) were able to go back in time more than 400 yr.

Most of the work on reef coral cores has emphasized temperature determinations, using the $\delta^{18}O$ paleothermometer. More than temperature can be determined from these cores, however, Fallon, White, and McCulloch (2002) used reef
coral cores to determine the impact of mining on the waters around Misima Island,
Papua New Guinea, and Heikoop, Tsujita, and Risk (1996) found the hydrothermal pulse associated with a volcanic eruption in the Banda islands, Indonesia,
recorded in corals near the eruption site. As useful and as flexible as the reef coral
record is, however, it suffers from severe geographical restrictions. To state the
obvious, reef coral records may only be obtained where reef corals grow — warm,
shallow waters. The most dynamic oceanographic processes are frequently located
deeper in the water column and further toward the poles.

The mid-90s saw increased interest in the records from deep-sea corals, sparked
by two key papers. Druffel et al. (1995) determined an age of approximately
1,800 yr for an individual *Gerardia* (zoanthid) collected off Little Bahama Bank,
suggesting that these organisms were perhaps the oldest living species in the ocean,

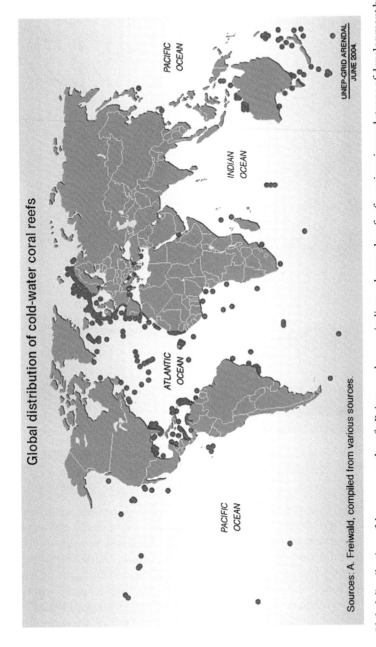

Figure 1 Global distribution of deep-sea coral reefs. Points on the map indicate observed reefs of varying size and stages of development but not the actual area covered. The high density of reefs shown in the North Atlantic reflects the intensity of research in this region. Further discoveries are expected worldwide, particularly in the deeper waters of subtropical and tropical regions. Figure from André Freiwald.

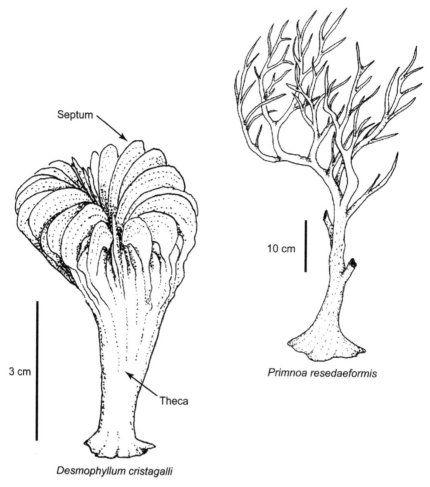

Figure 2   Two of the deep-sea scleractinian (*D. cristagalli*, a.k.a. *D. dianthus*; redrawn from Cairns, 1981) and horny corals (*P. resedaeformis*) discussed in this chapter.

rivaling Bristlecone Pines for the title of longest-lived organisms on the planet. They further suggested (p. 5031) "there is potential for *Gerardia* to serve as a millennial-scale integrator of upper ocean particle flux, and possibly reveal past changes in the productivity of the surface ocean." Smith, Risk, Schwarcz, and McConnaughey (1997) analyzed a suite of *Desmophyllum dianthus* from off Orphan Knoll, off the Newfoundland slope. The record of one pseudocolony captured a rapid cooling of intermediate waters at the onset of the Younger Dryas (~13 ka BP). Their analyses showed that this took place in as little as 5 yr, a startling find for the time.

The field has expanded dramatically in the past decade. At the Third International Symposium on Deep-Sea Corals, held in Miami in Nov–Dec 2005, there were climate-related papers from authors from almost 20 different countries.

1.2. Advantages of the Deep-Sea Coral Archive

To be accepted as dependable, any climate proxy extracted from organisms or sediments must be diagenetically stable, reproducible among specimens, and must faithfully record data of climatic significance. Advantages would be longevity, wide geographic and depth ranges, ease of analysis and abundance. Deep-sea corals fulfill all these criteria, with the possible exception of abundance. So-called deep-sea corals in fact range from depths of only a few meters (in places like Alaska and Chile), to >4 km, and in all oceans, but they are by no means abundant. In many cases, such as much of our own work, adventitious or opportunistic samples may be obtained from cultivating good relationships with fishermen, or from accidental trawl hauls. Samples specific to a given research project may be collected using deep submersibles, all of which are very expensive, and the use of which is mostly restricted to developed nations.

Although Scleractinians are relatively common and widespread in the deep-sea, obtaining long, coherent climate records from their skeletons can be challenging. They are usually small, less than 5 cm in height, and their growth banding is exceedingly narrow, *ca*. 10–100 μm (Lazier, Smith, & Risk, 1999; Sherwood, 2002). Recent technological advances, however, such as microsamplers and micro-probes, have made small sample sizes much less of a problem, and, whereas a century of reef-coral record could be carried on a strong shoulder (it could weigh ~100 kg), the same length of record in a deep-sea coral could be tucked in a shirt pocket. In the future, it is likely that more research will focus on horny corals, which have tree-like morphologies, and lifespans of several centuries (Druffel, King, Belastock, & Buesseler, 1990; Druffel et al., 1995; Risk, Heikoop, Snow, & Beukens, 2002; Sherwood, Scott, & Risk, 2006).

In summary, deep-sea corals provide a widespread archive of oceanographic processes, with annual- to decadal-resolution records spanning several centuries in length. They are ideally suited to the study of rapid changes, such as that which characterized the Pleistocene–Holocene transition.

2. Methods and Interpretations

2.1. Growth and Sclerochronology in Deep-Sea Scleractinians

Deep-sea scleractinians exhibit a wide variety of growth forms, from simple, solitary cup corals to complex, branching colonies (Figure 2). A brief introduction to skeletal nomenclature is provided here, to facilitate the discussion of sclerochro-nology. The animal itself, the polyp, ranges from a few mm to several cm in diameter. The polyp sits atop a vertical calcified tube, the corallite, supported by a thin horizontal plate called the dissepiment. The corallite is lengthened by periodic uplift of the polyp and formation of a new dissepiment. The wall of the corallite is called the theca. A series of thin vertical sheets, the septa, radiate from the theca into the center of the corallite.

At the microscopic scale, all the coral structures share a common mode of crystal nucleation and growth (see review in Cohen & McConnaughey, 2003). Crystals nucleate in so-called "centers of calcification" (COCs) producing randomly-oriented microgranules (Wainwright, 1964). In thin section, COCs appear opaque. Thickening of skeletal structures is achieved by the growth of crystal fibers outward from the COCs (Figure 3). Fan-shaped bundles of fibers (spherulites) take on an increasingly preferred orientation as they elongate (Barnes, 1970; Gladfelter, 1982). The fibrous region appears translucent in transmitted light, even though the crystals are more densely packed (Wainwright, 1964). Crystal fibers comprise the bulk of coral skeletons. Microbanding on the fibers occurs with a periodicity of ca. 1 μm (Sorauf & Jell, 1977). Cuif and Dauphin (2005) argued that fibers are not single crystals of aragonite, but rather composite structures built by the superposition of micron-thick layers, separated by an organic matrix.

Growth couplets consisting of alternating opaque-translucent band pairs (Figure 4) are found in many deep-sea scleractinians, ranging from 0.1 to 0.75 mm in width (Mortensen & Rapp, 1998; Lazier et al., 1999; Cheng, Adkins, Edwards, & Boyle, 2000; Adkins, Henderson, Wang, O'Shea, & Mokadem, 2004; Cohen, Gaetani, Lundälv, Corliss, & George, 2006). The delicate, sheet-like septae usually exhibit the clearest banding where they have not been secondarily thickened (Lazier et al., 1999). Growth bands are visible in transverse and longitudinal sections of the theca, but these may be susceptible to dissolution where the living tissue does not envelop the outer skeleton (Lazier et al., 1999). Cohen et al. (2006) relate opaque-translucent growth banding observed in *Lophelia pertusa* to the repetition of crystal nucleation and growth events. They suggest that the process of crystal nucleation and growth occurs quickly in succession, and that it may reflect a seasonal

Figure 3 SEM image of centers of calcification (COC) and crystal fibers (F) in the (shallow-water) scleractinian coral *Porites lutea*. Note occlusion of crystal fibers at points of lateral interference. Image from Cohen and McConnaughey (2003).

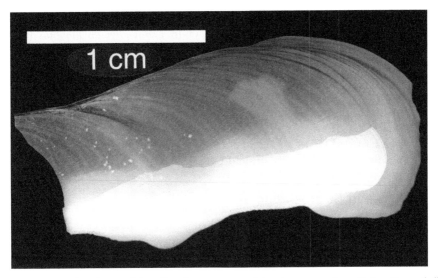

Figure 4 Example of growth banding in a septum of the deep-sea scleractinian *Desmophyllum dianthus*. White patch at bottom of image is where the theca was attached. Figure from Cheng et al. (2000), with permission.

growth spurt followed by a dormant period. Obviously, this has an important consequence for the continuity of geochemical proxy records.

The key utility of corals in paleoceanography is that long records of ambient conditions are locked into the growing skeleton in successive layers. When dealing with tropical reef scleractinians, massive colonies are selected for climate reconstructions because of the simplicity of their growth banding. Deep-sea scleractinians have more complicated cup and branching morphologies, making it difficult to visualize and sample the growth layers. The solitary cup corals are easier to deal with because they grow as one corallite over their 100+ yr lifespans (Smith et al., 1997; Adkins, Cheng, Boyle, Druffel, & Edwards, 1998; Cheng et al., 2000). Thus, the septa and theca exhibit the full complement of growth layers over the lifespan of the coral (Figure 4), provided that dissolution has not removed some of these layers (Lazier et al., 1999). Branching species are much more difficult to analyze because the individual corallites are relatively short lived. Although there are visible growth layers in the septa and theca of *Lophelia pertusa*, for example, they usually number <10 (Mortensen & Rapp, 1998; Cohen et al., 2006). Theoretically, extended chronologies could be developed by tracking successive growth layers into younger corallites, but this has not been demonstrated. Also, Pons-Branchu, Hillaire-Marcel, Deschamps, Ghaleb, and Sinclair (2005) raise the possibility that growth in *L. pertusa* occurs in relatively short bursts, separated by long hiatuses.

In most cases deep-sea corals cannot be observed growing in their natural habitat (but see Mortensen & Rapp, 1998), so growth rates must be determined with radiometric dating techniques (e.g., U/Th, ^{226}Ra/Ba, ^{210}Pb or ^{14}C). Vertical growth rates range between 0.1 and 3 mm/yr for solitary corals (Cheng et al., 2000; Adkins et al., 2002; Adkins et al., 2004) and 2 and 5 mm/yr for branching corals

(Mortensen & Rapp, 1998; Adkins et al., 2004). Corresponding periodicities of growth banding range from 0.3 to 3 bands/yr in *Desmophyllum dianthus* (Cheng et al., 2000) to ~0.1 bands/yr in *Enallopsammia rostrata* (Adkins et al., 2004). Banding periodicity probably varies within and among different species depending on depth and environmental conditions. For this reason, banding periodicity should be assessed on a case-by-case basis. Since they often live in near-constant physical conditions, the causes of growth banding in deep-sea scleractinians remains "quite literally, in the dark" (Lazier et al., 1999). Overall, it appears that growth banding in deep-sea scleractinians is not a reliable chronometer in most cases, and that skeletal chronologies must instead be established by radiometric dating methods.

2.2. Growth and Sclerochronology in Horny Corals

Several genera of deep-sea fans (Octocorallia: Order Gorgonacea) form durable, long-lived, tree-like skeletons (Figure 2). Compositionally, they differ from scleractinians in two key aspects. First, the carbonate phase is usually high magnesium calcite; some species deposit carbonate hydroxyapatite (Macintyre, Bayer, Logan, & Skinner, 2000). Relative to aragonite, the smaller lattice geometry of calcite favors co-precipitation of small cations such as Mg, and discriminates against larger cations, such as Sr and U. Second, the skeleton also contains a horny, fibrillar protein called gorgonin. The function of gorgonin is to lend flexibility to the otherwise stiff calcified skeleton (Grasshoff & Zibrowius, 1983; Lewis, Barnowski, & Telesnicki, 1992). Different regions of the skeleton may be composed of 'massive' (100%) calcite, 100% gorgonin, or a combination of both. For example, the Bamboo Corals (family Isididae) deposit gorgonin nodes, like the joints of a finger, between internodes of massive calcite. Red tree corals (family Primnoidae) deposit a two-part calcite-gorgonin "horny axis" towards the inner part of the axial skeleton, and a massive calcite cortex later on.

At the microscopic scale, massive calcite structures in octocorals appear almost identical to the aragonite in scleractinians (Lewis et al., 1992; Bond, Cohen, Smith, & Jenkins, 2005; Sherwood, 2002). Calcite fibers grow in spherulitic fashion, the fibers nucleating on gorgonin surfaces or COCs. Microbanding on the growing fibers occurs with a periodicity of *ca.* 1 μm. Where calcite is embedded with gorgonin, the calcite fibers take on stubbier morphologies, perhaps indicating a different process or rate of biomineralization.

Growth rings in octocorals may be observed in either the massive calcite or horny regions of the skeleton (Figure 5). In the massive calcite, the rings probably relate to the repetition of crystal growth and nucleation events (*sensu* Cohen et al., 2006). In the horny regions, the rings are produced by alternations in the ratio of gorgonin:calcite (Risk et al., 2002; Marschal, Garrabou, Harmelin, & Pichon, 2004; Sherwood, 2002), perhaps compounded by variations in the extent of protein tanning (Szmant-Froelich, 1974). Using bomb-[14]C (Sherwood, Scott, Risk, & Guilderson, 2005a) and ^{210}Pb-dating (Andrews et al., 2002) annual ring periodicity in the red tree coral *Primnoa resedaeformis* has been proven (Figure 6). Finer-scale growth rings, possibly lunar in origin, have also been described in *P. resedaeformis* (Risk et al., 2002; Sherwood, 2002) making this species perhaps the highest

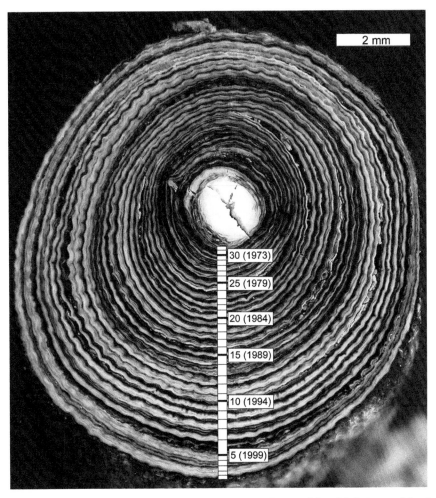

Figure 5   Example of the annual growth rings in a young (30-yr-old) axial skeleton of the deep-sea gorgonian *Primnoa resedaeformis*. Markings show individual rings isolated for analysis, with corresponding calendar age (in brackets), as determined by amateur growth ring counters. Figure from Sherwood et al. (2005a), with permission.

resolution deep water archive in the world. Thresher et al. (2004) verified annual ring formation in the calcite internodes of the bamboo coral *Keratoisis spp*. Other studies report more diffuse banding patterns, with ambiguous periodicities (Druffel et al., 1990; Roark et al., 2005; Andrews et al., 2005). As with scleractinians, it is highly likely that the appearance and timing of rings depends on local environmental factors, such as the downward flux of organic matter from spring plankton blooms (Sherwood, 2002). In regions of the ocean where seasonal variability is attenuated, growth rings may not be as prominently developed.

Lifespans of octocorals often exceed several hundreds of years (Druffel et al., 1990; Risk et al., 2002; Andrews et al., 2002; Thresher et al., 2004; Roark et al. 2005),

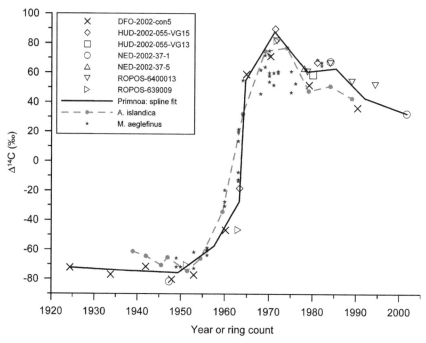

Figure 6 Timeseries $\Delta^{14}C$ of annual horny rings (after decalcification) of *Primnoa resedaeformis*. Data are from seven different colonies collected from 250 to 475 m in the NW Atlantic Ocean. Accurate reconstruction of the bomb ^{14}C signal proves annual ring periodicity. Reference curves for clam shell (*Arctica islandica*; Weidman & Jones, 1993) and haddock otoliths (*Melanogrammus aeglefinus*; Campana, 1997) also shown. Figure from Sherwood et al. (2005a), with permission.

the oldest reported lifespan being a 700 yr old specimen of *P. resedaeformis* (Sherwood et al., 2006). Compared with deep-sea scleractinian corals, obtaining centuries-long geochemical records from octocorals is simplified by the ability to sample across the growth rings of the axial skeletal.

Zoanthids and antipatharians (black corals) also show great promise as marine archives. These corals also form arborescent, concentrically banded skeletons composed of a horny protein much like the gorgonin found in octocorals (Goldberg, 1991; Druffel et al., 1995). The skeletons are not calcified, so the fossil record is sparse. However, these organisms have some of the longest lifespans of any corals. Druffel et al.'s (1995) 1,800-yr-old *Gerardia* is a case in point. Antipatharians may live for several hundreds of years (Williams, Risk, Sulak, Ross, & Stone, 2005), and their growth rings, demarcated by dark-colored protein "glue" (Goldberg, 1991), also appear to form annually (Williams et al., 2005).

2.3. Fossil Preservation

The degree of diagenetic stability of deep-sea corals seems excellent (Teichert, 1958). Repeated scanning electron microscope examination of deep-sea scleractinians from various depths (a few hundred meters to $>2\,km$) and various oceans shows that they remain the original aragonite, with virtually no recrystallization, since the early Pleistocene (Titschak & Freiwald, 2005; Remia & Taviani, 2004). The massive calcite structures in gorgonians have more durable and compact skeletons; cretaceous-aged specimens are excellently preserved (Grasshoff & Zibrowius, 1983). The organic phase of horny corals has poorer preservation potential, but amino acid and stable isotopic analyses suggest that it is stable over at least the last few thousand years (Goodfriend, 1997; Sherwood et al., 2006).

Both the reef coral record and the deep-sea coral record, at least in the case of the scleractinian record, may be compromised by bioerosion. All corals, from all depths, are bored by a variety of organisms, of which the most prevalent are sponges and, in shallow water, algae. In deep-sea corals, the major bioeroders are sponges, fungi, bryozoa and polychaetes (Boerboom, Smith, & Risk, 1998; Bromley, 2005; Beuck & Freiwald, 2005; Wisshak, Freiwald, Lundälv, & Gektidis, 2005). The effects of bioerosion may range from complete removal of the record to selective dissolution of aragonite COCs (Titschak & Freiwald, 2005) and subtle alterations of the trace element composition of the skeleton (Pons-Branchu et al., 2005; Robinson et al., 2006). Virtually nothing is known of the process by which climate records are distorted by bioerosion.

2.4. Sources of Carbon to Deep-Sea Corals

Based on radiocarbon measurements, Griffin and Druffel (1989) originally established that the carbonate fraction of deep-sea scleractinians and octocorals is mostly derived from ambient dissolved inorganic carbon (DIC) at depth. Other authors have noted a 1:1 relationship in the $\Delta^{14}C$ of coral carbonate and seawater DIC (Goldstein, Lea, Chakraborty, Kashgarian, & Murrell, 2001; Adkins et al., 2002; Frank et al., 2004; Roark et al., 2005). Radiocarbon evidence limits the amount of respiratory CO_2 incorporated into the carbonate to $<10\%$ (Adkins et al., 2002); however, $\delta^{13}C$ data indicate that the amount of respired CO_2 may be somewhat higher in select species (e.g., Figure 7).

In contrast to the carbonate, the organic fraction of horny corals is synthesized from food sources sinking out of surface waters, a conclusion based on the presence of bomb-^{14}C in deep-dwelling specimens (Figure 6; Griffin & Druffel, 1989; Druffel et al., 1995; Roark et al., 2005; Sherwood et al., 2005a). The actual food sources are mostly unknown because field monitoring is prohibitively expensive, and specimens usually die when raised to the surface. Studies on relatively shallow-dwelling octocorals consistently point to detrital particulate organic matter (POM) and zooplankton (in a $\sim$50:50 ratio) as the main food sources (Ribes, Coma, & Gili, 1999, 2003; Orejas, Gili, & Arntz, 2003). The sinking, rather than suspended fraction of detrital POM is likely to be more important to deeper-dwelling taxa, as the nutritional value of the latter decreases with depth. Our $\Delta^{14}C$ and $\delta^{15}N$ data support a sinking POM and/or zooplankton (as both have approximately the same $\Delta^{14}C$ and $\delta^{15}N$) diet in deep-sea gorgonians and antipatharians, while ruling out deep suspended POM and dissolved organic matter (DOM) as important food

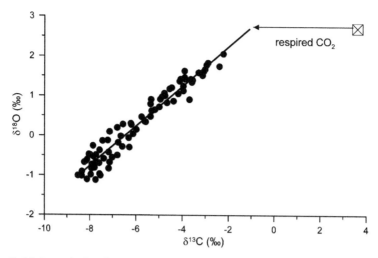

Figure 7   Stable isotopic data from a specimen of *Lophelia pertusa* collected from the NE Atlantic. Isotopic data plot along a straight line extending roughly to theoretical aragonite-seawater isotopic equilibrium (boxed cross). Horizontal offset from equilibrium is caused by skeletal incorporation of δ^{13}C-depleted respired CO_2 (i.e., metabolic effects). Data from Risk, Hall-Spenser, and Williams (2005), with permission.

sources (Heikoop, Hickmott, Risk, Shearer, & Atudorei, 2002; Sherwood et al., 2005a, 2005b; Williams et al., 2005).

2.5. Biocalcification Models

A debate on the mechanism of coral calcification has persisted for 40 yr (see review in Cohen & McConnaughey, 2003). A brief overview of this debate seems appropriate, as it frames our discussion of isotopic and trace elemental variability in coral carbonate. In the "organic matrix" model crystal growth is initiated, directed and terminated by organic secretions. Support for this idea is rooted in the detection of organic compounds within skeletal structures (Goreau, 1959; Johnston, 1980). More recently, the spatial distribution of sulfated polysaccharides within COCs and fibers has been worked out (Cuif & Dauphin, 2005). In the "physico-chemical" model (Barnes, 1970; Constanz, 1986) crystal growth occurs freely within pockets of enzymatically-modified seawater beneath the ectodermal tissue. This idea is rooted in the similarity of spherulitic crystal morphology in corals to that in mineral specimens (Bryan & Hill, 1941).

One of the advantages of the physicochemical model is that, being grounded in the principles of inorganic (thermodynamic and kinetic) solution/crystal chemistry, it may be quantitatively evaluated against a growing body of isotopic and trace elemental data from coral skeletons (Sinclair & Risk, 2006). These data indicate that isotopic and trace elemental compositions of corals almost always depart from thermodynamic equilibrium (i.e., "vital effects"). However, physicochemical-based models can account for the isotopic and trace elemental disequilibria in corals through biological manipulation of calcification fluids without appealing to an organic matrix (McConnaughey, 1989b; Adkins, Boyle, Curry, & Lutringer, 2003; Sinclair, 2005; Sinclair & Risk, 2006; Cohen et al., 2006).

In the classic model of calcification physiology (e.g., McConnaughey, 1989b; Cohen & McConnaughey, 2003) calcification occurs in a thin ($\sim$10 µm), extracellular calcifying fluid (ECF) between the coral tissue and the skeleton. The ECF is isolated from cell fluids by a membrane, which is permeable to small, uncharged molecules only. CO_2 passively diffuses across the cell membrane where it reacts with H_2O or OH^- to form HCO_3^-. The enzyme Ca-ATPase pumps Ca^{2+} into the ECF in exchange for two protons, to maintain charge neutrality. Proton removal increases the pH of the ECF, and converts HCO_3^- to CO_3^{2-}, which precipitates with Ca^{2+}. Small amounts of seawater HCO_3^- and $CO_3^=$, as well as trace elements, enter the ECF through leaky membranes or by passage of invaginated vacuoles across the cell membrane (Furla, Bénazet-Tambutté, Jaubert, & Allemand, 1998).

2.6. Stable Isotopic Disequilibria in Deep-Sea Corals

Oxygen and carbon isotope analysis of biogenic carbonates has been a cornerstone of paleoceanography for over 50 yr. The $\delta^{18}O$ of carbonate is used to infer past temperatures (Epstein, Buchsbaum, Lowenstam, & Urey, 1953) and the extent of continental glaciation (e.g., Shackleton, 1967), and $\delta^{13}C$ is used to identify water masses in the deep ocean (e.g., Sarnthein et al., 1994). Many biogenic carbonates that exhibit vital effects have consistent offsets from isotopic equilibrium. Deep-sea scleractinians are a special case; the extent of disequilibrium is highly variable within an individual coral (Emiliani et al., 1978; McConnaughey, 1989a; Smith, Schwarcz, Risk, McConnaughey, & Keller, 2000; Smith, Schwarcz, & Risk, 2002). When plotted in $\delta^{18}O$ vs. $\delta^{13}C$ space (Figure 7), isotopic data plot along straight lines extending roughly from equilibrium to a point up to 5‰ depleted for $\delta^{18}O$, and up to 15‰ depleted for $\delta^{13}C$. The slopes of these lines average $\sim$1/3. Octocoral calcites show similar patterns, though the extent of disequilibria is somewhat less (Druffel et al., 1990; Heikoop, Risk, Lazier, & Schwarcz, 1998).

According to McConnaughey's (1989b, 2003) kinetic model, strong isotopic depletions are mainly caused by hydration and hydroxylation of CO_2 within the ECF, although the pathways of $\delta^{18}O$ and $\delta^{13}C$ fractionation differ. For oxygen, both H_2O and OH^- are strongly depleted in $\delta^{18}O$: approximately -30‰ and -69‰, respectively. Since 1/3 of the oxygen atoms in HCO_3^- come from H_2O or OH^-, McConnaughey's (2003) calculations predict a DIC with a $\delta^{18}O$ composition well below that of aragonite in equilibrium with seawater. For carbon, the reaction of CO_2 with H_2O and OH^- strongly fractionates $\delta^{13}C$, approximately -7‰ and -27‰, respectively. Again, the resulting DIC takes on a $\delta^{13}C$ content much depleted compared with equilibrium aragonite. Hydroxylation of CO_2 predominates at pH > 8, where most calcification occurs; thus, the stronger $\delta^{18}O$ and $\delta^{13}C$ depletions associated with hydroxylation dominate the isotopic composition of the coral skeleton (McConnughey, 2003).

McConnaughey (1989b) calculated that HCO_3^- could precipitate with Ca^{2+} much faster than it can attain equilibrium with H_2O in the ECF, so the $\delta^{18}O$ depletion is buried in the precipitating aragonite. Furthermore, equilibration of

HCO_3^- occurs through a CO_2 intermediate, which diffuses back and forth across the cell membrane and equilibrates with cell H_2O at lower pH. Some of the carbon in the ECF is carried with the CO_2 molecule, allowing it to equilibrate within the cell also. Strong depletions in $\delta^{13}C$ incurred during hydration and hydroxylation are thereby partly "eroded" (Cohen & McConnaughey, 2003). The CO_2 molecule carries both C and O atoms across the cell membrane, allowing simultaneous equilibration of $\delta^{13}C$ and $\delta^{18}O$ in the cell. This accounts for the $\delta^{18}O-\delta^{13}C$ correlations in deep corals. In addition to "kinetic effects," a small amount of the CO_2 entering the ECF is derived from metabolism, and is $\sim 2\permil$ depleted in $\delta^{13}C$ (McConnaughey, 1989a). Thus, the lines in $\delta^{18}O-\delta^{13}C$ space extend to a point roughly $2\permil$ depleted in $\delta^{13}C$ compared to equilibrium.

In recent years, a new aspect of isotopic disequilibria in deep-sea corals has emerged. High resolution microsampling reveals that COCs are more isotopically depleted than crystal fibers (Figure 8; Adkins et al., 2003; Rollion-Bard, Blamart, & Cuif, 2003). Moreover, isotopic data from COCs may even fall off the $\delta^{18}O$ vs. $\delta^{13}C$ lines, where $\delta^{18}O$ continues to decrease but $\delta^{13}C$ does not. Such anomalous data went unnoticed in earlier studies because the dental drills used in sampling could not be used to separate fibers from COCs (Lutringer, Blamart, Frank, & Labeyrie, 2005). To account for these observations, Adkins et al. (2003) developed a "carbonate model" of isotopic disequilibrium. In their model, $\delta^{18}O$ and $\delta^{13}C$ disequilibria are driven by thermodynamic response to a biologically-induced pH gradient in the ECF. While McConnaughey's (1989b) kinetic model and

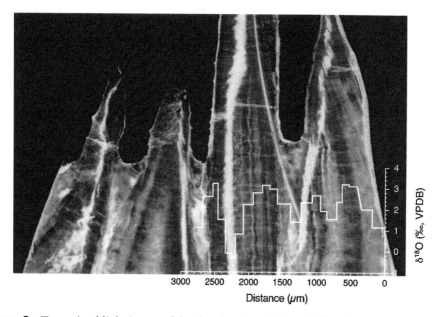

Figure 8 Transmitted light image of the thecal region of *Desmophyllum dianthus*, overlain with $\delta^{18}O$ results obtained with a high spatial resolution microsampler. Vertical line of COCs (lightest region) is associated with the lowest $\delta^{18}O$. Figure from Adkins et al. (2003), with permission.

Adkins et al.'s (2003) carbonate model are fundamentally different, both agree that isotopic depletions in COCs are an indication of a faster rate of calcification.

2.7. Overcoming Isotopic Disequilibria: The "Lines Technique"

Smith et al. (2000) proposed the "lines technique" as a way to overcome vital effects in deep-sea scleractinian corals. Their approach makes use of the fact that the lighter ends of $\delta^{18}O$–$\delta^{13}C$ regression lines for any one coral extend towards isotopic equilibrium (Figure 9). If $\delta^{13}C$ at equilibrium is known (from $\delta^{13}C$ of seawater or coeval benthic foraminifera), then, with a small correction for respiratory $\delta^{13}C$ (~2‰), the corresponding value of $\delta^{18}O$ at equilibrium, and hence temperature, can be calculated. Using 35 modern corals, Smith et al. (2000) showed that even if $\delta^{13}C$ is not known, the intercepts of $\delta^{18}O$–$\delta^{13}C$ regression lines were strongly correlated with temperature

$$(\delta^{18}O_c - \delta^{18}O_w) = -0.21 \ T(^\circ C) + 4.51 \tag{1}$$

Equation (1) is nearly identical to Grossman and Ku's (1986) experimentally-derived $\delta^{18}O$ vs. temperature equation. This suggests that setting $\delta^{13}C = 0$ is a reasonable approximation for seawater $\delta^{13}C$ minus the respiratory $\delta^{13}C$ correction.

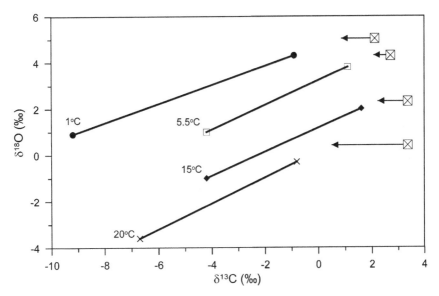

Figure 9 Isotopic data for four coral specimens collected from Antarctic shores, 410 m (solid circles); NW Atlantic slope, 1,025 m (open squares); NE Atlantic, 310 m (solid diamonds); Bahamas, 295 m (crosses). For clarity, only the least squares regression lines are shown. Crossed boxes represent theoretical isotopic equilibrium for each of the coral's datasets. Vertical offsets, reflecting temperature dependence of $\delta^{18}O$, are the basis for the 'lines technique'. Horizontal arrows point to intersection of regression lines with equilibrium $\delta^{18}O$; length of arrows reflects magnitude of theoretical metabolic effect. Figure modified from Smith et al. (2000), with permission.

Weaknesses to the lines technique stem from variability in metabolic effects (Adkins et al., 2003), or the slopes of $\delta^{18}O$ vs. $\delta^{13}C$ (Smith et al., 2000). Despite these limitations, Smith et al. (2000) could calculate present day temperatures to a precision of ± 0.36 to $\pm 1°C$ over a range of 1–28°C. The lines technique may not work as well with pre-Holocene corals, however, because seawater $\delta^{13}C$ was more variable, and coral metabolism may have been much different; this is apparent upon re-examination of their earlier results on *Desmophyllum* (Smith et al., 1997). In addition, Smith et al. (2000) used the classic dental drill sampling method, thus they did not take into account isotopic differences between fibers and COCs. Lutringer et al. (2005) analyzed fibers and COCs separately with an ion microprobe and essentially reproduced Smith et al.'s (2000) findings, albeit with slightly larger estimates of precision (± 0.7 to $\pm 1.5°C$). Overall, the lines technique may be useful in tracking relatively large changes in water temperature, such as that which occurs in upper intermediate waters.

2.8. Trace Element Vital Effects

A growing body of evidence shows that trace elemental vital effects are a ubiquitous feature of shallow- and deep-water scleractinians alike (Sinclair, Williams, & Risk, 2006). The mean compositions differ from inorganic carbonate, and the range of variability greatly exceeds that which can be accounted for by temperature-dependent partitioning alone. In scleractinians, different trace elements are strongly correlated with each other, much in the same way that $\delta^{18}O$ and $\delta^{13}C$ are correlated (Figure 10; Montagna, McCulloch, Taviani, Remia, & Rouse, 2005; Shirai et al., 2005; Sinclair et al., 2006). Moreover, elemental compositions are distinctly different in COCs than in surrounding fibers, with COCs being lower in Sr and U, and higher in Mg (Shirai et al., 2005; Gagnon & Adkins, 2005; Sinclair et al., 2006; Cohen et al., 2006). A recently developed trace element equilibrium/kinetic model

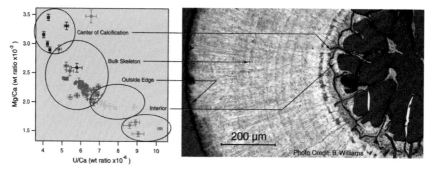

Figure 10 Negative correlation between Mg/Ca and U/Ca across the skeletal microstructures of *Lophelia pertusa*. Highest Mg and lowest U occurs in the optically dense COC. Conversely, the lowest Mg and highest U occurs in the optically light deposits that line the inside of the calyx. The bulk of the skeleton is a mixture between these two end-members. Figure from Sinclair et al. (2006), with permission.

of calcification can account for the apparent vital effects; specifically, Sr and U depletions are predicted at high calcification rates (Sinclair & Risk, 2006). Cohen et al. (2006) present evidence for a Sr/Ca paleothermometer in *Lophelia*; however, we are still a long way from producing reliable temperature reconstructions from scleractinian corals.

Following the extensive research on foraminiferal calcites, Mg/Ca in octocorals has been proposed as a potential paleothermometer (Weinbauer, Brandstätter, & Velimirov, 2000; Thresher et al., 2004; Bond et al., 2005; Sherwood et al., 2005c). Thresher et al. (2004) present 350 yr long records of Mg/Ca from Bamboo corals, which they interpret as evidence for recent cooling in the upper intermediate waters off Australia (Figure 11). More recent studies, however, suggest that Mg/Ca, as well as U/Ca and Sr/Ca, may be poorly reproducible along multiple transects of the same coral (Allard et al., 2005; Fallon, Roark, Guilderson, Dunbar, & Weber, 2005; Sinclair et al., 2005). Mg/Ca in particular may be strongly enriched in organic phases within the calcite, thus masking any temperature effect. On the other hand, Ba/Ca exhibits excellent reproducibility, and is therefore a good candidate as an environmental proxy (Allard et al., 2005; Fallon et al., 2005). Overall, further work is needed to assess the factors controlling elemental partitioning in octocoral calcite before reliable proxy data can be produced.

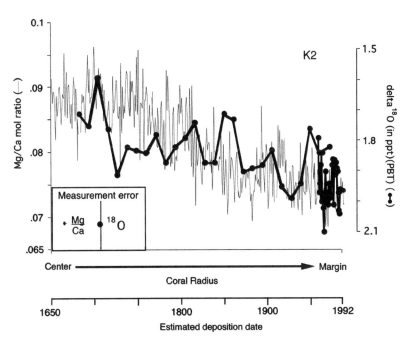

Figure 11 Records of Mg/Ca for a bamboo coral collected from ~1,000 m depth off Tasmania. Record of $\delta^{18}O$ is also shown. Both datasets indicate a gradual decline in water temperature over the last ~350 yrs. Figure from Thresher et al. (2004), with permission.

2.9. U-series Dating of Deep-Sea Corals

Deep-sea corals are excellent subjects for U-series dating. Scleractinians are easier to date because their aragonite skeletons contain higher levels of U (2–4 ppm) compared with the calcite skeletons of octocorals (3–4 ppb). With modern Multicollector ICP-MS, high precision U-series ages can be obtained on <1 g of sample. Here, we provide a brief overview of U-series dating as it applies to deep-sea corals. More detailed treatments are provided in Cheng et al. (2000) and Edwards, Gallup, and Cheng (2003). Both the ^{238}U–^{234}U–^{230}Th (U/Th) and the ^{235}U–^{231}Pa (U/Pa) decay schemes can be used. U/Th is the method of choice because it is somewhat more precise and covers a longer time range: 0–600 ka vs. 0–250 ka for U/Pa. However, U/Th dating can be used in concert with U/Pa (Goldstein et al., 2001), or even ^{226}Ra/Ba dating (Pons-Branchu et al., 2005), as a check on closed-system behavior and thus on dating accuracy.

In reef corals, it is normally the case that all of the measured ^{230}Th can be accounted for by radioactive in-growth. Deep-sea corals, on the other hand, live in a thorium-rich environment, since thorium increases with depth in the ocean. As a consequence, deep-sea corals may incorporate significant amounts of unsupported or non-radiogenic thorium, leading to older apparent ages. There are three sources of unsupported thorium (Cheng et al., 2000; Schröder-Ritzrau, Mangini, & Lomitschka, 2003): (1) a loosely attached detrital phase; (2) ferromanganese coatings, which adsorb thorium from seawater; and (3) an "initial", lattice-bound fraction incorporated during crystallization. The detritus and coatings can usually be removed by mechanical cleaning and by a series of chemical leaches, respectively (e.g., Shen & Boyle, 1988). Initial thorium cannot be separated from radiogenic thorium, so it must be accounted for in age calculations. The precision and accuracy of U/Th ages therefore depends on the precision and accuracy of the correction for unsupported thorium (Cheng et al., 2000).

The U/Th age equation, including the term for initial thorium is defined as (Edwards et al., 2003)

$$\{[^{230}\mathrm{Th}/^{238}\mathrm{U}] - [^{232}\mathrm{U}/^{238}\mathrm{U}][^{230}\mathrm{Th}/^{232}\mathrm{Th}]_i \, (e^{-\lambda_{230}t})\} - 1$$
$$= -e^{-\lambda_{230}t} + \{\delta^{234}\mathrm{U}_m/1000\}\{\lambda_{230}/(\lambda_{230} - \lambda_{234})\}\{1 - e^{-(\lambda_{230}-\lambda_{234})t}\}$$

$$(2)$$

where isotope ratios are activity ratios; λs are decay constants; t the ^{230}Th age; subscripts i and m represent initial and measured values, respectively; and $\delta^{234}\mathrm{U}_m$ the measured deviation in per mil of the $[^{234}\mathrm{U}/^{238}\mathrm{U}]$ ratio from secular equilibrium $\delta^{234}\mathrm{U} = ([^{234}\mathrm{U}/^{238}\mathrm{U}]-1) \times 1,000$. The measurement of ^{232}Th provides a check on the extent of unsupported thorium; thus, the only unknowns in equation (2) are the ^{230}Th age (t) and the initial ^{230}Th/^{232}Th activity ($[^{230}\mathrm{Th}/^{232}\mathrm{Th}]_i$).

There are two approaches to estimate $[^{230}\mathrm{Th}/^{232}\mathrm{Th}]_i$. In the first approach, a constant value is assumed, based on seawater measurements. This approach is suitable for relatively clean samples, with ^{232}Th contents of $ca.$ <20 ppb. Cheng et al. (2000) showed that $[^{230}\mathrm{Th}/^{232}\mathrm{Th}]$ measured in modern deep-sea scleractinians from between 400 and 2,000 m in the Atlantic and Pacific Oceans was indistinguishable, within error, from seawater measurements. Their estimate of $[^{230}Th/^{232}Th] = 14.8 +/- 14.8$ has been used as a "blanket" correction factor (Adkins et al., 1998; Cheng et al., 2000; Robinson et al., 2005). Alternatively, measurements of seawater $[^{230}Th/^{232}Th]$ specific to the area and depth of interest may be used to narrow the range of possible $[^{230}Th/^{232}Th]_i$ (Schröder-Ritzrau et al., 2003; Frank et al., 2004, 2005). Use of a constant $[^{230}Th/^{232}Th]_i$ correction assumes that seawater $[^{230}Th/^{232}Th]$ was constant in the past, which may not necessarily have been the case. Accordingly, errors in $[^{230}Th/^{232}Th]_i$, of $\pm 50\%$ to 100% are propagated through the age calculations to account for this uncertainty.

The second approach, using the so-called Rosholt-II isochron method (Ludwig & Titterington, 1994), assumes a two end-member mixing model between ^{232}Th-free carbonate and the ^{232}Th-rich contaminant (Lomitschka & Mangini, 1999; Schröder-Ritzrau et al., 2003, 2005). This method is suitable for heavily coated samples with ^{232}Th contents > 20 ppb (Schröder-Ritzrau et al., 2005). Two to four fractions of the same sample are prepared and analyzed separately. The first fraction consists of just the coating; the other fractions are subjected to cleanings of increasing intensity, from simple mechanical cleaning to strong chemical leaching in ascorbic acid and Na_2EDTA solution (Lomitschka & Mangini, 1999). When plotted in $[^{230}Th/^{232}Th]$ vs. $[^{238}Th/^{232}Th]$ space, data for the different fractions define an isochron, the y-intercept of which represents the $[^{230}Th/^{232}Th]_i$ correction factor for input to equation (2). A slightly different variation of the isochron approach is outlined in Goldstein et al. (2001).

The relationship between measured and initial $\delta^{234}U$ is defined as (Edwards et al., 2003)

$$\delta^{234}U_m = (\delta^{234}U_i)e^{-\lambda 234 t} \qquad (3)$$

If the ^{230}Th age can be solved independently, then Equation (3) can be used to calculate initial $\delta^{234}U$. Knowledge of $\delta^{234}U_i$ is a useful check on diagenetic alteration of the skeleton and hence open-system behavior, because marine $\delta^{234}U$ is fairly constant in time and space (Bard, Fairbanks, Hamelin, Zindler, & Chi Trach, 1991; Cheng et al., 2000; Henderson, 2002). If uranium is added to the skeleton by diagenetic processes (endolithic borings, secondary mineralization, microredox traps, pore water U; Swart & Hubbard, 1982; Pons-Branchu et al., 2005; Robinson et al., 2006), then the U/Th age equation may be invalidated. It is common practice to discard data when $\delta^{234}U_i$ falls outside a defined range of seawater $\delta^{234}U$ (e.g., 146 $\pm$ 7‰; Robinson et al., 2005).

Compilation of all published U/Th dates in deep-sea scleractinians provides an indication of the accuracy and precision of the technique (Figure 12). In general, $^{230}Th/^{232}Th$-corrected ages are younger than uncorrected ages by $< 3\%$. Thus, depending on the dating resolution required, a correction may not be necessary, provided that the ^{232}Th concentration is low. For uncorrected ages, the 2σ dating precision is better than 1%. For corrected data, the 2σ precision is larger, $\sim 5\%$, owing to uncertainty in the value of initial $[^{230}Th/^{232}Th]$. There are no apparent differences in the precision of the isochron vs. non-isochron dating techniques.

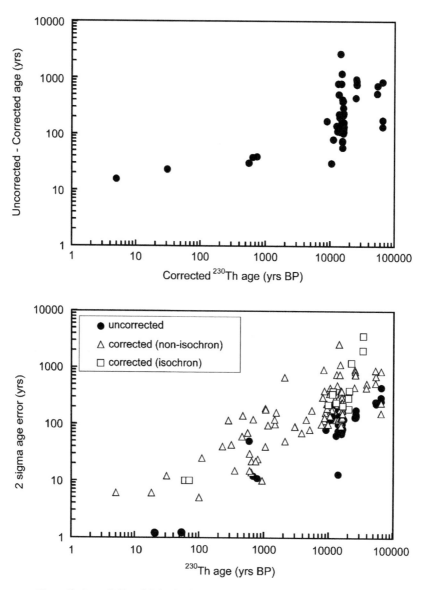

Figure 12 Compilation of all published U/Th dates from deep sea scleractinian corals. Upper panel: Difference between uncorrected and corrected ages averages <3%. Lower panel: 2σ age errors associated with uncorrected data, and corrected data using either the isochron or non-isochron approaches. See text for explanation.

In summary, U/Th dating of deep-sea corals is challenging because of the presence of unsupported Th, which leads to older apparent ages, and possible diagenetic incorporation of U, leading to younger apparent ages. With careful sample pre-treatment, correction for unsupported Th, and screening for diagenetically-altered samples, high precision U/Th dates may be obtained.

2.10. Radiocarbon Dating and Paleo-Δ^{14}C

Seawater Δ^{14}C has long been recognized as an important tool for tracking water mass circulation (Key, 2001). In the deep North Atlantic, for example, Δ^{14}C differentiates North Atlantic Deep Water (−65‰) from Antarctic Bottom Water (−165‰; Broecker, Gerard, Ewing, & Heezen, 1960; Stuiver, Quay, & Ostlund, 1983). The relative dominance of these water masses at any one location provides an indication of oceanic circulation patterns. Coral skeletons provide one of the few archives of past seawater Δ^{14}C variations, since their carbonate skeletons are derived from ambient DIC at depth (Griffin & Druffel, 1989; Adkins et al., 2002).

Coral skeletons contain enough carbonate material to perform both U/Th and ^{14}C dating on the same sample (Adkins et al., 1998; Mangini, Lomitschka, Eichstadter, Frank, & Vogler, 1998). Since the U/Th-dating can be used to calculate an absolute age, paired U/Th-^{14}C dating provides a direct measure of the ^{14}C age of the seawater in which the coral grew (Bard, Arnold, Fairbanks, & Hamelin, 1993; Edwards et al., 1993; Adkins et al., 2002). Adkins et al. (1998) and Mangini et al. (1998) first demonstrated this method on deep-sea corals, analogous to the method of differencing radiocarbon ages from pairs of benthic-planktic foraminifera sampled from the same core interval (B/P-dating; Broecker et al., 1990), but much improved by the fact that bioturbation and foraminiferal species effects are not at issue, and U/Th dating is model independent.

Paleo-Δ^{14}C may be calculated by (Adkins & Boyle, 1997)

$$\Delta^{14}C_{paleo} = \left\{ \frac{e(-^{14}C\ age/8033)}{e(-cal\ age/8266)} - 1 \right\} \times 1000 \qquad (4)$$

where ^{14}C age is the ^{14}C age measured on the coral, cal age is the U/Th age measured on coral, 8,033 is the Libby mean life and 8,266 is the true ^{14}C mean life in years. Knowledge of paleo-Δ^{14}C is powerful for tracking past variations in deep and intermediate water mass distributions (Frank et al., 2004, 2005; Robinson et al., 2005). In theory, by projecting paleo-Δ^{14}C values back in time to their intersection with the atmospheric record, the residence time or ventilation age of deep water masses may be calculated (Adkins et al., 1998; Mangini et al., 1998; Goldstein et al., 2001; Schröder-Ritzrau et al., 2003); however, this approach may be complicated by the mixing of multiple water masses with different convection histories.

2.11. Surface Signals from the Organic Skeletons of Horny Corals

Some of the more promising paleoceanographic information may found in the skeletons of horny corals. Druffel et al. (1995) originally hypothesized that certain aspects of export of POM may be monitored by studying the isotopic and amino acid chemistry of horny layers, in much the same way that down-core trends in sedimentary organic matter (SOM) is utilized in paleoceanography. In some species the annual rings may be decalcified and gently peeled apart and analyzed separately, without the need for expensive *in situ* microprobe analyses. Moreover, the amino acid composition of the horny material is stable over at least the last few thousand years, an indication that it is resistant to organic diagenesis (Goodfriend, 1997; Sherwood et al., 2006). Since the horny material is derived from sinking POM and/or zooplankton, annually banded horny corals provide a long-term, high resolution archive of biogeochemical processes occurring in surface waters (Sherwood et al., 2005b).

Accurate reconstruction of 20th century bomb radiocarbon from *Primnoa resedaeformis* (Figure 6) is a good example of the type of data that may be retrieved from horny corals (Sherwood et al., 2005a). As a variation on B/P dating, paired ^{14}C dating of both the carbonate and horny phases may prove to be useful in tracking ventilation rates of ambient water masses. The δ^{15}N and δ^{13}C of the horny phase may be another source of paleoceanographic data (Heikoop et al., 2002; Sherwood et al., 2005b; Ward-Paige, Risk, & Sherwood, 2005; Williams et al., 2005). The δ^{15}N and δ^{13}C in gorgonians and antipatharians is correlated with δ^{15}N and δ^{13}C in phytoplankton. Moreover, the intra- and inter-colony reproducibility of both isotopes appears to be excellent.

The δ^{15}N composition of horny corals is controlled by a variety of bottom-up processes, beginning with the isotopic composition of source nutrients, uptake by phytoplankton, and the length and complexity of the food web. The relative influence of denitrification vs. nitrification among different oceanographic regimes has a strong impact, for example, on the δ^{15}N of NO_3 (Cline & Kaplan, 1975; Altabet et al., 1999; Knapp, Sigman, & Lipschultz, 2005). Phytoplankton preferentially assimilate $^{14}NO_3$, leading to isotopically depleted biomass. In many regions of the ocean NO_3 is completely consumed, and the δ^{15}N of phytoplankton and sinking POM converges on the δ^{15}N of subsurface NO_3 (Altabet & McCarthy, 1985; Altabet, 1988; Altabet et al., 1999; Thunell, Sigman, Muller-Karger, Astor, & Varela, 2004). This is the basis for the interpretation of down-core sedimentary δ^{15}N as a proxy for the $\delta^{15}NO_3$ history of the oceans (Altabet, François, Murray, & Prell, 1995; Ganeshram, Pederson, Calvert, & Murray, 1995; Haug et al., 1998). The simple relation between subsurface NO_3 and sinking POM may be confounded by a variety of factors, including incomplete nutrient utilization (Altabet & François, 1994), alteration of sinking POM during transit through the water column (Saino & Hattori, 1987; Altabet, Deuser, & Honjo, 1991), and the complexity of the plankton food web (Altabet, 1988; Wu, Calvert, & Wong, 1999a). This latter factor presents the largest source of uncertainty in the interpretation of δ^{15}N data from horny corals, since trophic fractionation is large ($\sim$3.4‰ per trophic level; DeNiro & Epstein, 1981) relative to oceanic $\delta^{15}NO_3$ variability, and the corals may feed opportunistically on a wide range of plankton size classes (Ribes et al., 1999, 2003; Orejas et al., 2003). Future compound-specific isotope analyses will likely allow for the separation of trophic level and source nutrient isotope effects through separate analysis of different amino acids (McClelland & Montoya, 2002).

Specific processes that could be tracked with the δ^{15}N composition of horny corals include water mass movements, eutrophication, and trophic dynamics. Recent exploratory studies highlight some of this potential, particularly in assessing anthropogenic perturbations over the last few centuries. Sherwood (2006) presented a near 2,000-yr-long record of δ^{15}N from colonies of *P. resedaeformis* collected from the slope water region off eastern Canada (Figure 13). Anomalous

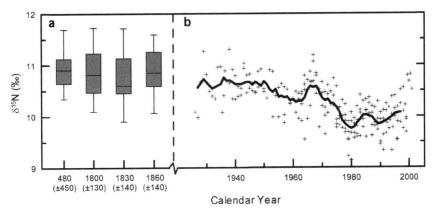

Figure 13 δ^{15}N measured in the annual horny rings of *Primnoa resedaeformis* (NW Atlantic; 250–475 m). (a) Data for subfossil colonies plotted as box-and-whisker plot because of uncertainties in radiocarbon age calibrations. Each box represents at least 20 annual rings. Numbers below boxes indicate mean calibrated ^{14}C age ($\pm 2\sigma$ range). (b) Data for six recent colonies plotted against calendar year, as dated by the bomb-^{14}C method and visual counting of annual rings. Bold line indicates 5 yr running mean. The 20th century decline in δ^{15}N is attributed to isotopic depletion of ambient NO$_3$, and/or trophic level effects. From Sherwood (2006), with permission.

decline in δ^{15}N over the 20th century was interpreted to reflect weakening of isotopically-heavier Labrador Current, or a change in the relative trophic level of the corals. Ward-Paige et al. (2005), using shallow gorgonians, and Williams et al. (2005), using deep antipatharians, presented records of long-term increases in δ^{15}N off SE USA, which they interpreted as evidence of increasing anthropogenic nutrient inputs.

The δ^{13}C of horny corals is controlled by a different assortment of bottom-up processes. Variability in the δ^{13}C of oceanic DIC is only 1–2‰ (Kroopnick, 1985), and trophic fractionation averages only ~1‰ per trophic level (DeNiro & Epstein, 1978). The δ^{13}C of SOM is often used to distinguish between marine and terrestrial inputs (Fry & Sherr, 1984); however, this generally does not apply to deep corals because they usually inhabit regions far from terrestrial influences. Most of the variability in δ^{13}C of phytoplankton and subsequent trophic levels arises from isotopic fractionation during phytoplankton growth. Earlier studies emphasized dissolved CO_2 in determining phytoplankton δ^{13}C, and the use of sedimentary δ^{13}C as a proxy for paleo-CO_2 (Jasper, Hayes, Mix, & Prahl, 1994; Rau, 1994). More recent studies point to growth rate, cell geometry, nutrient and light limitation and active carbon uptake (as opposed to diffusive uptake of CO_2) as equally, if not more important determinants of δ^{13}C (see review in Laws, Popp, Cassar, & Tanimoto, 2002). Long-term declines in δ^{13}C measured in octocorals and antipatharians appear to reflect the oceanic Suess effect, that is, the gradual depletion of oceanic δ^{13}C due to burning of isotopically-light fossil fuels (Williams et al., 2005; Sherwood, 2006). Over shorter time periods, the observed variations are yet to be explained. As with δ^{15}N, future compound-specific analysis may lead to a better understanding of δ^{13}C records obtained from horny corals.

3. Landmark Studies

Having explored the growth, geochemistry and radiometric dating of deep-sea corals, we now turn our attention to a few studies that have produced operationally useful paleoceanographic data. We focus on papers dealing with the rate of intermediate water mass variability across the Pleistocene–Holocene transition. The central theme of these papers is that climatic transitions occur over very short (decadal) timescales. Under most circumstances decadal-scale events are difficult to reconstruct from marine sediment records because of bioturbation.

Smith et al.'s (1997) landmark paper focused on *Desmophyllum* corals collected from the top of Orphan Knoll (1,800 m), an isolated seamount off the Newfoundland slope. The location and depth of Orphan Knoll allowed for the reconstruction of intermediate water mass variability in the Labrador Sea, an important component of the North Atlantic circulation. Stable isotopic data from these corals exhibited a major difference in the distribution of $\delta^{18}O$ vs. $\delta^{13}C$ lines between warm and cold periods spanning the last 15 ka (Figure 14). Foreshadowing the 'lines technique' published a few years later (Smith et al., 2000), they assumed that the lighter ends of their $\delta^{18}O$ vs. $\delta^{13}C$ lines approached isotopic equilibrium with seawater, an assumption supported by similarity of $\delta^{18}O$ values measured in contemporaneous foraminifera. One of their corals captured a rapid shift from warmer conditions at the base of the coral (13.400 ka BP), to cooler conditions at the top of the coral (13.322 ka), approximately co-incident with the

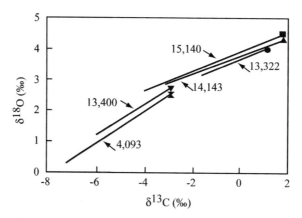

Figure 14 Isotopic data from *Desmophyllum dianthus* collected from Orphan Knoll (Labrador Sea). For clarity, only the least squares regression lines are shown. Symbols indicating isotopic maxima for each of the coral's datasets are assumed to approach isotopic equilibrium. Numbers indicate U/Th ages in years BP. The two oldest corals (15,140 and 14,143 yr) grew when the Labrador Sea retained much of its glacial character. Corresponding $\delta^{18}O$ maxima average 4.5‰. At the base of the next youngest coral (13,400 yr), $\delta^{18}O$ maxima (2.5‰) are much like that of a mid-Holocene coral (4,093 yr). Further up the same colony (13,322 yr), data return to that of the oldest-aged corals, indicating a rapid return to glacial-like conditions with the onset of the Younger Dryas. Figure from Smith et al. (1997), with permission.

accepted timing of the onset of the Younger Dryas event (12.9 ka BP). In this specimen, the shift from full warm to full cold conditions occurred between subsamples spaced 3 mm apart, an equivalent of 5 yr in time. This study was the first to demonstrate the utility of deep-sea corals in providing useful paleoceanographic data.

Shortly thereafter, a series of papers on paired U/Th-^{14}C dating of deep-sea corals came out, emphasizing rapid changes in intermediate water ventilation rates across the Pleistocene–Holocene transition (Adkins et al., 1998; Schröder-Ritzrau et al., 2003). The approach outlined in these studies culminated in the recent work of Robinson et al. (2005), with data from some 30 corals recovered from the New England Seamounts (33–39°N). Their reconstructions of paleo-Δ^{14}C over the interval 11–25 ka (Figure 15) document switches in the relative proportions of southern vs. northern source waters (a "bipolar seesaw"). Switches at intermediate depths occurred more frequently than they did in the abyss, and they coincided with climatic events documented in Greenland and Antarctic ice cores. The 15.4 ka event, in particular, saw a shift from full glacial to full modern-like conditions that took place in less than 100 yr.

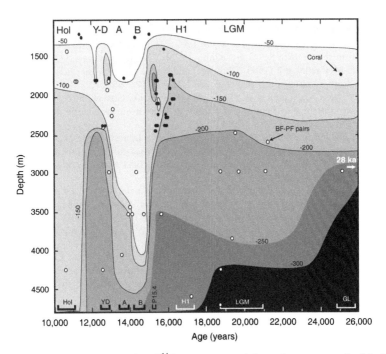

Figure 15   Mid-North Atlantic paleo-Δ^{14}C reconstructed from deep-sea corals (black circles) and foraminifera B/F pairs (white circles). During the last glacial period and Younger Dryas, the deep ocean was dominated by older, Δ^{14}C-depleted southern source waters, indicating cessation of deep water formation in the N. Atlantic. Younger, northern source waters characterize the warmer Bølling–Allerød and Holocene periods. Coral data indicate higher frequency variability in upper intermediate waters. Figure from Robinson et al. (2005), with permission.

REFERENCES

Adkins, J. F., & Boyle, E. A. (1997). Changing atmospheric $\Delta^{14}C$ and the record of deep water paleoventilation ages. *Paleoceanography, 12*, 337–344.

Adkins, J. F., Boyle, E. A., Curry, W. B., & Lutringer, A. (2003). Stable isotopes in deep-sea corals and a new mechanism for "vital effects". *Geochimica et Cosmochimica Acta, 67*, 1129–1143.

Adkins, J. F., Cheng, H., Boyle, E. A., Druffel, E. R. M., & Edwards, R. L. (1998). Deep-sea coral evidence for rapid change in ventilation of the deep North Atlantic 15,400 years ago. *Science, 280*, 725–728.

Adkins, J. F., Griffin, S., Kashgarian, M., Cheng, H., Druffel, E. R. M., Boyle, E. A., Edwards, R. L., & Shen, C.-C. (2002). Radiocarbon dating of deep-sea corals. *Radiocarbon, 44*, 567–580.

Adkins, J. F., Henderson, G. M., Wang, S.-L., O'Shea, S., & Mokadem, F. (2004). Growth rates of the deep-sea scleractinian *Desmophyllum cristagalli* and *Enallopsammia rostrata*. *Earth and Planetary Science Letters, 227*, 481–490.

Allard, G., Sinclair, D. J., Williams, B., Hillaire-Marcel, C., Ross, S., & Risk, M. (2005). Dendochronology in bamboo? Geochemical profiles and reproducibility in a specimen of the deep water bamboo coral *Keratoisis* spp. *Proceedings of 3rd International Symposium Deep-Sea Corals* (p. 204). Miami, USA.

Altabet, M. A. (1988). Variations in nitrogen isotopic composition between sinking and suspended particles: Implications for nitrogen cycling and particle transformation in the open ocean. *Deep Sea Research, 35*, 535–554.

Altabet, M. A., Deuser W.G., Honjo,S., & Honjo, S. (1991). Seasonal and depth-related changes in the source of sinking particles in the North Atlantic. *Nature, 354*, 136–139.

Altabet, M. A., & François, R. (1994). Sedimentary nitrogen isotopic ratio as a recorder for surface ocean nitrate utilization. *Global Biogeochemical Cycles, 8*, 103–116.

Altabet, M. A., François, R., Murray, D. W., & Prell, W. L. (1995). Climate-related variations in denitrification in the Arabian Sea from sediment $^{15}N/^{14}N$ ratios. *Nature, 373*, 506–508.

Altabet, M. A., & McCarthy, J. J. (1985). Temporal and spatial variations in the natural abundance of ^{15}N in PON from a warm-core ring. *Deep Sea Research, 32*, 755–772.

Altabet, M. A., Pilskaln, C., Thunell, R., Pride, C., Sigman, D., Chavez, F., & François, R. (1999). The nitrogen isotope biogeochemistry of sinking particles from the margin of the eastern North Pacific. *Deep Sea Research, 46*, 655–679.

Andrews, A. H., Cailliet, G. M., Kerr, L. A., Coale, K. H., Lundstrom, C., & DeVogelaere, A. P. (2005). Investigations of the age and growth for three deep-sea corals from the Davidson Seamount off central California. In: A. Freiwald & J. M. Roberts (Eds), *Cold-Water Corals and Ecosystems* (pp. 1021–1038). Berlin: Springer.

Andrews, A. H., Cordes, E. E., Mahoney, M. M., Munk, K., Coale, K. H., Cailliet, G. M., & Heifitz, J. (2002). Age, growth and radiometric age validation of a deep-sea, habitat-forming gorgonian (*Primnoa resedaeformis*) from the Gulf of Alaska. *Hydrobiologia, 471*, 101–110.

Bard, E., Arnold, M., Fairbanks, R. G., & Hamelin, B. (1993). ^{230}Th-^{234}U and ^{14}C ages obtained by mass spectrometry on corals. *Radiocarbon, 35*, 191–199.

Bard, E., Fairbanks, R. G., Hamelin, B., Zindler, A., & Chi Trach, H. (1991). Uranium-234 anomalies in corals older than 150000 years. *Geochimica et Cosmochimica Acta, 55*, 2385–2390.

Barnes, D. J. (1970). Coral skeletons: An explanation of their growth and structure. *Science, 170*, 1305–1308.

Beuck, L., & Freiwald, A. (2005). Bioerosion patterns in a deep-water Lophelia pertusa thicket (Propeller Mound, northern Porcupine Seabight). In: A. Freiwald & J. M. Roberts (Eds), *Cold-water corals and ecosystems* (pp. 915–936). Berlin: Springer.

Boerboom, C. M., Smith, J. E., & Risk, M. J. (1998). Bioerosion and micritisation in the deep-sea coral *Desmophyllum cristagalli*. *History of Biology, 13*, 53–60.

Bond, Z. A., Cohen, A. L., Smith, S. R., & Jenkins, W. J. (2005). Growth and composition of high-Mg calcite in the skeleton of a Bermudian gorgonian (*Plexaurella dichotoma*): Potential for paleothermometry. *Geochemistry, Geophysics, Geosystems, 6*, Q08010, doi:10.1029/2005GC000911.

Broecker, W. S., Gerard, R., Ewing, M., & Heezen, B. C. (1960). Natural radiocarbon in the Atlantic Ocean. *Journal of Geophysical Research*, 65(a), 2903–2931.

Broecker, W. S., Klas, M., Clark, E., Trumbore, S., Bonani, G., Wolfli, W., & Ivy, S. (1990). Accelerator mass spectrometric measurements on foraminifera shells from deep sea cores. *Radiocarbon*, 32, 119–133.

Bromley, R. G. (2005). Preliminary study of bioerosion in the deep-water coral Lophelia, Pleistocene, Rhodes, Greece. In: A. Freiwald & J. M. Roberts (Eds), *Cold-water Corals and Ecosystems* (pp. 895–914). Berlin: Springer.

Bryan, W. B., & Hill, D. (1941). Spherulitic crystallization as a mechanism of skeletal growth in the hexacorals. *Proceedings of Royal Society of Queensland*, 52, 78–91.

Cairns, S. D. (1981). Marine flora and fauna of the northeastern United States. Scleractinia. NOAA Tech. Rept. NMFS Circ. 438: 14pp., 16 figs.

Campana, S. E. (1997). Use of radiocarbon from nuclear fallout as a dated marker in the otoliths of haddock Melanogrammus aeglefinus. *Marine Ecology Progress Series*, 150, 49–56.

Carriquiry, J. D., & Risk, M. J. (1988). Timing and temperature record from stable isotopes of the 1982–1983 El Niño warming event in Eastern Pacific corals. *Palaios*, 3, 359–364.

Cheng, H., Adkins, J. A., Edwards, R. L., & Boyle, E. A. (2000). U-Th dating of deep-sea corals. *Geochimica et Cosmochimica Acta*, 64, 2401–2416.

Cline, J. D., & Kaplan, I. R. (1975). Isotopic fractionation of dissolved nitrate during denitrification in the eastern tropical north Pacific Ocean. *Marine Chemistry*, 3, 271–299.

Cohen, A. L., Gaetani, G. A., Lundälv, T., Corliss, B. H., & George, R. Y. (2006). Compositional variability in a cold-water scleractinian, *Lophelia pertusa*: New insights into "vital effects". *Geochemistry Geophysics Geosystems*, 7, Q12004, doi:10.1029/2006GC001354.

Cohen, A. L., & McConnaughey, T. A. (2003). Geochemical perspectives on coral mineralization. *Reviews in Mineralogy and Geochemistry*, 54, 151–187.

Constanz, B. R. (1986). Coral skeleton construction: A physiochemically dominated process. *Palaois*, 1, 152–157.

Cuif, J.-P., & Dauphin, Y. (2005). The two-step mode of growth in the scleractinian coral skeletons from the microscale to the overall scale. *Journal of Structural Biology*, 150, 319–331.

DeNiro, M. J., & Epstein, S. (1978). Influence of diet on the distribution of carbon isotopes in animals. *Geochimica et Cosmochimica Acta*, 42, 495–506.

DeNiro, M. J., & Epstein, S. (1981). Influence of diet on the distribution of nitrogen isotopes in animals. *Geochimica et Cosmochimica Acta*, 45, 341–351.

Druffel, E. R. M., Griffin, S., Witter, A., Nelson, E., Southon, J., Kashgarian, M., & Vogel, J. (1995). *Gerardia*: Bristlecone pine of the deep-sea? *Geochimica et Cosmochimica Acta*, 59, 5031–5036.

Druffel, E. R. M., King, L. L., Belastock, R. A., & Buesseler, K. O. (1990). Growth rate of a deep-sea coral using ^{210}Pb and other isotopes. *Geochimica et Cosmochimica Acta*, 54, 1493–1500.

Edwards, R. L., Beck, W. J., Burr, G. S., Donahue, D. J., Chappell, J. M. A., Bloom, A. L., Druffel, E. R. M., & Taylor, F. W. (1993). A large drop in atmospheric $^{14}C/^{12}C$ and reduced melting in the Younger Dryas, documented with ^{230}Th ages of corals. *Science*, 260, 962–967.

Edwards, R. L., Gallup, C. D., & Cheng, H. (2003). Uranium-series dating of marine and lacustrine carbonates. *Reviews in Mineralogy and Geochemistry*, 52, 363–405.

Emiliani, C., Hudson, J. H., Shinn, E. A., George, R. Y., & Lidz, B. (1978). Oxygen and carbon isotopic growth record in a reef coral from the Florida Keys and a deep-sea coral from Blake Plateau. *Science*, 202, 627–629.

Epstein, S., Buchsbaum, R., Lowenstam, H. A., & Urey, H. C. (1953). Revised carbonate-water isotopic temperature scale. *Bulletin of the Geological Society of America*, 64, 1315–1325.

Fallon, S. J., Roark, E. B., Guilderson, T. P., Dunbar, R. B., & Weber, P. (2005). Elemental imaging and proxy development in the deep sea coral, *Corallium secundrum*. Proceedings of 3rd International Symposium on Deep-Sea Corals (p. 187). Miami, USA.

Fallon, S. J., White, J. C., & McCulloch, M. T. (2002). Porites corals as recorders of mining and environmental impacts: Misima Island, Papua New Guinea. *Geochimica et Cosmochimica Acta*, 66, 45–62.

Frank, N., Lutringer, A., Paterne, M., Blamart, D., Henriet, J. P., van Rooij, D., & van Weering, T. C. E. (2005). Deep-water corals of the northeastern Atlantic margin: Carbonate mound evolution and upper intermediate water ventilation during the Holocene. In: A. Freiwald & J. M. Roberts (Eds), *Cold-Water Corals and Ecosystems* (pp. 113–133). Berlin: Springer.

Frank, N., Paterne, M., Ayliffe, L., van Weering, T., Henriet, J.-P., & Blamart, D. (2004). Eastern North Atlantic deep-sea corals: Tracing upper intermediate water $\Delta^{14}C$ during the Holocene. *Earth and Planetary Science Letters, 219,* 297–309.

Fry, B., & Sherr, E. B. (1984). $\delta^{13}C$ measurements as indicators of carbon flow in marine and freshwater ecosystems. *Contributions in Marine Science, 27,* 13–47.

Furla, P., Bénazet-Tambutté, S., Jaubert, J., & Allemand, D. (1998). Diffusional permeability of dissolved inorganic carbon through the isolated oral epithelial layers of the sea anemone, *Anemonia viridis. Journal of Experimental Marine Biology and Ecology, 230,* 71–88.

Gagnon, A. C., & Adkins, J. F. (2005). *Multiple proxy "vital effects" in a deep-sea coral.* Proceedings of 3rd International Symposium on Deep-Sea Corals (p. 188). Miami, USA.

Ganeshram, R. S., Pederson, T. F., Calvert, S. E., & Murray, J. W. (1995). Large changes in oceanic nutrient inventories from glacial to interglacial periods. *Nature, 376,* 755–757.

Gladfelter, E. H. (1982). Skeletal development in *Acropora cervicornis*: I. Patterns of calcium carbonate accretion in the axial corallite. *Coral Reefs, 1,* 45–51.

Goldberg, W. M. (1991). Chemistry and structure of skeletal growth rings in the black coral Antipathes fiordenis (Cnidaria, Antipatharia). *Hydrobiologia, 217/216,* 403–409.

Goldstein, S. J., Lea, D. W., Chakraborty, S., Kashgarian, M., & Murrell, M. T. (2001). Uranium-series and radiocarbon geochronology of deep-sea corals: Implications for Southern Ocean ventilation rates and the oceanic carbon cycle. *Earth and Planetary Science Letters, 193,* 167–182.

Goodfriend, G. A. (1997). Aspartic acid racemization and amino acid composition of the organic endoskeleton of the deep-water colonial anemone *Gerardia*: Determination of longevity from kinetic experiments. *Geochimica et Cosmochimica Acta, 61,* 1931–1939.

Goreau, T. F. (1959). The physiology of skeletal formation in corals I. A method for measuring the rate of calcium deposition by corals under different conditions. *Biological Bulletin, 116,* 59–75.

Grasshoff, M., & Zibrowius, H. (1983). Kalkkrusten auf Achsen von Hornkorallen. *Senckenbergiana Maritima, 15,* 111–145.

Griffin, S., & Druffel, E. R. M. (1989). Sources of carbon to deep-sea corals. *Radiocarbon, 31,* 533–543.

Grossman, E. L., & Ku, T.-L. (1986). Oxygen and carbon isotope fractionation in biogenic aragonite: Temperature effects. *Chemical Geology (Isotope Geoscience Section), 59,* 59–74.

Haug, G. H., Pederson, T. F., Sigman, D. M., Calvert, S. E., Nielsen, B., & Peterson, L. C. (1998). Glacial/interglacial variations in production and nitrogen fixation in the Cariaco Basin during the last 580 kyr. *Paleoceanography, 13,* 427–432.

Heikoop, J. M., Hickmott, D. D., Risk, M. J., Shearer, C. K., & Atudorei, V. (2002). Potential climate signals from the deep-sea gorgonian coral *Primnoa resedaeformis. Hydrobiologia, 471,* 117–124.

Heikoop, J. M., Risk, M. J., Lazier, A. V., & Schwarcz, H. P. (1998). $\delta^{18}O$ and $\delta^{13}C$ signatures of a deep-sea gorgonian coral from the Atlantic coast of Canada. *EOS, 79*(17, supplement), S179.

Heikoop, J. M., Tsujita, C. J., & Risk, M. J. (1996). Corals as proxy recorders of volcanic activity: Banda Api, Indonesia. *Palaios, 11,* 286–292.

Henderson, G. M. (2002). Seawater ($^{234}U/^{238}U$) during the last 800 thousand years. *Earth and Planetary Science Letters, 199,* 97–110.

Hendy, E. J., Gagan, M. K., Alibert, C. A., McCulloch, M. T., Lough, J. M., & Isdale, P. J. (2002). Abrupt decrease in tropical Pacific sea surface salinity at end of Little Ice Age. *Science, 295,* 1511–1514.

Jasper, J. P., Hayes, J. M., Mix, A. C., & Prahl, F. G. (1994). Photosynthetic fractionation of ^{13}C and concentrations of dissolved CO_2 in the central equatorial Pacific during the last 255,000 years. *Paleoceanography, 9,* 781–798.

Johnston, I. S. (1980). The ultrastructure of skeletogenesis in hermatypic corals. *International Review of Cytology, 67,* 171–214.

Key, R. M. (2001). Radiocarbon. In: J. Steele, S. Thorpe & K. Turekian (Eds), *Encyclopedia of Ocean Sciences* (pp. 2338–2353). London: Academic Press.

Knapp, A. N., Sigman, D. M., & Lipschultz, F. (2005). N isotopic composition of dissolved organic nitrogen and nitrate at the Bermuda Atlantic time-series site. *Global Biogeochemical Cycles, 19,* GB1018, doi:10.1029/2004GB002320.

Kroopnick, P. (1985). The distribution of ^{13}C of $\sum CO_2$ in the world oceans. *Deep-Sea Research I, 32,* 57–84.

Laws, E. A., Popp, B. N., Cassar, N., & Tanimoto, J. (2002). ^{13}C discrimination patterns in oceanic phytoplankton: Likely influence of CO_2 concentrating mechanisms, and implications for palaeo-reconstructions. *Functional Plant Biology, 29,* 323–333.

Lazier, A. V., Smith, J. E., & Risk, M. J. (1999). The skeletal structure of *Desmophyllum cristagalli*: The use of deep-water corals in sclerochronology. *Lethaia, 32,* 119–130.

Lewis, J. C., Barnowski, T. F., & Telesnicki, G. J. (1992). Characteristics of carbonates of gorgonian axes. *Biological Bulletin, 183,* 278–296.

Lomitschka, M., & Mangini, A. (1999). Precise Th/U-dating of small and heavily coated samples of deep sea corals. *Earth and Planetary Science Letters, 170,* 391–401.

Ludwig, K. R., & Titterington, D. M. (1994). Calculation of ^{230}Th/U isochrons, ages and errors. *Geochimica et Cosmochimica Acta, 58,* 5031–5042.

Lutringer, A., Blamart, D., Frank, N., & Labeyrie, L. (2005). Paleotemperatures from deep-sea corals scale effects. In: A. Freiwald & J. M. Roberts (Eds), *Cold-Water Corals and Ecosystems* (pp. 1081–1096). Berlin: Springer.

Macintyre, I. G., Bayer, F. M., Logan, M. A., & Skinner, H. C. W. (2000). Possible vestige of early phosphatic biomineralization in gorgonian octocorals (Coelenterata). *Geology, 28,* 455–458.

Mangini, A., Lomitschka, M., Eichstadter, R., Frank, N., & Vogler, S. (1998). Coral provides way to age deep water. *Nature, 392,* 347–348.

Marschal, C., Garrabou, J., Harmelin, J. G., & Pichon, M. (2004). A new method for measuring growth and age in the precious red coral *Corallium rubrum* (L.). *Coral Reefs, 23,* 423–432.

McClelland, J. W., & Montoya, J. P. (2002). Trophic relationships and the nitrogen isotopic composition of amino acids in plankton. *Ecology, 83,* 2173–2180.

McConnaughey, T. A. (1989a). ^{13}C and ^{18}O isotope disequilibria in biological carbonates. 1. Patterns. *Geochimica et Cosmochimica Acta, 53,* 151–162.

McConnaughey, T. A. (1989b). ^{13}C and ^{18}O isotope disequilibria in biological carbonates. 2. *In vitro* simulation of kinetic isotope effects. *Geochimica et Cosmochimica Acta, 53,* 163–171.

McConnughey, T. A. (2003). Sub-equilibrium oxygen-18 and carbon-13 levels in biological carbonates: carbonate and kinetic models. *Coral Reefs, 22,* 316–327.

Montagna, P., McCulloch, M., Taviani, M., Remia, A., & Rouse, G. (2005). High-resolution trace element compositions in deep-water scleractinian corals (*Desmophyllum dianthus*) from the Mediterranean Sea and the Great Australian Bight. In: A. Freiwald & J. M. Roberts (Eds), *Cold-Water Corals and Ecosystems* (pp. 1109–1126). Berlin: Springer.

Mortensen, P. B., & Rapp, H. T. (1998). Oxygen and carbon isotope ratios related to growth line patterns in skeletons of Lophelia pertusa (L.) (Anthozoa, Sclleractinia): Implications for determination of linear extension rates. *Sarsia, 83,* 433–446.

Orejas, C., Gili, J. M., & Arntz, W. (2003). Role of small-plankton communities in the diet of two Antarctic octocorals (*Primnoisis antarctica* and *Primnoella* sp.). *Marine Ecology Progress Series, 250,* 105–116.

Pons-Branchu, E., Hillaire-Marcel, C., Deschamps, P., Ghaleb, B., & Sinclair, D. J. (2005). Early diagenesis impact on precise U-series dating of deep-sea corals: Example of a 100–200-year old *Lophelia pertusa* sample from the northeast Atlantic. *Geochimica et Cosmochimica Acta, 69,* 4865–4879.

Rau, G. H. (1994). Variations in sedimentary organic δ^{13}C as a proxy for past changes in ocean and atmospheric [CO_2]. In: R. Zahn, M. Kamiski, L. D. Labeyrie & T. F. Pedersen (Eds), *Carbon Cycling in the Glacial Ocean Constraints on the Ocean's Role in Global Climate Change* (pp. 307–322). Berlin: Springer.

Remia, A., & Taviani, M. (2004). Shallow-buried Pleistocene Madrepora-coral mounds on a muddy continental slope. Tuscan Archipelago, NE Tyrrhenian Sea. Facies 50. doi:10.1007/s10347-004-0029-2.

Ribes, M., Coma, R., & Gili, J. M. (1999). Heterogeneous feeding in benthic suspension feeders: The natural diet and grazing rate of the temperate gorgonian *Paramuricea clavata* (Cnidaria: Octocorallia) over a year cycle. *Marine Ecology Progress Series, 183*, 125–137.

Ribes, M., Coma, R., & Rossi, S. (2003). Natural feeding of the temperate asymbiotic octocoral-gorgonian *Leptogorgia sarmentosa* (Cnidaria: Octocorallia). *Marine Ecology Progress Series, 254*, 141–150.

Risk, M. J., Hall-Spenser, J., & Williams, B. (2005). Climate records from the Faroe-Shetland Channel using *Lophelia pertusa* (Linnaeus, 1758). In: A. Freiwald & J. M. Roberts (Eds), *Cold-water corals and ecosystems* (pp. 1097–1108). Berlin: Springer.

Risk, M. J., Heikoop, J. M., Snow, M. G., & Beukens, R. (2002). Lifespans and growth patterns of two deep-sea corals: *Primnoa resedaeformis* and *Desmophyllum cristagalli*. *Hydrobiologia, 471*, 125–131.

Roark, E. B., Guilderson, T. P., Flood-Page, S., Dunbar, R. B., Ingram, B. L., Fallon, S. J., & McCulloch, M. (2005). Radiocarbon-based ages and growth rates of bamboo corals from the Gulf of Alaska. *Geophysical Research Letters, 32*, L04606, doi:10.1029/2004GL021919.

Robinson, L. F., Adkins, J. F., Fernandez, D. P., Burnett, D. S., Wang, S.-L., Gagnon, A. C., & Krakauer, N. (2006). Primary U distribution in scleractinian corals and its implications for U series dating. *Geochemistry, Geophysics, Geosystems, 7*, Q05022.

Robinson, L. F., Adkins, J. F., Keigwin, L. D., Southon, J., Fernandez, D. P., Wang, S.-L., & Scheirer, D. S. (2005). Radiocarbon variability in the Western North Atlantic during the Holocene. *Science, 310*, 1469–1473.

Rollion-Bard, C., Blamart, D., & Cuif, J.-P. (2003). Microanalysis of C and O isotopes of azooxanthellate and zooxanthellate corals by ion microprobe. *Coral Reefs, 22*, 405–415.

Saino, T., & Hattori, A. (1987). Geographical variation in the water column distribution of suspended particulate organic nitrogen and its ^{15}N natural abundance in the Pacific and its marginal seas. *Deep-Sea Research A, 34*, 807–827.

Sarnthein, M., Winn, K., Jung, S. J. A., Duplessy, J.-C., Labeyrie, L., Erlenkeuser, H., & Ganssen, G. (1994). Changes in east Atlantic deepwater circulation over the last 30,000 years: Eight time slice reconstructions. *Paleoceanography, 9*, 209–267.

Schröder-Ritzrau, A., Freiwald, A., & Mangini, A. (2005). U/Th-dating of deep-water corals from the eastern North Atlantic and the western Mediterranean Sea. In: A. Freiwald & J. M. Roberts (Eds), *Cold-Water Corals and Ecosystems* (pp. 157–172). Berlin: Springer.

Schröder-Ritzrau, A., Mangini, A., & Lomitschka, M. (2003). Deep-sea corals evidence periodic reduced ventilation in the North Atlantic during the LGM/Holocene transition. *Earth and Planetary Science Letters, 216*, 399–410.

Shackleton, N. J. (1967). Oxygen isotope analyses and Pleistocene temperatures reassessed. *Nature, 215*, 15–17.

Shen, G. T., & Boyle, E. A. (1988). Determination of lead, cadmium, and other trace metals in annually-banded corals. *Chemical Geology, 67*, 47–62.

Sherwood, O. A. (2002). *The deep-sea gorgonian Primnoa resedaeformis as an oceanographic monitor* (p. 65). M.Sc. Thesis. McMaster University, Hamilton, Canada.

Sherwood, O. A. (2006). *Deep-sea octocorals: Dating methods, stable isotopic composition, and proxy records of the slopewaters off Nova Scotia* (p. 240). Ph.D. Thesis. Dalhousie University, Halifax, Canada.

Sherwood, O. A., Heikoop, J. M., Scott, D. B., Risk, M. J., Guilderson, T. P., & McKinney, R. A. (2005b). Stable isotopic composition of deep-sea gorgonian corals *Primnoa* spp.: A new archive of surface processes. *Marine Ecology Progress Series, 301*, 135–148.

Sherwood, O. A., Heikoop, J. M., Sinclair, D. J., Scott, D. B., Risk, M. J., Shearer, C., & Azetsu-Scott, K. (2005c). Skeletal Mg/Ca in Primnoa resedaeformis: Relationship to temperature? In: A. Freiwald & J. M. Roberts (Eds), *Cold-water corals and ecosystems* (pp. 1061–1079). Berlin: Springer.

Sherwood, O. A., Scott, D. B., & Risk, M. J. (2006). Late Holocene radiocarbon and aspartic acid racemization dating of deep-sea octocorals. *Geochimica et Cosmochimica Acta, 70*, 2806–2814.

Sherwood, O. A., Scott, D. B., Risk, M. J., & Guilderson, T. P. (2005a). Radiocarbon evidence for annual growth rings in the deep-sea octocoral *Primnoa resedaeformis*. *Marine Ecology Progress Series*, *301*, 129–134.

Shirai, K., Kusakabe, M., Nakai, S., Ishii, T., Watanabe, T., Hiyagon, H., & Sano, Y. (2005). Deep-sea coral geochemistry: Implication for the vital effect. *Chemical Geology*, *224*, 212–222.

Sinclair, D. J. (2005). Correlated trace element "vital effects" in tropical corals: A new geochemical tool for probing biomineralization. *Geochimica et Cosmochimica Acta*, *69*, 3265–3284.

Sinclair, D. J., & Risk, M. J. (2006). A numerical model of trace-element coprecipitation in a physicochemical calcification system: Application to coral biomineralization and trace-element 'vital effects'. *Geochimica et Cosmochimica Acta*, *70*, 3855–3868.

Sinclair, D. J., Sherwood, O. A., Risk, M. J., Hillaire-Marcel, C., Tubrett, M., Sylvester, P., McCulloch, M., & Kinsley, L. (2005). Testing the reproducibility of Mg/Ca profiles in the deep-water coral *Primnoa resedaeformis*: Putting the proxy through its paces. In: A. Freiwald & J. M. Roberts (Eds), *Cold-water corals and ecosystems* (pp. 1039–1060). Berlin: Springer.

Sinclair D. J., Williams B., & Risk M. (2006). A biological origin for climate signals in corals — trace element "vital effects" are ubiquitous in scleractinian coral skeletons. *Geophysical Research Letters*, *33*, L17707, doi:10.1029/2006GL027183.

Smith, J. E., Risk, M. J., Schwarcz, H. P., & McConnaughey, T. A. (1997). Rapid climate change in the North Atlantic during the Younger Dryas recorded by deep-sea corals. *Nature*, *386*, 818–820.

Smith, J. E., Schwarcz, H. P., & Risk, M. J. (2002). Patterns of isotopic disequilibria in azooxanthellate coral skeletons. *Hydrobiologia*, *471*, 111–115.

Smith, J. E., Schwarcz, H. P., Risk, M. J., McConnaughey, T. A., & Keller, N. (2000). Paleotemperatures from deep-sea corals: Overcoming "vital effects". *Palaios*, *15*, 25–32.

Sorauf, J. E., & Jell, J. S. (1977). Structure and incremental growth in the ahermatypic coral *Desmophyllum cristagalli* from the North Atlantic. *Palaeontology*, *20*, 1–19.

Stuiver, M., Quay, P. D., & Ostlund, H. G. (1983). Abyssal C-14 distribution and the age of the world oceans. *Science*, *219*, 849–851.

Swart, P. K., & Hubbard, J. A. E. B. (1982). Uranium in scleractinian coral skeletons. *Coral Reefs*, *1*, 13–19.

Szmant-Froelich, A. (1974). Structure, iodination, and growth of the axial skeletons of *Muricea californica* and *M. fruticosa* (Coelenterata: Gorgonacea). *Marine Biology*, *27*, 299–306.

Teichert, C. (1958). Cold- and deep-water coral banks. *AAPG Bulletin*, *42*, 1064–1082.

Thresher, R., Rintoul, S. R., Koslow, J. A., Weidman, C., Adkins, J., & Proctor, C. (2004). Oceanic evidence of climate change in southern Australia over the last three centuries. *Geophysical Research Letters*, *31*, L07212, doi:10.1029/2003GL018869.

Thunell, R. C., Sigman, D. M., Muller-Karger, F., Astor, Y., & Varela, R. (2004). Nitrogen isotope dynamics of the Cariaco Basin, Venezuela. *Global Biogeochemical Cycles*, *18*, GB3001, doi:10.1029/2003GB002185.

Titschak, J., & Freiwald, A. (2005). Growth, deposition and faciesof Pleistocene bathyal coral communities from Rhodes, Greece. In: A. Freiwald & J. M. Roberts (Eds), *Cold-Water Corals and Ecosystems* (pp. 41–59). Berlin: Springer.

Tudhope, A. W., Chilcott, C. P., McCulloch, M. T., Cook, E. R., Chappell, J., Ellam, R. M., Lea, D. W., Lough, J. M., & Shimmield, G. B. (2001). Variability in the El Niño-Southern oscillation through a glacial-interglacial cycle. *Science*, *291*, 1511–1517.

Wainwright, S. (1964). Studies of the mineral phase of coral skeleton. *Experimental Cell Research*, *34*, 213–230.

Ward-Paige, C. A., Risk, M. J., & Sherwood, O. A. (2005). Reconstruction of nitrogen sources on coral reefs: $\delta^{15}N$ and $\delta^{13}C$ in gorgonians from the Florida Reef Tract. *Marine Ecology Progress Series*, *296*, 155–163.

Weber, J. N. (1973). Deep-sea scleractinian coral: Isotopic composition of skeleton. *Deep Sea Research*, *2*, 901–909.

Weidman, C. R., & Jones, G. A. (1993). A shell-derived time history of bomb ^{14}C on Georges Bank and its Labrador Sea implications. *Journal of Geophysics Research*, *98*(C8), 14577–14588.

Weinbauer, M. G., Brandstätter, F., & Velimirov, B. (2000). On the potential use of magnesium and strontium concentrations as ecological indicators in the calcite skeleton of the red coral (*Corallium rubrum*). *Marine Biology, 137*, 801–809.

Williams, B., Risk M. J., Sulak, K., Ross, S., & Stone, R. (2005). Deep-water Antipatharians and Gorgonians: Proxies of biogeochemical processes. *Proceedings of 3rd International Symposium on Deep-Sea Corals* (p. 70). Miami, USA.

Wisshak, M., Freiwald, A., Lundälv, T., & Gektidis, M. (2005). The physical niche of the bathyal *Lophelia pertusa* in a non-bathyal setting: environmental controls and paleoecological implications. In: A. Freiwald & J. M. Roberts (Eds), *Cold-Water Corals and Ecosystems* (pp. 979–1001). Berlin: Springer.

Wu, J., Calvert, S. E., & Wong, C. S. (1999). Carbon and nitrogen isotope ratios in sedimenting particulate organic matter at an upwelling site off Vancouver Island. *Estuarine Coastal and Shelf Science, 48*, 193–203.

CHAPTER THIRTEEN

TRANSFER FUNCTIONS: METHODS FOR QUANTITATIVE PALEOCEANOGRAPHY BASED ON MICROFOSSILS

Joël Guiot* *and* Anne de Vernal

Contents

1. Introduction	523
2. Methods Based on Calibration	527
2.1. Methods based on ordination	527
2.2. Generalized models and non-parametric methods	531
3. Methods Based on Similarity	533
3.1. Mutual range method	533
3.2. Modern analogues method	534
3.3. Constrained analogues	535
3.4. Response surfaces	536
4. Comparison of Methods with a Worked Example	537
5. Discussion and Future Developments	545
5.1. What are the best methods?	545
5.2. What is the effect of the correlations between climatic variables?	547
5.3. How many taxa can be used?	547
5.4. How independent is a cross-validation?	548
5.5. Bayesian methods	549
6. The applications of Transfer Functions *Sensu Lato* in Paleoceanography	550
6.1. The most common methods and reconstructed parameters	550
6.2. Multi-Method and multi-proxy approaches	552
7. Concluding Remarks	556
References	557

1. INTRODUCTION

One of the main avenues of research in paleoceanography is the use of microfossil assemblages for decoding or reconstructing climatic or oceanographic parameters. This requires appropriate statistical methods after the identification of biological remains and count proportions of each taxon, or presence/absence of specific taxa. The various statistical methods that have been developed for the

---

* Corresponding author.

Developments in Marine Geology, Volume 1
ISSN 1572-5480, DOI 10.1016/S1572-5480(07)01018-4

reconstruction of climate and ocean parameters based on microfossil assemblages are generally referred to as transfer functions in a very large sense. However, the term "transfer function" has a specific meaning from a mathematical point of view, and has different meanings depending on the context. In a paleoclimatic context, the term "transfer function" was apparently used for the first time by Webb and Bryson (1972). In paleoceanography, "transfer functions" may refer to the method developed by Imbrie and Kipp in the early 1970s for the reconstruction of sea-surface temperatures (SSTs) based on multi-variate analyses of planktonic foraminifer assemblages. According to Sachs, Webb, and Clark (1977), the transfer function procedure is defined as a method producing calibrated quantitative estimates of some parameters of the environment, such as seasonal or monthly air or ocean surface temperature, from proxy data such as biological assemblages. By extension, transfer functions may thus include all methods that permit quantitative estimates of ocean or climate parameters using paleontological data.

For a long time, Quaternary paleoecologists have used intuitive methods to reconstruct paleoclimates or paleoenvironments from biological data. The most common approach has been to compare the present-day distribution of selected species with the corresponding distribution of climate variables thought to be determinant for the distribution of the selected species. This method has been called "indicator-species". It is univariate because the species are analyzed separately and related to one climatic variable. Univariate methods can be the source of a number of problems. In particular, the species respond to a combination of climatic variables and their distribution is controlled by different climatic or hydrographic factors in different parts of their ranges. Moreover, climate parameters are often interrelated. Thus, it has been necessary to develop methods taking into account the ecological complexity of species and assemblages, and of their relationships with climatic factors.

A first attempt was made using a multi-variate indicator-species method, which takes into account several species (Iversen, 1944; Atkinson, Briffa, & Coope, 1987). However, this method still disregards much relevant information concerning species abundances and requires an accurate knowledge and understanding of the present distribution of the species. Such knowledge is possible on land from the observation of the modern vegetation and fauna, but it is much more difficult to develop in marine environments where *in situ* measurements and observations of the phytoplankton and zooplankton are limited in time and space.

A key solution was found with the development of response models and their inverse, the "transfer functions" (Imbrie & Kipp, 1971). These methods assume that each species lives in a given range of climatic and environmental conditions. The problem is having an adequate understanding of the response of plants, animals, or protists to climate and other abiotic and biotic factors. In the case of the response model, the assemblage $\mathbf{X}$ is expressed as a function $\mathbf{R}$ of climate $\mathbf{C}$ and non-climatic factors $\mathbf{D}$ as follows: $\mathbf{X} = \mathbf{R}(\mathbf{C},\mathbf{D})$. If climate is the main factor controlling the biological response, it is possible to approximate the response function by $\mathbf{X} = \mathbf{R_c}(\mathbf{C})$. The inverse of this response function $\mathbf{X}$ enables one to calculate climatic parameters from known biological responses: $\hat{\mathbf{C}} = \hat{\mathbf{R}}_\mathbf{c}^{-1}(\mathbf{X})$. With the exception of a situation characterized by one taxon responding to one climatic factor, the solution for $\hat{\mathbf{R}}_\mathbf{c}^{-1}$ can be approached only in a statistical way by minimizing the difference between **C** and its estimate. A backward approach is often more useful because it allows direct calibration of the climate on the assemblages, as follows: $\hat{C} = \hat{T}(X)$. Here, $\hat{T}$ is the transfer function, which can be empirically defined from calibrations using various statistical techniques.

Transfer function models and methods based on calibrations are based on a few assumptions:

(1) Climate is the ultimate cause of changes in the paleobiological data.
(2) The ecological properties of the species considered has not changed between the period analyzed and the present time, and the relationship between the species and the climate is thus uniform through time.
(3) The modern observations contain all the necessary information to interpret the fossil data.

The third assumption originates from the uniformatarian principle. It is implicit in most paleoecological studies, but constitutes an unprovable premise, especially when going back in time in the geological archives. One major consequence of the three assumptions mentioned above is the implicit condition of equilibrium between the present distribution of species and climate.

Since the early work of Imbrie and Kipp (1971), different types of approaches based on multi-variate statistical methods have been developed and widely applied in paleoceanography to reconstruct paleotemperatures. They used primarily planktonic foraminifera, diatoms, and radiolarian assemblages (e.g., CLIMAP, 1976, 1981; Pichon et al., 1992; Koç, Jansen, & Haflidason, 1993; Sabin & Pisias, 1996).

An alternative to the transfer function model and calibration approaches is the analogue-based approach, introduced for the first time in marine sciences by Hutson (1980). It fundamentally differs from transfer function models, as there is no strict calibration. It supposes that a given assemblage of taxa in the fossil record is most likely to have occurred under a combination of environmental conditions characterizing similar modern assemblages of taxa. This approach therefore relies on the comparison between assemblages rather than the direct relationship between taxa and climate or hydrographic parameters. When comparing the assemblages, it is necessary to define an adequate way to evaluate the distance, which quantitatively defines the dissimilarity between the spectra. The analogue-based approach requires the availability of extensive sets of modern assemblages for comparison, especially since there is no calibration allowing extrapolation. Adequate knowledge of the ecology and physiology of the taxa is necessary, not for calibration, but to avoid over-interpreting the data. Two advantages must be pointed out. First, several environmental variables can be reconstructed concurrently, thus taking into account their interactions. Second, no assumption is made about the form of the underlying relationship between taxon abundance or assemblage components and the environmental gradients, unlike calibration approaches that assume linearity or modal distributions. After an early use in paleoceanography by Hutson (1980) and Prell (1985), we had to wait until the 1990s (e.g., de Vernal, Rochon, Hillaire-Marcel, Turon, & Guiot, 1993; de Vernal, Turon, & Guiot, 1994; Waelbroeck et al., 1998) for a revival of the analogue-based approach in paleoceanography.

In this chapter, we first describe the principles of the different calibration- or similarity-based methods that have been developed for quantitative reconstruction of climate and ocean parameters using biological remains (Table 1). Subsequently, we try to compare the respective use and performance of these methods with an example using a North Atlantic foraminifer data set. In the discussion, we address

Table 1 Summary of the Various Transfer Function Methods used for Paleoclimate or Paleoceanographic Reconstructions Based on Microfossil Data (See Text for More Details).

Type of approach	Family	Sub-family	Method	Order
Calibration	Ordination techniques	Indirect gradient analyses	Regression on principal components (e.g., the Q-mode Imbrie and Kipp method — I&K)	Linear or quadratic
			Regression on correspondence factors (Roux, 1979)	Unimodal
		Direct gradient analyses	Partial least Square (PLS; ter Braak, Juggins, Birks, & Van der Voet, 1993)	Linear
			Canonical correlation (e.g., Guiot, Berger, Munaut, & Till, 1983)	Linear
			Canonical correspondence analyses (e.g., Gasse & Tekaia, 1983)	Unimodal
			Weighted averaging — Partial least-squares (WA-PLS; e.g., Korkhola & Weckström, 2000)	Unimodal
	Generalized models (GLM)		Logistic regression (e.g., ter Braak & Looman, 1986)	Unimodal
	Non-parametric methods		Generalized additive model (GAM; e.g., Gersonde, Crosta, Abelmann, & Armand, 2005)	Non-linear
			Artificial neural network (ANN; e.g., Malmgren & Nordlund, 1996)	Non-linear
Similarity	Presence/absence		Indicator species method (ISM; Iversen, 1944)	Non-linear
			Mutual climatic range method (MCR; Atkinson et al., 1987)	Non-linear
	Abundances	Unconstrained	Modern analogue technique (MAT; Hutson, 1980)	Non-linear
		Constrained	MAT constrained with external variable (e.g., Guiot et al., 1993)	Non-linear
			MAT constrained geographically (SIMMAX; Pflaumann, Duprat, Pujol, & Labeyrie 1996)	Non-linear
		Response surface	Response surface method (Bartlein, Prentice, & Webb, 1986)	Non-linear
			Revised analogue method (RAM; Waelbroeck et al., 1998)	Non-linear

key issues with respect to limitations and potential future developments. Finally, the applications of different methods using microfossil assemblages for the reconstruction of ocean parameters are briefly presented.

2. Methods Based on Calibration

The transfer function is a predictive method. Whereas response functions aim at describing the relationships between species and abiotic factors, transfer function models aim at making predictions that are as close as possible to reality. The error of prediction is, however, difficult to evaluate and is never known with certainty. It is generally larger than the calibration error, which is obtained by comparing the data used for calibration (also called "training data set") and the estimates. The values for the error of prediction can be evaluated by a validation procedure in which part of the modern data available is not used for calibration, and is reserved for prediction tests. This independent verification is nevertheless very limited (but better than nothing), as nobody knows for certain what is really happening when we move from the domain of modern data towards the domain of fossil data.

2.1. Methods Based on Ordination

Ordination is a general term for multi-variate techniques sorting multi-dimensional data points along one or several axes in order to visualize their structure. In paleo-climatology and paleoceanography, we hope that they are ordered along the environmental or climatic gradients we want to reconstruct. There are different ordination techniques (Table 1), all of which differ slightly with respect to the mathematical approach used to calculate species and sample similarity/dissimilarity. The first family of methods is the indirect gradient analysis, with ordination along hidden variables. It includes the principal component analysis and the correspondence analysis (CA). The second family of methods is the direct gradient analysis. It has the advantage of expressing explicitly the gradient in relation to environmental variables. It includes the canonical analyses and the methods derived from weighted averaging (WA) and partial least square.

Following ter Braak, Juggins, Birks, and Van der Voet (1993), the calibration is done between an environmental vector $\mathbf{y}$ (y_i, $i = 1\ldots n$) and a species matrix $\mathbf{X}$ (x_{ij}, $i = 1\ldots n$, $j = 1\ldots m$). The subscript i (from 1 to n) is the number of the sites where the m species are recorded (the columns of $\mathbf{X}$) and where the environmental variable are to be reconstructed.

2.1.1. Principal component analysis

The principal components analysis (PCA) is one of the earliest ordination techniques and is the basis of the Imbrie and Kipp method (I&K). It uses a rigid rotation to derive orthogonal axes, which maximize the variance in the species data set. However, it does not guarantee that the variance of the climatic variables is maximized. It is then accompanied by a multiple regression to project the climate on the species space.

The PCA searches for a weight vector **b**, which is the first component that maximizes the variance of a linear combination **z** of the species, subject to the condition that the norm of **b** is 1. The quantity V to be maximized is:

$$V = \mathbf{z'z}/\mathbf{b'b} \text{ where } \mathbf{z} = \mathbf{x\,b} \text{ (\textbf{z} being centered and normalized)} \tag{1}$$

In such an analysis, the species are first standardized, i.e., the columns of **X** have a zero mean and unit variance. The PCA can be done in R-mode, with **b** being an eigenvector of **X'X** (which is proportional to the correlation matrix of the species). Alternatively, the PCA can be made in a Q-mode, and **b** is then an eigenvector of **XX'** (which measures the association between the sites). In both modes, the maximum number of principal components is equal to the minimum of m (number of species) and n (number of sites). Equation (1) defines the first principal component. The second, third and following components are defined in the same way, with the condition that the components are all independent. The estimate of **y** is finally obtained by regression on these components **z**.

It is important to mention here that several terminologies are used for principal component analyses depending upon the application domain. Imbrie and Kipp (1971) used the term factor analysis in reference to the human sciences, which introduced the expression after identification of latent factors in data sets. In these analyses, a small number of factors yielding a clear interpretation are kept, whereas the remaining ones are considered to represent noise. In climatology, the term empirical orthogonal function (EOF) is often used because the focus is to detect independent patterns typical in climatic fields of research. All of these techniques use eigenvectors analyses of a covariance or correlation matrix and are mathematically identical. The term "factor analysis" has only been used in the recent literature to refer to the CA (Section 2.1.3). Here, we will continue to use PCA, although Imbrie and Kipp used the expression "factor analysis".

Imbrie and Kipp (1971) have developed a procedure in three steps. The first step consists of estimating the physical variables to describe the biotic assemblages in core-tops using a PCA, which is an objective way of clustering different species into assemblages of species. The PCA used is a Q-mode on non-centered percentages. To facilitate the interpretation, they used a varimax rotation, although it was not necessary to improve the performance of the method. A small number of components usually represent a large proportion of the variance. The components retained are related to typical assemblages, such as arctic, subarctic, subtropical and tropical. The second step leads to the derivation of a relationship between each environmental variable and these components using multiple (linear or quadratic) regressions (possibly stepwise regression). The third step projects the down-core data on the principal components and thus allows for the calculation of factor scores for each sample. In the final step, the scores are entered into the calibration equations to calculate paleotemperatures.

CLIMAP (1976, 1981) have adopted this method in their worldwide reconstruction of the last glacial maximum (LGM) SST. This reconstruction has been often used by climate modelers as a boundary condition for paleoclimate simulations of the LGM (e.g., Joussaume & Taylor, 1995; IPCC, 2001; Braconnot, 2004). Koç and Schrader (1990), Koç et al. (1993), Andersen, Koç, Jennings, and Andrews (2004) and De Sève (2005) have applied the method to marine diatoms in northern Atlantic regions. They have been able to interpret six factors explaining more than 90% of the species variance. Gersonde et al. (2005) have also used this technique, among two others, to estimate the climate at the LGM in the Circum-Antarctic Ocean based on siliceous microfossils. Sabin and Pisias (1996) and Pisias, Roelofs, and Weber (1997) have developed and applied the Imbrie and Kipp (1971) technique for the reconstruction of SSTs in the North Pacific based on radiolarian assemblages. Mix, Morey, and Pisias (1999) have modified the way with which to calculate the PCA in order to solve the problem of no-analogues in the LGM: instead of calculations based on modern data, they made calculations of the PCA on a large set of LGM data. Such a variant in the approach seems to be a suitable way to make progress in spite of the limitations resulting from a no-analogue situation.

2.1.2. Partial least square analysis and canonical correlation analysis

While the PCA does not try to fit at best the predictand $\mathbf{y}$ (recall that a predictand is a variable to predict, here the climate), the partial least square analysis (PLS) maximizes the covariance between the predictand and a linear combination of the species subjected to the condition that the norm of $\mathbf{b}$ is equal to 1. The quantity C to be maximized for the first component is

$$C = \mathbf{z}'\mathbf{y}/\mathbf{b}'\mathbf{b} \quad \text{where } \mathbf{z} = \mathbf{X}\mathbf{b} \tag{2}$$

$\mathbf{z}$ being a linear combination of the species abundance, the analysis aims at finding a linear combination to maximize the covariance with the environmental parameter. As in PCA, the following components must be orthogonal to the first one and together. In this method, the interpretation of the components is linked to the predictand to be reconstructed; whereas in the PCA, the components must be calibrated *a posteriori* on the predictor (a predictor is a variable used to predict an dependent variable, they are here the taxa percentages). There are two possibilities of generalization to several predictands:

(1) One is to maximize the sum of the correlations of each predictand. This is the principal component analysis of instrumental variables (Sabatier & Van Campo, 1984) or redundancy analysis (ter Braak & Prentice, 1988).

(2) The second possibility is to maximize the correlation between a linear combination of the predictors and a linear combination of the predictands. This allows for the deduction of the relationships of the predictands in relation to the predictors (Fritts, Blasing, Hayden, & Kutzbach, 1971; Guiot et al., 1983).

2.1.3. Correspondence analysis

With species assemblages, CA (Benzécri, 1973; Hill, 1974) is often more suitable than PCA because unimodal relationships replace linear relationships between species and climate parameters (ter Braak, 1985). However, the CA requires the introduction of norms. Let us define $\mathbf{R}$ as the diagonal matrix $(x_{1\cdot}, x_{2\cdot}, \ldots, x_{n\cdot})$

where x_i is the total abundance of all species at site i (it is equal to 100 in the case of percentages). K is the diagonal matrix $(x_{.1}, \; x_{.2}, \; \ldots, \; x_{.m})$ where $x_{.j}$ is the total abundance of species j at all sites. CA searches for a centered weight vector $\mathbf{u}$ that maximizes V:

$$V = \mathbf{z'Rz}/\mathbf{u'Ku} \quad \text{where} \quad \mathbf{z} = \mathbf{R}^{-1}\mathbf{Xu} \tag{3}$$

The following components are defined in the same way with the additional condition that they are orthogonal. The predictand $\mathbf{y}$ is estimated by regression on the components $\mathbf{z}$. Roux (1979) proved the superiority of this method by its application on the data set of Imbrie and Kipp (1971). As for PCA, this method imposes the projection of $\mathbf{y}$ on the components by regression. This two-step method is also known as an indirect gradient analysis (ter Braak, 1986). Detrended correspondence analysis (DCA) corresponds to CA corrected for the arch effect (when a distribution of sites along an arch is observed after plotting of their scores on axes 1 and 2).

2.1.4. Weighted averaging partial least square regression (WA-PLS)

The WA-PLS is related to the CA method as the PLS is related to the PCA (ter Braak et al., 1993). WA-PLS searches for a vector $\mathbf{u}$ with weighted norm equal to 1 that maximizes C:

$$C = \mathbf{z'Ry}/\mathbf{u'Ku} \quad \text{where} \quad \mathbf{z} = \mathbf{R}^{-1}\mathbf{Xu} \tag{4}$$

The following components are found in the same way as in the previous analyses, but with the additional condition that they are R-orthogonal. A particular case of this analysis is the WA method where only one component is retained. The score u_k is then given by the average of the environmental parameters on the n observations weighted by the abundance of the kth species (this corresponds to the first factor or component). The score obtained gives an indication of where the optimum of species k occurs in the environmental domain, and $\mathbf{y}$ is reconstructed by averaging the m values of u_k weighted by the species abundances in the analyzed sample (this corresponds to the regression). This method is efficient for reconstructing chemical parameters from lake diatoms (ter Braak & Juggins, 1993). The generalization of several predictands is provided by canonical correspondence analysis (CCA) (ter Braak, 1986, 1987; Gasse & Tekaia, 1983).

WA-PLS is commonly applied in paleolimnology and paleoclimatology by using lake diatoms (Racca, Gregory-Eaves, Pienitz, & Prairie, 2004; Korkhola & Weckström, 2000) and pollen (Seppä, Birks, Odland, Poska, & Veski, 2004). It has been applied in paleoceanography for the reconstruction of SSTs north of Iceland during the late Holocene based on diatoms (Jiang, Seidenkrantz, Knudsen, & Eiríksson, 2002). WA-PLS has also been used for SST reconstruction at LGM in the Pacific and Atlantic Oceans based on planktonic foraminifera by Morey, Mix, and Pisias (2005). In their paper, the authors provide a very careful description of the CCA approach showing its relationship with CA.

2.2. Generalized Models and Non-Parametric Methods

2.2.1. General linear model

The general linear model (GLM) can be seen as an extension of the linear multiple regression:

$$\mathbf{z} = \mathbf{Xb} = \sum_{j=1}^{m} b_j x_j$$

$$\mathbf{y} = \mathbf{z} + \mathbf{e} \qquad (5)$$

where $\mathbf{y}$ is the climatic variable, $\mathbf{z}$ its estimation, $\mathbf{X}$ the species matrix and $\mathbf{b}$ the vector of regression coefficients. A major limitation of the simple linear model is met when the $\mathbf{X}$ variables are not linearly independent. In such cases, the use of principal components of $\mathbf{X}$ solves the problem. An advantage of GLM is the possibility of using categorical variables and giving access to multi-variate tests for assessing the significance of the model. The basic idea of GLM is that the covariance matrix of the residuals is not required to be a diagonal matrix with all the terms equal to σ_e^2 with a Gaussian distribution. We may impose any sort of distribution, in particular a binomial distribution, and then define a logistic model (ter Braak & Looman, 1986). We may also weight the observations in the fitting process in order to emphasize some classes of observations.

The species assemblages have m variables, but the sum of the percentages is 100. Therefore, when $m-1$ species are known, the mth is fixed and one regressor is redundant, and we can remove it. However, because the aim is not only to reconstruct the climatic variable $\mathbf{y}$ but also to analyze how the species respond to climate, we do not want to remove any regressor arbitrarily. GLM allows the possibility of retaining all of the regressors in the analysis.

When categorical variables are added to the model, there are two ways of coding them. The first way is to code each category of the predictor by one numerical value (1, 2,...), thus adding only one predictor. The second way is to define a new binary predictor (taking the value of 1 or 0 depending on whether the category is realized or not) for each category, thus adding a large number of predictors for a single additional variable. GLM can be used with both coding methods. In transfer function models, a categorical variable could, for example, be the membership to a given biome (Guiot et al., 1993).

2.2.2. Generalized additive model

The generalized additive model (GAM) (Hastie & Tibshirani, 1990) extends traditional regression as effectively as GLM by replacing $\sum_{j=1}^{m} b_j x_j$ in Equation (5) by $\sum_{j=1}^{m} b_j f_j(x_j)$. Since the functions f_j can be non-linear, GAM belongs to a non-linear family of methods. It is also a non-parametric method in the sense that it is not necessary to know the analytical form of f_j. Usually, the function used is a spline function, which is very flexible and makes this model very efficient.

The non-linear modeling procedures are useful for two reasons. First, they help to prevent model misspecification, which can lead to incorrect conclusions regarding treatment efficacy. Second, they provide information about the relationships between predictors and predictands that is not shown by standard modeling techniques. Nevertheless, the counterpart is a risk of over-parameterization and the difficulty of understanding non-analytical functions.

The locally weighted regression proposed by Cleveland and Devlin (1988) fits a surface to the data in a multi-dimensional space by using the family of the spline functions. It has the advantage of being non-parametric and does not require formal specification of the surface function. Gersonde et al. (2005) used GAM to estimate the sea-ice concentration from diatom assemblages during the LGM. This family of methods is used to estimate the response of species to several environmental variables, as seen in the response surface section.

2.2.3. Artificial neural networks

The artificial neural network (ANN) seems to be an excellent approach for a large class of non-linear problems (Walczak & Wegscheider, 1994). ANN implies a black-box approach, which allows very close data fitting, although it is hard to analyze the causality between input and output variables. An ANN consists of a number of simple and interconnected processors called neurons. A signal coming from input variables, the species assemblages, passes through these neurons to reach the output variables, the environmental variables. The interconnections are defined by coefficients that are iteratively tuned to best fit the output variables. In the most popular case, the architecture of the ANN is composed of one hidden layer with a number of neurons, depending on the complexity of the problem. At each neuron, the incoming signal is transformed by a non-linear function (called activation function) and sent to the next layer. An example of possible activation function is the sigmoid function, which translates the input in the [0,1] domain and attenuates extreme values. This function introduces non-linearity in the model. The process of estimating the adjustable coefficients is called "training". Several iterations are required for this process. One of the most commonly used ANN methods is the feed-forward network trained with the back-propagation learning algorithm.

One major issue with ANN is the difficulty to find the proper structure of the network (number of neurons and hidden layers). There is no other method than to try different designs and to select the most satisfying one. Another issue is the number of iterations needed during the training for correct calibration and reconstruction. It is indeed important to stop the iterations before overestimation. Because of its non-linear property, ANN may fit the training data set very well, including the noise inherent to any data set, which can restrict the ability of making correct predictions outside of the range of the training. To avoid such a problem, it is necessary to divide the training data set into two subsets where one is used for calibration and the other is used for independent verification: when the verification error increases and becomes too large, the iterations have to be stopped. This is called early stopping.

This method has been introduced in continental paleoclimatology by Guiot, Cheddadi, Prentice, and Jolly (1996) and in paleoceanography by Malmgren and Nordlund (1996). Peyron et al. (1998) have shown its potential in reconstructing LGM climate from pollen data by means of reducing the number of predictors by grouping them into plant functional types. Such grouping is motivated by the fact that a large number of predictors leads to a much larger set of parameters and finally induces a risk of over-parameterization. Guiot and Tessier (1997) and Guiot et al. (2005) have shown the advantage of ANN for extracting climatic signals from tree-ring data. Racca, Philibert, Racca, and Prairie (2001) have shown the better predictive power of ANN over WA and WA-PLS with lake diatoms, but stated that a multi-model approach is better for capturing information more accurately from diatom assemblages.

 ## 3. Methods Based on Similarity

Instead of calibrating a relationship between climate and species, this family of methods (Table 1) is based on the comparison of past assemblages to modern assemblages. The mutual climatic range (MCR) methods (and their predecessor, the species indicator method) use maps of the modern distribution of species and consider only the occurrence in terms of presence/absence. The analogue methods directly or indirectly compare fossil assemblages to modern assemblages using a similarity index.

3.1. Mutual Range Method

The indicator species method (ISM) is based on a set of distribution maps showing the modern geographical range of taxa and records of the environmental or climatic variables that determine their range. Then, the distribution of the taxa is re-mapped into a space defined by axes, which are the environmental or climatic variables to be reconstructed. Iversen (1944) and Hintikka (1963) presented early examples of such re-mapping when they used records of the presence or absence of higher plant taxa in the proximity of meteorological stations to draw scatter plots showing the occurrence of taxa with respect to summer and winter temperatures.

Atkinson et al. (1987) generalized ISM to the MCR method dealing with a large number of taxa. Current reconstruction methods generally employ three or four environmental variables concurrently to define a three- or four-dimension environmental space. A regular grid is laid out in this space and the data are used to determine the status of the taxon at each grid point. From the re-mapped data, a climatic envelope that encompasses the environmental space of taxon occurrence is mathematically defined. In order to make reconstructions using MCRs, it is necessary to search for grid points at which the combination of taxa found in the fossil assemblage may co-occur. Although this method is straightforward in principle, problems arise from co-occurrence of taxa that do not have MCRs in the modern data set. In such circumstances, the best solution is to find the least dissimilar combination(s) of taxa. The reconstructed values given by the MCR method are ranges for each of the environmental variables. In general, these ranges are smaller for more diverse assemblages. However, in assemblages of low species diversity, the presence of a taxon characterized by an extremely narrow range of occurrence may reduce the range of possible environmental conditions.

This method has been criticized because it is based only on presence/absence of species, while abundances of taxa are often available. In some situations, when there are assemblages that no longer exist in present times, abundances cannot be used as with other methods. For example, in the Pliocene, fossil pollen assemblages from Mediterranean areas resemble modern ones but include Chinese and North American pollen taxa. In this case, the MCR approach is the most suitable. It has been used by Fauquette, Guiot, and Suc (1998) with some additional information such as absence/presence/strong presence. Klotz, Guiot, and Mosbrugger (2003) have improved the method by implementing a probability approach, which is more subtle than the common MCR based exclusively on binary information. With some proxies, such as plant macrofossil or mollusk remains, the abundance in the sediment is not necessarily a direct function of the actual abundance of the organism in the environment. In these cases, the MCR or its variants are more suitable for climate reconstruction than other methods (Moine, Rousseau, Jolly, & Vianey-Liaud, 2002; Pross, Klotz, & Mosbrugger, 2000). MRC has not been applied to marine data, but its recent development may offer sufficient improvement to motivate applications, notably for periods and regions where modern analogues do not exist.

3.2. Modern Analogues Method

The modern analogue technique (MAT) is based on the collection and analysis of a large number of modern or surface samples representing the same type(s) of sedimentary environment as the fossil samples from which microfossil assemblages have been recorded. These samples provide point data and thus contrast with the distribution maps forming the basis of the mutual range method, which gives the illusion of spatially continuous information. With MAT, environmental data must be interpolated at the sampling sites, as is the case for the calibration methods. Fossil assemblages may contain combinations of taxa that respond to different aspects of the environment. In such cases, it might be suitable to weight the taxa according to their sensitivity against the environmental variables to be reconstructed. The simplest weighting is the exclusion from the assemblages of taxa that do not display a response to any of the environmental variables being reconstructed. Another method of weighting is given by paleobioclimate indicator loadings (Guiot, 1990). The taxa weights are derived by means of an ordination of the fossil data: scores are obtained for a series of ordination axes and these axes are examined in order to determine which one is best related to the taxa showing a strong response to the environmental variables to be reconstructed. The scores for this axis are then used as weights when assessing the degree of analogy. Thus, the most responsive taxa have the greatest influence upon the choice of analogues.

An appropriate distance measure is required in order to evaluate the degree of analogy between fossil and modern samples. In most cases, MAT is applied to microfossil data that are expressed as percentage values. It is possible to use a variety of distance measures with percentage data, especially with data transformations reducing the influence of the ubiquitous taxa that are often more abundant in the assemblages. The most widely used distance measure is known as the "chord distance", which is particularly effective with percentage data (Overpeck, Webb, & Prentice, 1985) and may be computed using the formula below:

$$d_{ij}^2 = \sum_{k=1}^{m} \left(\sqrt{p_{ik}} - \sqrt{p_{jk}} \right)^2 \tag{6}$$

where m is the number of taxa, p_{ik} (resp. p_{jk}) is the proportion of taxon k in fossil assemblage i (resp. modern assemblage j) and d_{ij}^2 the squared chord distance between assemblages j and i.

The smaller the chord distance, the greater is the degree of analogy between the two samples. Thus, paleoenvironmental reconstructions are based on a small number of modern samples that are the most similar to the fossil sample. The dispersion of the analogues around the reconstructed value provides a partial estimate of the uncertainty, which is related to the tolerance of the proxy in a relatively large range of climatic or oceanographic conditions. However, this estimate does not include any uncertainty related to the fitting error between the climatic or oceanographic conditions. A solution has been proposed by Nakagawa, Tarasov, Nishidac, Gotandaa, and Yasudaa (2002), who applied MAT to meteorological station data and calculated a linear relationship between observations and reconstructions with confidence intervals. For a fossil assemblage, this relationship can be applied to the reconstruction from the MAT, providing the required error bar. Another issue in estimating the uncertainty depends upon the number of spatially close analogues available for a given sample. A way to remedy this problem is to do an even re-sampling of the modern assemblages similar to that of the response surface approach (cf. Section 3.3). The determination of the number of analogues to be taken is also critical. Waelbroeck et al. (1998) proposed selecting the set of analogues before a sharp increase of the distance; Gavin, Oswald, Wahl, and Williams (2003) proposed another statistical tool named the "receiver operating characteristic" curve which helps to decide the threshold between analogue and no-analogue spectra.

MAT was used for the first time in paleoceanography by Hutson (1980). It was applied by Prell (1985) on the CLIMAP (1976, 1981) data. de Vernal et al. (1994, 2001, 2005) have shown its ability to reconstruct past SST, salinity and ice cover in subpolar seas from dinocyst assemblages. It is now considered to be a standard method, which is often compared to other ones for assessing the reliability of reconstructions based on other methods (e.g., Kucera et al., 2005b; Barrows & Juggins, 2005; Chen, Huang, Pflaumann, Waelbroeck, & Kucera, 2005; Gersonde et al., 2005).

3.3. Constrained Analogues

A large range of analogues may reduce the possibility of the selection of "wrong analogues". However, the problem of wrong analogues remains acute in the case of fossil assemblages characterized by a limited number of taxa, which are ubiquitous and represent a wide range of environmental conditions. When additional qualitative or quantitative information is available, constrained-analogue methods have shown great promise with pollen data. Two main approaches can be used. One consists of applying a constraint for the selection of analogues in a given biome: all of the assemblages from other biomes are rejected even if the distance with fossil assemblage is very low. The second approach to constrain the choice of analogues uses paleoclimatic information derived from independent proxies. For example, fossil beetle assemblages (Guiot et al., 1993), sedimentological indices (Seret, Guiot, Wansard, Beaulieu, & Reille, 1992) and past lake levels (Cheddadi, Yu, Guiot, Harrison, & Prentice, 1997) have been used together to constrain the choice of analogues in pollen-based paleoclimate reconstructions. The advantage of this procedure is the possibility of reducing the confidence interval of the climatic parameter common to the various proxies and the opportunity to reconstruct additional climatic parameters. For example, constraints on pollen analogues derived from fossil beetle assemblages may be used to improve the precision of July temperature reconstructions, and the resulting constrained pollen analogues also yield more coherent winter temperature estimates. The principal disadvantage of the "multi-proxy constraint" approach is that independent paleoclimate inferences from alternate proxies are no longer available for comparison. Nevertheless, if the climatic control and sensitivity of the various proxies is well understood, the "multi-proxy constraint" approach can yield internally consistent reconstructions of paleoclimate parameters, including seasonality of temperature and precipitation.

In addition to the distance between spectra, Pflaumann et al. (1996) have introduced a constraint related to the geographical distance: the analogues chosen are those located at the smallest geographical distance from the fossil site (SIMMAX method). It is a MAT constrained by geography. Such an approach could be useful for warm periods characterized by fossil assemblages similar to the modern ones, but it is difficult to justify for glacial periods because the best analogues can be located at a very large distance from the study site. For example, Malmgren, Kucera, Nyberg, and Waelbroeck (2001) have shown that the results do not match the ranking obtained with the calibration data set when a geographical constraint is applied to an independent validation data set and to an additional fossil data set. Therefore, this approach is not further considered here.

3.4. Response Surfaces

The first step of the response surface analogue technique (RSAT) is a re-mapping of the taxa into an environmental space using the modern data set (and not biogeographical distribution of modern species as in the MRM approach). The main method of this type currently applied is the quantitative environmental response surfaces (Bartlein et al., 1986; Huntley, 1994). In this method, the surfaces are fitted to the abundance values of each taxon used for reconstructions. The abundance of the taxon at each grid point can be fitted either by a flexible extrapolation technique such as the locally weighted regression (Cleveland & Devlin, 1988) based on tri-cube weighting function (which is related to the GAM approach), or by an interpolation weighted inversely to the distance. The latter is the approach adopted by Waelbroeck et al. (1998) for SSTs estimates based on planktonic foraminifers.

Having fitted response surfaces for the suite of taxa used in reconstructions, the values for the taxa at each grid point can be combined to provide a series of assemblages that may be considered as potential analogues. Thereafter, the procedure is very similar to that of MAT, with 5–10 closest matching assemblages selected for each fossil assemblage and used to provide a reconstruction and a confidence interval. The degree of analogy is once again conventionally measured using the chord distance.

In the RSAT approach, the variation of taxa with distributions independent of climate is smoothed and thus removed by the procedure used to fit the surface. Therefore, if the principal cause of variability of a taxon is non-climatic, the response surface is flat and the taxon contributes little to the choice of analogues. As the new potential assemblages are sampled on a grid, each type of climate is evenly represented, which improves the value of the confidence intervals. These two features represent the greatest advantages of this method.

In paleoceanography, response surfaces have been used to generate new gridded assemblages and combined with original assemblages to fill the parts of the climatic space not covered by data. The method proposed by Waelbroeck et al. (1998), which is called the revised analogue method (RAM), uses gridded assemblages in addition to raw assemblages in order to provide a better shaped data set. The gridded samples are obtained by interpolation in the climatic space, which is more flexible than a surface (Figure 1B). Another modification is made by the introduction of a threshold for the selection of the analogues at the place where a sharp increase of the distance is observed (see Section 3.1). The use of RAM allows for improvements in the reconstruction of SSTs at cold and warm ends of the temperature range. It has been used together with MAT, SIMMAX and ANN in the Multiproxy Approach for the Reconstruction of the Glacial Ocean (MARGO) initiative to reconstruct LGM SST (Barrows & Juggins, 2005; Chen et al., 2005; Hayes, Kucera, Kallel, Sbaffi, & Rohling, 2005; Kucera et al., 2005b). A more formal approach, using a Bayesian probabilistic approach, is proposed by Haslett et al. (2006).

4. COMPARISON OF METHODS WITH A WORKED EXAMPLE

In this section, most of the methods reviewed above (*cf.* Table 1) are compared from their application to the same data set. For this purpose, we used the MARGO planktonic foraminifer core-top data set from the Atlantic Ocean (e.g., Kucera et al., 2004). It is composed of 862 spectra of 26 taxa (percentages) and the corresponding SSTs in winter (January-February-March) and in summer (July-August-September) is compiled from the 1998 version of the World Ocean Atlas (NODC, 1998). From this data set, we have removed the 26 spectra with winter SSTs $<-1°C$ representing very cold conditions in the North Atlantic, and the 23 spectra with summer SSTs $>29°C$ representing very warm conditions. These two subsets of data represent extreme conditions at the limits of the main data set and will be used to evaluate how each method performs in almost no-analogue conditions. The main data set contains 813 samples that are randomly divided in one calibration data set of

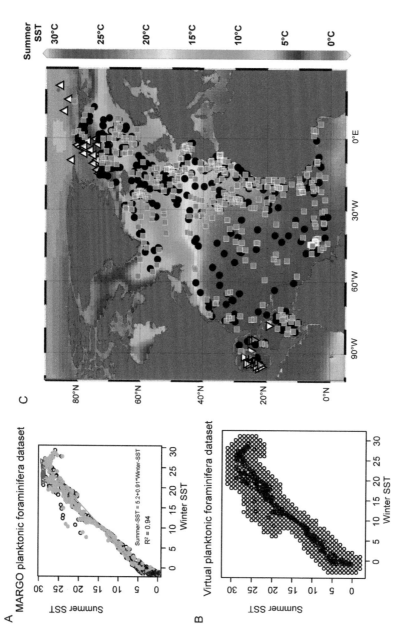

Figure 1 Hydrographical and geographical distribution of samples used in the worked example. (A) Summer vs. winter SSTs at selected sites: the calibration and verification data sets are shown by circles, and the cold and warm sub-sets of data are illustrated by blue and red triangles, respectively. (B) Summer vs. winter SSTs for the virtual planktonic foraminifer data set, established from the interpolation of the climatic space by defining a limit of 3 °C for the distance between grid points and raw data. (C) Map of the North Atlantic Ocean showing the summer SST and the location of surface sediment samples: grey squares and black dots are the calibration and verification samples, respectively; the open triangles correspond to the cold samples (Greenland Sea and Arctic Ocean) and the warm samples (Caribbean Sea).

407 samples and a verification data set of 406 samples. Figure 1A, C show the distribution of the samples in the hydrographic and geographic space, respectively. The distribution along the main diagonal is typical of open ocean where there is a high correlation ($r = 0.97$) between winter and summer SSTs.

To assess the prediction error of the various techniques, the root-mean-square error (RMSE) is computed as the square root of the sum of the squared differences between the observed and the predicted values divided by the number of observations. When it is applied to the verification set, the RMSE is called the root-mean-square error of prediction (RMSEP).

All the methods have been implemented in the *R environment for statistical computing and graphics* (http://www.r-project.org/). A script called "transfer.r" has been used to implement the eight methods on the worked example with text and graphics outputs. It can be found at the URL: http://servpal.cerege.fr/paleoceano_transfer/

Using the I&K method, we reduced the calibration matrix of species relative abundances to eight principal components. Imbrie and Kipp (1971) used the varimax technique to rotate the selected components in order to facilitate the interpretation of the variables. We did not apply this technique because it has no effect on the statistical performance of the reconstruction but only on the interpretation of the biogeographical patterns. The scores of the principal components were used as predictors of the winter and summer SSTs. A curvilinear multiple regression was performed between the two predictands and the eight principal components and their interactions (PC1 × PC2, PC1 × PC3 …, PC2 × PC3 …). Between 95% and 96% of the SST variance is explained by these predictors (Table 2). In parallel, verification samples were projected on the eight components and their scores were introduced into the calibrated regression. The RMSE was ~1.9°C, and the RMSEP on the verification data set was ~2.0°C and 2.2°C for summer and winter, respectively. The RMSEP on the "cold" samples is clearly larger (2.5°C and 4°C); the method cannot yield SST estimates lower than +1°C in winter and +4°C in summer (Figure 2). The performance for the "warm" samples was better, especially in winter (RMSEP = 1.5°C), but more dispersed values than found in reality were obtained for summer SSTs (RMSEP = 2.16°C; Figure 2b).

For the PLS method, we used the kernel algorithm with a leave-one-out cross-validation procedure to select the number of components to be included in the regression. The number of components selected is the one providing the minimum RMSEP on the cross-validation samples (it must be noted that this cross-validation is restricted to the 407 samples calibration data set, the verification data set being exclusively kept for an independent test). For winter SST, we used 13 components explaining 98% of the predictand variance, and for summer SST, we also used 13 components explaining 97% of the variance. The RMSE was lower than for the I&K method (Table 2). The RMSEP calculated on the 406 verification samples were also much lower. With respect to the samples from the cold and warm extremes, the performances were better than those of the I&K method; the winter SST estimates for the cold samples were better, and had values of ~0°C (Figure 2a), and the summer SST estimates for the warm samples were less dispersed, although slightly underestimated in comparison with the I&K estimates (Figure 2b).

Table 2 Performance of the Various Transfer Function Methods of Paleoclimate Reconstructions Applied to Planktonic Foraminifer Data.

Method	R^2 calib.	RMSE	RMSEP-406	RMSEP-cold	RMSEP-warm
Winter SST					
IK	0.96	1.88	2.05	2.48	1.49
PLS	0.98	1.38	1.48	2.05	1.54
WA-PLS	0.98	1.34	1.50	2.05	**0.99**
GLM	0.97	1.46	1.61	**1.07**	**1.24**
GAM	0.98	1.34	1.35	1.70	2.12
ANN	0.98	1.13	**1.29**	1.74	**1.08**
MAT	0.99	1.07	**1.06**	**1.43**	**1.26**
RAM	0.99	1.05	**1.16**	**1.24**	**0.82**
Std. dev.		8.95	8.43	0.32	1.14
Summer SST					
IK	0.95	1.95	2.22	3.98	2.16
PLS	0.97	1.49	1.57	3.48	1.80
WA-PLS	0.97	1.50	1.57	3.44	**1.07**
GLM	0.96	1.68	1.73	**2.05**	1.64
GAM	0.97	1.40	1.50	**3.00**	**1.16**
ANN	0.98	1.19	**1.28**	**3.00**	1.84
MAT	0.98	1.17	**1.14**	2.88	**1.01**
RAM	0.98	1.22	**1.22**	2.86	**0.81**
Std. dev.		8.37	7.51	1.75	0.07

Notes: The coefficient of determination (R^2) and the root square of the mean squared errors (RMSE) are calculated on the calibration data set (407 samples), the root square of the mean squared errors of the prediction (RMSEP) are calculated on the validation data set (406 samples), on the cold data set (26 samples with winter SST $< -1°$C) and on the warm data set (23 spectra with summer SST $> 29°$C). The best performing values are enhanced in bold.

For the WA-PLS method, we followed the method proposed by ter Braak et al. (1993) who demonstrated that the solution of WA-PLS can be obtained from a PLS algorithm by pre-processing the taxa and the environment matrices, and by applying the PLS algorithm to the transformed matrices. The leave-one-out cross-validation test led to the retention of 15 and 7 components for winter and summer, respectively. The percentage of variance reconstructed and the RMSE are quite similar to those obtained with the PLS approach, but the RMSEP on the warm samples gives much better results ($\sim 1°$C; Table 2), with estimates more closely concentrated along the diagonal than with the PLS method (Figure 2a, b).

For the GLM method, the predictors are the 26 taxa and the error distribution was assumed to follow a Gaussian law. Because the objective was to focus on the low temperatures, a weighting function equal to the square of the inverse of the SST (corrected to avoid zero values) was defined. For example, SSTs $< 0°$C are weighted > 1 and SSTs $> 20°$C are weighted < 0.0025. This arbitrary weighting procedure was applied to illustrate how a GLM performs by focusing on a given environment. The results were similar to those obtained with the WA-PLS, except that the cold samples yielded a RMSEP half as large, and allowed winter SST estimates of $-1°C$ for observations ranging from the freezing point ($-1.8°C$) to $-1°C$ (Figure 2). The improvement for summer SSTs was also significant, with estimates of $\sim2°C$ for observations ranging between $-1.8°C$ and $4.4°C$. For the warm samples, the performance was intermediate between those of the PLS and WA-PLS methods.

For the GAM method, the predictors are the first six factors CA extracted from the taxa, and Gaussian law is assumed for the residuals. The six predictors were smoothed by a thin plate regression spline. The multi-dimensional surface was fitted to SST in the six-dimension space. The performance on the calibration samples is similar to that of the previously described methods. The RMSEP of winter SST is larger than that of GLM for both cold and warm samples, but the performance for estimating summer SSTs of warm samples was excellent (Figure 2).

For the ANN method, the size of the species data set was also reduced by using the first ten correspondence factors. We used the architecture of a back-propagation network as described by Malmgren et al. (2001) and Peyron et al. (1998). The number of neurons was set at four. Thus, the size of the input layer was 10, that of the hidden layer size was equal to four and the size of the output layer corresponded to the number of predictands, i.e., two (winter and summer SSTs). During the iterative estimation of the ANN, we regularly calculated the calibration error (on the 407 samples of the calibration data set) and the verification error (on the 406 samples of the verification data set). As expected, the calibration error tended to continuously decrease but the verification error, after an initial decrease, tended to slightly increase. An increase that is too high indicates an over-parameterization. After $\sim1,000$ iterations, the convergence was considered satisfactory because the verification error remained similar to the calibration error. The RMSE was narrower than with the other methods and the RMSEP belongs to the group of methods that performed the best (Table 2). However, the performance remained slightly inferior to that of the GLM for the cold samples (Figures 2 and 3), which is to be expected as the GLM was designed to better reconstruct cold climates.

For the MAT method, we used the chord distance on the 26 taxa. A maximum of ten analogues was selected. A threshold for exclusion of the analogues was calculated based on a Monte-Carlo method. Distance vectors between each modern spectra and another modern spectrum randomly selected were calculated 1,000 times for the entire matrix, and the threshold was the first quartile for these 1,000 values. In the case of the calibration data set, the threshold was calculated to 1.9. The reference data set used as a reserve of analogues was the calibration data set of 407 assemblages. The reconstructed values for the spectra in the four subsets of data were obtained from the weighted average of the SSTs of selected analogues according to the inverse of their distance. This method provided the best RMSE (Table 2), due to frequent geographical proximity of numerous samples. The RMSEP on the 406 samples remained narrow for the same reasons. More interestingly, we also obtained the best results for the cold and warm samples at the extremities of the modern data set (Figure 2), which cannot be explained by geographical proximity. This indicates that the method remains robust in a situation where the analogues are limited.

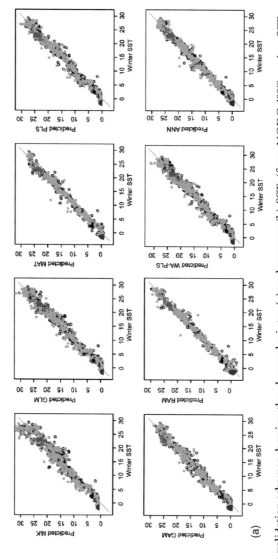

Figure 2 Cross-validation schemes showing the observed winter (a) and summer (b) SSTs (from NODC, 1998) against SSTs estimated by various methods (see calculated statistics in Table 2). The red and blue triangles correspond to the warm and cold samples, respectively; the open circles and grey circles correspond to the calibration and verification data sets, respectively.

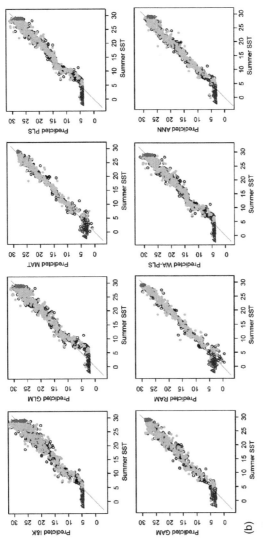

Figure 2 Continued

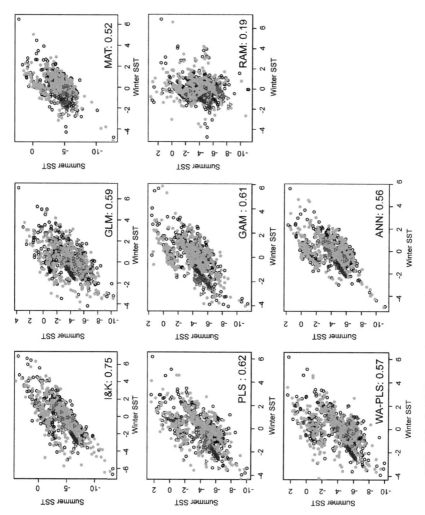

Figure 3 Winter *vs.* summer SST residuals (differences between observed and estimated values). The red and blue triangles correspond to the warm and cold samples, respectively; the open circles and grey circles correspond to the calibration and verification data sets, respectively.

For the RAM method, the modern data set was artificially expanded by remapping the original data onto a regular grid of SSTs (the climate of the new samples are represented by circles in Figure 1B). The 26 taxa were linked to the SST variables by a simple interpolation technique where the estimates of the taxa percentages at a grid point were weighted by the inverse of the distance between the grid point and the modern spectra in the climatic space. A maximum distance of $3°C$ was allowed. The ranges of SST variables were divided into 35 nodes spaced by $1°C$, and the percentages of the 26 taxa were estimated on this grid, yielding 405 virtual samples (after removing samples falling outside the filled climatic space). MAT was then applied on the new set of spectra (real and virtual, i.e., 812 samples). We did not implement the "jump" technique (Waelbroeck et al., 1998), which enables a better control on the selection of analogues. Instead, we used a threshold, as in the MAT calculated with the Monte-Carlo method (in our case the threshold was 1.2). A better performance was obtained with RAM than with MAT, except for summer SSTs of the warm samples, which yield comparable results (Table 1; Figure 2a, b).

5. DISCUSSION AND FUTURE DEVELOPMENTS

5.1. What are the Best Methods?

We have defined a general verification data set (406 samples) covering the same climatic space as the calibration data set. This was done to verify the risk of over-parameterization, which occurs when too many parameters are used, giving the impression that a large proportion of predictand variance is explained, while there is no real improvement of the prediction. In this respect, all of the methods provided reliable results, since the RMSEP on the general verification data set were close to the RMSE on the calibration data set. Because a clear spatial correlation exists between the assemblages, we cannot ascertain whether the RMSEP perfectly reflects the prediction capacity of each model. This is why we used cold and warm data sets. The cold data set represents the situation frequently met in the North Atlantic during glacial periods, with assemblages characterized by a low species diversity and the almost exclusive occurrence (>95%) of *Neogloboquadrina pachyderma* left-coiling (*cf.* Kucera et al., 2005b). Conversely, the warm data set contains assemblages characterized by a high diversity of species. Both data sets are at the extremes of the reference data set: the cold data set represents winter SSTs lower than $-1°C$, whereas the calibration data set contains samples with winter SSTs higher than $-1°C$. The warm data set contains samples with summer SSTs higher than $29°C$, whereas the calibration data set contains samples with summer SSTs lower than $29°C$. This definition allows us to limit the spatial correlation and to verify the behavior of the models at the limit of the extrapolation, i.e., when there are no good analogues.

Examining Table 2, which shows the respective performances of the methods, allows us to distinguish four categories:

(1) The first category contains the method which performed the least well (I&K) on all the data sets, including the modern and fossil data (as shown by Chen

et al., 2005). This was previously recognized by Pflaumann et al. (1996), Waelbroeck et al. (1998) and Malmgren et al. (2001).

(2) The second category contains PLS, WA-PLS and GAM. They have a good performance, except for the cold and warm data sets. WA-PLS is nevertheless better than the two other methods for the warm data set.

(3) The third category of methods is represented by GLM, which makes it possible to weight the observations and to emphasize the importance of some samples for the optimization of the method in a given environmental range. This is what we have shown by focusing on the cold samples, with the result of more accurate reconstructions at the cold end of the data set. However, the corollary is a lesser performance for the opposite extreme, i.e., at the warm end of the data set.

(4) The last category includes the analogue methods (MAT, RAM) and ANN, which provide acceptable results for both extremes of the data sets, even if the performance is lower than that of GLM for the cold samples. From the analyses of lake diatoms, Racca et al. (2001) have already shown the superiority of ANN to WA and WA-PLS, and its robustness for the selection of species. In this category of methods, RAM outperforms MAT on the extreme data sets. The performance obtained by Waelbroeck et al. (1998) is apparently even better than ours for the cold end of the data set. This can be explained by the fact that, here, the coldest samples have been excluded from the reference data set, thus limiting the possibility of creating adequate virtual spectra in the domain colder than $-1\,°C$, whereas Waelbroeck et al. (1998) used all of the samples of the reference data set. For the warm samples, which are characterized by higher species diversity, this limitation had a lower impact. It can be argued that RAM creates virtual samples for combinations of ecological parameters that do not exist today, making the bold assumption that the response of the species within this space is smooth and continuous. If done with caution, this method offers somewhat of an advantage, because the gaps in the climatic space are often due to a lack of data rather than to ecological impossibilities. Nevertheless, these gaps must not be too large and numerous to enable a reliable interpolation.

The comparison of MAT and RAM has been made in a few paleoceanographical studies based on planktonic foraminifers. For example, Barrows and Juggins (2005) applied both approaches for the reconstruction of LGM SSTs along the Australian margin and the Indian Ocean; Chen et al. (2005) also tested and used these methods to estimate LGM SSTs in the western Pacific. Both studies concluded that MAT and RAM yielded results of similar quality. Hayes et al. (2005) used RAM and ANN for the reconstruction of the LGM SSTs in the Mediterranean Sea, also based on planktonic foraminifera. They concluded that the two techniques perform similarly in the LGM but that RAM would benefit from larger calibration data sets. Peyron and de Vernal (2001) have applied MAT and ANN on dinocyst assemblages from the Arctic Seas and the northern North Atlantic. They did not find any significant differences in the performance of both methods from the calibration and validation exercises. However, they noticed that the application to late Quaternary series might yield different results in the case of distant analogue situations.

As pointed out by Kucera et al. (2005b), it is relevant to develop a multi-method approach, as it is not clear that any one method is indisputably better than the others. Kucera et al. (2005b) proposed a search for consistent patterns from the reconstructions based on different methods. A multi-method strategy was also suggested by Racca et al. (2001) who concluded that a better performance was obtained when estimates from WA-PLS and ANN are averaged, probably because the different methods capture and extract complementary information from assemblages. The comparison based on the example presented here shows that it is not necessary to try all of the methods, as some clearly underperform and others yield quite similar results. The best candidates for a multi-method approach are MAT, RAM, ANN and WA-PLS. GLM could be seen as an alternative to RAM.

5.2. What is the Effect of the Correlations between Climatic Variables?

The strong correlation between winter and summer SSTs ($r = 0.97$) in the reference data set makes it difficult to reconstruct seasonality patterns different from those occurring at present in the open North Atlantic Ocean, i.e., with a difference between summer and winter that is far from the mean present value of $3.9 \pm 0.15°C$. In particular, the methods based on calibration tend to reproduce the structure that dominates the calibration data set. In Figure 3, we have plotted the winter vs. summer SST residuals obtained with the eight methods. The calibration methods (I&K, GLM, CAM, PLS, WA-PLS, ANN) result in significant correlation between the residuals of winter and summer SSTs. This indicates that a non-negligible part of the correlation between the seasonal temperatures is not related to the species assemblages, with the consequence that the species distribution is more diversified than the temperature pattern. There is then a risk that the equations will not be able to reconstruct a different temperature pattern. The MAT also results in some correlation, but not the RAM. By creating virtual samples, RAM has extended the climatic space and defined a limit for the distance between grid points and raw data of $3°C$ (Figure 1B), which gives more flexibility in the relationship between summer and winter. Because the virtual assemblages were created by using response functions of the species, there is probably no artifact. By re-sampling the response domain of the species, RAM creates coherent assemblages and provides better results.

5.3. How Many Taxa can be Used?

Racca et al. (2004) have discussed the selection of relevant taxa for diatom-based transfer functions in paleolimnology, which is a particularly delicate matter due to the extremely large diversity of diatom species and the large number of potentially determining environmental parameters in lakes. In order to limit the number of taxa (>250 species retained for statistical analyses; 109 surface sediment samples), Racca et al. (2004) used a pruning algorithm of ANNs. This is equivalent to a stepwise selection of predictors in a regression model. The removal of species that corrupt the relationship between assemblages and environmental parameters make the reconstruction of the various parameters more independent. An approach like this may help to avoid a correlation structure such as the one found in Figure 3. However, the variables thus obtained independently could provide artificial combinations that do not exist in the present environment, which makes it difficult to provide a proper evaluation.

Generally speaking, some methods are more sensitive to large numbers of predictors. This is related to the parsimony principle, which stipulates that among several equivalent models, it is always better to choose the one having the lowest number of parameters. ANN is certainly the worst method in this respect. The first rule to follow is to remove the species that are proven not to be related to the variables to be reconstructed. The second rule is to preferably use orthogonal factors (principal components or correspondence factors) to restrict the number of parameters (this is preferable to the stepwise selection of predictors).

5.4. How Independent is a Cross-Validation?

As explained above, RMSE is not a proper measure of the quality of a method. It is simply a measure of the ability of the method to fit the training data set. Independent data must be used to independently calculate the RMSEP. However, the verification data set must be carefully defined. If the verification data set contains samples that are located geographically too close to the samples of the training data set, full independency of the data might not be achieved, which results in overestimation of the quality of the reconstruction. For example, ANN is often optimized on a test data set to avoid over-parametrization; consequently, this data set loses its independence. Telford, Andersson, Birks, and Juggins (2004) have shown that SIMMAX and RAM fail to ensure statistical independency of samples during cross validation. They have also shown that the RMSEP of the best ANN or MAT models, as defined on the basis of cross validation, is lower than the RMSEP calculated from a fully independent test set. Moreover, ANN does not outperform MAT when independent test sets are used. Based on the analyses of the common example presented above, we offer the same conclusions.

Telford and Birks (2005), using planktonic foraminifer data, examined the consequence of spatial autocorrelation on the performance evaluation of WA, WA-PLS, MAT, and ANN. Spatial autocorrelation, which is the tendency of sites that are close to each other to bear more similarities together than with randomly selected sites, is an intrinsic property of ecological data; in the case of performance testing, spatial autocorrelation presents a problem because autocorrelated data violate the assumption of independence that is a prerequisite for most standard statistical procedures (Legendre, 1993). Telford and Birks (2005) showed that the coefficient of determination (r^2) between observed and estimated values from a transfer function model based on an autocorrelated environment can be high even in the absence of relationships between the species and the environmental parameter reconstructed. They showed that MAT and ANN outperform, in calibration, transfer function methods based on unimodal species-environment response models (calibrated on North Atlantic data), but that all methods yield similar results when validated on spatially independent data (South Atlantic). They concluded that MAT and ANN could be misleading because of their incapacity to maintain a spatial autocorrelation structure. In our common example, truly independent data sets (the "cold" and the "warm" data sets) were used to take into account potential effects of spatial auto-correlation problems, but our results show that the conclusions based on these extreme data sets are not fundamentally different from the conclusions based on the validation data set obtained from random selection (the 406 samples data set in Table 2). Therefore, the conclusions of Telford and Birks (2005) are not unequivocal.

5.5. Bayesian Methods

The transfer function model uses a relationship which is not causal, as it expresses temperature as function of species abundances and vice versa. Recent development of Monte-Carlo-Markov-Chains (MCMC) methods has made accessible the use of causal models and their inversion in the context of Bayesian theory, which allows calculation of *a posteriori* probabilities for model parameters given *a priori* infor-mation and data. An example of application is given by Guiot et al. (2000) who used a mechanistic model of vegetation expressed as a function of climate. By inverting the model under the constraint that model outputs fit pollen data, they have reconstructed climate under various environmental conditions, such as low atmospheric CO_2 concentrations. The advantage of this approach is that obser-vations under all conditions are not necessary, thus making it possible to reconstruct situations without analogue under the modern climate. Moreover, as the same vegetation model also simulates other useful parameters as $\delta^{13}C$ in soils, the same approach can be used to constrain the type of vegetation from $\delta^{13}C$ in loess profiles (Hatté & Guiot, 2005). This way of proceeding is possible on continents because global process models are available. It is not yet the case in the oceans but offers very promising possibilities.

In another example of Bayesian approach, Toivonen, Mannila, Korhola, and Olander (2001) have developed a multi-nomial calibration model, Bummer, which was applied to lake diatoms. In this case, there is no process model but rather a statistical model that is calibrated in a different way. They compared the Baysian approach to a more standard approach called "frequentist approach". The major difference lies in the fact that frequentist methods assume that parameters being estimated have fixed values and that data are random observations from some population, whereas the Bayesian approach assumes that parameters are random variables and that the data are fixed. In the same way, Haslett et al. (2006) propose the use of Bayesian theory and its MCMC implementation with a probabilistic model, where the discrepancies between data and estimates are modeled in a more sophisticated way. *A priori* probabilities allow some degree of smoothing to be imposed on the modern data in the spatial domain (it is achieved by the response surfaces) and on the fossil data in the time domain (it is achieved by imposing a relationship between time-neighbor fossil samples). The *a posteriori* probabilities provide the distribution of the climate, taking into account the fossil pollen data as well, in addition to the smoothness constraints. This methodology offers the possi-bility of working on the basis of several proxies: one proxy could be used for defining *a priori* probabilities and another one to estimate the *a posteriori* probabilities. For example, in the context of paleoceanography, we may use the temperature reconstructed from planktonic foraminifera (with their confidence intervals) to define *a priori* probabilities, and the model calibrated on dinocysts or diatoms will be used to refine the reconstructions based on foraminifers. The final reconstruction is then compatible with both proxies. Such an approach has a great potential in paleoceanography for the development of integrated multi-proxy transfer functions.

6. THE APPLICATIONS OF TRANSFER FUNCTIONS *Sensu Lato* IN PALEOCEANOGRAPHY

6.1. The Most Common Methods and Reconstructed Parameters

In paleoceanography, the methods that have been the most commonly used are the transfer functions adapted from the Imbrie and Kipp (1971) method. They have been applied to planktonic foraminifer assemblages (e.g., CLIMAP, 1976, 1981; Ruddiman & Mix, 1993), radiolarians (e.g., Morley, 1979; Pisias et al., 1997; Abelmann, Brathauer, Gersonde, Sieger, & Zielinski, 1999), coccoliths (e.g., Molfino, Kipp, & Morley, 1982), and diatom assemblages from the Southern Ocean (e.g., Pichon et al., 1992; Zielinski, Gersonde, Sieger, & Fütterer, 1998; Gersonde et al., 2005) and the North Atlantic and Nordic seas (e.g., Koç & Schrader, 1990; Koç et al., 1993; Andersen et al., 2004). One advantage of I&K method was the possibility of developing transfer functions despite a limited number of modern spectra. This approach allowed important progress in paleoceanography during the 1970s and 1980s, when sedimentary cores were not very abundant. With the multiplication of coring expeditions, it has been possible to develop exhaustive databases of microfossil assemblages in surface sediment samples, and thus to adapt methods requiring large networks of reference samples, such as MAT and ANN. Since the early 1990s, MAT and methods derived from MAT (RAM and SIMMAX) are probably the most commonly used approaches; they were applied to foraminifer assemblages (e.g., Pflaumann et al., 2003; Chen et al., 2005; Kucera et al., 2005b), diatom assemblages (Crosta, Pichon, & Burckle, 1998; Crosta, Sturm, Armand, & Pichon, 2004; Gersonde et al., 2005), and dinocyst data (de Vernal et al., 1994, 2001, 2005; Marret, de Vernal, Benderra, & Harland, 2001). ANN is also becoming a method that is currently applied, but often as a complement of other methods (e.g., Kucera et al., 2005b; Chen et al., 2005).

The most common application of transfer functions using microfossil assemblages has been the reconstruction of the ocean surface temperatures of the past. Beyond general paleoclimate and paleoceanographic objectives, one of the objectives for SST reconstructions was to provide modelers with maps of the ocean temperature, which are as precise as possible, during past key intervals used in paleoclimate model experiments, notably within the context of COHMAP and PMIP (see COHMAP Members, 1988; Joussaume & Taylor, 1995, 1999). Since the pioneer work of CLIMAP (1976, 1981), the establishment of exhaustive "modern" databases from analyses of surface sediment samples, together with the development of various methods, based on calibrations or similarities, has allowed important improvements in the ability to reconstruct past temperatures and to assess the degree of reliability of reconstructions (e.g., from the RMSEP; Section 4). A comprehensive series of publications on the use of various methods, based on several proxies (including foraminifers, siliceous microfossils and dinocysts), has been produced within the context of the MARGO project (see Quaternary Science Reviews, Vol. 24, issues 7–9 published in 2005; http://www.pangaea.de/Projects/MARGO/). In this special MARGO issue, regional maps of the LGM Ocean have been published and constitute updates from the LGM maps produced by CLIMAP in 1976 and 1981.

As mentioned above, the majority of paleoceanographic applications of transfer functions *sensu lato* aim at reconstructing SSTs, which is unquestionably the most determinant parameter of the distribution of planktonic organisms in the ocean. In general, the thermal inertia of the ocean waters results in a relative uniformity of the temperature throughout the annual cycle, which is reflected by the strong correlation between winter SST and summer SST. However, the temperature in surface waters can be a complex parameter at regional scales, with different seasonal signals or variants in the structure of the upper water masses (thermocline depth), depending upon stratification and vertical mixing. Thus, the temperature of the upper water mass may show a heterogeneous distribution in time and space. Some studies have provided analyses of the distribution of planktonic foraminifer assemblages and their relationships with the vertical structure of the upper water masses, leading to propositions of transfer functions for the reconstruction of the depth of the thermocline or mixed layer (e.g., Andreasen & Ravelo, 1997; Chen & Prell, 1997; Watkins & Mix, 1998; Hüls & Zahn, 2000). Some studies based on the distribution of the mesopelagic radiolarians in the North Atlantic have led to the reconstruction of the thermocline structure from ANN (*cf.* Cortese, Dolven, Bjørklund, & Malmgren, 2005). Other researchers have examined the relationship between dinocyst assemblages and the SSTs of the warmest and coldest months of the year, which allowed them to propose reconstructions of various seasonal signals (*cf.* de Vernal & Hillaire-Marcel, 2000; de Vernal et al., 2001). These studies also suggested that the seasonal signal is intimately related to the stratification that is a function of salinity, and therefore led to the reconstruction of salinity and surface water density (e.g., Hillaire-Marcel, de Vernal, Bilodeau, & Weaver, 2001).

In addition to SSTs and related parameters, such as the depth of the thermocline or winter vs. summer temperatures, a few studies have examined the relationship between the density or seasonal duration of sea-ice cover and microfossil assemblages. They led to the application of MAT for the reconstruction of sea-ice cover at high latitudes of the Northern Hemisphere from dinocyst assemblages (e.g., de Vernal & Hillaire-Marcel, 2000; de Vernal et al., 1994, 2001, 2005), and in the Southern Ocean from diatom assemblages (Crosta et al., 1998, 2004; Gersonde et al., 2005).

Assuming a relationship between the trophic requirement of the species and primary productivity, diverse methods have been developed and tentatively applied to reconstruct biological productivity from planktonic foraminifer assemblages (e.g.,

Watkins & Mix, 1998; Cayre, Beaufort, & Vincent, 1999; Ivanova et al., 2003), dinocyst data (Radi & de Vernal, 2004) and diatoms (e.g., Schrader & Sorknes, 1991).

Attempts were also made to reconstruct diverse properties of sea water. In particular, Anderson and Archer (2002) proposed the reconstruction of carbonate saturation using modified MAT and planktonic foraminifer assemblages, which include some species more sensitive than others to calcium carbonate dissolution. Methods based on similarity have been applied to benthic ostracod data to tentatively reconstruct the bottom water mass properties, such as the pale temperature of shelf areas (e.g., Wood, Whatley, Cronin, & Holtz, 1993) or the origin of deep waters in the Arctic Ocean (e.g., Cronin, Holtz, & Whatley, 1994; Jones, Whatley, Cronin, Harry, & Dowsett, 1999). Many studies also explored the potential of benthic foraminifers to quantitatively reconstruct parameters such as productivity and carbon fluxes, bottom water quality and oxygenation, water depth, and deep current velocity, temperature and salinity. Because of the large number of species, complex ecological requirements and taphonomical processes, in addition to the typical mosaic pattern of the benthic foraminifer distribution on the seafloor, the development of transfer functions has been challenging (see Jorissen et al., this volume). To date, mostly indicator-species approaches have been used, but one can mention attempts at reconstructing paleo-water depth from MAT (Hayward, 2004), primary production and bottom water oxygenation using multivariate and regression methods (e.g., Loubere, 1994, 1996; Morigi, Jorissen, Gervais, Guichard, & Borsetti, 2001; Wollenburg, Kuhnt, & Mackensen, 2001).

6.2. Multi-Method and Multi-Proxy Approaches

Various methods and several proxies are now available for the reconstruction of paleoceanographical parameters, notably the SST. These methods include many techniques based on calibration or similarity, and various types of microfossils, but also molecular biomarkers (see Rosell-Melé & Clymont, this volume) and geochemical content of biogenic remains (see Rosenthal, this volume). Inter-comparisons of SSTs estimated from various methods are currently being made in order to improve the confidence level of paleoceanographical reconstructions.

An example of a down-core application of a multi-method approach is illustrated in Figure 4. It shows the results of MAT and ANN applied to dinocyst assemblages in a sequence of $\sim$25,000 yr collected in the northwest North Atlantic (Peyron & de Vernal, 2001). In this example, three distinct calibration data sets, that yielded comparable RMSE from ANN, were used to reconstruct the SST of August, which is the warmest month. The results show very large amplitudes of temperatures from glacial to interglacial, with a good record of climate oscillations related to Heinrich and Dansgaard–Oeschger events (cf. Hillaire-Marcel & Bilodeau, 2000). In general, there is a relatively good agreement between the estimates given the extreme variability of conditions at the core site (in the hydrography used for calibration the modern SST in August was $13.4\pm2.6°C$). The difference between the MAT and ANN estimates averages 1.5–2.6°C, which thus appear acceptable on the whole. However, SST discrepancies are particularly large for the early Holocene, which corresponds to an interval with relatively distant modern

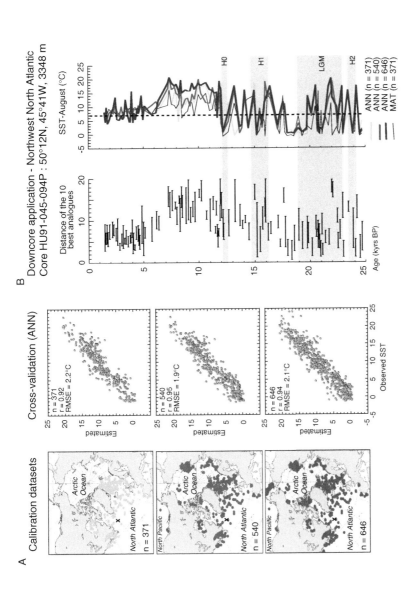

Figure 4 Data with permission from Peyron and de Vernal (2001) showing calibration results and down-core reconstructions of SST using ANN based on three dinocyst data sets, which represent the stages in the development of a single database that first covered the northern North Atlantic ($n = 371$) and progressively included the Arctic seas ($n = 540$) and the Northern North Pacific ($n = 646$). (A) Illustration of the calibration data sets in the left panel and of the cross-validation results for SST reconstruction in the right panel. The x on the maps show the location of the core analyzed. (B) The SST reconstructions from the three ANN calibrations are illustrated together with the results from MAT. The vertical dashed line corresponds to modern summer SST. It is noteworthy that most selected analogues are from the northern North Atlantic data set and that the MAT results are fairly similar among the three data sets since the selected analogues are approximately the same. The threshold for a no-analogue situation is ~40.

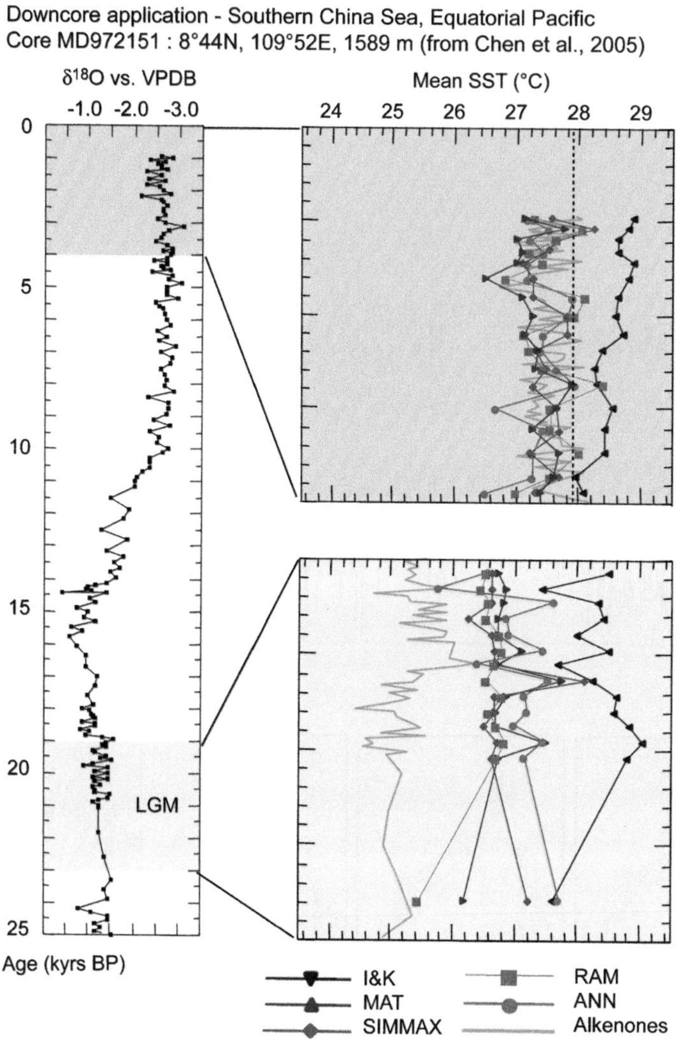

Downcore application - Southern China Sea, Equatorial Pacific
Core MD972151 : 8°44N, 109°52E, 1589 m (from Chen et al., 2005)

Figure 5 Data with permission from Chen et al. (2005, Copyright Elsevier) showing SST estimates in core MD972151 collected in the South China Sea (cf. Lee et al., 1999). The SST results are from five different transfer function methods using a western Pacific planktonic foraminifer data set that includes 694 samples. They are compared to SST estimates from alkenone data (Huang et al., 1999). The vertical dashed line in the Holocene part of the diagram corresponds to modern annual SST.

analogues (see Figure 5). From this example, it is clear that the definition of the calibration data set is determinant, since it allows for extrapolation and the divergence of the estimates. However, both calibration data sets yield satisfactory RMSE and the problem is to decide which one to retain.

Another example of a multi-method and multi-proxy approach, from a very different oceanographical setting, is shown in Figure 5. The data are from a sequence spanning the late Holocene and the LGM in the South China Sea, equatorial Pacific (*cf.* Lee, Wei, & Chen, 1999; Chen et al., 2005). In this core, SST estimates were obtained from the application of various methods (I&K, RAM, SIMMAX, MAT, ANN) on planktonic foraminifers; independent SSTs from alkenone measurements (*cf.* Huang, Wang, & Zhao, 1999) were used for comparison. The western Pacific modern data set ($n = 694$) covers a wide range of temperatures ($2–30°C$) with a high density of points in the subtropical-equatorial domain. The cross-validation of all methods yielded RMSE ranging from 0.85 to $1.27°C$, with the exception of I&K with a lower performance. Similarly, for the late Holocene interval, all methods provide consistent SST estimates in agreement with alkenone results, the only outsider being the I&K method. However, in this case, while SST estimates are consistent for recent time intervals, they yield discrepancies between alkenone and foraminifer proxies for the LGM and show a wide scatter of results in a few samples. This example illustrates that transfer functions, although well validated, cannot be widely extrapolated in space or time without some problems occurring.

The study of Chen et al. (2005) is one example, among others, that illustrates reconstructions of past intervals yielding significant discrepancies in SST estimates, i.e., with differences larger than the ranges of the RMSE. Many examples of differing reconstructions, or even conflicting SST estimates based on different proxies can be found in the literature (e.g., see Marret et al., 2001; Marchal et al., 2002). However, most often the comparison of the results is made on the basis of results obtained through individual projects, without standardization of the hydrographical database and transfer function methods. The formal inter-comparison exercise of the LGM that has been undertaken within the context of the MARGO project is probably a first in this direction. This inter-comparison has indeed implied first and foremost a standardization of the reference stratigraphy and of hydrographic data as input environmental variables (*cf.* Kucera, Rosell-Melé, Schneider, Waelbroeck, & Weinelt, 2005a). In the case of the MARGO inter-comparison exercise, the reconstructions have shown consistencies for many regions of the world oceans, which points to the adequacy of the methods used, and to the robustness of the quantitative results. However, in many regions of the ocean, in the northern North Atlantic, or along some continental margins marked by upwelling, the reconstructions show significant discrepancies (Kageyama et al., 2006; de Vernal et al., 2006).

Apparent inconsistencies recorded through multiple methods or multi-proxy approaches highlight the difficulty of making reconstructions in domains that represent extrapolation from the modern situation. Nevertheless, the inconsistencies themselves are important since they allow the identification of areas of uncertainties and lead to the identification of situations that are not well represented by the modern data sets used for paleoclimate reconstructions. SST is a relatively complex property of sea water that is intimately related to other parameters, including salinity, stratification and nutrients, pCO_2, etc. Although the temperature signal is most important, any change in the interaction among the properties of sea water may induce a bias in the SST record by a given proxy. Therefore, the discrepancies between SSTs estimated from various proxies could reflect the distinct response to changes in sea water properties that are interrelated in the modern data sets, but possibly not in the same way as in the past (e.g., Watkins & Mix, 1998; Mix et al., 2000). As expressed by Mix et al. (2000), "a multi-proxy strategy increases confidence in results and provides oceanographic context for more effective interpretation of processes".

7. CONCLUDING REMARKS

Transfer functions, in a broad sense, are powerful methods that allow complex biological assemblages analyzed by micropaleontologists to be translated into simple abiotic parameters for the use of the entire scientific community working in the fields of paleoceanography and paleoclimatology. They are extremely useful, but they imply a high degree of simplification of multiple biotic variables that are themselves dependent upon complex abiotic and biotic parameters. Various methods were developed over the last decades, based either on calibrations between assemblages and climate parameters or on the degree of similarity between assemblages (see Table 1). Among these methods, those yielding the most accurate reconstructions of the modern conditions are the weighting average — WA-PLS method, the ANN, the MATs and RAM. The artificial neuronal method is, however, very dependent upon the definition of the calibration data set, which makes it more risky than others for extrapolating outside its domain of training, because of its strong non-linear characteristic. The analogue methods do not necessarily provide accurate results outside the domain of calibration, but they allow reliability indices to be set based on the degree of analogy, as well as the identification of no-analogue situations. The RAM provides the possibility of improving the reference data sets in some parts of their range, but we must be cautious not to create virtual samples that are too distant from the actual climatic space. It is probably one of the most appropriate and reliable methods currently used in paleoceanography.

A multi-method approach has been proposed to obtain more robust reconstructions and to identify the assemblages for which the performance of the transfer function is low or problematic. Our comparison shows that it is not necessary to try all of the different methods, as some clearly underperform and others yield quite similar results. The best candidates for a multi-method approach are MAT, RAM, ANN and WA-PLS. The remaining problem is to make a decision from the various reconstructions. Should we select one, which is assumed to be the "best", or average all estimates? The first solution is difficult to apply, as we generally do not know which one is the "best" reconstruction. Averaging several reconstructions, however, may be biased by the inclusion of outliers, thus leading to an average different from the initial reconstructions. We can advise researchers to make at least three reconstructions, eventually remove possible outliers, and, finally, to use average values or a median, which is less sensitive to outliers.

Another classical problem of transfer functions comes from the possible correlation between the environmental variables to be reconstructed. It is necessary to collect surface samples representative of diverse environmental conditions in order to limit this problem; otherwise it is difficult to make an inference about a given parameter if conditions of the past were different from the present. In this respect, we have seen that the RAM method performs better than the other methods. Another strategy to evaluate is to select an appropriate subset of species for each environmental variable analyzed. Such a "stepwise" approach based on different predictors may help in producing calibration equations for each predictand that are more independent.

The respective performances of the transfer function methods can be evaluated from the cross-validation of modern data sets. For example, when tested on modern planktonic foraminifer data in the North Atlantic, the best methods produce temperature estimates within $\pm 1.3°C$, and those performing less well produce estimates within an error of $\pm 2.2°C$. These performances are possibly overestimated because of autocorrelation (cf. Telford & Birks, 2005), but contrary to what is claimed by these authors, MAT and RAM are not more affected by this problem than are other methods. In any case, a performance with a RMSEP of $\pm 1.5°C$ should be considered extremely satisfactory since the actual standard deviation of the "modern" mean SST is $>1.5°C$ in the largest part of the World Ocean, and throughout most of the North Atlantic (see World Ocean Atlas, 2001).

Beyond the evaluation of the performance of various methods, one important caveat is the fact that the parts of modern data sets that serve as analogues or as a basis for calibration do not necessarily provide the adequate environmental combinations of ocean surface properties for the past we seek to reconstruct. This issue has to be explored through multi-proxy strategies and would benefit from the development of Bayesian approaches, based on either process or stochastic models.

REFERENCES

Abelmann, A., Brathauer, U., Gersonde, R., Sieger, R., & Zielinski, U. (1999). A radiolarian-based transfer function for the estimation of summer sea-surface temperatures in the Southern Ocean (Atlantic sector). *Paleoceanography, 14,* 410–421.

Anderson, D. M., & Archer, D. (2002). Glacial-interglacial stability of ocean pH inferred from foraminifer dissolution rates. *Nature, 416,* 70–73.

Andersen, C., Koç, N., Jennings, A., & Andrews, J. T. (2004). Non-uniform response of the major surface currents in the Nordic Seas to insolation forcing: Implications for the Holocene climate variability. *Paleoceanography, 19,* PA2003, doi: 10.1029/2002PA000873.

Andreasen, D. G., & Ravelo, A. C. (1997). Tropical Pacific Ocean thermocline depth reconstructions for the last glacial maximum. *Paleoceanography, 12,* 395–413.

Atkinson, T. C., Briffa, K. R., & Coope, G. R. (1987). Seasonal temperatures in Britain during the past 22,000 years reconstructed using beetle remains. *Nature, 325,* 587–593.

Barrows, T. T., & Juggins, S. (2005). Sea-surface temperatures around the Australian margin and Indian Ocean during the Last Glacial Maximum. *Quaternary Science Reviews, 24,* 1017–1047.

Bartlein, P. J., Prentice, I. C., & Webb, T., III (1986). Climatic response surfaces from pollen data for some eastern North American taxa. *Journal of Biogeography, 13,* 35–57.

Benzécri, J. P. (1973). *L'analyse des données. II. l'analyse des correspondances.* Paris: Dunod.

ter Braak, C. J. F. (1985). Correspondence analysis of incidence and abundance data: Properties in terms of a unimodal response model. *Biometrics, 41,* 859–873.

ter Braak, C. J. F. (1986). Canonical correspondence analysis: A new eigenvector technique for multivariate direct gradient analysis. *Ecology, 67,* 1167–1179.

ter Braak, C. J. F. (1987). The analysis of vegetation-environment relationships by canonical correspondence analysis. *Vegetation, 69*, 69–77.

ter Braak, C. J. F., & Juggins, S. (1993). Weighted averaging partial least squares regression (WA-PLS): An improved method for reconstructing environmental variables from species assemblages. *Hydrobiologia, 1*, 1–18.

ter Braak, C. J. F., Juggins, S., Birks, H. J. B., & Van der Voet, H. (1993). Weighted averaging partial least squares regression (WA-PLS): Definition and comparison with other methods for species-environment calibration. In: G. P. Patil & C. R. Rao (Eds), *Multivariate environment statistics* (pp. 519–553). Amsterdam: Elsevier.

ter Braak, C. J. F., & Looman, C. W. N. (1986). Weighted averaging, logistic regression and the Gaussian response model. *Vegetation, 65*, 3–11.

ter Braak, C. J. F., & Prentice, I. C. (1988). A theory of gradient analysis. *Advances in Ecological Research, 18*, 271–317.

Braconnot, P. (2004). Modéliser le dernier maximum glaciaire et l'Holocène moyen. *C.R. Geosciences, 336*, 711–719.

Cayre, O., Beaufort, L., & Vincent, E. (1999). Paleoproductivity in the Equatorial Indian Ocean for the last 260,000 yr: A transfer function based on planktonic foraminifera. *Quaternary Science Reviews, 18*, 839–857.

Cheddadi, R., Yu, G., Guiot, J., Harrison, S. P., & Prentice, I. C. (1997). The climate 6000 years ago in Europe. *Climate Dynamics, 13*, 1–9.

Chen, M. T., Huang, C. C., Pflaumann, U., Waelbroeck, C., & Kucera, M. (2005). Estimating glacial western Pacific sea-surface temperature: Methodological overview and data compilation of surface sediment planktonic foraminifer faunas. *Quaternary Science Reviews, 24*, 1049–1062.

Chen, M.-T., & Prell, W. L. (1997). Reassessment of CLIMAP methods for estimating Quaternary sea-surface temperatures: Examination using Pacific core top data sets. *Terrestrial, Atmospheric and Oceanic Sciences, 8*, 111–139.

Cleveland, W. S., & Devlin, S. J. (1988). Locally weighted regression: An approach to regression analysis by local fitting. *Journal of the American Statistical Association, 83*, 596–610.

CLIMAP, Climate: Long-Range Investigation, Mapping, and Prediction Project Members (1976). The surface of the Ice Age Earth. *Science, 191*, 1131–1137.

CLIMAP, Climate: Long-Range Investigation, Mapping, and Prediction Project Members (1981). Seasonal reconstructions of the Earth's surface at the Last Glacial Maximum. *Geological Society of American Map Chart Series MC, 36*, 1–18.

COHMAP Members (1988). Climatic changes of the last 18,000 years: Observation and model simulations. *Science, 241*, 1043–1052.

Cortese, G., Dolven, J. K., Bjørklund, K. R., & Malmgren, B. A. (2005). Late Pleistocene-Holocene radiolarian paleotemperatures in the Norwegian Sea based on artificial neural networks. *Palaeogeography, Palaeoclimatology, Palaeoecology, 224*, 311–332.

Cronin, T. M., Holtz, T. R., Jr., & Whatley, R. C. (1994). Quaternary paleoceanography of the deep Arctic Ocean based on quantitative analysis of Ostracoda. *Marine Geology, 119*, 305–332.

Crosta, X., Pichon, J. J., & Burckle, L. H. (1998). Application of modern analog technique to marine Antarctic diatoms: Reconstruction of the maximum sea ice extent at the last glacial maximum. *Paleoceanography, 13*(3), 284–297.

Crosta, X., Sturm, A., Armand, L., & Pichon, J. J. (2004). Late Quaternary sea ice history in the Indian sector of the Southern Ocean as recorded by diatom assemblages. *Marine Micropaleontology, 50*, 209–223.

De Sève, M. A. (2005). Transfer function between surface sediment diatom assemblages and sea-surface temperature and salinity of the Labrador Sea. *Marine Micropaleontology, 36*, 249–267.

Fauquette, S., Guiot, J., & Suc, J. P. (1998). A method for climatic reconstruction of the Mediterranean Pliocene using pollen data. *Palaeogeography, Palaeoclimatology, Palaeoecology, 144*, 183–201.

Fritts, H. C., Blasing, T. J., Hayden, B. P., & Kutzbach, J. E. (1971). Multivariate techniques for specifying tree-growth and climate relationships and for reconstructing anomalies in paleoclimate. *Journal of Applied Meteorology, 10*, 845–864.

Gasse, F., & Tekaia, F. (1983). Transfer functions for estimating paleoecological conditions (pH) from East African diatoms. *Hydrobiologia*, *103*, 85–90.

Gavin, D. G., Oswald, W. W., Wahl, E. R., & Williams, J. W. (2003). A statistical approach to evaluating distance metrics and analog assignments for pollen records. *Quaternary Research*, *60*, 356–367.

Gersonde, R., Crosta, X., Abelmann, A., & Armand, L. (2005). Sea-surface temperature and sea ice distribution of the Southern Ocean at the EPILOG Last Glacial Maximum — a circum-Antarctic view based on siliceous microfossil records. *Quaternary Science Reviews*, *24*, 869–896.

Guiot, J. (1990). Methodology of the last climatic cycle reconstruction in France from pollen data. *Palaeogeography, Palaeoclimatology, Palaeoecology*, *80*, 49–69.

Guiot, J., Berger, A. L., Munaut, A. V., & Till, C. (1983). Some new mathematical procedures in dendroclimatology with examples for Switzerland and Morocco. *Tree-Ring Bulletin*, *42*, 33–48.

Guiot, J., Cheddadi, R., Prentice, I. C., & Jolly, D. (1996). A method of biome and land surface mapping from pollen data: Application to Europe 6000 years ago. *Palaeoclimates: Data and modelling*, *1*, 311–324.

Guiot, J., de Beaulieu, J. L., Cheddadi, R., David, F., Ponel, P., & Reille, M. (1993). The climate in Western Europe during the Last Glacial Interglacial Cycle derived from pollen and insect remains. *Palaeogeography, Palaeoclimatology, Palaeoecology*, *103*, 73–93.

Guiot, J., Nicault, A., Rathgeber, C., Edouard, J. L., Guibal, F., Pichard, G., & Till, C. (2005). Last-millennium summer-temperature variations in Western Europe based on proxy data. *The Holocene*, *15*, 489–500.

Guiot, J., & Tessier, L. (1997). Detection of pollution signals in tree-ring series using AR processes and neural networks. In: T.S. Rao, M. B. Priestley, & O. Lessi (Eds), *Applications of time-series analysis in astronomy and meteorology* (p. 413). London: Chapman & Hall.

Guiot, J., Torre, F., Jolly, D., Peyron, O., Boreux, J. J., & Cheddadi, R. (2000). Inverse vegetation modeling by Monte Carlo sampling to reconstruct palaeoclimates under changed precipitation seasonality and CO_2 conditions: Application to glacial climate in Mediterranean region. *Ecological Modelling*, *127*, 119–140.

Haslett, J., Salter-Townshend, M., Wilson, S. P., Bhattacharya, S., Whiley, M., Allen, J. R. M., Huntley, B., Mitchell, F. J. G. (2006) Bayesian palaeoclimate reconstruction. *Journal of the Royal Statistical Society. Series A*, *169*, 1–36, Part 3.

Hastie, T., & Tibshirani, R. (1990). *Generalized Additive Models*. New York: Chapman & Hall.

Hatté, C., & Guiot, J. (2005). Palaeoprecipitation reconstruction by inverse modelling using the isotopic signal of loess organic matter: application to the Nußlochloess sequence (Rhine Valley, Germany). *Climate Dynamics*, *25*, 315–327.

Hayes, A., Kucera, M., Kallel, N., Sbaffi, L., & Rohling, E. J. (2005). Glacial Mediterranean sea surface temperatures based on planktonic foraminiferal assemblages. *Quaternary Science Reviews*, *24*, 999–1016.

Hayward, B. W. (2004). Foraminifera-based estimates of paleobathymetry using Modern Analogue Technique, and the subsidence history of the early Miocene Waitemata Basin. *New Zealand Journal of Geology and Geophysics*, *47*, 749–767.

Hill, M. O. (1974). Correspondence analysis: A neglected multivariate method. *Applied Statistics*, *23*, 340–354.

Hillaire-Marcel, C., & Bilodeau, G. (2000). Instabilities in the Labrador Sea water mass structure during the last climatic cycle. *Canadian Journal of Earth Sciences*, *37*, 795–809.

Hillaire-Marcel, C., de Vernal, A., Bilodeau, G., & Weaver, A. J. (2001). Absence of deep water formation in the Labrador Sea during the last interglacial period. *Nature*, *410*, 1073–1077.

Hintikka, V. (1963). Über das Grossklima einiger Pflanzenareale in zwei Klimakoordinatensystemen dargestellt. *Annales Botanici Societatis Zoologicae Botanicae Fennicae 'Vanamo'*, *34*, 64.

Huang, C.-Y., Wang, C.-C., & Zhao, M. (1999). High-resolution carbonate stratigraphy of IMAGES Core MD972151 from South China Sea. *The Journal of Terrestrial, Atmospheric, and Oceanic Sciences*, *10*, 225–238.

Hüls, M., & Zahn, R. (2000). Millennial-scale sea surface temperature variability in the western tropical North Atlantic from planktonic foraminiferal census counts. *Paleoceanography*, *15*, 659–678.

Huntley, B. (1994). Late Devensian and Holocene palaeoecology and palaeoenvironments of the Morrone Birkwoods, Aberdeenshire, Scotland. *Journal of Quaternary Science*, 9, 311–336.

Hutson, W. H. (1980). The Agulhas current during the late Pleistocene: Analysis of modern faunal analogs. *Science*, 207, 64–66.

Imbrie, J., & Kipp, N. G. (1971). A new micropaleontological method for quantitative paleoclimatology: Application to a late Pleistocene Caribbean core. In: K. K. Turekian (Ed.), *The Late Cenozoïc Glacial Ages* (pp. 71–181). New Haven, CT: Yale University Press.

Intergovernmental Panel on Climate Change (IPCC) (2001). *Climate change 2001. The scientific basis*. Cambridge, UK: Cambridge University Press.

Ivanova, E., Schiebel, R., Deo Singh, A., Schmiedl, G., Niebler, S., & Hemleben, C. (2003). Primary production in the Arabian Sea during the last 135,000 years. *Palaeogeography, Palaeoclimatology, Palaeoecology*, 197, 61–82.

Iversen, J. (1944). *Viscum, Hedera* and *Ilex* as climatic indicators. *Geologiska Föreninger Förhandlingar*, 66, 463–483.

Jiang, H., Seidenkrantz, M.-S., Knudsen, K. L., & Eiríksson, J. (2002). Late-Holocene summer sea-surface temperatures based on a diatom record from the north Icelandic shelf. *The Holocene*, 12, 137–147.

Jones, R. L., Whatley, R. C., Cronin, T. M., Harry, J., & Dowsett, H. J. (1999). Reconstructing late Quaternary deep-water masses in the eastern Arctic Ocean using benthonic Ostracoda. *Marine Micropaleontology*, 37, 251–272.

Joussaume, S., & Taylor, K. E. (1995). Status of the Paleoclimate Modeling Intercomparison Project (PMIP). *Proceedings of the first international AMIP scientific conference*, Geneva, WCRP Report (pp. 425–430).

Joussaume, S., & Taylor, K. E. (1999). The Paleoclimate Modeling Intercomparison Project (PMIP). *Proceedings of the WCRP 111: Proceedings of the third PMIP Workshop* (pp. 1–10). http://www-lsce. cea.fr/pmip/

Kageyama, M., Laîná, A., Abe-Ouchi, A., Braconnot, P., Cortijo, E., Crucifix, M., de Vernal, A., Guiot, J., Hewitt, C. D., Kitoh, A., Marti, O., Ohgaito, R., Otto-Bliesner, B., Peltier, W. R., Rosell-Melé, A., Vettoretti, G., Weber, S.L., & MARGO Project Members. (2006). Last Glacial Maximum temperature over the North Atlantic, Europe and western Siberia: A comparison between PMIP models, MARGO sea-surface temperatures and pollen-based reconstructions. *Quaternary Science Reviews*, 25, 2082–2102.

Klotz, S., Guiot, J., & Mosbrugger, V. (2003). Continental European Eemian and early Würmian climate evolution: Comparing signals using different quantitative reconstruction approaches based on pollen. *Global and Planetary Change*, 789, 1–18.

Koç, N., Jansen, E., & Haflidason, H. (1993). Paleoceanographic reconstructions of surface ocean conditions in the Greenland, Iceland and Norwegian Seas through the last 14 ka based on diatoms. *Quaternary Science Reviews*, 12, 115–140.

Koç, N., & Schrader, H. (1990). Surface sediment diatom distribution and Holocene paleotemperature variations in the Greenland, Iceland and Norwegian Sea. *Paleoceanography*, 5, 557–580.

Korkhola, A., & Weckström, J. (2000). A quantitative Holocene climatic record from diatoms in Northern Fennoscandia. *Quaternary Research*, 54, 284–294.

Kucera, M., Rosell-Melé, A., Schneider, R., Waelbroeck, C., & Weinelt, M. (2005a). Multiproxy approach for the reconstruction of the glacial ocean surface (MARGO). *Quaternary Science Reviews*, 24, 813–819.

Kucera, M., Weinelt, M., Kiefer, T., Pflaumann, U., Hayes, A., Weinelt, M., Chen, M.-T., Mix, A.C., Barrows, T. T., Cortijo, E., Duprat, J., Juggins, S., & Waelbroeck, C. (2004). Compilation of planktic Foraminifera census data, modern from the Atlantic Ocean, *PANGAEA*, doi:10.1594/ PANGAEA.227322

Kucera, M., Weinelt, M., Kiefer, T., Pflaumann, U., Hayes, A., Weinelt, M., Chen, M. T., Mix, A. C., Barrows, T. T., Cortijo, E., Duprat, J., Juggins, S., & Waelbroeck, C. (2005b). Reconstruction of sea-surface temperatures from assemblages of planktonic foraminifera: Multi-technique approach based on geographically constrained calibration data sets and its application to glacial Atlantic and Pacific Oceans. *Quaternary Science Reviews*, 24, 951–998.

de Vernal, A., Eynaud, F., Henry, M., Hillaire-Marcel, C., Londeix, L., Mangin, S., Matthiessen, J., Marret, F., Radi, T., Rochon, A., Solignac, S., & Turon, J. L. (2005). Reconstruction of sea-surface conditions at middle to high latitudes of the Northern Hemisphere during the Last Glacial Maximum (LGM) based on dinoflagellate cyst assemblages. *Quaternary Science Reviews, 24,* 897–924.

de Vernal, A., Henry, M., Matthiessen, J., Mudie, P. J., Rochon, A., Boessenkool, K., Eynaud, F., Grøsfjeld, K., Guiot, J., Hamel, D., Harland, R., Head, M. J., Kunz-Pirrung, M., Levac, E., Loucheur, V., Peyron, O., Pospelova, V., Radi, T., Turon, J.-L., & Voronina, E. (2001). Dinoflagellate cyst assemblages as tracers of sea-surface conditions in the northern North Atlantic, Arctic and sub-Arctic seas: the new "*n* = 677" database and application for quantitative paleoceanographical reconstruction. *Journal of Quaternary Science, 16,* 681–699.

de Vernal, A., & Hillaire-Marcel, C. (2000). Sea-ice, sea-surface salinity and the halo/thermocline structure in the northern North Atlantic: Modern versus full glacial conditions. *Quaternary Science Reviews, 19,* 65–85.

de Vernal, A., Rosell-Melé, A., Kucera, M., Hillaire-Marcel, C., Eynaud, F., Weinelt, M., Dokken, T., & Kageyama, M. (2006). Multiproxy reconstruction of LGM sea-surface conditions in the northern North Atlantic. *Quaternary Science Reviews, 25,* 2820–2834.

de Vernal, A., Rochon, A., Hillaire-Marcel, C., Turon, J.-L., & Guiot, J. (1993). Quantitative reconstruction of sea-surface conditions, seasonal extent of sea-ice cover and meltwater discharges in high latitude marine environments from dinoflagellate cyst assemblages. Proceedings of the NATO Workshop on Ice in the Climate system. *NATO ASI Series, 112,* 611–621.

de Vernal, A., Rosell-Melé, A., Kucera, M., Hillaire-Marcel, C., Eynaud, F., Weinelt, M., Dokken, T., & Kageyama, M. (2006). Testing the validity of proxies for the reconstruction of LGM sea-surface conditions in the northern North Atlantic. *Quaternary Science Reviews* (in press).

de Vernal, A., Turon, J.-L., & Guiot, J. (1994). Dinoflagellate cyst distribution in high latitude environments and development of transfer function for the reconstruction of sea-surface temperature, salinity and seasonality. *Canadian Journal of Earth Sciences, 31,* 48–62.

Waelbroeck, C., Labeyrie, L., Duplessy, J.-C., Guiot, J., Labracherie, M., Leclaire, H., & Duprat, J. (1998). Improving past sea surface temperature estimates based on planktonic fossil faunas. *Paleoceanography, 13,* 272–283.

Walczak, B., & Wegscheider, W. (1994). Calibration of non-linear analytical systems by a Neuro-Fuzzy approach. *Chemometrics and Intelligent Laboratory Systems, 22,* 199–207.

Watkins, J. M., & Mix, A. C. (1998). Testing the effects of tropical temperature, productivity, and mixed-layer depth on foraminiferal transfer functions. *Paleoceanography, 13,* 96–105.

Webb, T., Jr., & Bryson, R. A. (1972). Late- and Postglacial climatic change in the northern Midwest, USA: Quantitative estimates derived from fossil pollen spectra by multivariate statistical analysis. *Quaternary Research, 2,* 70–115.

Wollenburg, J. E., Kuhnt, W., & Mackensen, A. (2001). Changes in Arctic Ocean paleoproductivity and hydrography during the last 145 kyr: The benthic foraminiferal record. *Paleoceanography, 16,* 65–77.

Wood, A. M., Whatley, R. C., Cronin, T. M., & Holtz, T. (1993). Pliocene palaeotemperature reconstruction for the southern North Sea based on Ostracoda. *Quaternary Science Reviews, 12,* 747–767.

World Ocean Atlas. (2001). http://www.nodc.noaa.gov/OC5/WOA01F/prwoa01f.html

Zielinski, U., Gersonde, R., Sieger, R., & Fütterer, D. K. (1998). Quaternary surface water temperature estimations — calibration of diatom transfer functions for the Southern Ocean. *Paleoceanography, 13,* 365–383.

PART 3: GEOCHEMICAL TRACERS

ELEMENTAL PROXIES FOR PALAEOCLIMATIC AND PALAEOCEANOGRAPHIC VARIABILITY IN MARINE SEDIMENTS: INTERPRETATION AND APPLICATION

Stephen E. Calvert* *and* Thomas F. Pedersen

Contents

1. Introduction	568
2. Sedimentary Components of Marine Sediments	569
3. Normalization of Elemental Data	569
4. Palaeoclimatic Records from the Sea Floor	571
4.1. Aeolian sediment inputs	571
4.2. Terrestrial runoff and precipitation records	573
4.3. Changes in lithogenous fluxes to the North Pacific during the cenozoic	580
5. Metalliferous Sedimentation in the Ocean	581
5.1. Hydrogenous deposits	581
5.2. Records of sea floor hydrothermal inputs to the ocean	583
6. Elemental Proxies for Palaeoproductivity	585
6.1. Barium	585
6.2. Phosphorus	594
6.3. Aluminium	597
7. Proxies for Redox Conditions at the Sea Floor and in Bottom Sediments	599
7.1. Proxies for oxic conditions of sedimentation	603
7.2. Proxies for suboxia	607
7.3. Proxies for anoxic conditions of sedimentation	611
7.4. Post-depositional migration or loss of authigenic metals	616
7.5. Distinguishing between suboxic and anoxic environments of sedimentation	617
7.6. History of anoxia in the Cariaco Basin	618
7.7. Sedimentary records of changes in ventilation and organic flux	619
8. Future Developments	621
9. Afterword	625
Acknowledgements	625
References	625

* Corresponding author.

Developments in Marine Geology, Volume 1
ISSN 1572-5480, DOI 10.1016/S1572-5480(07)01019-6

1. INTRODUCTION

The elemental composition of sediments and sedimentary rocks provides a wide range of information on sediment sources, the mode and direction of sediment transport, past climates, ocean circulation, organic production, sea-floor and sedimentary oxygenation and post-depositional changes in sediment sequences. The content of a given element in a sediment sample depends on the relative proportions of its constituent phases, which are ultimately derived from the continents, the sea floor and the water column. Physical and biological processes during deposition coupled with post-depositional chemical reactions yield a complex component mixture that can provide significant palaeoceanographic and palaeoclimatic information to complement and strengthen interpretations derived from the study of microfossils and the isotopic compositions of sedimentary components.

Early work on the construction of climate records from marine sediments focused primarily on stratigraphic studies of microfossil distributions and changes in assemblages over time (Schott, 1935; Cushman & Henbest, 1940; Phleger, Parker, & Peirson, 1953; Parker, 1962). But it was also realized that the lithology and mineralogy of deep-sea sediments could also identify the sequence of glacial–interglacial deposits on the sea floor (Bramlette & Bradley, 1940). Elemental data assembled in the reports of some of the classic deep-sea expeditions (Murray & Renard, 1891; Correns, 1937; Bramlette & Bradley, 1940; Arrhenius, 1952) amplified this information, and laid the groundwork for much of our current understanding of the geochemistry of deep-sea deposits and how geochemical data can be used to study the ocean system. One or two decades later information on elemental variability was developed to study marine geochemical cycles (Goldberg, 1954; Goldberg & Arrhenius, 1958; Wedepohl, 1960; Degens, 1965; Berner, 1971), and this greatly advanced understanding of ways in which the distributions, enrichments and depletions of elements can be used to interpret the sedimentary record.

In this chapter, we review the application of sedimentary geochemistry to the reconstruction of climatic and oceanographic changes over the Cenozoic, with emphasis on the Late Pleistocene. Records from both pelagic regimes and ocean margin provinces will be used to show how information on both ocean conditions and terrestrial climates can be assembled from the major, minor and trace element composition of sea-floor deposits. Processes governing the distribution and accumulation of organic carbon and nitrogen in marine sediments will not be specifically discussed in this chapter because they have been extensively reviewed in a number of publications (see for example, Emerson & Hedges, 1988; Pedersen & Calvert, 1990; Stein, 1991; Rullkötter, 2000). Carbon and nitrogen cycling in the ocean are fully discussed in Volume 2. Likewise, the application of rare earth element geochemistry to palaeoceanographic and palaeoclimatic problems is not treated here; the interested reader can refer to recent reviews by McLennan (1989), McLennan, Hemming, McDaniel, and Hanson (1993) and Holser (1997).

The next two sections present background sedimentary information and describe some common data manipulation techniques that are used to construct and study palaeoclimatic and palaeoceanographic records in marine sediments.

2. SEDIMENTARY COMPONENTS OF MARINE SEDIMENTS

Marine sediments and sedimentary rocks are mixtures of a number of components that are derived from different parts of the Geosphere (Goldberg, 1963). They comprise *lithogenous* components, crystalline and non-crystalline phases derived from continental surfaces, (soils and all rock types) and the crustal rocks of the sea floor, *biogenous* components, skeletal materials and the degraded tissues of marine and terrestrial organisms, *hydrogenous* components, inorganic phases that precipitate from seawater, *cosmogenous* components, meteoritic material from space, and *diagenetic* phases and constituents that form, or are precipitated, within accumulating sediments. The elemental composition of sediments reflects this admixture of components since virtually all sediments have contributions from more than one of these sediment sources. In the case where an element resides in only one of the components, its total content will be variably diluted by all other components. However, where an element is present in more than one component its concentration is determined by its amount in each separate component and the proportion of each component in the mixture. Considerable effort has been devoted to unraveling complex component mixtures in order to gain insight into the geochemistry of sedimentary deposits.

3. NORMALIZATION OF ELEMENTAL DATA

The lithogenous fraction of sediments carries information on the variable composition of Earth's crust, which can be modified by chemical erosion and sorting of materials on the land surface and on the sea floor, as well as transportation processes that abrade particles and sort them according to size. The concentration of aluminium is often used to estimate the total lithogenous content of a sediment because its concentration is very similar in acidic to basic extrusive and intrusive and most metamorphic rocks. This is in turn quite similar to its concentration in bulk upper crust and shales, the most common sedimentary rock (Turekian & Wedepohl, 1961; Wedepohl, Correns, Shaw, Turekian, & Zemann, 1969–1978; Wedepohl, 1971; Taylor & McLennan, 1985).

The conservative behaviour of Al is well known in weathering profiles and soil formation due to the *in situ* formation of clay minerals and the retention of hydroxides (Loughnan, 1969; Hem, 1969–1978; Bohn, O'Connor, & McNeal, 1979). This has led to its use in sediment studies as a normalizing parameter for the assessment of the relative degrees of enrichment or depletion of specific elements in a given sample, or to estimate the "background" contribution of an element derived from crustal sources. Data compilations of the World Shale Average (Wedepohl, 1971), the Post-Archaean Average Shale (Taylor & McLennan, 1985) and the North American Shale Composite (Gromet, Dymek, Haskin, & Korotev, 1984) have been used for this purpose. Van der Weijden (2002) recently reviewed the use of the element/Al ratio in sedimentary geochemistry and cautioned against its use for element correlations while confirming that it can provide useful qualitative information on element enrichments/depletions and in correcting for dilution.

An alternative method for normalizing elemental data utilizes the bulk Ti content of marine sediments. The Al/Ti ratio in carbonate-rich equatorial sediments of the central Pacific appears to be substantially higher than the crustal ratio, possibly due to adsorption of dissolved Al onto particle surfaces and its transport to the sea floor (Murray & Leinen, 1993; Murray, Leinen, & Isern, 1993). If correct, this discovery would obviate the use of Al as a conservative lithogenous proxy, but, on the other hand, would provide a useful tracer of variable particle accumulation on the sea floor, linked to biogenous fluxes in the dominantly carbonate-bearing sediments of remote pelagic regions of the sea (Section 6.3). For this reason, Murray and Leinen (1993) elected to use the bulk Ti content of their equatorial sediments as a general lithogenous proxy. This practice has been followed subsequently (Zabel, Bickert, Dittert, & Haese, 1999; Yarincik, Murray, & Peterson, 2000; Zabel et al., 2001; Mora & Martinez, 2005) in situations where the total biogenous fraction of the deposits is small and virtually all Al is contained in lithogenous aluminosilicate lattices.

The use of Ti as a lithogenous proxy is problematic because, although the inorganic fraction of North Pacific clays was found to have a remarkably uniform titanium concentration (Revelle, 1944) and the bulk Ti content of equatorial Pacific sediments has been used as a constant flux index (Arrhenius, Kjellberg, & Libby, 1951; Arrhenius, 1952), its content varies to a greater extent in different rock types compared with Al. Thus, Ti contents can range over a factor of five or more in silicic to mafic igneous rocks compared with a range of only *ca.* 10% for Al (Wedepohl et al., 1969–1978). As discussed in Section 4.1, Ti is hosted principally in discrete Ti-bearing heavy minerals in sediments and sedimentary rocks (Spears & Kanaris-Sotiriou, 1976). These are transported with the silt and fine sand fractions of marine sediments. Ti/Al ratios have consequently been used as a grain-size proxy (Boyle, 1983), suggesting that the assumption underlying the use of Ti as a general lithogenous index, namely its more or less uniform content in a wide range of crustal reservoirs, is more complicated than that of Al.

Discrete rutile and anatase crystals (TiO_2) were identified in Pacific pelagic sediments by Goldberg & Arrhenius (1958), who considered at least the anatase to be formed authigenically on the sea floor. This phase also forms authigenically in deeply weathered soils from dissolved Ti released from residual aluminosilicates (Walker, Sherman, & Katsura, 1969; Bain, 1976). Closer to volcanic islands and seamount chains in the North Pacific, Ti contents of the bulk sediments increase due to the derivation of augite, a well-known host for Ti, from basaltic rocks. Trace quantities of ilmenite ($FeTiO_3$) would also be expected from this source; and since Ti is hosted in biotite, amphiboles, pyroxenes and some clay minerals (Correns, 1969–1978) it is clear that Ti has multiple sources and hosts in marine sediments.

For these reasons, Al will be used as the lithogenous reference element in this chapter where data normalization is used to describe elemental variability.

4. PALAEOCLIMATIC RECORDS FROM THE SEA FLOOR

Information on past climate changes has often been difficult to obtain from continental records because of the poor preservation of proxies on land, which are subjected to severe alteration by chemical and physical weathering and erosion. Although sedimentary sequences on the sea floor can be affected by erosion and redeposition, the deep-sea repository is generally more complete, and judicious selection of core sites can produce high quality, high-resolution records of climatic variations. Elemental data have been used to glean information on changing terrestrial climates, for example whether the source regions were relatively arid, humid or ice covered, by identifying specific sedimentary components that have been transported to ocean core sites via the atmosphere, by river runoff and by ice transport, respectively. This type of information on provenance comes predominantly from the lithogenous fraction of marine sediments. Provenance studies using mainly mineralogical data have a long history (Pettijohn, 1957; Milner, 1962a, 1962b), but considerable progress has been made more recently using elemental data from sedimentary rock sequences to acquire information on likely sources of components (Nesbitt & Young, 1982; Argast & Donnelly, 1987; McCann, 1991; McLennan et al., 1993; Nesbitt & Young, 1996; Nesbitt, Young, McLennan, & Keays, 1996; Hurowitz & McLennan, 2005).

The identification of the primary source or sources — igneous, metamorphic or sedimentary rocks — using elemental data is made difficult by changes introduced by chemical weathering coupled with the sorting of granular sediment masses during physical erosion and transportation. Unless source material has been subjected only to mechanical disaggregation (Nesbitt & Young, 1996), the eroded material carried to the oceans, by whatever transport mode, reflects the composition of the weathered mantle of the land surface, not the unaltered primary rocks (Kronberg, Nesbitt, & Lam, 1986; Nesbitt et al., 1996; Nesbitt & Young, 1997). However, chemical weathering — its intensity and compositional trends — depends on terrestrial climate, and it is that fact that allows elemental compositions of marine sediments to be used as recorders of climatic changes over geological time (Stewart, 1990; Visser & Young, 1990).

4.1. Aeolian Sediment Inputs

The realization that records of aeolian sedimentation can be obtained from marine sediments began with the work of Radczewski (1939), following earlier reports of wind-blown dust at great distances from coastlines (Darwin, 1846). Early work on the distribution of quartz (Rex & Goldberg, 1958) revealed distinct distributional patterns of abundance that could be related to modern wind directions, and that study ushered in a large number of investigations of the abundance of aeolian materials in pelagic sediments. In addition to mineralogical data, the grain-size distribution of the lithogenous fraction of marine sediments can be used to derive relative wind strengths (Parkin & Shackleton, 1973) on the assumption that more vigorous atmospheric circulation will transport larger particles (mineral and rock grains) to a given site of deposition. Thus, both mineralogical and textural approaches yield information on changes in arid land areas and on the relative aridity in sediment source regions.

Aeolian inputs to marine sediments can be tracked using geochemical proxies for sediment grain-size. Among such proxies, Si/Al, Ti/Al and Zr/Al ratios have been successfully used. Silicon occurs both as the basic building block of all aluminosilicates and separately as quartz (SiO_2). Quartz most commonly occurs in the silt and sand fractions of soils and sediments and sedimentary rocks, reflecting the average sizes of quartz grains in source rocks (Feniak, 1944). Quartz is highly resistant to chemical leaching, which, together with its hardness (7 on the Mohs scale) and lack of cleavage, causes the mineral to survive multiple cycles of erosion and deposition (Rankama & Sahama, 1950). Its abundance in sediments is roughly the same as its abundance in common igneous and metamorphic rocks (Blatt, Middleton, & Murray, 1980), attesting to its longevity (Rex & Goldberg, 1958).

Normalization of bulk Si contents to Al (see above) in lithogenous materials yields values that vary widely depending on the ratio of quartz to aluminosilicate phases in the sediments, and since quartz is on average always coarser than the clay minerals in sediment, the Si/Al ratio identifies coarser and finer-grained sections of ocean cores. Silicon also occurs, of course, in the opaline skeletons of diatoms, silicoflagellates and radiolarians; thus, corrections must be made in samples containing appreciable biogenous Si, using direct measures of opal or assuming an average lithogenous Si/Al ratio, if the bulk sediment Si/Al ratio is used as a grain-size proxy.

As previously discussed, Ti resides in both aluminosilicates as a substituent for some of the major elements and as discrete Ti oxide and silicate mineral phases. It is enriched in soils, especially laterites and bauxites, due to the formation of Ti oxides during weathering (Raman & Jackson, 1965; Dolcaster, Syers, & Jackson, 1970; Milne & Fitzpatrick, 1977). In the form of rutile (TiO_2), sphene ($CaTiSiO_5$) and ilmenite ($FeTiO_3$), Ti is carried in the silt and fine-sand fractions during particle transportation, accompanying slightly coarser quartz grains. Correns (1954), for example, found that Ti was enriched in the coarser fractions of pelagic North Atlantic sediments. Spears and Kanaris-Sotiriou (1976) showed that the Ti/Al ratio correlates closely with the total quartz content of a series of Carboniferous sedimentary rocks, and that only a small amount of Ti resides in the clay fraction where it substitutes for Al, Fe, Mn and perhaps Si in the lattices of a wide range of aluminosilicates (Correns, 1969–1978), especially biotite, but also amphiboles, pyroxenes and some clay minerals. In general, but not invariably, therefore, the Ti/Al ratio can be used as a grain-size proxy because the heavy minerals are the main Ti carrier in many sediments.

Zirconium occurs in sediments almost exclusively as the mineral zircon ($ZrSiO_4$), which has a hardness close to that of quartz, has no cleavage and is chemically resistant in the weathering profile (Sudom & St. Arnaud, 1971; Milne & Fitzpatrick, 1977). In primary source rocks it occurs in the very fine sand and silt fractions (Feniak, 1944), and, due to its higher specific gravity than quartz, it is transported with the fine and medium sand quartz grains. Like quartz, therefore, it survives recycling and accumulates in soils (Mason & Jacobs, 1998) and sediments (Dypvik & Harris, 2001). The geochemical behaviour of Zr therefore suggests that the Zr/Al ratio can be used as a grain-size proxy, and it has been applied in this way in a number of studies (e.g. Pedersen et al., 1992; Ganeshram, Calvert, Pedersen, & Cowie, 1999).

An increase in aeolian flux to a site in the eastern equatorial Pacific during Marine Isotope Stage (MIS) 2–4 relative to MIS 1 and the MIS 5 interstadials was detected using changes in the Ti/Al ratio (Boyle, 1983). This is consistent with a source of dust from the arid areas of northern South America lying west of the Andes (Prospero & Bonatti, 1969). Likewise, Shimmield and Mowbray (1991) used variations in Ti/Al (and Zr/Al, Cr/Al) in ODP cores in the northwestern Arabian Sea to track changes in monsoonal wind strength during the Late Pleistocene.

Modern Saharan aerosols, which are currently derived from the extensive dry lake beds in northern Africa (Stuut et al., 2005), can be transported across the North Atlantic as far as the Caribbean and Bermuda (Delany et al., 1967; Prospero & Bonatti, 1969; Glaccum & Prospero, 1980; Denison, Koepnick, Burke, Hetherington, & Fletcher, 1994). Immediately downwind of the Sahara, such material is coarser grained than the enclosing clay-rich sediment (Parkin & Shackleton, 1973; Parkin, 1974; Parkin & Padgham, 1975; Sarnthein, 1978; Sarnthein, Tetzlaff, Koopmann, Wolter, & Pflaumann, 1981; Sarnthein et al., 1982; Grousset et al., 1989), and this can be observed by downcore changes in Si/Al, Ti/Al and Zr/Al (Grousset et al., 1989) (Figure 1). The supply of wind-blown dust appears to have increased under glacial conditions when the areal extent and the aridity of the Sahara increased significantly (Chylek, Lesins, & Lohmann, 2001).

Changes in aeolian fluxes to the Mediterranean Sea on glacial–interglacial timescales can be tracked by Si/Al and Zr/Al ratios (Figure 2), which peak in the marls that were deposited between the sapropel events (Calvert & Fontugne, 2001). This reflects the establishment of more arid conditions in the Mediterranean borderlands that alternated with more humid conditions that led to the formation of the sapropels on a precessional timescale. As in the case illustrated in Figure 1, Ti/Al does not follow the increases in quartz (monitored using Si/Al, as biogenous silica is virtually absent at this site) and zircon (Zr/Al) in the marls, suggesting that it is only partly a grain-size proxy in this basin.

The clay mineralogy of atmospheric dust varies latitudinally, a reflection of the climatic control of clay mineral formation. Thus, kaolinite is the dominant clay mineral in dusts collected from low latitudes in the Atlantic Ocean, whereas illite is the most abundant clay species at higher latitudes (Chester, Elderfield, Griffin, Johnson, & Padgham, 1972). Thus, the clay mineralogy of atmospheric dust carries information on terrestrial climates, which is preserved in ocean archives.

4.2. Terrestrial Runoff and Precipitation Records

Elemental indices of grain-size changes have also been used to infer changes in terrestrial runoff and weathering regimes. Schmitz (1987b), for example, interpreted the increase in Ti content of Bengal Fan sediments through the Cenozoic as reflecting an increase in the supply of coarse-grained material from the

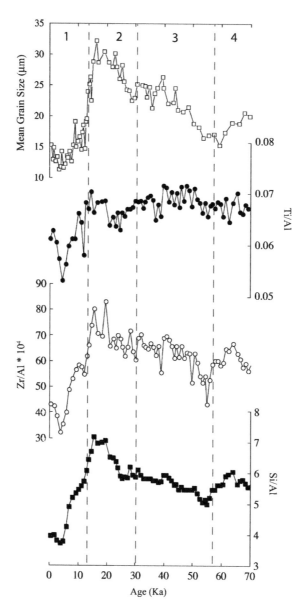

Figure 1 Temporal record of mean grain size (non-carbonate fraction), Ti/Al, Zr/Al & Si/Al changes over the last 70 kyr in core SEDORQUA 11 K (21.48°N, 17.95°W, 1,200 m depth) on the continental slope off Mauritania. Marine Isotope Stages (MIS) are indicated at the top of the upper panel. Data from Martinez et al. (1996). The larger mean grain-size in MIS 2 and the upper part of MIS 3 is closely mirrored by higher Zr/Al and Si/Al, reflecting higher contents of zircon and quartz, respectively, relative to aluminosilicate phases. Note that Ti/Al does not follow the grain-size trends as clearly, due to the location of Ti in aluminosilicates as well as in discrete titaniferous minerals.

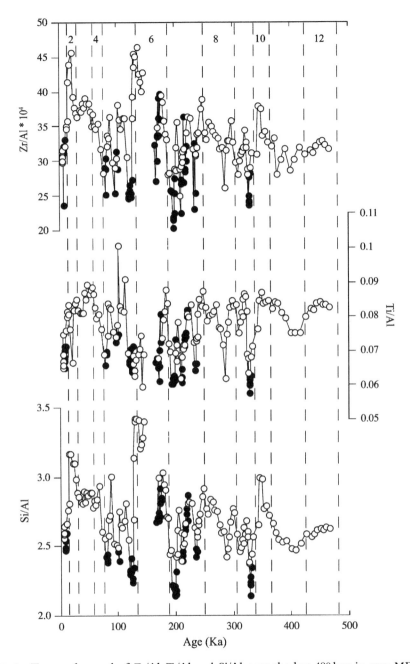

Figure 2 Temporal record of Zr/Al, Ti/Al and Si/Al over the last 480 kyr in core MD84641 (33.03°N, 32.63°E, 1,375 m depth) from the eastern Mediterranean. Data from Calvert and Fontugne (2001). Closed symbols are sapropel intervals, open symbols are marls. Zr/Al and Si/Al have similar downcore variations, reflecting the presence of coarser grained material in the marls and finer-grained sediment in the sapropels. Ti/Al is broadly similar, but there are significant deviations in MIS 2 and late Stage 6.

Ganges-Brahmaputra River system to the Bay of Bengal. This was supported by a strong curvilinear relationship between Ti content and grain size (settling velocity) in a size-fractionated Holocene varved clay. Thus, Schmitz used the Ti/Al ratio as a "palaeosteam indicator", interpreting the downcore changes in the ratio as recording the tectonic uplift of the Himalaya since the Late Miocene. The development of such a high mountain source region caused intensified monsoonal precipitation, led to the rejuvenation of river systems and intensified terrestrial denudation rates. Millennial-scale increases in Ti/Al and coherent increases in grain-size have also been identified in high-resolution cores from the NW European continental margin, reflecting in this case not tectonic effects but rather changing bottom water transport paths and/or winnowing (Kroon, Shimmield, & Shimmield, 2000).

Rubidium does not form discrete Rb mineral phases, but it substitutes readily for K in aluminosilicate minerals. The enrichment of Rb relative to K is greater in micas, especially biotite, compared with co-existing feldspars (Heier & Adams, 1963; Lange, Reynolds, & Lyons, 1966). In the absence of Al data, the K/Rb ratio has been used as a grain-size proxy in sedimentary rocks (Dypvik & Harris, 2001), Chinese loess sequences (Liu, Chen, Ji, & Chen, 2004) and in cores off north-west Africa (Matthewson, Shimmield, Kroon, & Fallick, 1995). Zr/Rb has likewise been used to track changes in sediment grain-sizes (Schneider, Price, Müller, Kroon, & Alexander, 1997), since on average Zr resides in coarser particles than does Rb.

Clay minerals are mainly formed by the hydrolytic decomposition of primary aluminosilicates during terrestrial weathering, which is principally controlled by the flow rate of water through the soil profile (Singer, 1979/80). This results in alteration of the primary minerals and the neoformation of new aluminosilicate frameworks. Different clay mineral species are therefore formed under different leaching regimes as governed by the prevailing precipitation. It follows that information on clay mineral assemblages could be used to reconstruct changes in terrestrial climate. Since clay mineral species have different elemental compositions, mineralogical changes will be reflected in the bulk lithogenous geochemistry of a given sediment.

The general sequence of clay mineral formation and stability has been summarized by Singer (1979/80), based on extensive earlier work. Given the same rock type, montmorillonite will form at the lowest precipitation rates, illite and vermiculite at intermediate rates and kaolinite/halloysite at the highest leaching intensities. This sequence reflects the extent of major cation (Mg, Fe, K, Na, etc.) retention (montmorillonite) or loss (kaolinite) from the weathering zone. The distribution of clay minerals in the Holocene sediments of the world ocean accords with latitudinal gradients in the intensity of land–surface weathering. Thus, kaolinite is the most abundant clay mineral in Holocene sediments of the eastern equatorial Atlantic (Griffin, Windom, & Goldberg, 1968) due to its derivation via river runoff from the deeply-weathered soils of the Niger and Zaire river basins. High kaolinite contents in Holocene sediments of the eastern Indian Ocean derive from the deflation of post-Miocene lateritic soils exposed in the interior of Australia. These "dead laterites" (Campbell, 1917) were formed under more humid conditions than those now prevailing. Illite is the most abundant clay mineral in the mid-latitudes of the Pacific and Atlantic Oceans, reflecting its derivation from the continents via rivers and the wind, where it forms both in soils and via long-term burial diagenesis in sedimentary sequences (Meunier & Velde, 2004). This material has K-Ar ages of hundreds of millions of years (Hurley, Heezen, Pinson, & Fairbairn, 1963), attesting to its strictly detrital nature. It is the dominant clay mineral in aerosols collected in the North Pacific (Arnold, Merrill, Leinen, & King, 1998). Montmorillonite is derived from lowland parts of river basins where leaching within the soil is minimal; its formation on the sea floor by the alteration of basaltic debris is inferred from its distribution and mineral associations in modern sediments (Griffin et al., 1968; Heath, 1969), but this has not been conclusively demonstrated. Iron-rich varieties of montmorillonite — nontronites — are probably formed on the sea floor from the reaction of dissolving biogenous silica and hydrothermal Fe oxyhydroxides (Hein, Yeh, & Alexander, 1979; Cole & Shaw, 1983; Cole, 1985). This is further discussed in Section 5.2.

Coherent downcore variations in Ti/Al, K/Al, Zr/Al and Rb/Al have been observed in cores located off large tropical drainage basins, where aeolian dust supply is minor. In cores off the Zaire River, for example (Schneider et al., 1997), changes in the relative abundances of kaolinite (with a high Al content) and feldspar (with a lower Al content) are thought to cause large variations in total Al contents, which drive the changes in the elemental ratios (Figure 3).

The mineralogical variations probably reflect past climatic changes in equatorial Africa, with kaolinite forming in soils during warm and humid periods when feldspar would also be less stable, and more feldspar surviving soil-forming processes during drier periods when kaolinite production decreased. The chemical and mineralogical data collectively suggest that African climate was warmer and more humid during the Holocene and MIS 5.1, 5.3, 5.5, 6.3 and 6.5. The Zr/Rb ratio in the same cores is highest when the other elemental ratios are lowest (thus, more Al-rich kaolinite) reflecting the enrichment of zircon relative to feldspar in deeply weathered soils. This climatic interpretation is supported by variations in clay mineral abundances off the Niger River (Pastouret, Chamley, Delibrias, Duplessy, & Thiede, 1978). Here, kaolinite is derived from well-drained lateritic soils in the humid equatorial climatic belt of Africa, whereas illite and smectite come from poorly-drained soils of the lower parts of the equatorial and tropical rivers; the relative abundance of these species in sediments off the Niger Delta records changes in the rainfall in this part of Africa over the last glacial–interglacial cycle. During MIS 2 and 3, the flow of the Niger River decreased and was able to deliver only illite- and smectite-rich material from its lower reaches to the offshore area, whereas during the Holocene increased precipitation augmented the supply of kaolinite from interior parts of the Niger drainage basin.

Climatically driven variations in river runoff from northern South America have been reconstructed from downcore changes in inferred major element contents of cores from the Cariaco Trench in the Caribbean. At ODP Site 1002 (Peterson, Haug, Hughen, & Rohl, 2000), downcore oscillations in the relative abundances of Ti and Fe, determined as count rates by a profiling X-ray fluorescence scanner (Jansen, Gaast, Koster, & Vaars, 1998), are considered to reflect changing input of fine-grained terrigenous siliciclastic sediment from the adjacent continental margin

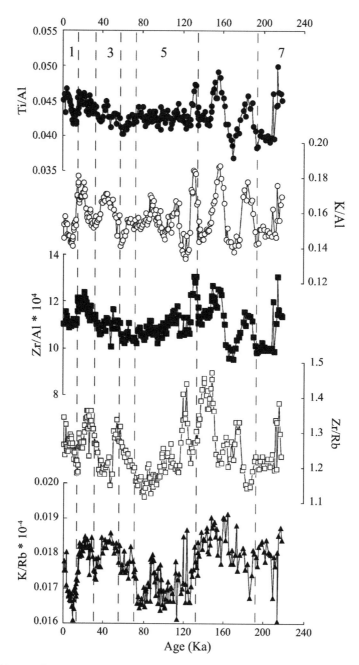

Figure 3 Temporal record of Ti/Al, K/Al, Zr/Al, Zr/Rb and K/Rb over the last 220 kyr in core GeoB 1083 in the eastern tropical Atlantic (6.58°S 10.32°E, 3,124 m water depth). Data from Schneider et al. (1997). Relative changes in Al, K and Rb at this site reflect the preferential enrichment of kaolinite (low K, high Al) in warm/humid intervals and the better survival of feldspar (high K/Al) during drier periods. Changes in Zr/Rb reflect the enrichment of zircon relative to feldspar in deeply weathered soils.

of Venezuela. Such oscillations are in turn likely to be driven by changes in rainfall and runoff from the watersheds of the local rivers, and are especially marked during MIS 3, when lower sea level and isolation of the Cariaco Basin from the Caribbean proper increased the importance of local sediment supply. Higher accumulation rates of the terrigenous indices (Ti and Fe) at Site 1002 coeval with the warm interstadial events in Greenland point to higher precipitation in northern South America at the same time.

A somewhat more complicated picture of sediment sources, and hence climatic changes, emerges from a more extensive set of quantitative elemental data from the same ODP site (Yarincik et al., 2000). Changes in K/Al and Ti/Al (used here rather than Al/Ti as in the original publication) are interpreted to reflect an increase in the relative proportions of kaolinite relative to illite (lower K/Al) during glacials, when local rivers supplied more kaolinite to a semi-isolated basin during lowered sea levels (Figure 4). At the same time, increased aeolian supply of Ti-rich dust from North Africa led to increases in Ti/Al. On the basis of this approach, the major element composition of the deposits has recorded changes in both riverine and

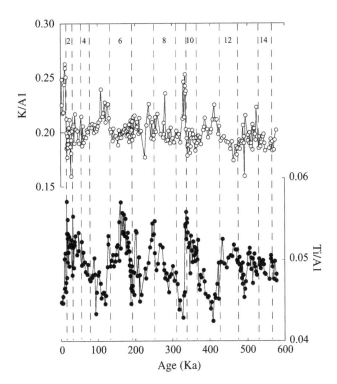

Figure 4 Temporal record of K/Al and Ti/Al over the last 590 kyr at ODP Site 1002, Cariaco Basin. Marine Isotope Stages are indicated at the top of the upper panel. Data from Yarincik et al. (2000). Changes in K/Al reflect an increase in the relative proportions of kaolinite relative to illite (lower K/Al) during glacials, when local rivers supplied more kaolinite to a semi-isolated basin during lowered sea-levels. Increases in Ti/Al point to increased aeolian supply of Ti-rich dust from North Africa led to increases in Ti/Al. These two proxies are anti-phased at this site.

aeolian sediment supply. The link to palaeoclimate is via increase in river flow and by implication increased rainfall, bringing more lithogenous detritus to the Cariaco Basin. In parallel, the enhanced atmospheric transport of dust implies greater aridity and/or increased wind speeds in the desert latitudes.

Similar changes in the proportions of lithogenous and biogenous components over the last 85 ka, possibly linked to changes in terrestrial climate, have been reported in cores recovered off central Brazil (Arz, Pätzold, & Wefer, 1998). As in the Cariaco Basin (Peterson et al., 2000), Ca abundances in the cores vary inversely with Fe and Ti abundances (as recorded by the XRF scanner technique), with higher lithogenous indices correlating with interstadial events in Greenland. The enrichments in lithogenous elements were interpreted as recording increased supplies of terrestrial detritus to the Brazilian margin during warmer periods due to an increase in precipitation in the Brazilian lowlands. However, ^{230}Th-normalized carbonate accumulation rates (François, Bacon, & Suman, 1990) in this part of the eastern Atlantic (Ruhlemann et al., 1996) varied by a factor of three over the last 180 kyr coherently with % $CaCO_3$, which is higher in MIS 1, 3, 5.1, 5.3, 5.5 and 6.5. Thus, variations in Fe and Ti are likely to reflect biogenous dilution and it will be necessary to obtain ^{230}Th-normalized lithogenous accumulation rates before it can be definitively concluded that detrital sediment supply from the adjacent land changed over this time period.

Higher resolution records of Fe and Ti abundances from the Cariaco Basin over the last 14 kyr (Haug, Hughen, Sigman, Peterson, & Rohl, 2001) show a marked decrease in inferred precipitation associated with the Younger Dryas (11.5 to 12.6 ka), smaller decreases during the Medieval Warm Period (0.7 to 1.05 ka), a large decrease during the "Little Ice Age" (0.55 to 0.2 ka), and a broad increase during the Holocene "thermal maximum" (5.4 to 10.5 ka). Such variability in the hydrological balance over northern South America, based on inferred runoff, is broadly coherent with similar variations throughout the tropical Atlantic (Hastenrath, 1985; Hodell et al., 1991; Hastenrath & Greishar, 1993; van der Hammen & Hoogheimstra, 1995; deMenocal, Ortiz, Guilderson, & Sarnthein, 2000). Latitudinal migration of the Intertropical Convergence Zone (ITCZ) appears to be the cause; this responds to seasonality of insolation at the precessional (21 kyr) periodicity (Haug et al., 2001).

4.3. Changes in Lithogenous Fluxes to the North Pacific during the Cenozoic

The history of lithogenous sediment supply to the North Pacific over the last 70 million years has been reconstructed from mineralogical and geochemical data obtained from a site (LL44–GPC) currently located in the pelagic clay province north of Hawaii (Leinen & Heath, 1981; Leinen, 1987, 1989; Kyte, Leinen, Heath, & Zhou, 1993). Two aeolian components have been identified on the basis of the mineralogy of this core, one originating from an easterly source by trade winds during the Palaeogene — possibly from North Africa — and the other originating from Asia via the westerlies during the Neogene. The change in composition reflects the progressive movement of the core site from low latitudes close to the East Pacific Rise to the central North Pacific Basin due to sea-floor spreading.

Further insights into the nature and sources of lithogenous supply to this site have been obtained from elemental data (Kyte et al., 1993). The low-latitude aeolian material was modeled as having an andesitic composition, whereas the more recent aeolian material is identical to quartz- and illite-rich wind-blown dust, and compositionally similar to average shale (Wedepohl, 1971). Figure 5 shows that Si/Al ratios are approximately constant over the last 30 Myr, whereas maxima in the ratio occur in the Late Cretaceous-Early Palaeocene and Eocene due to biogenic silica diagenesis (there are only rare radiolaria observed). Si/Al ratios in the post-Miocene deposits are lower than in Chinese loess (Taylor, McLennan, & McCulloch, 1983) because of loss of larger quartz grains during atmospheric transport from the source region(s); the ratio (mean $= 2.86 \pm 0.012$) reflects the admixture of clay minerals and quartz (Rahn, 1976). The Ti/Al ratio has a broad peak in the Oligo-Miocene when the site was closest to Hawaii during its movement north and westwards by sea-floor spreading. Ti concentrations in Hawaiian basalts can reach 12% by weight, and weathering of such rocks can account for the high ratio of Ti/Al in the mid-Tertiary deposits (Walker et al., 1969). K/Al and Rb/Al ratios show relatively high K and low Rb contents in the Cretaceous-Palaeocene, Early Eocene and Late Eocene-Early Oligocene sections of the core, implying silicic volcanism (Peterson & Goldberg, 1962). In the Miocene-Pliocene, Rb is enriched and K is depleted, possibly depicting a switch to higher illite contents (higher Rb, lower K) in Asian dust.

5. Metalliferous Sedimentation in the Ocean

Iron- and manganese-rich deposits accumulate on the sea floor by two processes. The first is the slow precipitation of poorly crystalline and amorphous oxyhydroxides from normal seawater, which yields *hydrogenous* deposits. These coat detrital and biogenous particles and are thoroughly dispersed within the bottom sediments, or they form discrete ferromanganese nodules and crusts on the sea floor that constitute large potential ore deposits. *Hydrothermal* processes comprise the second mechanism, and include the rapid deposition of sulphides, oxyhydroxides and the crystallization of authigenic clay minerals at mid-ocean ridge crests from hot, acidic waters that flow into bottom waters from hydrothermal vents (Boström & Peterson, 1966; Haymon & Kastner, 1981).

5.1. Hydrogenous Deposits

Hydrogenous ferromanganese oxyhydroxides occur as spheroidal and discoidal concretions that cover extensive areas of the ocean floor in all water depths, especially in areas of slow sedimentation, and as crusts and coatings on rock and mineral surfaces, especially on seamounts and similar topographic elevations throughout the ocean basins (Cronan, 1976). Both types of deposits are formed at

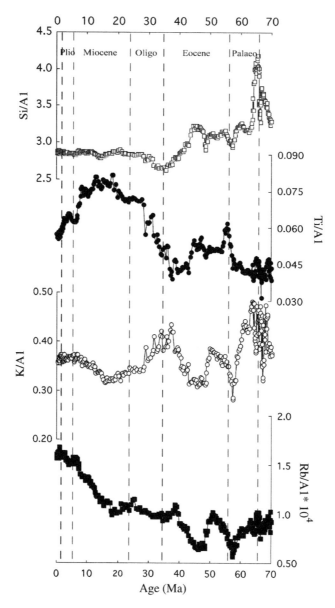

Figure 5 Temporal record of lithogenous proxies over the last 70 Myr in core LL44-GPC3, North Pacific. Data from Kyte et al. (1993). The roughly constant Si/Al ratios over the last 30 Myr record the supply of aeolian dust from the deserts of inland China, whereas maxima in the ratio in the Late Cretaceous-Early Palaeocene and Eocene probably reflect biogenic Si enrichment. Ti enrichment in the Oligo-Miocene has tracked the movement of the core site close to Hawaii during its movement north and westwards by sea-floor spreading. High K/Al and Rb/Al ratios in the Cretaceous-Palaeocene, Early Eocene and Late Eocene-Early Oligocene record silicic volcanism. In the Miocene-Pliocene, Rb enrichment and K depletion records a switch to higher illite contents in Asian dust.

rates ranging from <1 to >50 mm/Ma, significantly slower than the accumulation rate of the associated sediments (Ku, 1976). The nodules are therefore maintained at the sediment surface, probably by the activities of benthic organisms (Piper & Fowler, 1980), but they are sporadically found buried in the sedimentary section (Cronan & Tooms, 1967). Finely dispersed Fe and Mn oxyhydroxides are ubiquitous components of oxidized pelagic clays (Krishnaswami, 1976). Such phases are unstable under the reducing conditions normally encountered at depth in marginal and some deep-sea environments, so that they are only found in the sea-floor record in slowly accumulating deposits that have remained oxygenated since the deposits formed.

The abundance of ferromanganese nodules in the surface 2.5 m of the sediment column in gravity and piston cores from the central Pacific is roughly the same as the surface concentration (Cronan & Tooms, 1967). Deep-sea drilling has revealed that ferromanganese nodules, micro-nodules and crusts are common constituents of all slowly-accumulating pelagic sediments since the Late Cretaceous (Glasby, 1978, 1988), occurring preferentially on or close to hiatuses (Usui & Ito, 1994). They are mineralogical and chemically similar to modern nodules and crusts, and there is little evidence of post-depositional alteration of their compositions. This is in contrast to some nodules preserved in the geological record on land, where evidence of chemical and mineralogical alteration is found (Jenkyns, 1977; Margolis, Ku, Glasby, Fein, & Audley-Charles, 1978). The isotopic composition of Sr in the buried deep-sea deposits is closely similar to coeval seawater (Ito, Usui, Kajiwara, & Nakano, 1998), confirming the fossil nature of the nodules and that they are not slumped deposits from surface sediment horizons. In view of the fact that a decrease of dissolved oxygen content by 50–70% would have caused dissolution of the Mn phase of such deposits, the combined evidence suggests that the environmental conditions in most regions of the pelagic ocean, and in particular the oxygen content of bottom waters, have remained more or less constant for the last *ca.* 60 million years.

5.2. Records of Sea Floor Hydrothermal Inputs to the Ocean

Hydrothermal metalliferous sediments accumulate at mid-ocean spreading centres. The metals are derived from sub-sea floor rocks that are chemically altered by reaction with circulating seawater and are precipitated when hot solutions mix with cold, ambient seawater above the sea floor (Edmond et al., 1979a, 1979b). Sulphide chimneys and mounds form from sulphide- and metal-laden high-temperature hydrothermal waters close to vents (Bonatti, 1975; Haymon & Kastner, 1981). Fine-grained sulphide-rich particulates are deposited at greater distances from the ridge crests, and this material is subsequently oxidized when in contact with normal bottom waters (German et al., 1993). More widely dispersed poorly crystalline and amorphous Fe and Mn oxyhydroxides form when hydrothermal waters are cooled by rapid mixing with ambient oxygenated seawater (Boström & Peterson, 1966; Piper, 1973; Edmond et al., 1979a; Dymond, 1981). A wide range of metals and metalloids are coprecipitated or scavenged by the oxyhydroxides. In addition, poorly crystalline clay minerals form authigenically within the hydrothermal deposits (Dymond et al., 1973; McMurtry & Yeh, 1981; Haymon & Kastner, 1986). The degree of major element enrichment in ridge crest sediments can be seen in the exceptionally high Fe/Al, Mn/Al and Si/Al ratios in a transect across the East Pacific Rise (Figure 6). Similar relative enrichments of other metals and metalloids, including As, B, Cd, Cu, Ni, Pb, Tl and Zn, are found on other EPR transects (Piper, 1973). The enriched metals accumulate at significantly higher rates compared with hydrogenous sedimentation in the central ocean basins (Bender et al., 1971; Dymond & Veeh, 1975).

The hydrothermal sediments are buried beneath normal pelagic sediment when sea-floor spreading transports the newly-formed sea floor away from the ridge

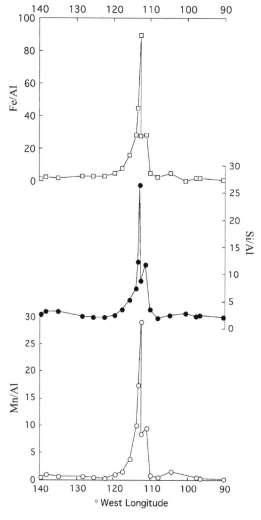

Figure 6   Longitudinal transect of metal enrichments in surface sediments across the East Pacific Rise. Data from Boström and Peterson (1969).

crests, so that all sedimentary sections in the ocean basins are floored by unique basal Fe- and Mn-rich deposits (Boström, Joensuu, Valdes, & Riera, 1972; Cronan et al., 1972; Dymond et al., 1973; Dymond & Corliss, 1977).

The history of metalliferous sedimentation in the North Pacific since the Late Cretaceous is shown by the record from core LL44-GPC discussed previously (Kyte et al., 1993). Fe/Al and Mn/Al maxima in the Cretaceous/Palaeocene and Eocene sections of the core reflect the accumulation of oxyhydroxides and Fe-rich smectite, as also recorded by maxima in Si/Al, formed when the site lay close to the East Pacific Rise (Figure 7). The smectites contain large amounts of Fe in octahedral positions, accounting for the low Al contents of the bulk sediments (Corliss, Lyle, & Dymond, 1978). Phosphorus maxima probably reflect adsorbed phosphate on volcanogenic Fe oxyhydroxides (Berner, 1973) as well as small amounts of insoluble fish debris (Doyle & Riedel, 1979). Maxima in As, Sb and V (not shown) coincide closely with the principal Fe/Al peaks, probably reflecting adsorption of the metal oxyanions on freshly precipitated Fe oxyhydroxides. The minor Mn maximum in the Oligocene section of the core records hydrogenous accumulation of oxyhydroxides, and is accompanied by high abundances of Co and Ni (not shown), two metals that are characteristically enriched in hydrogenous deposits throughout the Pacific.

6. ELEMENTAL PROXIES FOR PALAEOPRODUCTIVITY

A variety of sedimentary proxies for estimating marine palaeoproductivity have been developed in an effort to assess, through time, the contribution of export production in the sea to the global carbon cycle, as well as the factors that promote the preservation of organic matter in marine deposits. The methods utilize microfossil assemblages, stable isotope ratios, biomarker distributions, the concentrations and/or burial fluxes of organic carbon and biogenous silica (opal), as well as several major and minor elements. All of these approaches when used in concert can provide critical insights into the production and preservation of biogenous materials in marine sediments that can then be related to other information on climate states, ocean circulation and environmental conditions at the sea floor.

6.1. Barium

Pelagic sediments of the equatorial Pacific are enriched in barium relative to crustal abundances (Revelle, 1944), the additional Ba occurring as barite ($BaSO_4$) (Goldberg & Arrhenius, 1958). Although these authors were not able to identify an organism responsible for secreting the barite, they nevertheless concluded that the euhedral barite crystals, such as those illustrated by Arrhenius (1963), formed authigenically in the sediment following the release of Ba from benthonic faecal pellets. They further concluded that such Ba enrichment reflects the higher productivity characteristic of the equatorial Pacific (Graham, 1941; Barber et al., 1996). Goldberg and Arrhenius (1958) also showed that the Ba/Ti ratio increased

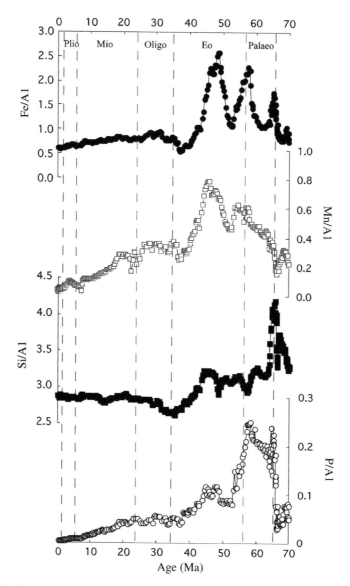

Figure 7 The Cenozoic history of metalliferous sedimentation in the North Pacific since the Late Cretaceous as recorded by the distribution of Fe/Al, Mn/Al, Si/Al and P/Al in core LL44-GPC from the North Pacific (30.32°N, 157.83°W, 5,705 m water depth). Data from Kyte et al. (1993). Hydrothermal inputs were high during the Cretaceous/Palaeocene and Eocene when there are Fe/Al, Mn/Al and Si/Al maxima. Phosphorus is enriched due to adsorption on volcanogenic Fe oxyhydroxides. The minor Mn maximum in the Oligocene section of the core records hydrogenous accumulation of oxyhydroxides.

during the last glacial in a Pacific core at 5°N, supporting the inferred glacial productivity increase (Arrhenius, 1952). Since these early discoveries, the variability of sedimentary Ba has become one of the most widely used proxies for palaeoproductivity.

Relatively high concentrations of discrete barite particles, on average 1 μm in diameter, occur in near-surface waters of the oceans (Dehairs, Chesselet, & Jedwab, 1980; Dehairs, Lambert, Chesselet, & Risler, 1987; Bishop, 1988; Dehairs et al., 1990; Dehairs, Stoobants, & Goeyens, 1991), and especially in areas of high new production (Dehairs, Baeyens, & Goeyens, 1992; Cardinal et al., 2005). The barite is mainly associated with biogenous aggregates in surface and near-surface waters, especially siliceous debris (Bishop, 1988), but occurs as free particles in deeper waters (Dehairs et al., 1990; Bertram & Cowen, 1997). Such barite particles constitute most of the suspended barite in the water column, although in some regions up to 50% of the standing stock of particulate Ba may be present in skeletal and organic debris. Since near-surface seawater is undersaturated with respect to barite (Monnin, Jeandel, Cattaldo, & Dehairs, 1999), precipitation of this phase requires the existence of microenvironments where concentrations of either dissolved Ba or sulphate are elevated, possibly in decaying organic matter (Chow & Goldberg, 1960). The possible direct biogenic formation of barite in the oceanic water column has prompted a search for the identity of Ba-secreting organisms. While some planktonic marine organisms do form barite intracellularly (Gayral & Fresnel, 1979; Rieder, Ott, Pfundstein, & Schoch, 1982), and xenophyophoria, giant benthic protozoa, also produce euhedral barite crystals within the protoplasm (Tendal, 1972; Gooday & Nott, 1982), no associations between barite and organisms has been described in plankton samples, and no excessive Ba enrichment has been found in modern plankton samples (Martin & Knauer, 1973). Laboratory culture experiments have shown that Ba can be removed from the solution during growth of a wide range of phytoplankton (Riley & Roth, 1971; Fisher, Guillard, & Bankston, 1991; Ganeshram, François, Commeau, & Brown-Leger, 2003) and a heterotrophic soil bacterium (Gonzales-Munoz et al., 2003) in proportion to the Ba concentration in the growth medium, but Ba is not incorporated in living cells (Sternberg, Tang, Ho, Jeandel, & Morel, 2005). Barite microcrystals are only formed during the decay of phytoplankton cultures in the dark (Ganeshram, Pedersen, & Murray, 1992), consistent with their formation in microenvironments within decaying masses of organic matter suspended in near-surface waters (Chow & Goldberg, 1960; Bishop, 1988). However, the culture conditions frequently employed in such experiments evidently cause Ba to be removed by Fe hydroxide scavenging on (organic) particle surfaces, rendering the interpretation of many laboratory results equivocal (Sternberg et al., 2005).

Barium is also incorporated into other biogenous phases in ocean waters, notably acantharian celestite ($BaSO_4$ is isostructural with $SrSO_4$) (Bernstein, Byrne, Betzer, & Greco, 1992) where it can constitute up to 0.5% by weight (Arrhenius, 1963). Acantharian tests are completely solubilized in the upper water column, however, rendering this vector an insignificant source of biogenic Ba in marine sediments.

The settling flux of Ba through the oceanic water column, as measured by moored sediment traps, is tightly correlated with organic carbon flux out of surface waters, changing seasonally with total primary production (Dymond, 1985; François, Honjo, Manganini, & Ravizza, 1995; Jeandel, Tachikawa, & Dehairs, 2000; Balakrishnan Nair, Ittekkot, Shankar, & Guptha, 2005). Below *ca.* 1,500 m water depth, the settling flux of Ba increases while that of organic carbon decreases due to the microbial degradation of organic matter and the additional formation of barite in settling particles. Away from regions of significant terrestrial organic matter supply, $C_{organic}$/Ba ratios decrease in a predictable manner in all ocean basins, providing a basis for using Ba fluxes for palaeoceanographic reconstructions of past productivity variations (Dymond, 1985; François et al., 1995; Fagel, Dehairs, Peinert, Andre, & Antia, 2004).

As well as barite microcrystals, Ba in marine sediments is hosted in several other phases, notably lithogenous material derived from crustal rocks, and biogenous debris, both skeletal material and degraded soft tissue produced in the ocean. The insoluble nature of barite leads to its high degree of preservation in slowly-accumulating sediments — constituting a "dissolution residue" (Dymond, 1981) — which have not been subjected to reducing conditions (see below). In alumino-silicates, Ba mainly occurs in K-feldspars and micas, where it substitutes isomor-phically for K (Puchelt, 1969–1978). Various attempts have been made to correct total Ba contents of sediments for the contribution from such lithogenous sources in order to arrive at estimates of the abundance of biogenous Ba. Normalization to Al (Section 3) is a common method, combined with estimates of the Ba/Al ratio of average crustal rocks or of the aluminosilicate debris in a specific sediment sample. Alternatively, a correction for aluminosilicate contributions has been made by ref-erence to the total K content (Schneider et al., 1997). A secure knowledge of the Ba content of associated mineral phases or the bulk lithogenous material is required in order to derive measures of excess or biogenous Ba (Reitz, Pfeifer, de Lange, & Klump, 2004). Direct determination of the barite content of sediments, either by means of chemical extraction (Paytan, Kastner, & Chavez, 1996) or by direct enumeration using electron microscopy/X-ray spectrometry (Robin et al., 2003), circumvents some of the problems inherent in correction procedures when the Ba content of the other carrier phases is poorly constrained.

Barium is enriched in Plio-Pleistocene sediments beneath the equatorial up-welling zone of the Indian Ocean, but Palaeocene-Eocene sections of deep sea drilling project (DSDP) cores lying north of this productive zone at present are similarly enriched, suggesting that the accumulation of excess or biogenous Ba has tracked the northward movement of the Indian Plate beneath the equatorial region during the Neogene (Schmitz, 1987a). Sedimentary opal concentrations are also significantly higher in the Ba-enriched sections of the cores, supporting the in-ference that Ba accumulates together with skeletal planktonic material in marine sediments. Similar Ba enrichments in higher-resolution sections from the Atlantic, Pacific and Southern Oceans that are also variably enriched in opal and organic carbon have also been identified, and used to study changes in ocean production on glacial–interglacial timescales. Thus, using the Ba proxy, productivity appears to have been higher during glacial (Matthewson et al., 1995; Gingele, Zabel, Kasten,

Bonn, & Nurnberg, 1999), interglacial (Shimmield, 1992; Shimmield, Derrick, Mackensen, Grobe, & Pudsey, 1994; Bonn, Gingele, Grobe, Mackensen, & Fütterer, 1998) or both glacial and interglacial intervals (Rutsch et al., 1995; Schneider et al., 1997; Thomson et al., 2000) of deep ocean cores, pointing to the modulation of ocean productivity by glacial–interglacial climate changes. In the Antarctic region, higher productivity is found during relatively ice-free interglacial stages, whereas in the central Arabian Sea, interglacial maxima in production record increased nutrient supplies via upwelling of more vigorously circulating intermediate and Antarctic Bottom Waters. These changes occur on the eccentricity and obliquity (ice-volume) frequencies rather than the precessional frequency that would be expected based on the low-latitude insolation forcing of coastal upwelling in the coastal Arabian Sea. In contrast, biogenous Ba enrichments coherent with organic carbon maxima are observed at approximately 21 kyr intervals on the Zaire Fan in the eastern equatorial Atlantic (Rutsch et al., 1995; Schneider et al., 1997), showing that productivity variations occur at the precessional frequency at this low-latitude site (Figure 8). This pattern is not replicated by the variations in opal content in the same core, suggesting that siliceous plankton is not the main carrier of the barite in this region.

Palaeoproductivity estimates for the equatorial Pacific using the Ba and other inorganic geochemical proxies (see Sections 6.2 and 6.3) have been controversial. The well-expressed cycles of $CaCO_3$ content in the Swedish Deep-Sea Expedition cores were initially taken to reflect changing glacial–interglacial productivity, with a superimposed effect of carbonate dissolution (Arrhenius, 1952, 1959). Although carbonate dissolution was thought to have increased during glacial stages due to enhanced flow of Antarctic Bottom Water, greatly increased production levels were invoked to explain the carbonate-rich glacial horizons in the equatorial cores. Subsequent studies (Berger, 1973) have emphasized variations in sea-floor preservation as controlling the carbonate cycles, and indeed it is now known that the depth of the lysocline has fluctuated markedly on glacial–interglacial timescales (Farrell & Prell, 1989). However, the change in carbonate saturation required to produce the observed carbonate cycles is quite large, and variable carbonate supply (production) appears to be a better explanation (Archer, 1991). In addition, invariant foraminiferal preservation indices in cores collected above the lysocline which also show marked carbonate cycles also argue for variations in production (Adelseck & Anderson, 1978). Similar sets of issues (production vs. preservation) have been encountered with the use of organic carbon (Emerson & Hedges, 1988) and biogenous opal (Archer, Lyle, Rodgers, & Froelich, 1993) for palaeoproductivity reconstructions, so that application of another palaeoproductivity proxy to this problem may provide independent evidence for changes in glacial–interglacial production in the oceans.

The settling flux of Ba intercepted by moored sediment traps is highest on average within five degrees of the equator at 140°W (Dymond & Collier, 1996), coincident with the highest primary production rates (Barber et al., 1996) and water column organic matter fluxes (Honjo, Dymond, Collier, & Manganini, 1995). Likewise, Ba is enriched in the surface sediment at the equator, although the peak in Ba content is much more narrowly centred on the equator compared with

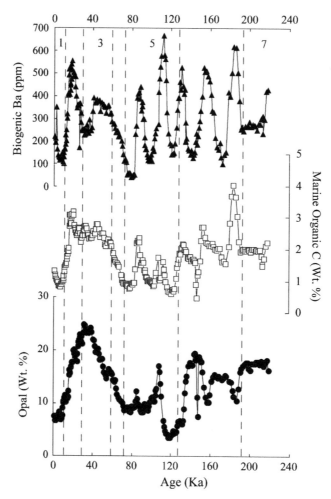

Figure 8 Temporal record of biogenous Ba, marine organic carbon (MOC) and opal over the last 230 kyr in core GeoB 1083 from the Zaire Fan, eastern tropical Atlantic. Data from Schneider et al. (1997). MOC was derived from total organic carbon values using the C isotopic composition of the bulk sediment and a mixing equation. Biogenous Ba, derived by subtracting aluminosilicate Ba from total Ba using the Ba/K ratio in crustal rocks, fluctuates coherently with MOC, with higher values in MIS 2, 3, 5 and 6. Opal values show increases in MIS 2, 3, 5 and 6, but they are not coherent with Ba or MOC.

the water column settling fluxes. The accumulation rate of barite in the surface sediment also peaks at the equator (Paytan et al., 1996) as does the bulk sediment Ba/Ti and Ba/Al ratio (Murray & Leinen, 1993; Murray, Knowlton, Leinen, Mix, & Polsky, 2000) (Figure 9). Thus, biogenic barium is enriched in Holocene equatorial sediments.

Estimates of the accumulation rate of Ba in surface sediments have been combined with sea surface primary production determinations to construct productivity algorithms for estimating palaeoproductivity (Dymond, Suess, & Lyle, 1992;

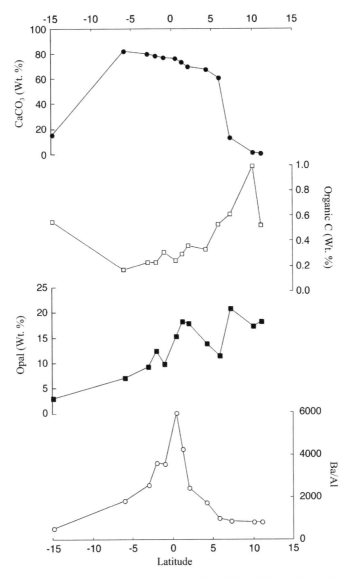

Figure 9 Distribution of CaCO₃, opal, Organic C and Ba/Al in surface sediments across a north-south transect in the central equatorial Pacific (data from Murray and Leinen, 1993).

François et al., 1995; Paytan et al., 1996), and these have been widely applied to long core sections recovered throughout the oceans (Gingele & Dahmke, 1994; Bonn et al., 1998; Gingele et al., 1999; Klump, Hebbeln, & Wefer, 2001). The widespread evidence for syndepositional focusing in deep-sea sediments (Marcantonio et al., 2001), including the Holocene sections of many equatorial Pacific cores (François, Frank, Rutgers van der Loeff, & Bacon, 2004; Loubere, Mekik, François, & Pichat, 2004; Kienast, Kienast, François, Mix, & Calvert, 2007, in press), renders such algorithms unreliable because the sediment-derived Ba accumulation rates (both barite and excess Ba) of many of the cores used for this estimate are too high by unknown factors.

This problem can be appreciated by examining the glacial–interglacial records of barite and bulk sediment barium enrichment and accumulation in piston cores from the central equatorial Pacific (Paytan et al., 1996; Murray et al., 2000). Barite concentrations are highest in both glacial and interglacial stages when $CaCO_3$ contents are lowest, reflecting variable carbonate dilution of insoluble barite (Figure 10). Conversion of these data to accumulation rates, using the stratigraphically derived linear sedimentation rates, reveals clear maxima in glacial stages MIS 2–8, and also in MIS 11, coherent with the barite content. The use of sediment accumulation rates derived in this way is likely to exaggerate these glacial increases in the accumulation rates if significant sediment focusing during the glacials has occurred (Frank et al., 1995; Frank, Gersonde, Rutgers van der Loeff, Kuhn, & Mangini, 1996; Marcantonio et al., 2001; François et al., 2004; Anderson & Winckler, 2005; Winckler, Anderson, & Schlosser, 2005). The barite maxima in low carbonate intervals (MIS 6, 8 and 10) due to lowered dilution by this major phase together with the inflated sedimentation rates caused by focusing in the glacial intervals produces sharp barite accumulation rate maxima that are more likely artifacts. Correction for focusing using $^{230}Th_{excess}$ shows that there has been no significant increase in glacial productivity in the equatorial Pacific west of the Panama Basin using the abundance of barite as a palaeoproductivity proxy (Marcantonio et al., 2001), consistent with observations using $^{231}Pa/^{230}Th$ (Pichat et al., 2004), benthic foraminiferal transfer functions (Loubere, 1999, 2000) and ^{230}Th-normalized sediment fluxes (Loubere et al., 2004).

Similar changes in glacial and interglacial palaeoproductivity in the Atlantic sector of the Southern Ocean south of the polar front but higher glacial Ba accumulation north of the polar front are also revealed by applying radionuclide constant flux tracers to derive true vertical fluxes of Ba (Nürnberg, Bohrmann, Schlüter, & Frank, 1997; Frank et al., 2000). In view of these considerations, the relative Ba enrichment in bulk sediments in the equatorial Pacific core appears to be a better proxy for changes in palaeoproductivity because both Ba and Al are diluted to the same extent when carbonate content changes. Broad maxima in Ba/Al ratios occur in MIS 2–3, 6, 7, 8, 9/10 and 11/12, which are not coherent with the sharp maxima in the barite accumulation rate, themselves artificially elevated by carbonate dissolution and sediment focusing.

Downcore Ba records can be compromised by the post-depositional loss of accumulated barite in suboxic sediments (McManus et al., 1998; McManus, Berelson, Hammond, & Klinkhammer, 1999) and especially in anoxic sediments in which either minimal (van Os, Middelburg, & de Lange, 1991; Schenau, Prins, deLange, & Monnin, 2001) or significant sulphate depletion (Brumsack & Gieskes, 1983; Brumsack, 1989; van Os et al., 1991; Von Breymann, Emeis, & Suess, 1992) has occurred. For this reason, barium records in many continental margin settings — where relatively high organic matter accumulation leads to lower redox potentials at shallow depth — do not record palaeoproductivity changes (Shimmield et al., 1994; Ganeshram et al., 1999). Direct measurements of Ba efflux from and

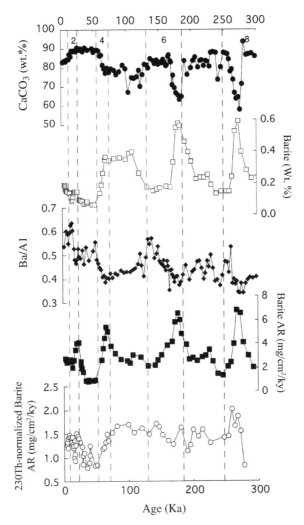

Figure 10 Temporal record over the last 280 kyr of % CaCO3, % barite, Ba/Al, and the barite accumulation rate, estimated using the stratigraphically derived mass accumulation rate (Paytan, 1995) and the ^{230}Th-normalized mass accumulation rate to correct for winnowing/focusing (François et al., 2004) in core TT013-PC72 from the central equatorial Pacific (0.11°N, 139.40°W, 4,298 m depth). Carbonate, Ba, and Al data from Murray et al. (2000), and barite data from Paytan (1995).

dissolved pore water Ba enrichment in sediments underlying the California Current (McManus, Berelson, Klinkhammer, Kilgore, & Hammond, 1994), the Arabian Sea and the equatorial Pacific (Paytan & Kastner, 1996; McManus et al., 1998) show that Ba is remobilized within suboxic sediments, as well as anoxic sediments, even though pore-water sulphate depletion is not detected. Evidently, sulphate reduction occurs in microenvironments in many oxic and suboxic settings (Berelson et al., 1996), and this leads to dissolution of some of the deposited barite. The concurrent determination of solid-phase U and Ba contents of a sediment core can be used to screen out those sites that have had significant post-depositional Ba remobilization (McManus et al., 1998) because U is known to become enriched in sediments under suboxic conditions (Anderson, 1982; Klinkhammer & Palmer, 1991). Uranium concentrations can therefore be used as a proxy for high rates of diagenetic organic matter oxidation and hence vertical organic flux to the sea floor (Kumar et al., 1995; Anderson et al., 1998). This aspect of U geochemistry is discussed in Section 7.2.1.

6.2. Phosphorus

Phosphorus is a limiting macro-nutrient for algal growth, along with fixed nitrogen, in many parts of the modern ocean, and its accumulation in marine sediments has been used as a palaeoproductivity proxy. Organically bound and skeletal P is exported from the surface ocean in biogenous particles and aggregates and is diagenetically recycled within accumulating sediments and, together with a relatively small amount of organic P, may become fixed in an inorganic phosphate mineral phase and adsorbed by Fe oxyhydroxides.

The relative abundance of different forms of sedimentary P has been estimated by selective chemical extractions (Ruttenberg & Berner, 1992). These show that in most deep-sea sediments the major phosphorus-bearing component is francolite (or carbonate fluor-apatite: $(Ca, Mg, Na)_5(PO_4, CO_3)_3(F, OH)$). This phase also accumulates within modern nearshore and continental shelf settings (Ruttenberg & Berner, 1993), and occasionally forms enriched deposits in highly productive regions (Burnett, 1977; Price & Calvert, 1978; Baturin & Bezrukov, 1979; Schuffert, Kastner, & Jahnke, 1998; Schenau, Slomp, & DeLange, 2000). Subsequent winnowing of the deposits on some shelves concentrates the phosphatic materials into concentrated phosphorite beds (Baturin, 1971), which resemble some of the giant phosphorite ores in the geological record. A minor amount of P in unconsolidated marine sediments is retained in refractory organic matter, fish debris (Suess, 1981) or is adsorbed by authigenic Fe oxyhydroxides (Froelich, 1988). The latter phase is reductively dissolved during burial under suboxic and anoxic conditions (see Section 7), the released P potentially precipitating as francolite.

The phosphorus originally delivered to sediments is evidently preserved — even though the organic material which it partly constitutes is largely degraded diagenetically — by "sink switching" (Ruttenberg & Berner, 1993), namely the loss of P from organic association and the subsequent precipitation of francolite. This phase, together with fish debris (biogenous P), is included in an operationally defined authigenic/biogenous fraction soluble in dilute acetic acid (pH 4). The accumulation of organic P can therefore be tracked by the distribution over time of organic and authigenic/biogenous P (Filippelli & Delaney, 1995) or total P (Murray et al., 2000), since these components together generally constitute >70% of total P.

A record of palaeoproductivity from the burial of P from the highly productive continental slope off northwestern Mexico (Figure 11) shows that phosphogenesis, and hence productivity, was higher in interglacial stages compared with the glacials over the last 130 kyr (Ganeshram, Pedersen, Calvert, & François, 2002).

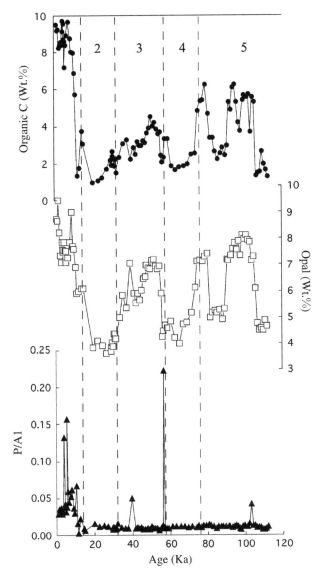

Figure 11 Temporal record of organic C, opal and P/Al over the last 114 kyr in core NH15P (22.68°N 106.48°W, 425 m water depth) from the northwestern Mexican slope (Ganeshram et al., 2002) show that phosphogenesis, and hence productivity, was higher in interglacial stages compared with the glacials over the last 130 kyr, as also recorded by organic C and biogenous silica contents.

Phosphorus enrichments at this site are coherent with other palaeoproductivity proxies, namely organic C and biogenous silica contents, and with the N isotopic proxy for higher water column denitrification. The accumulation of biogenous P is also higher during interglacials off the Peru margin (Burnett, 1980; Garrison &

Kastner, 1990) and in the Arabian Sea (Schenau et al., 2000) where the relationship with carbon burial and $\delta^{15}N$ is also similar, suggesting that the oceanic cycles of N and P are closely coupled on glacial–interglacial timescales (Ganeshram et al., 2002).

In contrast to the sedimentary records from productive continental margins, the record of P burial in the central equatorial Pacific appears to show increases in glacial stages compared to interglacial stages (Figure 12). The correspondence between the P/Al and Ba/Al ratios is strong and coherent with the $CaCO_3$ record, lending support to the suggestion that the fluctuations in $CaCO_3$ content at least partly reflect productivity variations.

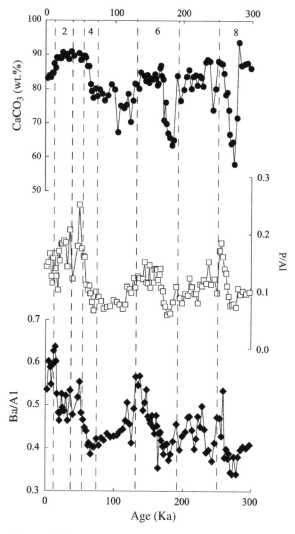

Figure 12 Temporal record of $CaCO_3$, P/Al and Ba/Al over the last 300 kyr in core TT013-72PC PC72 from the central equatorial Pacific (0.11 °N, 139.40 °W, 4,298 m water depth) showing P and Ba increases in glacial stages compared with interglacials. Data from Murray et al. (2000).

6.3. Aluminium

Surface sediments in the central equatorial Pacific are enriched in Al relative to Ti, Al/Ti ratios being significantly higher than whole crust ($\sim$16) or upper crustal values ($\sim$27) (Murray et al., 1993; Murray & Leinen, 1996). This enrichment parallels the markedly higher settling fluxes of biogenous debris on transects across the equator (Figure 13), and has been ascribed to preferential adsorption (scavenging) of Al relative to Ti by settling particles (Orians & Bruland, 1986), overwhelmingly comprising organic matter, $CaCO_3$ and opal in the equatorial Pacific (Honjo et al., 1995). Since the flux of particles is greatest close to the equatorial divergence due to the higher productivity (Chavez & Barber, 1987), the Al enrichment is directly related to organic particle flux, and has therefore been used as a proxy for productivity (Murray et al., 1993; Murray & Leinen, 1996).

Multivariate statistical analysis of the composition of the settling material on the 140°W transect suggests that the excess Al is largely associated with opal and that this component carries most of the Al to the sea floor (Dymond et al., 1997). This is consistent with reports of the biological removal of dissolved Al from seawater (Moran & Moore, 1988) and the presence of significant amounts of Al in diatom frustules in some bottom sediments (Van Bennekom, Jansen, Van der Gaast, van Iperen, & Pieters, 1989). Sedimentary Al/Ti ratios could therefore be used to track the accumulation of opal in the ocean, making this parameter a specific proxy for siliceous productivity.

Further investigation of the relationship between Al and Ti contents of the trapped particles yields a highly variable picture (Dymond et al., 1997). Although Al/Ti ratios are higher than upper crustal values close to the equator, the ratios fall below this value to the north and south (Figure 14). Moreover, significant non-zero intercepts on the Al axis in regressions of Al on Ti are only found at 5°N and 5°S; at the other stations on this transect, including moorings at the equator and 2°N and

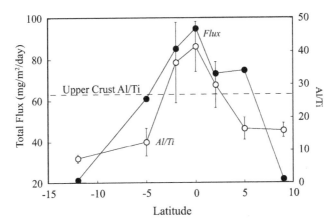

Figure 13 Total settling flux of particulate material (Honjo et al., 1995) and the Al/Ti ratio of bulk sediment trap material (Dymond, Collier, McManus, Honjo, & Manganini, 1997) at US JGOFS moorings across the equator at 140°W. Locations and depths of the traps are given in Table 1.

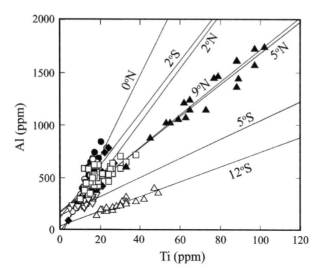

Figure 14   Regressions of Al on Ti for sediment trap samples from a trans-equatorial Pacific mooring transect at 140°W (Dymond et al., 1997). Model II regressions (Ricker, 1984). Intercepts are different from zero only for the moorings at 5°N and 5°S. Slopes of the regressions (Table 1) give the Al/Ti ratios of the settling material (aluminosilicate plus opal) that is added to the "excess"amounts of Al at these sites.

Table 1   Regressions Coefficients for Al and Ti Contents of Sediment Trap Samples from the US JGOFS Trans-Equatorial Mooring Transect at 140°W.

Latitude	Trap depth (m)	Slope	Intercept	r^2
9°N	2,250	16.03 ± 2.1	126.28 ± 156.0	0.93
5°N	2,100	16.63 ± 3.3	143.17 ± 79.6	0.81
2°N	2,200	29.84 ± 7.0	−74.77 ± 124.6	0.74
0°N	2,284	41.50 ± 7.6	−100.55 ± 106.7	0.85
0°N	3,650	45.56 ± 12.1	−134.32 ± 190.0	0.66
2°S	3,593	36.57 ± 12.2	−53.78 ± 182.8	0.46
5°S	2,316	12.55 ± 4.2	94.64 ± 54.3	0.52
12°S	1,292	7.50 ± 1.2	26.83 ± 38.2	0.87

Source: Dymond et al. (1997).

2°S, there is no significant amount of excess Al in the particles, but the slopes of the regressions are variable (Table 1). The nature of the regressions (a non-zero intercept at some stations and variable slopes) imply that there are at least two particle fractions in addition to any excess scavenged Al in the settling material (Timothy & Calvert, 1998), possibly aeolian dust (Murray et al., 1993) and opal (Dymond et al., 1997), and that the proportions of this additional material are site specific. This may be related to the position of the ITCZ at 4°N and the equatorial divergence where the highest production is observed.

Systematic downcore variations in Al/Ti close to the equator ranging from upper crustal values to values similar to the surface sediment values suggest that the Al enrichment carried by particles is preserved in the bottom sediment, even though some of the biogenous phases dissolve or are recycled (Murray et al., 1993). In the unique situation where the sediment is composed overwhelmingly (>99%) of biogenous phases, as in this particular case, Al is a poor proxy for aluminosilicate debris, and is the reason for the selection of Ti as a proxy for terrigenous aluminosilicate debris in this region.

As in the case of settling particulate material (Figure 14), downcore variations in Al and Ti contents are also strongly linearly correlated, with a variable non-zero intercept that represents the excess Al content (Timothy & Calvert, 1998). This relationship is best interpreted as the addition of aluminosilicate material with a more or less constant Al/Ti ratio to an aluminous sediment component that is devoid of Ti. This additional component could be scavenged Al, as originally hypothesized (Murray et al., 1993), Al originally incorporated in opaline shells (Van Bennekom et al., 1989; Dymond et al., 1997) or an aluminous phase that forms diagenetically in the sediment (Mackin & Aller, 1984). It is not possible to choose between these possibilities with our current understanding of Al geochemistry of biogenous sediments.

The downcore distribution of Al/Ti in the central equatorial Pacific over the last 300 kyr (Figure 15) shows that the ratios are generally higher where the $CaCO_3$ content is high (MIS 3–4, 6, 7), although the correlation is not strong. Such a postulated relationship suggests an increasing importance of excess Al in lithogenous-poor samples. However, regression analysis of the Ti and Al values shows that excess (or "scavenged") Al (Murray & Leinen, 1996) contributes 177 ± 58 ppm Al to the sediments at this site, a minor component at most core depths. The Al enrichments in the carbonate-rich sections of this core have been interpreted as support for the carbonate variations being due primarily to productivity fluctuations (Murray et al., 1993). However, the changes in Al/Ti correlate poorly with either the Ba/Al ratio or the ^{230}Th-normalized barite accumulation rate in the same core (Figure 15). Thus, the higher excess Al contents at this site do not correlate with other palaeoproductivity proxies.

7. PROXIES FOR REDOX CONDITIONS AT THE SEA FLOOR AND IN BOTTOM SEDIMENTS

Many trace and minor elements accumulate in the solid phase of marine sediments as a result of post-depositional precipitation or adsorption from the bottom waters or from pore waters. These processes are primarily controlled, in turn, by redox reactions in response to the oxidative decomposition of deposited organic matter (Froelich et al., 1979). The reactions proceed in a well-defined sequence (Figure 16) during which oxidants are consumed and reduced species accumulate in the pore waters. Some trace metals that have multiple valency states at Earth surface conditions may be enriched as a consequence of such reactions,

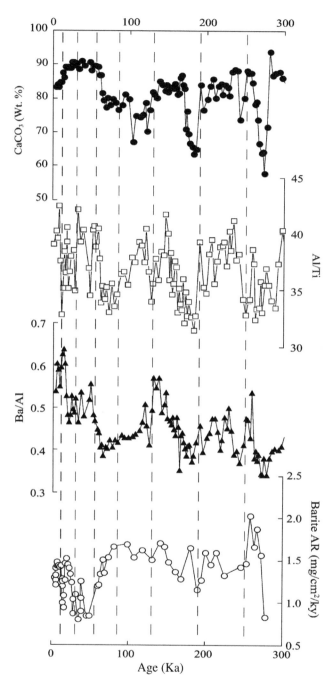

Figure 15 Temporal record of % CaCO$_3$, Al/Ti, Ba/Al (data from Murray et al., 2000) and the ^{230}Th-normalized accumulation rate of barite (Marcantonio et al., 2001) over the last 280 kyr in core TT013-PC72 (0.1°N, 139.4°W, 4,298 m water depth). Al/Ti is generally higher where the CaCO$_3$ content is high (MIS 3–4, 6, 7), but the changes in Al/Ti correlate poorly with either the Ba/Al ratio or the ^{230}Th-normalized barite accumulation rate in the same core.

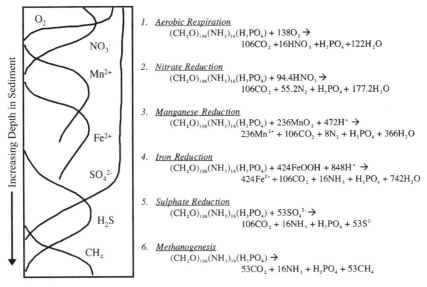

Figure 16 Schematic portrayal of the vertical sequence of oxidation-reduction reactions in marine sediments. Microbial consortia utilize the oxidants O_2, NO_3^-, Mn(IV), Fe(III) and SO_4^{2-} to metabolize buried organic matter, as shown by the idealized reactions on the right, formulated following Richards (1965) and Hartmann, Müller, Suess, and van der Weijden (1973) which are ordered sequentially (1–6) either by the decreasing free energy yield for each mole of carbon oxidized (Froelich et al., 1979) or by the physiological requirements of the respective microbial consortia (McCarty, 1972). Downcore profiles on the left depict the concentration gradients of the pore water constituents that are monitors of these reactions. Adapted from Eganhouse and Venkatesan (1993).

contributing to the *authigenic* fraction of the sediments. Such enrichment occurs either because the elements have different solubilities in oxygenated or oxygen-deficient seawater or are partitioned between the solid and solution phases to different extents under different redox conditions. Other metals are precipitated as sulphides within sediments in the presence of dissolved H_2S species produced by reduction of seawater sulphate when it is used as an oxidant during microbial organic matter degradation. Sedimentary records of relative enrichments and depletions of such authigenic metals and metalloids have been used to study both the geochemical behaviour of the elements in the ocean as well as the chemical state of ocean waters and bottom sediments in the past. This has been made possible by the rapid advances over the last two decades in our understanding of the biogeochemistry of trace and minor elements in the ocean, as summarized by Bruland & Lohan (2003).

Elements with variable valency can be added to accumulating sediments in three ways: (i) by diffusion from bottom waters and subsequent precipitation or adsorption; (ii) by release to pore waters from degrading organic matter followed by precipitation or adsorption; and (iii) by dissolution of a solid phase and subsequent fixation within the deposits. These vectors therefore contribute a "marine fraction" of elements to sediments (Piper, 1994) or "excess" metals over and above those supplied in detrital minerals and rock fragments. The proportion of such metals in a given sediment sample can be determined by correcting the total content for the likely contribution from lithogenous sources, taking into account the chemical composition of marine planktonic organic matter (Martin & Knauer, 1973; Collier & Edmond, 1984; Brumsack, 1986). The balance between organic sources and seawater sources will, of course, vary widely depending on the flux of organic matter to the sea floor, the sedimentation rate of the deposit and the redox conditions of the bottom water and the sediment. In many situations, organic sources of authigenic trace elements are dominant for Ag and Cd, whereas seawater itself dominates as a source for Mo, Re and U; Cr, V, Ni, Cu and Zn appear to fall between these two extremes (Piper, 1994; Piper & Medrano, 1994; Piper & Isaacs, 1995).

As indicated previously, trace and minor elements whose behaviours are most likely to be different under oxic and anoxic marine conditions fall into two categories. The first includes those elements whose valency can vary as a function of the prevailing redox potential. Included in this group are: Mn, which forms a highly insoluble oxyhydroxide where oxic conditions prevail; I, which, as the iodate ion, has a strong adsorptive affinity for organic matter in the presence of oxygen; and Cr, Mo, Re, U and V, which occur as highly soluble anionic species in oxic waters but are reduced to reactive or insoluble species of lower valency under anoxic conditions. The second category includes elements whose valency does not change, such as Ag, Cd, Cu, Ni and Zn, but which form highly insoluble sulphides and are usually removed from solution in the presence of H_2S. Study of the geochemistry of these groups of metals in sedimentary rocks has a long history, and has contributed significantly to the understanding of the environment of formation of sedimentary metalliferous ores (Maynard, 1983; Coveney & Glascock, 1989; Parnell, Ye, & Chen, 1990; Jones & Manning, 1994). The relative degrees of enrichment and depletion of the trace elements in oxic, reducing and anoxic deposits to be discussed in this section are summarized in Figure 17.

Because the authigenic accumulation of many metals occurs in reducing sediments, their enrichments have been used to infer conditions, both in the water column and in bottom sediments, that were more reducing in the past. Water column anoxia, and therefore surficial sedimentary anoxia, occurs where the rates of consumption of dissolved oxygen and other oxidants exceed their rates of replenishment and the large reservoir of sulphate in seawater becomes the dominant oxidant. This may occur in coastal basins where the ventilation of deep waters is restricted by a topographic barrier, or at intermediate water depths in the open ocean when ventilation is slow and oxidant demand is high due to a high settling flux of organic matter from the sea surface. Examples of the former category include the Black Sea, the Cariaco Trench and several high-latitude fjords (Richards, 1965), whereas larger areas of intermediate depth waters in the eastern Pacific and the Arabian Sea are examples of the second category (Brandhorst, 1959; Naqvi, Noronha, & Gangadhara Reddy, 1982; Naqvi, 1987). In addition, metals may also be sequestered authigenically where the rate of accumulation and burial of particulate organic matter (a higher organic rain rate) exceeds the rate of replenishment of dissolved oxygen from bottom waters by diffusion or irrigation, even

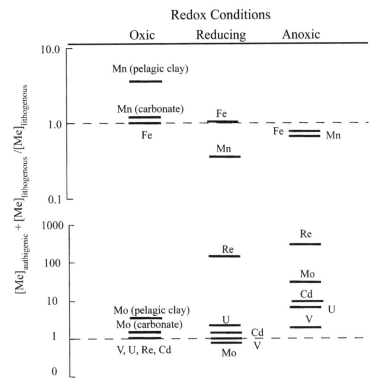

Figure 17 Enrichment and depletion factors for redox-sensitive metals in marine sediments (from Morford & Emerson, 1999). Iodine is omitted from this figure because its content in lithogenous detritus is poorly known (but probably very small).

though the bottom waters are fully oxygenated. This situation prevails in large areas along continental margins; the surficial sediments are oxic, but they become anoxic (sulphidic) at a variable depth below the sediment/water interface due to the microbial utilization of oxygen, nitrate and Fe and Mn oxyhydroxides during organic matter degradation, after which H_2S is produced as a byproduct of bacterial sulphate reduction (Figure 16).

Authigenic metal enrichments in sediments may therefore be used to infer anoxic bottom water conditions in the past, or a higher burial flux of organic matter, and hence higher productivity, in the past. In both cases, enrichments are directly due to the redox status of the depositional environment, but in the latter case the enrichments constitute an indirect proxy for productivity.

7.1. Proxies for Oxic Conditions of Sedimentation

7.1.1. Manganese

Manganese occurs as the Mn(II), Mn(III) and Mn(IV) valency states in Earth surface environments. The species Mn^{2+} and $MnCl^+$ are the principal soluble forms in seawater, but Mn(II) is thermodynamically unstable in the presence of oxygen and is sluggishly oxidized to insoluble Mn(III) and Mn(IV) oxides. Dissolved manganese is normally enriched in surface waters, reflecting the input of Mn as oxyhydroxide coatings on sediment particles and its partial release into solution. The element is scavenged throughout much of the oceanic water column below the mixed layer (Bender, Klinkhammer, & Spencer, 1977). An intermediate-depth Mn maximum occurs in the Oxygen Minimum Zone (OMZ) wherever oxygen concentrations fall below $100\,\mu M$ probably reflecting reductive dissolution of settling particulate Mn oxyhydroxides or the lateral transport of dissolved Mn that has diffused out of anoxic shelf and slope sediments (Klinkhammer & Bender, 1980).

Dissolved Mn(II) accumulates in the deep sulphidic waters of anoxic basins due to the reduction of Mn(IV) oxides that settle into the deeper waters from oxic waters above the redox boundary (Spencer & Brewer, 1971). A particulate concentration maximum is found immediately above the boundary, and this is formed by the oxidative precipitation of upward-diffusing Mn^{2+} from the anoxic waters. Manganese is thus actively recycled between oxidized and reduced forms across a redox boundary, whether that boundary be in a water column or in sea-floor deposits.

As discussed in Section 5.1, manganese is concentrated over its crustal abundance in sediments of the deep ocean (Murray & Irvine, 1895) and in surficial sediments on many continental margins (Figure 17), where it occurs as MnO_2 and MnOOH, collectively referred to as Mn oxyhydroxides (Grill, 1982; Kalhorn & Emerson, 1984; Murray, Balistrieri, & Paul, 1984). Its enrichment in pelagic sediments (Goldberg & Arrhenius, 1958) is due to the slow accumulation rates of terrestrial and biogenic detritus relative to the precipitation of authigenic Mn under oxic conditions (Krishnaswami, 1976). Manganese is preserved in pelagic sediments due to the very low settling fluxes of organic matter (Suess, 1980). The combination of the limited flux and the typically refractory nature of organic material at abyssal depths in the ocean ensures that oxic conditions are maintained to great depth in the bottom sediments.

In continental margin environments, where accumulation rates of organic matter are much higher, oxygen is completely consumed at relatively shallow depths during aerobic oxidation of labile organic matter, and under these conditions Mn recycles between surface and subsurface horizons (Figure 18). Burial of surface oxyhydroxides transports Mn(IV) into the subsurface reducing environment where it dissolves. Manganese(II) concentrations are therefore commonly much higher in the pore waters of such sediments immediately below the oxyhydroxide horizon. Dissolved Mn diffuses downwards into the sediment and upwards into the overlying oxic horizon where it is reprecipitated. This "zone refining" of Mn (Froelich et al., 1979) leads to surficial solid-phase Mn concentrations in hemipelagic and nearshore sediments that are often much higher than they are in pelagic sediments.

In spite of the accumulation of dissolved Mn in the bottom waters of anoxic basins, the concentrations of Mn, even in the presence of relatively high alkalinity, are insufficient for the precipitation of Mn(II) solid phase (i.e. $MnCO_3$) in the water column or underlying sediments. Hence, fully anoxic deposits — i.e. those that accumulate under oxygen-free bottom waters — characteristically have solid–phase Mn concentrations that are controlled entirely by the aluminosilicate fraction

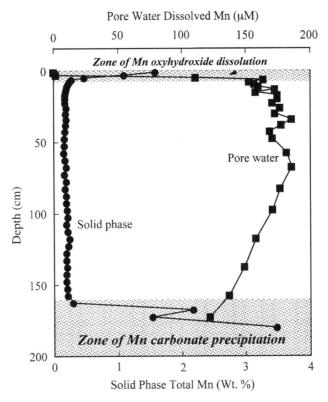

Figure 18 Downcore distribution of pore water and solid phase total Mn in core P8 from the Panama Basin (4.62°N, 82.99°W, 2,700 m water depth). Elevated solid phase Mn in the uppermost part of the core occurs as finely dispersed oxyhydroxides, which have accumulated from the oxidative precipitation of upwardly diffusing dissolved Mn, whereas the marked increase at the base of the core reflects the presence of authigenic mixed Mn carbonate $((Mn_{48}Ca_{47}Mg_8)CO_3)$. The dissolved Mn distribution reflects its production from buried surface oxyhydroxides and its downwards diffusion along the concentration gradient to the depth of Mn carbonate formation (Pedersen & Price, 1982).

(Figure 17). On the other hand, mixed Mn(II)–Ca–Mg carbonates are found in subsurface anoxic sections of marine sediments where surface horizons have high oxyhydroxide contents (Calvert & Pederson, 1996). The continual burial of surface oxyhydroxides as sedimentation proceeds increases the pore water concentration of dissolved Mn, which may reach levels where the ion activity product of Mn^{2+} and dissolved bicarbonate exceeds the solubility product of a carbonate phase (Figure 18). Even where the abundance of such authigenic carbonates is below typical analytical detection limits, pore water Mn profiles commonly show clear evidence of Mn removal within the sediments (Li, Bischoff, & Mathieu, 1969) that is almost certainly due to subsurface precipitation of a manganese carbonate phase.

The occurrence of Mn carbonate in marine sediments can be used as a proxy for original oxygenated bottom water conditions because the mechanism for increasing the concentration of dissolved Mn in sediment pore waters can only operate where the surface sediments accumulate Mn oxyhydroxides. These phases dissolve when they are buried below the oxygenated surface sediments and may saturate the interstitial waters. Such "manganese pumping" can only occur beneath oxygenated bottom waters. In contrast, the bottom sediments of anoxic basins or oxygen minima lack authigenic Mn enrichments because there is no Mn oxyhydroxide cap on the sediments, and therefore no effective mechanism for increasing Mn concentrations in the pore waters above the solubility product of a mixed carbonate. Thus, the 'manganese pump' only operates under oxic bottom waters, an observation that establishes the validity of sedimentary manganese carbonate occurrences as indicators of bottom water oxygenation history, as discussed by Calvert, Bustin, and Ingall (1996) and Calvert and Pederson (1996).

7.1.2. Iodine and bromine

The distribution and concentration of the halogens iodine and bromine in marine sediments are governed by the organic fraction of marine sediments and by diagenetic reactions involving organic matter degradation (Bojanowski & Paslawska, 1970; Price, Calvert, & Jones, 1970). Iodine occurs as I(V) in the form of the iodate (IO_3^-) ion in oxic seawater, with minor occurrences as the iodide ion (I^-) in some near-surface waters, presumably due to biochemical processes (Wong, 1980). In anoxic basins it occurs as the iodide ion (I^-) (Wong & Brewer, 1977; Emerson, Cranston, & Liss, 1979). In contrast, bromine behaves conservatively and occurs everywhere in marine waters as the bromide (Br^-) ion. The concentrations of both elements are closely related to the content of organic matter in oxic sediments, whereas under anoxic conditions iodine is depleted relative to bromine (Price et al., 1970; Price & Calvert, 1977; Shimmield & Pedersen, 1990). The enrichment of iodine in oxic sediments is due to the adsorption of the iodate ion by the high molecular weight organic fraction of the sediment, whereas iodide, the stable form in anoxic waters, is not adsorbed as effectively (François, 1987). Iodine is released from the organic matrix when surficial sediments with high adsorbed I contents are buried under anoxic conditions as a result of the reaction between dissolved pore water sulphide and sedimentary humic substances. Iodine contents therefore decrease with burial, whereas the concentration of bromine changes to a lesser degree, probably as a consequence of the degradation of the organic matrix itself (Pedersen & Price, 1980). Hence, I enrichments relative to organic carbon and Br contents in marine sediments can be used as proxies for oxygenated bottom water conditions in the past, as long as the sediments have remained oxic since deposition.

Using this approach, it was found that the organic-rich sediments deposited during the Last Glacial Maximum (LGM) in the Panama Basin have $I/C_{organic}$ ratios significantly greater than those characteristic of anoxic sediments, and, consequently, that the organic carbon maximum formed when the bottom waters of the basin were oxygenated (Pedersen, Pickering, Vogel, Southon, & Nelson, 1988). This conclusion was supported by micropalaeontological information and the behaviour of Mo, which is examined in Section 7.3.2. Similarly, I/Br ratios in a core from the central California continental slope (discussed in Section 7.7) indicate that bottom water conditions were probably oxic during stadial events and the Younger

Dryas during the last 60 kyr. This conclusion is consistent with information from trace metal proxies on the ventilation of bottom waters at this site.

7.2. Proxies for Suboxia

7.2.1. Chromium, vanadium, rhenium and uranium

These four trace elements occur in seawater in at least two oxidation states, and are probably removed to bottom sediments as the less soluble lower oxidation state. In oxic sea water, they occur predominantly as: chromate (CrO_4^{2-}) and also to a lesser extent as the cationic (III) aquahydroxy species $Cr(OH)_2^+(H_2O)_4$ (Elderfield, 1970); vanadate $(HVO_4^{2-}$ and $H_2VO_4^-$, which are probably hydrolyzed to $VO(OH)_3^-$ or $VO_2(OH)_3^{2-})$ (Turner et al., 1981; Sadiq, 1988; Wehrli and Stumm, 1989; Emerson and Huested, 1991); perrhenate (ReO_4^-) (Colodner, et al., 1993) and the uranyl (VI) carbonate complex $(UO_2(CO_3)_2^{2-})$ (Langmuir, 1978). The vertical distribution these metals in seawater show either very small surface depletions (Cr and V), signifying minor involvements in biogeochemical cycles (Murray et al., 1983; Collier, 1984), or are invariant (Re and U), pointing to their conservative behaviour in oxic waters (Ku et al., 1977; Anbar et al., 1992; Colodner et al., 1993).

In the eastern tropical Pacific, dissolved Cr(VI) exhibits a minimum at the depths of the OMZ $(O_2\ 10-40\ \mu M)$, while Cr(III) has a maximum concentration at the same depth, indicating the reductive transformation of Cr species (Murray et al., 1983; Rue, Smith, Cutter, & Bruland, 1997). This clearly shows that the reduction of Cr(VI) takes place under suboxic conditions as predicted thermodynamically. In anoxic basins, Cr(VI) is also removed from solution, probably by the adsorption of $Cr(OH)_2^+$ on particle surfaces (Emerson et al., 1979). Similarly, V is removed from the anoxic bottom waters of Lake Nitinat (British Columbia), the Black Sea and the Cariaco Trench, probably by surface adsorption of its reduced species VO^{2+} or $VO(OH)_3^-$ (Emerson & Huested, 1991), or by the precipitation of V_2O_3 or $V(OH)_3$. However, in Saanich Inlet (British Columbia) and Framvaren (Norway), V concentrations are higher in the anoxic water mass possibly due to the complexation of VO^{2+} by dissolved organic matter (Emerson & Huested, 1991).

Chromium, V, Re and U accumulate in marine sediments above their crustal abundances under suboxic conditions, that is in the absence of both oxygen and sulphide, as shown by their increases in concentration in pore waters at, or close to, the burial depths where Fe oxyhydroxides are also reduced. Where the identity of the predominant chemical species of the metals is known with some confidence, this behaviour is consistent with the thermodynamically predicted reactions that caused the conversion of the oxidized into the reduced forms of the metals. Where the chemical species have not been well established, the predicted reactions at suboxic redox potentials do not appear to be valid. On the basis of the observed behaviour of the metals, however, sedimentary enrichments of Cr, V, Re and U have been used as proxies for sedimentation under suboxic conditions.

In accordance with thermodynamic predictions, Cr and V are strongly removed from the pore waters of some continental margin sediments close to the depth of manganese oxide reduction, suggesting that they are diffusing into the sediments down a concentration gradient to be fixed at depth (Shaw, Gieskes, & Jahnke, 1990). The enrichment of the two metals in the solid phases of the sediments is difficult to discern in many cases due to their high background in the lithogenous fraction. Dissolved Cr and V also increase in concentration with depth in pore waters in organic-rich sediments in the poorly-ventilated Gulf of California (Brumsack & Gieskes, 1983) showing again that they are solubilized under low redox potentials. The solid phase V concentration increases with depth in these sediments, demonstrating that V is being removed from pore solutions (Brumsack, 1986).

Enrichments of Cr and V in the organic-rich Miocene Monterey Formation are much larger than those of Mo and U, suggesting that deposition of the sediments occurred under suboxic (denitrifying) conditions (Piper & Isaacs, 1995). This conclusion is in accord with the accepted palaeoenvironmental model for the formation of this deposit, which invokes deposition on an upwelling-intense continental margin (Isaacs, 2001).

Rhenium is greatly enriched in the suboxic sediments of nearshore and continental margin settings (Colodner et al., 1993; Crusius, Calvert, Pedersen, & Sage, 1996; Sundby, Martinez, & Gobeil, 2004) as well as in the sediments of anoxic basins (Koide, Hodge, Yang, Stallard, & Goldberg, 1986; Ravizza, Turekian, & Hay, 1991; Colodner, Edmond, & Boyle, 1995). In contrast, concentrations of the element in pelagic clays and ferromanganese nodules are less than or equal to crustal abundances. The degree of enrichment of Re in anoxic sediments is significantly larger (up to a factor of 300 relative to crustal abundances) than any other metal whose marine geochemistry is sufficiently well understood (Colodner et al., 1993). In these suboxic and anoxic environments, Re is delivered to the sediments via diffusion from bottom waters and is being removed from sediment pore waters by the reduction of Re(VII) to Re(IV), possibly ReO_2 or ReS.

Uranium has a constant concentration in seawater from all ocean basins (Ku et al., 1977), and is largely conservative in estuaries (Borole, Krishnaswami, & Somayajulu, 1982; Cochran, 1984). This conservative behaviour is due to the high degree of solubility of the uranyl(VI) carbonate complex $(UO_2(CO_3)_2^{2-})$ in seawater. In anoxic basins, U is removed from solution to the bottom sediments by reduction to the IV oxidation state (Anderson, 1987; Todd, Elsinger, & Moore, 1988; Anderson, Fleisher, & LeHuray, 1989a; Anderson, LeHuray, Fleisher, & Murray, 1989b; McKee & Todd, 1993). In spite of the evident removal of U in the deep water of these basins, it exists in solution in its higher oxidation state. Moreover, a little authigenic U is found in settling particulate matter collected by particle interceptor traps in Saanich Inlet (British Columbia) and the Black Sea (Anderson et al., 1989a; Anderson et al., 1989b). Its removal to the sediments must, therefore, occur by diffusion from the bottom water into the underlying deposits, and may involve the catalytic reduction of U(VI) to U(IV) on particle surfaces (Kochenov, Korolev, Dubinchuk, & Medvedev, 1977), as reflected in pore water and solid phase total U profiles in the Black Sea (Barnes & Cochran, 1991). Experimental work has also demonstrated bacterial reduction of U(VI) to U(IV) in the presence of hydrogen sulphide (Lovley, Phillips, Gorby, & Landa, 1991), a mechanism that would be favoured in sediments compared to the water column because of the higher concentrations of bacterial cells in sediments (Kriss, 1963).

Uranium also appears to be removed from bottom oxygenated waters by diffusion across the sediment/water interface into continental margin and some deep-sea sediments (Thomson, Wallace, Colley, & Toole, 1990; Klinkhammer & Palmer, 1991; Barnes & Cochran, 1993; McManus, Berelson, Klinkhammer, Hammond, & Holm, 2005). Such removal is demonstrated by the decrease with depth of dissolved pore water uranium and the concomitant increase in solid phase U in subsurface suboxic and anoxic sediments. The authigenic U at such depths has a $^{234}U/^{238}U$ ratio identical with that in normal seawater, attesting to its seawater source (Yamada & Tsunogai, 1984; Thomson et al., 1990). The sink is evidently below the depth where Mn and Fe oxyhydroxides are reduced, and probably lies at the depth where sulphate reduction begins. The largest of the global U sinks, accounting for the removal of at least 75% of the total dissolved U supplied to the ocean by rivers, evidently occurs in suboxic, continental margin sediments (Klinkhammer & Palmer, 1991).

The suboxic/anoxic sedimentary U sink is controlled by the settling flux of metabolizable organic matter to the sea floor. Under oxygenated bottom water conditions, oxygen is fully depleted in the pore waters at a sub-surface depth that depends on the organic content of the sediment, below which suboxic conditions (nitrate, Mn and Fe oxyhydroxide reduction) prevail (Figure 16). Increasing the organic flux to the sea floor — that is, increasing surface productivity — causes the oxic/suboxic boundary to shoal, steepening the concentration gradient of dissolved U from the bottom water to the boundary, thereby increasing the rate of fixation of U in the suboxic zone (Rosenthal, Lam, Boyle, & Thomson, 1995b). Changes in productivity can therefore be inferred from changes in U content of the bottom sediments, which constitutes a palaeoproductivity proxy. Uranium and several other trace metals are oxidized and re-precipitated close to the oxic/suboxic boundary in some deposits due to a decrease in bulk sedimentation rate in the Holocene compared with the LGM (Thomson et al., 1990; Thomson, Higgs, Croudace, Colley, & Hydes, 1993; Thomson, Higgs, & Colley, 1996; Mangini, Jung, & Laukenmann, 2001). This "burn-down" causes the formation of characteristic concentration peaks above and below the boundary between two sediment facies, one organic-poor and the other organic-rich. Where the Holocene sedimentation rate is < 2 cm/ky, authigenic U that accumulated in sediments during the glacial period can subsequently be removed during "burn-down diagenesis" (Mangini et al., 2001), so that periods which lack U enrichments do not necessarily signify periods of low productivity.

Changes in the drawdown of atmospheric CO_2 by processes in the high latitude regions of the ocean on glacial–interglacial timescales have been of considerable interest (Knox & McElroy, 1984; Sarmiento & Toggweiler, 1984; Siegenthaler & Wenk, 1984). The Southern Ocean region in particular has been identified as one of the systems where the efficiency of the biological pump might have changed (Sigman & Boyle, 2000), leading to some of the large changes in atmospheric pCO_2 recorded in Antarctic ice cores (Jouzel et al., 1987). This has focused interest on changes in the proportion of nutrients removed from the mixed layer and the rate of export of carbon to the deep sea via settling organic particles. Proxy records of palaeoproductivity using isotopic (François, Altabet, & Burckle, 1992; Shemesh,

Macko, Charles, & Rau, 1993; Mackensen, Grobe, Hubberten, & Kuhn, 1994) and elemental data (Kumar et al., 1995; Rosenthal et al., 1995b; Anderson et al., 1998) have been constructed for large regions of the Antarctic and sub-Antarctic circumpolar region to test this hypothesis.

The burial of organic carbon in Southern Ocean sediments appears to have changed drastically, but this change is areally variable. In the Subantarctic zone and in the Cape Basin of the South Atlantic, north of the Antarctic Polar Front, production appears to have been substantially higher during the LGM compared to the Holocene (Figure 19), as recorded by the U palaeoproductivity proxy (Rosenthal

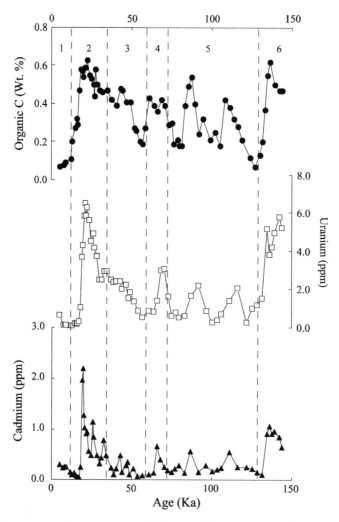

Figure 19   Downcore variations in organic C and authigenic U and Cd over the last 145 kyr in core MD88-769 in the Subantarctic Indian Ocean (46.07°S, 90.12°E, 3,420 m water depth). Data from Rosenthal, Boyle, Labeyrie, and Oppo (1995a). Organic carbon and the metal proxies point to substantially higher production during the LGM in the Subantarctic zone.

et al., 1995b). This may have been induced by a higher glacial supply of Fe by dust (Kumar et al., 1995). South of the Antarctic Polar Front, on the other hand, where sea ice was extensive during the LGM, changes in production show the inverse pattern. This has suggested that the modern belt of high productivity moved northwards during the LGM, and the degree of enrichment of U (and Cd) in glacial sediments suggests that the burial flux was higher in the Subantarctic zone possibly due to a greater partitioning of carbon into the deep sea, consistent with a more efficient biological pump. This conclusion is supported by sedimentary records of the $^{231}Pa/^{230}Th$ and $^{10}Be/^{230}Th$ particle flux proxies (Kumar et al., 1995), of opal and organic carbon burial (Charles, Froelich, Zibello, Mortlock, & Morley, 1991; Mortlock et al., 1991), of biogenous Ba enrichments (Figure 20) where U data show there has been no diagenetic loss of barite (Anderson et al., 1998; Anderson, Chase, Fleisher, & Sachs, 2002) and by diatom-bound $\delta^{15}N$ measurements (Robinson et al., 2005). Whether there was a net increase in export production and carbon burial in the glacial Southern Ocean or little change due to areal compensation of sea-surface production has not been fully resolved due to uncertainties in our understanding of Southern Ocean nutrient cycles, the phytoplankton groups responsible for production and the physical factors controlling export production during the Last Glacial (Anderson et al., 2002).

The enrichment of uranium in organic-rich sediments that accumulated during OIS 2 and 3 in the Arabian Sea has been ascribed to anoxic bottom water conditions at this time (Sarkar et al., 1993). However, the organic carbon flux appears to have been significantly higher where U is enriched, as is the case for cores in the Subantarctic region, suggesting that the oxic–anoxic boundary shoaled within the sediment, thereby increasing the authigenic fixation of U in the sediments. Although the oxygen content of the bottom waters could have decreased concomitantly because of the higher carbon flux into bottom waters, there is no requirement for the bottom waters to have been fully anoxic to explain the observed U enrichment. This conclusion is strongly supported by the areal distribution of glacial U enrichments in the Southern Ocean, and by the discovery of similar glacial enrichments in cores recovered from the depths of South Atlantic Intermediate Water, which could not have been oxygen deficient during this time period (Chase, Anderson, & Fleisher, 2001).

7.3. Proxies for Anoxic Conditions of Sedimentation

7.3.1. Silver, cadmium, copper, nickel and zinc

These five metals are chalcophile elements that, together with lead and mercury, are concentrated in sulphide deposits in the Earth's crust. They are commonly enriched in sedimentary rocks, especially black shales, where they occur mainly as disseminated sulphides. The range of metal contents in such shales is very wide, and some deposits contain ore-grade levels of some of the metals. The stable species of these metals in aqueous solution are the monovalent (Ag) and divalent forms, and their distributions in the ocean are dominated by biochemical reactions. Silver, Cd, Ni and Zn behave as micronutrients, being removed quantitatively in surface waters by plankton growth and liberated from settling organic debris in the upper part of the

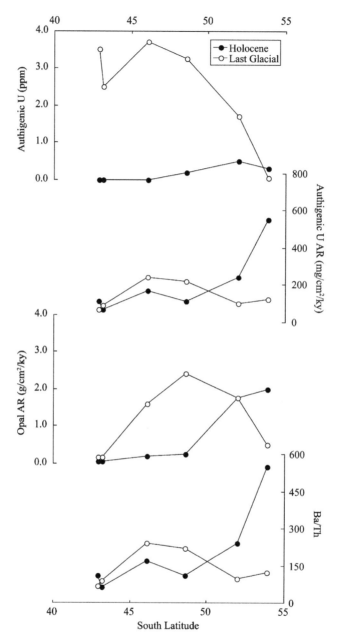

Figure 20 North-south transect of surface sediment authigenic U, authigenic U accumulation rate, opal accumulation rate and the Ba/Th ratio in the Atlantic sector of the Southern Ocean. Data from (Kumar, 1994). Accumulation rates are Th-normalized component accumulation rates.

oceanic water column. Concentration-depth profiles of Cd are very similar to those of phosphate, whereas distributions of Ag, Ni and Zn are closest to those of silicic acid, with surface minima, increasing concentrations with depth and substantial ocean–ocean fractionation (Bruland, 1983; Flegal, Sanudo-Wilhelmy, & Sceflo, 1995; Nozaki, 1997; Bruland & Lohan, 2003). Copper is somewhat unique among the trace metals in seawater in that it behaves partly like a micronutrient but is also scavenged from solution onto particle surfaces in deep water. Its depth distribution therefore shows surface depletions, more or less linear increases with depth in deep water and high concentrations in bottom waters, implying a sediment source (Boyle, Sclater, & Edmond, 1977). Cadmium can be used as a robust upwelling proxy because its concentration in surface waters responds in the same fashion as the macro-nutrients, especially phosphate, to upwelling events (van Geen & Husby, 1996). Assimilation of this upwelled Cd by phytoplankton that respond to the increased supply of nutrients is then transmitted to the sediment in the biogenous debris that settles into deeper waters.

In anoxic basins, such as the Cariaco Trench, Framvaren Fjord, Saanich Inlet and the Black Sea, the dissolved concentrations of Cd, Cu and Zn decrease by factors ranging from 2 to 10 from the upper oxic water mass into the underlying anoxic waters (Jacobs & Emerson, 1982; Jacobs, Emerson, & Skei, 1985; Jacobs, Emerson, & Huested, 1987; Haraldsson & Westerlund, 1988, 1991; Landing & Lewis, 1991). Equilibrium speciation calculations show that such decreases reflect the precipitation of the respective solid sulphides in the presence of dissolved sulphide in the anoxic waters. Silver is similarly stripped from solution in the anoxic waters of Saanich Inlet, with concentrations year-round being $<0.5\,pM$ in waters below the oxygen-zero boundary (Kramer, 2006). In contrast to the systematic behaviours of Ag, Cd, Cu and Zn, Ni does not show any decrease in concentration in anoxic waters even though a number of insoluble sulphides have the potential to limit its concentration in the presence of dissolved sulphide. Hence, we should expect that the bottom sediments of such basins would display enrichments in the concentrations of at least Ag, Cd, Cu and Zn above those characteristic of the detrital sediment input.

Cadmium, Cu, Ni and Zn are enriched in the bottom sediments of some modern anoxic basins, but these enrichments are highly variable (Jacobs et al., 1987). This is due to the relatively large contribution of the metals to the sediments in lithogenous phases, which dilute the authigenic fraction to different extents depending on their accumulation rates, Thus, Framvaren sediments are highly enriched in Cd, Cu, Ni and Zn because it is a sediment-starved basin that receives very little detrital supply. On the other hand, the fluxes of most of the metals in the Cariaco Trench are dominated by detrital input so that only in the case of Cd and Zn are significant enrichments found, and in Saanich Inlet, Cu is enriched whereas the Zn can be accounted for entirely by the supply of the metal from settling particulate organic matter (François, 1988). Paradoxically, Ni is significantly enriched in the bottom sediments of the Black Sea and Framvaren in spite of an evident lack of removal of the metal in the deep anoxic waters of the basin.

Authigenic Ag, Cd, Cu, Ni and Zn appear to be added to the solid fraction of many nearshore sediments by the diffusion of the dissolved metals from the overlying oxygenated waters or the pore waters into the subsurface anoxic horizons. Theoretical studies of the sedimentary removal of Cd and Cu show that the respective sulphide phases should control the concentration of both metals in the pore waters of estuarine sediments containing moderate concentrations of dissolved sulphide (Davies-Colley, Nelson, & Williamson, 1984, 1985). Where sulphide concentrations are higher ($\sim$1 mM), polysulphide formation may cause the metals to stabilize in solution. Gobeil, Silverberg, Sundby, and Cossa (1987) showed that authigenic Cd is added to the sediments of the Laurentian Trough by downward diffusion from a dissolved Cd maximum in the upper part of the sediment column. Likewise, Pedersen, Waters, and Macdonald (1989) suggested that Cd is markedly enriched in the organic-rich sediments of a small sediment-starved coastal embayment by the diffusion of dissolved Cd from the overlying waters and fixation as the sulphide within the sediments. In each of these cases, the sediments are accumulating under oxygenated bottom waters, but the metals are being precipitated as the sulphides under anoxic conditions at shallow depth in the sediment. The affinity of Cd for adsorption by organic matter is rather low (Guy & Chakrabarti, 1976) so that a high degree of correlation between the concentrations of Cd and organic carbon (Pedersen et al., 1989) reflect an indirect control of the production of dissolved sulphide by sulphate reduction that is fueled by the burial of large quantities of marine organic matter in this coastal setting. Evidently, wherever there is sufficient organic matter delivered to sediments and the redox boundary occurs at relatively shallow depth (itself due to the high settling flux of carbon), then Cd, Cu, Ni and Zn (and probably Ag) can be potentially sequestered by the sediments as the respective sulphides.

7.3.2. Molybdenum

The concentration of dissolved Mo is similar in the Atlantic and Pacific oceans (Sugawara, Okabe, & Tanaka, 1961; Morris, 1975; Collier, 1985), and it is slightly depleted in surface waters, reflecting a minor involvement in biogeochemical cycles. Molybdenum is removed from solution in anoxic basins (Berrang & Grill, 1974; Emerson & Huested, 1991), where concentrations observed vary widely, there being no relationship between the dissolved metal concentrations and the dissolved sulphide levels, which range over two orders of magnitude. This has suggested that there are no solubility equilibria involving Mo and that the metal is removed from the anoxic waters within the bottom sediments.

Molybdenum is enriched in both oxic and anoxic sediments. Thus, in spite of its occurrence as the stable soluble molybdate ion in oxygenated seawater (Turner, Whitfield, & Dickson, 1981), Mo is significantly enriched in ferromanganese nodules (Calvert & Price, 1977) and in surficial Mn oxyhydroxides in continental margin and nearshore sediments (Shimmield & Price, 1986). In the latter situation, Mo is released from the solid phases of surface sediments when the oxyhydroxides of Mn are reductively dissolved during burial into more reducing subsurface sediments (Figure 18). Where dissolved sulphide occurs in such sub-surface sediments, Mo may be removed from the pore waters to be immobilized in the solid phases.

Experimental and observational studies have revealed that a key step in Mo removal from seawater and pore waters is the conversion of inert MoO_4^{2-} to particle-reactive thiomolybdates ($MoO_xS_{4-x}^{2-}$), which may then be scavenged from solution by Fe sulphide phases and humic substances (Helz et al., 1996). Protonated mineral surfaces promote the interconversion of thiomolybdate species, which could lead to a higher rate of immobilization of Mo in sediments compared with overlying waters (Vorlicek & Helz, 2002). The relationship between Mo removal and aqueous sulphide concentration suggests that thiomolybdate formation increases markedly at an H_2S concentration of 11 μM at 25°C, an effective "switch point" (Helz et al., 1996), which enhances Mo immobilization on solid phases. Because of the kinetics of molybdate–thiomolybdate interconversion, a secondary control is the time of exposure of MoO_4^{2-} to sulphide (Erickson & Helz, 2000). Thus, Mo enrichment in marine sediments proceeds in the presence of sulphide, which can increase in concentration to relatively high levels in subsurface continental margin sediments as well as anoxic basins due to diagenetic sulphate reduction.

In Saanich Inlet and the Black Sea, Mo occurs at concentrations below the detection limit (1–2 ppm) in settling particulate matter in the deep anoxic waters (François, 1988; Crusius et al., 1996), whereas it is greatly enriched in the bottom anoxic sediments of both basins (90–100 ppm). This lends strong support to the mechanism of Mo enrichment under sulphidic conditions by the immobilization within the sediments, the decrease in concentration in the anoxic bottom waters being due to the diffusion of the dissolved metal into the sediment along a concentration gradient.

The largest sedimentary Mo enrichments are found in permanently anoxic basins, such as Framvaren (Piper, 1971), the Black Sea (Calvert, 1990), Saanich Inlet (François, 1988) and the Cariaco Basin (Piper & Dean, 2002). In continental margin sediments, on the other hand, where surface sediments are oxic, Mo concentrations are significantly lower (Pedersen, 1985; Shimmield & Price, 1986; Crusius et al., 1996; Zheng, Anderson, van Geen, & Kuwabara, 2000; Chaillou, Anschutz, Lavaux, Schafer, & Blanc, 2002; Sundby et al., 2004). This contrast may be produced by the ambient concentrations of dissolved sulphide in anoxic basins, the residence time of deep basin waters, as well as the kinetics of Mo species transformations. Thus, dissolved Mo depletion in the anoxic Black Sea waters, where H_2S concentration is ca. 400 μM and the deep water residence time is long, is >90% (relative to the overlying oxic waters), whereas it is only ca. 10% in Saanich Inlet where H_2S is ca. 10 μM and ventilation takes place on a quasi-annual time scale, with Lake Nitinat and the Cariaco Basin having intermediate values (Piper & Isaacs, 1996; Algeo & Lyons, 2006).

In spite of the lower degree of Mo enrichment in continental margin deposits, subsurface enrichments above crustal abundances are readily detected and have been used to hindcast past anoxic deep-water conditions and carbon fluxes to the sea floor. In such environments, pore water Mo is depleted and solid phase Mo concentrations increase below the zone of Mn oxyhydroxide recycling (Malcolm, 1985; Shaw et al., 1990; Zheng et al., 2000; Chaillou et al., 2002; Sundby et al., 2004). Because Mo is enriched in surficial horizons where Mn oxyhydroxides exist under surficial oxic conditions (Shimmield & Price, 1986), the metal is effectively pumped into the underlying anoxic zone by burial where it is released to solution. Molybdenum is also most likely fixed as Fe-Mo-S clusters, probably in solid solution in FeS phases (Helz, Vorlicek, & Kahn, 2004). The increasing concentration of solid phase Mo with depth in many situations suggests that the kinetics of Mo removal are sluggish (Sundby et al., 2004), and that the modest enrichment in continental margin settings is restricted by this kinetic hindrance, which lengthens the path over which dissolved Mo must diffuse before it is fixed. The metal begins to be removed from the Santa Barbara Basin sediment at dissolved sulphide levels of $\sim$0.1 μM, as shown by the decrease in dissolved Mo and the increase in solid phase concentrations, which reach 7 ppm (Zheng et al., 2000). This removal may involve scavenging by humic substances to form organic thiomolybdates (Adelson, Helz, & Miller, 2001). A second critical concentration of $\sim$100 μM total sulphide marks the further depletion of pore water Mo and the concomitant enrichment in solid phase Mo in this basin. It is here that the association with Fe sulphide phases is probably involved (Helz et al., 2004) rather than precipitation as a Mo sulphide. Pyrite (FeS_2), which forms from FeS during diagenetic sulphide reactions, is the host of significant Mo contents in anoxic sediments (Huerta-Diaz & Morse, 1992). These observations support the experimentally determined sulphide "switch point", which enhances the conversion rate of molybdate to thiomolybdate species in solution (Helz et al., 1996). In order to sustain H_2S concentrations above the switch point in near-surface sediment pore waters so that significant Mo fixation can occur, it has been argued that dissolved sulphide must be present in the bottom waters of a basin. Sulphide would otherwise be lost by reaction with O_2 diffusing downwards from ventilated bottom waters (Adelson et al., 2001). Data from the California margin discussed in Section 7.7 challenge this conclusion because Mo is enriched in sediments that have accumulated under oxic bottom water conditions (Hendy & Pedersen, 2005).

7.4. Post-Depositional Migration or Loss of Authigenic Metals

Several authigenic trace metals can be remobilized and relocated in buried sediments if they become exposed to oxygen after initial accumulation. This process has been reported in sediment sections of the Atlantic where a diminution in the Holocene sedimentation rate has allowed the diffusion of dissolved oxygen into deeper anoxic Late Glacial sediments that had built up relatively high concentrations of the metals due to a higher burial flux of organic matter at that time (Thomson, Higgs, Croudace et al., 1993; Thomson, Higgs, & Colley, 1996). The effect is to produce a redistribution of elements in a specific redox-controlled order above and below the boundary between an upper oxic section, which has not been altered, and an underlying section, which has been secondarily oxidized by O_2 diffusing downwards from oxygenated sediment. This so-called "burn-down" process causes the formation of sharp concentration peaks of, for example, Ag, Cd, Cu, V, Zn, V, I and U. Rhenium is also mobilized in the oxidized section and re-immobilized below the oxidation front, but its distribution extends over a broader depth range, and this distribution remains over at least 3.4 Ma (Crusius & Thomson, 2000). On the other hand, U forms sharp concentration peaks roughly 500 cm broad in the upper part of the glacial reducing sediments. Bioturbation can also lead to loss of U from sediments due to the post-depositional transport of reducing sediment into the near-surface, more oxidizing environment (Zheng, Anderson, van Geen, & Fleisher, 2002). Molybdenum can be lost from an anoxic sediment that has been subsequently exposed to O_2 unless it is associated with authigenic pyrite formation (Crusius & Thomson, 2000). Thus, Mo enrichments may be preserved in re-oxidized deposits if they were originally sufficiently anoxic for pyrite to have formed. Similar patterns of element remobilization and re-immobilization have been discovered around the sapropel horizons in the Mediterranean following re-ventilation of the basin after periods of increased production and bottom water anoxia (Pruysers, De Lange, Middelburg, & Hydes, 1993; Thomson et al., 1995; Crusius & Thomson, 2000).

These artifacts must be carefully evaluated when making palaeoceanographic deductions from authigenic metal distributions in sediment sections that have suffered post-depositional burn-down events. Authigenic metal records are well preserved in sediments that have high sedimentation rates and where the metal enrichments are permanently buried under suboxic or anoxic conditions.

7.5. Distinguishing between Suboxic and Anoxic Environments of Sedimentation

It is well established that Ag, Cd, Cr, Cu, Mo, Ni, Re, U, V and Zn are all removed from seawater under suboxic and/or anoxic bottom water conditions. Those metals that are readily fixed in solid phases under suboxic conditions (Cr, Re, U and V) without the requirement for the presence of dissolved sulphide will also be removed under anoxic conditions because the change in oxidation state will be promoted at lower redox potentials. A less than ideal method for distinguishing between suboxic and anoxic sedimentary environments could therefore involve the enrichment of Cr, Re, U or V accompanied by the absence of significant enrichments of Ag, Cd, Cu, Mo, Ni or Zn.

The contrasting geochemical behaviours of Re and Mo can provide a simple way to make this distinction (Crusius et al., 1996). As shown previously, these two metals show the largest degrees of enrichment above crustal abundances among the metals considered here. Rhenium is not associated with authigenic oxide phases, but is strongly removed to suboxic sediments. Molybdenum is adsorbed on Mn oxyhydroxides and is also removed to anoxic sediments at high dissolved sulphide levels, which are maintained in the surficial sediment by anoxic bottom waters. A ratio of total Re/Mo greater than that of seawater (0.4 mmol/mol) in a given deposit will be diagnostic of suboxic conditions, whereas a ratio close to that of seawater is expected under anoxic conditions. Under oxic conditions in the bottom water and the sediments, the ratio could be less than that in seawater due to the fixation of Mo on oxyhydroxides and the lack of uptake of Re on such phases.

7.6. History of Anoxia in the Cariaco Basin

The Cariaco Basin, lying on the Venezuelan continental shelf, is a 1,400 m deep permanently anoxic basin that is isolated from the Caribbean Sea by a sill with a maximum depth of 146 m. Sulphate reducing conditions prevail below 250 m depth, where dissolved sulphide concentrations are 40–70 µM (Richards & Vaccaro, 1956; Richards, 1975; Scranton, Astor, Bohrer, Ho, & Muller_karger, 2001), and the deep water residence time is approximately 100 years (Deuser, 1973). The bottom sediments of the basin are siliceous-calcareous organic-rich silty clays that are finely laminated, light-coloured diatom/silicoflagellate-rich laminae alternating with darker, clay-rich laminae (Hughen, Overpeck, Peterson, & Anderson, 1996). During the last glacial, the basin was further isolated from the Caribbean due to sea level lowering, when fully oxygenated conditions in the deep water prevailed. The sediments deposited during this period are moderately organic rich and weakly bioturbated (Peterson, Overpeck, Kipp, & Imbrie, 1991).

The history of environmental change in the Cariaco Basin has been recon-structed using a combination of foraminiferal census and isotopic data (Peterson et al., 1991; Lin, Peterson, Overpeck, Trumbore, & Murray, 1997) and a range of major and trace element proxies for sedimentation under oxic, suboxic and anoxic conditions (Dean, Piper, & Peterson, 1999; Piper & Dean, 2002). Production and sediment supply in the Cariaco Basin vary seasonally in response to changes in the position of the ITCZ in the tropical Atlantic. Between January and March, the ITCZ lies south of the equator, and strong trade winds blowing along the coast of Venezuela produce strong upwelling. Production in the Cariaco Basin is up to 20 times greater than in the open Caribbean to the north. This is also the dry season in the region, due to the location of the ITCZ far to the south. Beginning in June–July, the ITCZ moves northwards and lies near the Venezuelan coast; winds di-minish, upwelling weakens and the high rainfall results in increased river discharge.

Prior to 12.6 kyr ago, the foraminiferal faunas were dominated by species characteristic of the modern non-upwelling season, suggesting low productivity at this time. At 12.6 ka, a sharp increase in the abundance of species characteristic of upwelling regimes and high plankton productivity is found to be coincident with the first major rise in sea-level that accompanied the melt water pulse (1A) from the Laurentide ice sheet. Production is interpreted to have increased in response to an increase in water exchange between the basin and the Caribbean, and anoxia in deep waters was quickly established in response to the higher organic matter fluxes below sill depth. These changes in hydrography and production are coincident with the observed changes in the sedimentary record, from oxic, bioturbated clays to anoxic, micro-laminated clays.

Between 18 and 15 ka, when the basin was maximally isolated, and the most organic-lean sediment accumulated, the trace element proxies show, as expected from the bioturbated sediment structure, that the deep water basin waters were oxic; there are no enrichments of Cd, Cr or Mo (Figure 21), or of Cu, Ni, V and Zn (not shown) relative to average shale. Between 18 and 23 ka, however, when the organic C content of the sediment is higher, small enrichments of Cr and Mo (and Cu, Ni, V and Zn) may indicate modest uptake of the metals from bottom waters and from deposited organic materials by subsurface suboxic and anoxic sediment even though the surface sediments were oxic and bioturbated, in the same way that many continental margin sediments have small metal enrichments. Also note that the Ti/Al ratio is highest when the trace metals are lowest, possibly recording the delivery of coarser-grained material from local, exposed parts of the shelf and islands surrounding the basin.

The abrupt increase in organic C at 15 ka is followed shortly thereafter by very large enrichments in Cd, Cr, Mo (Figure 21) and Cu, Ni, V and Zn (not shown). Over a period of roughly 1.8 kyr, concentrations are maximal for the entire core. This interval, lasting from 14.6 to 12.8 ka, is coeval with the Bølling-Allerød interstadial event in the North Atlantic, which followed Heinrich Event 1 and marked the end of the last glacial period (Alley & Clark, 1999). Melt water pulse 1A (Fairbanks, 1989), triggered by abrupt melting of Antarctic ice (Weaver, Saenko, Clark, & Mitrovica, 2003), begins at 14.6 ka during which sea level rose by ~20 m in less than 500 years. The response to this dramatic event was clearly the rapid onset of anoxic bottom water conditions in the Cariaco Basin, due to the large increase in export production and the increase in water column stability induced by lower sea surface salinity. This set of conditions caused the largest Mo enrichments for the entire core. The state of the basin changed drastically once again at the beginning of the Younger Dryas event at 12.8 ka, when surface water temperatures decreased by 3–4°C (Lea, Pak, Peterson, & Hughen, 2003), and Mo and Cr contents decreased drastically; Cd, after an initial decrease, increased to levels observed during the B/A. Organic C contents also decreased, but biogenous carbonate and opal increased and diatom and zooplankton biomarkers reached a maxima for the last 12 ka (Werne, Hollander, Lyons, & Peterson, 2000), indicating an interval of increased productivity. Thus, Cd enrichments tracked the increase in export production, while the anoxia proxies suggest that deep water conditions were not as reducing as during the B/A. Following the end of the Younger Dryas, Mo and Cd increased markedly and Cr increased modestly, suggesting productivity continued to rise and intense anoxia returned to the deep waters. Conditions were wettest and river flow was highest in the Cariaco Basin borderlands between approximately 10.5 and 5.4 ka, the time interval of the Holocene "thermal maximum", as deduced from the record of Fe and Ti abundances. At ODP Site 1002 within the Basin, these track variations in lithogenous sediment supply (Haug et al., 2001). This is the same time interval of higher Cd, Cr and Mo enrichments in core 39PC (Figure 21) when the deep basin waters must have been more intensely anoxic probably due to increased water column stability induced by the higher freshwater flux into the basin. The time interval with the largest Mo enrichments (~15–10 ka) is also the period of accumulation of distinct laminations, whereas the laminations are weakly developed or less distinct after 10 ka (Peterson et al., 1991; Lin et al., 1997), when the sulphide level in deep waters decreased. Anoxic conditions persisted through to the Late Holocene, however.

7.7. Sedimentary Records of Changes in Ventilation and Organic Flux

As explained previously, the burial flux of metals sequestered in sediments under suboxic and anoxic conditions can be affected or controlled by the oxygen content

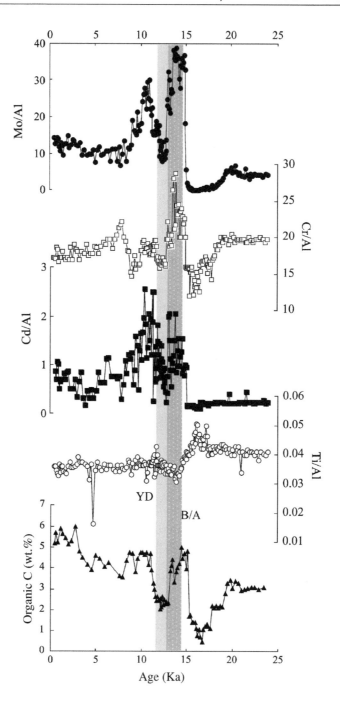

of bottom waters in contact with the sediments and by the accumulation rate of organic matter that is metabolized within the deposits, thereby changing the sedimentary redox conditions. Methods for unravelling these two controls are required in order to understand past changes in water column ventilation and productivity.

High-resolution sedimentary records of high-frequency palaeoclimatic and palaeoceanographic variability that address this problem have been obtained from the central California margin, where changes in bottom water oxygenation appear to be highly coherent with the Dansgaard-Oeschger climate oscillations in Greenland (Behl & Kennett, 1996; Cannariato & Kennett, 1999; Hendy & Kennett, 2000; Hendy, Kennett, Roark, & Ingram, 2002). Multi-element data have been used to discriminate between bottom water ventilation and organic flux variations as the main drivers of the oxygenation of the sea floor over the last 60 kyr (Ivanochko & Pedersen, 2004; Hendy & Pedersen, 2005). On the upper slope to the northwest of the Santa Barbara Basin, a core site located at the base of the OMZ shows enrichments of minor and trace element during Late Quaternary interstadials and the Holocene, consistent with reduced intermediate water ventilation (Figure 22). Marked enrichments in Ag, Cd and Mo indicate sulphate-reducing conditions in the pore waters during some of the major interstadials that are also coherent with proxies for productivity (organic C and $CaCO_3$ content, biogenic silica and planktonic foraminiferal assemblages) and nutrient content of source waters ($\delta^{15}N$), showing that the export flux of biogenous detritus as well as intermediate water ventilation played a role in modulating the intensity of the OMZ over the last 60 kyr. Such changes have been found in other parts of the central California margin from downcore records of Mo (Dean, Gardner, & Piper, 1997), and the benthic foraminiferal Cd/Ca palaeonutrient proxy (van Geen, Frairbanks, Dartnell, McGann, & Gardner, 1996). In the Santa Barbara Basin, on the other hand, which has very low dissolved oxygen contents (currently $< 2 \mu M$), a similar approach found that oxygenation changes were mainly controlled by basin ventilation, with a minor role for productivity variations over the last 50 kyr (Ivanochko & Pedersen, 2004).

8. FUTURE DEVELOPMENTS

In this chapter, we have shown how elemental abundance data have been applied to a variety of problems in palaeoclimatology and palaeoceanography. Continuing research will lead to the refinement of many of the proxies discussed, and

Figure 21 Temporal indices of changing bottom water oxygenation, sediment supply and productivity from core PL07-39PC (10.67°N, 64.94°W, 790 m water depth) in the Cariaco Basin. Data from Piper and Dean (2002). YD = Younger Dryas, B/A = Bølling-Allerød. Prior to the B/A, deep basin waters were oxic. Between 18 and 23 ka, minor Cr and Mo enrichments indicate modest uptake of the metals from bottom waters and from deposited organic materials by subsurface suboxic and anoxic sediment even though the surface sediments were oxic and bioturbated. Higher Ti/Al at this time records the delivery of coarser-grained material to the basin. Maximal enrichments in Cr and Mo and a peak in Cd are found in the B/A. All three metals are greatly depleted in the YD, and then increased after 11 ka.

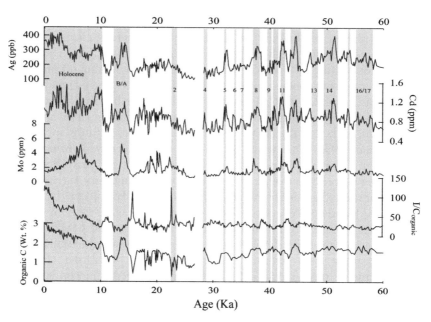

Figure 22   Comparison of temporal records of Ag, Cd, Mo, I/C$_{organic}$ and organic C at ODP Site 1017E on the central California continental slope (34.53°N, 121.1°W, 955 m water depth). Shading represents warm intervals as deduced from the ratio of dextral to sinistral *Neogloboquadrina pachyderma* (Hendy, Pedersen, Kennett, & Tada, 2004). Interstadial events are numbered beneath the Ag profile. YD = Younger Dryas, B/A = Bølling-Allerød. Ag, Cd and Mo enrichments show that many of the interstadials experienced sulphate-reducing conditions in the pore waters, suggesting that the export flux of biogenous detritus as well as intermediate water ventilation played a role in modulating the intensity of the OMZ over the last 60 kyr.

to the introduction of new proxies for tracking climatic and oceanographic variability in the past as more information on the geochemistry of marine sediments accumulates and as new requirements for insight into environmental changes are realized.

Further application of palaeoclimate proxies using major and minor element contents of lithogenous components, and cross validation of interpretations against other palaeoclimatic indices, e.g. terrestrial biomarkers (e.g. Zhao, Eglinton, & Zhang, 2000; Pancost & Boot, 2004), will increase our understanding of past variations in terrestrial climates and land–ocean interactions. The recent development of automated continuous scanning X-ray fluorescence determination of selected major element abundances (Jansen et al., 1998) promises to provide much-needed high-resolution sedimentary records, which could be used for these purposes. Calibration of the XRF response so that more quantitative information can be acquired, and extension of the method to include the determination of Al concentration, in addition to Fe, Ti and Ca reported so far, would mark a significant advance in the technique. In this way, true variability of the composition of the lithogenous component of sediments could be assessed via Al-normalization (Section 3), thereby avoiding the ambiguity of results currently available in the form of raw count rates.

Further work on the validation of palaeoproductivity proxies is urgently required. For the Ba proxy, experimental approaches using phytoplankton cultures need to be extended in view of the probability that the culture conditions themselves have compromised some earlier results, and that the precipitation of Fe hydroxide on particle surfaces leads to the removal of Ba from solution (Sternberg et al., 2005). The formation of hydroxide coatings on all particle surfaces could account for the presence of barite within masses of sinking planktonic debris (Dehairs et al., 1987), the association of barite with suspended diatom frustules (Bishop, 1988) and the precipitation of barite in cultures of both diatoms and coccolithophorids (Ganeshram et al., 2003), implying that barite may constitute an index of particle flux through the water column rather than productivity *per se*. The Ba or barite proxy also needs to be cross-checked against other palaeoproductivity proxies, e.g. phytoplankton group biomarkers (alkenones for coccoliths, sterols for diatoms (e.g. Engel & Macko, 1993).

Establishment of the degree of preservation of barite in bottom sediments is also an urgent requirement in view of conflicting results from oxic, suboxic and anoxic settings. This is especially puzzling in view of the fact that all deep ocean bottom waters are saturated with respect to barite (Monnin et al., 1999), and the observation that Ba is being diagenetically mobilized from fully oxic pelagic sediments that lack subsurface suboxia or anoxia (Paytan & Kastner, 1996; McManus et al., 1998). Assessment of the recommendation for the use of authigenic U enrichment in sub-surface sediments as an index of probable barite remobilization (McManus et al., 1998) should be carried out in parallel with this work.

Algorithms for palaeoproductivity using barite or excess Ba (Dymond et al., 1992; François et al., 1995) require re-evaluation in view of the evidence for significant focusing of Holocene sediment sections, as discussed later. This effect will have inflated some, if not many, of the Ba or barite accumulation rates estimated from conventional mass accumulation rates, which are required for the regressions of modern Ba accumulation rates on estimates of primary production at individual core sites.

Evidence for a large range of Al/Ti ratios and the highly variable amount of excess Al relative to Ti in settling particular material in the central equatorial Pacific (Dymond et al., 1997) suggests that the apparent excess Al in bottom sediments of this region is not related in a simple way to the particle flux to the sea floor. This will potentially compromise the use of excess Al as a particle flux, or palaeoproductivity, proxy as originally conceived (Murray et al., 1993; Murray & Leinen, 1996). More work on the geochemistry of the preponderantly biogenous sediments in this region is required in order to provide a means for distinguishing between adsorbed or scavenged Al and Al added to the sediment by authigenic formation of aluminosilicate phases or phases (Timothy & Calvert, 1998). Knowledge of the precise composition of these phases will then provide a means for distinguishing between authigenic and detrital (lithogenous) Al, which in turn will improve our ability to characterize the aluminosilicate contribution to sediments, for example in the form of aeolian dust (Anderson, Fleisher, & Lao, 2006), and the reliability of Al-normalization for assessing element enrichments and depletions over time.

Variations in the sedimentary concentrations of chalcophile elements that have a significant authigenic component can be valuable as indicators of bottom-water or upper-pore-water oxygenation. But there is a constraint on such inferences, in that the rate of sub-bottom precipitation as a metal sulphide, for example, is largely a function of the concentration gradient, $\partial C/\partial z$, which provides the primary control on the rate of molecular diffusion. A subtle change in precipitation depth of say, 1 to 2 mm, could halve the rate of precipitation of the authigenic fraction of a metal, without necessarily being directly related to the oxygen content at the sediment water interface. A short-term decline in sedimentation rate could yield such a change. Thus, there is considerable scope for high-spatial-resolution study of the distributions of dissolved chalcophile metals as a function of shallow sub-bottom depth in sedimentary pore waters, ideally as time-series measurements over the course of several seasons on selected continental margins. Such work would be expensive, technically challenging and time-consuming, but would provide data of considerable value in the calibration of trace-metal proxies.

Since the bulk composition of sediments and sedimentary rocks is determined by the accumulation rates of the individual components of the deposit, a better way to evaluate variations in the supply rates of proxies to the sea floor is to recast compositional information into component accumulation rates (Koczy, 1951). This approach avoids the constant sum problem when data are expressed as closed arrays (Rollinson, 1993). We have eschewed this approach here because reliable data on the vertical flux of materials, that is sediment flux to the sea floor through the seawater interface or from the sea surface, are scarce. The calculation of component accumulation rates — which relies on estimates of linear sedimentation rates determined as the sediment thickness deposited per unit time between two horizons or depths of known age — is unreliable because sediment winnowing and lateral transport, or focusing, on the sea floor appears to be pervasive (François et al., 2004). Thus, the thickness of a sediment section is not a reliable estimate of the mass of sediment supplied to the sea floor from the overlying water column, a requirement for assessing, for example, palaeoproductivity or aeolian dust fluxes. This is a controversial issue at the time of writing (Lyle et al., 2005), but if these effects are confirmed by on-going work, a great deal of effort will be required to re-evaluate the true vertical flux of materials to the sea floor by using a constant flux tracer. Two possibilities have been proposed, [230]Th (Bacon, 1984; François et al., 1990) and [3]He (Marcantonio, Anderson, & Mix, 1996), both of which have more or less constant fluxes to the sea floor due to their source functions — U decay in the first case and extra-terrestrial dust flux in the second. These tracers have the added advantage that they can provide normalized accumulation rates at high resolution, that is at each sampling point in a core profile (François et al., 2004), whereas conventional sedimentation rates can only yield average mass accumulation rates between more widely spaced sample depths, normally at much coarser resolution than the available compositional information. Modern mass spectrometric techniques are fully capable of generating the high-resolution data sets that will be required to generate reliable records of component accumulation rates throughout the ocean. The major limitation of the [230]Th normalization method is the relatively short half-life of the isotope (75.7 kyr), which restricts its application to the last

200–300 kyr. No such limitation exists for ^{3}He normalization, and we expect to see its wide application for this purpose in older sediment sections.

 ## 9. AFTERWORD

Nearly 150 years ago, natural philosophers began applying techniques of elemental analysis of sediments and sedimentary rocks in an effort to understand geological processes and to apply such knowledge to unravelling Earth history. The laborious techniques and crude detection limits of the past have been supplanted in the modern era by sophisticated, automated methods that generate large volumes of multi-element information. Although this wealth of data has provided an interpretative boon for modern sedimentary geochemists, challenges remain, none more compelling than developing a stronger understanding of climate history, particularly during the Quaternary Era.

This chapter has reviewed current approaches to palaeoclimate interpretation that are based on a wide range of analyses of marine sediments. Although single element distributions can serve as proxies for certain processes, such an approach is often insufficient to yield confident palaeoclimatic interpretations. A multiproxy approach — the collective interpretation of the typical multi-element assemblages that today can be gleaned from the analysis of sediment cores – presents a better option, as implied by the examples discussed in the preceding pages. We recommend that such an approach be applied wherever possible to circumvent some of the current proxy noise, for the additional investment in analytical effort pays disproportionately higher dividends in insight gained.

ACKNOWLEDGEMENTS

We thank Bob Anderson, Raja Ganeshram, Frank Kyte, David Piper and Ralph Schneider for generously supplying original data. We acknowledge the very useful anonymous review of the original manuscript. Preparation of this chapter was supported by research grants from the Natural Sciences and Engineering Research Council of Canada to SEC and TFP.

REFERENCES

Adelseck, C. G., Jr., & Anderson, T. F. (1978). The late Pleistocene record of productivity fluctuations in the eastern equatorial Pacific Ocean. *Geology, 6,* 388–391.

Adelson, J. M., Helz, G. R., & Miller, C. V. (2001). Reconstructing the rise of recent coastal anoxia: Molybdenum in Chesapeake Bay sediments. *Geochimica et Cosmochimica Acta, 65,* 237.

Algeo, T. J., & Lyons, T. W. (2006). Mo-total organic carbon covariation in modern anoxic marine environments: Implications for analysis of paleoredox and palaeohydrographic conditions. *Paleoceanography, 21,* doi:10.1029/2004PA001112.

Alley, R. B., & Clark, R. U. (1999). The deglaciation of the northern hemisphere: A global perspective. *Annual Review of Earth and Planetary Sciences, 27,* 149–182.

Anbar, A. D., Creaser, R. A., Papanastassiou, D. A., & Wasserburg, G. J. (1992). Rhenium in seawater: Confirmation of generally conservative behavior. *Geochimica et Cosmochimica Acta, 56,* 4099–4103.

Anderson, R. F. (1982). Concentration, vertical flux, and remineralization of particulate uranium in seawater. *Geochimica et Cosmochimica Acta, 46*, 1293–1299.

Anderson, R. F. (1987). Redox behavior of uranium in an anoxic marine basin. *Uranium, 3*, 145–164.

Anderson, R. F., Chase, Z., Fleisher, M. Q., & Sachs, J. (2002). The Southern Ocean's biological pump during the Last Glacial Maximum. *Deep Sea Research, 49*, 1909–1938.

Anderson, R. F., Fleisher, M. Q., & Lao, Y. (2006). Glacial-interglacial variability in the delivery of dust to the central equatorial Pacific Ocean. *Earth and Planetary Science Letters, 242*, 406–414.

Anderson, R. F., Fleisher, M. Q., & LeHuray, A. P. (1989a). Concentration, oxidation state, and particulate flux of uranium in the Black Sea. *Geochimica et Cosmochimica Acta, 53*, 2215–2224.

Anderson, R. F., Kumar, N., Mortlock, R. A., Froelich, P. N., Kubik, P., Dittrich-Hannen, B., & Suter, M. (1998). Late-Quaternary changes in productivity of the Southern Ocean. *Journal of Marine Systems, 17*, 497–514.

Anderson, R. F., LeHuray, A. P., Fleisher, M. Q., & Murray, J. W. (1989b). Uranium deposition in Saanich Inlet sediments, Vancouver Island. *Geochimica et Cosmochimica Acta, 53*, 2205–2213.

Anderson, R. F., & Winckler, G. (2005). Problems with paleoproductivity proxies. *Paleoceanography, 20*, doi:10.1029/2004PA001107.

Archer, D. (1991). Equatorial Pacific calcite preservation cycles: Production or dissolution? *Paleoceanography, 6*, 561–571.

Archer, D., Lyle, M., Rodgers, K., & Froelich, P. (1993). What controls opal preservation in tropical deep-sea sediments. *Paleoceanography, 8*, 7–21.

Argast, S., & Donnelly, T. W. (1987). The chemical dscirimination of clastic sedimentary components. *Journal of Sedimentary Petrology, 57*, 813–823.

Arnold, E., Merrill, J., Leinen, M., & King, J. (1998). The effect of source area and atmospheric transport on mineral aerosol collected over the North Pacific Ocean. *Global and Planetary Change, 18*, 137–159.

Arrhenius, G. (1952). Sediment cores from the East Pacific. In: H. Pettersson (Ed.), *Reports of the Swedish deep-sea expedition (1947–1948)* (p. 227). Goteborg: Elanders Boktryckeri Aktiebolag.

Arrhenius, G. (1959). Climatic records on the ocean floor. In: B. Bolin (Ed.), *Rossby memorial volume* (pp. 121–129). New York: Rockefeller Institute Press.

Arrhenius, G. (1963). Pelagic sediments. In: M. N. Hill (Ed.), *The sea* (pp. 655–727). New York: Wiley.

Arrhenius, G., Kjellberg, G., & Libby, W. F. (1951). Age determination of Pacific chalk ooze by radiocarbon and titanium content. *Tellus, 3*, 222–229.

Arz, H., Pätzold, J., & Wefer, G. (1998). Correlated millennial-scale changes in surface hydrography and terrigenous sediment yield inferred from last-glacial marine deposits off northeastern Brazil. *Quaternary Research, 50*, 157–166.

Bacon, M. P. (1984). Glacial to interglacial changes in carbonate and clay sedimentation in the Atlantic Ocean estimated from 230 Th measurements. *Isotope Geoscience, 2*, 97–111.

Bain, D. C. (1976). A titanium-rich soil clay. *Journal of Soil Science, 27*, 68–70.

Balakrishnan Nair, T. M., Ittekkot, V., Shankar, R., & Guptha, M. V. (2005). Settling barium fluxes in the Arabian Sea: Critical evaluation of relationship with export production. *Deep Sea Research, 52*, 1930–1946.

Barber, R. T., Sanderson, M. P., Lindley, S. T., Chai, F., Newton, J., Trees, C. C., Foley, D. G., & Chavez, F. (1996). Primary productivity and its regulation in the equatorial Pacific during and following the 1991–1922 El Niño. *Deep Sea Research II, 43*, 933–969.

Barnes, C. E., & Cochran, J. K. (1991). Geochemistry of uranium in Black Sea sediments. *Deep Sea Research, 38*(Suppl. 2), S1237–S1254.

Barnes, C. E., & Cochran, J. K. (1993). Uranium geochemistry in estuarine sediments: Controls on removal and release processes. *Geochimica et Cosmochimica Acta, 57*, 555–569.

Baturin, G. N. (1971). Stages of phosphorite formation on the ocean floor. *Nature, 232*, 61–62.

Baturin, G. N., & Bezrukov, P. L. (1979). Phosphorites on the sea floor and their origin. *Marine Geology, 31*, 317–332.

Behl, R. J., & Kennett, J. P. (1996). Brief interstadial events in the Santa Barbara basin, NE Pacific, during the last 60 kyr. *Nature, 379*, 243–246.

Bender, M. L., Broecker, W., Gornitz, V., Middel, V., Kay, R., Sun, S. S., & Biscaye, P. (1971). Geochemistry of three cores from the East Pacific Rise. *Earth and Planetary Science Letters, 12*, 425–433.

Bender, M. L., Klinkhammer, G. P., & Spencer, D. W. (1977). Manganese in seawater and the marine manganese balance. *Deep Sea Research, 24*, 799–812.

Van Bennekom, A. J., Jansen, J. H. F., Van der Gaast, S. J., van Iperen, J. M., & Pieters, J. (1989). Aluminium-rich opal: An intermediate in the preservation of biogenic silica in the Zaire (Congo) deep-sea fan. *Deep Sea Research, 36*, 173–190.

Berelson, W. M., McManus, J., Coale, K. H., Johnson, K. S., Kilgore, T., Burdige, D., & Pilskaln, C. (1996). Biogenic matter diagenesis on the sea floor: A comparison between two continental margin transects. *Journal of Marine Research, 54*, 731–762.

Berger, W. H. (1973). Deep-sea carbonates: Pleistocene dissolution cycles. *Journal of Foraminiferal Research, 3*, 187–195.

Berner, R. A. (1971). *Principles of chemical sedimentology* (p. 240). New York: McGraw-Hill.

Berner, R. A. (1973). Phosphate removal from seawater by adsorption on volcanogenic ferric oxides. *Earth and Planetary Science Letters, 18*, 77–86.

Bernstein, R. E., Byrne, R. H., Betzer, P. R., & Greco, A. M. (1992). Morphologies and transformations of celestite in seawaer: The role of acantharians in strontium and barium geochemistry. *Geochimica et Cosmochimica Acta, 56*, 3273–3279.

Berrang, P. G., & Grill, E. V. (1974). The effect of manganese oxide scavenging on molybdenum in Saanich Inlet, British Columbia. *Marine Chemistry, 2*, 125–148.

Bertram, M. A., & Cowen, J. P. (1997). Morphological and compositional evidence for biotic precipitation of marine barite. *Journal of Marine Research, 55*, 577–593.

Bishop, J. K. B. (1988). The barite-opal-organic carbon association in oceanic particulate matter. *Nature, 332*, 341–343.

Blatt, H., Middleton, G., & Murray, R. (1980). *Origin of sedimentary rocks*. Englewood Cliffs, New Jersy: Prentice-Hall.

Bohn, H. L., O'Connor, G. A., & McNeal, B. L. (1979). *Soil chemistry* (p. 329). New York: John Willey.

Bojanowski, R., & Paslawska, S. (1970). On the occurrence of iodine in bottom sediments and interstitial waters of the southern Baltic Sea. *Acta Geophysica Polonica, 18*, 277–286.

Bonatti, E. (1975). Metallogenesis at oceanic spreading centers. *Annual Review of Earth and Planetary Sciences, 3*, 401–431.

Bonn, W. J., Gingele, F. X., Grobe, H., Mackensen, A., & Fütterer, D. K. (1998). Paleoproductivity at the Antarctic continental margin: Opal and barium records for the last 400 ka. *Palaeogeography, Palaeoclimatology, Palaeoecology, 139*, 195–211.

Borole, D. V., Krishnaswami, S., & Somayajulu, B. L. K. (1982). Uranium isotopes in rivers, estuaries and adjacent coastal sediments of western India: Their weathering, transport and oceanic budget. *Geochimica et Cosmochimica Acta, 46*, 125–137.

Boström, K., Joensuu, O., Valdes, S., & Riera, M. (1972). Geochemical history of South Atlantic Ocean sediments since Late Cretaceous. *Marine Geology, 12*, 85–121.

Boström, K., & Peterson, M. N. A. (1966). Precipitates from hydrothermal exhalations of the East Pacific Rise. *Economic Geology, 61*, 1258–1265.

Boström, K., & Peterson, M. N. A. (1969). The origin of aluminum-poor ferromanganoan sediments in areas of high heat flow on the East Pacific Rise. *Marine Geology, 7*, 427–447.

Boyle, E. A. (1983). Chemical accumulation variations under the Peru Current during the past 130,000 years. *Journal of Geophysical Research, 88*, 7667–7680.

Boyle, E. A., Sclater, F. R., & Edmond, J. M. (1977). The distribution of dissolved copper in the Pacific. *Earth and Planetary Science Letters, 37*, 38–54.

Bramlette, M. N., & Bradley, W. H. (1940). Geology and biology of North Atlantic deep-sea cores between Newfoundland and Ireland. Part 1. Lithology and geologic interpretations. *US Geological Survey Professional Paper, 196A*, 1–34.

Brandhorst, W. (1959). Nitrification and denitrification in the eastern tropical north Pacific. *Journal du Conseil International Pour L'Exploration de la Mer, 25*, 187–193.

Von Breymann, M. T., Emeis, K.-C., & Suess, E. (1992). Water depth and diagenetic constraints on the use of barium as a paleoproductivity indicator. In: C. P. Summerhayes, W. L. Prell & K. C. Emeis (Eds), *Upwelling systems: Evolution since the early Miocene* (pp. 273–284). Geological Society Special Publication.

Bruland, K. W. (1983). Trace elements in sea-water. In: J. P. Riley & G. Skirrow (Eds), *Chemical oceanography* (2nd ed., Vol. 8, pp. 157–220). London: Academic Press.

Bruland, K. W., & Lohan, M. C. (2003). Controls of trace metals in seawater. In: H. Elderfield (Ed.), *Treatise on geochemistry* (pp. 23–47). New York: Elsevier.

Brumsack, H.-J. (1989). Geochemistry of recent TOC-rich sediments from the Gulf of California and the Black Sea. *Geologische Rundschau, 78,* 851–882.

Brumsack, H. J. (1986). The inorganic geochemistry of Cretaceous black shales (DSDP Leg 41) in comparison to modern upwelling sediments from the Gulf of California. In: C. P. Summerhayes & N. J. Shackleton (Eds), *North Atlantic palaeoceanography* (pp. 447–462). London, United Kingdom: Geological Society Special Publications, Geological Society of London.

Brumsack, H. J., & Gieskes, J. M. (1983). Interstitial water trace-element chemistry of laminated sediments from the Gulf of California, Mexico. *Marine Chemistry, 14,* 89–106.

Burnett, W. C. (1977). Geochemistry and origin of phosphorite deposits from off Peru and Chile. *Geological Society of America Bulletin, 88,* 813–823.

Burnett, W. C. (1980). Apatite-glauconite associations off Peru and Chile: Paleo-oceanographic implications. *Journal of the Geological Society of London, 137,* 757–764.

Calvert, S. E. (1990). Geochemistry and origin of the Holocene sapropel in the Black Sea. In: V. Ittekkot, S. Kempe, W. Michaelis & A. Spitzy (Eds), *Facets of modern biogeochemistry* (pp. 326–352). Berlin: Springer-Verlag.

Calvert, S. E., Bustin, R. M., & Ingall, E. D. (1996). Influence of water column anoxia and sediment supply on the burial and preservation of organic carbon in marine shales. *Geochimica et Cosmochimica Acta, 60,* 1577–1593.

Calvert, S. E., & Fontugne, M. R. (2001). On the late Pleistocene-Holocene sapropel record of climatic and oceanographic variability in the eastern Mediterranean. *Paleoceanography, 16,* 78–94.

Calvert, S. E., & Pederson, T. F. (1996). Sedimentary geochemistry of manganese: Implications for the environment of formation of manganiferous black shales. *Economic Geology, 91,* 36–50.

Calvert, S. E., & Price, N. B. (1977). Geochemical variation in ferromanganese nodules and associated sediments from the Pacific Ocean. *Marine Chemistry, 5,* 43–74.

Campbell, J. M. (1917). Laterite: Its origin, structure and minerals. *Mining Magazine, 17,* 67–77, 120–128, 171–179, 220–229.

Cannariato, K. G., & Kennett, J. P. (1999). Climatically related millennial-scale fluctuations in strength of California margin oxygen-minimum zone during the past 60 ky. *Geology, 27,* 975.

Cardinal, D., Savoye, N., Trull, T. W., Andre, L., Kopczynska, E. E., & Dehairs, F. (2005). Variations of carbon remineralisation in the Southern Ocean illustrated by the Baxs proxy. *Deep Sea Research, 52,* 355–370.

Chaillou, G., Anschutz, P., Lavaux, G., Schafer, J. M., & Blanc, G. (2002). The distribution of Mo, U and Cd in relation to major redox species in muddy sediments of the Bay of Biscay. *Marine Chemistry, 80,* 41–59.

Charles, C. D., Froelich, P. N., Zibello, M. A., Mortlock, R. A., & Morley, J. J. (1991). Biogenic opal in Southern Ocean sediments over the last 450,000 years: Implications for surface water chemistry and circulation. *Paleoceanography, 6,* 697–728.

Chase, Z., Anderson, R. F., & Fleisher, M. Q. (2001). Evidence from authigenic uranium for increased productivity of the glacial Subantarctic ocean (Paper 2000PA000542). *Paleoceanography, 16,* 468–478.

Chavez, F. P., & Barber, R. T. (1987). An estimate of new production in the equatorial Pacific. *Deep Sea Research, 34,* 1229–1243.

Chester, R., Elderfield, H., Griffin, J. J., Johnson, L. R., & Padgham, R. C. (1972). Eolian dust along the eastern margins of the Atlantic Ocean. *Marine Geology, 13,* 91–106.

Chow, T. J., & Goldberg, E. D. (1960). On the marine geochemistry of barium. *Geochimica et Comsochimica Acta, 20,* 192–198.

Chylek, P., Lesins, G., & Lohmann, U. (2001). Enhancement of dust source area during past glacial periods due to changes of the Hadley circulation. *Journal of Geophysical Research, 106,* 18477–18485.

Cochran, J. K. (1984). The fates of uranium and thorium decay series nuclides in the estuarine environment. In: V. S. Kennedy (Ed.), *The estuary as a filter* (pp. 179–219). Orlando: Academic Press.

Cole, T. G. (1985). Composition, oxygen isotope geochemistry, and origin of smectite in the metal-lioferous sediments of the Bauer Deep, southeast Pacific. *Geochimica et Comsochimica Acta, 49,* 221–235.

Cole, T. G., & Shaw, H. F. (1983). The nature and origin of authigenic smectites in some recent marine sediments. *Clay Minerals, 18,* 239–252.

Collier, R., & Edmond, J. (1984). The trace element geochemistry of marine biogenic particulate matter. *Progress in Oceanography, 13,* 113–199.

Collier, R. W. (1984). Particulate and dissolved vanadium in the North Pacific Ocean. *Nature, 309,* 441–444.

Collier, R. W. (1985). Molybdenum in the northeast Pacific Ocean. *Limnology and Oceanography, 30,* 1351–1354.

Colodner, D., Edmond, J., & Boyle, E. (1995). Rhenium in the Balck Sea: Comparison with molybdenum and uranium. *Earth and Planetary Science Letters, 131,* 1–15.

Colodner, D., Sachs, J., Ravizza, G., Turekian, K., Edmond, J., & Boyle, E. (1993). The geochemical cycle of rhenium: A reconnaissance. *Earth and Planetary Science Letters, 117,* 205–221.

Corliss, J. B., Lyle, M., & Dymond, J. (1978). The chemistry of hydrothermal mounds near the Galapagos Rift. *Earth and Planetary Science Letters, 40,* 12–24.

Correns, C. W. (1937). Die Sedimente des aquatorialen Atlantischen Ozeans. II. Geochemie der Sedimente. *Wissenschaftliche Ergebnisse der Deutsche Atlantische Expedition "Meteor", 1925–1927, 3,* 205–245.

Correns, C. W. (1954). Titan in Tiefseesedimenten. *Deep Sea Research, 1,* 78–85.

Correns, C. W. (1969–1978). Titanium. In: K. H. Wedepohl, C. W. Correns, D. M. Shaw, K. K. Turekian, J. Zemann (Eds), *Handbook of geochemistry* (pp. 22-D-1-22). Berlin: Springer.

Coveney, R. M., & Glascock, M. D. (1989). A review of the origins of metal-rich Pennsylvanian black shales, central U.S.A., with an inferred fole for basinal brines. *Applied Geochemistry, 4,* 347–367.

Cronan, D. S. (1976). Manganese nodules and other ferromanganese oxide deposits. In: J. P. Riley & R. Chester (Eds), *Chemical oceanography* (pp. 217–263). San Diego, CA: Academic.

Cronan, D. S., & Tooms, J. S. (1967). Sub-surface concentrations of manganese nodules in Pacific sediments. *Deep Sea Research, 15,* 215–223.

Cronan, D. S., Van Andel, T. H., Heath, G. R., Dinkelman, M. G., Bennett, R. H., Burkey, D., Charleston, S., Kaneps, H., Rudolfo, K. S., & Yeats, R. S. (1972). Iron-rich basal sediments from the Eastern Equatorial Pacific: Leg 16, Deep Sea Drilling Project. *Science, 175,* 61–63.

Crusius, J., Calvert, S. E., Pedersen, T. F., & Sage, D. (1996). Rhenium and molybdenum enrichments in sediments as indicators of oxic, suboxic and anoxic conditions of deposition. *Earth and Planetary Science Letters, 145,* 65–78.

Crusius, J., & Thomson, J. (2000). Comparative behavior of authigenic Re, U and Mo during reoxidation and subsequent long-term burial in marine sediments. *Geochimica et Comsochimica Acta, 64,* 2233–2242.

Cushman, J. A., & Henbest, L. G. (1940). Geology and biology of North Atlantic deep-sea cores between Newfoundland and Ireland. Part 2. Foraminifera. *US Geological Survey Professional Paper, 196A,* 35–55.

Darwin, C. (1846). An account of the fine dust which often falls on vessels in the Atlantic Ocean. *Quarterly Journal of the Geological Society of London, 2,* 26–30.

Davies-Colley, R. J., Nelson, P. O., & Williamson, K. J. (1984). Copper and cadmium uptake by estuarine sedimentary phases. *Environmental Science and Technology, 18,* 491–499.

Davies-Colley, R. J., Nelson, P. O., & Williamson, K. J. (1985). Sulfide control of cadmium and copper concentrations in an anaerobic estuarine sediments. *Marine Chemistry, 16,* 173–186.

Dean, W. A., Gardner, J. V., & Piper, D. Z. (1997). Inorganic geochemical indicators of glacial-interlgacial changes in productivity and anoxia on the California continental margin. *Geochimica et Cosmochimica Acta, 61,* 4507–4518.

Dean, W. A., Piper, D. Z., & Peterson, L. C. (1999). Molybdenum accumulation in Cariaco basin sediment over the past 24 k.y.: A record of water-column anoxia and climate. *Geology, 27,* 507–510.

Degens, E. T. (1965). *Geochemistry of sediments* (p. 342). Englewood Cliffs, NJ: Prentice-Hall.

Dehairs, F., Baeyens, W., & Goeyens, L. (1992). Accumulation of suspended barite at mesopelagic depths and export production in the Southern Ocean. *Science, 258,* 1332–1335.

Dehairs, F., Chesselet, R., & Jedwab, J. (1980). Discrete suspended particles of barite and the barium cycle in the open ocean. *Earth and Planetary Science Letters, 49,* 528–550.

Dehairs, F., Goeyens, L., Stoobants, N., Bernard, P., Goyet, C., Poisson, A., & Chesselet, R. (1990). On suspended barite and the oxygen minimum in the Southern Ocean. *Global Biogeochemical Cycles, 4,* 85–102.

Dehairs, F., Lambert, C. E., Chesselet, R., & Risler, N. (1987). The biological production of marine suspended barite and the barium cycle in the western Mediterranean Sea. *Biogeochemistry, 4,* 119–139.

Dehairs, F., Stoobants, N., & Goeyens, L. (1991). Suspended barite as a tracer of biological activity in the Southern Ocean. *Marine Chemistry, 35,* 399–410.

Delany, A. C., Delany, A. C., Parkin, D. W., Griffin, J. J., Goldberg, E. D., & Reimann, B. E. F. (1967). Airborne dust collected at Barbados. *Geochimica et Cosmochimica Acta, 31,* 885–909.

deMenocal, P. B., Ortiz, J. D., Guilderson, T., & Sarnthein, M. (2000). Coherent high- and low-latitude climate variability during the Holocene Warm Period. *Science, 288,* 2198–2202.

Denison, R. E., Koepnick, R. B., Burke, W. H., Hetherington, E. A., & Fletcher, A. (1994). Construction of the Mississippian, Pennsylvanian and Permain seawater 87Sr/86Sr curve. *Chemical Geology, 112,* 145–167.

Deuser, W. G. (1973). Cariaco Trench: Oxidation of organic matter and residence time of anoxic water. *Nature, 242,* 601–603.

Dolcaster, D. L., Syers, J. K., & Jackson, M. L. (1970). Titanium as free oxide and substituted forms in kaolinites and other soil minerals. *Clays and Clay Minerals, 18,* 71–79.

Doyle, P. S., & Riedel, W. R. (1979). Cretaceous to Neogene ichtioliths in a giant piston core from the central North Pacific. *Micropaleontology, 25,* 337–364.

Dymond, J. (1981). Geochemistry of Nazca plate surface sediments: An evaluation of hydrothermal, biogenic, detrital and hydrogenous sources. *Geological Society of America Memoir, 154,* 133–173.

Dymond, J. (1985). Particulate barium fluxes in the oceans: An indicator of new productivity. *Transactions of the American Geophysical Union, 66,* 1275.

Dymond, J., & Collier, R. (1996). Particulate barium fluxes and their relationships to biological productivity. *Deep Sea Research, 43,* 1283–1308.

Dymond, J., Collier, R., McManus, J., Honjo, S., & Manganini, S. (1997). Can the aluminium and titanium contents of ocean sediments be used to determine the paleoproductivity of the oceans? *Paleoceanography, 12,* 586–593.

Dymond, J., & Corliss, J. B. (1977). History of metalliferous sedimentation at Deep Sea Drilling Site 319 in the South Eastern Pacific. *Geochimica et Cosmochimica Acta, 41,* 741–753.

Dymond, J., Corliss, J. B., Heath, G. R., Field, C. W., Dasch, E. J., & Veeh, H. H. (1973). Origin of metalliferous sediments from the Pacific Ocean. *Geological Society of America Bulletin, 84,* 3355–3372.

Dymond, J., Suess, E., & Lyle, M. (1992). Barium in deep-sea sediments: A geochemical proxy for paleoproductivity. *Paleoceanography, 7,* 163–181.

Dymond, J., & Veeh, H. H. (1975). Metal accumulation rates in the Southeast Pacific and the origin of metalliferous sediments. *Earth and Planetary Science Letters, 28,* 13–22.

Dypvik, H., & Harris, N. B. (2001). Geochemical facies analysis of fine-grained siliciclastics using Th/U, Zr/Rb and (Zr+Rb)/Sr ratios. *Chemical Geology, 181,* 131–146.

Edmond, J. M., Measures, C., Mangum, B., Grant, B., Sclater, F. R., Collier, R., Hudson, A., Gordon, L. I., & Corliss, J. B. (1979a). On the formation of metal-rich deposits on ridge crests. *Earth and Planetary Science Letters, 46,* 19–30.

Edmond, J. M., Measures, C., McDuff, R. E., Chan, L. H., Collier, R., Grant, B., Gordon, L. I., & Corliss, J. B. (1979b). Ridge crest hydrothermal activity and the balances of the major and minor elements in the ocean: The Galapagos data. *Earth and Planetary Science Letters, 46,* 1–18.

Eganhouse, R. P., & Venkatesan, M. I. (1993). Chemical oceanography and geochemistry. In: M. D. Dailey, D. J. Reish & J. W. Anderson (Eds), *Ecology of the Southern California Bight: A synthesis and interpretation*. Berkely: University of California Press.

Elderfield, H. (1970). Chromium speciation in sea water. *Earth and Planetary Science Letters, 9*, 10–16.

Emerson, S., Cranston, R. E., & Liss, P. S. (1979). Redox species in a reducing fjord: Equilibrium and kinetic considerations. *Deep Sea Research, 26A*, 859–878.

Emerson, S., & Hedges, J. I. (1988). Processes controlling the organic carbon content of open ocean sediments. *Paleoceanography, 3*, 621–634.

Emerson, S. R., & Huested, S. S. (1991). Ocean anoxia and the concentrations of molybdenum and vanadium in seawater. *Marine Chemistry, 34*, 177–196.

Engel, M. H., & Macko, S. A. (Eds) (1993). *Organic geochemistry: Principles and applications* (p. 861). New York: Plenum.

Erickson, B. E., & Helz, G. R. (2000). Molybdenum (VI) speciation in sulfidic waters: Stability and lability of thiomolybdates. *Geochimica et Cosmochimica Acta, 64*, 1149–1158.

Fagel, N., Dehairs, F., Peinert, R., Andre, L., & Antia, A. N. (2004). Reconstructing export production at the NE Atlantic margin: Potential and limits of the Ba proxy. *Marine Geology, 204*, 11–25.

Fairbanks, R. G. (1989). A 17,000-year glacio-eustatic sea level record: Influence of glacial melting rates on the Younger Dryas event and deep-ocean circulation. *Nature, 342*, 637–642.

Farrell, J. W., & Prell, W. L. (1989). Climatic change and $CaCO_3$ preservation: An 800,000 year bathymetric reconstruction from the central equatorial Pacific Ocean. *Paleoceanography, 4*, 447–466.

Feniak, M. W. (1944). Grain sizes and shapes of various minerals in igneous rocks. *American Mineralogist, 29*, 415–421.

Filippelli, G. M., & Delaney, M. L. (1995). Phosphorus geochemistry and accumulation rates in the eastern Equatorial Pacific: Results from Leg 138. *Proceedings of the Ocean Drilling Program, Scientific Results, 138*, 757–767.

Fisher, N. S., Guillard, R. L., & Bankston, D. C. (1991). The accumulation of barium by marine phytoplankton grown in culture. *Journal of Marine Research, 49*, 339–354.

Flegal, A. R., Sanudo-Wilhelmy, S. A., & Sceflo, G. M. (1995). Silver in the eastern Atlantic Ocean. *Marine Chemistry, 49*, 315–320.

François, R. (1987). The influence of humic substances on the geochemistry of iodine in nearshore and hemipelagic marine sediments. *Geochimica et Cosmochimica Acta, 51*, 2417–2427.

François, R. (1988). A study on the regulation of the concentrations of some trace metals (Rb, Sr, Zn, Pb, Cu, V, Cr, Ni, Mn and Mo) in Saanich Inlet sediments, British Columbia, Canada. *Marine Geology, 83*, 285–308.

François, R., Altabet, M. A., & Burckle, L. H. (1992). Glacial to interglacial changes in surface nitrate utilization in the Indian sector of the Southern Ocean as recorded by sediment $\delta^{15}N$. *Paleoceanography, 7*, 589–606.

François, R., Bacon, M. P., & Suman, D. O. (1990). Thorium-230 profiling in deep-sea sediments: high-resolution records of flux and dissolution of carbonate in the equatorial Atlantic during the last 24,000 years. *Paleoceanography, 5*, 761–787.

François, R., Frank, M., Rutgers van der Loeff, M. M., & Bacon, M. P. (2004). ^{230}Th normalization: An essential tool for interpreting sedimentary fluxes during the late Quaternary. *Paleoceanography, 19*, PA1018, doi:10.1029/2003PA000939.

François, R., Honjo, S., Manganini, S. J., & Ravizza, G. E. (1995). Biogenic barium fluxes to the deep sea: Implications for paleoproductivity reconstruction. *Global Biogeochemical Cycles, 9*, 289–303.

Frank, M., Eisenhauer, A., Bonn, W. J., Walter, P., Grobe, H., Kubik, P. W., Dittrich-Hannen, B., & Mangini, A. (1995). Sediment redistribution versus paleoproductivity change: Weddell Sea margin sediment stratigraphy and biogenic particle flux of the last 250,000 years deduced from ^{230}Th$_{ex}$, ^{10}Be and biogenic barium profiles. *Earth and Planetary Science Letters, 136*, 559–573.

Frank, M., Gersonde, R., Rutgers van der Loeff, M. M., Bohrmann, G., Nurnberg, C., Kubik, P. W., Suter, M., & Mangini, A. (2000). Similar glacial and interglacial export bioproductivity in the Atlantic sector of the Southern Ocean: Multiproxy evidence and implications for glacial atmospheric CO_2. *Paleoceanography, 15*, 642–658.

Frank, M., Gersonde, R., Rutgers van der Loeff, M. M., Kuhn, G., & Mangini, A. (1996). Late Quaternary sediment dating and quantification of lateral sediment redistribution applying $^{230}Th_{ex}$: A study from the eastern Atlantic sector of the Southern Ocean. *Geologische Rundschau*, *85*, 554–566.

Froelich, P. N. (1988). Kinetic control of dissolved phosphate in natural rivers and estuaries: A primer on the phosphate buffer mechanism. *Limnology and Oceanography*, *33*, 649–668.

Froelich, P. N., Klinkhammer, G. P., Bender, M. L., Luedtke, N. A., Heath, G. R., Cullen, D., Dauphin, P., Hammond, D., Hartman, B., & Maynard, V. (1979). Early oxidation of organic matter in pelagic sediments of the eastern equatorial Atlantic: Suboxic diagenesis. *Geochimica et Cosmochimica Acta*, *43*, 1075–1090.

Ganeshram, R. S., Calvert, S. E., Pedersen, T. F., & Cowie, G. L. (1999). Factors controlling the burial of organic carbon in laminated and bioturbated sediments off NW Mexico: Implications for hydrocarbon preservation. *Geochimica et Cosmochimica Acta*, *63*, 1723–1734.

Ganeshram, R. S., François, R., Commeau, J., & Brown-Leger, S. L. (2003). An experimental investigation of barite formation in seawater. *Geochimica et Comsochimica Acta*, *67*, 2599–2605.

Ganeshram, R. S., Pedersen, T. F., Calvert, S. E., & François, R. (2002). Reduced nitrogen fixation in the glacial ocean inferred from changes in marine nitrogen and phosphorus inventories. *Nature*, *415*, 156–159.

Ganeshram, R. S., Pedersen, T. F., & Murray, J. W. (1992). The record of organic carbon burial in Holocene and LGM sediments in the oxygen minimum off northwestern Mexico. *Transactions – American Geophysical Union*, *73*, 309.

Garrison, R. E., & Kastner, M. (1990). Phosphatic sediments and rocks recovered from the Peru margin during ODP Leg 112. *Proceedings of the Ocean Drilling Program, Scientific Results*, *112*, 111–134.

Gayral, P., & Fresnel, J. (1979). *Exanthemachrysis gayraliae* Lepailleur: Ultrastucture et discussion. *Protistologica*, *15*, 271–282.

van Geen, A., Frairbanks, R. G., Dartnell, P., McGann, M., & Gardner, J. V. (1996). Ventilation changes in the northeast Pacific during the last deglaciation. *Paleoceanography*, *11*, 519–528.

van Geen, A., & Husby, D. M. (1996). Cadmium in the California Current system: Tracer of past and present upwelling. *Journal of Geophysical Research*, *101*, 3489–3507.

German, C. R., Higgs, N. C., Thomson, J., Mills, R., Elderfield, H., Blusztajn, J., Fleer, A. P., & Bacon, M. P. (1993). A geochemical study of metalliferous sediment from the TAG hydrothermal mound, 26°08′N, Mid-Atlantic Ridge. *Journal of Geophysical Research*, *98*, 9683–9692.

Gingele, F., & Dahmke, A. (1994). Discrete barite particles and barium as tracers of paleoproductivity in South Atlantic sediments. *Paleoceanography*, *9*, 151–168.

Gingele, F. X., Zabel, M., Kasten, S., Bonn, W. J., & Nurnberg, C. C. (1999). Biogenic barium as a proxy for paleoproductivity: Methods and limitations of application. In: G. Fischer & G. Wefer (Eds), *Use of proxies in paleoceanography: Examples from the South Atlantic* (pp. 345–364). Berlin: Springer-Verlag.

Glaccum, R. A., & Prospero, J. M. (1980). Saharan aerosols over the tropical North Atlantic – Mineralogy. *Marine Geology*, *37*, 295–321.

Glasby, G. P. (1978). Deep-sea manganese nodules in the stratigraphic record: Evidence from DSDP cores. *Marine Geology*, *28*, 51–64.

Glasby, G. P. (1988). Manganese deposition through geological time: Dominance of the post-Eocene deep-sea environment. *Ore Geology Reviews*, *4*, 135–144.

Gobeil, C., Silverberg, N., Sundby, B., & Cossa, D. (1987). Cadmium diagenesis in Laurentian Trough sediments. *Geochimica et Cosmochimica Acta*, *51*, 589–596.

Goldberg, E. D. (1954). Marine geochemistry, I. Chemical scavengers of the sea. *Journal of Geology*, *62*, 249–265.

Goldberg, E. D. (1963). Mineralogy and chemistry of marine sedimentation. In: F. P. Shepard (Ed.), *Submarine geology* (pp. 436–466). New York: Harper and Row.

Goldberg, E. D., & Arrhenius, G. O. S. (1958). Chemistry of Pacific pelagic sediments. *Geochimica et Cosmochimica Acta*, *13*, 153–212.

Gonzales-Munoz, M., Fernandez-Luque, B., Martinez-Ruiz, F., Ben Chekroun, K., Arias, J. M., Rodriguez-Gallego, M., Martinez-Canamero, M., de Linares, C., & Paytan, A. (2003). Precipitation of barite by *Myxococcus xanthus*: Possible implications for the biogeochemical cycle of barium. *Applied and Environmental Microbiology, 69,* 5722–5725.

Gooday, A. J., & Nott, J. A. (1982). Intracellular barite crystals in two xenophyophores, *Aschemonella ramuliformis* and *Galatheammina* sp. (Protozoa, Rhizopoda) with comments on the taxonomy of *A. ramuliformis. Journal of the Marine Biological Association of the U.K., 62,* 595–605.

Graham, H. W. (1941). Plankton production in relation to character of water in the open Pacific. *Journal of Marine Research, 4,* 189–197.

Griffin, J. J., Windom, H., & Goldberg, E. D. (1968). The distribution of clay minerals in the world ocean. *Deep Sea Research, 15,* 433–459.

Grill, E. V. (1982). Kinetic and thermodynamic factors controlling manganese concentrations in oceanic waters. *Geochimica et Cosmochimica Acta, 46,* 2435–2446.

Gromet, P. L., Dymek, P. F., Haskin, L. A., & Korotev, R. L. (1984). The "North American shale composite": Its composition, major and minor element characteristics. *Geochimica et Cosmochimica Acta, 48,* 2469–2482.

Grousset, F. E., Buat-Menard, P., Boust, D., Tian, R.-C., Baudel, S., Pujol, C., & Vergnaud-Grazzini, C. (1989). Temporal changes of aeolian Saharan input in the Capre Verde abyssal plain since that last Glacial period. *Oceanologica Acta, 12,* 177–185.

Guy, R. D., & Chakrabarti, C. L. (1976). Studies of metal-organic interactions in modern systems pertaining to natural waters. *Canadian Journal of Chemistry, 16,* 2600–2611.

van der Hammen, T., & Hoogheimstra, H. (1995). The El Abra Stadial, a Younger Dryas equivalent in Colombia. *Quaternary Science Reviews, 14,* 841–851.

Haraldsson, C., & Westerlund, S. (1988). Trace metals in the water columns of the Black Sea and Framvaren Fjord. *Marine Chemistry, 23,* 417–424.

Haraldsson, C., & Westerlund, S. (1991). Total and suspended cadmium, cobalt, copper, iron, lead, manganese, nickel and zinc in the water column of the Black Sea. In: J. W. Murray & E. Izdar (Eds), *Black Sea oceanography* (pp. 161–172). Dordrecht: Kluwer.

Hartmann, M., Müller, P. J., Suess, E., & van der Weijden, C. H. (1973). Oxidation of organic matter in recent marine sediments. *"Meteor" Forschungs Ergebnisse, C12,* 74–86.

Hastenrath, S. (1985). *Climate and circulation of the tropics* (pp. 184–187). Boston: D. Reidel.

Hastenrath, S., & Greishar, L. (1993). Circulation mechanisms related to northeast Brazil rainfall anomalies. *Journal of Geophysical Research, 98,* 5093–5102.

Haug, G. H., Hughen, K. A., Sigman, D. M., Peterson, L. C., & Rohl, U. (2001). Southward migration of the Intertropical Convergence Zone through the Holocene. *Science, 293,* 1304–1307.

Haymon, R. M., & Kastner, M. (1981). Hot spring deposits on the East Pacific Rise at 21°N: Preliminary description of mineralogy and genesis. *Earth and Planetary Science Letters, 53,* 363–381.

Haymon, R. M., & Kastner, M. (1986). The formation of high temperature clay minerals from basalt alteration during hydrothermal discharge on the East Pacific Rise axis at 21°N. *Geochimica et Cosmochimica Acta, 50,* 1933–1939.

Heath, H. R. (1969). Mineralogy of Cenozoic deep-sea sediments from the equatorial Pacific Ocean. *Geological Society of America Bulletin, 80,* 1997–2018.

Heier, K. S., & Adams, J. A. S. (1963). The geochemistry of the alkali metals. *Physics and Chemistry of the Earth, 5,* 253–381.

Hein, J. R., Yeh, H.-W., & Alexander, E. (1979). Origin of iron-rich montmorillouite from the manganese nodule belt of the North Equatorial Pacific. *Clays and Clay Minerals, 27,* 185–194.

Helz, G. R., Miller, C. V., Charnock, J. M., Mosselmans, J. F. W., Pattrick, R. A. D., Garner, C. D., & Vaughan, D. J. (1996). Mechanism of molybdenum removal from the sea and its concentration in black shales: EXAFS evidence. *Geochimica et Cosmochimica Acta, 60,* 3631–3642.

Helz, G. R., Vorlicek, T. P., & Kahn, M. D. (2004). Molybdenum scavenging by iron monosulfide. *Environmental Science and Technology, 38,* 4263–4268.

Hem, J. D. (1969–1978). Aluminium: Behavior during weathering and alteration of rocks. In: K. H. Wedepohl, C. W. Correns, D. M. Shaw, K. K. Turekian & J. Zemann (Eds), *Handbook of geochemistry* (pp. 13-G-1–13-G-6). Berlin: Springer.

Hendy, I. L., & Kennett, J. P. (2000). Dansgaard/Oeschger Cycles and the California Current system: Planktonic foraminiferal response to rapid climate change in Santa Barbara Basin, ODP Hole 893A. *Paleoceanography, 15*, 30–42.

Hendy, I. L., Kennett, J. P., Roark, E. B., & Ingram, B. L. (2002). Apparent synchroneity of submillennial scale climate events between Greenland and Santa Barbara Basin, California from 30–10ka. *Quaternary Science Reviews, 21*, 1167–1184.

Hendy, I. L., & Pedersen, T. F. (2005). Is pore water oxygen content decoupled from productivity on the California Margin? Trace element results from Ocean Drilling Program Hole 1017E, San Lucia slope, California. *Paleoceanography, 20*, doi:10.1029/2004PA001123.

Hendy, I. L., Pedersen, T. F., Kennett, J. P., & Tada, R. (2004). Intermittent existence of a southern Californian upwelling cell during submillennial climate change of the last 60 kyr. *Paleoceanography, 19*, PA3007, doi:10.1029/2003PA000965.

Hodell, D. A., Curtis, J. H., Jones, G. A., Higuera-Gundy, A., Brenner, M., Binford, M. W., & Dorsey, K. T. (1991). Reconstruction of Caribbean climate change over the past 10,500 years. *Nature, 352*, 790–793.

Holser, W. T. (1997). Evaluation of the application of rare-earth elements to paleoceanography. *Palaeogeography, Palaeoclimatology, Palaeoecology, 132*, 309–323.

Honjo, S., Dymond, J., Collier, R., & Manganini, S. J. (1995). Export production of particles to the interior of the equatorial Pacific Ocean during the EqPac experiment. *Deep Sea Research II, 42*, 831–870.

Huerta-Diaz, M. A., & Morse, J. W. (1992). Pyritization of trace metals in anoxic marine sediments. *Geochimica et Cosmochimica Acta, 56*, 2681–2702.

Hughen, K., Overpeck, J., Peterson, L., & Anderson, R. (1996). The nature of varved sedimentation in the Cariaco Basin, Venezuela, and its palaeoclimatic significance. In: A. E. S. Kemp (Ed.), *Palaeoclimatology and palaeoceanography from laminated sediments* (pp. 171–183). London: Geological Society of London.

Hurley, P. M., Heezen, B. C., Pinson, W. H., & Fairbairn, H. W. (1963). K-Ar age values in pelagic sediments of the North Atlantic. *Geochimica et Cosmochimica Acta, 27*, 393–399.

Hurowitz, J. A., & McLennan, S. M. (2005). Geochemistry of Cambro-Ordovician sedimentary rocks of the northeastern United States: Changes in sediment sources at the onset of Taconian Orogenesis. *Journal of Geology, 113*, 571–588.

Isaacs, C. M. (2001). Depositional framework of the Monterey Formation, California. In: C. M. Isaacs & J. Rullkötter (Eds), *The Monterey Formation: From rocks to molecules* (pp. 1–30). New York: Columbia University Press.

Ito, T., Usui, A., Kajiwara, Y., & Nakano, T. (1998). Strontium isotopic compositions and palaeoceanographic implication of fossil manganese nodules in DSDP/ODP cores, Leg 1-126. *Geochimica et Cosmochimica Acta, 62*, 1545–1554.

Ivanochko, T. S., & Pedersen, T. F. (2004). Determining the influences of Late Quaternary ventilation and productivity variations in Santa Barbara Basin sedimentary oxygenation: A multi-proxy approach. *Quaternary Science Reviews, 23*, 467–480.

Jacobs, L., & Emerson, S. (1982). Trace metal solubility in an anoxic fjord. *Earth and Planetary Science Letters, 60*, 237–252.

Jacobs, L., Emerson, S., & Huested, S. S. (1987). Trace metal geochemistry in the Cariaco Trench. *Deep Sea Research, 34*, 965–981.

Jacobs, L., Emerson, S., & Skei, J. (1985). Partitioning and transport of metals across the O_2/H_2S interface in a permanently anoxic basin: Framvaren Fjord, Norway. *Geochimica et Cosmochimica Acta, 49*, 1433–1444.

Jansen, J. H. F., Van der Gaast, S. J., Koster, B., & Vaars, A. J. (1998). CORTEX, a shipboard XRF-scanner for element analyses in split sediment cores. *Marine Geology, 151*, 143–153.

Jeandel, C., Tachikawa, K., & Dehairs, F. (2000). Biogenic bartum in suspended and traped material as a tracer of export production in the tropical NE Atlantic (EUMELI sites). *Marine chemistry, 71*, 125.

Jenkyns, H. C. (1977). Fossil nodules. In: G. P. Glasby (Ed.), *Marine manganese deposits* (pp. 87–108). Amsterdam: Elsevier.

Jones, B. F., & Manning, D. A. C. (1994). Comparison of geochemical indices used for the interpretation of palaeoredox conditions in ancient mudstones. *Chemical Geology, 111,* 111–129.

Jouzel, J., Lorius, C., Petit, J. R., Genthon, C., Barkov, N. I., Kotlyakov, V. M., & Petrov, V. M. (1987). Vostok ice core: A continuous isotope temperature record over the last climatic cycle (160,000 years). *Nature, 329,* 403–408.

Kalhorn, S., & Emerson, S. (1984). The oxidation state of manganese in surface sediments of the deep sea. *Geochimica et Cosmochimica Acta, 48,* 897–902.

Kienast, S. S., Kienast, M., François, R., Mix, A. C., & Calvert, S. E. (2007). [230]Th normalized particle flux and sediment focusing in the Panama Basin region during the last 30 kyrs. *Paleoceanography, 22,* in press.

Klinkhammer, G. P., & Bender, M. L. (1980). The distribution of manganese in the Pacific Ocean. *Earth and Planetary Science Letters, 46,* 361–384.

Klinkhammer, G. P., & Palmer, M. R. (1991). Uranium in the oceans: Where it goes and why. *Geochimica et Cosmochimica Acta, 55,* 1799–1806.

Klump, J., Hebbeln, D., & Wefer, G. (2001). High concentrations of biogenic barium in Pacific sediments after Termination I – a signal of changes in productivity and deep water chemistry. *Marine Geology, 177,* 1–11.

Knox, F., & McElroy, M. B. (1984). Changes in atmospheric CO_2: Influence of the marine biota at high latitude. *Journal of Geophysical Research, 89,* 4629–4637.

Kochenov, A. V., Korolev, K. G., Dubinchuk, V. T., & Medvedev, Y. L. (1977). Experimental data on the conditions of precipitation of uranium from aqueous solutions. *Geochemistry International, 14,* 82–87.

Koczy, F. F. (1951). Factors determining the element concentration in sediments. *Geochimica et Cosmochimica Acta, 1,* 73.

Koide, M., Hodge, V. F., Yang, J. S., Stallard, M., & Goldberg, E. G. (1986). Some comparative marine chemistries of rhenium, gold, silver and molybdenum. *Applied Geochemistry, 1,* 705–714.

Kramer, D. (2006). *The marine silver cycle in coastal and open ocean environments of the North Pacific.* Master's Thesis, University of Victoria, 113pp.

Krishnaswami, S. (1976). Authigenic transition elements in Pacific pelagic clays. *Geochimica et Cosmochimica Acta, 40,* 425–434.

Kriss, A. E. (1963). *Marine Microbiology* (p. 536). Edinburgh: Oliver and Boyd.

Kronberg, B. I., Nesbitt, H. W., & Lam, W. W. (1986). Upper Pleistocene Amazon deep-sea fan muds reflect intense chemical weathering of their mountainous source lands. *Chemical Geology, 54,* 283–294.

Kroon, D., Shimmield, G., & Shimmield, T. (2000). Century- to millennial-scale sedimentological-geochemical records of the glacial-Holocene sediment variations from the Barra Fan (NE Atlantic). *Journal of the Geological Society, 157,* 643–653.

Ku, T.-L., Knauss, K. G., & Mathieu, G. G. (1977). Uranium in the open ocean: Concentration and isotopic composition. *Deep Sea Research, 24,* 1005–1017.

Ku, T. L. (1976). Rates of accretion. In: G. P. Glasby (Ed.), *Marine manganese deposits* (pp. 249–267). New York: Elsevier.

Kumar, N. (1994). *Trace metals and natural radiounclides as tracers of ocean productivity.* Doctoral Thesis, Columbia University, New York, 317pp.

Kumar, N., Anderson, R. F., Mortlock, R. A., Froelich, P. N., Kubik, P., Dittrich-Hannen, B., & Suter, M. (1995). Increased biological productivity and export production in the glacial Southern Ocean. *Nature, 378,* 675–680.

Kyte, F. T., Leinen, M., Heath, G. R., & Zhou, L. (1993). Cenozoic sedimentation history of the central North Pacific: Inferences from the elemental geochemistry of core LL44-GPC3. *Geochimica et Cosmochimica Acta, 57,* 1719–1740.

Landing, W. M., & Lewis, B. L. (1991). Thermodynamic modeling of trace metal speciation in the Black Sea. In: J. W. Murray & E. Izdar (Eds), *Black Sea oceanography* (pp. 125–160). Dordrecht: NATO ASI Series, Kluwer.

Lange, I. M., Reynolds, R. C., & Lyons, J. B. (1966). K/Rb ratios in coexisting K-feldspars and biotites from some New England granites and metasediments. *Chemical Geology, 1,* 317–322.

Langmuir, D. (1978). Uranium solution-mineral equilibria at low temperatures with applications to sedimentary ore deposits. *Geochimica et Cosmochimica Acta, 42,* 547–569.

Lea, D. W., Pak, D. K., Peterson, L. C., & Hughen, K. A. (2003). Synchroneity of tropical and high-latitude Atlantic temperatures over the Last Glacial termination. *Science, 301,* 1361–1364.

Leinen, M. (1987). The origin of paleochemical signatures in North Pacific pelagic clays: Partitioning experiments. *Geochimica et Cosmochimica Acta, 51,* 305–319.

Leinen, M. (1989). The late Quaternary record of atmospheric transport to the northwest Pacific from Asia. In: M. Leinen & M. Sarnthein (Eds), *Paleoclimatology and paleometeorology: Modern and past patterns of global atmospheric transport* (pp. 693–732). Boston: NATO ASI Series, Kluwer.

Leinen, M., & Heath, R. (1981). Sedimentary indicators of atmospheric activity in the northern hemisphere during the Conozoic. *Palaeogeography, Palaeoclimatology, Palaeoecology, 36,* 1–21.

Li, Y.-H., Bischoff, J., & Mathieu, G. (1969). The migration of manganese in the Arctic basin sediment. *Earth and Planetary Science Letters, 7,* 265–270.

Lin, H.-L., Peterson, L. C., Overpeck, J. T., Trumbore, S. E., & Murray, D. W. (1997). Late Quaternary climate change from d18O records of multiple species of planktonic foraminifera: High-resolution records from the anoxic Cariaco Basin, Venezuela. *Paleoceanography, 12,* 415–427.

Liu, L., Chen, J., Ji, J., & Chen, Y. (2004). Comparison of paleoclimatic change from Zr/Rb ratios in Chinese loess with marine isotope records over the 2.6–1.2 Ma BP interval. *Geophysical Research Letters,* 31, L15204, doi:10.1029/2004GL019693.

Loubere, L. (1999). A multiproxy reconstruction of biological productivity and oceanography in the eastern equatorial Pacific for the past 30,000 years. *Marine Micropaleontology, 37,* 173–198.

Loubere, P. (2000). Marine control of biological production in the eastern equatorial Pacific. *Nature,* 406, 497–500.

Loubere, P., Mekik, F., François, R., & Pichat, S. (2004). Export fluxes of calcite in the eastern equatorial Pacific from the Last Glacial Maximum to present. *Paleoceanography, 19,* doi:10.1029/2003PA000986.

Loughnan, F. C. (1969). *Chemical weathering of the silicate minerals* (p. 154). New York: Elsevier.

Lovley, D. R., Phillips, E. J. P., Gorby, Y. A., & Landa, E. R. (1991). Microbial reduction of uranium. *Nature, 350,* 413–416.

Lyle, M., Mitchell, N., Pisias, N., Mix, A., Martinez, J. I., & Paytan, A. (2005). Do geochemical estimates of sediment focusing pass the sediment test in the equatorial Pacific? *Paleoceanography, 20,* PA1005, doi:10.1029/2004PA001019.

Mackensen, A., Grobe, H., Hubberten, H.-W., & Kuhn, G. (1994). Benthic foraminiferal assemblages and the δ^{13}C-signal in the Atlantic sector of the southern ocean. In: R. Zahn, T. F. Pedersen, M. A. Kaminski & L. D. Labeyrie (Eds), *Carbon cycling in the glacial ocean: Constraints on the ocean's role in global change* (pp. 105–144). Springer-Verlag.

Mackin, J. E., & Aller, R. C. (1984). Dissolved Al in sediments and waters of the East China Sea: Implications for authigenic mineral formation. *Geochimica et Cosmochimica Acta, 48,* 281–297.

Malcolm, S. J. (1985). Early diagenesis of molybdenum in estuarine sediments. *Marine Chemistry, 16,* 213–225.

Mangini, A., Jung, M., & Laukenmann, S. (2001). What do we learn from peaks of uranium and of manganese in deep sea sediments? *Marine Geology, 177,* 63–78.

Marcantonio, F., Anderson, R. F., Higgins, S. M., Stute, M., Schlosser, P., & Kubik, P. (2001). Sediment focusing in the central equatorial Pacific Ocean. *Paleoceanography, 16,* 260–267.

Marcantonio, F., Anderson, R. F., & Mix, A. (1996). Extraterrestrial 3He as a tracer of marine sediment transport and accumulation. *Nature, 383,* 705–707.

Margolis, S. V., Ku, T. L., Glasby, G. P., Fein, C. D., & Audley-Charles, M. G. (1978). Fossil manganese nodules from Timor: Geochemical and radiochemical evidence for deep-sea origin. *Chemical Geology, 21,* 185–198.

Martin, J. H., & Knauer, G. A. (1973). The elemental composition of plankton. *Geochimica et Cosmochimica Acta, 37,* 1639–1651.

Martinez, P., Bertrand, P., Bouloubassi, I., Bareille, G., Shimmield, G., Vautravers, B., Grousset, F., Guichard, S., Ternois, Y., & Sicre, M.-A. (1996). An integrated view of inorganic and organic biogeochemical indicators of paleoproductivity changes in a coastal upwelling area. *Organic Geochemistry*, *24*, 411–420.

Mason, J. A., & Jacobs, P. M. (1998). Chemical and particle-size evidence for addition of fine dust to soils of the midwestern United States. *Geology*, *26*, 1135–1138.

Matthewson, A. P., Shimmield, G. B., Kroon, D., & Fallick, A. E. (1995). A 300 kyr high-resolution aridity record of the North African continent. *Palaeoceanography*, *10*, 677–692.

Maynard, J. B. (1983). *Geochemistry of sedimentary ore deposits* (p. 305). New York: Springer-Verlag.

McCann, T. (1991). Petrological and geochemical discrimination of provenance in the southern Welsh Basin. In: A. C. Morton, S. P. Todd & P. D. W. Haughton (Eds), *Developments in sedimentary provenance studies* (pp. 215–230). London: Geological Society Special Publication No. 57.

McCarty, P. L. (1972). Energetics of organic matter degradation. In: R. Mitchell (Ed.), *Water pollution microbiology* (pp. 91–118). New York: J. Wiley.

McKee, B. A., & Todd, J. F. (1993). Uranium behavior in a permanently anoxic fjord: Microbial control?. *Limnology and oceanography*, *38*, 48–414.

McLennan, S. M. (1989). Rare earth elements in sedimentary rocks: Influence of provenance and sedimentary processes. *Reviews in Mineralogy*, *11*, 169–200.

McLennan, S. M., Hemming, S. R., McDaniel, D. K., & Hanson, G. N. (1993). Geochemical approaches to sedimentation, provenance, and tectonics. In: M. J. Johnsson & A. R. Basu (Eds), *Processes controlling the composition of clastic sediments* (pp. 21–40). Boulder, CO: Geological Society of America Special Paper 284.

McManus, J., Berelson, W. M., Hammond, D. E., & Klinkhammer, G. P. (1999). Barium cycling in the North Pacific: Implications for the utility as Ba as a paleoproductivity and paleoalkalinity proxy. *Paleoceanography*, *14*, 53–61.

McManus, J., Berelson, W. M., Klinkhammer, G. P., Hammond, D. E., & Holm, C. (2005). Authigenic uranium: Relationship to oxygen penetration depth and organic carbon rain. *Geochimica et Cosmochimica Acta*, *69*, 95–108.

McManus, J., Berelson, W. M., Klinkhammer, G. P., Johnson, K. S., Coale, K. H., Anderson, R. F., Kumar, N., Burdige, D. J., Hammond, D. E., Brumsack, H. J., McCorkle, D. C., & Rushdi, A. (1998). Geochemistry of barium in marine sediments: Implications for its use as a paleoproxy. *Geochimica et Cosmochimica Acta*, *62*, 3453–3473.

McManus, J., Berelson, W. M., Klinkhammer, G. P., Kilgore, T. E., & Hammond, D. E. (1994). Remobilization of barium in continental margin sediments. *Geochimica et Cosmochimica Acta*, *58*, 4899–4907.

McMurtry, G. M., & Yeh, H. W. (1981). Hydrothermal clay mineral formation of East Pacific Rise and Bauer Basin sediments. *Chemical Geology*, *32*, 189–205.

Meunier, A., & Velde, B. (2004). *Illite: Origins, evolution and metamorphism* (p. 286). Berlin: Springer-Verlag.

Milne, A. R., & Fitzpatrick, R. W. (1977). Titanium and zirconium minerals. In: J. B. Dixon & S. B. Weed (Eds), *Minerals in soil environments* (pp. 1131–1205). Madison, WI: Soil Science Society of America.

Milner, H. B. (1962a). *Sedimentary petrography, 1, methods in sedimentary petrography* (p. 643). New York: Macmillan.

Milner, H. B. (1962b). *Sedimentary petrography, 2, principles and applications* (p. 715). New York: Macmillan.

Monnin, C., Jeandel, C., Cattaldo, T., & Dehairs, F. (1999). The marine barite saturation state of the world's oceans. *Marine Chemistry*, *65*, 253–261.

Mora, G., & Martinez, J. I. (2005). Sedimentary metal ratios in the Colombia Basin as indicators for water balance change in northern South America during the past 400,000 years. *Palaeoceanography*, *20*, PA4013.

Moran, S. B., & Moore, R. M. (1988). Evidence from mesocosm studies for biological removal of dissolved aluminium from sea water. *Nature*, *335*, 706–708.

Morford, J. L., & Emerson, S. (1999). The geochemistry of redox sensitive trace metals in sediments. *Geochimica et Cosmochimica Acta*, *63*, 1735–1750.

Morris, A. W. (1975). Dissolved molybdenum and vanadium in the northeast Atlantic Ocean. *Deep Sea Research*, *22*, 49–54.

Mortlock, R. A., Charles, C. D., Froelich, P. N., Zibello, M. A., Saltzman, J., Hays, J. D., & Burckle, L. H. (1991). Evidence for lower productivity in the Antarctic Ocean during the last glaciation. *Nature*, *351*, 220–223.

Murray, J., & Irvine, R. (1895). On the manganese oxide and manganese nodules in marine deposits. *Transactions – Royal Society of Edinburgh*, *37*, 721–742.

Murray, J., & Renard, A. F. (1891). *Report on deep-sea deposits based on the specimens collected during the voyage of HMS Challenger in the years 1872–1876*. London: Her Majesty Stationary Office.

Murray, J. W., Balistrieri, L. S., & Paul, B. (1984). The oxidation state of manganese in marine sediments and ferromanganese nodules. *Geochimica et Cosmochimica Acta*, *48*, 1237–1248.

Murray, J. W., Spell, B., & Paul, B. (1983). The contrasting geochemistry of manganese and chromium in the eastern tropical Pacific Ocean. In: C. S. Wong, E. Boyle, K. W. Bruland & E. D. Goldberg (Eds), *Trace metals in sea water* (pp. 643–669). New York: Plenum Press.

Murray, R. W., Knowlton, C. W., Leinen, M., Mix, A. C., & Polsky, C. H. (2000). Export production and carbonate dissolution in the central equatorial Pacific over the past 1 Myr. *Paleoceanography*, *15*, 570–592.

Murray, R. W., & Leinen, M. (1993). Chemical transport to the seafloor of the equatorial Pacific Ocean across a latitudinal transect at 135°W: Tracking sedimentary major, trace, and rare earth element fluxes at the Equator and the Intertropical Convergence zone. *Geochimica et Cosmochimica Acta*, *57*, 4141–4163.

Murray, R. W., & Leinen, M. (1996). Scavenged excess aluminum and its relationship to bulk titanium in biogenic sediment from the central equatorial Pacific Ocean. *Geochimica et Cosmochimica Acta*, *60*, 3869–3878.

Murray, R. W., Leinen, M., & Isern, A. R. (1993). Biogenic flux of Al to sediment in the central equatorial Pacific Ocean: Evidence for increased productivity during glacial periods. *Paleoceanography*, *8*, 651–670.

Naqvi, S. W. A. (1987). Some aspects of the oxygen-deficient conditions and denitrification in the Arabian Sea. *Journal of Marine Research*, *45*, 1049–1072.

Naqvi, S. W. A., Noronha, R. J., & Gangadhara Reddy, C. V. (1982). Denitrification in the Arabian Sea. *Deep Sea Research*, *29*, 459–469.

Nesbitt, H. W., & Young, G. M. (1982). Early Proterozoic climates and plate motions inferred from major element chemistry of lutites. *Nature*, *299*, 715–717.

Nesbitt, H. W., & Young, G. M. (1996). Petrogenesis of sediments in the absence of chemical weathering: Effects of abrasion and sorting on bulk composition and mineralogy. *Sedimentology*, *43*, 341–358.

Nesbitt, H. W., & Young, G. M. (1997). Sedimentation in the Venezuelan Basin, Circulation in the Caribbean Sea and onset of northern hemisphere glaciation. *Journal of Geology*, *105*, 531–544.

Nesbitt, H. W., Young, G. M., McLennan, S. M., & Keays, R. R. (1996). Effects of chemical weathering and sorting on the petrogenesis of siliciclastic sediments, with implications for provenance studies. *Journal of Geology*, *104*, 525–542.

Nozaki, Y. (1997). A fresh look at element distribution in the North Pacific. *Transactions of the American Geophysical Union*, *78*, 221.

Nürnberg, C. C., Bohrmann, G., Schlüter, M., & Frank, M. (1997). Barium accumulation in the Atlantic sector of the Southern Ocean: Results from 190,000-year records. *Paleoceanography*, *12*, 594–603.

Orians, K. J., & Bruland, K. W. (1986). The biogeochemistry of aluminum in the Pacific Ocean. *Earth and Planetary Science Letters*, *78*, 397–410.

van Os, B. J. H., Middelburg, J. J., & de Lange, G. J. (1991). Possible diagentic mobilization of barium in sapropelic sediment from the eastern Mediterranean. *Marine Geology*, *100*, 125–136.

Pancost, R. D., & Boot, C. S. (2004). The palaeoclimatic utility of terrestrial biomarkers in marine sediments. *Marine Chemistry*, *92*, 239–261.

Parker, F. L. (1962). Planktonic foraminiferal species in Pacific sediments. *Micropaleontology*, *8*, 219–254.

Parkin, D. W. (1974). Trade-winds during glacial cycles. *Proceedings of the Royal Society of London, A337,* 73–100.

Parkin, D. W., & Padgham, R. C. (1975). Further studies on trade winds during the glacial cycles. *Proceedings of the Royal Society of London, 346,* 245–260.

Parkin, D. W., & Shackleton, N. J. (1973). Trade wind and temperature correlations down a deep-sea core off the Sahara Coast. *Nature, 245,* 455–457.

Parnell, J., Ye, L., & Chen, C. (Eds) (1990). *Sediment-hosted mineral deposits* (p. 227). Oxford: Blackwell.

Pastouret, L., Chamley, H., Delibrias, G., Duplessy, J. C., & Thiede, J. (1978). Late Quaternary climatic changes in western tropical Africa deduced from deep-sea sedimentation off the Niger delta. *Oceanologica Acta, 1,* 217–232.

Paytan, A. (1995). *Marine barite, a recorder of oceanic chemistry, productivity, circulation.* Doctoral Thesis, University of California, San Diego, 111pp.

Paytan, A., & Kastner, M. (1996). Benthic Ba fluxes in the central equatorial pacific, implications for the oceanic Ba cycle. *Earth and Planetary Science Letters, 142,* 439–450.

Paytan, A., Kastner, M., & Chavez, F. P. (1996). Glacial to interglacial fluctuations in productivity in the equatorial Pacific as indicated by marine barite. *Science, 274,* 1355–1357.

Pedersen, T. F. (1985). Early diagenesis of copper and molybdenum in mine tailings and natural sediments in Rupert and Holberg Inlets, British Columbia. *Canadian Journal of Earth Sciences, 22,* 1474–1484.

Pedersen, T. F., & Calvert, S. E. (1990). Anoxia vs. productivity: What controls the formation of organic-carbon-rich sediments and sedimentary rocks?. *American Association of Petroleum Geologists Bulletin, 74,* 454–466.

Pedersen, T. F., Pickering, M., Vogel, J. S., Southon, J., & Nelson, D. E. (1988). The response of benthic foraminifera to productivity cycles in the eastern equatorial Pacific: Faunal and geochemical constraints on glacial bottom-water oxygen levels. *Paleoceanography, 3,* 157–168.

Pedersen, T. F., & Price, N. B. (1980). The geochemistry of iodine and bromine in sediments of the Panama Basin. *Journal of Marine Research, 38,* 397–411.

Pedersen, T. F., & Price, N. B. (1982). The geochemistry of manganese carbonate in Panama Basin sediments. *Geochimica et Cosmochimica Acta, 46,* 59–68.

Pedersen, T. F., Shimmield, G. B., & Price, N. B. (1992). Lack of enhanced preservation of organic matter in sediments under the intense oxygen minimum on the Oman Margin. *Geochimica et Cosmochimica Acta, 56,* 545–551.

Pedersen, T. F., Waters, R. D., & Macdonald, R. W. (1989). On the natural enrichment of cadmium and molybdenum in the sediments of Ucluelet Inlet, British Columbia. *Science of the Total Environment, 79,* 125–139.

Peterson, L. C., Haug, G. H., Hughen, K. A., & Rohl, U. (2000). Rapid changes in the hydrologic cycle of the Tropical Atlantic during the last glacial. *Science, 290,* 1947.

Peterson, L. C., Overpeck, J. T., Kipp, N. G., & Imbrie, J. (1991). A high-resolution late Quaternary upwelling record from the anoxic Cariaco Basin, Venezuela. *Paleoceanography, 6,* 99–120.

Peterson, M. N. A., & Goldberg, E. D. (1962). Feldspar distributions in South Pacific pelagic sediments. *Journal of Geophysical Research, 67,* 3477–3492.

Pettijohn, F. J. (1957). *Sedimentary rocks* (p. 718). New York: Harper Bros.

Phleger, F. B., Parker, F. L., & Peirson, J. F. (1953). North Atlantic foraminifera. Reports of the Swedish deep-sea expedition 1947–1948, v. VIII, Sediment cores from the North Atlantic Ocean, 1. Elanders Boktryckeri Aktiebolag, Göteborg, 122pp.

Pichat, S., Sims, K. W. W., François, R., McManus, J. F., Leger, S. B., & Albarede, F. (2004). Lower export production during glacial periods in the equatorial Pacific derived from (231Pa/230Th)xs,0 measurements in deep-sea sediments. *Paleoceanography, 19,* PA4023, doi:10.1029/2003PA000994.

Piper, D. Z. (1971). The distribution of Co, Cr, Cu, Fe, Mn, Ni and Zn in Framvaren, a Norwegian anoxic fjord. *Geochimica et Cosmochimica Acta, 35,* 531–550.

Piper, D. Z. (1973). Origin of metalliferous sediments from the East Pacific Rise. *Earth and Planetary Science Letters, 19,* 75–82.

Piper, D. Z. (1994). Seawater as the source of minor elements in black shales, phosphorites and other sedimentary rocks. *Chemical Geology, 114,* 95–114.

Piper, D. Z., & Dean, W. E. (2002). *Trace-element deposition in the Cariaco Basin, Venezuela Shelf, under sulfate-reducing conditions – a history of the local hydrography and global climate, 20 ka to the present.* US Geological Survey Professional Paper 1670, Washington, DC, 41pp.

Piper, D. Z., & Fowler, B. (1980). New constraint on the maintenance of Mn nodules at the sediment surface. *Nature, 286,* 880–883.

Piper, D. Z., & Isaacs, C.M. (1995). *Geochemistry of minor elements in the Monterey Formation, CA: Seawater chemistry of deposition.* US Geological Survey Professional Paper No. 1566, Reston, VA, 41pp.

Piper, D. Z., & Isaacs, C. M. (1996). Instability of bottom-water redox conditions during accumulation of Quaternary sediment in the Japan Sea. *Paleoceanography, 11,* 171–190.

Piper, D. Z., & Medrano, M. D. (1994). *Geochemistry of the phosphoria formation at Montpelier Canyon, Idaho: Environment of Deposition.* US Geological Survey Bulletin 2023–B, B1–B28

Price, N. B., & Calvert, S. E. (1977). The contrasting geochemical behaviours of iodine and bromine in recent sediments from the Namibian shelf. *Geochimica et Cosmochimica Acta, 41,* 1769–1775.

Price, N. B., & Calvert, S. E. (1978). The geochemistry of phosphorites from the Namibian Shelf. *Chemical Geology, 23,* 151–170.

Price, N. B., Calvert, S. E., & Jones, P. G. W. (1970). The distribution of iodine and bromine in the sediments of the southwestern Barents Sea. *Journal of Marine Research, 28,* 22–34.

Prospero, J. M., & Bonatti, E. (1969). Continental dust in the atmosphere of the eastern equatorial Pacific. *Journal of Geophysical Research, 74,* 3362–3371.

Pruysers, P. A., De Lange, G. J., Middelburg, J. J., & Hydes, D. J. (1993). The diagenetic formation of metal-rich layers in sapropel-containing sediments in the eastern Mediterranean. *Geochimica et Cosmochimica Acta, 57,* 527–536.

Puchelt, H. (1969–1978). Barium: Abundance in rock-forming minerals. In: K. H. Wedepohl, C. W. Correns, D. M. Shaw, K. K. Turekian & J. Zemann (Eds), *Handbook of geochemistry* (pp. D1–D18). Berlin: Springer.

Radczewski, O. E. (1939). Eolian deposits in marine sediments. In: P. D. Trask (Ed.), *Recent marine sediments* (pp. 496–502). Tulsa: American Association of Petroleum Geologists.

Rahn, K. A. (1976). Silicon and aluminum in atmosheric aerosols: Crust-air fractionation?. *Atmospheric Environment, 10,* 597–601.

Raman, K. V., & Jackson, M. L. (1965). Rutile and anatase determination in soils and sediments. *American Mineralogist, 50,* 1086–1092.

Rankama, K., & Sahama, T. G. (1950). *Geochemistry.* Chicago: University of Chicago Press, 912pp.

Ravizza, G., Turekian, K. K., & Hay, B. J. (1991). The geochemistry of rhenium and osmium in recent sediments from the Black Sea. *Geochimica et Cosmochimica Acta, 55,* 3741–3752.

Reitz, A., Pfeifer, K., de Lange, G. J., & Klump, J. (2004). Biogenic barium and the detrital Ba/Al ratio: A comparison of their direct and indirect determination. *Marine Geology, 204,* 289–300.

Revelle, R. (1944). *Marine bottom samples collected in the Pacific by the Carnegie on its seventh cruise.* Carnegie Institution of Washington, Publication 556, 1–180, Washington, D.C.

Rex, R. W., & Goldberg, E. D. (1958). Quartz contents of pelagic sediments of the Pacific Ocean. *Tellus, 10,* 153–159.

Richards, F. A. (1965). Anoxic basins and fjords. In: J. P. Riley & G. Skirrow (Eds), *Chemical Oceanography* (pp. 611–645). New York: Academic Press.

Richards, F. A. (1975). The Cariaco Basin (Trench). *Oceanography and Marine Biology Annual Reviews, 13,* 11–67.

Richards, F. A., & Vaccaro, R. F. (1956). The Cariaco Trench, an anaerobic basin in the Caribbean Sea. *Deep Sea Research, 3,* 214–228.

Ricker, W. E. (1984). Computation and uses of central trend lines. *Canadian Journal of Zoology, 62,* 1897–1905.

Rieder, N., Ott, H. A., Pfundstein, P., & Schoch, R. (1982). X-ray microanalysis of the mineral content of some protozoa. *Journal of Protozoology, 29,* 15–18.

Riley, J. P., & Roth, I. (1971). The distribution of trace elements in some species of phytoplankton grown in culture. *Journal of Marine Biological Association of the UK, 51,* 63–72.

Robin, E., Rabouille, C., Martinez, G., Lefevre, I., Reyss, J. L., xvan Beek, P., & Jeandel, C. (2003). Direct barite determination using SEM/EDS-ACC system: Implication for constraining barium carriers and barite preservation in marine sediments. *Marine Chemistry, 82*, 289–306.

Robinson, R. S., Sigman, D. M., DiFiore, P. J., Rohde, M. M., Mashiotta, T. A., & Lea, D. W. (2005). Diatom-bound 15N//14N: New support for enhanced nutrient consumption in the ice age subantarctic. *Paleoceanography, 20*, doi:10.1029/2004PA001114.

Rollinson, H. R. (1993). *Using geochemical data: Evaluation, presentation, interpretation* (p. 352). Harlow: Longman.

Rosenthal, Y., Boyle, E.A., Labeyrie, L., & Oppo, D. (1995a). Glacial enrichments of authigenic Cd and U in Subantarctic sediments: A climatic control on the elements' oceanic budget? *Paleoceanography, 10*, 395–413.

Rosenthal, Y., Lam, P., Boyle, E. A., & Thomson, J. (1995b). Authigenic cadmium enrichments in suboxic sediments: Precipitation and postdepositional mobility. *Earth and Planetary Science Letters, 132*, 99–111.

Rue, E., Smith, G., Cutter, G., & Bruland, K. (1997). The response of trace element redox couples to suboxic conditions in the water column. *Deep Sea Research I, 44*, 113–134.

Ruhlemann, C., Frank, M., Hale, W., Mangini, A., Mulitza, S., Muller, P. J., & Wefer, G. (1996). Late Quaternary productivity changes in the western equatorial Atlantic: Evidence from ^{230}Th-normalized carbonate and organic carbon accumulation rates. *Marine Geology, 135*, 127–152.

Rullkötter, J. (2000). Organic matter: The driving force for early diagenesis. In: H. D. Schulz & M. Zabell (Eds), *Marine geochemistry* (pp. 129–172). Berlin: Springer.

Rutsch, H.-J., Mangini, A., Bonani, G., Dittrich-Hannen, B., Kubik, P. W., Suter, M., & Segl, M. (1995). ^{10}Be and Ba concentrations in West African sediments trace productivity in the past. *Earth and Planetary Science Letters, 133*, 129–143.

Ruttenberg, K. C., & Berner, R. A. (1992). Development of a sequential extraction method for different forms of phosphorus in marine sediments. *Limnology and Oceanography, 37*, 991–1007.

Ruttenberg, K. C., & Berner, R. A. (1993). Authigenic apatite formation and burial in sediments from non-upwelling, continental margin environments. *Geochimica et Cosmochimica Acta, 57*, 991–1007.

Sadiq, M. (1988). Thermodynamic solubility relationships of inorganic vanadium in the marine environment. *Marine Chemistry, 23*, 87–96.

Sarkar, A., Bhattacharya, S. K., & Sarin, M. M. (1993). Geochemical evidence for anoxic deep water in the Arabian Sea during the last glaciation. *Geochimica et Cosmochimica Acta, 57*, 1009–1016.

Sarmiento, J. L., & Toggweiler, J. R. (1984). A new model for the role of the oceans in determining atmospheric pCO2. *Nature, 308*, 621–624.

Sarnthein, M. (1978). Sand deserts during glacial maximum and climatic optimum. *Nature, 272*, 43–46.

Sarnthein, M., Tetzlaff, G., Koopmann, B., Wolter, K., & Pflaumann, U. (1981). Glacial and inter-glacial wind regimes over the eastern subtropical Atlantic and North-west Africa. *Nature, 293*, 193–195.

Sarnthein, M., Thiede, J., Pflaumann, U., Erlenkeuser, H., Fütterer, D., Koopman, B., Lange, H., & Seibold, E. (1982). Atmospheric and oceanic circulation patterns off Northwest Africa during the past 25 million years. In: U. von Rad, K. Hinz, M. Sarnthein & E. Seibold (Eds), *Geology of the Northwest African continental margin* (pp. 545–604). Berlin: Springer-Verlag.

Schenau, S. J., Prins, M. A., deLange, G. J., & Monnin, C. (2001). Barium accumulation in the Arabian Sea: Controls on barite preservation in marine sediments. *Geochimica et Cosmochimica Acta, 65*, 1545–1556.

Schenau, S. J., Slomp, C. P., & DeLange, G. J. (2000). Phosphogenesis and active phosphorite formation in sediments from the Arabian Sea oxygen minimum zone. *Marine Geology, 169*, 1–20.

Schmitz, B. (1987a). Barium, equatorial high productivity, and the northward wandering of the Indian continent. *Paleoceanography, 2*, 197–215.

Schmitz, B. (1987b). The TiO_2/Al_2O_3 ratio in the Cenozoic Bengal Abyssal Fan sediments and its use as a paleostream energy indicator. *Marine Geology, 76*, 195–206.

Schneider, R. R., Price, B., Müller, P. J., Kroon, D., & Alexander, I. (1997). Monsoon related variations in Zaire (Congo) sediment load and influence of fluvial silicate supply on marine productivity in the east equatorial Atlantic during the last 200,000 years. *Paleoceanography, 12*, 463–481.

Schott, W. (1935). Die Foraminiferen in dem äquatorialen Teil des Atlantischen Ozeans. Wissenschaftliche Ergebnisse der Deutsche Atlantische Expedition "Meteor" 1925–27, *3*(part 3), 43–134.

Schuffert, J. D., Kastner, M., & Jahnke, R. A. (1998). Carbon and phosphorus burial associated with modern phosphorite formation. *Marine Geology, 146*, 21–31.

Scranton, M. I., Astor, Y., Bohrer, R., Ho, T.-Y., & Muller_karger, F. (2001). Controls on temporal variability of the geochemistry of the deep Cariaco Basin. *Deep Sea Research II, 48*, 1605–1625.

Shaw, T. J., Gieskes, J. M., & Jahnke, R. A. (1990). Early diagenesis in differing depositional environments: The response of transition metals in pore water. *Geochimica et Cosmochimica Acta, 54*, 1233–1246.

Shemesh, A., Macko, S. A., Charles, C. D., & Rau, G. H. (1993). Isotopic evidence for reduced productivity in the glacial Southern Ocean. *Science, 262*, 407–410.

Shimmield, G., Derrick, S., Mackensen, A., Grobe, H., & Pudsey, C. (1994). The history of barium, biogenic silica and organic carbon accumulation in the Weddell Sea and Antarctic Ocean over the last 150,000 years. In: R. Zahn, T. F. Pedersen, M. A. Kaminski & L. Labeyrie (Eds), *Carbon cycling in the glacial ocean: Constraints on the ocean's role in global change* (pp. 555–574). Springer Verlag.

Shimmield, G. B. (1992). Can sediment geochemistry record changes in coastal upwelling palaeo-productivity? Evidence from Northwest Africa and the Arabian Sea. In: C. P. Summerhayes (Ed.), *Upwelling systems: Evolution since the Early Miocene* (pp. 29–46). London: Geological Society Special Publication 64.

Shimmield, G. B., & Mowbray, S. T. (1991). The inorganic geochemical record of the northwest Arabian Sea: A history of productivity variation over the last 400 K.Y. from sites 722 and 724. *Proceedings of the Ocean Drilling Program, Scientific Results, 117*, 409–419.

Shimmield, G. B., & Pedersen, T. F. (1990). The geochemistry of reactive trace elements and halogens in hemipelagic continental margin sediments. *Reviews in Aquatic Sciences, 3*, 255–279.

Shimmield, G. B., & Price, N. B. (1986). The behaviour of molybdenum and manganese during early sediment diagenesis — offshore Baja California, Mexico. *Marine Chemistry, 19*, 261–280.

Siegenthaler, U., & Wenk, T. (1984). Rapid atmospheric CO_2 variations and ocean circulation. *Nature, 308*, 624–626.

Sigman, D., & Boyle, E. A. (2000). Glacial/interglacial variations in atmospheric carbon dioxide. *Nature, 407*, 859–869.

Singer, A. (1979/80). The paleoclimatic interpretation of clay minerals in soils and weathering profiles. *Earth Science Reviews, 15*, 303–326.

Spears, D. A., & Kanaris-Sotiriou, R. (1976). Titanium in some Carboniferous sediments from Great Britain. *Geochimica et Cosmochimica Acta, 40*, 345–351.

Spencer, D. W., & Brewer, P. G. (1971). Vertical advection diffusion and redox potentials as controls on the distribution of manganese and other trace metals dissolved in waters of the Black Sea. *Journal of Geophysical Research, 76*, 5877.

Stein, R. (1991). *Accumulation of organic carbon in marine sediments*. Lecture Notes in Earth Sciences, 34. Springer, Heidelberg, 217pp.

Sternberg, E., Tang, D., Ho, T. Y., Jeandel, C., & Morel, F. (2005). Barium uptake and adsorption in diatoms. *Geochimica et Cosmochimica Acta, 69*, 2745–2752.

Stewart, A. D. (1990). Geochemistry, provenance and climate of the Upper Proterozoic Stoer Group in Scotland. *Scottish Journal of Geology, 26*, 89–97.

Stuut, J. B., Zabel, M., Ratmeyer, V., Helmke, P., Schefuss, E., Lavik, G., & Schneider, R. R. (2005). Provenance of present-day eolian dust collected off NW Africa. *Journal of Geophysical Research*, 110, doi:10.1029/2004JD005161.

Sudom, M. D., & St. Arnaud, R. J. (1971). Use of quartz, zirconium and titanium as indices in pedological studies. *Canadian Journal of Soil Science, 51*, 385–396.

Suess, E. (1980). Particulate organic carbon flux in the ocean—surface productivity and oxygen utilization. *Nature, 288*, 260–263.

Suess, E. (1981). Phosphate regeneration from sediments of the Peru continental margin by dissolution of fish debris. *Geochimica et Cosmochimica Acta, 45*, 577–588.

Sugawara, K., Okabe, S., & Tanaka, N. (1961). Geochemistry of molybdenum in natural waters (II). *Journal of Earth Sciences of Nagoya University, 9*, 114–128.

Sundby, B., Martinez, P., & Gobeil, C. (2004). Comparative geochemistry of cadmium, rhenium, uranium and molybdenum in continental margin sediments. *Geochimica et Cosmochimica Acta, 68*, 2485–2493.

Taylor, S. R., & McLennan, S. M. (1985). *The continental crust: Its composition and evolution.* Oxford: Blackwell Scientific, 312pp.

Taylor, S. R., McLennan, S. M., & McCulloch, M. T. (1983). Geochemistry of loess, continental crustal composition and crustal model ages. *Geochimica et Cosmochimica Acta, 47*, 1897–1905.

Tendal, O. S. (1972). A monograph of the xenophyophoria (Rhizopodea, Protozoa). *Galathea Report, 12*, 7–100.

Thomson, J., Higgs, N. C., & Colley, S. (1996). Diagenetic redistributions of redox-sensitive elements in northeast Atlantic glacial/interglacial transition sediments. *Earth and Planetary Science Letters, 139*, 365–377.

Thomson, J., Higgs, N. C., Croudace, I. W., Colley, S., & Hydes, D. J. (1993). Redox zonation of elements at an oxic/post-oxic boundary in deep-sea sediments. *Geochimica et Cosmochimica Acta, 57*, 579–595.

Thomson, J., Higgs, N. C., Wilson, T. R. S., Croudace, I. W., de Lange, G. J., & Santvoort, P. J. M. (1995). Redistribution and geochemical behaviour of redox-sensitive elements around S1, the most recent eastern Mediterranean sapropel. *Geochimica et Cosmochimica Acta, 59*, 3487–3501.

Thomson, J., Nixon, S., Summerhayes, C. P., Rohling, E. J., Schonfeld, J., Zahn, R., Grootes, P., Abrantes, F., Gaspar, L., & Vaqueiro, S. (2000). Enhanced productivity on the Iberian margin during glacial/interglacial transitions revealed by barium and diatoms. *Journal of the Geological Society, 157*, 667–677.

Thomson, J., Wallace, H. E., Colley, S., & Toole, J. (1990). Authigenic uranium in Atlantic sediments of the last glacial stage — A diagenetic phenomenon. *Earth and Planetary Science Letters, 98*, 222–232.

Timothy, D. A., & Calvert, S. E. (1998). Systematics of variations in excess Al and Al/Ti in sediments from the central equatorial Pacific. *Paleoceanography, 13*, 127–130.

Todd, J. F., Elsinger, R. J., & Moore, W. S. (1988). The distributions of uranium, radium and thorium isotopes in two anoxic fjords; Framvaren Fjord (Norway) and Saanich Inlet (British Columbia). *Marine Chemistry, 23*, 393–415.

Turekian, K. K., & Wedepohl, L. H. (1961). Distribution of the elements in some major units of the Earth's crust. *Bulletin of the Geological Society of America, 72*, 175–192.

Turner, D. R., Whitfield, M., & Dickson, A. G. (1981). The equilibrium speciation of dissolved components in freshwater and seawater at 25°C and 1 atm pressure. *Geochimica et Cosmochimica Acta, 45*, 855–881.

Usui, A., & Ito, T. (1994). Fossil manganese deposits buried within DSDP/ODP cores, Legs 1–126. *Marine Geology, 119*, 111–136.

Van der Weijden, C. H. (2002). Pitfalls of normalization of marine geochemical data using a common divisor. *Marine Geology, 184*, 167–187.

Visser, J. N. J., & Young, G. M. (1990). Major element geochemistry and paleoclimatology of the Permo-Carboniferous glacigene Dwyka Formation and post-glacial mudrocks in southern Africa. *Palaeogeography, Palaeoclimatology, Palaeoecology, 81*, 49–57.

Vorlicek, T. P., & Helz, G. R. (2002). Catalysis by mineral surfaces: Implications for Mo geochemistry in anoxic environments. *Geochimica et Cosmochimica Acta, 66*, 3679–3692.

Walker, J. L., Sherman, G. D., & Katsura, T. (1969). The iron and titanium minerals in the titaniferous ferruginous latosols of Hawaii. *Pacific Science, 23*, 291–304.

Weaver, A. J., Saenko, O. A., Clark, P. U., & Mitrovica, J. X. (2003). Meltwater pulse 1A from Antarctica as a trigger of the Bolling-Allerod Warm Interval. *Science, 299*, 1709–1712.

Wedepohl, K. H. (1960). Spurenanalytische Untersuchungen an Tiefseetonen aus dem Atlantik. *Geochimica et Cosmochimica Acta, 18*, 200–231.

Wedepohl, K. H. (1971). Environmental influences on the chemical composition of shales and clays. In: L. H. Ahrens, F. Press, S. K. Runcorn & H. C. Urey (Eds), *Physics and Chemistry of the Earth* (pp. 307–331). Oxford: Pergamon.

Wedepohl, K. H., Correns, C. W., Shaw, D. M., Turekian, K. K., & Zemann, J. (Eds) (1969–1978). *Handbook of geochemistry, I, II-1 through II-5*. Berlin: Springer.

Wehrli, B., & Stumm, W. (1989). Vanadyl in natural waters: Adsorption and hydrolysis promote oxygenation. *Geochimica et Cosmochimica Acta, 53*, 69–77.

Werne, J. P., Hollander, D. J., Lyons, T. W., & Peterson, L. C. (2000). Climate-induced variations in productivity and planktonic ecosystem structure from the Younger Dryas to Holocene in the Cariaco Basin, Venezuela. *Paleoceanography, 15*, 19–29.

Winckler, G., Anderson, R. F., & Schlosser, P. (2005). Equatorial Pacific poductivity and dust flux during the mid-Pleistocene climate transition. *Paleoceanography, 20*, doi:1029/2005PA001177.

Wong, G. T. F. (1980). The stability of dissolved inorganic species of iodine in seawater. *Marine Chemistry, 9*, 13–24.

Wong, G. T. F., & Brewer, P. G. (1977). The marine chemistry of iodine in anoxic basins. *Geochimica et Cosmochimica Acta, 41*, 151–159.

Yamada, M., & Tsunogai, S. (1984). Postdepositional enrichment of uranium in sediment from the Bering Sea. *Marine Geology, 54*, 263–276.

Yarincik, K. M., Murray, R. W., & Peterson, L. C. (2000). Climatically sensitive eolian and hemipelagic deposition in the Cariaco Basin, Venezuela, over the past 578,000 years: Results from Al/Ti and K/Al. *Paleoceanography, 15*, 210–228.

Zabel, M., Bickert, T., Dittert, L., & Haese, R. R. (1999). Significance of the sedimentary Al:Ti ratio as an indicator for variations in the circulation patterns of the equatorial North Atlantic. *Paleoceanography, 14*, 789–799.

Zabel, M., Schneider, R. R., Wagner, T., Adegbie, A. T., deVries, U., & Kolonic, S. (2001). Late Quaternary climate changes in central Africa as inferred from terrigenous input to the Niger Fan. *Quaternary Research, 56*, 207–217.

Zhao, M., Eglinton, G., & Zhang, Z. (2000). Marine and terrestrial biomarker records for the last 35,000 years at ODP site 658C off NW Africa. *Organic Geochemistry, 31*, 919–930.

Zheng, Y., Anderson, R. F., van Geen, A., & Fleisher, M. Q. (2002). Remobilization of authigenic uranium in marine sediments by bioturbation. *Geochimica et Cosmochimica Acta, 66*, 1759–1772.

Zheng, Y., Anderson, R. F., van Geen, A., & Kuwabara, J. (2000). Authigenic molybdenum formation in marine sediments: A link to pore water sulfide in the Santa Barbara Basin. *Geochimica et Cosmochimica Acta, 64*, 4165.

CHAPTER FIFTEEN

Isotopic Tracers of Water Masses and Deep Currents

Christelle Claude* *and* Bruno Hamelin

Contents

1. Introduction   645
2. Present State of Methodological Approaches and Interpretations   648
   2.1. The overall general circulation   648
   2.2. Analytical techniques   650
   2.3. Sources and fates of metals, residence time in the ocean, and
        present-day distribution of the different radiogenic tracers   650
   2.4. Chronometers and archives   659
3. Examples of Applications   664
   3.1. The sedimentary record of radiogenic tracers with long residence
        times (Sr and Os)   664
   3.2. Examples using tracers with short and intermediate residence times   665
4. Conclusion and Perspectives   670
References   671

1. Introduction

The main topic of this chapter is a review of the applications of "radiogenic" isotopes in chemical oceanography and of their potential use as tracers for paleoceanographic research on timescales going back to $\sim$60 Ma. Radiogenic isotopes, such as ^{87}Sr, ^{187}Os, ^{143}Nd, ^{206}Pb, ^{207}Pb, ^{208}Pb, and ^{176}Hf, are the stable (i.e., non-radioactive) end-products of radioactive isotopes of various trace elements (^{87}Rb, ^{187}Re, ^{147}Sm, ^{238}U, ^{235}U, ^{232}Th, and ^{176}Lu, respectively). These isotopes are characterized by huge variations in their relative abundances between different natural reservoirs of the Earth, due to both large fractionation effects between

* Corresponding author.

Developments in Marine Geology, Volume 1
ISSN 1572-5480, DOI 10.1016/S1572-5480(07)01020-2

parent and daughter chemical elements through geochemical reactions, and progressive ingrowth of daughter isotopes integrated over the Earth's history. As usual in radiochronology, we will use hereafter the adjectives "radiogenic" and "unradiogenic" to designate samples with isotopic compositions relatively enriched or depleted in radiogenic isotopes.

As with all other chemical elements, radiogenic isotopes are introduced into the ocean as weathering and erosional products of the Earth's crust, transported via aeolian, riverine, or hydrothermal circulations (see Table 1). The isotopic composition of the corresponding elements, i.e., the relative abundance of their radiogenic isotopes, depends upon the geological history of bedrock sources. This isotopic composition is preserved and remains normally unaltered during biogeochemical reactions in the oceanic water column. Mass-dependent isotopic fractionations occurring during such reactions are generally negligible compared to radiogenic variations, and are moreover filtered out by analytical standard normalization procedures. The relative abundance of radiogenic isotopes in the ocean may thus be seen as an isotopic "signature" of chemical element sources and of their subsequent transport by oceanic currents. While these elements have mostly low solubility and are highly reactive, their isotopic composition depends only on mixing between different source-rocks, and behave as conservative tracers of carrier water masses in the ocean.

Unlike other conservative tracers of oceanic circulation, radiogenic isotopes have the immense advantage of being trapped in solid phases precipitated in chemical equilibrium with seawater. Thus, they have the ability to record some aspects of past ocean circulation properties.

The pioneer study in this field was published by Chow and Patterson (1959), who measured lead isotopes in Fe-Mn nodules and showed that their isotopic compositions differed significantly between the Atlantic and Pacific oceans. More than 20 years elapsed before the full potential of that approach was really understood and began to be systematically exploited. A second benchmark study on radioactive/radiogenic isotope oceanography (Brass, 1976) demonstrated that strontium isotopes, although homogeneous at any given time in the ocean, had suffered extremely large variations over the Phanerozoic period. Again, this study initiated a large number of subsequent studies and an ongoing debate as to the causes of the variations. These two early studies had also established the basic distinction between soluble tracers, with a long-residence time in the ocean and sensitivity to long-term changes in the global geochemical cycles, and short residence-time tracers, which respond both to the regional changes in the sources of input to the ocean and to the changes in the internal mixing rates and circulation patterns within the ocean. We illustrate below various examples of different isotopic tracers and different case studies. We focus specifically on the stable "radiogenic" isotopes. Short-lived radioactive isotopes are continuously produced as intermediate nuclides in the natural decay series of uranium and thorium and cosmogenic nuclides by cosmic rays are not included in this chapter, except ^{10}Be, which will be discussed for its radiochronological application. The behavior of these elements is examined in Chapters 5 and 16.

Table 1 Radiogenic Isotopes Used for Reconstructing Past–Ocean Circulation.

Radiogenic tracers and isotope ratios[a]	Residence time in the ocean (years)	Input sources	Stable isotopes	Parent isotope	Half-life (Gyr)
Lead isotopes					
^{206}Pb, ^{207}Pb, **^{208}Pb**[b], $^{206}Pb/^{204}Pb$, $^{207}Pb/^{204}Pb$, $^{208}Pb/^{204}Pb$	~5–50	Erosion of continental crust and hydrothermal input	^{204}Pb	^{238}U, ^{235}U, ^{232}Th	4.47, 0.704, 14.01
Neodymium isotopes					
^{143}Nd, $^{143}Nd/^{144}Nd$	600–2000	Erosion of continental crust (riverine and eolian particles)	**^{142}Nd**, ^{144}Nd, ^{145}Nd, ^{146}Nd, ^{148}Nd, ^{150}Nd	^{147}Sm	106
Hafnium isotopes					
^{176}Hf, $^{176}Hf/^{176}Hf$	~2000	Erosion of continental crust and hydrothermal input	^{174}Hf, ^{176}Hf, ^{178}Hf, ^{179}Hf, **^{180}Hf**	^{176}Lu	15.7
Osmium isotopes					
^{187}Os, $^{187}Os/^{188}Os$	10000–20000	Erosion of continental crust , weathering of abyssal peridotite, cosmic dusts	^{184}Os, ^{186}Os[c], ^{188}Os, ^{189}Os, ^{190}Os, **^{192}Os**	^{187}Re	42.3
Strontium isotopes					
^{87}Sr, $^{87}Sr/^{86}Sr$	Several millions	Weathering of continental crust, hydrothermal input, dissolution of marine carbonates	^{84}Sr, ^{86}Sr, **^{88}Sr**	^{87}Rb	48.8

[a] Isotope ratios commonly used as paleoceanographic tracers.
[b] In bold are the major isotopes.
[c] Previously used for the isotope ratio.

2. Present State of Methodological Approaches and Interpretations

2.1. The Overall General Circulation

The global ocean deep circulation is characterized primarily by the sinking of dense water masses at high latitudes and slow ventilation of the deep basins forming a global loop known as the conveyor belt (Broecker, 1991). Radiogenic isotopes are used to complement the standard hydrographic tracers in order to distinguish the different water masses and characterize their mixing, both in the present-day ocean and in past geological periods. The principal water masses and their acronyms, which will appear later in the discussion, are briefly summarized below.

The production of ~20 sverdrups (Sv) (Ganachaud & Wunsch, 2000) of North Atlantic Deep Water (NADW) is a prominent feature of the global thermohaline circulation. It originates from the joining of the Norwegian Sea Overflow Water (cf. North East Atlantic Deep Water; Iceland-Scotland Ridge Overflow Water) and the Denmark Strait Overflow Water (DSOW) (Figure 1). NADW, characterized by distinctive temperature and salinity, flows southward and leaves the South Atlantic where it joins the Antarctic Circumpolar Current (ACC) in its main water mass, the Circumpolar Deep Water (CDW). The eastward flowing ACC is the largest current of the ocean in terms of volume transport (140–150 Sv).

The Antarctic Bottom Waters (AABW) and Antarctic Intermediate Waters (AAIW) are formed at the second main source of deep-water formation, the Antarctic shelf. AABW directly results from the sinking of cold dense waters left after the seasonal melting of Antarctic ice cap. AABW is denser than NADW and flows northward along with other water masses, mainly the CDW (21 ± 6 Sv). AAIW flows above, as it is less dense than NADW (Figure 1).

In the Southern Atlantic Ocean, a total flow of ~24 Sv resulting from the contribution of NADW and AABW is conveyed northward into the Atlantic basin (6 ± 1.3 Sv) and eastward into the Indian (11 ± 4 Sv) and Pacific (7 ± 2 Sv) oceans (Ganachaud & Wunsch, 2000).

These water masses form the global thermohaline circulation system, also known as conveyor belt, which is mainly driving the oceanic transport of heat and

Figure 1 The thermohaline circulation (reproduced from Ganachaud & Wunsch, 2000). Numbers give the zonally integrated layer mass transport in sverdrups (Sv), calculated by an inverse model based on physical oceanographic data gathered during the World Ocean Circulation Experiment (WOCE) international program. Three different density classes of water are considered, bounded by neutral surfaces given in the insert. Arrows indicate upwelling and downwelling fluxes, with the same grey-scale as the horizontal fluxes, depending on the layer from which the water is coming. Radiogenic isotopes can be used as geochemical tracers to help understand mixing processes. These isotopes have the great interest that they are recorded through time in various sedimentary phases. On the other hand, they have the serious drawback of being affected by various solid-solution interactions, either at their continental sources during weathering, within the oceanic water column during scavenging, or at the sediment water interface during diagenesis. Possible non-conservative behavior resulting from those interactions must, therefore, be investigated carefully in any attempt of quantitative modeling of their distribution.

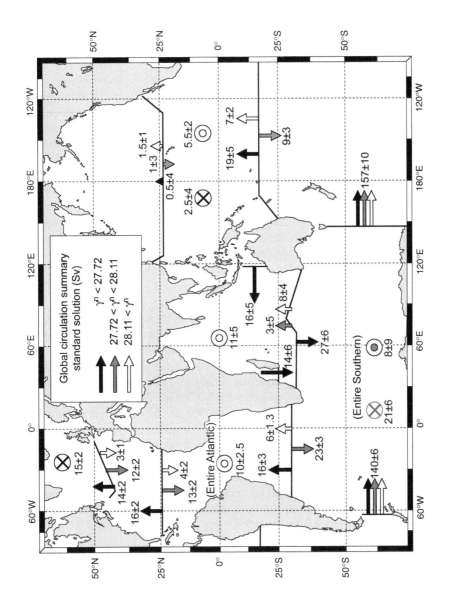

moisture from low to high latitudes and is therefore one important factor in controlling the Earth's climate.

These deep currents are balanced by intermediate and surface water masses flowing back westward at low latitudes, such as the Pacific-Indonesian throughflow (16 ± 5 Sv) and Southern Ocean waters, the AAIW and the Gulf Stream surface current (16 ± 3 Sv).

2.2. Analytical Techniques

Radiogenic isotopes, used as tracers of deep currents, are present at trace levels (ppm to ppt) in sediments and ferromanganese crusts, which are the two main geological sinks of the corresponding elements in the ocean. Measuring the isotopic ratios of such elements (see Table 1) generally requires their chemical separation and purification from major elements of the sample matrix (Birck, 1986; David, Birck, Telouk, & Allègre, 1999). This is followed by their analysis using some form of mass spectrometry. Thermal ionization mass spectrometry (TIMS) was the pioneering technique and is still used commonly for Nd and Sr isotopes, with a typical level of precision of ~ 20 ppm for both elements.

For lead, the precision obtained with conventional TIMS is limited to the per mil level due to difficulty in correcting precisely mass fractionation biases occurring during thermal emission in the TIMS source. However, this problem has been circumvented by using the double (or triple) spike method (Doucelance & Manhès, 2001; Galer, 1999; Thirlwall, 2000).

The rapid analytical developments of multiple-collector-inductively coupled plasma mass spectrometry (MC-ICPMS) has led to considerable improvement in the precision of lead isotope measurements, as well as in that of other metals particularly difficult to ionize, such as Zr, Hf, and Th (Albarède et al., 2004; Halliday et al., 1998; Rehkämper & Halliday, 1998).

Recently, laser-ablation systems coupled to MC-ICPMS have been used to analyze the isotopic composition of trace metals directly from untreated samples (Jochum et al., 2005).

Cosmogenic ^{10}Be displays extremely low isotopic abundance (on the order of 10^{-15}), which cannot be measured by TIMS or MC-ICPMS. In this case, Accelerator mass spectrometry (AMS) has been successfully used, although some measurements have also been obtained in ferromanganese crusts with secondary ion mass spectrometry (SIMS) coupled to a double-focusing mass spectrometer (Belshaw, O'Nions, von Blanckenburg, 1995).

2.3. Sources and Fates of Metals, Residence Time in the Ocean, and Present-Day Distribution of the Different Radiogenic Tracers

2.3.1. Strontium

Strontium (Sr^{2+}) has a conservative behavior in the ocean, with a residence time of several million years. Therefore, its concentration and isotopic composition are homogeneous in the different ocean basins ($7.65 \, \mu g \, g^{-1}$, $^{87}Sr/^{86}Sr = 0.70918$ (Goldstein & Jacobsen, 1987). The only significant sink for Sr is its incorporation into biogenic carbonates and subsequent burial in marine sediments. Sr-isotopic composition results from a mass balance between continental (riverine and aeolian) and mantle-derived (hydrothermal) inputs. These two sources (continental and mantle) are characterized by very distinct Sr isotope ratios (on average 0.7119 and 0.7035, respectively; Palmer & Edmond, 1989). This major contrast is the consequence of the significantly distinct solid-liquid partition coefficients of Rb (the parent element of ^{87}Sr) and Sr during magmatic processes ($D_{Rb} < D_{Sr}$). The internal differentiation of the Earth resulted in a major export of Rb into the new forming continental crust, and a relative Rb-depletion of the residual mantle. Over billions of years, the continental rocks thus developed very high (i.e., radiogenic) ^{87}Sr/^{86}Sr ratios, compared to those of mid-ocean ridge basalts and mantle rocks. The present-day inventory of Sr in the ocean is 125×10^{15} mol. The dominant input (*ca.* 75%) is the fluvial discharge (33×10^9 mol yr^{-1}).

The importance of the hydrothermal flux has been actively disputed (Goldstein & Jacobsen, 1987; Morton & Sleep, 1985; Palmer & Edmond, 1989). In principle, estimating the Sr hydrothermal flux at mid-Ocean ridges should provide an independent assessment of hydrothermal water fluxes through the crust, to be compared with other direct (heat flux) or indirect (helium isotope budget) approaches. Successive attempts gave contrasting results, ranging from 2×10^9 to 15×10^9 mol yr^{-1} for hydrothermal Sr inputs. Palmer and Edmond (1989) favored a value in the high range, based on a revised value of the world average isotopic composition of rivers (0.7119) calculated from 76 river data corresponding to 47% of the global runoff.

The diagenesis of deep-sea carbonate sediments constitutes another source of Sr (3.4×10^9 mol yr^{-1}; Palmer & Edmond, 1989). However, pore waters have on an average an Sr-isotope composition close to that of seawater (0.7084). Therefore, such diagenetic input acts essentially as a buffer of the isotopic variations of Sr from other sources. So far, sources such as aeolian inputs, off-ridge oceanic crust weathering, and direct exchange with continental groundwaters in coastal areas, although potentially significant, have been considered as minor.

2.3.2. Osmium

Thermodynamic data suggest that the main chemical species of Os in the ocean are H_2OsO_5, $H_3OsO_6^-$ (+8 oxidation state) and organic complexes (Levasseur, Birck, & Allègre, 1998). Precise determinations of the concentration (10.87 ± 0.07 fg g^{-1}) and isotopic composition (^{187}Os/^{188}Os $= 1.059 \pm 0.008$) have been measured only recently (Levasseur et al., 1998), using a negative ion TIMS technique. This has also shown that the Os concentration is homogeneous in the different ocean basins.

As far as Sr is concerned, the main control on the isotopic composition of Os in the ocean is the balance between continental and oceanic crust weathering. However, the difference in isotopic composition between these two end-members is much larger than it is for Sr because of the very strong partitioning between rhenium (magmaphile) and osmium (compatible) during the formation of the continental crust. In addition, for Os, the cosmic dust fallout, which is enriched in siderophile elements, must be taken into account in the budget calculations. The isotopic composition of this input is unradiogenic (primitive) and indistinguishable from that of the mantle ($^{187}Os/^{188}Os = 0.126$; Pegram, Krishnaswami, Ravizza, & Turekian, 1992; Ravizza, 1993). Assuming that the Os cycle is at steady state, Levasseur et al. (1998) calculated that the continental runoff ($^{187}Os/^{188}Os = 1.54$) represents 70% of the total Os input to the ocean. Cosmic dust amounts to 14% of the unradiogenic sources.

However, this assumption has been questioned recently. While early studies (Ravizza & Turekian, 1992; Sharma, Papanastassoiu, & Wasserburg, 1997) suggested an Os residence time much longer than the mixing time of the ocean, more recent studies (Oxburgh, 1998), (Ravizza et al., 2001) and (Oxburgh, 2001) have reported Os isotopic variations on a 2–3 kyr timescale, suggesting that the residence time should be less than 10 kyr.

Moreover (Woodhouse, Ravizza, Falkner, Statham, & Peucker-Ehrenbrink, 1999; Sharma, Wasserburg, Hofmann, & Butterfield, 2000a) suggested a slight Os depletion in the oxygen minimum zone of the Pacific Ocean, which is compatible with a reactive behavior for this element in the water column.

2.3.3. Neodymium

The present-day isotopic composition of Nd ($^{143}Nd/^{144}Nd$) is a distinctive characteristic of various continental landmasses, depending on their age and lithology. It is usually expressed in conventional ε-units, with:

$$\varepsilon_{Nd} = \left[\frac{(^{143}Nd/^{144}Nd)_{Sample} - (^{143}Nd/^{144}Nd)_{CHUR}}{(^{143}Nd/^{144}Nd)_{CHUR}} \right] \times 10^4$$

where "CHUR" stands for "Chondritic Uniform Reservoir" (Goldstein et al., 1984), i.e., the global mean isotopic composition deduced from measurements in chondrites (($^{143}Nd/^{144}Nd)_{CHUR} = 0.512638$).

The possibility of using this tracer for fingerprinting water masses has attracted a lot of interest since the early measurements of (Piepgras, Wasserburg, & Dash, 1979). Indeed, numerous studies of different depth profiles in the world oceans have clearly established that ε_{Nd} can be considered to some extent as a conservative tracer (Amakawa, Alibo, Alibo, & Nosaki, 2000; Bertram & Elderfield, 1993; Jeandel, 1993; Jeandel, Bishop, & Zindler, 1995; Jeandel, Thouron, & Fieux, 1998; Piepgras & Jacobsen, 1988; Piepgras & Wasserburg, 1980; Shimizu, Tachikawa, Masuda, & Nozaki, 1994; Spivack & Wasserburg, 1988; Stordal & Wasserburg, 1986; Tachikawa, Jeandel, & Roy-Barman, 1999).

The main features of the global distribution of Nd isotopes in the ocean are now well established. Ocean basins surrounded by Proterozoic or Archean continental crust are characterized by very low ε_{Nd} values (resulting from the low time-integrated Sm/Nd ratios in the parent rocks), such as in Baffin Bay, Northwestern Atlantic ($\varepsilon_{Nd} = -26$; Stordal & Wasserburg, 1986). Despite its mixing with other water masses, NADW keeps a distinctly unradiogenic composition ($\varepsilon_{Nd} = -13.5$) that can be followed along the track of the conveyor belt, all the way to the deep Indian Ocean, or even further into the deep Southern Pacific.

At the other extreme, Pacific intermediate and surface waters are much more radiogenic, with ε_{Nd} values between 0 and -4, due to the dominance of young mantle-derived volcanic terrains around that ocean. Other water masses encountered in between those two end-members have intermediate Nd compositions, increasing progressively along the track of the conveyor belt.

The Mediterranean Outflow Water (MOW) is characterized by an ε_{Nd} value of -9.4 (Spivack & Wasserburg, 1988). The water masses formed in the South Atlantic and in the Southern Ocean (AAIW, CDW, AABW) display ε_{Nd} values between -7 and -9, because they receive some contribution from deep Pacific waters ($\varepsilon_{Nd} = -3$ to -6; Jeandel, 1993; Piepgras & Jacobsen, 1988; Piepgras & Wasserburg, 1982; Piepgras & Wasserburg, 1987; Shimizu et al., 1994).

The Nd isotopic composition of the Indian Ocean is intermediate between the Atlantic and the Pacific values, with a relatively homogeneous ε_{Nd} value of -7 (Bertram & Elderfield, 1993), whereas in the Indonesian Throughflow, water masses originating from the Pacific Ocean are traceable by ε_{Nd} values between -5 and -3 (Jeandel et al., 1998).

However, in spite of this seemingly straightforward pattern, extending this tracing technique to the past ocean circulation by using sedimentary records requires a detailed understanding of the geochemical mechanisms through which seawater acquires its ε_{Nd} isotopic "label".

Dissolved neodymium (Nd^{3+}) is mainly present in the ocean as carbonate and sulfate complexes (Bruland, 1983). Similar to the other rare earth elements, Nd is particle-reactive and is thus rapidly removed from the upper water column by scavenging. Contrary to Sr and Os, the vertical distribution of dissolved Nd is therefore depleted in the surface layers and increases with depth due to desorption processes and remineralization of biogenic particulate matter. On average, the Nd concentration is ~4 ppt in deep waters. Although still debated, the residence time of Nd in the ocean is between 600 and 2,000 years (Jeandel, 1993; Jeandel et al., 1995; Tachikawa et al., 1999). Nd is dominantly introduced to the ocean from continental sources, with no significant hydrothermal contribution at ocean ridges. The debate about the relative importance of aeolian vs. riverine inputs is still going on. Several studies of Nd isotopes in sediment traps, desert-derived dusts, and surface seawater from the Atlantic and the Pacific oceans reached the conclusion that there is a significant contribution from aeolian dusts (upto 20% of the total input) (Goldstein & Jacobsen, 1987; Greaves, Elderfield, & Sholkovitz, 1999; Tachikawa, Jeandel, & Dupré, 1997; Tachikawa et al., 1999). On the other hand, Jones, Halliday, Rea, & Owen (1994) found no indication of the Asian dust plume impact on the isotopic composition of the Pacific deep waters.

Notwithstanding the general agreement that the main source of dissolved Nd to the ocean is the riverine input, the exact processes at the ocean/continent interface are still largely debated. Most of the Nd delivered by rivers is very efficiently trapped in the estuarine sediments. However, several studies have shown that Nd can be remobilized from sediments by early diagenetic processes and returned to seawater (Elderfield & Sholkovitz, 1987; Goldstein, O'Nions, & Hamilton, 1984; Sholkovitz, Elderfield, Szymczak, & Casey, 1999). For instance, (Sholkovitz et al., 1999) have demonstrated a clear similarity between the rare earth elements (REE) patterns of the Equatorial Undercurrent (EUC) waters in the Pacific Ocean and Papua New Guinea rivers. Lacan & Jeandel, 2001 confirmed this influence by measuring the Nd isotopic composition of the EUC waters before and after in-teraction with the continental slope off Papua New Guinea.

A final problem in trying to relate the Nd isotopic composition of ocean water masses to specific source areas on land is linked to possible isotopic fractionation during rock weathering and dissolution of detrital materials (Goldstein et al., 1984). A recent study performed on glacial tills has shown that unradiogenic Nd is prefe-rentially released during weathering (Öhlander, Ingri, Land, & Schöberg, 2000). An unradiogenic signature was also identified in boreal rivers from Northern Scandinavia and interpreted to reflect partial dissolution of the watershed basement rocks (Andersson, Dahlqvist, Ingri, & Gustafsson, 2001). These studies are supported by experimental data obtained from acid leaching of Greenland river sediment (von Blanckenburg & Nägler, 2001). Until now, however, the significance of such isotopic fractionation effects on the seawater Nd composition has not been quantified.

2.3.4. Lead

Dissolved lead is mainly present in the ocean as Pb^{2+} and carbonate complexes, such as $Pb(CO_3)$ and $Pb(CO_3)_2^{2-}$ (Bruland, 1983). Lead is highly reactive with marine particles (Cochran et al., 1990), as demonstrated by the strong removal of ^{210}Pb from surface waters. This natural radioactive isotope (half-life $= 22.6\,yr$) is produced from the decay of ^{222}Rn, a daughter product of the ^{238}U-series. Measurements of its deficit with respect to the parent isotope ^{226}Ra allowed Craig et al. (1973) to give the first estimate of its residence time in the ocean, which is in the order of 50 yr. However, large differences are suspected between oceans, depending on regional variations in particle fluxes. For instance, estimates as high as several hundred years have been recently proposed for the Pacific ocean (Henderson & Maier-Reimer, 2002).

Since the work of Schaule and Patterson (1981), it is well established that anthropogenic emissions have completely overwhelmed the natural background of detrital lead in the ocean over the last century. Typical present-day vertical profiles of Pb concentrations are enriched in surface waters, due to the atmospheric fallout, and decrease with depth. Even the very low concentrations found at great depths (in the order of $1\,ng\,L^{-1}$) are probably contaminated by anthropogenic Pb because of its efficient vertical transfer by scavenging. Nevertheless, studies of stable Pb isotopes in contaminated surface, thermocline, and deep waters have shown that the element is also advected with water masses (Alleman, Véron, Church, Flegal, & Hamelin, 1999; Boyle, Chapnick, Shen, & Bacon, 1986; Shen & Boyle, 1988; Véron et al., 1998). Therefore, much like Nd, Pb can be used as a circulation tracer. However, due to the anthropogenic interference, the pre-anthropogenic Pb cycle is more conjectural than it is for most other tracers.

Nonetheless, Pb would be a good paleo-circulation tracer if appropriate natural authigenic phases could be identified in cored sequences. Information on the pre-anthropogenic seawater Pb isotope composition can only be derived from records preserved in authigenic phases of sediments and in ferromanganese crusts and nodules (Abouchami & Goldstein, 1995; Chow & Patterson, 1959; Chow & Patterson, 1962). Present-day NADW is characterized by a $^{206}Pb/^{204}Pb$ atomic ratio of 19.3 and the deep Pacific water mass by a value ranging between 18.5 and 18.8. Water masses in the Indian and Southern Oceans are intermediate between those values. Unfortunately, surface abrasion of Mn nodules yields Pb-isotope compositions that are averaged over tens to hundreds of thousand years (Figure 2), smoothing out glacial–interglacial variations.

Pb is strongly leached from basalts during hydrothermal weathering of mid-ocean ridges (Chen, Waserburg, Von Damm, & Edmond, 1986). However, the impact of this source on a global ocean scale is restricted due to immediate precipitation when the hydrothermal fluid mixes with seawater, forming metal-liferous deposits such as those found in ophiolites. The fate of hydrothermal Pb-inputs is strongly coupled to that of manganese. The mantle Pb-isotope signatures can be identified off-ridge in metalliferous sediments (Barrett, Taylor, & Lugowski, 1987), and probably constitutes one end-member of Pb mixing trends in some parts of the ocean, such as the central Indian Ocean. Nevertheless, mantle-source Pb probably represents less than a few percent of the total oceanic Pb-budget, com-pared to continental sources.

Aeolian inputs have been estimated at $\sim 10\%$ of the pre-anthropogenic Pb budget in the ocean (Chow & Patterson, 1962). This corresponds to the dissolution of 8% of the atmospheric dust deposited onto the ocean (Henderson & Maier-Reimer, 2002), and seems thus compatible with experimental determinations of partition coefficients between seawater and particulate matter (Duce et al., 1991). Nevertheless, it must be kept in mind that aeolian dust may be a major input in remote regions of the ocean with low riverine inputs (Jones, Halliday, Rea, & Owen, 2000).

As for Nd, estuarine and shelf processes affecting Pb-transfer from continents to the ocean are still poorly understood. We know that riverine Pb is efficiently precipitated in estuaries, and that oceanic ^{210}Pb is subject to boundary scavenging on shelf regions. However, reversible exchange between those two sources ("boundary exchange" such as for Nd) has yet to be demonstrated.

Finally, incongruent weathering of continental rocks must be taken into account for mass balance reconstructions. Some minor mineral phases that are most easily weathered are also enriched in radiogenic isotopes (^{206}Pb, ^{207}Pb, and ^{208}Pb) (Erel et al., 1994; Jones et al., 2000). The isotopic composition of dissolved Pb in rivers is thus more radiogenic than the basement rocks and therefore does not reflect the bulk composition of the continental terrains. An example of such incongruent weathering has been demonstrated recently in the North Atlantic (von Blanc-kenburg & Nägler, 2001).

2.3.5. Hafnium

Hafnium (Hf $^{4+}$) is present in seawater as hydroxides (Hf(OH)$_5^-$, Hf(OH)$_4$) (Bruland, 1983; Turner, Whitfiels, & Dickson, 1981). Hf has a nutrient-type ver-tical depth profile like Nd (Godfrey, White, & Salters, 1996; McKelvey & Orians, 1998) and is thus a reactive element like the other REE. Hf concentrations are very low (0.04–0.2 pg g^{-1}; Godfrey et al., 1996; McKelvey & Orians, 1998) and avail-able data were scarce until the recent improvements were made to the ionization efficiency with MC-ICPMS instruments. The Hf residence time has been

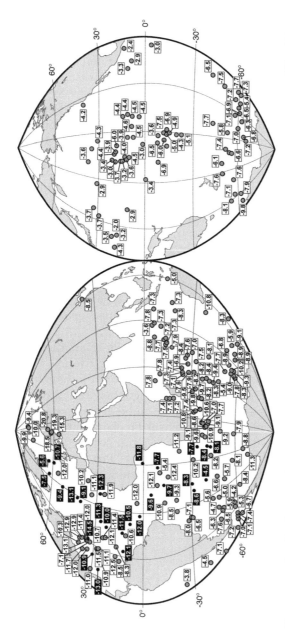

Figure 2 The present-day distribution of Nd and lead isotopes in the ocean (reproduced from Frank, 2002), ε_{Nd} (2a), and $^{206}Pb/^{204}Pb$ (2b) ratios of deep-water masses have been obtained from measurements in surface scrapings of ferromanganese crusts and nodules. Black dots associated to numbers in white indicate measurements of ferromanganese coating of foraminifera in surface sediments. The main feature in both figures is the clear contrast between the Atlantic and Pacific oceans, and the progressive transition of both tracers along the track of the global conveyor belt.

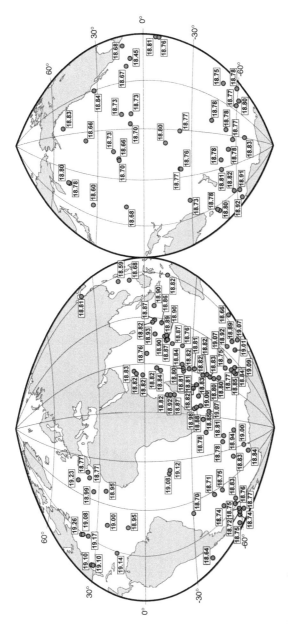

Figure 2 (*Continued*).

estimated as about twice that of Nd (1,500 to 2,000 yr) based on water column studies and Hf isotope measurements in Fe-Mn crusts and nodules (David et al., 2001; Godfrey et al., 1997; Lee et al., 1999).

The Hf isotopic composition (^{176}Hf/^{177}Hf) is also commonly expressed with the ε-notation as defined above for Nd, with (^{176}Hf/^{177}Hf)$_{CHUR}$ = 0.282769. The global variability of ε_{Hf} values in crustal rocks is in general well correlated with ε_{Nd}, along a mantle-crust correlation known as the Hf-Nd "global array" (Figure 3). ε_{Hf} displays a wider range of variation than ε_{Nd} does, from -30 in old continental terrains to $+25$ in mantle rocks. The present-day distribution of Hf in seawater has been inferred, as for lead, from measurements of Fe-Mn crusts and Mn nodules. NADW ε_{Hf} values range from -2 to $+3$, while Pacific deep waters are generally between $+3$ and $+9$. Indian Ocean values are intermediate ($+2$ to $+5.5$) (Godfrey et al., 1997; Lee et al., 1999; David et al., 2001). The Hf isotopic composition in deep waters is controlled mainly by the continental inputs, although the hydrothermal contribution cannot be neglected (Godfrey et al., 1997; White, Patchett, & Ben Othman, 1986).

A prominent feature of Hf in the ocean is that ε_{Hf} values measured in ferromanganese deposits are higher than expected based on their ε_{Nd} values, with respect to the mantle crust "global array" (Figure 3). This anomalous composition measured in the deposits has been interpreted as reflecting zircon enrichment in the sandy fraction of sediments. Indeed, zircons have a very low Lu/Hf ratio, and consequently, a low time-integrated ^{176}Hf/^{177}Hf ratio. Zircons are also very

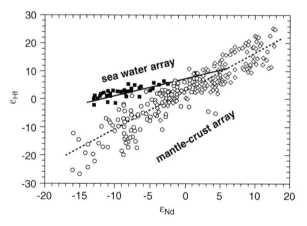

Fig. 3 ε_{Nd} vs. ε_{Hf} in sea water, the "zircon paradox" (reproduced from Albarède et al., 1998). The "mantle-crust array" is a compilation of hafnium and neodymium isotope measurements on terrestrial rocks. The dispersion and the overall correlation of the two tracers result from their parallel behavior during the differentiation of the Earth's internal reservoirs, and from the chemical fractionation between radioactive parents (lutetium, samarium) and radiogenic daughters (hafnium, neodymium) during these magmatic processes. By comparison, the data obtained in Fe-Mn nodules are offset toward higher ε_{Hf}. This is attributed to the low ^{176}Hf/^{177}Hf ratios in weathering-resistant minerals, like zircon, which escape dissolution during continental meteoric alteration, and thus leave a more "radiogenic" hafnium (i.e., relatively enriched in the radiogenic ^{176}Hf isotope) entering the ocean.

resistant to weathering. Therefore, the sandy fraction of sediments constitutes a "missing" refractory reservoir of unradiogenic Hf, while the more labile fraction, which can undergo weathering and chemical exchange, bears the more radiogenic Hf component found in seawater. This interpretation has not yet been verified by direct measurements in the river load. This is nevertheless one of the best examples of the influence of incongruent weathering on the composition of radiogenic isotopes dissolved in seawater.

2.4. Chronometers and Archives

2.4.1. Fe-Mn crusts and nodules

Except for Sr, which is incorporated in carbonates, radiogenic trace metals dissolved in the ocean have a strong affinity with Fe and Mn and are scavenged mainly in association with these two elements. Therefore, hydrogeneous or hydrothermal Fe-Mn crusts, manganese nodules, and manganese coatings on foraminifera are major carriers of trace metals to the deep seafloor. Crusts, nodules and coatings precipitate directly from seawater as iron hydroxides and manganese oxides, and they are thus considered as reliable recorders of the seawater composition. Fe-Mn nodules cover the sea floor in some regions far away from hydrothermal areas. In these nodules, Fe hydroxide and Mn oxide are added in thin concentric layers forming nodules that can reach diameters of several centimeters. In contrast, Fe-Mn crusts precipitate at the top of seamounts or directly on the basaltic seafloor, in the deep ocean (>1,000 m depth), where the deposition of sediments is precluded by erosive bottom currents. Fe-Mn crusts can reach more than 10 cm thick.

The chemical composition of Fe-Mn crusts is controlled by complex interactions between different factors (e.g., the metal supply available in seawater at the site, the hydrographic conditions, and the chemical conditions in the water column, in relation to biological productivity). Most metal concentrations, including those of Mn and Fe, often relatively increase as reduced ions within the oxygen minimum zone (OMZ) in the Atlantic (at $\sim$1,000 m depth). Re-oxidation and precipitation occurs when this OMZ water mixes with oxygen-rich deeper waters, such as NADW and AABW. This mixing occurs typically around seamounts where turbulent bottom eddies may be generated.

The chemistry of Fe-Mn crusts is considered as highly sensitive to the variations of local physical and chemical conditions (Halbach & Puteanus, 1984; Hein, Bohrson, Schulz, Noble, & Clague, 1992). Such perturbation may occur for example in the OMZ, which is subject to periodic intensification or weakening, directly linked to productivity changes in surface waters. The depth of the lysocline, the rate of $CaCO_3$ dissolution as well as the strength of deep currents may also intervene. Koschinsky and Halbach, 1995 proposed a genetic model of the formation of hydrogenous Fe-Mn crusts. Fe-bearing vernadite (δ-MnO_2), the most abundant mineral phase in hydrogenous Fe-Mn crusts, is the main carrier of Co and Ni while other metals like Al, Ti, and Si form oxide and hydroxide colloids bound to the amorphous FeO-OH phase.

The growth rates of Fe-Mn crusts and nodules are very slow (less than 10 mm kyr^{-1}). This is a disadvantage for the reconstruction of marine records.

Nevertheless, this disadvantage is compensated by: (1) the very high concentrations of the radiogenic trace metals allowing mg-size samples to be analyzed, (2) the recent development of very precise analytical techniques, and (3) the fact that Fe-Mn hydroxides and oxides also strongly scavenge ^{230}Th and ^{10}Be elements that are reliable chronometers on timescales of thousands to millions of years. Thus, while precise growth rates cannot be obtained for Fe-Mn nodules (see below), surface samples obtained by scrapping the outer micro-layer of nodules can provide us with data for the average isotopic composition of Hf, Nd, Os, and Pb over the last million years.

It is well established since the early work by Ku (1976) and Edwards, Chen, and Wasserburg (1987) that ferromanganese deposits can be dated by the radioactive disequilibrium of ^{234}U and ^{230}Th in the ^{238}U decay series. $^{230}Th_{ex}$ is ^{230}Th activity "in excess" of that resulting from the radioactive decay of uranium in the sample. This excess is due to the extremely reactive character of thorium, resulting in preferential removal of that element from the dissolved phase and incorporation in solid phases, during scavenging or precipitation of authigenic phases on the sea-floor. $^{234}U_{ex}$ is the "excess" of the short-lived isotope 234 over the major isotope 238, due to the preferential enrichment of that isotope during weathering, and thus in the fresh water input to the ocean. Both $^{230}Th_{ex}$ and $^{234}U_{ex}$ decay exponentially with time once incorporated in a closed-system solid phase. Mass spectrometric techniques have considerably improved the resolution of this method, compared to the earlier measurements by alpha counting spectroscopy (Claude-Ivanaj, Abouchami, Galer, Hofmann, & Koschinsky, 1998; Eisenhauer, Gögen, Pernicka, & Mangini, 1992; Neff, Bollhöffer, Frank, & Mangini, 1999). Excellent analytical precision can be achieved over the entire range of the chronometer, from a few hundred years to 4–5 hundred kyr. On the other hand, several stringent conditions must be satisfied to obtain accurate ages. In order to obtain smooth decay profiles with depth in the sample, the surface activity of the radio-nuclide must have remained constant through time, i.e., the flux of deposition, or the isotopic ratio when the activity is normalized to a stable isotope for reference (^{230}Th to ^{232}Th, ^{10}Be to ^{9}Be) must have been constant. Furthermore, closed-system conditions are required, with negligible diffusion and no dissolution/re-precipitation nor exchange with seawater. The role of ^{230}Th and ^{10}Be diffusion has been a focus of discussion since the early studies. Ku, Omura, and Chen (1979) and Krishna-swami et al. (1982) have shown that diffusion may be a serious problem in the case of Mn-nodules. Growth rates calculated from $^{234}U_{ex}$ (over ^{238}U) decay profiles are generally higher than those derived from ($^{230}Th_{ex}$) profiles (Claude, Suhr, Koschinsky, & Hofmann, 2005; Henderson & Burton, 1999; Neff et al., 1999). This discordance between the observed ($^{234}U/^{238}U$) profiles and those expected from ($^{230}Th_{ex}$) enabled Henderson and Burton (1999) to calculate an effective uranium diffusivity of 1.2×10^{-6} to $4.7 \times 10^{-8} \, cm^2 \, yr^{-1}$ for crusts of different porosities. Claude et al. (2005) have examined in detail the relationships between the texture and diffusion processes in Fe-Mn crusts, and suggested the existence of local closed-system domains related to textural heterogeneities (laminated vs. botryoidal) of the crusts. In regions of strong scavenging, such as continental margins, ^{232}Th can be used as an index of thorium contamination of continental detrital origin, and the ^{230}Th/^{232}Th ratio appears to be more reliable than ^{230}Th$_{ex}$ activity alone (Claude et al., 2005).

Beyond ages accessible to ^{230}Th dating, reliable chronologies have been established for timescales up to 10 Myr by using cosmogenic ^{10}Be (Krishnaswami, Somayajulu, & Moore, 1972; Ku et al., 1979; Ling et al., 1997; McMurtry, Von der Haar, Eisenhauer, Mahonery, & Yeh, 1994; Segl, Mangini, & Bonani, 1984; Sharma & Somayajulu, 1982). In the case of ^{230}Th, normalization of ^{10}Be concentrations to the stable isotope ^{9}Be can be used in order to filter out possible variations in deposition flux due to changes in hydrographic conditions.

For older periods, no other radiometric dating method is available. Since the longest records cover the last 60 Ma, chronologies of the oldest sections have been either extrapolated from ^{10}Be growth rates, or estimated by assuming constant rates of incorporation of Co into the ferromanganese crust (Frank, O'Nions, Hein, & Banakar, 1999; Halbach, Segl, Puteanus, & Mangini, 1983; Manheim & Lane-Bostwick, 1986; Puteanus & Halbach, 1988). Magnetostratigraphy (Joshima & Usui, 1998) and biostratigraphy using microfossils, which are sometimes incorporated into the crusts, have also been attempted (Harada & Nishida, 1976; Kadko & Burckle, 1980).

2.4.2. Other archives

High-resolution records of seawater isotopic variations should be attainable from marine sediments on much shorter timescales than for ferromanganese deposits. Therefore, great efforts have been devoted to isolating a "pure" seawater record from foraminiferal calcite, especially for Nd (Burton & Vance, 2000; Palmer & Elderfield, 2000, 1986; Pomiès, Davies, & Conan, 2002; Vance & Burton, 1999). However, REE concentrations are very low in calcite tests, contrary to the very high concentrations found in Mn hydroxide coatings that are ubiquitous on all grain surfaces, including calcite tests, within the sediment.

Specific cleaning methods based on redox chemistry were developed to remove these coatings (Boyle & Keigwin, 1986; Palmer & Elderfield, 1985). Although a reduction of the REE concentrations by one order of magnitude could be obtained by these techniques, compared to uncleaned samples, Sholkovitz (1989) and Sholkovitz et al. (1999) showed that the highly reactive REEs were in fact re-adsorbed onto foraminiferal surfaces during cleaning.

Vance and Burton (1999) and Burton and Vance (2000) revisited this issue, and succeeded for the first time in measuring reliable Nd isotopic compositions from foraminifera. They showed that the Nd isotopic composition of planktonic foraminifera from the Bay of Bengal exhibited a clear climatic signal over glacial–interglacial cycles, without evidence of diagenetic interference. However, the Nd/Ca ratios measured in these studies is still 500 times higher than in seawater, at odds with previous Nd measurements in calcite (Show & Wasserburg, 1988) and with the calcite/seawater partition coefficient obtained for other trace elements in foraminifera (Havach, Chancler, Wilson-Finelli, & Shaw, 2001; McCorkle, Martin, Lea, & Klinkhammer, 1995).

More recent studies have suggested that Nd might in fact be stored in another minor phase intermingled with the calcite, for example barite. Haley and

Klinkhammer (2002) developed a flow-through system including a DTPA (diethylene triamine pentaactic acid) step, based on the original cleaning method of Lea and Boyle (1991) to dissolve barite. Finally, a study by Vance et al. (2004) confirmed that Nd is isotopically identical in core-top foraminifera and local surface seawater, without any perturbation due to secondary incorporation deeper in the water column or within the sediment.

Alternatively to assuming that the Fe-Mn coatings are contaminants and removing them, it can be assumed that this phase is authigenic and that it records the deep-water composition, thus possibly providing high-resolution records of marine-Nd changes in sediment cores. This approach has been used both for long-term records (Palmer & Elderfield, 1986) and for studies of glacial–interglacial variations (Bayon et al., 2002; Rutberg, Hemming, & Goldstein, 2000). However, potential problems may arise due to the diagenetic mobility of Mn and REE in the sediments under reducing conditions. These coatings form mainly by remobilization and upward diffusion of the reduced species (Mn^{2+}, Fe^{2+}) into oxic pore waters, where they re-precipitate. Diagenetic evolution may thus alter the authigenic isotopic signatures. Nevertheless, Palmer and Elderfield (1985), and Rutberg et al. (2000) reported ε_{Nd} variations of diagenetic coatings consistent with a decrease in the influence of NADW in the South Atlantic at the Last Glacial Maximum-Holocene transition.

It is important to emphasize that, in parallel with the quest for a paleo-seawater record, useful complementary information can be obtained from radiogenic isotopes (Sr, Nd, and Pb) measured in the detrital fraction of marine sediments. This fraction is composed of lithogenic material of continental origin, transported to the ocean by rivers and wind, and then redistributed by ocean currents. A number of studies (Bout-Ramazeilles, Davies, & Labeyrie, 1998; Dia, Dupré, & Allègre, 1992; Fagel, Innocent, Stevenson, & Hillaire-Marcel, 1999; Grousset, Biscaye, Zindler, Prospero, & Chester, 1988; Hemming et al., 1998; Innocent, Fagel, & Hillaire-Marcel, 2000; Revel, Cremer, Grousset, & Labeyrie, 1996) have reported glacial–interglacial variations of Sr, Pb, and Nd isotopes in the detrital fraction, which have been interpreted in terms of changes in the sediment sources and mixing of particulate matter transported by different water masses (Figure 4). For example, Innocent et al. (2000) showed that analyses of the clay fraction in cores from the region of NADW may help to reconstruct changes in deep current strength, because of a longer residence time for the fine particles in the water column, whereas the coarse fraction is more representative of local inputs from turbidites and gravity flows.

Finally, it must be mentioned that various other archives have been investigated as possible authigenic records of deep-water Nd isotope composition, such as fish teeth (Martin & Haley, 2000; Staudigel, Doyle, & Zindler, 1985; Thomas, 2004; Thomas, Bralower, & Jones, 2003), shark teeth (Vennemann & Hegner, 1998), marine phosphates, carbonates, and other fish remains (Staudigel et al., 1985; Stille, 1992; Stille & Fisher, 1990; Stille, Steinmann, & Riggs, 1996). Although temporally discontinuous by nature, these samples gave useful complementary information on long-term variations of the marine isotopic composition.

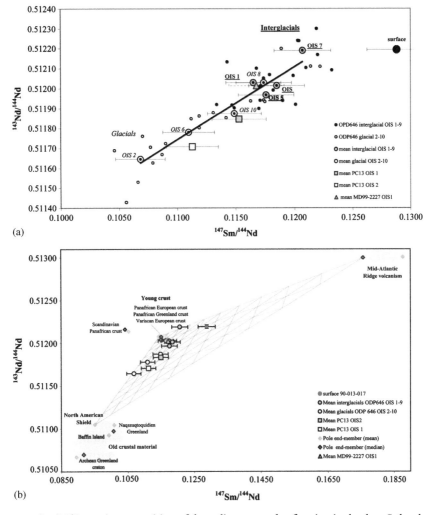

Figure 4   Sm/Nd isotopic composition of the sedimentary clay fraction in the deep Labrador Sea (from Fagel & Hillaire-Marcel, in press). (a) ^{143}Nd/^{144}Nd vs. ^{147}Sm/^{144}Nd diagram of the sediment samples from the Site 646, Leg ODP 105, representative of the Western Boundary Undercurrent (WBUC). Open symbol: samples from glacial stages; filled symbol: samples from interglacial intervals. Mean value for each Oxygen Isotope Stage (OIS) is plotted as a larger circle. Previous data from previous cores PC13 (HU90-013-013, square) and MD99-2227 (triangle), both drilled at close site location are indicated for comparison. (b) Comparison between the data from Site 646, Leg ODP 105 and those of the potential regional end-members. To take into account the scattering of the Sm/Nd composition of the sources, each end-member is characterized by a mean and a median value calculated from literature database. The North American Shield (NAS) end-member corresponds to a composite ("*Canadian Shield Composite*") value proposed by Mc Culloch and Wasserburg (1978). Mixing calculation is based on three end-members: (1) an old craton represent by the North American Shield composition (^{143}Nd/^{144}Nd = 0.51105, ^{147}Sm/^{144}Nd = 0.095, [Nd] = 25 ppm); (2) a young crust characterized by a Pan-African and/or Variscan signature from Europe or Greenland (^{143}Nd/144Nd = 0.51208, ^{147}Sm/^{144}Nd = 0.1145, [Nd] = 33 ppm); (3) a volcanic source defined by mantel-like isotope signature (^{143}Nd/^{144}Nd = 0.51300, ^{147}Sm/^{144}Nd = 0.1756, [Nd] = 14.5 ppm). For details and literature references see Innocent et al. (1997), Fagel et al. (1999).

3. EXAMPLES OF APPLICATIONS

The different case studies briefly described below are not intended to be an exhaustive compilation of all the studies that have been performed on various parts of the world ocean. Instead, we have selected a few specific regions to illustrate some of the main methodological benchmarks, as well as some intrinsic limitations of the isotopic tracers.

3.1. The Sedimentary Record of Radiogenic Tracers with Long Residence Times (Sr and Os)

The Sr isotopic composition of seawater has been reconstructed for the Phanerozoic using marine carbonate shells (foraminifera, brachiopods, belemnites, conodonts (Hodell, Mueller, & Garrido, 1991; Veizer et al., 1999), marine barites (Paytan, Kastner, & Martin, 1993), fish teeth (Staudigel et al., 1985), and phosphatic pelloids (Show & Wasserburg, 1988) (Figure 5). The marine Sr isotope curve is characterized by an asymmetric trough shape over the Phanerozoic, starting from a composition close to present-day seawater (Figure 5), followed by several large oscillations until the Cenozoic period. Then, the $^{87}Sr/^{86}Sr$ ratio remained fairly constant from the beginning of the Cenozoic until ~35 Ma ago, followed by a steep increase from 35 Ma to present (Figure 5). This steep and well-defined rise in the Sr isotope ratio has been used as a chronostratigraphic tool, with a resolution of about ± 1 Myr, in regions or core sections devoid of the usual microfaunal biostratigraphic markers (Hodell et al., 1991).

The changes in the $^{87}Sr/^{86}Sr$ ratio of Sr dissolved in seawater respond to variations in the relative contributions from continental and oceanic rocks at the global scale. Increased contribution from continental weathering (raising the $^{87}Sr/^{86}Sr$) may be due either to climate change or to enhanced uplift during phases of continental collisions. In contrast, increased hydrothermal exchanges at mid-ocean ridges, during phases of continental breakup or enhanced rifting, tend to lower the isotopic ratio. The variation in the non-radiogenic input originating from the mantle is generally considered as minor compared to continental sources (Palmer & Elderfield, 1986). Major plate tectonic and paleogeographical changes have been invoked as key factors in controlling the ocean Sr budget (Richter, Rowley, & DePaolon, 1992). In particular, the pronounced shift since 35 Ma has been attributed to Himalayan uplift (Richter et al., 1992) increasing the contribution of very radiogenic sediment transported by the Ganges and Brahmaputra rivers and inherited from metamorphic complexes present in preexisting orogenic terrains (Edmond, 1993). The exact balance between the changes in river runoff (climatic signal) and the changes in the isotopic composition of the rivers due to exposure of fresh bedrock through uplifting (tectonic signal) remains unanswered for now.

Several Os oceanic records have been reported for the past 70 Ma (Pegram et al., 1992; Pegram & Turekian, 1999; Peucker-Ehrenbrink & Ravizza, 2000; Peucker-Ehrenbrink, Ravizza, & Hofmann, 1995; Ravizza, 1993) (Figure 5). The Os isotope marine record shows a general increase since ~65 Ma to present, quite similar

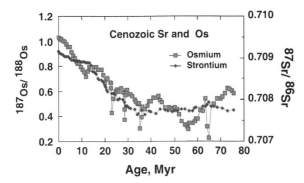

Figure 5 Cenozoic evolution of the Sr and Os isotope ratio in seawater (figure kindly provided by Burton, 2005). Strontium and osmium dissolved in seawater are incorporated in calcium carbonate shells, and in the authigenic fraction in sediments and ferromanganese deposits, respectively. Due to the very long residence time of Strontium, the $^{87}Sr/^{86}Sr$ ratio is homogeneous in the ocean at any given period. By contrast, this isotopic ratio has varied widely with time, depending on the global balance between strontium coming from continental rocks weathering (relatively enriched in ^{87}Sr) and from the oceanic crust hydrothermal sources (relatively depleted in ^{87}Sr). The long-term regular increase in $^{87}Sr/^{86}Sr$ over the Cenozoic has been strongly influenced by the Himalayan orogeny, due to the combination of increased weathering fluxes caused by climate change, and also changes in the isotopic composition of the altered rocks caused by the uplift and exposition of more radiogenic lithologies. Short-term discontinuities in the Sr and Os isotope records (i.e., abrupt changes in the rate of variation) have also attracted a lot of interest, since they must reflect huge perturbations of the continental alteration regime. However, their definite interpretation is still debated.

to Sr. In contrast, high values dominated prior to 65 Ma. In general, the good match between the Sr and Os marine records suggests that inputs of Sr and Os to the ocean are related to the same processes, in particular to the Himalayan orogeny for the most recent period of the record. However, Sharma, Wasserburg, Hofmann, and Chakrapani (2000b) pointed out that the Os concentrations are too low in the Himalayan rivers to have any significant impact on the Os marine budget. Global variations in the weathering regime and intensity must then be advocated.

The Os marine record is also characterized by several prominent excursions toward low $^{187}Os/^{188}Os$ values. The most distinctive excursion occurred at the Cretaceous/Tertiary boundary. It has been attributed to the addition of a huge quantity of unradiogenic Os of extraterrestrial origin that spread over the Earth after the K/T meteorite impact(s) (Peucker-Ehrenbrink et al., 1995). The slow increase in $^{187}Os/^{188}Os$ above the boundary could be explained by increased weathering of excavated continental material. The other minimum at 33 Ma (Eocene/Oligocene boundary) is much less constrained as it has been documented only in one core.

3.2. Examples using Tracers with Short and Intermediate Residence Times

3.2.1. Evolution of NADW in the North Atlantic

The isotope composition of the North Atlantic Deep Water (NADW) and of its different components has been studied with particular interest because the formation of this water mass is critical to global thermohaline circulation. Reference records have been obtained from three Fe-Mn crusts from the North Atlantic region (Burton, Bourdon, Birck, Allègre, & Hein, 1999; Burton, Lee, Christensen, Halliday & Hein, 1999; Burton, Ling, & O'Nions et al., 1997; O'Nions, Frank, Blanckenburg, & Ling et al., 1998; Reynolds, Frank, & O'Nions et al., 1999) covering the last 60 Ma (Figure 6).

Pb and Nd isotopes show similar patterns, with a rather constant composition prior to 3 Ma and then a sharp change afterward, with decreasing ε_{Nd}, $^{207}Pb/^{206}Pb$, $^{208}Pb/^{206}Pb$ and increasing $^{206}Pb/^{204}Pb$ (Burton et al., 1999; Burton et al., 1997; O'Nions et al., 1998; Reynolds et al., 1999). Hf isotopes also change in parallel during the same period (Piotrowski, 2000; van de Flierdt, Frank, Lee, & Halliday et al., 2002). By contrast, prior to 3 Ma, the NADW-Hf isotope record is decoupled from those of Nd and Pb, with a broad minimum between 12 and 23 Ma. This discrepancy is not yet explained, but a number of theories have been proposed.

It has been suggested that the drastic change after 3 Ma occurs because of an increase in the production of deep water in the Labrador Sea. This is supported mainly by the high Nd concentrations and very low ε_{Nd}, which reflects supplies from the surrounding Archean terrains. As discussed below, this major change in the configuration of oceanic circulation may have its ultimate origin in geodynamics, and possibly related to the closure of the Isthmus of Panama at ~3.5 Ma (Keigwin, 1982).

However, it has been shown recently that the sharp increase in Pb isotopes and decrease in ε_{Hf} could result also from enhanced erosion of the Canadian Shield and Greenland, related to the onset of Northern Hemisphere glaciations at ~2.7 Ma. Indeed, Von Blanckenburg and Nägler (2001) demonstrated that weathering of the old continental terrains in the North Atlantic region is not congruent, leading to the preferential release of trace metals with more radiogenic signatures than the exposed whole rock. This process could also explain the negative correlation observed between ε_{Nd} and $^{206}Pb/^{204}Pb$ in the time series records from the Atlantic Ocean. Interestingly, this correlation is not found in the Pacific Ocean, where the main terrains contributing to erosion are mostly young volcanic island arcs, for which weathering is essentially congruent. The strong decrease seen in the Hf isotope record also supports a change in weathering regime in the North Atlantic following the onset of the glaciations, since very low ε_{Hf} are provided to the LSW and NADW by the abrasion of old zircons from the eroded terrains around the Labrador Sea (van de Flierdt et al., 2002).

Finally, Pb and Nd isotopes measured by Winter, Johnson, and Clark (1997) in Mn-micronodules of sediment cores from the Arctic Ocean also corroborate the predominance of erosion processes rather than changes in ocean circulation to explain the trends observed in time series from the North Atlantic Ocean.

3.2.2. Central atlantic

Present-day deep waters in the eastern Atlantic basin (ENADW — Eastern North Atlantic Deep Water) result from vigorous vertical mixing of the northern flowing branch of NADW with AABW. This confluence occurs mainly through the Romanche fracture zone (1.4 Sv; Mercier & Morin, 1997) and the Vema fracture

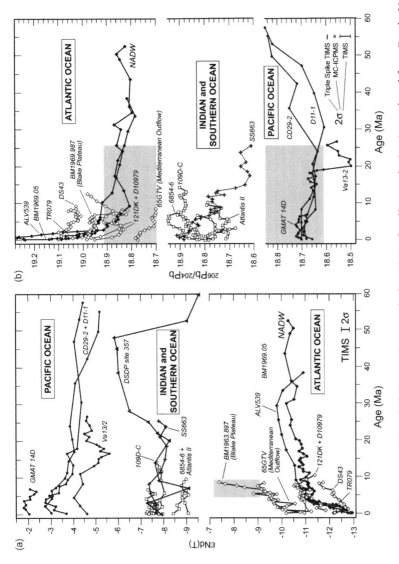

Fig. 6   Comparison of the deep-water Nd and Pb time series over the last 70 Ma for all ocean basins (reproduced from Frank, 2002). These figures show that the interbasin differences observed in the present-day distribution of Nd (6a) and Pb (6b) isotopes have persisted over the entire Cenozoic, but have been strongly amplified since the late Cenozoic. Various factors that may have contributed to this evolution are still debated, including climate change (initiation of the Northern Hemisphere Glaciation, which changed continental alteration regimes) and changes in the deep ocean circulation (installation of the global conveyor belt in its present-day configuration).

zone (2–2.4 Sv; Fisher, Rhein, Schott, & Stramma, 1996; McCartney, Bennett, & Woodgate-Jones, 1991). ENADW has an ε_{Nd} value between 11.6 and 11.9 (Tachikawa et al., 1999), resulting from the mixing of NADW ($\varepsilon_{Nd} = -13.5$) with AABW ($\varepsilon_{Nd} = 7$ to -9) (Piepgras & Wasserburg, 1982, 1987; Jeandel, 1993; Bertram & Elderfield, 1993) within the fracture zones.

The Nd and Pb isotope time series of ENADW derived from ferromanganese crusts show similar patterns as NADW over the past 4 Myr, although with a smaller amplitude of variation (Abouchami, Galer, & Koschinsky et al., 1999; Reynolds et al., 1999). This is consistent with continuous advection of NADW and AABW into the eastern Atlantic basin through the Romanche and the Vema fracture zones. However, Claude-Ivanaj, Hofmann, Vlastélic, and Koschinsky et al. (2001) demonstrated that the Pb isotopic record of ENADW for the past 400 kyr cannot be explained only by water mass advection and mixing of NADW and AABW. A significant addition of Pb from the Amazon or the Orinoco rivers to the NADW is required before it enters the Romanche fracture zone. This has been confirmed by Pb and Nd isotope time series for the last 100 kyr, obtained from ferromanganese crusts located in the Romanche fracture zone (Frank et al., 2003). However, the exact mechanism of Pb transfer from the river load to the deep ocean water is not well understood.

Prior to 3 Myr, comparison of the isotopic evolution of the deep waters in the fracture zones with the time series in western and eastern North Atlantic suggests that the isotopic composition of the deep eastern North Atlantic was strongly influenced by the direct supply of Labrador Sea Water via a northern route. In addition, the Mediterranean Outflow Water may have played a more important role in the past than at present (Abouchami et al., 1999; Frank et al., 2003).

3.2.3. NADW export in the Southern Ocean

In the present-day Southern Ocean, the Antarctic Circumpolar Current (ACC) is composed of 50% NADW. The evolution of this NADW export in the past can be estimated by comparing the Nd isotope evolution of NADW with that of the Southern Ocean. All the available archives (five Fe-Mn crusts from the South Atlantic and the Southern Ocean, together with cores from the Rio Grande Rise and the deep Cape Basin) are consistent with each other and do not reveal any resolvable long-term variability in ε_{Nd} (Frank, Whiteley, Kasten, Hein, & O'Nions et al., 2002; Palmer & Elderfield, 1985; Rutberg et al., 2000). This is in marked contrast with the NADW records (Figure 6). The explanation that has been proposed is a progressive reduction of NADW export since the onset of the northern hemisphere glaciations, ~3 Ma ago. Such a diminution of NADW export has also been documented during the glacial periods of the late Quaternary by carbon isotopes (Charles & Fairbanks, 1992; Raymo & Ruddiman, 1992; Raymo et al., 1990; Charles et al., 1996) and by Nd isotopes (Rutberg et al., 2000). Mass balance calculations based on the assumptions of a Pacific-dominated end-member with a ε_{Nd} value of -4 to -6, and no systematic difference in Nd concentration between NADW and the Circumpolar Deep Water are consistent with a present-day NADW contribution of 45–55% at the latitude of Walvis Ridge (Piepgras & Wasserburg, 1982; Döös, 1995). The same calculation applied to the time series suggests a 15 to

35% larger contribution of NADW at 3.5 Ma than at present (Frank et al., 2002). For the period from 14 Ma to present, the data are still consistent with a strong and continuous export of a NADW-type water mass from the North Atlantic.

3.2.4. Consequences of the Panama gateway closure on the general circulation

Burton et al. (1997) showed that the isotopic signal associated with NADW strengthened around 3 to 4 million years ago, due to the enhanced contribution of Labrador Sea Water. This seemed in good agreement with the idea that the closure of the Isthmus of Panama played a key role in setting today's general pattern of ocean circulation, as initially suggested by (Keigwin, 1982). A cessation of the deep water exchange between the two oceans is indeed suggested at ~4–6 Ma by sedimentological observations (Haug & Tiedemann, 1998), and by the comparison of faunal evolution of deep dwelling microorganisms. Before the closure, the direct flow of low-salinity water from the Pacific to the Atlantic ocean would have led to a smaller NADW flow, as corroborated by general circulation models simulations (Maier-Reimer, Mikolajewicks, & Crowley, 1990). The results from Burton et al. (1997) were obtained from measurements of Nd and Pb isotopes in a hydrogenous ferromanganese crust (BM 1969.05) from the San Pablo seamount in the western North Atlantic. The Pb and Nd isotope time series in crust BM 1963.897 from the Blake Plateau (850 m, now within Gulf Stream waters), published by Reynolds et al. (1999), shows a shift of both tracers between 2 and 3 Ma, comparable with those observed in the deeper western North Atlantic crusts (Burton et al., 1997; O'Nions et al., 1998; Reynolds et al., 1999). Frank (2002) interprets this shift as a consequence of within-basin mixing of NADW. As mentioned above, two crusts from the eastern Atlantic basin also show similar variations (Abouchami et al., 1999).

Over the period between 8 and 5 Ma, the Nd isotope composition decreased by 2.5 ε_{Nd} units and $^{206}Pb/^{204}Pb$ increased from 18.83 to 18.97 (this later change was complete by 6.5 Ma). These patterns are interpreted as an indication of a progressive decrease in the advection of Pacific water masses into the Atlantic Ocean. This implies that the closure of the Panama gateway could not have been the direct cause but only one of the prerequisites for the onset of Northern Hemisphere glaciations. The immediate consequence of the Panama gateway closure was the early Pliocene climate warming rather than a major cooling.

3.2.5. Indian Ocean deep water and consequences of Himalayan erosion

The eastern Indian Ocean is another region of particular interest when investigating the mechanism of geochemical exchange between the continents and the ocean, because of the huge detrital input derived from the Himalayas and the Indonesian volcanic arc.

Burton and Vance (2000) have analyzed Nd isotopes in planktonic foraminifera for the past 150 kyr from ODP site 758, on the southernmost reaches of the Bay of Bengal (2,925 m depth, 1,000 m above the Bengal fan). They found core-top ε_{Nd} values of −10.12 and −10.28, indistinguishable from present-day seawater. They also demonstrated systematic variations ($\Delta\varepsilon_{Nd} = 3$) on glacial–interglacial timescales, in remarkable correspondence with the global oxygen isotope record. Their interpretation was that reduced monsoonal rainfall over India and reduced runoff to the Bay of Bengal at the Last Glacial Maximum led to decreasing amounts of unradiogenic Nd delivered by the Ganges-Brahmaputra rivers during glacial intervals. This resulted in an increased proportion of radiogenic Nd derived from western Indian thermocline waters. Increased upwelling (minimal today, at that site) may have also played a role, as well as increased aeolian deposition from the Arabian Desert and Persian Gulf (ε_{Nd} of -6 and -9, respectively).

Colin, Turpin, Bertaux, Desprairies, & Kissel (1999) have studied gravity cores from the Bay of Bengal (MD77-180), the Andaman Sea (MD77-169), and 15 other core-top samples, to map the relative contributions of Ganges-Brahmaputra and Irrawaddy rivers, and the western part of the Indo-Burman ranges. The samples showed increased $^{87}/^{86}Sr$ in the non-carbonate fraction of the sediment during cold intervals and were interpreted to reflect a decrease in chemical weathering (that preferentially released the radiogenic minerals). Indeed, reduced precipitation is likely to lead to lower degrees of chemical weathering in river plains, with a relative increase in the proportion of material derived from physical erosion.

On longer timescales, ferromanganese crust SS663 from the deep central Indian Ocean (O'Nions et al., 1998; Frank & O'Nions, 1998) revealed a strikingly constant ε_{Nd} (within the narrow range of present-day Indian Ocean water, -7.4 to -7.8) for the past 26 Myr. This suggests that the huge input of Himalayan weathering (ε_{Nd} of -16 ± 2, Derry & France-Lanord, 1996) had a minimal influence on the Nd isotope budget of the Indian Ocean deep water. In contrast, a significant increase of $^{206}Pb/^{204}Pb$ from 18.6 to 18.9 has been observed on the same timescale, attributed to the Himalayan erosion by Frank and O'Nions (1998), probably originating mainly from the High Himalayan gneisses and leucogranites. The decoupling of Nd and Pb isotopes in that crust is intriguing and it is certainly linked to the difference in reactivity of Pb and Nd with particles. Nd is less efficiently scavenged than Pb, and the riverine Himalayan input may have been advected away and diluted, while local sediments recorded a clear signal for Pb isotopes.

4. CONCLUSION AND PERSPECTIVES

Radiogenic isotopes are now well established as important geochemical tracers for chemical oceanographers. The large number of Nd isotope studies carried out both on the modern ocean and on various sedimentary archives have played a major role in demonstrating the potential of this tracing method. To date, the variations of the other radiogenic isotopes remain less documented, but it is clearly demonstrated that each of them has the capability of capturing specific aspects of the ocean circulation and of land–sea interactions, depending on its mineral sources and chemical properties, and thus on its residence time in the ocean.

It is clear that the recent improvement of the analytical techniques, in both higher precision and sample throughput of the MC-ICPMS, has opened new perspectives of development of this method. It is now possible to obtain complete maps of the distribution of each of these isotopes, at the scale of the world ocean as well as in key regions of interest. High-resolution time records as well as detailed geographical investigations are now possible with a multi-element perspective.

A major step forward will be to integrate the observations obtained from these tracers as parameters in quantitative geochemical models of the ocean circulation. Numerous examples of box model calculations have already been published, for both long and short residence time isotopic tracers. A significant breakthrough has been reached recently by simulating Pb isotopes variations within a GCM model. This approach needs to be extended to the different isotopes. Input functions to the ocean must be documented precisely, each isotope having its own sources on continental surfaces, and its specific availability to dissolution processes. Possible complexities resulting, for instance, from incongruent weathering on land, or from precipitation/remobilization processes in estuaries and continental platforms must be further documented, and their impact on ocean-scale isotopic patterns must be more thoroughly evaluated.

As seen in the examples given above, the main difficulty in interpreting the isotopic time records stems from the persistent ambiguity between changes in hydrographic circulation (leading to changes in mixing proportions between different water masses), and changes in the strength and/or isotopic composition of the inputs. It is not yet clear to what extent unequivocal solutions can be found to this ambiguity. However, significant progress on that issue may be expected from: (1) systematic inter-element correlations; (2) increased spatial resolution of sample collection in regions of identified point-sources of input (e.g., major rivers and estuaries); and (3) comparison between different types of archives (i.e., authigenic vs. detrital phases).

On this last issue, it must be emphasized that the high-time-resolution records embedded in deep-sea sediment cores may ultimately reveal more information than medium- to low-resolution records in ferromanganese crusts, although the latter are less prone to diagenetic perturbation. So far, much more effort has been devoted to long-term studies based on ferromanganese deposits, because they seem more appealing as undisturbed seawater recorders. However, regional scale studies of glacial–interglacial variations, including short-term events where the radiogenic isotopes can be systematically compared with traditional paleocirculation and paleoproductivity tracers are needed to improved resolution.

REFERENCES

Abouchami, W., Galer, S. J. G., & Koschinsky, A. (1999). Pb and Nd isotopes in NE Atlantic Fe-Mn crusts: Proxies for trace metal paleosources and paleocean circulation. *Geochimica et Cosmochimica Acta, 63*, 1489–1505.

Abouchami, W., & Goldstein, S. L. (1995). A lead isotopic study of circum—Antarctic manganese nodules. *Geochimica et Cosmochimica Acta, 59*, 1809–1820.

Albarède, F., Simonetti, A., Vervoort, J. D., Blichert-Toft, J., & Abouchami, W. (1998). A Hf-Nd correlation in ferromanganese nodules. *Geophysical Research Letters, 25*, 3895–3898.

Albarède, F., Telouk, P., Blichert-Toft, J., Boyet, M., Agranier, A., & Nelson, B. (2004). Precise and accurate isotopic measurements using multiple-collector ICPMS. *Geochimica et Cosmochimica Acta, 68*, 2725–2744.

Alleman, L. Y., Véron, A. J., Church, T. M., Flegal, A. R., & Hamelin, B. (1999). Invasion of the abyssal North Atlantic by modern anthropogenic lead. *Geophysical Research Letters, 26*, 1477–1480.

Amakawa, H., Alibo, D. S., & Nosaki, Y. (2000). Nd isotopic composition and REE pattern in the surface waters of the eastern Indian Ocean and its adjacent seas. *Geochimica et Cosmochimica Acta, 64*, 1715–1727.

Andersson, P., Dahlqvist, R., Ingri, J., & Gustafsson, Ö. (2001). The isotopic composition of Nd in a boreal river: A reflexion of selective weathering and colloïdal transport. *Geochimica et Cosmochimica Acta, 65*, 521–527.

Barrett, T. J., Taylor, P. N., & Lugowski, J. (1987). Metalliferous sediments from DSDP Leg 92: The East Pacific Rise transect. *Geochimica et Cosmochimica Acta, 51*, 2241–2253.

Bayon, G., German, C. R., Boella, R. M., Milton, J. A., Taylor, R. N., & Nesbitt, R. W. (2002). An improved method for extracting marine sediments fractions and its application to Sr and Nd isotopic analysis. *Chemical Geology, 187*, 179–199.

Belshaw, N. S., O'Nions, R. K., & von Blanckenburg, F. (1995). A SIMS method for $^{10}Be/^{9}Be$ ratio measurement in environmental materials. *International Journal of Mass Spectrometry and Ion Physics, 142*, 55–67.

Bertram, C. J., & Elderfield, H. (1993). The chemical balance of rare earth elements and Nd isotopes in the oceans. *Geochimica et Cosmochimica Acta, 57*, 1957–1986.

Birck, J.-L. (1986). Precision K-Rb-Sr isotopic analysis: Application to Rb-Sr chronology. *Chemical Geology, 56*, 73–83.

von Blanckenburg, F., & Nägler, T. F. (2001). Weathering versus circulation-controlled changes in radiogenic isotope tracer composition in Labrador Sea and Northern Atlantic Deep Water. *Paleoceanography, 16*, 424–434.

Bout-Ramazeilles, V., Davies, G., & Labeyrie, L. (1998). Nd-Sr-Pb evidence of glacial-interglacial variations in clay provenance and transport in the North Atlantic Ocean. *Mineralogical Magazine, 62A*, 1443–1444.

Boyle, E. A., Chapnick, S. D., Shen, G. T., & Bacon, M. P. (1986). Temporal variability of lead in the western North Atlantic. *Journal of Geophysical Research, 91*, 8573–8593.

Boyle, E. A., & Keigwin, L. D. (1986). Comparison of Atlantic and Pacific paleochemical records for the last 215,000 years: Changes in deep ocean circulation and chemical inventories. *Earth and Planetary Science Letters, 76*, 135–150.

Brass, G. W. (1976). The variation of the marine $^{86}Sr/^{87}Sr$ during the Phanerozoic time: Interpretation using a flux model. *Geochimica et Cosmochimica Acta, 40*, 721–730.

Broecker, W. S. (1991). The great ocean conveyor. *Oceanography, 4*, 79–89.

Bruland, K. W. (1983). Trace elements in seawater. In: J. P. Riley & R. Chester (Eds), *Chemical Oceanography* (Vol. 8, pp. 175–220). San Diego, CA: Academic Press.

Burton, K. W., Bourdon, B., Birck, J.-L., Allègre, C. J., & Hein, J. (1999). Osmium isotope variations in the oceans recorded by Fe-Mn crusts. *Earth and Planetary Science Letters, 171*, 185–197.

Burton, K. W., Lee, D. C., Christensen, J. N., Halliday, A. N., & Hein, J. R. (1999). Actual timing of neodymium isotopic variations recorded by Fe-Mn crusts in the western North Atlantic. *Earth and Planetary Science Letters, 171*, 149–156.

Burton, K. W., Ling, H.-F., & O'Nions, R. K. (1997). Closure of the central american isthmus and its effect on deep-water formation in the North-Atlantic. *Nature, 386*, 382–385.

Burton, K. W., & Vance, D. (2000). Glacial-interglacial variations in the neodymium isotope composition of seawater in the Bay of Bengal recorded by planktonic foraminifera. *Earth and Planetary Science Letters, 176*, 425–441.

Charles, C. D., & Fairbanks, R. G. (1992). Evidence from Southern Ocean sediments for the effect of North Atlantic Deep water flux on climate. *Nature, 355*, 416–419.

Charles, C. D., Lynch-Stieglitz, J., Ninnemann, U. S., & Fairbanks, R. G. (1996). Climate connections between the hemisphere revealed by deep sea sediment core/ice core correlations. *Earth and Planetary Science Letters, 142*, 19–27.

Chen, J. H., Waserburg, G. J., Von Damm, K. L., & Edmond, J. M. (1986). The U-Th-Pb systematics in hot springs on the East Pacific Rise at 21°N and Guyamas Basin. *Geochimica et Cosmochimica Acta, 50*, 2467–2479.

Chow, T. J., & Patterson, C. C. (1959). Lead isotopes in manganese nodules. *Geochimica et Cosmochimica Acta, 17*, 21–31.

Chow, T. J., & Patterson, C. C. (1962). The occurrence and significance of lead isotopes in pelagic sediments. *Geochimica et Cosmochimica Acta, 26*, 263–308.

Claude-Ivanaj, C., Abouchami, W., Galer, S. J. G., Hofmann, A. W., & Koschinsky, A. (1998). High resolution ^{230}Th/^{232}Th and ^{234}U/^{238}U chronology of a hydrogenous Fe-Mn crust from the NE Atlantic. *Mineralogical Magazine, 62*, 335–336.

Claude-Ivanaj, C., Hofmann, A. W., Vlastélic, I., & Koschinsky, A. (2001). Recording changes in ENADW composition over the last 340 Ka using high-precision lead isotopes in a Fe-Mn crust. *Earth and Planetary Science Letters, 188*, 73–89.

Claude, C., Suhr, G., Koschinsky, A., & Hofmann, A. W. (2005). U-Th chronology and paleoceanographic record in a Fe-Mn crust from the eastern NE Atlantic over the last 700 Ka. *Geochimica et Cosmochimica Acta, 69*, 4845–4854.

Cochran, J. K., McKibbin-Vaughan, T., Dornblaser, M. M., Hirshberg, D., Livingstone, H. D., & Buesseler, K. O. (1990). ^{210}Pb scavenging in the North Atlantic and the North Pacific Oceans. *Earth and Planetary Science Letters, 97*, 332–352.

Colin, C., Turpin, L., Bertaux, J., Desprairies, A., & Kissel, C. (1999). Erosional history of the Himalayan and Burman ranges during the last two glacial-interglacial cycles. *Earth and Planetary Science Letters, 171*, 647–660.

Craig, H., Krishnaswami, S., & Somayajulu, B. L. K. (1973). ^{210}Pb-^{226}Ra: Radioactive disequilibrium in the deep ocean. *Earth and Planetary Science Letters, 17*, 295–305.

David, K., Birck, J.-L., Telouk, P., & Allègre, C. J. (1999). Application of isotope dilution for precise measurement of Zr/Hf and ^{176}Hf/^{177}Hf mass spectrometry (ID-TIMS, ID-MC-ICPMS). *Chemical Geology, 157*, 1–12.

David, K. M., Frank, M., O'Nions, R. K., Belshaw, N. S., Arden, J. W., & Hein, J. R. (2001). The Hf isotope composition of global seawater and the evolution of Hf isotopes in the deep Pacific Ocean from Fe-Mn crusts. *Chemical Geology, 178*, 23–42.

Derry, L. A., & France-Lanord, C. (1996). Neogene Himalayan weathering history and river ^{87}Sr/^{86}Sr: impact on the marine Sr record. *Earth and Planetary Science Letters, 142*, 59–74.

Dia, A., Dupré, B., & Allègre, C. J. (1992). Nd isotopes in Indian Ocean sediments used as tracers of supply to the ocean and circulation paths. *Marine Geology, 103*, 349–359.

Döös, K. (1995). Interocean exchange of water masses. *Journal of Geophysical Research, 100*, 13499–13514.

Doucelance, R., & Manhès, G. (2001). Reevaluation of precise lead isotope measurements by thermal ionization mass spectrometry: Comparison with determinations by plasma source mass spectrometry. *Chemical Geology, 176*, 361–377.

Duce, R. A., Liss, P. S., Merrill, J. T., Atlas, E. L., Buat-Menard, P., Hicks, B. B., Miller, J. M., Prospero, J. M., Arimoto, R., Church, T. M., Ellis, W., Galloway, J. N., Hansen, L., Jickells, T. D., Knap, A. H., Reinhardt, K. H., Schneider, B., Soudine, A., Wollast, R., & Zhou, M. (1991). The atmospheric input of trace species to the world ocean. *Global Biogeochemical Cycles, 5*, 193–259.

Edmond, J. M. (1993). Himalayan tectonics, weathering processes and the Sr isotope record in marine limestones. *Science, 258*, 1594–1597.

Edwards, R. L., Chen, J. H., & Wasserburg, G. J. (1987). ^{238}U-^{234}U-^{230}Th-^{232}Th systematics and the precise measurement of time over the past 500,000 years. *Earth and Planetary Science Letters, 81*, 175–192.

Eisenhauer, A., Gögen, K., Pernicka, E., & Mangini, A. (1992). Climatic influences on the growth rates of Mn crusts during the late Quaternary. *Earth and Planetary Science Letters, 109*, 25–36.

Elderfield, H., & Sholkovitz, E. R. (1987). Rare earth elements in the pore waters of reducing nearshore sediments. *Earth and Planetary Science Letters, 82*, 280–288.

Erel, Y., Harlavan, Y., & Blum, J. D. (1994). Lead isotope systematics of granitoid weathering. *Geochimica et Cosmochimica Acta, 58*, 5299–5306.

Fagel, N., & Hillaire-Marcel, C. (2006). Glacial/interglacial instabilities of the Western Boundary Under Current during the last 360 kyr from Sm/Nd signatures of sedimentary clay fractions at ODP-Site 646 (Labrador Sea). *Marine Geology, 232*, 87–99.

Fagel, N., Innocent, C., Stevenson, R. K., & Hillaire-Marcel, C. (1999). Deep circulation changes in the Labrador Sea since the last Glacial Maximun: New constraints from Sm-Nd data on sediments. *Paleoceanography, 14*, 777–788.

Fisher, J., Rhein, M., Schott, F., & Stramma, L. (1996). Deep water masses and transports in the Vema Fracture Zone. *Deep Sea Research, 43*, 1067–1074.

van de Flierdt, T. M., Frank, M., Lee, D.-C., & Halliday, A. N. (2002). Glacial weathering and the hafnium isotope composition of seawater. *Earth and Planetary Science Letters, 198*, 167–175.

Frank, M. (2002). Radiogenic isotopes: tracers of past ocean circulation and erosional input. *Review of Geophysics, 40*, 1001.

Frank, M., van de Flierdt, T., Halliday, A. N., Kubik, P. W., Hattendork, B., & Günther, D. (2003). Evolution of deepwater mixing and weathering inputs in the central Atlantic Ocean over the past 33 Myr. *Paleoceanography, 18*, 1029–1046.

Frank, M., & O'Nions, R. K. (1998). Sources of Pb for Indian Ocean ferromanganese crusts: A record of Himalayan erosion? *Earth and Planetary Science Letters, 158*, 121–130.

Frank, M., O'Nions, R. K., Hein, J. R., & Banakar, V. K. (1999). 60 Myr records of major elements and Pb-Nd isotopes from hydrogenous ferromanganese crusts: Reconstruction of seawater paleochemistry. *Geochimica et Cosmochimica Acta, 63*, 1689–1708.

Frank, M., Whiteley, N., Kasten, S., Hein, J. R., & O'Nions, R. K. (2002). North Atlantic Deep Water export to the Southern Ocean over the past 14 Ma: Evidence from Nd and Pb isotopes in ferromanganese crusts. *Paleoceanography, 17*, 1022, doi: 10.1029/ 2000PA 000606.

Galer, S. J. G. (1999). Optimal double and triple spiking for high precision lead isotopic measurement. *Chemical Geology, 157*, 255–274.

Ganachaud, A., & Wunsch, C. (2000). Improved estimates of global ocean circulation, heat transport and mixing from hydrographic data. *Nature, 408*, 453–457.

Godfrey, L. V., Lee, D.-C., Sangrey, W. S., Halliday, A. N., Salters, V. J. M., Hein, J. R., & White, W. M. (1997). The Hf isotopic composition of ferromanganese nodules and crusts and hydrothermal manganese deposits: Implications for seawater Hf. *Earth and Planetary Science Letters, 151*, 91–105.

Godfrey, L. V., White, W. M., & Salters, V. J. M. (1996). Dissolved zirconium and hafnium across a shelf break in the NE Atlantic. *Geochimica et Cosmochimica Acta, 60*, 3995–4006.

Goldstein, S. J., & Jacobsen, S. B. (1987). The Nd and Sr isotopic systematics of river-water dissolved material: Implications for the sources of Nd and Sr in seawater. *Chemical Geology, 66*, 245–272.

Goldstein, S. J., O'Nions, R. K., & Hamilton, P. J. (1984). A Sm-Nd isotopic study of atmospheric dusts and particulates from major river system. *Earth and Planetary Science Letters, 70*, 221–236.

Greaves, M. J., Elderfield, H., & Sholkovitz, E. R. (1999). Eolian sources of rare earth elements to the Western Pacific ocean. *Marine Chemistry, 68*, 31–38.

Grousset, F. E., Biscaye, P. E., Zindler, A., Prospero, J., & Chester, R. (1988). Neodymium isotopes as tracers in marine sediments and aerosols: North Atlantic. *Earth and Planetary Science Letters, 87*, 367–378.

Halbach, P., & Puteanus, D. (1984). The influence of the carbonate dissolution rate on the growth and composition of Co-rich ferromanganese crusts from central Pacific seamounts areas. *Earth and Planetary Science Letters, 68*, 73–87.

Halbach, P., Segl, M., Puteanus, D., & Mangini, A. (1983). Co-fluxes and growth rates in ferromanganese deposits from central Pacific seamount areas. *Nature, 304*, 719–722.

Haley, B. A., & Klinkhammer, G. P. (2002). Development of a flowthrough system for cleaning and dissolving foraminiferal tests. *Chemical Geology, 185*, 51–69.

Halliday, A. N., Lee, D.-C., Christensen, J. N., Rehkämper, M., Yi, W., Luo, X., Hall, C. M., Ballentine, C. J., Pettke, T., & Stirling, C. (1998). Applications of multiple collector-ICPMS to cosmochemistry, geochemistry, and paleoceanography. *Geochimica et Cosmochimica Acta, 62*, 919–940.

Harada, K., & Nishida, S. (1976). Biostratigraphy of some marine manganese nodules. *Nature, 260*, 770–771.

Haug, G. H., & Tiedemann, R. (1998). Influence of Panamian isthmus formation on Atlantic Ocean thermohaline circulation. *Nature, 393*, 673–676.

Havach, S. M., Chancler, G. T., Wilson-Finelli, A., & Shaw, T. J. (2001). Experimental partition coefficient in cultured benthic foraminifera. *Geochimica et Cosmochimica Acta, 65*, 1277–1283.

Hein, J. R., Bohrson, W. A., Schulz, M. S., Noble, M., & Clague, D. A. (1992). Variations in the fine-scale composition of a central Pacific ferromanganese crust: Paleoceanographic implications. *Paleoceanography, 7*, 63–77.

Hemming, S. R., Biscaye, P. E., Broecker, W. S., Hemming, N. G., Klas, M., & Hajdas, I. (1998). Provenance change coupled with increased clay flux during deglacial times in the western equatorial Atlantic. *Palaeogeography, Palaeoclimatology, Palaeoecology, 142*, 217–230.

Henderson, G. M., & Burton, K. W. (1999). Using ($^{234}U/^{238}U$) to assess diffusion rates of isotope tracers in ferromanganese crusts. *Earth and Planetary Science Letters, 170*, 169–179.

Henderson, G. M., & Maier-Reimer, E. (2002). Advection and removal of ^{210}Pb and stable leéd isotopes in the oceans: A general circulation model study. *Geochimica et Cosmochimica Acta, 66*, 257–272.

Hodell, D. A., Mueller, P. A., & Garrido, J. R. (1991). Variations in the strontium isotopic compositions of seawater during the Neogene. *Geology, 19*, 24–27.

Innocent, C., Flagel, N., Stevenson, R. K., & Hillaire-Marcel, C. (1997). Sm-Nd signature of modern and late Quaternary sediments from the northwest North Atlantic: Implications for past current changes since the Last glacial Maximum. *Earth and Planetary Science Letters, 146*, 607–625.

Innocent, C., Fagel, N., & Hillaire-Marcel, C. (2000). Sm-Nd isotope systematics in deep-sea sediments: Clay-size vs. coarser fractions. *Marine Geology, 168*, 79–87.

Jeandel, C. (1993). Concentration and isotopic composition of Nd in the south Atlantic ocean. *Earth and Planetary Science Letters, 117*, 581–591.

Jeandel, C., Bishop, J. K., & Zindler, A. (1995). Exchange of neodymium and its isotope between seawater and small and large particles in the Sargasso Sea. *Geochimica et Cosmochimica Acta, 59*, 537–545.

Jeandel, C., Thouron, D., & Fieux, M. (1998). Concentrations and isotopic compositions of neodymium in the eastern Indian Ocean and Indonesian straits. *Geochimica et Cosmochimica Acta, 62*, 2597–2607.

Jochum, K. P., Stoll, B., Herwig, K., Amini, M., Abouchami, W., & Hofmann, A. W. (2005). Lead isotope ratio measurements in geological glasses by laser ablation-sector field-ICP mass spectrometry (LA-SF-ICPMS). *International Journal of Mass Spectrometry, 242*, 281–289.

Jones, C. E., Halliday, A. N., Rea, D. K., & Owen, R. M. (1994). Neodymium isotopic variations in North Pacific modern silicate sediment and the insignificance of detrital REE contributions to seawater. *Earth and Planetary Science Letters, 127*, 55–66.

Jones, C. E., Halliday, A. N., Rea, D. K., & Owen, R. M. (2000). Eolian inputs of lead to the North Pacific. *Geochimica et Cosmochimica Acta, 64*, 1405–1416.

Joshima, M., & Usui, A. (1998). Magnetostratigraphy of hydrogenetic manganese crusts from Northwestern Pacific seamounts. *Marine Geology, 146*, 53–62.

Kadko, D., & Burckle, L. H. (1980). Manganese nodule growth rates determined by fossil diatoms dating. *Nature, 287*, 725–726.

Keigwin, L. D. (1982). Isotopic paleoceanography of the Caribbean and East Pacific: Role of Panama uplift in late neogene time. *Science, 217*, 350–352.

Koschinsky, A., & Halbach, P. (1995). Sequential leaching of marine ferromanganese precipitates: Genetic implications. *Geochimica et Cosmochimica Acta, 59*, 5113–5132.

Krishnaswami, S., Mangini, A., Thomas, J. H., Sharma, P., Cochran, J. K., Turekian, K. K., & Parker, P. D. (1982). ^{10}Be and Th isotopes in manganese nodules and adjacent sediments: Nodule growth histories and nuclide behavior. *Earth and Planetary Science Letters, 59*, 217–234.

Krishnaswami, S., Somayajulu, B. L. K., & Moore, W. S. (1972). Dating of manganese nodules using beryllium-10. In: D. R. Horn (Ed.), *Ferromanganese deposits on the ocean floor* (pp. 117–121). Washington D.C.: National Science Foundation.

Ku, T. L. (1976). The uranium-series methods of age determination. *Annual Review of Earth and Planetary Science, 4*, 347–379.

Ku, T. L., Omura, A., & Chen, P. S. (1979). *Marine geology and oceanography of the Pacific manganese province.* In: J. L. Bischoff & Z. Piper (Eds) (pp. 791–814). New York: Plenum.

Lacan, F., & Jeandel, C. (2001). Tracing Papua New Guinea imprint on the central Equatorial Pacific Ocean using neodymium isotopic compositions and rare earth element patterns. *Earth and Planetary Science Letters, 186*, 457–512.

Lea, D. W., & Boyle, E. A. (1991). Barium in planktonic foraminifera. *Geochimica et Cosmochimica Acta, 55*, 3321–3331.

Lee, D.-C., Halliday, A. N., Hein, J. R., Burton, K. W., Christensen, J. N., & Günther, D. (1999). Hafnium isotope stratigraphy of ferromanganese crusts. *Science, 285*, 1052–1054.

Levasseur, S., Birck, J.-L., & Allègre, C. J. (1998). Direct measurement of femtomoles of osmium and $^{187}Os/^{186}Os$ ratio in seawater. *Science, 282*, 272–274.

Ling, H.-F., Burton, K. W., O'Nions, R. K., Kamber, B. S., von Blanckenburg, F., Gibb, A. J., & Hein, J. R. (1997). Evolution of Nd and Pb isotopes in central Pacific seawater from ferromanganese crusts. *Earth and Planetary Science Letters, 146*, 1–12.

Maier-Reimer, E., Mikolajewicks, U., & Crowley, T. J. (1990). Ocean General Circulation Model sensitivity experiment with an open central America seaway. *Paleoceanography, 5*, 349–366.

Manheim, F. T., & Lane-Bostwick, C. M. (1986). Cobalt in ferromanganese crusts as a monitor of hydrothermal discharge on the Pacific seafloor. *Nature, 335*, 59–62.

Martin, E. E., & Haley, E. A. (2000). Fossil fish teeth as proxies for seawater Sr and Nd isotopes. *Geochimica et Cosmochimica Acta, 64*, 835–847.

McCartney, M. S., Bennett, S. L., & Woodgate-Jones, M. E. (1991). Eastward flow through the mid-Atlantic ridge at 11°N and its influence on the abyss of the eastern basin. *Journal of Physical Oceanography, 21*, 1089–1121.

McCorkle, D. C., Martin, P. A., Lea, D. W., & Klinkhammer, G. P. (1995). Evidence of a dissolution effect on benthic foraminiferal shell chemistry: $\delta^{13}C$, Cd/Ca, Ba/Ca and Sr/Ca results from the Ontong-Java Plateau. *Paleoceanography, 10*, 699–714.

McCulloch, M. T., & Wasserburg, G. J. (1978). Sm-Nd and Rb-Sr Chronology of Continental Crust Formation. *Science, 200*, 1003–1011.

McKelvey, B. A., & Orians, K. J. (1998). The determination of dissolved zirconium and hafnium from seawater using isotope dilution inductively coupled plasma mass spectrometry. *Marine Chemistry, 60*, 245–255.

McMurtry, G. M., Von der Haar, D. L., Eisenhauer, A., Mahonery, J. J., & Yeh, H.-W. (1994). Cenozoic accumulation history of a Pacific ferromanganese crust. *Earth and Planetary Science Letters, 125*, 105–118.

Mercier, H., & Morin, P. (1997). Hydrography of the Romanche and Chain Fracture Zones. *Journal of Geophysical Research, 102*, 10373–10389.

Morton, J. L., & Sleep, N. H. (1985). A mid-ocean ridge thermal model: Constraints on the volume of axial hydrothermal heat flux. *Journal of Geophysical Research, 90*, 11353–11354.

Neff, U., Bollhöffer, A., Frank, N., & Mangini, A. (1999). Explaining discrepant profiles of $^{234}U/^{238}U$ and $^{230}Th_{ex}$ in Mn-crusts. *Geochimica et Cosmochimica Acta, 63*, 2211–2218.

O'Nions, R. K., Frank, M., von Blanckenburg, F., & Ling, H.-F. (1998). Secular variation of Nd and Pb isotopes in ferromanganese crusts from the Atlantic, Indian and Pacific oceans. *Earth and Planetary Science Letters, 155*, 15–28.

Öhlander, B., Ingri, J., Land, M., & Schöberg, H. (2000). Change of Sm-Nd isotopic composition during weathering of till. *Geochimica et Cosmochimica Acta, 64*, 813–820.

Oxburgh, R. (1998). Variations in the osmium isotope composition of seawater over the past 200,000 years. *Earth and Planetary Science Letters, 159*, 183–191.

Oxburgh, R. (2001). Residence time of osmium in the ocean. *Geochemistry, Geophysics, and Geosystems, 2*, paper number 2000GC000104.

Palmer, M. R., & Edmond, J. M. (1989). The strontium isotope budget of the modern ocean. *Earth and Planetary Science Letters, 92*, 11–26.

Palmer, M. R., & Elderfield, H. (1985). Variations in the Nd isotopic of foraminifera from Atlantic Ocean sediments. *Earth and Planetary Science Letters, 73*, 299–305.

Palmer, M. R., & Elderfield, H. (1986). Rare earth elements and neodymium isotopes in ferromanganese oxide coatings and cenozoic foraminifera from the Atlantic Ocean sediments. *Geochimica et Cosmochimica Acta, 50*, 409–417.

Paytan, A., Kastner, M., & Martin, E. E. (1993). Marine barite as a monitor of seawater strontium isotope composition. *Nature, 366*, 445–449.

Pegram, W. J., Krishnaswami, S., Ravizza, G., & Turekian, K. K. (1992). The record of seawater $^{187}Os/^{186}Os$ variation through the cenozoic. *Earth and Planetary Science Letters, 113*, 569–576.

Pegram, W. J., & Turekian, K. K. (1999). The osmium isotopic composition change of cenozoic seawater as inferred from a deep-sea core corrected for meteoritic contributions. *Geochimica et Cosmochimica Acta, 63*, 4053–4058.

Peucker-Ehrenbrink, B., & Ravizza, G. (2000). The marine osmium isotope record. *Terra Nova, 12*, 205–219.

Peucker-Ehrenbrink, B., Ravizza, G., & Hofmann, A. W. (1995). The marine $^{187}Os/^{186}Os$ record of the past 80 millions years. *Earth and Planetary Science Letters, 130*, 155–167.

Piepgras, D. J., & Jacobsen, S. B. (1988). The isotopic composition of neodymium in the North Pacific. *Geochimica et Cosmochimica Acta, 52*, 1373–1381.

Piepgras, D. J., & Wasserburg, G. J. (1980). Neodymium isotopic variations in seawater. *Earth and Planetary Science Letters, 50*, 128–138.

Piepgras, D. J., & Wasserburg, G. J. (1982). Isotopic composition of neodymium in the waters from the Drake Passage. *Science, 217*, 207–214.

Piepgras, D. J., & Wasserburg, J. G. (1987). Rare earth element transport in the western north Atlantic inferred from Nd isotopic observations. *Geochimica et Cosmochimica Acta, 51*, 1257–1271.

Piepgras, D. J., Wasserburg, G. J., & Dash, E. J. (1979). The isotopic composition of Nd in different water masses. *Earth and Planetary Science Letters, 45*, 223–236.

Piotrowski, A. M., Lee, D.-C., Christensen, J. N., Burton, K. W., Halliday, A. N., Hein, J. R., & Günther, D. (2000). Changes in erosion ocean circulation circulation recorded in the Hf isotopic compositions of North Atlantic and Indian Ocean ferromanganese crusts. *Earth and Planetary Science Letters, 181*, 315–325.

Pomiès, C., Davies, G. R., & Conan, S. M.-H. (2002). Neodymium isotopes in modern foraminifera from the Indian Ocean: Implication for the use of foraminiferal Nd isotope compositions in paleo-oceanography. *Earth and Planetary Science Letters, 203*, 1031–1045.

Puteanus, D., & Halbach, P. (1988). Correlation of Co concentration and growth rate: A method for age determination of ferromanganese crusts. *Chemical Geology, 69*, 73–85.

Ravizza, G. (1993). $^{187}Os/^{186}Os$ ratios of seawater over the past 28 Ma as inferred from metalliferous carbonates. *Earth and Planetary Science Letters, 118*, 335–348.

Ravizza, G., Norris, G. N., Blusztajn, J., & Aubry, M. P. (2001). An osmium isotope excursion associated with the late paleocene thermal maximum: Evidence of intensified chemical weathering. *Paleoceanography, 16*, 155–163.

Ravizza, G., & Turekian, K. K. (1992). The osmium isotopic composition of organic-rich marine sediments. *Earth and Planetary Science Letters, 110*, 1–6.

Raymo, M. E., & Ruddiman, W. F. (1992). Tectonic forcing of late cenozoic climate. *Nature, 359*, 117–122.

Raymo, M. E., Ruddiman, W. F., Shackelton, N. J., & Oppo, D. W. (1990). Evolution of Atlantic-Pacific $\delta^{13}C$ gradients over the last 2.5 m y. *Earth and Planetary Science Letters, 97*, 353–368.

Rehkämper, M., & Halliday, A. N. (1998). Accuracy and long-term reproducibility of lead isotopic measurements by multiple-collector inductively coupled plasma mass spectrometry using an external method for correction of mass discrimination. *International Journal of Mass Spectrometry, 181*, 123–133.

Revel, M., Cremer, M., Grousset, F. E., & Labeyrie, L. (1996). Grain size and Sr-Nd isotopes as tracer of paleo-bottom current strength, Northeast Atlantic Ocean. *Marine Geology, 131*, 233–249.

Reynolds, B. C., Frank, M., & O'Nions, R. K. (1999). Nd- and Pb-isotope time series from Atlantic ferromanganese crusts: Implications for changes in provenance and paleocirculation over the last 8 Myr. *Earth and Planetary Science Letters, 173*, 381–396.

Richter, F., Rowley, D. B., & DePaolon, D. J. (1992). Sr isotope evolution of seawater: The role of tectonics. *Earth and Planetary Science Letters, 109*, 11–23.

Rutberg, R. L., Hemming, S. R., & Goldstein, S. L. (2000). Reduced North Atlantic deep water flux to the glacial southern ocean inferred from neodymium isotope ratios. *Nature, 405*, 935–938.

Schaule, B. K., & Patterson, C. C. (1981). Lead concentrations in the North Pacific: Evidence for global anthropogenic perturbations. *Earth and Planetary Science Letters, 54*, 97–116.

Segl, M., Mangini, A., Bonani, G., Hofmann, H. J., Nessi, M., Suter, M., Wölfli, W., Friedrich, G., Plüger, W. L., Wiechowski, A., & Beer, J. (1984). [10]Be dating of manganese crust from Central North Pacific and implications for oceanic paleocirculation. *Nature, 309*, 540–543.

Sharma, M., Papanastassoiu, D. A., & Wasserburg, G. J. (1997). The concentration and isotopic composition of osmium in the oceans. *Geochimica et Cosmochimica Acta, 61*, 3287–3299.

Sharma, P., & Somayayulu, B. L. K. (1982). [10]Be dating of large manganese nodules from world oceans. *Earth and Planetary Science Letters, 59*, 235–244.

Sharma, M., Wasserburg, G. J., Hofmann, A. W., & Butterfield, D. A. (2000a). Osmium isotopes in hydrothermal fluids from the Juan de Fuca Ridge. *Earth and Planetary Science Letters, 179*, 139–152.

Sharma, M., Wasserburg, G. J., Hofmann, A. W., & Chakrapani, G. J. (2000b). Himalayan uplift and osmium isotopes in ocean and rivers. *Geochimica et Cosmochimica Acta, 63*, 4005–4012.

Shen, G. T., & Boyle, E. A. (1988). Thermocline ventilation of anthropogenic lead in the western North Atlantic. *Journal of Geophysical Research, 93*, 15715–15732.

Shimizu, H. K., Tachikawa, K., Masuda, A., & Nozaki, Y. (1994). Cerium and Nd isotope ratios and REE patterns in seawater from the North Pacific Ocean. *Geochimica et Cosmochimica Acta, 58*, 323–333.

Sholkovitz, E. R. (1989). Artifacts associated with the chemical leaching of sediments for rare-earth elements. *Chemical Geology, 77*, 47–51.

Sholkovitz, E. R., Elderfield, H., Szymczak, R., & Casey, K. (1999). Island weathering: River sources of rare earth elements to the western Pacific Ocean. *Marine Chemistry, 68*, 39–57.

Show, H. F., & Wasserburg, G. J. (1988). Sm-Nd in marine carbonates and phosphates: Implications for Nd in seawater and crustal ages. *Geochimica et Cosmochimica Acta, 49*, 503–518.

Spivack, A. J., & Wasserburg, G. J. (1988). Neodymium isotopic composition of the Mediterranean outflow and the eastern North Atlantic. *Geochimica et Cosmochimica Acta, 48*, 2267–2773.

Staudigel, H., Doyle, P., & Zindler, A. (1985). Sr and Nd systematics of fish teeth. *Earth and Planetary Science Letters, 76*, 45–56.

Stille, P. (1992). Nd-Sr evidence for dramatic changes of paleocurrents in the Atlantic Ocean during the past 80 my. *Geology, 20*, 387–390.

Stille, P., & Fisher, H. (1990). Secular variation in the isotopic of Nd in Tethys seawater. *Geochimica et Cosmochimica Acta, 54*, 3139–3145.

Stille, P., Steinmann, M., & Riggs, S. R. (1996). Nd isotope evidence for the evolution of the paleocurrents in the Atlantic and Tethys oceans during the past 180 Ma. *Earth and Planetary Science Letters, 144*, 9–19.

Stordal, M. C., & Wasserburg, G. J. (1986). Neodymium isotopic study of Baffin Bay water: Source of REE from very old terranes. *Earth and Planetary Science letters, 77*, 259–272.

Tachikawa, K., Jeandel, C., & Dupré, B. (1997). Distribution of rare earth elements and neodymium isotopes in settling particulate material of the tropical Atlantic Ocean (EUMELI site). *Deep Sea Research, 44*, 1769–1792.

Tachikawa, K., Jeandel, C., & Roy-Barman, M. (1999). A new approach to the Nd residence time in the ocean: The role of atmospheric inputs. *Earth and Planetary Science Letters, 170*, 433–446.

Thirlwall, M. F. (2000). Inter-laboratory and other errors in Pb isotope analyses investigated using a [207]Pb-[204]Pb double spike. *Chemical Geology, 163*, 299–322.

Thomas, D. J. (2004). Evidence for deep-water production in the North Pacific Ocean during the early cenozoic warm interval. *Nature, 430*, 65–68.

Thomas, D. J., Bralower, T. J., & Jones, C. E. (2003). Neodymium isotopic reconstruction of late paleocene-early eocene thermohaline circulation. *Earth and Planetary Science Letters, 209*, 209–322.

Turner, D. R., Whitfiels, M., & Dickson, A. G. (1981). Equilibrium speciation of dissolved components in freshwater and seawater at 25°C and 1 atm pressure. *Geochimica et Cosmochimica Acta, 45*, 855–881.

Vance, D., & Burton, K. W. (1999). Neodymium isotopes in planktonic foraminifera: A record of the response of continental weathering and ocean circulation rates to climate change. *Earth and Planetary Science Letters, 173*, 365–379.

Vance, D., Scrivner, A. E., Beney, P., Staubwasser, M., Henderson, G. M., & Slowey, N. C. (2004). The use of foraminifera as a record of the past neodymium isotope composition of seawater. *Paleoceanography, 19,* PA2009, doi: 10.1029/2003 PA 00957.

Veizer, J., Ala, D., Azmy, K., Bruckschen, P., Buhl, D., Bruhn, F., Carden, G. A. F., Diener, A., Ebneth, S., Godderis, Y., Jasper, T., Korte, C., Pawellek, F., Podlaha, O. G., & Strauss, H. (1999). ^{87}Sr/^{86}Sr, δ^{13}C, and δ^{18}O evolution of Phanerozoic seawater. *Chemical Geology, 161,* 59–88.

Véron, A. J., Church, T. M., & Flegal, A. R. (1998). Lead isotopes in the western North Atlantic: Transient tracers of pollutant lead inputs. *Environmental research, 78,* 104–111.

Vennemann, T. W., & Hegner, E. (1998). Oxygen, strintium and neodymium isotope composition of fossil shark teeth as a proxy for the paleoceanography and paleoclimatology of the miocene northern Alpine Paratethys. *Paleogeography, Paleoclimatologie, Palaoecology, 142,* 107–121.

White, W., Patchett, P. J., & Ben Othman, D. (1986). Hf isotope ratios of marine sediments and Mn nodules. *Earth and Planetary Science Letters, 79,* 46–54.

Winter, B. L., Johnson, C. M., & Clark, D. L. (1997). Strontium, neodymium, and lead isotope variations of authigenic and silicate sediment components from the late cenozoic Arctic ocean: Implications for sediment provenance and the source of trace metals in seawater. *Geochimica et Cosmochimica Acta, 61,* 4181–4200.

Woodhouse, O. B., Ravizza, G. E., Falkner, K. K., Statham, P. J., & Peucker-Ehrenbrink, B. (1999). Osmium in sea water: Concentration and isotopic composition vertical profiles in the eastern Pacific Ocean. *Earth and Planetary Science Letters, 173,* 223–233.

CHAPTER SIXTEEN

PALEOFLUX AND PALEOCIRCULATION FROM SEDIMENT ^{230}Th AND ^{231}Pa/^{230}Th

Roger François

Contents

1. Introduction	681
2. Factors Controlling the Distribution of ^{230}Th and ^{231}Pa in the Ocean	684
2.1. The removal term: particle scavenging and the distribution of ^{230}Th and ^{231}Pa in the water column	684
2.2. The transport term: boundary scavenging and meridional overturning	686
3. Paleoceanographic Applications	698
3.1. Reconstructing past changes in the preserved vertical rain rate of particles and particle constituents by ^{230}Th-normalization	698
3.2. Reconstructing past changes in the vertical rain rate of particles from sediment ^{231}Pa/^{230}Th	704
3.3. Reconstructing past changes in the rate of the atlantic meridional overturning circulation from sediment ^{231}Pa/^{230}Th	707
4. Conclusions	712
References	712

1. INTRODUCTION

The distribution of trace elements and isotopes in the ocean is often difficult to explain precisely because it is controlled by the complex interactions between localized sources, sinks, and transport by mixing and ocean currents, all of which often vary considerably in space and time. Understanding the distribution of two of these trace isotopes, ^{230}Th (half-life: 75.69 ± 0.23 ka; Cheng et al., 2000) and ^{231}Pa (half-life: 32.71 ± 0.11 ka; Robert, Miranda, & Muxard, 1969), is significantly simplified, however, because their source is constant and well established (Figure 1). Both are produced from the decay of uranium isotopes dissolved in seawater (^{234}U and ^{235}U, respectively). All dissolved uranium isotopes are added to the oceans by rivers and ground waters and reside in the ocean for $\sim$400 ka before removal in anoxic sediments, biogenic carbonates and hydrothermal systems (Dunk, Mills, & Jenkins, 2002). With such a long residence time, uranium concentration in seawater

Developments in Marine Geology, Volume 1
ISSN 1572-5480, DOI 10.1016/S1572-5480(07)01021-4

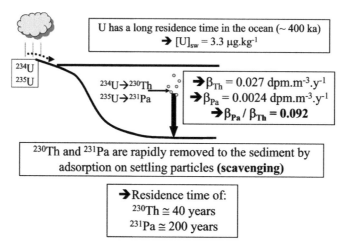

Figure 1 ^{234}U and ^{235}U are added to the oceans by rivers and ground waters and reside in the ocean for ~400 ka before removal. Since the residence time of uranium is much larger than the ocean mixing time (~1,000 yr), the uranium concentration and isotopic composition of seawater is uniform. The two uranium isotopes decay to ^{230}Th and ^{231}Pa at rates (β_{Th} and β_{Pa}) which are also uniform and well constrained. The two isotopes are thus produced at a uniform rate ratio (β_{Pa}/β_{Th}) of 0.092. After production, both ^{230}Th and ^{231}Pa are quickly adsorbed on settling particles and removed to the underlying sediments, a process called "particle scavenging". As a result, the residence time of ^{230}Th and ^{231}Pa in the water column is short. Because Th has a higher affinity for marine particles than Pa, the residence time of ^{230}Th in deep water (~30–40 yr) is shorter than that of ^{231}Pa (~200 yr).

is nearly uniform and proportional to salinity. At 35‰ salinity, the concentration of the most abundant uranium isotope (^{238}U) has been estimated at 3.23 ppb (Delanghe et al., 2002) and 3.33 ppb (Robinson, Henderson, Hall, & Matthews, 2004), corresponding to an activity ranging from 2.41 to 2.49 dpm kg^{-1}. The uranium isotopic composition of seawater is also uniform. The ^{234}U/^{238}U activity ratio has been measured at 1.1496 (Delanghe et al., 2002) and 1.1466 (Robinson et al., 2004). Using these values and natural abundance ratio (0.04604) for ^{235}U/^{238}U, the range of ^{234}U and ^{235}U activities in seawater at 35‰ salinity (density at $1°C = 1,028$ kg·m^{-3}) is thus 2,850–2,935 and 114–118 dpm·m^{-3}, respectively, resulting in constant production rates of ^{230}Th (β_{Th}) and ^{231}Pa (β_{Pa}):

$$\beta_{Th} = \lambda_{230}A_{234} = 9.20 \times 10^{-6} \times \sim 2,900 = \sim 0.0267 \text{ dpm m}^{-3} \text{ yr}^{-1}$$

$$\beta_{Pa} = \lambda_{231}A_{235} = 2.12 \times 10^{-5} \times \sim 116 = \sim 0.00246 \text{ dpm m}^{-3} \text{ yr}^{-1}$$

where λ_{230} is the ^{230}Th decay constant (9.20×10^{-6} yr^{-1}), λ_{231} the ^{231}Pa decay constant (2.12×10^{-5} yr^{-1}), A_{234} and A_{235} are the activities of ^{234}U and ^{235}U (dpm m^{-3}), respectively.

The two radioisotopes are thus produced at a constant production rate ratio (β_{Pa}/β_{Th}) of ~0.092. Thorium is present in seawater at the +IV oxidation state (Santschi et al., 2006), while protactinium is found at the +V oxidation state. After production, both ^{230}Th and ^{231}Pa are quickly adsorbed on settling particles and removed to the underlying sediments, a process called "particle scavenging" (it has been suggested Pa(V) may have to be reduced to Pa(IV) before scavenging). The rapid removal of these two isotopes from seawater is reflected by their very low seawater activities compared with that of their parent uranium isotopes (Moore & Sackett, 1964). In the absence of removal by scavenging or any other processes besides radioactive decay, the activity of the daughters (^{230}Th; ^{231}Pa) in seawater would be equal to the activity of the parents (^{234}U; ^{235}U). Instead, seawater activities of the former are $\sim$1,000 times lower than the activities of the latter, indicating rapid removal by a process other than radioactive decay. This process is clearly particle scavenging since the missing daughters are found in the underlying sediments.

A simple mass balance equation can be used to calculate the residence time of the daughters in the water column. The rate at which atoms of the daughters are produced (i.e., the activity of the parent isotope; A_U) must be equal to the rate at which they are removed by radioactivity decay (i.e., the activity of the daughters; $A_{Th;Pa}$) and by scavenging. The latter is taken to be proportional to the concentration of the daughters in seawater ($\kappa N_{Th;Pa}$). Hence:

$$A_U = A_{Th;Pa} + \kappa N_{Th;Pa} \tag{1}$$

where κ is the scavenging rate constant, which is equivalent to the inverse of the residence time of the daughters in seawater ($1/\tau_{scav}$), and $N_{Th;Pa}$ is the number of atoms of the daughters per unit volume of seawater.

Multiplying by the decay constant of the daughter ($\lambda_{Th;Pa}$) to express the scavenging rate in activity units, we obtain:

$$\lambda_{Th;Pa} A_U = \lambda_{Th;Pa} A_{Th;Pa} + \frac{A_{Th;Pa}}{\tau_{scav}} \tag{2}$$

$$\tau_{scav} \cong \frac{1}{\lambda_{Th;Pa}} \frac{A_{Th;Pa}}{A_U} \tag{3}$$

Seawater ^{230}Th activities vary from $<0.1\,\text{dpm}\,\text{m}^{-3}$ in surface waters to $\sim$1.5 dpm m^{-3} in the deep Pacific Ocean (Nozaki, Yang, & Yamada, 1987; Roy-Barman, Chen, & Wasserburg, 1996), resulting in residence times ranging from a few months in surface waters to $\sim$50 yr in Pacific bottom waters. Seawater ^{231}Pa activities vary from $<0.1\,\text{dpm}\,\text{m}^{-3}$ in surface waters to $\sim$0.6 dpm m^{-3} at mid depth in the North Pacific (Nozaki & Nakanishi, 1985) resulting in residence times ranging from a few months to $\sim$200 yr. Thus, as ^{230}Th is more rapidly removed from the water column than ^{231}Pa, it is as if nature were providing us with the perfect "dye experiment", adding uniformly in space and time two similar isotopes with slightly different chemical properties. Consequently, the difference in the distribution of these two isotopes in marine sediments can be more clearly linked to processes associated with their transport and removal, which gave rise to three important tools for paleoceanographic studies covering the last few hundred thousand years of the Quaternary: (1) normalization to scavenged ^{230}Th concentration in marine sediment to estimate past variations in the vertical rain rate of sedimentary constituents, (2) variations in scavenged ^{231}Pa/^{230}Th in sediment to inform us on past changes in particle rain rates and (3) variations in scavenged ^{231}Pa/^{230}Th in sediments to constrain past changes in the rate of the Atlantic meridional overturning circulation (MOC). In this chapter, we review how particle scavenging and ocean circulation control the vertical profiles and lateral transport of ^{230}Th and ^{231}Pa in the water column. We then review the principles, underlying assumptions, advantages and limitations involved in each of the three paleoceanographic applications outlined above, while providing a few examples of paleoceanographic records that illustrate the merit of these approaches.

2. FACTORS CONTROLLING THE DISTRIBUTION OF ^{230}Th AND ^{231}Pa IN THE OCEAN

2.1. The Removal Term: Particle Scavenging and the Distribution of ^{230}Th and ^{231}Pa in the Water Column

The first vertical profiles of particulate and total ^{230}Th measured in the water column clearly indicated a gradual increase in concentration with depth (Krishnaswami, Lal, Somayajulu, Weiss, & Craig, 1976; Nozaki, Horibe, & Tsubota, 1981; Bacon & Anderson, 1982; Anderson, Bacon, & Brewer, 1983a, 1983b; Nozaki & Nakanishi, 1985). This observation was taken to imply that scavenging must occur as a result of reversible exchanges between settling particles and the dissolved pool of ^{230}Th in the water column, and prompted the development of simple scavenging models (Bacon & Anderson, 1982; Bacon, Huh, Fleer, & Deuser, 1985; Nozaki et al., 1987). In the simplest version of these models (Nozaki et al., 1981; Bacon & Anderson, 1982), dissolved ^{230}Th $\left(^{230}\text{Th}_\text{d}\right)$ is produced at a known rate $\left(\beta_\text{Th}\right)$ from the decay of ^{234}U and adsorbed reversibly on particles, producing particulate ^{230}Th $\left(^{230}\text{Th}_\text{p}\right)$ that sinks to the seafloor at a mean velocity S $\left(\text{m\,yr}^{-1}\right)$:

$$^{234}\text{U} \xrightarrow{\beta_\text{Th}} {}^{230}\text{Th}_\text{d} \underset{k_{-1}}{\overset{k_1}{\rightleftarrows}} {}^{230}\text{Th}_\text{p} \downarrow_S$$

where k_1 and k_{-1} are the adsorption and desorption rate constants. If we assume that adsorption and desorption follow first order kinetics and neglect radioactive decay, which is justified as the residence time of ^{230}Th in the water column is much shorter than its half-life of 75,690 yr, we can write the following steady-state equations:

$$\frac{\partial[\text{Th}]_\text{d}}{\partial t} = \beta_\text{Th} + k_{-1}[^{\text{xs}}\text{Th}]_\text{p} - k_1[\text{Th}]_\text{d} = 0 \tag{4}$$

$$\frac{\partial[\text{Th}]_\text{p}}{\partial t} = k_1[\text{Th}]_\text{d} - k_{-1}[^{\text{xs}}\text{Th}]_\text{p} - \frac{S\partial[^{\text{xs}}\text{Th}]_\text{p}}{\partial Z} = 0 \tag{5}$$

For total ^{230}Th concentration $([\text{Th}]_\text{t} = [\text{Th}]_\text{d} + [^{\text{xs}}\text{Th}]_\text{p})$, we can write:

$$\frac{\partial[\text{Th}]_\text{t}}{\partial t} = \beta_\text{Th} - \frac{S\partial(K[\text{Th}]_\text{t})}{\partial Z} = 0 \tag{6}$$

where $[\text{Th}]_\text{t}$, $[\text{Th}]_\text{d}$ and $[^{\text{xs}}\text{Th}]_\text{p}$ = seawater total, dissolved and "excess" particulate ^{230}Th concentrations; $K = [^{\text{xs}}\text{Th}]_\text{p}/[\text{Th}]_\text{t}$, and Z = water depth. Particulate ^{230}Th consists of a detrital fraction locked in the mineral lattices of lithogenic constituents and a scavenged (or "excess") fraction adsorbed from seawater. $[^{xs}Th]_p$ refers to the latter fraction only.

These equations state that production of dissolved ^{230}Th by ^{234}U decay and desorption from particles must be balanced by its removal by adsorption on particles (equation 4), and production of particulate ^{230}Th by adsorption of dissolved ^{230}Th on particles must be balanced by its removal by desorption and sinking (equation 5). Further assuming constant S, k_1 and k_{-1} with depth, this model predicts a simple linear increase in concentrations versus depth with a slope inversely proportional to the sinking velocity of the particles (S):

$$[Th]_p = \left[\frac{\beta_{Th}}{S}\right] Z \tag{7}$$

$$[Th]_d = \left[\frac{\beta_{Th}}{k_1}\right] + \left[\frac{k_{-1}\beta_{Th}}{k_1 S}\right] Z \tag{8}$$

In this simpler model, the processes of particle aggregation, disaggregation and remineralization are implicit. ^{230}Th is initially adsorbed on small particles as a result of their comparatively higher surface areas. These small particles have very slow sinking rates and are not effectively removed from the water column by direct gravitational settling. Instead, they are packaged into larger, faster sinking particles by processes mostly mediated by the biota through formation of fecal pellets or marine snow. As these larger aggregates sink rapidly to the seafloor, they undergo disaggregation, releasing fine particles back to suspension in the water column. These fine particles remain in suspension until re-capture by large sinking particles which bring then further down through successive cycles of aggregation and disaggregation before reaching the seafloor (Bacon et al., 1985). This process results in a mean sinking velocity (S), which is typically ranging between 300 and 1,000 m yr^{-1}. Considering the similarity in its mode of addition and removal, one could expect that such model would be equally valid to describe the seawater profiles of ^{231}Pa.

Although more detailed models, explicitly considering the aggregation/disaggregation cycles between fine and large particles, have been proposed to describe scavenging in seawater (Bacon et al., 1985; Nozaki et al., 1987; Clegg & Whitfield, 1990), the simpler model described above provides a good first-order description of the behavior of particle-reactive elements in regions where the effects of water movement are not important. Seawater profiles of ^{230}Th that are nearly linear with depth have been found in some regions of the ocean (Figure 2). However, a rapidly expanding database, made possible by the recent development of mass spectrometric techniques to analyze ^{230}Th and ^{231}Pa in seawater (Choi et al., 2001; Shen et al., 2002, 2003), indicates that linear ^{230}Th profiles are the exception rather than the rule and most ^{230}Th seawater profiles measured to date display significant deviations from the linear profiles predicted by these scavenging models. In the case of ^{231}Pa, no linear profiles have thus far been documented. The main reason for the failure of these simple models

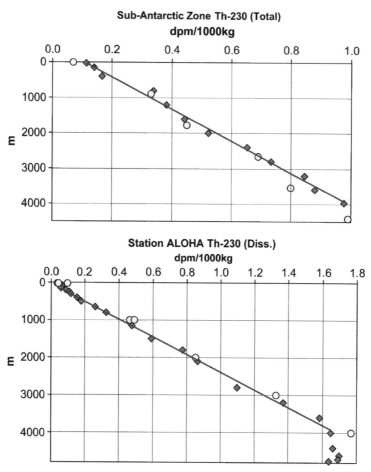

Figure 2 [230]Th seawater profiles close to linearity are found in the subantarctic Ocean (diamonds: 47°S–140°E, François et al., unpublished; circles: 53°S–175°W, Chase et al., 2003) and station ALOHA (22°45′N–158°W) in the North Pacific (circles: Roy-Barman et al., 1996; diamonds: François & Bacon, unpublished).

at describing accurately the vertical distribution of [230]Th and [231]Pa in seawater in most oceanic regions is that the effect of water mass circulation can rarely be neglected.

2.2. The Transport Term: Boundary Scavenging and Meridional Overturning

2.2.1. Lateral transport by turbulent diffusion: boundary scavenging

According to Equation (8), higher particle sinking rates (S) or higher adsorption rate constant (k_1) must decrease the concentration of dissolved [230]Th and [231]Pa at a given depth. Geographic variations in these two parameters, reflecting changes in particle flux and particle composition, thus generate lateral concentration gradients and induce a lateral transport along isopycnals by eddy diffusion from zones of high

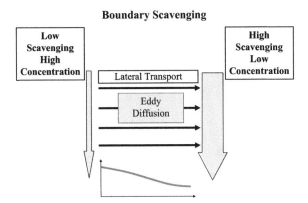

Figure 3 Geographic variations in scavenging intensity creates lateral gradients in the concentration of scavenged elements which drives lateral transport by eddy diffusion along isopycnals from regions of low scavenging intensity, where dissolved concentrations are relatively high to regions of high scavenging intensity, where concentrations are lower. The latter are found primarily at ocean margins and other regions of high marine productivity. This mode of lateral transport of scavenged elements is called "boundary scavenging".

concentration, where particle flux and/or affinity is smaller, to zones of low concentration where particle flux and/or affinity is larger (Figure 3). This effect is clearly illustrated in the Pacific Ocean when comparing ²³⁰Th concentration profiles from the central North Pacific (station ALOHA) and the eastern equatorial Pacific (Figure 4). This mode of lateral transport by eddy diffusion is called "boundary scavenging" (Spencer, Bacon, & Brewer, 1981; Bacon, 1988), reflecting the fact that regions of higher scavenging intensity are often found at ocean margins. However, this same label is also used to describe transport by eddy diffusion toward open ocean regions with high particle scavenging (e.g., equatorial upwelling zones). [It has been suggested that enhanced scavenging by particles resuspended and slowly transported down continental slopes could also contribute to boundary scavenging (Lao, Anderson, Broecker, Hofmann, & Wolfli, 1993), but the significance of this potential mechanism remains to be established.]

The extent of lateral transport by eddy diffusion depends on the residence time of the scavenged elements or isotopes in the water column (Bacon, 1988; Anderson et al., 1990). Elements and isotopes with shorter residence times can only be transported laterally over short distances from the site where they were added to seawater prior to removal by scavenging. The very short residence time of ²³⁰Th in the water column limits severely the extent to which this isotope is affected by boundary scavenging (Bacon, 1984). On the other hand, with its significantly longer residence times, ²³¹Pa is more effectively transported to and preferentially removed in regions of higher scavenging intensity where particle flux or particle composition enhances adsorption of ²³¹Pa from seawater (Figure 5). A larger fraction of the ²³¹Pa produced in the water column is thus transported laterally by eddy diffusion to regions with higher scavenging intensity than for ²³⁰Th. As a result, the ²³¹Pa/²³⁰Th of settling particles and sediments is

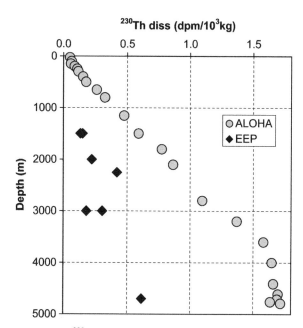

Figure 4 Seawater dissolved ^{230}Th profiles at station ALOHA in the central North Pacific (22°45′N–158°W; François & Bacon, unpublished) and two stations from the eastern equatorial Pacific (st 1110, 5°6′N, 81°32′W; st 1114, 11°9′N, 87°35′W; Bacon & Anderson, 1982).

higher than the production rate ratio (>0.092) in regions of higher scavenging intensity and lower (<0.092) in regions of lower scavenging intensity (Bacon, 1988; Anderson et al., 1990).

Scavenging intensity increases with particle flux. Particle fluxes are higher near continental margins and in upwelling regions, where we expect to find ^{231}Pa/^{230}Th>0.092. The distribution of ^{231}Pa/^{230}Th measured in the surface sediments of the Pacific Ocean clearly meets this expectation (Yang, Nozaki, Sakai, & Masuda, 1986; Lao et al., 1992; Walter, Rutgers van der Loeff, & François, 1999; Pichat et al., 2004; Figure 6). However, changes in particle composition also play an important role in controlling boundary scavenging. In particular, experimental (Guo, Chen, & Gueguen, 2002) and field (Anderson et al., 1983b; Shimmield & Price, 1988; Walter, van der Loeff, & Hoeltzen, 1997; Chase, Anderson, Fleisher, & Kubik, 2002) observations clearly indicate that ^{231}Pa has a much higher affinity for biogenic silica and MnO_2 than for the other major constituents of marine particles. On the other hand, Th appears to have a higher affinity for carbonate (Chase et al., 2002). Two regions with similar particle flux but different opal or MnO_2 concentration must therefore also generate boundary scavenging and lateral transport of ^{231}Pa from the low opal or MnO_2 to the high opal or MnO_2 region.

The effect of particle composition is best expressed by the "fractionation factor" (F(Th/Pa); Anderson et al., 1983b), which is the ratio of the distribution coefficients for Th ($^{Th}K_d$) and Pa ($^{Pa}K_d$). Distribution coefficients refer to the concentration of

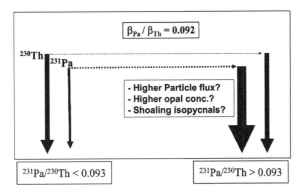

Figure 5 ^{230}Th and ^{231}Pa are produced uniformly in seawater at a constant rate ratio of 0.093. Both are removed rapidly to the sediment by adsorption on settling particles. The very short residence time of ^{230}Th limits its lateral transport before removal and its flux to the seafloor always approximate its production rate in the overlying water column. In contrast, ^{231}Pa has a longer residence time in the water column and can be more effectively transported toward areas where particle flux or composition results in higher scavenging intensity. This differential transport between the two radioisotopes creates variations in the ^{231}Pa/^{231}Th of the underlying sediment.

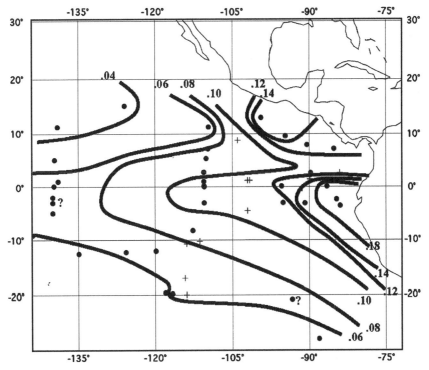

Figure 6 Surface ^{231}Pa/^{230}Th in the Eastern Equatorial Pacific. Data are tabulated in Pichat et al. (2004). Crosses are samples taken in the vicinity of mid ocean ridges with anomalous high values suspected to be influenced by ^{231}Pa scavenging by manganese oxides.

scavenged ^{230}Th and ^{231}Pa per mass of marine particles divided by the concentration of dissolved ^{230}Th and ^{231}Pa in seawater:

$$F\left(\frac{\text{Th}}{\text{Pa}}\right) = \frac{^{\text{Th}}K_\text{d}}{^{\text{Pa}}K_\text{d}} \tag{9}$$

$$^{\text{Th}}K_\text{d} = \frac{[^{\text{xs}}\text{Th}]_\text{p}}{[p][\text{Th}]_\text{d}} \tag{10}$$

$$^{\text{Pa}}K_\text{d} = \frac{[^{\text{xs}}\text{Pa}]_\text{p}}{[p][\text{Pa}]_\text{d}} \tag{11}$$

where $[^{\text{xs}}\text{Th}]_\text{p}$ and $[\text{Th}]_\text{d}$ are as defined above, $[^{\text{xs}}\text{Pa}]_\text{p}$ and $[\text{Pa}]_\text{d}$ are the equivalent for ^{231}Pa, and $[p]$ is the concentration of particles in seawater (g m^{-3}). $^{\text{Th}}K_\text{d}$ typically ranges between 10^6 and 10^7, while $^{\text{Pa}}K_\text{d}$ is somewhat lower $(10^5–10^6)$ reflecting the higher particle reactivity of Th for most constituents of marine particles (see Henderson & Anderson, 2003 for a recent review).

$F(\text{Th/Pa})$ varies with particle composition (Chase et al., 2002). For particles with high opal or MnO_2 concentrations, $F(\text{Th/Pa})$ approaches 1. These minerals have similar affinity for the two radioisotopes and fractionate them very little (i.e., Th/Pa measured on particles is very close to $[\text{Th}]_\text{d}/[\text{Pa}]_\text{d}$ in seawater). On the other hand, marine particles dominated by carbonate minerals have $F(\text{Th/Pa})$ that can exceed 10. These particles have thus a higher affinity for Th and their Th/Pa exceeds $[\text{Th}]_\text{d}/[\text{Pa}]_\text{d}$ in seawater. Opal and MnO_2 are often found at higher concentration in particles settling at ocean margins, reflecting the dominance of diatoms and the diagenetic remobilization of MnO_2 from sediments (Anderson et al., 1983b). Therefore, both enhanced particle flux and differences in particle composition may contribute to the boundary scavenging of ^{231}Pa. There is a stronger correlation between ^{231}Pa/^{230}Th and opal/carbonate ratio in particles collected with sediment traps than with particle flux measured with these same sediment traps, suggesting that particle composition plays a predominant role in controlling the fractionation of ^{231}Pa and ^{230}Th (Chase et al., 2002).

While the effect of boundary scavenging is clearly expressed in the surface sediments of the Pacific Ocean (Figure 6), it appears much more muted in the Atlantic Ocean (Yu, François, & Bacon, 1996; Walter et al., 1999). It has been proposed that this contrast between the two oceans is a direct consequence of the abyssal circulation (Yu, François, Bacon, & Fleer, 2001). In the Atlantic Ocean, deep waters have comparatively short residence times because of the well developed deep MOC and the continuous "flushing" of this basin by the North Atlantic Deep Water (NADW) (Talley, 2003). The mean residence time of deep water in the Atlantic, which was estimated at $\sim$100 yr (Broecker, 1979) or $\sim$275 yr (Stuiver, Quay, & Ostlund, 1983), is nearly equivalent to the time required for basin-wide lateral mixing (Anderson et al., 1990). As a result, the lateral gradients of dissolved ^{231}Pa concentration that are needed to drive transport by eddy diffusion along isopycnals cannot be fully established in this ocean and boundary scavenging in not fully expressed.

2.2.2. Lateral transport by the meridional overturning circulation

In addition to eddy diffusion, lateral transport of ^{230}Th and ^{231}Pa can also occur as a result of the MOC of the ocean. In contrast to boundary scavenging, lateral transport by overturning moves the two radioisotopes from zones of low seawater concentrations toward zones of high seawater concentrations. Here, however, the contrast in radioisotope concentrations does not arise from differential scavenging intensity, but is created by ocean circulation (Figure 7). This effect is particularly well documented in the Atlantic Ocean where there is a well developed deep overturning which initiates the global "conveyor belt" circulation of the ocean (Talley, 2003).

^{230}Th profiles typically display two kinds of deviation from the linear increase with depth predicted by the scavenging models. In some profiles, deep water concentrations are lower than expected from the models (Vogler, Scholten, van der Loeff, & Mangini, 1998; Moran et al., 2001, 2002). In other profiles, we find ^{230}Th concentrations that are in excess to those expected from the model at mid-depth (Rutgers van der Loeff & Berger, 1993; Chase, Anderson, Fleisher, & Kubik, 2003). These deviations and their geographic distribution can be readily explained when taking into account the meridional overturning of the ocean (Figure 7). Shallow waters entering the Nordic Seas to produce new deep water have low ^{230}Th and ^{231}Pa concentrations. Deep winter convective mixing thus results

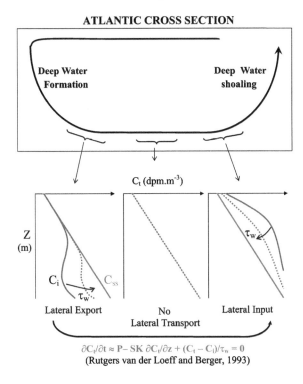

ATLANTIC CROSS SECTION

$$\partial C_i / \partial t \approx P - SK\ \partial C_i / \partial z + (C_t - C_i)/\tau_w = 0$$
(Rutgers van der Loeff and Berger, 1993)

Figure 7 The effect of the meridional overturning circulation on dissolved ^{230}Th and ^{231}Pa vertical seawater profiles (see text for explanation).

in low and nearly constant concentration profiles in the water column of the Norwegian (Moran, Hoff, Buesseler, & Edwards, 1995) and Labrador Sea (Moran, Charette, Hoff, Edwards, & Landing, 1997). Concentrations are higher than predicted by the scavenging model at shallow depths and lower than predicted in deep waters. The concentration deficit in deep waters spreads southward with the Deep Western Boundary Current and the NADW. During transit to the southern ocean, the newly formed deep water is continuously subjected to the particle rain that originates from surface waters and which scavenges the ^{230}Th and ^{231}Pa produced in the water column. When particles reach the depth of the newly formed deep water where the ^{230}Th and ^{231}Pa seawater concentrations are below equilibrium concentrations dictated by scavenging, desorption from particles is enhanced and a fraction of the ^{230}Th and ^{231}Pa scavenged at shallower depths is released to the deep waters instead of being removed into the underlying sediments. Thus, the ^{230}Th and ^{231}Pa concentrations in deep waters gradually increase during transit to the southern ocean until the concentration at steady-state with respect to scavenging is regained, at which point the water column profiles have relaxed back to linearity (Figure 7). As deep water concentrations increase, more ^{230}Th and ^{231}Pa is exported "downstream" than is imported from "upstream" and therefore the fraction of ^{230}Th and ^{231}Pa produced in the water column which contributes to the buildup of deep water dissolved ^{230}Th and ^{231}Pa concentrations is laterally transported with the water and does not reach the underlying sediment. Once the profile linearity is regained, this mode of lateral transport ceases and all the ^{230}Th and ^{231}Pa produced in the water column is either removed into the underlying sediments or is transported laterally by eddy diffusion (boundary scavenging). When the deep waters reach the upwelling limb of the meridional overturning cell, such as in the Circum Polar Deep Water (CPDW) around Antarctica, shoaling isopycnals bring deep waters with relatively high ^{230}Th and ^{231}Pa concentrations to mid-depth where concentrations exceed the concentrations predicted by the scavenging models. When particles reach these shoaling waters with relatively high dissolved ^{230}Th and ^{231}Pa concentrations, adsorption on particles is enhanced, which tends to gradually trim the mid-water concentration excess (Figure 7). Zones of shoaling isopycnals are thus characterized by a net lateral addition of ^{230}Th and ^{231}Pa, which compensates for the lateral export from zones of deep water formation. The shapes of the ^{230}Th profiles measured in the Atlantic and the southern ocean are in agreement with this simple conceptual model. For instance, water column profiles from the Atlantic Ocean clearly show a ^{230}Th deficits when intercepting the Deep Western Boundary Current, while profiles measured south of the Antarctic Polar Front display the curvature predicted for regions of shoaling isopycnals (Figure 8).

The gradual relaxation of the profiles to linearity can be described for each isopycnal by adding a lateral transport term to the initial scavenging model, as first proposed by Rutgers van der Loeff and Berger (1993):

$$\frac{\partial [\text{Th}]_t}{\partial t} = \beta_{\text{Th}} - \frac{S \partial (K[\text{Th}]_t)}{\partial Z} + \frac{[\text{Th}]_t - {}^i[\text{Th}]_t}{\tau_{\text{w}}} = 0 \qquad (12)$$

where ${}^i[\text{Th}]_t$ and $[\text{Th}]_t$ are total ^{230}Th (or ^{231}Pa) concentration measured at two locations on the same isopycnal and τ_{w} is the "transit time" of water between these

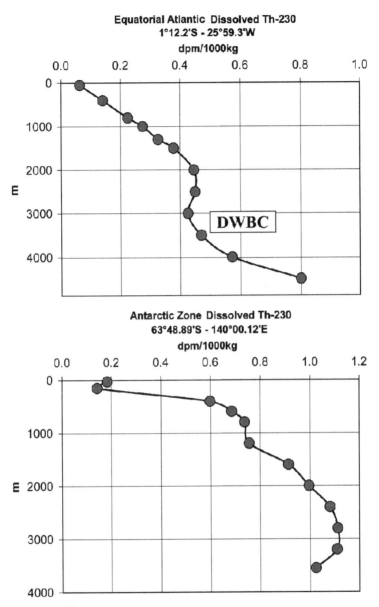

Figure 8 Dissolved ^{230}Th seawater profile from the equatorial Atlantic and the southern ocean (François et al., unpublished) showing deviations from linearity consistent with the conceptual model reported in Figure 7.

two sites. In deep waters, K ($[^{xs}Th]_p/[Th]_t$) is nearly constant. Integrating equation (12) thus gives:

$$[Th]_t \cong (\beta_{Th}\tau_w + {}^i[Th]_t)(1 - e^{-Z/\tau_w SK}) \qquad (13)$$

$$[Th]_t \cong (\beta_{Th}\tau_w + {}^i[Th]_t)(1 - e^{-\tau_{scav}/\tau_w}) \qquad (14)$$

These equations predict that the radioisotope profiles relax back to linearity with an e-folding time equivalent to the residence time of the radioisotope, which depends on water depth Z ($\tau_{scav} = Z/SK$). Profile linearity is thus regained faster at shallower depths, where residence times are shorter, and [230]Th regains linearity faster than [231]Pa because of its shorter residence time in the water column. For instances, at 4,500 m depth, [230]Th virtually regains its steady-state linear profile within ~200 yr, while it takes ~1,000 yr for [231]Pa to reach the same stage. This explains why we never find linear profiles for [231]Pa, since it necessitates a water column that has been vertically stable for a period equivalent to the mixing time of the ocean.

This gradual relaxation of [230]Th and [231]Pa profiles is clearly illustrated when comparing profiles measured in different regions of the Atlantic Ocean (Figure 9). For instance, [230]Th concentration in the Norwegian Sea is nearly constant with depth as a result of intensive vertical mixing (Moran et al., 1997). At 46°N, 21°W, we find slightly lower concentrations in the upper water column and slightly higher concentrations in deep water. The relatively small concentration changes in deep waters suggest a rapid transit time between the two sites. As we move further south however, deep water [230]Th concentrations rapidly increase, indicating much longer transit times. The dissolved [230]Th seawater concentrations decrease toward the seafloor suggesting increased ventilation below 3,500 m. This is the depth coinciding with the core of the Deep Western Boundary Current that flows along the western margin of the North Atlantic. The clear eastward increase in [230]Th concentrations below 3,000 m (Figure 10) is consistent with the more rapid ventilation of the Northwestern Atlantic basin, and documents the propagation of the DWBC ventilation signal to the east and the gradual relaxation of

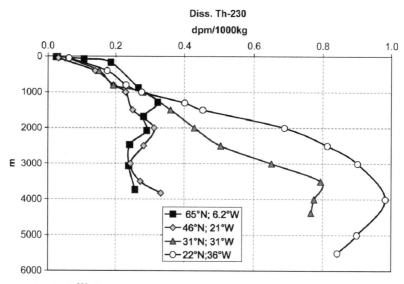

Figure 9   Dissolved [230]Th profiles from the Northeastern Atlantic Ocean (65°N profile from Moran et al., 1997; 46°–22° profiles from François & Bacon, unpublished).

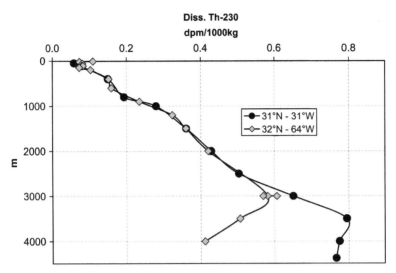

Figure 10 Dissolved ^{230}Th concentration profiles in the western (BATS; Buesseler & Edwards, unpublished) and eastern (François & Bacon, unpublished) North Atlantic.

the ^{230}Th profiles toward linearity. ^{231}Pa profiles show similar trends (François & Bacon, unpublished data). In the South Atlantic, the ^{230}Th profiles have nearly regain linearity at all depths (Figure 11a). This is not the case for ^{231}Pa, however, consistent with its much longer response time (Figure 11b). Linear profiles for ^{230}Th are found in the subantarctic zone of the southern ocean (Figure 2), but further south, as we enter the Circumpolar Deep Water, we find the expected convex curvature created by shoaling isopycnals (Figure 8). In the North Pacific, ^{231}Pa documents the deep overturning associated with Deep Pacific Water produced by the upwelling of the Lower Circumpolar Deep Water (Nozaki & Nakanishi, 1985; Figure 12). This overturning is however too slow to substantially distort the ^{230}Th profile, which is nearly linear over the entire North Pacific water column (Figure 2).

Because the residence time of ^{230}Th and ^{231}Pa increases with depth, the water column profiles regain linearity faster at shallower depths. A clear illustration of the difference in response time with depth is provided by comparing two stations in the central equatorial Atlantic. In this region, one interpretation of the general flow pattern at 1,800 m deduced from SOFAR floats (Schmitz, 1996) suggests that a fraction of the Deep Western Boundary Current diverges from the western margin at the equator and follows a west–east–west equatorial loop (Figure 13). This schematic flow pattern is supported by tritium measurements in the equatorial Atlantic (Jenkins, unpublished), which also reveals the presence of two distinct cores of water ventilation, one centered at 1,500 m and the other at 4,000 m (Figure 14). Dissolved ^{230}Th and ^{231}Pa were measured at two stations from this region (Figure 15). Based on the flow diagram deduced from the SOFAR floats (Figure 13), station 4 is "downstream" from station 3, which is consistent with the ^{230}Th and ^{231}Pa profiles

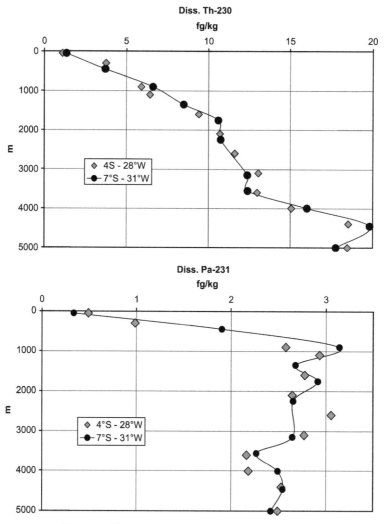

Figure 11   dissolved ^{230}Th and ^{231}Pa in the South Atlantic at 4 °S, 28 °W (grey diamonds) and 7 °S, 31 °W (black circles) (François & Bacon, unpublished).

measured at these two stations. The two dissolved ^{230}Th profiles measured in the upper 2,000 m overlap perfectly, showing no overturning signal associated with the shallow branch of the equatorial loop identified with tritiated water (Figure 14). At 1,500 m, the residence time of ^{230}Th is ~10 yr (~$[1/9.2 \times 10^{-6}] \times [0.3/2,930]$). Linearity should thus be regained within 40–50 yr, which is about the minimum transit time that would be needed for tritiated water added to surface waters in the North Atlantic in the early 1960s from nuclear bomb testing to reach the equatorial region in the late 1990s. In the deeper branch of the equatorial extension of the Deep Western Boundary Current, however, ^{230}Th residence times are about three

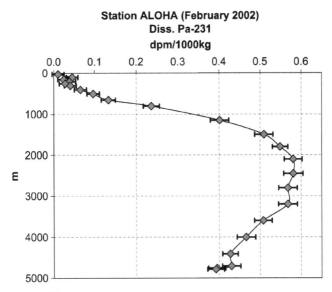

Figure 12 Dissolved ^{231}Pa profile at Hawaii Ocean Time-series (HOTS) Station ALOHA (22.75°N, 158°W) (François and Bacon, unpublished).

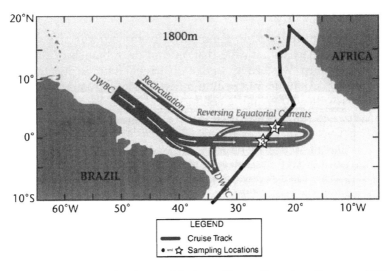

Figure 13 Apparent equatorial pathway of the Deep Western Boundary Current at 1,800 m based on SOFAR float trajectories (with permission from Schmitz, 1996). The position of stations 3 and 4 are indicated by stars.

times as long, so that dissolved ^{230}Th concentrations in these waters is still far removed from equilibrium with particles, as indicated by the clear deviation from linearity found within this depth interval (Figure 15). We can also clearly observe the gradual build up of ^{230}Th in this water mass from station 3 to 4, consistent

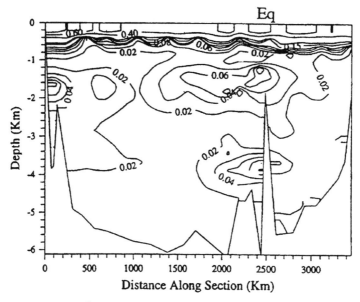

Figure 14 Distribution of ^{3}H on a SW–NE section across the equatorial Atlantic (Jenkins, unpublished).

with the flow diagram reported in Figure 13. On the other hand, ^{231}Pa exhibits deviations from linearity in both the shallow and deep branch of the equatorial extension of the Deep Western Boundary Current. This is due to the longer residence time of ^{231}Pa in the water column. Even at a depth of 1,500 m, the ^{231}Pa residence time is still ~50 yr and it would take ~200 yr for the ^{231}Pa profile to relax to linearity. As for ^{230}Th, we clearly see the gradual build-up of dissolved ^{231}Pa concentration as the water moves from station 3 to 4. One of the consequences for the difference in response time with depth is that deeper overturning exports ^{230}Th and ^{231}Pa from a wider region than shallow overturning. At shallower depth, profile linearity is more quickly regained, at which point horizontal transport as a result of overturning ceases.

 3. PALEOCEANOGRAPHIC APPLICATIONS

3.1. Reconstructing Past Changes in the Preserved Vertical Rain Rate of Particles and Particle Constituents by ^{230}Th-Normalization

Estimating past changes in particle flux from the sediment record is an essential part of paleoceanographic reconstructions. The main goal is often to gain information on past vertical flux of particles originating from surface waters, which can be interpreted in terms of export production of biogenic material or supply rate of lithogenic particles. Traditionally, fluxes of material to the seafloor were

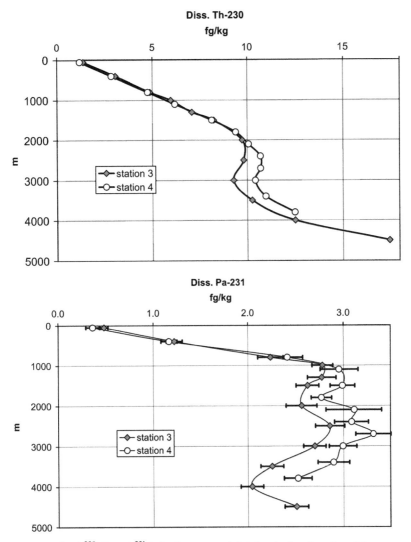

Figure 15 Dissolved ^{230}Th and ^{231}Pa in the equatorial Atlantic (stations 3 and 4 are reported in Figure 12) (François & Bacon, unpublished).

reconstructed from the burial flux of sedimentary constituents obtained by calculating mass accumulation rates based on core chronology. Core chronology provides the mean linear sedimentation rate (LSR$_{i,j}$) between successively dated sediment horizons i, j:

$$\text{LSR}_{i,j} = \frac{z_j - z_i}{t_j - t_i} \tag{15}$$

where $z_{i,j}$ are sediment depths and $t_{i,j}$ are the corresponding calendar ages. The latter is generally obtained from ^{14}C dating or δ^{18}O stratigraphy. Mass accumulation rates ($MAR_{i,j}$) are then calculated by multiplying $LSR_{i,j}$ by the mean dry bulk density of the sediment within this core section ($\rho_{i,j}$):

$$MAR_{i,j} = \rho_{i,j} LSR_{i,j} \tag{16}$$

Mass accumulation rates of individual sedimentary constituents ($^{\alpha}MAR_{i,j}$) are then obtained by multiplying total mass accumulation rates by the mean sediment concentration of the constituents of interest between horizons i and j ($[\alpha]_{i,j}$). This approach is still widely used, but before interpreting these results in terms of vertical rain rates of sedimentary constituents reaching the seafloor from the overlying surface, we must carefully consider several factors. Burial efficiency must evidently be taken into account. Preservation before burial for most biogenic constituents of marine sediments is low and is often difficult to quantify. We must also consider the possibility that sediments may be redistributed on the seafloor post- or syndepositionally by bottom currents from zones of sediment winnowing to zones of sediment focusing. Syndepositional redistribution is particularly difficult to identify, since it does not disturb sediment stratigraphy. Finally, the resolution of this approach is inherently limited by the need of having to take a difference between two dates ($\Delta t_{i,j} = t_j - t_i$). Documenting changes in $MAR_{i,j}$ at higher resolution requires increasingly smaller $\Delta t_{i,j}$. Since the error of individual ages is determined by the chronological method used, higher resolution automatically results in higher uncertainties on $\Delta t_{i,j}$ and $^{\alpha}MAR_{i,j}$. Here, it is important to recognize that while concentrations of sedimentary constituents ($[\alpha]_z$) can often be measured at resolutions that are much higher than for the core chronology, they cannot be translated into high resolution $^{\alpha}MAR_{i,j}$ changes by multiplying $[\alpha]_z$ by $MAR_{i,j}$. Yet, it is not uncommon to find in the literature changes in $^{\alpha}MAR_{i,j}$ reported at a resolution that appears to exceed the resolution of the core chronology (e.g., Lyle et al., 1988; Murray, Leinen, & Isern, 1993; Paytan, Kastner, & Chavez, 1996; Harris et al., 1996; Schneider et al., 1996). Calculating variations in $^{\alpha}MAR_{i,j}$ within chronological tie-points by multiplying $MAR_{i,j}$ by variations in concentration between depths i and j implies that we know that sediment accumulation rate is constant over this depth interval. If we have sediment consisting of a biogenic and lithogenic fraction, that would also imply that the flux of these two constituents have changed in perfect opposition to keep the overall sedimentation rate constant. However, we could equally assume that the flux of either of these sedimentary constituents stays constant between tie-points. None of these assumptions can be justified and all we can do is report average $^{\alpha}MAR_{i,j}$ between chronological tie-points by multiplying $MAR_{i,j}$ by the mean concentration of constituent α ($[\alpha]_{i,j}$) in the sediment section between z_i and z_j.

Some of these problems can be greatly alleviated by ^{230}Th normalization. The approach was first proposed by Bacon (1984) who pointed out that, because of its very short residence time in the water column and its limited lateral transport prior to removal by scavenging, the vertical flux of ^{230}Th scavenged to the seafloor (V_{Th}) must always approximate the rate at which it is produced in the overlying water column (P_{Th}). The latter can be easily quantified by multiplying the rate of ^{230}Th production per unit volume of seawater (β_{Th}) by the height of the overlying water column (Z; m):

$$V_{Th} \cong P_{Th} = Z(m) \times 0.0267 \ (dpm \ m^{-3} \ y^{-1}) \tag{17}$$

This approximately known flux of ^{230}Th can then be used as a reference against which the vertical flux of particle reaching the seafloor (V_p) can be estimated. A constant V_{Th} automatically generates a simple inverse relationship between particle flux and the concentration of scavenged ^{230}Th in settling particles reaching the seafloor ($[^{xs}Th]_{part}$):

$$[^{xs}Th]_{part} = \frac{V_{Th}}{V_p} \qquad (18)$$

We can thus estimate V_p by measuring $[^{xs}Th]_{part}$, which is obtained by measuring total ^{230}Th in the particles and subtracting the contribution from ^{230}Th present in the lithogenic constituents.

$$V_p \cong \frac{P_{Th}}{[^{xs}Th]_{part}} \qquad (19)$$

Before burial, a fraction of the biogenic material reaching the seafloor is remineralized. Because of the very high particle reactivity of thorium, the ^{230}Th scavenged to the seafloor by the remineralized fraction of settling particles remains in the sediment and the ^{230}Th concentration of the residual particles that eventually get buried as marine sediment increases. Downcore variations in excess ^{230}Th concentration decay-corrected since the time of deposition ($^{o}[^{xs}Th]_{sed}$) thus reflect changes in the vertical rain rate of particles reaching the seafloor minus the dissolution rate of labile constituents on the seafloor, a.k.a. the "preserved rain rate" ($^{pr}V_p$):

$$^{pr}V_p \cong \frac{P_{Th}}{^{o}[^{xs}Th]_{sed}} \qquad (20)$$

In addition to correcting for ^{230}Th present in the lithogenic fraction, evaluating $^{o}[^{xs}Th]_{sed}$ in marine sediments may require correcting for ingrowth from authigenic uranium, which is added to the sediment from seawater, either because of adsorption on oxides or precipitation in anoxic and suboxic sediments (François, Frank, Rutgers van der Loeff, & Bacon, 2004). This correction can sometimes be problematic as authigenic uranium can be remobilized after its initial emplacement (Zheng, Anderson, van Geen, & Fleisher, 2002).

How does this method improve on the shortcomings of the MAR approach?

It does not help with the preservation problem. If we want to estimate past changes in the vertical rain rate of material reaching the seafloor, whether we use MAR or ^{230}Th-normalization, we must find ways of quantifying %preservation or %dissolution of sediment constituents on the seafloor to estimate their vertical rain rates or export flux from surface waters. Increasingly reliable methods are being used to estimate carbonate dissolution (e.g., Lohmann, 1995; Broecker & Clark, 2001; Mekik, Loubere, & Archer, 2002; Loubere, Mekik, François, & Pichat, 2004; Marchitto, Lynch-Stieglitz, & Hemming, 2005) but no ready solutions for this problem are yet available for organic matter and biogenic silica.

^{230}Th-normalization provides considerable improvements over mass accumulation rate calculations, however, in providing a means to correct for sediment redistribution on the seafloor and by increasing the resolution of flux reconstructions (see François et al., 2004 for a recent review). If sediment winnowing and focusing occur in the same general area subjected to a similar particle flux at about the same depth, and if there is no fractionation between ^{230}Th and sediment during redistribution, then $^{o}[^{xs}$Th$]_{sed}$ and the rain rate information that it provides are conserved. In addition, the accumulation rate of ^{230}Th between two dated sediment horizons (MAR$_{i,j} \times {}^{o}[^{xs}Th]_{i,j}$) at a given site divided by the amount of ^{230}Th produced in the overlying water column during the corresponding time interval ($P_{Th} \times \Delta t_{i,j}$) also provides a good approximation of the ratio of vertically and horizontally transported sediments at this site. This ratio is called the focusing factor (Ψ; Suman & Bacon, 1989):

$$\Psi_{i,j} = \text{MAR}_{i,j} \frac{{}^{o}[^{xs}\text{Th}]_{i,j}}{P_{Th}(t_j - t_i)} \tag{21}$$

Quantification is less accurate when laterally transported sediments originates from shallower depths and have very different chemical composition (François, Bacon, & Suman, 1990; François et al., 2004), but lateral transport via bottom nepheloid layers typically produces small uncertainties. For instance, if we assume uniform lateral transport within the bottom 1,000 m of a 5,000 m water column that triples sediment accumulation rates to 30 g m^{-2} yr^{-1} from a $^{pr}V_p$ of 10 g m^{-2} yr^{-1}, it can be calculated that ^{230}Th-normalization overestimates preserved rain rate by 23% compared with 200% for chronology-based mass accumulation rates, and the focusing factor is underestimated by 19% (François et al., 2004).

The ^{230}Th-normalization method also affords flux reconstructions at much higher temporal resolution than can be achieved by the chronology-based mass accumulation rate method. This is because flux estimations by ^{230}Th-normalization rely on point measurements instead of age differences between sediment horizons. This eliminates the compromise between resolution and precision facing mass accumulation rate (and sediment focusing) estimates. The resolution of ^{230}Th-normalized flux measurements that can be achieved is only limited by bioturbation and its precision is dictated by the precision of $^{o}[^{xs}$Th$]$ measurements. The latter depends on water depth, sediment composition (fraction of lithogenic material, presence of authigenic uranium) and sediment age (François et al., 2004).

Although the method is gradually gaining wider acceptance to reconstruct paleoflux from the late Quaternary marine sediment records, some still question its validity and contend that the scavenged flux of ^{230}Th to the seafloor is much more variable and deviates widely from the production rate (Lyle et al., 2005). Extensive lateral transport of dissolved ^{230}Th, however, would require a much longer residence time of ^{230}Th in the water column and an activity in seawater that would be severalfold higher than has been measured so far. Lateral transport of ^{230}Th adsorbed on fine particles (Lyle et al., 2005) is equally impossible considering that the mean residence time of suspended particles in the water column is even shorter than for dissolved ^{230}Th (5–10 yr; Bacon & Anderson, 1982). Modeling (Bacon, 1988; Henderson, Heinze, Anderson, & Winguth, 1999; Marchal, François, Stocker, & Joos, 2000) and sediment trap studies (Yu et al., 2001; Scholten et al., 2001) have confirmed that lateral transport of ^{230}Th in the water column is limited and its flux to the seafloor is always relatively close (typically within 30–50%) to its rate of production in the water column. Although it is recognized that the assumption of a constant flux of ^{230}Th to the seafloor exactly equal to its production rate is an approximation which produces systematic errors (i.e., the method underestimates paleoflux in higher flux regions and overestimates paleoflux in low flux regions and thus underestimates flux variability), these errors are small compared with the extent of sediment focusing deduced by ^{230}Th-normalization. There are numerous examples where $\Psi > 5$, indicating that chronology-based mass accumulation rate overestimate vertical rain rate by more than fivefold.

Increasingly widespread application of this method has revealed that sediment redistribution on the seafloor is rather common. In particular, core collections tend to be biased toward regions of sediment focusing, which is not surprising since cores are not collected from regions of sediment winnowing. Applying ^{230}Th-normalization has also revealed that sediment focusing has often been overlooked (Pondaven et al., 2000; Marcantonio et al., 2001; Loubere et al., 2004; Kienast, Kienast, François, Mix, & Calvert, in press) and has resulted in misleading interpretations of the sedimentary flux record. There are sites where sediment redistribution is negligible and chronology-based accumulation rates provide the correct preserved rain rates (but with a lower resolution; Figure 16). However, whether sediment redistribution is insignificant cannot be known before measuring ^{230}Th, and in many instances ^{230}Th-normalization reveals significant sediment focusing. One particularly important and contentious example comes from the equatorial Pacific (Paytan et al., 1996; Marcantonio et al., 2001; Loubere et al., 2004; Lyle et al., 2005; Kienast et al., in press). Many of the cores collected in this oceanic region have been affected by sediment focusing, particularly during the Last Glacial Maximum. This results in higher glacial mass accumulation rates, which have often been interpreted as reflecting higher export of biogenic material from the overlying water (e.g., Lyle, Prahl, & Sparrow, 1992; Lyle, Mix, & Pisias, 2002; Paytan et al., 1996). When normalizing to ^{230}Th, however, this trend totally disappears (Marcantonio et al., 2001; Loubere et al., 2004; Kienast et al., in press). Combining ^{230}Th-normalized carbonate fluxes from four eastern equatorial Pacific cores with estimates of carbonate dissolution based on the recently developed Menardii fragmentation index (MFI; Mekik et al., 2002) reveals a consistent increase in carbonate export during the last 25 ka at the four sites, which is in sharp contrast to the much higher and erratic fluxes obtained by applying the same dissolution corrections to carbonate accumulation rates obtained from core chronology (Figure 17).

Admittedly, the ^{230}Th-normalized method could be refined. In particular, a means of assessing the recognized but limited bias produced by the effect of boundary scavenging would increase the accuracy of the method, which in its present form (assuming $V_{Th} = P_{Th}$) somewhat overestimates low fluxes and underestimates high fluxes. Sediment ^{231}Pa/^{230}Th may eventually provide a means of bringing to bear such a correction, once we can better constrain the relative importance of the factors (particle flux versus particle composition) that control the contrasting boundary scavenging of these two radioisotopes. It is possible that in some localized region of the ocean, deviations exceeding 30% may be found, particularly in areas with sharp horizontal gradients in particle flux. These

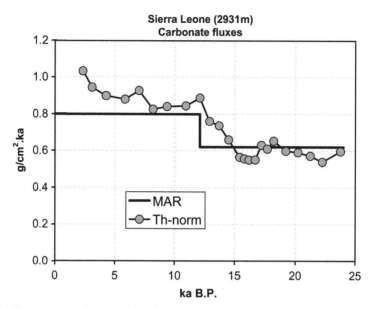

Figure 16   Comparison of preserved carbonate flux based on mass accumulation rates (MAR) derived from $\delta^{18}O$ chronology and ^{230}Th normalization on Sierra Leone Rise (EN066-38GGC; 4°55'N, 20°30'W), a site with little or no sediment redistribution (François et al., 1990). In this situation, the agreement between the two approaches is good. While the resolution of the MAR record could be improved by ^{14}C dating, it would be impossible to match the resolution afforded by ^{230}Th-normalization without significantly compromising the precision of the flux measurements.

refinements will require precise and high-resolution determination of the distribution of ^{230}Th and ^{231}Pa in the water column, especially in the vicinity of productivity fronts. Nonetheless, our present understanding of the behavior of ^{230}Th in the ocean is sufficient to warrant the systematic use of ^{230}Th-normalization for paleoflux reconstructions. In most situations, the method in its present form clearly provides important new insights into past changes in particle export from surface water and corrects the most serious errors associated with earlier interpretations of sediment mass accumulation rates.

3.2. Reconstructing Past Changes in the Vertical Rain Rate of Particles from Sediment $^{231}Pa/^{230}Th$

This is arguably the least developed of the paleoceanographic applications based on sediment ^{230}Th and ^{231}Pa. Its potential has not yet been fully established but results from a recent study (Chase et al., 2002) warrants further examination of the paleoflux information imbedded in this ratio. The method is based on the effect of boundary scavenging on the distribution of the two isotopes in marine sediments (Figure 5). If scavenging intensity and fractionation between ^{231}Pa and ^{230}Th were only controlled by particle flux, sediment $^{231}Pa/^{230}Th$ (corrected for their lithogenic fractions and radioactive decay) would provide a relatively simple tool

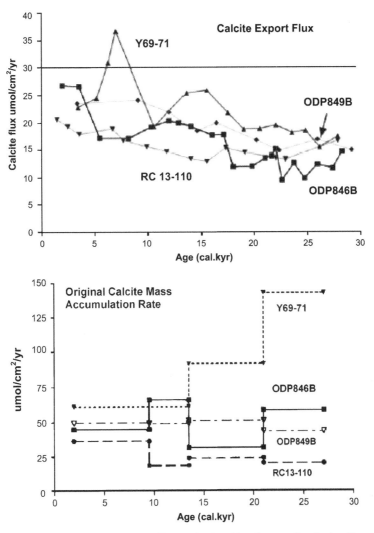

Figure 17 Calcite export fluxes estimated by combining % carbonate dissolution (based on MFI; Mekik et al., 2002) and (a) ^{230}Th-normalized carbonate fluxes (b) chronology-based carbonate mass accumulation rates (with permission from Loubere et al., 2004).

informing us on past changes in the geographic distribution of particle flux (Kumar, Gwiazda, Anderson, & Froelich, 1993; Kumar et al., 1995; François, Bacon, Altabet, & Labeyrie, 1993). Correct interpretation would require a synoptic database, because ^{231}Pa/^{230}Th at a given site does not only depend on particle flux at this site but also in surrounding regions which act as sources or sinks for the laterally transported ^{231}Pa. One particularly appealing characteristic of this proxy is that it is not affected by post-depositional particle remineralization or dissolution and it could potentially provide direct information on the vertical rain rate of material reaching the seafloor. The synoptic distribution of surface sediment

^{231}Pa/^{230}Th in Eastern Equatorial Pacific (Figure 6) lends support to the idea that, at least in this region, downcore changes in sediment ^{231}Pa/^{230}Th may record changes in export production. However, the correlation between export production and surface sediment ^{231}Pa/^{230}Th breaks down in the western equatorial Pacific (Pichat et al., 2004). Complications arise from the fact that sediment ^{231}Pa/^{230}Th depends not only on particle flux, but also on particle composition (Walter et al., 1997; Chase et al., 2002) and on the rate of the deep overturning circulation of the ocean (Yu et al., 1996; Marchal et al., 2000). The effect of particle scavenging and circulation dominates in different oceans, leading to different applications and interpretations of past changes in sediment ^{231}Pa/^{230}Th depending on location. In particular, the dominant effect of the MOC in the Atlantic reduces boundary scavenging and the sensitivity of ^{231}Pa/^{230}Th to changes in particle scavenging in this ocean (Yu et al., 2001), and changes in sediment ^{231}Pa/^{230}Th are thus interpreted as mainly reflecting changes in deep water circulation (Yu et al., 1996; Marchal et al., 2000; McManus, François, Gherardi, Keigwin, & Brown-Leger, 2004; Gherardi et al., 2005), after due consideration for a possible particle scavenging overprint (see Section 3.3). In the other oceanic regions, particle scavenging becomes the dominant effect, but it appears to be more strongly driven by particle composition than particle flux (Chase et al., 2002). In particular, opal concentration appears to exert a strong effect on the ^{231}Pa/^{230}Th ratio of marine particles due to its high affinity for ^{231}Pa (Walter et al., 1997; Chase et al., 2002). Because diatoms dominate the more productive regions of the ocean, opal concentration and particle flux often covary and it is difficult to unravel their respective influence.

As we gain further insights into the factors controlling ^{231}Pa/^{230}Th in marine particles, the prospect of reconstructing past changes in the geographical distribution of particle rain rate from synoptic maps of sediment ^{231}Pa/^{230}Th becomes increasingly tenuous. On the other hand, another application starts to emerge, which could be equally important. Compiling data obtained from settling particles intercepted by sediment traps from a wide range of locations, Chase et al. (2002) showed that ^{231}Pa/^{230}Th is tightly correlated to the carbonate to opal ratio of the settling particles. This ratio is an important variable which is controlled by the relative dominance of coccolithophorid and diatom production in marine ecosystems. In turn, it exerts a primary control on the organic carbon to carbonate "rain ratio" which affects atmospheric CO_2 by changing seawater alkalinity (Archer & Maier-Reimer, 1994; Archer, Winguth, Lea, & Mahowald, 2000), and possibly on the sequestration organic carbon to the deep sea through the ballasting effect of carbonate minerals (Klass & Archer, 2002; François, Honjo, Krishfield, & Manganini, 2002). Reconstructing the opal/carbonate rain ratio from the accumulation rates or ^{230}Th-normalized fluxes of sedimentary opal and carbonate is problematic because of poor and variable preservation on the seafloor, which is difficult to quantify. Sediment ^{231}Pa/^{230}Th thus offers the alluring prospect of providing information on past changes in this critical rain ratio without having to develop means of accurately assessing the preservation of biogenic silica and carbonate in sediments. There are clear and climatically coherent variations in ^{231}Pa/^{230}Th in the sedimentary record of the equatorial Pacific (Figure 18), showing higher values and suggesting higher

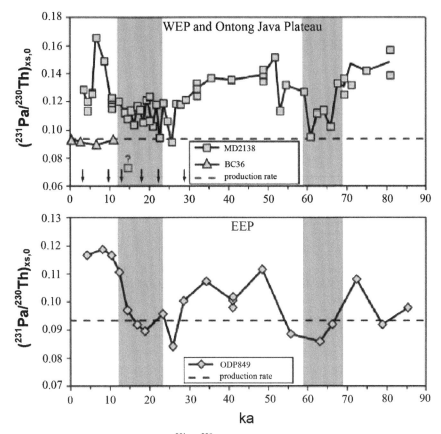

Figure 18 Variations in sediment ^{231}Pa/^{230}Th in the western (WEP) and eastern (EEP) equatorial Pacific during the last 85 ka (with permission from Pichat et al., 2004). Glacial periods (isotopic stages 2 and 4) are shown in light gray. The vertical arrows in the upper panel represent the depth of the ^{14}C ages that were used to establish the age model of core MD2138.

opal/carbonate rain rates during interglacial periods (Pichat et al., 2004). However, quantification of such data set requires systematic water column and modeling studies to gain a better understanding of the processes giving rise to the striking correlation between opal/carbonate and ^{231}Pa/^{230}Th reported by Chase et al. (2002).

3.3. Reconstructing past changes in the rate of the Atlantic meridional overturning circulation from sediment ^{231}Pa/^{230}Th

As already indicated, water column profiles clearly document the influence of the Atlantic MOC in controlling the distribution of ^{231}Pa and ^{230}Th in this ocean (Section 2.2.2). The export of ^{231}Pa from the Atlantic by the MOC was first revealed by a clear ^{231}Pa deficit in surface sediments (Yu et al., 1996). Today, the mean residence time of deep water in the Atlantic is equivalent to the mean residence time of ^{231}Pa in seawater (~200 yr). As a result, we would expect that

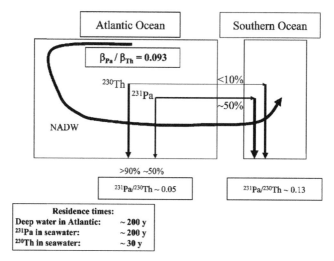

Figure 19 Conceptual diagram of the principle behind the use of sediment $^{231}Pa/^{230}Th$ in Atlantic sediment to estimate the rate of the Atlantic meridional overturning circulation (see text for explanation).

nearly half of the ^{231}Pa produced in the Atlantic as newly formed deep water transits through this basin would be exported to the southern ocean (Figure 19). On the other hand, because ^{230}Th has a much shorter residence time, most of the ^{230}Th produced in this water should be removed in Atlantic sediments. Reflecting this differential export, the mean $^{231}Pa/^{230}Th$ should only be about half the ratio of the rate of production (0.092) in Atlantic sediments and significantly higher in the sediments of the southern ocean, which is consistent with surface sediment data (Yu et al., 1996). If the rate of the Atlantic MOC were reduced to such an extent as to double the transit time of deep water in the Atlantic, we would expect only $\sim 1/4$ of the ^{231}Pa produced in the Atlantic to be exported and the $^{231}Pa/^{230}Th$ of Atlantic sediments would increase correspondingly. Lower rates of NADW formation and export to the southern ocean should thus result in systematically higher $^{231}Pa/^{230}Th$ in Atlantic sediments. This was confirmed by adding particle scavenging into intermediate complexity ocean circulation models (Marchal et al., 2000; Siddall et al., 2005). As the rate of the Atlantic MOC was reduced, the mean $^{231}Pa/^{230}Th$ of surface sediments in the North Atlantic increased and approached the production rate ratio of 0.092 for a total shutdown (Marchal et al., 2000). In contrast, the $^{231}Pa/^{230}Th$ of sediments in the South Atlantic (Marchal et al., 2000) and the southern ocean (Asmus et al., 1999) appears much less sensitive to changes in MOC.

Considering the cycle of aggregation and disaggregation between fine suspended and large sinking particles (see Section 2.1), Thomas, Henderson, and Robinson (2006) also pointed out that particles reaching the seafloor last equilibrate with the overlying 1,000 m of the water column. Consequently, $^{231}Pa/^{230}Th$ ratios measured in deep sea sediments are more sensitive to deep meridional overturning. The signal is muted when the overturning occurs at depth much shallower than the recording sediments. In this situation, the water column ^{231}Pa deficit generated by the overturning (Figure 7) gradually propagates into deeper waters. As particles sinking in the overturning waters adsorb less ^{231}Pa, they can adsorb more ^{231}Pa in deep waters and depress dissolved ^{231}Pa concentrations. It is as if the shallow overturning signal is "spread thinner" into the deep water below the section of the water column water that is directly affected by the overturning. Since the particles reaching the seafloor ultimately equilibrate with these deeper waters, they only record a fraction of the shallow overturning signal.

The results from the initial study comparing ^{231}Pa/^{230}Th in sediments from the Holocene and the last glacial maximum (Yu et al., 1996) indicated little difference between the mean ^{231}Pa/^{230}Th of the two time periods, suggesting that the rates of MOC were comparable. Moreover, as benthic foraminifera Cd/Ca and δ^{13}C clearly indicate that the Atlantic overturning cell was shallower during the last glacial period (Boyle, 1992), and considering that shallower overturnings are less effective at exporting ^{231}Pa (see Section 2.2.2) and only partially recorded in deep sea sediments, these results also pointed to the possibility that the shallow glacial overturning might have actually been faster that the deep overturning of the Holocene. However, the relatively large error bars on the ^{231}Pa/^{230}Th average for the entire Atlantic basin calculated by Yu et al. (1996) prevented a definitive conclusion, and Marchal et al. (2000) indicated that these data could not rule out the possibility of a 30% weaker glacial overturning.

We have now measured ^{230}Th and ^{231}Pa in several high sedimentation rate cores from the North Atlantic, which confirms the high circulation rates in the shallower glacial overturning cell (McManus et al., 2004; Gherardi et al., 2005; Gherardi, 2006). Analyzing sediment from Bermuda Rise deposited at 4,550 m depth, McManus et al. (2004) found a glacial ^{231}Pa/^{230}Th slightly higher (0.068 ± 0.010) than during the Holocene (0.055 ± 0.006; Figure 20), suggesting at face value a ~30% slowdown of the glacial MOC. However, in a shallower core taken from the Iberian margin (3,135 m; Gherardi et al., 2005), the ^{231}Pa/^{230}Th of LGM sediments is significantly lower (0.049 ± 0.001; Figure 21), indicating a faster overturning. The contrast between the two cores is in part due to the difference in ventilation between the eastern and western North Atlantic and in part due to the difference in water depth at the two coring sites. While the signal of the shallow glacial overturning is muted in the deep sediments deposited at 4,550 m, it is more fully recorded in the shallower core of the Iberian margin.

These high resolution records also revealed abrupt changes in the rate of MOC associated with all major climatic shifts of the last deglaciation (Figure 20). In particular, ^{231}Pa/^{230}Th approaches 0.092 during the early stages of the last deglaciation, suggesting a nearly total shutdown of the Atlantic MOC. The timing of this rapid change is coincident with that of a massive iceberg discharge known as Heinrich Event 1 (H1) and suggests that the MOC has been strongly affected by icebergs melting directly in the area of deep water formation (McManus et al., 2004). This was followed by a rapid resumption of the MOC at the beginning of the Bolling-Allerod warm period (14.5 ka B.P.), also clearly recorded as a sharp air temperature increase in the δ^{18}O of Greenland ice (Figure 20). Following this relatively warm period, the MOC slowed down again during the cold Younger

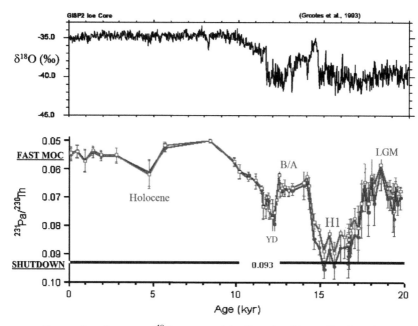

Figure 20 Comparison between $\delta^{18}O$ measured in Greenland ice reflecting air temperature (Grootes, Stuiver, White, Johnsen, & Jouzel, 1993) and sediment $^{231}Pa/^{230}Th$ measured in a high-resolution core from Bermuda Rise (OCE326-GGC5; $33°42'N; 57°35'W$; 4,550 m) showing abrupt changes in the rate of the Atlantic MOC from the last glacial maximum (LGM), to Henrich Event H1, the Bolling-Allerod warm period (B/A), the Younger Dryas (YD) and the Holocene (with permission from McManus et al., 2004).

Dryas period before a final and gradual acceleration to modern overturning rates toward the beginning of the Holocene. The $^{231}Pa/^{230}Th$ record from the Iberian margin shares similar features but with some differences (Figure 21), revealing some level of differentiation between the ventilation history of the western and eastern North Atlantic basins and the shallower and deeper overturning (Gherardi et al., 2005). In particular, the near collapse of the MOC at the beginning of the deglaciation recorded in the deep western basin was delayed by $\sim$1,000 yr in the eastern basin, suggesting the possibility that the shallow glacial overturning originated from two different sites (south Labrador Sea and west of Rockall Plateau) which may have been differently affected by iceberg discharges from the Laurentide and Fennoscandian ice sheets.

The major concern that we must address when interpreting sediment $^{231}Pa/^{230}Th$ in Atlantic sediments as a tracer of changes in the rate of the MOC is the possibility that a change in scavenging intensity, resulting from a change in particle flux or composition, could independently alter sediment $^{231}Pa/^{230}Th$. According to Figure 19, for a given rate of MOC, the mean $^{231}Pa/^{230}Th$ of Atlantic sediment will increase if the residence time of ^{231}Pa in the water column decreases. To address this concern, it must be established whether increases in sediment $^{231}Pa/^{230}Th$ are concurrent with increases in the preserved rain rates of particles obtained by ^{230}Th-normalization (see Section 3.1) and/or increased in sediment

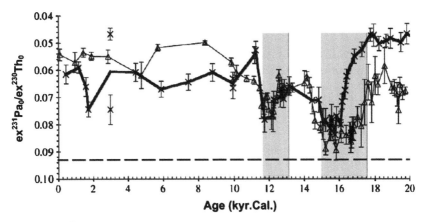

Figure 21 ^{231}Pa/^{230}Th profile from core SU81-18 (thick line with crosses) taken from the Iberian margin (37°46'N, 10°11'W, 3,135 m) and core EN GGC5 (thin line with open triangle) taken from Bermuda Rise (OCE326-GGC5; 33°42'N, 57°35'W, 4,550 m) versus calendar year (with permission from Gherardi et al., 2005).

opal concentration (see Section 2.2.1). Should it be the case, interpreting higher ^{231}Pa/^{230}Th as reflecting a slowdown in MOC would be unwarranted. This approach was used in the studies of McManus et al. (2004) and Gherardi et al. (2005) and confirmed that high ^{231}Pa/^{230}Th ratios recorded in these cores were not associated with increased particle flux or opal concentration, thereby lending credence to the interpretation of the record as a circulation signal. Nonetheless, we are still faced with the lingering problem that opal in settling particles may not always be preserved in sediments and measuring sedimentary opal may not be an indubitable argument against changes in particle scavenging, since its absence in sediment does not necessarily means that it was not present in the initial settling flux. The relatively high opal concentrations that have to be reached to significantly affect the ^{231}Pa/^{230}Th of particles (Walter et al., 1997; Chase et al., 2002) would likely result in some level of opal preservation in sediments, but this is a question that still needs to be addressed quantitatively. In the meantime, coherent changes in ^{231}Pa/^{230}Th from different Atlantic cores with distinct sedimentary regimes lend further support to the interpretation of the signal as recording basin-wide changes in circulation (Gherardi et al., 2005; Gherardi, 2006).

The weak expression of boundary scavenging in the surface sediments of the Atlantic Ocean is believed to be a direct consequence of the rapid meridional overturning found today in this ocean basin (see Section 2.2.1). Surface sediments with ^{231}Pa/^{230}Th higher than the production rate ratio have only been found in restricted regions off N.W. Africa, where primary production is high as a result of upwelling (Yu et al., 1996), and south of Iceland, where sediments have higher opal concentration (Anderson, unpublished; François et al., unpublished). During periods with significantly slower MOC, a stronger expression of boundary scavenging should be expected in Atlantic sediments, which should result in a more pronounced increase in ^{231}Pa/^{230}Th at the margins.

4. CONCLUSIONS

The uniquely uniform source of ^{230}Th and ^{231}Pa in seawater greatly facilitates our understanding of the factors that control the distribution of these two isotopes in the marine environment and has led to three important paleoceanographic applications which inform us on the past variability of several key oceanic processes.

As a result of its very short residence time in the water column, ^{230}Th provides a quasi constant flux tracer which can be used to estimate past changes in the preserved vertical rain rate of particles to the seafloor during the last several hundred thousand years (Henderson & Anderson, 2003; François et al., 2004). The approach is quantitative to the extent that the underlying assumption ($V_{Th} \cong P_{Th}$) is verified and will be further refined as future water column and modeling studies provide the necessary constraints to estimate the limited lateral transport of this isotope in the water column as a result of eddy diffusion and ocean circulation.

Changes in sediment ^{231}Pa/^{230}Th reflect different processes depending on geographic location. Outside the Atlantic, this ratio seems to mostly record past changes in scavenging intensity driven either by changes in particle flux or changes in particle composition. This application may ultimately develop as a powerful synoptic tool to reconstruct past changes in particle flux field or opal/carbonate rain ratio (Chase et al., 2002), but further progress in that direction needs additional water column and modeling studies to clearly identify and quantify the processes that control boundary scavenging. A synoptic database will also be required for quantitative interpretation. In the Atlantic Ocean, ^{231}Pa/^{230}Th appears to mostly record changes in the rate of the meridional overturning and has already provided semi-quantitative estimates of past circulation changes (McManus et al., 2004). Analyzing cores from different locations and depths is providing further insights into the geometry and strength of the meridional circulation in relation to climatic variability (Gherardi et al., 2005). Further development of sediment ^{231}Pa/^{230}Th as a quantitative tracer for paleocirculation or paleoflux will require a better understanding of the interaction between particle scavenging and ocean circulation in controlling the distribution of ^{231}Pa/^{230}Th in sediments, which can only be achieved by increasing the database in the water column and modern sediments, and by imbedding ^{231}Pa and ^{230}Th scavenging into general circulation models (Marchal et al., 2000; Siddall et al., 2005). Much work still needs to be done, but the potential of these approaches is becoming increasingly evident as we learn more about the behavior of ^{230}Th and ^{231}Pa in the marine environment.

REFERENCES

Anderson, R. F., Bacon, M. P., & Brewer, P. G. (1983a). Removal of Th-230 and Pa-231 from the open ocean. *Earth and Planetary Science Letters, 62*, 7–23.

Anderson, R. F., Bacon, M. P., & Brewer, P. G. (1983b). Removal of Th-230 and Pa-231 at ocean margins. *Earth and Planetary Science Letters, 66*, 73–90.

Anderson, R. F., Lao, Y., Broecker, W. S., Trumbore, S., Hofmann, H. J., & Wolfli, W. (1990). Boundary scavenging in the Pacific Ocean: A comparison of Be-10 and Pa-231. *Earth and Planetary Science Letters, 96,* 287–304.

Archer, D., & Maier-Reimer, E. (1994). Effect of deep-sea sedimentary calcite preservation on atmospheric CO_2 concentration. *Nature, 367,* 260–263.

Archer, D., Winguth, A.M.E., Lea, D.W., Mahowald, N. (2000). What caused the glacial/interglacial atmospheric pCO_2 cycles? *Review of Geophysics, 38,* 159, 189

Asmus, T., Frank, M., Koschmeider, C., Frank, N., Gersonde, R., & Mangini, A. (1999). Variations of the biogenic particle flux at the Southern Atlantic Section of the Subantarctic Front during the late Quaternary: Evidence from sedimentary ^{231}Pa$_{ex}$ and ^{230}Th$_{ex}$. *Marine Geology, 159,* 63–78.

Bacon, M. P. (1984). Glacial to interglacial changes in carbonate and clay sedimentation in the Atlantic Ocean estimated from ^{230}Th measurements. *Isotope Geoscience, 2,* 97–111.

Bacon, M. P. (1988). Tracers of chemical scavenging in the ocean: Boundary effects and large-scale chemical fractionation. *Philosophical Transactions of the Royal Society of London A, 325,* 147–160.

Bacon, M. P., & Anderson, R. F. (1982). Distribution of thorium isotopes between dissolved and particulate forms in the deep sea. *Journal of Geophysical Research, 87,* 2045–2056.

Bacon, M. P., Huh, C.-A., Fleer, A. P., & Deuser, W. G. (1985). Seasonality in the flux of natural radionuclides and plutonium in the deep Sargasso Sea. *Deep-Sea Research, 32,* 273–286.

Boyle, E. A. (1992). Cadmium and δ^{13}C paleochemical ocean distributions during the Stage 2 Glacial Maximum. *Annual Reviews of Earth Planetary Sciences, 20,* 245–287.

Broecker, W. S. (1979). A revised estimate for the radiocarbon age of North Atlantic Deep Water. *Journal of Geophysical Research, 84,* 3218–3226.

Broecker, W. S., & Clark, E. (2001). An evaluation of Lohmann's foraminifera weight dissolution index. *Paleoceanography, 16,* 431–434.

Chase, Z., Anderson, R. F., Fleisher, M. Q., & Kubik, P. W. (2002). The influence of particle composition and particle flux on scavenging of Th, Pa and Be in the ocean. *Earth and Planetary Science Letters, 204,* 215–229.

Chase, Z., Anderson, R. F., Fleisher, M. Q., & Kubik, P. W. (2003). Scavenging of ^{230}Th, ^{231}Pa and ^{10}Be in the southern ocean (SW Pacific sector): The importance of particle flux, particle composition and advection. *Deep-Sea Research II, 50,* 739–768.

Cheng, H., Edwards, R. L., Hoff, J., Gallup, C. D., Richards, D. A., & Asmerom, Y. (2000). The half-lives of uranium-234 and thorium-230. *Chemical Geology, 169,* 17–33.

Choi, M.-S., François, R., Sims, K., Bacon, M. P., Brown-Leger, S., Fleer, A. P., Ball, L., Schneider, D., & Pichat, S. (2001). Rapid determination of ^{230}Th and ^{231}Pa in seawater by desolvated-micron-ebulization inductively-coupled magnetic sector mass spectrometry. *Marine Chemistry, 76,* 99–112.

Clegg, S. L., & Whitfield, M. (1990). A generalized model for the scavenging of trace metals in the open ocean—I. Particle cycling. *Deep-Sea Research I, 37,* 809–832.

Delanghe, D., Bard, E., & Hamelin, B. (2002). New TIMS constraints on the uranium-238 and uranium-234 in seawaters from the main ocean basins and the Mediterranean Sea. *Marine Chemistry, 80,* 79–93.

Dunk, R. M., Mills, R. A., & Jenkins, W. J. (2002). A reevaluation of the oceanic uranium budget for the Holocene. *Chemical Geology, 190,* 45–67.

François, R., Bacon, M. P. B., Altabet, M. A., & Labeyrie, L. D. (1993). Glacial/interglacial changes in sediment rain rate in the S.W. Indian sector of subantarctic waters as recorded by ^{230}Th, ^{231}Pa, U and δ^{15}N. *Paleoceanography, 8,* 611–629.

François, R., Bacon, M. P., & Suman, D. O. (1990). Th-230 profiling in deep-sea sediments: High-resolution records of flux and dissolution of carbonate in the equatorial Atlantic during the last 24,000 years. *Paleoceanography, 5,* 761–787.

François, R., Frank, M., Rutgers van der Loeff, M. M., & Bacon M. P. (2004). ^{230}Th-normalization: An essential tool for interpreting sedimentary fluxes during the late Quaternary. *Paleoceanography, 19*(1), PA1018, doi:10.1029/2003PA000939.

François, R., Honjo, S., Krishfield, R., & Manganini, S. (2002). Factors controlling the flux of organic carbon to the bathypelagic zone of the ocean. *Global Biogeochemical Cycles, 16,* 1087, doi:10.1029/2001GB001722.

Gherardi, J. -M. (2006). Changements de la circulation océanique au cours de la déglaciation : Apport des traceurs [231]Pa/[230]Th du sédiment et Mg/Ca des foraminifères benthiques. Ph.D. thesis, Université Paris 6.

Gherardi, J.-M., Labeyrie, L. D., McManus, J., François, R., Skinner, L. C., & Cortijo, E. (2005). Evidence from the North Eastern Atlantic Basin for variability in the rate of the meridional overturning circulation through the last deglaciation. *Earth and Planetary Science Letters*, *240*, 710–723.

Grootes, P. M., Stuiver, M., White, J. W. C., Johnsen, S., & Jouzel, J. (1993). Comparison of oxygen isotope records from the GISP2 and GRIP Greenland ice cores. *Nature*, *366*, 552–554.

Guo, G., Chen, M., & Gueguen, C. (2002). Control of Pa/Th ratio by particulate chemical composition in the ocean. *Geophysical Research Letters*, *29*(20), 1961, doi:10.1029/2002GL015666.

Harris, P. G., Zhao, M., Rosell-Melé, A., Tiedemann, R., Sarnthein, M., & Maxwell, J. R. (1996). Chlorin accumulation rat as a proxy for Quaternary marine primary productivity. *Nature*, *383*, 63–65.

Henderson, G. M., & Anderson, R. F. (2003). The U-series toolbox for paleoceanography. *Reviews in Mineralogy and Geochemistry*, *52*, 493–531.

Henderson, G. M., Heinze, C., Anderson, R. F., & Winguth, A. M. E. (1999). Global distribution of the [230]Th flux to ocean sediments constrained by GCM modeling. *Deep Sea Research Part I*, *46*, 1861–1893.

Kienast, S. S., Kienast, M., Mix, A., Calvert, S. E., & François, R. (in press). [230]Thorium normalized particle flux and sediment focusing in the Panama Basin region during last 30,000 years. *Paleoceanography*.

Klass, C., & Archer, D. (2002). Association of sinking organic matter with various types of mineral ballast in the deep sea: Implication for the rain ratio. *Global Biogeochemical Cycles*, *16*, 1116, doi:10.1029/2001GB001765.

Krishnaswami, S., Lal, D., Somayajulu, B. L. K., Weiss, R., & Craig, H. (1976). Large-volume *in-situ* filtration of deep Pacific waters: Mineralogical and radioisotope studies. *Earth and Planetary Science Letters*, *32*, 420–429.

Kumar, N., Anderson, R. F., Mortlock, R. A., Froelich, P. N., Kubik, P., Dittrich-Hannen, B., & Suter, M. (1995). Increased biological productivity and export production in the glacial Southern Ocean. *Nature*, *378*, 675–680.

Kumar, N., Gwiazda, R., Anderson, R. F., & Froelich, P. N. (1993). [231]Pa/[230]Th ratios in sediments as a proxy for past changes in Southern Ocean productivity. *Nature*, *362*, 45–48.

Lao, Y., Anderson, R. F., Broecker, W. S., Hofmann, H. J., & Wolfli, W. (1993). Particle fluxes of Th-230, Pa-231 and Be-10 in the Northeastern Pacific Ocean. *Geochimica et Cosmochimica Acta*, *50*, 205–217.

Lao, Y., Anderson, R. F., Broecker, W. S., Trumbore, S. E., Hofmann, H. J., & Wolfli, W. (1992). Increased productivity of cosmogenic [10]Be during the Last Glacial Maximum. *Nature*, *357*, 576–578.

Lohmann, G. P. (1995). A model for variation in the chemistry of planktonic foraminifera due to secondary calcification and selective dissolution. *Paleoceanography*, *10*, 445–457.

Loubere, P., Mekik, F., François, R., & Pichat, S. (2004). Export fluxes of calcite in the eastern equatorial Pacific from the Last Glacial Maximum to present. *Paleoceanography*, *19*, PA2018, doi:10.1029/2003PA000986.

Lyle, M., Mitchell, N., Pisias, N., Mix, A., Martinez, J. I., & Paytan A. (2005). Do geochemical estimates of sediment focusing pass the sediment test in the equatorial Pacific? *Paleoceanography*, *20*, PA1005, doi:10.1029/2004PA001019.

Lyle, M., Mix, A., & Pisias, N. (2002). Patterns of $CaCO_3$ deposition in the eastern tropical Pacific Ocean for the last 150 kyr: Evidence for a southeast Pacific depositional spike during marine isotope stage (MIS) 2. *Paleoceanography*, *17*, 1013, doi:10.1029/2000PA000538.

Lyle, M., Murray, D. W., Finney, B. P., Dymond, J., Robbins, J. M., & Brooksforce, K. (1988). The record of Late Pleistocene biogenic sedimentation in the Eastern Tropical Pacific Ocean. *Paleoceanography*, *3*, 39–59.

Lyle, M., Prahl, F. G., & Sparrow, M. A. (1992). Upwelling and productivity changes inferred from a temperature record in the central Equatorial Pacific. *Nature*, *355*, 812–815.

Marcantonio, F., Anderson, R. F., Higgins, S., Stute, M., Schlosser, P., & Kubik, P. W. (2001). Sediment focusing in the central equatorial Pacific Ocean. *Paleoceanography, 16*, 260–267.

Marchal, O., François, R., Stocker, T. F., & Joos, F. (2000). Ocean thermohaline circulation and sedimentary ^{231}Pa/^{230}Th ratio. *Paleoceanography, 15*, 625–641.

Marchitto, T. M., Lynch-Stieglitz, J., & Hemming, S. (2005). Deep Pacific CaCO$_3$ compensation and glacial–interglacial atmospheric CO2. *Earth and Planetary Science Letters, 231*, 317–336.

McManus, J. F., François, R., Gherardi, J.-M., Keigwin, L. D., & Brown-Leger, S. (2004). Collapse and rapid resumption of Atlantic meridional circulation linked to deglacial climate changes. *Nature, 428*, 834–837.

Mekik, F., Loubere, P., & Archer, D. (2002). Organic carbon flux and organic carbon to calcite flux ratio recorded in deep sea carbonates: Demonstration and a new proxy. *Global Biogeochemical Cycles, 16*(3), 1052, doi:10.1029/2001GB001634.

Moore, W. S., & Sackett, W. M. (1964). Uranium and thorium series inequilibrium in sea water. *Journal of Geophysical Research, 69*, 5401–5405.

Moran, S. B., Charette, M. A., Hoff, J. A., Edwards, R. L., & Landing, W. M. (1997). Distribution of ^{230}Th in the Labrador Sea and its relation to ventilation Processes: Constraints on ^{231}Pa/^{230}Th as a Paleocirculation Tracer in the Deep Atlantic. *Earth and Planetary Science Letters, 150*, 151–160.

Moran, S. B., Hoff, J. A., Buesseler, K. O., & Edwards, R. L. (1995). High Precision ^{230}Th and ^{232}Th in the Norwegian Sea and Denmark by thermal ionization mass spectrometry. *Geophysical Research Letters, 22*, 2589–2592.

Moran, S. B., Shen, C.-C., Edmonds, H. N., Weinstein, S. E., Smith, J. N., & Edwards, R. L. (2002). Dissolved and particulate ^{231}Pa and ^{230}Th in the Atlantic Ocean: Constraints on intermediate/deep water age, boundary scavenging, and ^{231}Pa/^{230}Th fractionation. *Earth and Planetary Science Letters, 203*, 999–1014.

Moran, S. B., Shen, C.-C., Weinstein, S. E., Hettinger, L. H., Hoff, J. H., Edmonds, H. N., & Edwards, R. L. (2001). Constraints on deep water age and particle flux in the Equatorial and South Atlantic Ocean based on seawater Pa-231 and Th-230 data. *Geophysical Research Letters, 28*, 3437–3440.

Murray, R. W., Leinen, M., & Isern, A. R. (1993). Biogenic fluxes of Al to sediment in the central equatorial Pacific Ocean: Evidence for increased productivity during glacial periods. *Paleoceanography, 8*, 651–670.

Nozaki, Y., Horibe, Y., & Tsubota, H. (1981). The water column distributions of thorium isotopes in the western North Pacific. *Earth and Planetary Science Letters, 54*, 203–216.

Nozaki, Y., & Nakanishi, T. (1985). ^{231}Pa and ^{230}Th profiles in the open ocean water column. *Deep-Sea Research, 32*, 1209–1220.

Nozaki, Y., Yang, H.-S., & Yamada, M. (1987). Scavenging of thorium in the ocean. *Journal of Geophysical Research, 92*, 772–778.

Paytan, A., Kastner, M., & Chavez, F. P. (1996). Glacial to interglacial fluctuations in productivity in the equatorial Pacific as indicated by marine barite. *Science, 274*, 1355–1357.

Pichat, S., Sims, K. W., François, R., McManus, J. F., Brown Leger, S., & Albarede, F. (2004). Lower biological productivity during glacial periods in the Equatorial Pacific as derived from (231Pa/230Th)xs,0 measurements in deep-sea sediments. *Paleoceanography, 19*, PA4023, doi:10.1029/2003PA000994.

Pondaven, P., Ragueneau, O., Treguer, P., Hauvespre, A., Dezileau, L., & Reyss, J.-L. (2000). Resolving the "opal paradox" in the southern ocean. *Nature, 405*, 168–172.

Robinson, L. F., Henderson, G. M., Hall, L., & Matthews, I. (2004). Climatic control of riverine and seawater uranium isotope ratios. *Science, 205*, 851–854.

Robert, P. J., Miranda, C. F., & Muxard, R. (1969). Mesure de la période du protactinium 231 par microcalorimétrie. *Radiochimica Acta, 11*, 104–108.

Roy-Barman, M., Chen, J. H., & Wasserburg, G. J. (1996). ^{230}Th-^{232}Th systematics in the central Pacific Ocean: The sources and the fates of thorium. *Earth and Planetary Science Letters, 139*, 351–363.

Rutgers van der Loeff, M. M., & Berger, G. W. (1993). Scavenging of ^{230}Th and ^{231}Pa near the Antarctic Polar Front in the South Atlantic. *Deep-Sea Research I, 40*, 339–357.

Santschi, P. H., Murray, J. W., Baskaran, M., Benitez-Nelson, C. R., Guo, L. D., Hung, C.-C., Lamborg, C., Moran, S. B., Passow, U., & Roy-Barman, M. (2006). Thorium speciation in seawater. *Marine Chemistry, 100,* 250–268.

Schmitz, W. J., Jr. (1996). *On the world ocean circulation: Volume I. Some global features/North Atlantic circulation.* Woods Hole Oceanographic Technical Report. WHOI-96-03.

Schneider, R. R., Muller, P. J., Ruhland, G., Meinecke, G., Schmidt, H., & Wefer, G. (1996). Late Quaternary surface temperatures and productivity in the east-equatorial South Atlantic: Response to changes in trade/monsoon wind forcing and surface water advection. In: G. Wefer, W. H. Berger & D. J. Webb (Eds), *The South Atlantic: Present and past circulation* (pp. 527–551). Berlin: Springer-Verlag.

Scholten, J. C., Fietzke, J., Vogler, S., Rutgers van der Loeff, M. M., Mangini, A., Koeve, W., Waniek, J., Stoffers, P., Antia, A., & Kuss, J. (2001). Trapping efficiencies of sediment traps from the deep Eastern North Atlantic: The ^{230}Th calibration. *Deep-Sea Research II, 48,* 2383–2408.

Shen, C.-C., Cheng, H., Edwards, R. L., Moran, S. B., Edmonds, H. N., Hoff, J. A., & Thomas, R. B. (2003). Measurement of attogram quantities of ^{231}Pa in dissolved and particulate fractions of seawater by isotope dilution thermal ionization mass spectroscopy. *Analytical Chemistry, 75,* 1075–1079.

Shen, C.-C., Edwards, R. L., Cheng, H., Dorale, J. A., Thomas, R. B., Moran, S. B., Weinstein, S. E., & Edmonds, H. N. (2002). Uranium and thorium isotopic and concentration measurements by magnetic sector inductively coupled plasma mass spectrometry. *Chemical Geology, 185,* 165–178.

Shimmield, G. B., & Price, N. B. (1988). The scavenging of U, ^{230}Th and ^{231}Pa during pulsed hydrothermal activity at 20°S, East Pacific Rise. *Geochimica Cosmochimica Acta, 52,* 669–677.

Siddall, M., Stocker, T., Henderson, G., Edwards, N. R., Muller, S. A., Joos, F., & Frank, M. (2005). ^{231}Pa/^{230}Th fractionation by ocean transport, biogenic particle flux and particle type. *Earth and Planetary Science Letters, 237,* 135–155.

Spencer, D. W., Bacon, M. P., & Brewer, P. G. (1981). Models of the distribution of ^{210}Pb in a section across the North Equatorial Atlantic Ocean. *Journal of Marine Research, 39,* 119–137.

Stuiver, M., Quay, P. D., & Ostlund, H. G. (1983). Abyssal water carbon-14 distribution and the age of the world ocean. *Science, 219,* 849–851.

Suman, D. O., & Bacon, M. P. (1989). Variations in Holocene sedimentation in the North American basin determined from ^{230}Th measurements. *Deep-Sea Research, 36,* 869–878.

Talley, L. D. (2003). Shallow, intermediate, and deep overturning components of the global heat budget. *Journal of Physical Oceanography, 33,* 530–560.

Thomas, A. L., Henderson, G. M., & Robinson, L. F. (2006). Interpretation of the ^{231}Pa/^{230}Th paleocirculation proxy: New water-column measurements from the southwest Indian Ocean. *Earth and Planetary Science Letters, 241,* 493–504.

Vogler, S., Scholten, J. C., van der Loeff, M. M. R., & Mangini, A. (1998). ^{230}Th in the eastern North Atlantic: The importance of water mass ventilation in the balance of ^{230}Th. *Earth and Planetary Science Letters, 156,* 61–74.

Walter, H.-J., Rutgers van der Loeff, M. M., & François, R. (1999). Reliability of the ^{231}Pa/^{230}Th activity ratio as a tracer for bioproductivity of the ocean. In: G. Fischer & G. Wefer (Eds), *Proxies in paleoceanography* (pp. 393–408). Breman, Germany: University of Bremen.

Walter, H.-J., Rutgers van der Loeff, M. M., & Hoeltzen, H. (1997). Enhanced scavenging of ^{231}Pa relative to ^{230}Th in the south Atlantic south of the polar front: Implications for the use of the ^{231}Pa/^{230}Th ratio as a paleoproductivity proxy. *Earth and Planetary Science Letters, 149,* 85–100.

Yang, H.-S., Nozaki, Y., Sakai, H., & Masuda, A. (1986). The distribution of Th-230 and Pa-231 in the deep-sea surface sediments of the Pacific Ocean. *Geochimica et Cosmochimica Acta, 50,* 81–89.

Yu, E.-F., François, R., & Bacon, M. P. (1996). Similar rates of modern and last-glacial ocean thermohaline circulation inferred from radiochemical data. *Nature, 379,* 689–694.

Yu, E.-F., François, R., Bacon, M. P., & Fleer, A. P. (2001). Fluxes of ^{230}Th and ^{231}Pa to the deep sea: Implications for the interpretation of excess ^{230}Th and ^{231}Pa/^{230}Th profiles in sediments. *Earth and Planetary Science Letters, 191,* 219–230.

Zheng, Y., Anderson, R. F., van Geen, A., & Fleisher, M. Q. (2002). Remobilization of authigenic uranium in marine sediments by bioturbation. *Geochimica Cosmochimica Acta, 66*(10), 1759–1772.

CHAPTER SEVENTEEN

Boron Isotopes in Marine Carbonate Sediments and the pH of the Ocean

Nicholas Gary Hemming* *and* Bärbel Hönisch

Contents

1. Introduction	717
2. Empirical Observations and Theoretical Background	718
2.1. Synthetic carbonate studies	720
2.2. Culture experiments	721
3. Caveats and Complications	721
3.1. Isotope fractionation factor	721
3.2. Foraminifera size and dissolution effects	723
3.3. Analytical challenges	723
3.4. Secular variation in seawater $\delta^{11}B$	725
4. Applications of the Boron Isotope Paleo-pH Proxy	726
5. Summary and Conclusion	730
Acknowledgments	731
References	731

1. Introduction

Until recently, our understanding of ancient seawater pH has relied on model predictions, despite its being an important chemical parameter of the oceans. Exchange between the surface ocean and the atmosphere means that changes in PCO_2 (ocean) and pCO_2 (atmosphere) are coupled, so knowing one can provide information on the other. If we could measure ancient ocean pH, then constraints can be placed on natural, pre-industrial pCO_2. This may help us better understand these natural fluctuations and controls on atmospheric chemistry so that we may better understand the recent anthropogenic perturbations and their future implications. One of the most critical environmental threats today is global warming, attributed to the industrial increase in atmospheric CO_2, making the development of a proxy for paleo–pH both timely and necessary.

* Corresponding author.

Developments in Marine Geology, Volume 1
ISSN 1572-5480, DOI 10.1016/S1572-5480(07)01022-6

A promising candidate for a paleo-pH proxy is the boron isotope composition of marine calcium carbonate. Since the first published analysis of the boron isotope composition of modern marine carbonates (Hemming & Hanson, 1992; Vengosh, Kolodny, Starinsky, Chivas, & McCulloch, 1991), there has been a steady increase in our knowledge of this proxy. Although complications have been identified, confidence in the proxy has strengthened as our understanding of the mechanisms and controls on boron isotope uptake and fractionation in carbonates has increased. The goal of this paper is to present a history of the development of the boron isotope paleo-pH proxy as well as the theory behind it, identify the caveats, complications, and assumptions, and briefly review the applications of the proxy.

2. EMPIRICAL OBSERVATIONS AND THEORETICAL BACKGROUND

Our understanding of the boron isotope composition of seawater begins with an early study by Schwarcz, Agyei, and McMullen (1969). They proposed that the isotopically heavy value of seawater ($\sim+40\permil$ relative to the NIST standard SRM951 boric acid) is due to the fact that boron is adsorbed onto clay particles as they enter the ocean. The clays flocculate and are deposited, thus removing some boron from the marine system. The boron that is removed is isotopically light, leaving seawater isotopically heavy. Normally one would expect the adsorbed species to be isotopically heavier than the water, as the lower vibration energy of the heavier isotope is typically favored in adsorption reactions. However, because boron does not occur as a metal in most natural systems, the aqueous boron species must be considered. Boron is present in seawater primarily as the trigonal $B(OH)_3$ species and the tetrahedral $B(OH)_4^-$ species (~80 and 20%, respectively, in seawater). The distribution of these species is pH controlled (Figure 1a). Because of coordination-controlled vibration differences between the species, there is an isotopic offset. The isotope exchange reaction

$$^{11}B(OH)_3 + {}^{10}(BOH)_4^- \leftrightarrow {}^{10}B(OH)_3 + {}^{11}B(OH)_4^-$$

has a equilibrium constant <1, which means the trigonal species is isotopically heavy relative to the tetrahedrally coordinated species. Thus, as the distribution of aqueous species changes with pH, so does the isotopic composition of the species (Figure 1b). The suggestion by Schwarcz et al. (1969) that adsorbed boron on clays is isotopically light gave the first clue that there could be a pH control on the isotope composition of adsorbed boron. Subsequent work on boron isotopes focused on distinguishing marine vs. non-marine origins of borate deposits (see Swihart, Moore, & Callis, 1986) until Spivack, Palmer, and Edmond (1987) presented their study of the sedimentary cycle of boron isotopes. This study represented a great leap in our understanding of boron isotope systematics. This was soon followed by an experimental study of boron adsorption on clays (Palmer, Spivack, & Edmond, 1987), which supported the early work of Schwarcz et al. (1969). At about this same time, two groups were beginning the first analyses on marine carbonates. Vengosh et al. (1991) and Hemming and Hanson (1992)

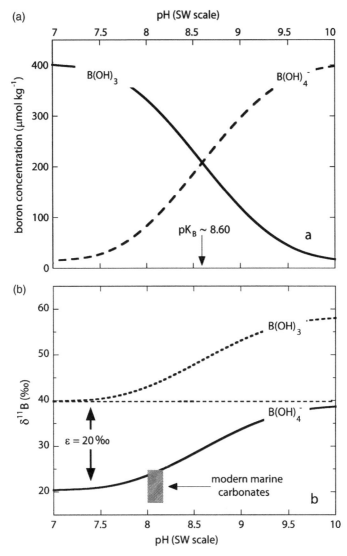

Figure 1 (a). The pH controlled distribution of the two dominant aqueous species of boron in seawater (pKB value from Dickson, 1990). (b). The isotopic composition of the two dominant aqueous species of boron in seawater (calculated using the theoretical fractionation factor of Kakihana et al., 1977). The box represents the range of measured modern carbonates analyzed by Hemming and Hanson (1992) and include bivalves, corals, echinoderms, calcareous algae, and ooids. Since the exact growth pH for those samples is not known, data are plotted at typical surface seawater-pH (after Hemming & Hanson, 1992).

pursued an improved negative thermal ionization method (NTIMS) of boron isotope analysis following initial work on this method by Duchateau and de Bièvre (1983) and Heumann and Zeininger (1985). Both the Hemming and Hanson (1992) and Vengosh et al. (1991) studies took the survey approach, measuring a variety of marine carbonates, and both found significant offsets from seawater. The Hemming and Hanson (1992) study reported a relatively narrow range in boron isotope compositions considering the wide variety of samples analyzed (corals, echinoderms, brachiopods, ooids, bivalves, etc.), representing calcite, high-Mg calcite, and aragonite (Figure 1b). Hemming and Hanson (1992) proposed that the range in boron isotope composition was due to the fact that only the $B(OH)_4^-$ aqueous species is involved in reactions incorporating boron into the carbonate structure, as the survey data plotted remarkably close to the isotopic composition of the tetrahedral species in seawater as calculated using the only available estimate of the fractionation factor at that time (Kakihana, Kotaka, Satoh, Nomura, & Okamoto, 1977). In contrast, Vengosh et al. (1991) reported a much larger range in boron isotope values for modern carbonates, from +13.3‰ for a foraminifera sample to +31.9‰ for a coral sample. It is not clear why there is a difference between these two studies, but a later study by Gaillardet and Allègre (1995) reported modern coral analyses that overlap within analytical uncertainty of the Hemming and Hanson (1992) data (+24 to +27‰). This study used a positive ion technique similar to that used by Spivack and Edmond (1986), and gave support to the Hemming and Hanson (1992) interpretation.

2.1. Synthetic Carbonate Studies

The next step in the development of the proxy was to determine if the interpretation of Hemming and Hanson (1992) was robust. The uncertainties of the survey study (the exact pH at sample localities, as well as salinity, temperature, and other factors, was not known), did not allow quantitative analysis of the controls on boron uptake and fractionation. Therefore, a laboratory-based study was initiated to test the Hemming and Hanson (1992) hypothesis. In that study, synthetic calcite was grown from solutions with similar ionic strength to seawater, and a range of boron concentrations (Hemming, Reeder, & Hanson, 1995). The boron isotope standard, NIST SRM951, was used so the starting composition of the fluid was known. The results showed a remarkable agreement to the $B(OH)_4^-$ isotope composition of the solution, and thus lent strong support for the initial hypothesis (Figure 1b). There was no indication from that study that mineralogy (calcite, high-Mg calcite, or aragonite) played a significant role in the isotopic fractionation. This set the stage for the use of boron isotopes as a paleo-pH proxy, as it was now clear that the boron isotope composition of the precipitated carbonate matched perfectly with the known pH of the experimental solution. A second experimental study of inorganic precipitates confirmed the tight coupling between boron isotopes in synthetic carbonates and the pH of the growth solution over a range of pH conditions (Sanyal, Nugent, Reeder, & Bijma, 2000). However, there was an offset of approximately −2‰ between the growth solution $B(OH)_4^-$ composition and the carbonate. It is not known why there is a discrepancy between the Hemming et al. (1995) and Sanyal et al. (2000) studies, but the experimental setup was quite different. Whereas Hemming et al. (1995) adjusted pH by controlling pCO_2 in the experiment chamber, Sanyal et al. (2000) controlled pH by the addition of salts, which also altered alkalinity. Perhaps the differences in the experimental controls had an influence on the results, but to date these experiments have not been replicated. Regardless, both studies show the strong dependence of boron isotope compositions of calcium carbonate with pH.

2.2. Culture Experiments

While it is impossible to recreate natural conditions in the laboratory, coral and foraminifera culture experiments come closest to simulating true conditions, while allowing control of environmental parameters. Coral and foraminifera experiments over a range of pH values (Hönisch et al., 2004; Reynaud, Hemming, Juillet-Leclerc, & Gattuso, 2004; Sanyal, Bijma, Spero, & Lea, 2001; Figure 2, Sanyal et al., 1996) all faithfully reproduce the shape of the Kakihana et al. (1977) $B(OH)_4^-$ curve, although they are variably offset from the curve (see discussion on this offset below). Hönisch et al. (2003, 2004) further evaluated the effects of variable light intensity and respiration on the boron isotope composition of cultured foraminifers and corals, respectively. Whereas planktic foraminifers quantitatively record the influence of those physiological parameters on pH in their microenvironment (Hönisch et al., 2003), corals are not strongly affected by these controls, thus reflecting the different calcification mechanism of the corals (Hönisch et al., 2004).

3. CAVEATS AND COMPLICATIONS

3.1. Isotope Fractionation Factor

There has been a flurry of activity recently to recalculate and measure the isotope fractionation between aqueous $B(OH)_3$ and $B(OH)_4^-$ (Byrne, Yao, Klochko, Tossell, & Kaufman, 2006; Liu & Tossell, 2005; Oi, 2000; Pagani, Lemarchand, Spivack, & Gaillardet, 2005a; Sanchez-Valle et al., 2005; Zeebe, 2005). Whereas Pagani et al. (2005a) merely fitted a curve to the synthetic calcite calibration established by Sanyal et al. (2000), most of these studies attempt to refine the fractionation factor (α) using *ab initio* calculations of the theoretical isotope effect based on the vibrational energies of the two coordinations. All but one of these studies results in α significantly greater than that determined by Kakihana et al. (1977). The range in α determined in these studies is from 1.0176 to 1.030 (see discussion in Zeebe et al., in press). This large range is an indication of the problems with determining α that stems from uncertainties in the vibration frequency data and assumptions used in the calculations. Byrne et al. (2006) provide the first true experimental estimate of $\alpha = 1.0285$. Although this estimate is consistent with other theoretical estimates, the shape of the empirical calibration curves discussed above (Figure 3, Hönisch et al., 2004; Reynaud et al., 2004; Sanyal et al., 2001; Sanyal et al., 1996, 2000) is remarkably consistent with a fractionation factor of ~1.020, close to the Kakihana et al. (1977) fractionation factor. The discrepancy between the empirically determined and the experimentally determined fractionation factors is not understood at this time. This is an unsatisfying aspect of the boron isotope paleo-pH proxy. It should be noted that a larger fractionation factor

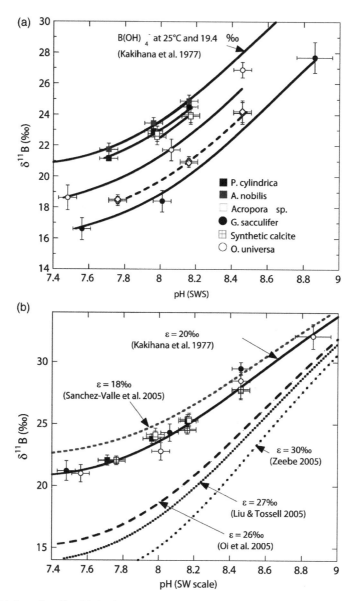

Figure 2 (a) Data for all published calibration studies (solid and shaded squares after Hönisch et al., 2004; open squares after Reynaud et al., 2004; open circles after Sanyal et al., 2001; open diamond after Sanyal et al., 2000; closed circles after Sanyal et al., 1996). Note that all empirical curves are parallel to each other but show constant offsets relative to each other and to the theoretical borate curve by Kakihana et al. (1977). (b) Empirical data from Figure 2a corrected for their specific offsets, and five theoretical borate curves. Empirical data match the shape of the borate curve by Kakihana et al. (1977). Note that minor variability between theoretical and empirical curves may be due to differences in experimental temperature and salinity.

cannot explain the observed $\delta^{11}B$ offsets from the Kakihana et al. (1977) $B(OH)_4^-$ curve, as vital effects would move the curves left or right on Figure 2. Further, it is unlikely to be a problem with the empirical fractionation factor, as other values for α change the shape of the curve and therefore do not fit the calibration data. Zeebe (2005) and (Zeebe et al., in press) recognize the importance of the empirical calibrations, stating that the current uncertainty in α does not preclude the use of the proxy, and that the proxy is robust as long as single species, empirical calibrations are used. Further study, particularly with synthetic carbonate approaches, may help to solve this conundrum.

3.2. Foraminifera Size and Dissolution Effects

In an effort to evaluate all potential complications behind the boron isotope paleo-pH proxy, Hönisch and Hemming (2004) analyzed natural foraminifers from core top samples over a depth range representing variable calcite saturation (Ontong-Java Plateau in the Pacific and the $90°$ East Ridge in the Indian Ocean). It was found that both shell size and shell weight influence the boron isotope values measured. Shell size relates to the depth of calcification of the species analyzed (the symbiont-bearing *Globigerinoides sacculifer*). Smaller shells are consistent with the foraminifers residing deeper in the water column, where light levels are lower and the pH-elevating effect of symbiont photosynthesis is reduced (Figure 3). The lighter $\delta^{11}B$ values of smaller individuals are also in good agreement with Mg/Ca measurements on these same individuals and the expected cooler temperatures at depth. Shells from deeper location core tops have lower size-normalized shell weights, indicative of the amount of dissolution the foraminifera test has experienced in the corrosive bottom waters. The isotopic composition of those shells picked from deeper sediment cores is offset from pristine, unaltered samples, towards lighter boron isotope values (Figure 3). The conclusion of this study is that, in order to make robust interpretations of paleo-pH using the boron isotope proxy in foraminifera, one must select large ($>400\,\mu m$), pristine shells of single calibrated species. Previous authors were not aware of these potential complications so those studies need to be interpreted with caution (see application discussion below).

3.3. Analytical Challenges

Perhaps the biggest obstacle that must be overcome in order to bring the boron isotope paleo-pH proxy into widespread use is the analytical challenge. Early work relied on positive ion methods, beginning with procedures that measured the $Na_2BO_2^+$ ion (Oi et al., 1989; Swihart et al., 1986). This method was capable of measuring geologic samples to a precision of $\pm 2‰$ and required large amounts of boron ($>100\,\mu g$), but was adequate for early studies of borate minerals where sample size was not a problem. However, the method also required high sample purity, and was sensitive to ionization temperature, as well as fractionation during the analysis. The introduction of the $Cs_2BO_2^+$ method (Leeman, Vocke, Beary, & Paulsen, 1991; Nakamura, Ishikawa, Birck, & Allegre, 1992; Ramakumar, Khodade, Parab, Chitambar, & Jain, 1985; Spivack & Edmond, 1986; Xiao, Beary,

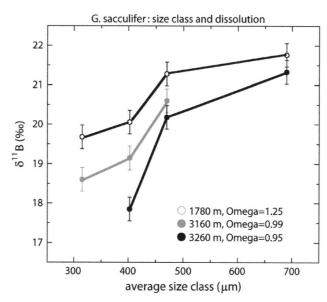

Figure 3 The δ^{11}B of G. *sacculifer* shells from the Ontong-Java Plateau decrease with shell size and with increasing dissolution (with permission from Hönisch & Hemming, 2004). Note that δ^{11}B of the large shells show less difference between shallow and deep sites, and the authors recommend only selecting large shells with no evidence of dissolution for paleo-reconstructions.

& Fassett, 1988) improved on the previous methods by a reduction in sample size to 0.5–1 μg, and the heavier compound reduced problems with in-run fractionation. However, sample purity is still a problem with this method.

In order for the boron isotope paleo-pH proxy to have wide application to marine chemistry and climate change studies, a method that would allow the analysis of very small samples is necessary. Foraminifer shells are currently the most useful archive for paleo-climate proxies, but only contain on the order of 10–15 ppm boron. An unreasonably large sample size would be necessary in order to analyze foraminifers from marine sediment cores using the positive ionization methods. Improvements to NTIMS first developed by Duchateau and de Bièvre (1983), Heumann and Zeininger (1985), and Vengosh, Chivas and McCulloch (1989) opened the door to analysis of very small amounts of boron with high precision (Hemming & Hanson, 1992, 1994). Because the negative ion species BO_2^- ionizes extremely efficiently, sub-nanogram quantities of boron could be analyzed with precision of better than 0.7‰. Recent refinements of this method now result in reported precision of < 0.3‰, on par with the positive ion method (see Hönisch & Hemming, 2005; Pelejero et al., 2005).

Obtaining this precision comes at a price. Each sample must be replicated at least three times (Hönisch & Hemming, 2004, 2005) and extreme care must be taken in order to reproduce running conditions. Because internal normalization is not possible as with stable isotope analysis, mass fractionation during acquisition must be corrected by repeated analysis of standards using exactly the same analytical protocol as the samples. The protocol we have developed includes consistent sample loading procedures, filament heating procedures, strict acceptance criteria for in-run fractionation, and a minimum of three replicate analyses. Some samples must be analyzed 10 times or more in order to obtain three runs that meet all of these criteria. While we think this protocol is essential in order to obtain high precision data, few labs apply these criteria. This makes it difficult to compare results from one lab to another. One lab that has adhered to these guidelines has produced perhaps the best data published to date (Pelejero et al., 2005). Following is a brief review of our procedure.

Samples for data published by Hemming, Sanyal, and Hönisch were measured a minimum of three times, and a minimum of 50–60 isotope ratios are collected over a period of 20–30 min. Data are rejected if the increase in the measured ratio during that time interval exceeds 1‰. Data are also rejected if a signal is observed at mass 26 ($^{12}C^{14}N$ ions), an indication that an isobaric interference may be present. NTIMS measures BO_2^- ions on masses 43 and 42 ($^{11}B^{16}O_2^-$ and $^{10}B^{16}O_2^-$) and organic matter contamination interferes on mass 42 ($^{12}C^{14}N^{16}O$), which results in an underestimate of the original $\delta^{11}B$ of the carbonate (Hemming & Hanson, 1992). As a consequence, samples need to be treated with an oxidizing solution to remove organic matter. Samples are crushed and bleached overnight in commercial bleach or NaOCl. We refrain from H_2O_2 because it is too aggressive and may cause partial dissolution during the cleaning process. The oxidizing solution, as well as potentially adsorbed boron and clay particles, are then removed by repeated rinsing (10 times) and ultrasonication in distilled water. The cleaned carbonate sample is then dissolved in distilled HCl and measured at approximately 980°C on an out-gassed Re filament. The filament temperature and heat-up time are critical for the analyses. Experience has shown that it is best to heat the filament slowly, over 20–30 min. Our goal is to collect three analyses that fall within ± 0.5‰ for every sample, but excessive fractionation and/or isobaric interference through organic matter contamination sometimes require up to ten or more analyses of a single sample solution to reach that goal. Over the years we have optimized heat up time, analysis temperature and we now refrain completely from waiting after filament heat up. We can measure boron amounts as small as 1 ng and obtain 2σ analytical uncertainties of ± 0.3‰ or better. Daily analysis of SRM NBS 951 and/or a seawater standard allows for monitoring of the mass spectrometer performance and data reference relative to SRM NBS 951. Sample cleaning and analysis are always done in exactly the same manner and resulting data for monospecific samples of similar age are internally consistent.

3.4. Secular Variation in Seawater $\delta^{11}B$

The boron isotope paleo-pH proxy requires the knowledge of the isotope composition of the parent fluid in order for it to work. Since boron has a relatively long residence time in seawater, (~3–5 Myr, Lemarchand, Gaillardet, Lewin, & Allègre, 2000, 2002; ~16 Myr, Taylor & McLennan, 1985), it is reasonable to assume there has not been significant variation in the isotopic composition of seawater over those time scales. However, there has been some concern regarding studies that report pH interpretations going back tens of millions of years (Palmer, Pearson, & Cobb, 1998;

Pearson & Palmer, 1999, 2000; Spivack, You, & Smith, 1993). In response to this concern Lemarchand et al. (2000) modeled the expected limits of boron isotope composition of seawater going back ~120 Ma. Their model predicts boron isotope fluctuations over a range of ~6‰, related primarily to continental discharge, alteration of oceanic crust, and mechanical erosion of continents as it supplies a source of sediment for adsorption of boron. They suggest that the relatively large pH changes reported by Pearson and Palmer (2000) and Spivack et al. (1993) may therefore be incorrect. If pH is calculated using the boron isotope data in fora-minifers and the modeled boron isotope composition of seawater, little change in pH is seen over that time period. A secular variation of 0.1‰/Ma is estimated by the Lemarchand et al. (2000) model. However, evidence from past seawater Ca^{2+} and Mg^{2+} concentration and calcite compensation depth suggests that pH has probably increased over the past 100 Ma and that variations have occurred over the Cenozoic (Tyrrell & Zeebe, 2004). Given this independent evidence, an almost constant seawater pH based on the estimated $\delta^{11}B_{seawater}$ (Lemarchand et al., 2000) appears unrealistic.

Clearly, for paleo-pH studies of deep time, a boron isotope secular variation curve must be constructed similar to the Sr isotope curve. There are several potential approaches to this goal:

1) analysis of a deep marine carbonate that grew at low pH values (<7.8) which would represent the maximum offset from the seawater value and allow calculation of the bulk seawater composition; problems with this method include uncertainties in the habitats of benthic foraminifers and ancient species and species-specific offsets as reported in Figure 2;
2) analysis of an extant species (such as brachiopods) that have spanned a large part of the Phanerozoic; problems with this approach again include uncertainties in species-specific offsets (although less likely for this particular organism), the habitat of ancient species, and the ocean pH, increasing the uncertainty of the ocean isotope composition estimate;
3) pristine seawater fluid inclusion analysis of calcium carbonate or other marine mineral precipitates; problems with this method include the difficulty in extraction of the fluid for analysis, and uncertainties in determining whether the inclusion is pristine;
4) analysis of two minerals with known (and different) fractionation from seawater would allow calculation of the seawater isotope composition, however a second suitable mineral (besides $CaCO_3$) has not been identified.

4. APPLICATIONS OF THE BORON ISOTOPE PALEO-PH PROXY

To date, there are surprisingly few published applications of the boron isotope paleo-pH proxy. This is most likely due to the fact that the proxy is still in development, as well as the fact that the analyses are difficult. However, some interesting records are now available, and it is expected that, with the recent increase in our knowledge of the proxy (Hönisch et al., 2003, 2004; Hönisch & Hemming, 2004, 2005), more application studies will be forthcoming.

The first attempt at applying the proxy was the study of Spivack et al. (1993), who completed down-core analyses of mixed species of foraminifers and associated pore water going back approximately 21 Ma. This early attempt suffered from several problems not recognized at the time, including potential species effects, shell size effects, and dissolution effects which could be present in a bulk sample. Further, the interpretation that seawater $\delta^{11}B$ was 5‰ lighter than today prior to ~7.5 Ma is unlikely, and the interpretation that seawater pH was as low as 7.4 is highly suspect.

The second attempt at applying the proxy addressed many of these potential problems by analyzing mono-specific samples, and restricting the study to glacial–interglacial time scales, thus avoiding potential complications with seawater secular variation and reducing the potential for diagenetic effects (Sanyal, Hemming, Hanson, & Broecker, 1995). In this study, the pH of glacial surface water was calculated to be ~0.3 pH units more basic than present-day seawater based on the boron isotope composition of glacial age planktonic foraminifers. The authors showed this was reasonable by calculating the expected surface ocean pH difference based on the known atmospheric pCO_2 of glacial times as measured in air trapped in ice cores. Whereas the authors used a single species for the surface water pH part of this study, and met many of the analytical criteria described previously, the determination of bottom water pH was based on analysis of bulk, mixed benthic foraminifer species. As with the Spivack et al. (1993) study, interpretations based on mixed species analysis are suspect, and the very high bottom water pH (~0.3 pH units greater than Holocene) was not accepted by the paleoceanographic community. The mismatch between the boron isotope estimate and reconstructions based on carbonate preservation (Anderson & Archer, 2002) and Zn/Ca ratios in benthic foraminifers (Marchitto et al., 2005), which suggested a much smaller increase in deepwater saturation, had the unfortunate side effect of making people suspicious of the proxy. Although we still do not know the reason for the heavy boron isotope values measured in that study (perhaps resulting from using mixed benthic species), a recent application of the boron isotope pH proxy using *Cibicidoides wuellerstorfi* has shown that the glacial ocean was at most ~0.1 pH units more basic than today (Hönisch, unpublished data). Sanyal, Hemming, and Broecker (1997) followed up their first study with an application of the boron isotope paleo-pH proxy to a core in the eastern equatorial Pacific. In that study, no surface water change was seen through the marine isotope stage 5–6 transition, which they interpreted as a result of higher glacial upwelling intensity at that site. As with the Sanyal et al. (1995) study, mixed benthic species were analyzed for bottom water pH determinations. The absolute isotope values obtained from those samples were ~2‰ lower, and the relative difference between glacial and Holocene samples was ~1‰ greater as compared to the earlier study, indicating the poor consistency and large uncertainty of the mixed benthic species approach. So again caution should be exercised in interpreting the high bottom water pH values reported there.

Two studies used the boron isotope pH proxy to interpret changes in upwelling over glacial–interglacial time scales (Palmer & Pearson, 2003; Sanyal & Bijma, 1999). The basis for using boron isotopes in this context is that increased upwelling at a particular site will decrease the surface water pH at the site, which should be reflected in the boron isotope composition of foraminifers. Sanyal and Bijma (1999) compared glacial–interglacial changes in pH at two locations, off northwest Africa and in the eastern equatorial Pacific. They interpret the minimal change in pH of surface water in the Pacific core as an indication this area was a significantly greater source of CO_2 to the atmosphere than was the northwest Africa region, which showed a 0.2 pH unit higher value than today. Similarly, the Palmer and Pearson (2003) study of a western equatorial Pacific core interprets the calculated low surface water pH values as disequilibrium between the surface ocean and atmosphere. According to that study, the deglacial surface Pacific had excess CO_2 relative to the atmosphere, and thus the tropical oceans were a source of CO_2 to the atmosphere, so may have triggered or amplified Holocene warming.

Palmer et al. (1998) used the boron isotope composition of various foraminifer species that represent a range of depth of calcification and thus can be used to create a pH-depth profile. Using the same $\delta^{11}B/pH$ relationship for all species investigated, similar pH-depth profiles at five time slices going back ~16 Ma are obtained and suggest little change in the ocean pH structure. Although this approach provides potential for reconstructing the boron isotope composition of seawater through time, it does not take into account specific differences in the isotopic composition recorded by foraminifer species uncalibrated for their $\delta^{11}B$-pH relationship. So again, caution should be applied in interpreting those data.

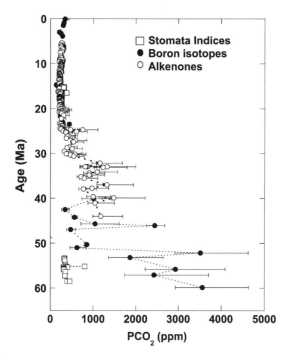

Figure 4 Estimates of pCO_2 through the Cenozoic using boron isotopes (Pearson & Palmer, 2000), alkenones (Pagani et al., 1999, 2005), and stomata indices (Royer et al., 2001).

Pearson and Palmer (1999) and Pearson and Palmer (2000) represent attempts to take the proxy back to deep time. As with the Spivack et al. (1993) study, there is concern whether there has been secular variation of the boron isotope composition of seawater as suggested by models (Lemarchand et al., 2000, 2002). Pearson and Palmer (1999) measured pH–depth profiles from planktic foraminifers over the past 43 Ma and concluded pCO₂ was similar or only slightly elevated relative to pre-industrial levels. The Pearson and Palmer (2000) study analyzed foraminifer samples going back 60 Ma. They construct a surface ocean pH curve that shows relatively small deviations going back ∼20 Ma, but then erratic fluctuations and significantly more acidic surface ocean pH conditions from 40 to 60 Ma (there is a gap in the time series between 20 and 40 Ma). These data can be compared with other proxies for pCO₂ (Figure 4), such as stomata indices (Royer et al., 2001) and the photo-synthetic carbon isotope fractionation as reported in alkenones (Pagani, Freeman, & Arthur, 1999; Pagani, Zachos, Freeman, Tipple, & Bohaty, 2005b). As discussed above, Lemarchand et al. (2002, 2000) reinterpret the Pearson and Palmer (2000) data in light of their model secular variation calculations and find that pH variations over the past 60 Ma may be minor.

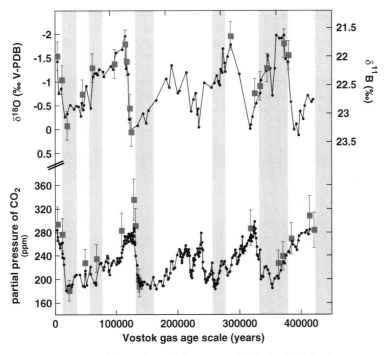

Figure 5   Boron isotope evidence for pH changes at ODP site 668B in the eastern equatorial Atlantic in synchrony with atmospheric pCO₂ variations recorded in the Vostok ice core (Hönisch & Hemming, 2005). In the top panel the black curve is the $\delta^{18}O$ of G. *ruber* from the sediment core, and the red symbols are the measured $\delta^{11}B$ of G. *sacculifer*. In the lower panel, the black curve is the pCO₂ record from Vostok, and the solid symbols are the calculated PCO₂ based on the $\delta^{11}B$ as well as temperature, salinity and alkalinity estimates (for a detailed description see Hönisch & Hemming, 2005).

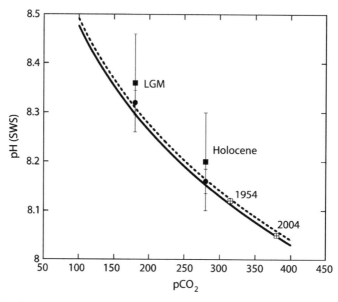

Figure 6 Plot of calculated surface ocean pH for a given atmospheric pCO_2. The two curves are calculated for salinities and temperatures estimated for open tropical waters in the Holocene (34.8 psu, 27°C, 2,300 μmol kg^{-1} alkalinity), and the LGM (35.8 psu, 25°C, 2,370 μmol kg^{-1}). Also shown are reconstructed pH data from planktic foraminifers (black squares: Sanyal et al., 1995, and black circles: Hönisch & Hemming, 2005). The error bars for both are propagated according to the description in the publications. Offsets for data from Sanyal et al. (1995) and Hönisch and Hemming (2005) are due to differences in the analytical protocols used for those two studies. Also shown are the atmospheric pCO_2 data for 1954 and 2004 from Keeling and Whorf (2005), plotted along the calculated pH profiles.

Finally, the most recent study by Hönisch and Hemming (2005) represents a rigorous test of the boron isotope paleo-pH proxy by using boron isotope data of mono-specific foraminifers representing two glacial cycles, calculating aqueous PCO_2 from the pH data, and comparing the results directly with the known atmospheric pCO_2 as measured in air trapped in ice cores. The results are a remarkable success, with PCO_2 values from the proxy matching the ice core data to within $+/-20$ ppmv (Figure 5). Glacial/Holocene studies from ocean areas where atmospheric pCO_2 and aqueous PCO_2 are in equilibrium (Hönisch & Hemming, 2005; Sanyal et al., 1995, 1997) can also be used to assess the robustness of the proxy, as ice core pCO_2 data are available for this time period. Figure 6 compares these studies and demonstrates the good agreement between boron isotope-pH reconstructions and measured changes in atmospheric pCO_2.

 ## 5. SUMMARY AND CONCLUSION

While the development of the boron isotope paleo-pH proxy continues, the general understanding of the proxy allows it to be applied if done carefully. The observations and lessons learned from calibration studies and first attempts at applying the proxy include:

1) data quality is of utmost importance, so any application of the proxy must include evidence that strict analytical protocol has been adhered too, and include replicate analysis of all samples;
2) core locations must be selected carefully, depending on the question being pursued with the proxy (i.e., evidence that upwelling or other major oceanographic changes have not occurred at the location; water depths are above the lysocline);
3) calibrated, mono-specific foraminifera samples of the appropriate size range and with no signs of dissolution or recrystallization must be chosen; extinct species must be calibrated to extant species;
4) absolute pH values can only be reconstructed within the residence time of boron in the oceans, as secular variation of the boron isotope composition of seawater cannot yet be determined beyond this time; until past variations of $\delta^{11}B$ seawater are satisfactorily determined, older samples can only be interpreted with regard to relative pH changes within 3–5 Myr time intervals (i.e., intervals similar to the residence time of boron in seawater).

These observations allow application of the proxy to an important time window that could provide invaluable information on the controls on climate, and thus addresses ocean–atmosphere interactions resulting from recent anthropogenic perturbations.

ACKNOWLEDGMENTS

Discussions, suggestions, and editing by Sidney Hemming, as well as thoughtful suggestions by an anonymous reviewer, greatly improved this manuscript. This research was supported by NSF grants OCE-0326952 and OCE-0083061.

REFERENCES

Anderson, D. M., & Archer, D. (2002). Glacial-interglacial stability of ocean pH inferred from foraminifer dissolution rates. *Nature, 416,* 70–72.

Byrne, R. H., Yao, W., Klochko, K., Tossell, J. A., & Kaufman, A. J. (2006). Experimental evaluation of the isotopic exchange equilibrium $10B(OH)_3 + 11B(OH)_4^- = 11B(OH)_3 + 10B(OH)_4^-$ in aqueous solution. *Deep Sea Research Part I: Oceanographic Research Papers, 53,* 684–688.

Dickson, A. G. (1990). Thermodynamics of the dissociation of boric acid in synthetic seawater from 273.15 to 318.15 K. *Deep-Sea Research, 37,* 755–766.

Duchateau, N. L., & de Bièvre, P. (1983). Boron isotopic measurements by thermal ionization mass spectrometry using the negative BO_2^- ion. *International Journal of Mass Spectrometry and Ion Processes, 54,* 289–297.

Gaillardet, J., & Allègre, C. J. (1995). Boron isotopic compositions of corals: Seawater or diagenesis record? *Earth and Planetary Science Letters, 136,* 665–676.

Hemming, N. G., & Hanson, G. N. (1992). Boron isotopic composition and concentration in modern marine carbonates. *Geochimica et Cosmochimica Acta, 56,* 537–543.

Hemming, N. G., & Hanson, G. N. (1994). A procedure for the isotopic analysis of boron by negative thermal ionization mass spectrometry. *Chemical Geology, 114*, 147–156.

Hemming, N. G., Reeder, R. J., & Hanson, G. N. (1995). Mineral-fluid partitioning and isotopic fractionation of boron in synthetic calcium carbonate. *Geochimica et Cosmochimica Acta, 59*, 371–379.

Heumann, K. G., & Zeininger, H. (1985). Boron trace determination in metals and alloys by isotope dilution mass spectrometry with negative thermal ionization. *International Journal of Mass Spectrometry and Ion Processes, 67*, 237–252.

Hönisch, B., Bijma, J., Russell, A. D., Spero, H. J., Palmer, M. R., Zeebe, R. E., & Eisenhauer, A. (2003). The influence of symbiont photosynthesis on the boron isotopic composition of foraminifera shells. *Marine Micropaleontology, 49*, 87–96.

Hönisch, B., & Hemming, N. G. (2004). Ground-truthing the boron isotope paleo-pH proxy in planktonic foraminifera shells: Partial dissolution and shell size effects. *Paleoceanography, 19*, doi:10.1029/2004PA001026.

Hönisch, B., & Hemming, N. G. (2005). Surface ocean pH response to variations in pCO$_2$ through two full glacial cycles. *Earth and Planetary Science Letters, 236*, 305–314.

Hönisch, B., Hemming, N. G., Grottoli, A. G., Amat, A., Hanson, G. N., & Bijma, J. (2004). Assessing scleractinian corals as recorders for paleo-pH: Empirical calibration and vital effects. *Geochimica et Cosmochimica Acta, 68*, 3675–3685.

Kakihana, H., Kotaka, M., Satoh, S., Nomura, M., & Okamoto, M. (1977). Fundamental studies on the ion-exchange of boron isotopes. *Bulletin of the Chemical Society of Japan, 50*, 158–163.

Keeling, C. D., & Whorf, T. P. (2005). Atmospheric CO$_2$ records from sites in the SIO sampling network, in trends: A compendium of data on global change. C. d. i. a. center, Oak Ridge National Laboratory, TN: U.S. Department of Energy.

Leeman, W. P., Vocke, R. D., Beary, E. S., & Paulsen, P. J. (1991). Precise boron isotopic analysis of aqueous samples: Ion exchange extraction and mass spectrometry. *Geochimica et Cosmochimica Acta, 55*, 3901–3907.

Lemarchand, D., Gaillardet, J., Lewin, É., & Allègre, C. J. (2000). The influence of rivers on marine boron isotopes and implications for reconstructing past ocean pH. *Nature, 408*, 951–954.

Lemarchand, D., Gaillardet, J., Lewin, E., & Allegre, C. J. (2002). Boron isotope systematics in large rivers: Implications for the marine boron budget and paleo-pH reconstruction over the Cenozoic. *Chemical Geology, 190*, 123–140.

Liu, Y., & Tossell, J. A. (2005). *Ab initio* molecular orbital calculations for boron isotope fractionations on boric acids and borates. *Geochimica et Cosmochimica Acta, 69*, 3995–4006.

Marchitto, T. M., Lynch-Stieglitz, J., & Hemming, S. R. (2005). Deep Pacific CaCO$_3$ compensation and glacial-interglacial atmospheric CO$_2$. *Earth and Planetary Science Letters, 231*, 317–336.

Nakamura, E., Ishikawa, T., Birck, J.-L., & Allegre, C. J. (1992). Precise boron isotopic analysis of natural rock samples using a boron-mannitol complex. *Chemical Geology, 94*, 193–204.

Oi, T. (2000). *Ab initio* molecular orbital calculations of reduced partition function ratios of polyboric acids and polyborate anions. *Zeitschrift fuer Naturforschung, 55*, 623–628.

Oi, T., Nomura, M., Musashi, M., Ossaka, T., Okamoto, M., & Kakihana, H. (1989). Boron isotopic compositions of some boron minerals. *Geochimica et Cosmochimica Acta, 53*, 3189–3195.

Pagani, M., Freeman, K. H., & Arthur, M. A. (1999). Late Miocene atmospheric CO$_2$ concentrations and the expansion of C4 grasses. *Science, 285*, 876–879.

Pagani, M., Lemarchand, D., Spivack, A., & Gaillardet, J. (2005a). A critical evaluation of the boron isotope-pH proxy: The accuracy of ancient ocean pH estimates. *Geochimica et Cosmochimica Acta, 69*, 953–961.

Pagani, M., Zachos, J. C., Freeman, K. H., Tipple, B., & Bohaty, S. (2005b). Marked decline in atmospheric carbon dioxide concentrations during the Paleogene. *Science, 309*, 600–603.

Palmer, M. R., & Pearson, P. N. (2003). A 23,000-year record of surface water pH and PCO$_2$ in the Western Equatorial Pacific Ocean. *Science, 300*, 480–482.

Palmer, M. R., Pearson, P. N., & Cobb, S. J. (1998). Reconstructing past ocean pH-depth profiles. *Science, 282*, 1468–1471.

Palmer, M. R., Spivack, A. J., & Edmond, J. M. (1987). Temperature and pH controls over isotopic fractionation during adsorption of boron on marine clay. *Geochimica et Cosmochimica Acta, 51*, 2319–2323.

Pearson, P. N., & Palmer, M. R. (1999). Middle Eocene seawater pH and atmospheric carbon dioxide concentrations. *Science, 284*, 1824–1826.

Pearson, P. N., & Palmer, M. R. (2000). Atmospheric carbon dioxide concentrations over the past 60 million years. *Nature, 406*, 695–699.

Pelejero, C., Calvo, E., McCulloch, M. T., Marshall, J. F., Gagan, M. K., Lough, J. M., & Opdyke, B. N. (2005). Preindustrial to modern interdecadal variability in coral reef pH. *Science, 309*, 2204–2207.

Ramakumar, K. L., Khodade, P. S., Parab, A. R., Chitambar, S. A., & Jain, H. C. (1985). Application of dicesium metaborate ion for the determination of boron by thermal ionization mass spectrometry. *Journal of Radioanalytical and Nuclear Chemistry, 107*, 215–223.

Reynaud, S., Hemming, N. G., Juillet-Leclerc, A., & Gattuso, J.-P. (2004) Effect of pCO_2 and temperature on the boron isotopic composition of a zooxanthellate coral: *Acropora* sp. *Coral Reefs, 23*, 539–546. doi:10.1007/s00338-00004-00399.

Royer, D. L., Wing, S. L., Beerling, D. J., Jolley, D. W., Koch, P. L., Hickey, L. J., & Berner, R. A. (2001). Paleobotanical evidence for near present-day levels of atmospheric CO_2 during part of the tertiary. *Science, 292*, 2310–2313.

Sanchez-Valle, C., Reynard, B., Daniel, I., Lecuyer, C., Martinez, I., & Chervin, J.-C. (2005). Boron isotopic fractionation between minerals and fluids: New insights from *in situ* high pressure-high temperature vibrational spectroscopic data. *Geochimica et Cosmochimica Acta, 69*, 4301–4313.

Sanyal, A., & Bijma, J. (1999). A comparative study of northwest Africa and eastern equatorial Pacific upwelling zones as sources of CO_2 during glacial periods based on boron isotope paleo-pH estimation. *Paleoceanography, 14*, 753–759.

Sanyal, A., Bijma, J., Spero, H. J., & Lea, D. W. (2001). Empirical relationship between pH and the boron isotopic composition of G. *sacculifer*: Implications for the boron isotope paleo-pH proxy. *Paleoceanography, 16*, 515–519.

Sanyal, A., Hemming, N. G., & Broecker, W. S. (1997). Changes in pH in the eastern equatorial Pacific across stage 5–6 boundary based on boron isotopes in foraminifera. *Global Biogeochemical Cycles, 11*, 125–133.

Sanyal, A., Hemming, N. G., Broecker, W. S., Lea, D. W., Spero, H. J., & Hanson, G. N. (1996). Oceanic pH control on the boron isotopic composition of foraminifera: Evidence from culture experiments. *Paleoceanography, 11*, 513–517.

Sanyal, A., Hemming, N. G., Hanson, G. N., & Broecker, W. S. (1995). Evidence for a higher pH in the glacial ocean from boron isotopes in foraminifera. *Nature, 373*, 234–236.

Sanyal, A., Nugent, M., Reeder, R. J., & Bijma, J. (2000). Seawater pH control on the boron isotopic composition of calcite: Evidence from inorganic calcite precipitation experiments. *Geochimica et Cosmochimica Acta, 64*, 1551–1555.

Schwarcz, H. P., Agyei, E. K., & McMullen, C. C. (1969). Boron isotopic fractionation during clay adsorption from sea-water. *Earth and Planetary Science Letters, 6*, 1–5.

Spivack, A. J., & Edmond, J. M. (1986). Determination of boron isotope ratios by thermal ionization mass spectrometry of the dicesium metaborate cation. *Analytical Chemistry, 58*, 31–35.

Spivack, A. J., Palmer, M. R., & Edmond, J. M. (1987). The sedimentary cycle of the boron isotopes. *Geochimica et Cosmochimica Acta, 51*, 1939–1949.

Spivack, A. J., You, C.-F., & Smith, H. J. (1993). Foraminiferal boron isotope ratios as a proxy for surface ocean pH over the past 21 Myr. *Nature, 363*, 149–151.

Swihart, G. H., Moore, P. B., & Callis, E. L. (1986). Boron isotopic composition of marine and nonmarine evaporite borates. *Geochimica et Cosmochimica Acta, 50*, 1297–1301.

Taylor, S. R., & McLennan, S. M. (1985). The continental crust: Its composition and evolution (pp. 312). Oxford: Blackwell.

Tyrrell, T., & Zeebe, R. E. (2004). History of carbonate ion concentration over the last 100 million years. *Geochimica Cosmochimica Acta, 68*, 3521–3530.

Vengosh, A., Chivas, A. R., & McCulloch, M. T. (1989). Direct determination of boron and chlorine isotopes in geological materials by negative thermal-ionization mass spectrometry. *Chemical Geology (Isotope Geoscience Section)*, *79*, 333–343.

Vengosh, A., Kolodny, Y., Starinsky, A., Chivas, A. R., & McCulloch, M. T. (1991). Coprecipitation and isotopic fractionation of boron in modern biogenic carbonates. *Geochimica et Cosmochimica Acta*, *55*, 2901–2910.

Xiao, Y.-K., Beary, E. S., & Fassett, J. D. (1988). An improved method for the high-precision isotopic measurement of boron by thermal ionization mass spectrometry. *International Journal of Mass Spectrometry*, *85*, 203–213.

Zeebe, R. E. (2005). Stable boron isotope fractionation between dissolved $B(OH)_3$ and $B(OH)_4^-$. *Geochimica et Cosmochimica Acta*, *69*, 2753–2766.

Zeebe, R. E., Bijma, J., Hönisch, B., Sanyal, A., Spero, H. J., & Wolf-Gladrow, D. A. (in press). Vital effects and beyond: A modelling perspective on developing paleoceanographic proxy relationships in foraminifera. In R. James et al. (Eds), *Biogeochemical controls on palaeoceanographic proxies*. London: The Geological Society.

CHAPTER EIGHTEEN

The Use of Oxygen and Carbon Isotopes of Foraminifera in Paleoceanography

Ana Christina Ravelo* *and* Claude Hillaire-Marcel

Contents

1. Introduction	735
2. Notation and Standards	736
3. Stratigraphic and Paleoecological Use of Foraminifera	738
4. Foraminiferal Oxygen Isotopes as Environmental Proxies	740
4.1. Factors that influence the $\delta^{18}O$ of seawater	741
4.2. Factors that influence the $\delta^{18}O$ of foraminifera	746
5. Foraminiferal Carbon Isotopes as Environmental Proxies	751
5.1. Factors that influence the $\delta^{13}C$ of DIC of seawater	752
5.2. Factors that influence the $\delta^{13}C$ of foraminifera	756
6. Conclusion and Summary	759
References	760

1. Introduction

Oxygen and carbon isotopes of foramininfera have been used in paleoceanography studies for decades. Urey (1947) laid the foundations for stable isotope geochemistry when he described and calculated the thermodynamic properties and fractionation of isotopes. Simultaneously, Alfred Nier (1947) developed mass spectrometry techniques allowing the measurement of minute differences in isotopic compositions between natural compounds. Urey's calculations and Nier's technological developments thus opened a large array of applications in stable isotopes geochemistry. Among these, and following McCrea's (1950) early work, Urey and his graduate students developed the use of oxygen isotope composition of calcite as a paleothermometer (Epstein, Buchsbaum, Lowenstam, & Urey, 1953; Urey, Epstein, Lowenstam, & McKinney, 1951). Cesare Emiliani, one of Urey's students, was the first to use oxygen isotope paleothermometry to reconstruct the glacial–interglacial swings in climate of the late Pleistocene using fossil foramininfera shells from deep-sea

* Corresponding author.

Developments in Marine Geology, Volume 1
ISSN 1572-5480, DOI 10.1016/S1572-5480(07)01023-8

sediments (Emiliani, 1955). Although he initially overestimated the glacial–interglacial temperature changes by not adequately taking into account the large changes in the oxygen isotopic composition of seawater, he was the first to use oxygen isotopic records in support of the Milankovitch theory (Emiliani & Geiss, 1959), and was responsible for the initial use of Marine Isotope Stages (MIS) (Emiliani, 1955).

In the early 1960s, major studies about isotope fractionation processes within the hydrological cycle were published by Craig (1961) and Dansgaard (1964). They led Shackleton (1967) to revise Emiliani's interpretation and to conclude that glacial/interglacial variations in foraminiferal oxygen isotope records were primarily influenced by changes in the oxygen isotopic composition of seawater, rather than temperature. In subsequent years, MIS finally emerged as a basic stratigraphic tool (Shackleton & Opdyke, 1973). Thus, the work of Urey, his students, Shackleton and others, developed fundamental concepts that launched the field of paleoceanography, and for the last 50 yr or so, measuring the oxygen and carbon isotopic composition of fossil foraminiferal calcite has been one of the most effective techniques for reconstructing ocean and climate conditions of past times.

Over the last few decades, oxygen and carbon isotopic records derived from measurements of fossil foraminiferal shells have been used to address a large range of questions regarding the evolution and history of the ocean and climate. However, these discoveries were not possible without parallel studies aimed at understanding biological factors, or 'vital effects', that cause some species of foraminifera to calcify out of equilibrium with seawater. In addition, the species-specific ecology of planktonic and benthic foraminifera has been studied to understand how isotopic records might be interpreted in light of, for example, the seasonality and depth of calcification in the water column (for planktonic species) and in the sediment (for benthic species). Overviews of oxygen isotope paleo-thermometry techniques and the 'vital effects' and ecological factors that impact the oxygen and carbon isotopes of biogenic carbonates have been recently published (Hoefs, 2004; Rohling & Cooke, 1999; Sharp, 2006). As such, this chapter is not intended to be a comprehensive review of foraminiferal oxygen and carbon stable isotope biogeochemistry, rather it is intended to be a practical introduction to the factors that must be taken into account when interpreting paleoceanographic records derived from measurements of the oxygen and carbon isotopic composition of foraminifera.

2. NOTATION AND STANDARDS

The stable isotopes of oxygen used in paleoceanographic studies are ^{16}O and ^{18}O, which comprise 99.63 and 0.1995% of the oxygen on Earth, respectively (Faure, 1986). The stable isotopes of carbon are ^{12}C and ^{13}C, which comprise 98.89 and 1.11% of the stable carbon on Earth, respectively (Faure, 1986). Accurate quantification of low abundances of the rare isotopes (^{18}O and ^{13}C) is possible only as ratios to the more common isotopes ($^{18}O/^{16}O$ and $^{13}C/^{12}C$) in the sample, as expressed in comparison with the ratios of a known standard. The difference in the ratio of the sample compared with the standard is expressed as a delta (δ) value:

$$\delta^{18}O = \frac{^{18}O/^{16}O_{sample} - ^{18}O/^{16}O_{standard}}{^{18}O/^{16}O_{standard}} \times 1000$$

$$\delta^{13}C = \frac{^{13}C/^{12}C_{sample} - ^{13}C/^{12}C_{standard}}{^{13}C/^{12}C_{standard}} \times 1000$$

The $\delta^{18}O$ and $\delta^{13}C$ values have concentration units of per thousand, or 'per mil' (‰) relative to the standard. For example, a $\delta^{18}O$ value of 1.0‰ means that the sample has an $^{18}O/^{16}O$ ratio that is 0.1% greater than the standard, or a $\delta^{13}C$ value of -25‰ means that the sample has a $^{13}C/^{12}C$ ratio that is 2.5% lower than that of the standard.

The first step in analyzing the carbon and oxygen isotopic composition of most substrates is to produce carbon dioxide (CO_2) gas that has no offset or a known isotopic offset, relative to the sample; this is done by various methods depending on the substrate. To determine the $\delta^{18}O$ and $\delta^{13}C$ values of foraminifera shells, the shells are dissolved at a given temperature in orthophosphoric acid (see Bowen, 1966; Burman, Gustafsson, Segl, & Schmitz, 2005) to produce CO_2. To determine the $\delta^{18}O$ of seawater, CO_2 gas is isotopically equilibrated with seawater at a constant temperature following the original procedure of Epstein and Mayeda (1953). To determine the $\delta^{13}C$ value of dissolved inorganic carbon (DIC) of seawater, CO_2 is stripped from the seawater by acidification (e.g., St-Jean, 2003). Once CO_2 is isolated, a gas-source stable isotope mass spectrometer with three collectors is typically used to ionize the CO_2 gas and to quantify the isotopic ratios ($^{18}O/^{16}O$ and $^{13}C/^{12}C$) of the sample CO_2 and of an aliquot of CO_2 reference gas. Until recently, variations of natural isotope abundances were usually measured using dual inlet mass spectrometers derived from the original design of Alfred Nier, allowing the introduction and comparison of, alternatively, a reference gas and the sample gas, in the mass spectrometer ionization chamber (i.e., source). Recent improvements in mass spectrometers now allow similar measurements in continuous flow mode, using a carrier gas (helium) for CO_2 and introducing sequentially standard and sample gases. The $\delta^{18}O$ and $\delta^{13}C$ values can then be calculated.

Various standards are used in different laboratories, but all lab standards are calibrated to international reference standards. Originally, they were the "historical" Pee Dee Belemnite (*Belemnitella americana*) shell from the Cretaceous Pee Dee formation (PDB) and/or the Standard Mean Ocean Water (SMOW). The need for clarification arose in the 1990s, notably due to the exhaustion of the original PDB standard and because the definition of the "Standard Mean Ocean Water" was unclear. Coplen (1996) clarified guidelines to report isotopic compositions against standards of the *International Atomic Energy Agency* of Vienna, the VPDB and VSMOW (i.e., the "Vienna" PDB and SMOW). These guidelines are now largely adopted. Worthy of mention is the fact that most laboratories doing stable isotope studies on foraminifera also use the "Carrara Marble" standard as calibrated by

M. Hall (University of Cambridge) against VPDB ($\delta^{13}C = 2.25‰$; $\delta^{18}O = -1.27‰$). It should be noted that the reported carbon isotope composition of this "Carrara Marble" is slightly distinct from the value listed in IAEA documents for the Carrara Marble-C1 radiocarbon reference material ($\delta^{13}C = 2.42‰$).

All $\delta^{18}O$ and $\delta^{13}C$ values of carbonates, and all $\delta^{13}C$ values of dissolved inorganic carbon (DIC) of seawater are thus reported relative to VPDB, and all $\delta^{18}O$ values of water (snow, ice, rain, groundwater, seawater) are reported relative to VSMOW. Typically the analytical errors, or external precision, that are reported in the literature are based on the long term (month to year) reproducibility of a lab standard, and, for calcite are approximately 0.05‰ for $\delta^{13}C$ and 0.08% for $\delta^{18}O$ (both $\pm 1\sigma$), or slightly better.

3. STRATIGRAPHIC AND PALEOECOLOGICAL USE OF FORAMINIFERA

The use of foraminiferal stable isotope data from deep-sea sediment studies requires minimum knowledge of both sediment-sample properties and analytical procedures. Due to benthic organism mobility, and given subsequent mixing of sediment, mixing of foraminiferal populations has differential impacts according to shell size (Bard, 2001) and sediment-sample thickness, for example. Mixing results in smoothed isotope records, adding to biases in peak-abundances for any given shell fluxes to the sea-floor (e.g., Guinasso & Shink, 1975), thus in the recording of isotopic "excursions" and shifts. The interpretation of foraminiferal isotopic data has thus intrinsic limitations in terms of temporal resolution, absolute timing, and sensitivity. Unfortunately, mixing is not a constant function, as briefly explained in Chapter 6, and foraminiferal fluxes also vary in time, thus making this caveat even more critical.

Any gradual change of isotopic composition observed in a given core where some "bioturbation" of the sediment has occurred can thus be interpreted either as the reflection of gradual changes in paleoenvironmental conditions, or as a smoothing effect through a sharp transition resulting from a rapid paleoenvironmental shift. In such situations, comparative measurements of single-shells from each sediment sample may provide a better insight into the modal distribution of isotopic compositions. Thomas, Zachos, Bralower, Thomas, and Bohaty (2002) used this approach to document a paleoclimate event in the Early Eocene. Improvements in analytical techniques notably with respect to sample size, means this approach can be used to examine critical isotopic transitions.

Another critical aspect when analyzing foraminiferal populations concerns the size dependence of isotopic compositions, notably in planktic foraminifers, a feature that was recognized long ago (see examples in Berger, Be, & Vincent, 1981 and Section 4.2, below). The usual procedure to obtain isotopic records thus consists of making isotopic measurements on specimens of a given size range (e.g., 150–250 µm for most planktic foraminifera). However, this size-selection may also result in paleoecologically biased reconstruction when large interannual and/or seasonal

variability in hydrographic conditions and primary productivity have been responsible for variable growth rates and variable size distributions of shells of a given
species or morphotype. Size-selection may then results in records biased towards
optimum environmental conditions and the smoothing out of departures from such
conditions.

Recent progress in the genetics of foraminiferal populations has shed new light
on some of the aforementioned problems, but they have also shown that some
genera could present morphological convergences between species or sub-species
with drastically distinct ecological requirements, thus leading to potential biases in
isotopic records if samples composed of a mixed assemblage of foraminifera are used
for isotopic analyses. A morphotype commonly referred to as *Neogloboquadrina
pachyderma* dextral (Npd) constitutes an example. Np is the most common species in
sub-Arctic basins and the only planktic foraminifer present at very high latitudes.
The left-coiled morphotype of this species (Npl) seems to represent a relatively
homogeneous genetic population, whereas Npd may either correspond to a small
percentage of specimens with true Np affinities, or to what Darling, Kucera,
Kroon, and Wade (2006) identify now as a distinct species (*N. incompta*). This later
taxon has ecological requirements and isotopic compositions not unlike those of the
more temperate species *Globigerina bulloides* (e.g., de Vernal & Hillaire-Marcel,
2006). However, the dextral form of Np *stricto sensu* has isotopic compositions
and likely ecological behaviors like those of Npl (e.g., Hillaire-Marcel, de Vernal,
Polyak, & Darby, 2004).

Benthic foraminifers are less sensitive to most of the limitations discussed above,
but they also require some analytical precautions. Most species show a well-
developed ^{13}C-depleted organic lining. This requires an analytical pretreatment
prior to CO_2 extraction with orthophosphoric acid to avoid contamination of the
calcite-derived CO_2 by the isotopically light CO_2 from the lining. Pretreatment
typically involves heating the sample for one to one and a half hours, in a furnace set
at 250–300°C, under a flux of helium or under vacuum. As a matter of fact, the
relative abundance of benthic foraminiferal linings vs. shells, in the sediment,
can provide an index for quantifying carbonate dissolution (de Vernal, Bilodeau,
Hillaire-Marcel, & Kassou, 1992) and thus for discarding isotopic records potentially biased by selective dissolution of lighter shells.

Within the above intrinsic limitations, the oxygen isotope composition of
foraminiferal calcite is linked to environmental conditions through the so-called
"paleotemperature equation" that represents a second order polynomial approximation of the thermodependent fractionation factor between calcite and water in
the following (simplified) isotopic equilibrium:

$$1/3CaC^{16}O_3 + H_2^{18}O \rightarrow 1/3CaC^{18}O_3 + H_2^{16}O$$

Actually, isotopic exchanges between water and calcite occur through DIC species, and the final contribution of CO_3^{2-} or HCO_3^- ions to the calcite of the shell
can be responsible of significant departures from the theoretical equilibrium, as
discussed in Section 4.2 below. Many scientists working with foraminiferal calcite
still use the 1953 equation of Epstein et al., slightly modified by Shackleton (1974),
using notably experimental data from O'Neil that provide better constrains in the

low temperature domain ($<10°C$) lacking in the original Epstein and others'
dataset:

$$t = 16.9 - 4.38\,(\delta_c - A) + 0.10\,(\delta_c - A)^2$$
$$A = \delta_w - 0.27\%$$

where t represents the temperature (in °C), δ_c and δ_w, the isotopic composition in
δ-units (i.e., ‰ deviation of $^{18}O/^{16}O$ ratio vs. standard value) of, respectively,
calcite (measured against VPDB), and ambient water (measured against VSMOW).
The offset value of -0.27% allows for conversion to the VPDB scale after Coplen
(1988). This equation is valid for calcite precipitation in seawater with near-normal
salinity (as well as in freshwater), but may require adjustments in high salinity
environments, where the oxygen-content of DIC species cannot be considered
negligible with respect to that of ambient water. This somewhat complicated expression of the "classical" paleotemperature equation, linking oxygen compositions
of biogenic calcite to that of its ambient water, is directly inherited from historical
analytical procedures for $\delta^{18}O-CO_2$ and $\delta^{18}O-H_2O$ measurements developed
in the 1950s by Urey and his students. Later on, several authors tried to better
document the carbonate–water paleotemperature equation based on experimental
growth of foraminifera shells (Erez & Luz, 1983; Bemis et al., 1988). These equations have small offsets but generally provide clustered slopes with $d\delta/dt$ relationships ranging from -0.21 to $-0.23\%/°C$ within the foraminiferal growth
temperature domain.

Whatever its mathematical expression, the "paleotemperature equation" thus
links the isotopic composition of calcite to temperature during calcite precipitation,
and to the isotopic composition of ambient water. The latter depends in turn on
a large array of variables, salinity being indirectly a primary one. In subsequent
sections we will examine to what extent these environmental parameters can be
quantitatively reconstructed from foraminiferal isotopic compositions.

4. Foraminiferal Oxygen Isotopes as Environmental Proxies

As discussed in the above section, the oxygen isotopic composition of a
foraminifera shell reflects the oxygen isotopic composition of the seawater in which
the shell calcifies, but the offset of the $\delta^{18}O$ of a foraminiferal shell from the $\delta^{18}O$ of
seawater also depends on temperature assuming thermodynamic isotopic equili-
brium between seawater and calcite (Figure 1). In the temperature range of biogenic
calcite precipitation (i.e., ~40 to $-2°C$), calcite is enriched in ^{18}O by $30-35\%$, in
comparison with ambient water, when both compositions are expressed against the
same standard reference value. Within the same temperature range, the expression
(δ_c-A) in the paleotemperature equation which is not much different from
$(\delta_c-\delta_w)$ with δ_c and δ_w expressed against VPDB and VSMOW, respectively, varies
between ~0 and $+5\%$.

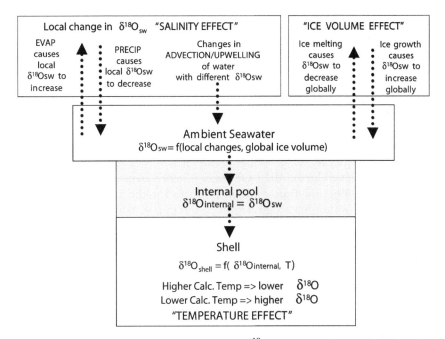

Figure 1 Environmental factors that influence the $\delta^{18}O$ of foraminifera shells. This schematic assumes that thermodynamic equilibrium fractionation occurs, and does not depict potential vital effects.

Many species produce shells that are in oxygen isotopic equilibrium with seawater, however some do not notably because of biological 'vital' effects. This section includes discussions of how downcore foraminiferal $\delta^{18}O$ records are used, in light of such effects, to record changes in the $\delta^{18}O$ of seawater and changes in the temperature of calcification.

4.1. Factors that Influence the $\delta^{18}O$ of Seawater

The $\delta^{18}O$ of seawater can vary with time due to several different processes, some which influence the $\delta^{18}O$ of the global ocean, and some which influence the local $\delta^{18}O$ of seawater. The $\delta^{18}O$ of the global ocean is primarily influenced by changes in the amount of water stored as ice on land which influences the $\delta^{18}O$ of the global ocean on the timescale of the mixing time of the ocean (millennium) (Shackleton, 1967), and by changes in temperature-dependent isotopic exchange with the oceanic crust (Gregory & Taylor, 1981; Muehlenbachs & Clayton, 1976) on the timescale that hydrothermal fluxes influence ocean chemistry (hundreds of millions of years). For the purposes of understanding late Cenozoic climate change, paleoceanographers typically focus on ice volume as the dominant factor influencing the $\delta^{18}O$ of the global ocean. Due to differences in vapor pressure, $H_2^{16}O$ evaporates more readily than $H_2^{18}O$, and therefore, the $\delta^{18}O$ value of water vapor, cloud droplets, and precipitation are low compared with that of seawater. Because

$H_2^{18}O$ condenses more readily than $H_2^{16}O$, as water is removed from clouds through precipitation, the remaining cloud water vapor becomes increasingly depleted in ^{18}O with increasing latitude and altitude, and decreasing temperature, in a Rayleigh distillation process. Thus, high latitude cloud water vapor and therefore snowfall derived from those clouds, is strongly depleted in ^{18}O, and therefore ice sheets are a reservoir of water with low $\delta^{18}O$ values, in the range of −30 to −50‰ (e.g., IAEA, 2000).

During glacial periods, when low $\delta^{18}O$ water is stored in ice sheets, the mean $\delta^{18}O$ value of the world's oceans was relatively high. For example, in the most recent glacial period ∼21 ka ago, or the Last Glacial Maximum (LGM), the average $\delta^{18}O$ of the global ocean was ∼1.1‰ higher than today (Adkins, McIntyre, & Schrag, 2002). Sea level during the LGM was at least 120 m below present day sea level (Fairbanks, 1989), and thus a rough estimate for the change in the average $\delta^{18}O$ of the global ocean due to glaciation is: 0.09‰/−10 m of sea level.

However, for any other time period, the precise relationship would depend on the average $\delta^{18}O$ of the ice sheets during that time period (which is dependent on their geographical location and the climatic conditions when they formed).

Foraminiferal $\delta^{18}O$ records do not only reflect changes in global ice volume, but they also reflect changes in the $\delta^{18}O$ of seawater due to local processes. As discussed above, $H_2^{16}O$ evaporates more readily than $H_2^{18}O$, and $\delta^{18}O$ values of precipitation are low compared with that of seawater. Consequently, evaporation causes surface water salinity and $\delta^{18}O$ values to increase, and precipitation causes surface water salinity and $\delta^{18}O$ values to decrease. Because salinity and $\delta^{18}O$ are both influenced by the balance of evaporation relative to precipitation, they are highly correlated in the surface ocean with higher values at low latitudes and lower values at high latitudes. Building on a dataset from the GEOSECS expedition, analyzed by Harmon Craig, LeGrande, and Schmidt (2006) and Schmidt, Bigg, and Rohling (1999) provided a larger $\delta^{18}O$ database with which to assess the distribution of $\delta^{18}O$ throughout the modern ocean (Figure 2). Taking into account the global distribution of surface water salinity and $\delta^{18}O$, the relationship between geographical variations in $\delta^{18}O$ and salinity is roughly: 0.5‰/1.0 psu. This relationship is approximately what is expected due to Rayleigh distillation processes, with high-latitude surface water being influenced by excess precipitation that has low values, and low-latitudes being dominated by excess evaporation which causes relatively high $\delta^{18}O$ values of surface water. However, regional relationships of $\delta^{18}O$ to salinity are dominated by the effects of mixing between regional precipitation (fresh water) and seawater, and as such, the slope of regional mixing lines deviate significantly from the global relationship given above. As a consequence, in tropical regions, where the $\delta^{18}O$ of rainfall is relatively high (0 to −10‰) the slope of the mixing line is less steep and the relationship between surface water $\delta^{18}O$ and salinity is less than the global average of 0.5‰/psu. For example in the eastern equatorial Atlantic, the western equatorial Atlantic, and the eastern equatorial Pacific, the relationship between $\delta^{18}O$ and salinity of surface water is only 0.08, 0.18, and 0.26‰/psu, respectively (Fairbanks, Charles, & Wright, 1992). As a result, the $\delta^{18}O$ values of tropical and sub-tropical surface water, away from coastal regions where rivers enter the ocean, are fairly uniform, ranging from ∼0.9 to 1.2‰ in the tropical Atlantic, and ∼0.2–0.5‰ in the tropical Pacific.

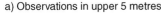

a) Observations in upper 5 metres

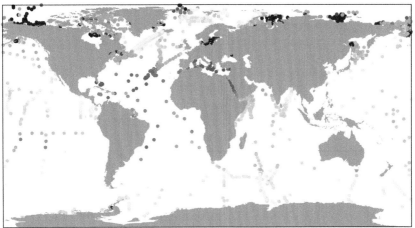

b) Gridded Surface Dataset

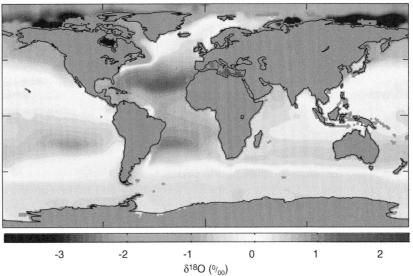

$\delta^{18}O$ (‰)

Figure 2   (a) All $\delta^{18}O$ measurements from the upper 5 meters of the water column (Schmidt et al., 1999) illustrates good data coverage in the Arctic and Atlantic Oceans, but very sparse data in areas such as the Pacific and Southern Oceans; (b) the 1 X 1 gridded data set of surface $\delta^{18}O$. Figure is from LeGrande and Schmidt (2006). Reproduced by permission of American Geophysical Union.

At high latitudes, where precipitation has relatively low values and falls directly on to the ocean surface or where large amounts of freshwater from Arctic rivers and snow and ice melt enters the ocean, the relationship between $\delta^{18}O$ and salinity of surface water is greater than 0.5‰/psu. The situation is slightly more complicated in the North Atlantic where water mass advection linked to the Atlantic Meridional Overturning component of the general thermohaline circulation tightly controls the salinity relationships in the surface water layer. A strong longitudinal regionalism superimposes the latitudinal gradients (Figure 3) with $\delta^{18}O$ vs. salinity gradients that vary from 0.4‰/psu northeastward, under the influence of the evaporated waters of the North Atlantic Drift, to 0.6‰/psu, northwestward, where Arctic rivers constrain the freshwater end-member of the relationship ($\delta^{18}O$ ranging from ~−19 to −22‰ for most Arctic rivers, except the Ob that is slightly less ^{18}O-depleted; Cooper et al., 2005). As a matter of fact, Arctic surface waters significantly depart from simple mixing patterns between two end-members (Bédard, Hillaire-Marcel, & Pagé, 1981) (Figure 3). There, sea-ice growth from low salinity surface waters is responsible for the production of isotopically light brines that sink and mixes with the subsurface more saline North Atlantic Water mass. When sea-ice melts, it releases high $\delta^{18}O$-low salinity waters that mix with the surface layer resulting in a large scatter in $\delta^{18}O$ vs. salinity data (Figure 3).

Thus, in addition to the precipitation/evaporation balance, processes linked to sea-ice growth and melting as well as advection or upwelling of seawater can impact local $\delta^{18}O$ of surface water significantly. The influence of advection is clear for high latitude regions (Rohling & Bigg, 1998), but it is likely to be a less significant factor at low latitude in localities away from rivers and upwelling regions.

In sum, variations in the $\delta^{18}O$ of surface water at any one location reflects changes in the $\delta^{18}O$ of the global ocean due to changes in ice volume, changes in the local precipitation/evaporation balance, changes in the input of freshwater (if close to continental margins) through rivers or snow/ice melt which may reflect regional continental climate change, changes in contribution of advected or upwelled water with a different $\delta^{18}O$ value to that location and, at high latitudes, disturbances due to sea-ice growth and melting processes.

The $\delta^{18}O$ of seawater in the deep ocean reflects primarily the distribution of subsurface water masses. Like salinity, the $\delta^{18}O$ of subsurface water is basically conservative, and therefore the distribution of $\delta^{18}O$ in the deep ocean reflects water mass flow and mixing. The relationship between $\delta^{18}O$ and salinity in the deep ocean is different than that in the surface ocean mainly because the $\delta^{18}O$ values of deep Southern Ocean water are low relative to the $\delta^{18}O$ values of surface water of the same salinity (Figure 4). Around Antarctica, sea-ice formation also causes brine rejection increasing the salinity and density of the deep Southern Ocean Water. However, sea-ice formation only very slightly fractionates oxygen isotopes (Tan & Fraser, 1976; Tan & Strain, 1999), although on purely thermodynamics ground, ice should be ~3‰ enriched in ^{18}O vs. liquid water (O'Neil, 1968). Thus, brine rejection causes increased salinity, but practically no increase in $\delta^{18}O$ of surface water. As a result, deep Southern Ocean water has relatively low $\delta^{18}O$ values (−0.3‰) given its relatively high salinity (~34.7 psu). In regions where deep water forms, such as the North Atlantic, the $\delta^{18}O$ of deep water is relatively high (+0.3‰), as expected for water with relatively high salinity (~35.0 psu). In past times, changes in the circulation of deep water masses and/or in the contribution of brines to deep water masses, could have had an impact on the $\delta^{18}O$ values of deep water recorded at a single location (Adkins et al., 2002) (Figure 4). The implications are that the relationship between salinity and $\delta^{18}O$ of seawater is different in the deep ocean compared with the surface ocean, and that temporal variations in the

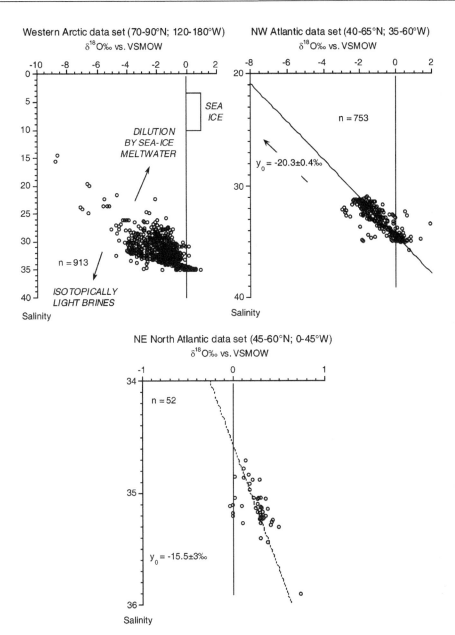

Figure 3   Salinity vs. $\delta^{18}O–H_2O$ relationship in the Arctic vs. North Atlantic Oceans (surface layer: 0–500 m). The scatter of values, in the Arctic, illustrate processes linked to sea-ice growth and melting. In the North Atlantic, a strong contrast exists between the NW sector, where Arctic rivers constrain the $\delta^{18}O$ value of the freshwater end-member of the mixing system, and the NE sector, where the influence of mixing with evaporated waters dominates. Data from Schmidt et al. (1999).

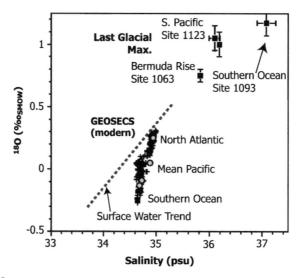

Figure 4 The $\delta^{18}O$ and salinity of deep-water masses from Adkins et al. (2002) deviates from the surface water relationship between $\delta^{18}O$ and salinity. The North Atlantic end-member has values close to the surface water line because it is derived from surface water in the North Atlantic. However, the Southern Ocean end-member has relatively high salinity because it is derived partially from sea-ice formation and brine rejection which enriches the salinity but not the ^{18}O. In the last glacial maximum, $\delta^{18}O$ and salinity of all water masses are enriched because fresh low $\delta^{18}O$ water was stored in ice sheets, however, the relationship between the deep-water masses was different, indicating that the signature of deep-water sources was different.

$\delta^{18}O$ of deep water in any one location reflect several indicators of climate change: global ice volume and changes in the $\delta^{18}O$ signatures and mixing of deep water masses.

A final caveat about $\delta^{18}O$-salinity relationships mostly concerns the North Atlantic and the Arctic. It relates to meltwater pulses and has been relatively well documented for the deglacial period (e.g., Flower, Hastings, Hill, & Quinn, 2004). Inland storage of water in the glacier and surrounding glacial lakes may result in the delayed release of ^{18}O-depleted freshwaters into the ocean at the occasion of major drainage events. This pattern may slightly bias the coupling of the $\delta^{18}O$ of the ocean to salinity and sea level changes.

4.2. Factors that Influence the $\delta^{18}O$ of Foraminifera

The interpretation of a foraminiferal $\delta^{18}O$ record must take into account all processes described above which alter the $\delta^{18}O$ value of seawater where the foraminifera calcifies its shell. In addition, the $\delta^{18}O$ of the shell is offset from the seawater $\delta^{18}O$ values because of the thermodynamic isotopic fractionation that occurs during calcite precipitation (Figure 1). In many species of foraminifera, the shell calcite is in oxygen isotopic equilibrium with seawater, and thus the isotopic separation factor between calcite and seawater ($\varepsilon \sim \delta^{18}O_{CaCO_3} - \delta^{18}O_{H_2O}$) is inversely related to calcification temperature (for equilibrium fractionation, the $\delta^{18}O$ of calcite decreases by ~ 0.21–$0.23\permil$ for a $1°C$ increase in temperature for a given $\delta^{18}OH_2O$). As such, if the $\delta^{18}O$ of seawater is known, foraminiferal calcite can be used as a paleothermometer to reconstruct past ocean temperature. Whatever the paleotemperature equation used (see Section 3 above), given the uncertainty in the $\delta^{18}O$ of seawater in the past and the lack of absolute precision on the temperature dependence of the $CaCO_3$–H_2O isotopic equilibrium in the low temperature range, it is most common to use the slopes of these equations to quantify relative changes in temperature after the potential influences of ice volume and local to regional impacts on the $\delta^{18}O$ of seawater are constrained.

While it is common to apply oxygen isotope paleotemperature equations, there is strong indication that foraminifera sometimes do not calcify in oxygen isotopic thermodynamic equilibrium with seawater notably because of biological 'vital' effects. Such effects were identified very early, notably by Duplessy, Lalou, and Vinot (1970). Some of these 'vital effects' seem to be related to the photosynthetic activity of algal symbionts. In contradiction to the explanation proposed by Duplessy et al. (1970), there is no firm evidence for the incorporation of low $\delta^{18}O$ metabolic CO_2 during calcite precipitation. Species that contain algal symbionts probably have higher calcification rates (through CO_2 consumption by algae), which induce a kinetic fractionation resulting in a 0.35–$0.5\permil$ depletion in $\delta^{18}O$ of large (adult) shells relative to equilibrium values [see Ravelo & Fairbanks, 1992; Spero, 1992; Spero, Bijma, Lea, & Bemis, 1997; Spero & Lea, 1993, and references therein].

The fact that the $\delta^{18}O$ of a foraminifera shell changes with size in some species indicates that there are ontogenetic effects probably related to changes in the intensity of photosynthesis and/or changes in depth habitat as a foraminifera matures from juvenile to adult (Williams, Bé, & Fairbanks, 1979; Spero & Lea, 1996). The absence of size-dependent changes in $\delta^{18}O$ of calcite for many non-symbiont bearing species confirms the idea that changes in photosynthetic rates can drive some ontogenic effects (Ravelo & Fairbanks, 1995). However, in some cases, the $\delta^{18}O$ of calcite increases with shell size, even in non-symbiont bearing species, due to the addition of gametogenic calcite as the foraminifera sinks at the end of its lifecycle, as has been clearly documented for *Globorotalia truncatulinoides* (Lohmann, 1990).

The other factor that may influence the $\delta^{18}O$ of foramininfera is carbonate ion concentrations ($[CO_3^{2-}]$), which causes decreasing $\delta^{18}O$ of calcite with increasing $[CO_3^{2-}]$ (Spero et al., 1997). The $[CO_3^{2-}]$ may influence calcification rates and induce kinetic fractionation effects, with competitive incorporation of bicarbonate and dissolved carbonate ions in calcite. This could explain why even non–symbiont bearing planktonic and benthic foraminifera sometimes do not calcify in oxygen isotopic equilibrium with seawater (Bemis, Spero, Bjima, & Lea, 1998; Spero et al., 1997). However, the potential role of dissolved carbonate ions and pH on the isotopic composition of foraminiferal calcite is still controversial (Deines, 2005; Zeebe, 2005) and will require further examination.

Overall, there are documented effects of photosynthesis, $[CO_3^{2-}]$, and gametogenic calcification which influence the $\delta^{18}O$ of planktonic foramininfera causing slight $\delta^{18}O$ depletion in symbiont-bearing species, and $\delta^{18}O$ enrichment in species that add a lot of gametogenic calcite in the subsurface. For benthic foramininfera, interspecies differences in δ^{18}O are probably due to the $[CO_3^{2-}]$ of the microenvironment, with the lower pH and $[CO_3^{2-}]$ of porewaters causing slight δ^{18}O enrichment in infaunal species (e.g., *Uvigerina* spp.) compared with epifaunal species (*Cibicidoides* spp.) which calcify in oxygen isotopic equilibrium with ambient water (Bemis et al., 1988). Once the potential biological and $[CO_3^{2-}]$ effects are accounted for, the paleotemperature estimates must be put into proper oceanographic context. For example, many planktonic foraminifera species have strong seasonal or depth habitat preferences (Figure 5), and the interpretation of δ^{18}O records must take into account these preferences (see Section 6).

In the above context, one specific feature of Arctic specimens of the planktic foraminifera *N. pachyderma* (*stricto sensu*), left as well as right-coiled (Npl and Npd), must be examined. As early as the sixties, Van Donk and Matthieu (1969) had noticed large departure from isotopic equilibrium with ambient Arctic waters in such shells. Later on, several authors further documented this offset (Kohfeld et al., 1996; Bauch et al., 1997; Hillaire-Marcel et al., 2004) which varies from −1 to −3‰ relative to equilibrium conditions for a calcite precipitated at mid-thermocline depth (Figure 6). Interestingly, despite this offset, the shells still present a size-dependence in their δ^{18}O values, but with a reverse trend, in comparison with their North Atlantic counterpart (Figure 7). Hillaire-Marcel et al. (2004) have interpreted this feature as a response to the reverse temperature gradient of the Arctic

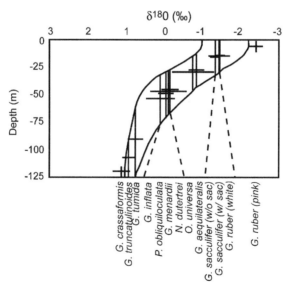

Figure 5 The δ^{18}O of tropical Atlantic species of foraminifera from a core top (V29-144) from Ravelo and Fairbanks (1992). The profiles are the envelope around the monthly predicted δ^{18}O of calcite profile predicted for the upper 125 m of the water column overlying the core top. Core top foraminifera values are plotted to overlap with the predicted profiles in order to infer calcification depth in the water column. The vertical lines are the average δ^{18}O values for each species, and the horizontal lines crossing each vertical line are the standard deviation of the measurements which are made on different size fractions.

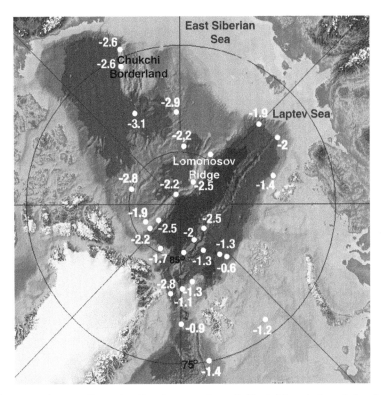

Figure 6 Compilation of isotopic offsets between foraminiferal (*N. pachyderma* left coiled – Npl) calcite δ^{18}O-values and equilibrium conditions for a calcite precipitated at mid-depth along the pycnocline between the surface cold and dilute water layer and the underlying North Atlantic Water (from literature; see refs. in Hillaire-Marcel et al., 2004). Npl populations are from surface sediment samples. These offsets are tentatively linked to the rate of production and accumulation of isotopically light brines along this pycnocline, due to sea-ice formation. Arctic topography from the GEBCO (1979) and IBCAO (2000) maps.

Ocean thermocline, where cold low salinity waters overly an intermediate more saline and warmer water mass originating from the North Atlantic. Due to the low salinity prevailing in the surface water layer, Npl cannot occupy this layer. It has been shown to occupy essentially the 35–36 psu domain with few questionable excursions below this minimum threshold (e.g., Hilbrecht, 1996). Thus Arctic Npl populations are restricted to the more saline (and slightly warmer) interface with the underlying North Atlantic subducted water that makes a large gyre into the Arctic Ocean, as an intermediate water mass. As a matter of fact, experiments by Spero and Lea (1996) on the more temperate species *G. bulloides*, have also shown that despite an overall offset with equilibrium conditions, isotopic shifts between shells were still preserving temperature differences.

With respect to the overall -1 to -3% isotopic offset in Arctic Npl shells, the temperature along the thermocline cannot be the cause. The only correlation that has been put forth by several authors is an apparent increase of this offset with the

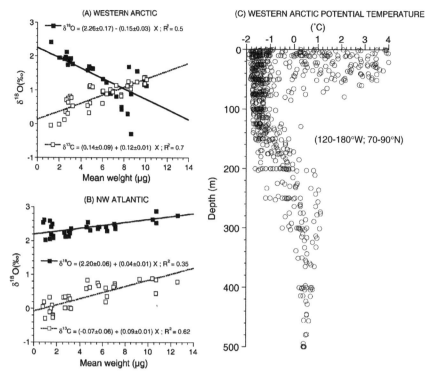

Figure 7 Example of size dependence of isotopic compositions in the deep-dwelling *N. pachyderma* left coiled (Npl) assemblages from Arctic (A) and North Atlantic (B) surface sediments interpreted as a consequence of Npl-calcite precipitation along the thermocline with either negative (North Atlantic) or positive (western Arctic) gradients within the underlying water mass (from Hillaire-Marcel et al., 2004). Note that in the Arctic Npl cannot develop in the very shallow dilute layer due to its low salinity (C). Data from Candon (2000).

duration/intensity of the sea-ice cover. From this viewpoint, Npl presents another peculiarity as well: its tolerance to very high salinity conditions. In brines linked to Antarctic sea-ice formation Spindler (1996) identified live specimens in water of up to 82 psu, and Npl growth in water with up to 58 psu. Given their tolerance to high salinity conditions and their growth at the top of the pycnocline above the underlying North Atlantic Water mass, Npl are likely to grow in high-salinity, low-$\delta^{18}O$ droplets (or thin layers) in regions where sinking isotopically depleted brines form in the Arctic (see Section 4.1 above). As such, the isotopic offset of Npl relative to equilibrium would indeed show some proportionality with the intensity of sea-ice (and brine) formation.

Whatever the cause for the modern behavior of Npl in Arctic environments, its abundance in the glacial North Atlantic Ocean, where it often represented most of planktic foraminifers, has led to intensive investigations. The interpretation of its isotopic composition changes, often linked to salinity changes in surface waters (e.g., Duplessy et al., 1992), is a challenge in view of all parameters that may account for its $\delta^{18}O$ values. In principle, salinity and temperature changes should be the most important factors. Independent estimates of temperature, such as Mg/Ca ratios in Npl shells (see Chapter 19), could help to constrain surface salinity. Unfortunately, the fact that Npl lives in the low-temperature range of the Mg/Ca calibration curve results in equivocal interpretations as demonstrated by Meland et al. (2006).

Attempts at directly linking $\delta^{18}O$ values to potential density values (σ_θ), which also depends on salinity and temperature, have been made (e.g., Hillaire-Marcel, de Vernal, Bilodeau, & Stoner, 2001). Indeed, both δ and σ_θ expressions show analogies and are directly proportional to salinity and inversely proportional to temperature:

$$\delta = \left(\frac{R_i}{R_s} - 1\right) \times 10^3 \text{ and } \sigma_\theta = \left(\frac{\rho_i}{\rho_m} - 1\right) \times 10^3$$

where R_i and R_s stand for $^{18}O/^{16}O$ ratios in the shell and reference materiel, ρ_i and ρ_m, for the density of the ambient water and that of pure water (at the same temperature). The rationale is to obtain information on the density structure of the upper water column, at sites of the North Atlantic where deep convection may or may not occur. Unfortunately, this approach requires an independent calibration of the σ_θ vs. δ relationship. This relationship is nearly linear, at sites where salinity is the prominent parameter, and polynomial, where temperature also plays an important role. Thus, solving the basic "paleotemperature" equation remains largely an issue dependent on the site, on the time and the foraminiferal species considered. Some hopes are seen ahead, in the development of new isotopic measurement techniques. Recently, it has been shown that in addition to the measurements of CO_2 masses 44, 45, 46, representing various combinations of O and C isotopes, it is possible to measure mass 47 (proportional to $^{13}C-^{18}O$ bonds in carbonate minerals, a direct function of the temperature during crystal growth), thus providing an access to the equilibrium temperature, independent of changes in the $^{18}O/^{16}O$ of environmental water (Ghosh et al., 2006). The analytical precision achieved is still insufficient to consider immediate applications in paleoceanography, but there may be hope to improve it in near future.

5. FORAMINIFERAL CARBON ISOTOPES AS ENVIRONMENTAL PROXIES

The $\delta^{13}C$ of a foraminifera shell reflects the carbon isotopic composition of the DIC in seawater in which the shell calcified, but it is not in isotopic equilibrium with seawater. The main reason that it is not equilibrium is because biogenic calcification is relatively rapid, resulting in kinetic isotope fractionation, and because of strong biological 'vital' effects. Kinetic fractionation for C-isotopes does not imply similar effects for O-isotopes, since the "equilibrating" pools of oxygen from seawater is considerably larger than that of carbon. In addition, the $\delta^{13}C$ of DIC in seawater ($\delta^{13}C_{DIC}$) is not uniform throughout the world's ocean, nor is the average $\delta^{13}C_{DIC}$ of the ocean constant with time. Thus, paleoceanographic records

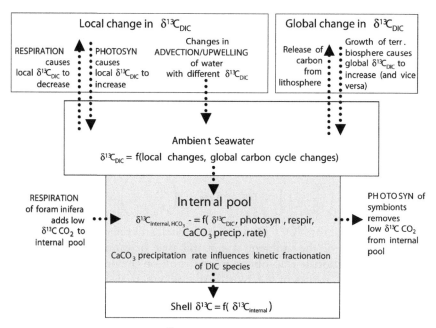

Figure 8   Factors that influence the δ^{13}C of foraminifera shells.

of foraminaferal δ^{13}C reflect multiple parameters (Figure 8). This section includes discussions of how downcore foraminiferal δ^{13}C records are used, in light of vital effects, to monitor changes in δ^{13}C$_{DIC}$.

5.1. Factors that Influence the δ^{13}C of DIC of seawater

The average δ^{13}C$_{DIC}$ of the whole ocean is influenced by the global carbon cycle, specifically the partitioning of carbon between the ocean, atmosphere, and terrestrial biosphere reservoirs. Because photosynthesis discriminates against ^{13}C and in favor of ^{12}C, the terrestrial biosphere (living vegetation and soil organic matter) is a reservoir of ^{13}C-depleted carbon (with a major mode of δ^{13}C values of around $-25\permil$ due to the photosynthetic cycle of C3-plants; Park & Epstein, 1960; Bender, 1971). If the size of the terrestrial biosphere increases, carbon with low δ^{13}C values is sequestered, and the ocean and atmosphere become enriched in ^{13}C. The average ocean and atmosphere δ^{13}C value can also be influenced by changes in geological sources and sinks of carbon (e.g., gas hydrates, volcanic outgassing, organic matter accumulation and sedimentation, alteration of ophiolites, etc.).

Thermodynamic (temperature-dependent) equilibration between the surface ocean DIC and the atmospheric CO_2 influences their isotopic separation factor $\varepsilon \cong \delta^{13}C_{CO_2} - \delta^{13}C_{DIC}$, which increases by $0.1\permil$ with a $1°$ decrease in surface ocean temperature (Mook, Bommerson, & Staverman, 1974; Zhang, Quay, & Wilbur, 1995; Szaran, 1997; see also Siegenthaler & Münnich, 1981). However, because the ocean contains more than 50 times more inorganic carbon than the atmosphere, changes in surface temperature have a large influence on the δ^{13}C value of the atmospheric CO_2 and a significant but reduced influence on the $\delta^{13}C$ of the surface ocean, but a negligible effect on the $\delta^{13}C_{DIC}$ of the ocean as a whole. Therefore, changes in the $\delta^{13}C_{DIC}$ of the mean ocean primarily respond to changes in the size of the terrestrial biosphere and/or to a perturbation in the carbon cycle due to a geological sink/source. Because the carbon in the biosphere and in organic carbon sinks is usually characterized by very low $\delta^{13}C$ values compared with those of the ocean DIC, the transfer of carbon between these reservoirs and the ocean DIC can be quantified (e.g., Emerson & Hedges, 1988; Buchmann, Brooks, Flanagan, & Ehleringer, 1998).

As in the case of foraminiferal $\delta^{18}O$ paleoceanographic records, foraminiferal $\delta^{13}C$-records can be influenced by other factors besides changes in the isotopic composition of the mean ocean. Specifically, as outlined below, the $\delta^{13}C_{DIC}$ of seawater at any one location can be influenced by local changes in the balance between photosynthesis and respiration, changes in the relative mixture of water masses, and changes in the original $\delta^{13}C_{DIC}$ signature of water masses reaching the site location. In the surface ocean, where photosynthesis dominates over respiration, [DIC] is relatively low and $\delta^{13}C_{DIC}$ is relatively high reflecting the net export of low $\delta^{13}C$ carbon (in the form of organic particulate matter with values around −19 to −23‰; Deines, 1980; Macko, Engle, & Quian, 1994; see also Burdige, 2006) out of the surface water. In the deep ocean, where respiration of organic particulate matter dominates, [DIC] is relatively high and $\delta^{13}C_{DIC}$ is relatively low (Figure 9). Thus, changes in photosynthesis or respiration will readily impact the $\delta^{13}C_{DIC}$ of seawater. For example, increases in photosynthesis that lead to greater export of carbon from the surface to the deep ocean, thus a stronger 'biological pump', would result in a larger surface to deep water $\delta^{13}C_{DIC}$ difference. Because photosynthesis and respiration also influence major nutrient distributions in the ocean, the concentrations of, for example, dissolved phosphate or nitrate in the ocean are inversely related to the $\delta^{13}C_{DIC}$ (e.g., Ortiz, Wheeler, Mix, & Key, 2000), and thus changes in the foraminiferal $\delta^{13}C$ are often interpreted as changes in the $\delta^{13}C_{DIC}$ once the impact of global carbon cycle changes on the mean ocean $\delta^{13}C_{DIC}$ values is taken into account. This relationship between foraminifera $\delta^{13}C$ values and photosynthetic activity in the surface ocean has been used for quite some time as a proxy for quantifying paleoproductivity changes between glacial and interglacial periods (e.g., Zahn, Winn, & Sarnthein, 1986; Sarnthein, Winn, Duplessy, & Fontugne, 1988).

Although changes in photosynthesis/respiration strongly influence major nutrient concentrations and $\delta^{13}C_{DIC}$, there are other possible causes of local changes in the $\delta^{13}C_{DIC}$ of seawater at a given location linked to water mass formation and circulation. As deep water forms in high latitude regions where surface-water cools and sinks, it obtains its $\delta^{13}C_{DIC}$ signature from its surface water sources. North Atlantic Deep Water (NADW) has relatively high $\delta^{13}C_{DIC}$ values owing to its North Atlantic surface water source, and Antarctic Bottom Water (AABW) has relatively low $\delta^{13}C_{DIC}$ values because it is comprised of Southern Ocean surface water and deep water from all basins, all of which are sources with relatively low $\delta^{13}C_{DIC}$ values (Figure 9). Today, in the Atlantic Ocean, mixing between nutrient-depleted, high $\delta^{13}C_{DIC}$ NADW flowing from north to south and denser

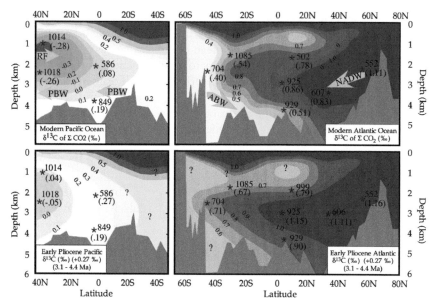

Figure 9 Upper panel: The $\delta^{13}C_{DIC}$ distribution throughout the western Atlantic and Pacific oceans, redrawn from Kroopnick (1985). The $\delta^{13}C_{DIC}$ is relatively high in the surface ocean where photosynthesis dominates, and is relatively low in the deep ocean where respiration dominates. The $\delta^{13}C_{DIC}$ of the North Atlantic Deep Water (NADW) is relatively high, reflecting the North Atlantic surface water source of NADW. The $\delta^{13}C_{DIC}$ distribution in the Atlantic mainly reflects mixing between different water masses with different $\delta^{13}C_{DIC}$ signatures. In the Pacific, the $\delta^{13}C_{DIC}$ distribution reflects the 'aging' of Pacific Bottom Water (PBW) and the mid-depth return flow (RF). The asterisks are plotted at the depth and latitude of some Ocean Drilling Program (ODP) sites. ODP site numbers are given next to the average values (in parentheses) of benthic foraminifera (*Cibicidoides* spp.) $\delta^{13}C$ from the interglacial periods of the last 400,000 yr reproduce the modern distribution of $\delta^{13}C_{DIC}$ fairly well (Ravelo & Andreasen, 2000). Lower panel: The distribution of $\delta^{13}C$ values of water masses during an interval of time within the warm Pliocene period reconstructed using the average $\delta^{13}C_{cal}$ values of benthic foraminifera (*Cibicidoides* spp.) at each site; 0.27 is added to all values to account for the global $\delta^{13}C$ offset between the Pliocene and late Pleistocene (Ravelo & Andreasen, 2000).

nutrient-enriched, low $\delta^{13}C_{DIC}$ AABW flowing south to north results in a strong horizontal nutrient gradient in the deep Atlantic basin (Figure 9). The $\delta^{13}C_{DIC}$ values in epibenthic foraminifera provide a way to reconstruct past deep-water mass distribution and changes in the relative strength of deep-water masses (i.e., the North Atlantic component of the thermohaline circulation). For example, the distribution of water masses has been reconstructed for the Early Pliocene warm period indicating that the mixing zone between NADW and AABW was shifted south due to enhanced flux of NADW relative to AABW (Figure 9) (Ravelo & Andreasen, 2000). In contrast, the mixing zone shifted northward and to shallower depths during cold periods [Pleistocene glaciations and Heinrich events (Zahn et al., 1997)] in response to a strong reduction in the production rate of NADW relative to AABW (Figure 10) (e.g., Duplessy et al., 1984, 1988; Zahn et al., 1986;

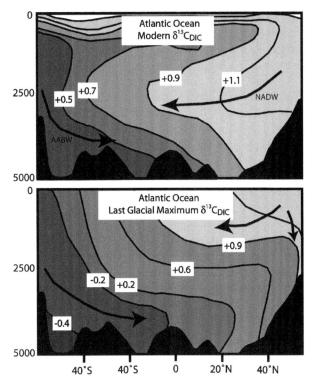

Figure 10   LGM vs. modern conditions in the Atlantic Ocean, based on δ^{13}C values in benthic foraminifera from variable depths (Duplessy et al., 1988; Labeyrie et al., 1992). A much shallower distribution of the LGM NADW is apparent.

Sarnthein et al., 1988, 1994; Curry, Duplessy, Labeyrie, & Shackleton, 1988; Boyle, 1997).

In the Pacific Ocean, AABW enters the south Pacific and flows northward, becoming more nutrient-enriched and ^{13}C-depleted as it returns at mid-depth ($\sim$2 km). This mid-depth water, found throughout the North Pacific, is the most nutrient-enriched and oldest (radiocarbon depleted) water in the global ocean. Reconstruction of Pacific δ^{13}C distributions using benthic foraminifera can be used to monitor the aging of mid-depth water in the North Pacific. For example, in the Pliocene warm period, the mid-depth water of the Pacific had higher δ^{13}C values probably indicating enhanced ventilation (less aging) of the interior of the Pacific Ocean (Figure 9). In sum, deep-water nutrient distributions and gradients are reflected in the distribution of δ^{13}C$_{DIC}$ (Kroopnick, 1985). In past times, either a change in the circulation pattern or the relative strength of deep-water masses can impact the δ^{13}C$_{DIC}$ of deep water at a sediment core location.

Surface water δ^{13}C$_{DIC}$ at a given location can also be influenced by changes in the mixing of water masses, especially the vertical upwelling of deeper (low δ^{13}C$_{DIC}$) water into the surface. As described above, changes in the δ^{13}C$_{DIC}$ at any one location is likely to reflect both global average δ^{13}C$_{DIC}$ changes due to perturbations in the global carbon cycle and changes in the local $\delta^{13}C_{DIC}$ due to processes that also influence the nutrient distribution in the ocean: photosynthesis, respiration, and variations in the advected or upwelled relative contributions of water masses with different $\delta^{13}C_{DIC}$ signatures to the region. An additional mechanism that impacts $\delta^{13}C_{DIC}$, and that is unrelated to processes that influence the nutrient distributions in the ocean, is isotopic thermodynamic equilibration between the atmospheric CO_2 and oceanic DIC. Although the average $\delta^{13}C_{DIC}$ of the surface ocean reflects isotopic equilibrium with the atmosphere globally, many regions are not locally in isotopic equilibrium. The locations where the surface ocean $\delta^{13}C_{DIC}$ does approaches equilibrium are in regions with high air–sea CO_2 fluxes (e.g., the Southern Ocean) and regions with long surface water residence times (e.g., sub-tropical gyres). In past times, the $\delta^{13}C_{DIC}$ of the surface ocean could vary if there was locally more or less isotopic equilibration with the atmosphere (due to changes in residence time or changes in air–sea CO_2 fluxes). Alternatively, in regions where equilibration tends to have a large impact on the $\delta^{13}C_{DIC}$ of surface water, then $\delta^{13}C_{DIC}$ could vary with a change in temperature (since air–sea carbon isotopic fractionation is temperature dependent). Since the source of deep water is high latitude surface water, then changes in the thermodynamic equilibration of surface waters can have an impact on the $\delta^{13}C_{DIC}$ of deep-water masses (Charles, Wright, & Fairbanks, 1993). The thermodynamic isotopic fractionation between the atmosphere and the ocean is a potential mechanism by which changes in the $\delta^{13}C_{DIC}$ of the ocean and the concentration of nutrients can be decoupled in past times. Thus, to reconstruct nutrient distributions accurately requires other proxies that are not influenced by air–sea exchange. However, regardless of the causes of changes in the $\delta^{13}C_{DIC}$ signature of water masses, the mixing zone between water masses can be reconstructed and can provide tremendous insight into circulation patterns of past times as illustrated by Duplessy et al. (1988) and Labeyrie et al. (1992) in the case of the relative strength of NADW.

In sum, $\delta^{13}C_{DIC}$ is influenced by many processes: The average $\delta^{13}C_{DIC}$ of the global oceans is influenced by changes in the amount of carbon stored on land, and by changes in the sink/source of carbon from geological reservoirs. The $\delta^{13}C_{DIC}$ at any one location can be related to the balance of photosynthesis vs. respiration, the relative mixture of water masses with different $\delta^{13}C_{DIC}$ signatures, and changes in the $\delta^{13}C_{DIC}$ of source waters from which those water masses are derived. The $\delta^{13}C$ analyses of plantkonic and benthic foraminifera are used to reconstruct changes in $\delta^{13}C_{DIC}$, but foraminiferal calcite is not in isotopic equilibrium with DIC and is readily affected by microhabitat conditions.

5.2. Factors that Influence the $\delta^{13}C$ of Foraminifera

Differences of isotopic composition between small (possibly juvenile) and large (possibly mature) specimens may depend on (i) distinct "vital effects", (ii) changes of habitat during the life cycle and/or seasonal shifts in environmental conditions, (iii) the deposition of secondary calcite in large sinking shells, and (iv) differential dissolution of shells depending on the thickness of their walls during early diagenetic processes (e.g., Lohmann, 2006).

The $\delta^{13}C$ of foraminiferal calcite is not what would be expected if calcite were in isotopic thermodyamic isotopic equilibrium with the DIC of seawater due to both abiotic kinetic fractionation and biological vital effects. Abiotic kinetic fractionation results in a 1.0 ± 0.2‰ (Romanek, Grossman, & Morse, 1992) enrichment of $\delta^{13}C$ in calcite relative to bicarbonate (HCO_3^-), which comprises $\sim 95\%$ of the DIC at the pH of the modern ocean. Over the range of expected foraminiferal calcification rates and temperatures, this enrichment is nearly constant (Romanek et al., 1992; Turner, 1982); this abiotic kinetic fractionation effect causes foraminiferal calcite $\delta^{13}C$ to be 1.0 ± 0.2‰ enriched relative to $\delta^{13}C_{DIC}$. However, the observed $\delta^{13}C$ offset between foraminiferal calciate and $\delta^{13}C_{DIC}$ varies widely depending on species and ontogenetic stage because of biological vital effects in planktonic (Ravelo & Fairbanks, 1995) and benthic (Grossman, 1987) species.

Like $\delta^{18}O$, the $\delta^{13}C$ of calcite is influenced by photosynthesis of algal symbionts and respiration, however the impact of these biological processes on the $\delta^{13}C$ of foraminiferal shells is much more severe. Both photosynthesis and respiration have an impact on the microenvironment (Figure 8), or 'internal carbon pool' of the foraminifera which can have a different pH, DIC concentration, and $\delta^{13}C_{DIC}$ than the surrounding ambient seawater (Zeebe, Bijma, & Wolf-Gladrow, 1999). The $\delta^{13}C_{DIC}$ of the internal carbon pool can be enriched in ^{13}C when CO_2 with low $\delta^{13}C$ values is sequestered by algae during photosynthesis, and can be depleted in ^{13}C when contaminated by low $\delta^{13}C$ metabolic CO_2 during respiration. Since the foraminifera draws on this internal pool, the $\delta^{13}C$ of the foraminiferal test can be influenced by the rate of photosynthesis, respiration, and calcite precipitation relative to the turnover rate of DIC in the internal pool. In addition, the rate of isotopic equilibration during CO_2 hydroxylation and hydration within the internal pool is slow relative to calcite precipitation, and thus the chemistry of the internal pool (its pH and CO_3^{2-} concentration) and the rate of precipitation which influences the degree of isotopic equilibration between species of carbon, can influence the $\delta^{13}C$ of the test (McConnaughey, 1989a, 1989b). The chemistry of the internal pool, and therefore the degree of kinetic fractionation as described by McConnaughey (1989a, 1989b), can be influenced both by biological processes as well as by the ambient seawater chemistry, specifically the pH and concentration CO_3^{2-} of seawater (Spero et al., 1997).

Given the multiple potential factors that determine the offset between the $\delta^{13}C$ of foraminiferal calcite and $\delta^{13}C_{DIC}$, there are significant species-specific differences in this offset. In general, foraminiferal $\delta^{13}C$ values tend to be lower than what would be expected by abiotic kinetic fractionation (calcite $\delta^{13}C$ enrichment of 1.0 ± 0.2‰ relative to $\delta^{13}C_{DIC}$) alone, probably due to the contamination of the internal pool with CO_2 from respiration possibly combined with the kinetic fractionation effects described by McConnaughey (1989a, 1989b). This is true for benthic foraminiferal species (Grossman, 1987; Mackensen, Schumacher, Radke, & Schmidt, 2000; McCorkle, Corliss, & Farnham, 1997) and many planktonic species, particularly those without algal symbionts (Ravelo & Fairbanks, 1995). In addition, algal-bearing planktonic species typically have strong size-dependent $\delta^{13}C$ values, with increasing $\delta^{13}C$ values with size, probably due to increases in the rate of photosynthesis (Ravelo & Fairbanks, 1995; Spero & Deniro, 1987) (Figure 11),

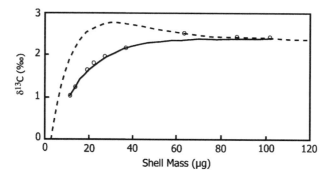

Figure 11 The δ^{13}C of *Globigernoides sacculifer* (a symbiont-bearing planktonic foraminifera) from the core top of V28-122 from Ravelo and Fairbanks (1995). Each circle is the average δ^{13}C values of 10–15 shells vs. the average mass of an individual shell. The increase in δ^{13}C values with size reflects the increasing effect of photosynthesis on the δ^{13}C of the shell as the foraminifera grows. The dashed curve is the calculated δ^{13}C of calcite being added at a given shell mass. See Ravelo and Fairbanks (1995) for a full explanation.

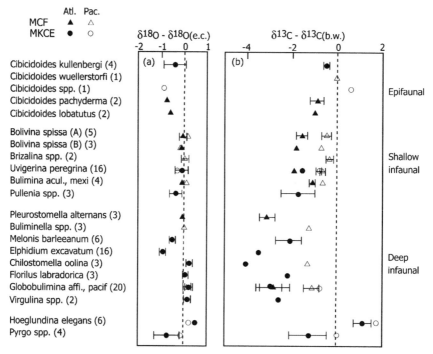

Figure 12 Average δ^{18}O offsets from calcite in equilibrium with bottom water (δ^{18}O e.c.) (a) and average δ^{13}C (b) offsets from bottom water DIC (δ^{13}Cb.w.) (b) were calculated for each species in each core. The averages and standard deviations of single-core values are shown here. The values in parentheses give the number of cores in which each species was analyzed. Triangles denote data from McCorkle et al. (1997) ("MCF") and circles show data from McCorkle et al. (1990) ("MKCE"); solid symbols show the averages of Atlantic Ocean cores and open symbols show Pacific Ocean averages. The species are grouped according to their stained abundance patterns with epibenthic and epifaunal species at the top (five *Cibicidoides* spp.), shallow infaunal species in the second group (*B. spissa* to *Pulknia* spp.), and intermediate and deep infaunal species in the third group (*P. alternans* to *Virgulina* spp.). Figure is redrawn from McCorkle et al. (1997).

however the $\delta^{13}C$ enrichment relative to $\delta^{13}C_{DIC}$ often remains below $1.0 \pm 0.2‰$ indicating that other factors (e.g., respiration) continue to influence the $\delta^{13}C$ of the test throughout the foraminifera lifecycle. In large adult shells, the $\delta^{13}C$ size-dependence decreases, and paleoceanographic studies that utilize symbiont-bearing planktonic foraminifera almost always rely on records produced from shells from the largest size fraction in which abundant shells are found.

As described above, species-specific $\delta^{13}C$ offsets from the $\delta^{13}C_{DIC}$ are attributed to the combination of abiotic kinetic fractionation, biological vital effects of photosynthesis and respiration on the internal carbon pool, and kinetic fractionation due to the slow isotopic equilibration of carbon species within the internal pool. However, it is also important to account for the fact that there can be large species-specific differences in the habitat and ecological preferences, and therefore in the $\delta^{13}C_{DIC}$ of the ambient seawater in which the foraminifera live. For example, different species of planktonic foraminifera have different temperature, nutrient, and light requirements and therefore can live and calcify at different vertical depths and seasons. As such their shells will reflect the vertical and seasonal variations in $\delta^{13}C_{DIC}$ within the upper ocean. Benthic foraminifera have different oxygen and food requirements and therefore can live and calcify at different depths in the sediments. Since the $\delta^{13}C_{DIC}$ of porewaters can be quite depleted in sediment with high rates of respiration of sedimentary organic matter, infaunal species tend to have lower $\delta^{13}C$ values than epifaunal species growing at the same time at the same location (Mackensen et al., 2000; McCorkle et al., 1997) (Figure 12).

6. CONCLUSION AND SUMMARY

Paleoceanographic records derived by analyzing the $\delta^{13}O$ and $\delta^{13}C$ of foraminiferal shells must obviously be generated and interpreted with caution and with close consideration of a wide array of factors. The common strategies are to generate records using only one species in order to minimize the variability due to specific biological effects. Furthermore, it is common to select species that have well-known ecological preferences that are fairly predictable across regions with different oceanographic conditions in the modern ocean. Otherwise, it is difficult to interpret records accurately with respect to understanding how isotopic records might be biased by the potential effects of changes in seasonality, nutrient levels, and depth habitat, not to mention interannual variability (with exceptional blooms) and early diagenetic effects (mixing, selective dissolution). It is also common to use shells from one size fraction of the sediment; usually the largest abundant size fraction in which shells are abundant is chosen since the ontogenetic isotope fractionation effects are more severe during early stages of growth, although this may result in a bias toward optimal conditions for the given species, and not necessarily the mean conditions of the time interval considered.

These practices are thought to minimize variability due to biological factors, and the resulting $\delta^{18}O$ downcore record can then be interpreted as primarily a reflection of changes in the $\delta^{18}O$ of seawater, the temperature of calcification, and possibly ocean carbon chemistry ([CO_3^{2-}]). The resulting $\delta^{13}C$ downcore record will reflect changes in the $\delta^{13}C_{DIC}$ and [CO_3^{2-}] The isotopic composition ($\delta^{18}O$

and $\delta^{13}C_{DIC}$) of seawater, the temperature of calcification, and the seawater carbon chemistry can each vary as a function of multiple processes as discussed above. For example, changes in $\delta^{13}C_{DIC}$ could be related to changes in the global carbon cycle, regional changes in water mass mixing and sources, photosynthesis/respiration processes, and changes in $[CO_3^{2-}]$. However, in some cases, isotopic records are predominantly influenced by single oceanographic/climate process. Perhaps the best example of this is the dominance of ice volume changes, relative to changes in calcification temperature and $[CO_3^{2-}]$, on most foraminiferal $\delta^{18}O$ records. For this reason, $\delta^{13}O$ records remain one of the best stratigraphic tools available for correlating records across the globe.

In many cases, because records from around the globe all have some shared variance due to whole ocean changes in $\delta^{18}O$ and $\delta^{13}C_{DIC}$, then comparisons between records (differences) can be used to isolate changes in local conditions in one location relative to another. If the $\delta^{13}O$ and $\delta^{13}C_{DIC}$ characteristics of water masses can be reconstructed, then changes in relative flux of each water mass can be quantified using cores in the mixing zone between end-members. However, in most cases, to isolate absolute changes in local oceanographic conditions and to separate the influence of multiple effects, other paleoceanographic proxies in the same core must be analyzed, and to understand the context (e.g., the geographic extent) of a single record, comparisons with records from other regions must be made.

REFERENCES

Adkins, J. F., McIntyre, K., & Schrag, D. P. (2002). The salinity, temperature, and $\delta^{18}O$ of the Glacial Deep Ocean. *Science, 298*, 1769–1773.

Bard, E. (2001). Paleoceanographic implications of the difference in deep-sediment mixing between large and fine particles. *Paleoceanography, 16*, 235–239.

Bédard, P., Hillaire-Marcel, C., & Pagé, P. (1981). ^{18}O-modelling of freshwater inputs in Baffins Bays and Canadian Arctic coastal waters. *Nature, 293*, 287–289.

Bemis, B. E., Spero, H. J., Bjima, J., & Lea, D. W. (1998). Reevaluation of the oxygen isotopic composition of planktonic foraminifera: Experimental results and revised paleotemperature equations. *Paleoceanography, 13*, 150–160.

Bender, M. M. (1971). Variations in the $^{13}C/^{12}C$ ratios of plants in relation to the pathway of photosynthetic carbon dioxide fixation. *Phytochemistry, 10*, 1239–1244.

Berger, W. H., Be, A. W. H., & Vincent, E. (Eds). (1981). Oxygen and carbon isotopes in foraminifera. *Palaeogeography, Palaeoclimatology, Palaeoecology, 33*, 1–277.

Bowen, R. (1966). *Paleotemperature analysis. Methods in geochemistry and geophysics* (Vol. 2, p. 265) Amsterdam: Elsevier.

Boyle, E. A. (1997). Characteristics of the deep ocean carbon system during the past 150,000 years: Sigma CO_2 distributions, deep water flow patterns, and abrupt climate change. *Proceedings of National Academy of Sciences, USA, 94*, 8300–8307.

Buchmann, N., Brooks, J. R., Flanagan, L. B., & Ehleringer, J. R. (1998). Carbon isotope discrimination of terrestrial ecosystems. In: H. Griffiths (Ed.), *Stable isotopes integration of biological, ecological, and geochemical processes* (pp. 203–221). Oxford, UK: BIOS Scientific Publications.

Burdige, D.J. (2006). *The geochemistry of marine sediments* (p. 630). Princeton University Press.

Burman, J., Gustafsson, O., Segl, M., & Schmitz, B. (2005). A simplified method of preparing phosphoric acid for stable isotope analyses of carbonates. *Rapid Communications in Mass Spectrometry, 19*, 3086–3088.

Candon, L. (2000). *Structure des populations et composition isotopique des foraminifères planctoniques dans le nord-ouest de l'Atlantique Nord (actuel versus dernier maximum glaciaire)*. M.Sc. thesis, Université du Québec à Montréal, 154 pp.

Charles, C. D., Wright, J. D., & Fairbanks, R. G. (1993). Thermodynamic influences on the marine carbon isotope record. *Paleoceanography, 8*, 691–699.

Cooper, L. W., Benner, R., McClelland, J. W., Peterson, B. J., Holmes, R. M., Raymond, P. A., Hansell, D. A., Grebmeier, J. M., & Codispoti, L. A. (2005). Linkages among runoff, dissolved organic carbon, and the stable oxygen isotope composition of seawater and other water mass indicators in the Arctic Ocean. *Journal of Geophysical Research, 110*, G02013, doi:10.1029/2005JG000031.

Coplen, T. B. (1988). Normalization of oxygen and hydrogen isotope data. *Chemical Geology, 72*, 293–297.

Coplen, T. B. (1996). New guidelines for the reporting of stable hydrogen, carbon, and oxygen isotope ratio data. *Geochimica et Cosmochimica Acta, 60*, 3359.

Craig, H. (1961). Isotopic variations in meteoric waters. *Science, 133*, 1702–1703.

Curry, W. B., Duplessy, J. C., Labeyrie, L. D., & Shackleton, N. J. (1988). Changes in the distribution of C13 of deep water CO_2 between the last glaciation and the Holocene. *Paleoceanography, 3*, 317–342.

Dansgaard, W. (1964). Stable isotopes in precipitation. *Tellus, 16*, 436–468.

Darling, K. F., Kucera, M., Kroon, D., & Wade, C. M. (2006). A resolution for the coiling direction paradox in *Neogloboquadrina pachyderma*. *Paleoceanography, 21*, PA2011, doi:10.1029/2005PA001189.

Deines, E. T. (1980). The isotopic composition of reduced organic carbon. In: P. Fritz & J. Ch. Fontes (Eds), *Handbook of environmental isotope geochemistry* (Vol. 1A, pp. 329–406). New York: Elsevier.

Deines, P. (2005). Comment on "An explanation of the effect of seawater carbonate concentration on foraminiferal oxygen isotopes," by R. E. Zeebe (1999). *Geochimica et Cosmochimica Acta, 69*, 787.

Duplessy, J.-C., Labeyrie, L., Arnold, M., Paterne, M., Duprat, J., & van Weering, T. C. E. (1992). Changes in surface salinity of the North Atlantic Ocean during the last deglaciation. *Nature, 358*, 485–488.

Duplessy, J.-C., Lalou, C., & Vinot, A. C. (1970). Differential isotopic fractionation in benthic foraminifera and paleotemperatures reassessed. *Science, 168*, 250–251.

Duplessy, J. C., Matthews, R. K., Prell, W., Ruddiman, W. F., Caralp, M., & Hendy, C. H. (1984). C-13 record of benthic foraminifera in the last interglacial ocean: Implications for the carbon cycle and global deep water circulation. *Quaternary Research, 21*, 225–243.

Duplessy, J. C., Shackleton, N. J., Fairbanks, R. G., Labeyrie, L., Oppo, D., & Kallel, N. (1988). Deepwater source variations during the last climatic cycle and their impact on the global deepwater circulation. *Paleoceanography, 3*, 343–360.

Emerson, S., & Hedges, J. I. (1988). Processes controlling the organic carbon content of open ocean sediments. *Paleoceanography, 3*, 621–634.

Emiliani, C. (1955). Pleistocene temperatures. *Journal of Geology, 63*, 538–578.

Emiliani, C., & Geiss, J. (1959). On glaciations and their causes. *International Journal of Earth Sciences, 46*, 576–601.

Epstein, S., Buchsbaum, H. A., Lowenstam, H. A., & Urey, H. C. (1953). Revised carbonate-water isotopic temperature scale. *Geological Society of America Bulletin, 64*, 1315–1326.

Epstein, S., & Mayeda, T. (1953). Variation of O-18 content of water from natural sources. *Geochimica et Cosmochimica Acta, 4*, 213–224.

Erez, J., & Luz, B. (1983). Experimental paleotemperature equation for planktonic foraminifera. *Geochemica et Cosmochimica Acta, 47*, 1025–1031.

Fairbanks, R. G. (1989). A 17,000-year glacio-eustatic sea level record: Influence of glacial melting rates on the Younger Dryas event and deep-ocean circulation. *Nature, 342*, 637.

Fairbanks, R. G., Charles, C. D., & Wright, J. D. (1992). Origin of global meltwater pulses. In: R. E. Taylor (Ed.), *Radiocarbon after four decades* (pp. 473–500). Springer-Verlag.

Faure, G. (1986). *Principles of isotope geology* (p. 335). Wiley.

Flower, B. J., Hastings, D. W., Hill, H. W., & Quinn, T. M. (2004). Phasing of deglacial warming and Laurentide Ice Sheet meltwater in the Gulf of Mexico. *Geology, 32*, 597–600.

GEBCO. (1979). General Bathymetric Chart of the Oceans. www.ngdc.noaa.gov/mgg/gebco

Ghosh, P., Adkins, J., Affek, H., Balta, B., Guo, G., Schauble, E. A., Schrag, D., & Eiler, J. M. (2006). $^{13}C-^{18}O$ bonds in carbonate minerals: A new kind of paleothermometer. *Geochimica et Cosmochimica Acta, 70,* 1439–1456.

Gregory, R. T., & Taylor, H. P. (1981). An oxygen isotopic profile in a section of cretaceaous oceanic crust, Samali ophiolite: Evidence for $\delta^{18}O$ buffering of the oceans by deep (>5 km) seawater-hydrothermal circulation at midocean ridges. *Journal of Geophysical Research, 86,* 2737–2755.

Grossman, E. L. (1987). Stable isotopes in modern benthic foraminifera: A study of vital effects. *Journal of Foraminiferal Research, 17,* 48–61.

Guinasso, N. L., & Shink, D. R. (1975). Quantitative estimates of biological mixing rates in abyssal sediments. *Journal of Geophysical Research, 80,* 3032–3043.

Hilbrecht, H. (1996). Planktic foraminifera and the physical environment in the Atlantic and Indian Oceans. http://www.ngdc.noaa.gov/mgg/geology/hh1996/pachy.html.

Hillaire-Marcel, C., de Vernal, A., Bilodeau, G., & Stoner, J. (2001). Changes of potential density gradients in the northwestern North Atlantic during the last climatic cycle based on a multiproxy approach. In: D. Seidov et al. (Eds), *The oceans and rapid climate changes: Past, present and future* (pp. 83–100). Geophysical Monograph Series 126 .

Hillaire-Marcel, C., de Vernal, A., Polyak, L., & Darby, D. (2004). Size-dependent isotopic composition of planktic foraminifers from Chukchi Sea vs. NW Atlantic sediments: Implications for the Holocene paleoceanography of the western Arctic. *Quaternary Science Reviews, 23,* 245–260.

Hoefs, J. (2004). *Stable isotope geochemistry* (p. 244). Berlin: Springer-Verlag.

IAEA. (2000). Stable isotope processes in the water cycle, International Atomic Energy Agency, Vienna. www.iaea.or.at/programmes/ripc/ih/volumes/vol_two/cht_ii_03.pdf

IBCAO. (2000). International bathymetric chart of the Arctic Ocean. http://www.ngdc.noaa.gov/mgg/bathymetry/arctic/provisionalmap.html

Kroopnick, P. M. (1985). The distribution of ^{13}C of ΣCO_2 in the world oceans. *Deep-Sea Research, 32,* 57–84.

Labeyrie, L., Duplessy, J. C., Duprat, J., Juillet-Leclerc, A. J., Moyes, J., Michel, E., Kallel, N., & Shackleton, N. J. (1992). Changes in the vertical structure of the north Atlantic ocean between glacial and modern times. *Quaternary Science Reviews, 11,* 401–413.

LeGrande, A. N., & Schmidt, G. A. (2006). Global gridded data set of the oxygen isotopic composition in seawater. *Geophysics Research Letter, 33,* L12604, doi:10.1029/2006GL026011.

Lohmann, G. P. (1990). *Globorotalia truncatulinoides*'s growth and chemistry as probes of the past thermocline. I, Shell size. *Paleoceanography, 5,* 55–75.

Lohmann, G. P. (2006). A model for variation in the chemistry of planktonic foraminifera due to secondary calcification and selective dissolution. *Paleoceanography, 10,* 445–458.

Mackensen, A., Schumacher, S., Radke, J., & Schmidt, D. N. (2000). Microhabitat preferences and stable carbon isotopes of endobenthic foraminifera: Clue to quantitative reconstruction of oceanic new production?. *Marine Micropaleontology, 40,* 233.

Macko, S. A., Engle, M. H., & Quian, Y. (1994). Early diagenesis and organic matter preservation: A molecular stable isotope perspective. *Chemical Geology, 114,* 365–379.

McConnaughey, T. (1989a). ^{13}C and ^{18}O isotopic disequilibrium in biological carbonates: II. In vitro simulation of kinetic isotope effects. *Geochimca et Cosmochimca Acta, 53,* 163–171.

McConnaughey, T. (1989b). ^{13}C and ^{18}O isotopic disequilibrium in biological carbonates: I. Patterns. *Geochimca et Cosmochimca Acta, 53,* 151–162.

McCorkle, D. C., Corliss, B. H., & Farnham, C. A. (1997). Vertical distributions and stable isotopic compositions of live (stained) benthic foraminifera from the North Carolina and California continental margins.. *Deep Sea Research Part I: Oceanographic Research Papers, 44,* 983.

McCrea, J. M. (1950). On the isotopic chemistry of carbonates and a paleotemperature scale. *The Journal of Chemical Physics, 18,* 849–857.

Meland, M. Y., Jansen, E., Elderfield, H., Dokken, T. M., Olsen, A., & Bellerby, R. G. J. (2006). Mg/Ca ratios in the planktonic foraminifer *Neogloboquadrina pachyderma* (sinistral) in the northern North Atlantic/Nordic Seas. *Geochemistry, Geophysics, Geosystems, 7,* Q06P14, doi:10.1029/2005GC001078.

Mook, W. G., Bommerson, J. C., & Staverman, W. H. (1974). Carbon isotope fractionation between dissolved bicarbonate and gaseous carbon dioxide. *Earth and Planetary Science Letters, 22,* 169–176.

Muehlenbachs, K., & Clayton, R. N. (1976). Oxygen isotope composition of the oceanic crust and its bearing on seawater. *Journal of Geophysical Research, 81,* 4365–4369.

O'Neil, J. R. (1968). Hydrogen and oxygen isotope fractionation between ice and water. *Journal of Physical Chemistry, 72,* 3683–3684.

Ortiz, J. D., Wheeler, P. A., Mix, A. C., & Key, R. M. (2000). Anthropogenic CO_2 invasion into the northeast Pacific based on concurrent [13]CDIC and nutrient profiles from the California Current. *Global Biogeochemical Cycles, 14,* 917–929.

Park, R., & Epstein, S. (1960). Carbon isotope fractionation during photosynthesis. *Geochimica et Cosmochimica Acta, 21,* 110–126.

Ravelo, A. C., & Fairbanks, R. G. (1992). Oxygen isotopic composition of multiple species of planktonic foraminifera: Recorders of the modern photic zone temperature gradient. *Paleoceanography, 7,* 815–832.

Ravelo, A. C., & Fairbanks, R. G. (1995). Carbon isotopic fractionation in multiple species of planktonic-foraminifera from core-tops in the tropical Atlantic. *Journal of Foraminiferal Research, 25,* 53–74.

Ravelo, A. C., & Andreasen, D. H. (2000). Enhanced circulation during a warm period. *Geophysical Research Letters, 27,* 1001–1004.

Rohling, E. J., & Bigg, G. R. (1998). Paleosalinity and $\delta^{18}O$: A critical assessment. *Journal of Geophysical Research, 103,* 1307–1318.

Rohling, E. J., & Cooke, S. (1999). Stable oxygen and carbon isotope ratios in foraminiferal carbonate. In: B. K. Sen Gupta (Ed.), *Modern foraminifera* (pp. 239–258). Dordrecht, The Netherlands: Kluwer Academic.

Romanek, C. S., Grossman, E. L., & Morse, J. W. (1992). Carbon isotopic fractionation in synthetic aragonite and calcite: Effects of temperature and precipitation rate. *Geochimica et Cosmochimica Acta, 56,* 419.

Sarnthein, M. K., Winn, K., Duplessy, J.-C., & Fontugne, M. R. (1988). Global variations of surface ocean productivity in low and mid-latitudes: Influence on CO_2. *Paleoceanography, 3,* 361–399.

Sarnthein, M. K., Winn, K., Jung, S. J. A., Duplessy, J.-C., Labeyrie, L., Erlenkeuser, H., & Ganssen, G. (1994). Changes in east Atlantic deepwater circulation over the last 30,000 years: Eight time slices reconstruction. *Paleoceanography, 9,* 209–267.

Schmidt, G. A., Bigg, G. R., & Rohling, E. J. (1999). Global seawater oxygen-18 database, http://data.giss.nasa.gov/o18data/

Shackleton, N. J. (1967). Oxygen isotope analyses and Pleistocene temperatures reassessed. *Nature, 215,* 15–17.

Shackleton, N. J. (1974). Attainment of isotopic equilibrium between ocean water and the benthic foraminifera genus *Uvigerina*: Isotopic changes in the ocean during the last glacial. In: J. Labeyrie (Ed.), *Méthodes quantitatives d'étude des variations du climat au cours du Pléistocène* (pp. 203–209). France: Editions du C.N.R.S.

Shackleton, N. J., & Opdyke, N. D. (1973). Oxygen isotope and palaeomagnetic stratigraphy of equatorial Pacific core V28-238: Oxygen isotope temperatures and ice volumes on a 105 and 10 year scale. *Quaternary Research, 3,* 39–55.

Sharp, Z. (2006). *Principles of stable isotope geochemistry* (p. 360). New Jersey: Prentice Hall.

Siegenthaler, U., & Münnich, K. O. (1981). HHH ^{13}C/^{12}C fractionation during CO_2 transfer from air to sea. In: B. Bolin (Ed.), *Carbon cycle modelling*. SCOPE Report 16 (pp. 249–257). Chichester, UK: Wiley.

Spero, H. J. (1992). Do planktic foraminifera accurately record shifts in the carbon isotopic composition of seawater sigma-CO_2. *Marine Micropaleontology, 19,* 275–285.

Spero, H. J., Bijma, J., Lea, D. W., & Bemis, B. E. (1997). Effect of seawater carbonate concentration on foraminiferal carbon and oxygen isotopes. *Nature, 390,* 497–500.

Spero, H. J., & Deniro, M. J. (1987). The influence of symbiont photosynthesis on the Delta-O-18 and Delta-C-13 values of planktonic foraminiferal shell calcite. *Symbiosis, 4,* 213–228.

Spero, H. J., & Lea, D. W. (1993). Intraspecific stable-isotope variability in the planktic foraminifera *Globigerinoides sacculifer*: Results from laboratory experiments. *Marine Micropaleontology, 22,* 221–234.

Spero, H. J., & Lea, D. W. (1996). Experimental determination of stable-isotope variability in *Globigerinoides bulloides*: Implications for paleoceanographic reconstructions. *Marine Micropaleontology, 28,* 221–234.

Spindler, M. (1996). On the salinity tolerance of the planktonic foraminifer *Neogloboquadrina pachyderma* from Antarctic sea ice. National Institute for Polar Research symposium. *Polar Biology, 9,* 85–91.

St-Jean, G. (2003). Automated quantitative and isotopic (C-13) analysis of dissolved inorganic carbon and dissolved organic carbon in continuous-flow using a total organic carbon analyser. *Rapid Communications in Mass Spectrometry, 17,* 419–428.

Szaran, J. (1997). Achievement of carbon isotope equilibrium in the system HCO_3^- (solution) CO_2 (gas). *Chemical Geology, 142,* 79–86.

Thomas, D. J., Zachos, J. C., Bralower, T. J., Thomas, E., & Bohaty, S. (2002). Warming the fuel for the fire: Evidence for the thermal dissociation of methane hydrate during the Paleocene–Eocene thermal maximum. *Geology, 30,* 1067–1070.

Tan, F. C., & Fraser, W. D. (1976). Oxygen isotope studies in the Gulf of St. Lawrence. *Journal of the Fisheries Research Board of Canada, 33,* 1397–1401.

Tan, F. C., & Strain, P. M. (1999). Sea ice and oxygen isotopes in Foxe Basin, Hudson Bay and Hudson Strait, Canada. *Journal of Geophysical Research, 101,* 20869–20876, 1996.

Turner, J. V. (1982). Kinetic fractionation of carbon-13 during calcium carbonate precipitation. *Geochemica et Cosmochimica Acta, 46,* 1183–1191.

Urey, H. C. (1947). The thermodynamic properties of isotopic substances. Liversidge lecture of 1946. *Journal of the Chemical Society,* 562–581.

Urey, H. C., Epstein, S., Lowenstam, H. A., & McKinney, C. R. (1951). Measurement of paleo-temperatures and temperatures of the upper Cretaceous of England, Denmark, and the south-eastern United States. *Geological Society of America Bulletin, 62,* 399–416.

de Vernal, A., Bilodeau, G., Hillaire-Marcel, C., & Kassou, N. (1992). Quantitative assessment of carbonate dissolution in marine sediments from foraminifer linings vs. shell ratios: Davis Strait, northwest North Atlantic. *Geology, 20,* 527–530.

de Vernal, A., & Hillaire-Marcel, C. (2006). Provincialism in trends and high frequency changes in the northwest North Atlantic during the Holocene. *Global and Planetary Change, 54,* 263–290.

Williams, D. F., Bé, A. W. H., & Fairbanks, R. G. (1979). Seasonal oxygen isotopic variations in living planktonic foraminifera off Bermuda. *Science, 206,* 447–449.

Zahn, R., Schönfeld, J., Kudrass, H.-R., Park, M.-H., Erlenkeuser, H., & Grootes, P. (1997). Thermohaline instability in the North Atlantic during meltwater events: Stable isotope and ice-rafted detritus records from core SO75-26KL, Portuguese margin. *Paleoceanography, 5,* 696–710.

Zahn, R., Winn, K., & Sarnthein, M. (1986). Benthic foraminiferal ^{13}C and accumulation rates of organic carbon: *Uvigerina peregrina* group and *Cibicidoides wuellerstorfi. Paleoceanography, 1,* 27–42.

Zeebe, R. E. (2005). Reply to the comment by P. Deines on "An explanation of the effect of seawater carbonate concentration on foraminiferal oxygen isotopes," by R. E. Zeebe (1999). *Geochimica et Cosmochimica Acta, 69,* 789–790.

Zeebe, R. E., Bijma, J., & Wolf-Gladrow, D. A. (1999). A diffusion-reaction model of carbon isotope fractionation in foraminifera. *Marine Chemistry, 64,* 199–227.

Zhang, J., Quay, P. D., & Wilbur, D. O. (1995). Carbon isotope fractionation during gas–water exchange and dissolution of CO_2. *Geochimica et Cosmochimica Acta, 59,* 107–114.

CHAPTER NINETEEN

ELEMENTAL PROXIES FOR RECONSTRUCTING CENOZOIC SEAWATER PALEOTEMPERATURES FROM CALCAREOUS FOSSILS

Yair Rosenthal

Contents

1. Introduction   765
2. Thermodynamic Effects on Mg Co-Precipitation in Calcites   766
3. Foraminiferal Mg/Ca Paleothermometry   767
   3.1. Temperature calibrations in planktonic foraminifera   767
   3.2. Inter-species and intra-species variability   769
   3.3. Intra-test variability   772
   3.4. Methodology   773
   3.5. Secondary non-temperature effects   774
   3.6. Temperature calibrations of Mg/Ca in benthic foraminifera   776
4. Ostracode Mg/Ca Paleothermometry   777
5. Coralline Sr/Ca Paleothermometry   780
   5.1. Secondary non-temperature effects   782
   5.2. Mg/Ca, U/Ca in corals   784
6. Contributions to Cenozoic Climate History   784
   6.1. The value of pairing $\delta^{18}O$ and Mg/Ca records   784
   6.2. Quaternary records of orbitally paced variability   786
   6.3. Tertiary climate evolution   787
   6.4. Seasonal to inter-annual variability   789
References   790

1. INTRODUCTION

The distribution of sea surface and bottom water temperatures and salinity is perhaps the best representation of the state of the climate system, as the ocean plays a fundamental role in the evolution of the Earth's climate. Therefore, determining the past temperature evolution of the ocean is key to understanding Earth's history. Oxygen isotope ($\delta^{18}O$) measurements in foraminifera, the standard tool in paleoceanography, have provided seminal information about the evolution of Cenozoic climate. However, their dependence on both temperature and the isotopic composition of the water — the $\delta^{18}O$ of seawater ($\delta^{18}O$ water) on both the extent of continental ice sheets and the local salinity — limits their utility for reconstructing

Developments in Marine Geology, Volume 1
ISSN 1572-5480, DOI 10.1016/S1572-5480(07)01024-X

past ocean temperatures, and hence the need for independent proxies for temperature. Elemental ratios in biogenic carbonates are particularly useful as they are measured on the same phase as $\delta^{18}O$ and thus reduce the uncertainties associated with the use of different proxy carriers. Although considerable research on magnesium (Mg) and strontium (Sr) geochemistry of marine calcareous fossils has been carried out over the past several decades, it was only recently that their potential for reconstructing ocean paleotemperatures has been fully realized. Among the new elemental proxies are Mg/Ca in planktonic foraminifera and Sr/Ca in corals as recorders of sea surface temperatures (SSTs) and Mg/Ca in benthic foraminifera and ostracodes as proxies of bottom water temperatures. Mg, Sr and Ca have relatively long oceanic residence times, $\sim$13 My for Mg, 5 My for Sr and 1 My for Ca, implying nearly constant Mg-, and Sr- to Ca-ratios in seawater on timescales of $<1 \times 10^6$ yr. A clear advantage of these carbonate-based thermometers is that coupling $\delta^{18}O$ and Mg/Ca measurements in foraminifera or Sr/Ca in corals, provides potentially a novel way to adjust for the temperature-dependency of $\delta^{18}O$ and isolate the record of $\delta^{18}O_{water}$, which can then be used to reconstruct local changes in evaporation–precipitation (and by inference salinity) and provide valuable information about changes in continental ice volume.

2. THERMODYNAMIC EFFECTS ON MG CO-PRECIPITATION IN CALCITES

Among the alkaline earth elements, Mg^{2+} and Sr^{2+} form the most important solid solutions with carbonates. Because of the small ionic radii of Mg^{2+}, $MgCO_3$ is isostructural with calcite (rhombohederal). Element-to-calcium ratios in $CaCO_3$ minerals depend on two factors: the corresponding element-to-calcium activity ratios of ocean water, and the distribution coefficients of these elements between the carbonate mineral and seawater, respectively. These relationships are expressed as $D_{El} = (El/Ca)_{mineral}/(El/Ca)_{seawater}$, where D_{El}, is the empirical homogeneous distribution coefficient calculated based on the molar concentration ratios of, for example, Mg/Ca in calcite and seawater. At equilibrium, the partitioning constant between the two pure mineral phases depends on temperature, as the substitution of Mg into calcite is associated with a change in enthalpy or heat of reaction, which is sensitive to temperature. As the substitution of Mg into calcite is an endothermic reaction, the Mg/Ca ratio of calcite is expected to increase with increasing temperature (see discussions by Mucci & Morse, 1990; Rosenthal, Boyle, & Slowey, 1997b).

Based on thermodynamic considerations, Rosenthal et al. (1997b) and Lea, Mashiotta, and Spero (1999) proposed an exponential temperature dependence of Mg uptake into calcite of $\sim$3% per °C. This prediction is consistent with experiments of Mg incorporation into inorganically precipitated calcites (Burton & Walter, 1991; Katz, 1973; Mucci, 1987; Oomori, Kameshima, Maezato, & Kitano, 1987) and see also reviews by Morse and Bender (1990) and Mucci and Morse (1990). These inorganic experiments also suggest that the distribution coefficient is essentially independent of precipitation rate, thus supporting a dominant temperature control. As instructive as these data may be, however, biologically mediated processes are rarely at thermodynamic equilibrium. For example, fora-minifera contain 1–2 orders of magnitude lower Mg than found in marine inor-ganic calcites. Such offsets suggest that biological processes exert a major influence on the co-precipitation of metals in biogenic carbonates, thus highlighting the need for species-specific empirical calibrations.

3. FORAMINIFERAL MG/CA PALEOTHERMOMETRY

3.1. Temperature Calibrations in Planktonic Foraminifera

The temperature dependence of Mg uptake into planktonic foraminiferal tests has been determined using three different approaches:

(i)   *Culture-based calibrations*: Planktonic foraminifera grown under controlled laboratory conditions in which the temperature is fixed independently of other environmental parameters (e.g., light, salinity and pH) (e.g., Lea et al., 1999; Russell, Hoenisch, Spero, & Lea, 2004). Also, in these experiments the environmental parameters are accurately determined and replication is possible, thus making temperature a true independent variable. Samples are kept under the same conditions throughout the experiment, as opposed to the variable environment encountered by foraminifera under natural conditions. A potential disadvantage is, however, that test growth under laboratory conditions may be significantly different from the natural environment. For example, cultured foraminifera do not go through gametogenesis, and there-fore lack the gametogenic calcite that in some species (e.g., *Globogerinoides sacculifer*) constitute a major part of their tests (Bè, 1980).

(ii) *Sediment trap calibrations*: Measurements of planktonic foraminifera from sediment trap time-series in sites characterized by significant seasonal SST variability, offer the closest settings to culture experiments under natural conditions (Anand, Elderfield, & Conte, 2003). Sediment trap samples have gone through complete, natural life cycle including gametogenesis, the tests are well preserved, and the conditions under which the foraminifera calcified are well constrained.

(iii) *Core top calibrations*: The analysis of fossil foraminifera obtained from surface sediments is valuable because it is based on the same material used for down-core studies (e.g., Elderfield & Ganssen, 2000; Lea, Pak, & Spero, 2000). Whereas, calibrations of live foraminifera provide a direct test for the tem-perature sensitivity of foraminiferal Mg/Ca, it is known that post-depositional dissolution can significantly alter the original test composition (Brown & Elderfield, 1996; Rosenthal, Lohmann, Lohmann, & Sherrell, 2000). In such cases, applying water-based calibrations to sedimentary samples may lead to substantial inaccuracies in temperature estimates. Thus, core-top calibrations provide a critical evaluation to the accuracy by which the surface signal is transferred to, and preserved in the sediment.

A summary of published multiple- and single-species calibrations is given in Table 1. At present, Mg/Ca calibrations are expressed as an exponential dependence of temperature in the form:

$$\frac{Mg}{Ca} \,(\text{mmol mol}^{-1}) = Be^{AT} \tag{1}$$

where A and B are the exponential and pre-exponential constants, respectively and T is temperature in °C. Justification for the choice of exponential fit to the empirical data comes primarily from the thermodynamic prediction of exponential response as described above. In this relationship, the exponential constant reflects the Mg/Ca response to a given temperature change (in mmol mol^{-1} per °C), implying increased sensitivity with temperature. The pre-exponential constant determines the absolute temperature and is species specific. It is important to note that whereas the exponential relationship provides the best fit to the multi-species data, most single-species data sets can be described by linear relationships with the same degree of confidence as the exponential equations given here. Single-species calibrations cover, however, only a relatively limited temperature range. Therefore, within the limited temperature ranges, estimates derived either from an exponential or linear regression of the same single-species data sets are typically in a good agreement.

Culture calibrations are available for a limited number of species including, *Globogerinoides sacculifer* (Nürnberg et al., 1996a; Nürnberg, Bijma, & Hemleben, 1996b), *Orbulina universa* (Lea et al., 1999; Russell et al., 2004), *Globoratalia bulloides* (Lea et al., 1999; Mashiotta et al., 1999) and *Neogloboquadrina pachyderma* (d) (Langen et al., 2005) (Figure 1A: Table 1). Combined, the available experiments suggest a temperature sensitivity of $9.7 \pm 0.9\%$ change in Mg/Ca per °C, thus providing the strongest direct evidence for temperature control on planktonic foraminiferal Mg/Ca, much in the same way inorganic experiments support the thermodynamic expectations. The sediment-trap based multi-species calibration of Anand et al. (2003) from the Sargasso Sea is in good agreement with these culturing results, suggesting a temperature sensitivity of $9 \pm 0.3\%$ per °C (Figure 1B). Anand et al. (2003) derived calcification temperatures for the different planktonic species from paired $\delta^{18}O$ measurements rather than applying estimates from the local hydrography because the latter approach might be associated with significant errors due to the uncertainty about the exact calcification depth. The multiple-species core top calibration of Elderfield and Ganssen (2000), re-evaluated by Rosenthal and Lohmann (2002) suggests temperature sensitivity of $9.5 \pm 0.5\%$ per °C (Figure 1C). The good agreement between the multi-species calibrations provides robust evidence that the temperature signal imprinted during the test formation is reliably transferred into the sediment, thus supporting the use of planktonic foraminifera Mg/Ca for paleotemperature reconstructions, despite evidence for significant post-depositional diagenetic alteration (Dekens, Lea, Pak, and Spero, 2002; Rosenthal et al., 2000).

Offsets in Mg/Ca among individual species observed in both culture and field studies stress the need for single-species calibrations. Although there is significant variability among the different equations, the median exponential value of the

14 currently available calibrations is $8.6 \pm 1.6\%$ per °C (average $8.3 \pm 1.6\%$). In a few cases, however, the estimated temperature sensitivity seems to be significantly lower (Table 1). The pre-exponential constants (B) are more variable. Differences in B primarily reflect inter-species variability (Anand et al., 2003; Elderfield & Ganssen, 2000), but are also affected by diagenetic overprints (Rosenthal & Lohmann, 2002). Given the latter biases, relative changes in seawater temperatures can be estimated using Mg/Ca thermometry with greater accuracy than absolute temperatures.

3.2. Inter-Species and Intra-Species Variability

Inter-species variability in Mg/Ca is generally correlated with calcification depth. Accordingly, shallow mixed-layer dwellers (e.g., G. ruber and G. sacculifer) have high Mg/Ca, whereas deep dwellers (e.g., G. tumida and G. dutertrei), are characterized by relatively low Mg/Ca (Bender, Lorens, & Williams, 1975; Rosenthal & Boyle, 1993). This decreasing trend is consistent with the enrichment in $\delta^{18}O$, reflecting the depth-dependent decrease in calcification temperature (Fairbanks, Sverdlove, Free, Wiebe, & Bè, 1982; Fairbanks, Wiebe, & Bè, 1980). However, although it is common to assign planktonic foraminifera specific calcification depths, it is also well known that many species continue to calcify while migrating vertically through the water column (Hemleben, Spindler, & Anderson, 1989) and some species such as G. sacculifer, also add an outer calcite crust prior to reproduction (aka "gametogenic crust") at depths significantly deeper (and colder) than their principal habitat depth (Bè, 1980). These processes are likely to cause significant variability in the Mg distribution within, and among foraminifera tests, which should be linked to changes in calcification temperature. It follows, that the "whole test" Mg/Ca composition of the test, most commonly used for temperature reconstructions, represents a weighted average of calcite layers formed at different depths/temperatures (Benway, Haley, Klinkhammer, & Mix, 2003; Lohmann, 1995). This is less of an issue for the shallow, mixed-layer species of G. ruber and G. sacculifer and more critical for deep species such as N. dutertrei, G. tumida and G. trancatulinoides. The fact that the latter may also change their preferred depth habitat in response to climate/oceanographic changes, requires caution in paleoceanographic reconstructions of the thermocline from planktonic foraminifera (Skinner & Elderfield, 2005; Spero, Mielke, Kalve, Lea, & Pak, 2003).

There is also significant variability in test composition among specimens of the same species. In particular, there is significant dependence of Mg/Ca ratios on the shell size. In most species Mg/Ca increases with increasing test size, possibly reflecting variability in calcification rate with smaller individuals calcifying faster than larger individuals (Elderfield, Vautravers, & Cooper, 2002). In addition, there is evidence for Mg/Ca variability among different morphotypes of the same species, which maybe related to differences in either growth season (e.g., white vs. pink variety of G. ruber; Anand et al., 2003) or depth habitat (e.g., G. ruber s.s. vs. G. ruber s.l.; Steinke et al., 2005). The intra-species variability implies that Mg/Ca analyses should be performed on the same morphotypes and within a single size fraction and that the calibration should be carried out on the same morphotypes and size

Table 1 Summary of Published Mg/Ca-Temperature Calibrations for Multiple, and Single Species of Planktonic Foraminifera (Mg/Ca in mmol mol^{-1} and T in °C).

Species	Mg/Ca $= Be^{AT}$		Material	Region	Temperature range	Reference
	B	A				
Multiple-species equations						
Two planktonic species	0.47	0.082	Core tops	Atlantic and Pacific	0–30	1
Eight planktonic species[a]	0.52	0.100	Core tops	North Atlantic	8–22	2
Eight planktonic species	0.78	0.095	Core tops	North Atlantic	8–22	3
Ten planktonic species	0.38	0.090	Sediment trap	North Atlantic	15–28	4
Single-species equations						
G. sacculifer	0.39	0.089	Culture		19–30	1
O. universa	1.36	0.085	Culture		16–25	5
O. universa	0.85	0.096	Culture		15–25	6
G. bulloides	0.53	0.100	Culture		16–25	5
G. bulloides	0.47	0.107	Culture/core tops		10–25	7
N. pachyderma (d)	0.51	0.104	Culture		9–19	8
G. ruber (white; 250–350 μm)[b]	0.34	0.102	Sediment trap	North Atlantic	22–28	4
G. ruber (white; 350–500 μm)	0.48	0.085	Sediment trap	North Atlantic	22–28	4
G. ruber (white; 212–355 μm)	0.69	0.068	Sediment trap	Gulf of California	20–33	9

G. sacculifer with sac (350–500 μm)	0.67	0.069	Sediment trap	North Atlantic	22–28	4
G. sacculifer without sac (350–500 μm)	1.06	0.048	Sediment trap	North Atlantic	22–28	4
G. bulloides (212–355 μm)	1.20	0.057	Sediment trap	Gulf of California	16–31	9
G. ruber (white; 250–350 μm)	0.30	0.089	Core tops	Equatorial Pacific	24–29	10
G. ruber (white; 250–350 μm)	0.38	0.090	Core tops	Atlantic and Pacific	21–29	11
O. universa[c]	0.95	0.086	Core tops	North Atlantic	8–22	12
G. bulloides	0.56	0.100	Core tops	North Atlantic	8–22	2
G. bulloides	0.81	0.081	Core tops	North Atlantic	8–22	2
N. pachyderma (s)	0.55	0.099	Core tops	Norwegian Sea	1–12	13
N. pachyderma (s)	0.41	0.083	Core tops	South Atlantic	0–15	13
G. trancatulinoides (r)	0.36	0.098	Core tops	Indian Ocean	8–25	14

References: 1. Nürnberg, Bijma, and Hemleben (1996a); 2. Elderfield and Ganssen (2000); 3. Rosenthal and Lohmann (2002) based on data from Elderfield and Ganssen (2000); 4. Anand et al. (2003); 5. Lea et al. (1999); 6. Russell et al. (2004); 7. Mashiotta, Lea, and Spero (1999); 8. Langen, Pak, Spero, and Lea (2005); 9. McConnell and Thunell (2005); 10. Lea et al. (2000); 11. Dekens et al. (2002); 12. Hathorne, Alard, James, and Rogers (2003); 13. Nürnberg (1995); 14. McKenna and Prell (2005).

[a]Used an exponential constant of 0.1 based on Lea et al. (1999).

[b]Size fraction.

[c]Based on micro-analysis methods.

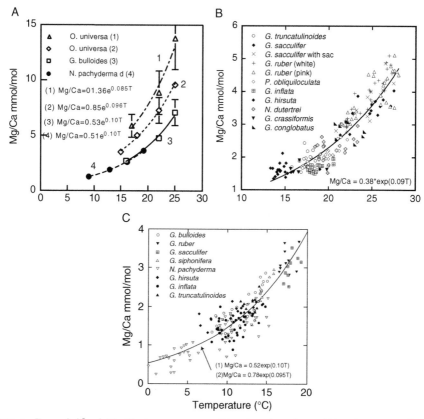

Figure 1 Foraminiferal Mg/Ca-temperature calibrations based on: (A) culture experiments. Data from: (1) Lea et al. (1999), (2) Russell et al. (2004), (3) Lea et al. (1999), (4) Langen et al. (2005); (B) sediment traps from the Sargasso Sea (with permission from Anand et al., 2003); (C) core tops (Elderfield & Ganssen, 2000); regression line 1 from Elderfield and Ganssen (2000) with permission, and regression line 2 from Rosenthal and Lohmann (2002) based on the same data set.

fraction. The specific size fraction is often chosen to match that used for isotope analysis.

3.3. Intra-Test Variability

Studies using high spatial resolution analytical methods, reveal large compositional variability within individual chambers and within the test as a whole (Anand & Elderfield, 2005; Benway et al., 2003; Brown & Elderfield, 1996; Duckworth, 1977; Gehlen, Bassinot, Beck, & Khodja, 2004; McKenna & Prell, 2005). Compositional differences in Mg/Ca, observed in sequentially precipitated test layers, are apparently consistent with the extent of vertical migration experienced by the different species during their adult life; shallow dwelling species (e.g., *G. ruber*) tend to have a more homogeneous composition than deep-dwelling species (e.g., *N. dutertrei*) thus suggesting that temperature exerts significant control on intra-test variability (Eggins, De Deckker, & Marshall, 2003). Likewise, the observation of low Mg/Ca outer-wall layers on *G. sacculifer* tests (Eggins et al., 2003) is consistent with the calcification of gametogenic calcite at deeper and colder depths, in contrast with previous results (Nürnberg et al., 1996a). The pattern of Mg/Ca variation is, however, significantly different between symbiont-bearing and symbionet-free species. Tests of symbiont-bearing foraminifera show cyclic variations between high and low Mg/Ca layers within individual chambers, which are largely absent from symbiont-free species (Eggins, Sadekov, & De Deckker, 2004; Sadekov, Eggins, & De Deckker, 2005). This variability cannot simply be attributed to changes in ambient temperature, but is more likely due to biological effects on the co-precipitation of Mg in foraminiferal calcite (Sadekov et al., 2005). Evidence for systematic Mg/Ca variations in different layers within the tests of shallow water, high-Mg calcite benthic species (Erez, 2003; Toyofuku & Kitazato, 2005) provide further evidence for the role of biological processes on Mg uptake during the biomineraliztion of foraminifera tests (Bentov & Erez, 2005; Bentov & Erez, 2006; Erez, 2003).

3.4. Methodology

The two most commonly used cleaning protocols for Mg/Ca analysis of fora-minifera are variants of the Cd cleaning method outlined in (Boyle & Keigwin, 1985) (but see modification in Rosenthal, Boyle, & Labeyrie, 1997a). The method includes multiple rinses with ddH$_2$O and methanol in an ultrasonic bath followed by a reduction and oxidation steps devised to remove contaminants associated with adherent sediment, secondary Mn- and Fe-oxides, and organic matter, respectively. A final, dilute acid leaching step is often added to remove contamination from secondary MnCO$_3$ overgrowths on the foraminiferal tests (Boyle, 1983). The "Cd cleaning method" is often used by laboratories measuring trace elements in foraminifera (e.g., Cd/Ca, Ba/Ca and U/Ca in addition to Mg/Ca). Other laboratories, use a shorter version (aka "Mg cleaning"), which does not include the reductive step (Barker, Greaves, & Elderfield, 2003). The removal of adherent sediment by multiple washes is a most critical step in the preparation for Mg/Ca analysis as some clays contain high levels of Mg and thus may lead to significant bias in measured Mg/Ca ratios (Barker et al., 2003; Lea, 2003). Another potentially significant source for Mg contamination is from secondary Mg-rich Mn-CO$_3$ overgrowths on the foraminiferal test surfaces. Evidence for Mg-Mn-CO$_3$ rich phases comes from micro-analytical investigations (Gehlen et al., 2004; Sadekov et al., 2005), XRD (Pena, Calvo, Cacho, Eggins, & Pelejero, 2005) and flow-through leaching of foraminiferal tests (Benway et al., 2003). Adding the reduction step and weak acid leaches can potentially minimize this contamination (Barker et al., 2003; Pena et al., 2005). However, as the solubility of Mn-CO$_3$ phases is lower than that of CaCO$_3$, the efficacy of these steps might be limited. Likewise, contamination from sediment debris cannot always be eliminated (Lea, 2003). Therefore, Al/Ca, Fe/Ca and Mn/Ca are routinely measured with Mg/Ca, to monitor for potential contamination by secondary mineral phases.

An inter-laboratory comparison study suggests that differences in cleaning methods used by different laboratories constitute a significant source of inter-laboratory variability in Mg/Ca analysis (Rosenthal et al., 2004). The results show that foraminiferal Mg/Ca ratios obtained by the "Cd cleaning" protocol are on average 10–15% lower than those obtained by the "Mg cleaning" method (Rosenthal et al., 2004). This offset is attributable to the addition of reductive step (Barker et al., 2003; Martin & Lea, 2002). The average inter-laboratory Mg/Ca variance is $\sim \pm 8\%$ (%RSD) for Mg/Ca measurements, which translates to $\sim \pm 1°C$ uncertainty in Mg-based temperature estimates. In practice, however, the difference is significantly smaller as some of the offsets imparted by the application of different cleaning techniques is corrected for by using the appropriate calibration obtained using the same cleaning protocol and the inter-laboratory bias in temperature should be better than $1°C$ (Rosenthal et al., 2004).

Among several instruments capable of measuring minor element to calcium ratios in marine biogenic carbonates either inductively coupled plasma-optical emission spectrophotometry (ICP-OES) or inductively coupled plasma-mass spectrometry (ICP-MS) is used. The ICP- MS offers a versatile combination of very high sensitivity simultaneously across a wide isotopic range and is mainly used by researchers, who are also measuring trace metals in foraminifera (e.g., Cd/Ca and Zn/Ca) and need its very high sensitivity (Lea & Martin, 1996; Rosenthal, Field, & Sherrell, 1999; Yu, Day, Greaves, & Elderfield, 2005). However, because of their relative economy, streamlined operation and relative ease of use — less sensitive ICP-OES instruments have gained favor for routine minor element analysis of marine biogenic carbonates (De Villiers, Greaves, & Elderfield, 2002; Green, Cooper, German, & Wilson, 2003; Schrag, 1999; Wara et al., 2003). Both instruments offer comparable external precision for Mg/Ca analysis ($\pm 0.5\%$) (Rosenthal et al., 2004) and exhibit similar Ca matrix effects. After accounting for matrix effects, data generated on either instrument are essentially interchangeable (Andreasen et al., 2006). An alternative approach to bulk test analysis is the use of micro-analytical techniques including electron microprobe (e.g., Anand & Elderfield, 2005), laser ablation ICP-MS (Eggins et al., 2003) and secondary ion mass spectrometry (SIMS) (Curry & Marchitto, 2005). To date, these methods have mainly been used to investigate the distribution of minor elements within foraminiferal tests. In the future, however, these methods may become crucial for obtaining good elemental data from cores where the samples are compromised by diagenetic artifacts (e.g., secondary overgrowths and/or partial re-crystallization) that cannot be rid of by the cleaning methods described above.

3.5. Secondary Non-Temperature Effects

3.5.1. Salinity and pH effect

Culture studies show a weak positive relationship between planktonic foraminiferal Mg/Ca and seawater salinity (Lea et al., 1999; Nürnberg et al., 1996a) and an inverse dependence on pH (Lea et al., 1999; Russell et al., 2004). Over a large salinity gradient, the effect is $\sim 7 \pm 4\%$ change in Mg/Ca per salinity unit (SU), for *G. sacculifer* and *O. universa* (Lea, 2003). In both species, Mg/Ca ratios decrease by $\sim 7 \pm 6\%$ per 0.1 pH (Lea et al., 1999; Russell et al., 2004). Given a temperature sensitivity of $\sim 9\%$ per $^{\circ}C$ and based on results from culture experiments, the salinity effect translates to an uncertainty of ~ 0.3–$1^{\circ}C$ for a change of 1 SU and -0 to $-1.2^{\circ}C$ per 0.1 pH unit. However, caution must be taken when applying these relationships. First, no discernible changes are seen at salinity changes of <3 (Nürnberg et al., 1996a). And second, the errors in both relationships are relatively large.

3.5.2. Dissolution effects

Early studies of core-top sediments have found that planktonic foraminifera Mg/Ca is correlated with the depth of the seafloor from which the tests were collected, independently of the overlying sea surface temperatures, thereby suggesting that foraminiferal Mg/Ca is altered by post-depositional dissolution on the seafloor (Lorens, Williams, & Bender, 1977; Savin & Douglas, 1973). The degree of alteration increases with water depth and the concomitant decrease in carbonate saturation levels (Rosenthal & Boyle, 1993; Russell, Emerson, Nelson, Erez, & Lea, 1994). The systematic decrease in Mg/Ca ratios of planktonic foraminifera is likely caused by preferential dissolution of Mg-rich calcite, which might be more susceptible to dissolution (Bè, Morse, & Harrison, 1975; Brown & Elderfield, 1996; Hecht, Eslinger, & Garmon, 1975). Indeed, Rosenthal et al. (2000) suggest that selective dissolution of Mg-rich chamber calcite, formed in deeper and warm surface waters, shift the test's bulk Mg/Ca toward the composition acquired in colder, thermocline waters. The decrease in Mg/Ca with depth is due to changes in each individual test, rather than due to the progressive loss of more dissolution-prone tests, which might have calcified in warmer water (Rosenthal et al., 2000). The decrease in Mg/Ca with depth is higher in non-spinose, thermocline dwelling species (e.g., *G. tumida and N. dutertrei*) than in spinose, shallow dwelling species (*G. ruber* and *G. sacculifer*), in accord with the larger range of temperatures over which the deep dwellers calcify (Brown & Elderfield, 1996; Dekens et al., 2002).

Besides temperature, dissolution is the main cause for the observed variance in Mg/Ca data. Various approaches have been offered to account for dissolution effects on Mg/Ca. Lea et al. (2000) and Dekens et al. (2002) proposed that combining core top measurements of the magnitude of the depth-dependent decrease of Mg/Ca with independent estimates for past shifts in lysocline depth, can provide an estimate as to the potential error in Mg/Ca-based temperature estimates due to dissolution effects. Applying this approach on two Pacific records, they suggest a potential error of $0.5^{\circ}C$ in their estimates of the tropical Pacific glacial–interglacial temperature amplitude. This approach, however, does not quantify the dissolution effects on specific samples, which might respond more to changes in pore-water chemistry than shifts in lysocline depth. Taking a different approach, Rosenthal and Lohmann (2002), suggested that the relationship between size-normalized test weight and the dissolution driven decrease in Mg/Ca, could be used for correcting down core Mg-based temperature estimates for dissolution effects. While this method is very attractive, as it directly addresses dissolution effects experienced by a particular sample, further studies have demonstrated that the test size-weight relationship is not strictly a function of dissolution, as initially thought, but also depends on the carbonate ion $[CO_3^{2-}]$ content of seawater in which they grow (Barker & Elderfield, 2002). Thus, the initial test weight must be known in order to correct Mg/Ca ratios for dissolution effects. Therefore, assessing and quantifying dissolution effects on Mg/Ca based temperature estimates from planktonic foraminifera remain problematic, especially for time intervals characterized by major changes in the ocean carbonate system (e.g., Eocene/Oligocene transitions), and new innovative approaches are needed. One such approach is, for example, the use of planktonic foraminifera test crystallinity (inferred from the full width at half maximum of calcite (104) X-ray diffraction peak) as a proxy for post-burial dissolution. Core-top depth transects show that foraminiferal crystallinity and size-normalized test weight co-vary with depth and the degree of calcite saturation, thereby suggesting that combining the two proxies might offer a way to quantify and correct for variable dissolution imprints on the Mg/Ca record (Bassinot, Melieres, & Labeyrie, 2004).

3.6. Temperature Calibrations of Mg/Ca in Benthic Foraminifera

The first calibration is based on the species *Cibicidoides pachyderma* from a shallow bathymetric transect in Little Bahama Banks (LBB) spanning a temperature range of 5–18°C (Rosenthal et al., 1997b). These initial results have been confirmed by Lear, Rosenthal, and Slowey (2002) and Martin, Lea, Rosenthal, Papenfuss, and Sarnthein (2002) who argued that the temperature sensitivity of *Cibicidoides* Mg/Ca is essentially identical with the ~10% increase in Mg/Ca per °C observed in planktonic foraminifera (Table 2). A similar temperature sensitivity is observed in other benthic species including *Planulina* spp., *Oridorslis umbunatus* and *Melonis* spp. (Lear et al., 2002; Rathmann et al., 2004) whereas other species *Uvigerina* spp. and *Planulina aremenensis*, exhibit significantly lower sensitivity of ~6% per °C.

While these observations lend further support to the dominant effect of temperature on foraminiferal Mg/Ca, there are still a few important questions about the accuracy of these calibrations. Most importantly, it has been questioned whether samples from LBB have been compromised by diagenetic calcite overgrowths enriched in Mg (Marchitto & deMenocal, 2003). Ongoing research efforts, at the time of this writing (April 2006), provide new insights on benthic foraminiferal Mg/Ca. Two new calibration studies of *C. pachyderma* from depth transects in the Florida Strait and Great Bahama Banks, where the problem of diagenetic calcite overgrowth is rather minimal, suggest a relatively low, linear dependence of Mg/Ca over a temperature range of 6–19°C (Marchitto, Bryan, Curry, & McCorkle, 2007; Curry & Marchitto, 2005). These results are consistent with a new calibration, proposed by Rosenthal et al. (2005), which is based on a new compilation of core-top data from the Arctic Ocean, LBB, Hawaii and Indonesia. Thus, there is a growing body of evidence to suggest that the temperature sensitivity of benthic foraminiferal Mg/Ca is significantly lower than initially thought, and more similar with that reported for *Uvigerina* spp. (Table 2). A few papers explore the potential of the aragonitic benthic foraminifer *Hoeglundina elegans* for reconstructing bottom water temperatures (Reichart, Jorissen, Anschutz, & Mason, 2003; Rosenthal et al., 2006). This species has a widespread distribution, occurring in the Atlantic, Pacific and Indian Oceans. However, because of its high dissolution susceptibility it does not occur continuously through the geological record, except for shallow water Cenozoic sections. Recently, based on core top calibrations from both Atlantic and Pacific locations, Rosenthal et al. (2006) proposed that Mg/Ca and Sr/Ca in *H. elegans* tests can be used to reconstruct thermocline temperatures. The application of Sr/Ca ratios seems especially promising as it avoids many of the problems associated with contamination with high-Mg overgrowths.

Another potentially confounding issue in the calibration of benthic foraminfera, is the evidence of non-temperature related effects, which may exert additional control on Mg/Ca in foraminifera tests. For example, Mg/Ca in dead calcitic benthic foraminifera from deep bathymetric transects in the Atlantic and Pacific Oceans decreases with depth, significantly beyond the change predicted from the global calibration when applied to the *in situ* temperature (Martin et al., 2002; Russell et al., 1994). The similarity with trends observed in other trace metals, raises the possibility that dissolution might preferentially remove metals from the calcitic tests (McCorkle, Martin, Lea, & Klinkhammer, 1995), however, more recent studies suggest that these trends reflect a primary saturation effect on trace metal co-precipitation in foraminiferal tests. A $[CO_3]$ saturation effect has already been suggested for Zn and Cd in calcitic (Marchitto, Curry, & Oppo, 2000), Mg and Sr in aragonitic foraminifera (Marchitto et al., 2000; Rosenthal et al., 2006) and Mg in calcific formainifera (Elderfield, Yu, Anand, Kiefer, & Nyland, 2006). While these secondary effects may typically have only small influence on bottom water temperature estimates, they can lead to a large bias during times of substantial change in the oceanic calcite compensation depth (CCD) (Lear, Rosenthal, Coxall, & Wilson, 2004).

4. OSTRACODE MG/CA PALEOTHERMOMETRY

Studies over the past decade have provided strong evidence for the dependence of ostracode Mg/Ca on temperature (Dwyer, Cronin, & Baker, 2002) and references therein) and demonstrated its utility for paleoceanography (e.g., Dwyer et al., 1995). Ostracodes are bivalved crustaceans that grow via a process of molting. Ostracodes excrete an exosekelton known as a carapace that in some taxa consists of 80–90% calcite and the rest of chitin and proteins. Tests of adult specimens, used in paleoceanography, are flat, typically 0.5–1 mm in length and weigh between 20 and 200 µg, thus providing a good target for geochemical analyses. Temperature calibrations are currently available for two marine, epifaunal benthic ostracodes, the deep-sea genus *Krithe* and shallow marine/estuarine genus *Loxoconcha* (Cronin, Dwyer, Baker, Rodriguez-Lazaro, & Briggs, 1996; Dwyer et al., 1995, 2002), both long-lived genera known since the Cretaceous. Mg/Ca ratios in these genera are significantly higher than in deep-sea benthic foraminifera. Based on core top calibrations, Dwyer proposed a linear relationship between *Krithe* Mg/Ca and temperature, with a slope of $0.95 \pm 0.15 \, \text{mmol} \, \text{mol}^{-1}$ per °C which can also be described by an exponential dependence of 6% per °C (Figure 2).

Table 2   Summary of Published Mg/Ca-Temperature Calibrations for Benthic Foraminifera (Mg/Ca in mmol mol^{-1} and T in °C).

Species	B	A	Material	Region	Temperature range	Method	Reference
Exponential equations Mg/Ca $= B^*_{exp}(AT)$							
C. pachyderma	1.36	0.100	Low–Mg calcite	Little Bahama Banks, Hawaii	4.5–18	a	1
C. pachyderma, C. wuellerstorfi	0.85	0.11	Low–Mg calcite	Atlantic and Pacific	0–18	a	2
C. pachyderma, C. wuellerstorfi, C. compressus	0.87	0.11	Low–Mg calcite	Atlantic and Pacific	1–18	a	3
Planulina spp.	0.79	0.120	Low–Mg calcite	Pacific	2–12	a	3
O. umbonatus	1.01	0.114	Low–Mg calcite	Atlantic and Pacific	1–10	a	3
O. umbonatus	1.53	0.09	Low–Mg calcite	Atlantic	3–10	b	4
Melonis spp.	0.98	0.101	Low–Mg calcite	Atlantic and Pacific	1–18	a	3
P. ariminensis	0.91	0.062	Low–Mg calcite	Pacific	3–15	a	3
Uvigerina spp.	0.92	0.061	Low–Mg calcite	Atlantic and Pacific	2–18	a	3
Linear equations Mg/Ca $= B+AT$							
C. pachyderma	0.35	0.25	Low–Mg calcite	Little Bahama Banks, Hawaii	4–18	a	5
C. pachyderma, C. wuellerstorfii	1.28	0.16	Low–Mg calcite	Little Bahama Banks, Hawaii and Arctic Ocean	−1–18	a	6

C. pachyderma	1.2	0.12	Low-Mg calcite	Florida Straits	6–19	a	7
C. pachyderma	0.7	0.14	Low-Mg calcite	Great Bahama Banks and Florida Straits	6–19	c	8
Planoglabratella opercularis	89.7	2.22	High-Mg calcite	Culture	10–25	a	9
Quinqueloculina yabei	66.0	2.9	High-Mg calcite	Culture	10–25	a	9
Planoglabratella opercularis	81.5	1.6	High-Mg calcite	Culture	10–25	d	10
Hoeglundona elegans	0.96	0.034	Aragonite	Core tops	2–18	a	11
Linear Sr/Ca = B+AT							
Hoeglundona elegans	1.53	0.060	Aragonite	Core tops	2–18	a	11

Method: (a) Whole test dissolution; (b) micro-analysis by laser ablation ICP-MS; (c) micro-analysis by secondary ion mass spectrometry; (d) micro-analysis by electron microprobe.
References: 1. Rosenthal et al. (1997b); 2. Martin et al. (2002); 3. Lear et al. (2002); 4. Rathmann, Hess, Kuhnert, and Mulitza, 2004; 5. Marchitto and deMenocal (2003); 6. Rosenthal, Lear, Oppo, and Linsley (2005); 7. Marchitto et al. (2007); 8. Curry and Marchitto (2007); 9. Toyofuku, Kitazato, Kawahata, Tsuchiya, and Nohara (2000); 10. Toyofuku and Kitazato (2005); 11. Rosenthal, Lear, Oppo, and Linsley (2006).

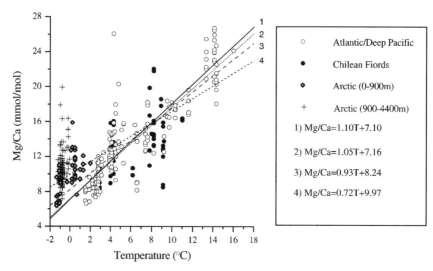

Figure 2 Ostracod Mg/Ca-temperature calibrations for adult shells of genus *Krithe* from different ocean regions. Regression lines are based on samples from (1) Atlantic and central Pacific core tops; (2) data from regression 1 and core tops from Chilean Fjords; (3) same as 2 with Arctic core tops between 0 and 900 m; (4) same as 3 with Arctic core tops deeper than 900 m. Figure modified with permission from Dwyer et al. (2002).

In principle, the factors controlling the co-precipitation of Mg and its preservation in ostracode tests are similar to those discussed for foraminifera. There are however, significant differences in both Mg/Ca ratios, and the temperature sensitivity between the two ostracode genera and among species (Dwyer et al., 2002). There are also significant ontogenic effects, which can be minimized by studying only tests of the adult stage (Dwyer et al., 2002). Culture studies show no discernible salinity effects, and dissolution experiments suggest that the effect of post-depositional dissolution on the test composition are minimal (Dwyer et al., 2002). Ostracode Mg/Ca has been used successfully for reconstructing the history of bottom water temperatures through the Pliocene and Pleistocene (e.g., Cronin, Dowsetta, Dwyerb, Baker, & Chandler, 2005; Dwyer et al., 1995; Dwyer, Cronin, Baker, & Rodriguez-Lazaro, 2000).

5. CORALLINE SR/CA PALEOTHERMOMETRY

Because of the large ionic radius of Sr^{2+}, $SrCO_3$ is iso-structural with the aragonite crystal cell (orthorombic). The principles governing the Sr/Ca ratios in inorganically precipitated aragonite are similar to those discussed above for the incorporation of Mg into calcite. However, in contrast with Mg co-precipitation in calcite, the substitution of Sr into aragonite is an exothermic reaction, thus the Sr/Ca ratio of aragonite is expected to decrease with increasing temperature. This prediction is borne out by inorganic co-precipitation experiments of Sr with aragonite showing a negative, linear dependence with a slope of $-0.039\,\mathrm{mmol\,mol^{-1}}$ per $1\,^\circ\mathrm{C}$ (Kinsman & Holland, 1969). Subsequent research, has shown that the variability of Sr/Ca in hermatypic coral skeletons is highly correlated with monthly or seasonal variations in water temperature, thus suggesting that the Sr/Ca in coral skeletal aragonite is negatively correlated with water temperature (Beck et al., 1992; Smith, Buddemeier, Redalje, & Houck, 1979). The relationship obtained in these studies is of the form:

$$\left(\frac{Sr}{Ca}\right)_{coral} (\mathrm{mmol\,mol^{-1}}) = B + A(\mathrm{SST}) \qquad (2)$$

where SST is given in $^\circ\mathrm{C}$, and the slope $A = -0.062\,\mathrm{mmol\,mol^{-1}}$ per $^\circ\mathrm{C}$ (Beck et al., 1992). These initial observations are supported by more recent studies showing strong correlations between near-monthly coral Sr/Ca and surface water temperature (e.g., Alibert & McCulloch, 1997; Linsley et al., 2004a, 2006; Marshall & McCulloch, 2002) (Figure 3). Generally the slope of the Sr/Ca–SST relationship is similar for near-monthly coral Sr/Ca and monthly SST (Table 3). Although

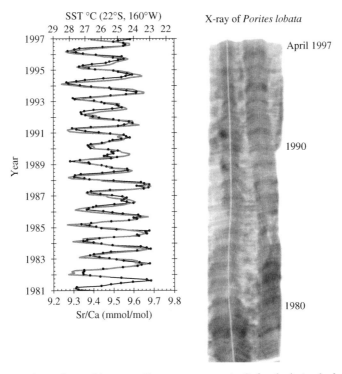

Figure 3 Comparison of monthly sea surface temperature (red) for the latitude–longitude grid including Rarotonga in the South Pacific (1° by 1°; centered at 22°S, 160°W) and coralline Sr/Ca measurements (black) analyzed at 1 mm intervals in a *Porites lutea* coral core from Rarotonga. The X-ray positive of the coral core shows the sampling track and annual growth density bands. Figure modified with permission from Schrag and Linsley (2002).

Table 3 Several Published Monthly Sr/Ca-Temperature Calibrations for *Porites* Corals from the Pacific Ocean (Sr/Ca in mmol mol^{-1} and T in °C).

Site	Species	Sr/Ca = $B+AT$		Reference	Method
		B	A		
New Caledonia	*Porites lobata*	10.479	−0.062	Beck et al. (1992)	TIMS
Hawaii	*P. lobata*	10.956	−0.079	De Villiers et al. (1994)	TIMS
Taiwan	*P. lobata*	10.286	−0.051	Shen et al. (1996)	TIMS
Great Barrier Reef	*Porites* sp.	10.38	−0.055	Alibert and McCulloch (1997)	TIMS
Great Barrier Reef	*Porites* sp.	10.4	−0.059	Marshall and McCulloch (2002)	TIMS
Fiji	*P. lutea*	10.65	−0.053	Linsley et al. (2004a)	ICP-AES
Rarotonga	*P. lutea*	11.12	−0.065	Linsley, Wellington, and Schrag (2000)	ICP-AES

Note: TIMS, thermal ion mass spectrometry; ICP-AES, inductively coupled plasma atomic emission spectrometer.

temperature is the dominant control on Sr/Ca in coral aragonite, there are additional potentially complicating factors that can limit the use of this tracer.

5.1. Secondary Non-Temperature Effects

5.1.1. Disequilibrium offset

Although in *Porites* corals the slope of the Sr/Ca-temperature relationships, derived at different sites with different individual corals, is generally the same, the intercepts of the linear relationship are significantly different among sites. These offsets are interpreted as reflecting disequilibrium effects on Sr incorporation in corals (Marshall & McCulloch, 2002), similar to those proposed to explain the $^{18}O/^{16}O$ signal in corals (McConnaughey, 1989). The different intercepts for coral Sr/Ca-SST relationships, thus, reflect the fact that the magnitude of disequilibrium offset varies among corals of the same species even at the same site. The similarity among the slopes of the Sr/Ca-SST relationships in the genus *Porites* (Table 3), suggests that Sr/Ca can be used to accurately reconstruct seawater paleotemperature as long as adequate site and coral specific temperature calibration is carried out. However, for fossil corals lacking a site-specific calibration, the consistent Sr/Ca-SST slope, but potentially different intercept implies that only relative changes in SST can accurately be reconstructed (±0.5°C). In all cases, however, for an individual coral Sr/Ca record to be an absolute temperature proxy the degree of disequilibrium offset must be constant in that coral. This is generally the case for corals sampled along the maximum growth axis.

5.1.2. Growth effects

The effects of metabolic processes such as growth and calcification rate on coralline Sr/Ca are somewhat still controversial. De Villiers, Shen, and Nelson (1994) argued that Sr/Ca ratios in *Pavona clavus* depend on growth rate, with higher ratios associated with slower skeletal extension rates. These researchers argued that the difference between skeleton accreting at 12 and 6 mm/yr is equivalent to 1–2°C based on near monthly calibration with monthly SST. In contrast, Shen et al. (1996) and Alibert and McCulloch (1997) find little growth rate effect on Sr/Ca in *Porites lutea* skeletons accreted in the maximum growth axis and Shen et al. (1996) found no perceptible growth rate effect in *Porites* at accretion rates of 18–23 mm/yr. The discrepancy between these two studies, might be due the fact that De Villiers et al. (1994) did not sample exclusively along the corals maximum growth axis, which according to Marshall and McCulloch (2002) invalidates their calcification rate argument since it is widely known that samples from off-axis sections of *Porites* skeletons show higher Sr/Ca ratios, apparently due to smaller polyp size and lower skeletal density.

5.1.3. Photosynthetic effects
Microanalysis of coral's skeleton showed that temporally equivalent skeletal elements in *Porites* can have different Sr/Ca ratios (Cohen, Layne, Hart, & Lobel, 2001). Some of the skeletal Sr/Ca heterogeneity is apparently due to the presence of night-time and day-time deposited skeletal elements (Cohen et al., 2001). During the night, small micron-size crystals are deposited at the axial spines of the corallites (COC: centers of calcification), whereas larger needle shaped crystals are formed during the day along all surfaces of the precipitating skeleton. It was further shown that the night-time and day-time skeleton have significantly different Sr/Ca (Cohen et al., 2001). Subsequent ion probe analyses on symbiont-bearing and symbiont-free corals suggested that much of the sensitivity of Sr/Ca in corals might be related to the symbionts' photosynthetic activity (Cohen, Layne, & Hart, 2002). Nonetheless, the smaller amplitude response of Sr/Ca in corals without symbionts, demonstrates the important role of temperature in controlling the skeletal Sr/Ca.

5.1.4. Diagenetic alteration
Diagenesis of coral aragonite can compromise Sr/Ca-SST records. In submerged corals, typically used for studying recent climate variability, there is evidence for precipitation of inorganic aragonite in the pores of older parts of the coral skeleton. The secondary aragonite fillings have higher Sr/Ca (and also U/Ca) ratios than in the primary biogenic aragonite, which can potentially alter the bulk skeleton elemental composition (Enmar et al., 2000; Lazar et al., 2004; Muller, Gagan, & McCulloch, 2001). These effects, which are not easily detected, may lead to significant bias in SST estimates. Diagenetic effects are more severe in uplifted, sub-aerially exposed corals, as the interaction with freshwater may lead to alteration of skeletal aragonite into calcite, which tends to lower skeletal Sr, as it is less compatible with the calcite than aragonite crystal lattice (Bar-Matthews, Wasserburg, & Chen, 1993). The problem can potentially be mitigated by avoiding the recrystalized parts of the skeleton, which can be identified by the presence of calcite using a combination of X-ray diffraction analysis and thin sections (Bar-Matthews et al., 1993; McGregor & Gagan, 2003). The application of micro-analytical techniques (e.g., laser ablation ICP or microprobe) allows for targeting highly preserved parts of the skeleton in high spatial resolution (Sinclair, Kinsley, & McCulloch, 1998).

5.2. Mg/Ca, U/Ca in Corals

Studies have shown that like Sr/Ca, also Mg/Ca (Mitsuguchi, Matsumoto, Abe, Uchida, & Isdale, 1996) and U/Ca (Min, Edwards, Taylor, Gallup, & Beck, 1995; Shen & Dunbar, 1995) in hermatypic coral skeletons are highly correlated with seasonal variations in water temperature, thus suggesting that coralline Mg/Ca and U/Ca may serve as temperature proxies. Despite the covariance with temperature, there are still questions whether temperature directly controls the co-precipitation of Mg and Sr in corals. High-resolution studies, using laser ablation ICP-MS show high variability in Mg/Ca and Sr/Ca, which accounts for significant fraction of the total variance, yet is not correlated with the *in situ* temperature cycle (Sinclair et al., 1998; Sinclair, 2005). This and the fact that there are significant differences among elemental ratio records suggest that although temperature might play a key role in determining the Mg/Ca and U/Ca composition of corals, other factors might also be important (Fallon, McCulloch, Woesik, & Sinclair, 1999). Given the covariance between several elemental ratios it is quite likely that metabolic effects, which may be temperature related, have a significant influence on the skeletal composition. At least in the case of U/Ca, the possibility that the skeletal composition is more related to the variability in $[CO_3]$ ion concentration than temperature has been raised (Shen & Dunbar, 1995). A carbonate ion control is consistent with the hypothesis put forward by Russell et al. (2004) to explain U/Ca variability in foraminifera, though it does not preclude the possibility of more than one controlling variable.

6. Contributions to Cenozoic Climate History

6.1. The Value of Pairing $\delta^{18}O$ and Mg/Ca Records

Since the publication of the first calibrations for planktonic and benthic foraminifera (Nürnberg et al., 1996a; Rosenthal et al., 1997b), coupling Mg/Ca and $\delta^{18}O$ measurements has become a standard practice in many laboratories as it offers distinct advantages that were not available to the paleoceanographic community in the past. Previously, the interpretations of foraminiferal oxygen isotope records, while of major importance on their own, were limited because the $\delta^{18}O$ composition of carbonates depends on both the temperature and the isotopic composition of the water in which the test was formed. The $\delta^{18}O$ of surface water reflects the hydrological balance between evaporation and precipitation (E/P), i.e., the surface salinity. Over longer time scales seawater $\delta^{18}O$ is also a function of the extent of continental ice sheets. The effect of each of these variables could not, however, be determined without an independent proxy for the actual temperature at which the test precipitated. Thus, the development of Mg/Ca thermometry in foraminifera provides the necessary means for adjusting $\delta^{18}O_{calcite}$ for the temperature-dependency and calculating seawater $\delta^{18}O_{water}$. For example, applying the calibration of (Bemis, Spero, & Lea, 1998), which is widely used for tropical surface dwelling planktonic species (e.g., *G. ruber* and *G. sacculifer*) the isotopic composition of the seawater can be calculated from foraminiferal $\delta^{18}O_{calcite}$ using the following equation:

$$\delta^{18}O_{water} = 0.27 + \frac{T - 16.5 + 4.8 \times \delta^{18}O_{calcite}}{4.8} \tag{3}$$

where T is the Mg/Ca based temperature in °C and the factor 0.27 is used to convert from calcite standard units of Pee Dee Belemnite (PDB) to water units based on the mean ocean water standard (SMOW). Given a strong covariance between $\delta^{18}O_{water}$ and surface salinity (Craig & Gordon, 1965), changes in E/P can be constructed and by inference salinity (e.g., Mashiotta et al., 1999; Stott et al., 2004). Caution should be taken, however, when interpreting $\delta^{18}O_{water}$ variations in terms of surface salinity. First, the choice of the pre-exponent constant in the Mg/Ca-temperature calibration has significant influence on the estimate of absolute temperatures, which in turn affects the values of reconstructed $\delta^{18}O_{water}$. Secondly, the relationship between $\delta^{18}O_{water}$ and salinity may not be spatially and temporally constant and thus potentially introducing a large uncertainty in paleo-salinity reconstructions (Schmidt, 1999). This is especially critical in the tropical ocean where the $\delta^{18}O_{water}$ and salinity relationship is highly variable (Fairbanks et al., 1997). Another advantage of paired measurements on the same samples is that it allows for precise determination of temporal leads and lags between variations in seawater temperature and the build-up or erosion of continental ice sheets, which is fundamental to understanding climate change (e.g., Lea et al., 2000; Rosenthal, Oppo, & Linsley, 2003).

On time scales longer than the oceanic residence times of Mg and Ca (13 and 1 My for Mg and Ca, respectively), secular variations in seawater Mg/Ca need to be considered before interpreting foraminiferal Mg/Ca records strictly in terms of temperature. In this case, the Mg/Ca paleotemperature equations include a term accounting for the relative difference between the seawater Mg/Ca ratio at a specific time interval and in the modern ocean:

$$\left(\frac{Mg}{Ca}\right)_{Foram} = \left[\frac{(Mg/Ca)_{sw-t}}{(Mg/Ca)_{sw-0}}\right] \times Be^{AT} \tag{4}$$

Estimates of secular variations in seawater Mg/Ca come primarily from three sources: (1) studies of the composition of fluid inclusions in evaporate minerals (Lowenstein, 2001); (2) Mg/Ca analysis of fossil echinoderms (Dickson, 2002); and (3) modeling of the processes that may change Mg and Ca concentrations in the ocean. (Demicco, Lowenstein, Hardie, & Spencer, 2005; Wilkinson & Algeo, 1989). Among the most important processes are hydrothermal alteration of basaltic crust, calcification of marine organisms, dolomite formation and changes in weathering rates and thus the riverine flux of these elements. While these reconstructions vary considerably, there is an agreement that seawater Mg/Ca generally increased by two to threefold throughout the Cenozoic, reaching the modern value of $(Mg/Ca)_{sw-0} = 5.5\,mol\,mol^{-1}$.

Below we discuss several studies that illustrate the great potential of combining elemental proxies of paleotemperature with oxygen isotope measurements for addressing questions in Cenozoic climate history. This is by no means a comprehensive list of all the studies employing this method to date, but rather a "sampler" of a few important ones.

6.2. Quaternary Records of Orbitally Paced Variability

Since its introduction less than a decade ago, the new approach of combining Mg/ Ca and $\delta^{18}O$ records has already helped in addressing some critical questions in paleoceanography and paleoclimatology. Perhaps the most important to date are records from the equatorial ocean that shed new light on the controversial question of the magnitude of surface water cooling of the tropical warm pools during glacial intervals (Crowley, 2000).

Results from early studies, based on faunal assemblages (CLIMAP, 1984) (Moore et al., 1980) and $\delta^{18}O$ composition (Broecker, 1986) of tropical planktonic foraminifera, suggest that the change in tropical SSTs during the last glacial maximum (LGM) was minimal. These suggestions are apparently at odds with terrestrial records of temperature and climatological expectations from the lower glacial atmospheric carbon dioxide (pCO_2) (Broecker & Denton, 1989). The new Mg/Ca records, though not without uncertainty, suggest that the tropical seas cooled by $\sim 3°C$ during the LGM (Hastings, Russell, & Emerson, 1998; Lea et al., 2000; Rosenthal et al., 2003), an estimate consistent with expectations from changes in continental temperatures and the inferred change in lapse rate at that time (Pierrehumbert, 1999). Moreover, the combined Mg/Ca and $\delta^{18}O$ records suggest a substantial shift in tropical Pacific hydrology, which attests to reorganization of the major tropical climate systems (e.g., El Niño Southern-Oscillation; East Asian Monsoon) in the past (Dannenmann, Linsley, Oppo, Rosenthal, & Beaufort, 2003; Koutavas, Lynch-Stieglitz, Marchitto, & Sachs, 2002; Lea et al., 2000; Stott et al., 2004; Stott, Poulsen, Lund, & L., 2002; Visser, Thunell, & Stott, 2003).

The inferred freshening of the western equatorial Pacific during the LGM, explains the low LGM-Holocene $\delta^{18}O$ amplitude in planktonic foraminiferal records, which erroneously was interpreted as suggesting minimal SST change (Broecker & Denton, 1989), thus underscoring the advantage of pairing Mg/Ca and $\delta^{18}O$ records. However, paired records not only allow for the determination of hydrographic changes, but also provide hitherto unavailable information about the temporal relationships among different climate variables. Such records provide new insights about the thermal and hydrological evolution of the tropical Pacific during the Pleistocene (Figure 4), revealing that during glacial terminations, changes in equatorial SST led by $\sim 3,000$ yr the northern hemisphere deglaciation (de Garidel, Rosenthal, Bassinot, & Beaufort, 2005; Medina-Elizalde & Lea, 2005). The records also show that throughout the Pleistocene, changes in the tropical Pacific hydrography were tightly linked to the orbitally-paced changes in the northern high latitude. However, in contrast with the eastern equatorial pacific, the western Pacific warm pool SST and salinity remain, on average, stable for the past ~ 2 million years, at the time of major Northern Hemisphere glaciation. This type of records offer valuable perspective on the climate state of the tropical ocean, which is critically needed for the current discussion on future climate change.

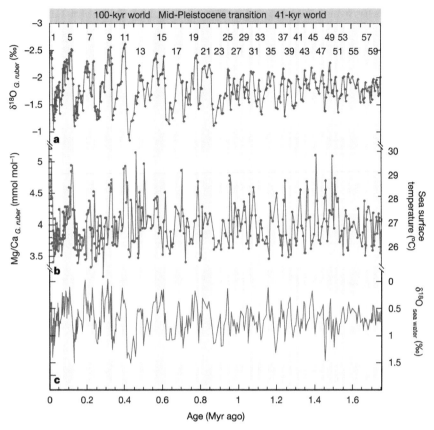

Figure 4 Planktonic foraminiferal records (*G. ruber*) from core MD97-2140 in the western equatorial Pacific (Eauripik rise 2° 02N, 141° 46E, 2,547 m) (with permission from de Garidel et al., 2005). (A) $\delta^{18}O$; (B) Mg/Ca-derived SST; and (C) calculated $\delta^{18}O_{sea-water}$. Interglacial periods are numbered at the top. Note the relatively large glacial–integlacial SST variability (°C), the long-term stability in SST and the shift from 41 to 100 ky dominated periodicity during the mid-Pleistocene.

6.3. Tertiary Climate Evolution

Paired $\delta^{18}O$ and Mg/Ca measurements in benthic foraminifera have been used to infer changes in whole ocean temperature and salinity. Combined, these records provide the best proxy for constraining the Cenozoic climate history and the evolution of the cryosphere. Low-resolution benthic foraminiferal Mg/Ca record offers new insights on the evolution of Cenozoic climate. The Mg/Ca record depicts a generally decreasing trend of global temperature throughout the Cenozoic, with the most enhanced cooling occurring during the Eocene and the Plio-Pleistocene (Figure 5) (Billups, 2003; Billups & Schrag, 2002; Lear, Elderfield, & Wilson, 2000). The inferred seawater $\delta^{18}O$ record suggests that the Cenozoic expansion of continental ice sheets occurred primarily in three steps, during the Eocene-Oligocene (E/O) transition, the middle Miocene and Plio-Pleistocene.

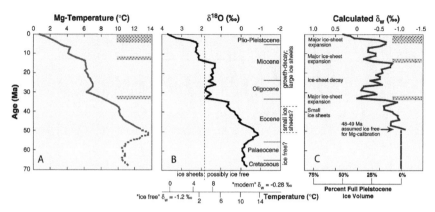

Figure 5 (A) Bottom water temperature record obtained from benthic composite foraminiferal Mg/Ca. Broken line indicates temperatures calculated from the $\delta^{18}O$ record assuming an ice-free world; (B) composite benthic foraminfera $\delta^{18}O$ record normalized to *Cibicidoides* spp.; (C) seawater $\delta^{18}O$ record, a measure of global ice volume, calculated from the Mg/Ca and $\delta^{18}O$ records. Blue areas indicate periods of substantial ice-sheet growth. Figure from Lear et al. (2000) with permission.

Interestingly, only the latter two climate transitions appear to be associated with cooling of deep-sea and polar surface temperatures (Billups & Schrag, 2002; Lear et al., 2000; Lear, Rosenthal, & Wright, 2003; Shevenell, Kennett, & Lea, 2004). In contrast, the Mg/Ca record shows no evidence of cooling during the E/O transition (Lear et al., 2000, 2004). At face value, the lack of cooling of polar surface water suggests that the initiation of the Antarctic ice sheet might be attributable to increased moisture flux to the Antarctic continent (Lear et al., 2000). Alternatively, it is possible that the lack of cooling is due to an artifact related to the effect of increased ocean calcite saturation on benthic foraminiferal Mg/Ca (Coxall, Wilson, Palike, Lear, & Backman., 2005; Lear et al., 2004; Rea & Lyle, 2005). Given this uncertainty, the temperature history across E/O climate transition is currently poorly constrained.

Although not many Mg/Ca-based SST records are available for the Tertiary, they make an important contribution, thereby highlighting the potential of the new approach beyond the Quaternary. One such example is the magnitude of warming during the Late Paleocene Thermal Maximum (LPTM). Although an increase in atmospheric pCO_2 has been implicated in causing this major warming event its magnitude is still debated. Estimating the magnitude of global warming provides an important constraint on the magnitude of this perturbation. Planktonic foraminifera $\delta^{18}O$ from Ocean Drilling Program (ODP) Site 1209 in the Pacific Ocean suggests a 2–3°C SST warming. Mg/Ca measurements of the same samples suggest, however, a 4–5°C warming; the difference is attributable to an increase in surface salinity and its effect on the $\delta^{18}O$ signal (Zachos et al., 2003). These two estimates have very different implications for our understanding of the causes and forcings behind the LPTM warming. For instance, only the high estimate of low latitude increase in surface temperature is consistent with model predictions of 8–10°C higher temperatures in Antarctica in response to about four times increase in atmospheric pCO_2 (Shellito, Sloan, & Huber, 2003).

Another interval that attracted much attention is the middle Miocene climate transition ~14 million years ago, which is marked by one of the three large "steps" in the Cenozoic benthic foraminiferal oxygen isotope record (Mi-1 *sensu* Miller, Fairbanks, & Mountain, 1987). Lear et al. (2000) suggest, based on benthic foraminiferal Mg/Ca, that roughly two-thirds of the $\delta^{18}O$ increase can be attributed to an increase in Antarctic ice cover, with the rest representing a 2–3°C cooling of deep-sea temperatures (Lear et al., 2000). But what caused this climate deterioration? Previously, it has been suggested that the mid-Miocene cooling was a triggered by a decrease in atmospheric pCO_2 (Flower & Kennett, 1993). A recent high-resolution planktonic foraminiferal Mg/Ca and $\delta^{18}O$ records from a Southern Ocean site indicates a ~7°C cooling of sea surface temperatures associated with significant surface freshening (Shevenell et al., 2004). Based on the new record, Shevenell et al. (2004) suggest that strengthening of the Antarctic Circumpolar Current in response to changes in the Earth's orbital parameters, isolated the Antarctic continent from tropical heat and moisture sources, thereby leading to enhanced cooling and ice sheet build up.

6.4. Seasonal to Inter-Annual Variability

Corals are the most important archive for resolving climate variability in the tropical ocean on sub- to inter-annual time scales. Although not without complications, the high correlation between skeletal Sr/Ca and the seasonal variability in water temperature at many sites, offers a potentially very useful tracer in paleoceanographic studies. Indeed, the high temporal resolution obtainable from corals allows us to extend the record of tropical SST beyond that of the instrumental period, thus providing a better perspective for understanding tropical climate dynamics (e.g., Hendy et al., 2002; Linsley et al., 2000). For instance, the method has been employed successfully by Linsley et al. who generated sub-annual SST records for the past ~250 yr from the Raratonga and Fiji Islands in the South Pacific (Linsley et al., 2000, 2004a, 2004b). These long, high-resolution records allow the researchers to assess the decadal and inter-decadal climate variability in the south Pacific and its links to extra-tropical climate systems. Generating surface salinity records from paired $\delta^{18}O$ and Sr/Ca records further allows them to track movements of the South Pacific Convergence Zone (SPCZ) over the past >300 yr, possibly in response to a change in the mean climate state of the tropical Pacific (Linsley et al., 2006). Furthermore, paired measurements of Sr/Ca and $\delta^{18}O$ in corals can potentially provide information about high temporal-resolution variations in surface-ocean hydrologic balance, thereby offer insights into the seasonal balance between precipitation and evaporation (Gagan et al., 2000). This approach has been successfully used to reconstruct changes in tropical Pacific surface salinity over the past few centuries (Hendy et al., 2002). However, such salinity estimates suffer from the same caveats discussed above for foraminifera based reconstructions as they are sensitive to both the choice of Sr/Ca-temperature calibration and the $\delta^{18}O$ water and salinity relationship. A similar approach has been be used for reconstructing

ENSO-related variations in runoff in the Great Barrier Reef (Australia) (McCulloch, Gagan, Mortimer, Chivas, & Isdale, 1994).

Currently there are only a few Sr/Ca records from submerged fossil corals in the south Pacific (Beck, Recy, Taylor, & Cabloch, 1997), Caribbean Sea (Guilderson, Fairbanks, & Rubenstone, 1994) and western Pacific (McCulloch et al., 1999) that extend through the Holocene. Sr/Ca measurements from these sites suggest significant ($\sim$5–6°C) tropical Pacific and Caribbean cooling prior to 10,000 yr B.P., a significantly larger estimate than the $\sim$3°C inferred from foraminiferal Mg/Ca. However, as noted above, there are still questions/concerns about the accuracy of the coral Sr/Ca thermometer due to secondary effects. However, modeling studies suggest that associated with glacial–interglacial sea level changes, there were $\sim$1–2% variations in seawater Sr/Ca due to shifts in carbonate deposition between continental shelves, dominated by Sr-rich aragonite, and the deep ocean, dominated by Sr-poor calcite (Stoll & Schrag, 1998). Somewhat larger shifts in seawater Sr/Ca ($\sim$3%) are apparently recorded in foraminifera, suggesting secular shifts (Martin, Lea, Mashiotta, Papenfuss, & Sarnthein, 1999; Shen et al., 2001). These secular shifts in seawater composition can account for half of the signal observed in the corals' Sr/Ca record, thus explaining the discrepancy between coral-, foraminiferal Mg/Ca- and alkenone-based estimates of the glacial cooling of the tropical ocean.

REFERENCES

Alibert, C., & McCulloch, M. T. (1997). Strontium/calcium ratios in modern Porites corals from the Great Barrier Reef as a proxy for sea surface temperature: Calibration of the thermometer and monitoring of ENSO. *Paleoceanography, 12*, 345–363.

Anand, P., & Elderfield, H. (2005). Variability of Mg/Ca and Sr/Ca between and within the planktonic foraminifers *Globigerina bulloides* and *Globorotalia truncatulinoides*. *Geochemistry, Geophysics, and Geosystems, 6*, 15, doi:10.1029/2004GC000811.

Anand, P., Elderfield, H., & Conte, M. H. (2003). Calibration of Mg/Ca thermometry in planktonic foraminifera from a sediment trap time-series. *Paleoceanography, 18*(2), 15, doi:10.1029/2002PA000846.

Andreasen, H. D., Sosdian, S., Perron-Cashman, S., Lear, C. H., de Garidel Thoron, T., & Rosenthal Y. (2006). Fidelity of ICP-OES and ICP-MS measurement of Mg/Ca and Sr/Ca ratios in marine biogenic carbonates: Are they trustworthy together? *Geochemistry, Geophysics, and Geosystems, 7*(10), 15.

Bar-Matthews, M., Wasserburg, G. J., & Chen, J. H. (1993). Diagenesis of fossil coral skeletons: Correlations between trace elemental, textures and 234U/238U. *Geochimica et Cosmochimica Acta, 37*, 257–276.

Barker, S., & Elderfield, H. (2002). Foraminiferal calcification response to glacial–interglacial changes in atmospheric CO_2. *Science, 297*, 833–836.

Barker, S., Greaves, M., & Elderfield, H. (2003). A study of cleaning procedures used for foraminiferal Mg/Ca paleothermometry. *Geophysics, Geochemistry, and Geosystems, 4*(9), 20, doi:10.1029/2003GC000559.

Bassinot, F. C., Melieres, F., & Labeyrie, L. (2004). Crystallinity of foraminifera shells: A proxy to reconstruct past bottom water CO_3 changes? *Geochemistry, Geophysics, and Geosystems, 5*(8), 12, doi:10.1029/2003GC000668.

Bè, A. W. H. (1980). Gametogenic calcification in a spinose planktonic foraminifer, *Globogerinoides sacculifer*. *Marine Micropaleontology, 5*, 283–310.

Bè, A. W. H., Morse, J. W., & Harrison, S. M. (1975). Progressive dissolution and ultrastructural breakdown in planktonic foraminifera. In: W. V. Sliter, A. W. H. Bè & W. H. Berger (Eds), *Dissolution of deep-sea carbonates*. Special Publications of Cushman Foundation for Foraminiferal Research (pp. 27–55).

Beck, W. J., Edwards, L. R., Ito, E., Taylor, F. W., Recy, J., Rougerie, F., Joannot, P., & Henin, C. (1992). Sea-surface temperature from coral skeletal strontium/calcium ratios. *Science, 257*, 644–647.

Beck, J. W., Recy, J., Taylor, F. W., & Cabloch, G. (1997). Abrupt changes in early Holocene tropical sea surface temperature derived from coral records. *Nature, 385*, 705–707.

Bemis, B. E., Spero, H. J., & Lea, D. W. (1998). Reevaluation of the oxygen isotopic composition of planktonic foraminifera: Experimental results and revised paleotemperature equations. *Paleoceanography, 13*, 150–160.

Bender, M. L., Lorens, R. B., & Williams, F. D. (1975). Sodium, magnesium and strontium in the tests of planktonic foraminifera. *Micropaleontology, 21*, 448–459.

Bentov, S., & Erez, J. (2005). Novel observations on biomineralization processes in foraminifera and implications for Mg/Ca ratio in the shells. *Geology, 33*, 841–844.

Bentov, S., & Erez, J. (2006). Impact of biomineralization processes on the Mg content of foraminiferal shells: A biological perspective. *Geochemistry, Geophysics, and Geosystems, 7*(1), 11, doi:10.1029/2005GC001015.

Benway, H. M., Haley, B. A., Klinkhammer, G. P., & Mix, A. C. (2003). Adaptation of a flow-through leaching procedure for Mg/Ca paleothermometry. *Geochemistry, Geophysics, and Geosystems, 4*(2), 15, doi:8410.1029/2002GC000312.

Billups, K. (2003). Application of benthic foraminiferal Mg/Ca ratios to questions of early Cenozoic climate change. *Earth Planetary Science Letters, 209*, 181–195.

Billups, K., & Schrag, D. P. (2002). Paleotemperatures and ice volume of the past 27 Myr revisited with paired Mg/Ca and $^{18}O/^{16}O$ measurements on benthic foraminifera. *Paleoceanography, 17*, 1–11.

Boyle, E. A. (1983). Manganese carbonate overgrowths on foraminifera tests. *Geochimica et Cosmochimica Acta, 47*, 1815–1819.

Boyle, E. A., & Keigwin, L. D. (1985). Comparison of Atlantic and Pacific paleochemical records for the last 250,000 years: Changes in deep ocean circulation and chemical inventories. *Earth Planetary Science Letters, 76*, 135–150.

Broecker, W. S. (1986). Oxygen isotope constraints on surface ocean temperature. *Quaternary Research, 26*, 121–134.

Broecker, W. S., & Denton, G. H. (1989). The role of ocean-atmosphere reorganizations in glacial cycles. *Geochimica et Cosmochimica Acta, 53*, 2465–2501.

Brown, S. J., & Elderfield, H. (1996). Variations in Mg/Ca and Sr/Ca ratios of planktonic foraminifera caused by postdepositional dissolution: Evidence of shallow Mg-dependent dissolution. *Paleoceanography, 11*, 543–551.

Burton, E. A., & Walter, L. M. (1991). The effects of pCO_2 and temperature on magnesium incorporation in calcite in seawater and $MgCl_2$–$CaCl_2$ solutions. *Geochimica et Cosmochimica Acta, 55*, 775–785.

CLIMAP (1984). The last interglacial ocean. *Quaternary Research, 21*, 123–124.

Cohen, A. L., Layne, G. D., & Hart, S. R. (2002). The effect of algal symbionts on the accuracy of Sr/Ca paleotemperatures from coral. *Science, 296*, 331–333.

Cohen, A. L., Layne, G. D., Hart, S. R., & Lobel, P. S. (2001). Kinetic control of skeletal Sr/Ca in a symbiotic coral: Implications for the paleoceanographic proxy. *Paleoceanography, 16*, 20–26.

Coxall, H. K., Wilson, P. A., Palike, H., Lear, C. H., & Backman, J. (2005). Rapid stepwise onset of Antarctic glaciation and deeper calcite compensation in the Pacific Ocean. *Nature, 433*, 53–57.

Craig, H., & Gordon, L. I. (1965). Deuterium and oxygen-18 variations in the ocean and the atmosphere: Stable isotopes in oceanographic studies and paleotemperatures. In: E. Tongiori (Ed.), *Proceedings of the Third Spoleto Conference*, Spoleto, Italy. Sischi and Figli, Pisa (pp. 9–130).

Cronin, T. M., Dowsetta, H. J., Dwyerb, G. S., Baker, P. A., & Chandler, M. A. (2005). Mid-Pliocene deep-sea bottom-water temperatures based on ostracode Mg/Ca ratios. *Marine Micropaleontology, 54*, 249–261.

Cronin, T. M., Dwyer, G. S., Baker, P. A., Rodriguez-Lazaro, J., & Briggs, W. M. (Eds) (1996). *Deep-Sea ostracode shell chemistry (Mg:Ca ratios) and Late Quaternary Arctic Ocean history*. *Late Quaternary*

Paleoceanography of the North Atlantic margins (Spl. Publn. 111, pp. 117–134). London: Geological Society Publications.

Curry, W. B., & Marchitto, T. M. (2005). A SIMS calibration of benthic foraminiferal Mg/Ca. *EOS, Transactions, of American Geophysical Union, Fall Meeting, 86*(52), PP51A-0583.

Dannenmann, S., Linsley, B. K., Oppo, D. W., Rosenthal, Y., & Beaufort, L. (2003). East Asian monsoon forcing of suborbital variability in the Sulu Sea during marine isotope stage 3: Link to Northern Hemisphere climate. *Geochemistry, Geophysics, and Geosystems, 4*(1), 13.

De Villiers, S., Greaves, M., & Elderfield, H. (2002). An intensity ratio calibration method for the accurate determination of Mg/Ca and Sr/Ca of marine carbonates by ICP-AES. *Geochemistry, Geophysics, and Geosystems, 3*, 14.

De Villiers, S., Shen, G. T., & Nelson, B. K. (1994). The Sr/Ca-temperature relationship in coralline aragonite: Influence of variability in (Sr/Ca)seawater and skeletal growth parameters. *Geochimica et Cosmochimica Acta, 58*, 197–208.

Dekens, P. S., Lea, D. W., Pak, D. K., & Spero, H. J. (2002). Core top calibration of Mg/Ca in tropical foraminifera: Refining paleo-temperature estimation. *Geochemistry, Geophysics, and Geosystems, 3*(4), 29.

Demicco, R. V., Lowenstein, T. K., Hardie, L. A., & Spencer, R. J. (2005). Model of seawater composition for the Phanerozoic. *Geology, 33*, 877–880.

Dickson, J. A. D. (2002). Fossil echinoderms as monitor of the Mg/Ca ratio of Phanerozoic oceans. *Science, 298*, 1222–1224.

Duckworth, D. L. (1977). Magnesium concentration in the tests of the planktonic foraminifer *Globorotalia truncatulinoides. Journal of Foraminiferal Research, 4*, 304–312.

Dwyer, G. S., Cronin, T. M., & Baker, P. A. (2002). Trace elements in marine ostracodes, the Ostracoda: Applications in Quaternary Research. Washington DC: American Geophysical Union (pp. 205–224).

Dwyer, G. S., Cronin, T. M., Baker, P. A., Raymo, M. E., Buzas, J. S., & Correge, T. (1995). North Atlantic deepwater temperature change during late Pliocene and late Quaternary climatic cycles. *Science, 270*, 1347–1351.

Dwyer, G. S., Cronin, T. M., Baker, P. A., & Rodriguez-Lazaro, J. (2000). Changes in North Atlantic deep-sea temperature during climatic fluctuations of the last 25,000 years based on ostracode Mg/Ca ratios. *Geochemistry Geophysics Geosystems, 1*, 17, doi:10.1029/2000GC000046.

Eggins, S., De Deckker, P., & Marshall, J. (2003). Mg/Ca variation in planktonic foraminifera tests: Implications for reconstructing palaeo-seawater temperature and habitat migration. *Earth Planetary Science Letters, 212*, 291–306.

Eggins, S., Sadekov, A., De Deckker, P. (2004). Modulation and daily banding of Mg/Ca in *Orbulina universa* tests by symbiont photosynthesis and respiration: A complication for seawater thermometry? *Earth Planetary Science Letters, 225*, 411–419.

Elderfield, H., & Ganssen, G. (2000). Past temperature and d^{18}O of surface ocean waters inferred from foraminiferal Mg/Ca ratios. *Nature, 405*, 442–445.

Elderfield, H., Vautravers, M., & Cooper, M. (2002). The relationship between shell size and Mg/Ca, Sr/Ca, d18O, and d13C of species of planktonic foraminifera. *Geochemistry, Geophysics, and Geosystems, 3*(8), 13, doi:10.1029/2001GC000194.

Elderfield, H., Yu, J., Anand, P., Kiefer, T., & Nyland, B. (2006). Calibrations for benthic foraminiferal Mg/Ca paleothermometry and the carbonate ion hypothesis. *Earth Planetary Science Letters*, 1–17.

Enmar, R., Stein, M., Bar-Matthews, M., Sass, E., Katz, A., & Lazar, B. (2000). Diagenesis in live corals from the Gulf of Agaba: I. The effect on paleo-oceanography tracers. *Geochimica et Cosmochimica Acta, 64*(18), 3123–3132.

Erez, J. (2003). The source of ions for biomineralization in foraminifera and their implications for paleoceanographic proxies. In: M. P. Dove, D. Y. J. & S. Weiner (Eds), Biomineralization. *Reviews in Mineralogy and Geochemistry* (Vol. 54, pp. 115–149).

Fairbanks, R. G., Evans, M. N., Rubenstone, J. L., Mortlock, R. A., Broad, K., Moore, M. D., & Charles, C. D. (1997). Evaluating climate indices and their geochemical proxies measured in corals. *Coral Reefs, 16*(Suppl), S93–S100.

Fairbanks, R. G., Sverdlove, M., Free, R., Wiebe, P. H., & Bè, A. W. H. (1982). Vertical distribution and isotopic fractionation of living planktonic foraminifera in the Panama Basin. *Nature, 298*, 841–844.

Fairbanks, R. G., Wiebe, P. H., & Bè, A. W. H. (1980). Vertical distribution and isotopic composition of living planktonic foraminifera in the western North Atlantic. *Science, 207*, 61–63.

Fallon, S. J., McCulloch, M. T., Woesik, R. V., & Sinclair, D. J. (1999). Corals at their latitudinal limits: laser ablation trace element systematics in Porites from Shirigai Bay, Japan. *Earth Planetary Science Letters, 172*, 221–238.

Flower, B. P., & Kennett, J. P. (1993). Relations between Monterey Formation deposition and middle Miocene global cooling: Naples Beach section, California. *Geology, 21*, 877–880.

Gagan, M. K., Ayliffe, L. K., Beck, J. W., Cole, J. E., Druffel, E. R. M., Dunbar, R. B., & Schrag, D. P. (2000). New views of tropical paleoclimates from corals. *Quaternary Science Reviews, 19*, 45–64.

de Garidel, T., Rosenthal, Y., Bassinot, F., & Beaufort, L. (2005). Stable sea surface temperatures in the western Pacific warm pool over the past 1.75 million years. *Nature, 433*, 294–298.

Gehlen, M., Bassinot, F., Beck, L., & Khodja, H. (2004). Trace element cartography of *Globigerinoides ruber* shells using particle-induced X-ray emission. *Geochemistry, Geophysics, and Geosystems, 5*, 9, doi:10.1029/2004GC000822.

Green, D. R., Cooper, M. J., German, C. R., & Wilson, P. A. (2003). Optimization of an inductively coupled plasma-optical emission spectrometry method for the rapid determination of high-precision Mg/Ca and Sr/Ca in foraminifera calcite. *Geochemistry, Geophysics, and Geosysystems, 4*(6), 11.

Guilderson, T. P., Fairbanks, R. G., & Rubenstone, J. L. (1994). Tropical temperature variations since 20,000 years ago: Modulating interhemispheric climate change. *Science, 263*, 663–665.

Hastings, D. W., Russell, A. D., & Emerson, S. R. (1998). Foraminiferal magnesium in *Globeriginoides sacculifer* as a paleotemperature proxy. *Paleoceanography, 13*, 161–169.

Hathorne, E. C., Alard, O., James, R. H., & Rogers, N. W. (2003). Determination of intratest variability of trace elements in foraminifera by laser ablation inductively coupled plasma-mass spectrometry. *Geochemistry, Geophysics, and Geosystems, 4*, 14, doi:10.1029/2003GC000539.

Hecht, A. D., Eslinger, E. V., & Garmon, L. B. (1975). Experimental studies on the dissolution of planktonic foraminifera. Dissolution of deep-sea carbonates. Cushman Foundation for Foraminiferal Research, Special Publications, pp. 56–69.

Hemleben, C., Spindler, M., & Anderson, O. R. (1989). *Modern planktonic foraminifera* (p. 363). New York: Springer-Verlag.

Hendy, E. J., Gagan, M. K., Alibert, C. A., McCulloch, M. T., Lough, J. M., & Isdale, P. J. (2002). Abrupt decrease in tropical Pacific sea surface salinity at end of the Little Ice Age. *Science, 295*, 1511–1514.

Katz, A. (1973). The interaction of magnesium with calcite during crystal growth at 25–90°C and one atmosphere. *Geochimica et Cosmochimica Acta, 37*, 1563–1586.

Kinsman, D. J. J., & Holland, H. D. (1969). The co-precipitation of cations with $CaCO_3$-IV. The co-precipitation of Sr with aragonite between 16 and 96 degrees. *Geochimica et Cosmochimica Acta, 33*, 1–18.

Koutavas, A., Lynch-Stieglitz, J., Marchitto, T. M., & Sachs, J. P. (2002). El Niño: Like pattern in Ice Age tropical Pacific Sea surface temperature. *Science, 297*, 226–230.

Langen, P. J. v., Pak, D. K., Spero, H. J., & Lea, D. W. (2005). Effects of temperature on Mg/Ca in neogloboquadrinid shells determined by live culturing. *Geochemistry, Geophysics, and Geosystems, 6*, 11, doi:10.1029/2005GC000989.

Lazar, B., Enmar, R., Schossberger, M., Bar-Matthews, M., Halicz, L., & Stein, M. (2004). Diagenetic effects on the distribution of uranium in live and Holocene corals from the Gulf of Aqaba. *Geochimica et Cosmochimica Acta, 68*, 4583–4593.

Lea, D. W. (2003). Elemental and isotopic proxies of pastocean temperatures. In: K. Turkien (Ed.), *Treatise on geochemistry*. Amsterdam: Elsevier.

Lea, D. W., & Martin, P. A. (1996). A rapid mass spectrometric method for the analysis of barium, cadmium, and strontium in foraminifera shells. *Geochimica et Cosmochimica Acta, 60*, 3143–3149.

Lea, D. W., Mashiotta, T. A., & Spero, H. J. (1999). Controls on magnesium and strontium uptake in planktonic foraminifera determined by live culturing. *Geochimica et Cosmochimica Acta, 63*, 2369–2379.

Lea, D. W., Pak, D. K., & Spero, H. J. (2000). Climate impact of late Quaternary equatorial Pacific sea surface temperature variations. *Science, 289*, 1719–1724.

Lear, C. H., Elderfield, H., & Wilson, P. A. (2000). Cenozoic deep-sea temperatures and global ice volumes from Mg/Ca in benthic foraminiferal calcite. *Science, 287,* 269–272.

Lear, C. H., Rosenthal, Y., Coxall, H. K., & Wilson, P. A. (2004). Late Eocene to early Miocene ice sheet dynamics and the global carbon cycle. *Paleoceanography, 19,* 11, doi:10.1029/2004PA001039.

Lear, C. H., Rosenthal, Y., & Slowey, N. (2002). Benthic foraminiferal Mg/Ca-paleothermometry: A revised core-top calibration. *Geochimica et Cosmochimica Acta, 66,* 3375–3387.

Lear, C. H., Rosenthal, Y., & Wright, J. D. (2003). The closing of a seaway: Ocean water masses and global climate change. *Earth Planetary Science Letters, 210,* 425–436.

Linsley, B. K., Kaplan, A., Gouriou, Y., Salinger, J., deMenocal, P. B., Wellington, G. M., & Howe, S. S. (2006). Tracking the extent of the South Pacific Convergence Zone since the early 1600s. *Geochemistry, Geophysics, and Geosystems, 7,* Q05003, doi:10.1029/2005GC001115.

Linsley, B. K., Wellington, G. M., & Schrag, D. P. (2000). Decadal sea surface temperature variability in the subtropical South Pacific from 1726 to 1997 AD. *Science, 290,* 1145–1148.

Linsley, B. K., Wellington, G. M., Schrag, D. P., Ren, L., Salinger, M. J., & Tudhope, A. W. (2004a). Coral evidence for changes in the amplitude and spatial pattern of South Pacific interdecadal climate variability over the last 300 years. *Climate Dynamics, 22,* 1–11, doi:10.1007/s00382-003-0364-y.

Linsley, B. K., Wellington, G. M., Schrag, D. P., Ren, L., Salinger, M. J., & Tudhope, A. W. (2004b). Geochemical evidence from corals for changes in the amplitude and spatial pattern of South Pacific interdecadal climate variability over the last 300 years. *Climate Dynamics, 22,* 1–11.

Lohmann, G. P. (1995). A model for variation in the chemistry of planktonic foraminifera due to secondary calcification and selective dissolution. *Paleoceanography, 10,* 445–457.

Lorens, R. B., Williams, D. F., & Bender, M. L. (1977). The early nonstructural chemical diagenesis of foraminiferal calcite. *Journal of Sedimentary Petrology, 47,* 1602–1609.

Lowenstein, T. K. (2001). Oscillations in Phanerozoic seawater chemistry: Evidence from fluid inclusions. *Science, 294,* 1086–1088.

Marchitto, T. M., Bryan, S. P., Curry, W. B., & McCorkle, D. C. (2007). Mg/Ca temperature calibration for the benthic foraminifer cibicidoides pachyderma. *Paleoceanography, 22*(PA1203), 1–9.

Marchitto, T. M., Curry, W. B., & Oppo, D. W. (2000). Zinc concentrations in benthic foraminifera reflect seawater chemistry. *Paleoceanography, 15,* 299–306.

Marchitto, T. M., & deMenocal, B.P. (2003). Late Holocene variability of upper North Atlantic Deep Water temperature and salinity. *Geochemistry, Geophysics, and Geosystems, 4,* 1100, doi:1010.1029/2003GC000598.

Marshall, J. F., & McCulloch, M. T. (2002). An assessment of the Sr/Ca ratio in shallow water hermatypic corals as a proxy for sea surface temperature. *Geochimina et Cosmochimica Acta, 66*(18), 3263–3280.

Martin, P. A. & Lea, D. W. (2002). A simple evaluation of cleaning procedures on fossil benthic foraminifera. *Geophysics Geochemistry Geosystems, 3*(1), doi:10.1029/2001GC000280.

Martin, P. A., Lea, D. W., Mashiotta, T. A., Papenfuss, T., & Sarnthein, M. (1999). Variation of foraminiferal Sr/Ca over Quaternary glacial–interglacial cycles: Evidence for changes in mean ocean Sr/Ca. *Geochemistry, Geophysics, and Geosystems, 1,* 19, doi:1999GC000006.

Martin, P. A., Lea, D. W., Rosenthal, Y., Papenfuss, T. P., & Sarnthein, M. (2002). Late Quaternary Deep-Sea temperatures inferred from benthic foraminiferal magnesium. *Earth Planetary Science Letters, 198,* 193–209.

Mashiotta, T. A., Lea, D. W., & Spero, H. J. (1999). Glacial–interglacial changes in subantarctic sea surface temperature and d^{18}O-water using foraminiferal Mg. *Earth Planetary Science Letters, 170,* 417–432.

McConnaughey, T. (1989). ^{13}C and ^{18}O isotopic disequilibrium in biological carbonates: I. Patterns. *Geochimica et Cosmochimica Acta, 53,* 151–162.

McConnell, M. C., & Thunell, R. C. (2005). Calibration of the planktonic foraminiferal Mg/Ca paleothermometer: Sediment trap results from the Guaymas Basin, Gulf of California. *Paleoceanography, 20,* 18, doi:10.1029/2004PA001077.

McCorkle, D. C., Martin, P. A., Lea, D. W., & Klinkhammer, G. P. (1995). Evidence of a dissolution effect on benthic foraminiferal shell chemistry: Delta-C-13, Cd/Ca, Ba/Ca, and Sr/Ca results from the Ontong Java Plateau. *Paleoceanography, 10*(4), 699–714.

McCulloch, M. T., Gagan, M. K., Mortimer, G. E., Chivas, A. R., & Isdale, P. J. (1994). A high-resolution Sr/Ca and d^{18}O coral record from the Great Barrier Reef, Australia, and the 1982–1983 El Nino. *Geochimica et Cosmochimica Acta, 58*, 2747–2754.

McCulloch, M. T., Tudhope, A. W., Esat, T. M., Mortimer, G. E., Chappell, J., Pillans, B., Chivas, A. R., & Omura, A. (1999). Coral record of equatorial sea-surface temperatures during the penultimate deglaciation at Huon peninsula. *Science, 283*, 202–204.

McGregor, H. V., & Gagan, M. K. (2003). Diagenesis and geochemistry of Porites corals from Papua New Guinea: Implications for paleoclimate reconstruction. *Geochimica et Cosmochimica Acta, 67*(12), 2147–2156.

McKenna, V. M., & Prell, W. L. (2005). Calibration of the Mg/Ca of *Globorotalia truncatulinoides* (R) for the reconstruction of marine temperature gradients. *Paleoceanography, 19*, 12, doi:10.1029/2000PA000604.

Medina-Elizalde, M., & Lea, D. W. (2005). The mid pleistocene transition in the tropical Pacific. *Science, 310*, 1009–1012.

Miller, K. G., Fairbanks, R. G., & Mountain, G. S. (1987). Tertiary oxygen isotope synthesis, sea level history, and continental margin erosion. *Paleoceanography, 2*, 1–9.

Min, G., Edwards, R., Taylor, F., Gallup, C., & Beck, J. (1995). Annual cycles of U/Ca in coral skeletons and U/Ca thermometry. *Geochimica et Cosmochimica Acta, 59*(10), 2025–2042.

Mitsuguchi, T., Matsumoto, E., Abe, O., Uchida, T., & Isdale, P. J. (1996). Mg/Ca thermometry in coral skeletons. *Science, 274*, 961–963.

Moore, Jr., T. C., Burckle, L. H., Geitzenauer, K., Luz, B., Molina-Cruz, A., Robertson, J. H., Sachs, H., Sancetta, C., Thiede, J., Thompson, P., & Wenkam, C. (1980). The reconstruction of sea surface temperatures in the Pacific Ocean of 18,000 BP. *Marine Micropalaeontology, 5*, 215–247.

Morse, J. W., & Bender, M. L. (1990). Partition coefficient in calcite: Examination of factors influencing the validity of experimental results and their application to natural systems. *Chemical Geology, 82*, 265–277.

Mucci, A. (1987). Influence of temperature on the composition of magnesian calcite overgrowths precipitated from seawater. *Geochimica et Cosmochimica Acta, 51*, 1977–1984.

Mucci, A., & Morse, J. W. (1990). Chemistry of low-temperature abiotic calcites: experimental studies on coprecipitation, stability, and fractionation. *Reviews in Aquatic Sciences, 3*, 217–254.

Muller, A., Gagan, M. K., & McCulloch, M. T. (2001). Early marine diagenesis in corals and geochemical consequences for paleoceanographic reconstructions. *Geophysical Research Letters, 28*(23), 4471–4474.

Nürnberg, D. (1995). Magnesium in tests of *Neogloboquadrina pachyderma* sinisteral from high northern and southern latitudes. *Journal of Foraminiferal Research, 25*, 350–368.

Nürnberg, D., Bijma, J., & Hemleben, C. (1996a). Assessing the reliability of magnesium in foraminiferal calcite as a proxy for water mass temperature. *Geochimica et Cosmochimica Acta, 60*, 803–814.

Nürnberg, D., Bijma, J., & Hemleben, C. (1996b). Erratum: Assessing the reliability of magnesium in foraminiferal calcite as a proxy for water mass temperature. *Geochimica et Cosmochimica Acta, 60*, 2483–2484.

Oomori, T., Kameshima, H., Maezato, Y., & Kitano, Y. (1987). Distribution coefficient of Mg^{2+} ions between calcite and solution at 10–50°C. *Marine Chemistry, 20*, 327–336.

Pena, L. D., Calvo, E., Cacho, I., Eggins, S., & Pelejero, C. (2005). Identification and removal of Mn-Mg-rich contaminant phases on foraminiferal tests: Implications for Mg/Ca past temperature reconstructions. *Geochemistry, Geophysics, and Geosystems, 6*(9), 25, doi:10.1029/2005GC000930.

Pierrehumbert, R. (1999). Huascaran delta O-18 as an indicator of tropical climate during the last glacial maximum. *Geophysical Research Letters, 26*(9), 1345–1348.

Rathmann, S., Hess, S., Kuhnert, H., & Mulitza, S. (2004). Mg/Ca ratios of the benthic foraminifera *Oridorsalis umbonatus* obtained by laser ablation from core top sediments: Relationship to bottom water temperature. *Geochemistry, Geophysics, and Geosystems, 5*, 10, doi:10.1029/2004GC000808.

Rea, D. K., & Lyle, M. (2005). Paleogene calcite compensation depth in the eastern subtropical Pacific: Answers and questions. *Paleoceanography, 20*, 9, doi:10.1029/2004PA001064.

Reichart, G.-J., Jorissen, F., Anschutz, P., & Mason, P. R. (2003). Single foraminiferal test chemistry records the marine environment. *Geology, 31*, 355–358.

Rosenthal, Y., & Boyle, E. A. (1993). Factors controlling the fluoride content of planktonic foraminifera: An evaluation of its paleoceanographic applicability. *Geochimica et Cosmochimica Acta*, *57*, 335–346.

Rosenthal, Y., Boyle, E. A., & Labeyrie, L. (1997a). Last glacial paleochemistry and deep water circulation in the Southern Ocean: Evidence from foraminiferal cadmium. *Paleoceanography*, *12*, 778–787.

Rosenthal, Y., Boyle, E. A., & Slowey, N. (1997b). Environmental controls on the incorporation of Mg, Sr, F and Cd into benthic foraminiferal shells from Little Bahama Bank: Prospects for thermocline paleoceanography. *Geochimica et Cosmochimica Acta*, *61*, 3633–3643.

Rosenthal, Y., Field, F., & Sherrell, R. M. (1999). Precise determination of element/calcium ratios in calcareous samples using Sector Field Inductively Coupled Plasma Mass Spectrometry. *Analytical Chemistry*, *71*, 3248–3253.

Rosenthal, Y., Lear, C. H., Oppo, D. W., & Linsley, B. K. (2005). Temperature and [CO$_3$] effects on Mg/Ca and Sr/Ca in the benthic foraminifera species of *Hoeglundina Elegans* and Cibicidoides. *EOS, Transactions, of American Geophysical Union, Fall Meeting*, *86*(52), PP33A-1544.

Rosenthal, Y., Lear, C. H., Oppo, D. W., & Linsley, B. K. (2006). Temperature and carbonate ion effects on Mg/Ca and Sr/Ca ratios in benthic foraminifera: The aragonitic species *Hoeglundina elegan*. *Paleoceanography*, *21*, 14, doi:10.1029/2005PA001158.

Rosenthal, Y., & Lohmann, G. P. (2002). Accurate estimation of sea surface temperatures using dissolution-corrected calibrations for Mg/Ca paleothermometry. *Paleoceanography*, *17*(4), 6.

Rosenthal, Y., Lohmann, G. P., Lohmann, K. C., & Sherrell, R. M. (2000). Incorporation and preservation of Mg in *G. sacculifer*: Implications for reconstructing sea surface temperatures and the oxygen isotopic composition of seawater. *Paleoceanography*, *15*, 135–145.

Rosenthal, Y., Oppo, D. W., & Linsley, B. K. (2003). The amplitude and phasing of climate change during the last deglaciation in the Sulu Sea, western equatorial Pacific. *Geophysical Research Letters*, *30*(8), 4.

Rosenthal, Y., Perron-Cashman, S., Lear, C. H., Bard, E., Barker, S., Billups, K., Bryan, M., Delaney, M. L., DeMenocal, P. B., Dwyer, G. S., Elderfield, H., German, C. R., Greaves, M., Lea, D. W., Marchitto, Jr, T. M., Pak, D. P., Paradis, G. L., Russell, A. D., Schneider, R. R., Scheindrich, K., Stott, L., Tachikawa, K., Tappa, E., Thunell, R., Wara, M., Weldeab, S., & Wilson, P. A. (2004). Laboratory inter-comparison study of Mg/Ca and Sr/Ca measurements in planktonic foraminifera for paleoceanographic research. *Geochemistry, Geophysics, and Geosystems*, *5*(4), 29, doi:10.1029/2003GC000650.

Russell, A. D., Emerson, S., Nelson, B., Erez, J., & Lea, D. W. (1994). Uranium in foraminiferal calcite as a recorder of seawater uranium concentrations. *Geochimica et Cosmochimica Acta*, *58*, 671–681.

Russell, A. D., Hoenisch, B., Spero, H. J., & Lea, D. W. (2004). Effects of seawater carbonate ion concentration and temperature on shell U. Mg, and Sr in cultured planktonic foraminifera. *Geochimica et Cosmochimica Acta*, *68*, 4347–4361.

Sadekov, A. Y., Eggins, S. M., & De Deckker, P. (2005). Characterization of Mg/Ca distributions in planktonic foraminifera species by electron microprobe mapping. *Geochemistry, Geophysics, and Geosystems*, *6*, 14, doi:10.1029/2005GC000973.

Savin, S. M., & Douglas, R. G. (1973). Stable isotope and magnesium geochemistry of recent planktonic foraminifera from the South Pacific. *Geological Society of American Bulletin*, *84*, 2327–2342.

Schmidt, G. A. (1999). Error analysis of paleosalinity calculations. *Paleoceanography*, *14*, 422–429.

Schrag, D. P. (1999). Rapid analysis of high-precision Sr/Ca ratios in corals and other marine carbonates. *Paleoceanography*, *14*, 97–102.

Schrag, D. P., & Linsley, B. K. (2002). Paleoclimate: Corals, chemistry, and climate. *Science*, *296*(5566), 277–278.

Shellito, C. J., Sloan, L. C., & Huber, M. (2003). Climate model sensitivity to atmospheric CO$_2$ levels in the Early-Middle Paleogene. *Palaeogeography, Palaeoclimatology, Palaeoecology*, *193*, 113–123.

Shen, G. T., & Dunbar, R. B. (1995). Environmental controls on uranium in reef corals. *Geochimica et Cosmochimica Acta*, *59*(10), 2009–2024.

Shen, C. C., Lee, T., Chen, C. Y., Wang, C. H., Dai, C. F., & Li, L. A. (1996). The calibration of D[Sr/Ca] versus sea surface temperature relationship for Porites corals. *Geochimica et Cosmochimica Acta*, *60*, 3849–3858.

Shen, C.-C., Hastings, D. W., Lee, T., Chiu, C.-H., Lee, M.-Y., Wei, K.-Y., & Edwards, R. L. (2001). High precision glacial–interglacial benthic foraminiferal Sr/Ca records from the eastern Equatorial Atlantic Ocean and Caribbean Sea. *Earth Planetary Science Letters, 190*(3–4), 197–209.

Shevenell, A., Kennett, J. P., & Lea, D. W. (2004). Middle Miocene Southern Ocean cooling and Antarctic cryosphere expansion. *Science, 305*, 1766–1770.

Sinclair, D. J. (2005). Correlated trace element "vital effects" in tropical corals: A new geochemical tool for probing biomineralization. *Geochimica et Cosmochimica Acta, 69*(13), 3265–3284.

Sinclair, D. J., Kinsley, L. P. J., & McCulloch, M. T. (1998). High resolution analysis of trace elements in corals by laser ablation ICP-MS. *Geochimica et Cosmochimica Acta, 62*(11), 1889–1901.

Skinner, L. C., & Elderfield, H. (2005). Constraining ecological and biological bias in planktonic foraminiferal Mg/Ca and d^{18}Occ: A multispecies approach to proxy calibration testing. *Paleoceanography, 20*(8), 15, doi:10.1029/2004PA00105.

Smith, S. V., Buddemeier, R. W., Redalje, R. C., & Houck, J. E. (1979). Strontium–calcium thermometry in coral skeletons. *Science, 204*, 404–407.

Spero, H. J., Mielke, K. M., Kalve, E. M., Lea, D. W., & Pak, D. K. (2003). Multispecies approach to reconstructing eastern equatorial Pacific thermocline hydrography during the past 360 kyr. *Paleoceanography, 18*(1), 16, doi:10.1029/2002PA000814.

Steinke, S., Chiu, H. -Y., Yu, P. -S., Shen, C. -C., Lowemark, L., Mii, H. -S., & Chen, M. -T. (2005). Mg/Ca ratios of two *Globigerinoides ruber* (white) morphotypes: Implications for reconstructing past tropical/subtropical surface water conditions. *Geochemistry, Geophysics, and Geosystems, 6*, 13, doi:10.1029/2005GC000926.

Stoll, H. M., & Schrag, D. P. (1998). Effects of Quaternary sea level changes on strontium in seawater. *Geochimica et Cosmochimica Acta, 62*, 1107–1118.

Stott, L., Cannariato, K., Thunell, R., Haug, G. H., Koutavas, A., & Lund, S. (2004). Decline of surface temperature an salinity in the western tropical Pacific Ocean in the Holocene epoch. *Nature, 431*, 56–59.

Stott, L., Poulsen, C., Lund, S., & Thunell, R. (2002). Super ENSO and Global Climate Oscillations at Millennial Time Scales. *Science, 297*, 222–226.

Toyofuku, T., & Kitazato, H. (2005). Micromapping of Mg/Ca values in cultured specimens of the high-magnesium benthic foraminifera. *Geochemistry, Geophysics, and Geosystems, 6*, 10, doi:10.1029/2005GC009619.

Toyofuku, T., Kitazato, H., Kawahata, H., Tsuchiya, M., & Nohara, M. (2000). Evaluation of Mg/Ca thermometry in foraminifera: Comparison of experimental results and measurements in nature. *Paleoceanography, 7155*, 456–464.

Visser, K., Thunell, R. C., & Stott, L. (2003). Magnitude and timing of temperature change in the Indo-Pacific warm pool during deglaciation. *Nature, 421*, 152–155.

Wara, M. W., Delaney, M. L., Bullen, T. D., & Ravelo, A. C. (2003). Application of a radially viewed inductively coupled plasma-optical emission spectrophotomer to simultaneous measurement of Mg/Ca, Sr/Ca, and Mn/Ca ratios in marine biogenic carbonates. *Geochemistry, Geophysics, and Geosystems, 4*(8), 14, doi:10.1029/2003GC000525.

Wilkinson, B. H., & Algeo, T. J. (1989). Sedimentary carbonate record of calcium–magnesium cycling. *American Journal of Science, 289*, 1158–1194.

Yu, J., Day, J., Greaves, M., & Elderfield, H. (2005). Determination of multiple element/calcium ratios in foraminiferal calcite by quadrupole ICP-MS. *Geochemistry, Geophysics, and Geosystems, 6*(8), 9, doi:10.1029/2005GC000964.

Zachos, J. C., Wara, M. W., Bohaty, S., Delaney, M. L., Petrizzo, M. R., Brill, A., Bralower, T. J., Premoli-Silva, I., et al. (2003). A transient rise in tropical sea surface temperature during the Paleocene–Eocene thermal maximum. *Science, 302*, 1551–1554.

RECONSTRUCTING AND MODELING PAST OCEANS

Katrin J. Meissner

Contents

1. A Brief Historical Overview	800
2. Classification of Climate Models	801
2.1. Conceptual models	802
2.2. Numerical models	802
3. Models and Proxy Data	804
3.1. Data assimilation	805
3.2. Integrating proxy data into models	806
3.3. Testing and evaluating models	806
4. International Programs	807
5. Conclusion	808
References	809

The science of reconstructing past climates has evolved rapidly over the past three decades. Having started as a subdiscipline of geology and geochemistry (Imbrie & Imbrie, 1979), it is today an interdisciplinary research field that unifies a large international scientific community. Some of the most innovative and surprising research in this field has important implications for future climate change. As such, paleoscience is today recognized as being of great relevance to societal concerns. For example, ice core analyses revealed that greenhouse gas levels are higher today than they have been for hundreds of thousands of years. In other words, anthropogenic impacts are pushing the climate system towards a state for which there is no "reference climate" in paleo records of the Quaternary. Therefore, a better understanding of the climate system and feedbacks within the climate system is crucial for our future. Another disconcerting finding of paleoresearch is the fact that the climate system is highly non-linear: In the past, rapid and large-amplitude climate change has occurred in response to slowly varying, small-amplitude forcing. It is critical that policy makers understand what paleoscience has to say about rapid climate change in order to fully understand the significance of future rapid climate change. The UN-sanctioned Intergovernmental Panel on Climate Change (IPCC) recognizes the importance of understanding past climate change to predict future climate scenarios and has devoted an entire chapter in the fourth assessment report (AR4) to paleoclimates.

Developments in Marine Geology, Volume 1
ISSN 1572-5480, DOI 10.1016/S1572-5480(07)01025-1

The ocean's high heat capacity and its ability to transport energy and to sequester and release greenhouse gases give it an important role in helping to determine the state of the planet's climate. Compared to the atmosphere, the ocean has a long response time to perturbations and this makes it a key player in climate change on timescales of hundreds to ten thousands of years. Thus, paleoceanography is a crucial component of paleoclimatic research.

1. A Brief Historical Overview

The reconstruction of paleocean characteristics and dynamics requires climatic detective work. It involves the dating and interpretation of paleoclimatic records as well as the definition of physical and dynamical constraints, which specify possible circulation patterns and characteristics. As a single drill core, analysis method or numerical model cannot possibly address fundamental scientific questions regarding past climate change, international cooperative efforts are necessary to construct multi-proxy records and evaluate climate models. The first interdisciplinary project to reconstruct climate began in the 1970s and was led by the marine geologists John Imbrie and Jim Hayes and the geochemist Nick Shackelton. The 'Climate: Long-range Investigation, Mapping, and Prediction' project (CLIMAP) reconstructed the surface of the Earth at the Last Glacial Maximum (LGM) at 21,000 yr BP [CLIMAP Project Members, 1976, 1981]. Although the reconstruction was based on ocean sediment proxy data, the expertise of scientists from other disciplines was also sought after including researchers from the still very new field of climate modeling, for example.

The first climate models evolved from numerical weather prediction models and were developed during the early 1960s. These atmosphere general circulation models (AGCMs) had become the central tool of climate science by the 1970s. The initial focus of modeling past climates was the LGM (since it represented a very different climate from today and there was a relatively rich amount of proxy data) at 21,000 BP. Some of the first studies were simulations by Williams, Barry, and Washington (1974); Gates (1976a, 1976b) and Hansen et al. (1984), amongst others. As atmosphere-only models, these simulations relied on boundary conditions (e.g., CLIMAP sea surface temperatures). An exhaustive overview of the results of these early model studies can be found in Crowley and North (1991).

Meanwhile, ocean modelers were beginning to build similar models of the ocean (ocean general circulation models or OGCMs). As soon as the first ocean models became operational, it became apparent that the poleward ocean heat transport plays a key factor in climate change and variability (Manabe & Bryan, 1985; Bryan, 1986). The studies of Bryan (1986) and Manabe and Stouffer (1988) should be highlighted in this context as these early OGCM simulations showed for the first time that multiple equilibria of the thermohaline circulation could exist in more explicit models under identical boundary conditions. This finding resulted in an exhaustive and still ongoing study of the hysteresis behavior of the thermohaline circulation in response to buoyancy forcing (see Weaver & Hughes, 1992 for a thorough early overview paper or some of the more recent studies: Rahmstorf,

1995; Weaver, 1995; Ganopolski & Rahmstorf, 2001; Rahmstorf, 2002; Rahmstorf et al., 2005 among many others).

In the late 1980s, the Cooperative Holocene Mapping Project (COHMAP) took the place of CLIMAP in again successfully combining skills of many disciplines into a single interdisciplinary effort. The objective of the COHMAP project was to reconstruct climate conditions at successive 3,000-yr intervals from 18,000 BP to present. As climate models had evolved drastically in the 1980s, a variety of proxy data sources were used this time in combination with model simulations to reconstruct paleo environments. Therefore, COHMAP was the first interdisciplinary effort to promote the interaction between climate modelers and Quaternary scientists. Broad data sets were used to define model boundary conditions, in order to test model accuracy and improve our understanding of external and internal climate forcing mechanisms (COHMAP Members, 1988).

The next deciding step forward in the history of climate modeling of past environments was the consideration of biogeochemical cycles as a crucial component of the climate system. Until the early 1990s, the main focus of climate modeling was on physical feedbacks between climate components (mainly the ocean, atmosphere, and sea ice), but a few researchers had begun to investigate the importance of biogeochemical cycles as early as the mid-1980s. These first simulations were carried out using 3-box models (Knox & McElroy, 1984; Sarmiento & Toggweiler, 1984; Siegenthaler & Wenk, 1984) and highlighted an important aspect of the marine carbon cycle: The CO_2 content of the warm (low latitude) surface ocean and hence also of the atmosphere is regulated by the surface waters of the cold high-latitude box. This finding is referred to as the Harvardton-Bear effect after the authors' three institutions (Havard, Princeton and Bern) and has important implications for paleoclimate conditions with reduced atmospheric CO_2 (Broecker et al., 1999). The growing evidence over the last 20 yr is that a better quantification of biogeochemical interactions is critical in the understanding of climate change and variability, and this has led to the development of biogeochemical processes in box models (Stephens & Keeling, 2000; Toggweiler, 1999; Gildor, Tziperman, & Toggweiler, 2002) and OGCMs. As a result, the first paleoclimate model studies including oceanic and/or global biogeochemical processes in OGCMs started to appear in the early 1990s (Heinze, Maier-Reimer, & Winn, 1991) and are still being developed (Schmittner (2005) among others).

Today there are several interdisciplinary alliances that aim to catalyze and compare studies from a wide range of different research groups with a diverse set of research tools. These groups include the Coupled Carbon Cycle Climate Model Intercomparison Project (C4MIP), Past Global Changes (PAGES), Rapid Climate Change (RAPID) and Paleoclimate Modeling Intercomparison Project (PMIP), which will be discussed in Section 4.

2. CLASSIFICATION OF CLIMATE MODELS

The theoretical approach to paleoceanography uses quantitative ocean and climate models to reconstruct paleoconditions of the ocean and interpret observations. A number of different ocean and coupled climate models have been used in the past, and a brief overview of the hierarchy and use of these models in the broad research area of paleoceanography will now be given (for an extensive overview of numerical climate models, the reader is referred to McGuffie & Henderson-Sellers, 2001).

2.1. Conceptual Models

Climate models can be split into two categories: conceptual and numerical models. Whereas conceptual models provide non- or semi-quantitive, non-dynamic simulations of paleoclimate, numerical models are quantitative and can be dynamic (Parrish, 1998). One of the first paleoclimate studies using a conceptual model is described in Nairn and Smithwick (1976). Other purely conceptual models of the atmosphere (Parrish, 1982) and ocean (Ziegler et al., 1981) were used for paleoclimate studies in the late 1970s and early 1980s. The next step in the hierarchy of climate models are the numerical conceptual models. These models are still very simple and take only a few processes into account, but make use of computer power. To understand the usefulness of these simple models, consider the following example.

The 100,000-yr rhythm of the major glacial–interglacial cycles during the past 700,000 yr is thought to be tied to the changing eccentricity of the Earth's orbit (which has characteristic periods of 95,000, 131,000, 413,000, and 2.1 million yr) in what is commonly referred to as Milankovitch theory. Challengers of this theory argue that these changes result in very small changes in insolation and are unlikely to drive the climate system. Also, the 413,000 period of eccentricity resulted in minimal insolation changes during the warm interglacial episode about 400,000 yr ago. The proxy records show an especially warm interglacial during that time contrary to all predictions solely based on insolation in the Northern Hemisphere. Paillard (1998) developed two simple multi-equilibria ice models, which were able to give more insight into Milankovitch forcing and the warm interglacial episode. With his two very simple multi-equilibria models, Paillard (1998) successfully simulated both — the transition from a 41 to 100 kyr period in the glacial–interglacial cycle as well as a warm interglacial 400,000 yr ago.

2.2. Numerical Models

It is, of course, impossible to build an exact model of the climate system; for the ocean alone, the position and momentum of approximately 5×10^{46} molecules would have to be calculated at each instant of time. Instead, the ocean, atmosphere, sea ice, or land surface is split into discrete macroscopic elements with measurable characteristics such as temperature, density, velocity, etc. The state of and exchange between these discrete elements follow physical laws and can therefore be determined with numerical models. Small-scale processes within each element can also influence the large-scale pattern and therefore have to be parameterized. Existing ocean (and climate) models differ in regards to:

- their temporal and spatial resolution (resolution is defined as the spatial scale which defines the boundary between processes that are resolved by the model and those which are parameterized)
- the nature of processes which are resolved (e.g., some models include biogeo-chemical cycles whereas other models resolve physical processes only)
- the number of subsystems taken into account (e.g., an ocean model needs boundary conditions at the surface which can be provided either by data or by an atmosphere model that is physically interacting with the ocean model)

As the different subsystems (ocean, atmosphere, sea ice, continental ice sheets, vegetation, etc.) of the climate system interact with each other in a complex manner and on a very broad range of timescales, climate modeling reduces to the process of identifying isolable subsystems and processes that are relevant to the problem at hand. While identifying these subsystems and processes, the researcher has also to keep in mind that these processes have to be suitable to be simulated by limited mathematical models and will provide results in a reasonable computational time (Crowley & North, 1991).

The simplest class of climate models includes one-dimensional Energy Balance Models which were first developed by Budyko (1969) and Sellers (1969). It is interesting to note that these simple models yield two stable solutions under present-day boundary conditions: the present-day climate and a completely frozen Earth (also called "Snowball Earth").

2.2.1. General circulation models

Because the boundary data for paleoclimatic simulations tend to contain large uncertainties, global models are better suited for this research area than regional models. The most comprehensive results are of course given by global general circulation models (GCMs). GCMs consist of a three-dimensional representation of the ocean (or atmosphere) and are to date the most complex models available. They are built on the conservation equations of momentum, mass, and tracers, and can include parameterizations of the carbon cycle or atmospheric chemistry. When run in an uncoupled mode (atmosphere or ocean model only), they rely on recon-structed data specifying the boundary conditions at the bottom of the atmosphere (or surface of the ocean). For example, as already mentioned in Section 1, nu-merous modeling studies restored the ocean surface characteristics to the CLIMAP data set (CLIMAP Project Members, 1976) for simulations of the LGM (e.g., Fichefet et al., 1994).

A better approach is to use coupled ocean–atmosphere models; by computing the surface boundary conditions, one can bypass the data problem. For example, AGCMs coupled to a mixed-layer ocean are still commonly used. This type of model predicts changes in sea surface temperatures and sea ice by treating the ocean as though it were a layer of water of constant depth (typically 50 m), heat transports within the ocean being specified and remaining constant while climate changes. With a constant heat transport in the ocean, these models are quite limited for simulations under very different climatic conditions (Bice, Scotese, Seidov, & Barron, 2000). The most complex models in use today are coupled atmosphere–ocean general circulation models (AOGCMs, e.g. Bush, & Philander, 1998). Some recent models include the biosphere, carbon cycle, and atmospheric chemistry as well. However, coupled ocean–atmosphere GCMs often need flux adjustments. Flux adjustments balance surface fluxes at the ocean–atmosphere interface to avoid a numerical drift of the coupled system. As flux adjustments have been "tuned" to the present-day climate, the use of these adjustments to simulate past climates is not very reliable. However, some recent studies use coupled atmosphere–ocean GCMs that do not need artificial flux adjustments (e.g., LeGrande et al., 2006).

2.2.2. Earth System Models of Intermediate Complexity
Earth System Models of Intermediate Complexity (EMICs, Claussen et al., 2002) have been developed recently to bridge the gap between simple, conceptual models (such as box models or EBMs) and comprehensive GCMs. EMICs include most of the processes described in comprehensive models, but in a more reduced (parameterized) form. As climate subsystems other than the atmosphere and ocean also play an important role in climate change (e.g., continental ice sheets, sea ice, and land surface processes), the class of Earth System Models explicitly simulates the interactions among several components of the natural Earth system. The advantage of EMICs lies in their computational efficiency, which allows for long-term climate simulations over several tens to thousands of years. Also, the large number of processes described in EMICs enables the user to investigate interactions and feedbacks within the climate system in a broad range of sensitivity experiments (Kubatzki & Claussen, 2003). EMICs are used to explore the parameter space, develop analysis methods, and analyze the response of climate over long timescales. The modeling community is continuously integrating new processes and subsystems in their models to obtain a better representation of the climate system dynamics. For example, there is growing evidence that biogeochemical interactions between the subsystems are more important than initially thought (carbon cycle, nitrogen cycle, methane, etc.) and most of the EMICs used as of today include some parameterization of oceanic and even global biogeochemical cycles (Bern 2.5D climate model, CLIMBER-2, CLIMBER 3α, GENIE, ISAM-2, LOVE-CLIM, MIT Integrated Global System Model, MoBidiC, UVic Earth System Model).

EMICs are today widely used for paleoceanographic and paleoclimate simulations.

3. MODELS AND PROXY DATA

The interpretation of paleoproxy data is an ongoing challenge for paleoclimate scientists. As a striking example, one could compare the studies of Clark,

Pisias, Stocker, and Weaver (2002) and Bond et al. (2001). Both papers are highly regarded, yet draw opposite conclusions from the atmospheric $\delta^{14}C$ record. Whereas Bond et al. (2001) relate the variability of atmospheric $\delta^{14}C$ to changes in solar radiation, Clark et al. (2002) interpret the same type of record as a signature of variability in the thermohaline circulation and ocean heat transport. On the other hand, model simulations of past climates depend strongly on boundary conditions, assumptions, and the model used for the study. The simulated climate for a certain time span can be radically different depending on the model and boundary conditions used. Interpretation of measured paleoclimate data is thus urgently needed through collaboration between modelers and observers.

3.1. Data Assimilation

One possible approach to combine proxy data with climate models is that of "data assimilation". Data assimilation involves the construction of a field that accommodates best the information obtained from paleoproxies with the physical (and dynamical) constraints of the climate system using coupled climate models (e.g., Paul & Schäfer-Neth, 2005). Thus, the models help to physically interpret the rather local proxy data in a global context. Inverse models, for example, seek for a compromise between the hydrographic and paleoproxy data. They determine circulation and biogeochemical characteristics by fitting the model to datasets of temperature, salinity, oxygen, nutrients, radiocarbon, etc. For example, Winguth, Archer, Maier-Reimer, and Mikolajewicz (2000) used an inverse method (Wunsch, 1997) by which an OGCM (the Hamburg large-scale geostrophic OGCM LSG) can be adjusted to fit the paleonutrient data in order to reproduce ocean circulation and surface boundary conditions for the LGM. The results of this assimilation study confirm that the North Atlantic Deep Water was shallower and the North Atlantic Intermediate Water stronger during the LGM compared to present day. The optimized glacial circulation patterns in the Southern Ocean are consistent with the Cd/Ca measurements but cannot explain changes in $\delta^{13}C$.

Other modelers use a more empirical approach by realizing several model simulations and comparing the simulated characteristics to paleodata. For example, Schmittner, Meissner, Eby, and Weaver (2002) compared simulated changes in sea surface properties for a large variety of overturning strengths with different reconstruction sets. In this study, the results were found to depend strongly on the data set used. Sea surface temperature reconstructions from CLIMAP (CLIMAP Project Members, 1976, 1981) and earlier salinity reconstructions based on planktonic foraminifera (Seidov, Sarnthein, Stattegger, Prien, & Weinelt, 1996; Duplessy et al., 1991; Sarnthein et al., 1995) are most consistent with a significant reduction of the circulation, while recent reconstructions using dynocyst assemblages (de Vernal, Hillaire-Marcel, Turon, & Matthiessen, 2000) allow no unequivocal conclusion.

The disconcerting results of these two examples point to one of the most important challenges one encounters when combining paleoproxy data and climate models: the scale difference. Climate models reproduce large-scale features (reflecting global differential heating, the impact of large-scale topography, etc.) whereas small-scale features (secondary mountain ranges, marginal seas, etc.) are not well described because of insufficient resolution (Renssen, Braconnot, Tett, von Storch, & Weber, 2004). On the other hand, the spatial and temporal sampling of proxy data is sparse. Therefore, the local character of proxy records might show the signature of local events that are not resolved in a global model with coarse resolution. Also, the proxy data usually reflects seasonal phenomena, which might be misinterpreted as representing annual averages.

The uncertainties of paleoproxy reconstructions combined with the highly nonlinear nature of the climate system make a complete agreement between models and proxy data impossible. Therefore, palaeodata assimilation schemes may be seen as an "estimate of the probability distribution functions of past climatic features and model parameters" [Crucifix, 2005b].

3.2. Integrating Proxy Data into Models

Another possible approach is to incorporate paleoproxy data ($\delta^{18}O$, δD, $\delta^{14}C$, $\delta^{13}C$, $\delta^{10}Be$, etc.) as prognostic active tracers in climate models. These climate models can then be run under constant forcing to reproduce pattern of proxy data for a certain equilibrium climate state (e.g., the LGM), which can be compared to observed proxy data obtained from ice cores, marine sediments, and other records. Additionally, time-dependent (or transient) experiments can be performed under time-varying forcings (such as meltwater events or changing solar activity) to produce time series, which can be directly compared to the proxy record. This has been done with uncoupled ocean (or atmosphere) GCMs (e.g., Schmidt, 1999; Werner, Mikolajewicz, Hoffmann, & Heimann, 2000; Butzin, Prange, & Lohmann, 2005), vegetation models (Kaplan, Prentice, Knorr, & Valdes, 2002), and continental ice sheet models (e.g. Clarke, Lhomme, & Marshall, 2005). However, the importance of interactions between atmosphere, oceans, and other systems such as the biosphere and the cryosphere points to the necessity of using coupled models. To date, there have been only a few studies simulating paleoproxy data with either coupled ocean–atmosphere GCMs or Earth System Models (e.g. LeGrande et al., 2006; Crucifix, 2005a; Roche, Paillard, Ganopolski, & Hoffman, 2004; Meissner, Schmittner, Weaver, & Adkins, 2003; Stocker & Wright, 1996). These quantitative comparisons between climate models and paleoproxies are one of the key foci in paleoclimate modeling as they will allow for improved interpretation of the paleoproxy records.

A third way to bridge the gap between the modeling and proxy data approach is to find locations of proxy data records of special interest with the use of climate models. A simulation including prognostic paleoproxy tracers can determine the geographical region of greatest impact on a given paleoproxy data during a given climate event (e.g., Meissner et al., 2003).

3.3. Testing and Evaluating Models

Finally, geological data provide us with the only way of verifying climate models. Even though paleoproxy data are point measurements whereas climate models operate on large grid cells, paleoproxy time series offer anchor points to test and evaluate models. In addition, sensitivity experiments can be performed to describe mechanisms that assess the model response. Thus, numerical simulations of past climates constitute an independent test of the performance and reliability of the climate models that are used for future climate change scenarios (Crucifix, Braconnot, Harrison, & Otto-Bliesner, 2005).

In conclusion, large amounts of paleoproxy data have been retrieved from various types of archives, but attempts to use numerical models for verification and interpretation of this data are sparse. The science of using three-dimensional climate models to interpret paleo records is still in its infancy.

4. INTERNATIONAL PROGRAMS

Today, there are several international and interdisciplinary programs to co-ordinate and encourage the systematic study of past climates. For example, the Paleoclimate Modeling Intercomparison Project (PMIP) was launched in 1992 and is endorsed by the World Climate Research Programme (WCRP) and the International Geosphere-Biosphere Programme (IGBP). Its first phase focused on the LGM and the mid-Holocene (6 kyr BP) and was designed to test and compare atmospheric components of climate models (Harrison, Braconnot, Joussaume, Hewitt, & Stouffer, 2002). Phase I of PMIP met with considerable success, with 67 model experiments archived in a central database and over 70 papers published in the refereed literature (Crucifix et al., 2005). One of the most important conclusions arising from this initial phase of PMIP was the recognition of the importance of ocean and land surface (vegetation) feedbacks in the climate system. In 2002, the PMIP steering committee decided to launch PMIP Phase II that focuses on coupled ocean–atmosphere (OAGCM) and ocean–atmosphere–vegetation models (OAVGCM). It also expanded the set of standard experiments to include simulations of the previous interglacial/glacial transition (glacial inception at 115 kBP), the Younger Dryas (12.7–11.7 kBP), the early Holocene (9 kBP) and the abrupt cooling event 8.2 kBP (Crucifix et al., 2005).

Rapid Climate Change (RAPID), a 6-yr program (2001–2007) of the Natural Environment Research Council aims to investigate the causes of rapid climate change, with a main focus on the role of the thermohaline circulation. Using a combination of present-day observations, paleoproxy data, and climate models, RAPID seeks to assess the probability and magnitude of future climate change (Rapid Climate Science Plan, available at http://www.noc.soton.ac.uk/rapid/rapid.php).

PAst Global changES (PAGES) is a core project of the IGBP and is funded by the U.S. and Swiss National Science Foundations, and the National Oceanic and Atmospheric Administration (NOAA). It acts as an umbrella organization to facilitate international collaborations and interdisciplinary science. PAGES does not fund individual research, but it sponsors workshops, symposia, and conferences. PAGES's scope of interest includes the physical climate system, biogeochemical cycles, ecosystem processes, biodiversity, and human dimensions, on different time scales such as the Pleistocene, the Holocene, the last millennium and the recent past. Over 3,800 scientists in more than 100 countries around the world currently subscribe to PAGES (http://www.pages.unibe.ch/).

The Global Analysis, Interpretation and Modeling Task Force (GAIM) is a component of the International Geosphere Biosphere Programme (IGBP) of the International Council of Scientific Unions (ICSU). GAIM aims to advance the study of the coupled dynamics of Earth Systems with emphasis on the biogeochemical processes using both data and models. Modeling projects are structured by topic (CO_2, trace gases and climate–vegetation interactions) and time period (paleo (from 20 kBP), fossil fuel (from 200 BP), contemporary (from 20 BP) and future). For example, the Ocean Carbon Cycle Model Intercomparison Project (OCMIP) and the Coupled Carbon Cycle Climate Model Intercomparison Project (C4MIP) are two of the GAIM continuing current projects (http://gaim.unh.edu).

The commission on Palaeoclimate (PALCOMM) is one of the five commissions of the International Union for Quaternary Research (INQUA). Although mostly data-oriented, PALCOMM also assists the comparisons of paleo data and models. Its overall objective is a better understanding of Quaternary climatic perturbations, transitions and abrupt events.

Finally, DEKLIM (German Climate Research Programme), a 37 million Euro project funded by the Federal Ministry of Education and Research, dedicated one of its four major research areas to Paleoclimate Research. The program seeks to (a) enhance our understanding of the climate system, (b) support young scientists, and (c) invigorate international cooperation.

5. CONCLUSION

Understanding the Earth's past environments is not only a fascinating research field satisfying human curiosity, but also essential in order to make predictions for the future. Today, several international and interdisciplinary collaborations help to develop common international science directions and ensure that important scientific questions are addressed in a coherent manner. Scientists in each of the two classic schools of paleoceanography (reconstructions through proxy data and theoretical quantitative analysis using models) are more and more exchanging expertise and working together for a better understanding of the past oceanographic environment. However, inherent problems associated with the poor spatial and temporal resolution of proxy data as well as uncertainties related to the proxy data themselves make their interpretation difficult and sometimes impossible. There is an urgent need for the creation of new well-calibrated and time-resolved paleo data records. At the same time, models are limited by resolved processes, resolution, and the quality of boundary conditions. Overall, there is much progress to be made in both fields. In spite of these problems, the geological data as well as model simulations provide a substantial set of results which gives us some insight into how the ocean and the whole climate system functions. This knowledge is of ultimate importance to understand and predict future climate changes due to anthropogenic perturbations.

REFERENCES

Bice, K. L., Scotese, C. R., Seidov, D., & Barron, E. J. (2000). Quantifying the role of geographic change in Cenozoic ocean heat transport using uncoupled atmosphere and ocean models. *Palaeogeography, Palaeoclimatology, Palaeoecology, 161*, 295–310.

Bond, G., Kromer, B., Beer, J., Muscheler, R., Evans, M., Showers, W., Hoffmann, S., Lotti-Bond, R., Hajdas, I., & Bonani, G. (2001). Persistent solar influence on North Atlantic surface circulation during the Holocene. *Science, 294*, 2130–2136.

Broecker, W., Lynch-Stieglitz, J., Archer, D., Hofmann, M., Maier-Reimer, E., Marchal, O., Stocker, T., & Gruber, N. (1999). How strong is the Harvardton-Bear constraint? *Global Biogeochemical Cycles, 13*(4), 817–820.

Bryan, F. (1986). High-latitude salinity effects and interhemispheric thermohaline circulations. *Nature, 323*, 301–304.

Budyko, M. I. (1969). The effect of solar radiation variations on the climate of the earth. *Tellus, 21*, 611–619.

Bush, A. B., & Philander, S. G. H. (1998). The role of ocean–atmosphere interactions in tropical cooling during the Last Glacial Maximum. *Science, 297*, 1341–1344.

Butzin, M., Prange, M., & Lohmann, G. (2005). Radiocarbon simulations for the glacial ocean: The effects of wind stress, Southern Ocean sea ice and Heinrich events. *Earth and Planetary Science Letters, 235*(1–2), 45–61.

Clarke, G. K. C., Lhomme, N., & Marshall, S. J. (2005). Tracer transport in the Greenland ice sheet: Three-dimensional isotopic stratigraphy. *Quaternary Science Reviews, 24*, 155–171.

Clark, P. U., Pisias, N. G., Stocker, T. F., & Weaver, A. J. (2002). The role of the thermohaline circulation in abrupt climate change. *Nature, 415*, 863–869.

Claussen, M., Mysak, L. A., Weaver, A. J., Crucifix, M., Fichefet, T., Loutre, M. F., Weber, S. L., Alcamo, J., Alexeev, V. A., Berger, A., Calov, R., Ganopolski, A., Goosse, H., Lohmann, G., Lunkeit, F., Mokhov, I. I., Petoukhov, V., Stone, P., & Wang, Z. (2002). Earth system models of intermediate complexity: Closing the gap in the spectrum of climate models. *Climate Dynamics, 18*, 579–586.

CLIMAP Project Members (1976). The surface of the Ice-Age Earth. *Science, 191*, 1131–1137.

CLIMAP Project Members (1981). Seasonal reconstructions of the earth's surface at the Last Glacial Maximum. *Geological Society of America*, Map Chart Ser, MC-36.

COHMAP Members (1988). Climatic changes of the last 18,000 years: Observations and model simulations. *Science, 241*, 1043–1052.

Crowley, T.J., & North, G.R. (1991). *Paleoclimatology, Oxford Monographs on Geology and Geophysics* (Vol. 18). New York: Oxford University Press.

Crucifix, M. (2005a). Distribution of carbon isotopes in the glacial ocean: A model study. *Paleoceanography, 20*(PA4020), 1–18.

Crucifix, M. (2005b). *Palaeoclimate modeling: Achievements and perspectives*. Predictability of the evolution and variation of the multi-scale earth system, 21st Century Earth Science COE Workshop, University of Tokyo, Tokyo.

Crucifix, M., Braconnot, P., Harrison, S. P., & Otto-Bliesner, B. (2005). Second phase of Paleoclimate Modelling Intercomparison Project. *EOS, 86*(28).

Duplessy, J.-C., Labeyrie, L., Juillet-Leclerc, A., Maître, F., Duprat, J., & Sarnthein, M. (1991). Surface salinity reconstruction of the North Atlantic Ocean during the Last Glacial Maximum. *Oceanology Acta, 14*, 311–324.

Fichefet, T., Hovine, S., & Duplessy, J.-C. (1994). A model study of the Atlantic thermohaline circulation during the Last Glacial Maximum. *Nature, 372*(6503), 252–255.

Ganopolski, A., & Rahmstorf, S. (2001). Rapid changes of glacial climate simulated in a coupled climate model. *Nature, 409*, 153–158.

Gates, W. L. (1976a). Modeling the ice-age climate. *Science, 191*, 1138–1144.

Gates, W. L. (1976b). The numerical simulation of ice-age climate with a global general circulation model. *Journal of the Atmospheric Sciences, 33*, 1844–1873.

Gildor, H., Tziperman, E., & Toggweiler, R. J. (2002). The sea-ice switch mechanism and glacial interglacial CO_2 variations. *Global Biogeochemical Cycles, 16*(10.1029/2001GB001446).

Hansen, J., Lacis, A., Rind, D., Russell, G., Stone, P., Fung, I., Ruedy, R., & Lerner, J. (1984). Climate sensitivity: Analysis of feedback mechanisms. Climate Processes and Climate Sensitivity. Geophysical Monograph (Vol. 29, pp. 130–163). Washington, DC: AGU.

Harrison, S. P., Braconnot, P., Joussaume, S., Hewitt, C., & Stouffer, R. J. (2002). Fourth international workshop of the Paleoclimate Intercomparison Project (PMIP): Launching PMIP Phase II. EOS.

Heinze, C., Maier-Reimer, E., & Winn, K. (1991). Glacial pCO_2 reduction by the world ocean: Experiments with the Hamburg carbon cycle model. Paleoceanography, 6(4), 395–430.

Imbrie, J., & Imbrie, K. P. (1979). Ice ages, solving the mystery. Cambridge, Massachusetts and London, England: Harvard University Press.

Kaplan, J. O., Prentice, I. C., Knorr, W., & Valdes, P. J. (2002). Modeling the dynamics of terrestrial carbon storage since the Last Glacial Maximum. Geophysical Research Letters, 29(22), 31: 1–4.

Knox, F., & McElroy, M. B. (1984). Changes in atmospheric CO_2: Influence of the marine biota at high latitude. Journal of Geophysical Research, 89, 4629–4637.

Kubatzki, C., & Claussen, M. (2003). Modelers and geologists join forces at workshop. EOS, 84(9), 79.

LeGrande, A. N., Schmidt, G. A., Shindell, D. T., Field, C. V., Miller, R. L., Koch, D. M., Faluvegi, G., & Hoffmann, G. (2006). Consistent simulations of multiple proxy responses to an abrupt climate change event. Proceedings of the National Academy of Sciences of the United States of America, 103(4), 837–842.

Manabe, S., & Bryan, K. (1985). CO_2-induced change in a coupled ocean-atmosphere model and its paleoclimatic implications. Journal of Geophysical Research, 90, 11689–11708.

Manabe, S., & Stouffer, R. J. (1988). Two stable equilibria of a coupled ocean–atmosphere model. Journal of Climate, 1(9), 841–866.

McGuffie, K., & Henderson-Sellers, A. (2001). Forty years of numerical climate modelling. International Journal of Climatology, 21, 1067–1109.

Meissner, K. J., Schmittner, A., Weaver, A. J., & Adkins, J. F. (2003). The ventilation of the North Atlantic Ocean during the Last Glacial Maximum — a comparison between simulated and observed radiocarbon ages. Paleoceanography, 18(2), 1: 1–13.

Nairn, A. E. M., & Smithwick, M. E. (1976). The continental Permian in Central, West, and South Europe. Permian paleogeography and climatology. Boston, MA: D. Reidel.

Paillard, D. (1998). The timing of Pleistocene glaciations from a simple multi-state climate model. Nature, 391, 378–381.

Parrish, J. T. (1982). Upwelling and petroleum source beds, with reference to the Paleozoic. American Association of Petroleum Geologists Bulletin, 66, 750–774.

Parrish, J. T. (1998). Interpreting pre-Quaternary climate from the geologic record. New York: Columbia University Press.

Paul, A., & Schäfer-Neth, C. (2005). How to combine sparse proxy data and coupled climate models. Quaternary Science Reviews, 24, 1095–1107.

Rahmstorf, S. (1995). Bifurcations of the Atlantic thermohaline circulation in response to changes in the hydrological cycle. Nature, 378, 145–149.

Rahmstorf, S. (2002). Ocean circulation and climate during the past 120,000 years. Nature, 419, 207–214.

Rahmstorf, S., Crucifix, M., Ganopolski, A., Goosse, H., Kamenkovich, I. V., Knutti, R., Lohmann, G., Marsh, R., Mysak, L., Wang, Z., & Weaver, A. J. (2005). Thermohaline circulation hysteresis: A model intercomparison. Geophysical Research Letters, 32, doi:10.1029/2005GL023655.

Renssen, H., Braconnot, P., Tett, S. F. B., von Storch, H., & Weber, S. L. (2004). Past climate variability through Europe and Africa. Recent developments in Holocene climate modelling. Dordrecht, The Netherlands: Kluwer Academic Publishers.

Roche, D., Paillard, D., Ganopolski, A., & Hoffman, G. (2004). Oceanic oxygen-18 at the present day and LGM: Equilibrium simulations with a coupled climate model of intermediate complexity. Earth and Planetary Science Letters, 218, 317–330.

Sarmiento, J. L., & Toggweiler, J. R. (1984). A new model for the role of the oceans in determining atmospheric pCO_2. Nature, 308, 621–624.

Sarnthein, M., Jansen, E., Weinelt, M., Arnold, M., Duplessy, J.-C., Erlenkeuser, H., Flatoy, A., Johannessen, G., Johannessen, T., Jung, S., Koc, N., Labeyrie, L., Maslin, M., Pflaumann, U., & Schulz, H. (1995). Variations in Atlantic surface ocean paleoceanography, 50°–80° N: A time-slice record of the last 30,000 years. *Paleoceanography, 10*(6), 1063–1094.

Schmidt, G. A. (1999). Forward modelling of carbonate proxy data from planktonic foraminifera using oxygen isotope tracers in a global ocean model. *Paleoceanography, 14*, 482–497.

Schmittner, A. (2005). Decline of the marine ecosystem caused by a reduction in the Atlantic overturning circulation. *Nature, 434*, 628–633.

Schmittner, A., Meissner, K. J., Eby, M., & Weaver, A. J. (2002). Forcing of the deep ocean circulation in simulations of the Last Glacial Maximum. *Paleoceanography, 17*(2), 1–16.

Seidov, D., Sarnthein, M., Stattegger, K., Prien, R., & Weinelt, M. (1996). North Atlantic ocean circulation during the Last Glacial Maximum and subsequent meltwater event: A numerical model. *Journal of Geophysical Research, 101*(C7), 16305–16332.

Sellers, W. D. (1969). A global climatic model based on the energy balance of the earth-atmosphere system. *Journal of Applied Meteorology, 8*, 392–400.

Siegenthaler, U., & Wenk, T. (1984). Rapid atmospheric CO_2 variations and ocean circulation. *Nature, 308*, 624–626.

Stephens, B. B., & Keeling, R. F. (2000). The influence of Antarctic sea ice on glacial-interglacial CO_2 variations. *Nature, 404*, 171–174.

Stocker, T. F., & Wright, D. G. (1996). Rapid changes in ocean circulation and atmospheric radiocarbon. *Paleoceanography, 11*(6), 773–795.

Toggweiler, J. R. (1999). Variation of atmospheric CO_2 by ventilation of the ocean's deepest water. *Paleoceanography, 14*(5), 571–588.

de Vernal, A., Hillaire-Marcel, C., Turon, J. L., & Matthiessen, J. (2000). Reconstruction of sea surface temperature, salinity and sea ice cover in the northern North Atlantic during the Last Glacial Maximum. *Canadian Journal of Earth Sciences, 37*, 725–750.

Weaver, A. J. (1995). Driving the ocean conveyor. *Nature, 378*, 135.

Weaver, A., & Hughes, T. M. C. (1992). Stability and variability of the thermohaline circulation and its link to climate. *Trends in Physical Oceanography Research Trends Series Council of Scientific Research Integration Trivandrum India, 1*, 15–70.

Werner, M., Mikolajewicz, U., Hoffmann, G., & Heimann, M. (2000). Possible changes of $\delta^{18}O$ in precipitation caused by a meltwater event in the North Atlantic. *Journal of Geophysical Research, 105*(D8), 10161–10168.

Williams, J., Barry, R. G., & Washington, W. M. (1974). Simulation of the atmospheric circulation using the NCAR global circulation model with ice age boundary conditions. *Journal of Applied Meteorology, 13*, 305–317.

Winguth, A. M. E., Archer, D., Maier-Reimer, E., & Mikolajewicz, U. (2000). Inverse methods in global biogeochemical cycles. Paleonutrient data analysis of the glacial Atlantic using an adjoint ocean general circulation model (pp. 171–183). Geophysical Monograph Series. Washington, DC: AGU.

Wunsch, C. (Ed.) (1997). *The ocean circulation inverse problem.* Cambridge: Cambridge University Press.

Ziegler, A. M., Bambach, R. K., Parrish, J. T., Barrett, S. F., Gierlowski, E. H., Parker, W. C., Raymond, A., Sepkoski, J. J. J. (1981). *Paleobotany, paleoecology, and evolution. Paleozoic biogeography and climatology* (Vol. 2). New York: Praeger.

INDEX OF TAXA

PLANKTONIC FORAMINIFERA SPECIES

Globigerina bulloides, 200, 216, 220, 240, 739
Globigerinella adamsi, 221
Globigerinella siphonifera, 219–220
Globigerinita glutinata, 220, 223
Globigerinita uvula, 220
Globigerinoides ruber (pink), 220
Globigerinoides ruber (white), 220
Globigerinoides sacculifer, 216, 220, 723
Globoquadrina conglomerata, 221
Globorotalia hirsuta, 220, 240
Globorotalia inflata, 220
Globorotalia menardii, 220, 223, 229, 236, 243, 748
Globorotalia scitula, 220, 229
Globorotalia truncatulinoides, 216, 220, 240, 747
Globorotalia tumida, 220
Globorotaloides hexagonus, 221
Hastigerina pelagica, 216
Neogloboquadrina dutertrei, 200, 220, 225
Neogloboquadrina incompta, 220
Neogloboquadrina pachyderma, 216, 220, 240, 545, 622, 739, 768
Orbulina universa, 220, 244, 768
Pulleniatina obliquiloculata, 220
Turborotalia quinqueloba, 220

BENTHIC FORAMINIFERA TAXA

Alabaminella weddellensis, 301
Ammonia batavus, 276
Angulogerina angulosa, 304–305
Anomalina sp., 276
Bolivina cf. *B. dilatata*, 281
Bulimina aculeata, 270
Bulimina exilis, 270
Cancris auriculus, 276
Cancris inaequalis, 276
Cancris oblongus, 276
Cancris panamensis, 276
Cancris sagra, 276
Cassidulina carinata, 276
Cassidulina crassa, 276

Cassidulina cushmani, 276
Cassidulina delicata, 276, 280, 284
Cassidulina depressa, 276
Cassidulina laevigata, 276, 304
Cassidulina limbata, 276
Cassidulina minuta, 276
Cassidulina oblonga, 276
Cassidulina sgarellae, 276
Cassidulina subcarinata, 276
Cassidulina subglobosa, 270, 276, 301, 310–312
Cassidulina teretis, 276, 309
Cassidulina tumida, 276
Chilostomella ovoidea, 270
Chilostomella spp., 280, 282, 284
Cibicides bradyi, 276
Cibicides fletcheri, 276
Cibicides lobatulus, 305–306, 309, 311
Cibicides refulgens, 276, 309
Cibicides ungerianus, 276
Cibicidoides spp., 306, 748, 754, 758, 788
Cibicidoides pachyderma, 776
Cibicidoides wuellerstorfi, 276, 302, 312, 727
Discanomalina spp., 305
Elphidium albiumbilicatulum, 276
Elphidium excavatum, 276
Elphidium incertum, 276, 311
Elphidium tumidum, 276
Epistominella decorata, 276
Epistominella exigua, 270, 276, 287, 301–302, 304–306, 309–312
Epistominella pusilla, 301
Epistominella smithi, 276, 280, 284
Epistominella vitrea, 276
Eponides antillarum, 276
Eponides leviculus, 276
Eponides regularis, 276
Fontbotia wuellerstorfi, 304
Fursenkoina spp., 301
Gavelinopsis lobatulus, 276
Gavelinopsis translucens, 276
Globobulimina affinis, 270
Globobulimina spp., 270, 280, 282, 284, 312

Globocassidulina subglobosa, 270, 301,
    310–312
Gyroidina altiformis, 287, 310
Gyroidina io, 276
Gyroidina lamarckiana, 276, 311
Gyroidina multilocula, 276
Gyroidina parva/pulchra, 276
Gyroidina rotundimargo, 276
Gyroidina umbonata, 276
Gyroidinoides neosoldanii, 276
Gyroidinoides soldanii, 276, 310
Hanzawaia boueana, 276
Hanzawaia concentrica, 276
Hoeglundina elegans, 276–277, 310–312,
    776
Hyalinea balthica, 276
Islandiella sp., 276
Islandiella subglobosa, 276
Lenticulina articulata, 276
Melonis barleeanus, 270, 311
Melonis spp., 287, 776, 778
Nonionella stella, 280, 284
Nuttallides umboniferus, 302, 304, 306
Oridorsalis tener, 306, 310–311
Oridorsalis umbonatus, 276, 309, 312
Osangularia culter, 276
Osangularia rugosa, 276
Planulina ariminensis, 276, 305–306
Planulina exorna, 276
Planulina limbata/ornata, 276
Planulina spp., 776, 778
Pullenia sp., 276
Rhabdammina spp., 305
Stainforthia fusiformis, 313
Textularia kattegatensis, 270
Uvigerina mediterranea, 287, 301
Uvigerina perigrina, 270, 287, 301, 304, 309, 310,
    311, 312
Uvigerina spp., 288, 301, 310, 748, 776,
    778
Valvulineria araucana, 276
Valvulineria javana, 276
Valvulineria oblonga, 276

DIATOM TAXA

Actinocyclus, 335
Azpeitia tabularis, 336–337
Bacillariineae, 328
Biddulphiniineae, 328

Chaetoceros, 335–336, 339
Coscinodiscineae, 328
Coscinodiscus, 335
Eucampia Antarctica, 339
Fragilariinea, 328
Fragilariopsis curta, 337
Fragilariopsis cylindrus, 338
Fragilariopsis doliolus, 336
Fragilariopsis kerguelensis, 330, 337
Fragilariopsis obliquescostata, 352
Hyalochaete, 335
Navicula, 335
Phaeoceros, 335
Rhizosolenia, 334, 335
Rhizosoleniineae, 328
Roperia tessalata, 336
Thalassionema, 335
Thalassiosira lentiginosa, 338
Thalassiothrix, 334, 335
Triceratium, 335

DINOFLAGELLATE TAXA

Alexandrium catenella, 395
Alexandrium minutum, 395
Alexandrium tamarense, 395
Coolia monotis, 395
Dinophysis acuminata, , 395
Dinophysis acuta, 395
Dinophysis fortii, 395
Dinophysis norvegica, 395
Gambierdiscus toxicus, 395
Gonyaulax baltica, 386
Gonyaulax digitale, 381
Gonyaulax scrippsae, 381
Gonyaulax sp., 378
Gonyaulax spinifera, 378, 380, 381
Gymnodinium catenatum, 394–395
Karenia bicuneiformis, 395
Karenia papilionaceae, 395
Karenia selliformis, 395
Karenia brevis, 395
Lingulodinium polyedrum, 394
Noctiluca, 394
Ostreopsis siamensis, 395
Pentapharsodinium dalei, 383, 389–390
Peridiniella catenata, 388
Pheopolykrikos hartmannii, 380
Polarella glacialis, 388
Polykrikos kofoidii, 392

Polykrikos schwartzii, 378
Polykrikos sp., 387–389
Preperidinium meunieri, 380
Prorocentrum lima, 395
Protoceratium reticulatum, 392
Protoceratium sp., 380
Protoperidinium americanum, 392
Protoperidinium avellanum, 380
Protoperidinium conicoides, 380
Protoperidinium conicum, 381
Protoperidinium leone, 380
Protoperidinium nudum, 381
Protoperidinium sp., 378
Protoperidinium stellatum, 381
Protoperidinium subinerme, 381
Pyrodinium bahamense, 395
Scrippsiella trifida, 381
Symbiodinium, 372

DINOFLAGELLATE CYST TAXA

Achomosphaera ramulifera, 12
Ataxiodinium choane, 380, 388
Batiacasphaera sphaerica, 12
Bitectatodinium spongium, 380, 388
Bitectatodinium tepikiense, 387
Brigantedinium cariacoense, 380
Brigantedinium simplex, 12, 380
Brigantedinium spp., 387, 389, 392
Corrudinium harlandii, 12
Corrudinium sp. I, 12
Corrudinium sp. II, 12
Cymatiosphaera sp., 12
Cyst of *Polykrikos* sp. Arctic morphotype, 387
Cyst type I, 12
Cysts of cf. *Scrippsiella trifida*, 381
Cysts of *Gonyaulax baltica*, 386
Cysts of *Gymnodinium catenatum*, 394
Cysts of *Pentapharsodinium dalei*, 383, 390
Cysts of *Pheopolykrikos hartmannii*, 380, 388
Cysts of *Polykrikos kofoidii*, 380, 388, 391, 392
Cysts of *Protoperidinium americanum*, 381, 388, 391, 392
Cysts of *Protoperidinium nudum*, 381
Cysts of *Protoperidinium stellatum*, 381, 388
Cysts of *Polykrikos schwartzii*, 378
Dalella chathamensis, 388
Dubridinium caperatum, 380, 388

Echinidinium aculeatum, 380, 388
Echinidinium bispiniformum, 380, 388
Echinidinium delicatum, 380, 388
Echinidinium granulatum, 380, 388, 391
Echinidinium karaense, 380, 388
Echinidinium transparantum, 380, 388
Filisphaera filifera, 12
Impagidinium aculeatum, 387
Impagidinium japonicum, 12,
Impagidinium pacificum, 380
Impagidinium paradoxum, 12, 380, 383, 388, 391
Impagidinium patulum, 12, 380, 383, 388, 391
Impagidinium plicatum, 380, 383, 388
Impagidinium sphaericum, 12, 380, 384, 388, 391
Impagidinium strialatum, 380, 388, 391
Impagidinium velorum, 12, 380
Impagidinium pallidum, 387, 389
Impagidinium variaseptum, 388
Impletosphaeridium sp., 12
Incertae sedis sp. I, 12
Incertae sedis sp. II, 12
Invertocycta sp. I, 12
Invertocysta lacrymosa, 12
Islandinium brevispinosum, 380, 388
Islandinium minutum, 12, 380, 388, 391
Islandinium? cezare, 380, 388, 391
Leipokatium invisitatum, 380
Lejeunecysta cf. *fallax*, 12
Lejeunecysta cf. *paratenella*, 12
Lejeunecysta oliva, 380
Lejeunecysta sabrina, 380, 391
Lingulodinium machaerophorum, 383
Nematosphaeropsis labyrinthus, 383, 399
Nematosphaeropsis rigida, 380
Nematosphaeropsis sp. I, 12
Operculodinium centrocarpum, 383, 386–387, 389–390
Operculodinium crassum, 12
Operculodinium israelianum, 383
Operculodinium janduchenei, 380
Operculodinium longispinigerum, 12, 380
Operculodinium centrocarpum-short processes form, 380
Operculodinium sp. I, 12
Polysphaeridium zoharyi, 383–384, 395
Pyxidinopsis psilata, 381
Pyxidinopsis reticulata, 381, 383, 388, 391
Quinquecuspis concreta, 378
Selenopemphix antarctica, 387–388, 399

Selenopemphix cf. *armata*, 12
Selenopemphix nephroides, 12, 381, 388, 391, 392
Selenopemphix quanta, 381, 388, 391, 392
Spiniferites alaskensis, 381
Spiniferites belerius 381
Spiniferites bentorii, 381, 383, 384, 388
Spiniferites bulloideus, 381
Spiniferites cruciformis, 386
Spiniferites delicatus, 381, 388, 392
Spiniferites elongatus, 385, 388–389
Spiniferites frigidus, 388
Spiniferites hyperacanthus, 381, 388
Spiniferites lazus, 381, 388
Spiniferites membranaceus, 381, 388, 392
Spiniferites mirabilis, 387, 399
Spiniferites pachydermus, 381, 383, 388
Spiniferites ramosus, 12, 381, 383, 388
Stelladinium reidii, 381, 388
Stelladinium robustum, 381
Tectatodinium pellitum, 12, 381, 388
Tectatodinium sp. I, 12
Tectatodinium sp. II, 12
Trinovantedinium applanatum, 381, 388, 392
Trinovantedinium variabile, 388
Tuberculodinium vancampoae, 384
Votadinium calvum, 381, 388, 391
Votadinium spinosum, 381, 388, 391
Xandarodinium variabile, 12
Xandarodinium xanthum, 381, 388

COCCOLITHOPHORID TAXA

Calcidiscus leptoporus, 423
Calcidiscus macintyrei, 450
Coccolithaceae, 412
Coccolithus pelagicus, 411
Emiliania huxleyi, 410, 446
Florisphaera profunda, 427
Gephyrocapsa oceanica, 412, 425, 429, 430, 446,
　450, 458, 461
Helicosphaera carteri, 412
Hellicosphaera selli, 450
Noelataerhabdaceae, 412
Phaeocystis, 411, 430
Prymnesiophyceae, 411, 445
Prymnesium, 411

Rhabdosphaera clavigera, 412
Rhabdosphaeracea, 412
Reticulofenestra asanoi, 450
Syracosphaera pulchra, 276
Syracosphaeracea, 412
Umbelicosphaera sibogae, 412

CORAL TAXA

Acropora cervicornis, 518
Anemonia viridis, 518
Anthozoa, 519
Antipatharia, 492, 500, 501, 512, 513
Antipathes fiordenis, 518
Arctica islandica, 500
Corallium rubrum, 519, 522
Corallium secundrum, 517
Desmophyllum cristagalli, 494
Desmophyllum dianthus, 494, 497–498, 504,
　514
Enallopsammia rostrata, 498
Gerardia, 492, 494, 500
Gorgonacea, 498
Isididae, 498
Keratoisis spp., 499
Leptogorgia sarmentosa, 520
Lophelia pertusa, 496–497, 502, 506
Madrepora, 520
Melanogrammus aeglefinus, 500
Muricea californica, 521
Octocorallia, 498
Paramuricea clavata, 520
Pavona clavus, 782
Plexaurella dichotoma, 516
Porites lobata, 782
Porites lutea, 496, 781, 783
Porites sp., 782
Primnoa resedaeformis, 498–500, 512–513
Primnoidae, 498
Primnoisis antarctica, 519
Scleractinia, 492

OSTRACOD TAXA

Krithe, 599, 602
Loxoconcha, 599

SUBJECT INDEX

Abiotic parameters, 9, 376, 425, 428, 556
Abyssal currents, 149
Acantharia, 587
Acantharian tests, 587
Accelerator Mass Spectrometry (AMS), 189, 650
Accretion rates, 783
Accumulation rates, 41, 145, 166, 292, 294,
    298, 353, 443, 452, 459–460, 472, 474,
    579–580, 592, 604, 612–613, 623–624,
    699–700, 702–706
Acetolysis, 382
Acoustic
    impedance, 66, 67
    profiles, 21
    reflectivity, 40
    velocity, 67
Adsorption, 570, 585–586, 597, 599, 601,
    606–607, 614, 684–687, 689, 692, 701,
    718, 726
Aegean Sea, 453
Aeolian
    dust, 577, 582, 598, 623–624, 653, 655
    inputs, 572, 651, 655
    sediment, 29, 567, 571, 580
    transport, 466
Aerobic
    oxidation, 456, 604
    remineralization, 269
Aerosols, 148, 465–466, 573, 577
African margin, 390, 425
Age
    control, 266, 293–295, 331
    model, 20, 196, 294, 300, 443, 707
    modeling, 20, 196, 294, 300, 443, 707
Aggregates, 24–25, 29–32, 34–36, 286, 381,
    455, 587, 594, 685
Aggregation, 24, 28, 31, 455, 685, 708
Agulhas Current, 167, 230, 232
Air-sea CO_2 fluxes, 756
Al/Ti ratio, 570, 597, 599, 623
Alboran Sea, 468
Algal
    biomarkers, 460
    blooms, 371, 394

culture, 448
symbionts, 215, 218, 747, 757
Algorithms, 225, 231, 234, 342, 421, 590, 592,
    623
Aliphatic hydrocarbons, 441
Alkalinity, 4, 6, 14, 272, 303, 604, 706, 720,
    729–730
Alkenone, 5, 48–49, 442–452, 456–459, 461,
    463, 471–473, 623, 728–729
    fluxes, 447
    index, 445
    paleothermometry, 467
    producers, 449, 458
    production rate, 447
Allerød, 197, 619
ALOHA (A Long Term Oligotrophic Habitat
    Assessment), 455, 686, 687, 688, 697
Alpha Counting Spectrometry, 660
Alumina, 144
Aluminosilicate, 570, 572, 574, 576, 588, 590,
    598–599, 604, 623
Aluminum, 71
American margin, 49, 174
Amino acids, 441, 512
Amirante Passage, 22, 43–45
AMO (Atlantic meridional overturning), 14
Amphiboles, 570, 572
AMS, 189–192, 194, 351, 355, 650
Anaerobic
    conditions, 287, 300
    oxidation, 456
Analogue, 80, 272, 373, 395, 397, 525–526,
    529, 533–537, 546–547, 549, 553, 556
Analogy, 103, 534–535, 537, 556
Anatase, 570
Angola
    Basin, 274, 453
    Current, 470
Anhysteretic remanent magnetization, 109–110
Anisotropy, 24–26, 170
Annual
    cycles, 91, 379
    rings, 511, 513

Anoxia, 229, 275, 280, 445, 452, 567, 602,
  617–619, 623
Anoxic
  conditions, 11, 275, 280, 394, 463, 473, 567,
    594, 602, 606, 611, 614, 617–619
  environments, 268–269, 273, 567, 608, 617
  pore water, 605
  sediments, 67, 269, 454, 592–593, 606,
    608–609, 614–617, 681
Antarctic
  Bottom Water (AABW), 149, 308, 341, 511,
    589, 648, 753
  Circumpolar Current, 28, 167, 355, 648,
    668, 789
  ice cores, 28, 470, 515, 609
  ice sheet, 159, 301, 788
  Intermediate Water, 308, 648
  Ocean, 158–159, 330, 373, 389, 529, 686
  Polar Front, 347–348, 356, 610–611, 692
  surface water, 355
Anthropogenic
  impact, 799
  perturbations, 512, 717, 731, 808
Antipatharians, 492, 500–501, 512–513
Apatite, 594
Aphotic zone, 454
Apparent ages, 192, 508, 510
Arabian Sea, 45, 147, 151–152, 155–156, 160,
    230, 235, 237, 375, 390, 392, 447, 453,
    455, 466, 472, 573, 589, 593, 596, 602,
    611
Aragonite, 193, 496, 498, 501–503, 508, 720,
    779–783, 790
Aragonitic corals, 193
Archaea, 453–456
Arctic
  Bottom Water, 149, 308, 341, 511, 589, 648,
    753
  Ocean, 103, 139, 150, 158–159, 162, 169,
    171–172, 229, 289, 291, 309, 330, 373,
    389, 529, 538, 552–553, 666, 686, 749,
    776, 778
  Polar front, 49, 347–348, 356, 610–611,
    692
  seas, 157, 387–388, 396, 400, 546, 553
  surface waters, 355, 744
  taxa, 387
Argentine Basin, 24, 39, 48
Artificial neural networks, 396, 421, 532
Asian seas, 386

Astronomical
  theory, 13
  tuning, 122, 415
Atlantic
  meridional overturning circulation, 681, 684,
    707–708
  Ocean, 145–148, 150–151, 154, 156–157,
    161–163, 197, 201, 223, 231, 290–291,
    302, 304, 309, 311–312, 331, 373, 375,
    383, 385, 390, 392, 396–397, 413, 426,
    454, 473, 500, 530, 537–538, 547, 573,
    577, 648, 666, 669, 690–692, 694,
    711–712, 743, 745, 750, 753, 755, 758
  Ridge, 148–149, 154, 166
Atmospheric
  ^{14}C concentration, 188, 197
  circulation, 392, 475, 572
  CO_2, 251, 353, 355, 460, 467, 471, 549, 609,
    706, 717, 752–753, 756, 801
  dust, 466, 573, 655
  General Circulation Models (AGCM), 800,
    804
  nitrogen, 187
  pCO_2, 6, 609, 727, 729–730, 788–789
  radiocarbon, 187
  transport, 580–581
Attenuation coefficient, 73, 88
Augite, 570
Australian
  monsoon, 171
  shelf, 175
Authigenic
  carbonates, 605
  clays, 144, 146–147
  derived chlorite 144, 145, 146, 147
  metals, 567, 601, 616
  pyrite, 617
  trace elements, 602
Automated coccolithophorids analysis, 420
Automatic coccolith recognition, 421
Autotrophic, 8, 372, 377, 379, 390–391

Ba/Ti ratio, 585
Backpropagation learning algorithm, 532
Bacteria, 31, 451, 455
Baffin Bay, 652
Bahama Outer Ridge, 39–40
Baie des
  Chaleurs, 74
  Ha!Ha!, 84

Baltic Sea, 386, 409, 456
Bamboo corals, 498, 507
Banda islands, 492
Barchan, 20
Barents Sea, 151
Barite, 585, 587–590, 592–593, 599–600, 611, 623, 661–662
Barium, 567, 585, 587–588, 590, 592
Basalts, 146, 158, 166, 581, 651, 655
Bathymetry, 8, 23, 148, 289, 302, 308
Bathypelagic, 455
Bauxites, 572
Bay of
    Bengal, 152, 155, 576, 661, 669–670
    Biscay, 82–83, 301–302, 305, 309
Bayesian approaches, 557
Beaufort Sea, 116, 157
Bedforms, 20–22
Beidellite, 143
Benguela upwelling system, 330, 452
Benthic
    foraminifera, 26, 43, 199, 216–217, 248–249, 263–267, 269–275, 278–279, 284–285, 287–288, 290–292, 294, 296, 298–299, 301, 303–308, 505, 592, 621, 709, 736, 739, 747, 754–757, 759, 765–766, 776–778, 784, 787–789
    foraminiferal accumulation rate, 288, 292, 298
    oxygen isotopes, 54
Bering Sea, 157
Bermuda Rise, 40, 128, 443, 709–711
Bicarbonate, 191, 460–461, 605, 747, 757
Bio-calcification, 272
    models, 491, 502
Biocenoses, 371, 377
Bioclimatic domains, 388
Biodiversity, 271, 280, 807
Bioerosion, 501
Biogenic
    calcite, 740
    carbonates, 11, 203, 503, 651, 681, 736, 766–767, 774
    magnetite, 118
    minerals, 3–5, 10
    production, 9
    silica, 349–350, 353, 581, 621, 688, 701, 706
Biogenous dilution, 580

Biogeochemical
    cycles, 215, 475, 607, 614, 801, 803–804, 807
    provinces, 428–430
Biogeographic distribution, 221, 426
Biogeographical
    provinces, 246
    reconstructions, 384
    zonation, 383
Biogeography, 28–29, 337, 343, 371, 379, 382, 409, 413, 424–425, 429, 431
Biohorizons, 415
Biological
    productivity, 3, 7, 472, 551, 659
    pump, 470–471, 609, 611, 753
Biomarker, 3, 5, 7, 10, 194, 211, 415, 441–443, 445, 447, 449, 451, 453, 455, 457–461, 463–469, 471–473, 475, 552, 619, 622–623
    degradation rates, 459
    oxidation rates, 459
Biomass, 291, 329, 345, 425–428, 445, 512
Biomineralization, 14, 253, 332, 346, 357, 498
Biostratigraphic
    markers, 266, 664
    zonation, 415
Biostratigraphy, 5, 12, 213, 240, 331, 373, 409, 415–416, 424, 450, 661
Biosynthesis, 441, 464
Biotic parameters, 9, 376, 425, 428, 556
Biotite, 570, 572, 576
Bioturbation, 20, 70, 117, 185–186, 198–199, 202, 293, 337, 443, 511, 514, 617, 702, 738
Biozones, 100
Bipolar minerals, 156
Black Sea, 76, 386, 453, 602, 607–608, 613, 615
Blake
    Basin, 28
    Outer Ridge, 21, 24, 49
    Plateau, 28, 492, 669
Blooms, 7, 216, 224, 246, 301, 371, 377, 394, 410, 447, 456, 499, 759
Bølling-Allerød, 619
Borate minerals, 723
Boric acid, 718
Boron, 5–6, 717–721, 723–731
    isotopes, 5, 717–721, 723, 725, 727–729

Bottom
  current, 9, 29, 149, 154, 165, 166, 167, 168,
    170, 174, 273, 278, 304, 305, 306, 382,
    659, 700
  flow, 26, 43, 48
  oxygen, 609
  sediment transport, 26
  water circulation, 170, 303
  water density, 44
  water oxygenation, 263, 267, 269,
    272–275, 277, 279–280, 282–285, 290,
    304, 306–307, 552, 606, 621
  water temperature, 272, 765–766, 776–777,
    780, 788
Boundary
  conditions, 352, 357, 800–801, 803, 805, 808
  layer, 19, 31–34, 36, 50
  scavenging, 655, 681, 686–688, 690–692,
    703–704, 706, 711–712
Brackish environments, 386
Brassicasterol, 473
Brazil Current, 470
Brazilian
  continental slope, 171
  shield, 163
British Ice Sheet, 75
Bromine, 606
Brunhes, 100–101, 122, 126–127
Buffer layer, 32, 34, 36

Cabot Strait, 385
Cadmium, 611, 613
Calcareous
  cysts, 372
  dinoflagellates, 373
  fossils, 765–766
  nannofossils, 3, 417, 419–420
  nannoplankton, 411
  ooze, 228
  phytoplankton, 415, 419
Calcification, 216–218, 244–245, 268, 272,
    414, 496, 502–503, 505, 507, 721, 723,
    728, 736, 741, 746–747, 751, 757,
    759–760, 768–769, 773, 782–783, 785
  depth, 245, 268, 768–769
  rates, 507, 747, 757
  temperature, 746, 760, 768–769
Calcite, 6, 193, 213–215, 217–219, 228, 233,
    241–245, 247–252, 272, 411, 422–424,
    462, 471, 498–499, 501, 507–508, 661,

705, 720–721, 723, 726, 735–736,
    738–740, 746–750, 756–758, 766–767,
    769, 773, 775–780, 783, 785, 788, 790
  compensation depth, 228, 247, 726, 777
  dissolution, 213, 243–244, 247–249, 251–252
  overgrowth, 776
  precipitation, 740, 746–747, 750, 757
  saturation, 233, 249–250, 723, 776, 788
Calcium carbonate, 3, 110, 247, 382, 552, 665,
    718, 721, 726
Calendar age, 194–196, 203–204, 351, 355,
    499, 699
CALIB, 192, 205, 351, 355, 540
Calibration
  dataset, 189, 195, 196, 198, 204, 205, 206,
    227–229, 235, 344, 397, 427, 576, 537,
    539, 540, 541, 545, 546, 547, 552, 553,
    554, 556,
  equation, 344, 448–449, 528, 557
  of ^{14}C ages, 186, 189, 193, 194, 195, 196,
    204, 513
  methods, 534, 547
  programs, 185, 204–205
  techniques, 204
California
  Current, 593
  margin, 228, 616, 621
CalPal, 205
Calvin-Benson, 463
Canadian margin, 165, 173, 398
Canary Current, 469
Canonical
  correlation analysis, 529
  correspondence analysis, 235, 530
Cap Ferret Canyon, 269, 277
Cape Breton Canyon, 82, 83
Carbohydrates, 461
Carbon
  cycle, 1, 185–187, 194–197, 278, 384,
    422, 431, 441, 455, 466, 470, 475, 585,
    752–753, 756, 760, 801, 803–804, 808
  cycling, 186, 194, 201, 248, 460
  isotopes, 42, 52, 188, 272, 668, 735–737,
    739, 741, 743, 745, 747, 749, 751, 753,
    755, 757, 759
  isotope fractionation, 187, 461, 729
  Preference Index, 464–465
  reservoir, 186
  sequestration, 471

Carbonate, 3, 6, 13, 21, 28, 31, 52, 67, 70,
    110, 150, 186, 191–194, 213, 216, 236,
    243, 245–248, 250–251, 271–272, 286,
    292, 303–304, 330, 344, 349, 355, 382,
    409–410, 414, 417, 419–420, 422–424,
    442, 461, 475, 498, 501–506, 509, 511–512,
    552, 570, 574, 580, 589, 592–594, 599,
    605–608, 619, 651, 653–654, 664–665,
    670, 688, 690, 701, 703–707, 712, 717–721,
    723, 725–727, 729, 739–740, 747, 751, 766,
    775–776, 784, 790
  dissolution, 247, 292, 303, 382, 420, 552,
    589, 592, 701, 703, 705, 739
  flux, 245, 414, 422–423, 703
  minerals, 13, 186, 690, 706, 751
  pump, 410
  saturation, 236, 251, 271–272, 304,
    552, 589, 775
Carbonic acid, 191
Cariaco Basin, 195–198, 201–205, 230, 453,
    473, 567, 579–580, 615, 618–619, 621
Caribbean Sea, 154, 538, 618, 790
Carotenoids, 441
Carrara Marble, 737–738
Causal models, 549
CCD, 69–70, 247–248, 272, 304, 777
Celadonite, 144
Celestite, 587
Cenozoic, 1, 3, 5–7, 9, 11–14, 100, 122, 139,
    141, 147, 158–159, 167, 169, 187, 213,
    219, 227, 240, 331–332, 373, 415, 419,
    449, 466–467, 567–568, 573, 580, 586,
    664–665, 667, 726, 728, 741, 765, 767,
    769, 771, 773, 775, 777, 779, 781,
    783–787, 789
  biostratigraphy, 213
Centennial scale oscillations, 91
Central
  Arctic, 151, 172, 309
  Pacific, 29, 310, 466, 570, 583
Centrales, 327–328
Centrifugation technique, 334
Chalcophile elements, 611, 624
Champlain Sea, 200
Chatham
  Drift, 42
  Rise, 28, 42–43
Chemical
  erosion, 569
  tracers, 6, 10, 12, 565, 648, 670

  weathering, 141, 144–145, 154–156, 159,
    571, 670
Chemocline, 472
Chemolysis, 452
Chilean fjords, 780
China Sea, 160, 171, 291, 443, 554–555
Chlorin, 459, 472
  Index, 459
Chlorite, 140–147, 150, 152–158, 160, 162,
    164–168, 170–172, 175–176
Chlorophyll
  concentration, 456
  maximum, 447
Chondrites, 652
Chord distance, 343, 535, 537, 541
Chromatography, 194, 475
Chromium, 607
Chronology, 10, 28, 124, 128, 195, 331, 351,
    699–700, 702–705
Chronostratigraphical markers, 11
Chrysophyta, 327
Chukchi Sea, 151
Circum Antarctic Deep Water, 165
Circumpolar Deep Water, 42, 162, 308, 355,
    648, 695
Clades, 219
Clay, 4, 10–11, 13, 24, 34, 37–38, 42,
    45–48, 139–176, 247, 414, 569–570,
    572–573, 576–577, 580–581, 583, 618,
    662–663, 718, 725
  distribution, 139, 145, 150–151, 155–157,
    162, 171–172, 174–175
  flux, 173
  mineral assemblages, 141, 150, 152, 158–160,
    163, 165–166, 169–175, 576
  mineral provinces, 148, 160, 175
  mineralogy, 48, 139, 141, 158, 160, 162–163,
    166–168, 176, 573
  minerals, 4, 11, 13, 34, 139, 141–145,
    147–151, 153–161, 163, 165, 167–176,
    569–570, 572, 576, 581, 583
  particles, 148, 167, 718, 725
  sedimentation, 169
CLIMAP, 7, 214, 228, 233, 254, 352–354,
    398, 424, 525, 528, 535, 550–551, 786,
    800–801, 803, 805
Climate
  change, 140, 142, 168–170, 193, 219, 285,
    341, 347, 352, 356–357, 397–398, 414,
    467–468, 475, 491, 571, 589, 664–665,

667, 724, 741, 744, 746, 785–786,
    799–801, 804, 807–808
dynamics, 468, 789
events, 40
forcing mechanisms, 801
Model Intercomparison Project (CMIP),
    801, 808
modeling, 357, 800–801, 803, 806–807
shifts, 199, 473
transition, 199, 202, 470, 788–789
vegetation interactions, 808
warming, 394, 669
zonation, 141
Climatic
    events, 515
    oscillations, 88
    regimes, 199
    variables, 19, 475, 523–524, 527,
        533, 547
Clinoptilolite, 158
Coarse fraction, 248, 662
Coastal
    ecosystems, 266
    upwelling, 223, 294, 390, 392, 413,
        425, 447, 589
Coccolith
    assemblages, 413, 450
    weight, 423–424
Coccolithales, 411
Coccolithophore biogeography, 424
Coccolithophorids, 3, 8, 372, 377, 623
Coccoliths, 5, 9, 373, 400, 411, 413–415, 417,
    420–424, 426, 428, 446, 550, 623
Coccosphere, 411
Coelenterate, 492
Coercivity ratios, 112
Cohesive silt, 34, 163
COHMAP, 550, 801
Coiling, 200, 219, 238–242, 545
Cold periods, 43, 170, 347, 514, 754
Color
    reflectance, 63–64, 67–68
    variability, 69
Combinable Magnetic Resonance, 79
Compaction, 70, 72, 117
Computerized axial tomography, 73, 106
Computing Guided Nannofossil Identification
    System, 421
Conceptual models, 799, 802, 804
Confocal scanning, 79, 81

Conservative tracers, 52, 646
Constrained-analogue methods, 535
Continental
    ice sheet, 765, 784–785, 787, 803–804,
        806
    ice volume, 766
    margin, 6, 9, 22, 27–29, 40–41, 48–49,
        146–148, 159, 278, 290, 294, 297, 299,
        302, 305, 373, 379, 384, 386, 398, 400,
        414, 425, 443–444, 464, 555, 576–577,
        592, 596, 603–604, 607–609, 614–616,
        619, 624, 660, 688, 744
    rise, 29, 174
    shelve, 216, 373, 376, 491, 790
    slope, 10, 148, 159, 171, 274, 292, 301, 311,
        574, 594, 606, 622, 654, 687
    source, 141, 145, 153, 162, 168, 170, 648,
        653, 655, 664
    weathering, 141, 156, 160–161, 166, 168,
        664
Continentality Index, 383, 385
Contourites, 20, 23
Convection, 2, 49, 419, 511, 751
Conventional radiocarbon age, 188
Conveyor-belt, 197
Copper, 611, 613
Coral, 3–4, 11, 193, 195–197, 203, 205–206,
        372, 491–495, 497–515, 719–721,
        765–766, 782–784, 789–790
    aragonite, 193, 782–783
    calcification, 502
    growth rings, 2
    metabolism, 506
    skeletons, 193, 492, 496, 502, 511,
        781, 784
Coralline
    Mg/Ca, 784
    Sr/Ca paleothermometry, 765, 780
Corallites, 497, 783
Core-top
    calibration, 251, 446, 448, 767
    samples, 50, 175, 234, 248, 250, 263, 288,
        340, 342–343, 350, 670
Coriolis force, 40, 355
Correlation matrix, 528
Correspondence Analysis, 235, 291, 527,
        529–530
Cosmic
    dust, 647, 651–652
    radiation, 187

Cosmogenic
   isotopes, 124, 126–127, 204
   nuclides, 646
   ^{14}C production, 186
Coulter counter, 35, 46–47, 90
Coupled
   Carbon Cycle Climate Model
      Intercomparison Project, 801, 808
   climate models, 802, 805
   ocean-atmosphere models, 803
Covariance, 344, 528–529, 531, 784–785
Crag-and-tail, 20
Crenarchaeota, 454–455
Cretaceous, 122, 449, 463, 501, 581–583,
   585–586, 665, 737, 777
Cross-validation, 523, 539, 542, 548, 553, 555,
   557
Cryogenic magnetometer, 66, 87, 108–109, 114
Cryosphere, 787, 806
Cryptic species, 219, 227, 242
Crystal
   fibers, 496, 504
   growth, 13, 498, 502, 751
Crystallinity, 75, 776
Current
   flow, 21, 149, 169, 305
   index, 34
   intensity, 20, 45
   redistribution, 29
   regime, 45, 162, 305–306
   strength, 25, 40, 163, 662
   velocity, 3–4, 161, 263, 273, 290, 303–307,
      552
   zonation, 21
Cyst
   assemblages, 12, 371, 373, 377, 379,
      382–384, 386–387, 390–391, 394,
      396–397, 400, 535, 546, 551–552, 805
   forming dinoflagellates, 379
   production, 379, 387

Dansgaard-Oeschger, 467, 552, 621
Databases, 92, 307, 342–344, 395–397, 455,
   550–551
Dating techniques, 123, 186, 331, 497, 509
Dauwe Index, 459
Decadal
   scale events, 514
   variability, 7
Decanting, 419–420

Decarbonation, 346
Decay constant, 187–188, 203, 508, 682–683
Declination, 101–102, 104, 110, 115,
   119, 129
Deconvolution techniques, 114
Deep
   circulation, 28, 49, 139, 141–143, 145,
      147, 149, 151, 153, 155, 157, 159,
      161–163, 165, 167, 169, 171–173, 175,
      248, 648
   convection, 751
   dwelling species, 772
   ocean flow, 26
   ocean ventilation, 193–194, 203–204
   sea corals, 491, 492, 494, 495, 497, 500, 501,
      503, 504, 508, 510, 511, 515
   sea currents, 3, 29, 30, 38
   species, 769
   ventilation, 201–202
   water flux, 192
   water formation, 192, 515, 692, 709
   water masses, 141, 166, 511, 744, 746
   Western Boundary Current, 23–24, 42, 692,
      694–696, 698
Deglaciation, 54, 161, 200, 202–205, 301, 350,
   709–710, 786
Demagnetization, 87, 99, 108–109, 112–113,
   115–116, 119–121
Dendrochronologies, 195
Denitrification, 348, 512, 595
Denmark Strait Overflow Water, 648
Density, 2–3, 21, 30–31, 44–45, 49, 52, 63–67,
   69–70, 73, 79, 84–85, 87, 89, 91, 106, 198,
   244–245, 269, 271–272, 281–282, 293,
   334, 382, 386, 417, 420, 423, 493, 551,
   555, 648, 682, 700, 744, 751, 781, 783,
   802
Depositing flow, 20, 23
Deposition, 19, 21, 23–24, 29–30, 32–35,
   37–40, 46, 48–50, 74, 82–83, 101, 103,
   117, 145–146, 149, 158, 160–161, 163,
   168–170, 174, 185, 191, 213, 226, 229,
   243, 246, 266, 281, 294, 296, 301,
   303–304, 355, 441–442, 451, 459, 472,
   568, 572, 581, 606, 608, 659–661, 670,
   701, 756, 790
   rate, 34, 37, 40, 49, 174
Desert-derived dust, 653
Desorption, 653, 684–685, 692
Detrended Correspondence Analysis, 530

Detrital
    clays, 140–141, 144–146, 148, 154, 158, 161
    microcalcite, 422
    minerals, 602
    particles, 162
    remanent magnetization, 119
Deuterium, 457–458
Dextral coiling, 240
Diagenesis, 118, 141–142, 161, 168, 197, 278, 332, 346–347, 355, 441, 451–452, 459, 512, 577, 581, 609, 648, 651, 783
Diagenetic
    bias, 451–452
    effects, 193, 203, 727, 759, 783
    processes, 10, 117, 146, 271, 277–278, 286, 509, 653, 756
Diatom
    assemblages, 263, 330, 332, 335, 340–341, 343, 350, 353, 532–533, 550–551
    biogeography, 337
    biostratigraphy, 331
    distribution, 329, 333–334, 338
    ecology, 330, 335, 337, 357
    flux, 330
    intrinsic organic matter, 332
    oozes, 354
    productivity, 341
    silica, 349
    taxonomy, 328, 331, 342
    zonation, 330
Diatomite, 329
Diatoms, 3–4, 8–9, 31, 215, 270, 327–339, 341, 343–349, 351, 353, 355–357, 372, 377, 382, 390, 461, 471, 473, 525, 529–530, 533, 546, 549–550, 552, 572, 623, 690, 706
Differential dissolution, 199, 756
Diffuse reflectance, 67, 85
Diffusivity, 346, 660
Digital
    imaging, 92, 420
    X-radiography, 70, 85
Dilution, 155, 160, 170, 174, 188, 333, 384, 419, 452, 459, 570, 580, 592
Dimorphic, 411

Dinocyst, 5, 9, 11, 371, 373, 375–377, 379, 381–384, 386–388, 390, 392, 394–398, 400, 550–551
    assemblages, 12, 371, 373, 382, 384, 386–387, 390–391, 394, 396–397, 400, 535, 546, 551–552
    distribution, 397
    fluxes, 390
Dinoflagellate cysts, 3, 371, 373, 375, 377, 379, 381–383, 385, 387, 389, 391, 393, 395, 397, 399
Dinoflagellates, 215, 371–373, 376–377, 379, 387, 390, 394, 398, 473
Dinosporin, 373
Dinosterol, 473
Direct gradient analysis, 527, 530
Disaggregation, 571, 685, 708
Disequilibrium
    effects, 782
    offset, 782
Dissimilarity coefficient, 343
Dissolution, 10, 21, 118, 144, 193, 199–200, 213, 220, 243–244, 246–252, 292, 301, 303, 330, 337–339, 343, 348, 357, 373, 382, 398, 410, 420, 422, 424, 496–497, 501, 552, 583, 588–589, 592–593, 601, 604, 647, 654–655, 658–660, 671, 701, 703, 705, 717, 723–725, 727, 731, 739, 756, 759, 767, 775–777, 779–780
    effects, 717, 723, 727, 775–776
    experiments, 780
    rate, 247, 701
Dissolved
    carbon dioxide, 441, 460
    inorganic carbon, 187, 191, 460, 501, 737–738
    nutrients, 345
    organic matter, 501, 607
    oxygen, 4, 304, 583, 602, 616, 621
Diversity index, 280, 284, 307
Dolomite formation, 785
Down-core
    records, 337–339, 346, 417
    variability, 39
Downwelling, 648
Drainage events, 746
Drammensfjord, 453
DSDP (Deep Sea Drilling Project), 146, 158, 169, 331, 398, 399, 588
Dysaerobic mineralization, 269
Dysoxic indicators, 279

Early diagenesis, 278, 347
Earth System Models, 804, 806
Earth's orbital parameters, 789
Earthquakes, 84, 87
East
    American margin, 174
    Asian monsoon, 171, 786
    Canadian Arctic, 77
    Greenland Current, 351
Eastern
    margins, 145
    pacific, 155, 352–353, 392, 602
    south Atlantic, 470, 473
Eccentricity, 13, 589, 802
Echo sounder, 21, 23
Ecophenotypy, 241
Eddy diffusion, 355, 686–687, 690–692, 712
Eigenvector, 528
Ekman divergence, 430
El Nino Southern Oscillation (ENSO), 431
Electrical
    conductivity, 46
    resistivity, 64
Electronic Microfossil Image Database System,
    253
Embayment, 614
Empirical
    calibration, 226–227, 231, 235, 242, 250,
        721, 723, 767
    equations, 291
    relationships, 67, 251, 264
ENSO, 204, 431, 492, 790
Environmental factors, 2, 120, 240, 266, 290,
    306, 345, 386, 499, 741
Eocene, 24, 449, 470, 581–582, 585–586,
    588, 665, 738, 776, 787
Eocene/Oligocene transition, 776
Epibenthic, 26, 52, 193, 305, 754, 758
    carbon isotope, 52
    foraminifera, 26, 305, 754
Epicontinental, 373, 376, 396, 398
    seas, 373, 376, 396, 398
Epipelagic, 7
Episodic events, 270, 272
Equatorial
    Atlantic, 76, 151, 154, 162, 165, 330,
        340–341, 375, 392, 427, 576, 589, 693,
        695, 698–699, 729, 742
    Indo-Pacific Ocean, 409, 431–432

Pacific, 127, 274, 288, 292, 311, 331,
    430, 447, 468, 470, 555, 570, 573,
    585, 589, 591–593, 596–597, 599,
    623, 687, 703, 706, 727–728, 742, 771,
    786
Undercurrent, 653
upwelling zone, 588, 687
Equilibrium, 13, 192, 345, 419, 492, 502–506,
    508, 514, 525, 613, 646, 692, 697, 718,
    730, 736, 739–741, 746–751, 756–757,
    766–767, 806
temperature, 13, 751
Esters, 463
Estuaries, 266, 373, 376, 386, 394, 396, 454,
    608, 655, 671
Euphotic zone, 376–377, 446–447, 451, 455
Euryarchaeota, 454
Euryhaline, 386
Eutrophic waters, 9
Eutrophication, 282, 371, 373, 392–394,
    512
Evaporate minerals, 785
Evaporation–precipitation, 766
Export productivity, 265, 277, 285, 392–393,
    441, 459, 466
Extracellular calcifying fluid, 503

Factor analysis, 332, 339, 341, 425, 528
Farquhar Ridge, 43
Fast-settling particles, 246
Fast-sinking specimens, 246
Fecal pellets, 10, 286, 381–382, 414, 685
Feldspars, 142, 576, 588
Ferrimagnetic minerals, 65
Ferromanganese
    coatings, 508
    crusts, 650, 654, 656, 668, 671
    nodules, 581, 583, 608, 614
Filtration technique, 417, 419
Fine-grained particles, 141, 148
Fjords, 337, 357, 392–393, 602, 780
Flood layer, 84
Florida Current, 28
Flow rates, 52, 54, 576
Fluorescence in situ hybridization, 455
Fluorescent radiation, 75
Fluvial
    input, 464
    source, 47
    transport, 194

Flux
    equation, 286, 289–290, 292–293, 297
    Index, 570
Focusing, 14, 191, 540, 546, 591–593,
    623–624, 650, 700, 702–703
Food
    availability, 8, 237, 245, 250, 268, 282
    web, 286, 394, 455, 512
Foraminifera, 6, 13, 20, 26, 31, 43, 192–193,
    195, 197, 199, 201, 213–217, 219–221,
    223–225, 227–229, 231–237, 239–254,
    264, 266–269, 272–273, 275, 277–278,
    281, 285, 288, 290, 292–294, 297–299,
    301–305, 332, 349–350, 357, 372–373,
    382, 397–400, 426, 443–444, 505, 511,
    514–515, 524–526, 530, 537–538, 540,
    546, 548, 550–552, 554–555, 557, 656,
    659, 661–662, 664, 669, 709, 717,
    720–721, 723–724, 727–729, 731,
    735–741, 743, 745–749, 751–759,
    765–770, 773–778, 780, 784, 786–790,
    805
Foraminiferal
    assemblages, 228, 231, 236, 263, 266,
        272–274, 277, 287, 302–303, 305–306,
        308, 392, 398, 621
    calcite, 6, 214, 218, 241–242, 244–245,
        251, 507, 661, 736, 739, 747, 756–757, 773
    ecology, 263–267, 305
    fluxes, 198–199, 738
    Gametogenetic calcite, 219
    Mg/Ca, 765, 767–768, 772, 774–776, 785,
        787–790
    niche, 268
    stratigraphy, 28
Fossil
    assemblages, 7, 11, 213–214, 301, 332, 337,
        341, 343, 350, 397, 409, 411, 413–415,
        417, 419, 421, 423–425, 427, 429, 431,
        450, 523–524, 527, 533–536, 550–551,
        585
    hydrocarbons, 464
Fractionated settling technique, 333–334
Fractionation, 6, 163, 187–190, 192, 203,
    334, 345–348, 443, 457–461, 463, 503,
    512–513, 613, 645, 650, 654, 658, 688,
    690, 702, 704, 717–721, 723–726, 729,
    735–736, 739, 741, 746–747, 751,
    756–757, 759

factor, 345, 347, 688, 717,
    719–721, 723, 739
Fragmentation, 226, 249, 251–252, 338, 703
    Index, 338, 703
Fram Strait, 51, 229
Framvaren, 607, 613, 615
Francolite, 594
French Riviera, 83
Freshwater, 9, 147, 376, 384, 386, 619, 740,
    743–745, 783
    discharge, 384
Frontal zones, 384
Frustules, 270, 329, 334, 344, 357, 597, 623
Functional morphology, 241, 253
Future climate changes, 808

GAIM, 808
Gametogenic calcification, 747
Gamma ray, 66, 106
    attenuation density, 106
Gardar Drift, 23, 40
Gas bubbles, 31
Gaussian distribution, 531
GCM, 671
General
    Circulation Models (GCM), 669, 712, 800,
        803–804
    Linear Model, 531
Generalized Additive Model, 526, 531
Genetic
    diversity, 241–242, 244
    types, 219–222, 240–242, 244–245,
        253
Genotypic variability, 423
Geochemical
    cycles, 215, 475, 568, 607, 614, 646, 801,
        803–804, 807
    proxies, 67, 215, 224, 233, 241–242,
        244–245, 252–253, 264–265, 272, 308,
        341, 415, 572, 589
Geochemistry, 6, 327, 329, 331–333, 335, 337,
    339, 341, 343–345, 347, 349, 351, 353,
    355, 357, 420, 428, 445, 491–492, 514,
    568–569, 576, 594, 599, 602, 608,
    622–623, 735, 766, 799
Geographical North, 102
Geomagnetic
    equator, 103
    excursions, 126–127

field, 100–102, 108, 110, 115, 117, 120, 123,
    126–128, 130, 194–195
North Pole, 102
paleointensity, 119
Polarity Time Scale, 99–100, 121
Geomagnetism, 99–101
Geostrophic
    current, 22
    flow, 26, 32, 50
GeP & CO (Geochemistry, phytoplankton and
    Color of Ocean project), 428, 430
Gibbs Fracture, 170
Gibbsite, 144, 151, 153–154, 156
GISP, 124
Glacial, 7, 12, 28–29, 40, 42–45, 49, 54, 139,
    156–157, 159, 161–162, 165, 167–176,
    197, 199–200, 204, 214, 225, 227, 229,
    231, 233–235, 237–238, 246, 251, 254,
    291, 296, 338, 340–341, 347–350,
    353–355, 375, 392, 397–398, 431–432,
    460, 466–468, 470–471, 473, 492,
    514–515, 528, 536–537, 545, 552, 568,
    573, 577, 587–589, 592, 596, 606,
    609, 611, 616–619, 654–655, 661–663,
    668–671, 703, 707, 709–710, 727–728,
    730, 735–736, 742, 746, 750, 753, 775,
    786–787, 790, 800, 802, 805, 807
    cycles, 29, 165, 170–172, 214, 296, 398,
        467–468, 470, 661, 730, 802
    deposits, 157, 568
    erosion, 169
    lakes, 746
    pCO₂, 786
    periods, 42, 162, 169–171, 173, 238, 294,
        296, 350, 466, 470, 536, 545, 668, 707,
        742, 753–754, 787
    productivity, 348, 587, 589, 592
    terminations, 468, 786
Glacial/Holocene transition, 173–174
Glacial-interglacial
    cycles, 29, 40, 43, 291, 398, 577, 661, 802
    time scales, 231, 573, 588–589, 596, 609,
        669, 727
Glaciomarine sediments, 89, 91, 159
Global
    carbon cycle, 186–187, 194–195, 278, 384,
        422, 585, 752–753, 756, 760
    high resolution paleointensity stack, 121
    ice volume, 468–469, 474, 742, 746, 788
    ocean flow, 19, 26

thermohaline circulation, 195, 648, 666
warming, 6, 350, 717, 788
Glycerol dialkyl glycerol tetraethers, 452
Goethite, 67
Gonyaulacales, 378–379
Gorgonians, 501, 512–513
Gorgonin, 498, 500
Grain size, 3, 19, 21, 23, 28, 33–34, 39–40,
    42, 45, 47, 49, 52–54, 65–66, 70, 75, 82,
    84–85, 87, 90–91, 110–113, 117–118, 120,
    144, 150, 304–306, 574, 576
    analysis, 45, 82, 305
    proxies, 49, 570
Granulometry, 304
Gravity
    cores, 670
    currents, 149
    flow, 29, 442, 662
Gray level, 70, 73, 424
Greenhouse gas, 799–800
Greenland, 14, 25, 41, 44, 52, 127, 129, 170,
    173–174, 195, 289, 311, 331, 351, 468,
    515, 538, 579–580, 621, 654, 663, 666,
    709–710
    ice cores, 127, 195
    ice sheet, 44
Grenarchaeota, 454, 456
GRIP, 124
Growth
    effects, 782
    rate, 7, 245, 346, 455, 458, 460–463, 470–471,
        497, 513, 659–661, 739, 782–783
Guinea Basin, 289
Gulf of
    Aden, 147, 156
    Alaska, 157
    Cadiz, 26, 305
    California, 76, 447, 608, 770–771
    Guinea, 289, 393
    Maine, 451
    Mexico, 152, 154, 302, 375
    St. Lawrence, 386
Gulf Stream, 28, 170, 650, 669
Guyana Basin, 149
Gymnodiniales, 378–379

Hafnium, 647, 655, 658
Half-life, 12, 82–83, 188–189, 624, 647, 654,
    681, 684
Halley Bay, 453

Halloysite, 576
Halocline, 8
Haptophyceae, 445
Haptophytes, 411, 430, 449, 461–463,
    471, 473
Harbours, 215, 237, 392
Harmful Algal Blooms, 371, 394
Harvardton-Bear effect, 801
Hatch-Slack, 463
Heat transfer, 193
Heavy
    isotopes, 344–345
    minerals, 346, 349, 570, 572
Hectorite, 143
Heinrich
    events, 65, 467–468, 754
    layers, 41
Helium, 107, 651, 737, 739
Hematite, 67, 110–111
Henry's Law, 460–461
Heterococcoliths, 411
Heterotrophs/Autotrophs ratio, 8
Heterotrophic, 8, 215, 372, 377, 379, 381,
    390–391, 587
High frequency
    oscillations, 91
    variability, 123
High latitudes, 148, 156, 169, 197,
    301, 387–388, 395, 397–398,
    445, 449, 551, 648, 650, 739,
    742–744
High nutrient low chlorophyll, 428
High resolution
    biostratigraphy, 331
    images, 64
    records, 49, 123, 709
    sampling, 415
High temporal resolution, 789
High-productivity regimes, 223, 229, 237
Holocene, 40–41, 54, 75, 119, 128–130,
    139, 160–161, 168, 172–174, 176, 202,
    204, 230, 237, 240, 278, 284, 294,
    296–297, 305, 338–339, 350–351,
    355, 357, 413, 426, 431, 467, 469,
    495, 506, 514–515, 530, 552, 554–555,
    576–577, 580, 590–591, 609–610, 616,
    619, 621, 623, 662, 709–710, 727–728,
    730, 786, 790, 801, 807–808
    climate optimum, 351
    thermal maximum, 580, 619

Holococcoliths, 411
Horny corals, 491, 494–495, 498, 501,
    511–513
Hydraulic permeability, 79
Hydrodynamics, 24, 156, 271
Hydrogen, 457–458, 608
Hydrogenous deposits, 567, 581, 585
Hydrographic parameters, 371, 373, 375,
    395–396, 525
Hydrography, 26, 162, 241, 552, 618,
    768, 786
Hydrological
    balance, 580, 784
    changes, 170, 430
    fronts, 336, 382
    models, 225
Hydrology, 332, 467, 786
Hydropyrolysis, 452
Hydrothermal, 144, 146, 147, 278, 285, 492,
    577, 581, 583, 584, 586, 646, 647, 651,
    653, 655, 658, 659, 664, 665, 681, 741,
    785
    activity, 146
    deposits, 583–584
    environments, 147
    flux, 651, 741
    input, 567, 583, 586, 647
    vents, 278, 285, 581
    weathering, 655
Hydroxides, 569, 655, 659–660
Hydroxylation, 503–504, 757
Hyperpycnal
    flow, 84
    sequences, 84
    turbidite, 83
Hypersaline environments, 386
Hyperthermophilic, 454
Hyposaline conditions, 225
Hypoxia, 273, 275, 282
Hypoxic
    bottom water, 273, 282
    conditions, 273, 275, 277, 279–280, 282
    event, 282
Hystrichospheres, 397

Iberian margin, 40, 42, 195, 453, 709–711
Ice
    age, 40, 83, 580
    cores, 28, 126–127, 195, 357, 432, 457, 470,
        515, 609, 727, 730, 806

rafted detritus, 28-29, 48
rafted layers, 161
sediment transport, 157
sheets, 710, 742, 746, 765, 784–785, 787,
    803–804
Iceberg discharge, 709–710
Iceland
    Scotland Overflow Water, 40
    Sea, 195
Ice-ocean interaction, 475
Igneous rocks, 101, 570
Illite, 34, 140–145, 147, 149–150, 152–157,
    159–160, 165–168, 170–176, 573,
    576–577, 579, 581–582
Ilmenite, 570, 572
Image analysis, 70–71, 73
IMAGES, 14, 64, 68–73, 76–77, 79–81,
    84–86, 88–89, 91–92, 253, 410, 412,
    420–421, 423
Imbrie and Kipp technique, 396
Inclination, 85, 101–103, 110, 114–115, 119–120
Incubation, 447, 456
Index species, 229, 303, 415
Indian
    Basin, 148, 151
    Bottom Water, 308
    Deep Water, 308
    Ocean, 20, 22, 43–44, 139, 147,
        150–152, 155–156, 160, 163, 165–167,
        171, 175, 237, 241, 250, 254, 290–291,
        310–312, 347, 427, 430–431, 447, 451,
        468, 546, 576, 588, 610, 652–653, 655,
        658, 669–670, 723, 771, 777
    Ridge, 27, 164
Indicator species, 229, 284, 301, 390, 526, 533
    methods, 524, 533
Indirect gradient analysis, 527, 530
Indonesian
    Archipelago, 168, 175
    Islands Arc, 175
    Throughflow, 168, 175–176, 650, 653
Inductively
    coupled plasma-mass spectrometry (ICPMS),
        774
    coupled plasma-optical emission
        spectrophotometry, 774
Infaunal, 268, 270, 273, 275, 277, 279–284,
    288, 299, 306, 748, 758–759
    microhabitat, 275, 279–280, 288, 306
    taxa, 268, 270, 273, 277, 281–284, 299

Inorganic
    aragonite, 783
    carbon, 13, 186–187, 191, 410, 460, 501,
        506, 737–738, 752
    phosphate, 594
Insolation
    cycles, 428, 467
    forcing, 351, 428, 468, 589
Integrated Ocean Drilling Program (IODP), 64,
    105
Interannual variability, 204, 759
Interdecal variability, 7
Interglacials, 6, 42–43, 49, 162, 170–171, 432,
    467, 595–596
Intermediate
    nepheloid layers, 29–30
    water circulation, 28
    water flow, 28
International
    code of Botanical Nomenclature, 411
    Geosphere-Biosphere Program (IGBP), 807
    Marine Past Global Change Study
        (IMAGES), 64
    Union for Quaternary Research,
        808
Interocean connections, 168
Interpolation technique, 545
Interspecific variability, 423
Interstadials, 573, 621–622
Interstitial
    pore pressure, 83
    waters, 304, 606
Intertropical convergence zone (ITCZ), 580
Intraspecific variability, 242, 421
Intra-test variability, 765, 772
Intrinsic limitations, 11, 664, 738–739
Inverse models, 805
Iodine, 603, 606
IODP, 14, 64, 72, 105
Ionian Basin, 165
Irminger Basin, 167, 173
Iron, 67, 100, 287, 348–349, 353, 355, 577,
    581, 659
    fertilization, 348–349
    supply, 355
Isobathyal species, 266
Isochrysidales, 411
Isopycnals, 166, 686–687, 690, 692, 695
Isothermal Remanent Magnetization,
    109–110

Isotherms, 354

Isotope, 4–6, 10–13, 42, 52, 54, 68, 82, 101, 124, 126–127, 161–163, 188–190, 204, 225, 264, 272, 327, 332, 344–345, 347–348, 353, 356–357, 415, 419, 467, 475, 512, 645–648, 650, 652–656, 659, 662, 666–671, 681–683, 687, 704, 712, 717–721, 723, 725, 727–729, 735–737, 739–741, 743–745, 747, 749, 751, 753, 755, 757, 759

  fractionation, 187–188, 419, 461, 717, 721, 729, 736, 751, 759

  fractionation factor, 717, 721

  geochemistry, 327, 329, 331–333, 335, 337, 339, 341, 343–345, 347, 349, 351, 353, 355, 357, 420, 735

Isotopic

  composition, 6, 162–163, 190, 218, 251, 349, 443, 462–463, 465, 474, 503, 512, 568, 583, 590, 646, 650–655, 658, 660–665, 668, 671, 682, 718–720, 723, 725, 728, 735–740, 747, 750–751, 753, 756, 759, 765, 784

  depletion, 458, 503, 505, 513

  disequilibrium, 504

  enrichment, 345, 472

  equilibrium, 502–503, 505, 514, 739–741, 746–748, 751, 756–757

  ratio, 188, 332, 345, 349, 355, 458, 475, 650, 660, 664–665, 737

  tracers, 6, 48–49, 645–647, 649, 651, 653, 655, 657, 659, 661, 663–665, 667, 669, 671

Isthmus of Panama, 666, 669

Japan Sea, 413, 473

Jaramillo, 122

Jet Stream, 29

JGOFS (Joint Global Ocean Flux Study), 422

JOIDES Resolution, 104, 108

Kaolinite, 140, 142–146, 148, 150–157, 162–169, 171–172, 175–176, 573, 576–579

Kaolinite/chlorite ratio, 157, 162, 165–167, 171

Kara Sea, 151, 172

Kau Bay, 453

Kinematic

  properties, 190

  viscosity, 30, 36

Kinetic

  effects, 504

  energy, 36, 38, 52

  fractionation, 747, 751, 757, 759

  model, 503–504, 506

Koljö Fjord, 394

Kuroshio, 28

Labile organic matter, 269–270, 287, 289, 299–300, 346, 604

Labrador

  Current, 149, 513

  Sea, 149, 161, 164–165, 167, 173–174, 514, 663, 666, 668–669, 692, 710

Laflamme Sea, 85

Lagoons, 464

Lake Agassiz, 79

Laminated sediments, 228, 230, 275, 280, 357

Land-ocean

  correlations, 441, 463

  interaction, 384, 475, 622

Laptev Sea, 151, 172, 291

Laser

  ablation ICP-MS, 774, 779, 784

  secondary neutral mass spectrometry, 13

  size analyser, 54

Last

  deglaciation, 161, 202–205, 301, 350, 709

  Glacial Maximum, 7, 162, 214, 231, 233–234, 254, 338, 354, 375, 467, 528, 606, 662, 670, 703, 709–710, 742, 746, 786, 800

  glacial period, 168, 348–349, 355, 515, 619, 709

  glacial-interglacial cycle, 40, 43, 291, 470, 577

  interglacial, 12, 43

Late

  Cenozoic, 1, 3, 5–7, 9, 11–14, 167, 187, 219, 240, 373, 667, 741

  Holocene, 351, 530, 555, 619

  Paleocene thermal maximum, 788

  Quaternary, 75, 162, 167, 170, 175, 238, 242, 280, 373, 395, 397–398, 412–413, 416, 468, 547, 621, 668, 702

Lateral

  advection, 246, 286, 342, 443

  dispersion, 382

  fluxes, 10

transport, 167, 275, 330, 381–382, 414, 604, 624, 684, 686–689, 691–692, 700, 702, 712
Laterites, 572, 576
Latitudinal zonation, 157–158, 175
Laurentian Channel, 149
Laurentide ice sheet, 618
Lead, 11, 40, 44, 49, 70–71, 83, 92, 161, 196, 199, 237, 265, 291, 294, 330, 338, 443, 452, 459, 513, 531, 555, 611, 615, 617, 621, 646–647, 650, 654, 656, 658, 670, 753, 767, 773, 777, 783
Lead-210, 82, 293, 294, 305, 497, 498, 654, 655
Leeuwin current, 167–168, 175–176
LGM, 49, 162, 167, 201, 338, 341, 350, 352–353, 355, 375, 397–398, 528–530, 532, 537, 546, 551, 553, 555, 606, 609–611, 709–710, 730, 742, 755, 786, 800, 803, 805–807
Libby's age, 204
Light isotopes, 344
Light
  level, 329, 723
  limitation, 513
  microscopy, 417
Limiting factors, 271, 306
Lineage, 227
Linear
  regression, 343–344, 352, 457, 768
  relationships, 529, 768
Lines technique, 491, 505–506, 514
Liquid scintillation counting, 189
Lithogenic tracers, 354
Lithogenous
  index, 570
  proxy, 570
Lithology, 64–66, 70, 158, 568, 652
Lithostatic pressure, 66
Little
  Bahama Banks, 776, 778
  Ice Age, 40, 580
Loess, 47, 169, 549, 576, 581
Logarithmic transformation, 340, 343, 397
Logistic model, 531
Low oxygen indicators, 280
Lutetium, 658
Lysocline, 247–248, 252, 303–304, 311–312, 382, 398, 414, 589, 659, 731, 775
  depth, 775

Macroaggregates, 414
Madeira Abyssal Plain, 451
Magnesium, 498, 766
  calcite, 498
Magnetic
  anomalies, 100, 122
  deviation, 101
  field, 100–104, 108, 110, 115, 117, 120, 123–124, 126–128, 130, 194–195
  grains, 25, 52, 117–118
  induction, 103–104
  meridian, 102
  mineralogy, 99, 111–113, 118–119
  minerals, 77, 118
  oxides, 118
  polarity, 99–101, 105, 121–122
  properties, 3–4, 65, 76, 105, 120, 123
  resonance, 63, 79–80
  Resonance Imaging, 63, 79–80
  stratigraphy, 65, 99, 101, 103, 105, 107, 109, 111, 113, 115, 117, 119, 121, 123, 125, 127, 129–130, 331
  susceptibility, 24, 26, 50, 63–66, 70, 79–80, 85, 87, 103–104, 109–110, 112, 121, 170
Magnetism, 99, 103, 117
Magnetite, 25, 52, 65, 110–113, 118–121, 170
Magnetization, 99, 101, 103–104, 107–110, 112–113, 115, 117–120
Magnetometers, 99, 104, 107, 109
Magnetostratigraphy, 4, 661
Manganese, 3, 272, 287, 581, 603–607, 655, 659, 689
  coating, 508, 656, 659
  nodules, 3, 272, 581, 583, 608, 614, 659
  oxides, 659, 689
  pump, 606
Marginal seas, 157, 449, 467, 805
MARGO, 254, 354, 537, 551, 555
Marine
  ^{14}C reservoir age, 185–186, 196
  isotopic stages, 295
  reservoir correction, 192–193, 197, 205
  snow, 10, 286, 382, 442, 685
Marker
  pigment, 430
  species, 272, 280, 284, 290, 307
Mass
  accumulation rates, 41, 460, 623–624, 699–700, 702–705

spectrometry, 13, 189, 475, 650, 735, 774, 779, 782
Matuyama, 101, 118, 122, 126
Maximum
    angular deviation, 115
    chlorophyll zone, 377
MC-ICPMS, 650, 655, 670
Mechanistic model, 549
Medieval warm period, 580
Mediterranean, 26, 76, 83, 151, 158, 160, 165, 197, 229, 234, 266, 269, 280, 282, 289, 305, 375, 447, 454, 458, 472, 534, 546, 573, 575, 617, 653, 668
    outflow, 26, 305, 653, 668
    sapropels, 229, 266, 280, 458
Meltwater, 350, 746, 806
Membrane lipids, 446, 454–456
Menardii fragmentation index, 703
Mercury, 611
Meridional overturning circulation, 195, 681, 684, 691, 707–708
Mesopelagic, 455, 551
Mesotrophic, 268, 292
Mesozoic, 141, 159, 169, 373, 415, 419
Metabolic
    effects, 506, 784
    fractionation, 245
Metalliferous sedimentations, 581, 585
Metamorphism, 147
Methanogenic conditions, 451
Methanotrophic archaea, 456
Mg/Ca paleothermometry, 765, 767, 777
Micas, 144, 150, 576, 588
Microbial activity, 451
Microfossil assemblages, 7, 11, 523–524, 527, 534, 550–551, 585
Microfossils, 7, 254, 379, 386, 398, 417, 443, 523, 525, 527, 529, 531, 533, 535, 537, 539, 541, 543, 545, 547, 549, 551–553, 555, 568, 661
Microhabitat, 245, 253, 267–268, 275, 279, 284, 288, 306, 756
Micronutrients, 611
Micropaleontology, 327, 329, 331–333, 335, 337, 339, 341, 343, 345, 347, 349, 351, 353, 355, 357, 414, 420
Mid-Atlantic Ridge, 148–149, 154
Mid-Miocene cooling, 789
Mid-ocean canyon, 170
Mid-Pleistocene climate transition, 470

Milankovitch, 230, 339, 357, 467, 736, 802
Millennial time scales, 13, 469
Mineral assemblages, 113, 141, 150, 152, 158–160, 163, 165–166, 169–175, 576
Mineralogical provinces, 160, 171
Mineralogy, 3–4, 47–48, 66–67, 73, 99, 110–113, 118–120, 139, 141, 158, 160, 162–163, 166–168, 176, 253, 568, 573, 580, 720
Miocene, 301, 331, 452, 470, 492, 576, 581–582, 608, 787, 789
Mississippi delta, 266, 269
Mixed layer, 49, 117–118, 142–143, 149, 216, 428, 430, 447, 462, 551, 604, 609
Mixotrophic, 8, 372, 377, 381
Modal distributions, 525, 738
Model simulations, 801, 805, 808
Modeling, 6, 14, 118, 128, 357, 475, 531–532, 648, 702, 707, 712, 785, 790, 799–801, 803–808
Modern
    analogue technique (MAT), 272, 373, 397, 526, 534
    assemblages, 335, 343–344, 525, 533, 535
    atmospheric carbon, 193
    datasets, 352, 450, 533, 541, 544–545, 557
Molecular
    approaches, 470
    compounds, 194
    diffusion, 624
    isotopic composition, 463
    viscosity, 30
Molybdate, 614–616
Molybdenum, 75, 284, 614, 616–617
Monophyletic, 219
Monsoon, 160, 171, 230, 447, 473, 786
    wind, 160
Monte-Carlo-Markov-Chain Techniques, 549
Monte-Carlo method, 541, 545
Montmorillonite, 34, 142, 150, 155, 160, 576–577
Morphological variation, 241–242
Morphology, 213, 217, 219, 225–227, 235, 241–242, 248, 253, 267, 271, 275, 279, 284, 288–289, 329, 376, 378–379, 502
Morphometric analyses, 420, 422
Morphospecies, 219, 221, 241–242
Morphotypes, 242, 288, 386, 397, 426, 769
MSCL, 64–69, 72, 87, 89, 91–92

MST, 64
Mudwaves, 39
Multisensor
    Core Logger (MSCL), 64
    Track (MST), 64
Multichannel seismic profile, 24
Multi-cores, 286
Multi-nominal calibration model, 549
Multi-technique approach, 258, 403, 560
Multiple regression functions, 288
Multiple regressions, 288, 290–291, 426, 527,
    530, 539
Multiproxy approaches, 176, 537, 625
Multivariate analyses, 386–387, 424
Mutual range method, 523, 533–534

n-alkanes, 452, 463–466
n-alkanoic acids, 463–464, 466
n-alkanols, 463–464, 466
n-alkyl lipids, 442, 463–464, 466, 473
Namibian slope, 443
Nannofossil biostratigraphy, 416
Nannoplankton, 411
Nansen Basin, 172
Natural
    gamma radiation, 63–64, 68
    Remanent Magnetization, 99, 101, 109–110,
    113, 117
Negative thermal ionization method, 719
Neodymium, 52, 652–653, 658
    isotope, 52, 658
Neogene, 11, 100, 159, 227, 237, 240, 243,
    245, 331, 420, 473, 580, 588
    climate oscillations, 243
Nepheloid layer, 10, 14, 29–30, 149, 163, 702
Neptune database, 254
Neritic environments, 376
New production, 587
Newfoundland Basin, 174
NGRIP, 44, 195
Nickel, 611
Nitrate, 287, 347–349, 355, 394, 461, 603, 609,
    753
Nitrification, 512
Nitrogen, 187, 332, 347–348, 355–356, 568,
    594, 804
    isotopes, 347
NOAA, 8, 205, 254, 429, 807
Non destructive physical methods, 121
Non-analogue situations, 397

Non-linear function, 532
Non-radiogenic thorium, 508
Non-temperature effects, 765, 774, 782
Nontronite, 143–144
Nordic Seas, 76, 231, 336, 352, 448, 457, 550,
    691
Normalizing parameter, 569
North
    American Shield, 663
    Atlantic Deep Water (NADW), 25, 162, 201,
        204,
        303, 308, 511, 648, 665–666, 690,
        753–754, 805
    Atlantic drift, 744
    Atlantic Intermediate Water, 805
    Atlantic Oscillation (NAO), 351
    Atlantic paleointensity stack, 121, 124
    East Atlantic basins, 149, 167, 170
    Eastern Atlantic Deep Water (NEADW), 40,
        308, 648
    Pacific, 48, 76, 146, 152, 155, 157, 201, 302,
        330–331, 340, 350, 390, 396, 447, 455,
        466, 529, 567, 570, 577, 580–581, 585,
        683, 687, 695, 755
    Pacific subtropical gyre, 447, 455
    Western Pacific, 466
Northeastern Atlantic, 149, 694
Northern Hemisphere, 102, 154, 157, 171, 200,
    331, 351, 375, 387–389, 430, 468, 470,
    551, 666–669, 786, 802
Northwest
    Atlantic, 167, 170, 174, 456
    Pacific, 301, 302, 466, 468
Norwegian
    Bottom Currents, 167
    deep-sea water, 170
    Sea, 174, 233, 288, 308, 350, 447, 648, 694,
    771
    Sea Overflow Water, 308, 648
Norwegian-Greenland Sea, 289, 311, 331
Nova Scotian Rise, 21, 35
Nuclear Magnetic Resonance, 63, 79
Nucleation, 216, 496, 498
Numerical models, 799, 802, 807
Nutricline, 427
Nutrient
    availability, 8, 376, 384, 396, 461
    content, 336, 394, 425, 430, 463, 621
    cycling, 332, 347, 349, 357
    enrichment, 384

fractionation, 345
limitation, 396, 428
sources, 347
utilization, 512

Obliquity, 339, 432, 589
Ocean
  Carbon cycle Model Intercomparison
    Project, 808
  Drilling Program (ODP), 14, 64, 104–105,
    415, 754, 788
  dynamics, 2–3
  general circulation models, 800, 804
  ridges, 651, 653, 655, 664, 689
  ventilation, 189, 193–194, 203–204
Oceanic
  crust, 144, 147, 152, 651, 665, 726, 741
  provinces, 302
  residence time, 766, 785
Oceanographic parameters, 226, 264, 306, 308,
  357, 523
Octocorals, 492, 498–501, 507–508, 513
Okinawa Trough, 28
Oligocene, 38, 159, 301, 470, 581–582,
  585–586, 665, 776, 787
Oligotrophic
  conditions, 223, 306
  ecosystems, 267, 269
  Open Ocean, 237
  subtropical gyre, 221
Oman margin, 288, 292
Online calibration, 205
Ontogenic effects, 747, 780
Ontogenetic growth, 235
Opal, 10, 158, 247, 382, 398, 572, 585,
  588–591, 595, 597–598, 611–612, 619,
  688, 690, 706–707, 711–712
Ophiolites, 655, 752
Opportunistic species, 271, 301, 430
Orbital parameters, 789
Orbitally-paced variability, 786
Ordination techniques, 526–527
Ordovician, 492
Organic
  carbon, 8, 13, 67, 117, 186–187, 191,
    193–194, 247, 278, 285–289, 291–292,
    294–295, 297–300, 304, 353, 410, 414,
    444, 451, 460, 470, 501, 506, 568, 585,
    588–590, 606, 610–611, 614, 706,
    737–738, 752–753

carbon flux, 117, 285–287, 291, 295, 300,
  304, 588, 611
flux, 264, 267, 269, 271–272, 277, 280,
  282–294, 296–297, 299–300, 306–307,
  385, 567, 594, 609, 619, 621
lining, 739
matter, 3, 10, 31, 67, 75, 187, 193, 247,
  266, 268–271, 277, 284–287, 289–294,
  297, 299–301, 304–305, 307, 332–333,
  344–347, 349, 373, 382, 398, 414, 419,
  443, 448, 452, 455, 459, 461, 464, 466,
  472, 499, 501, 511, 585, 587–589, 592,
  594, 597, 599, 601–604, 606–607, 609,
  613–614, 616, 618, 621, 701, 725, 752,
  759, 773
matter degradation, 287, 601, 603, 606
matter oxidation, 594
molecules, 194, 441, 475
rich sediments, 303, 606, 608, 611,
  614
walled dinoflagellate cysts, 371, 373, 375,
  377, 379, 381–383, 385, 387, 389, 391,
  393, 395, 397, 399
Orphan Knoll, 494, 514
Orthophosphoric acid, 737, 739
Osmium, 647, 651, 665
Ostracodes, 766, 777
OxCal, 204–205
Oxic layer, 451–452
Oxidation, 193, 346, 382, 398, 419, 456, 459,
  594, 601, 604, 607–608, 616–617, 651,
  659, 682, 773
Oxiphylic taxa, 279
Oxygen
  availability, 6, 382–383
  concentration, 267, 269, 271–280, 282–284,
    287, 290–291, 303, 604
  depletion, 267
  index, 279–280
  isotope composition, 349, 735, 739
  isotope stratigraphy, 332
  isotopes, 54, 101, 357, 467, 735, 740, 744
  isotopic equilibrium, 741, 746–748
  minimum zones, 274
Oxyhydroxides, 577, 581, 583, 585–586, 594,
  603–607, 609, 614–615, 617

Pacific
  Bottom Water, 308, 683
  Deep Water, 308, 653, 658

Equatorial Divergence, 428
North Equatorial Countercurrent, 428
Ocean, 127, 146, 151, 155–157, 160,
    171, 249, 290, 350, 373, 375, 392, 409,
    427–428, 431–432, 454, 508, 614, 646,
    652–654, 666, 683, 687–688, 690, 755,
    777, 782, 788
PAGES, 801, 807–808
Paleo seafloors, 82
Paleobathymetry, 266
Paleocene, 449, 788
Paleocirculation, 139, 171, 671, 681, 683, 685,
    687, 689, 691, 693, 695, 697, 699, 701,
    703, 705, 707, 709, 711–712
Paleoclimate, 141, 159, 176, 185–186, 193,
    338, 357, 526, 528, 536, 540, 550, 555,
    738, 801–802, 804–808
  modeling, 357, 801, 806–807
  Modeling Intercomparison Project (PMIP),
    801, 807
  simulations, 804
Paleoclimatic models, 332, 341, 352–353, 357
Paleoclimatology, 7, 205, 527, 530, 532, 556,
    786
Paleocurrent, 56, 179, 678
Paleoecological
  equation, 343, 424–425, 428
  model, 426
Paleo-export production, 287
Paleoflux, 681, 683, 685, 687, 689, 691,
    693, 695, 697, 699, 701–705, 707, 709,
    711–712
Paleogene, 11, 159, 279, 331, 449
Paleohydrological reconstructions, 159
Paleointensity, 3, 77, 99, 115, 119, 121,
    123–124, 126–127
Paleomagnetic
  inclinations, 85
  records, 99, 109, 118
Paleomagnetic secular variations, 115, 128
Paleomagnetism, 100–101, 103, 107–108,
    123
Paleo-nutrient, 486, 805, 811
Paleo-oxygenation, 274
Paleo-pH, 6, 717–718, 720–721, 723–727,
    730
  proxy, 717–718, 720–721, 723–727, 730
Paleoproductivity, 6, 235, 263, 269, 285,
    287–288, 290–293, 299, 346, 348, 671,
    753

Paleosalinities, 225
Paleosols, 473
Paleotemperature
  equation, 5, 349, 739–740, 747, 785
  reconstruction, 7, 413, 426, 768
Palygorskite, 142–144, 147, 156, 160
Palynology, 373
Palynomorphs, 10, 384, 386
Panama Basin, 592, 605–606
Partial
  dissolution, 410, 654, 725
  Least Square analysis, 529
Particle
  aggregation, 24, 28, 685
  flux, 285, 494, 597, 611, 623, 654, 686–688,
    690, 698, 701–706, 710–712
  scavenging, 681–684, 687, 706, 708,
    711–712
  sinking rates, 686
Particulate organic
  carbon, 117, 285–286, 414
  matter, 286, 345, 448, 501, 602, 613
PDB, 189, 729, 737, 785
Pee Dee
  Belemnitella (PDB), 737
  formation, 737
Pelagic
  flux, 10, 28
  oozes, 248
  settling, 48
Pennales, 327–328
Perennial sea ice cover, 353
Peridiniales, 378–379
Persian Gulf, 670
Peruvian margins, 390
Phanerozoic, 646, 664, 726
Phosphate, 52, 394, 461–462, 585, 594,
    613, 753
  proxy, 52
Phosphogenesis, 594
Phosphorite, 594
Phosphorus, 567, 585–586, 594–595
Photic zone, 8, 215, 329, 376–377, 445–447,
    451, 454–455, 472
Photodegradation, 451
Photoluminescence, 79, 81
Photosynthesis, 7, 187, 414, 461, 463, 723, 747,
    752–754, 756–760
Photosynthetic effects, 783
Phyllosilicate, 142

Physical
    properties, 3–4, 63–65, 67–69, 71,
        73–75, 77, 79, 81, 83, 85, 87, 89, 91,
        225–226, 235, 246
    weathering, 159, 571
Physico-chemical parameters, 271, 301, 306,
        333, 341
Phytoplankton, 215, 224, 285, 300–301, 339,
        345–346, 355, 377, 415, 419, 425–428,
        430, 441, 445, 456, 459–461, 470–471,
        475, 512–513, 524, 587, 611, 613, 623
Picoplankton, 455
Piston cores, 88, 161, 214, 291, 294, 583, 592
Plankton samples, 229, 377, 587
Planktonic foraminifera, 6, 41, 193, 197, 201,
        213–221, 223–225, 227–229, 231–254,
        263, 294, 373, 397–398, 400, 426, 443,
        525,
        530, 546, 550, 621, 661, 669, 748, 758–
        759, 765–770, 774–776, 786–789, 805
Plant functional types, 475, 532
Pleistocene, 11, 13, 50, 122, 160–161, 214, 229,
        331, 409, 413, 415–416, 422–424, 450,
        452, 466, 468–470, 474, 495, 501,
        514–515, 568, 573, 588, 735, 754, 780,
        786–787, 808
Pliocene, 11, 50, 452, 468–469, 472, 534,
        581–582, 669, 754–755, 780
    warm period, 754–755
Polar
    front, 49, 174, 301, 341, 347–348, 356, 399,
        592, 610–611, 692
    Regions, 154, 330, 466
    water, 167, 221, 239, 246, 336, 457
Polarity chrons, 122
Poles, 100, 103, 156, 229, 236, 386, 492
Pollen, 373, 384–385, 463, 466, 530, 532, 534,
        536, 549
Pollen/dinocyst ratio, 385
Polynyas, 371, 390, 396
Polyp size, 783
Pore water, 193, 247, 273, 277, 280, 287,
        509, 593, 599, 601, 604–609, 614–616,
        621–622, 624, 651, 662, 727
    oxygenation, 280
    chemistry, 775
Porosity, 21, 66, 70, 73, 75, 79–80, 91,
        243–244, 304, 333
Porphyrins, 463, 470
Portuguese margin, 40, 305

Positive ionization methods, 724
Post-burial dissolution, 776
Postglacial sediments, 89, 91
Precession, 431–432, 473
Precipitation, 143–144, 147, 457, 498, 536,
        567, 573, 576–577, 579–581, 587,
        594, 599, 601, 604–605, 607, 613, 616,
        623–624, 655, 659–660, 670–671, 701,
        740–744, 746–747, 750, 757, 765–767,
        777, 780, 783–784, 789
Precipitation/evaporation balance, 744
Prediction error, 235, 344, 397, 539
Preferential
    adsorption, 597
    dissolution, 246, 775
Preservation index, 338
Primary
    producers, 372, 390, 409
    productivity, 224, 237, 247, 289, 384, 391,
        396, 413, 425, 427–428, 459, 551, 739
Principal component Analysis, 87, 115, 233,
        339, 527, 529
Pristine, 64, 71, 105, 723, 726
Productivity, 3–5, 7, 10, 14, 48, 194, 223–224,
        229–230, 237, 247, 265, 272, 277–278,
        280, 284–285, 287–290, 292, 307, 332,
        336, 339–341, 347–348, 353, 357, 371,
        373, 382, 384, 390–398, 400, 413, 417,
        425, 427–428, 441, 459, 463, 466–467,
        471–473, 494, 551–552, 585, 587–590,
        592, 594–597, 599, 603, 609, 611,
        618–619, 621, 623, 659, 687, 704, 739
    proxies, 263, 285, 287–288, 290, 292,
        347–348, 427, 472, 595, 599, 623
Proterozoic, 652
Protista, 327, 411
Protozoa, 587
Provincialism, 8–9
Prymnesiophyceae, 411, 445
P-wave velocity, 63–64, 66–67
Pycnocline, 216, 377, 749–750
Pyrite, 616–617
Pyroxenes, 570, 572

Q-mode
    factor analysis, 339, 425
    principal component analysis, 233
Qualitative
    proxy, 264
    reconstructions, 231

Quantitative
  analyses, 228, 419
  approaches, 352, 371, 395
Quartz, 29, 31, 38, 142, 166–167, 174,
  571–574, 581
Quartz/feldspar ratio, 167
Quaternary, 6, 19, 38, 75, 100–101, 108, 115,
  119, 127, 159, 162, 167, 170, 175, 214,
  219, 222, 238, 241–243, 245, 280, 301,
  303, 373, 379, 395, 397–398, 412–413,
  416, 466–468, 524, 547, 551, 621, 625,
  668, 683, 702, 765, 786, 788, 799, 801,
  808
  climate change, 219

Radioactive
  decay, 187–189, 191, 660, 683–684, 704
  disequilibrium, 660
  isotopes, 68, 645–646
  tracers, 28
Radiocarbon, 3, 13, 128, 185–189, 191–195,
  197, 199, 201, 203–206, 357, 444, 460,
  491, 501, 511–513, 738, 755, 805
  activity, 3
  age, 188, 194, 199, 444, 511, 513
  calibration, 205
  dating, 128, 185, 187, 189, 191, 193,
    195, 197, 199, 201, 203, 205, 460,
    491, 511
  timescale, 185, 194
Radiochronology, 646
Radiogenic
  isotopes, 5, 10–11, 82, 645–648,
    650, 655, 659, 662, 670–671
  thorium, 508
  tracers, 645, 647, 650, 664
Radiolarians, 8, 346, 372, 550–551, 572
Radiometric techniques, 189
Radionuclide, 592, 660
Radioscopy, 70
Random settling technique, 417
Rapid climate change, 352, 467, 799,
  801, 807
Rapidly deposited layers, 65, 86–87
Rare
  earth elements, 653
  species, 228, 343
Rayleigh distillation processes, 742
Rayleigh's model, 345
Recrystallization, 193, 501, 731

Red
  Sea, 44, 76, 147, 156, 225, 357, 579, 787
  tides, 371, 377, 394
Redistribution, 29, 159, 199, 224, 248, 616,
  700–704
Redox
  conditions, 266, 268, 284, 347, 451, 459,
    567, 599, 601–602, 621
  potentials, 592, 607–608, 617
Redundancy analysis, 529
Reef corals, 492, 508
Reference database, 342, 396–397
Reflectance, 63–64, 67–68, 85
Refractory organic matter, 3, 269, 299, 398,
  594
Regression functions, 288, 291
Relative
  abundance, 145, 150, 152, 156, 199, 226,
    228, 237, 282, 334, 336–338, 340–341,
    343, 350, 383–384, 387, 392, 395, 417,
    420, 427, 450, 455, 464, 539, 577, 594,
    645–646, 739
  paleointensity, 77, 99, 115, 119,
    123–124
Remineralisation, 268–269, 286–287
Reservoir
  age, 185–186, 196–198, 201–202, 205, 351,
    355
  correction, 192–193, 197, 202, 205
  variability, 197–198
Residence time, 12, 147, 186, 197, 245–247,
  511, 615, 618, 645–647, 650, 652–655,
  662, 664–665, 670–671, 681–684, 687,
  689–690, 694–696, 698, 700, 702,
  707–708, 710, 712, 725, 731, 756, 766,
  785
Resistant taxa, 273, 277, 279, 282–284
Response surface, 523, 526, 532, 535–537, 549
Resuspended particles, 149
Reversals, 99–101, 115, 118–119, 121–123,
  196
Revised analogue method, 526, 537
Reworked sediments, 158
Reworking, 49, 157, 159, 166, 169, 337–339,
  343, 382, 386, 442
Reykjanes Ridge, 27, 45–46, 170
Rhenium, 607–608, 616–617, 651
Rio Grande Rise, 165–166, 171, 668
Ripples, 20–21, 32, 37, 84
River discharges, 398

R-mode, 528
Root mean
  square error, 343, 352
  square error of prediction, 343
Rose Bengal, 268, 276, 304
Rosholt-II isochron method, 509
Rubidium, 576
Rutile, 570, 572
Ryuku Ridge, 28

Saanich Inlet, 77–78, 453, 607–608, 613, 615
Sagami Bay, 447
Saguenay Fjord, 63, 82–87
Salinity, 2–8, 26, 163, 175, 225, 240–241,
    266, 271–272, 285, 303–304, 306–307,
    309–310, 312, 329, 346, 349, 371, 373,
    376, 379, 382, 384–386, 388, 395–398,
    400, 425, 441, 456–458, 460, 475, 535,
    551–552, 555, 619, 648, 669, 682, 720,
    722, 729, 740, 742–746, 749–751,
    765–767, 774–775, 780, 784–789, 805
  tolerance, 225, 388
Samarium, 658
Sangamonian, 91
Santa
  Barbara basin, 280, 616, 621
  Monica Basin, 453
Sao Paolo Plateau, 166
Saponite, 143–144
Sapropels, 229, 266, 280–282, 458, 573, 575
Sargasso Sea, 41, 447, 768, 772
Satellite imagery, 431
Saturation
  effect, 777
  magnetization, 112–113
Scandinavian coast, 398
Scanning electron microscope (SEM), 411, 420,
    500
Scavenging, 49, 587, 597, 616, 648, 653–655,
    660, 681–692, 700, 703–704, 706, 708,
    710–712
  intensity, 687–689, 691, 704, 710, 712
  models, 684–685, 691–692
Scleractinia, 492
Sclerochronology, 491, 495, 498
Scotia Sea, 28, 375
Sea
  floor, 11, 118, 141, 193, 199, 214, 216,
    228, 245–247, 269, 277, 285, 293–295,
    297–300, 306–307, 373, 379, 382, 400,

    567–571, 577, 581, 583–585, 594,
    597, 599, 602, 609, 615, 621, 623–624,
    659
  ice, 4, 173, 197, 216, 263, 611, 801–804
  level, 30, 43–44, 159, 168, 170–171,
    383–384, 475, 579, 618–619, 742,
    746, 790
  level change, 170, 746, 790
  surface conditions, 1, 242, 382, 383, 398
  surface salinity, 456, 619
  surface temperature, 28, 41, 234, 263,
    441, 445, 766, 775, 781, 789, 800,
    803, 805
  water density, 2, 31
Sea-ice, 5, 148, 172, 216, 225, 231, 327,
    329, 336–339, 341, 348, 350–354,
    357, 371, 373, 375–376, 382, 387–389,
    395–398, 400, 443, 532, 551, 744–746,
    749–750
  cover, 336, 350, 353, 357, 371, 373,
    375–376, 382, 387–389, 395–396, 398,
    400, 551, 750
  formation, 172, 744, 746, 749–750
  melting, 348
  model, 357
Seasonal
  cycles, 91
  scale, 357
  sea ice, 357, 388
  variability, 285, 294, 447, 499, 789
Seasonality, 3–4, 264, 285, 291, 301, 371,
    386–387, 395, 398, 428, 447–448, 536,
    547, 580, 736, 759
Sea-to-air flux, 470
Secondary ion mass spectrometry, 650,
    774, 779
Secular
  equilibrium, 508
  variations, 102, 128, 785
Sedigraph, 45–46
Sediment
  color, 68
  density, 21, 65–66, 69–70
  focusing, 592, 700, 702–703
  redistribution, 159, 701, 703–704
  reworking, 442
  transport, 25–26, 30, 45, 148, 157, 161, 172,
    568, 664
  trap calibration, 767

traps, 246, 289, 419, 422, 588–589, 653, 690, 706, 772
Sedimentary
    fluxes, 174, 379
    records, 6, 11, 28, 169, 353, 451, 567, 596, 601, 611, 619, 621–622, 653
    sources, 170
Sedimentation rates, 11, 19, 38, 43, 48, 127, 146, 198–199, 278, 382, 592, 617, 624
Sedimentology, 70, 75
Sediment-water interface, 3, 268, 274–275, 277, 279, 282, 284, 286, 299, 303, 305–306
Seismic
    reflection, 20–22, 38
    reflectors, 67
Selective
    degradation, 382
    deposition, 34–35, 37–38
    dissolution, 144, 193, 501, 739, 759, 775
    erosion, 36, 149
    transport, 160–161
Semi-quantitative
    estimates, 172, 712
    index, 251, 279
Sepiolite, 142–144, 147, 155
Sequence boundary, 71
Settling
    flux, 34, 588–590, 597, 602, 604, 609, 614, 711
    velocity, 23, 30–31, 34–35, 38, 45–46, 245–246, 576
Shallow
    dwelling species, 772, 775
    water, 35, 197, 492, 501, 691, 773, 777
Shatsky Rise, 146
Shell
    morphology, 213, 227, 235, 241–242
    porosity, 243–244
    size effects, 727
    weight, 243, 249–251, 723
Si/Al ratio, 572, 581–582, 584
Siberian Sea, 151, 172
Siderophile elements, 651
Sierra Leone Rise, 704
Silica, 3, 70, 110, 144, 328, 331–332, 346, 348–350, 353, 357, 382, 394, 398, 475, 573, 577, 581, 585, 595, 621, 688, 701, 706
    cycle, 357
    diagenesis, 581

Siliceous
    microfossils, 398, 529, 551
    plankton, 589
    productivity, 332, 339, 597
Silicic acid, 330, 348, 355, 613
Silicoflagellates, 572
Silicon, 332, 348–349, 355–357, 572
    fractionation, 348
    isotopes, 348, 356–357
    uptake, 349
Silt, 10, 19, 23–24, 34–35, 37–40, 42, 45–48, 50, 52, 54, 83, 149, 163, 414, 570, 572
Silver, 611, 613
SIMMAX, 526, 536–537, 548, 550, 555
Simple linear model, 531
Sinistral coiling, 240–241
Sinking velocity, 685
Size dependence, 25, 31, 119, 163, 738, 750
Skagerrak, 394, 453, 456
Skagerrak-Kattegat, 394
Skan Bay, 453
Slope, 23, 28–30, 39, 41, 47, 149–150, 159, 166, 171, 175, 274, 286, 292, 301, 304, 311, 343, 352, 383, 443, 448, 494, 505, 512, 514, 574, 594–595, 598, 604, 606, 621–622, 654, 685, 742, 777, 781–782
Smectite, 140, 142–144, 146, 150, 152–159, 161, 165, 167, 170–175, 577, 585
Smectite/illite ratio, 167
Solar
    activity, 195, 339, 806
    variability, 126, 194, 202
    wind, 195
Somali Basins, 43
Sortable silt, 19, 24, 34, 37, 40, 47, 52, 54
South
    Atlantic, 114, 121–122, 124, 127, 146, 151–152, 154, 158, 161–167, 171, 304, 338, 352–353, 390, 448, 461–462, 470, 473, 549, 610–611, 648, 653, 662, 668, 695–696, 708, 771
    Atlantic Intermediate Water, 611
    Atlantic paleointensity stack, 121, 124
    Atlantic tropical gyre, 470
    East Atlantic, 449, 450
    East Indian Ocean, 151
    Java Current, 171
    Pacific, 146, 152, 155, 241, 243, 428, 755, 789–790

Pacific convergence zone, 789
Pacific subtropical gyre, 428
Southeast Pacific, 162
Southeastern Canadian margin, 398
Southern
    Hemisphere, 102, 156–157, 171, 187, 205,
        224, 387
    Ocean, 76, 103, 202, 241, 303, 327, 330, 332,
        336–339, 341, 343, 347–350, 352–353,
        355, 375, 388, 396, 398, 447–448, 469,
        550–551, 588, 592, 609–612, 650, 653,
        655, 668, 692–693, 695, 708, 743–744,
        746, 753, 756, 789, 805
    Subtropical Convergence, 428
    subtropical Front, 355–356
Southwest African margin, 425
Southwestern Pacific, 155
Spatial
    autocorrelation, 397, 548–549
    variability, 20, 40, 172, 270
SPECMAP, 42, 124
Spectral analysis, 91, 121, 432
Spectrometer, 190, 650, 725, 737, 782
Spectrometric methods, 422
Spectrophotometer, 67, 87
Spectroscopy, 660
Speleothems, 80, 195, 457
Sphene, 572
Spherical harmonic analyses, 102
Spherulitic crystal, 502
Spiking and spraying, 418
Spiking with microbeads, 417
Spring
    bloom, 447
    flux, 225
Square chord distance, 343
Sr/Ca ratio, 462, 777, 780, 782–783
St. Lawrence Estuary, 88, 126, 129, 385
Stable
    carbon isotope, 188
    isotopes, 225, 264, 415, 419, 647, 735–736
    isotopic composition, 218, 251
    isotopic disequilibria, 491, 503
Stadials, 167
Standard Mean Ocean Water, 737
Standardized database, 395
Statistical
    analyses, 11, 115, 424, 547
    reproducibility, 228, 335
    techniques, 525

Stenohaline, 373, 400
Sterols, 623
Stevensite, 143
Stochastic models, 557
Stratification, 7–8, 221, 236, 272, 275, 282,
    336, 348–349, 353, 376, 384, 396, 398,
    400, 431, 468, 551, 555
Stratified surface water, 197
Stratigraphic
    markers, 266, 331–332, 664
    resolution, 20, 32, 416
Stratigraphy, 4, 28, 48, 64–65, 74, 99–101,
    103, 105, 107, 109, 111, 113, 115,
    117–119, 121–123, 125, 127–130,
    204–205, 331–332, 351, 355, 555,
    699–700
Strontium, 646–647, 650, 665, 766
Subarctic Zone, 95, 367, 368
Subantarctic
    Front, 356
    Mode Water, 355
Sub-millennial time-scales, 410
Suboxia, 567, 607, 623
Suboxic
    indicators, 279
    sediments, 592–593, 608, 617, 701
    water, 451
Subsurface water masses, 744
Subtropical Pacific, 430
Sudoite, 143
Suess effect, 513
Sulphate, 273, 287, 451, 587, 592–593,
    601–603, 609, 614–615, 618, 621–622
Sulphides, 452, 581, 601–602, 611, 613–614
Sulu Sea, 276
Sunda Strait, 171
Surface
    current transport, 172
    currents, 160, 167, 171, 246
    dwelling taxa, 277
    sediments, 139, 146, 149–152, 155, 160,
        165–168, 171–173, 175, 241, 243, 247,
        327, 330, 332, 336, 340, 374–375, 388,
        424, 447–452, 455–456, 459, 461–462,
        464, 466, 584, 590–591, 597, 606,
        614–615, 619, 621, 623, 656, 688, 690,
        707–708, 711, 750, 767
    water, 27, 29, 196–197, 199, 201–202,
        216, 225, 231, 251, 285–286, 288, 290,
        292–293, 297, 304, 329–330, 332,

335–337, 339, 344–349, 353, 355, 377,
379, 382, 394, 396, 400, 413–414,
424–425, 427–429, 431, 449, 456, 461,
469, 501, 512, 551, 587–588, 604, 606,
611, 613–614, 619, 650, 652, 654, 659,
683, 692, 696, 698, 701, 704, 727–728,
742–744, 746, 749–750, 753–756, 775,
781, 784, 786, 788, 801
Surface western boundary currents, 28
Suspended sediment, 29, 33, 38, 47, 84, 141,
163
Svalbard, 172
Symbiont-bearing foraminifera, 773
Symbionts, 215, 218, 220, 245, 372, 747, 757,
783
Syndepositional redistribution, 700
Synthetic
    calcite approaches, 720-721
    seismic profiles, 67
Système de Reconnaissance Automatique de
    Coccolithes, 421

Taphonomical processes, 265, 270, 284, 294,
303, 371, 377, 552
Taphonomy, 226, 265, 270, 278, 284, 294, 303,
305, 308, 552
Tasmanian Basin, 165
Taxonomy, 219, 235, 253, 328, 331, 342, 395,
397, 409–411
Teleconnections, 468
Temperature-dependency, 766
Temporal variability, 47, 108, 409, 428,
449
Tephra, 106, 331, 351
    chronology, 331
    layers, 106
Terrestrial
    biomarkers, 466, 622
    organic matter, 269, 466, 588
Terrigeneous
    input, 75
    sediments, 75, 144, 147, 162, 167
Tertiary, 331–332, 373, 581, 665, 765,
787–788
Tetra-unsaturated alkenone, 445, 448
TEX_{86} index, 452–453
Thanathocoenoses, 377
Thermal
    alteration, 144, 146, 452, 785
    ionization mass spectrometry, 650

Thermocline, 204, 216, 229, 432, 551, 654,
670, 748–750, 769, 775, 777
    dwelling species, 775
Thermodynamic equilibrium, 502, 741,
747, 767
Thermohaline circulation, 1–2, 6, 12, 14, 26,
192, 195, 350, 467, 648, 666, 743, 754,
800, 805, 807
Thermophilic taxa, 387
Thiomolybdates, 615–616
Thorium, 508, 646, 660, 682, 701
Thulean Rise, 53
Ti/Al ratio, 570, 572–573, 576, 581, 619
Timor Sea, 168, 175
Titanium, 570
Tomography, 63–64, 73, 106
Total
    Clay Index, 173–174
    organic carbon, 194, 297, 299, 444, 451, 590
Toxic species, 394
Trace
    element fractionation, 6
    elements, 225, 251, 415, 419, 503, 506, 602,
    607, 645, 661, 681, 773
    metals, 353, 599, 609, 613, 616, 619, 650,
    659–660, 666, 774, 777
Trade winds, 28, 430, 466, 580, 618
Transfer functions, 3, 7, 215, 228–229,
231, 233, 235, 246, 332, 341–342,
350, 352–353, 409, 413, 424, 523–525,
527, 529, 531, 533, 535, 537, 539,
541, 543, 545, 547, 549–553, 555–556,
592
Transit time, 193, 692, 694, 696, 708
Triassic, 415, 492
Tritium, 695
Trophic
    behaviour, 215, 218
    fractionation, 512–513
Tropical
    climate dynamics, 789
    Ocean, 216, 224–225, 236–237, 428, 431,
    728, 785–786, 789–790
    Pacific, 204, 409, 428, 430–431, 466, 607,
    742, 775, 786, 789–790
TROX model, 267–268
Turbid
    bottom nepheloid layers, 29
    layer, 29

Turbidity, 22, 29–30, 47, 49, 75, 87,
    147–149, 170, 172, 174, 266, 286
Turbulence, 6, 8, 31, 36, 376, 384, 386
Turbulent boundary layers, 19, 33

U_{37}^k index, 445, 446, 448, 451, 452, 467
Ubiquitous taxa, 387, 389, 395, 397, 534
U-channel method, 99, 105
Ultrasonication, 417, 725
Unimodal relationships, 529
Unsupported thorium, 508
Upwelling, 14, 169, 192, 196–197, 201–202,
    204, 223, 229–230, 237, 267, 280, 289, 294,
    296–297, 300, 330, 336–339, 355–356,
    371, 390–392, 396–397, 400, 413, 425, 430,
    447, 452, 467, 469–470, 555, 588–589, 608,
    613, 618, 648, 670, 687–688, 692, 695, 711,
    727, 731, 744, 755
    intensity, 229, 237, 390–391, 400, 469, 727
    proxy, 613
    zones, 223, 687
Uranium, 13, 509, 594, 607–609, 611, 646,
    660, 681–683, 701–702
    isotopes, 681–683
U-series
    dating methods, 193
    isotopes, 6, 12–13

Validation data set, 536
Vanadium, 607
Var Canyon, 83
Varimax technique, 539
Variscan signature, 663
Varved sediments, 77, 195
Vema Channel, 24–25, 45, 164–166, 171
Ventilation, 3–5, 52, 189, 193–194, 201–204,
    511–512, 515, 567, 602, 607, 615, 617,
    619, 621–622, 648, 694–695, 709–710,
    755
    rates, 3, 193, 202, 512, 515
Vermiculite, 142, 149, 165, 576
Vertical
    convection, 2
    distribution, 607, 653, 686
    flux, 10, 38, 382, 592, 624, 698, 700–701
    gradient, 221, 236
    rain, 681, 683, 698, 700–701, 703–705, 712
    stratification, 7
Viscosity, 30–32, 36

Vital effects, 491, 502–503, 505–507, 723, 736,
    747, 752, 756–757, 759
Volcanogenic sediments, 146, 158
Volumetric Magnetic susceptibility, 65, 87,
    109–110
Vostok, 346, 355, 470–471, 729
    ice core, 470–471, 729

Wadden Sea, 453
Walvis Ridge, 164, 668
Warm periods, 40–41, 170, 536
Washington margin, 453
Water
    column, 1–3, 8, 10, 38, 49, 147–148, 213,
        216, 244–245, 247, 275, 282, 285–286,
        291, 330, 377, 379, 382, 413, 417, 422,
        441, 446–448, 451, 454–456, 459, 461,
        475, 492, 512, 568, 587–590, 595, 602,
        604, 608, 613, 619, 621, 623–624, 646,
        648, 652–653, 658–659, 662, 681–685,
        687, 689, 692, 694–695, 698, 700,
        702–704, 707–710, 712, 723, 736, 743,
        748, 751, 769
    column stability, 619
    corrosiveness, 303, 304, 306, 723
    depth, 28, 30, 35, 40, 42–43, 82, 117, 146,
        149, 165, 167–168, 170, 173, 175, 264,
        266, 271, 282, 285, 288–293, 295, 297,
        301–302, 329, 332, 384, 422, 552, 578,
        581, 586, 588, 595–596, 600, 602, 605,
        610, 621–622, 684, 694, 702, 709, 731,
        775
    mass advection, 668, 743
    masses, 6, 8, 28, 49, 52, 141, 148, 156, 163,
        166–167, 170, 238, 266, 272, 278, 285,
        301–303, 307–308, 376, 413, 424, 445,
        457, 460, 503, 511–512, 551, 645–657,
        659, 661–663, 665, 667, 669, 671, 744,
        746, 753–756, 760
    samples, 428, 447, 455
    stratification, 221, 349, 353, 431
Wavelet analysis, 121
Weddell Sea, 149, 274, 304, 338, 353
Weighted
    averaging, 341, 526–527, 530
    averaging partial least square, 341, 530
West
    Atlantic, 162, 167, 170, 174, 456
    Australian Currents, 175
    Equatorial Pacific, 274

Indian Ocean, 151, 171, 311–312
Pacific, 42, 146, 151, 160, 163, 165, 468
Westerlies, 466, 580
Western
boundary Undercurrent (WBUC), 49, 161, 167, 663
South Atlantic, 166
tropical Atlantic, 375
WinCal25, 205
Wind transport, 148, 154–155
Wind-driven circulation, 28
Winnowing, 21, 36–37, 50, 53, 163, 303, 305, 337, 576, 593–594, 624, 700, 702–703
Winter sea-ice, 338, 341, 353–354, 389
World
Climate Research Programme (WCRP), 807
magnetic model, 102
Ocean Atlas, 8, 537, 557
Ocean Circulation Experiment (WOCE), 648

X-ray, 63–65, 69–78, 82–85, 92, 106, 139–140, 142, 150, 161, 176, 305, 577, 588, 622, 776, 781, 783
analysis, 75, 83
attenuation, 70, 73

density, 106
diffraction, 139, 142, 150, 176, 776, 783
fluorescence, 63, 74–75, 92, 577, 622
fluorescence spectrometry, 63, 74–75
imaging, 64–65, 70, 84, 92
microfluorescence, 75, 106
radiography, 69

Yermak Plateau, 51
Younger Dryas, 50, 197, 202, 204, 398, 467, 494, 514–515, 580, 619, 621–622, 710, 807

Zaire River plume, 225
Zeolites, 146, 154–155
Zijderveld diagrams, 115
Zinc, 611
Zircon, 572–574, 577–578, 658
Zirconium, 572
Zoanthids, 492, 500
Zooplankton, 8, 215, 286, 414, 463, 501, 512, 524, 619
Zr/Al ratio, 572–573

Printed and bound by CPI Group (UK) Ltd, Croydon, CR0 4YY

08/05/2025

01864806-0006